T0217238

Lecture Notes in Computer Science   9815

Commenced Publication in 1973
Founding and Former Series Editors:
Gerhard Goos, Juris Hartmanis, and Jan van Leeuwen

More information about this series at http://www.springer.com/series/7410

Matthew Robshaw · Jonathan Katz (Eds.)

Advances in Cryptology – CRYPTO 2016

36th Annual International Cryptology Conference
Santa Barbara, CA, USA, August 14–18, 2016
Proceedings, Part II

 Springer

Editors
Matthew Robshaw
Impinj, Inc.
Seattle, WA
USA

Jonathan Katz
University of Maryland
College Park, MD
USA

ISSN 0302-9743   ISSN 1611-3349   (electronic)
Lecture Notes in Computer Science
ISBN 978-3-662-53007-8   ISBN 978-3-662-53008-5   (eBook)
DOI 10.1007/978-3-662-53008-5

Library of Congress Control Number: 2016945783

LNCS Sublibrary: SL4 – Security and Cryptology

Printed on acid-free paper

This Springer imprint is published by Springer Nature
The registered company is Springer-Verlag GmbH Berlin Heidelberg

Preface

The 36th International Cryptology Conference (Crypto 2016) was held at UCSB, Santa Barbara, CA, USA, during August 14–18, 2016. The workshop was sponsored by the International Association for Cryptologic Research.

Crypto continues to grow. This year the Program Committee evaluated a record 274 submissions out of which 70 were chosen for inclusion in the program. Each paper was reviewed by at least three independent reviewers, with papers from Program Committee members receiving at least five reviews. Reviewers with potential conflicts of interest for specific papers were excluded from all discussions about those papers, and this policy was extended to the program chairs as well.

The 44 members of the Program Committee were aided in this complex and time-consuming task by many external reviewers. We would like to thank them all for their service, their expert opinions, and their spirited contributions to the review process. It was a tremendously difficult task to choose the program for this conference, as the quality of the submissions was very high. It was even harder to identify a single best paper, but our congratulations go to Elette Boyle, Niv Gilboa, and Yuval Ishai from IDC Herzliya, Ben Gurion University, and the Technion, respectively, whose paper "Breaking the Circuit Size Barrier for Secure Computation Under DDH" was awarded Best Paper. Our congratulations also go to Mark Zhandry of MIT and Princeton University who won the award for the Best Student Paper "The Magic of ELFs."

The invited speakers at Crypto 2016 were Brian Sniffen, Chief Security Architect at Akamai Technologies, Inc., and Paul Kocher, founder of Cryptography Research. Brian's presentation cast a fascinating light on the issues of real-world cryptographic deployment while Paul's presentation, a joint invitation from the program co-chairs of both Crypto 2016 and CHES 2016, marked 20 years since his publication of the first paper on side-channel attacks at Crypto 1996.

We are, of course, indebted to Brian LaMacchia, the general chair, as well as the local Organizing Committee, who together proved ideal liaisons for establishing the layout of the program and for supporting the speakers. Our job as program co-chairs was made much easier by the excellent tools developed by Shai Halevi; both Shai and Brian were always available at short notice to answer our queries. Finally, we would like to thank all the authors who submitted their work to Crypto 2016. Without you the conference would not exist.

August 2016

Matthew Robshaw
Jonathan Katz

Crypto 2016

The 36th IACR International Cryptology Conference

University of California, Santa Barbara, CA, USA
August 14–18, 2016

Sponsored by the *International Association for Cryptologic Research*

General Chair

Brian LaMacchia   Microsoft

Program Chairs

Matthew Robshaw   Impinj, USA
Jonathan Katz   University of Maryland, USA

Program Committee

Alex Biryukov   University of Luxembourg, Luxembourg
Anne Canteaut   Inria, France
Dario Catalano   Università di Catania, Italy
Nishanth Chandran   Microsoft Research, India
Melissa Chase   Microsoft Research, USA
Joan Daemen   STMicroelectronics, Belgium and Radboud University,
                         The Netherlands
Martin Van Dijk   University of Connecticut, USA
Itai Dinur   Ben-Gurion University, Israel
Pierre-Alain Fouque   Université Rennes 1, France
Steven Galbraith   Auckland University, New Zealand
Sanjam Garg   University of California, Berkeley, USA
S. Dov Gordon   George Mason University, USA
Jens Groth   University College London, UK
Sorina Ionica   Université de Picardie, France
Tetsu Iwata   Nagoya University, Japan
Aggelos Kiayias   National and Kapodistrian University of Athens,
                         Greece
Gregor Leander   Ruhr Universität Bochum, Germany
Shengli Liu   Shanghai Jiao Tong University, China
Alexander May   Ruhr Universität Bochum, Germany
Willi Meier   FHNW, Switzerland
Payman Mohassel   Visa Research, USA

Elke De Mulder	Cryptographic Research, France
Steven Myers	Indiana University, USA
Phong Nguyen	Inria, France and CNRS/JFLI and University of Tokyo, Japan
Kaisa Nyberg	Aalto University, Finland
Kenny Paterson	Royal Holloway University of London, UK
Thomas Peyrin	Nanyang Technological University, Singapore
Benny Pinkas	Bar-Ilan University, Israel
David Pointcheval	École Normale Supérieure, France
Manoj Prabhakaran	University of Illinois, USA
Bart Preneel	KU Leuven, Belgium
Mariana Raykova	Yale University, USA
Christian Rechberger	TU-Graz, Austria and DTU, Denmark
Mike Rosulek	Oregon State University, USA
Rei Safavi-Naini	University of Calgary, Canada
Alessandra Scafuro	Boston University and Northeastern University, USA
Patrick Schaumont	Virginia Tech, USA
Dominique Schröder	Saarland University, Germany
Jae Hong Seo	Myongji University, Korea
Yannick Seurin	ANSSI, France
Abhi Shelat	University of Virginia, USA
Nigel Smart	University of Bristol, UK
Ron Steinfeld	Monash University, Australia
Mehdi Tibouchi	NTT Secure Platform Laboratories, Japan

Additional Reviewers

André Chailloux
Jie Chen
Céline Chevalier
Chongwon Cho
Seung Geol Choi
Ashish Choudhury
Sherman Chow
Kai-Min Chung
Michele Ciampi
Michael Clear
Ran Cohen
Geoffroy Couteau
Dana Dachman-Soled
Deepesh Data
Jean Paul Degabriele
David Derler
Daniel Dinu
Christoph Dobraunig
Yevgeniy Dodis
Nico Döttling
Natnatee Dokmai
Leo Ducas
Tuyet Duong
Keita Emura
Frederic Ezerman
Pooya Farshim
Sebastian Faust
Dario Fiore
Marc Fischlin
Joe Fitzsimons
Nils Fleischhacker
Emmanuel Fouotsa
Georg Fuchsbauer
Eiichiro Fujisaki
Martin Gagne
François Le Gall
Chaya Ganesh
Juan Garay
Christina Garman
Romain Gay
Essam Ghadafi
Benedikt Gierlichs
Niv Gilboa
Vipul Goyal
Frédéric Grosshans
Aurore Guillevic

Divya Gupta
Felix Günther
Shai Halevi
Mike Hamburg
Shuai Han
Helena Handschuh
Christian Hanser
Carmit Hazay
Ethan Heilman
Ryan Henry
Gottfried Herold
Felix Heuer
Viet Tung Hoang
Dennis Hofheinz
Ziyuan Hu
Yan Huang
Michael Hutter
Malika Izabachene
Håkon Jacobsen
Mahavir Jhawar
Dingding Jia
Keting Jia
Thomas Johansson
Aaron Johnson
Kimmo Järvinen
Yael Tauman Kalai
Bhavana Kanukurthi
Petteri Kaski
Marcel Keller
Nathan Keller
Carmen Kempka
Iordanis Kerenidis
Dmitry Khovratovich
Dakshita Khurana
Eike Kiltz
Jinsu Kim
Taechan Kim
Paul Kirchner
Elena Kirshanova
Susumu Kiyoshima
Simon Knellwolf
Stefan Koelbl
Vlad Kolesnikov
Takeshi Koshiba
Luke Kowalczyk
Thorsten Kranz

Daniel Kraschewski
Anna Krasnova
Hugo Krawczyk
Fernando Krell
Stephan Krenn
Ranjit Kumaresan
Alptekin Kupcu
Fabien Laguillaumie
Virginie Lallemand
Enrique Larraia
Changmin Lee
Hyung Tae Lee
Kwangsu Lee
Nikos Leonardos
Tancrède Lepoint
Anthony Leverrier
Benoit Libert
Fuchun Lin
Rachel Lin
Yehuda Lindell
Feng-Hao Liu
Yi-Kai Liu
Patrick Longa
Steve Lu
Stefan Lucks
Atul Luykx
Anna Lysyanskaya
Lin Lyu
Vadim Lyubashevsky
Mohammad Mahmoody
Hemanta Maji
Giulio Malavolta
Tal Malkin
Alex Malozemoff
Mark Marson
Daniel Masny
Takahiro Matsuda
Florian Mendel
Bart Mennink
Thyla van der Merwe
Peihan Miao
Christof Michel
Ian Miers
Andrew Miller
Brice Minaud
Kazuhiko Minematsu

Ilya Mironov
Ameer Mohammad
Amir Moradi
Tal Moran
Nicky Mouha
Pratyay Mukherjee
Jörn Müller-Quade
Valérie Nachef
Michael Naehrig
Maria Naya-Plasencia
Soheil Nemati
Khoa Nguyen
Ivica Nikolic
Ventzi Nikov
Ryo Nishimaki
Anca Nitulescu
Adam O'Neill
Miyako Ohkubo
Go Ohtake
Tatsuaki Okamoto
Ozgur Oksuz
Cristina Onete
Claudio Orlandi
Elisabeth Oswald
Léo Paul Perrin
Jiaxin Pan
Giorgos Panagiotakos
Omkant Pandey
Kostas
  Pappagiannopoulos
Anat Paskin-Cherniavsky
Rafael Pass
Valerio Pastro
Arpita Patra
Souradyuti Paul
Christopher Peikert
Rene Peralta
Trevor Perrin
Giuseppe Persiano
Christophe Petit
Rafael Del Pino
Oxana Poburinnaya
Antigoni Polychroniadou
Orazio Puglisi
Baodong Qin
Max Rabkin

Carla Rafols
Srinivasan Raghuraman
Vanishree Rao
Manuel Reinert
Oscar Reparaz
Silas Richelson
Thomas Ristenpart
Damien Robert
Alon Rosen
Adeline Roux-Langlois
Arnab Roy
Tim Ruffing
Hansol Ryu
Sondre Rønjom
Akshayaram Srinivasan
Amin Sakzad
Katerina Samari
Ruediger Schack
Christian Schaffner
John Schanck
Thomas Schneider
Peter Scholl
Peter Schwabe
Sven Schäge
Adam Sealfon
Setareh Sharifian
Tom Shrimpton
Sandeep Shukla
Siang Meng Sim
Luisa Siniscalchi
Daniel Slamanig
Yongsoo Song
Kannan Srinathan
Akshayaram Srinivasan
Douglas Stebila
Damien Stehlé
John Steinberger
Marc Stevens
Valentin Suder
Willy Susilo
Björn Tackmann
Katsuyuki Takashima
Qiang Tang
Stefano Tessaro
Aishwarya
  Thiruvengadam

Jean-Pierre Tillich
Yosuke Todo
Yiannis Tselekounis
Michael Tunstall
Himanshu Tyagi
Aleksei Udovenko
Jon Ullman
Dominique Unruh
Prashant Vasudevan
Vesselin Velichkov
Muthu
  Venkitasubramaniam
Frederik Vercauteren
Damien Vergnaud
Jorge Villar
Dhinakaran
  Vinayagamurthy
Ivan Visconti
Michael Walter
Pengwei Wang
Qingju Wang
Xiao Wang
Hoeteck Wee
Mor Weiss
Yunhua Wen
Carolyn Whitnall
Daniel Wichs
Xiaodi Wu
Keita Xagawa
Sophia Yakoubov
Shota Yamada
Kan Yasuda
Arkady Yerukhimovich
Ouyang Yingkai
Thomas Zacharias
Mark Zhandry
Bingsheng Zhang
Liang Feng Zhang
Xiao Zhang
Yupeng Zhang
Hong-Sheng Zhou
Vassilis Zikas
Dionysis Zindros

Contents – Part II

Secure Computation and Protocols I

Obfuscation

Asymmetric Cryptography and Cryptanalysis II

Asymmetric Cryptography

Adversary-Dependent Lossy Trapdoor Function from Hardness of Factoring Semi-smooth RSA Subgroup Moduli

Takashi Yamakawa[1,2]($\boxtimes$), Shota Yamada[2], Goichiro Hanaoka[2], and Noboru Kunihiro[1]

[1] The University of Tokyo, Chiba, Japan
yamakawa@it.k.u-tokyo.ac.jp, kunihiro@k.u-tokyo.ac.jp
[2] National Institute of Advanced Industrial Science and Technology (AIST), Tokyo, Japan
{yamada-shota,hanaoka-goichiro}@aist.go.jp

Abstract. Lossy trapdoor functions (LTDFs), proposed by Peikert and Waters (STOC'08), are known to have a number of applications in cryptography. They have been constructed based on various assumptions, which include the quadratic residuosity (QR) and decisional composite residuosity (DCR) assumptions, which are factoring-based *decision* assumptions. However, there is no known construction of an LTDF based on the factoring assumption or other factoring-related search assumptions. In this paper, we first define a notion of *adversary-dependent lossy trapdoor functions* (ad-LTDFs) that is a weaker variant of LTDFs. Then we construct an ad-LTDF based on the hardness of factorizing RSA moduli of a special form called semi-smooth RSA subgroup (SS) moduli proposed by Groth (TCC'05). Moreover, we show that ad-LTDFs can replace LTDFs in many applications. Especially, we obtain the first factoring-based deterministic encryption scheme that satisfies the security notion defined by Boldyreva et al. (CRYPTO'08) without relying on a decision assumption. Besides direct applications of ad-LTDFs, by a similar technique, we construct a chosen ciphertext secure public key encryption scheme whose ciphertext overhead is the shortest among existing schemes based on the factoring assumption w.r.t. SS moduli.

1 Introduction

1.1 Background

In modern cryptography, constructing provably secure cryptographic primitives is an important research topic. In this line of researches, Peikert and Waters [27] proposed *lossy trapdoor functions* (LTDFs) and constructed a number of cryptographic primitives such as a collision resistant hash function, a chosen plaintext (CPA) and chosen ciphertext (CCA) secure public key encryption (PKE) schemes and an oblivious transfer scheme based on LTDFs. Following the work,

The first author is supported by a JSPS Fellowship for Young Scientists.

© International Association for Cryptologic Research 2016
M. Robshaw and J. Katz (Eds.): CRYPTO 2016, Part II, LNCS 9815, pp. 3–32, 2016.
DOI: 10.1007/978-3-662-53008-5_1

it is also shown that LTDFs can be used for constructing a deterministic encryption (DE) scheme [5] and a selective opening attack (SOA) secure PKE scheme [3]. As seen above, LTDFs have many applications, and therefore it is important to research concrete constructions of LTDFs.

As concrete constructions of LTDFs, Peikert and Waters [27] constructed schemes based on the decisional Diffie-Hellman (DDH) and learning with errors (LWE) assumptions. After that, many constructions of LTDFs have been proposed thus far. Among them, LTDFs related to the factoring are based on the quadratic residuosity (QR) [11], decisional composite residuosity (DCR) [11], Φ-hiding [20], or general class of subgroup decision assumptions [36], all of which are decision assumptions. On the other hand, there is no known construction of an LTDF based on the factoring assumption or a factoring-related search assumption. In general, search assumptions are rather weaker than decision assumptions. Thus it is important to research the possibility of constructing LTDFs based on a search assumption.

1.2   Our Result

In this paper, though we do not construct LTDFs based on the factoring assumption, we construct an *adversary dependent lossy trapdoor function* (ad-LTDF), which is a new notion we introduce, based on the factoring assumption w.r.t. semi-smooth RSA subgroup (SS) moduli, which are RSA moduli of a special form [13]. Then we show that ad-LTDFs can replace LTDFs in many applications. As a result, we immediately obtain factoring-based cryptographic primitives including a hash function, PKE scheme and DE scheme. Besides direct applications of ad-LTDFs, by using similar technique, we construct CCA secure PKE scheme with compact ciphertext based on the factoring assumption w.r.t. SS moduli. More details are given in the following.

Adversary-Dependent Lossy Trapdoor Function. We first reconsider the definition of LTDFs, and introduce a notion of an ad-LTDF, which is a weaker variant of an LTDF. Intuitively, an LTDF is a computationally indistinguishable pair of an injective and lossy functions. Here, the description of lossy functions should be fixed by the scheme. On the other hand, for ad-LTDFs, we allow a description of lossy function to depend on an adversary. That is, we only require that for any efficient adversary $\mathcal{A}$ there exists a lossy function that $\mathcal{A}$ cannot distinguish from an injective function. We observe that this significant relaxation does not harm the security of many LTDF-based cryptographic constructions. This is because in many LTDF-based schemes, lossy functions are used only in security proofs and they do not appear in the real scheme. This means that even if lossy functions depend on an adversary, we can still prove the security of the scheme. By this observation, we can see that ad-LTDFs can replace LTDFs in many applications.

Moreover, we construct an ad-LTDF based on the factoring assumption w.r.t. SS moduli, which is introduced by Groth [13]. As a result, we can instantiate many LTDF-based constructions based on the factoring assumption w.r.t. SS moduli. The intuition of the construction of the ad-LTDF is given in Sect. 1.3.

Applications of ad-LTDFs. As stated above, ad-LTDFs can replace LTDFs in many applications, and we give a construction of an ad-LTDF under the factoring assumption w.r.t. SS moduli. Thus we immediately obtain new factoring-based constructions of many cryptographic primitives such as a collision resistant hash function, CPA secure PKE scheme and a DE scheme. Among them, the DE scheme obtained by this way is the first factoring-based scheme that satisfies the PRIV security for block-sources, which is defined in [5], without relying on any decision assumption.

Table 1. Comparison among CCA secure PKE schemes based on the factoring assumption: ℓ_N is the bit-length of an underlying composite number N, ℓ_{MAC} denotes the bit-length of a message authentication code, Factoring SS denotes the factoring assumption w.r.t. SS moduli, and we assume that an exponentiation with an exponent of length ℓ can be computed by 1.5ℓ multiplications.

Schemes	Ciphertext overhead (bit)	Public key size (bit)	Computational cost for		Assumption
			encryption (mult)	decryption (mult)	
HK09 [18]	$2\ell_N$	$3\ell_N$	$3\ell_N + 3.5\lambda$	$1.5\ell_N + 10.5\lambda$	Factoring
MLLJ11 [25]	$2\ell_N$	$3\ell_N$	18.5λ	18λ	Factoring SS
Ours	$\ell_N + \ell_{MAC}$	$O(\lambda^2 \ell_N / \log \lambda)$	$O(\lambda \ell_N^2 / \log \lambda)$	$O(\lambda \ell_N^2 / \log \lambda)$	Factoring SS

CCA Secure PKE with Short Ciphertext. Besides direct applications of ad-LTDFs, we construct a CCA secure PKE scheme whose ciphertext overhead is the shortest among schemes based on the factoring assumption w.r.t. SS moduli. Table 1 shows the efficiency of CCA secure PKE schemes based on the factoring assumption. Among existing schemes, the scheme proposed by Hofheinz and Kiltz [18] is one of the best in regard to the ciphertext overhead, which consists of 2 elements of $\mathbb{Z}_N^*$. Mei et al. [25] improved the efficiency of the Hofheinz-Kiltz scheme [18] in regard to encryption and decryption costs by using SS moduli. However, they did not improve the ciphertext overhead. In contrast, the ciphertext overhead of our scheme consists of only 1 element of $\mathbb{Z}_N^*$ and a message authentication code (MAC), whose bit-length can be much smaller than that of N. By giving a concrete parameter, the ciphertext overhead of our scheme is 1360-bit for 80-bit security whereas that of [18] is 2048-bit. On the other hand, the public key size of our scheme is much larger than that of [18], and an encryption and decryption are much less efficient than those in [18]. We note that the reduction from the CCA security of our scheme to the factoring assumption w.r.t. SS moduli is quite loose, but all known CCA secure PKE scheme based on the factoring assumption (including [18,25]) also require loose reductions because they require Blum-Blum-Shub pseudo-random number generator [4].

We note that there is a strong negative result for a CCA secure PKE scheme whose ciphertext overhead is less than 2 group elements in a prime order setting [14]. Even in a composite order setting, there are only a few CCA secure PKE schemes whose ciphetext overhead is less than 2 group elements, all of which rely on a subgroup decision assumption [16,17,21] or an interactive assumption [19] stronger than the factoring assumption. Ours is the first scheme to overcome this bound based solely on the factoring assumption (though our assumption is the factoring assumption w.r.t. SS moduli, which may not be considered standard).

1.3 Our Technique

Difficulty of Constructiing LTDFs Based on a Search Assumption. Before explaining our technique, we first explain why it is difficult to construct LTDFs based on a search assumption. Recall that an LTDF is a computationally indistinguishable pair of injective and lossy functions. Apparently, the definition of LTDFs itself requires the hardness of a decision problem. Thus for constructing LTDFs based on a search assumption, we have to rely on some "search-to-decision" reduction. As a general technique for such a reduction, there is the Goldreich-Levin hardcore theorem [12], which enables us to extract "pseudorandomness" from hardness of any search problem. However, the Goldreich-Levin hardcore bit destroys algebraic structures of original problems. On the other hand, considering existing constructions of LTDFs, algebraic structures of underlying problems are crucial for constructing LTDFs. Thus, for constructing LTDFs based on search assumptions, we have to establish another "search-to-decision" reduction technique that does not hurt underlying algebraic structures. In the context of lattice problems, this has been already done. Namely, it is shown that search-LWE and decision-LWE assumptions are equivalent [33]. Thus LTDFs can be constructed based on the search-LWE assumption. However, there is no known such a reduction in the context of the factoring problem. Namely, we have no reduction from decision assumptions such as QR, DCR, or more general subgroup decision assumptions to the factoring assumption.

New Search-to-Decision Reduction Technique. The core of this work is to give a new search-to-decision reduction technique in the context of factoring w.r.t. SS moduli. Namely, we introduce a new decision assumption that we call the adversary-dependent decisional RSA subgroup (ad-DRSA) assumption, and reduce the ad-DRSA assumption to the factoring assumption w.r.t. SS moduli. In the following, we explain the technique in more detail.

We say that a composite number N is an SS modulus if it can be written as $N = PQ = (2pp' + 1)(2qq' + 1)$, where P and Q are primes with the same length, p and q are "smooth" numbers (i.e., products of distinct small primes) and p' and q' are relatively large primes. Then the group of quadratic residues $\mathbb{QR}_N$ is a cyclic group of order $pqp'q'$, and has many subgroups since pq is smooth. With respect to SS moduli, Groth [13] proposed the decisional RSA subgroup (DRSA) assumption, which claims that any PPT adversary cannot distinguish a random element of G from that of $\mathbb{QR}_N$ where G is the unique subgroup of $\mathbb{QR}_N$ of order $p'q'$.

Our first observation is that if there exists an algorithm that breaks the DRSA assumption, then one can find at least one small prime that divides $\Phi(N)$. This can be seen by the following argument: Assume that all prime factors of pq are of ℓ_B-bit length. (Since pq is smooth, ℓ_B is relatively small. Especially, we set $\ell_B = O(\log \lambda)$.) Recall that the DRSA assumption claims that any PPT algorithm cannot distinguish a random element of G from that of $\mathbb{QR}_N$. This is equivalent to that the distributions of $g^{p_1 \cdots p_M}$ and g are indistinguishable where $g \xleftarrow{\$} \mathbb{QR}_N$ and $p_1, \ldots, p_M$ are the all ℓ_B-bit primes (and thus M is the number of the all ℓ_B-bit primes). If there exists an algorithm $\mathcal{A}$ that breaks the DRSA assumption, then it distinguishes these two distributions. Thus, by the hybrid argument, there exists $j \in [M]$ such that $\mathcal{A}$ distinguish the distribution of $g^{p_1 \cdots p_{j-1}}$ from $g^{p_1 \cdots p_j}$. By using $\mathcal{A}$, one can find this p_j by the exhaustive search since M is polynomial in the security parameter in our parameter setting. (See Sect. 2.4 for more detail.) For this p_j, we have $p_j | \Phi(N)$ (with overwhelming probability) since otherwise p_j-th power on $\mathbb{QR}_N$ is a permutation on the group and thus distributions of $g^{p_1 \cdots p_{j-1}}$ and $g^{p_1 \cdots p_j}$ are completely identical. The above argument proves that if there exists an algorithm that breaks the DRSA assumption, then one can find at least one small prime that divides $\Phi(N)$. However, this fact states nothing about the reduction from the DRSA assumption to the factoring assumption since even if one can find one small prime p that divides $\Phi(N)$, we do not know how to factorize N.

Here, we relax the DRSA assumption to define the *adversary-dependent decisional RSA subgroup* (ad-DRSA) assumption. Intuitively, the ad-DRSA assumption claims that for any PPT adversary $\mathcal{A}$, there exists a subgroup $S_{\mathcal{A}}$ of $\mathbb{QR}_N$ such that $\mathcal{A}$ does not distinguish a random element of $S_{\mathcal{A}}$ from that of $\mathbb{QR}_N$. More precisely, the ad-DRSA assumption is parametrized by an integer $m \leq M$, and m-ad-DRSA assumption claims that for any PPT algorithm $\mathcal{A}$, there exists at least one choice of $p_1, \ldots p_m$ out of all ℓ_B-bit primes such that $\mathcal{A}$ cannot distinguish $g^{p_1 \cdots p_m}$ from g where $g \xleftarrow{\$} \mathbb{QR}_N$. By this definition, if there exists a PPT algorithm $\mathcal{A}$ that breaks the m-ad-DRSA assumption, then $\mathcal{A}$ distinguishes $g^{p_1 \cdots p_m}$ from g for *all* choices of $p_1, \ldots, p_m$. If m is sufficiently smaller than M, then there exists "many" choices of $p_1, \ldots, p_m$ and thus one can find "many" primes that divides $\Phi(N)$: One can find at least one such prime for each choice of $p_1, \ldots, p_m$ by the similar method as in the case of the DRSA assumption. Then the product of these primes is a large divisor of $\Phi(N)$ and thus one can factorize N by using the Coppersmith theorem [6], which claims that if one is given a "large" divisor of $\Phi(N)$, then one can factorize N efficiently. Thus, the m-ad-DRSA assumption is reduced to the factoring assumption.

Remark 1. *We remark that if m is so small that there exists a choice of $p_1, \ldots, p_m$, all of which are coprime to $\Phi(N)$, then the m-ad-DRSA assumption is trivial since in that case g and $g^{p_1 \cdots p_m}$ are distributed identically. We show that there exists a parameter choice such that m-ad-DRSA assumption is non-trivial and it can be reduced to the factoring assumption simultaneously.*

How to Use the ad-DRSA Assumption. As explained above, we show a reduction from the ad-DRSA assumption, which is a certain type of a subgroup assumption, to the factoring assumption. However, the ad-DRSA assumption is not an ordinary subgroup decision assumption: Roughly speaking, it only claims that for any PPT adversary $\mathcal{A}$, there exists a subgroup $S_{\mathcal{A}} \in \mathbb{QR}_N$ such that $\mathcal{A}$ cannot distinguish random elements of $S_{\mathcal{A}}$ from $\mathbb{QR}_N$. Therefore, it cannot be used for constructions where elements of a subgroup are used in the real descriptions of the scheme. On the other hand, if elements of a subgroup are used only in the security proof, the ad-DRSA assumption suffices. We give two examples of such cases.

One is ad-LTDFs. As explained in Sect. 1.2, ad-LTDFs is a relaxation of LTDFs such that descriptions of lossy functions can depend on an adversary. For constructing ad-LTDFs based on the ad-DRSA assumption, we simply imitate the construction by Xu et al. [36], who constructed LTDFs based on the (standard) DRSA assumption. We observe that in their construction, the descriptions of injective functions consist only of elements of $\mathbb{QR}_N$, and elements of its subgroup are used only in the descriptions of lossy functions. Therefore even if we replace the DRSA assumption with the ad-DRSA assumption, only lossy functions depend on an adversary. This meets the definition of the ad-LTDFs.

The other is the hash-proof system-based CCA secure public key encryption. Hofheniz and Kiltz [16] introduced the concept of constraind CCA (CCCA) security, and showed efficient constructions of CCA secure public key encryption schemes based on a hash proof system, which can be constructed from any subgroup decision assumption [8]. Though elements of a subgroup are used in the real protocol of their original construction, it is easy to see that even if elements of a subgroup are replaced with those of a larger group, the scheme is still secure because they are indistinguishable by the assumption. Thus that scheme can be instantiated based on the ad-DRSA assumption.

1.4   Discussion

Plausibility of the Factoring Assumption w.r.t. SS Moduli. Here, we discuss the plausibility of the assumption we used. SS moduli was first introduced by Groth [13] in 2005 and they have been used in some works [25,36,37]. All of these works assume the factoring assumption w.r.t. SS moduli (or more stronger assumptions). On the other hand, in 2011, Coron et al. [7] gave a cryptanalysis against the Groth's work [13]. However, they did not improve attacks against SS moduli. Thus, we can say that SS moduli has attracted a certain amount of attention in the sense of both constructions and cryptanalysis, but no fatal attack is found thus far. Therefore we believe that the hardness of factoring SS moduli is rather reliable.

Interpretation of Our Result. In this paper, we constructed a weaker variant of LTDF (ad-LTDF) based on the factoring assumption w.r.t. SS moduli. One may wonder how meaningful our result is since an SS modulus is not an RSA modulus of a standard form. We believe that our result is meaningful in terms of that we constructed an "LTDF-like primitive" (ad-LTDF), which can replace LTDFs in many applications, based on a search assumption (factoring w.r.t. SS moduli) rather than a decision assumption. Although the application given in this paper is limited to the case of SS moduli, we hope that our new search-to-decision reduction technique can be extended to other general settings.

Limitation of ad-LTDFs. Though ad-LTDFs can replace LTDFs in many cases, there exist some LTDF-based primitives that cannot be obtained from ad-LTDFs. A typical example is the oblivious transfer protocol proposed by Peikert and Waters [27]. The reason why we cannot construct the scheme based on ad-LTDFs is that in the scheme, a lossy function is explicitly required. Specifically, a receiver sends a pair of injective and lossy functions to a sender. Since we cannot specify a lossy function before fixing an adversary, we cannot instantiate this scheme based on ad-LTDFs.

1.5   Related Work

Deterministic Encryption. Bellare et al. [1] initiated the study of DE, defined the security notion of DE called the PRIV security, and gave constructions of PRIV secure DE schemes in the random oracle model. Boldyreva et al. [5] slightly weakened the PRIV security to what they call the PRIV security for block-sources, and constructed DE schemes with this security in the standard model based on LTDFs. Bellare et al. [2] showed that DE scheme with a weaker security notion (where messages are uniformly random) can be constructed from any one-way trapdoor permutation. In this paper, we only consider the PRIV security for block-sources as defined in [5].

Factoring Based CCA Secure PKE Schemes. In 2009, Hofheinz and Kiltz [18] proposed the first practical CCA secure PKE scheme under the factoring assumption in the standard model. After that, many variants of the scheme are proposed thus far [23–25,37]. However, none of them improve the ciphertext overhead of the scheme. On the other hand, the ciphretext overhead of our proposed scheme is shorter than those of them.

2   Preliminaries

Here we review some basic notations and definitions.

2.1   Notations

We use $\mathbb{N}$ to denote the set of all natural numbers and $[n]$ to denote the set $\{1, \ldots n\}$ for $n \in \mathbb{N}$. If S is a finite set, then we use $x \xleftarrow{\$} S$ to denote that x is chosen uniformly at random from S. If $\mathcal{A}$ is an algorithm, we use $x \leftarrow \mathcal{A}(y)$ to denote that x is output by $\mathcal{A}$ whose input is y. For a finite set S, $|S|$ denotes the cardinality of S. For a real number x, $\lceil x \rceil$ denotes the smallest integer not smaller than x and $\lfloor x \rfloor$ denotes the largest integer not larger than x. For a bit string a, ℓ_a denotes the length of a. For a function f in λ, we often denote f to mean $f(\lambda)$ for notational simplicity. We say that a function $f(\cdot) : \mathbb{N} \to [0,1]$ is negligible if for all polynomials $p(\cdot)$ and all sufficiently large $\lambda \in \mathbb{N}$, we have $f(\lambda) < |1/p(\lambda)|$. We say f is overwhelming if $1 - f$ is negligible. We say that a function $f(\cdot) : \mathbb{N} \to [0,1]$ is noticeable if there exists a polynomial p such that for all sufficiently large λ, we have $f(\lambda) > |1/p(\lambda)|$. We say that an algorithm $\mathcal{A}$ is probabilistic polynomial time (PPT) if there exists a polynomial p such that running time of $\mathcal{A}$ with input length λ is less than $p(\lambda)$. We use $a|b$ to mean that a is a divisor of b. For a natural number N, $\Phi(N)$ denote the number of natural numbers smaller than N that are coprime to N. For random variables X and Y, $\Delta(X,Y)$ denote the statistical distance between them. We use the fact that for any (probabilistic) function f, $\Delta(f(X), f(Y)) \leq \Delta(X,Y)$ holds, and that $\Delta((X_1, Z), (Y_1, Z)) = \mathbb{E}_Z[\Delta(X_1, Y_1)]$ where $\mathbb{E}$ denotes the expected value. For random variables X and Y, we define min-entropy of X as $H_\infty(X) := -\log(\max_x \Pr[X = x])$ and average min-entropy of X given Y as $\tilde{H}_\infty(X|Y) := -\log(\Sigma_y \Pr[Y = y] \max_x \Pr[X = x|Y = y])$. We use λ to denote the security parameter.

2.2 Syntax and Security Notions

Here, we review definitions of cryptographic primitives. We omit the definitions of a collision resistant hash function and a CPA/CCA secure PKE scheme due to the page limitation. They are standard and can be found in the full version.

Key Encapsulation Mechanism. Here, we review the definition of key encapsulation mechanism (KEM) and its security. It is shown that a CCA secure PKE scheme is obtained by combining a constrained CCA (CCCA) secure KEM and a CCA secure authenticated symmetric key encryption scheme [16]. In the following, we recall the definitions of KEM and its CCCA security.

A KEM consists of three algorithms (Gen, Enc, Dec). Gen takes a security parameter 1^λ as input and outputs (PK, SK), where PK is a public key and SK is a secret key. Enc takes a public key PK as input and outputs (C, K), where C is a ciphertext and K is a symmetric key. Dec takes a secret key SK and a ciphertext C as input and outputs a key K with length ℓ_K or $\bot$. We require that for all (PK, SK) output by Gen and all (C, K) output by Enc(PK), we have Dec$(SK, C) = K$.

To define the CCCA security of KEM $=$ (Gen, Enc, Dec), we consider the following game between an adversary $\mathcal{A}$ and a challenger $\mathcal{C}$. First, $\mathcal{C}$ generates $(PK, SK) \leftarrow$ Gen(1^λ) and $(C^*, K) \leftarrow$ Enc(PK), chooses a random bit $b \xleftarrow{\$} \{0,1\}$, and sets $K^* := K$ if $b = 1$ and otherwise $K^* \xleftarrow{\$} \{0,1\}^{\ell_K}$. Then (PK, C^*, K^*) is given to the adversary $\mathcal{A}$. In the game, $\mathcal{A}$ can query pairs of ciphertexts and predicates any number of times. When $\mathcal{A}$ queries (C, pred), $\mathcal{C}$ computes $K \leftarrow$ Dec(SK, C) and returns K to $\mathcal{A}$ if $C \neq C^*$ and $\mathsf{pred}(K) = 1$, and otherwise $\bot$. Finally, $\mathcal{A}$ outputs a bit b'. We define the CCCA advantage of $\mathcal{A}$ as $\mathsf{Adv}_{\mathcal{A},\mathsf{KEM}}^{\mathsf{CCCA}}(\lambda) := |\Pr[b = b'] - 1/2|$. We say that KEM is CCCA secure if $\mathsf{Adv}_{\mathcal{A},\mathsf{KEM}}^{\mathsf{CCCA}}(\lambda)$ is negligible for any PPT *valid* adversary $\mathcal{A}$, where "valid" is defined below.

Before defining "valid", we prepare two definitions. We say that a predicate pred is *non-trivial* if $\Pr[\mathsf{pred}(K) = 1 : K \xleftarrow{\$} \{0,1\}^{\ell_K}]$ is negligible. We say that an algorithm $\mathcal{C}'$ is an *alternative challenger* if it has the same syntax as the real challenger $\mathcal{C}$. We say that an adversary $\mathcal{A}$ is valid if for any PPT alternative challenger $\mathcal{C}'$, all predicates pred queried by $\mathcal{A}$ in the game between $\mathcal{A}$ and $\mathcal{C}'$ are non-trivial.

Though the above definition of the CCCA security slightly differs from the original definition given in [16], we can easily prove that our definition still yields the "hybrid encryption theorem" that a CCA secure PKE scheme can be obtained by a CCCA secure KEM and authenticated symmetric key encryption.

Deterministic Encryption. A deterministic encryption scheme consists of three algorithms (Gen, Enc, Dec). Gen takes a security parameter 1^λ as input and outputs (PK, SK), where PK is a public key and SK is a secret key. Enc is a deterministic algorithm that takes a public key PK and a message msg as input and outputs a ciphertext C. Dec takes a secret key SK and a ciphertext C as input and outputs a message msg or $\bot$. We require that for all msg, (PK, SK) output by Gen and C output by $\mathsf{Enc}(PK, msg)$, we have $\mathsf{Dec}(SK, C) = msg$.

We recall security notions for deterministic encryption following [5]. In [5], the authors considered three security notions called PRIV, PRIV1 and PRIV1-IND, and proved all of them are equivalent. Therefore we consider only the simplest security definition PRIV1-IND in this paper. A random variable X over $\{0,1\}^n$ is called a (u, n)-source if $H_\infty(X) \geq u$. For ATK $\in$ {CPA, CCA}, a deterministic encryption scheme DE = (Gen, Enc, Dec) for ℓ-bit message is PRIV1-IND-ATK secure for (t, n)-sources if for any (t, n)-sources M_0 and M_1 and all PPT adversaries $\mathcal{A}$, $\mathsf{Adv}^{\mathsf{PRIV1-IND-ATK}}_{\mathcal{A}, M_0, M_1, \mathsf{DE}}(\lambda) := |\Pr[b = b' : (PK, SK) \leftarrow \mathsf{Gen}(1^\lambda); b \xleftarrow{\$} \{0,1\}; msg^* \xleftarrow{\$} M_b; C^* \leftarrow \mathsf{Enc}(PK, msg^*); b' \leftarrow \mathcal{A}^{\mathcal{O}}(PK, C^*)] - 1/2|$ is negligible where if ATK = CPA, then $\mathcal{O}$ is an oracle that always returns $\bot$, and if ATK = CCA, then $\mathcal{O}$ is an decryption oracle that is given a ciphertext C and returns $\mathsf{Dec}(SK, C)$ if $C \neq C^*$ and otherwise $\bot$.

2.3 Known Lemmas

Here, we review three known lemmas used in this paper. First, we review a simple variant of the Hoeffding inequality [15].

Lemma 1 *(Hoeffding inequality).* *Let $\mathcal{D}_1$ and $\mathcal{D}_2$ be probability distributions over $\{0,1\}$. Let $X_1, \ldots, X_K$ be K independent random variables with the distribution $\mathcal{D}_1$ and $Y_1, \ldots, Y_K$ be K independent random variables with the distribution $\mathcal{D}_2$. If we define $\epsilon := |\Pr[X = 1 : X \xleftarrow{\$} \mathcal{D}_1] - \Pr[Y = 1 : Y \xleftarrow{\$} \mathcal{D}_2]|$, then $\Pr[|\frac{\Sigma^K_{k=1} X_k - \Sigma^K_{i=k} Y_k}{K}| - \epsilon| \geq \delta] \leq 4e^{-\delta^2 K/2}$ holds.*

The following is the generalized leftover hash lemma [10].

Lemma 2 *(Generalized leftover hash lemma).* *Let $X \in \{0,1\}^{n_1}$ and Y be random variables. Let $\mathcal{H}$ be a family of pairwise independent hash function from $\{0,1\}^{n_1}$ to $\{0,1\}^{n_2}$. Then we have $\Delta((H(X), H, Y), (U, H, Y)) \leq \delta$ where $H \xleftarrow{\$} \mathcal{H}$ as long as $\tilde{H}_\infty(X|Y) \geq n_2 + 2\log(1/\delta)$.*

The following is the "crooked version" of the above lemma proven by Boldyreva et al. [5].

Lemma 3 *(Generalized crooked leftover hash lemma [5, Lemma 7.1]). Let $X \in \{0,1\}^n$ and Y be random variables. Let $\mathcal{H}$ be a family of pairwise independent hash function from $\{0,1\}^n$ to R and f be a function from R to S. Then for $H \xleftarrow{\$} \mathcal{H}$, we have $\Delta((f(H(X)), H, Y), (f(U), H, Y)) \leq \delta$ as long as $\tilde{H}_\infty(X|Y) \geq \log|S| + 2\log(1/\delta) - 2$.*

Finally, we review the Coppersmith theorem about bivariate integer equations. The following lemma is a special case of [6, Theorem 3].

Lemma 4. *Let $p(x,y) = a + bx + cy$ be a polynomial over $\mathbb{Z}$. For positive integers X, Y and $W = \max\{a, bX, cY\}$, if $XY < W$ holds, then one can find all solutions (x_0, y_0) such that $p(x_0, y_0) = 0$, $|x_0| < X$ and $|y_0| < Y$ in time polynomial in $\log_2 W$.*

2.4   Semi-smooth RSA Subgroup Modulus

For integers ℓ_B, t_p and t_q, We say that $N = PQ = (2pp' + 1)(2qq' + 1)$ is an (ℓ_B, t_p, t_q)-semi-smooth RSA subgroup $((\ell_B, t_p, t_q)$-SS) modulus if the following conditions hold.

– P and Q are distinct prime numbers with the same length that satisfy $\gcd(P - 1, Q - 1) = 2$.
– p' and q' are distinct primes larger than 2^{ℓ_B}.
– p and q are products of t_p and t_q distinct ℓ_B-bit primes. Here, an ℓ_B-bit prime means a prime number between $2^{\ell_B - 1}$ and 2^{ℓ_B}. We note that we have $\gcd(p, q) = 1$ since we have $\gcd(P - 1, Q - 1) = 2$.

We define $t := t_p + t_q$. Let $\mathcal{P}_{\ell_B}$ be the set of all ℓ_B-bit primes, and $M_{\ell_B} := |\mathcal{P}_{\ell_B}|$. We define the group of quadratic residues as $\mathbb{QR}_N := \{u^2 : u \in \mathbb{Z}_N^*\}$. This is a subgroup of $\mathbb{Z}_N^*$, and a cyclic group of order $pqp'q'$. Then there exists unique subgroups of order $p'q'$ and pq, and we denote them by G and $G^\perp$ respectively. Then we have $\mathbb{QR}_N = G \times G^\perp$. That is, for any element $g \in \mathbb{QR}_N$, we can uniquely represent $g = g(G)g(G^\perp)$ by using $g(G) \in G$ and $g(G^\perp) \in G^\perp$. Moreover, if the factorization of N is given, then we can compute $g(G)$ and $g(G^\perp)$ from g efficiently.

When N is an SS modulus, we cannot say that a random element g of $\mathbb{QR}_N$ is a generator (i.e., $\mathrm{ord}(g) = pqp'q'$) with overwhelming probability. However, we can prove that g has an order larger than a certain value with overwhelming probability.

Lemma 5 *([13, Lemma 2]). Let N be an (ℓ_B, t_p, t_q)-SS modulus. For any integer $d < t$ if $\frac{(t2^{1-\ell_B})^{d+1}}{(1 - t2^{1-\ell_B})(d+1)!}$ is negligible, then $\Pr[\mathrm{ord}(g) \geq p'q'2^{(t-d)(\ell_B - 1)} : g \xleftarrow{\$} \mathbb{QR}_N]$ is overwhelming. Especially, $\Pr[\mathrm{ord}(g(G^\perp)) \geq 2^{(t-d)(\ell_B - 1)} : g \xleftarrow{\$} \mathbb{QR}_N]$ is overwhelming.*

When ℓ_B is small, $\mathrm{ord}(G^\perp)$ is smooth, and therefore the discrete logarithm on the group can be solved efficiently by the Pohlig-Hellman algorithm [28].

Lemma 6 ([13]). *If $\ell_B = O(\log \lambda)$, then the discrete logarithm problem on $G^\perp$ can be solved efficiently. More precisely, there exists a PPT algorithm that, given an (ℓ_B, t_p, t_q)-SS modulus N, $g \in G^\perp$ and g^x, outputs $x \mod \mathrm{ord}(g)$.*

By combining the above lemmas, we obtain the following lemma.

Lemma 7. *Let N be an (ℓ_B, t_p, t_q)-SS modulus and we assume $\ell_B = O(\log(\lambda))$. If $\frac{(t2^{1-\ell_B})^{d+1}}{(1-t2^{1-\ell_B})(d+1)!}$ is negligible and $x \le 2^{(t-d)(\ell_B-1)}$ holds, then there exists a PPT algorithm PLog that, given P, Q, g, g^x, outputs x with overwhelming probability where $g \xleftarrow{\$} \mathbb{QR}_N$.*

Hardness Assumptions. Here, we give definitions of two hardness assumptions. Let IGen be an algorithm that is given the security parameter 1^λ and outputs an (ℓ_B, t_p, t_q)-SS modulus with its factorization. We first define the factoring assumption.

Definition 1. *We say that the factoring assumption holds with respect to IGen if for any PPT algorithm $\mathcal{A}$, $\Pr[\mathcal{A}(N) \in \{P, Q\} : (N, P, Q) \leftarrow \mathsf{IGen}(1^\lambda)]$ is negligible.*

Next, we define the *decisional RSA subgroup* (DRSA) assumption proposed by Groth [13]. This assumption claims that any PPT algorithm cannot distinguish a random element of G from that of $\mathbb{QR}_N$. We note that actually we do not use this assumption in this paper. We include this only for the information of the reader.

Definition 2. *We say that the decisional RSA subgroup (DRSA) assumption holds with respect to IGen if for any PPT algorithm $\mathcal{A}$, $|\Pr[1 \leftarrow \mathcal{A}(N, g) : (N, P, Q) \leftarrow \mathsf{IGen}(1^\lambda); g \xleftarrow{\$} \mathbb{QR}_N] - \Pr[1 \leftarrow \mathcal{A}(N, g) : (N, P, Q) \leftarrow \mathsf{IGen}(1^\lambda); g \xleftarrow{\$} G]|$ is negligible.*

Attacks. We review factoring attacks against SS moduli as discussed in [13]. As shown in [13], by using Pollard's ρ-method [30], we can factorize an SS modulus in time $\tilde{O}(\min(\sqrt{p'}, \sqrt{q'}))$. As another method, by using Naccache et al.'s algorithm [26], if a divisor of $P-1$ or $Q-1$ larger than $N^{1/4}$ is given, then N can be factorized efficiently. Thus ℓ_B should be large enough so that it is difficult to guess a significant portion of factors of p or q. In 2011, Coron et al. [7] proposed a new factoring algorithm for a certain class of RSA moduli that includes SS moduli. For the case of SS moduli, their algorithm work in time $\tilde{O}(\min(\sqrt{p'}, \sqrt{q'}))$, which matches the time complexity of Pollard's ρ-method. As observed in [13], other methods such as the baby-step giant-step algorithm [34], Pollard's λ-method [31] or Pollard's $p-1$ method [29] require $O(\min(p', q'))$ time.

The above attacks use the structure of SS moduli. On the other hand, there are algorithms such as the elliptic curve method [22] or the general number field sieve [9], which can be applied to general RSA moduli. Among these algorithms, general number field sieve is asymptotically the most efficient and its heuristic running time is $\exp((1.92 + o(1)) \ln(N)^{1/3} \ln\ln(N)^{2/3})$.

Parameter Settings. Here, we discuss parameter settings of SS moduli. We have to set parameters to avoid the above attacks. We first give an asymptotic parameter setting. We set $\ell_{p'} = \ell_{q'} = O(\lambda)$, $\ell_B = \lfloor 4 \log \lambda \rfloor$ and $t_p = t_q = O(\lambda^3 / \log \lambda)$ (then we have $\ell_N \approx \ell_{p'} + \ell_{q'} + t\ell_B = O(\lambda^3)$). In this setting, we have $M_{\ell_B} = O(\lambda^4 / \log \lambda)$ by the prime number theorem and thus there exists exponentially many choices of p and q. If we set $d := \lfloor t/4 \rfloor$, then $\frac{(t2^{1-\ell_B})^{d+1}}{(1-t2^{1-\ell_B})(d+1)!}$ is negligible[1]. We use the fact that in this parameter setting, given N, $g \in \mathbb{QR}_N$ and $p_1, \ldots, p_m$ for $m \leq M_{\ell_B}$, $g^{p_1 \cdots p_m}$ can be computed in polynomial time in λ. This is because we have $m \leq M_{\ell_B} = O(\lambda^4 / \log(\lambda))$ and $p_1 \ldots p_m \leq 2^{\ell_B M_{\ell_B}} = 2^{O(\lambda^4)}$, and thus $p_1 \ldots p_m$-th power can be computed by $O(\lambda^4)$ multiplications. We use this asymptotic parameter setting throughout the paper. As a concrete parameter, Groth [13] proposed to set $\ell'_p = \ell'_q = 160$, $\ell_B = 15$, $t_p = t_q = 32$ and $d = 7$ for 80-bit security (then we have $\ell_N = 160 \cdot 2 + 15 \cdot 2 \cdot 32 = 1280$). We use this parameter for the construction of CCA secure PKE scheme with compact ciphertext (Sect. 6). However, this parameter does not give us enough lossiness in the construction of ad-LTDFs. Thus we propose to set $\ell'_p = \ell'_q = 160$, $\ell_B = 15$, $t_p = t_q = 70$ and $d = 8$ (then we have $\ell_N = 160 \cdot 2 + 15 \cdot 2 \cdot 70 = 2420$) for 80-bit security for other applications (Sects. 4 and 5). We note that the number of $\ell_B = 15$-bit primes is 1612. Therefore the possible choice of $t = 64$ or 140 primes out of them is much larger than 2^{80} and thus it is hard to guess the significant portion of their factors.

3   Adversary-Dependent Decisional RSA Subgroup Assumption

In this section, we generalize the DRSA assumption. Specifically, we define the m-adversary-dependent decisional RSA subgroup (m-ad-DRSA) assumption for any integer $m \leq M_{\ell_B}$ with respect to (ℓ_B, t_p, t_q)-SS moduli. Intuitively, this assumption claims that for any PPT algorithm $\mathcal{A}$, there exist distinct ℓ_B-bit primes $p_1, \ldots, p_m$ such that $\mathcal{A}$ does not distinguish g from $g^{p_1 \cdots p_m}$ where g is a random element of $\mathbb{QR}_N$. We prove that under a certain condition, the m-ad-DRSA assumption holds under the factoring assumption.

First we give the precise definition of the m-ad-DRSA assumption.

Definition 3. *Let* IGen *be a PPT algorithm that generates an (ℓ_B, t_p, t_q)-SS RSA modulus. We say that for any integer $m \leq M_{\ell_B}$, the m-adversary-dependent decisional RSA subgroup (m-ad-DRSA) assumption holds with respect to* IGen *if for any noticeable function ϵ and PPT algorithm $\mathcal{A}$, there exists a PPT algorithm $\mathcal{S}_{\mathcal{A},\epsilon}$ that is given (ℓ_B, t_p, t_q)-SS RSA modulus N and outputs distinct ℓ_B-bit primes $p_1, \ldots, p_m$, such that the following is satisfied. If we let*

---

[1] In fact, d can be set as $d := \lfloor ct \rfloor$ for any small enough constant c.

$$P_0 := \Pr \left[1 \leftarrow \mathcal{A}(N, g) : \begin{array}{c} (N, P, Q) \leftarrow \mathsf{IGen}(1^\lambda) \\ g \xleftarrow{\$} \mathbb{QR}_N \end{array} \right]$$

$$P_1 := \Pr \left[1 \leftarrow \mathcal{A}(N, g^{p_1 \cdots p_m}) : \begin{array}{c} (N, P, Q) \leftarrow \mathsf{IGen}(1^\lambda) \\ g \xleftarrow{\$} \mathbb{QR}_N \\ \{p_1, \ldots, p_m\} \leftarrow \mathcal{S}_{\mathcal{A}, \epsilon}(N) \end{array} \right]$$

then we have $|P_0 - P_1| \leq \epsilon(\lambda)$ for sufficiently large λ.

Remark 2. *One may think that the above defined assumption cannot be used for proving security of any cryptographic scheme since ϵ is noticeable. However, an important remark here is that ϵ can be an arbitrary noticeable function. Thus, in security proofs, we can set ϵ depending on an adversary $\mathcal{A}$'s advantage against the scheme that we want to prove secure, such that ϵ is smaller than the advantage of $\mathcal{A}$ (for infinitely many security parameters). This can be done if $\mathcal{A}$ breaks the security of the scheme since in these cases, the advantage of $\mathcal{A}$ should be non-negligible. See security proofs in Sects. 5 and 6 to see this argument indeed works.*

Remark 3. *In the above definition, if m is so small that there exists a choice of $p_1, \ldots, p_m$, all of which are coprime to $\Phi(N)$, then $g^{p_1 \cdots p_m}$ is distributed uniformly on $\mathbb{QR}_N$. In this case, m-ad-DRSA assumption is trivial. This occurs if and only if we have $M_{\ell_B} - m \geq t$. In this paper, we set m to be relatively large so that m-ad-DRSA assumption is non-trivial. (See Remark 4.)*

The following theorem claims that the m-ad-DRSA assumption holds under the factoring assumption if m is small enough.

Theorem 1. *Let IGen be a PPT algorithm that generates an (ℓ_B, t_p, t_q)-SS RSA modulus where $\ell_B = O(\log \lambda)$. If the factoring assumption holds with respect to IGen and there exists a constant c such that $(M_{\ell_B} - m + 1)(\ell_B - 1) \geq (1/2 + c)\ell_N$ holds, then the m-ad-DRSA assumption holds with respect to IGen.*

Remark 4. *If we set $m := \lfloor M_{\ell_B} + 1 - (1/2 + c)\ell_N/(\ell_B - 1) \rfloor$ for sufficiently small c, then by the above theorem, the m-ad-DRSA assumption holds under the factoring assumption. Moreover, by setting the parameter as given in Sect. 2.4, we have $M_{\ell_B} - m \approx (1/2 + c)\ell_N/(\ell_B - 1) \approx (1/2 + c)(\ell_{p'} + \ell_{q'} + t\ell_B)/\ell_B \leq t$ for sufficiently large λ if $c < 1/2$ since $t = O(\lambda^3/\log \lambda)$ and $\ell_{p'} = \ell_{q'} = O(\lambda)$. Thus the m-ad-DRSA assumption is non-trivial.*

Before proving the theorem, we prepare a lemma related to the Coppersmith attack. Though a heuristic proof appeared in [26], to the best of our knowledge, this has not been proven rigorously in the literature.

Lemma 8. *Let P and Q be primes with the same length and $N = PQ$. Let e be a divisor of $\Phi(N) = (P-1)(Q-1)$. If there exists a positive constant c such that $e > N^{1/2+c}$ holds, then there exists a polynomial time algorithm that is given N and e, and factorizes N.*

Proof. We define e_1 and e_2 such that $e = e_1e_2$, $e_1|P-1$ and $e_2|Q-1$. (Note that we cannot always compute e_1 and e_2 from e.) Then we can write $P = e_1k_1 + 1$ and $Q = e_2k_2 + 1$ by using integers k_1 and k_2. Then we have $N = PQ = (e_1k_1 + 1)(e_2k_2 + 1) = ek_1k_2 + e_1k_1 + e_2k_2 + 1$. Therefore if we define $p(x, y) = N + ex + y$, then $p(x, y) = 0$ has a solution $(x_0, y_0) = (-k_1k_2, -(e_1k_1 + e_2k_2 + 1))$. Let $X := N^{1/2-c}$, $Y := 3N^{1/2}$ and $W := \max(N, eX, Y)$. One can see that $|x_0| < X$, $|y_0| < Y$ and $XY = 3N^{1-c} < N \leq W$ hold (for sufficiently large N). Therefore one can compute the solution $(x_0, y_0) = (-k_1k_2, -(e_1k_1 + e_2k_2 + 1))$ in polynomial time in $\log N$ by Lemma 4. Then one can compute $P + Q = e_1k_1 + e_2k_2 + 2 = -y + 1$ and factorize N. $\square$

Intuition for the Proof of Theorem 1. Here, we give an intuition for the proof of Theorem 1. We remark that the following argument is not a rigorous one. What we have to do is to construct a PPT algorithm $\mathcal{S}_{\mathcal{A},\epsilon}$ that is given N and outputs $\{p_1, \ldots, p_m\}$ such that $\mathcal{A}$'s advantage to distinguish g from $g^{p_1 \cdots p_m}$ is smaller than ϵ where $g \xleftarrow{\$} \mathbb{QR}_N$. Let list $L := \mathcal{P}_{\ell_B}$, which is the set of all ℓ_B-bit primes. First, $\mathcal{S}_{\mathcal{A},\epsilon}$ randomly chooses m distinct primes $\{p_1, \ldots, p_m\}$ from L and test whether $\mathcal{A}$'s advantage to distinguish g from $g^{p_1 \cdots p_m}$ is smaller than ϵ or not. More precisely, $\mathcal{S}_{\mathcal{A},\epsilon}$ approximates $\mathcal{A}$'s advantage by iterating the execution of $\mathcal{A}(g)$ and $\mathcal{A}(g^{p_1 \cdots p_m})$ for independently random $g \xleftarrow{\$} \mathbb{QR}_N$ a number of times and counting the number that each of them outputs 1. We denote the approximated advantage by ϵ'. Due to the Hoeffding inequality [15], the approximation error can be made smaller than $\epsilon/4$ by polynomial times iterations since ϵ is noticeable. If $\epsilon' < \epsilon/2$, then $\mathcal{A}$'s real advantage is smaller than $3\epsilon/4 < \epsilon$ and thus $\mathcal{S}_{\mathcal{A},\epsilon}$ outputs $\{p_1, \ldots, p_m\}$ and halts. Otherwise, $\mathcal{A}$'s advantage to distinguish g from $g^{p_1 \cdots p_m}$ is larger than $\epsilon/4$. Then there exists p_j such that $\mathcal{A}$'s advantage to distinguish $g^{p_1 \cdots p_{j-1}}$ from $g^{p_1 \cdots p_j}$ is larger than $\epsilon/(4m)$ by the hybrid argument. $\mathcal{S}_{\mathcal{A},\epsilon}$ can find this p_j in polynomial time since $\epsilon/(4m)$ is noticeable. We remark that we have $p_j|\Phi(N)$. This is because, otherwise $\mathcal{A}$'s advantage to distinguish $g^{p_1 \cdots p_{j-1}}$ from $g^{p_1 \cdots p_j}$ is 0 since their distributions are completely identical and thus ϵ' should be smaller than $\epsilon/2$. Then $\mathcal{S}_{\mathcal{A},\epsilon}$ removes p_j from L. Then it randomly chooses m distinct primes $\{p_1, \ldots, p_m\}$ from L again, and do the same as the above. Then it outputs $\{p_1, \ldots, p_m\}$ and halts if approximated $\mathcal{A}$'s advantage to distinguish g from $g^{p_1 \cdots p_m}$ is smaller than $\epsilon/2$, or otherwise removes some $p_{j'}|\Phi(N)$ from L. $\mathcal{S}_{\mathcal{A},\epsilon}$ repeat this procedure many times. Assume that $\mathcal{S}_{\mathcal{A},\epsilon}$ does not halts by the time it cannot choose m distinct primes from L. By that time, $M_{\ell_B} - m + 1$ distinct ℓ_B-bit primes that divide $\Phi(N)$ are removed from L. Let e be the product of them. Then we have $e|\Phi(N)$ and $e \geq 2^{(\ell_B-1)(M_{\ell_B}-m+1)} \geq N^{1/2+c}$. Therefore if e is given, then we can factorize N efficiently by Lemma 8. Thus under the factoring assumption, $\mathcal{S}_{\mathcal{A},\epsilon}$ must output some $\{p_1, \ldots, p_m\}$ before $|L|$ becomes smaller than m with overwhelming probability, and $\mathcal{A}$'s advantage to distinguish g from $g^{p_1 \cdots p_m}$ is smaller than $3\epsilon/4 < \epsilon$.

Now we give the full proof of Theorem 1.

Proof (of Theorem 1). First, we prove the following two claims.

Claim 1. *For any PPT algorithm $\mathcal{A}$ and a noticeable function δ, there exists a PPT algorithm $\mathsf{Approx}_{\mathcal{A},\delta}$ that satisfies the following. Let $\mathcal{D}_1$ and $\mathcal{D}_2$ be descriptions of distributions that are samplable in polynomial time in λ, and $\epsilon := |\Pr[1 \leftarrow \mathcal{A}(X) : X \xleftarrow{\$} \mathcal{D}_1] - \Pr[1 \leftarrow \mathcal{A}(X) : X \xleftarrow{\$} \mathcal{D}_2]|$. Then $\mathsf{Approx}_{\mathcal{A},\delta}(1^\lambda, \mathcal{D}_1, \mathcal{D}_2)$ outputs ϵ' such that $|\epsilon' - \epsilon| \leq \delta(\lambda)$ with overwhelming probability. (We say that $\mathsf{Approx}_{\mathcal{A},\delta}$ succeeds if it outputs such ϵ'.)*

Proof. The construction of $\mathsf{Approx}_{\mathcal{A},\delta}$ is as follows.

$\mathsf{Approx}_{\mathcal{A},\delta}(1^\lambda, \mathcal{D}_1, \mathcal{D}_2)$: For $i = 1$ to K where $K := \lambda/\delta(\lambda)^2$, choose X_i and Y_i according to $\mathcal{D}_1$ and $\mathcal{D}_2$, respectively, and run $\mathcal{A}(X_i)$ and $\mathcal{A}(Y_i)$ for each i. Let k_1 be the number of the event that $\mathcal{A}(X_i)$ outputs 1 and k_2 be the number of the event that $\mathcal{A}(Y_i)$ outputs 1. Output $|k_1 - k_2|/K$.

Since δ is noticeable, K is polynomial in λ and therefore $\mathsf{Approx}_{\mathcal{A},\delta}$ is a PPT algorithm. It can be seen by Lemma 1 that $\mathsf{Approx}_{\mathcal{A},\delta}$ satisfies the desired property. $\qed$

Claim 2. *For any PPT algorithm $\mathcal{A}$ and a noticeable function ϵ, there exists a PPT algorithm $\mathsf{Find}_{\mathcal{A},\epsilon}$ that satisfies the following. For any (ℓ_B, t_p, t_q)-SS RSA modulus N and a set $I = \{p_1, \ldots, p_m\}$ of distinct ℓ_B-bit primes, if $|\Pr[1 \leftarrow \mathcal{A}(N, g) : g \xleftarrow{\$} \mathbb{QR}_N] - \Pr[1 \leftarrow \mathcal{A}(N, g^{p_1 \cdots p_m}) : g \xleftarrow{\$} \mathbb{QR}_N]| > \epsilon(\lambda)$ holds, then $\mathsf{Find}_{\mathcal{A},\epsilon}(N, I)$ outputs $p_j \in I$ that divides $\Phi(N)$ with overwhelming probability. (We say that $\mathsf{Find}_{\mathcal{A},\epsilon}$ succeeds if it outputs such p_j or the inequality assumed is false.)*

Proof. The construction of $\mathsf{Find}_{\mathcal{A},\epsilon}$ is as follows.

$\mathsf{Find}_{\mathcal{A},\epsilon}(N, I = \{p_1, \ldots, p_m\})$: Define distributions $\mathcal{D}_0 := \{(N, g) : g \xleftarrow{\$} \mathbb{QR}_N\}$ and $\mathcal{D}_j := \{(N, g^{p_1 \cdots p_j}) : g \xleftarrow{\$} \mathbb{QR}_N\}$ $(j = 1, \ldots, m)$. For $j := 1$ to m, repeat the following.

   Compute $\epsilon' \leftarrow \mathsf{Approx}_{\mathcal{A},\epsilon/(2m)}(1^\lambda, \mathcal{D}_{j-1}, \mathcal{D}_j)$.

   If $\epsilon' > \epsilon/(2m)$, then output p_j and halt.

If it does not halt by the time the above loop is finished, then output $\bot$.

First, we show $\mathsf{Find}_{\mathcal{A},\epsilon}$ is a PPT algorithm. Since $m \leq M_{\ell_B}$ is polynomial in λ and thus $\epsilon/(2m)$ is noticeable, $\mathsf{Approx}_{\mathcal{A},\epsilon/(2m)}$ is a PPT algorithm. Therefore $\mathsf{Find}_{\mathcal{A},\epsilon}$ is a PPT algorithm. We prove that $\mathsf{Find}_{\mathcal{A},\epsilon}$ satisfies the desired property. First, we assume that all executions of $\mathsf{Approx}_{\mathcal{A},\epsilon/(2m)}$ called by $\mathsf{Find}_{\mathcal{A},\epsilon}$ succeed. The probability that this assumption is satisfied is overwhelming since the number of executions of $\mathsf{Approx}_{\mathcal{A},\epsilon/(2m)}$ is polynomial in λ and each execution succeeds with overwhelming probability. First, we prove that $\mathsf{Find}_{\mathcal{A},\epsilon}$ outputs any prime $p_j \in I$ if we have $|\Pr[1 \leftarrow \mathcal{A}(N, g) : g \xleftarrow{\$} \mathbb{QR}_N] - \Pr[1 \leftarrow \mathcal{A}(N, g^{p_1 \cdots p_m}) : g \xleftarrow{\$} \mathbb{QR}_N]| > \epsilon$. By the hybrid argument, there exists $j \in [m]$ such that $|\Pr[1 \leftarrow$

$\mathcal{A}(X) : X \xleftarrow{\$} \mathcal{D}_{j-1}] - \Pr[1 \leftarrow \mathcal{A}(X) : X \xleftarrow{\$} \mathcal{D}_j]| > \epsilon/m$ holds. For such j, if we let $\epsilon' := \mathsf{Approx}_{\mathcal{A},\epsilon/(2m)}(\mathcal{D}_{j-1}, \mathcal{D}_j)$, then we have $\epsilon' > \epsilon/m - \epsilon/(2m) = \epsilon/(2m)$ and thus p_j is output. Then we prove that if p_j is output by $\mathsf{Find}_{\mathcal{A},\epsilon}$, then $p_j|\Phi(N)$ holds. If p_j does not divide $\Phi(N)$, then p_j is coprime to $\mathrm{ord}(\mathbb{QR}_N)$, and especially p_j-th power is a permutation on the group $\{g^{p_1 \cdots p_{j-1}} : g \in \mathbb{QR}_N\}$. Therefore $\mathcal{D}_{j-1}$ and $\mathcal{D}_j$ are completely the identical distributions. Therefore we have $|\Pr[1 \leftarrow \mathcal{A}(X) : X \xleftarrow{\$} \mathcal{D}_{j-1}] - \Pr[1 \leftarrow \mathcal{A}(X) : X \xleftarrow{\$} \mathcal{D}_j]| = 0$. Thus if we let $\epsilon' := \mathsf{Approx}_{\mathcal{A},\epsilon/(2m)}(\mathcal{D}_{j-1}, \mathcal{D}_j)$, then we have $\epsilon' < \epsilon/(2m)$, and thus such p_j cannot be output. □

Then we go back to the proof of Theorem 1. For any PPT algorithm $\mathcal{A}$ and a noticeable function ϵ, we construct a PPT algorithm $\mathcal{S}_{\mathcal{A},\epsilon}$ such that $\Pr[1 \leftarrow \mathcal{A}(N,g) : (N,P,Q) \leftarrow \mathsf{IGen}(1^\lambda); g \xleftarrow{\$} \mathbb{QR}_N] - \Pr[1 \leftarrow \mathcal{A}(N, g^{p_1 \cdots p_m}) : (N,P,Q) \leftarrow \mathsf{IGen}(1^\lambda); g \xleftarrow{\$} \mathbb{QR}_N; \{p_1, \ldots, p_m\} \leftarrow \mathcal{S}_{\mathcal{A},\epsilon}(N)] \leq \epsilon(\lambda)$ holds for sufficiently large λ. The construction of $\mathcal{S}_{\mathcal{A},\epsilon}$ is as follows.

$\mathcal{S}_{\mathcal{A},\epsilon}(N)$: Let $L := \mathcal{P}_{\ell_B}$. (Recall that $\mathcal{P}_{\ell_B}$ is the set of all ℓ_B-bit primes.)
  While $|L| \geq m$, repeat the following.
    Choose distinct ℓ_B-bit primes $p_1, \ldots, p_m$ from L randomly, and let $I := \{p_1, p_2, \ldots, p_m\}$, $\mathcal{D}_0 := \{(N,g) : g \xleftarrow{\$} \mathbb{QR}_N\}$ and $\mathcal{D}_m := \{(N, g^{p_1 \cdots p_m}) : g \xleftarrow{\$} \mathbb{QR}_N\}$. Compute $\epsilon' \leftarrow \mathsf{Approx}_{\mathcal{A},\epsilon/4}(1^\lambda, \mathcal{D}_0, \mathcal{D}_m)$. If $\epsilon' < \epsilon/2$, then output I and halts. Otherwise run $\tilde{p} \leftarrow \mathsf{Find}_{\mathcal{A},\epsilon/4}(N, I)$. If $\tilde{p} \in L$ then remove $\tilde{p}$ from L, otherwise remove a random element from L.
  If it does not halt by the time the above loop finishes, then it outputs $\bot$.

First, we prove that $\mathcal{S}_{\mathcal{A},\epsilon}(N)$ is a PPT algorithm. Since ϵ is noticeable, $\mathsf{Approx}_{\mathcal{A},\epsilon/4}$ and $\mathsf{Find}_{\mathcal{A},\epsilon/4}$ are PPT algorithms. Moreover the number of repeat is at most $M_{\ell_B} - m + 1 \leq M_{\ell_B}$, which is polynomial in λ. Therefore $\mathcal{S}_{\mathcal{A},\epsilon}(N)$ is a PPT algorithm.

Then we prove that $\mathcal{S}_{\mathcal{A},\epsilon}(N)$ satisfies the desired property. In the following, we assume that all executions of $\mathsf{Approx}_{\mathcal{A},\epsilon/4}$ and $\mathsf{Find}_{\mathcal{A},\epsilon/4}$ called by $\mathcal{S}_{\mathcal{A},\epsilon}(N)$ succeed. The probability that the above assumption holds is overwhelming since the number of executions is polynomial and each execution succeeds with overwhelming probability. If $\mathcal{S}_{\mathcal{A},\epsilon}$ outputs some $I = \{p_1, \ldots, p_m\}$, then we have $\Pr[1 \leftarrow \mathcal{A}(N,g) : g \xleftarrow{\$} \mathbb{QR}_N] - \Pr[1 \leftarrow \mathcal{A}(N, g^{p_1 \cdots p_m}) : g \xleftarrow{\$} \mathbb{QR}_N]| < \epsilon' + \epsilon/4 < \epsilon/2 + \epsilon/4 = 3\epsilon/4$. Next, we prove that for overwhelming fraction of N generated by IGen, the probability that $\mathcal{S}_{\mathcal{A},\epsilon}(N)$ outputs $\bot$ is negligible. First, we prove that in each repeat, $\tilde{p}$ that is removed from L divides $\Phi(N)$. We let $\epsilon' \leftarrow \mathsf{Approx}_{\mathcal{A},\epsilon/4}(1^\lambda, \mathcal{D}_0, \mathcal{D}_m)$. If $\epsilon' \geq \epsilon/2$, then we have $|\Pr[1 \leftarrow \mathcal{A}(N,g) : g \xleftarrow{\$} \mathbb{QR}_N] - \Pr[1 \leftarrow \mathcal{A}(N, g^{p_1 \cdots p_m}) : g \xleftarrow{\$} \mathbb{QR}_N]| > \epsilon' - \epsilon/4 \geq \epsilon/2 - \epsilon/4 = \epsilon/4$. Therefore $\mathsf{Find}_{\mathcal{A},\epsilon/4}(N, I)$ outputs $p_j \in I$ that divides $\Phi(N)$ since it succeeds. Thus if $\mathcal{S}_{\mathcal{A},\epsilon}(N)$ outputs $\bot$, then one ℓ_B-bit prime factor of $\Phi(N)$ is removed from L in each repeat, and the repeat is done $M_{\ell_B} - m + 1$ times. Therefore throughout the execution of $\mathcal{S}_{\mathcal{A},\epsilon}(N)$, $M_{\ell_B} - m + 1$ distinct ℓ_B-bit prime factors of $\Phi(N)$ are removed from L. If we let e be the product of these primes, then we have $e > (2^{\ell_B - 1})^{M_{\ell_B} - m + 1} \geq 2^{(1/2+c)\ell_N} > N^{1/2+c}$ and $e|\Phi(N)$. By Lemma 8, we can factorize N efficiently by using e. Therefore for overwhelming fraction of N generated by IGen, the probability that $\mathcal{S}_{\mathcal{A},\epsilon}(N)$ outputs $\bot$ is negligible under the factoring assumption. Therefore for overwhelming fraction of N generated by IGen, we have $\Pr[1 \leftarrow \mathcal{A}(N,g) : g \xleftarrow{\$} \mathbb{QR}_N] - \Pr[1 \leftarrow \mathcal{A}(N, g^{p_1 \cdots p_m}) : \{p_1, \ldots, p_m\} \leftarrow \mathcal{S}_{\mathcal{A},\epsilon}(N); g \xleftarrow{\$} \mathbb{QR}_N]| < 3\epsilon/4$ with overwhelming probability over the randomness of $\mathcal{S}_{\mathcal{A},\epsilon}$. Since ϵ is noticeable, by the averaging argument, $\Pr[1 \leftarrow \mathcal{A}(N,g) : (N,P,Q) \leftarrow \mathsf{IGen}(1^\lambda); g \xleftarrow{\$} \mathbb{QR}_N] - \Pr[1 \leftarrow \mathcal{A}(N, g^{p_1 \cdots p_m}) : (N,P,Q) \leftarrow \mathsf{IGen}(1^\lambda); g \xleftarrow{\$} \mathbb{QR}_N; \{p_1, \ldots, p_m\} \leftarrow \mathcal{S}_{\mathcal{A},\epsilon}(N)] \le \epsilon(\lambda)$ holds for sufficiently large λ.

4 Adversary-Dependent Lossy Trapdoor Function

In this section, we define ad-LTDFs. Then we give a construction of an ad-LTDF based on the m-ad-DRSA assumption, which can be reduced to the factoring assumption by Theorem 1.

4.1 Definition

Here we define ad-LTDFs. Intuitively, ad-LTDFs are defined by weakening LTDFs so that descriptions of lossy functions that cannot be distinguished from those of injective functions may depend on a specific distinguisher. Namely, the algorithm that generates lossy functions takes a "lossy function index" I as well as a public parameter as input, and we require that for any PPT algorithm $\mathcal{A}$, there exists at least one I such that $\mathcal{A}$ does not distinguish lossy functions generated with index I from injective functions. Moreover, we require that such I can be efficiently computed given $\mathcal{A}$. The precise definition is as follows. For integers n and k such that $0 < k < n$, an (n,k)-sd LTDF consists of 5 algorithms $(\mathsf{ParamsGen}, \mathsf{SampleInj}, \mathsf{SampleLossy}, \mathsf{Evaluation}, \mathsf{Inversion})$ with a family $\{\mathcal{I}(\lambda)\}_{\lambda \in \mathbb{N}}$ of lossy function index sets.

$\mathsf{ParamsGen}(1^\lambda) \to (PP, SP)$: It takes a security parameter 1^λ as input, and outputs a public parameter PP and a secret parameter SP.

$\mathsf{SampleInj}(PP) \to \sigma$: It takes a public parameter PP as input, and outputs a function description σ, which specifies an injective function f_σ over the domain $\{0,1\}^n$.

$\mathsf{SampleLossy}(PP, I) \to \sigma$: It takes a public parameter PP and a lossy function index $I \in \mathcal{I}(\lambda)$ as input, and outputs a function index σ, which specifies a "lossy" function f_σ over the domain $\{0,1\}^n$.

$\mathsf{Evaluation}(PP, \sigma, x) \to f_\sigma(x)$: It takes a public parameter PP, function description σ and $x \in \{0,1\}^n$ as input, and outputs $f_\sigma(x)$

$\mathsf{Inversion}(SP, \sigma, y) \to f_\sigma^{-1}(y)$: It takes a secret parameter SP, a function description σ and y and outputs $f_\sigma^{-1}(y)$.

We require ad-LTDFs to satisfy the following three properties.

Correctness: For all $x \in \{0,1\}^n$, we have $\mathsf{Inversion}(SP, \sigma, \mathsf{Evaluation}(PP, \sigma, x)) = x$ with overwhelming probability where $(PP, SP) \leftarrow \mathsf{ParamsGen}(1^\lambda)$ and $\sigma \leftarrow \mathsf{SampleInj}(PP)$.

Lossiness: For all $\lambda \in \mathbb{N}$, $(PP, SP) \leftarrow \mathsf{ParamsGen}(1^\lambda)$, $I \in \mathcal{I}(\lambda)$ and $\sigma \leftarrow \mathsf{SampleLossy}(PP, I)$, the image of f_σ has size at most 2^{n-k}.

Indistinguishability Between Injective and Lossy Functions. Intuitively, we require that for any PPT adversary $\mathcal{A}$, there exists at least one lossy function index $I \in \mathcal{I}(\lambda)$ such that $\mathcal{A}$ cannot distinguish an injective function from a lossy function with the lossy function index I.

The more precise definition is as follows. For any PPT adversary $\mathcal{A}$ and noticeable function $\epsilon(\lambda)$, there exists a PPT algorithm $\mathcal{S}_{\mathcal{A},\epsilon}$ that takes a public parameter PP as input and outputs $I \in \mathcal{I}(\lambda)$ such that the following is satisfied. If we let

$$P_{\mathsf{inj}} := \Pr\left[1 \leftarrow \mathcal{A}(PP, \sigma) : \begin{array}{c} (PP, SP) \leftarrow \mathsf{ParamsGen}(1^\lambda) \\ \sigma \leftarrow \mathsf{SampleInj}(PP) \end{array}\right]$$

$$P_{\mathsf{lossy}} := \Pr\left[1 \leftarrow \mathcal{A}(PP, \sigma) : \begin{array}{c} (PP, SP) \leftarrow \mathsf{ParamsGen}(1^\lambda) \\ I \leftarrow \mathcal{S}_{\mathcal{A},\epsilon}(PP) \\ \sigma \leftarrow \mathsf{SampleLossy}(PP, I) \end{array}\right]$$

then we have $|P_{\mathsf{inj}} - P_{\mathsf{lossy}}| \leq \epsilon(\lambda)$ for sufficiently large λ.

As mentioned in Remark 2, though ϵ must be noticeable in the above definition, ad-LTDFs can be used for many cryptographic applications. This is because ϵ can be set depending on the advantage of an adversary in security reductions.

Remark 5. *Besides what is explained above, there is a minor difference between the definition of ad-LTDFs and that of LTDFs. In the definition of ad-LTDFs, ParamsGen is explicitly separated from SampleInj or SampleLossy, whereas there is no separation between them in the definition of LTDFs [27]. This is only for simplifying the presentation, and there is no significant difference here.*

4.2   Construction

We construct an ad-LTDF based on the m-ad-DRSA assumption. Let IGen be an algorithm that generates an ℓ_N-bit (ℓ_B, t_p, t_q)-SS RSA modulus with the parameter given in Sect. 2.4 and $n := (t - d)(\ell_B - 1)$.

Definition of $\mathcal{I}(\lambda)$: $\mathcal{I}(\lambda)$ is defined as the set of all m-tuple of distinct primes of length ℓ_B. That is, we define $\mathcal{I}(\lambda) := \{\{p_1, \ldots, p_m\} : p_1, \ldots, p_m \text{ are distinct } \ell_B \text{ bit primes}\}$.

$\mathsf{ParamsGen}(1^\lambda) \to (PP, SP)$: Generate $(N, P, Q) \leftarrow \mathsf{IGen}(1^\lambda)$, set $PP := N$ and $SP := (P, Q)$, and output (PP, SP).

$\mathsf{SampleInj}(PP = N) \to \sigma$: Choose $g \xleftarrow{\$} \mathbb{QR}_N$ and output $\sigma := g$.

$\mathsf{SampleLossy}(PP = N, I = \{p_1, \ldots, p_m\}) \to \sigma$: Choose $g \xleftarrow{\$} \mathbb{QR}_N$ and output $\sigma := g^{p_1 \cdots p_m}$.

Evaluation($PP = N, \sigma = g, x \in \{0,1\}^n$) $\to f_\sigma(x)$: Interpret x as an element of $[2^n]$ and output g^x.
Inversion($SP = (P,Q), \sigma = g, y$) $\to f_\sigma^{-1}(y)$: Compute $x = \mathsf{PLog}(P,Q,g,y)$ and output x where PLog is the algorithm given in Lemma 7.

Theorem 2. *If the m-ad-DRSA assumption holds with respect to IGen, then the above scheme is an $(n, n - (\ell_{p'} + \ell_{q'} + (M_{\ell_B} - m)\ell_B))$-ad-LTDF.*

Then the following corollary follows by combining the above theorem and Theorem 1.

Corollary 1. *If the factoring assumption holds with respect to IGen for the parameter setting given in Sect. 2.4, then there exists an ad-LTDF.*

Proof. Recall that we set $\ell_{p'} = \ell_{q'} = O(\lambda)$, $\ell_B = \lfloor 4\log\lambda \rfloor$, $t_p = t_q = O(\lambda^3/\log\lambda)$ (then we have $\ell_N \approx \ell_{p'} + \ell_{q'} + t\ell_B = O(\lambda^3)$) and $d := t/4$. We let $m := \lfloor M_{\ell_B} + 1 - (1/2 + c)\frac{\ell_N}{(\ell_B - 1)} \rfloor$ for a constant $c < 1/4$. Then we have $(M_{\ell_B} - m + 1)(\ell_B - 1) \geq (1/2 + c)\ell_N$ and therefore the m-ad-DRSA assumption holds under the factoring assumption by Theorem 1. Then we prove that the above ad-LTDF for this m is non-trivial, i.e., we have $n - (\ell_{p'} + \ell_{q'} + (M_{\ell_B} - m)\ell_B) > 0$. Since we have $m \approx M_{\ell_B} - (1/2 + c)\frac{\ell_N}{(\ell_B - 1)}$, we have $n - (\ell_{p'} + \ell_{q'} + (M_{\ell_B} - m)\ell_B) \approx (t - d)\ell_B - (\ell_{p'} + \ell_{q'} + (1/2 + c)\frac{\ell_N \ell_B}{(\ell_B - 1)}) \approx (1/4 - c)t\ell_B - (3/2 + c)(\ell_{p'} + \ell_{q'}) > 0$ for sufficiently large λ since $t\ell_B = O(\lambda^3)$ and $\ell_{p'} + \ell_{q'} = O(\lambda)$. Thus the obtained ad-LTDF for this m is non-trivial.

Remark 6. *If we set $\ell_{p'} = \ell_{q'} = 160$, $\ell_B = 15$, $t = 64$, $d = 7$ and $\ell_N = 2420$ as given in Sect. 2.4, and $c = 1/10$ then by setting $m := \lfloor M_{\ell_B} + 1 - (1/2 + c)\frac{\ell_N}{(\ell_B - 1)} \rfloor$, the obtained scheme is a $(1848, 103)$-ad-LTDF. If better lossiness is required, then one may set t larger (as long as factorizing N is hard).*

Then we prove Theorem 2.

Proof (of Theorem 2).

Correctness. If g is generated by $\mathsf{SampleInj}$, then it is a random element of $\mathbb{QR}_N$. Thus $\mathsf{Inversion}((P,Q), g, \mathsf{Evaluation}(N, \sigma, x)) = \mathsf{Inversion}((P,Q), g, g^x) = x$ holds by the correctness of PLog given in Lemma 7.

Lossiness. Next, we prove that the above construction satisfies $(n, n - (\ell_{p'} + \ell_{q'} + (M_{\ell_B} - m)\ell_B))$-lossiness. Let σ be a function description generated by $\mathsf{SampleLossy}(N, I = \{p_1, \ldots, p_m\})$. What we should prove is that the image size of f_σ is at most $2^{\ell_{p'} + \ell_{q'} + (M_{\ell_B} - m)\ell_B}$. There exists $g \in \mathbb{QR}_N$ such that $\sigma = g^{p_1 \cdots p_m}$, and thus any output of f_σ is an element of the group $S := \{h^{p_1 \cdots p_m} : h \in \mathbb{QR}_N\}$. We consider the order of S. S is a subgroup of $\mathbb{QR}_N = G \times G^\perp$ and $p_1 \ldots p_m$ is coprime to $\mathrm{ord}(G) = p'q'$. Therefore there exists a subgroup $S^\perp$ of $G^\perp$ such that $S = G \times S^\perp$. We can see that $\mathrm{ord}(S^\perp)$ is the product of some distinct ℓ_B-bit primes and coprime to $p_1 \ldots p_m$ by the definition. Therefore that is the product of at most $M_{\ell_B} - m$ such primes, and can be bounded by $2^{(M_{\ell_B} - m)\ell_B}$. Therefore the order of S is at most $2^{\ell_{p'} + \ell_{q'} + (M_{\ell_B} - m)\ell_B}$.

Indistinguishability Between Injective and Lossy Functions. This immediately follows from the m-ad-DRSA assumption. Indeed, clearly we have $P_{\mathsf{inj}} = P_0$ and $P_{\mathsf{lossy}} = P_1$ where P_0 and P_1 are defined in Definition 3, and the m-ad-DRSA assumption requires $|P_0 - P_1| < \epsilon(\lambda)$ for sufficiently large λ.

As an analogue of all-but-one lossy trapdoor function defined in [27], we can define adversary-dependent all-but-one lossy trapdoor functions (ad-ABO). The definition and constructions are given in the full version.

5   Applications

Here we discuss applications of ad-LTDFs. As mentioned before, ad-LTDFs can replace LTDFs in many applications. Informally, ad-LTDFs can replace LTDFs if a lossy function is used only in the security proof and not used in the real protocol. In such cases, a lossy function may depend on an adversary that tries to distinguish it from an injective function since an adversary is firstly fixed in security proofs. As a result, we can immediately obtain a collision resistant hash function [27], a CPA secure PKE scheme [27] and a DE scheme [5] based on ad-LTDFs by simply replacing LTDFs by ad-LTDFs. Among them, by using our ad-LTDF based on the factoring assumption given in Sect. 4, we obtain the first DE scheme that satisfies the PRIV security for block-sources defined in [5] under the factoring assumption.

5.1   Collision Resistant Hash Function

Here, we give an analogue of the collision resistant hash function in [27] based on ad-LTDFs. In fact, our scheme is obtained by simply replacing LTDFs in the scheme in [27] by ad-LTDFs. The concrete construction is as follows. Let $(\mathsf{ParamsGen}, \mathsf{SampleInj}, \mathsf{SampleLossy}, \mathsf{Evaluation}, \mathsf{Inversion})$ be an (n, k)-ad-LTDF and $\mathcal{H}$ be a family of pairwise independent hash functions from $\{0,1\}^k$ to $\{0,1\}^{\kappa n}$ where $\kappa := 2\rho + \delta$, $\rho < 1/2$ is a constant that satisfies $n - k \leq \rho n$ and δ is some constant in $(0, 1 - 2\rho)$,

$\mathsf{Gen}_{\mathsf{crh}}(1^\lambda)$: Run $(PP, SP) \leftarrow \mathsf{ParamsGen}(1^\lambda)$ and $\sigma \leftarrow \mathsf{SampleInj}(PP)$, and choose $H \xleftarrow{\$} \mathcal{H}$. Output a function description $h := (H, PP, \sigma)$.
$\mathsf{Evaluation}_{\mathsf{crh}}((H, PP, \sigma), x)$: Compute $H(\mathsf{Evaluation}(PP, \sigma, x))$ and output it.

Theorem 3. *The above hash function is collision resistant.*

We omit the proof since this can be proven by modifying the proof in [27] in a similar way as in Sect. 5.3.

5.2 CPA Secure Public Key Encryption

Here, we give an analogue of the CPA secure PKE scheme in [27] based on ad-LTDFs. In fact, our scheme is obtained by simply replacing LTDFs in the scheme in [27] by LTDFs. The concrete construction is as follows.

Let (ParamsGen, SampleInj, SampleLossy, Evaluation, Inversion) be an (n, k)-ad-LTDF and $\mathcal{H}$ be a family of pairwise independent hash functions from $\{0,1\}^n$ to $\{0,1\}^\ell$, where $\ell \leq k - 2\log(1/\delta)$ for some negligible δ. The construction of our scheme PKE = (Gen, Enc, Dec) is as follows.

Key Generation: $\mathsf{Gen}(1^\lambda)$ generates $(PP, SP) \leftarrow \mathsf{ParamsGen}(1^\lambda)$ and $\sigma \leftarrow \mathsf{SampleInj}(PP)$. It also chooses a hash function $H \xleftarrow{\$} \mathcal{H}$. It outputs a public key $PK = (PP, \sigma, H)$ and a secret key $SK = (SP, H)$.

Encryption: Enc takes as input a public key $PK = (PP, \sigma, H)$ and a message $msg \in \{0,1\}^\ell$. It chooses $x \xleftarrow{\$} \{0,1\}^n$, sets $C_1 := \mathsf{Evaluation}(PP, \sigma, x)$ and $C_2 := msg \oplus H(x)$ and outputs $C = (C_1, C_2)$

Decryption: Dec takes as input a secret key $SK = (SP, H)$ and a ciphertext $C = (C_1, C_2)$, computes $x := \mathsf{Inversion}(SP, \sigma, C_1)$ and $msg := C_2 \oplus H(x)$, and outputs msg.

Theorem 4. *The above scheme is CPA secure.*

This can be proven by modifying the proof in [27] in a similar way as in Sect. 5.3. The proof is given in the full version.

Remark 7. *If we use ad-ABO given in the full version, we can construct CCA secure PKE scheme similarly as in [27].*

5.3 Deterministic Encryption

Here, we construct a DE scheme based on ad-LTDFs. The construction is a simple analogue of the scheme in [5] based on LTDFs. Indeed, our scheme is obtained by simply replacing LTDFs by ad-LTDFs in their scheme. The concrete construction is as follows. Let (ParamsGen, SampleInj, SampleLossy, Evaluation, Inversion) be an (n, k)-ad-LTDF and $\mathcal{H}$ be a family of pairwise independent permutations on $\{0,1\}^n$, where $u \geq n - k + 2\log(1/\delta) - 2$ holds for some negligible δ. The construction of our scheme DE = (Gen, Enc, Dec) is as follows.

$\mathsf{Gen}(1^\lambda)$: Generate $(PP, SP) \leftarrow \mathsf{ParamsGen}(1^\lambda)$ and $\sigma \leftarrow \mathsf{SampleInj}(PP)$ and choose $H \xleftarrow{\$} \mathcal{H}$. Output a public key $PK = (PP, \sigma, H)$ and a secret key $SK = (SP, \sigma)$.

$\mathsf{Enc}(PK = (PP, \sigma, H), msg)$: Compute $C \leftarrow \mathsf{Evaluation}(PP, \sigma, H(msg))$ and output C.

$\mathsf{Dec}(SK, C)$: Compute $msg' \leftarrow \mathsf{Inversion}(SP, \sigma, C)$ and $msg := H^{-1}(msg')$ and output msg.

Theorem 5. *The above scheme is PRIV1-IND-CPA secure deterministic encryption for (u, n)-sources.*

Proof. Assume that the above scheme is not PRIV1-IND-CPA secure. There exists (u, n)-sources M_0, M_1 and a PPT adversary $\mathcal{A}$ such that $\mathsf{Adv}_{\mathcal{A},\mathsf{DE}}^{\mathsf{PRIV-IND-CPA}}(\lambda)$ is non-negligible. Then there exist a polynomial poly such that for infinitely many λ, $\mathsf{Adv}_{\mathcal{A},\mathsf{DE}}^{\mathsf{PRIV-IND-CPA}}(\lambda) > 1/\mathsf{poly}(\lambda)$ holds. We consider the following sequence of games.

Game 1: This game is the original PRIV1-IND-CPA game with respect to M_0, M_1 and $\mathcal{A}$. That is, a challenger computes $(PP, SP) \leftarrow \mathsf{ParamsGen}(1^\lambda)$ and $\sigma \leftarrow \mathsf{SampleInj}(PP)$, chooses $H \leftarrow \mathcal{H}$, sets $PK := (PP, \sigma, H)$, chooses $b \xleftarrow{\$} \{0,1\}$, $msg^* \xleftarrow{\$} M_b$ and computes $C^* \leftarrow \mathsf{Evaluation}(PP, \sigma, H(msg^*))$. $\mathcal{A}$ is given (PK, C^*) and outputs b'.

Game 2: This game is the same as the previous game except that σ is generated by $\mathsf{SampleLossy}(PP, I)$, where intuitively, I is an index such that "it is difficult to distinguish an injective function from a lossy function with index I for $\mathcal{A}$". To describe this precisely, we consider the following PPT algorithm $\mathcal{B}$.
$\mathcal{B}(PP, \sigma)$: Choose $H \xleftarrow{\$} \mathcal{H}$, $b \xleftarrow{\$} \{0,1\}$, $msg^* \xleftarrow{\$} M_b$, set $PK := (PP, \sigma, H)$, compute $C^* \leftarrow \mathsf{Evaluation}(PP, \sigma, H(msg^*))$, run $b' \leftarrow \mathcal{A}(PK, C^*)$ and output 1 if $b = b'$, and otherwise 0.
Let $\mathcal{S}_{\mathcal{B}, 1/(2\mathsf{poly})}$ be the algorithm that is assumed to exists in the definition of ad-LTDFs. (Note that $\mathcal{B}$ is a PPT algorithm and $1/(2\mathsf{poly})$ is noticeable.) In this game, we let $I \leftarrow \mathcal{S}_{\mathcal{B}, 1/(2\mathsf{poly})}(PP)$ and $\sigma \leftarrow \mathsf{SampleLossy}(PP, I)$.

Game 3: This game is the same as the previous game except that a challenge ciphertext is set as $C^* \leftarrow \mathsf{Evaluation}(PP, \sigma, H(U))$ where $U \in \{0,1\}^n$ is a uniformly random string.

Let T_i be the event that $b = b'$ in Game i. Clearly we have $|\Pr[T_1] - 1/2| = \mathsf{Adv}_{\mathcal{A},\mathsf{DE}}^{\mathsf{PRIV-IND-CPA}}(\lambda)$. Then we prove the following lemmas.

Lemma 9. *For sufficiently large any λ, we have $|\Pr[T_2] - \Pr[T_1]| \leq 1/(2\mathsf{poly}(\lambda))$.*

Proof. By the definition of an adversary-dependent lossy trapdoor function, if we let

$$P_{\mathsf{inj}} := \Pr\left[1 \leftarrow \mathcal{B}(PP, \sigma) : \begin{array}{l}(PP, SP) \leftarrow \mathsf{ParamsGen}(1^\lambda) \\ \sigma \leftarrow \mathsf{SampleInj}(PP)\end{array}\right]$$

$$P_{\mathsf{lossy}} := \Pr\left[1 \leftarrow \mathcal{B}(PP, \sigma) : \begin{array}{l}(PP, SP) \leftarrow \mathsf{ParamsGen}(1^\lambda) \\ I \leftarrow \mathcal{S}_{\mathcal{B}, 1/(2\mathsf{poly})}(PP) \\ \sigma \leftarrow \mathsf{SampleLossy}(PP, I)\end{array}\right]$$

then we have $|P_{\mathsf{inj}} - P_{\mathsf{lossy}}| \leq 1/(2\mathsf{poly})$ for sufficiently large λ. It is clear that $P_{\mathsf{inj}} = \Pr[T_1]$ and $P_{\mathsf{lossy}} = \Pr[T_2]$ holds. Therefore the lemma follows.

Lemma 10. *We have $|\Pr[T_3] - \Pr[T_2]| \leq \delta$.*

In Lemma 3, we let $f := \mathsf{Evaluation}(PP, \sigma, \cdot)$, $X := msg^*$ and $Y := (PP, \sigma)$. Then by the lossiness, $|S| \leq 2^{n-k}$ holds where S is the range of f. By the definition of (u, n)-sources, we have $\tilde{H}_\infty(X|Y) \geq u$ and $u \geq n - k + 2\log(1/\delta) - 2 \geq |S| + 2\log(1/\delta) - 2$. By Lemma 3, the statistical distance between $(C^*, H, (PP, \sigma))$ in Game 2 and that in Game 3 is at most δ. Thus the lemma follows.

Lemma 11. *We have* $\Pr[T_3] = 1/2$.

Proof. In Game 3, $\mathcal{A}$ is given no information about b. Therefore the probability that $\mathcal{A}$ can correctly guess b is $1/2$.

By combining these lemmas, for all sufficiently large λ, we have $|\Pr[T_1] - 1/2| \leq 1/(2\mathsf{poly}(\lambda)) + \delta$, equivalently, $\mathsf{Adv}_{\mathcal{A},\mathsf{DE}}^{\mathsf{PRIV-IND-CPA}}(\lambda) \leq 1/(2\mathsf{poly}(\lambda)) + \delta$. On the other hand, we assumed, $\mathsf{Adv}_{\mathcal{A},\mathsf{DE}}^{\mathsf{PRIV-IND-CPA}}(\lambda) > 1/\mathsf{poly}(\lambda)$ for infinitely many λ. Combining these two inequalities, we have $1/(2\mathsf{poly}(\lambda)) < \delta$ for infinitely many λ, which contradicts to that δ is negligible. Therefore there does not exist a PPT adversary that breaks the scheme.

Remark 8. *If we use ad-ABO given in the full version, we can construct PRIV1-IND-CCA secure DE scheme similarly as in [5].*

6   CCA Secure PKE with Short Ciphertext

In this section, we construct a CCCA secure KEM under the m-ad-DRSA assumption. By Theorem 1, under certain condition, this scheme is CCCA secure under the factoring assumption w.r.t. SS moduli. By setting a parameter appropriately, we obtain a PKE scheme whose ciphertext overhead is minimum among schemes that are CCA secure under the factoring assumption by combining our KEM and an authenticated symmetric key encryption scheme.

6.1   Construction

Idea of Our Construction. Since the m-ad-DRSA assumption is a type of subgroup decision assumptions, we can consider an "adversary-dependent version" of hash proof systems as in [8], where it is shown that a hash proof system can be constructed based on any subgroup decision assumption. Then we construct a KEM similarly as in [16], where the authors constructed a CCCA secure KEM based on a hash proof system. Though our construction is based on the above idea, for clarity, we give a direct construction of our KEM rather than defining the "adversary-dependent version" of hash proof systems.

The construction of our scheme $\mathsf{KEM}_{\mathsf{CCCA}}$ is as follows. Let IGen be a PPT algorithm that generates (ℓ_B, t_p, t_q)-SS RSA modulus, $\mathcal{H}$ be a family of pairwise independent hash functions from $(\mathbb{Z}_N^*)^n$ to $\{0,1\}^\lambda$ where $n := \lceil \frac{(2\ell_N+1)\lambda}{\ell_B-1} \rceil$, and $h : G \to \{0,1\}^\lambda$ be a target collision resistant hash function. For simplicity, we assume that the KEM key length is equal to the security parameter λ.

$\mathsf{Gen}(1^\lambda)$: Generate $(N, P, Q) \leftarrow \mathsf{IGen}(1^\lambda)$. Choose $H \stackrel{\$}{\leftarrow} \mathcal{H}$, $g \stackrel{\$}{\leftarrow} \mathbb{QR}_N$ and $x_{i,j}^{(k)} \stackrel{\$}{\leftarrow} [(N-1)/4]$ and set $X_{i,j}^{(k)} := g^{x_{i,j}^{(k)}}$ for $i = 1, \ldots, \lambda$, $j = 1, \ldots, n$ and $k = 0, 1$. Output $PK := (N, h, H, \{X_{i,j}^{(k)}\}_{i \in [\lambda], j \in [n], k \in \{0,1\}})$ and $SK := (\{x_{i,j}^{(k)}\}_{i \in [\lambda], j \in [n], k \in \{0,1\}}, PK)$.

Enc(PK): Choose $r \xleftarrow{\$} [(N-1)/4]$, compute $C := g^r$, $t := h(C)$ and $K := H((\prod_{i=1}^{\lambda} X_{i,1}^{(t_i)})^r, \ldots, (\prod_{i=1}^{\lambda} X_{i,n}^{(t_i)})^r)$ where t_i denotes the i-th bit of t. Output (C, K).

Dec(SK, C): Compute $t := h(C)$ and $K := H(C^{\sum_{i=1}^{\lambda} x_{i,1}^{(t_i)}}, \ldots, C^{\sum_{i=1}^{\lambda} x_{i,n}^{(t_i)}})$ where t_i denotes the i-th bit of t, and output K.

6.2 Security

Theorem 6. *If m-ad-DRSA assumption holds with respect to IGen and $(\ell_B - 1)(t_p + t_q + m - M_{\ell_B}) \geq \lambda$ holds, then $\mathsf{KEM}_{\mathsf{CCCA}}$ is CCCA secure.*

Corollary 2. *If the factoring assumption holds with respect to IGen for the parameter setting given in Sect. 2.4, then $\mathsf{KEM}_{\mathsf{CCCA}}$ is CCCA secure for $n = O(\lambda^4/\log(\lambda))$.*

Proof (of Corollary 2). Let $m := \lfloor M_{\ell_B} + 1 - (1/2 + c)\frac{\ell_N}{(\ell_B - 1)} \rfloor$ for a constant $c < 1/4$. Then we have $(M_{\ell_B} - m + 1)(\ell_B - 1) \geq (1/2 + c)\ell_N$ and therefore the m-ad-DRSA assumption holds under the factoring assumption by Theorem 1. Moreover, we have $(\ell_B - 1)(t_p + t_q + m - M_{\ell_B}) = O(\lambda^3)$ if we use the parameter setting given in Sect. 2.4. Thus the obtained scheme is CCCA secure under the factoring assumption. □

Theorem 6 can be proven almost similarly as the security of the CCCA secure KEM based on a hash proof system in [16]. However, for a technical reason, we need the following variant of the leftover hash lemma unlike in [16]. Specifically, in the leftover hash lemma (Lemma 2), a random variable X should be independent from H. On the other hand, in our proof, we need a variant in which a random variable X may depend on H. The following lemma states that this is possible with the loss of the number of possible random variables X. We note that this idea is already used in some existing works [32,35].

Lemma 12. *Let $\mathcal{X}$ be a set of random variables X on $\{0,1\}^{n_1}$ such that $H_\infty(X) \geq n_2 + 2\log(1/\delta)$, and $\mathcal{H}$ be a family of pairwise independent hash functions from $\{0,1\}^{n_1}$ to $\{0,1\}^{n_2}$. Then for any computationally unbounded algorithm $\mathcal{F}$, which is given $H \in \mathcal{H}$ and outputs a description of a distribution $X \in \mathcal{X}$, we have $\Delta((H(X), H), (U, H)) \leq |\mathcal{X}|\delta$ where $H \xleftarrow{\$} \mathcal{H}$ and $X \leftarrow \mathcal{F}(H)$.*

The proof is given in the full version.

Then we give the proof of Theorem 6.

Proof (of Theorem 6). Assume that there exists a valid PPT adversary $\mathcal{A}$ that breaks the CCCA security of the above scheme. Then there exists a polynomial poly such that $\mathsf{Adv}_{\mathcal{A},\mathsf{KEM}_{\mathsf{CCCA}}}^{\mathsf{CCCA}}(\lambda) > 1/\mathsf{poly}(\lambda)$ for infinitely many λ. We consider the following sequence of games.

Game 1: This game is the original CCCA game of $\mathsf{KEM}_{\mathsf{CCCA}}$ for $\mathcal{A}$. That is, a challenger $\mathcal{C}$ generates $(N, P, Q) \leftarrow \mathsf{IGen}(1^\lambda)$, chooses $H \xleftarrow{\$} \mathcal{H}$, $g \xleftarrow{\$} \mathbb{QR}_N$ and

$x_{i,j}^{(k)} \xleftarrow{\$} [(N-1)/4]$ and sets $X_{i,j}^{(k)} := g^{x_{i,j}^{(k)}}$ for $i = 1, \ldots, \lambda$, $j = 1, \ldots, n$ and $k = 0, 1$ and sets $PK := (N, h, H, \{X_{i,j}^{(k)}\}_{i \in [\lambda], j \in [n], k \in \{0,1\}})$. Then it chooses $b \xleftarrow{\$} \{0,1\}$ and $r^* \xleftarrow{\$} [(N-1)/4]$, and computes $C^* := g^{r^*}$, $t^* := h(C^*)$ and $K^* := H((\prod_{i=1}^{\lambda} X_{i,1}^{(t_i^*)})^{r^*}, \ldots, (\prod_{i=1}^{\lambda} X_{i,n}^{(t_i^*)})^{r^*})$ where t_i^* denotes the i-th bit of t^* if $b = 1$ and $K^* \xleftarrow{\$} \{0,1\}^\lambda$ otherwise. Then it gives (PK, C^*, K^*) to $\mathcal{A}$. In the game, $\mathcal{A}$ can query pairs of ciphertexts and predicates to an oracle $\mathcal{O}_{\mathsf{Dec}}$. When $\mathcal{A}$ queries (C, pred), $\mathcal{O}_{\mathsf{Dec}}$ computes $K \leftarrow \mathsf{Dec}(SK, C)$ and returns K to $\mathcal{A}$ if $C \neq C^*$ and $\mathsf{pred}(K) = 1$, and otherwise $\bot$. Finally, $\mathcal{A}$ outputs a bit b'.

Game 2: This game is the same as the previous game except that K^* is set differently if $b = 1$. Specifically, it is set as $K^* := H(C^{* \sum_{i=1}^{\lambda} x_{i,1}^{(t_i^*)}}, \ldots, C^{* \sum_{i=1}^{\lambda} x_{i,n}^{(t_i^*)}})$ if $b = 1$.

Game 3: This game is the same as the previous game except that C^* is set differently. Specifically, it is uniformly chosen from $\mathbb{QR}_N$.

Game 4: This game is the same as the previous game except that g is uniformly chosen from a subgroup S of $\mathbb{QR}_N$, which is defined as follows. First, we define a PPT algorithm $\mathcal{B}$ as follows.

$\mathcal{B}(N, g)$: Choose $H \xleftarrow{\$} \mathcal{H}$, $x_{i,j}^{(k)} \xleftarrow{\$} [(N-1)/4]$ and set $X_{i,j}^{(k)} := g^{x_{i,j}^{(k)}}$ for $i \in [\lambda], j \in [n]$ and $k = 0, 1$ and $PK := (N, h, H, \{X_{i,j}^{(k)}\}_{i \in [\lambda], j \in [n], k \in \{0,1\}})$, choose $C^* \xleftarrow{\$} \mathbb{QR}_N$ and $b \xleftarrow{\$} \{0,1\}$, and set $K^* := H(C^{* \sum_{i=1}^{\lambda} x_{i,1}^{(t_i^*)}}, \ldots, C^{* \sum_{i=1}^{\lambda} x_{i,n}^{(t_i^*)}})$ where $t^* := h(C^*)$ and t_i^* is the i-th bit of t^* if $b = 1$, and $K^* \xleftarrow{\$} \{0,1\}^\ell$ otherwise. Run $b' \leftarrow \mathcal{A}^{\mathcal{O}_{\mathsf{Dec}}}(PK, C^*, K^*)$ and output b'. We note that $\mathcal{B}$ can simulate $\mathcal{O}_{\mathsf{Dec}}$ for $\mathcal{A}$ since it knows $SK = (\{x_{i,j}^{(k)}\}_{i \in [\lambda], j \in [n], k \in \{0,1\}}, PK)$.

Let $\mathcal{S}_{\mathcal{B}, \mathsf{poly}/2}$ be the algorithm that is assumed to exist in the definition of m-ad-DRSA assumption. Note that this algorithm actually exists since $\mathcal{B}$ is a PPT algorithm and $\mathsf{poly}/2$ is noticeable. Then we define the subgroup S as follows: We run $\{p_1, \ldots, p_m\} \leftarrow \mathcal{S}_{\mathcal{B}, \mathsf{poly}/2}$ and define $S := \{h^{p_1, \ldots, p_m} : h \in \mathbb{QR}_N\}$.

Game 5: This game is the same as the previous game except that the decryption oracle $\mathcal{O}_{\mathsf{Dec}}$ is replaced with an alternative decryption oracle $\mathcal{O}_{\mathsf{Dec}'}$ that works as follows: $\mathcal{O}_{\mathsf{Dec}'}$, given C and pred, computes $t := h(C)$ and returns $\bot$ if $t = t^*$. Otherwise it computes $K := H(C^{\sum_{i=1}^{\lambda} x_{i,1}^{(t_i)}}, \ldots, C^{\sum_{i=1}^{\lambda} x_{i,n}^{(t_i)}})$ and outputs K if $\mathsf{pred}(K) = 1$, and otherwise $\bot$.

Game 6: This game is the same as the previous game except that $x_{i,j}^{(k)}$ is set differently. Specifically, it is uniformly chosen from $\mathsf{ord}(\mathbb{QR}_N)$ instead of from $[(N-1)/4]$ for $i = 1, \ldots, \lambda$, $j = 1, \ldots, n$ and $k = 0, 1$.

Game 7: This game is the same as the previous game except that the decryption oracle $\mathcal{O}_{\mathsf{Dec}'}$ is replaced with an alternative decryption oracle $\mathcal{O}_{\mathsf{Dec}''}$ that works as follows: $\mathcal{O}_{\mathsf{Dec}''}$, given C and pred, computes $t := h(C)$ and returns $\bot$ if $t = t^*$ or $C \notin S$, where S is the group defined in Game 4. Otherwise it computes $K := H(C^{\sum_{i=1}^{\lambda} x_{i,1}^{(t_i)}}, \ldots, C^{\sum_{i=1}^{\lambda} x_{i,n}^{(t_i)}})$ and outputs K if $\mathsf{pred}(K) = 1$, and otherwise $\bot$.

Game 8: This game is the same as the previous game except that K^* is always an independently random string.

Let T_i be the event that $b = b^*$ holds in Game i. Then clearly we have $\mathrm{Adv}_{\mathcal{A},\mathsf{PKE}_{\mathsf{CCCA}}}^{\mathsf{CCCA}} = |\Pr[T_1] - 1/2|$. First, we prove that the group S defined in Game 4 is a proper subgroup of $\mathbb{QR}_N$. Moreover, we prove that $\mathrm{ord}(S)/\mathrm{ord}(\mathbb{QR}_N) \leq 2^{-\lambda}$ holds. By the definition of SS moduli, $\mathrm{ord}(\mathbb{QR}_N)$ has $t_p + t_q$ distinct prime factors $p'_1, \ldots, p'_{t_p+t_q}$ of ℓ_B-bit. Since the number of the all ℓ_B-bit primes is M_{ℓ_B}, there exist at least $t_p + t_q + m - M_{\ell_B}$ distinct primes contained in both $\{p'_1, \ldots, p'_{t_p+t_q}\}$ and $\{p_1, \ldots, p_m\}$. We denote those primes by $p''_1, \ldots p''_{t_p+t_q+m-M_{\ell_B}}$. Those primes cannot be a factor of $\mathrm{ord}(S)$ since $S = \{h^{p_1 \cdots p_m} : h \in \mathbb{QR}_N\}$ by the definition whereas they are a factor of $\mathrm{ord}(\mathbb{QR}_N)$. Thus $\mathrm{ord}(S)/\mathrm{ord}(\mathbb{QR}_N) \leq \frac{1}{p''_1 \cdots p''_{t_p+t_q+m-M_{\ell_B}}} \leq 2^{-(t_p+t_q+m-M_{\ell_B})(\ell_B-1)} \leq 2^{-\lambda}$. Then we prove the following lemmas.

Lemma 13. $\Pr[T_2] = \Pr[T_1]$ *holds.*   □

Proof. The modification between Game 1 and 2 is only conceptual.   □

Lemma 14. $|\Pr[T_3] - \Pr[T_2]|$ *is negligible.*

Proof. This follows from the fact that the statistical distance between the uniform distributions on $[(N-1/4)]$ and $[\mathrm{ord}(\mathbb{QR}_N)]$ are negligible.   □

Lemma 15. *We have* $|\Pr[T_4] - \Pr[T_3]| \leq 1/(2\mathsf{poly})$ *for sufficiently large* λ.

Proof. This follows immediately from the definition of m-ad-DRSA assumption.   □

Lemma 16. *If* h *is collision resistant, then* $|\Pr[T_5] - \Pr[T_4]|$ *is negligible.*

Proof. From the view of $\mathcal{A}$, Game 4 and 5 may differ only if $\mathcal{A}$ makes a query (C, pred) such that $h(C) = t^*$. If $\mathcal{A}$ makes such a query, then this means that it finds a collision of h.   □

Lemma 17. $|\Pr[T_6] - \Pr[T_5]|$ *is negligible.*

Proof. This follows from the fact that the statistical distance between the uniform distributions on $[(N-1/4)]$ and $[\mathrm{ord}(\mathbb{QR}_N)]$ are negligible.   □

Lemma 18. $|\Pr[T_7] - \Pr[T_6]|$ *is negligible.*

Proof. Let q be an upper bound of the number of decryption queries $\mathcal{A}$ makes. We consider hybrids $\mathsf{H}_0, \ldots, \mathsf{H}_q$ that are defined as follows. A hybrid H_ℓ is the same as Game 6 except that the oracle to which $\mathcal{A}$ accesses works similarly as $\mathcal{O}_{\mathsf{Dec}''}$ for the first ℓ queries, and similarly as $\mathcal{O}_{\mathsf{Dec}'}$ for the rest of queries. Let $T_{6,\ell}$ be the event that $b = b'$ holds in the hybrid H_ℓ. Clearly, We have $\Pr[T_{6,0}] = \Pr[T_6]$ and $\Pr[T_{6,\ell}] = \Pr[T_7]$. Let F_ℓ be the event that $\mathcal{O}_{\mathsf{Dec}''}$ returns $\bot$ for $\mathcal{A}$'s ℓ-th query $(C_\ell, \mathsf{pred}_\ell)$ but $\mathcal{O}_{\mathsf{Dec}'}$ does not return $\bot$ for it. That is, F_ℓ is the event

that $C_\ell \in \mathbb{QR}_N \setminus S$, $t \neq t^*$ and $\mathsf{pred}(K_\ell) = 1$ hold where $K_\ell := H(C_\ell^{\sum_{i=1}^\lambda x_{i.1}^{(t_i)}},$
$\ldots, C_\ell^{\sum_{i=1}^\lambda x_{i.n}^{(t_i)}})$, $t := h(C_\ell)$ and t_i denotes the i-th bit of t. Unless F_ℓ occurs, hybrids H_ℓ and $\mathsf{H}_{\ell-1}$ are exactly the same from the view of $\mathcal{A}$. Therefore we have $|\Pr[T_{6,\ell}] - \Pr[T_{6,\ell-1}]| \leq \Pr[F_\ell]$. Let view_ℓ be the view from $\mathcal{A}$ before it is given the response for its ℓ-th query. That is, view_ℓ consists of PK, C^*, K^*, C_ℓ and decryption queries and responses for them before the ℓ-th query. We prove the following claim.

Claim 3. *If $C_\ell \in \mathbb{QR}_N \setminus S$ and $t \neq t^*$, then K_ℓ is distributed almost uniformly on $\{0,1\}^\lambda$ from the view of $\mathcal{A}$ in the hybrids $\mathsf{H}_{\ell-1}$ and H_ℓ. More precisely, we have $\Delta((K_\ell, \mathsf{view}_\ell), (U, \mathsf{view}_\ell)) \leq 2^{-\lambda}$ where $U \xleftarrow{\$} \{0,1\}^\lambda$.*

Assume this claim is true. Then we prove that $\Pr[F_\ell]$ is negligible for any $\ell \in [q]$. Since $\mathcal{A}$ is valid, pred is non-trivial. That is, for independently uniform U, $\Pr[\mathsf{pred}_i(U) = 1]$ is negligible. By Claim 3, if $C_\ell \in \mathbb{QR}_N \setminus S$ and $t \neq t^*$, then we have $\Delta((K_\ell, \mathsf{view}), (U, \mathsf{view})) \leq 2^{-\lambda}$ where $U \xleftarrow{\$} \{0,1\}^\lambda$. Therefore $\Pr[\mathsf{pred}(K_\ell) = 1]$ is negligible and thus $\Pr[F_\ell]$ is negligible. Thus $|\Pr[T_7] - \Pr[T_6]|$ is negligible by the hybrid argument. What is left is to prove Claim 3.

Proof (of Claim 3). Since we assumed $t \neq t^*$, there exists $i \in [\lambda]$ such that $t_i \neq t_i^*$. We denote minimum such i by i^*. Since $C \in \mathbb{QR}_N \setminus S$ and S is a proper subgroup of $\mathbb{QR}_N$, there exists an ℓ_B-bit prime $\bar{p}$ that divides $\mathsf{ord}(C)$ but does not divide $\mathsf{ord}(S)$. Here, we claim that the decryption oracle before ℓ-th query can be simulated by using $\{x_{i,j}^{(k)} \bmod \mathsf{ord}(S)\}_{i \in [\lambda], j \in [n], k \in \{0,1\}}$ and PK. This can be seen by that the oracle immediately returns $\bot$ for a query (C, pred) such that $C \notin S$. If we define $\mathsf{view}_\ell' := (\overline{PK}, C^*, K^*, C_\ell, \{x_{i,j}^{(k)} \bmod \mathsf{ord}(S)\}_{i \in [\lambda], j \in [n], k \in \{0,1\}})$ where $\overline{PK}$ denotes a public key except H, then we have $\Delta((K_\ell, \mathsf{view}_\ell), (U, \mathsf{view}_\ell)) \leq \Delta((K_\ell, H, \mathsf{view}_\ell'), (U, H, \mathsf{view}_\ell'))$. Thus it suffices to show that conditioned on any fixed value of view_ℓ', $\Delta((K_\ell, H), (U_\ell, H)) \leq 2^{-\lambda}$ holds. One can see that view_ℓ' does not depend on $(x_{i^*,j}^{(t_{i^*})} \bmod \bar{p})$ at all for $j \in [n]$: $\overline{PK}$ does not depend on $(x_{i^*,j}^{(t_{i^*})} \bmod \bar{p})$ since $g \in S$ by the modification from Game 3 to 4. (C^*, K^*) does not depend on $(x_{i^*,j}^{(t_{i^*})} \bmod \bar{p})$ since we assumed $t_{i^*} \neq t_{i^*}^*$ and thus $x_{i^*,j}^{(t_{i^*})}$ is not used for generating K^*. $\{x_{i,j}^{(k)} \bmod \mathsf{ord}(S)\}_{i \in [\lambda], j \in [n], k \in \{0,1\}}$ does not depend on $(x_{i^*,j}^{(t_{i^*})} \bmod \bar{p})$ since $\mathsf{ord}(S)$ is coprime to $\bar{p}$. Thus conditioned on any value of view_ℓ', $(x_{i^*,j}^{(t_{i^*})} \bmod \bar{p})$ is distributed uniformly for all $j \in [n]$. Therefore we have $H_\infty(C_\ell^{\sum_{i=1}^\lambda x_{i.1}^{(t_i)}}, \ldots, C_\ell^{\sum_{i=1}^\lambda x_{i.n}^{(t_i)}} | \mathsf{view}_\ell') \geq n \log \bar{p} \geq n(\ell_B - 1) \geq \lambda + 2\ell_N \lambda$. Here, we use Lemma 12: We set $\mathcal{X} := \{X_C\}_{C \in \mathbb{QR}_N \setminus S}$ where X_C denotes a random variable that is distributed as $(C^{\sum_{i=1}^\lambda x_{i.1}^{(t_i)}}, \ldots, C^{\sum_{i=1}^\lambda x_{i.n}^{(t_i)}})$ conditioned on view_ℓ', $\delta := 2^{-\ell_N \lambda}$, and $\mathcal{F}$ as an algorithm that simulates the game between $\mathcal{A}$ and the challenger and outputs X_{C_ℓ} where C_ℓ is $\mathcal{A}$'s ℓ-th decryption query. Then we have $\Delta((K_\ell, H), (U, H)) \leq |\mathbb{QR}_N \setminus S| 2^{-\ell_N \lambda} \leq 2^{-\lambda}$ where $U \xleftarrow{\$} \{0,1\}^\lambda$, conditioned on any fixed value of view_ℓ'. Thus the proof of Claim 3 is completed. $\square$

This concludes the proof of Lemma 18. $\square$

Lemma 19. $|\Pr[T_8] - \Pr[T_7]|$ *is negligible.*

Proof. Since we have $\Pr[C^* \in S : C^* \xleftarrow{\$} \mathbb{QR}_N] \leq 2^{-\lambda}$, in the following, we assume $C^* \notin S$. Then there exists $\bar{p}$ that divides $\mathrm{ord}(C^*)$ but does not divide $\mathrm{ord}(S)$. Let view be the view from $\mathcal{A}$ in Game 8 except K^*, and $\mathsf{view}' := \{\overline{PK}, C^*, \{x_{i,j}^{(k)}$ mod $\mathrm{ord}(S)\}_{i \in [\lambda], j \in [n], k \in \{0,1\}}\}$. By a similar argument as in the proof of Claim 3, we have $\Delta((K^*, \mathsf{view}), (U, \mathsf{view})) \leq \Delta((K^*, H, \mathsf{view}'), (U, H, \mathsf{view}'))$ and $\tilde{H}_\infty(C^{*\sum_{i=1}^{\lambda} x_{i,1}^{(t_i)}}, \ldots, C^{*\sum_{i=1}^{\lambda} x_{i,n}^{(t_i)}} |\mathsf{view}') \geq n \log \bar{p} \geq n(\ell_B - 1) \geq (2\ell_N + 1)\lambda$. If we let $X := (C^{*\sum_{i=1}^{\lambda} x_{i,1}^{(t_i)}}, \ldots, C^{*\sum_{i=1}^{\lambda} x_{i,n}^{(t_i)}})$, $Y := \mathsf{view}$ and $\delta := 2^{-\ell_N \lambda}$ in Lemma 2, then we have $\Delta((K^*, H, \mathsf{view}'), (U, H, \mathsf{view}')) \leq 2^{-\ell_N \lambda}$ where $K^* = H(C^{*x_1 + t^* y_1}, \ldots, C^{*x_n + t^* y_n})$ and $U \xleftarrow{\$} \{0,1\}^k$. Thus the lemma follows. $\square$

By the above lemmas, we have $\mathsf{Adv}_{\mathcal{A}, \mathsf{KEM}_{\mathsf{CCCA}}}^{\mathsf{CCCA}}(\lambda) = |\Pr[T_1] - \Pr[T_8]| \leq \mathsf{negl}$ $(\lambda) + 1/(2\mathsf{poly}(\lambda))$ for sufficiently large λ where negl is some negligible function. On the other hand, we assumed that there are infinitely many λ such that $\mathsf{Adv}_{\mathcal{A}, \mathsf{KEM}_{\mathsf{CCCA}}}^{\mathsf{CCCA}}(\lambda) > 1/\mathsf{poly}(\lambda)$. Therefore for infinitely many λ, we have $1/(2\mathsf{poly}(\lambda)) < \mathsf{negl}(\lambda)$, which contradicts to that $\mathsf{negl}(\lambda)$ is negligible. Thus there does not exist a valid PPT adversary that breaks the CCCA security of the scheme. $\square$

Discussion. Here, we discuss the efficiency of the CCA secure PKE scheme that is obtained by combining the above KEM and an authenticated symmetric key encryption scheme. Table 1 shows the efficiency and hardness assumption of CCA secure PKE schemes based on the factoring in the standard model. Among existing schemes, the scheme proposed by Hofheinz and Kiltz [18] is one of the best in regard to the ciphertext overhead, which consists of 2 elements of $\mathbb{Z}_N^*$. In contrast, the ciphertext overhead of our scheme consists of only 1 element of $\mathbb{Z}_N^*$ plus a MAC. By giving a concrete parameter ($\ell_p' = \ell_q' = 160$, $\ell_B = 15$, $t_p = t_q = 32$ and $\ell_N = 1280$), the ciphertext overhead of our scheme is 1360-bit for 80-bit security whereas that of [18] is 2048-bit. On the other hand, the public key size of our scheme is much larger than that of [18], and an encryption and decryption are much less efficient than those in [18].

Acknowledgment. We would like to thank the anonymous reviewers and members of the study group "Shin-Akarui-Angou-Benkyou-Kai" for their helpful comments. Especially, we would like to thank the reviewer of EUROCRYPT 2016 who suggested to use the term "adversary-dependent" instead of "generalized", and Atsushi Takayasu for giving us useful comments on the Coppersmith theorem. This work was supported by CREST, JST and JSPS KAKENHI Grant Number 14J03467.

References

1. Bellare, M., Boldyreva, A., O'Neill, A.: Deterministic and efficiently searchable encryption. In: Menezes, A. (ed.) CRYPTO 2007. LNCS, vol. 4622, pp. 535–552. Springer, Heidelberg (2007)

2. Bellare, M., Fischlin, M., O'Neill, A., Ristenpart, T.: Deterministic encryption: definitional equivalences and constructions without random oracles. In: Wagner, D. (ed.) CRYPTO 2008. LNCS, vol. 5157, pp. 360–378. Springer, Heidelberg (2008)
3. Bellare, M., Hofheinz, D., Yilek, S.: Possibility and impossibility results for encryption and commitment secure under selective opening. In: Joux, A. (ed.) EUROCRYPT 2009. LNCS, vol. 5479, pp. 1–35. Springer, Heidelberg (2009)
4. Blum, L., Blum, M., Shub, M.: A simple unpredictable pseudo-random number generator. SIAM J. Comput. **15**(2), 364–383 (1986)
5. Boldyreva, A., Fehr, S., O'Neill, A.: On notions of security for deterministic encryption, and efficient constructions without random oracles. In: Wagner, D. (ed.) CRYPTO 2008. LNCS, vol. 5157, pp. 335–359. Springer, Heidelberg (2008)
6. Coppersmith, D.: Finding a small root of a bivariate integer equation; factoring with high bits known. In: Maurer, U.M. (ed.) EUROCRYPT 1996. LNCS, vol. 1070, pp. 178–189. Springer, Heidelberg (1996)
7. Coron, J.-S., Joux, A., Mandal, A., Naccache, D., Tibouchi, M.: Cryptanalysis of the RSA subgroup assumption from TCC 2005. In: Catalano, D., Fazio, N., Gennaro, R., Nicolosi, A. (eds.) PKC 2011. LNCS, vol. 6571, pp. 147–155. Springer, Heidelberg (2011)
8. Cramer, R., Shoup, V.: Universal hash proofs and a paradigm for adaptive chosen ciphertext secure public-key encryption. In: Knudsen, L.R. (ed.) EUROCRYPT 2002. LNCS, vol. 2332, pp. 45–64. Springer, Heidelberg (2002)
9. Crandall, R., Pomerance, C.: Prime Numbers: A Computational Perspective, 2nd edn. Springer, New York (2005)
10. Dodis, Y., Ostrovsky, R., Reyzin, L., Smith, A.: Fuzzy extractors: how to generate strong keys from biometrics and other noisy data. SIAM J. Comput. **38**(1), 97–139 (2008)
11. Freeman, D.M., Goldreich, O., Kiltz, E., Rosen, A., Segev, G.: More constructions of lossy and correlation-secure trapdoor functions. In: Nguyen, P.Q., Pointcheval, D. (eds.) PKC 2010. LNCS, vol. 6056, pp. 279–295. Springer, Heidelberg (2010)
12. Goldreich, O., Levin, L.A.: A hard-core predicate for all one-way functions. In: STOC, pp. 25–32 (1989)
13. Groth, J.: Cryptography in subgroups of Z_n^*. In: Kilian, J. (ed.) TCC 2005. LNCS, vol. 3378, pp. 50–65. Springer, Heidelberg (2005)
14. Hanaoka, G., Matsuda, T., Schuldt, J.C.N.: On the impossibility of constructing efficient key encapsulation and programmable hash functions in prime order groups. In: Safavi-Naini, R., Canetti, R. (eds.) CRYPTO 2012. LNCS, vol. 7417, pp. 812–831. Springer, Heidelberg (2012)
15. Hoeffding, W.: Probability inequalities for sums of bounded random variables. J. Am. Stat. Assoc. **58**(301), 13–30 (1963)
16. Hofheinz, D., Kiltz, E.: Secure hybrid encryption from weakened key encapsulation. In: Menezes, A. (ed.) CRYPTO 2007. LNCS, vol. 4622, pp. 553–571. Springer, Heidelberg (2007)
17. Hofheinz, D., Kiltz, E.: The group of signed quadratic residues and applications. In: Halevi, S. (ed.) CRYPTO 2009. LNCS, vol. 5677, pp. 637–653. Springer, Heidelberg (2009)
18. Hofheinz, D., Kiltz, E.: Practical chosen ciphertext secure encryption from factoring. In: Joux, A. (ed.) EUROCRYPT 2009. LNCS, vol. 5479, pp. 313–332. Springer, Heidelberg (2009)
19. Kiltz, E., Mohassel, P., O'Neill, A.: Adaptive trapdoor functions and chosen-ciphertext security. In: Gilbert, H. (ed.) EUROCRYPT 2010. LNCS, vol. 6110, pp. 673–692. Springer, Heidelberg (2010)

20. Kiltz, E., O'Neill, A., Smith, A.: Instantiability of RSA-OAEP under chosen-plaintext attack. In: Rabin, T. (ed.) CRYPTO 2010. LNCS, vol. 6223, pp. 295–313. Springer, Heidelberg (2010)
21. Kiltz, E., Pietrzak, K., Stam, M., Yung, M.: A new randomness extraction paradigm for hybrid encryption. In: Joux, A. (ed.) EUROCRYPT 2009. LNCS, vol. 5479, pp. 590–609. Springer, Heidelberg (2009)
22. Lenstra Jr., H.W.: Factoring integers with elliptic curves. Ann. Math. **126**(3), 649–673 (1987)
23. Lu, X., Li, B., Liu, Y.: How to remove the exponent GCD in HK09. In: Susilo, W., Reyhanitabar, R. (eds.) ProvSec 2013. LNCS, vol. 8209, pp. 239–248. Springer, Heidelberg (2013)
24. Lu, X., Li, B., Mei, Q., Liu, Y.: Improved efficiency of chosen ciphertext secure encryption from factoring. In: Ryan, M.D., Smyth, B., Wang, G. (eds.) ISPEC 2012. LNCS, vol. 7232, pp. 34–45. Springer, Heidelberg (2012)
25. Mei, Q., Li, B., Lu, X., Jia, D.: Chosen ciphertext secure encryption under factoring assumption revisited. In: Catalano, D., Fazio, N., Gennaro, R., Nicolosi, A. (eds.) PKC 2011. LNCS, vol. 6571, pp. 210–227. Springer, Heidelberg (2011)
26. Naccache, D., Stern, J.: A new public key cryptosystem based on higher residues. In: ACM Conference on Computer and Communications Security, pp. 59–66 (1998)
27. Peikert, C., Waters, B.: Lossy trapdoor functions and their applications. In: STOC, pp. 187–196 (2008)
28. Pohlig, S.C., Hellman, M.E.: An improved algorithm for computing logarithms over gf(p) and its cryptographic significance (corresp.). IEEE Trans. Inf. Theor. **24**(1), 106–110 (1978)
29. Pollard, J.M.: Theorems of factorization and primality testing. In: Proceedings of the cambridge philosophical society, vol. 76, pp. 521–528 (1974)
30. Pollard, J.M.: A monte carlo method for factorization. BIT **15**, 331–334 (1975)
31. Pollard, J.M.: Monte Carlo methods for index computation (mod p). Math. Comput. **32**, 918–924 (1978)
32. Raghunathan, A., Segev, G., Vadhan, S.: Deterministic public-key encryption for adaptively chosen plaintext distributions. In: Johansson, T., Nguyen, P.Q. (eds.) EUROCRYPT 2013. LNCS, vol. 7881, pp. 93–110. Springer, Heidelberg (2013)
33. Regev, O.: On lattices, learning with errors, random linear codes, and cryptography. In: STOC 2005, pp. 84–93 (2005)
34. Shanks, D.: Class number, a theory of factorization, and genera. In: 1969 Number Theory Institute (Proceedings of the Symposium Pure Mathematics, vol. XX, State University New York, Stony Brook, N.Y., 1969), pp. 415–440, Providence, R.I (1971)
35. Trevisan, L., Vadhan, S.P.: Extracting randomness from samplable distributions. In: 41st Annual Symposium on Foundations of Computer Science, FOCS 2000, 12–14 November 2000, Redondo Beach, California, USA, pp. 32–42 (2000)
36. Xue, H., Li, B., Lu, X., Jia, D., Liu, Y.: Efficient lossy trapdoor functions based on subgroup membership assumptions. In: Abdalla, M., Nita-Rotaru, C., Dahab, R. (eds.) CANS 2013. LNCS, vol. 8257, pp. 235–250. Springer, Heidelberg (2013)
37. Yamakawa, T., Yamada, S., Nuida, K., Hanaoka, G., Kunihiro, N.: Chosen ciphertext security on hard membership decision groups: the case of semi-smooth subgroups of quadratic residues. In: Abdalla, M., De Prisco, R. (eds.) SCN 2014. LNCS, vol. 8642, pp. 558–577. Springer, Heidelberg (2014)

Optimal Security Proofs for Signatures from Identification Schemes

Eike Kiltz[1], Daniel Masny[1], and Jiaxin Pan[1,2(✉)]

[1] Ruhr-Universität Bochum, Bochum, Germany
{eike.kiltz,daniel.masny,jiaxin.pan}@rub.de
[2] Karlsruher Institut Für Technologie, Karlsruhe, Germany

Abstract. We perform a concrete security treatment of digital signature schemes obtained from canonical identification schemes via the Fiat-Shamir transform. If the identification scheme is random self-reducible and satisfies the weakest possible security notion (hardness of key-recoverability), then the signature scheme obtained via Fiat-Shamir is unforgeable against chosen-message attacks in the multi-user setting. Our security reduction is in the random oracle model and loses a factor of roughly Q_h, the number of hash queries. Previous reductions incorporated an additional multiplicative loss of N, the number of users in the system. Our analysis is done in small steps via intermediate security notions, and all our implications have relatively simple proofs. Furthermore, for each step, we show the optimality of the given reduction in terms of model assumptions and tightness.

As an important application of our framework, we obtain a concrete security treatment for Schnorr signatures in the multi-user setting.

Keywords: Signatures · Identification · Schnorr · Tightness

1 Introduction

CANONICAL IDENTIFICATION SCHEMES AND THE FIAT-SHAMIR TRANSFORM. A canonical identification scheme ID as formalized by Abdalla et al. [1] is a three-move public-key authentication protocol of a specific form. The prover (holding the secret-key) sends a commitment R to the verifier. The verifier (holding the public-key) returns a random challenge h, uniformly chosen from a set ChSet (of exponential size). The prover sends a response s. Finally, using the verification algorithm, the verifier publicly checks correctness of the transcript (R, h, s). There is a large number of canonical identification schemes known (e.g. [13,15,20,28,29,31,34,36,38,39,42], the most popular among them being the scheme by Schnorr [42]. The Fiat-Shamir method [20] transforms any such

E. Kiltz—Supported in part by ERC Project ERCC (FP7/615074).
D. Masny—Supported by the DFG Research Training Group GRK 1817/1.
J. Pan—Supported by the DFG Research Training Group GRK 1817/1 and by the DFG grant HO 4534/4-1.

© International Association for Cryptologic Research 2016
M. Robshaw and J. Katz (Eds.): CRYPTO 2016, Part II, LNCS 9815, pp. 33–61, 2016.
DOI: 10.1007/978-3-662-53008-5_2

canonical identification scheme into a digital signature scheme SIG[ID] using a hash function.

DIGITAL SIGNATURES IN THE MULTI-USER SETTING. When it comes to security of digital signature schemes, in the literature almost exclusively the standard security notion of unforgeability against chosen message attacks (UF-CMA) [30] is considered. This is a *single-user setting*, where an adversary obtains one single public-key and it is said to break the scheme's security if he can produce (after obtaining Q_s many signatures on messages of his choice) a valid forgery, i.e. a message-signature pair that verifies on the given public-key. However, in the real world the attacker is usually confronted with many public-keys and presumably he is happy if he can produce a valid forgery under any of the given public-keys. This scenario is captured in the *multi-user setting* for signatures schemes. Concretely, in multi-user unforgeability against chosen message attacks (MU-UF-CMA) the attacker obtains N independent public-keys and is said to break the scheme's security if he can produce (after obtaining Q_s many signatures on public-keys of his choice) a valid forgery that verifies under any of the public-keys.

There are essentially two reasons why one typically only analyzes signatures in the single-user setting. First, the single-user security notion and consequently their analysis are simpler. Second, there exists a simple generic security reduction [25] between multi-user security and standard single-user security. Namely, for any signature system, attacking the scheme in the multi-user setting with N public-keys cannot increase the attacker's success ratio (i.e., the quotient of its success probability and its running time) by a factor more than N compared to attacking the scheme in the single-user setting. As the number of public-keys N is bounded by a polynomial, asymptotically, the single-user and the multi-user setting are equivalent. However, the security reduction is not tight: it has a loss of a non-constant factor N. This is clearly not satisfactory as in complex environments one can easily assume the existence of at least $N = 2^{30}$ ($\approx$ 1 billion) public-keys, thereby increasing the upper bound on the attacker's success ratio by a factor of 2^{30}. For example, if we assume the best algorithm breaking the single-user security having success ratio $\rho = 2^{-80}$, then it can only be argued that the best algorithm breaking the multi-user security has success ratio $\rho' = 2^{-80} \cdot 2^{30} = 2^{-50}$, which is not a safe security margin that defends against today's attackers.

TIGHTNESS. Generally, we call a security implication between two problems *tight* [9], if the success ratio ρ of any adversary attacking the first problem cannot decease by more than a small constant factor compared to the success ratio ρ' of any adversary attacking the second problem [7,26]. Here the success ratio ρ is defined as the quotient between the adversary's success probability and its running time. We note that this notion of tightness is slightly weaker than requiring that both, success probability and running time, cannot decrease by more than a small constant factor (called strong tightness in [26]). However, the main goal of a concrete security analysis is to derive parameters provably guaranteeing *k-bit security*. As the term *k-bit security* is commonly defined as

PIMP-KOA and IMP-KOA security is a generalization of Seurin's impossibility result to canonical identification schemes [43]; Lemmas 11 and 12 proving the impossibility of a reduction in the non-programmable random oracle model between PIMP-KOA, UF-KOA, and UF-CMA can be considered as a fine-grained version of a general impossibility result by Fukumitsu and Hasegawa [24] who only consider the implication IMP-PA → UF-CMA; All our impossibility results assume the reductions to be key-preserving [40] and are conditional in the sense that the existence of a reduction would imply that ID does not satisfy some other natural security property (that is believed to hold).

FROM SINGLE-USER TO MULTI-USER SECURITY FOR SIGNATURES. Our second main theorem can be informally stated as follows.

Theorem 2. *If* ID *is* UF-KOA*-secure against any adversary having success ratio* ρ*, then it is* MU-UF-CMA*-secure in the random oracle model against any adversary having success ratio* $\rho' \approx \rho/4$*, independent of the number of users* N *in the multi-user scenario.*

This theorem improves the bound implied by previous generic reductions [25] by a factor of N. Following our modular approach, the theorem is proved in two steps via Lemmas 7 and 8. Lemma 7 proves that UF-KOA tightly implies MU-UF-KOA. Tightness stems from the RSR property, meaning that from a given public key pk we can derive properly distributed $pk_1, \ldots, pk_N$ such that any signature σ which is valid under pk can be transformed into a signature σ_i which is valid under pk_i and vice-versa.

Lemma 8 is our main technical contribution and proves MU-UF-KOA → MU-UF-CMA in the programmable random oracle model, again with a tight reduction. One is tempted to believe that it can be proved the same way as in the single user setting (i.e., the same way as UF-KOA → UF-CMA). In the single user setting, the reduction simulates signatures on m_j using the HVZK property to obtain a valid transcript (R_j, h_j, s_j) and programs the random oracle as $H(R_j, m_j) := h_j$. However, in the MU-UF-KOA experiment an adversary can ask for a signature under pk_1 on message m which makes the reduction program the random oracle $H(R_1, m) := h_1$. Now, if the adversary submits a forgery (R_1, s_2) under pk_2 on the same message m, the reduction cannot use this forgery to break the MU-UF-KOA experiment because the random oracle $H(R_1, m)$ was externally defined by the reduction. Hence, for the MU-UF-KOA experiment, $m, (R_1, s_2)$ does not constitute a valid forgery. In order to circumvent the above problem we make a simple probabilistic argument. In our reduction, about one half of the multi-user public-keys are coming from the MU-UF-KOA experiment, for the other half the reduction knows the corresponding secret-keys. Which secret-keys are known is hidden from the adversary's view. Now, if the multi-user adversary first obtains a signature on message m under pk_1 and then submits a forgery on the same message m under pk_2, the reduction hopes for the good case that one of the public-keys comes from the MU-UF-KOA experiment and the other one is known. This happens with probability $1/4$ which is precisely the loss of our new reduction.

1.2   Example Instantiations

SCHNORR SIGNATURES. One of the most important and signature schemes in the discrete logarithm setting is the Schnorr signature scheme [42]. It is obtained via the Fiat-Shamir transform applied to the Schnorr identification protocol. The recent expiry of the patent in 2008 has triggered a number of initiatives to obtain standardized versions of it.

Theorems 1 and 2 can be used to derive a concrete security bound for strong multi-user MU-UF-CMA-security of Schnorr signatures in the random oracle model from the DLOG problem.[2] Our reduction loses a factor of roughly Q_h, the number of random oracle queries. This improves previous bounds by a factor of N, the number of users in the system. We derive concrete example parameters for a provably secure instantiation. Figure 1 shows that DLOG is tightly equivalent to IMP-KOA-security and PIMP-KOA-security is tightly equivalent to MU-UF-CMA-security, meaning the tightness barrier for Schnorr lies precisely between IMP-KOA and PIMP-KOA security.

KATZ-WANG SIGNATURES. The Chaum-Pedersen identification scheme [19] is a double-generator version of Schnorr. It is at least as secure as Schnorr which means one cannot hope for a tight proof under the DLOG assumption. However, we can use a simple argument from [29,34] for a tight security proof of its PIMP-KOA security under the (stronger) Decision Diffie-Hellman Assumption. The resulting signature scheme is known as the Katz-Wang signature scheme [34] and our framework yields a tight proof of its strong MU-UF-CMA-security. Again, this improves previous bounds by a factor of N, the number of users in the system.

GUILLOU-QUISQUATER SIGNATURES. Another canonical identification scheme of interest with the required properties is the one by Guillou-Quisquater [31]. Similar to Katz-Wang, for the Guillou-Quisquater scheme, we can use an argument from [2] for a tight proof of PIMP-KOA security under the Phi-hiding assumption. Alternatively, we can give a proof with loss Q_h under the Factoring assumption. Our framework also shows that this loss is unavoidable. Details are shown in the full version [35].

1.3   Related Work

SINGLE-USER SECURITY. There have been many different works addressing the single-user security of Fiat-Shamir based signature schemes SIG[ID]. In pioneering work, Pointcheval and Stern [41] introduced the Forking Lemma as a tool to prove UF-CMA security of SIG[ID] from HVZK, SS and KR-KOA-security. Ohta and Okamoto [37] gave an alternative proof from IMP-KOA security and HVZK. Abdalla et al. [1] prove the equivalence of IMP-PA-security of ID and UF-CMA security of SIG[ID] in the random oracle model. All above results incorporate a

---

[2] We can even prove *strong* MU-UF-CMA security of Schnorr signatures in the sense that a new signature on a previously signed message already counts as a valid forgery.

security loss of at least Q_h and can be seen as a special case of our framework. Furthermore, [6] consider stronger security notions (e.g., IMP-AA and man-in-the middle security) for the Schnorr and GQ identification schemes. Abdalla et al. [3] show that lossy identification schemes tightly imply UF-CMA-secure signatures in the random oracle model from decisional assumptions. Our Multi-Instance Reset Lemma (Lemma 1) is a generalization to the Reset Lemma of Bellare and Palacio [6].

MULTI-USER SECURITY. To mitigate the generic security loss problem in the multi-user setting for the special case of Schnorr's signature scheme, Galbraith, Malone-Lee, and Smart (GMLS) proved [25] a tight reduction, namely that attacking the Schnorr signatures in the multi-user setting with N public-keys provably cannot decrease (by more than a small constant factor) the attacker's success ratio compared to attacking the scheme in the single-user setting. Unfortunately, Bernstein [11] recently pointed out an error in the GMLS proof leaving a tight security reduction for Schnorr signatures as an open problem. Even worse, Bernstein identifies an "apparently insurmountable obstacle to the claimed [GMLS] theorem". Section 4.3 of [11] further expands on the insurmountable obstacle. Our Theorem 2 shows there is such a tight security reduction for Schnorr signatures if one is willing to rely on the random oracle model. Additionally, in [35] we also prove an alternative tight reduction in the standard model which assumes *strong* UF-CMA security. (Schnorr is generally believed to be strongly UF-CMA secure and this is provably equivalent to UF-CMA security in the random oracle model.) Proving the original GMLS theorem (i.e., without random oracles and from standard UF-CMA security) remains an open problem.

IMPOSSIBILITY RESULTS. In terms of impossibility results, Seurin [43], building on earlier work of [27,40], proves that there is no tight reduction from the (one-more) discrete logarithm assumption to UF-KOA-security of Schnorr signatures. A more recent result by [23] even excludes a reduction from any non-interactive assumption.[3] Fukumitsu and Hasegawa [24], generalizing earlier work on Schnorr signatures [21,40], prove that SIG[ID] cannot be proved secure in the non-programmable random oracle model only assuming IMP-PA security of ID.

SCHNORR SIGNATURES VS. KEY-PREFIXED SCHNORR SIGNATURES. After identifying the error in the GMLS proof, Bernstein [11] uses the lack of a tight security reduction for Schnorr's signature scheme as a motivation to promote a "key-prefixed" modification to Schnorr's signature scheme which includes the verifier's public-key in the hash function. The EdDSA signature scheme by Bernstein et al. [12] is essentially a key-prefixing variant of Schnorr's signature scheme. (In the context of security in a multi-user setting, key-prefixing was considered before, e.g., in [14].) In [12] key-prefixing is advertized as "an inexpensive way to alleviate concerns that several public keys could be attacked simultaneously." Indeed, Bernstein [11] proves that single-user security of the original

---

[3] The main result of the published paper [23] even excludes reduction from any *interactive* assumption (with special algebraic properties), but the proof turned out to be flawed.

Schnorr signatures scheme tightly implies multi-user security of the key-prefixed variant of the scheme. That is, the key-prefixed variant has the advantage of a standard model proof of its tight multi-user security, whereas for standard Schnorr signatures one has to assume strong security or rely on the random oracle model.

The TLS standard used to secure HTTPS connections is maintained by the Internet Engineering Task Force (IETF) which delegates research questions to the Internet Research Task Force (IRTF). Cryptographic research questions are usually discussed in the Crypto Forum Research Group (CFRG) mailing list. In the last months the CFRG discussed the issue of key-prefixing.

Key-prefixing comes with the disadvantage that the entire public-key has to be available at the time of signing. Specifically, in a CFRG message from September 2015 Hamburg [32] argues "having to hold the public key along with the private key can be annoying" and "can matter for constrained devices". Independent of efficiency, we believe that a cryptographic protocol should be as light as possible and prefixing (just as any other component) should only be included if its presence is justified. Naturally, in light of the GMLS proof, Hamburg [32] and Struik [44] (among others) recommended against key prefixing for Schnorr. Shortly after, Bernstein [10] identifies the error in the GMLS theorem and posts a tight security proof for the key-prefixed variant of Schnorr signatures. In what happens next, the participant of the CFRG mailing list switched their minds and mutually agree that key-prefixing should be preferred, despite of its previously discussed disadvantages. Specifically, Brown writes about Schnorr signatures that "this justifies a MUST for inclusion of the public key in the message of the classic signature" [16]. As a consequence, key-prefixing is contained in the current draft for EdDSA [33]. In the light of our new results, we recommend to reconsider this decision.

2   Definitions

2.1   Preliminaries

For an integer p, define $[p] := \{1, \ldots, p\}$ and $\mathbb{Z}_p$ as the residual ring $\mathbb{Z}/p\mathbb{Z}$. If A is a set, then $a \xleftarrow{\$} A$ denotes picking a from A according to the uniform distribution. All our algorithms are probabilistic polynomial time unless stated otherwise. If A is an algorithm, then $a \xleftarrow{\$} A$ denotes the random variable which is defined as the output of $\mathcal{A}$ on input b. To make the randomness explicit, we use the notation $a := (A)(b; \rho)$ meaning that the algorithm is executed on input b and randomness ρ. Note that A's execution is now deterministic.

2.2   Canonical Identification Schemes

A canonical identification scheme ID is a three-move protocol of the form depicted in Fig. 2. The prover's first message R is called *commitment*, the verifier selects a uniform *challenge* h from set ChSet, and, upon receiving a *response* s from the prover, makes a deterministic decision.

Definition 1 (Canonical Identification Scheme). *A canonical identification scheme* ID *is defined as a tuple of algorithms* ID := (IGen, P, ChSet, V).

- *The key generation algorithm* IGen *takes system parameters* par *as input and returns public and secret key* (pk, sk)*. We assume that* pk *defines* ChSet*, the set of challenges.*
- *The prover algorithm* P $= (P_1, P_2)$ *is split into two algorithms.* P_1 *takes as input the secret key* sk *and returns a commitment* R *and a state* St*;* P_2 *takes as input the secret key* sk*, a commitment* R*, a challenge* h*, and a state* St *and returns a response* s*.*
- *The verifier algorithm* V *takes the public key* pk *and the conversation transcript as input and outputs a* deterministic *decision,* 1 *(acceptance) or* 0 *(rejection).*

We require that for all $(pk, sk) \in$ IGen(par)*, all* $(R, St) \in P_1(sk)$*, all* $h \in$ ChSet *and all* $s \in P_2(sk, R, h, St)$*, we have* V$(pk, R, h, s) = 1$*.*

We make a couple of useful definitions. An identification scheme ID is called *unique* if for all $(pk, sk) \in$ IGen(par), $(R, St) \in P_1(sk)$, $h \in$ ChSet, there exists at most one response $s \in \{0, 1\}^*$ such that V$(pk, R, h, s) = 1$. A *transcript* is a three-tuple (R, h, s). It is called *valid* (with respect to public-key pk) if V$(pk, R, h, s) = 1$. Furthermore, it is called *real*, if it is the output of a real interaction between prover and verifier as depicted in Fig. 2. A canonical identification schemes ID has α *bits of min-entropy*, if for all $(pk, sk) \in$ IGen(par), the commitment generated by the prover algorithm is chosen from a distribution with at least α bits of min-entropy. That is, for all strings R' we have $\Pr[R = R'] \le 2^{-\alpha}$, if $(R, St) \xleftarrow{\$} P_1(sk)$ was honestly generated by the prover.

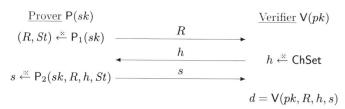

Fig. 2. A canonical identification scheme and its transcript (R, h, s).

We now define (parallel) impersonation against key-only attack (KOA), passive attack (PA), and active attack (AA).

Definition 2 ((Parallel) Impersonation). *Let* YYY $\in \{$KOA, PA, AA$\}$*. A canonical identification* ID *is said to be* $(t, \varepsilon, Q_{\mathrm{CH}}, Q_{\mathrm{O}})$*-PIMP-YYY secure (parallel impersonation against* YYY *attacks) if for all adversaries* $\mathcal{A}$ *running in time at most* t *and making at most* Q_{CH} *queries to the challenge oracle* CH *and* Q_{O} *queries to oracle* O,

$$\Pr\left[V(pk, R_{i^*}, h_{i^*}, s_{i^*}) = 1 \wedge i^* \in [Q_{\mathrm{CH}}] \;\middle|\; \begin{array}{l} (pk, sk) \xleftarrow{\boxtimes} \mathsf{IGen}(par) \\ St \xleftarrow{\boxtimes} \mathcal{A}^{\mathrm{O}(\cdot)}(pk) \\ (i^*, s_{i^*}) \xleftarrow{\boxtimes} \mathcal{A}^{\mathrm{CH}(\cdot)}(pk) \end{array}\right] \le \varepsilon,$$

where on the i-th query $\mathrm{CH}(R_i)$ $(i \in [Q_{\mathrm{CH}}])$, the challenge oracle returns $h_i \xleftarrow{\boxtimes} \mathsf{ChSet}$ to $\mathcal{A}$.[4] Depending on YYY, oracle O is defined as follows.

- If YYY = KOA (key-only attack), then O always returns $\perp$.
- If YYY = PA (passive attack), then $\mathrm{O} := \mathrm{TRAN}$, where on the j-th empty query $\mathrm{TRAN}(\epsilon)$ $(j \in Q_{\mathrm{O}})$, the transcript oracle returns a real transcript (R'_j, h'_j, s'_j) to $\mathcal{A}$, where $(R'_j, St'_j) \xleftarrow{\boxtimes} \mathsf{P}_1(sk)$, $h'_j \xleftarrow{\boxtimes} \mathsf{ChSet}$; $s'_j \xleftarrow{\boxtimes} \mathsf{P}_2(sk, R'_j, h'_j, St'_j)$.
- If YYY = AA (active attack), then $\mathrm{O} := \mathrm{PROVER} = (\mathrm{PROVER}_1, \mathrm{PROVER}_2)$, where on the j-th query $\mathrm{PROVER}_1(\epsilon)$ $(j \in Q_{\mathrm{O}})$, the prover oracle returns R'_j for $(R'_j, St'_j) \xleftarrow{\boxtimes} \mathsf{P}_1(sk)$ to $\mathcal{A}$; on query $\mathrm{PROVER}_2(j, h'_j)$, the oracle returns $s'_j \xleftarrow{\boxtimes} \mathsf{P}_2(sk, R'_j, h'_j, St'_j)$, if R'_j is already defined (and $\perp$ otherwise).

If YYY = KOA, then the parameter Q_{O} is not used and we simply speak of $(t, \varepsilon, Q_{\mathrm{CH}})$-PIMP-KOA. Moreover, $(t, \varepsilon, Q_{\mathrm{O}})$-IMP-YYY (impersonation against YYY attack) security is defined as $(t, \varepsilon, 1, Q_{\mathrm{O}})$-PIMP-YYY security, i.e., the adversary is only allowed $Q_{\mathrm{CH}} = 1$ query to the CH oracle.

Definition 3 (Key-recovery). Let YYY $\in \{\mathsf{KOA}, \mathsf{PA}, \mathsf{AA}\}$. A canonical identification ID is said to be (t, ε)-KR-YYY secure (key recovery under YYY attack) if for all adversaries $\mathcal{A}$ running in time at most t,

$$\Pr\left[(sk^*, pk) \in \mathsf{IGen}(par) \;\middle|\; \begin{array}{l} (pk, sk) \xleftarrow{\boxtimes} \mathsf{IGen}(par) \\ sk^* \xleftarrow{\boxtimes} \mathcal{A}^{\mathrm{O}(\cdot)}(pk) \end{array}\right] \le \varepsilon,$$

where depending on YYY oracle O is defined as in Definition 2. The winning condition $(sk^*, pk) \in \mathsf{IGen}(par)$ means that the tuple (sk^*, pk) is in the support of $\mathsf{IGen}(par)$, i.e., that $\mathcal{A}$ outputs a valid secret-key sk^* with respect to pk.

Definition 4 (Special Soundness). A canonical identification ID is said to be SS (special sound) if there exists an extractor algorithm Ext such that, for all $(pk, sk) \in \mathsf{IGen}(par)$, given any two accepting transcripts (R, h, s) and (R, h', s') (where $h \ne h'$), we have $\Pr[(sk^*, pk) \in \mathsf{IGen}(par) \mid sk^* \xleftarrow{\boxtimes} \mathsf{Ext}(pk, R, h, s, h', s')] = 1$.

Definition 5 (Random Self-reducibility). A canonical identification ID is said to be RSR (random self-reducible) if there is an algorithm Rerand and two deterministic algorithms Tran and Derand such that, for all $(pk, sk) \in \mathsf{IGen}(par)$:

- pk' and pk'' have the same distribution, where $(pk', \tau') \xleftarrow{\boxtimes} \mathsf{Rerand}(pk)$ is the rerandomized key-pair and $(pk'', sk'') \xleftarrow{\boxtimes} \mathsf{IGen}(par)$ is a freshly generated key-pair.

---

[4] On two queries $\mathrm{CH}(R_i)$ and $\mathrm{CH}(R_{i'})$ with the same input $R_i = R_{i'}$ the oracle returns two independent random challenges $h_i \xleftarrow{\boxtimes} \mathsf{ChSet}$ and $h_{i'} \xleftarrow{\boxtimes} \mathsf{ChSet}$.

- *For all* $(pk', \tau') \in \mathsf{Rerand}(pk)$, *all* $(pk', sk') \in \mathsf{IGen}(\mathsf{par})$, *and* $sk^* = \mathsf{Derand}(pk,$ $pk', sk', \tau')$, *we have* $(pk, sk^*) \in \mathsf{IGen}(\mathsf{par})$, *i.e.,* Derand *returns a valid secret-key* sk^* *with respect to* pk, *given any valid* sk' *for* pk'.
- *For all* $(pk', \tau') \in \mathsf{Rerand}(pk)$, *all transcripts* (R', h', s') *that are valid with respect to* pk', *the transcript* $(R', h', s := \mathsf{Tran}(pk, pk', \tau', (R', h', s')))$ *is valid with respect to* pk.

Definition 6 (Honest-verifier Zero-knowledge). *A canonical identification* ID *is said to be (perfect)* HVZK *(honest-verifier zero-knowledge) if there exists an algorithm* Sim *that, given public key* pk, *outputs* (R, h, s) *such that* (R, h, s) *is a real (i.e., properly distributed) transcript with respect to* pk.

2.3   Digital Signatures

We now define syntax and security of a digital signature scheme. Let par be common system parameters shared among all participants.

Definition 7 (Digital Signature). *A digital signature scheme* SIG *is defined as a triple of algorithms* $\mathsf{SIG} = (\mathsf{Gen}, \mathsf{Sign}, \mathsf{Ver})$.

- *The key generation algorithm* $\mathsf{Gen}(\mathsf{par})$ *returns the public and secret keys* (pk, sk).
- *The signing algorithm* $\mathsf{Sign}(sk, m)$ *returns a signature* σ.
- *The deterministic verification algorithm* $\mathsf{Ver}(pk, m, \sigma)$ *returns 1 (accept) or 0 (reject).*

We require that for all $(pk, sk) \in \mathsf{Gen}(\mathsf{par})$, *all messages* $m \in \{0, 1\}^*$, *we have* $\mathsf{Ver}(pk, m, \mathsf{Sign}(sk, m)) = 1$.

Definition 8 (Multi-user Security). *A signature scheme* SIG *is said to be* (t, ε, N, Q_s)-MU-SUF-CMA *secure (multi-user strongly unforgeable against chosen message attacks) if for all adversaries* $\mathcal{A}$ *running in time at most* t *and making at most* Q_s *queries to the signing oracle,*

$$\Pr\left[\begin{array}{l} \mathsf{Ver}(pk_{i^*}, m^*, \sigma^*) = 1 \\ \wedge \, (i^*, m^*, \sigma^*) \notin \{(i_j, m_j, \sigma_j) \mid j \in [Q_s]\} \end{array} \middle| \begin{array}{l} \textit{For } i = 1, \dots, N : (pk_i, sk_i) \stackrel{\$}{\leftarrow} \mathsf{Gen}(\mathsf{par}) \\ (i^*, m^*, \sigma^*) \stackrel{\$}{\leftarrow} \mathcal{A}^{\mathrm{SIGN}(\cdot, \cdot)}(pk_1, \dots, pk_N) \end{array} \right] \leq \varepsilon,$$

where on the j-*th query* $(i_j, m_j) \in [N] \times \{0, 1\}^*$ $(j \in [Q_s])$ *the signing oracle* SIGN *returns* $\sigma_j \stackrel{\$}{\leftarrow} \mathsf{Sign}(sk_{i_j}, m_j)$ *to* $\mathcal{A}$, *i.e., a signature on message* m_j *under public-key* pk_{i_j}.

We stress that an adversary in particular breaks multi-user security if he asks for a signature on message m under pk_1 and submits a valid forgery on the same message m under pk_2.

The first condition in the probability statement of Definition 8 is called the correctness condition, the second condition is called the freshness condition. Definition 8 covers *strong* security in the sense that a new signature on a previously queried message is considered as a fresh forgery. For standard (non-strong) MU-UF-CMA security (multi-user unforgeablility against chosen message attack) we modify the freshness condition in the experiment to $(i^*, m^*) \notin \{(i_j, m_j,) \mid j \in [Q_s]\}$, i.e., to break the scheme the adversary has to come up with a signature on a message-key pair which has not been queried to the signing oracle. We also define (t, ε, N)-MU-UF-KOA security (multi-user unforgeability against key only attack) as $(t, \varepsilon, N, 0)$-MU-UF-CMA security, i.e. $Q_s = 0$, the adversary is not allowed to make any signing query.

Definition 9 (Single-user Security). *In the single-user setting, i.e. $N = 1$ users, (t, ε, Q_s)-SUF-CMA security (strong unforgeablility against chosen message attacks) is defined as $(t, \varepsilon, 1, Q_s)$-MU-SUF-CMA security. Similarly, standard (non-strong) (t, ε, Q_s)-UF-CMA security (unforgeablility against chosen message attack) is defined as $(t, \varepsilon, 1, Q_s)$-MU-UF-CMA security. Further, (t, ε)-UF-KOA security (unforgeablility against key-only attack) is defined as $(t, \varepsilon, 1, 0)$-MU-SUF-CMA security, i.e., $N = 1$ users and $Q_s = 0$ signing queries.*

SECURITY IN THE RANDOM ORACLE MODEL. The security of identification and signature schemes containing a hash function can be analyzed in the random oracle model [8]. In this model hash values can only be accessed by an adversary through queries to an oracle H. On input x this oracle returns a uniformly random output $H(x)$ which is consistent with previous queries for input x. Using the random oracle model, the maximal number of queries to H becomes a parameter in the concrete security notions. For example, for $(t, \varepsilon, N, Q_s, Q_h)$-MU-SUF-CMA security we consider all adversaries making at most Q_h queries to the random oracle. We make the convention that each query to the random oracle made during a signing query is counted as the adversary's random oracle query, meaning $Q_h \geq Q_s$.

2.4   Signatures from Identification Schemes

Let $\mathsf{ID} := (\mathsf{IGen}, \mathsf{P}, \mathsf{ChSet}, \mathsf{V})$ be a canonical identification scheme. By the generalized Fiat-Shamir transformation [6], the signature scheme $\mathsf{SIG}[\mathsf{ID}] := (\mathsf{Gen}, \mathsf{Sign}, \mathsf{Ver})$ from ID is defined as follows. par contains the system parameters of ID and a hash function $H : \{0, 1\}^* \rightarrow \mathsf{ChSet}$.

$\mathsf{Gen}(\mathsf{par})$:	$\mathsf{Sign}(sk, m)$:	$\mathsf{Ver}(sk, m, \sigma)$:
$(pk, sk) \overset{\boxtimes}{\leftarrow} \mathsf{IGen}(\mathsf{par})$	$(R, St) \overset{\boxtimes}{\leftarrow} \mathsf{P}_1(sk)$	Parse $\sigma = (R, s)$
Return (pk, sk)	$h = H(R, m)$	$h = H(R, m)$
	$s \overset{\boxtimes}{\leftarrow} \mathsf{P}_2(sk, R, h, St)$	Return $\mathsf{V}(pk, R, h, s)$
	Return $\sigma = (R, s)$	

In some variants of the Fiat-Shamir transform, the hash additionally inputs some public parameters, for example $h = H(pk, R, m)$.

We call ID *commitment-recoverable*, if $\mathsf{V}(pk, R, h, s)$ first recomputes $R' = \mathsf{V}'(pk, h, s)$ and then outputs 1 iff $R' = R$. For commitment-recoverable

ID, we can define an alternative Fiat-Shamir transformation $\mathsf{SIG}'[\mathsf{ID}] := (\mathsf{Gen}, \mathsf{Sign}', \mathsf{Ver}')$, where Gen is as in $\mathsf{SIG}[\mathsf{ID}]$. Algorithm $\mathsf{Sign}'(sk, m)$ is defined as $\mathsf{Sign}(sk, m)$ with the modified output $\sigma' = (h, s)$. Algorithm $\mathsf{Ver}'(pk, m, \sigma')$ first parses $\sigma' = (h, s)$, then recomputes the commitment as $R' := \mathsf{V}'(pk, h, s)$, and finally returns 1 iff $H(R', m) = h$.

Since $\sigma = (R, s)$ can be publicly transformed into $\sigma' = (h, s)$ and vice-cersa, $\mathsf{SIG}[\mathsf{ID}]$ and $\mathsf{SIG}'[\mathsf{ID}]$ are equivalent in terms of security. On the one hand, the alternative Fiat-Shamir transform yields shorter signatures if $h \in \mathsf{ChSet}$ has a smaller representation size than response s. On the other hand, signatures of the Fiat-Shamir transform maintain their algebraic structure which in some cases enables useful properties such as batch verification.

3 Security Implications

In this section we will prove the following two main results.

Theorem 3 (Main Theorem 1). *Suppose* ID *is* SS, HVZK, RSR *and has* α *bit min-entropy. If* ID *is* (t, ε)-KR-KOA *secure then* $\mathsf{SIG}[\mathsf{ID}]$ *is* $(t', \varepsilon', Q_s, Q_h)$-UF-CMA-*secure and* $(t'', \varepsilon'', N, Q_s, Q_h)$-MU-UF-CMA-*secure in the programmable random oracle model, where*

$$\frac{\varepsilon'}{t'} \le 6(Q_h + 1) \cdot \frac{\varepsilon}{t} + \frac{Q_s}{2^\alpha} + \frac{1}{|\mathsf{ChSet}|},$$

$$\frac{\varepsilon''}{t''} \le 24(Q_h + 1) \cdot \frac{\varepsilon}{t} + \frac{Q_s}{2^\alpha} + \frac{1}{|\mathsf{ChSet}|}.$$

The proof of Theorem 3 is obtained by combining Lemmas 3–8 below and using $Q_h \le t' - 1$.

Theorem 4 (Main Theorem 2). *Suppose* $\mathsf{SIG}[\mathsf{ID}]$ *is* HVZK, RSR *and has* α *bit min-entropy. If* $\mathsf{SIG}[\mathsf{ID}]$ *is* $(t, \varepsilon, Q_h + Q_s)$-UF-KOA *secure then* $\mathsf{SIG}[\mathsf{ID}]$ *is* $(t', \varepsilon', N, Q_s, Q_h)$-MU-UF-CMA *secure in the programmable random oracle model, where*

$$\varepsilon' \le 4\varepsilon + \frac{Q_h Q_s}{2^\alpha}, \qquad t' \approx t$$

and Q_s, Q_h *are upper bounds on the number of signing and hash queries in the* MU-UF-CMA *experiment, respectively.*

The proof of Theorem 4 is obtained by combining Lemmas 7 and 8 below.

Here we present the proofs of Lemmas 1 and 3 (a new Multi-Instance Reset Lemma and an application of it), Lemmas 7 and 8 (the implication of "UF-KOA → MU-UF-CMA"), which are the main contributions of this paper. All remaining proofs are deferred to [35].

3.1   Multi-Instance Reset Lemma

We first state a new reset lemma that we will later use in the proof of Theorem 3. It is presented in the style of Bellare and Neven's General Forking Lemma [5] and does not talk about signatures or identification protocols. It is a generalization to many parallel instances of the Reset Lemma [6], which is obtained by setting $N = 1$.

Lemma 1 (Multi-Instance Reset Lemma). *Fix an integer $N \geq 1$ and a non-empty set H. Let $\mathcal{C}$ be a randomized algorithm that on input (I, h) returns a pair (b, σ), where b is a bit and σ is called the side output. Let IG be a randomized algorithm that we call the input generator. The accepting probability of $\mathcal{C}$ is defined as*

$$\mathsf{acc} := \Pr[b = 1 \mid I \xleftarrow{\boxtimes} \mathsf{IG}; h \xleftarrow{\boxtimes} H; (b, \sigma) \xleftarrow{\boxtimes} \mathcal{C}(I, h)]$$

The (multi-instance) reset algorithm $\mathcal{R}_{\mathcal{C}}$ associated to $\mathcal{C}$ is the randomized algorithm that takes input $I_1, \ldots, I_N$ and proceeds as follows.

Algorithm $\mathcal{R}_{\mathcal{C}}(I_1, \ldots, I_N)$:

For $i \in [N]$:
  Pick random coins ρ_i
  $h_i \xleftarrow{\boxtimes} H$
  $(b_i, \sigma_i) \xleftarrow{\boxtimes} \mathcal{C}(I_i, h_i; \rho_i)$
If $b_1 = \ldots = b_N = 0$ then return $(0, \epsilon, \epsilon)$ // Abort in Phase 1
Fix $i^* \in [N]$ such that $b_{i^*} = 1$
For $j \in [N]$:
  $h'_j \xleftarrow{\boxtimes} H$
  $(b'_j, \sigma'_j) \xleftarrow{\boxtimes} \mathcal{C}(I_{i^*}, h'_j; \rho_{i^*})$
If $\exists j^* \in [N] : (h_{i^*} \neq h'_{j^*}$ and $b'_{j^*} = 1)$ then return $(i^*, \sigma_{i^*}, \sigma'_{j^*})$
Else return $(0, \epsilon, \epsilon)$ // Abort in Phase 2

Let $\mathsf{res} := \Pr[i^* \geq 1 \mid I_1, \ldots, I_N \xleftarrow{\boxtimes} \mathsf{IG}; (i^*, \sigma, \sigma') \xleftarrow{\boxtimes} \mathcal{R}_{\mathcal{C}}(I_1, \ldots, I_N)]$. *Then*

$$\mathsf{res} \geq \left(1 - \left(1 - \mathsf{acc} + \frac{1}{|H|}\right)^N\right)^2.$$

Proof. For fixed instance I and coins ρ, we define the probabilities

$$\mathsf{acc}(I, \rho) := \Pr_{h \xleftarrow{\boxtimes} H} [b = 1 \mid (b, \sigma) \xleftarrow{\boxtimes} \mathcal{A}(I, h; \rho)],$$

$$\mathsf{res}(I, \rho) := \Pr_{h, h' \xleftarrow{\boxtimes} H} \left[b = 1 \wedge b' = 1 \wedge h \neq h' \,\middle|\, \begin{matrix} (b, \sigma) \xleftarrow{\boxtimes} \mathcal{A}(I, h; \rho); \\ (b', \sigma') \xleftarrow{\boxtimes} \mathcal{A}(I, h'; \rho) \end{matrix}\right].$$

As for fixed I, ρ, the two events $b = 1$ and $b' = 1$ are independent and we obtain

$$\mathsf{res}(I, \rho) \geq \mathsf{acc}(I, \rho) \cdot \left(\mathsf{acc}(I, \rho) - \frac{1}{|H|} \right), \tag{1}$$

where the additive factor $\frac{1}{|H|}$ accounts for the fact that $\Pr[h' = h] = 1/|H|$. With the expectation taken over $I \xleftarrow{\$} \mathsf{IG}$ and random coins ρ, we bound

$$\mathsf{E}_{I, \rho} \left[\mathsf{res}(I, \rho) \right] \geq \mathsf{E}_{I, \rho} \left[\mathsf{acc}(I, \rho) \cdot \left(\mathsf{acc}(I, \rho) - \frac{1}{|H|} \right) \right]$$

$$\geq \mathsf{E}_{I, \rho} [\mathsf{acc}(I, \rho)] \cdot \left(\mathsf{E}_{I, \rho} [\mathsf{acc}(I, \rho)] - \frac{1}{|H|} \right)$$

$$= \mathsf{acc} \left(\mathsf{acc} - \frac{1}{|H|} \right).$$

Above, we used (1), Jensen's inequality[5] applied to the convex function $\varphi(X) := X \cdot (X - 1/|H|)$, and the fact that $\mathsf{acc} = \mathsf{E}_{I, \rho} [\mathsf{acc}(I, \rho)]$.

Next, consider the random variables b_{i^*} and b'_j ($j \in [N]$) as defined during in the execution of $\mathcal{R}_\mathcal{A}(I_1, \ldots, I_N)$. Using $\mathsf{acc} = \Pr[b_{i^*} = 1]$ and $\Pr[b'_j = 1 \wedge b_{i^*} = 1] = \mathsf{E}_{I_{i^*}, \rho_{i^*}} [\mathsf{res}(I_{i^*}, \rho_{i^*})]$, we obtain

$$\Pr[b'_j = 1 \mid b_{i^*} = 1] = \frac{\Pr[b'_j = 1 \wedge b_{i^*} = 1]}{\Pr[b_{i^*} = 1]} \geq \mathsf{acc} - \frac{1}{|H|}.$$

Finally, we bound

$$\Pr[\text{no abort in phase 2} \mid \text{no abort in phase 1}] = 1 - \prod_{j=1}^{N} (1 - \Pr[b'_j = 1 \mid b_{i^*} = 1])$$

$$\geq 1 - \left(1 - \mathsf{acc} + \frac{1}{|H|} \right)^N,$$

and

$$\Pr[\text{no abort in phase 1}] = 1 - \prod_{i=1}^{N} (1 - \Pr[b_i = 1]) = 1 - (1 - \mathsf{acc})^N$$

to establish

$$\mathsf{res} = \Pr[\text{no abort in phase 1} \wedge \text{no abort in phase 2}] \geq (1 - (1 - \mathsf{acc} + \frac{1}{|H|})^N)^2.$$

This completes the proof. $\qquad\qquad\qquad\qquad\qquad\qquad\qquad\qquad\qquad\square$

---

[5] Jensen's inequality states that if φ is a convex function and X is a random variable, then $\mathsf{E}[\varphi(X)] \geq \varphi(\mathsf{E}[X])$.

3.2   Proof of the Main Theorems

Lemma 2 (XXX-KOA → XXX-PA). *Let* XXX ∈ {KR, IMP, PIMP}. *If* ID *is* $(t, \varepsilon, Q_{\text{CH}})$-XXX-KOA *secure and* HVZK, *then* ID *is* $(\approx t, \varepsilon, Q_{\text{CH}}, Q_O)$-XXX-PA *secure.*

The proof is given in the full version [35].

Lemma 3 below proving that KR-KOA tightly implies IMP-KOA uses the Multi-Instance Reset Lemma and that takes advantage of ID's random self-reducibility (RSR).

Lemma 3 (KR-KOA $\xrightarrow{\text{rewinding}}$ IMP-KOA). *If* ID *is* (t, ε)-KR-KOA *secure,* SS *and* RSR, *then* ID *is* (t', ε')-IMP-KOA *secure, where for any* $N \geq 1$,

$$\varepsilon \geq (1 - (1 - \varepsilon' + \frac{1}{|\text{ChSet}|})^N)^2, \quad t \approx 2Nt'. \tag{2}$$

In particular, the two success ratios are related as

$$\frac{\varepsilon'}{t'} - \frac{1}{t'|\text{ChSet}|} \leq 6 \cdot \frac{\varepsilon}{t}. \tag{3}$$

We remark that without RSR, we can still obtain the weaker bounds $\varepsilon \geq \varepsilon'(\varepsilon' - \frac{1}{|\text{ChSet}|})$, $t \approx 2t'$.

Proof. We first show how to derive (3) from (2). If $\varepsilon' \leq 1/|\text{ChSet}|$, then (3) holds trivially. Assuming $\varepsilon' > 1/|\text{ChSet}|$, we set $N := (\varepsilon' - 1/|\text{ChSet}|)^{-1}$ to obtain $t \approx 2t'/(\varepsilon' - 1/|\text{ChSet}|)$ and $\varepsilon \geq (1 - \frac{1}{e})^2 \geq \frac{1}{3}$. Dividing ε by t yields (3).

To prove (2), let $\mathcal{A}$ be an adversary against the (t', ε')-IMP-KOA-security of ID. We now build an adversary $\mathcal{B}$ against the (t, ε)-KR-KOA security of ID, with (t, ε) as claimed in (2).

We use the Multi-Instance Reset Lemma (Lemma 1), where $H := \text{ChSet}$ and IG runs $(pk, sk) \xleftarrow{\text{\tiny{\$}}} \text{IGen}$ and returns pk as instance I. We first define adversary $\mathcal{C}(pk, h; \rho)$ that executes $\mathcal{A}(pk; \rho)$, answers $\mathcal{A}$'s single query R with h, and finally receives s from $\mathcal{A}$. If transcript (R, h, s) is valid with respect to pk (i.e., $\text{V}(pk, R, h, s) = 1$), $\mathcal{C}$ returns $(b = 1, \sigma = (R, h, s))$; otherwise, it returns $(b = 0, \epsilon)$. By construction, $\mathcal{C}$ returns $b = 1$ iff $\mathcal{A}$ is successful: $\text{acc} = \varepsilon'$.

Adversary $\mathcal{B}$ is defined as follows. For each $i \in [N]$, it uses the RSR property of ID to generate a fresh public key/trapdoor pair $(pk_i, \tau_i) \xleftarrow{\text{\tiny{\$}}} \text{Rerand}(pk)$. Next, it runs $(i^*, \sigma, \sigma') \xleftarrow{\text{\tiny{\$}}} \mathcal{R}_{\mathcal{C}}(pk_1, \ldots, pk_N)$, with $\mathcal{C}$ defined above. If $i^* \geq 1$, then both transcripts $\sigma = (R, h, s)$ and $\sigma' = (R, h', s')$ are valid with respect to pk_{i^*} and $h \neq h'$. $\mathcal{B}$ uses the SS property of ID and computes $sk_{i^*} \leftarrow \text{Ext}(pk_{i^*}, R, h, s, h', s')$. Finally, using the RSR property of ID, it returns $sk = \text{Derand}(pk_{i^*}, sk_{i^*}, \tau_{i^*})$ and terminates. By construction, $\mathcal{B}$ is successful iff $\mathcal{R}_{\mathcal{C}}$ is. By Lemma 1 we can bound $\mathcal{B}$'s success probability as

$$\varepsilon = \text{res} \geq (1 - (1 - \varepsilon' + \frac{1}{|\text{ChSet}|})^N)^2.$$

The running time t of $\mathcal{B}$ is that of $\mathcal{R}_\mathcal{C}$, meaning $2Nt'$ plus the N times the time to run the Rerand and Derand algorithms of RSR plus the time to run the Ext algorithm of SS. We write $t \approx 2Nt'$ to indicate that this is the dominating running time of $\mathcal{B}$. $\qquad\square$

Lemma 4 (IMP-KOA $\xrightarrow{\text{loss } Q}$ PIMP-KOA). *If* ID *is* (t, ε)-IMP-KOA *secure, then* ID *is* $(t', \varepsilon', Q_{\text{CH}})$-PIMP-KOA *secure, where*

$$\varepsilon' \le Q_{\text{CH}} \cdot \varepsilon, \quad t' \approx t.$$

The proof is given in the full version [35].

Lemma 5 (PIMP-KOA $\xrightarrow{\text{PRO}}$ UF-KOA). *If* ID *is* $(t, \varepsilon, Q_{\text{CH}})$-PIMP-KOA *secure, then* SIG[ID] *is* (t', ε', Q_h)-UF-KOA *secure in the programmable random oracle model, where*

$$\varepsilon' = \varepsilon, \quad t' \approx t, \quad Q_h = Q_{\text{CH}} - 1.$$

The proof is given in the full version [35].

The following lemma is a special case of Lemma 8 (with a slightly improved bound).

Lemma 6 (UF-KOA $\xrightarrow{\text{PRO}}$ UF-CMA). *Suppose* ID *is* HVZK *and has* α *bit min-entropy. If* SIG[ID] *is* (t, ε, Q_h)-UF-KOA *secure, then* SIG[ID] *is* $(t', \varepsilon', Q_s, Q_h)$-UF-CMA *secure in the programmable random oracle model, where*

$$\varepsilon' \le \varepsilon + \frac{Q_h Q_s}{2^\alpha}, \quad t' \approx t,$$

and Q_s, Q_h *are upper bounds on the number of signing and hash queries in the* UF-CMA *experiment, respectively.*

Lemma 7 (UF-KOA $\xrightarrow{\text{RSR}}$ MU-UF-KOA). *Suppose* ID *is* RSR. *If* SIG[ID] *is* (t, ε)-UF-KOA *secure, then* SIG[ID] *is* (t', ε', N)-MU-UF-KOA *secure, where*

$$\varepsilon' = \varepsilon, \quad t' \approx t.$$

Note that without the RSR property one can use the generic bounds from [25] to obtain a non-tight bound with a loss of N.

Proof. Let $\mathcal{A}$ be an algorithm that breaks (t', ε', N)-MU-UF-KOA security of SIG[ID]. We will describe an adversary $\mathcal{B}$ invoking $\mathcal{A}$ that breaks (t, ε)-UF-KOA security of SIG[ID] with (t, ε) as stated in the lemma. Adversary $\mathcal{B}$ is executed in the UF-KOA experiment and obtains a public-key pk.

SIMULATION OF PUBLIC-KEYS INPUT TO $\mathcal{A}$. For each $i \in [N]$, $\mathcal{B}$ generates $(pk_i, \tau_i) \xleftarrow{\text{\$}} \textsf{Rerand}(pk)$ by using the RSR property of ID. Then $\mathcal{B}$ runs $\mathcal{A}$ on input $(pk_1, \ldots, pk_N)$.

FORGERY. Eventually, $\mathcal{A}$ will submit its forgery $(i^*, m^*, \sigma^* := (R^*, s^*))$ in the MU-UF-KOA experiment. $\mathcal{B}$ computes $h^* = H(m^*, R^*)$ and runs

$s \xleftarrow{\$} \mathsf{Tran}(pk, pk_{i^*}, \tau_{i^*}, (R^*, h^*, s^*))$. By the RSR property of ID, the random variables (pk, R^*, h^*, s) and $(pk_{i^*}, R^*, h^*, s^*)$ are identically distributed. If σ^* is a valid signature on message m^* under pk_{i^*}, then (R^*, s) is also a valid signature on m^* under pk. Thus, we have $\varepsilon = \varepsilon'$. The running time t of $\mathcal{B}$ is t' plus the N times the time to run the Rerand and Tran algorithms of RSR. We again write $t \approx t'$.   □

Lemma 8 (MU-UF-KOA $\xrightarrow{\text{PRO}}$ MU-UF-CMA). *Suppose* ID *is* HVZK *and has* α *bit min-entropy. If* SIG[ID] *is* (t, ε, N, Q_h)-MU-UF-KOA *secure, then* SIG[ID] *is* $(t', \varepsilon', N, Q_s, Q_h)$-MU-UF-CMA *secure in the programmable random oracle model, where*

$$\varepsilon' \leq 4\varepsilon + \frac{Q_h Q_s}{2^\alpha}, \quad t' \approx t,$$

and N *is the number of users and* Q_s *and* Q_h *are upper bounds on the number of signing and hash queries in the* MU-UF-CMA *experiment, respectively.*

Proof. Let $\mathcal{A}$ be an algorithm that breaks $(t', \varepsilon', N, Q_s, Q_h)$-MU-UF-CMA security of SIG[ID]. We will describe an adversary $\mathcal{B}$ invoking $\mathcal{A}$ that breaks (t, ε, N, Q_h)-MU-UF-KOA security of SIG[ID] with (t, ε) as stated in the lemma. Adversary $\mathcal{B}$ is executed in the MU-UF-KOA experiment and obtains public-keys $(pk_1, \ldots, pk_N)$, and has access to a random oracle H.

PREPARATION OF PUBLIC-KEYS. For each $i \in [N]$, adversary $\mathcal{B}$ picks a secret bit $b_i \xleftarrow{\$} \{0, 1\}$. If $b_i = 1$ then $\mathcal{B}$ defines $pk'_i := pk_i$, else $\mathcal{B}$ generates the key-pair $(pk'_i, sk'_i) \xleftarrow{\$} \mathsf{Gen}(\mathsf{par})$ itself. We note that all simulated public-keys are correctly distributed.

Adversary $\mathcal{B}$ runs $\mathcal{A}$ on input $(pk'_1, \ldots, pk'_N)$ answering hash queries to random oracle H' and signing queries as follows.

SIMULATION OF HASH QUERIES. A hash query $H'(R, m)$ is answered by $\mathcal{B}$ by querying its own hash oracle $H(R, m)$ and returning its answer.

SIMULATION OF SIGNING QUERIES. On $\mathcal{A}$'s j-th signature query (i_j, m_j), $\mathcal{B}$ returns a signature σ_j on message m_j under pk_{i_j} according to the following case distinction.

– Case A: $b_{i_j} = 0$. In that case sk'_{i_j} is known to $\mathcal{B}$ and the signature is computed as $\sigma_j := (R_j, s_j) \xleftarrow{\$} \mathsf{Sign}(sk'_{i_j}, m_j)$. Note that this involves $\mathcal{B}$ making a hash query and defining $H'(R_j, m_j) := H(R_j, m_j)$.
– Case B: $b_{i_j} = 1$. In that case sk'_{i_j} is unknown to $\mathcal{B}$ and the signature is computed using the HVZK property of ID. Concretely, $\mathcal{B}$ runs $(R_j, h_j, s_j) \xleftarrow{\$} \mathsf{Sim}(pk'_{i_j})$. If hash value $H'(R_j, m_j)$ was already defined (via one of $\mathcal{A}$'s hash/signing queries) and $H'(R_j, m_j) \neq h_j$, $\mathcal{B}$ aborts. Otherwise, it defines the random oracle

$$H'(R_j, m_j) := h_j \tag{4}$$

and returns $\sigma_j := (R_j, s_j)$, which is a correctly distributed valid signatures on m_j under pk_{i_j}. Note that by (4), $\mathcal{B}$ makes H and H' inconsistent, i.e., we have $H(R_j, m_j) \neq H'(R_j, m_j)$ with high probability. Also note that for each signing query, $\mathcal{B}$ aborts with probability at most $Q_h/2^\alpha$ because R_j has min-entropy α. Since the number of signing queries is bounded by Q_s, $\mathcal{B}$ aborts overall with probability at most $Q_h Q_s/2^\alpha$.

FORGERY. Eventually, $\mathcal{A}$ will submit its forgery $(i^*, m^*, \sigma^* := (R^*, s^*))$. We assume that it is a valid forgery in the MU-UF-CMA experiment, i.e., for $h^* = H'(R^*, m^*)$ we have $\mathsf{V}(pk'_{i^*}, R^*, h^*, s^*) = 1$. Furthermore, it satisfies the freshness condition, i.e.,

$$(i^*, m^*) \notin \{(i_j, m_j) : j \in [Q_s]\}. \tag{5}$$

After receiving $\mathcal{A}$'s forgery, $\mathcal{B}$ computes a forgery for the MU-UF-KOA experiment according to the following case distinction.

– Case 1: There exists a $j \in [Q_s]$ such that $(m^*, R^*) = (m_j, R_j)$. (If there is more than one j, fix any of them.) In that case we have and $h^* = h_j$ and furthermore $i^* \neq i_j$ by the freshness condition (5).
  - Case 1a: $(b_{i^*} = 1)$ and $(b_{i_j} = 0)$. Then the hash value $h^* = H'(R^*, m^*)$ was not programmed by $\mathcal{B}$ in (4). That means $h^* = H'(R^*, m^*) = H(R^*, m^*)$ and $\mathcal{B}$ returns $(i^*, m^*, (R^*, s^*))$ as a valid forgery to its MU-UF-KOA experiment.
  - Case 1b: $(b_{i^*} = b_{i_j})$ or $(b_{i^*} = 0 \wedge b_{i_j} = 1)$. Then $\mathcal{B}$ aborts.
  Note that in case 1 we always have $i^* \neq i_j$ and therefore $\mathcal{B}$ does not abort with probability $1/4$ in which case it outputs a valid forgery.
– Case 2: For all $j \in [Q_s]$ we have: $(m^*, R^*) \neq (m_j, R_j)$.
  - Case 2a: $b_{i^*} = 1$. Then the hash value $h^* = H'(R^*, m^*)$ was not programmed by $\mathcal{B}$ in (4). That means $h^* = H'(R^*, m^*) = H(R^*, m^*)$ and $\mathcal{B}$ returns $(i^*, m^*, (R^*, s^*))$ as a valid forgery to its MU-UF-KOA experiment.
  - Case 2b: $b_{i^*} = 0$. Then $\mathcal{B}$ aborts.
  Note that in case 2, $\mathcal{B}$ does not abort with probability $1/2$ in which case it outputs a valid forgery.

Overall, $\mathcal{B}$ returns a valid forgery of MU-UF-KOA experiment with probability

$$\varepsilon \geq \min\left\{\frac{1}{4}, \frac{1}{2}\right\} \cdot \left(\varepsilon' - \frac{Q_h Q_s}{2^\alpha}\right) = \frac{1}{4}\left(\varepsilon' - \frac{Q_h Q_s}{2^\alpha}\right).$$

The running time of $\mathcal{B}$ is that of $\mathcal{A}$ plus the Q_s executions of Sim. We write $t' \approx t$. This completes the proof.   □

If s in ID is uniquely defined by (pk, R, h) (e.g., as in the Schnorr identification scheme), then one can show the above proof even implies MU-SUF-CMA security of SIG[ID]. The simulation of hash and signing queries is the same as in the above proof. Let (i^*, m^*, R^*, s^*) be $\mathcal{A}$'s forgery. The freshness condition of the MU-SUF-CMA experiment says that $(i^*, m^*, R^*, s^*) \notin \{(i_j, m_j, R_j, s_j) : j \in [Q_s]\}$. Together with the uniqueness of ID, this implies $(i^*, m^*, R^*) \notin \{(i_j, m_j, R_j) : j \in [Q_s]\}$. If $(i^*, m^*) \notin \{(i_j, m_j) : j \in [Q_s]\}$, then $\mathcal{B}$ can break MU-UF-KOA security by the same case distinction as in the proof above. Otherwise, we have $R^* \notin \{R_j : j \in [Q_s]\}$, in which case we can argue as in case 2.

4   Impossibility Results

In this section, we show that Theorems 3 and 4 from the previous section are optimal in the sense that the security reduction requires: rewinding (Lemma 9), security loss of at least $O(Q)$ (Lemma 10) and programmability of random oracles (Lemmas 11 and 12).

Let X and Y be some hard cryptographic problems, defined through a (possibly) interactive experiment. A black-box reduction $\mathcal{R}$ from X to Y is an algorithm that, given black-box access to an adversary $\mathcal{A}$ breaking problem Y, breaks problem X. If X and Y are security notions for identification or signatures schemes, then a reduction $\mathcal{R}$ is called key-preserving, if $\mathcal{R}$ only makes calls to $\mathcal{A}$ with the same pk that it obtained by its own problem X. All our reductions considered in this section are key-preserving. All proofs from this section are given in the full version [35].

Lemma 9 (KR-KOA $\xrightarrow{\text{non-rewind.}}$ IMP-KOA). *If there is a key-preserving reduction $\mathcal{R}$ that $(t_\mathcal{R}, \varepsilon_\mathcal{R})$-breaks KR-KOA security of ID with one-time black-box access to an adversary $\mathcal{A}$ that $(t_\mathcal{A}, \varepsilon_\mathcal{A})$-breaks IMP-KOA security of ID, then there exists an algorithm $\mathcal{M}$ that $(t_\mathcal{M}, \varepsilon_\mathcal{M}, Q_O)$-breaks IMP-AA security of ID, where*

$$\varepsilon_\mathcal{M} \geq \varepsilon_\mathcal{R} - \frac{1}{|\mathsf{ChSet}|}, \quad t_\mathcal{M} \approx t_\mathcal{R}, \quad Q_O = 1.$$

For our next impossibility result, we will require the following definition for identification schemes.

Definition 10 (Concurrent (Weak) Impersonation against Man-in-the-Middle Attacks). *A canonical identification ID is said to be $(t, \varepsilon, Q_{\mathrm{CH}}, Q_O)$-IMP-MIM secure (impersonation against man-in-the-middle attacks) if for all adversaries $\mathcal{A}$ running in time at most t and adaptively making at most Q_O queries to the prover oracle PROVER and Q_{CH} queries to the challenge oracle CH,*

$$\Pr\left[\begin{array}{l}\mathsf{V}(pk, R_{i^*}, h_{i^*}, s_{i^*}) = 1 \wedge (i^* \in [Q_{\mathrm{CH}}]) \\ \wedge (R_{i^*}, h_{i^*}, s_{i^*}) \notin \{(R'_j, h'_j, s'_j) \mid j \in [Q_O]\}\end{array} \middle| \begin{array}{l}(pk, sk) \xleftarrow{\boxtimes} \mathsf{IGen}(par) \\ (i^*, s_{i^*}) \xleftarrow{\boxtimes} \mathcal{A}^{\mathrm{PROVER}(\cdot), \mathrm{CH}(\cdot)}(pk)\end{array}\right] \leq \varepsilon,$$

where oracles PROVER and CH are defined as in Definition 2. We define weak impersonation against man-in-the-middle attack (wIMP-MIM) by restricting $R_{i^*} \in \{R'_1, \ldots, R'_{Q_O}\}$.

We remark that wIMP-MIM is a non-standard definition without any practical relevance, but it will only be used for showing negative results. The following generalizes a result by Seurin [43] to canonical identification schemes.

Lemma 10 (IMP-KOA $\xrightarrow{\text{loss}<\mathbf{Q}}$ PIMP-KOA). *Suppose that ID has α bit min-entropy and there is a key-preserving reduction $\mathcal{R}$ that $(t_\mathcal{R}, \varepsilon_\mathcal{R})$-breaks IMP-KOA* *security of* ID *with* n-time black-box *access to an adversary* $\mathcal{A}$ *that* $(t_{\mathcal{A}}, \varepsilon_{\mathcal{A}}, Q_{\mathrm{Ch}})$-*breaks* PIMP-KOA *security of* ID. *Then there exists an algorithm* $\mathcal{M}$ *that* $(t_{\mathcal{M}}, \varepsilon_{\mathcal{M}}, 1, Q_{\mathrm{O}} = nQ_{\mathrm{Ch}})$-*breaks* IMP-MIM *security of* ID, *where*

$$\varepsilon_{\mathcal{M}} \geq \varepsilon_{\mathcal{R}} - \frac{n \ln \left((1 - \varepsilon_{\mathcal{A}})^{-1} \right)}{Q_{\mathrm{Ch}}} - \frac{n}{|\mathsf{ChSet}|} - \frac{n}{2^{\alpha}}, \quad t_{\mathcal{M}} \approx t_{\mathcal{R}}.$$

For a precise analysis of the function $\ln \left((1 - \varepsilon_{\mathcal{A}})^{-1} \right)$, we refer to [43]. For our purpose, it is sufficient that for a concrete choice of $\varepsilon_{\mathcal{A}}$, there is a constant c such that $c \cdot \varepsilon_{\mathcal{A}} = \ln \left((1 - \varepsilon_{\mathcal{A}})^{-1} \right)$. Hence Lemma 10 gives roughly $\varepsilon_{\mathcal{M}} \geq \varepsilon_{\mathcal{R}} - (c \cdot n / Q_{\mathrm{Ch}}) \cdot \varepsilon_{\mathcal{A}}$ for a suitable choice of $\varepsilon_{\mathcal{A}}$. Therefore $\varepsilon_{\mathcal{R}}$ can be at most $(c \cdot n / Q_{\mathrm{Ch}}) \cdot \varepsilon_{\mathcal{A}}$. Otherwise $\mathcal{M}$ would break IMP-MIM security of ID with $\varepsilon_{\mathcal{M}} > 0$.

In the proof of Lemma 10 (cf. [35]), the meta-reduction just forwards all $R_{j,i}$ received during the Man-in-the-Middle attack and R sent by $\mathcal{R}$. So if $\mathcal{R}$ is furthermore randomness-preserving, i.e., it chooses $R \in \{R_{1,1}, \ldots, R_{n,Q_{\mathrm{Ch}}}\}$, then $\mathcal{M}$ attacks wIMP-MIM-security of ID. This observation (formalized in the following corollary) is important since the Schnorr identification scheme is wIMP-MIM but not IMP-MIM-secure.

Corollary 1. *If* ID *has* α *bit min-entropy and there exists a key- and randomness-preserving reduction* $\mathcal{R}$ *that* $(t_{\mathcal{R}}, \varepsilon_{\mathcal{R}})$-*breaks* IMP-KOA *security of* ID *with* n-time black-box *access to an adversary* $\mathcal{A}$ *that* $(t_{\mathcal{A}}, \varepsilon_{\mathcal{A}}, Q_{\mathrm{Ch}})$-*breaks* PIMP-KOA *security of* ID, *then there exists an algorithm* $\mathcal{M}$ *that* $(t_{\mathcal{M}}, \varepsilon_{\mathcal{M}}, 1, Q_{\mathrm{O}} = nQ_{\mathrm{Ch}})$-*breaks* wIMP-MIM *security of* ID, *where*

$$\varepsilon_{\mathcal{M}} \geq \varepsilon_{\mathcal{R}} - \frac{n \ln \left((1 - \varepsilon_{\mathcal{A}})^{-1} \right)}{Q_{\mathrm{Ch}}} - \frac{n}{|\mathsf{ChSet}|} - \frac{n}{2^{\alpha}}, \quad t_{\mathcal{M}} \approx t_{\mathcal{R}}.$$

Lemma 11 (IMP-KOA $\xrightarrow{\text{NPRO}}$ UF-KOA). *If there exists a key-preserving reduction* $\mathcal{R}$ *in the non-programmable random oracle (NPRO) model that* $(t_{\mathcal{R}}, \varepsilon_{\mathcal{R}})$-*breaks* IMP-KOA *security of* ID *with* n-time black-box access to an adversary $\mathcal{A}$ that $(t_{\mathcal{A}}, \varepsilon_{\mathcal{A}}, Q_h)$-breaks UF-KOA security of SIG[ID], then there exists an algorithm $\mathcal{M}$ that $(t_{\mathcal{M}}, \varepsilon_{\mathcal{M}}, 1)$-breaks IMP-AA-security of ID, where

$$\varepsilon_{\mathcal{M}} \geq \varepsilon_{\mathcal{R}} - \frac{1}{|\mathsf{ChSet}|}, \quad t_{\mathcal{M}} \approx t_{\mathcal{R}}.$$

By Lemmas 4 and 11 implies that there is no reduction from PIMP-KOA to UF-KOA in the non-programmable random oracle model.

The following simple lemma actually holds for any signature scheme SIG.

Lemma 12 (UF-KOA $\xrightarrow{\text{NPRO}}$ UF-CMA). *Suppose that there is a key-preserving reduction* $\mathcal{R}$ *in the non-programmable random oracle (NPRO) model that* $(t_{\mathcal{R}}, \varepsilon_{\mathcal{R}}, Q_h)$-*breaks* UF-KOA *security of* SIG *with* n-time black-box *access to an adversary* $\mathcal{A}$ *that* $(t_{\mathcal{A}}, \varepsilon_{\mathcal{A}}, Q_s, Q_h)$-*breaks* UF-CMA *security of* SIG. *Then there exists an algorithm* $\mathcal{M}$ *that* $(t_{\mathcal{M}}, \varepsilon_{\mathcal{M}})$-*breaks* UF-KOA *security of* SIG, *where*

$$\varepsilon_{\mathcal{M}} \geq \varepsilon_{\mathcal{R}}, \quad t_{\mathcal{M}} \approx t_{\mathcal{R}}.$$

Remark 1. All the reductions considered in this section are key-preserving which is the main restriction of our results. If pk and R are elements from some multiplicative group $\mathbb{G}$ of prime order p, then we can extend our previous techniques to exclude the larger class of algebraic reductions. A reduction is algebraic, if for all group elements h output by the reduction, their respective representation is known. That is, if at some point of its execution the reduction holds group elements $g_1, \ldots, g_n \in \mathbb{G}$ and outputs a new group element h, then it also knows it representation meaning it also outputs $(\alpha_1, \ldots, \alpha_n) \in \mathbb{Z}_p^n$ satisfying $h = \prod g_i^{\alpha_i}$. Note that key-preserving and randomness-preserving reductions are a special case of algebraic reductions.

5   Instantiations

In this section we consider two important identification schemes, namely the ones by Schnorr [42] and by Katz-Wang [19,34]. We use our framework to derive tight security bounds and concrete parameters for the corresponding Schnorr/Katz-Wang signature schemes. In the full version [35] we discuss one more identification scheme, namely the one by Guillou-Quisquater [31].

5.1   Schnorr Identification/Signature Scheme

Schnorr's Identification Scheme. The well-known Schnorr's identification scheme is one of the most important instantiations of our framework. For completeness we show that Schnorr's identification has large min-entropy, special soundness (SS), honest-verifier zero-knowledge (HVZK), random-self reducibility (RSR) and key-recovery security (KR-KOA) based on the discrete logarithm problem (DLOG). Moreover, based on the one-more discrete logarithm problem (OMDL), Schnorr's identification is actively secure (IMP-AA) and weakly secure against man-in-the-middle attack (wIMP-MIM).

Let $\mathsf{par} := (p, g, \mathbb{G})$ be a set of system parameters, where $\mathbb{G} = \langle g \rangle$ is a cyclic group of prime order p with a hard discrete logarithm problem. Examples of groups $\mathbb{G}$ include appropriate subgroups of certain elliptic curve groups, or subgroups of $\mathbb{Z}_q^*$. The Schnorr identification scheme $\mathsf{ID_S} := (\mathsf{IGen}, \mathsf{P}, \mathsf{ChSet}, \mathsf{V})$ is defined as follows.

$\mathsf{IGen}(\mathsf{par})$:	$\mathsf{P}_1(sk)$:
$sk := x \xleftarrow{\boxtimes} \mathbb{Z}_p$	$r \xleftarrow{\boxtimes} \mathbb{Z}_p;\ R = g^r$
$pk := X = g^x$	$St := r$
$\mathsf{ChSet} := \{0,1\}^n$	Return (R, St)
Return (pk, sk)	
	$\mathsf{P}_2(sk, R, h, St)$:
	Parse $St = r$
$\mathsf{V}(pk, R, h, s)$:	Return $s = x \cdot h + r \bmod p$
If $R = g^s \cdot X^{-h}$ then return 1	
Else return 0	

We recall the DLOG assumption.

Definition 11 (Discrete Logarithm Assumption). *The discrete logarithm problem* DLOG *is* (t, ε)-*hard in* par $= (p, g, \mathbb{G})$ *if for all adversaries* $\mathcal{A}$ *running in time at most* t, $\Pr\left[g^x = X \mid X \xleftarrow{\pm} \mathbb{G}; x \xleftarrow{\pm} \mathcal{A}(X)\right] \leq \varepsilon$.

Lemma 13. $\mathsf{ID_S}$ *is a canonical identification with* $\alpha = \log p$ *bit min-entropy and it is unique, has special soundness* (SS), *honest-verifier zero-knowledge* (HVZK) *and is random-self reducible* (RSR). *Moreover, if* DLOG *is* (t, ε)-*hard in* par $= (p, g, \mathbb{G})$ *then* $\mathsf{ID_S}$ *is* (t, ε)-KR-KOA *secure.*

Proof. The correctness of $\mathsf{ID_S}$ is straightforward to verify. We note that R in $(R, St) \xleftarrow{\pm} \mathsf{P}_1(sk)$ is uniformly random over $\mathbb{G}$. Hence, $\mathsf{ID_S}$ has $\log |\mathbb{G}| = \log p$ bit min-entropy. We show the other properties as follows.

UNIQUENESS. For all $(X, x) \in \mathsf{IGen}(\mathsf{par})$, $(R := g^r, St := r) \in \mathsf{P}_1(sk)$ and $h \in \{0,1\}^n$, the value $s \in \mathbb{Z}_p$ satisfying $g^s = X^h R \Leftrightarrow s = xh + r$ is uniquely defined.

SPECIAL SOUNDNESS (SS). Given two accepting transcripts (R, h, s) and (R, h', s') with $h \neq h'$, we define an extractor algorithm $\mathsf{Ext}(X, R, h, s, h', s') := x^* := (s - s')/(h - h')$ such that, for all $(X := g^x, x) \in \mathsf{IGen}(\mathsf{par})$, we have $\Pr[g^{x^*} = X] = 1$, since we have $R = g^s X^{-h} = g^{s'} X^{-h'}$ and then $X = g^{(s-s')/(h-h')}$.

HONEST-VERIFIER ZERO-KNOWLEDGE (HVZK). Given public key X, we let $\mathsf{Sim}(X)$ first sample $h \xleftarrow{\pm} \{0,1\}^n$ and $s \xleftarrow{\pm} \mathbb{Z}_p$ and then output $(R := g^s X^{-h}, h, s)$. Clearly, (R, h, s) is a real transcript, since s is uniformly random over $\mathbb{Z}_p$ and R is the unique value satisfying $R = g^s X^{-h}$.

RANDOM-SELF REDUCIBILITY (RSR). Algorithm Rerand and two deterministic algorithm Derand and Tran are defined as follows:

- $\mathsf{Rerand}(X)$ chooses $\tau' \xleftarrow{\pm} \mathbb{Z}_p$ and outputs $(X' := X \cdot g^{\tau'}, \tau')$. We have that, for all $(X, x) \in \mathsf{IGen}(\mathsf{par})$, X' is uniform and has the same distribution as X'', where $(X'', x'') \xleftarrow{\pm} \mathsf{IGen}(\mathsf{par})$.
- $\mathsf{Derand}(X, X', x', \tau')$ outputs $x^* = x' - \tau'$. We have, for all $(X', \tau') \xleftarrow{\pm} \mathsf{Rerand}(X := g^x)$ and $(X', x') \in \mathsf{IGen}(\mathsf{par})$, $X' = g^{x'}$ and $x' = x + \tau'$ and thus $x^* = x$.
- $\mathsf{Tran}(X, X', \tau', (R', h', s'))$ outputs $s = s' - \tau' \cdot h'$. We have, for all $(X', \tau') \in \mathsf{Rerand}(X := g^x)$, if (R', h', s') is valid with respect to $X' := g^{x+\tau'}$ then $s = s' - \tau' \cdot h' = (x + \tau')h' + r - \tau' \cdot h' = xh' + r$ and (R', h', s) is valid with respect to X.

KEY-RECOVERY AGAINST KEY-ONLY ATTACK (KR-KOA). KR-KOA-security for ID is exactly the DLOG assumption. $\qquad\square$

Under the one-more discrete logarithm assumption [4], $\mathsf{ID_S}$ is IMP-AA secure [6] and in the full version [35] we show that $\mathsf{ID_S}$ is weakly IMP-MIM secure.

We now define the Q-interactive discrete-logarithm problem which precisely models PIMP-KOA-security for $\mathsf{ID_S}$, where $Q = Q_O$ is the number of parallel impersonation rounds.

Definition 12 (Q-IDLOG). *The interactive discrete-logarithm assumption Q-IDLOG is said to be (t, ε)-hard in* par $= (p, g, \mathbb{G})$ *if for all adversaries $\mathcal{A}$ running in time at most t and making at most Q queries to the challenge oracle* CH,

$$\Pr\left[s \in \{xh_i + r_i \mid i \in [Q]\} \;\middle|\; \begin{array}{l} x \xleftarrow{\boxtimes} \mathbb{Z}_p; X = g^x \\ s \xleftarrow{\boxtimes} \mathcal{A}^{\text{CH}(\cdot)}(X) \end{array}\right] \leq \varepsilon,$$

where on the i-th query $\text{CH}(g^{r_i})$ *$(i \in [Q])$, the challenge oracle returns $h_i \xleftarrow{\boxtimes} \mathbb{Z}_p$ to $\mathcal{A}$.*

In [35] we prove that in the generic group model, the Q-IDLOG problem in groups of prime-order p is at least $(t, 2t^2/p)$-hard. Note that the bound is independent of Q.

Schnorr's Signature Scheme. Let $H : \{0,1\}^* \to \{0,1\}^n$ be a hash function with $n < \log_2(p)$. As IDs is commitment-recoverable we can use the alternative Fiat-Shamir transformation to obtain the Schnorr signature scheme Schnorr := (Gen, Sign, Ver).

Gen(par):	Sign(sk, m):	Ver(sk, m, σ):
$sk := x \xleftarrow{\boxtimes} \mathbb{Z}_p$	$r \xleftarrow{\boxtimes} \mathbb{Z}_p;\ R = g^r$	Parse $\sigma = (h, s) \in \{0,1\}^n \times \mathbb{Z}_p$
$pk := X = g^x$	$h = H(R, m)$	$R = g^s X^{-h}$
Return (pk, sk)	$s = x \cdot h + r \bmod p$	If $h = H(R, m)$ then return 1
	$\sigma = (h, s) \in \{0,1\}^n \times \mathbb{Z}_p$	Else return 0.
	Return σ	

The DLOG problem is tightly equivalent to the 1-IDLOG problem by Lemma 3. Assuming the OMDL problem is hard, Schnorr is wIMP-MIM-secure and by Corollary 1 there cannot exist a tight implication 1-IDLOG → Q-IDLOG meaning the bound from Lemma 4 is optimal. By Lemmas 5 and 6, the Q-IDLOG problem is tightly equivalent to SUF-CMA-security of Schnorr in the programmable ROM. The latter is only tightly equivalent to MU-SUF-CMA-security in the programmable ROM (via Lemmas 7 and 8). In the full version [35] we improve this by proving that SUF-CMA security is tightly equivalent to MU-SUF-CMA-security in the standard model. Figure 3 summarizes the modular security implications for Schnorr.

We derive the following concrete security implications.

Lemma 14. *If* DLOG *is* (t, ε)-hard *in* par $= (p, g, \mathbb{G})$ *then* Schnorr *is* $(t', \varepsilon', Q_s, Q_h)$-SUF-CMA *secure and* $(t'', \varepsilon'', N, Q_s, Q_h)$-MU-SUF-CMA *secure in the programmable random oracle model, where*

$$\frac{\varepsilon'}{t'} \leq 6(Q_h + 1) \cdot \frac{\varepsilon}{t} + \frac{Q_s}{p} + \frac{1}{2^n},$$

$$\frac{\varepsilon''}{t''} \leq 12(Q_h + 1) \cdot \frac{\varepsilon}{t} + \frac{Q_s}{p} + \frac{1}{2^n}.$$

Lemma 15. *If Q_h-IDLOG is (t, ε)-hard in* par *then* Schnorr *is* $(t', \varepsilon', N, Q_s, Q_h)$-MU-SUF-CMA *secure in the programmable random oracle model, where*

$$\varepsilon' \leq 2\varepsilon + \frac{Q_h Q_s}{p}, \qquad t' \approx t.$$

We leave it an open problem to come up with a more natural hard problem over par that tightly implies Q-IDLOG (and hence MU-SUF-CMA-security of Schnorr). Note that according to [23], the hard problem has to have at least one round of interaction.

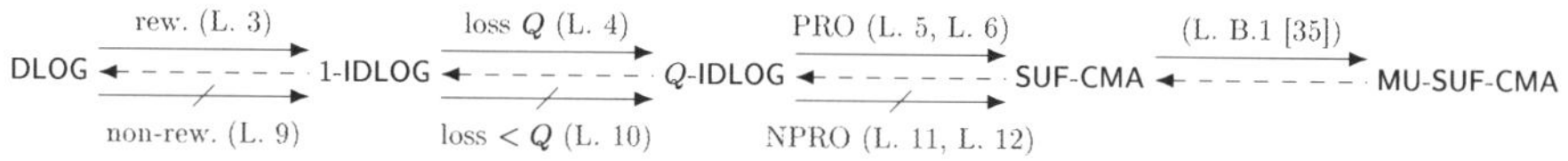

Fig. 3. Security relations for the Schnorr signature scheme. All implications except "1-IDLOG → Q-IDLOG" are tight.

The interpretation for the multi-user security of Schnorr over elliptic-curve groups is as follows. It is well-known that a group of order p providing k-bits security against the DLOG problem requires $\log p \geq 2k$. If one requires provable security guarantees for Schnorr under DLOG, then one has to increase the group size by $\approx \log(Q_h)$ bits. Reasonable upper bounds for $\log Q_h$ are between 40 and 80. However, the generic lower bound for the Q-IDLOG problem indicates that the only way to attack Schnorr in the sense of UF-KOA (and hence to attack Q-IDLOG) is to break the DLOG problem. In that case using groups with $\log p \approx 2k$ already gives provable security guarantees for Schnorr.

5.2   Chaum-Pedersen Identification/Katz-Wang Signature Scheme

Chaum-Pedersen Identification Scheme. Let par $:= (p, g_1, g_2, \mathbb{G})$ be a set of system parameters, where $\mathbb{G} = \langle g_1 \rangle = \langle g_2 \rangle$ is a cyclic group of prime order p. The Chaum-Pedersen identification scheme $\mathsf{ID}_{\mathsf{CP}} := (\mathsf{IGen}, \mathsf{P}, \mathsf{ChSet}, \mathsf{V})$ is defined as follows.

$\mathsf{IGen}(\mathsf{par})$:	$\mathsf{P}_1(sk)$:
$sk := x \xleftarrow{\$} \mathbb{Z}_p$	$r \xleftarrow{\$} \mathbb{Z}_p; \; R = (R_1, R_2) = (g_1^r, g_2^r)$
$pk := (X_1, X_2) = (g_1^x, g_2^x)$	$St := r$
$\mathsf{ChSet} := \{0,1\}^n$	Return (R, St)
Return (pk, sk)	
	$\mathsf{P}_2(sk, R, h, St)$:
$\mathsf{V}(pk, R = (R_1, R_2), h, s)$:	Parse $St = r$
If $R_1 = g^s \cdot X_1^{-h}$ and $R_2 = g^s \cdot X_2^{-h}$	Return $s = x \cdot h + r \bmod p$
then return 1	
Else return 0	

We recall the DDH assumption.

Definition 13 (Decision Diffie-Hellman Assumption). *The Decision Diffie-Hellman problem* DDH *is* (t, ε)*-hard in* par $= (p, g_1, g_2, \mathbb{G})$ *if for all adversaries* $\mathcal{A}$ *running in time at most* t,

$$\left| \Pr\left[1 \xleftarrow{\$} \mathcal{A}(g_1^x, g_2^x) \mid x \xleftarrow{\$} \mathbb{Z}_p \right] - \Pr\left[1 \xleftarrow{\$} \mathcal{A}(g_1^{x_1}, g_2^{x_2}) \mid x_1 \xleftarrow{\$} \mathbb{Z}_p; x_2 \xleftarrow{\$} \mathbb{Z}_p \setminus \{x_1\} \right] \right| \le \varepsilon.$$

Clearly, all security results of Schnorr carry over to the Chaum-Pedersen identification scheme, i.e., $\mathsf{ID_{CP}}$ is at least as secure as $\mathsf{ID_S}$. That also means that we cannot hope for tight PIMP-KOA security from the DLOG assumption. Instead, for the Chaum-Pedersen identification scheme, we give a direct tight proof of PIMP-KOA security under the DDH assumption which we extracted from [34].

Lemma 16. $\mathsf{ID_{CP}}$ *is a canonical identification scheme with* $\alpha = \log p$ *bit min-entropy and it is unique, has special soundness (SS), honest-verifier zero-knowledge (HVZK) and is random-self reducible (RSR). Moreover, if* DDH *is* (t, ε)*-hard in* par $= (p, g_1, g_2, \mathbb{G})$ *then* $\mathsf{ID_{CP}}$ *is* $(t', \varepsilon', Q_{\mathrm{CH}})$*-PIMP-KOA secure, where* $t \approx t'$ *and* $\varepsilon \ge \varepsilon' - Q_{\mathrm{CH}}/2^n$.

Proof. The proof of SS, HVZK, uniqueness, and RSR is the same as in $\mathsf{ID_S}$.

To prove PIMP-KOA-security under DDH, let $\mathcal{A}$ be an adversary that $(t', \varepsilon', Q_{\mathrm{CH}})$-breaks PIMP-KOA security. We build an adversary $\mathcal{B}$ against the (t, ε)-hardness of DDH as follows. Adversary $\mathcal{B}$ inputs (X_1, X_2) and defines $pk := (X_1, X_2)$. On the i-th challenge query $\mathrm{CH}(R_{i,1}, R_{i,2})$, it returns $h_i \xleftarrow{\$} \mathbb{Z}_p$. Eventually, $\mathcal{A}$ returns $i^* \in [Q_{\mathrm{CH}}]$ and s_{i^*} and terminates. Finally, $\mathcal{B}$ outputs $d := \mathsf{V}(pk, R_{i^*}, h_{i^*}, s_{i^*})$.

ANALYSIS OF $\mathcal{B}$. If $(X_1, X_2) = (g_1^x, g_2^x)$, then $\mathcal{B}$ perfectly simulates the PIMP-KOA game and hence $\Pr[d = 1 \mid (X_1, X_2) = (g_1^x, g_2^x)] = \varepsilon'$. If $(X_1, X_2) = (g_1^{x_1}, g_2^{x_2})$ with $x_1 \ne x_2$, then we claim that even a computationally unbounded $\mathcal{A}$ can only win with probability $Q_{\mathrm{CH}}/2^n$, i.e., $\Pr[d = 1 \mid (X_1, X_2) = (g_1^{x_1}, g_2^{x_2})] \le Q_{\mathrm{CH}}/2^n$.

It remains to prove the claim. For each index $i \in [Q_{\mathrm{CH}}]$, $\mathcal{A}$ first commits to $R_{i,1} = g_1^{r_{i,1}}$ and $R_{i,2} = g_2^{r_{i,2}}$ (for arbitrary $r_{i,1}, r_{i,2} \in \mathbb{Z}_p$) and can only win if there exists an $s_i \in \mathbb{Z}_p$ such that

$$r_{i,1} + h_i x_1 = s_i = r_{i,2} + h_i x_2$$
$$\Leftrightarrow h_i = \frac{r_{i,2} - r_{i,1}}{x_1 - x_2}$$

where $h_i \xleftarrow{\$} \{0, 1\}^n$ is chosen independently of $r_{i,1}, r_{i,2}$. This happens with probability at most $1/2^n$, so by the union bound we obtain the bound $Q_{\mathrm{CH}}/2^n$, as claimed. $\qquad\square$

Katz-Wang Signature Scheme. Let $H : \{0, 1\}^* \to \{0, 1\}^n$ be a hash function with $n < \log_2(p)$. As $\mathsf{ID_{CP}}$ is commitment-recoverable we can use the alternative Fiat-Shamir transformation to obtain a signature scheme which is known as the Katz-Wang signature scheme $\mathsf{KW} := (\mathsf{Gen}, \mathsf{Sign}, \mathsf{Ver})$.

Gen(par):	Sign(sk, m):	Ver(sk, m, σ):
$sk := x \xleftarrow{\$} \mathbb{Z}_p$	$r \xleftarrow{\$} \mathbb{Z}_p$	Parse $\sigma = (h, s) \in \{0,1\}^n \times \mathbb{Z}_p$
$(X_1, X_2) = (g_1^x, g_2^x)$	$R = (R_1, R_2) = (g_1^r, g_2^r)$	$R = g^s X^{-h}$
$pk := (X_1, X_2)$	$h = H(R, m)$	If $h = H(R, m)$ then return 1
Return (pk, sk)	$s = x \cdot h + r \bmod p$	Else return 0.
	$\sigma = (h, s) \in \{0,1\}^n \times \mathbb{Z}_p$	
	Return σ	

By our results we obtain the following concrete security statements, where the first bound matches [34, Theorem1].

Lemma 17. *If* DDH *is* (t, ε)-*hard in* par $= (p, g_1, g_2, \mathbb{G})$ *then* KW *is* $(t', \varepsilon', Q_s, Q_h)$-SUF-CMA *secure and* $(t'', \varepsilon'', N, Q_s, Q_h)$-MU-SUF-CMA *secure in the programmable random oracle model, where*

$$\frac{\varepsilon'}{t'} \leq \frac{\varepsilon}{t} + \frac{Q_s}{p} + \frac{1}{2^n},$$

$$\frac{\varepsilon''}{t''} \leq 4 \cdot \frac{\varepsilon}{t} + \frac{Q_s}{p} + \frac{1}{2^n}.$$

References

1. Abdalla, M., An, J.H., Bellare, M., Namprempre, C.: From identification to signatures via the Fiat-Shamir transform: minimizing assumptions for security and forward-security. In: Knudsen, L.R. (ed.) EUROCRYPT 2002. LNCS, vol. 2332, pp. 418–433. Springer, Heidelberg (2002)
2. Abdalla, M., Ben Hamouda, F., Pointcheval, D.: Tighter reductions for forward-secure signature schemes. In: Kurosawa, K., Hanaoka, G. (eds.) PKC 2013. LNCS, vol. 7778, pp. 292–311. Springer, Heidelberg (2013)
3. Abdalla, M., Fouque, P.-A., Lyubashevsky, V., Tibouchi, M.: Tightly-secure signatures from lossy identification schemes. In: Pointcheval, D., Johansson, T. (eds.) EUROCRYPT 2012. LNCS, vol. 7237, pp. 572–590. Springer, Heidelberg (2012)
4. Bellare, M., Namprempre, C., Pointcheval, D., Semanko, M.: The one-more-RSA-inversion problems and the security of Chaum's blind signature scheme. J. Cryptology **16**(3), 185–215 (2003)
5. Bellare, M., Neven, G.: Multi-signatures in the plain public-key model and a general forking lemma. In: Juels, A., Wright, R.N., Vimercati, S. (eds.) ACM CCS 2006, pp. 390–399. ACM Press, October/November 2006
6. Bellare, M., Palacio, A.: GQ and schnorr identification schemes: proofs of security against impersonation under active and concurrent attacks. In: Yung, M. (ed.) CRYPTO 2002. LNCS, vol. 2442, pp. 162–177. Springer, Heidelberg (2002)
7. Bellare, M., Ristenpart, T.: Simulation without the artificial abort: simplified proof and improved concrete security for Waters' IBE scheme. In: Joux, A. (ed.) EUROCRYPT 2009. LNCS, vol. 5479, pp. 407–424. Springer, Heidelberg (2009)
8. Bellare, M., Rogaway, P.: Random oracles are practical: a paradigm for designing efficient protocols. In: Ashby, V. (ed.) ACM CCS 1993, pp. 62–73. ACM Press, November 1993
9. Bellare, M., Rogaway, P.: The exact security of digital signatures - how to sign with RSA and Rabin. In: Maurer, U.M. (ed.) EUROCRYPT 1996. LNCS, vol. 1070, pp. 399–416. Springer, Heidelberg (1996)

10. Bernstein, D.: [Cfrg] key as message prefix => multi-key security. https://mailarchive.ietf.org/arch/msg/cfrg/44gJyZlZ7-myJqWkChhpEF1KE9M, 2015
11. Bernstein, D.J.: Multi-user Schnorr security, revisited. Cryptology ePrint Archive, Report 2015/996, 2015. http://eprint.iacr.org/
12. Bernstein, D.J., Duif, N., Lange, T., Schwabe, P., Yang, B.-Y.: High-speed high-security signatures. In: Preneel, B., Takagi, T. (eds.) CHES 2011. LNCS, vol. 6917, pp. 124–142. Springer, Heidelberg (2011)
13. Beth, T.: Efficient zero-knowledged identification scheme for smart cards. In: Günther, C.G. (ed.) EUROCRYPT 1988. LNCS, vol. 330, pp. 77–84. Springer, Heidelberg (1988)
14. Boneh, D., Gentry, C., Lynn, B., Shacham, H.: Aggregate and verifiably encrypted signatures from bilinear maps. In: Biham, E. (ed.) EUROCRYPT 2003. LNCS, vol. 2656, pp. 416–432. Springer, Heidelberg (2003)
15. Brickell, E.F., McCurley, K.S.: An interactive identification scheme based on discrete logarithms and factoring. In: Damgård, I.B. (ed.) EUROCRYPT 1990. LNCS, vol. 473, pp. 63–71. Springer, Heidelberg (1991)
16. Brown, D.: [Cfrg] key as message prefix => multi-key security. http://www.ietf.org/mail-archive/web/cfrg/current/msg07336.html, 2015
17. Canetti, R., Goldreich, O., Halevi, S.: The random oracle methodology, revisited (preliminary version). In: 30th ACM STOC, pp. 209–218. ACM Press, May 1998
18. Chatterjee, S., Koblitz, N., Menezes, A., Sarkar, P.: Another look at tightness II: practical issues in cryptography. Cryptology ePrint Archive, Report 2016/360 (2016). http://eprint.iacr.org/
19. Chaum, D., Pedersen, T.P.: Wallet databases with observers. In: Brickell, E.F. (ed.) CRYPTO 1992. LNCS, vol. 740, pp. 89–105. Springer, Heidelberg (1993)
20. Fiat, A., Shamir, A.: How to prove yourself: practical solutions to identification and signature problems. In: Odlyzko, A.M. (ed.) CRYPTO 1986. LNCS, vol. 263, pp. 186–194. Springer, Heidelberg (1987)
21. Fischlin, M., Fleischhacker, N.: Limitations of the meta-reduction technique: the case of schnorr signatures. In: Johansson, T., Nguyen, P.Q. (eds.) EUROCRYPT 2013. LNCS, vol. 7881, pp. 444–460. Springer, Heidelberg (2013)
22. Fischlin, M., Lehmann, A., Ristenpart, T., Shrimpton, T., Stam, M., Tessaro, S.: Random oracles with(out) programmability. In: Abe, M. (ed.) ASIACRYPT 2010. LNCS, vol. 6477, pp. 303–320. Springer, Heidelberg (2010)
23. Fleischhacker, N., Jager, T., Schröder, D.: On tight security proofs for schnorr signatures. In: Sarkar, P., Iwata, T. (eds.) ASIACRYPT 2014. LNCS, vol. 8873, pp. 512–531. Springer, Heidelberg (2014)
24. Fukumitsu, M., Hasegawa, S.: Black-box separations on Fiat-shamir-type signatures in the non-programmable random oracle model. In: López, J., Mitchell, C.J. (eds.) ISC 2015. LNCS, vol. 9290, pp. 3–20. Springer, Heidelberg (2015)
25. Galbraith, S.D., Malone-Lee, J., Smart, N.P.: Public key signatures in the multi-user setting. Inf. Process. Lett. 83(5), 263–266 (2002)
26. Galindo, D.: The exact security of pairing based encryption and signature schemes. Based on a talk at Workshop on Provable Security, INRIA, Paris (2004). http://www.dgalindo.es/galindoEcrypt.pdf
27. Garg, S., Bhaskar, R., Lokam, S.V.: Improved bounds on security reductions for discrete log based signatures. In: Wagner, D. (ed.) CRYPTO 2008. LNCS, vol. 5157, pp. 93–107. Springer, Heidelberg (2008)
28. Girault, M.: An identity-based identification scheme based on discrete logarithms modulo a composite number. In: Damgård, I.B. (ed.) EUROCRYPT 1990. LNCS, vol. 473, pp. 481–486. Springer, Heidelberg (1991)

29. Goh, E.-J., Jarecki, S., Katz, J., Wang, N.: Efficient signature schemes with tight reductions to the Diffie-Hellman problems. J. Cryptology **20**(4), 493–514 (2007)

30. Goldwasser, S., Micali, S., Rivest, R.L.: A digital signature scheme secure against adaptive chosen-message attacks. SIAM J. Comput. **17**(2), 281–308 (1988)

31. Guillou, L.C., Quisquater, J.-J.: A "Paradoxical" identity-based signature scheme resulting from zero-knowledge. In: Goldwasser, S. (ed.) CRYPTO 1988. LNCS, vol. 403, pp. 216–231. Springer, Heidelberg (1990)

32. Hamburg, M.: Re: [Cfrg] EC signature: next steps (2015). https://mailarchive.ietf. org/arch/msg/cfrg/af170b6OrLyNZUHBMOPWxcDrVRI

33. Josefsson, S., Liusvaara, I.: Edwards-curve digital signature algorithm (EdDSA), 7 October 2015. https://tools.ietf.org/html/draft-irtf-cfrg-eddsa-00

34. Katz, J., Wang, N.: Efficiency improvements for signature schemes with tight security reductions. In: Jajodia, S., Atluri, V., Jaeger, T. (eds.) ACM CCS 2003, pp. 155–164. ACM Press, October 2003

35. Kiltz, E., Masny, D., Pan, J.: Optimal security proofs for signatures from identification schemes. Cryptology ePrint Archive, Report 2016/191 (2016). http://eprint.iacr.org/

36. Micali, S., Shamir, A.: An improvement of the Fiat-Shamir identification and signature scheme. In: Goldwasser, S. (ed.) CRYPTO 1988. LNCS, vol. 403, pp. 244–247. Springer, Heidelberg (1990)

37. Ohta, K., Okamoto, T.: On concrete security treatment of signatures derived from identification. In: Krawczyk, H. (ed.) CRYPTO 1998. LNCS, vol. 1462, pp. 354–369. Springer, Heidelberg (1998)

38. Okamoto, T.: Provably secure and practical identification schemes and corresponding signature schemes. In: Brickell, E.F. (ed.) CRYPTO 1992. LNCS, vol. 740, pp. 31–53. Springer, Heidelberg (1993)

39. Ong, H., Schnorr, C.-P.: Fast signature generation with a Fiat-Shamir-like scheme. In: Damgård, I.B. (ed.) EUROCRYPT 1990. LNCS, vol. 473, pp. 432–440. Springer, Heidelberg (1991)

40. Paillier, P., Vergnaud, D.: Discrete-log-based signatures may not be equivalent to discrete log. In: Roy, B. (ed.) ASIACRYPT 2005. LNCS, vol. 3788, pp. 1–20. Springer, Heidelberg (2005)

41. Pointcheval, D., Stern, J.: Security arguments for digital signatures and blind signatures. J. Cryptology **13**(3), 361–396 (2000)

42. Schnorr, C.-P.: Efficient signature generation by smart cards. J. Cryptology **4**(3), 161–174 (1991)

43. Seurin, Y.: On the exact security of schnorr-type signatures in the random oracle model. In: Pointcheval, D., Johansson, T. (eds.) EUROCRYPT 2012. LNCS, vol. 7237, pp. 554–571. Springer, Heidelberg (2012)

44. Struik, R.: Re: [Cfrg] EC signature: next steps (2015). https://mailarchive.ietf.org/arch/msg/cfrg/TOWH1DSzB-PfDGK8qEXtF3iC6Vc

FHE Circuit Privacy Almost for Free

Florian Bourse[(✉)], Rafaël Del Pino, Michele Minelli, and Hoeteck Wee

ENS, CNRS, INRIA and PSL Research University, Paris, France
{fbourse,delpino,minelli,wee}@di.ens.fr

Abstract. Circuit privacy is an important property for many applications of fully homomorphic encryption. Prior approaches for achieving circuit privacy rely on superpolynomial noise flooding or on bootstrapping. In this work, we present a conceptually different approach to circuit privacy based on a novel characterization of the noise growth amidst homomorphic evaluation. In particular, we show that a variant of the GSW FHE for branching programs already achieves circuit privacy; this immediately yields a circuit-private FHE for NC^1 circuits under the standard LWE assumption with polynomial modulus-to-noise ratio. Our analysis relies on a variant of the discrete Gaussian leftover hash lemma which states that $\mathbf{e}^\mathsf{T}\mathbf{G}^{-1}(\mathbf{v}) + small\ noise$ does not depend on $\mathbf{v}$. We believe that this result is of independent interest.

Keywords: Homomorphic encryption · Circuit privacy · Branching program · Noise flooding · Learning with errors · Rerandomization

1 Introduction

A fully homomorphic encryption (FHE) scheme is an encryption scheme which supports computation on encrypted data: given a ciphertext that encrypts some data μ, one can compute a ciphertext that encrypts $f(\mu)$ for any efficiently computable function f, without ever needing to decrypt the data or know the decryption key. FHE has numerous theoretical and practical applications, the canonical one being to the problem of outsourcing computation to a remote server without compromising one's privacy. In 2009, Gentry put forth the first candidate construction of FHE based on ideal lattices [Gen09]. Since then, substantial progress has been made [vDGHV10, SS10, SV10, BV11a, BV11b, BGV12, GHS12, GSW13, BV14, AP14], offering various improvements in conceptual and

F. Bourse—Supported by the European Research Council under the European Community's Seventh Framework Programme (FP7/2007-2013 Grant Agreement no. 339563 CryptoCloud)

R. Del Pino—Supported by SAFEcrypto (H2020 ICT-644729)

M. Minelli was supported by European Union's Horizon 2020 research and innovation programme under grant agreement No H2020-MSCA-ITN-2014-643161 ECRYPT-NET.

H. Wee—Columbia University and CQT, NUS. Supported in part by the ERC Project aSCEND (H2020 639554) and NSF Award CNS-1445424.

© International Association for Cryptologic Research 2016
M. Robshaw and J. Katz (Eds.): CRYPTO 2016, Part II, LNCS 9815, pp. 62–89, 2016.
DOI: 10.1007/978-3-662-53008-5_3

technical simplicity, efficiency, security guarantees, assumptions, etc.; in particular, Gentry, Sahai and Waters presented a very simple FHE (hereafter called the GSW cryptosystem) based on the standard learning with errors (LWE) assumption.

Circuit Privacy. An additional requirement in many FHE applications is that the evaluated ciphertext should also hide the function f, apart from what is inevitably leaked through the outcome of the computation $f(\mu)$; we refer to this requirement as *circuit privacy* [SYY99, IP07]. In the context of outsourcing computation, a server may wish to hide its proprietary algorithm from the client. Circuit privacy is also a requirement when we use FHE for low-communication secure two-party computation. In all existing FHE schemes, there is a "noise" term in the ciphertext, which is necessary for security. The noise grows and changes as a result of performing homomorphic operations and, in particular, could leak information about the function f. The main challenge for achieving FHE circuit privacy lies precisely in avoiding the leakage from the noise term in the evaluated ciphertext.

Prior Works. Prior works achieve circuit privacy by essentially canceling out the noise term in the evaluated ciphertext. There are two main approaches for achieving this. The first is "noise flooding" introduced in Gentry's thesis, where we add a much larger noise at the end of the computation; in particular, the noise that is added needs to be super-polynomially larger than the noise that accumulates amidst homomorphic operations, which in turn requires that we start with a super-polynomial modulus-to-noise ratio.[1] This is a fairly mild assumption for the early constructions of FHE schemes, which required a quasi-polynomial modulus-to-noise ratio just to support homomorphic operations for circuits in NC^1 (i.e., circuits of logarithmic depth). The second is to decrypt and re-encrypt the evaluated ciphertext, also known as bootstrapping in the FHE literature. This can be achieved securely without having to know the secret key in the clear in one of two ways: (i) with the use of garbled circuits [OPP14, GHV10], and (ii) via homomorphic evaluation of the decryption circuit given an encryption of the secret key under itself [DS16], which requires the additional assumption of circular security.

Both of the prior approaches have some theoretical and practical draw-backs, if we consider FHE for NC^1 circuits (the rest of the discussion also applies to leveled FHE for general circuits). First, recall that we now have FHE for NC^1 circuits under the LWE assumption with a polynomial modulus-to-noise ratio [BV14, AP14], and we would ideally like to achieve circuit privacy under the same assumption. Relying on noise flooding for circuit privacy would require quantitatively stronger assumptions with a super-polynomial modulus-to-noise ratio, which in turn impacts practical efficiency due to the use of larger parameters. Similarly, the use of bootstrapping for circuit privacy can also be computationally expensive (indeed, the bootstrapping operation is the computational

---

[1] Recall that LWE hardness depends on the modulus-to-noise ratio: the smaller the ratio, the harder the problem.

bottleneck in existing FHE schemes, cf. [DM15, HS15]). Moreover, realizing bootstrapping via an encryption of the secret key requires an additional circular security assumption, which could in turn also entail the use of larger parameters in order to account for potential weaknesses introduced by circular security. Realizing bootstrapping via garbled circuits avoids the additional assumption, but is theoretically and practically unsatisfying as it requires encoding the algebraic structure in existing FHEs as boolean computation, and sacrifices the multi-hop property in that we can no longer perform further homomorphic computation on the evaluated ciphertexts.

1.1 Our Results

Our main result is a circuit-private FHE for NC^1 circuits – and a circuit-private leveled FHE for general circuits – under the LWE assumption with a polynomial modulus-to-noise ratio, and whose efficiency essentially matches that of existing variants of the GSW cryptosystem in [BV14, AP14]; in other words, we avoid noise flooding or bootstrapping and obtain circuit privacy almost for free!

We obtain our main result via a conceptually different approach from prior works: instead of canceling out the noise term in the evaluated ciphertext, we directly analyze the *distribution* of the noise term (prior works on FHE merely gave a bound on the noise term). Concretely, we show that adding a small noise in each step of homomorphic evaluation in the GSW cryptosystem already hides the computation itself which yields circuit privacy. Along the way, we gain better insights into the algebraic structure and the noise distribution in GSW scheme and provide new tools for analyzing noise randomization which we believe could be of independent interest.

As an immediate corollary, we obtain a two-party protocol for secure function evaluation where Alice holds x, Bob holds a branching program f, and we want Alice to learn $f(x)$ while protecting the privacy of x and f to the largest extent possible, that is, Bob learns nothing about x and Alice learns nothing about f (apart from a bound on the size of f). Our protocol achieves semi-honest security under the standard LWE assumption with polynomial hardness, and where the total communication complexity and Alice's computation are poly-logarithmic in the size of f.

The core of our analysis is a variant of the Gaussian leftover hash lemma [AGHS13, AR13]: given a "small" vector $\mathbf{e}$ and any vector $\mathbf{v}$, we have

$$\mathbf{e}^{\mathsf{T}} \cdot \mathbf{G}_{\mathrm{rand}}^{-1}(\mathbf{v}) + y \approx_s e'$$

where

- $\mathbf{G}_{\mathrm{rand}}^{-1}(\mathbf{v})$ outputs a random short vector $\mathbf{x}$ satisfying $\mathbf{G}\mathbf{x} = \mathbf{v} \mod q$ according to a discrete Gaussian with parameter $r = \tilde{O}(1)$;
- both y and e' are drawn from discrete Gaussians with parameter $O(r \cdot \|\mathbf{e}\|)$ (the norm of e' will be slightly larger than that of y).

We stress that the distribution of e' is independent of $\mathbf{v}$ and that the norm of y, e' are polynomially related to that of $\|\mathbf{e}\|$. Indeed, a similar statement is true via noise flooding, where we pick y, e' to have norm super-polynomially larger than that of $\|\mathbf{e}\|$. Using this leftover hash lemma to hide the argument of $\mathbf{G}_{\mathrm{rand}}^{-1}(\cdot)$ is new to this work and will be crucial in proving circuit privacy. In Table 1 we show a comparison with previous works on how to perform a step of computation for branching program evaluation.

1.2 Technical Overview

We proceed with a technical overview of our construction. We build up to our main construction in three steps.

Generating Fresh LWE Samples. How do we generate a fresh LWE sample from a large but bounded number of samples? That is, we need to randomize $(\mathbf{A}, \mathbf{s}^\mathsf{T}\mathbf{A} + \mathbf{e}^\mathsf{T})$. The first idea, going back to [Reg05, GPV08, ACPS09] is to choose $\mathbf{x}$ according to a discrete Gaussian with parameter $r = \tilde{O}(1)$ and a small "smoothing" noise y from a discrete Gaussian with parameter $O(r \cdot \|\mathbf{e}\|)$ and output

$$\mathbf{A}\mathbf{x}, (\mathbf{s}^\mathsf{T}\mathbf{A} + \mathbf{e}^\mathsf{T})\mathbf{x} + y$$

The vector $\mathbf{A}\mathbf{x}$ is statistically close to uniform (by leftover hash lemma), and the error $\mathbf{e}^\mathsf{T}\mathbf{x} + y$ in the resulting sample is statistically close to a discrete Gaussian with parameter $O(r \cdot \|\mathbf{e}\|)$. We stress that the norm of y is polynomially related to that of $\mathbf{e}$, which is better than naive noise flooding. One draw-back compared to noise flooding is that the error in the new sample leaks $\|\mathbf{e}\|$. In the case of generating fresh LWE samples, we just need to repeat the process to generate many more samples than what we started out with.

Randomizing GSW Ciphertexts. Next, we note that the above idea can also be used to randomize GSW ciphertexts. Recall that a GSW encryption of a message μ is of the form

$$\mathbf{C} = \begin{pmatrix} \mathbf{A} \\ \mathbf{s}^\mathsf{T}\mathbf{A} + \mathbf{e}^\mathsf{T} \end{pmatrix} + \mu\mathbf{G} \in \mathbb{Z}_q^{n \times (n \log q)}$$

where $\mathbf{s} \in \mathbb{Z}_q^n$ is the secret key and $\mathbf{G}$ is the "powers of 2" gadget matrix. We can randomize $\mathbf{C}$ to be a fresh encryption of μ by computing

$$\mathbf{C} \cdot \mathbf{G}_{\mathrm{rand}}^{-1}(\mathbf{G}) + \begin{pmatrix} \mathbf{0} \\ \mathbf{y}^\mathsf{T} \end{pmatrix}$$

where $\mathbf{G}_{\mathrm{rand}}^{-1}(\mathbf{G})$ is chosen according to a discrete Gaussian of parameter r satisfying $\mathbf{G} \cdot \mathbf{G}_{\mathrm{rand}}^{-1}(\mathbf{G}) = \mathbf{G}$ and $\mathbf{y}$ is again a small smoothing noise vector. Here, we need an extension of the previous lemma showing that each coordinate in $\mathbf{e}^\mathsf{T} \cdot \mathbf{G}_{\mathrm{rand}}^{-1}(\mathbf{G}) + \mathbf{y}^\mathsf{T}$ is statistically close to a discrete Gaussian; this in turn follows from an extension of the previous lemma where the vector $\mathbf{x}$ is drawn from

Table 1. The first row of the table shows the plaintext computation that happens at each step of the computation for evaluating a branching program (cf. Sect. 5.1). The next three rows describe how this computation is carried out homomorphically on ciphertexts $\mathbf{V}_0, \mathbf{V}_1, \mathbf{C}$ corresponding to encryptions of the input bits v_0, v_1, x. In the [GSW13,BV14] FHE schemes, homomorphic evaluation is deterministic, whereas in [AP14] and this work, homomorphic evaluation is randomized. In particular, our construction introduces an additional small Gaussian shift on top of [AP14].

Plaintext	$v_{\text{out}} = v_x = xv_1 + (1-x)v_0$
[GSW13, BV14]	$\mathbf{V}_{\text{out}} = \mathbf{C} \cdot \mathbf{G}_{\text{det}}^{-1}(\mathbf{V}_1) + (\mathbf{G} - \mathbf{C}) \cdot \mathbf{G}_{\text{det}}^{-1}(\mathbf{V}_0)$
[AP14]	$\mathbf{V}_{\text{out}} = \mathbf{C} \cdot \mathbf{G}_{\text{rand}}^{-1}(\mathbf{V}_1) + (\mathbf{G} - \mathbf{C}) \cdot \mathbf{G}_{\text{rand}}^{-1}(\mathbf{V}_0)$
[This work]	$\mathbf{V}_{\text{out}} = \mathbf{C} \cdot \mathbf{G}_{\text{rand}}^{-1}(\mathbf{V}_1) + (\mathbf{G} - \mathbf{C}) \cdot \mathbf{G}_{\text{rand}}^{-1}(\mathbf{V}_0) + \begin{pmatrix} \mathbf{0} \\ \hline \mathbf{y}^{\mathsf{T}} \end{pmatrix}$

discrete Gaussian over the coset of a lattice (cf. Lemma 3.6). And again, the norm of $\mathbf{y}$ is polynomially related to that in $\mathbf{e}$, which is better than naive noise flooding.

Scaling GSW Ciphertexts. More interesting, given a constant $a \in \{0, 1\}$, we can scale a GSW encryption of μ to obtain a fresh encryption of $a \cdot \mu$ while revealing no information about a beyond what is leaked in $a \cdot \mu$. In particular, if $\mu = 0$, then the resulting ciphertext should completely hide a. To achieve this, we simply proceed as before, except we use $\mathbf{G}_{\text{rand}}^{-1}(a \cdot \mathbf{G})$ so that $\mathbf{G} \cdot \mathbf{G}_{\text{rand}}^{-1}(a \cdot \mathbf{G}) = a \cdot \mathbf{G}$. Here, we crucially rely on the fact that the error $\mathbf{e}^{\mathsf{T}} \cdot \mathbf{G}_{\text{rand}}^{-1}(a \cdot \mathbf{G}) + \mathbf{y}^{\mathsf{T}}$ in the resulting ciphertext is independent of a.

Circuit-Private Homomorphic Evaluation. The preceding construction extends to the setting where we are given a GSW encryption $\mathbf{C}'$ of a instead of a itself, so that we output

$$\mathbf{C} \cdot \mathbf{G}_{\text{rand}}^{-1}(\mathbf{C}') + \begin{pmatrix} \mathbf{0} \\ \hline \mathbf{y}^{\mathsf{T}} \end{pmatrix}$$

We can handle homomorphic encryption as in GSW; this then readily extends to a circuit-private homomorphic evaluation for branching programs, following [BV14, AP14].

Branching programs are a relatively powerful representation model. In particular, any logarithmic space or NC^1 computation can be carried out by a family of polynomial-size branching programs. Branching programs can also directly capture several representation models often used in practice such as decision trees, OBDDs, and deterministic finite automaton.

The key insight from Brakerski and Vaikuntanathan [BV14] is that when homomorphically evaluating a branching program, we will only need to perform homomorphic additions along with homomorphic multiplications of ciphertexts

$\mathbf{V}_j, \mathbf{C}_i$ where $\mathbf{V}_j$ is the encryption of an intermediate computation and $\mathbf{C}_i$ is an encryption of the input variable x_i. To obtain decryption correctness with polynomial noise growth, they computed the product as

$$\mathbf{C}_i \cdot \mathbf{G}_{\det}^{-1}(\mathbf{V}_j),$$

where $\mathbf{G}_{\det}^{-1}(\cdot)$ denotes the deterministic binary decomposition, cleverly exploiting the asymmetric noise growth in GSW ciphertexts and the fact that the noise in $\mathbf{C}_i$ is smaller than that in $\mathbf{V}_j$. To obtain circuit privacy, we will compute the product as

$$\mathbf{C}_i \cdot \mathbf{G}_{\mathrm{rand}}^{-1}(\mathbf{V}_j) + \begin{pmatrix} \mathbf{0} \\ \mathbf{y}_j^{\mathsf{T}} \end{pmatrix}.$$

Note that we made two modifications:

- First, we switched to a randomized $\mathbf{G}_{\mathrm{rand}}^{-1}(\cdot)$. The use of a randomized $\mathbf{G}_{\mathrm{rand}}^{-1}(\cdot)$ for homomorphic evaluation was first introduced in [AP14], but for the very different purpose of a mild improvement in the noise growth (i.e. *efficiency*); here, we crucially exploit randomization for *privacy*.
- Next, we introduced an additional Gaussian shift $\mathbf{y}_j^{\mathsf{T}}$.

Interestingly, it turns out that computing the product as $\mathbf{C}_i \cdot \mathbf{G}_{\mathrm{rand}}^{-1}(\mathbf{V}_j)$ instead of $\mathbf{V}_j \cdot \mathbf{G}_{\mathrm{rand}}^{-1}(\mathbf{C}_i)$ is useful not only for polynomial noise growth, but also useful for circuit privacy. Roughly speaking, the former hides which $\mathbf{V}_j$ is used, which corresponds to hiding the intermediate states that lead to the final output state, which in turn hides the branching program.

We highlight a subtlety in the analysis: $\mathbf{V}_j$ could in principle encode information about $\mathbf{C}_i$, if the variable x_i has been read prior to reaching the intermediate state encoded in $\mathbf{V}_j$, whereas to apply our randomization lemma, we crucially rely on independence between $\mathbf{C}_i$ and $\mathbf{V}_j$. The analysis proceeds by a careful induction argument showing that $\mathbf{V}_j$ looks like a fresh GSW ciphertext independent of input ciphertexts $\mathbf{C}_1, \ldots, \mathbf{C}_\ell$ apart from some dependencies on the norm of the noise terms in the input ciphertexts (see Lemma 5.4 for a precise statement). These dependencies mean that homomorphic evaluation leaks the number of times each variable appears in the branching program, but that can be easily fixed by padding the branching program.

1.3   Discussions

One draw-back of our approach is that it is specific to the GSW cryptosystem and variants there-of, whereas previous approaches based on noise flooding and bootstrapping are fairly generic; another is that we need to pad the branching program so that each variable appears the same number of times. Nonetheless, we stress that the GSW cryptosystem turns out to be ubiquitous in many applications outside of FHE, including attribute-based encryption and fully homomorphic signatures [BGG+14, GVW15]. We are optimistic that the additional insights we gained into the noise distributions of GSW ciphertexts in this work will find applications outside of FHE.

We conclude with several open problems pertaining to FHE circuit privacy. The first is to achieve circuit privacy against malicious adversaries [OPP14]: namely, the result of a homomorphic evaluation should leak no information about the circuit f, even if the input ciphertexts are maliciously generated. Our analysis breaks down in this setting as it crucially uses fresh uniform randomness in the input ciphertexts for left-over hash lemma, and the fact that the noise in the input ciphertexts are small (but does not need to be discrete Gaussian). Another is to achieve circuit-private CCA1-secure FHE [LMSV12]; here, the technique that [DS16] uses to achieve circuit privacy cannot obtain such a result since giving out an encryption of the secret key violates CCA1-security. A third open problem is to extend the techniques in this work to other FHE schemes, such as those in [BV11a,DM15,HS15].

2　Preliminaries

In this section we clarify our notation and recall some definitions, problems and lemmas that we are going to use throughout the paper.

Notation. We denote the real numbers by $\mathbb{R}$, the integers by $\mathbb{Z}$, the integers modulo some q by $\mathbb{Z}_q$, and let $[N]$ indicate the integer numbers $\{1, \ldots, N\}$. Throughout the paper we use λ to denote the security parameter. We say that a function is *negligible* in λ, and we denote it by $\mathrm{negl}\,(\lambda)$, if it is a $f\,(\lambda) = o\,(\lambda^{-c})$ for every fixed constant c. We also say that a probability is *overwhelming* if it is $1 - \mathrm{negl}\,(\lambda)$.

Vectors are denoted by lower-case bold letters (e.g., $\mathbf{v}$) and are always in column form ($\mathbf{v}^\mathsf{T}$ is a row vector), while matrices are indicated by upper-case bold letters. We let $(\mathbf{a}, \mathbf{b})$ denote the vector obtained by concatenating the two vectors, i.e. $\begin{pmatrix} \mathbf{a} \\ \mathbf{b} \end{pmatrix}$. We also write $(\mathbf{v}_1 \mid \mathbf{v}_2 \mid \ldots \mid \mathbf{v}_k)$ to denote the matrix whose columns are the vectors $\mathbf{v}_i$. Unless otherwise stated, the norm $\|\cdot\|$ considered in this paper is the ℓ_2 norm and log denotes the base-2 logarithm, while ln denotes the natural logarithm.

Given two distributions X, Y over a finite or countable domain D, their statistical distance is defined as $\Delta\,(X, Y) = \frac{1}{2} \sum_{v \in D} |X\,(v) - Y\,(v)|$. We say that two distributions are *statistically close* (denoted by $\approx_s$) if their statistical distance is $\mathrm{negl}\,(\lambda)$. Given a set A, we will write $a \xleftarrow{\$} A$ to indicate that a is sampled from A uniformly at random. If $\mathcal{D}$ is a probability distribution, we will write $d \leftarrow \mathcal{D}$ to indicate that d is sampled according to the distribution $\mathcal{D}$. Following [MP12], we denote by $\mathbf{G}$ the gadget matrix, i.e. $\mathbf{G} = \mathbf{g}^\mathsf{T} \otimes \mathbf{I}_n$, where $\mathbf{g}$ is the vector $\left(1, 2, 4, \ldots, 2^{\lceil \log q \rceil - 1}\right)$, for given parameters n, q.

Lattices. A m-dimensional lattice Λ is a discrete additive subgroup of $\mathbb{R}^m$. For an integer $k < m$ and a rank k matrix $\mathbf{B} \in \mathbb{R}^{m \times k}$, $\Lambda\,(\mathbf{B}) = \left\{\mathbf{Bx} \in \mathbb{R}^m \mid \mathbf{x} \in \mathbb{Z}^k\right\}$ is the lattice generated by the columns of $\mathbf{B}$. We will let $\Lambda_q^\perp\,(\mathbf{B})$ denote $\{\mathbf{v} \in \mathbb{Z}^m \mid \mathbf{B}^\mathsf{T}\mathbf{v} = \mathbf{0} \mod q\}$.

Gaussian Function. For any $\alpha > 0$, the spherical Gaussian function with parameter α (omitted if 1) is defined as $\rho_\alpha(\mathbf{x}) = \exp\left(-\pi\|\mathbf{x}\|^2/\alpha^2\right)$, for any $\mathbf{x} \in \mathbb{R}^m$. Given a lattice $\Lambda \subseteq \mathbb{R}^m$, a parameter $r \in \mathbb{R}$ and a vector $\mathbf{c} \in \mathbb{R}^m$ the spherical Gaussian distribution with parameter r and support $\Lambda + \mathbf{c}$ is defined as

$$\mathcal{D}_{\Lambda+\mathbf{c},r}(\mathbf{x}) = \frac{\rho_r(\mathbf{x})}{\rho_r(\Lambda+\mathbf{c})}, \quad \forall \mathbf{x} \in \Lambda + \mathbf{c}$$

where $\rho_r(\Lambda + \mathbf{c})$ denotes $\sum_{\mathbf{x}\in\Lambda+\mathbf{c}} \rho_r(\mathbf{x})$. Note that $\rho_r(\mathbf{x}) = \rho\left(r^{-1}\mathbf{x}\right)$.

We now give an algorithm for the randomized bit decomposition $\mathbf{G}_{\text{rand}}^{-1}(\cdot)$.

Definition 2.1 (The $\mathbf{G}_{rand}^{-1}(\cdot)$ algorithm, adapted from [MP12], [AP14, Claim 3.1]). *There is a randomized, efficiently computable function $\mathbf{G}_{rand}^{-1}(\cdot) : \mathbb{Z}_q^n \to \mathbb{Z}^m$, where $m = n\lceil\log q\rceil$ such that $\mathbf{x} \leftarrow \mathbf{G}_{rand}^{-1}(\mathbf{v})$ is drawn from a distribution close to a Gaussian with parameter $r = \tilde{O}(1)$ conditioned on $\mathbf{Gx} = \mathbf{v}$ mod q, i.e. $\mathbf{G}_{rand}^{-1}(\mathbf{v})$ outputs a sample from the distribution $\mathcal{D}_{\Lambda_q^\perp(\mathbf{G}^\intercal)+\mathbf{G}_{det}^{-1}(\mathbf{v}),r}$ where $\mathbf{G}_{det}^{-1}(\cdot)$ denotes (deterministic) bit decomposition. We will also write $\mathbf{X} \leftarrow \mathbf{G}_{rand}^{-1}(\mathbf{M})$ to denote that the columns of the matrix $\mathbf{X} \in \mathbb{Z}^{m\times p}$ are obtained by applying the algorithm separately to each column of a matrix $\mathbf{M} \in \mathbb{Z}_q^{n\times p}$.*

In particular, using the exact sampler in [BLP+13, Sect. 5] (which is a variant of the algorithm presented in [GPV08]), $\mathbf{G}_{\text{rand}}^{-1}(\mathbf{v})$ outputs a sample from the discrete Gaussian

$$\mathcal{D}_{\Lambda_q^\perp(\mathbf{G}^\intercal)+\mathbf{G}_{\text{det}}^{-1}(\mathbf{v}),r}$$

Next, we recall the definition of the *smoothing parameter* of a lattice from [MR04]. Intuitively, this parameter provides the width beyond which the discrete Gaussian measure on a lattice behaves like a continuous one.

Definition 2.2 (Smoothing parameter). *For a lattice $\Lambda \subseteq \mathbb{Z}^m$ and positive real $\varepsilon > 0$, the smoothing parameter $\eta_\varepsilon(\Lambda)$ is the smallest real $r > 0$ such that $\rho_{1/r}(\Lambda^* \setminus \{\mathbf{0}\}) \leq \varepsilon$, where $\Lambda^* := \{\mathbf{x} \in \mathbb{R}^m \mid \mathbf{x}^\intercal\Lambda \subseteq \mathbb{Z}\}$.*

We will also need the following probability results.

Lemma 2.3 (Simplified version of [Pei10, Theorem 3.1]). *Let $\varepsilon > 0$, $r_1, r_2 > 0$ be two Gaussian parameters, and $\Lambda \subseteq \mathbb{Z}^m$ be a lattice. If $\frac{r_1 r_2}{\sqrt{r_1^2+r_2^2}} \geq \eta_\varepsilon(\Lambda)$, then*

$$\Delta(\mathbf{y}_1 + \mathbf{y}_2, \mathbf{y}') \leq 8\varepsilon$$

where $\mathbf{y}_1 \leftarrow \mathcal{D}_{\Lambda,r_1}$, $\mathbf{y}_2 \leftarrow \mathcal{D}_{\Lambda,r_2}$, and $\mathbf{y}' \leftarrow \mathcal{D}_{\Lambda,\sqrt{r_1^2+r_2^2}}$.

Lemma 2.4 [AP14, Lemma 2.1]. *There exists a universal constant $C > 0$, such that*

$$\Pr\left[\|\mathbf{x}\| > Cr\sqrt{m}\right] \leq 2^{-\Omega(m)}$$

where $\mathbf{x} \leftarrow \mathcal{D}_{\mathbb{Z}^m,r}$.

Next, we recall the LWE problem and its hardness assumption.

The LWE Problem and Assumption. The learning with errors (LWE) problem was introduced by Regev in [Reg05] as a generalization of "learning parity with noise". Let $q \geq 2$, n and $m = \text{poly}(n)$ be positive integers, and let χ be a probability distribution over $\mathbb{Z}_q$. We define the following advantage function for an adversary $\mathcal{A}$:

$$\text{Adv}_{\mathcal{A}}^{\text{LWE}_{n,q,\chi}} := \left| \Pr\left[\mathcal{A}\left(\mathbf{A}, \mathbf{s}^\mathsf{T}\mathbf{A} + \mathbf{e}^\mathsf{T}\right) = 1\right] - \Pr\left[\mathcal{A}\left(\mathbf{A}, \mathbf{u}\right) = 1\right] \right|$$

where $\mathbf{A} \overset{\$}{\leftarrow} \mathbb{Z}_q^{n \times m}$, $\mathbf{s} \overset{\$}{\leftarrow} \mathbb{Z}_q^n$, $\mathbf{e} \leftarrow \chi$ and $\mathbf{u} \overset{\$}{\leftarrow} \mathbb{Z}_q^m$. The LWE assumption asserts that for any PPT adversary $\mathcal{A}$, the advantage $\text{Adv}_{\mathcal{A}}^{\text{LWE}_{n,q,\chi}}$ is $\text{negl}(n)$.

Finally, we recall the definition of a homomorphic encryption scheme, evaluation correctness and semantic security.

Homomorphic Encryption Scheme. A homomorphic (secret-key) encryption scheme $\mathcal{E} = (\mathcal{E}.\text{Setup}, \mathcal{E}.\text{Encrypt}, \mathcal{E}.\text{Decrypt}, \mathcal{E}.\text{Eval})$ is a quadruple of PPT algorithms as follows:

- $\mathcal{E}.\text{Setup}\left(1^\lambda\right)$: given the security parameter λ, outputs a secret key sk and an evaluation key evk
- $\mathcal{E}.\text{Encrypt}\left(sk, \mu\right)$: using the secret key sk, encrypts a message $\mu \in \{0,1\}$ into a ciphertext c and outputs c
- $\mathcal{E}.\text{Decrypt}\left(sk, c\right)$: using the secret key sk, decrypts a ciphertext c to recover a message $\mu \in \{0,1\}$
- $\mathcal{E}.\text{Eval}\left(evk, f, c_1, \ldots, c_\ell\right)$: using the evaluation key evk, applies a function $f : \{0,1\}^\ell \to \{0,1\}$ to ciphertexts $c_1, \ldots, c_\ell$ and outputs a ciphertext c_f

Evaluation Correctness. We say that the $\mathcal{E}.\text{Eval}$ algorithm correctly evaluates all functions in $\mathcal{F}$ if, for any function $f \in \mathcal{F} : \{0,1\}^\ell \to \{0,1\}$ and respective inputs $x_1, \ldots, x_\ell \in \{0,1\}$ it holds that

$$\Pr\left[\mathcal{E}.\text{Decrypt}\left(sk, \mathcal{E}.\text{Eval}\left(evk, f, c_1, \ldots, c_\ell\right)\right) = f\left(x_1, \ldots, x_\ell\right)\right] = 1 - \text{negl}(\lambda)$$

where $sk \leftarrow \mathcal{E}.\text{Setup}\left(1^\lambda\right)$ and $c_i \leftarrow \mathcal{E}.\text{Encrypt}\left(sk, x_i\right)$.

Semantic Security. A secret key encryption scheme $\mathcal{E}$ is said to be semantically secure (or IND-CPA secure) if any PPT adversary $\mathcal{A}$ cannot distinguish between encryptions of two known plaintexts. More formally, let $sk \leftarrow \mathcal{E}.\text{Setup}(1^\lambda)$ and $\mathcal{O}_b\left(\mu_0, \mu_1\right) = \mathcal{E}.\text{Encrypt}\left(sk, \mu_b\right)$ for $b \in \{0,1\}$. Then $\mathcal{E}$ is IND-CPA secure if

$$\left| \Pr\left[\mathcal{A}^{\mathcal{O}_0}\left(1^\lambda\right) = 1\right] - \Pr\left[\mathcal{A}^{\mathcal{O}_1}\left(1^\lambda\right) = 1\right] \right| = \text{negl}(\lambda)$$

where the probability is taken over the internal coins of $\mathcal{E}.\text{Setup}$, $\mathcal{E}.\text{Encrypt}$ and $\mathcal{A}$.

3 Core Randomization Lemma

Note that throughout the rest of the paper we set q to be a power of 2, and $m = n \log q$. We discuss the use of a modulus q that is not a power of 2 in Sect. 5.4.

The goal of this Section is to establish the following lemma:

Lemma 3.1 (Core randomization lemma). *Let* $\varepsilon, \varepsilon' > 0$, $r > \eta_\varepsilon(\Lambda_q^\perp(\mathbf{G}^\mathsf{T}))$ *be a Gaussian parameter. For any* $\mathbf{e} \in \mathbb{Z}_q^m$, $\mathbf{v} \in \mathbb{Z}_q^n$, *if* $r \geq$ $\max\left(4\left((1-\varepsilon)(2\varepsilon')^2\right)^{-\frac{1}{m}}, \sqrt{5}(1+\|\mathbf{e}\|)\sqrt{\frac{\ln(2m(1+1/\varepsilon))}{\pi}}\right)$, *then*

$$\Delta\left((\mathbf{A}, \mathbf{Ax}, \mathbf{e}^\mathsf{T}\mathbf{x} + y), (\mathbf{A}, \mathbf{u}, e')\right) < \varepsilon' + 2\varepsilon$$

where $\mathbf{x} \leftarrow \mathbf{G}_{rand}^{-1}(\mathbf{v})$, $\mathbf{A} \xleftarrow{\$} \mathbb{Z}_q^{(n-1)\times m}$, $\mathbf{u} \xleftarrow{\$} \mathbb{Z}_q^{n-1}$, $y \leftarrow \mathcal{D}_{\mathbb{Z},r}$ *and* $e' \leftarrow \mathcal{D}_{\mathbb{Z},r\sqrt{1+\|\mathbf{e}\|^2}}$.

Asymptotically, $r = \tilde{\Theta}(\|\mathbf{e}\|\sqrt{\lambda})$ *is enough to obtain negligible statistical distance.*

Remark 1 (on the necessity of randomization). We note here that the use of randomization in $\mathbf{G}_{rand}^{-1}(\cdot)$ and the shift are both necessary.

First, the shift is necessary for both distributions to have the same support. For example, $\mathbf{e}^\mathsf{T}\mathbf{G}_{rand}^{-1}((1, 0, \ldots, 0))$ and $\mathbf{e}^\mathsf{T}\mathbf{G}_{rand}^{-1}(\mathbf{0})$ might lie in two different cosets of the lattice $\mathbf{e}^\mathsf{T}\Lambda_q^\perp(\mathbf{G}^\mathsf{T})$, depending on the value of $\mathbf{e}$: if the first coordinate of $\mathbf{e}$ is odd and all the others are even, then $\mathbf{e}^\mathsf{T}\mathbf{G}_{rand}^{-1}((1, 0, \ldots, 0))$ will be odd, while $\mathbf{e}^\mathsf{T}\mathbf{G}_{rand}^{-1}(\mathbf{0})$ will be even, for a q even. The shift by a Gaussian over $\mathbb{Z}$ ensures that the support of the two distributions is $\mathbb{Z}$. Proving that $\mathbf{e}^\mathsf{T}\Lambda_q^\perp(\mathbf{G}^\mathsf{T}) = \mathbb{Z}$ with overwhelming probability over the choice of $\mathbf{e}$ is still an open question that would remove the necessity of the shift, thus proving circuit privacy for standard GSW only using randomized $\mathbf{G}_{rand}^{-1}(\cdot)$.

Finally, the randomization of $\mathbf{G}_{rand}^{-1}(\cdot)$ is necessary for both distributions to have the same center. Using the same example, $\mathbf{e}^\mathsf{T}\mathbf{G}_{det}^{-1}((1, 0, \ldots, 0)) + y$ and $\mathbf{e}^\mathsf{T}\mathbf{G}_{det}^{-1}(\mathbf{0}) + y$ would be two Gaussians, centered respectively on e_1 (the first coordinate of $\mathbf{e}$) and on 0. Instead, using the randomized algorithm $\mathbf{G}_{rand}^{-1}(\cdot)$, the center of both distributions will be 0.

3.1   Additional Preliminaries

Before proving Lemma 3.1, we need to recall some additional results.

Lemma 3.2 [MR07, Lemma 3.3]. *Let* Λ *be any rank-m lattice and* ε *be any positive real. Then*

$$\eta_\varepsilon(\Lambda) \leq \lambda_m(\Lambda) \cdot \sqrt{\frac{\ln(2m(1+1/\varepsilon))}{\pi}}$$

where $\lambda_m(\Lambda)$ *is the smallest* R *such that the ball* $\mathcal{B}_R$ *centered in the origin and with radius* R *contains* m *linearly independent vectors of* Λ.

Lemma 3.3 [GPV08, Corollary 2.8]. *Let* $\Lambda \subseteq \mathbb{Z}^m$ *be a lattice,* $0 < \varepsilon < 1$, $r > 0$. *For any vector* $\mathbf{c} \in \mathbb{R}^m$, *if* $r \geq \eta_\varepsilon(\Lambda)$, *then we have*

$$\rho_r(\Lambda + \mathbf{c}) \in \left[\frac{1-\varepsilon}{1+\varepsilon}, 1\right] \cdot \rho_r(\Lambda).$$

Lemma 3.4 [Reg05, Claim 3.8]. *Let $\Lambda \subseteq \mathbb{Z}^m$ be any lattice, $\mathbf{c} \in \mathbb{R}^m$, $\varepsilon > 0$ and $r \geq \eta_\varepsilon(\Lambda)$. Then*

$$\rho_r\left(\Lambda + \mathbf{c}\right) \in \frac{r^m}{\det\left(\Lambda\right)}\left(1 \pm \varepsilon\right).$$

Generalized Leftover Hash Lemma. We state here a simplified version of the generalized leftover hash lemma which is sufficient for our use. The min-entropy of a random variable X is defined as

$$H_\infty\left(X\right) = -\log\left(\max_x \Pr\left[X = x\right]\right).$$

Lemma 3.5 (Generalized leftover hash lemma [DRS04]**).** *Let $\mathbf{e}$ be any random variable over $\mathbb{Z}_q^m$ and $f : \mathbb{Z}_q^m \to \mathbb{Z}_q^k$. Then*

$$\Delta((\mathbf{Xe}, \mathbf{X}, f(\mathbf{e})), (\mathbf{r}, \mathbf{X}, f(\mathbf{e}))) \leq \frac{1}{2}\sqrt{q^{n+k} \cdot 2^{-H_\infty(\mathbf{e})}}.$$

where $\mathbf{X} \xleftarrow{\$} \mathbb{Z}_q^{n \times m}$ and $\mathbf{r} \xleftarrow{\$} \mathbb{Z}_q^n$.

3.2 Proof of Lemma 3.1

We first prove that given $\mathbf{e}$, the new error term $\mathbf{e}^\intercal \mathbf{x} + y$ is indeed a Gaussian with parameter $r\sqrt{1 + \|\mathbf{e}\|^2}$. This proof is inspired by [AR13], which in turn is an improvement of [AGHS13], but it is different in two aspects: on one hand, in [AR13] the proof is done for the specific case where $\mathbf{x}$ is drawn from a Gaussian over a coset of $\mathbb{Z}^m$; on the other hand, they consider the more general case of an ellipsoidal Gaussian distribution.

Lemma 3.6 (adapted from [AR13, Lemma 3.3]**).** *Let $\varepsilon, r > 0$. For any $\mathbf{e} \in \mathbb{Z}^m$, $\mathbf{c} \in \mathbb{R}^m$, if $r \geq \sqrt{5}(1 + \|\mathbf{e}\|) \cdot \sqrt{\frac{\ln(2m(1+1/\varepsilon))}{\pi}}$, then*

$$\Delta\left(\mathbf{e}^\intercal \mathbf{x} + y, e'\right) < 2\varepsilon$$

where $\mathbf{x} \leftarrow \mathcal{D}_{\Lambda_q^\perp(\mathbf{G}^\intercal) + \mathbf{c}, r}$, $y \leftarrow \mathcal{D}_{\mathbb{Z}, r}$, and $e' \leftarrow \mathcal{D}_{\mathbb{Z}, r\sqrt{1+\|\mathbf{e}\|^2}}$.

Asymptotically, $r = \tilde{\Theta}(\|\mathbf{e}\|\sqrt{\lambda})$ is enough to obtain negligible statistical distance. We stress that the distribution of e' does not depend on the coset $\mathbf{c}$.

Proof. Let $\widehat{\mathbf{e}} = (\mathbf{e}, 1) \in \mathbb{Z}^{m+1}, \widehat{\mathbf{c}} = (\mathbf{c}, 0) \in \mathbb{Z}^{m+1}$ and $\widehat{\Lambda} = \Lambda_q^\perp(\mathbf{G}^\intercal) \times \mathbb{Z}$, we want to show that

$$\Delta\left(\widehat{\mathbf{e}}^\intercal \mathcal{D}_{\widehat{\Lambda} + \widehat{\mathbf{c}}, r}, \mathcal{D}_{\mathbb{Z}, \|\widehat{\mathbf{e}}\|r}\right) \leq 2\varepsilon$$

The support of $\widehat{\mathbf{e}}^\intercal \mathcal{D}_{\widehat{\Lambda} + \widehat{\mathbf{c}}, r}$ is $\widehat{\mathbf{e}}^\intercal \widehat{\Lambda} + \widehat{\mathbf{e}}^\intercal \widehat{\mathbf{c}} = \mathbf{e}^\intercal \Lambda_q^\perp(\mathbf{G}^\intercal) + \mathbb{Z} + \mathbf{e}^\intercal \mathbf{c} = \mathbb{Z}$. Fix some $z \in \mathbb{Z}$. The probability mass assigned to z by $\widehat{\mathbf{e}}^\intercal \mathcal{D}_{\widehat{\Lambda} + \widehat{\mathbf{c}}, r}$ is proportional to $\rho_r(\mathcal{L}_z)$, where

$$\mathcal{L}_z = \left\{\mathbf{v} \in \widehat{\Lambda} + \widehat{\mathbf{c}} : \widehat{\mathbf{e}}^\intercal \mathbf{v} = z\right\}$$

We define the lattice $\mathcal{L} = \left\{ \mathbf{v} \in \widehat{\Lambda} : \widehat{\mathbf{e}}^{\mathsf{T}}\mathbf{v} = 0 \right\}$; note that $\mathcal{L}_z = \mathcal{L} + \mathbf{w}_z$ for any $\mathbf{w}_z \in \mathcal{L}_z$. Let $\mathbf{u}_z = \frac{z}{\|\widehat{\mathbf{e}}\|^2 r}\widehat{\mathbf{e}}$, then $\mathbf{u}_z$ is clearly proportional to $\widehat{\mathbf{e}}$. Observe that $\mathbf{u}_z$ is orthogonal to $r^{-1}\mathcal{L}_z - \mathbf{u}_z$, indeed for any $\mathbf{t} \in r^{-1}\mathcal{L}_z$ we have $\widehat{\mathbf{e}}^{\mathsf{T}}(\mathbf{t} - \mathbf{u}_z) = 0$. From this we have $\rho(\mathbf{t}) = \rho(\mathbf{u}_z) \cdot \rho(\mathbf{t} - \mathbf{u}_z)$, and by summing for $\mathbf{t} \in r^{-1}\mathcal{L}_z$:

$$\rho(r^{-1}\mathcal{L}_z) = \rho(\mathbf{u}_z) \cdot \rho(r^{-1}\mathcal{L}_z - \mathbf{u}_z)$$

Observe that we have $r^{-1}\mathcal{L}_z - \mathbf{u}_z = r^{-1}(\mathcal{L} - \mathbf{c}')$ for some $\mathbf{c}'$ in the vector span of the lattice $\mathcal{L}$ (because $\mathcal{L}_z - r\mathbf{u}_z = \mathcal{L} + \mathbf{w}_z - r\mathbf{u}_z$ and $\widehat{\mathbf{e}}^{\mathsf{T}}(\mathbf{w}_z - r\mathbf{u}_z) = 0$). Thus using Lemmas 3.3 and 3.7 with $r \geq \sqrt{5}(1 + \|\mathbf{e}\|) \cdot \sqrt{\frac{\ln(2m(1+1/\varepsilon))}{\pi}} \geq \eta_\varepsilon(\mathcal{L})$, we obtain

$$\rho(r^{-1}\mathcal{L}_z) = \rho(\mathbf{u}_z) \cdot \rho_r(\mathcal{L} - \mathbf{c}')$$

$$\in \left[\frac{1-\varepsilon}{1+\varepsilon}, 1\right] \cdot \rho_r(\mathcal{L}) \cdot \rho(\mathbf{u}_z)$$

$$= \left[\frac{1-\varepsilon}{1+\varepsilon}, 1\right] \cdot \rho_r(\mathcal{L}) \cdot \rho\left(\frac{z}{\|\widehat{\mathbf{e}}\|^2 r}\widehat{\mathbf{e}}\right)$$

$$= \left[\frac{1-\varepsilon}{1+\varepsilon}, 1\right] \cdot \rho_r(\mathcal{L}) \cdot \rho_{\|\widehat{\mathbf{e}}\|r}(z)$$

This implies that the statistical distance between $\widehat{\mathbf{e}}^{\mathsf{T}}\mathcal{D}_{\widehat{\Lambda}+\widehat{\mathbf{c}},r}$ and $\mathcal{D}_{\mathbb{Z},\|\widehat{\mathbf{e}}\|r}$ is at most $1 - \frac{1-\varepsilon}{1+\varepsilon} \leq 2\varepsilon$. □

In order to conclude the previous proof, we now give a bound on the smoothing parameter of the lattice $\mathcal{L}$.

Lemma 3.7. *Let $\varepsilon > 0$. For any $\mathbf{e} \in \mathbb{Z}^m$, let $\mathcal{L}$ be as defined in Lemma 3.6. Then we have:*

$$\eta_\varepsilon(\mathcal{L}) \leq \sqrt{5}(1 + \|\mathbf{e}\|) \cdot \sqrt{\frac{\ln\left(2m\left(1 + 1/\varepsilon\right)\right)}{\pi}}.$$

Proof. We use Lemma 3.2 to bound the smoothing parameter of $\mathcal{L}$. Since $\widehat{\Lambda} = \Lambda_q^\perp(\mathbf{G}^{\mathsf{T}}) \times \mathbb{Z}$ is of dimension $m + 1$ and $\mathcal{L}$ is the sublattice of $\widehat{\Lambda}$ made of the vectors that are orthogonal to $\mathbf{e}$, we have that $\mathcal{L}$ is of dimension m. We thus exhibit m independent short vectors of $\mathcal{L}$ to obtain an upper bound on $\lambda_m(\mathcal{L})$. We first define the matrix

$$\overline{\mathbf{B}} = \begin{pmatrix} 2 & & & \\ -1 & \ddots & & \\ & \ddots & \ddots & \\ & & -1 & 2 \end{pmatrix} \in \mathbb{Z}^{(\log q) \times (\log q)}$$

and remark that it is a basis for the lattice $\Lambda_q^\perp(\mathbf{g}^{\mathsf{T}})$. The lattice $\widehat{\Lambda}$ is then generated by the columns of the matrix:

$$\mathbf{B} = (\mathbf{b}_1 \mid \ldots \mid \mathbf{b}_{m+1}) = \left(\begin{array}{c|c} \mathbf{I}_n \otimes \overline{\mathbf{B}} & \mathbf{0} \\ \hline \mathbf{0}^{\mathsf{T}} & 1 \end{array}\right) \in \mathbb{Z}^{(m+1) \times (m+1)}$$

For $k \leq m$ let $\mathbf{u}_k = \mathbf{b}_k - \mathbf{b}_{m+1} \cdot \hat{\mathbf{e}}^{\mathsf{T}} \mathbf{b}_k$, since $\hat{\mathbf{e}}^{\mathsf{T}} \mathbf{b}_{m+1} = 1$ we directly have $\hat{\mathbf{e}}^{\mathsf{T}} \mathbf{u}_k = 0$ and thus $\mathbf{u}_k \in \mathcal{L}$. The vectors $\mathbf{u}_1, \ldots, \mathbf{u}_m$ are linearly independent since $\mathrm{span}\,(\mathbf{u}_1, \ldots, \mathbf{u}_m, \mathbf{b}_{m+1}) = \mathrm{span}\,(\mathbf{b}_1, \ldots, \mathbf{b}_m, \mathbf{b}_{m+1}) = \mathbb{R}^{m+1}$ (which comes from the fact that $\mathbf{B}$ is a basis of an $(m+1)$-dimensional lattice). We now bound the norm of $\mathbf{u}_k$:

$$\|\mathbf{u}_k\| \leq \|\mathbf{b}_k\| + \|\mathbf{b}_{m+1}\| \|\mathbf{e}\| \|\mathbf{b}_k\|$$
$$= \sqrt{5}(1 + \|\mathbf{e}\|)$$

Note that $|\hat{\mathbf{e}}^{\mathsf{T}} \mathbf{b}_k| \leq \|\mathbf{e}\| \|\mathbf{b}_k\|$ since the last coefficient of $\mathbf{b}_k$ is 0. Finally we obtain $\lambda_m(\mathcal{L}) \leq \max_{k \leq m} \|\mathbf{u}_k\| \leq \sqrt{5}(1 + \|\mathbf{e}\|)$ and the result. $\qquad \square$

The final proof of Lemma 3.1 will necessitate a call to the leftover hash lemma, so before continuing we analyze the min-entropy of $\mathbf{x} \leftarrow \mathcal{D}_{\Lambda_q^\perp(\mathbf{G}^{\mathsf{T}}) + \mathbf{c}, r}$.

Lemma 3.8. *Let $\varepsilon > 0$, $r \geq \eta_\varepsilon\left(\Lambda_q^\perp(\mathbf{G}^{\mathsf{T}})\right)$. For any $\mathbf{c} \in \mathbb{R}^m$, we have*

$$H_\infty\left(\mathcal{D}_{\Lambda_q^\perp(\mathbf{G}^{\mathsf{T}}) + \mathbf{c}, r}\right) \geq \log\,(1 - \varepsilon) + m \log\,(r) - m.$$

Proof. For any $\mathbf{v} \in \Lambda_q^\perp(\mathbf{G}^{\mathsf{T}}) + \mathbf{c}$

$$\mathcal{D}_{\Lambda_q^\perp(\mathbf{G}^{\mathsf{T}}) + \mathbf{c}, r}(\mathbf{v}) \leq \mathcal{D}_{\Lambda_q^\perp(\mathbf{G}^{\mathsf{T}}) + \mathbf{c}, r}(\mathbf{v}_0), \text{ for } \mathbf{v}_0 \text{ the point of } \Lambda_q^\perp(\mathbf{G}^{\mathsf{T}}) + \mathbf{c} \text{ closest to } \mathbf{0}$$

$$= \frac{\rho_r(\mathbf{v}_0)}{\rho_r(\Lambda_q^\perp(\mathbf{G}^{\mathsf{T}}) + \mathbf{c})}$$

$$\leq \frac{1}{\rho_r(\Lambda_q^\perp(\mathbf{G}^{\mathsf{T}}) + \mathbf{c})}, \text{ since } \rho_r(\mathbf{v}_0) < 1$$

$$\leq (1 - \varepsilon)\frac{r^m}{\det\left(\Lambda_q^\perp(\mathbf{G}^{\mathsf{T}})\right)}, \text{ by Lemma 3.4 since } r \geq \eta_\varepsilon\left(\Lambda_q^\perp(\mathbf{G}^{\mathsf{T}})\right)$$

The lattice $\Lambda_q^\perp(\mathbf{G}^{\mathsf{T}})$ is generated by the basis $\mathbf{I}_n \otimes \overline{\mathbf{B}}$, with $\overline{\mathbf{B}}$ defined as above, which has determinant $\left(2^{\log q}\right)^n = 2^m$. The result follows:

$$H_\infty\left(\mathcal{D}_{\Lambda_q^\perp(\mathbf{G}^{\mathsf{T}}) + \mathbf{c}, r}\right) \geq \log\,(1 - \varepsilon) + m \log\,(r) - m \qquad \square$$

We are now ready to prove Lemma 3.1.

Proof. The proof is done in two steps. First, by Lemma 3.8, we know that $\mathbf{x}$ has min entropy at least $\log(1 - \varepsilon) + m \log(r) - m \geq (n + 1) \log(q) - 2 \log(\varepsilon') - 2$. Moreover, $\mathbf{e}^{\mathsf{T}} \mathbf{x} + y$ is in $\mathbb{Z}_q$. Applying the leftover hash lemma 3.5, we obtain

$$\Delta\left((\mathbf{A}, \mathbf{A}\mathbf{x}, \mathbf{e}^{\mathsf{T}} \mathbf{x} + y), (\mathbf{A}, \mathbf{u}, \mathbf{e}^{\mathsf{T}} \mathbf{x} + y)\right) < \varepsilon'$$

where $\mathbf{u} \xleftarrow{\$} \mathbb{Z}_q^{n-1}$. Now, using Lemma 3.6, we know that

$$\Delta\left(\mathbf{e}^{\mathsf{T}} \mathbf{x} + y, e'\right) < 2\varepsilon$$

Summing the two statistical distances concludes the proof. $\qquad \square$

3.3 Rerandomizing LWE samples

We finally describe a simple application of Lemma 3.1. Generating fresh LWE samples for a fixed secret $\mathbf{s}$ from a bounded number of samples is very useful, for example to build a public key encryption scheme from a symmetric one. It has already been shown in the succession of papers [Reg05, GPV08, ACPS09] that multiplying a matrix of m LWE samples $(\mathbf{A}, \mathbf{s}^{\mathsf{T}}\mathbf{A} + \mathbf{e}^{\mathsf{T}})$ by a discrete Gaussian $\mathbf{x} \leftarrow \mathcal{D}_{\mathbb{Z}^m, r}$ and adding another Gaussian term $y \leftarrow \mathcal{D}_{\mathbb{Z}, r}$ to the error part yields a fresh LWE sample $(\mathbf{a}', \mathbf{s}^{\mathsf{T}}\mathbf{a}' + e')$ with a somewhat larger Gaussian noise e'. Here we have shown that picking $\mathbf{x}$ according to a discrete Gaussian distribution over a coset $\mathbf{c}$ of $\Lambda_q^{\perp}(\mathbf{G}^{\mathsf{T}})$ is enough for this rerandomization process. Moreover, we show that the distribution of the final error is independent of the coset $\mathbf{c}$, which will come in handy for hiding homomorphic evaluations. We note that this could be extended to any other lattice with a small public basis (see the last paragraph of Sect. 5), but we mainly focus on $\Lambda_q^{\perp}(\mathbf{G}^{\mathsf{T}})$ because this is sufficient for our use.

4 Basic GSW Cryptosystem

In this section, we present the Homomorphic Encryption scheme introduced by [GSW13], with notation inspired by [AP14]. We defer setting the parameters to Sect. 5.3. The scheme is composed of the following algorithms:

- Setup (1^{λ}): samples $\bar{\mathbf{s}} \xleftarrow{\$} \mathbb{Z}_q^{n-1}$ and returns the secret key $\mathbf{s} = (-\bar{\mathbf{s}}, 1) \in \mathbb{Z}_q^n$.
- Encrypt$(\mathbf{s}, \mu)$: given the secret key $\mathbf{s} = (-\bar{\mathbf{s}}, 1)$ and a message $\mu \in \{0, 1\}$, samples a matrix $\mathbf{A} \xleftarrow{\$} \mathbb{Z}_q^{(n-1)\times m}$ and $\mathbf{e} \leftarrow \mathcal{D}_{\mathbb{Z}^m, \alpha}$. The algorithm then returns
$$\mathbf{C} = \begin{pmatrix} \mathbf{A} \\ \bar{\mathbf{s}}^{\mathsf{T}}\mathbf{A} + \mathbf{e}^{\mathsf{T}} \end{pmatrix} + \mu\,\mathbf{G} \in \mathbb{Z}_q^{n\times m}$$
as the ciphertext. Notice that $\mathbf{s}^{\mathsf{T}}\mathbf{C} = \mathbf{e}^{\mathsf{T}} + \mu\mathbf{s}^{\mathsf{T}}\mathbf{G}$, the last column of which is close to $\mu \frac{q}{2}$.
- Decrypt$(\mathbf{s}, \mathbf{C})$: given a ciphertext $\mathbf{C}$ and the secret key $\mathbf{s}$, computes the inner product of $\mathbf{s}^{\mathsf{T}}$ and the last column of $\mathbf{C}$, and finally returns 0 if the norm of the result is smaller than $\frac{q}{4}$, otherwise it returns 1.

We omit the original Eval algorithm since our modified version, which guarantees circuit privacy, is presented in Sect. 5.1.

The IND-CPA security of this scheme comes directly from [GSW13] and the LWE assumption.

In order to shorten several formulas in the rest of the paper, we slightly abuse the notation and define a modified version of the encryption algorithm Encrypt$_{\gamma}(\mathbf{s}, \mu)$, which is exactly the same as the previously defined Encrypt$(\mathbf{s}, \mu)$, except that $\mathbf{e} \leftarrow \mathcal{D}_{\mathbb{Z}^m, \gamma}$. We implicitly use Encrypt$(\mathbf{s}, \mu)$ to denote Encrypt$_{\alpha}(\mathbf{s}, \mu)$.

Extension to Public Key Setting. This scheme can be easily adapted to the public key setting. We now describe Setup$_{\mathsf{pub}}$ and Encrypt$_{\mathsf{pub}}$, as the other algorithms are identical to the private key setting.

- $\mathsf{Setup_{pub}}\left(1^\lambda\right)$: given the security parameter λ, samples $\bar{\mathbf{s}} \xleftarrow{\$} \mathbb{Z}_q^{n-1}$, $\mathbf{A} \xleftarrow{\$}$ $\mathbb{Z}_q^{(n-1)\times m}$, $\mathbf{e} \leftarrow \mathcal{D}_{\mathbb{Z}^m,\alpha}$. The algorithm returns the secret key $\mathbf{s} = (-\bar{\mathbf{s}}, 1) \in \mathbb{Z}_q^n$ and the public key $\widehat{\mathbf{A}} = \begin{pmatrix} \mathbf{A} \\ \bar{\mathbf{s}}^{\mathsf{T}}\mathbf{A} + \mathbf{e}^{\mathsf{T}} \end{pmatrix}$.

- $\mathsf{Encrypt_{pub}}\left(\widehat{\mathbf{A}}, \mu\right)$: given the public key $\widehat{\mathbf{A}}$ and a message $\mu \in \{0, 1\}$, samples a matrix $\mathbf{R} \xleftarrow{\$} \{-1, 0, 1\}^{m \times m}$. The algorithm then sets $\mathbf{C} = \widehat{\mathbf{A}}\mathbf{R} + \mu\,\mathbf{G}$ and returns $\mathbf{C} \in \mathbb{Z}_q^{n \times m}$ as the ciphertext. Notice that $\mathbf{s}^{\mathsf{T}}\mathbf{C} = \mathbf{e}^{\mathsf{T}}\mathbf{R} + \mu\mathbf{s}^{\mathsf{T}}\mathbf{G}$ the last column of which is close to $\mu\frac{q}{2}$.

Basic Homomorphic Operations. The homomorphic operations are done as follows:

- Homomorphic addition: $\mathbf{C}_1 \boxplus \mathbf{C}_2 = \mathbf{C}_1 + \mathbf{C}_2$
- Homomorphic multiplication: $\mathbf{C}_1 \boxdot \mathbf{C}_2 \leftarrow \mathbf{C}_1 \cdot \mathbf{G}_{\mathrm{rand}}^{-1}\left(\mathbf{C}_2\right)$

where the $\mathbf{G}_{\mathrm{rand}}^{-1}\left(\cdot\right)$ algorithm is the randomized bit decomposition described in Definition 2.1.

From now on and for readability, we will assume a correct choice of parameters has been made. This setting is discussed in Sect. 5.3.

4.1   Rerandomizing and Scaling GSW Ciphertexts

Here we describe our new technique to rerandomize GSW ciphertexts. This method allows the scaling of GSW ciphertexts, which will be used in our circuit evaluation procedure.

We recall the form of a GSW ciphertext

$$\mathbf{C} = \begin{pmatrix} \mathbf{A} \\ \bar{\mathbf{s}}^{\mathsf{T}}\mathbf{A} + \mathbf{e}^{\mathsf{T}} \end{pmatrix} + \mu\mathbf{G}$$

Using the rerandomization of LWE samples presented in Sect. 3, it is possible to generate a fresh encryption of 0 by computing $\mathbf{C} \cdot \mathbf{G}_{\mathrm{rand}}^{-1}\left(\mathbf{V}\right)$, where $\mathbf{C}$ is an encryption of 0 and $\mathbf{V}$ is any matrix in $\mathbb{Z}_q^{n \times m}$.

Lemma 4.1. *Let $r > 0$. For any $\mathbf{V} \in \mathbb{Z}_q^{n \times m}$, if $r = \Omega\left(\alpha\sqrt{\lambda m \log m}\right)$, with α being the Gaussian parameter of fresh encryptions, then*

$$\left(\mathbf{C} \cdot \mathbf{G}_{rand}^{-1}\left(\mathbf{V}\right) + \begin{pmatrix} \mathbf{0} \\ \mathbf{y}^{\mathsf{T}} \end{pmatrix}, \mathbf{C}\right) \approx_s \left(\mathbf{C}', \mathbf{C}\right)$$

where $\mathbf{C} = \begin{pmatrix} \mathbf{A} \\ \bar{\mathbf{s}}^{\mathsf{T}}\mathbf{A} + \mathbf{e}^{\mathsf{T}} \end{pmatrix} \leftarrow \mathsf{Encrypt}\left(\mathbf{s}, 0\right)$, $\mathbf{C}' \leftarrow \mathsf{Encrypt}_\gamma\left(\mathbf{s}, 0\right)$, with $\gamma = r\sqrt{1 + \|\mathbf{e}\|^2}$.

Proof. Fix $\mathbf{v} \in \mathbb{Z}_q^m$ and $\mathbf{e}$ such that $\|\mathbf{e}\| \leq C\alpha\sqrt{m}$, where C is as in Lemma 2.4. Then by applying Lemma 3.1 with $r = \Omega\left(\alpha\sqrt{\lambda m \log m}\right)$ and $\varepsilon' = \varepsilon = 2^{-\lambda}$ we have

$$\Delta\left((\mathbf{A}, \mathbf{A}\mathbf{x}, \mathbf{e}^\mathsf{T}\mathbf{x} + y), (\mathbf{A}, \mathbf{u}, e')\right) < 3 \cdot 2^{-\lambda}$$

where $\mathbf{A} \xleftarrow{\$} \mathbb{Z}_q^{(n-1)\times m}$, $\mathbf{x} \leftarrow \mathbf{G}_{\mathrm{rand}}^{-1}(\mathbf{v})$ and $y \leftarrow \mathcal{D}_{\mathbb{Z},r}$. From this we obtain that for $\mathbf{e} \leftarrow \mathcal{D}_{\mathbb{Z}^m,\alpha}$:

$$\begin{aligned}
&\Delta\left((\mathbf{A}, \mathbf{e}, \mathbf{A}\mathbf{x}, \mathbf{e}^\mathsf{T}\mathbf{x} + y), (\mathbf{A}, \mathbf{e}, \mathbf{u}, e')\right) \\
&= \sum_{\mathbf{w} \in \mathbb{Z}^m} \Delta\left((\mathbf{A}, \mathbf{A}\mathbf{x}, \mathbf{w}^\mathsf{T}\mathbf{x} + y), (\mathbf{A}, \mathbf{u}, w')\right) \cdot \Pr\left[\mathbf{e} = \mathbf{w}\right] \\
&\leq \sum_{\|\mathbf{w}\| < C\alpha\sqrt{m}} 3 \cdot 2^{-\lambda} \Pr\left[\mathbf{e} = \mathbf{w}\right] + \sum_{\|\mathbf{w}\| \geq C\alpha\sqrt{m}} \Pr\left[\mathbf{e} = \mathbf{w}\right] \\
&\leq 3 \cdot 2^{-\lambda} + \Pr\left[\|\mathbf{e}\| \geq C\alpha\sqrt{m}\right] \\
&\leq 3 \cdot 2^{-\lambda} + 2^{-\Omega(\lambda)}
\end{aligned}$$

In the left operand of the third equation we bound the statistical distance by $3 \cdot 2^{-\lambda}$ and in the right operand we bound it by 1. To obtain the last inequality we use Lemma 2.4 and have $\Pr\left[\|\mathbf{e}\| > C\alpha\sqrt{m}\right] \leq 2^{-\Omega(m)} \leq 2^{-\Omega(\lambda)}$ since $m \geq \lambda$. By rewriting this distance we have for any $\mathbf{v} \in \mathbb{Z}_q^m$

$$\left(\mathbf{C} \cdot \mathbf{G}_{\mathrm{rand}}^{-1}(\mathbf{v}) + \begin{pmatrix} \mathbf{0} \\ y \end{pmatrix}, \mathbf{C}\right) \approx_s \left(\begin{pmatrix} \mathbf{u} \\ \bar{\mathbf{s}}^\mathsf{T}\mathbf{u} + e' \end{pmatrix}, \mathbf{C}\right)$$

By writing $\mathbf{V} = (\mathbf{v}_1 \mid \ldots \mid \mathbf{v}_m)$ and $\mathbf{y} = (y_1, \ldots, y_m)$, we have

$$\mathbf{C} \cdot \mathbf{G}_{\mathrm{rand}}^{-1}(\mathbf{V}) + \begin{pmatrix} \mathbf{0} \\ \mathbf{y}^\mathsf{T} \end{pmatrix} = \left(\mathbf{C} \cdot \mathbf{G}_{\mathrm{rand}}^{-1}(\mathbf{v}_1) + \begin{pmatrix} \mathbf{0} \\ y_1 \end{pmatrix} \mid \ldots \mid \mathbf{C} \cdot \mathbf{G}_{\mathrm{rand}}^{-1}(\mathbf{v}_m) + \begin{pmatrix} \mathbf{0} \\ y_m \end{pmatrix}\right)$$

We define the distributions $(D_i)_{0 \leq i \leq m}$ in which the first i columns of $\mathbf{C} \cdot \mathbf{G}_{\mathrm{rand}}^{-1}(\mathbf{V}) + \begin{pmatrix} \mathbf{0} \\ \mathbf{y}^\mathsf{T} \end{pmatrix}$ are replaced with "fresh" $\begin{pmatrix} \mathbf{u} \\ \bar{\mathbf{s}}^\mathsf{T}\mathbf{u} + e' \end{pmatrix}$ and we obtain through a hybrid argument that

$$\Delta\left(\left(\mathbf{C} \cdot \mathbf{G}_{\mathrm{rand}}^{-1}(\mathbf{V}) + \begin{pmatrix} \mathbf{0} \\ \mathbf{y}^\mathsf{T} \end{pmatrix}, \mathbf{C}\right), \left(\begin{pmatrix} \mathbf{A}' \\ \bar{\mathbf{s}}^\mathsf{T}\mathbf{A}' + \mathbf{e}'^\mathsf{T} \end{pmatrix}, \mathbf{C}\right)\right) \leq m(3 \cdot 2^{-\lambda} + 2^{-\Omega(\lambda)}) \qquad \square$$

As a direct corollary we remark that the scaling of a GSW encryption $\mathbf{C}$ of μ by a bit a, defined as $\mathbf{C} \cdot \mathbf{G}_{\mathrm{rand}}^{-1}(a \cdot \mathbf{G}) + \begin{pmatrix} \mathbf{0} \\ \mathbf{y}^\mathsf{T} \end{pmatrix}$, where $\mathbf{y} \leftarrow \mathcal{D}_{\mathbb{Z}^m,r}$, does not depend on a, but only on $a\mu$.

5   Our Scheme: Circuit-Private Homomorphic Evaluation for GSW

In this section, we prove that a slight modification of the GSW encryption scheme is enough to guarantee circuit privacy, i.e. that an evaluation of any branching program does not reveal anything more than the result of the computation and the length of the branching program, as long as the secret key holder is honest.

First, we state our definition of circuit privacy, similar to [IP07, Definition 7], which is stronger than the one given in [Gen09, Definition 2.1.6] in the sense that it is simulation based, but weaker in the sense that we leak information about the length of the branching program.

Definition 5.1 (Simulation-based circuit privacy). *We say that a homomorphic encryption scheme $\mathcal{E}$ is circuit private if there exists a PPT algorithm* Sim *such that for any branching program Π of length $L = \text{poly}(\lambda)$ on ℓ variables, any $x_1, \ldots, x_\ell \in \{0, 1\}$, the following holds:*

$$\left(\mathcal{E}.\text{Eval}\left(evk, \Pi, (\mathbf{C}_1, \ldots, \mathbf{C}_\ell)\right), \mathbf{C}_1, \ldots, \mathbf{C}_\ell, 1^\lambda, \mathbf{s}\right)$$
$$\approx_s \left(\text{Sim}\left(1^\lambda, \Pi(x_1, \ldots, x_\ell), 1^L, (\mathbf{C}_1, \ldots, \mathbf{C}_\ell)\right), \mathbf{C}_1, \ldots, \mathbf{C}_\ell, 1^\lambda, \mathbf{s}\right)$$

where $\mathbf{s} \leftarrow \mathcal{E}.\text{Setup}\left(1^\lambda\right)$, $\mathbf{C}_i \leftarrow \mathcal{E}.\text{Encrypt}(\mathbf{s}, x_i)$ for $i \in [\ell]$.

We can now state our main theorem:

Theorem 5.2 (Main theorem). *There exists a fully homomorphic encryption scheme for branching programs that is circuit private and whose security is based on the* LWE *assumption with polynomial noise-to-modulus ratio.*

Remark 2. The aforementioned scheme is also multi-hop (see definition in [GHV10]) for branching programs, as long as the noise does not grow beyond $q/4$. This means that the output of an evaluation can be used as input for further computation, while the property of circuit privacy is maintained for every hop. More in detail, the evaluation can be carried out by multiple parties and any subset of these parties is not able to gain information about the branching program applied by an evaluator which is not in the subset, beside its length, input and output, even given access to the secret key.

5.1   Homomorphic Evaluation for Branching Programs

We first recall the branching program evaluation algorithm given in [BV14] and describe our modified version.

Permutation Branching Programs. A permutation branching program Π of length L and width W with input space $\{0, 1\}^\ell$ is a sequence of L tuples of the form $\left(\text{var}(t), \pi_{t,0}, \pi_{t,1}\right)$ where

- var $: [L] \to [\ell]$ is a function that associates the t-th tuple with an input bit $x_{\text{var}(t)}$

– $\pi_{t,0}, \pi_{t,1} : [W] \rightarrow [W]$ are permutations that dictate the t-th step of the computation.

On input $(x_1, \ldots, x_\ell)$, Π outputs 1 iff

$$\pi_{L,x_{\mathrm{var}(L)}}(\cdots(\pi_{1,x_{\mathrm{var}(1)}}(1))\cdots) = 1.$$

Following [BV14, IP07], we will evaluate Π recursively as follows. We associate each $t \in [L]$ with the characteristic vector $\mathbf{v}_t \in \{0,1\}^W$ of the current "state", starting with $\mathbf{v}_0 = (1, 0, \ldots, 0)$. We can then compute the w-th entry of $\mathbf{v}_t$ (denoted by $\mathbf{v}_t[w]$) as follows: for all $t \in [L], w \in [W]$,

$$
\begin{aligned}
\mathbf{v}_t[w] &= \mathbf{v}_{t-1}\left[\pi_{t,x_{\mathrm{var}(t)}}^{-1}(w)\right] \\
&= x_{\mathrm{var}(t)} \cdot \mathbf{v}_{t-1}\left[\pi_{t,1}^{-1}(w)\right] + \left(1 - x_{\mathrm{var}(t)}\right) \cdot \mathbf{v}_{t-1}\left[\pi_{t,0}^{-1}(w)\right]. \quad (5.1)
\end{aligned}
$$

Our Branching Program Evaluation. Here we present our $\mathsf{Eval}(\Pi, (\mathbf{C}_1, \ldots, \mathbf{C}_\ell))$ algorithm (note that it does not require any evaluation key), which homomorphically evaluates a branching program Π over ciphertexts $\mathbf{C}_1, \ldots, \mathbf{C}_\ell$. The first state vector is encrypted without noise: the initial encrypted state vector is $\mathbf{V}_0 = (\mathbf{G}, \mathbf{0}, \ldots, \mathbf{0})$, i.e. $\mathbf{V}_0[1] = \mathbf{G}$ and $\mathbf{V}_0[w] = \mathbf{0}$, for $2 \leq w \leq W$. Note that $\mathbf{G}$ and $\mathbf{0}$ are noiseless encryptions of 1 and 0, respectively. The encrypted state vector is then computed at each step by homomorphically applying (5.1) and adding a noise term: for $t \in [L]$ and $w \in [W]$

$$
\begin{aligned}
\mathbf{V}_t[w] \leftarrow &\mathbf{C}_{\mathrm{var}(t)} \cdot \mathbf{G}_{\mathrm{rand}}^{-1}\left(\mathbf{V}_{t-1}\left[\pi_{t,1}^{-1}(w)\right]\right) \\
&+ \left(\mathbf{G} - \mathbf{C}_{\mathrm{var}(t)}\right) \cdot \mathbf{G}_{\mathrm{rand}}^{-1}\left(\mathbf{V}_{t-1}\left[\pi_{t,0}^{-1}(w)\right]\right) + \boxed{\begin{pmatrix}\mathbf{0} \\ \mathbf{y}_{t,w}^{\mathsf{T}}\end{pmatrix}}
\end{aligned}
\quad (5.2)
$$

where $\mathbf{y}_{t,w} \leftarrow \mathcal{D}_{\mathbb{Z}^m, r\sqrt{2}}$. The output of the evaluation algorithm is $\mathbf{V}_L[0] \in \mathbb{Z}_q^{n \times m}$.

Remark 3 (Comparison with [BV14, AP14]. Cf. also Table 1). The differences between our homomorphic evaluation procedure and the previous ones are as follows:

– We added an additional Gaussian noise to the computation, as captured in the boxed term;
– [BV14] uses the deterministic $\mathbf{G}_{\mathrm{det}}^{-1}(\cdot)$ whereas [AP14] introduced the randomized $\mathbf{G}_{\mathrm{rand}}^{-1}(\cdot)$ for efficiency. Here, we crucially exploit the randomized $\mathbf{G}_{\mathrm{rand}}^{-1}(\cdot)$ for privacy.

Simulator. Towards proving circuit privacy, we need to specify a simulator Sim. We first describe a simulator that is given access to the number of times each variable is used and prove that its output distribution is statistically close to the result of Eval (Lemma 5.5). We can then pad the branching program so that each variable is used the same number of times. Given the security parameter λ, the length L of the branching program Π, the number of times τ_i that Π uses the i-th variable, the final value x_f of the evaluation of Π on input $(x_1, \ldots, x_\ell)$, the ciphertexts $\mathbf{C}_i$ encrypting x_i for $i \in [\ell]$, Sim mimics the way error grows in the states of Eval by doing τ_i dummy steps of computation with the i-th variable. This gives a new encryption $\widehat{\mathbf{A}}_f$ of 0 with the same noise distribution as the ciphertext output by the Eval procedure. Sim then adds the message part x_f to this ciphertext and outputs $\mathbf{C}_f = \widehat{\mathbf{A}}_f + x_f \mathbf{G}$.

In other words,

$$\mathsf{Sim}\left(1^\lambda, x_f, (1^{\tau_1}, \ldots, 1^{\tau_\ell}), (\mathbf{C}_1, \ldots, \mathbf{C}_\ell)\right)$$
$$\leftarrow \sum_{i=1}^{\ell} \sum_{t=1}^{\tau_i} \left(\mathbf{C}_i \cdot \left(\mathbf{G}_{\mathrm{rand}}^{-1}(\mathbf{0}) - \mathbf{G}_{\mathrm{rand}}^{-1}(\mathbf{0}) \right) + \begin{pmatrix} \mathbf{0} \\ \mathbf{y}_t^\mathsf{T} \end{pmatrix} \right) + x_f \mathbf{G}$$

where $\mathbf{y}_t \leftarrow \mathcal{D}_{\mathbb{Z}^m, r\sqrt{2}}$ for $t \in [L]$.

We note that the sum of $2\tau_i$ samples $\mathbf{G}_{\mathrm{rand}}^{-1}(\mathbf{0})$ can be sampled at once using the $\mathbf{G}_{\mathrm{rand}}^{-1}(\cdot)$ algorithm with a larger parameter $r\sqrt{2\tau_i}$, and the sum of τ_i samples from $\mathcal{D}_{\mathbb{Z}^m, r\sqrt{2}}$ is close to a sample from $\mathcal{D}_{\mathbb{Z}^m, r\sqrt{2\tau_i}}$.

5.2   Proof of Circuit Privacy

We proceed to establish circuit privacy in two steps. We first analyze how the ciphertext distribution changes in a single transition, and then proceed by induction to reason about homomorphic evaluation of the entire branching program.

Step 1. We begin with the following lemma, which is useful for analyzing the output of (5.2). Roughly speaking, this lemma says that if at step t, the state vector consists of fresh GSW encryptions with some noise parameter ζ, then at step $t+1$, the state vector is statistically close to fresh GSW encryptions with a somewhat larger noise which depends on the error in the input ciphertext and on ζ.

Lemma 5.3 *For any* $x, v_0, v_1 \in \{0, 1\}$ *and* $\mathbf{s} = (-\bar{\mathbf{s}}, 1) \leftarrow \mathsf{Setup}(1^\lambda)$, *the following holds:*

$$\left(\mathbf{C} \cdot \mathbf{G}_{\mathrm{rand}}^{-1}(\mathbf{V}_1) + (\mathbf{G} - \mathbf{C}) \cdot \mathbf{G}_{\mathrm{rand}}^{-1}(\mathbf{V}_0) + \begin{pmatrix} \mathbf{0} \\ \mathbf{y}^\mathsf{T} \end{pmatrix}, \mathbf{C} \right) \approx_s (\mathbf{V}'_x, \mathbf{C})$$

where $\mathbf{V}_b \leftarrow \mathsf{Encrypt}_\gamma(\mathbf{s}, v_b)$ *for* $b \in \{0, 1\}$, $\mathbf{C} = \begin{pmatrix} \mathbf{A} \\ \bar{\mathbf{s}}^\mathsf{T} \mathbf{A} + \mathbf{e}^\mathsf{T} \end{pmatrix} + x\mathbf{G} \leftarrow$ $\mathsf{Encrypt}(\mathbf{s}, x)$, $\mathbf{y} \leftarrow \mathcal{D}_{\mathbb{Z}^m, r\sqrt{2}}$ *and* $\mathbf{V}'_x \leftarrow \mathsf{Encrypt}_\zeta(\mathbf{s}, v_x)$, *with* $\zeta = \sqrt{\gamma^2 + 2r^2(1 + \|\mathbf{e}\|^2)}$.

Proof. We begin with a simple identity which is useful in the remainder of the proof:

$$\mathbf{C} \cdot \mathbf{G}_{\mathrm{rand}}^{-1}(\mathbf{V}_1) + (\mathbf{G} - \mathbf{C}) \cdot \mathbf{G}_{\mathrm{rand}}^{-1}(\mathbf{V}_0) = \widehat{\mathbf{A}} \cdot \left(\mathbf{G}_{\mathrm{rand}}^{-1}(\mathbf{V}_1) - \mathbf{G}_{\mathrm{rand}}^{-1}(\mathbf{V}_0) \right) + \mathbf{V}_x$$

where $\widehat{\mathbf{A}} = \begin{pmatrix} \mathbf{A} \\ \bar{\mathbf{s}}^\mathsf{T}\mathbf{A} + \mathbf{e}^\mathsf{T} \end{pmatrix}$ and $\mathbf{V}_0, \mathbf{V}_1, \mathbf{C}$ are as defined in the statement of the Lemma. Showing this identity is correct just requires performing the calculations:

$$\mathbf{C} \cdot \mathbf{G}_{\mathrm{rand}}^{-1}(\mathbf{V}_1) + (\mathbf{G} - \mathbf{C}) \cdot \mathbf{G}_{\mathrm{rand}}^{-1}(\mathbf{V}_0)$$

$$= \left(\widehat{\mathbf{A}} + x\mathbf{G}\right) \cdot \mathbf{G}_{\mathrm{rand}}^{-1}(\mathbf{V}_1) + \left((1-x)\,\mathbf{G} - \widehat{\mathbf{A}}\right) \cdot \mathbf{G}_{\mathrm{rand}}^{-1}(\mathbf{V}_0)$$

$$= \widehat{\mathbf{A}} \cdot \left(\mathbf{G}_{\mathrm{rand}}^{-1}(\mathbf{V}_1) - \mathbf{G}_{\mathrm{rand}}^{-1}(\mathbf{V}_0)\right) + x\mathbf{V}_1 + (1-x)\,\mathbf{V}_0$$

$$= \widehat{\mathbf{A}} \cdot \left(\mathbf{G}_{\mathrm{rand}}^{-1}(\mathbf{V}_1) - \mathbf{G}_{\mathrm{rand}}^{-1}(\mathbf{V}_0)\right) + \mathbf{V}_x$$

Then we observe that by applying Lemma 2.3 we have

$$\begin{pmatrix} \mathbf{0} \\ \mathbf{y}^\mathsf{T} \end{pmatrix} \approx_s \begin{pmatrix} \mathbf{0} \\ \mathbf{y}_1^\mathsf{T} \end{pmatrix} - \begin{pmatrix} \mathbf{0} \\ \mathbf{y}_0^\mathsf{T} \end{pmatrix}$$

where $\mathbf{y}_b \leftarrow \mathcal{D}_{\mathbb{Z}^m,r}, b \in \{0,1\}$. Lemma 4.1 also gives

$$\left(\widehat{\mathbf{A}} \cdot \mathbf{G}_{\mathrm{rand}}^{-1}(\mathbf{V}_b) + \begin{pmatrix} \mathbf{0} \\ \mathbf{y}_b^\mathsf{T} \end{pmatrix}, \mathbf{C}\right) \approx_s (\mathbf{C}_b, \mathbf{C})$$

where $\mathbf{C}_b \leftarrow \mathsf{Encrypt}_{\zeta'}(\mathbf{s},0)$, for $b \in \{0,1\}$, with $\zeta' = r\sqrt{1 + \|\mathbf{e}\|^2}$. We now have

$$\left(\mathbf{C} \cdot \mathbf{G}_{\mathrm{rand}}^{-1}(\mathbf{V}_1) + (\mathbf{G} - \mathbf{C}) \cdot \mathbf{G}_{\mathrm{rand}}^{-1}(\mathbf{V}_0) + \begin{pmatrix} \mathbf{0} \\ \mathbf{y}^\mathsf{T} \end{pmatrix}, \mathbf{C}\right) \approx_s (\mathbf{C}_1 - \mathbf{C}_0 + \mathbf{V}_x, \mathbf{C})$$

By additivity of variance on independent variables, we obtain that $\mathbf{C}_1 - \mathbf{C}_0 + \mathbf{V}_x = \mathbf{V}_x'$ looks like a fresh encryption of $0 - 0 + v_x = v_x$ with parameter $\sqrt{\gamma^2 + 2r^2(1 + \|\mathbf{e}\|^2)}$. $\qquad \square$

Step 2. We now prove that, at each step of the evaluation, each entry of the encrypted state $\mathbf{V}_t$ looks like a fresh GSW encryption of the corresponding entry of the state $\mathbf{v}_t$, even given the GSW encryptions of the input bits, except for a small correlation in the noise.

Lemma 5.4 (Distribution of the result of Eval). *For any branching program Π of length L on ℓ variables, we define $\tau_{t,i}$ to be the number of times the i-th variable has been used after t steps of the evaluation, i.e. $\tau_{t,i} = |var^{-1}(i) \cap [t]|$, for i in $[\ell]$ and $t \in [L]$.*

For any $x_1, \ldots, x_\ell \in \{0,1\}$, any $\mathbf{s} = (-\bar{\mathbf{s}}, 1) \leftarrow \mathsf{Setup}(1^\lambda)$, at each step $t \in [L]$, for all indexes $w \in [W]$, the following holds:

$$\left(\mathbf{V}_t[w], (\mathbf{C}_i)_{i\in[\ell]}\right) \approx_s \left(\mathbf{C}_{t,w}', (\mathbf{C}_i)_{i\in[\ell]}\right)$$

where $\mathbf{C}_i = \begin{pmatrix} \mathbf{A}_i \\ \bar{\mathbf{s}}^\mathsf{T}\mathbf{A}_i + \mathbf{e}_i^\mathsf{T} \end{pmatrix} + x_i\mathbf{G} \leftarrow \mathsf{Encrypt}(\mathbf{s}, x_i)$ *for* $i \in [\ell]$, $\mathbf{C}_{t,w}' \leftarrow$
$\mathsf{Encrypt}_{r_t}(\mathbf{s}, \mathbf{v}_t[w])$ *for* $(t,w) \in [L] \times [W]$ *and* $r_t = r\sqrt{2 \sum_{i=1}^{\ell} \tau_{t,i}(1 + \|\mathbf{e}_i\|^2)}$.

Proof. We prove this lemma by induction on $t \in [L]$. At step $t > 1$, for index $w \in [W]$ we use a series of hybrid distributions $\mathcal{H}_{t,w,k}$ for $0 \leq k \leq 2$ to prove that $\left(\mathbf{V}_t\,[w], (\mathbf{C}_i)_{i \in [\ell]}\right) \approx_s \left(\mathbf{C}'_{t,w}, (\mathbf{C}_i)_{i \in [\ell]}\right)$. In particular $\mathcal{H}_{t,w,0} = \left(\mathbf{V}_t\,[w], (\mathbf{C}_i)_{i \in [\ell]}\right)$, and $\mathcal{H}_{t,w,2} = \left(\mathbf{C}'_{t,w}, (\mathbf{C}_i)_{i \in [\ell]}\right)$.

Hybrid $\mathcal{H}_{t,w,0}$. Let $w_b = \pi^{-1}_{t,b}(w)$ for $b \in \{0,1\}$. We write w_β to denote $w_{x_{\mathrm{var}(t)}}$, i.e. w_0 or w_1, depending on the value of the variable which is used at time t.

$$\mathcal{H}_{t,w,0} = \left(\mathbf{V}_t\,[w], (\mathbf{C}_i)_{i \in [\ell]}\right)$$

$$= \left(\mathbf{C}_{\mathrm{var}(t)} \cdot \mathbf{G}^{-1}_{\mathrm{rand}}\left(\mathbf{V}_{t-1}\,[w_1]\right) + \left(\mathbf{G} - \mathbf{C}_{\mathrm{var}(t)}\right) \cdot \mathbf{G}^{-1}_{\mathrm{rand}}\left(\mathbf{V}_{t-1}\,[w_0]\right)\right.$$

$$\left. + \begin{pmatrix} \mathbf{0} \\ \mathbf{y}^{\mathsf{T}}_{t,w} \end{pmatrix}, (\mathbf{C}_i)_{i \in [\ell]}\right)$$

where $\mathbf{C}_i \leftarrow \mathsf{Encrypt}\,(\mathbf{s}, x_i)$ and $\mathbf{y}_{t,w} \leftarrow \mathcal{D}_{\mathbb{Z}^m, r\sqrt{2}}$.

Hybrid $\mathcal{H}_{t,w,1}$. We set

$$\mathcal{H}_{t,w,1} = \left(\mathbf{C}_{\mathrm{var}(t)} \cdot \mathbf{G}^{-1}_{\mathrm{rand}}\left(\mathbf{C}'_{t-1,w_1}\right) + \left(\mathbf{G} - \mathbf{C}_{\mathrm{var}(t)}\right) \cdot \mathbf{G}^{-1}_{\mathrm{rand}}\left(\mathbf{C}'_{t-1,w_0}\right)\right.$$

$$\left. + \begin{pmatrix} \mathbf{0} \\ \mathbf{y}^{\mathsf{T}}_{t,w} \end{pmatrix}, (\mathbf{C}_i)_{i \in [\ell]}\right)$$

where $\mathbf{C}_i \leftarrow \mathsf{Encrypt}\,(\mathbf{s}, x_i)$, $\mathbf{y}_{t,w} \leftarrow \mathcal{D}_{\mathbb{Z}^m, r\sqrt{2}}$ and $\mathbf{C}'_{t-1,w_b} \leftarrow \mathsf{Encrypt}_{r_{t-1}}(\mathbf{s}, \mathbf{v}_{t-1}[w_b])$ for $b \in \{0,1\}$.

By induction hypothesis we have $\mathcal{H}_{t-1,w_b,0} \approx_s \mathcal{H}_{t-1,w_b,2}$ for $b \in \{0,1\}$, i.e.

$$\left(\mathbf{V}_{t-1}\,[w_b], (\mathbf{C}_i)_{i \in [\ell]}\right) \approx_s \left(\mathbf{C}'_{t-1,w_b}, (\mathbf{C}_i)_{i \in [\ell]}\right)$$

where $\mathbf{C}_i \leftarrow \mathsf{Encrypt}\,(\mathbf{s}, x_i)$ and $\mathbf{C}'_{t-1,w_b} \leftarrow \mathsf{Encrypt}_{r_{t-1}}(\mathbf{s}, \mathbf{v}_{t-1}[w_b])$ for $b \in \{0,1\}$. We use the fact that applying a function to two distributions does not increase their statistical distance to obtain $\mathcal{H}_{t,w,0} \approx_s \mathcal{H}_{t,w,1}$.

Hybrid $\mathcal{H}_{t,w,2}$. Let

$$\mathcal{H}_{t,w,2} = \left(\mathbf{C}', (\mathbf{C}_i)_{i \in [\ell]}\right)$$

with $\mathbf{C}_i \leftarrow \mathsf{Encrypt}\,(\mathbf{s}, x_i)$, $\mathbf{C}' \leftarrow \mathsf{Encrypt}_\zeta(\mathbf{s}, \mathbf{v}_{t-1}[w_\beta])$ and $\zeta = \sqrt{r^2_{t-1} + 2r^2\left(1 + \|\mathbf{e}_{\mathrm{var}(t)}\|^2\right)}$.

By Lemma 5.3 we have:

$$\left(\mathbf{C}_{\mathrm{var}(t)} \cdot \mathbf{G}^{-1}_{\mathrm{rand}}\left(\mathbf{C}'_{t-1,w_1}\right) + \left(\mathbf{G} - \mathbf{C}_{\mathrm{var}(t)}\right) \cdot \mathbf{G}^{-1}_{\mathrm{rand}}\left(\mathbf{C}'_{t-1,w_0}\right)\right.$$

$$\left. + \begin{pmatrix} \mathbf{0} \\ \mathbf{y}^{\mathsf{T}}_{t,w} \end{pmatrix}, (\mathbf{C}_i)_{i \in [\ell]}\right) \approx_s \left(\mathbf{C}', (\mathbf{C}_i)_{i \in [\ell]}\right)$$

where $\mathbf{C}_i \leftarrow \mathsf{Encrypt}(\mathbf{s}, x_i)$, $\mathbf{y}_{t,w} \leftarrow \mathcal{D}_{\mathbb{Z}^m, r\sqrt{2}}$, $\mathbf{C}'_{t-1,w_b} \leftarrow \mathsf{Encrypt}_{r_{t-1}}(\mathbf{s}, \mathbf{v}_{t-1}[w_b])$ for $b \in \{0,1\}$ and $\mathbf{C}' \leftarrow \mathsf{Encrypt}_\zeta(\mathbf{s}, \mathbf{v}_{t-1}[w_\beta])$. Note that $\mathbf{v}_{t-1}[w_\beta] = \mathbf{v}_t[w]$ and $r_t = \sqrt{r_{t-1}^2 + 2r^2\left(1 + \|\mathbf{e}_{\mathrm{var}(t)}\|^2\right)} = \zeta$ from which we have that $\mathbf{C}'$ and $\mathbf{C}'_{t,w}$ are identically distributed, and directly $\mathcal{H}_{t,w,1} \approx_s \mathcal{H}_{t,w,2}$.

We note that this recursive formula does not apply to step $t = 0$, we thus use $t = 1$, $w \in [W]$ as the base case. We only describe the steps that differ from the case $t > 1$.

Hybrid $\mathcal{H}_{1,w,1}$. We have $\mathbf{G}_{\mathrm{rand}}^{-1}(\mathbf{V}_0[w_b]) = \mathbf{G}_{\mathrm{rand}}^{-1}(\mathbf{v}_0[w_b] \cdot \mathbf{G})$ for $b \in \{0,1\}$. Notice that we now have exactly $\mathcal{H}_{1,w,1} = \mathcal{H}_{1,w,0}$.

Hybrids $\mathcal{H}_{1,w,2}$. The proof for $\mathcal{H}_{1,w,1} \approx_s \mathcal{H}_{1,w,2}$ is identical to the one of Lemma 5.3 except for the fact that the ciphertext $\mathbf{V}_x$ from the proof is now of the form $\mathbf{v}_0[w_\beta]\mathbf{G}$. The resulting ciphertext $\mathbf{C}'_{1,w}$ is now only the sum of two encryptions of 0 and $\mathbf{v}_0[w_\beta]$ and has a Gaussian parameter $r\sqrt{2\left(1 + \|\mathbf{e}_{\mathrm{var}(1)}\|^2\right)} = r_1$. This implies $\mathcal{H}_{1,w,1} \approx_s \mathcal{H}_{1,w,2}$. $\qquad\square$

We now proceed to prove circuit privacy. We will first prove the following lemma, which states that the Eval algorithm presented in Sect. 5.1 only leaks the final result of the evaluation and the number of times each variable is used.

Lemma 5.5. *Let $\mathcal{E}$ be the scheme defined in Sect. 4 with evaluation defined as in this section, and Sim be the corresponding simulator. Then for any branching program Π of length $L = \mathrm{poly}(\lambda)$ on ℓ variables, such that the i-th variable is used τ_i times, and any $x_1, \ldots, x_\ell \in \{0,1\}$, the following holds:*

$$\left(\mathcal{E}.\mathsf{Eval}\left(\Pi, (\mathbf{C}_1, \ldots, \mathbf{C}_\ell)\right), \mathbf{C}_1, \ldots, \mathbf{C}_\ell, 1^\lambda, \mathbf{s}\right)$$
$$\approx_s \left(\mathsf{Sim}\left(1^\lambda, \Pi\left(x_1, \ldots, x_\ell\right), (1^{\tau_1}, \ldots, 1^{\tau_\ell}), (\mathbf{C}_1, \ldots, \mathbf{C}_\ell)\right), \mathbf{C}_1, \ldots, \mathbf{C}_\ell, 1^\lambda, \mathbf{s}\right)$$

where $\mathbf{s} \leftarrow \mathcal{E}.\mathsf{Setup}\left(1^\lambda\right)$, $\mathbf{C}_i \leftarrow \mathcal{E}.\mathsf{Encrypt}(\mathbf{s}, x_i)$ for i in $[\ell]$.

Proof. As shown in Lemma 5.4, the final result of the homomorphic evaluation of the branching program Π is of the form

$$\mathbf{V}_L[0] \approx_s \left(\genfrac{}{}{0pt}{}{\mathbf{A}}{\bar{\mathbf{s}}^\mathsf{T}\mathbf{A} + \mathbf{f}^\mathsf{T}}\right) + x_f\mathbf{G}$$

where $\mathbf{A} \xleftarrow{\$} \mathbb{Z}_q^{(n-1)\times m}$, $\mathbf{f} \leftarrow \mathcal{D}_{\mathbb{Z}^m, r_L}$ and $r_L = r\sqrt{2\sum_{i=1}^\ell \left(1 + \|\mathbf{e}_i\|^2\right)\tau_i}$.

Now we prove that the output of Sim is statistically close to the same distribution. This proof follows from the fact that scaling GSW ciphertexts yields a result which is independent of the argument of $\mathbf{G}_{\mathrm{rand}}^{-1}(\cdot)$. Let $\mathbf{A}_{i,t}, \mathbf{A}'_{i,t} \xleftarrow{\$} \mathbb{Z}_q^{(n-1)\times m}$, $\mathbf{f}_{i,f}, \mathbf{f}'_{i,t} \leftarrow \mathcal{D}_{\mathbb{Z}^m, r\sqrt{1+\|\mathbf{e}_i\|}}$, then the joint distribution of the output of Sim and ciphertexts $(\mathbf{C}_i)_{i\in[\ell]}$ is

$$(\mathcal{S}, (\mathbf{C}_i)_{i\in[\ell]}) = \left(\sum_{i=1}^{\ell} \mathbf{C}_i \sum_{t=1}^{\tau_i} \left(\mathbf{G}_{\mathrm{rand}}^{-1}(\mathbf{0}) - \mathbf{G}_{\mathrm{rand}}^{-1}(\mathbf{0}) \right) + \begin{pmatrix} \mathbf{0} \\ \mathbf{y}_t^\mathsf{T} \end{pmatrix} + x_f \mathbf{G}, (\mathbf{C}_i)_{i\in[\ell]} \right)$$

$$\approx_s \left(\sum_{i=1}^{\ell} \sum_{t=1}^{\tau_i} \begin{pmatrix} \mathbf{A}_{i,t} \\ \bar{\mathbf{s}}^\mathsf{T} \mathbf{A}_{i,t} + \mathbf{f}_{i,t} \end{pmatrix} + \begin{pmatrix} \mathbf{A}'_{i,t} \\ \bar{\mathbf{s}}^\mathsf{T} \mathbf{A}'_{i,t} + \mathbf{f}'_{i,t} \end{pmatrix}, (\mathbf{C}_i)_{i\in[\ell]} \right)$$

by Lemma 3.1

$$\approx_s \left(\begin{pmatrix} \mathbf{A} \\ \bar{\mathbf{s}}^\mathsf{T} \mathbf{A} + \mathbf{f}^\mathsf{T} \end{pmatrix}, (\mathbf{C}_i)_{i\in[\ell]} \right)$$

by Lemma 2.3 and summing uniform variables.

The result is the same as the joint distribution of the output of Eval and ciphertexts $(\mathbf{C}_i)_{i\in[\ell]}$, thus concluding the proof. □

We are now ready to prove Theorem 5.2.

Proof (Main theorem). Theorem 5.2 follows from Lemma 5.5 by tweaking the Eval algorithm of $\mathcal{E}$: it is sufficient that this algorithm pads the branching program $\varPi$ so that each variable is used L times. This padding is done by using the identity permutation for all steps after the L-th. After this proof, we discuss more efficient ways to pad branching program evaluations. It is easy to see that this step is enough to reach the desired circuit privacy property: the only information leaked besides the final result is $\tau_i = L$. □

Padding Branching Program Evaluations. In order to pad a branching program $\varPi$ that uses the i-th variable τ_i times to one that uses the i-th variable L times, we add $L - \tau_i$ steps, using the identity permutation at each one of these. Given $\mathbf{V}_L[0]$ the final result of the computation, this padding corresponds to steps $t \in [L+1, 2L - \tau_i]$ defined as follows:

$$\mathbf{V}_t[0] \leftarrow \mathbf{V}_{t-1}[0] + \mathbf{C}_i \left(\mathbf{G}_{\mathrm{rand}}^{-1}(\mathbf{V}_{t-1}[0]) - \mathbf{G}_{\mathrm{rand}}^{-1}(\mathbf{V}_{t-1}[0]) \right) + \begin{pmatrix} \mathbf{0} \\ \mathbf{y}_{t,0}^\mathsf{T} \end{pmatrix}$$

Using the same proof as Lemma 5.5 the final output will be

$$\mathbf{V}_{2L-\tau_i}[0] \leftarrow \mathbf{V}_L[0] + \mathbf{C}_i \sum_{t=L}^{2L-\tau_i-1} \left(\mathbf{G}_{\mathrm{rand}}^{-1}(\mathbf{V}_t[0]) - \mathbf{G}_{\mathrm{rand}}^{-1}(\mathbf{V}_t[0]) \right) + \begin{pmatrix} \mathbf{0} \\ \mathbf{y}_{t,0}^\mathsf{T} \end{pmatrix}$$

$$\approx_s \mathbf{V}_L[0] + \mathbf{C}_i \sum_{t=L}^{2L-\tau_i-1} \left(\mathbf{G}_{\mathrm{rand}}^{-1}(\mathbf{0}) - \mathbf{G}_{\mathrm{rand}}^{-1}(\mathbf{0}) \right) + \begin{pmatrix} \mathbf{0} \\ \mathbf{y}_{t,0}^\mathsf{T} \end{pmatrix}$$

Observe that by using Lemma 2.3 we have that

$$\sum_{t=L}^{2L-\tau_i-1} \left(\mathbf{G}_{\mathrm{rand}}^{-1}(\mathbf{0}) - \mathbf{G}_{\mathrm{rand}}^{-1}(\mathbf{0}) \right) \approx_s \mathcal{D}_{\Lambda_q^\perp(\mathbf{G}^\mathsf{T}), r_f}$$

$$\sum_{t=L}^{2L-\tau_i-1} \begin{pmatrix} \mathbf{0} \\ \mathbf{y}_{t,0}^\mathsf{T} \end{pmatrix} \approx_s \mathcal{D}_{\mathbb{Z}^m, r_f}$$

where $r_f = r\sqrt{2(L - \tau_i)}$. We can thus do all the steps at once by outputting $\mathbf{V}_L[0] + \mathbf{C}_i \cdot \mathbf{X} + \begin{pmatrix} \mathbf{0} \\ \mathbf{y}_f^\mathsf{T} \end{pmatrix}$, where $\mathbf{X} \leftarrow \mathcal{D}^m_{\Lambda_q^\perp(\mathbf{G}^\mathsf{T}),r_f}$ and $\mathbf{y}_f \leftarrow \mathcal{D}_{\mathbb{Z}^m,r_f}$. We note that $\mathbf{X}$ can be sampled using the $\mathbf{G}^{-1}_{\mathrm{rand}}(\cdot)$ algorithm with parameter r_f instead of r.

5.3   Setting the Parameters

In this section we show that, for appropriate values of the parameters, the output of the homomorphic evaluation $\mathbf{V}_L[0]$ decrypts to $\Pi(x_1, \dots, x_\ell)$ with overwhelming probability and guarantees circuit privacy.

We first recall the bounds on the parameters needed for both correctness and privacy. Let $n = \Theta(\lambda)$, $q = \mathrm{poly}(n)$, $m = n \log q$, α be the Gaussian parameter of fresh encryptions, r be the parameter of $\mathbf{G}^{-1}_{\mathrm{rand}}(\cdot)$. Let $B = \Theta(\alpha\sqrt{m})$ be a bound on the norm of the error in fresh encryptions (using a tail cutting argument we can show that $B = C\alpha\sqrt{m}$ is sufficient to have a bound with overwhelming probability), $L_{\max} = \mathrm{poly}(n)$ be a bound on the size of the branching programs we consider and $\ell_{\max} = \mathrm{poly}(n)$ an upper bound on their number of variables. Let $\varepsilon = O(2^{-\lambda})$ and $\varepsilon' = O(2^{-\lambda})$.

We have the following constraints:

- $\alpha = \Omega(\sqrt{m})$ for the hardness of $\mathsf{LWE}_{n-1,q,\mathcal{D}_{\mathbb{Z},\alpha}}$
- $r \geq \sqrt{\dfrac{5\ln(2m(1+1/\varepsilon))}{\pi}}$ for the correctness of $\mathbf{G}^{-1}_{\mathrm{rand}}(\cdot)$ sampling
- $r \geq 4\left((1 - \varepsilon)(2\varepsilon')^2\right)^{-\frac{1}{m}}$ for the leftover hash lemma
- $r \geq \sqrt{5}(1 + B)\sqrt{\dfrac{\ln(2m(1+1/\varepsilon))}{\pi}}$ for Lemma 3.7
- $q = \Omega\left(\sqrt{m}r\alpha(m L_{\max} \ell_{\max})^{1/2}\right)$ for the correctness of decryption

We can thus set the parameters as follows:

- $n = \Theta(\lambda)$,
- $L_{\max} = \mathrm{poly}(n)$,
- $\ell_{\max} = \mathrm{poly}(n)$,
- $\alpha = \Theta(\sqrt{n})$,
- $r = \tilde{\Theta}(n)$,
- $q = \tilde{\Theta}\left(n^{5/2} \cdot L_{\max} \cdot \ell_{\max}\right)$, a power of 2.

Note that the ciphertext size grows with $\log L_{\max}$. Correctness follows directly.

Lemma 5.6 (Correctness). *For any branching program Π of length L on ℓ variables, any $x_1, \dots, x_\ell \in \{0, 1\}$, the result of the homomorphic evaluation $\mathbf{C}_f = \mathsf{Eval}(\Pi, (\mathbf{C}_1, \dots, \mathbf{C}_\ell))$ decrypts to $\Pi(x_1, \dots, x_\ell)$ with overwhelming probability, where $\mathbf{C}_i \leftarrow \mathsf{Encrypt}(\mathbf{s}, x_i)$ for $i \in [\ell]$ and $\mathbf{s} \leftarrow \mathsf{Setup}(1^\lambda)$.*

Proof. Lemma 5.4 shows that the noise distribution of the output $\mathbf{C}_f$ of Eval has parameter $r_f = r\sqrt{2\sum_{i=1}^{\ell} \tau_i (1 + \|\mathbf{e}_i\|^2)}$, that is $r\sqrt{2L\sum_{i=1}^{\ell} (1 + \|\mathbf{e}_i\|^2)}$ because of the padding we applied to Π. We have $r_f \leq r\sqrt{2L\ell(1 + C^2\alpha^2 m)}$ with C the universal constant defined in Lemma 2.4, Using the bounds $L_{\max}$ and $\ell_{\max}$ we have $r_L = \tilde{O}(r\alpha(m\,L_{\max}\,\ell_{\max})^{1/2})$. Finally, by a tail cutting argument, $q = \tilde{\Theta}(r_L\sqrt{n}) = \tilde{\Theta}(n^{5/2}L_{\max}\ell_{\max})$ is enough for decryption to be correct with overwhelming probability. □

5.4   Arbitrary Modulus and Random Trapdoor Matrix

In this paragraph we show how to instantiate our proofs in a more generic setting.

Our GSW ciphertext rerandomization can be straightforwardly adapted to any matrix $\mathbf{H}$ and modulus q, as long as the lattice $\Lambda_q^\perp(\mathbf{H}^\mathsf{T})$ has a small public basis, i.e. a small public trapdoor. Observe that the conditions needed to apply GSW ciphertext rerandomization are given in Lemma 3.7, which bounds the smoothing parameter of the lattice

$$\mathcal{L} = \{\mathbf{v} \in \Lambda_q^\perp(\mathbf{H}^\mathsf{T}) \times \mathbb{Z} : \widehat{\mathbf{e}}^\mathsf{T}\mathbf{v} = 0\}$$

and in Lemma 3.8 which gives the min-entropy of a Gaussian over $\Lambda_q^\perp(\mathbf{H}^\mathsf{T})$.

Let $\beta \geq \|\mathbf{t}_i\|$, where $\mathbf{T} = \{\mathbf{t}_1, ..., \mathbf{t}_m\}$ is the public trapdoor of $\mathbf{H}$ (i.e. $\mathbf{T}$ is a small basis of $\Lambda_q^\perp(\mathbf{H}^\mathsf{T})$), we show that the previous two lemmas can be proven for $\mathbf{H}$ and the parameter r only grows by a factor β.

First, observe that Lemma 3.7 aims to find m small independent vectors in $\mathcal{L}$. By noticing that

$$\mathcal{L} = \{(\mathbf{v}, -\mathbf{v}^\mathsf{T}\mathbf{e}) : \mathbf{v} \in \Lambda_q^\perp(\mathbf{H}^\mathsf{T})\}$$

we can exhibit m small vectors $\mathbf{u}_i = (\mathbf{t}_i, -\mathbf{t}_i^\mathsf{T}\mathbf{e}), i \in [m]$ which are of norm

$$\|\mathbf{u}_i\| \leq \|\mathbf{t}_i\|(1 + \|\mathbf{e}\|) \leq \beta(1 + \|\mathbf{e}\|)$$

This bound is the one we obtain in Lemma 3.7 for $\Lambda_q^\perp(\mathbf{G}^\mathsf{T})$ where $\|\mathbf{T}\| = \sqrt{5}$.

Second, we show that the bound on the min-entropy of Lemma 3.8 can be expressed as a function of β, simply by using the fact that $\det(\mathbf{T}) \leq \|\mathbf{T}\|^m = \beta^m$. From this we have the following bound on the min-entropy:

$$H_\infty\left(\mathcal{D}_{\Lambda_q^\perp(\mathbf{H}^\mathsf{T})+\mathbf{c},r}\right) \geq \log(1 - \varepsilon) + m\log(r) - m\log(\beta)$$

This bound is slightly worse that the one we obtain in Lemma 3.8 for $\mathbf{G}$ (where we had 2 instead of β). However this is not a problem as it is a weaker bound than the one obtained in Lemma 3.7.

By using these two lemmas we can rerandomize GSW ciphertexts and ensure circuit privacy for arbitrary modulus q, and any matrix $\mathbf{H}$ with public trapdoor by setting the Gaussian parameter of $\mathbf{H}^{-1}(\cdot)$ to $r = \tilde{\Theta}(\beta n)$.

5.5   Extension to General Circuits

We can realize circuit-private FHE for general circuits via bootstrapping using the technique of [OPP14] by combining a compact FHE for general circuits with decryption in NC^1 with our circuit-private FHE for NC^1 circuits as follows: the server receives a ciphertext under the first FHE scheme, evaluates its circuit and bootstraps to the second (circuit hiding) FHE scheme. The ensuing scheme however will not satisfy the multi-hop requirement. Nevertheless, by using the construction given in [GHV10] it is possible to reach i-hop circuit private FHE for any a priori chosen i by giving out i pairs of switching keys to bootstrap from one scheme to the other and vice versa.

Acknowledgements. We thank Vinod Vaikuntanathan for insightful discussions, as well as Damien Stehlé and the organizers of the HEAT summer school where this research started.

References

[ACPS09] Applebaum, B., Cash, D., Peikert, C., Sahai, A.: Fast cryptographic primitives and circular-secure encryption based on hard learning problems. In: Halevi, S. (ed.) CRYPTO 2009. LNCS, vol. 5677, pp. 595–618. Springer, Heidelberg (2009)

[AGHS13] Agrawal, S., Gentry, C., Halevi, S., Sahai, A.: Discrete gaussian leftover hash lemma over infinite domains. In: Sako, K., Sarkar, P. (eds.) ASIACRYPT 2013, Part I. LNCS, vol. 8269, pp. 97–116. Springer, Heidelberg (2013)

[AP14] Alperin-Sheriff, J., Peikert, C.: Faster bootstrapping with polynomial error. In: Garay, J.A., Gennaro, R. (eds.) CRYPTO 2014, Part I. LNCS, vol. 8616, pp. 297–314. Springer, Heidelberg (2014)

[AR13] Aggarwal, D., Regev, O.: A note on discrete gaussian combinations of lattice vectors. CoRR, abs/1308.2405 (2013)

[BGG+14] Boneh, D., Gentry, C., Gorbunov, S., Halevi, S., Nikolaenko, V., Segev, G., Vaikuntanathan, V., Vinayagamurthy, D.: Fully key-homomorphic encryption, arithmetic circuit ABE and compact garbled circuits. In: Nguyen, P.Q., Oswald, E. (eds.) EUROCRYPT 2014. LNCS, vol. 8441, pp. 533–556. Springer, Heidelberg (2014)

[BGV12] Brakerski, Z., Gentry, C., Vaikuntanathan, V.: (Leveled) fully homomorphic encryption without bootstrapping. In: Goldwasser, S. (ed.) ITCS 2012, pp. 309–325. ACM, January 2012

[BLP+13] Brakerski, Z., Langlois, A., Peikert, C., Regev, O., Stehlé, D.: Classical hardness of learning with errors. In: Boneh, D., Roughgarden, T., Feigenbaum, J. (eds.) 45th ACM STOC, pp. 575–584. ACM Press, June 2013

[BV11a] Brakerski, Z., Vaikuntanathan, V.: Efficient fully homomorphic encryption from (standard) LWE. In: Ostrovsky, R. (ed.) 52nd FOCS, pp. 97–106. IEEE Computer Society Press, October 2011

[BV11b] Brakerski, Z., Vaikuntanathan, V.: Fully homomorphic encryption from Ring-LWE and security for key dependent messages. In: Rogaway, P. (ed.) CRYPTO 2011. LNCS, vol. 6841, pp. 505–524. Springer, Heidelberg (2011)

[BV14] Brakerski, Z., Vaikuntanathan, V.: Lattice-based FHE as secure as PKE. In: Naor, M. (ed.) ITCS 2014, pp. 1–12. ACM, January 2014

[DM15] Ducas, L., Micciancio, D.: FHEW: Bootstrapping homomorphic encryption in less than a second. In: Oswald, E., Fischlin, M. (eds.) EURO-CRYPT 2015. LNCS, vol. 9056, pp. 617–640. Springer, Heidelberg (2015)

[DRS04] Dodis, Y., Reyzin, L., Smith, A.: Fuzzy extractors: how to generate strong keys from biometrics and other noisy data. In: Cachin, C., Camenisch, J.L. (eds.) EUROCRYPT 2004. LNCS, vol. 3027, pp. 523–540. Springer, Heidelberg (2004)

[DS16] Ducas, L., Stehlé, D.: Sanitization of FHE ciphertexts. In: Fischlin, M., Coron, J.-S. (eds.) EUROCRYPT 2016. LNCS, vol. 9665, pp. 294–310. Springer, Heidelberg (2016). doi:10.1007/978-3-662-49890-3_12

[Gen09] Gentry, C.: A fully homomorphic encryption scheme. Ph.D. thesis, Stanford University (2009). http://crypto.stanford.edu/craig

[GHS12] Gentry, C., Halevi, S., Smart, N.P.: Homomorphic evaluation of the AES circuit. In: Safavi-Naini, R., Canetti, R. (eds.) CRYPTO 2012. LNCS, vol. 7417, pp. 850–867. Springer, Heidelberg (2012)

[GHV10] Gentry, C., Halevi, S., Vaikuntanathan, V.: i-Hop homomorphic encryption and rerandomizable yao circuits. In: Rabin, T. (ed.) CRYPTO 2010. LNCS, vol. 6223, pp. 155–172. Springer, Heidelberg (2010)

[GPV08] Gentry, C., Peikert, C., Vaikuntanathan, V.: Trapdoors for hard lattices and new cryptographic constructions. In: Ladner, R.E., Dwork, C. (eds.) 40th ACM STOC, pp. 197–206. ACM Press, May 2008

[GSW13] Gentry, C., Sahai, A., Waters, B.: Homomorphic encryption from learning with errors: conceptually-simpler, asymptotically-faster, attribute-based. In: Canetti, R., Garay, J.A. (eds.) CRYPTO 2013, Part I. LNCS, vol. 8042, pp. 75–92. Springer, Heidelberg (2013)

[GVW15] Gorbunov, S., Vaikuntanathan, V., Wichs, D.: Leveled fully homomorphic signatures from standard lattices. In: Servedio, R.A., Rubinfeld, R. (eds.) 47th ACM STOC, pp. 469–477. ACM Press, June 2015

[HS15] Halevi, S., Shoup, V.: Bootstrapping for HElib. In: Oswald, E., Fischlin, M. (eds.) EUROCRYPT 2015. LNCS, vol. 9056, pp. 641–670. Springer, Heidelberg (2015)

[IP07] Ishai, Y., Paskin, A.: Evaluating branching programs on encrypted data. In: Vadhan, S.P. (ed.) TCC 2007. LNCS, vol. 4392, pp. 575–594. Springer, Heidelberg (2007)

[LMSV12] Loftus, J., May, A., Smart, N.P., Vercauteren, F.: On CCA-Secure somewhat homomorphic encryption. In: Miri, A., Vaudenay, S. (eds.) SAC 2011. LNCS, vol. 7118, pp. 55–72. Springer, Heidelberg (2012)

[MP12] Micciancio, D., Peikert, C.: Trapdoors for lattices: simpler, tighter, faster, smaller. In: Pointcheval, D., Johansson, T. (eds.) EUROCRYPT 2012. LNCS, vol. 7237, pp. 700–718. Springer, Heidelberg (2012)

[MR04] Micciancio, D., Regev, O.: Worst-case to average-case reductions based on Gaussian measures. In: 45th FOCS, pp. 372–381. IEEE Computer Society Press, October 2004

[MR07] Micciancio, D., Regev, O.: Worst-case to average-case reductions based on gaussian measures. SIAM J. Comput. **37**(1), 267–302 (2007)

[OPP14] Ostrovsky, R., Paskin-Cherniavsky, A., Paskin-Cherniavsky, B.: Maliciously circuit-private FHE. In: Garay, J.A., Gennaro, R. (eds.) CRYPTO 2014, Part I. LNCS, vol. 8616, pp. 536–553. Springer, Heidelberg (2014)

[Pei10] Peikert, C.: An efficient and parallel gaussian sampler for lattices. In: Rabin, T. (ed.) CRYPTO 2010. LNCS, vol. 6223, pp. 80–97. Springer, Heidelberg (2010)

[Reg05] Regev, O.: On lattices, learning with errors, random linear codes, and cryptography. In: Gabow, H.N., Fagin, R. (eds.) 37th ACM STOC, pp. 84–93. ACM Press, May 2005

[SS10] Stehlé, D., Steinfeld, R.: Faster fully homomorphic encryption. In: Abe, M. (ed.) ASIACRYPT 2010. LNCS, vol. 6477, pp. 377–394. Springer, Heidelberg (2010)

[SV10] Smart, N.P., Vercauteren, F.: Fully homomorphic encryption with relatively small key and ciphertext sizes. In: Nguyen, P.Q., Pointcheval, D. (eds.) PKC 2010. LNCS, vol. 6056, pp. 420–443. Springer, Heidelberg (2010)

[SYY99] Sander, T., Young, A., Yung, M.: Non-interactive cryptocomputing for NC1. In: 40th FOCS, pp. 554–567. IEEE Computer Society Press, October 1999

[vDGHV10] van Dijk, M., Gentry, C., Halevi, S., Vaikuntanathan, V.: Fully homomorphic encryption over the integers. In: Gilbert, H. (ed.) EUROCRYPT 2010. LNCS, vol. 6110, pp. 24–43. Springer, Heidelberg (2010)

Symmetric Cryptography

Cryptanalysis of a Theorem: Decomposing the Only Known Solution to the Big APN Problem

Léo Perrin[1][(⊠)], Aleksei Udovenko[1][(⊠)], and Alex Biryukov[1,2][(⊠)]

[1] SnT, University of Luxembourg, Luxembourg City, Luxembourg
{leo.perrin,aleksei.udovenko}@uni.lu
[2] CSC, University of Luxembourg, Luxembourg City, Luxembourg
alex.biryukov@uni.lu

Abstract. The existence of Almost Perfect Non-linear (APN) permutations operating on an even number of bits has been a long standing open question until Dillon et al., who work for the NSA, provided an example on 6 bits in 2009.

In this paper, we apply methods intended to reverse-engineer S-Boxes with unknown structure to this permutation and find a simple decomposition relying on the cube function over $GF(2^3)$. More precisely, we show that it is a particular case of a permutation structure we introduce, the *butterfly*. Such butterflies are $2n$-bit mappings with two CCZ-equivalent representations: one is a quadratic non-bijective function and one is a degree $n + 1$ permutation. We show that these structures always have differential uniformity at most 4 when n is odd. A particular case of this structure is actually a 3-round Feistel Network with similar differential and linear properties. These functions also share an excellent non-linearity for $n = 3, 5, 7$.

Furthermore, we deduce a bitsliced implementation and significantly reduce the hardware cost of a 6-bit APN permutation using this decomposition, thus simplifying the use of such a permutation as building block for a cryptographic primitive.

Keywords: Boolean functions · APN · Butterfly structure · S-Box decomposition · CCZ-equivalence · Feistel Network · Bitsliced implementation

1 Introduction

When designing a symmetric primitive, it is common to use functions operating on a small part of the internal state to provide non-linearity. These are called

The work of Léo Perrin is supported by the CORE ACRYPT project (ID C12-15-4009992) funded by the *Fonds National de la Recherche* (Luxembourg). The work of Aleksei Udovenko is supported by the *Fonds National de la Recherche*, Luxembourg (project reference 9037104).

© International Association for Cryptologic Research 2016
M. Robshaw and J. Katz (Eds.): CRYPTO 2016, Part II, LNCS 9815, pp. 93–122, 2016.
DOI: 10.1007/978-3-662-53008-5_4

S-Boxes and their properties can be leveraged to justify security against differential [1] and linear [2] attacks using for example a wide-trail argument, as was done for the AES [3].

A popular strategy for choosing S-Boxes with desirable cryptographic properties is to use mathematical construction based for example on the inverse in a finite field [4]. A function with optimal differential property (in a sense that we will define later) is called *Almost Perfect Non-linear* or *APN*. While it is easy to find functions with this property, permutations are more rare. Many monomials are known to be APN permutations in finite fields of size 2^n for n odd (for example the cube function), but whether there even exists APN permutations operating on an even number of bits is still an important research area.

In this context, the 6-bit APN permutation described by a team of mathematicians from the NSA (Dillon *et al.*) in [5] is of great theoretical importance: it is the only known APN permutation for even n so far. Furthermore, it has already been used to design an authenticated cipher: Fides [6]. However, the method used by the Dillon et al. to find it relies on sophisticated considerations related to error correcting codes and no generalization of their results has been published to the best of our knowledge. In their paper, the authors state the "big APN problem" and it is, 6 years later, still as much of an open question:

(STILL) The Big APN Problem: Does there exist an APN permutation on $GF(2^m)$ if m is EVEN and GREATER THAN 6?

Our Contribution. By applying methods designed by Biryukov *et al.* to reverse-engineer the S-Box of the last Russian cryptographic standards [7], we show the existence of a much simpler expression of the 6-bit APN permutation. This is stated in Theorem 3 which we reproduce here.

Main Theorem (A Family of 6-bit APN Permutations). The 6-bit permutation described by Dillon *et al.* in [5] is affine equivalent to any involution built using the structure described in Fig. 1, where $\odot$ denotes multiplication in the finite field $GF(2^3)$, $\alpha \neq 0$ is such that $\text{Tr}(\alpha) = 0$ and $\mathcal{A}$ denotes any 3-bit APN permutation.

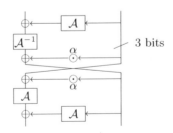

Fig. 1. Some S-Boxes affine-equivalent to the Dillon APN permutation.

We study extensively this structure, both experimentally and mathematically, and derive in particular new families of differentially 4-uniform permutations of $2n$ bits for n odd.

Outline. This paper is devoted to first deriving this theorem and then exploring its consequences. Section 2 describes how the cryptanalysis strategy described in [7] can be successfully applied to the 6-bit APN permutation to identify a highly structured decomposition. We then study this structure in Sect. 3. Next, we show in Sect. 4 that the same structure can be used to build differentially 4-uniform permutations with algebraic degree at least n in fields of size $2n$ for odd n. Finally, we use our results on the decomposition of 6-bit APN permutations to describe efficient bit-sliced and hardware implementation of some of them in Sect. 5.

Notations and Definitions

We use common definitions and notations throughout this paper. For the sake of clarity, we list them here. First, we describe the notations related to finite field:

- $\mathbb{F}_{2^n}$ is a finite field of size 2^n,
- for any x in $\mathbb{F}_{2^n}$, the trace of x is $\mathrm{Tr}(x) = \sum_{i=0}^{n-1} x^{2^i}$,

The differential properties of an S-Box $f : \mathbb{F}_2^n \to \mathbb{F}_2^m$ are studied using its Difference Distribution Table (DDT), the $2^n \times 2^m$ matrix $\mathcal{D}(f)$ such that $\mathcal{D}(f)[\delta, \Delta] = \#\{x \in \mathbb{F}_{2^n}, f(x + \delta) + f(x) = \Delta\}$. The maximum coefficient[1] in $\mathcal{D}(f)$ is the differential uniformity of f and, if it is equal to u, then we say that f is differentially u-uniform. A differentially 2-uniform function is called Almost Perfect Nonlinear (APN).

Similarly, security against linear attacks can be justified using the Linear Approximation Table (LAT)[2] of f. It is the $2^n \times 2^m$ matrix $\mathcal{L}(f)$ such that $\mathcal{L}(f)[a, b] = \#\{x \in \mathbb{F}_{2^n}, a \cdot x = b \cdot y\} - 2^{n-1}$ (where "$\cdot$" denotes the scalar product). The *non-linearity* of a $f : \mathbb{F}_2^n \to \mathbb{F}_2^m$ is $\mathcal{NL}(f) = 2^{n-1} - \max(|\mathcal{L}(f)[a, b]|)$ where the maximum is taken over all non-zero line and column indices a and b.

Finally, we also consider algebraic decompositions of the functions we study using the following tools:

- if x and u are vectors of $\mathbb{F}_2^n$, then $x^u = \prod_{i=0}^{n-1} x_i^{u_i}$ so that $x^u = 1$ if and only if $x_i = 1$ for all i such that $u_i = 1$,
- the Algebraic Normal Form (ANF) of a Boolean function f is its unique expression $f(x) = \bigoplus_{u \in \mathbb{F}_2^n} a_u x^u$ where all a_u are in $\{0, 1\}$,
- the algebraic degree of a Boolean function f is denoted $\deg(f)$ and is equal to the maximum Hamming weight of u such that $a_u = 1$ in the ANF of f,

---

[1] The maximum is taken over all non-zero line indices.

[2] This object is also sometimes referred to as the "correlation matrix". Up to a multiplication by a constant factor, the coefficients in the LAT of a function also form its Walsh Spectrum.

– the field polynomial representation of f mapping $\mathbb{F}_{2^n}$ to itself is its unique expression as a univariate polynomial of $\mathbb{F}_{2^n}$, so that $f(x) = \sum_{i=0}^{2^n-1} c_i x^i$ with c_i in $\mathbb{F}_{2^n}$. It can be obtained using Lagrange interpolation.

Note that the algebraic degree of a polynomial of $\mathbb{F}_{2^n}$ is equal to the maximum Hamming weight of the binary expansions of the exponents in its field polynomial representation. For example, the algebraic degree of the cube function $x \mapsto x^3$ in $\mathbb{F}_{2^n}$ is equal to 2.

Two functions f and g are *affine equivalent* if there exist affine permutations A and B such that $g = B \circ f \circ A$. If we also add an affine function C to the output, that is, $g = B \circ f \circ A + C$, then f and g are *extended affine-equivalent (EA-equivalent)*.

Finally, we denote the concatenation of two binary variables using the symbol "$||$". In particular, we will often interpret bit-strings of length $2n$ as $x||y$, where x and y are in $\mathbb{F}_2^n$.

2   A Decomposition of the 6-Bit APN Permutation

In this section, we identify a decomposition of the Dillon APN permutation. We denote this permutation $S_0 : \mathbb{F}_2^6 \rightarrow \mathbb{F}_2^6$ and give its look-up table in Table 1. As we are interested only in its being an APN permutation, we allow ourselves to compose it with affine permutations as such transformations preserve this property. We will omit the respective inverse permutations to simplify our description.

Table 1. The Dillon permutation S_0 in hexadecimal (e.g. $S_0(\texttt{0x10}) = \texttt{0x3b}$).

	.0	.1	.2	.3	.4	.5	.6	.7	.8	.9	.a	.b	.c	.d	.e	.f
0	00	36	30	0d	0f	12	35	23	19	3f	2d	34	03	14	29	21
1	3b	24	02	22	0a	08	39	25	3c	13	2a	0e	32	1a	3a	18
2	27	1b	15	11	10	1d	01	3e	2f	28	33	38	07	2b	2c	26
3	1f	0b	04	1c	3d	2e	05	31	09	06	17	20	1e	0c	37	16

Our strategy is identical to the one used to recover the structure of the S-Box of the last Russian cryptographic standards described in [7]. First, we obtain a high level decomposition of the permutation relying on two distinct but closely related 3-bit keyed permutations (the "TU-decomposition") in Sect. 2.1. Then, we decompose these keyed permutations in Sects. 2.2. Finally, we provide the complete decomposition of an S-Box affine-equivalent to S_0 in Sect. 2.3.

2.1   High-Level TU-Decomposition

As suggested in [7,8], we looked at the "Jackson Pollock" representation of the absolute value of the LAT of the S-Box (see Fig. 2a). We can see some patterns, namely columns and aligned short vertical segments of black and white colors within a grey rectangle (white is 0, grey is 4 and black is 8). The black-and-white columns also have the 8 topmost coefficients equal to zero. Moreover, their horizontal coordinates form a linear subspace of $\mathbb{F}_2^6$.

Therefore, as was done in [7], we compose the S-Box with a particular linear permutation chosen so that these particular columns are clustered to the left of the picture, i.e. their abscissa become $[0, 7]$. The black-and-white columns have coordinates $\{0, 4, 10, 14, 16, 20, 26, 30\}$ and the binary expansion of these numbers form a linear subspace of $\mathbb{F}_2^6$ spanned by the binary expansions of $\{4, 10, 16\}$. We thus construct a permutation η, linear over $GF(2)$, such that $\eta : 1 \mapsto 4, 2 \mapsto 10, 4 \mapsto 16$ and then we complete it by setting $\eta : 8 \mapsto 1, 16 \mapsto 2, 32 \mapsto 32$ so that η is a permutation. By Theorem 1 from [7], the composition $\eta^t \circ S_0$ of such mapping with the S-Box will group the black-and-white columns in the LAT. The Jackson Pollock representation of $\eta^t \circ S_0$ is given in Fig. 2b.

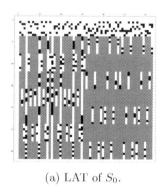 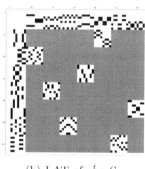

(a) LAT of S_0.   (b) LAT of $\eta^t \circ S_0$.

Fig. 2. The Jackson Pollock representation of the LAT of two permutations (absolute value). Row/column indices correspond to input/output linear approximation masks respectively. White pixels correspond to 0, grey to 4 and black to 8.

As we can see the columns are now aligned, as was our goal, and the short segments became grouped into small squares, thus making the whole picture more structured. Doing this also caused the appearance of a "white-square" in the top-left square $[0, 7] \times [0, 7]$. This last pattern is a known side effect of the existence of specific integral properties (see Lemma 2 of [7] which is itself derived from [9]). Hence, we checked for integral/multiset properties as defined in [10] and identified the following property: fixing the last 3 bits of the input and letting the first 3 take all possible values leads to the last 3 bits of the output taking all possible values.

We keep following the blueprint laid out in [7] and investigate the consequences of this integral distinguisher. In fact we generalize their next step, which consists in providing a high level decomposition of the S-Box, by describing the *TU-decomposition*.

Lemma 1. *Let f be a function mapping $\mathbb{F}_2^n \times \mathbb{F}_2^n$ to itself such that fixing the right input to any value and letting the left one take all 2^n possible values leads to the left output taking all 2^n possible values. Then f can be decomposed using a keyed n-bit permutation T and a keyed n-bit function U (see Fig. 3a):*

$$f(x, y) = \big(T_y(x), U_{T_y(x)}(y)\big),$$

Besides, if f is a permutation then U is a keyed permutation.

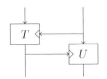

(a) Basic TU-decomposition.

(b) TU-decomposition composed with a swap.

Fig. 3. Principle of the TU-decomposition.

Proof. We simply define $T_y(x)$ to be the left side of $f(x, y)$. Because of the multiset property, T_y is a permutation for all y. We then define U to be such that $U_k(y)$ is the right side of $f\big(T_y^{-1}(k), y\big)$.

If f is a permutation then $(x, y) \mapsto f\big(T_y^{-1}(x), y\big)$ is a permutation equal to $(x, y) \mapsto (x, U_x(y))$. In particular, it holds that U_x is a permutation for all x, making it a keyed permutation.   □

We apply Lemma 1 to $\eta^t \circ S_0$ and deduce its TU-decomposition. We actually have the output halves swapped so we may draw the structure in a more symmetric fashion (see Fig. 3b). The corresponding keyed permutations T and U are given in Table 2.

Table 2. The keyed permutations T and U. T_k and U_k denote the permutations corresponding to the key k.

	0	1	2	3	4	5	6	7
T_0	0	6	4	7	3	1	5	2
T_1	7	5	1	6	4	2	0	3
T_2	4	3	2	0	5	6	1	7
T_3	3	5	2	1	4	6	7	0
T_4	1	2	0	6	4	3	7	5
T_5	6	5	2	4	7	0	1	3
T_6	5	2	6	4	0	3	1	7
T_7	2	0	1	6	5	3	4	7

(a) T.

	0	1	2	3	4	5	6	7
U_0	0	3	6	4	2	7	1	5
U_1	7	4	0	2	3	6	1	5
U_2	1	4	2	6	3	0	5	7
U_3	7	2	5	1	3	0	4	6
U_4	7	3	4	1	0	2	6	5
U_5	3	7	1	4	2	0	5	6
U_6	1	3	7	4	6	2	5	0
U_7	4	6	3	0	5	1	7	2

(b) U.

The degree of T as a 6-bit permutation is equal to 3 and that of U is equal to 2. However the degree of T^{-1} is equal to 2 as well. One may think that T^{-1} and U are somehow related and we indeed found that T^{-1} and U are linearly equivalent using the algorithm by Biryukov *et al.* from [11]. The linear equivalence of T^{-1} and U is given by:

$$U(x) = M'_U \circ T^{-1} \circ M_U(x),$$

where T and U are considered as 6-bit permutations and the linear permutations M_U and M'_U are given as the following binary matrices:

$$M_U = \begin{bmatrix} 0\,1\,1\,0\,1\,0 \\ 1\,0\,0\,0\,1\,0 \\ 0\,0\,1\,0\,1\,1 \\ 0\,0\,0\,1\,0\,0 \\ 0\,0\,0\,0\,1\,0 \\ 0\,0\,0\,0\,1\,1 \end{bmatrix}, M'_U = \begin{bmatrix} 1\,1\,0\,0\,0\,0 \\ 0\,1\,0\,0\,1\,0 \\ 1\,0\,1\,1\,1\,1 \\ 0\,0\,0\,1\,0\,0 \\ 0\,0\,0\,0\,1\,0 \\ 0\,0\,0\,0\,1\,1 \end{bmatrix}.$$

2.2 Decomposing T

As we applied a linear mapping on the output of the S-Box, we might have scrambled the initial structure of U. Hence, we choose the decomposition of T^{-1} as our main target. We start by composing it with a Feistel round to ensure that 0 is mapped to itself for all keys. Again, this simplification was performed while reverse-engineering the GOST S-Box. If we apply such an appropriate Feistel round before or after T^{-1}, the corresponding Feistel function is always a permutation. Moreover, in the case when the Feistel function is used between T and U, the Feistel function is linear[3] so we choose this side. We define $t(k) = T_k(0)$ and $T'_k(x) = T_k(x) \oplus t(k)$ so that $T'_k(0) = T'^{-1}_k(0) = 0$ for all k (see Fig. 4a). The linear permutation t is given by $t(x) = (0, 7, 4, 3, 1, 6, 5, 2)$.

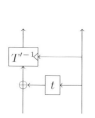 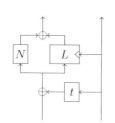 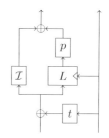

(a) Detaching a linear Feistel round.

(b) Splitting T'^{-1} into N and L.

(c) Simplifying N into $\mathcal{I}$ and linear functions.

Fig. 4. Simplifying the keyed permutation T'^{-1}.

---

[3] If we had attacked U instead of T^{-1}, then detaching a Feistel function in this way leads only to a nonlinear Feistel function (regardless of the side), which supports our choice of T'^{-1} as an easier target.

We then check the existence of particular algebraic structure in T'. We choose the irreducible polynomial $X^3 + X + 1$ to represent elements of $\mathbb{F}_{2^3}$ as binary strings and, furthermore, we represent these binary strings as integers. In equations we represent such constants in italic. Note that this representation was motivated by convenience reasons for working in Sage [12] and we are using it only in this section for describing the decomposition process.

Now we use Lagrange interpolation to represent each T'^{-1}_k as a polynomial over $\mathbb{F}_{2^3}$. The result is given in Table 3. Interestingly, the coefficients of the non-linear terms x^6, x^5, x^3 are key-independent. We therefore decompose T'^{-1} as a sum of its non-linear part N and its key-dependent linear part L_k so that $T'^{-1}_k(x) = N(x) + L_k(x)$, where $N(x) = 3x^6 + 2x^5 + 5x^3$ and $L_k(x)$ is linear for any k (see Fig. 4b).

Table 3. The values and polynomial interpolation of each T'^{-1}_k.

	0	1	2	3	4	5	6	7	Interpolation polynomial
T'^{-1}_0	0	5	7	4	2	6	1	3	$3x^6 + 2x^5 + 3x^4 + 5x^3 + 2x^2 + 0x$
T'^{-1}_1	0	3	1	4	7	5	2	6	$3x^6 + 2x^5 + 1x^4 + 5x^3 + 4x^2 + 2x$
T'^{-1}_2	0	4	5	7	3	6	2	1	$3x^6 + 2x^5 + 0x^4 + 5x^3 + 0x^2 + 0x$
T'^{-1}_3	0	2	3	7	6	5	1	4	$3x^6 + 2x^5 + 2x^4 + 5x^3 + 6x^2 + 2x$
T'^{-1}_4	0	2	5	1	7	4	6	3	$3x^6 + 2x^5 + 3x^4 + 5x^3 + 0x^2 + 5x$
T'^{-1}_5	0	4	3	1	2	7	5	6	$3x^6 + 2x^5 + 1x^4 + 5x^3 + 6x^2 + 7x$
T'^{-1}_6	0	3	7	2	6	4	5	1	$3x^6 + 2x^5 + 0x^4 + 5x^3 + 2x^2 + 5x$
T'^{-1}_7	0	5	1	2	3	7	6	4	$3x^6 + 2x^5 + 2x^4 + 5x^3 + 4x^2 + 7x$

We now simplify N by applying a linear function of our choice after T'^{-1} (see Fig. 4c). We allow ourselves to do this because this side corresponds to the input of the S-Box on which, as we said before, we may apply any affine layer as those would preserve the differential uniformity of the whole permutation. Choosing this side also prevents the need for a corresponding modification of U. We choose $p(x) = 4x^4 + x^2 + x$ because $(p \circ N)(x) = x^6$ is the inverse function in $\mathbb{F}_{2^3}$, denoted $\mathcal{I}$.

We further remark that $p \circ L_k$ is simpler than L_k too: there are nonzero coefficients only at x^2 and x^4 (see Table 4). Note also that $p \circ L_2 = 0$ so we add 2 to k to obtain these linear layers:

$$(p \circ L_k)(x) = l_2(k+2)x^2 + l_4(k+2)x^4,$$

where $l_2(x) = 2x^4 + 4x^2 + x$ and $l_4(x) = x^4 + 6x^2 + 2x$ are obtained from the Lagrange interpolations of $p \circ L_k$ given in Table 4.

In our effort to simplify the structure, we search for a linear permutation q such that both $l_2 \circ q$ and $l_4 \circ q$ have a simpler form and find that $q(x) = 3x^4 + 7x^2 + 3x$ is such that $(l_2 \circ q)(x) = x^4$ and $(l_4 \circ q)(x) = x^2$. Therefore,

Table 4. The interpolation polynomials of each $p \circ L_k$.

Function	Polynomial		Function	Polynomial
$p \circ L_0$	$7x^4 + 3x^2$		$p \circ L_4$	$4x^4 + 6x^2$
$p \circ L_1$	$2x^4 + 4x^2$		$p \circ L_5$	$1x^4 + 1x^2$
$p \circ L_2$	$0x^4 + 0x^2$		$p \circ L_6$	$3x^4 + 5x^2$
$p \circ L_3$	$5x^4 + 7x^2$		$p \circ L_7$	$6x^4 + 2x^2$

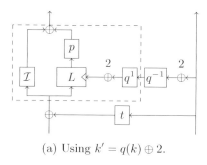

(a) Using $k' = q(k) \oplus 2$.

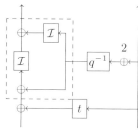

(b) Using Equation (1).

Fig. 5. Simplifying $p \circ L$ and thus T'^{-1}. The dashed area corresponds to the equivalence given by Eq. 1.

we can write $(p \circ L_k)(x) = k'^4 x^2 + k'^2 x^4$, where $k' = q^{-1}(k + 2)$. We deduce a representation of the whole structure of $p \circ T'^{-1}$ depending only on linear functions and the inverse function which we describe in Eq. (1) and Fig. 5.

$$(p \circ T'^{-1}_k)(x) = x^6 + x^2 k'^4 + x^4 k'^2 = (x + k')^6 + k'^6, \text{ with } k' = q^{-1}(k + 2). \quad (1)$$

Then, we replace the application of $x \mapsto q^{-1}(x+2)$ on the horizontal branch in Fig. 5b by its application on the right vertical branch followed by its inverse (see Fig. 6a; note that $q^{-1}(2) = 5$). By then discarding the affine permutation applied on the top of the right branch (we omit the affine layers applied to the outside of the complete permutation), we obtain the equivalent structure shown in Fig. 6b. Finally, we merge the two linear Feistel functions into $z(x) = t(q(x)) \oplus x$ to obtain our final decomposition of T^{-1}:

$$T^{-1}(\ell \| r) = \mathcal{I}(\ell + z(q^{-1}(r)) + 5) + \mathcal{I}(q^{-1}(r) + 5) \| (q^{-1}(r) + 5),$$

which is also is described in Fig. 6c. Now that we have found a decomposition of T, we shall use it to express a whole permutation affine-equivalent to S_0.

2.3 Joining the Decompositions of T and U

Let us now join the decomposition of T and U together, that of U being obtained using that $\mathcal{U}(x) = M'_U \circ T^{-1} \circ M_U(x)$. The affine transformations applied on

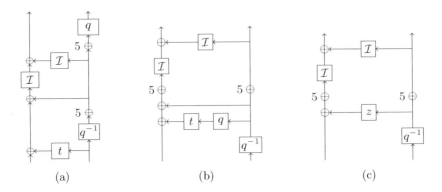

Fig. 6. Finishing the decomposition of T^{-1}: moving q, q^{-1} and $x \mapsto x + 2$ around, removing the outer affine layer and merging the Feistel linear rounds.

the top of T'^{-1} make the relation between T^{-1} and U affine instead of linear on one side. This side corresponds to the output of the S-Box and we omit this transformation. The other linear mapping M_U connecting T^{-1} and U merges with the linear part of T^{-1} and its symmetric copy from U into the linear mapping M (see Fig. 7a and b). The linear permutation M is given by the following matrix over $\mathbb{F}_2$:

$$M = \begin{bmatrix} 1 & 0 & 1 & 1 & 1 & 1 \\ 1 & 1 & 0 & 0 & 1 & 0 \\ 0 & 0 & 1 & 1 & 1 & 0 \\ 1 & 1 & 0 & 1 & 0 & 1 \\ 1 & 1 & 1 & 1 & 1 & 0 \\ 1 & 0 & 1 & 0 & 0 & 1 \end{bmatrix}.$$

In order to further improve our decomposition, we studied how each component of this structure could be modified so as to preserve the APN property of the permutation. We investigated both the replacement of the linear and non-linear permutations used and describe our findings in Sect. 3.3. In particular, we found that we could modify the central affine layer in the following fashions while still keeping the APN property of the permutation (see Theorem 2):

– changing the xor constants to any value, in particular 0;
– inserting two arbitrary 3-bit linear permutations a and b as shown in Fig. 7c.

Thus, we remove the xors from the structure and exhaustively check all linear permutations a, b such that the resulting linear layer from Fig. 7c has the simplest form. We found that for $a(x) = 2x^4 + 2x^2 + 4x$ and $b(x) = 2x^4 + 3x^2 + 2x$ the resulting matrix can be represented as the following matrix M' over $\mathbb{F}_{2^3}$:

$$M' = \begin{bmatrix} 2 & 5 \\ 1 & 2 \end{bmatrix}.$$

Interestingly, M' is an involution which, because of the symmetry of our decomposition, makes the whole S-Box involutive too! The matrix M' can more-

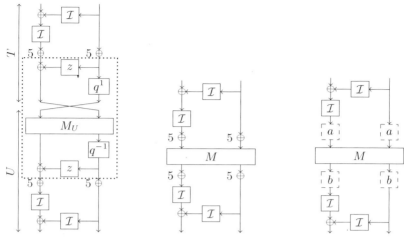

(a) Joining the decompo-   (b) Merging linear layers.   (c) Allowed transforma-
sitions of T and U.   tions.

Fig. 7. Simplifying the middle affine layer. The linear mappings in the dotted area in
Fig. 7a form the linear layer M.

over be decomposed into a 2-round Feistel Network with finite field multiplications by 2 as Feistel functions. We deduce the final decomposition from this final
observation and describe it in the following theorem.

Theorem 1. *There exist linear bijections A and B such that the Dillon 6-bit
permutation is equal to*

$$S_0(x) = B(S_{\mathcal{I}}(A(x) \oplus 9) \oplus 4,$$

*where the output of $S_{\mathcal{I}}(\ell||r)$ is the concatenation of two bivariate polynomials of
$\mathbb{F}_2[X]/(X^3 + X + 1)$, namely $S_{\mathcal{I}}^L(\ell, r)$ and $S_{\mathcal{I}}^R(\ell, r)$. These are equal to*

$$\begin{cases} S_{\mathcal{I}}^R(\ell||r) = (r^6 + \ell)^6 + 2r, \\ S_{\mathcal{I}}^L(\ell||r) = \left(r + 2S_{\mathcal{I}}^R(\ell||r)\right)^6 + S_{\mathcal{I}}^R(\ell||r)^6. \end{cases}$$

A picture representing a circuit computing $S_{\mathcal{I}}$ is given Fig. 8.

For the sake of completeness, we give the matrices of the linear permutations
A and B:

$$A = \begin{bmatrix} 1 & 1 & 0 & 1 & 0 & 1 \\ 1 & 1 & 1 & 1 & 0 & 0 \\ 1 & 0 & 0 & 0 & 0 & 0 \\ 0 & 0 & 0 & 1 & 0 & 1 \\ 0 & 0 & 0 & 1 & 0 & 0 \\ 0 & 0 & 0 & 1 & 1 & 0 \end{bmatrix}, \quad B = \begin{bmatrix} 0 & 1 & 1 & 1 & 0 & 1 \\ 0 & 0 & 0 & 0 & 0 & 1 \\ 0 & 0 & 1 & 1 & 1 & 0 \\ 0 & 0 & 0 & 1 & 1 & 1 \\ 0 & 0 & 1 & 0 & 1 & 0 \\ 1 & 0 & 1 & 1 & 0 & 1 \end{bmatrix}.$$

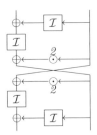

Fig. 8. The APN involution $S_\mathcal{I}$, where $\mathcal{I}$ denotes the inverse in the finite field $\mathbb{F}_{2^3}$ with the irreducible polynomial $X^3 + X + 1$, i.e. the monomial $x \mapsto x^6$.

3   Analysing Our Decomposition

In this section, we study the structure of the 6-bit APN permutation we derived from the Dillon S-Box in Sect. 2. We start with a description of its cryptographic properties in Sect. 3.1. Then, we generalize this structure into the *Butterfly structure* (see Sect. 3.2). We investigate how 3-bit affine permutations propagate through the different components of our decomposition in Sect. 3.3 and then we use this information to deduce how much freedom we have when choosing the different components of the permutation (see Sect. 3.4).

We discover some new relations between the APN permutation, the Kim function and the cube mapping over $\mathbb{F}_{2^6}$ in Sect. 3.5. Furthermore, we describe some simple univariate representations of the structure in Sect. 3.6. We have also noticed that $S_\mathcal{I}$ is CCZ-equivalent to the concatenation of two bent functions. However, because it could not produce any new 6-bit APN permutations, we discuss this in the full version of this paper [13].

3.1   Cryptographic Properties

The first consequence of our decomposition is the surprising observation that the 6-bit APN permutation is affine-equivalent to an involution. To the best of our knowledge, this was not known.

The permutation $S_\mathcal{I}$ is obviously APN due to how it was obtained, so that the highest differential probability is equal to $2/64 = 2^{-5}$. The Jackson Pollock representation of the DDT of $\mathsf{Swap} \circ S_\mathcal{I} \circ \mathsf{Swap}$, where Swap is a simple branch swap, is provided in Fig. 9a. The LAT of $S_\mathcal{I}$ contains[4], in absolute value, only 3 different coefficients: 945 occurrences of 0, 2688 occurrences of 4 and 336 occurrences of 8 (see Fig. 9b). Its maximum linear bias is thus $8/32 = 2^{-2}$. The left half of its output bits have algebraic degree 4 and those on the right half have algebraic degree 3.

---

[4] As $S_\mathcal{I}$ is a permutation, we ignore the first line and the first column of its LAT.

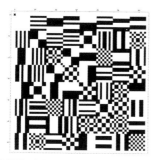
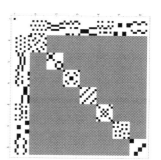

(a) DDT of Swap $\circ\, S_{\mathcal{I}} \circ$ Swap (white: 0,   (b) LAT of $S_{\mathcal{I}}$ (white: 0, grey: 4, black: 8).
black: 2).

Fig. 9. The Jackson Pollock representation of the DDT and LAT of $S_{\mathcal{I}}$.

3.2 The Butterfly Structure

As described above, the output of our 6-bit APN permutation $S_{\mathcal{I}}$ is the concatenation of two bivariate polynomials of $\mathbb{F}_{2^3}$. We define the keyed permutation R_k of $\mathbb{F}_{2^3}$ with a key in $\mathbb{F}_{2^3}$ as

$$R_k(x) = (x + 2k)^6 + k^6,$$

where R_k is indeed a permutation affine equivalent to the inverse function $x \mapsto x^6$. In fact, its inverse R_k^{-1} such that $R_k^{-1}(R_k(x)) = x$ is equal to $R_k^{-1} = (x + k^6)^6 + 2k$. Using this keyed permutation and its inverse, it is easy to express $S_{\mathcal{I}}$ (see also Fig. 10a):

$$S_{\mathcal{I}}(\ell || r) \;=\; R_{R_r^{-1}(\ell)}(r) \;||\; R_r^{-1}(\ell).$$

Using this representation, we show that $S_{\mathcal{I}}$ is CCZ-equivalent to a quadratic function with a very similar structure. First, we recall the definition of CCZ-equivalence (where CCZ stands for Carlet-Charpin-Zinoviev [14]) as it is defined e.g. in [15].

Definition 1 (CCZ-equivalence). *Let f and g be two functions mapping $\mathbb{F}_{2^n}$ to itself. They are said to be CCZ-equivalent if the sets $\{(x, f(x)) \mid x \in \mathbb{F}_{2^n}\}$ and $\{(x, g(x)) \mid x \in \mathbb{F}_{2^n}\}$ are affine equivalent. In other words, they are CCZ-equivalent if and only if there exists a linear permutation L of $(\mathbb{F}_{2^n})^2$ such that*

$$\big\{(x, f(x)), \forall x \in \mathbb{F}_{2^n}\big\} \;=\; \big\{L\,(x, g(x))\,, \forall x \in \mathbb{F}_{2^n}\big\}.$$

For example, a permutation is CCZ-equivalent to its inverse. As is shown in Proposition 2 of [16], CCZ-equivalence preserves both the differential uniformity and the Walsh spectrum (i.e. the distribution of the coefficients in the LAT).

Lemma 2. *The permutation $S_{\mathcal{I}}$ is CCZ-equivalent to the quadratic function $Q_{\mathcal{I}} : \mathbb{F}_2^6 \to \mathbb{F}_2^6$ obtained by concatenating two bivariate polynomials of $\mathbb{F}_{2^3}$:*

$$Q_{\mathcal{I}}(\ell || r) \;=\; R_r(\ell) || R_\ell(r).$$

A representation of $Q_{\mathcal{I}}$ is given Fig. 10b.

(a) The permutation $S_{\mathcal{I}}$.   (b) The function $Q_{\mathcal{I}}$.

Fig. 10. Two CCZ-equivalent APN functions of $\mathbb{F}_2^6$.

Proof. The functional graph of the function $Q_{\mathcal{I}}$ is the following set:

$$\{(x||y,\ R_y(x)||R_x(y)),\ \forall x||y \in \mathbb{F}_2^6\},$$

in which we can replace the variable x by $z = R_y(x)$ so that $x = R_y^{-1}(z)$ as R_k is invertible for all k. We obtain a new description of the same set:

$$\{(R_y^{-1}(z)||y,\ z||R_{R_y^{-1}(z)}(y)),\ \forall z||y \in \mathbb{F}_2^6\}.$$

As the function $\mu : (\mathbb{F}_2^6)^2 \to (\mathbb{F}_2^6)^2$ with $\mu(x||y,a||b) = (a||y,b||x)$ is linear, this graph is linearly equivalent to the following one:

$$\{(z||y,\ R_{R_y^{-1}(z)}(y))||R_y^{-1}(z),\ \forall z||y \in \mathbb{F}_2^6\},$$

which is the functional graph of $S_{\mathcal{I}}$: the two functions are CCZ-equivalent. □

Definition 2 (Butterfly Structure). *Let α be in $\mathbb{F}_{2^n}$, e be an integer such that $x \mapsto x^e$ is a permutation of $\mathbb{F}_{2^n}$ and $R_k[e,\alpha]$ be the keyed permutation*

$$R_k[e,\alpha](x) = (x + \alpha k)^e + k^e.$$

We call Butterfly Structures *the functions of $(\mathbb{F}_{2^n})^2$ defined as follows:*

- *the* Open Butterfly *with branch size n, exponent e and coefficient α is the permutation denoted H_e^α defined by:*

$$\mathsf{H}_e^\alpha(x,y) = \left(R_{R_y[e,\alpha](x)}^{-1}(y),\ R_y[e,\alpha](x)\right),$$

- *the* Closed Butterfly *with branch size n, exponent e and coefficient α is the function denoted V_e^α defined by:*

$$\mathsf{V}_e^\alpha(x,y) = \left(R_y[e,\alpha](x),\ R_x[e,\alpha](y)\right).$$

Furthermore, the permutation H_e^α and the function V_e^α are CCZ-equivalent.

Pictures representing such functions are given in Fig. 11. Our decomposition of the 6-bit APN permutation and its CCZ-equivalent function have butterfly structures: $S_{\mathcal{I}} = \mathsf{H}_6^2$ and $Q_{\mathcal{I}} = \mathsf{V}_6^2$. In fact, the proof of the CCZ-equivalence of open and closed butterfly is identical to that of Lemma 2. The properties of such structures for $n > 3$ are studied in Sect. 4.1, in particular in Theorem 4. In this section, we focus on the case $n = 3$.

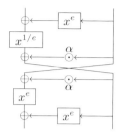

(a) Open (bijective) butterfly H_e^α.

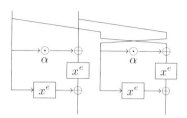

(b) Closed (non-bijective) butterfly V_e^α.

Fig. 11. The two types of butterfly structure with coefficient α and exponent e.

3.3   Propagation of Affine Mappings Through the Components

As we have seen, affine-equivalence and CCZ-equivalence are key concepts in our analysis of $S_{\mathcal{I}}$. In this context, it is natural to extend our analysis not only to outer affine layers applied before and after the permutation but also to the inner affine permutation itself: what modifications can we make to this function while preserving the APN property of the structure? In this section, we study the "propagation" of affine layers in the sense defined below. Our study will show some interesting properties of the structure and why changing some components can lead to an affine equivalent structure.

Definition 3 (Propagation of Affine Layers). *We say that an affine transformation A propagates through a component C if there exists an affine transformation A' such that $C \circ A = A' \circ C$.*

Note that this definition is another way of looking at self-equivalence: indeed, $C \circ A = A' \circ C$ is equivalent to $C = A'^{-1} \circ C \circ A$.

Theorem 2. *Consider the two permutations of $\mathbb{F}_2^6$ with structures shown in Fig. 12, where $A, B : \mathbb{F}_2^3 \to \mathbb{F}_2^3$ are some linear bijections,*

$$M = \begin{bmatrix} p & q \\ r & s \end{bmatrix}$$

is an invertible matrix operating on column-vectors, p, q, r, s are 3×3 sub-matrices over $\mathbb{F}_2$ and a, b, c, d are constants of $\mathbb{F}_{2^3}$. Assume also that q is invertible. Then both structures are affine-equivalent for any choice of M (with q invertible) and constants. As a consequence, all such structures are in the same affine-equivalence class.

Proof. We start by proving that adding constants a, b, c, d as described in Fig. 12 leads to affine-equivalent permutations. For now, we assume that A and B are the identity. First, we modify the constants without modifying the function to move them to the right branches only. To do this, we move a through the linear

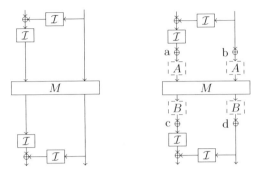

Fig. 12. Affine equivalent structures.

layer M and modify b in such a way that c cancels out. The difference required, $x = b' \oplus b$, is a solution to the equation $p(a) \oplus q(x) = c$, so that $x = q^{-1}(p(a) \oplus c)$ and x always exists since q is invertible. Thus, for

$$b' = b \oplus x = b \oplus q^{-1}(p(a) \oplus c),$$
$$d' = d \oplus r(a) \oplus s(x) = d \oplus r(a) \oplus s(q^{-1}(p(a) \oplus c)),$$

constructions with the structure described in Fig. 13a and b are functionally equivalent.

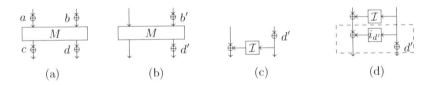

Fig. 13. How the xors around the central linear layer are affine equivalent to outer linear layers.

The xors remaining on the right branches propagate through the Feistel function $\mathcal{I}$ and are equivalent to particular outer affine transformations. Note that in $\mathbb{F}_{2^3}$ we have[5]

$$\mathcal{I}(x + d') = (x + d')^6 = x^6 + d'^2 x^4 + d'^4 x^2 + d'^6 = \mathcal{I}(x) + i_{d'}(x),$$

where $i_{d'}(x) = d'^2 x^4 + d'^4 x^2 + d'^6$ is an affine function and can be seen as an additional Feistel round. The propagation of the xor with d' is illustrated on

---

[5] For larger fields the inverse function does not satisfy the property and therefore such propagation is impossible. An anonymous reviewer pointed out that this works in $\mathbb{F}_{2^3}$ because the inverse function there has boolean algebraic degree 2 and therefore its derivative is linear.

Fig. 13c and d: the functions described on both figures are functionally equivalent. The case with b' is symmetrical.

We have now showed that the xors a, b, c, d can be removed and the resulting S-Box stays in the same affine equivalence class. Since the equivalence relation is symmetric, we can also modify the constants to arbitrary values. We now move on to studying the impact of branch-wise affine permutations.

It is sufficient to show how the two applications of B propagate through the bottom field inverses, the case of A being symmetric. We start by analyzing propagation through a single inverse function (see Fig. 14).

In the case when the input transformation is linear (when $c = 0$), it is easy to see that if the equivalent output transformation is affine, then it is actually linear, since $B(0) = \mathcal{I}(0) = 0$. By exhaustively checking all linear 3-bit permutations B we found that the only functions which propagate in such way are 21 functions of the form $x \mapsto \lambda x^{2^e}$, where $e \in \{0, 1, 2\}, \lambda \in \mathbb{F}_{2^3}, \lambda \neq 0$. This propagation is quite obvious since $(\lambda x^{2^e})^6 = \lambda^6 (x^6)^{2^e}$.

The more interesting case is when the input transformation is affine. By exhaustive search we found that *any* linear bijection B propagates through the field inverse in $\mathbb{F}_{2^3}$, but only *together* with a particular B-dependent xor constant. That is, for any linear bijection B there exists a constant c such that $\mathcal{I}(B(x) + c) = B'(\mathcal{I}(x)) + c'$ for some linear bijection B' and constant c', i.e. the affine function $B(x) + c$ propagates through the inverse function in the affine way (see Fig. 14b).

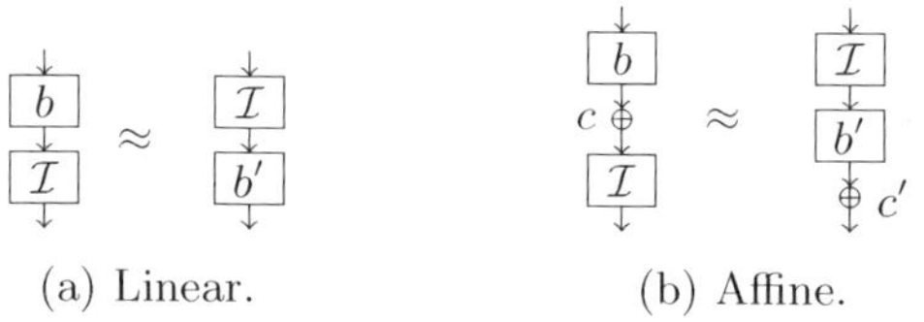

(a) Linear.   (b) Affine.

Fig. 14. Propagation of linear/affine permutation through the field inverse.

Note that after applying the linear bijections A and B the top right submatrix of M becomes $B \times q \times A$ and is still invertible, therefore the part of theorem about constant addition, which we already proved, is still applicable. Hence for any linear mappings A, B we can add the xor constants required for propagation of A, B. Let x, y be the values on the left and right branches respectively after applying the linear layer M. Then the left half of the output is equal to

$$x' = \mathcal{I}(B(x) + c) + \mathcal{I}(B(y) + c) = B'(\mathcal{I}(x)) + c' + B'(\mathcal{I}(y)) + c' = B'(\mathcal{I}(x) + \mathcal{I}(y)),$$

and the right half is simply $y' = B(x) + c$. The procedure is shown in Fig. 15. $\square$

Theorem 2 shows an interesting property of the field inverse in $\mathbb{F}_{2^3}$: all linear bijections propagate through it together with some xor constant. We have checked all nonlinear exponent functions in $\mathbb{F}_{2^n}$ for $n = 4, 5, 6, 7$ and none of

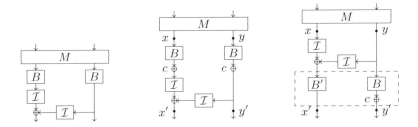

Fig. 15. Propagation of affine mappings through the inverses. The dashed area contains the outer affine parts.

them has this property. By using self-equivalence algorithm from [11] we found that in these fields the only affine transformations which propagate through such nonlinear monomial functions are the linear mappings of the form $x \mapsto \lambda x^{2^e}$, where $e \in [0, n-1], \lambda \in \mathbb{F}_{2^n}, \lambda \neq 0$.

In our decomposition the central linear layer is a 2-round Feistel Network where the round function σ is multiplication by 2 in the finite field defined by a particular polynomial (see Fig. 16a). By applying linear transformations around as in Theorem 2 we obtain an affine equivalent S-Box. We can move the linear functions a through the linear Feistel network, such that the round functions are modified and the linear functions a merge with the linear functions b as shown in Figs. 16b and c. Since by Theorem 2 the outer linear function $b \circ a$ can be omitted, we conclude that σ may be replaced by $a^{-1} \circ \sigma \circ a$ for arbitrary linear permutation a. By exhaustively checking $a^{-1} \circ \sigma \circ a$ for all a we found that there are 24 unique variants of σ. In particular, in the field defined by the irreducible polynomial $X^3 + X + 1$ the allowed multiplications by a constant α are when $\alpha \in \{2, 4, 6\}$, where the latter two are obtained from $\sigma(x) = 2x$ by setting $a(x) = x^2$ and $a(x) = x^4$. In the field defined by the other irreducible polynomial $X^3 + X^2 + 1$ such constants become $\alpha \in \{3, 5, 6\}$. We note that all these elements can be unambiguously defined by the conditions $\mathrm{Tr}(\alpha) = 0, \alpha \neq 0$ in both fields.

3.4   Replacing Components

It is natural to ask how unique are the components of the decomposition; can we get a different APN permutation by changing the central linear layer or the inverse functions?

We made an exhaustive[6] search for an invertible matrix such that when it is used as the middle linear layer in our decomposition, the resulting S-Box is an APN permutation. All the APN permutations we found are CCZ-equivalent to the original S-Box. However not all of them are affine-equivalent to it. By studying the new matrices we found that all of them can be obtained by using

---

[6] Actually we optimized the search by utilizing the equivalence classes given by Theorem 2.

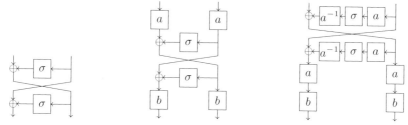

(a) The linear layer from the decomposition

(b) Applying arbitrary linear bijections a and b.

(c) Moving the linear functions a down.

Fig. 16. Propagation of the linear function a through the middle linear layer.

transformations from Theorem 2 together with swaps applied before and/or after the linear layer. All four different combinations of swaps result in four S-Boxes from distinct affine-equivalence classes (see Fig. 17). However they form two pairs of EA-equivalent S-Boxes: Fig. 17a and c, Fig. 17b and d. The proof for EA-equivalence is given in the full version of this paper [13]. Note that the function shown in Fig. 17c is the inverse of the function from Fig. 17b and both functions from Figs. 17a and d are involutions. Whether all four functions are EA-equivalent remains an open question.

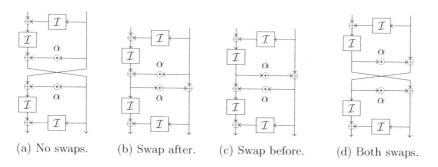

(a) No swaps.   (b) Swap after.   (c) Swap before.   (d) Both swaps.

Fig. 17. Four APN permutations from different affine-equivalence classes, obtained by adding swaps before and/or after the central linear layer.

We also made an exhaustive search of all 3-bit permutations and tried to use them instead of the field inverses. A non-involutive function has to be inverted in one of the places, as in the butterfly construction we introduced in Sect. 3.2. It turns out that the set of all 3-bit permutations for which the respective S-Box is an APN permutation is exactly the set of all 3-bit APN permutations. It is not surprising because all 3-bit APN permutations are in the same affine equivalence class. By using Theorem 2 and by applying some outer affine transformations we can easily replace the field inverses with arbitrary affine-equivalent functions and therefore with arbitrary 3-bit APN permutation. It follows that the two APN permutations at the top and the two APN permutations at the bottom may be different and the resulting S-Box will still be an APN permutation. We also note that one of the 3-bit APN permutations is such that its DDT and LAT are identical up to the signs in the LAT. It is the S-Box used in the block cipher 3-way [17].

As a summary of our observations we give the following theorem:

Theorem 3 (A Family of 6-bit APN Permutations). *The 6-bit permutation described by Dillon et al. in [5] is affine equivalent to the involution built using the structure described in Fig. 1, where $\odot$ denotes multiplication in the finite field $GF(2^3)$, $\alpha \neq 0$ is such that $Tr(\alpha) = 0$ and $\mathcal{A}$ denotes any 3-bit APN permutation.*

3.5 Relations with the Kim and the Cube Functions

It is suggested in [11] to count the number of pairs of affine permutations A, B such that $S_{\mathcal{I}} = B \circ S_{\mathcal{I}} \circ A$ as a measure of the symmetries inside $S_{\mathcal{I}}$. An algorithm performing this task is also provided. Using it, we have found that there are only 7 such pairs (including the pair of identity mappings). This property is preserved by affine transformations and the number could therefore be obtained without our decomposition. However, for the S-Box $S_{\mathcal{I}}$, these 7 pairs of transformations have a simple description:

$$S_{\mathcal{I}}(\lambda x, \lambda^{-1} y) = (\lambda, \lambda^{-1}) \otimes S_{\mathcal{I}}(x, y) \text{ for all } \lambda \in \mathbb{F}_{2^3}^*, \tag{2}$$

where "$\otimes$" is such that $(a, b) \otimes (c, d) = (ac, bd)$. In other words, multiplying the inputs by λ and λ^{-1} is equivalent to multiplying the outputs by the same values. As we have shown in Sect. 3.3, there are more symmetries *inside* the structure.

An anonymous reviewer pointed out that the observed property is quite similar to that of "Kim mapping", a non-bijective quadratic APN function from which Dillon *et al.* [5] obtained the APN permutation by applying transformations preserving CCZ-equivalence. The Kim function is defined over $\mathbb{F}_{2^6}$ as $k(x) = x^3 + x^{10} + ux^{24}$, where u is some primitive element of $\mathbb{F}_{2^6}$. It is pointed in [5] that the following holds:

$$k(\lambda x) = \lambda^3 k(x) \text{ for all } \lambda \in \mathbb{F}_{2^3}. \tag{3}$$

We found experimentally that the Kim mapping is actually affine-equivalent to all Closed Butterflies V_e^α with $n = 3, e \in \{3, 5, 6\}, Tr(\alpha) = 0$ and $\alpha \neq 0$. In particular, it is affine-equivalent to the function $Q_{\mathcal{I}} = \mathsf{V}_6^2$ described before.

The property that $k(\lambda x) = \lambda^3 k(x)$ for all $\lambda \in \mathbb{F}_{2^3}$ can be nicely translated to V_e^α structure (when $\alpha \neq 0$). Indeed, it is easy to see that the following holds:

$$\mathsf{V}_e^\alpha(\lambda x, \lambda y) = (\lambda^e, \lambda^e) \otimes \mathsf{V}_e^\alpha(x, y) \text{ for all } \lambda \in \mathbb{F}_{2^3}. \tag{4}$$

In particular, setting $e = 3$ and α such that V_e^α is affine-equivalent to the Kim mapping leads to a branch-wise variant of the property from Eq. 3.

Similarly, the Open Butterflies H_e^α exhibit the following property:

$$H_e^\alpha(\lambda^e x, \lambda y) = (\lambda^e, \lambda) \otimes H_e^\alpha(x, y) \text{ for all } \lambda \in \mathbb{F}_{2^3}. \tag{5}$$

While V_e^α is an interesting decomposition of the Kim function (when $Tr(\alpha) = 0, \alpha \neq 0$), we also found a very similar decomposition for the cube function over $\mathbb{F}_{2^6}$, which is also a quadratic APN function. Recall that the closed butterfly V_3^α maps (x, y) to $R_{kim}(x, y) \| R_{kim}(y, x)$, where $R_{kim}(x, y) = (x + \alpha y)^3 + y^3$. We have found that changing R_{kim} to $R_{cube}(x, y) = (x + \alpha y)^3 + x^3 + \alpha y^3$ leads to a function affine-equivalent to the cube function over $\mathbb{F}_{2^6}$. We describe the way we found this decomposition in the full version of this paper [13].

3.6   Univariate Polynomial Representations

In this section we describe several univariate polynomial representations of APN permutations from the affine-equivalence classes described in Sect. 3.4. We obtained them by interpolating the structures from previous sections in various bases relying on the field decomposition $\mathbb{F}_{2^6} \simeq (\mathbb{F}_{2^3})^2$. All polynomials described in this section are specified over $\mathbb{F}_{2^6}$ and w is a primitive element such that $w = X$ in $\mathbb{F}_2[X]/(X^6 + X^4 + X^3 + X + 1)$.

In [5], Dillon *et al.* represented the APN permutation as a univariate polynomial over $\mathbb{F}_{2^6}$ with 52 nonzero coefficients. Using our decomposition, we managed to find an APN permutation whose univariate polynomial has only 25 terms. Due to lack of space we give the polynomial in the full version of this paper [13].

Originally, the APN permutation was obtained as a composition $g = f_2 \circ f_1^{-1}$, where $f_1(x)$ and $f_2(x)$ contain 18 monomials each (as given in [5]). We have found a variant with much simpler polynomials. The function g is still an APN permutation if f_1 and f_2 as defined in [5] are replaced by the following two functions:

$$f_1(x) = w^{11} x^{34} + w^{53} x^{20} + x^8 + x,$$
$$f_2(x) = w^{28} x^{48} + w^{61} x^{34} + w^{12} x^{20} + w^{16} x^8 + x^6 + w^2 x.$$

Additionally, we found a few other simple representations relying on a composition of simple polynomials. Let $g(x) = i \circ m \circ i^{-1}(x)$, then g is an APN permutation when

$$i(x) = w^{21} x^{34} + x^{20} + x^8 + x, \quad m(x) = w^{52} x^8 + w^{36} x$$

or when

$$i(x) = w^{37} x^{48} + x^{34} + w^{49} x^{20} + w^{21} x^8 + w^{30} x^6 + x, \quad m(x) = x^8.$$

In these representations, i corresponds to the sum of the two inverse functions $\mathcal{I}$ so that i and i^{-1} are the non-linear parts of the open butterfly. The function m corresponds to the central linear layer (including possible branch swaps).

4   Differentially 4-Uniform Permutations of Larger Blocks

An up to date overview of known APN functions can be found in [15]. As APN functions operating on an even number of bits are still to be found for even block sizes larger than 6, differentially 4-uniform permutations have received a lot of attention from researchers. An obvious example is the inverse function $x \mapsto x^{2^n - 2}$ of $\mathbb{F}_{2^n}$ studied in the seminal work of Nyberg [4].

However, security against differential cryptanalysis is not sufficient and linear attack need to be taken into account too. The search can thus be focused on differentially 4-uniform permutations of $2n$ bits with non-linearity $2^{2n-1} - 2^n$ which is, as far as we know, the best that can be achieved. Whether there exists functions improving this bound is an open problem (Open Problem 2 in [18]). The same paper also states Open Problem 1: we must find other highly non-linear differentially 4-uniform functions operating on fields of even degree. Several papers have then presented constructions for such permutations, for example using binomials [19] or an APN permutation on $\mathbb{F}_{2^{n+1}}$ for even n [20].

In this section, we study the butterfly structure. In Sect. 4.1, we study butter-flies with $\alpha \neq 0, 1$ and, in Sect. 4.2, the case $\alpha = 1$ in which the open butterfly is functionally equivalent to a 3-round Feistel Network. We show that these structures are always differentially 4-uniform for block sizes $2n$ (n odd) and have algebraic degree $n + 1$ (when $\alpha \neq 1$) and n (when $\alpha = 1$) in the bijective case, 2 otherwise. While we could not prove it in the general case, we conjecture that they both have non-linearity $2^{2n-1} - 2^n$.

4.1   Butterfly with Non-Trivial α

Theorem 4 (Properties of the Butterfly Structure). *Let* V_e^α *and* H_e^α *respectively be the closed and open $2n$-bit butterflies with exponent $e = 3 \times 2^t$ for some t, coefficient α not in $\{0, 1\}$ and n odd. Then:*

- *the differential uniformity of both* H_e^α *and* V_e^α *is at most 4,*
- V_e^α *is quadratic, and*
- *half of the coordinates of* H_e^α *have algebraic degree n, the other half have algebraic degree $n + 1$.*

Proof. In this proof, we rely a lot on the *univariate degree* of a polynomial of $\mathbb{F}_2^n$. It is different from the algebraic degree: the cube function has *univariate degree* 3 and *algebraic degree* 2.

Differential Properties. As V_e^α and H_e^α are CCZ-equivalent, they have the same differential uniformity. It is thus sufficient to prove that the one of V_e^α is at most 4. First, note that the functions V_e^α with exponent 3×2^t is affine equivalent to V_3^α which uses the exponent 3 as V_3^α can be obtained simply by applying the linear permutation $x \mapsto x^{2^{n-t}}$ on each half of the output of V_e^α. Thus, it is sufficient to study the case where the exponent is equal to 3.

Let T_α be the linear permutation of $\mathbb{F}_2^n \times \mathbb{F}_2^n$ defined by the matrix

$$T_\alpha = \begin{bmatrix} 1 & \alpha \\ \alpha & 1 \end{bmatrix}.$$

As affine equivalence preserves differential uniformity, we will prove that the differential uniformity of $P = T_\alpha \circ V_3^\alpha$ is at most equal to 4 and deduce that V_3^α has the same property. The left side of the output of P is equal to

$$
\begin{aligned}
P_L(x,y) &= R(x,y) + \alpha R(y,x) \\
&= (x+\alpha y)^3 + y^3 + \alpha\big((y+\alpha x)^3 + x^3\big) \\
&= x^3(1+\alpha+\alpha^4) + y^3(1+\alpha+\alpha^3) + x^2 y(\alpha+\alpha^3)
\end{aligned}
$$

and the right side to

$$
\begin{aligned}
P_R(x,y) &= R(y,x) + \alpha R(x,y) \\
&= y^3(1+\alpha+\alpha^4) + x^3(1+\alpha+\alpha^3) + xy^2(\alpha+\alpha^3).
\end{aligned}
$$

To simplify expressions, we use the notation $\beta = \alpha^3 + \alpha$. Note that for the values of α we are interested in, namely $\alpha \neq 0, 1$, it holds that $\beta \neq 0$.

By definition of differential uniformity, the differential uniformity of P is at most 4 if and only if the following system has at most 4 solutions for any a, b, c, d (unless $a = b = 0$):

$$
\begin{cases}
P_L(x,y) + P_L(x+a, y+b) = c \\
P_R(x,y) + P_R(x+a, y+b) = d,
\end{cases}
$$

which is equivalent to

$$
\begin{cases}
(ax^2 + a^2 x)(1+\alpha+\alpha^4) + (by^2 + b^2 y)(1+\beta) + (bx^2 + a^2 y)\beta = c + P_L(a,b) \\
(by^2 + b^2 y)(1+\alpha+\alpha^4) + (ax^2 + a^2 x)(1+\beta) + (b^2 x + ay^2)\beta = d + P_R(a,b).
\end{cases}
$$

If $a = 0$ then the second line of the system yields the sum of a univariate degree 2 polynomial in y with $b^2 \beta x$. As $b \neq 0$ (recall that $a = b = 0$ is impossible), we deduce that x is equal to a univariate degree 2 polynomial in y and replace it by this expression in the first equation. We obtain an equation with univariate degree 4 only in y with at most 4 solutions, for each of which we deduce a unique value x. Hence, the system has at most 4 solutions. The case $b = 0$ is treated similarly.

We now suppose $a \neq 0$ and $b \neq 0$. We replace the left side of the first line ℓ_1 by a linear combination of the left sides of the two equations: $\ell_1 := ab^2\ell_1 + a^2 b\ell_2$. This quantity is a degree one bivariate polynomial with variables $X = ax^2 + a^2 x$ and $Y = by^2 + b^2 y$ so that we can write $\ell_1 = \gamma_0 X + \gamma_1 Y = \epsilon$, where ϵ is obtained by computing the same linear combination on the right side of the equations. If $\gamma_0 = 0$ then ℓ_1 actually is a degree 2 equation in y. For each of its at most 2 solutions, we obtain a degree 2 equation in x in ℓ_2 with at most 2 solutions. Hence, the total number of solutions is at most equal to 4. The case $\gamma_1 = 0$ is identical.

We now suppose $\gamma_0 \neq 0$ and $\gamma_1 \neq 0$. Using that $\gamma_0 X + \gamma_1 Y = \epsilon$, we deduce that $(ax^2 + a^2 x) = \big(\epsilon + (by^2 + b^2 y)\gamma_1\big)/\gamma_0$. We can therefore replace $(ax^2 + a^2 x)$ by this quantity in the second equation which becomes the sum of a degree 2 equation in y with a degree 1 term in x. As before, we deduce an expression of x as a degree 2 polynomial in y and replace it by this polynomial in the other equation. Hence, the initial system has as many solutions as an equation with univariate degree 4, i.e. at most 4.

Therefore, $P(x, y) + P(x + a, y + b) = (c, d)$ has at most 4 solutions, meaning that the differential uniformity of P is at most 4.

Algebraic Degrees. As the left and right side of $\mathsf{V}_e^\alpha(x, y)$ are equal to, respectively, $(x + \alpha y)^3 + y^3$ and $(y + \alpha x)^3 + x^3$, it is obvious that it is quadratic (recall that the algebraic degree of the univariate polynomial $x \mapsto x^e$ of $\mathbb{F}_2^n$ is the Hamming weight of the binary expansion of e).

Consider now the open butterfly H_e^α. For the sake of simplicity, we treat the case $e = 3$; other cases yield identical proofs. The right side of the output of such an open butterfly is equal to $(x + \alpha y^3)^{1/3} + \alpha y$, where $x\|y$ is the input. We deduce from Theorem 1 of [21] (or equivalently from Proposition 5 of [4]) that the inverse of 3 modulo $2^n - 1$ for odd n is

$$1/3 \equiv \sum_{i=0}^{(n-1)/2} 2^{2i} \mod 2^n - 1,$$

which implies in particular why the algebraic degree of $x \mapsto x^{1/3}$ is equal to $(n+1)/2$. We deduce from this expression that $(x + \alpha y^3)^{1/3}$ is equal to $\prod_{i=0}^{(n-1)/2}(x + \alpha y^3)^{2^{2k}}$. This sum can be developed as follows:

$$(x + \alpha y^3)^{1/3} = \sum_{J \subseteq [0, (n-1)/2]} \underbrace{\prod_{j \in J} \alpha^{2^{2j}} y^{3 2^{2j}}}_{\deg < 2|J|} \underbrace{\prod_{j \in \overline{J}} x^{2^{2j}}}_{\deg = (n+1)/2 - |J|},$$

where $\overline{J}$ is the complement of J in $[0, (n-1)/2]$, i.e. $J \cap \overline{J} = \emptyset$ and $J \cup \overline{J} = [0, (n-1)/2]$. The algebraic degree of each term in this sum is at most equal to $|J| + (n+1)/2$. If $\overline{J} = \emptyset$ then x is absent from the term so that the maximum algebraic degree is n. If $\overline{J} = \{j\}$ for some j, then the term is equal to $(xy^{-1})^{2^{2j}}$ (we omit the constant factor) which has algebraic degree $1 + (n - 1) = n$. If $|J| < (n-1)/2$, then the whole degree is smaller than n. Thus, the right side of the output has an algebraic degree equal to n.

The left side is equal to

$$\left(y + \alpha\left((x + \alpha y^3)^{1/3} + \alpha y\right)\right)^3 + \left((x + \alpha y^3)^{1/3} + \alpha y\right)^3.$$

The terms of highest algebraic degree in this equation are of the shape $y^2(x + \alpha y^3)^{1/3}$ and $y(x + \alpha y^3)^{2 \times 1/3}$. Because of what we established above, we have:

$$y^2(x + \alpha y^3)^{1/3} = \sum_{J \subseteq [0, (n-1)/2]} \underbrace{y^2 \times \prod_{j \in J} \alpha^{2^{2j}} y^{3 \times 2^{2j}}}_{\deg < 2|J| + 1} \underbrace{\prod_{j \in \overline{J}} x^{2^{2j}}}_{\deg = (n+1)/2 - |J|},$$

so that the algebraic degree of this term is at most equal to $|J| + (n+1)/2 + 1 \leq n + 1$. If $J = [0, (n-1)/2] \setminus \{j\}$ for some j, then the algebraic degree of the expression is $(1 + (n-1)/2) + (n+1)/2 = n+1$, meaning that this bound is reached. The terms $y(x + \alpha y^3)^{2/3}$ are treated similarly. Hence, the left side of the output has algebraic degree $n+1$.   □

This proof lead us to some interesting observations.

Remark 1. The proof relies on the study of $T_\alpha \circ V_e^\alpha$ which, for $n = 3$, has as its output the concatenation of $b(x, y)$ and $b(y, x)$ for a bent function b with a Maiorana-MacFarland structure. We provide further analysis for this observation in the full version of this paper [13]. We also note that the idea of building APN or differentially 4-uniform functions by concatenating two functions, at least one of which is bent, was discussed by Carlet in [22].

We have also studied the butterfly structure experimentally. While we could not find a pair (e, α) for which a butterfly is APN for $n > 3$, we did notice a variation in the distribution of 0, 2 and 4 in their DDT. It is therefore possible that APN butterflies exist but not for $n = 5, 7$. Moreover, butterflies are never differentially 4-uniform for $n = 4, 8, 10$. However, the case $n = 6$ yields the following proposition.

Proposition 1. *If $n = 6$, then there exists α such that H_5^α is a 12-bit permutation that is differentially 4-uniform. In fact, all of the coefficients in its DDT are in $\{0, 4\}$. Its non-linearity is $1920 = 2^{2n-1} - 2^{n+1}$.*

A natural generalization would be to have the same result for $e = 5$ whenever $x \mapsto x^5$ is a permutation. However, we found experimentally that this result does not hold for $n = 10$, although $x \mapsto x^5$ is a permutation of $\mathbb{F}_{2^{10}}$. We note also that, unlike in Theorem 4, Proposition 1 does not hold for all values of α but only for few of those.

We also found experimentally that the maximum LAT coefficient of a butterfly structure operating on $2n$ bits is equal to 2^n for $n = 3, 5, 7$. This implies that the non-linearity of the butterfly structure is "optimal" in the sense that no known permutations of a field of size $2n$ have a non-linearity higher than $2^{2n-1} - 2^n$. It is however not known if this bound holds for all permutations (see Open Problem 2 in [18]).

Proposition 2. *The non-linearity of a butterfly structure operating on $2n$ bits is equal to $2^{2n-1} - 2^n$ for $n = 3, 5, 7$.*

We conjecture that this proposition is true for every odd n.

4.2   Feistel Network ($\alpha = 1$)

If we set $\alpha = 1$ in an open butterfly structure, the resulting permutation is functionally equivalent to a 3-round Feistel Network with round functions $x \mapsto x^e$, $x \mapsto x^{1/e}$ and $x \mapsto x^e$, as described in Fig. 18. We denote such a Feistel

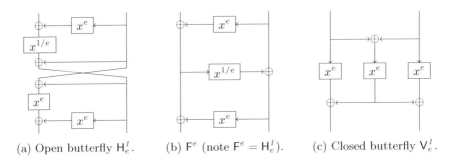

(a) Open butterfly H_e^I.   (b) F^e (note $\mathsf{F}^e = \mathsf{H}_e^I$).   (c) Closed butterfly V_e^I.

Fig. 18. The equivalence between H_e^I and F^e.

Network F^e. We note that the closed butterfly V_e^I has a structure reminiscent of a Lai-Massey round (see Fig. 18c).

In [23], Li and Wang proved that the $2n$-bit Feistel Networks F^e with $e = 2^k + 1$ and odd n such that $gcd(n, k) = 1$ have very good cryptographic properties:

1. the differential spectrum of F^e is equal to $\{0, 4\}$;
2. the non-linearity of F^e is the best known and is equal to $2^{2n-1} - 2^n$;
3. the algebraic degree of F^e is equal to n.

Note that the butterfly structures from Theorem 4 have degree $n + 1$ on half of the coordinates. We have proved that F^3 has degree n on all coordinates. The proof is given in the full version of this paper [13].

Remark 2. The proof for the algebraic degree of the left side of F^3 relies on particular cancellation occurring in the sum $y^2(x + y^3)^{1/3} + y(x + y^3)^{2/3}$. Such cancellations do not occur when $\alpha \neq 1$ as the terms in the corresponding sum are preceded by different coefficients which are both functions of α. This explains why the algebraic degree of F^3 and the open butterfly structure with $\alpha \neq 1$ are different.

We also note that the monomial $x \mapsto x^5$ in $\mathbb{F}_{2^{2n}}$ shares the same differential and linear properties. In [23] it is mentioned that for $n = 3$ the Feistel Network F^3 is CCZ-equivalent to the monomial $x \mapsto x^5$. We observe that the closed butterfly V_5^I, which is CCZ-equivalent to F^5, is actually linear-equivalent to the monomial $x \mapsto x^5$ over $\mathbb{F}_{2^{2n}}$ for all odd $n \geq 3$. We state the generalized result in the following theorem.

Theorem 5. *Let $n \geq 3$ be an odd integer and $e = 2^{2k} + 1$ for some positive integer k. Then the closed $2n$-bit butterfly V_e^1 is linear-equivalent to the monomial $x \mapsto x^e$ of $\mathbb{F}_{2^{2n}}$.*

Corollary 1. *Let $n \geq 3$ be an odd integer and $e = 2^{2k} + 1$ for some positive integer k, such that the monomial $x \mapsto x^e$ defines a permutation of $\mathbb{F}_{2^{2n}}$. Then the $2n$-bit Feistel Network F^e is CCZ-equivalent to this permutation.*

The proof is based on the field decomposition $\mathbb{F}_{2^{2n}} \simeq (\mathbb{F}_{2^n})^2$ and is given it in the full version of this paper [13].

5   Implementing 6-Bit APN Permutations

We can use the open butterfly structure to efficiently implement 6-bit APN permutations in both a bit-sliced fashion for use in software and in hardware. In this section, we explore this idea and provide an S-Box A_o which is affine equivalent to H_3^2 and for which there exists such efficient implementation.

5.1   Efficient Bit-Sliced Implementations

Starting from the algebraic normal forms of the operations used to compute H_3^2, it is easy to write a first naive bitsliced implementation (see full version [13]).

This implementation can be optimized by using Boolean algebra and removing the linear component of $x \mapsto x^3$ in the first and last steps. Doing this is equivalent to applying an affine permutation before and after the H_3^2 to obtain a new permutation A_o. This operation preserves the differential and linear property of the permutation while also keeping the property that $A_o^{-1} = \mathsf{Swap}_6 \circ A_o \circ \mathsf{Swap}_6$, where Swap_6 simply swaps the two 3-bit branches. The bitsliced implementation of this simplified S-Box is given in Algorithm 1 and its look-up table in Table 5.

Table 5. The look-up table of A_o in hexadecimal, e.g. $A_o(\mathtt{0x32}) = \mathtt{0x21}$.

	.0	.1	.2	.3	.4	.5	.6	.7	.8	.9	.a	.b	.c	.d	.e	.f
0.	0	1d	6	3f	3c	3b	31	12	22	35	17	2c	16	33	30	39
1.	2d	a	38	2b	1	4	2f	1e	3	34	2e	25	27	1a	29	28
2.	2a	7	14	3d	36	19	b	20	3e	d	37	8	1b	2	9	1c
3.	10	1f	21	3a	26	13	24	5	c	f	11	e	23	32	15	18

5.2   Hardware Implementation

Our decompositions also eases the hardware implementation of these S-Boxes. To illustrate this, we simulated the circuit computing these functions in three different ways. First, we simply gave the look-up table to the software[7] and let it find the best implementation it could (*no decomposition* case). Then, we fed it our decomposition of the different structures (*decomposed* case).

The optimization performed by the software is done for two competing criteria. The first is the area which simply corresponds to the physical space needed to implement the circuit using the logical gates available. The second is the propagation time, i.e. the delay necessary for the electronic signal to go through the circuit implementing the S-Box and to stabilize itself to the output value.[8]

---

[7] We used the digital cell library `SAED90n-1P9M` in the "normal V_t, high temperature, nominal voltage" corner.

[8] We also considered implementing the cube function using finite field arithmetic but could not easily improve our results.

Algorithm 1. An optimised bitsliced implementation of an S-Box affine-equivalent to the open butterfly with $\alpha = 2$, $e = 3$.

function $A_o(X_0, ..., X_5)$

1 . $t = (X_5 \wedge X_3)$
2 . $X_0 \oplus= t \oplus (X_5 \wedge X_4)$
3 . $X_1 \oplus= t$
4 . $X_2 \oplus= (X_4 \vee X_3)$
5 . $t = (X_1 \vee X_0)$
6 . $X_0 \oplus= (X_2 \wedge X_1) \oplus X_4$
7 . $X_1 \oplus= (X_2 \wedge X_0) \oplus X_5 \oplus X_3$
8 . $X_2 \oplus= t \oplus X_3$
9 . $X_3 \oplus= X_1$
10 . $X_4 \oplus= X_2 \oplus X_0$
11 . $X_5 \oplus= X_0$

12 . $u = X_3$
13 . $t = X_4$
14 . $X_3 \oplus= t$
15 . $X_3 = X_3 \wedge X_5 \oplus t$
16 . $X_4 \oplus= ((\neg X_5) \wedge u)$
17 . $X_5 \oplus= (t \vee u)$
18 . $t = (X_2 \wedge X_0)$
19 . $X_3 \oplus= t \oplus (X_2 \wedge X_1)$
20 . $X_4 \oplus= t$
21 . $X_5 \oplus= (X_1 \vee X_0)$

end function

For each function, we repeated the experience several times using different periods for the clock cycles: when the period is maximum, priority is given to optimizing the area and, as the period decreases, the priority shifts toward the propagation time. The results are given in Table 6.

Table 6. Results on the hardware implementation of our S-Boxes. The area a is in $(\mu m)^2$, the delay d is in ns and $a \times d$ is their product.

S-Box	Period (ns)	Base			Decomposed		
		a	d	$a \times d$	a	d	$a \times d$
H_3^2	100	799	56.42	45079.58	414	39.31	16274.34
	20	827	19.75	16333.25	404	18.7	7554.8
	10	928	9.81	9103.68	431	9.76	4206.56
	5	1062	4.81	5108.22	569	4.81	2736.89
A_o	100	774	53.13	41122.62	384	42.01	16131.84
	20	812	19.3	15671.6	384	15.43	5925.12
	10	869	9.63	8368.47	382	9.77	3732.14
	6	1041	5.8	6037.8	464	5.8	2691.2

As we can see, the knowledge of the decompositions always allows a more efficient implementation: regardless of what the main optimisation criteria is, both the area and the delay are decreased.

6   Conclusion

We have identified a decomposition of the 6-bit APN permutation published by Dillon *et al.* [5] and found it to be affine equivalent to an involution. We generalized the structure found to larger block sizes, although we could only prove its being differentially 4-uniform in those cases. We also deduced efficient implementation of 6-bit APN permutations in both a bit-sliced fashion and in hardware.

Our work also raised the following open questions.

Open Problems (On the properties of Butterfly Structures).

1. Is there a tuple n, e, α where $n > 3$ and e are integers, and α is a finite field element such that H_e^α operating on $(\mathbb{F}_{2^n})^2$ is APN?
2. Is it true that the non-linearity of a butterfly structure on $2n$ bits with $\alpha \neq 0, 1$ and n odd is always $2^{2n-1} - 2^n$?

Acknowledgements. We thank the anonymous reviewers for their helpful comments. We also thank Yann Le Corre for studying the hardware implementation of the permutation. The work of Léo Perrin is supported by the CORE ACRYPT project (ID C12-15-4009992) funded by the *Fonds National de la Recherche* (Luxembourg). The work of Aleksei Udovenko is supported by the *Fonds National de la Recherche*, Luxembourg (project reference 9037104).

References

1. Biham, E., Shamir, A.: Differential cryptanalysis of DES-like cryptosystems. J. Cryptol. **4**(1), 3–72 (1991)
2. Matsui, M.: Linear cryptanalysis method for DES cipher. In: Helleseth, T. (ed.) EUROCRYPT 1993. LNCS, vol. 765, pp. 386–397. Springer, Heidelberg (1994)
3. Daemen, J., Rijmen, V.: The Design of Rijndael: AES - The Advanced Encryption Standard. Springer, Heidelberg (2002)
4. Nyberg, K.: Differentially uniform mappings for cryptography. In: Helleseth, T. (ed.) EUROCRYPT 1993. LNCS, vol. 765, pp. 55–64. Springer, Heidelberg (1994)
5. Browning, K., Dillon, J., McQuistan, M., Wolfe, A.: An APN permutation in dimension six. Finite Fields Theory Appl. **518**, 33–42 (2010)
6. Bilgin, B., Bogdanov, A., Knežević, M., Mendel, F., Wang, Q.: Fides: lightweight authenticated cipher with side-channel resistance for constrained hardware. In: Bertoni, G., Coron, J.-S. (eds.) CHES 2013. LNCS, vol. 8086, pp. 142–158. Springer, Heidelberg (2013)
7. Biryukov, A., Perrin, L., Udovenko, A.: Reverse-engineering the S-box of Streebog, Kuznyechik and STRIBOBr1. In: Fischlin, M., Coron, J.-S. (eds.) EUROCRYPT 2016. LNCS, vol. 9665, pp. 372–402. Springer, Heidelberg (2016). doi:10.1007/978-3-662-49890-3_15
8. Biryukov, A., Perrin, L.: On reverse-engineering S-Boxes with hidden design criteria or structure. In: Gennaro, R., Robshaw, M. (eds.) CRYPTO 2015. LNCS, vol. 9215, pp. 116–140. Springer, Berlin Heidelberg (2015)
9. Bogdanov, A., Leander, G., Nyberg, K., Wang, M.: Integral and multidimensional linear distinguishers with correlation zero. In: Wang, X., Sako, K. (eds.) ASIACRYPT 2012. LNCS, vol. 7658, pp. 244–261. Springer, Heidelberg (2012)
10. Biryukov, A., Shamir, A.: Structural cryptanalysis of SASAS. In: Pfitzmann, B. (ed.) EUROCRYPT 2001. LNCS, vol. 2045, pp. 394–405. Springer, Heidelberg (2001)

11. Biryukov, A., De Cannière, C., Braeken, A., Preneel, B.: A toolbox for cryptanalysis: linear and affine equivalence algorithms. In: Biham, E. (ed.) EUROCRYPT 2003. LNCS, vol. 2656, pp. 33–50. Springer, Heidelberg (2003)

12. Developers, T.S.: SageMath, the Sage Mathematics Software System (Version 7.1) (2016). http://www.sagemath.org

13. Perrin, L., Udovenko, A., Biryukov, A.: Cryptanalysis of a Theorem: Decomposing the Only Known Solution to the Big APN Problem (Full Version). Cryptology ePrint Archive, Report 2016/539 (2016). http://eprint.iacr.org/

14. Carlet, C., Charpin, P., Zinoviev, V.: Codes, bent functions and permutations suitable for DES-like cryptosystems. Des. Codes Crypt. $15(2)$, 125–156 (1998)

15. Blondeau, C., Nyberg, K.: Perfect nonlinear functions and cryptography. Finite Fields Appl. 32, 120–147 (2015). Special Issue: Second Decade of FFA

16. Budaghyan, L., Carlet, C., Pott, A.: New classes of almost bent and almost perfect nonlinear polynomials. IEEE Trans. Inf. Theory $52(3)$, 1141–1152 (2006)

17. Daemen, J., Govaerts, R., Vandewalle, J.: A new approach to block cipher design. In: Anderson, R. (ed.) FSE 1993. LNCS, vol. 809, pp. 18–32. Springer, Heidelberg (1994)

18. Bracken, C., Leander, G.: A highly nonlinear differentially 4 uniform power mapping that permutes fields of even degree. Finite Fields Appl. $16(4)$, 231–242 (2010)

19. Bracken, C., Tan, C.H., Tan, Y.: Binomial differentially 4 uniform permutations with high nonlinearity. Finite Fields Appl. $18(3)$, 537–546 (2012)

20. Li, Y., Wang, M.: Constructing differentially 4-uniform permutations over $GF(2^{2m})$ from quadratic APN permutations over $GF(2^{2m+1})$. Des. Codes Crypt. $72(2)$, 249–264 (2014)

21. Kyureghyan, G.M., Suder, V.: On inverses of APN exponents. In: 2012 IEEE International Symposium on Information Theory Proceedings (ISIT), pp. 1207–1211. IEEE (2012)

22. Carlet, C.: Relating three nonlinearity parameters of vectorial functions and building APN functions from bent functions. Des. Codes Crypt. $59(1)$, 89–109 (2011)

23. Li, Y., Wang, M.: Constructing S-Boxes for lightweight cryptography with Feistel structure. In: Batina, L., Robshaw, M. (eds.) CHES 2014. LNCS, vol. 8731, pp. 127–146. Springer, Heidelberg (2014)

The SKINNY Family of Block Ciphers and Its Low-Latency Variant MANTIS

Christof Beierle[1], Jérémy Jean[2], Stefan Kölbl[3], Gregor Leander[1], Amir Moradi[1(✉)], Thomas Peyrin[2], Yu Sasaki[4], Pascal Sasdrich[1], and Siang Meng Sim[2]

[1] Horst Görtz Institute for IT Security, Ruhr-Universität Bochum, Bochum, Germany
{Christof.Beierle,Gregor.Leander,Amir.Moradi,Pascal.Sasdrich}@rub.de
[2] School of Physical and Mathematical Sciences, Nanyang Technological University, Singapore, Singapore
Jean.Jeremy@gmail.com, Thomas.Peyrin@ntu.edu.sg, SSIM011@e.ntu.edu.sg
[3] DTU Compute, Technical University of Denmark, Kongens Lyngby, Denmark
stek@dtu.dk
[4] NTT Secure Platform Laboratories, Tokyo, Japan
Sasaki.Yu@lab.ntt.co.jp

Abstract. We present a new tweakable block cipher family SKINNY, whose goal is to compete with NSA recent design SIMON in terms of hardware/software performances, while proving in addition much stronger security guarantees with regards to differential/linear attacks. In particular, unlike SIMON, we are able to provide strong bounds for all versions, and not only in the single-key model, but also in the related-key or related-tweak model. SKINNY has flexible block/key/tweak sizes and can also benefit from very efficient threshold implementations for side-channel protection. Regarding performances, it outperforms all known ciphers for ASIC round-based implementations, while still reaching an extremely small area for serial implementations and a very good efficiency for software and micro-controllers implementations (SKINNY has the smallest total number of AND/OR/XOR gates used for encryption process).

Secondly, we present MANTIS, a dedicated variant of SKINNY for low-latency implementations, that constitutes a very efficient solution to the problem of designing a tweakable block cipher for memory encryption. MANTIS basically reuses well understood, previously studied, known components. Yet, by putting those components together in a new fashion, we obtain a competitive cipher to PRINCE in latency and area, while being enhanced with a tweak input.

Keywords: Lightweight encryption · Low-latency · Tweakable block cipher · MILP

Updated information on SKINNY will be made available via https://sites.google.com/site/skinnycipher/.

M. Robshaw and J. Katz (Eds.): CRYPTO 2016, Part II, LNCS 9815, pp. 123–153, 2016.
DOI: 10.1007/978-3-662-53008-5_5

1   Introduction

Due to the increasing importance of pervasive computing, lightweight cryptography is currently a very active research domain in the symmetric-key cryptography community. In particular, we have recently seen the apparition of many (some might say too many) lightweight block ciphers, hash functions and stream ciphers. While the term *lightweight* is not strictly defined, it most often refers to a primitive that allows compact implementations, i.e. minimizing the area required by the implementation. While the focus on area is certainly valid with many applications, most of them require additional performance criteria to be taken into account. In particular, the throughput of the primitive represents an important dimension for many applications. Besides that, power (in particular for passive RFID tags) and energy (for battery-driven device) may be major aspects.

Moreover, the efficiency on different hardware technologies (ASIC, FPGA) needs to be taken into account, and even micro-controllers become a scenario of importance. Finally, as remarked in [3], software implementations should not be completely ignored for these lightweight primitives, as in many applications the tiny devices will communicate with servers handling thousands or millions of them. Thus, even so research started by focusing on chip area only, lightweight cryptography is indeed an inherent multidimensional problem.

Investigating the recent proposals in more detail, a major distinction is eye-catching and one can roughly split the proposals in two classes. The first class of ciphers uses very strong, but less efficient components (like the Sbox used in PRESENT [5] or LED [15], or the MDS diffusion matrix in LED or PICCOLO [31]). The second class of designs uses very efficient, but rather weak components (like the very small KATAN [9] or SIMON [2] round function)[1].

From a security viewpoint, the analysis of the members of the first class can be conducted much easily and it is usually possible to derive strong arguments for their security. However, while the second class strategy usually gives very competitive performance figures, it is much harder with state-of-the-art analysis techniques to obtain security guarantees even with regards to basic linear or differential cryptanalysis. In particular, when using very light round functions, bounds on the probabilities of linear or differential characteristics are usually both hard to obtain and not very strong. As a considerable fraction of the lightweight primitives proposed got quickly broken within a few months or years from their publication date, being able to give convincing security arguments turns out to be of major importance.

Of special interest, in this context, is the recent publication of the SIMON and SPECK family of block ciphers by the NSA [2]. Those ciphers brought a huge leap in terms of performances. As of today, these two primitives have an important efficiency advantage against all its competitors, in almost all implementation scenarios and platforms. However, even though SIMON or SPECK are quite

---

[1] Actually, this separation is not only valid for lightweight designs. It can well be extended to more classical ciphers or hash functions as well.

elegant and seemingly well-crafted designs, these efficiency improvements came at an essential price. Echoing the above, since the ciphers have a very light round function, their security bounds regarding classical linear or differential cryptanalysis are not so impressive, quite difficult to obtain or even non-existent. For example, in [22] the authors provide differential/linear bounds for SIMON, but, as we will see, one needs a big proportion of the total number of rounds to guarantee its security according to its block size. Even worse, no bound is currently known in the related-key model for any version of SIMON and thus there is a risk that good related-key differential characteristics might exist for this family of ciphers (while some lightweight proposals such as LED [15], PICCOLO [31] or some versions of TWINE [33] do provide such a security guarantee). One should be further cautious as these designs come from a governmental agency which does not provide specific details on how the primitives were built. No cryptanalysis was ever provided by the designers. Instead, the important analysis work was been carried out by the research community in the last few years and one should note that so far SIMON or SPECK remain unbroken.

It is therefore a major challenge for academic research to design a cipher that can compete with SIMON's performances and additionally provides the essential strong security guarantees that SIMON is clearly lacking. We emphasize that this is both a research challenge and, in view of NSA's efforts to propose SIMON into an ISO standard, a challenge that has likely a practical impact.

Lightweight Tweakable Block Ciphers and Side-Channel Protected Implementations. We note that tiny devices are more prone to be deployed into insecure environments and thus side-channel protected implementations of lightweight encryption primitives is a very important aspect that should be taken care of. One might even argue that instead of comparing performances of unprotected implementations of these lightweight primitives, one should instead compare protected variants (this is the recent trend followed by ciphers like ZORRO [14] or PICARO [28] and has actually already been taken into account long before by the cipher NOEKEON [13]). One extra protection against side-channel attacks can be the use of leakage resilient designs and notably through an extra tweak input of the cipher. Such tweakable block ciphers are rather rare, the only such candidate being Joltik-BC [18] or the internal cipher from SCREAM [34]. Coming up with a tweakable block cipher is indeed not an easy task as one must be extremely careful how to include this extra input that can be fully controlled by the attacker.

Low-Latency Implementations for Memory Encryption. One very interesting field in the area of lightweight cryptography is memory encryption (see e.g. [16] for an extensive survey of memory encryption techniques). Memory encryption has been used in the literature to protect the memory used by a process domain against several types of attackers, including attackers capable of monitoring and even manipulating bus transactions. Examples of commercial uses do not abound, but there are at least two: IBM's SecureBlue++ [36] and

Intel's SGX whose encryption and integrity mechanisms have been presented by Gueron at RWC 2016[2]. No documentation seems to be publicly available regarding the encryption used in IBM's solution, while Intel's encryption method requires additional data to be stored with each cache line. It is optimal in the context of encryption with memory overhead, but if the use case does not allow memory overhead then an entirely different approach is necessary.

With a focus on data confidentiality, a tweakable block cipher in ECB mode would then be the natural, straightforward solution. However, all generic methods to construct a tweakable block cipher from a block cipher suffer from an increased latency. Therefore, there is a clear need for lightweight tweakable block ciphers which do not require whitening value derivation, have a latency similar to the best non-tweakable block ciphers, and that can also be used in modes of operation that do not require memory expansion and offer beyond-birthday-bound security.

While being of great practical impact and need, it is actually very challenging to come up with such a block cipher. It should have three main characteristics. First, it must be executed within a single clock cycle and with a very low latency. Second, a tweak input is required, which in the case of memory encryption will be the memory address. Third, as one necessarily has to implement encryption and decryption, it is desirable to have a very low overhead when implementing decryption on top of encryption. The first and the third characteristics are already studied in the block cipher PRINCE [7]. However, the second point, i.e. having a tweak input, is not provided by PRINCE. It is not trivial to turn PRINCE into a tweakable block cipher, especially without increasing the number of rounds (and thereby latency) significantly.

Our Contributions. Our contributions are twofold. First, we introduce a new lightweight family of block ciphers: SKINNY. Our goal here is to provide a competitor to SIMON in terms of hardware/software performances, while proving in addition much stronger security guarantees with regard to differential/linear attacks. Second, we present MANTIS, a dedicated variant of SKINNY that constitutes a very efficient solution to the aforementioned problem of designing a tweakable block cipher for memory encryption.

Regarding SKINNY, we have pushed further the recent trend of having a SPN cipher with locally non-optimal internal components: SKINNY is an SPN cipher that uses a compact Sbox, a new very sparse diffusion layer, and a new very light key schedule. Yet, by carefully choosing our components and how they interact, our construction manages to retain very strong security guarantees. For all the SKINNY versions, we are able to prove using mixed integer linear programming (MILP) very strong bounds with respect to differential/linear attacks, not only in the single-key model, but also in the much more involved related-key model. Some versions of SKINNY have a very large key size compared to its block size and this theoretically renders the bounds search space huge. Therefore, the MILP

---

[2] The slides can be found here.

methods we have devised to compute these bounds for a SKINNY-like construction can actually be considered a contribution by itself. As we will see later, compared to SIMON, in the single-key model SKINNY needs a much lower proportion of its total number of rounds to provide a sufficient bound on the best differential/linear characteristic. In the related-key model, the situation is even more at SKINNY's advantage as no such bound is known for any version of SIMON as of today.

With regard to performance, SKINNY reaches very small area with serial ASIC implementations, yet it is actually the very first block cipher that leads to better performances than SIMON for round-based ASIC implementations, arguably the most important type of implementation since it provides a very good throughput for a reasonably low area cost, in contrary to serial implementations that only minimizes area. We also exhibit ASIC threshold implementations of our SKINNY variants that compare for example very favourably to AES-128 threshold implementations. As explained above, this is an integral part of modern lightweight primitives.

Regarding software, our implementations outperform all lightweight ciphers, except SIMON which performs slightly faster in the situation where the key schedule is performed only once. However, as remarked in [3], it is more likely in practice that the key schedule has to be performed everytime, and since SKINNY has a very lightweight key schedule we expect the efficiency of SKINNY software implementations to be equivalent to that of SIMON. This shows that SKINNY would perfectly fit a scenario where a server communicate with many lightweight devices. These performances are not surprising, in particular for bit-sliced implementations, as we show that SKINNY uses a much smaller total number of AND/NOR/XOR gates compared to all known lightweight block ciphers. This indicates that SKINNY will be competitive for most platforms and scenarios. Micro-controllers are no exception, and we show that SKINNY performs extremely well on these architectures.

We further remark that the decryption process of SKINNY has almost exactly the same description as the encryption counterpart, thus minimizing the decryption overhead.

We finally note that similarly to SIMON, SKINNY very naturally encompasses 64- or 128-bit block versions and a wide range of key sizes. However, in addition, SKINNY provides a tweakable capability, which can be very useful not only for leakage resilient implementations, but also to be directly plugged into higher-level operating modes, such as SCT [27]. In order to provide this tweak feature, we have generalized the STK construction [17] to enable more compact implementations while maintaining a high provable security level.

The SKINNY specifications are given in Sect. 2. The rationale of our design as well as various theoretical security and efficiency comparisons are provided in Sect. 3. Finally, we conducted a complete security analysis in Sect. 4 and we exhibit our implementation results in Sect. 5 (all the details are provided in the full version of the paper).

Regarding MANTIS, we propose in Sect. 6 a low-latency tweakable block cipher that reuses some design principles of SKINNY[3]. It represents a very efficient solution to the aforementioned problem of designing a tweakable block cipher tailored for memory encryption.

The main challenge when designing such a cipher is that its latency is directly related to the number of rounds. Thus, it is crucial to find a design, i.e. a round function and a tweak-scheduling, that ensures security already with a minimal number of rounds. Here, components of the recently proposed block ciphers PRINCE and MIDORI [1] turn out to be very beneficial.

The crucial step in the design of MANTIS was to find a suitable tweak-scheduling that would ensure a high number of active Sboxes not only in the single-key setting, but also in the setting where the attacker can control the difference in the tweak. Using, again, the MILP approach, we are able to demonstrate that a rather small number of rounds is already sufficient to ensure the resistance of MANTIS to differential (and linear) attacks in the related-tweak setting.

Besides the tweak-scheduling, we emphasize that MANTIS basically reuses well understood, previously studied, known components. It is mainly putting those components together in a new fashion, that allows MANTIS to be very competitive to PRINCE in latency and area, while being enhanced with a tweak. Thus, compared to the performance figures of PRINCE, we get the tweak almost for free, which is the key to solve the pressing problem of memory encryption.

2   Specification of SKINNY

Notations and SKINNY Versions. The lightweight block ciphers of the SKINNY family have 64-bit and 128-bit block versions and we denote n the block size. In both $n = 64$ and $n = 128$ versions, the internal state is viewed as a 4×4 square array of cells, where each cell is a nibble (in the $n = 64$ case) or a byte (in the $n = 128$ case). We denote $IS_{i,j}$ the cell of the internal state located at Row i and Column j (counting starting from 0). One can also view this 4×4 square array of cells as a vector of cells by concatenating the rows. Thus, we denote with a single subscript IS_i the cell of the internal state located at Position i in this vector (counting starting from 0) and we have that $IS_{i,j} = IS_{4 \cdot i + j}$.

SKINNY follows the TWEAKEY framework from [17] and thus takes a tweakey input instead of a key or a pair key/tweak. The user can then choose what part of this tweakey input will be key material and/or tweak material (classical block cipher view is to use the entire tweakey input as key material only). The family of lightweight block ciphers SKINNY have three main tweakey size versions: for a block size n, we propose versions with tweakey size $t = n$, $t = 2n$ and $t = 3n$ (versions with other tweakey sizes between n and $3n$ are naturally obtained from these main versions) and we denote $z = t/n$ the tweakey size to block size ratio. The tweakey state is also viewed as a collection of z 4×4 square arrays of cells of s bits each. We denote these arrays $TK1$ when $z = 1$, $TK1$ and $TK2$ when

---

[3] For the genesis of the cipher MANTIS, we acknowledge the contribution of Roberto Avanzi, as specified in Sect. 6.

$z = 2$, and finally $TK1$, $TK2$ and $TK3$ when $z = 3$. Moreover, we denote $TKz_{i,j}$ the cell of the tweakey state located at Row i and Column j of the z-th cell array. As for the internal state, we extend this notation to a vector view with a single subscript: $TK1_i$, $TK2_i$ and $TK3_i$. Moreover, we define the adversarial model **SK** (resp. **TK1**, **TK2** or **TK3**) where the attacker cannot (resp. can) introduce differences in the tweakey state.

Initialization. The cipher receives a plaintext $m = m_0\|m_1\|\cdots\|m_{14}\|m_{15}$, where the m_i are s-bit cells, with $s = n/16$ (we have $s = 4$ for the 64-bit block SKINNY versions and $s = 8$ for the 128-bit block SKINNY versions). The initialization of the cipher's internal state is performed by simply setting $IS_i = m_i$ for $0 \le i \le 15$:

$$IS = \begin{bmatrix} m_0 & m_1 & m_2 & m_3 \\ m_4 & m_5 & m_6 & m_7 \\ m_8 & m_9 & m_{10} & m_{11} \\ m_{12} & m_{13} & m_{14} & m_{15} \end{bmatrix}$$

This is the initial value of the cipher internal state and note that the state is loaded row-wise rather than in the column-wise fashion we have come to expect from the AES; this is a more hardware-friendly choice, as pointed out in [24].

The cipher receives a tweakey input $tk = tk_0\|tk_1\|\cdots\|tk_{30}\|tk_{16z-1}$, where the tk_i are s-bit cells. The initialization of the cipher's tweakey state is performed by simply setting for $0 \le i \le 15$: $TK1_i = tk_i$ when $z = 1$, $TK1_i = tk_i$ and $TK2_i = tk_{16+i}$ when $z = 2$, and finally $TK1_i = tk_i$, $TK2_i = tk_{16+i}$ and $TK3_i = tk_{32+i}$ when $z = 3$. We note that the tweakey states are loaded row-wise.

The Round Function. One encryption round of SKINNY is composed of five operations in the following order: SubCells, AddConstants, AddRoundTweakey, ShiftRows and MixColumns (see illustration in Fig. 1). The number r of rounds to perform during encryption depends on the block and tweakey sizes. The actual values are summarized in Table 1. Note that no whitening key is used in SKINNY. Thus, a part of the first and last round do not add any security. We motivate this choice in Sect. 3.

SubCells. A s-bit Sbox is applied to every cell of the cipher internal state. For $s = 4$, SKINNY cipher uses a Sbox $\mathcal{S}_4$ very close to the PICCOLO Sbox [31]. The action of this Sbox in hexadecimal notation is given by the following Table 2.

Table 1. Number of rounds for SKINNY-n-t, with n-bit internal state and t-bit tweakey state.

Block size n	Tweakey size t		
	n	$2n$	$3n$
64	32	36	40
128	40	48	56

Table 2. 4-bit Sbox $\mathcal{S}_4$ used in SKINNY when $s = 4$.

x	0	1	2	3	4	5	6	7	8	9	a	b	c	d	e	f
$\mathcal{S}_4[x]$	c	6	9	0	1	a	2	b	3	8	5	d	4	e	7	f
$\mathcal{S}_4^{-1}[x]$	3	4	6	8	c	a	1	e	9	2	5	7	0	b	d	f

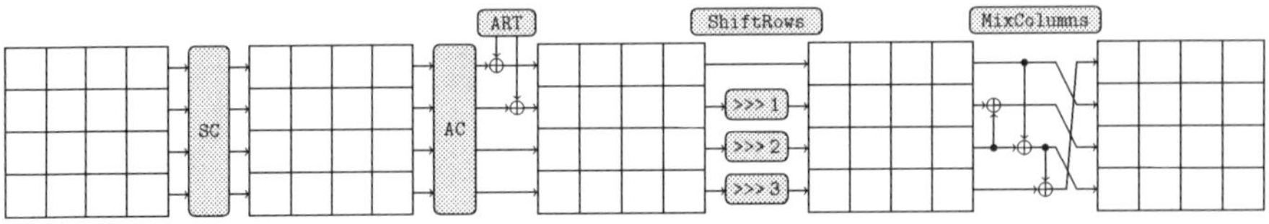

Fig. 1. The SKINNY round function applies five different transformations: SubCells (SC), AddConstants (AC), AddRoundTweakey (ART), ShiftRows (SR) and MixColumns (MC).

Note that $\mathcal{S}_4$ can also be described with four NOR and four XOR operations, as depicted in Fig. 2. If x_0, x_1, x_2 and x_3 represent the four inputs bits of the Sbox (x_0 being the least significant bit), one simply applies the following transformation:

$$(x_3, x_2, x_1, x_0) \rightarrow (x_3, x_2, x_1, x_0 \oplus (\overline{x_3 \vee x_2})),$$

followed by a left shift bit rotation. This process is repeated four times, except for the last iteration where the bit rotation is omitted.

For the case $s = 8$, SKINNY uses an 8-bit Sbox $\mathcal{S}_8$ that is built in a similar manner as for the 4-bit Sbox $\mathcal{S}_4$ described above. The construction is simple and is depicted in Fig. 3. If x_0, ..., x_7 represent the eight inputs bits of

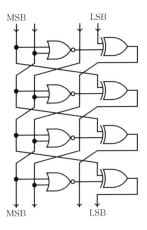

Fig. 2. Construction of the Sbox $\mathcal{S}_4$.

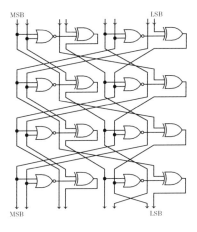

Fig. 3. Construction of the Sbox $\mathcal{S}_8$.

the Sbox (x_0 being the least significant bit), it basically applies the below transformation on the 8-bit state:

$$(x_7, x_6, x_5, x_4, x_3, x_2, x_1, x_0) \rightarrow (x_7, x_6, x_5, x_4 \oplus (\overline{x_7 \vee x_6}), x_3, x_2, x_1, x_0 \oplus (\overline{x_3 \vee x_2})),$$

followed by the bit permutation:

$$(x_7, x_6, x_5, x_4, x_3, x_2, x_1, x_0) \longrightarrow (x_2, x_1, x_7, x_6, x_4, x_0, x_3, x_5),$$

repeating this process four times, except for the last iteration where there is just a bit swap between x_1 and x_2.

AddConstants. A 6-bit affine LFSR, whose state is denoted (rc_5, rc_4, rc_3, rc_2, rc_1, rc_0) (with rc_0 being the least significant bit), is used to generate round constants. Its update function is defined as:

$$(\text{rc}_5 \| \text{rc}_4 \| \text{rc}_3 \| \text{rc}_2 \| \text{rc}_1 \| \text{rc}_0) \rightarrow (\text{rc}_4 \| \text{rc}_3 \| \text{rc}_2 \| \text{rc}_1 \| \text{rc}_0 \| \text{rc}_5 \oplus \text{rc}_4 \oplus 1).$$

The six bits are initialized to zero, and updated *before* use in a given round. The bits from the LFSR are arranged into a 4×4 array (only the first column of the state is affected by the LFSR bits), depending on the size of internal state:

$$\begin{bmatrix} c_0 & 0 & 0 & 0 \\ c_1 & 0 & 0 & 0 \\ c_2 & 0 & 0 & 0 \\ 0 & 0 & 0 & 0 \end{bmatrix},$$

with $c_2 = \texttt{0x2}$ and

$$(c_0, c_1) = (\text{rc}_3 \| \text{rc}_2 \| \text{rc}_1 \| \text{rc}_0, \ 0 \| 0 \| \text{rc}_5 \| \text{rc}_4) \text{ when s} = 4$$

$$(c_0, c_1) = (0 \| 0 \| 0 \| 0 \| \text{rc}_3 \| \text{rc}_2 \| \text{rc}_1 \| \text{rc}_0, \ 0 \| 0 \| 0 \| 0 \| 0 \| 0 \| \text{rc}_5 \| \text{rc}_4) \text{ when s} = 8.$$

The round constants are combined with the state, respecting array positioning, using bitwise exclusive-or.

AddRoundTweakey. The first and second rows of all tweakey arrays are extracted and bitwise exclusive-ored to the cipher internal state, respecting the array positioning. More formally, for $i = \{0, 1\}$ and $j = \{0, 1, 2, 3\}$, we have:
- $IS_{i,j} = IS_{i,j} \oplus TK1_{i,j}$ when $z = 1$,
- $IS_{i,j} = IS_{i,j} \oplus TK1_{i,j} \oplus TK2_{i,j}$ when $z = 2$,
- $IS_{i,j} = IS_{i,j} \oplus TK1_{i,j} \oplus TK2_{i,j} \oplus TK3_{i,j}$ when $z = 3$.

Then, the tweakey arrays are updated as follows (this tweakey schedule is illustrated in Fig. 4). First, a permutation P_T is applied on the cells positions of all tweakey arrays: for all $0 \leq i \leq 15$, we set $TK1_i \leftarrow TK1_{P_T[i]}$ with

$$P_T = [9, 15, 8, 13, 10, 14, 12, 11, 0, 1, 2, 3, 4, 5, 6, 7],$$

and similarly for $TK2$ when $z = 2$, and for $TK2$ and $TK3$ when $z = 3$. This corresponds to the following reordering of the matrix cells: $(0, \ldots, 15) \overset{P_T}{\longmapsto} (9, 15, 8, 13, 10, 14, 12, 11, 0, 1, 2, 3, 4, 5, 6, 7)$, indices being taken row-wise.

Table 3. The LFSRs used in SKINNY to generate the round constants. The TK parameter gives the number of tweakey words in the cipher, and the s parameter gives the size of cell in bits.

TK	s	LFSR
$TK2$	4	$(x_3\|\|x_2\|\|x_1\|\|x_0) \rightarrow (x_2\|\|x_1\|\|x_0\|\|x_3 \oplus x_2)$
	8	$(x_7\|\|x_6\|\|x_5\|\|x_4\|\|x_3\|\|x_2\|\|x_1\|\|x_0) \rightarrow (x_6\|\|x_5\|\|x_4\|\|x_3\|\|x_2\|\|x_1\|\|x_0\|\|x_7 \oplus x_5)$
$TK3$	4	$(x_3\|\|x_2\|\|x_1\|\|x_0) \rightarrow (x_0 \oplus x_3\|\|x_3\|\|x_2\|\|x_1)$
	8	$(x_7\|\|x_6\|\|x_5\|\|x_4\|\|x_3\|\|x_2\|\|x_1\|\|x_0) \rightarrow (x_0 \oplus x_6\|\|x_7\|\|x_6\|\|x_5\|\|x_4\|\|x_3\|\|x_2\|\|x_1)$

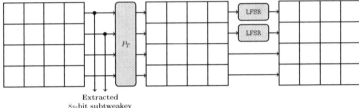

Fig. 4. The tweakey schedule in SKINNY. Each tweakey word $TK1$, $TK2$ and $TK3$ (if any) follows a similar transformation update, except that no LFSR is applied to $TK1$.

Finally, every cell of the first and second rows of $TK2$ and $TK3$ (for the SKINNY versions where $TK2$ and $TK3$ are used) are individually updated with an LFSR. The LFSRs used are given in Table 3 (x_0 stands for the LSB of the cell).

ShiftRows. As in AES, in this layer the rows of the cipher state cell array are rotated, but they are to the right. More precisely, the second, third, and fourth cell rows are rotated by 1, 2 and 3 positions to the right, respectively. In other words, a permutation P is applied on the cells positions of the cipher internal state cell array: for all $0 \le i \le 15$, we set $IS_i \leftarrow IS_{P[i]}$ with

$$P = [0, 1, 2, 3, 7, 4, 5, 6, 10, 11, 8, 9, 13, 14, 15, 12].$$

MixColumns. Each column of the cipher internal state array is multiplied by the following binary matrix $\mathbf{M}$:

$$\mathbf{M} = \begin{pmatrix} 1 & 0 & 1 & 1 \\ 1 & 0 & 0 & 0 \\ 0 & 1 & 1 & 0 \\ 1 & 0 & 1 & 0 \end{pmatrix}.$$

The final value of the internal state array provides the ciphertext with cells being unpacked in the same way as the packing during initialization. Note that decryption is very similar to encryption as all cipher components have very simple inverse (SubCells and MixColumns are based on a generalized Feistel structure, so their respective inverse is straightforward to deduce and can be implemented with the exact same number of operations).

Extending to Other Tweakey Sizes. The three main versions of SKINNY have tweakey sizes $t = n$, $t = 2n$ and $t = 3n$, but one can easily extend this to any size[4] of tweakey $n \leq t \leq 3n$:

- for any tweakey size $n < t < 2n$, one simply uses exactly the $t = 2n$ version but the last $2n - t$ bits of the tweakey state are fixed to the zero value. Moreover, the corresponding cells in the tweakey state $TK2$ will not be updated throughout the rounds with the LFSR.
- for any tweakey size $2n < t < 3n$, one simply uses exactly the $t = 3n$ version but the last $3n - t$ bits of the tweakey state are fixed to the zero value. Moreover, the corresponding cells in the tweakey state $TK3$ will not be updated throughout the rounds with the LFSR.

We note that some of our 64-bit block SKINNY versions allow small key sizes (down to 64-bit). We emphasize that we propose these versions mainly for simplicity in the description of the SKINNY family of ciphers. Yet, as advised by the NIST [26], one should not to use key sizes that are smaller than 112 bits.

Instantiating the Tweakey State with Key and Tweak Material. Following the TWEAKEY framework [17], SKINNY takes as inputs a plaintext or a ciphertext and a tweakey value, which can be used in a flexible way by filling it with key and tweak material. Whatever the situation, the user must ensure that the key size is always at least as big as the block size.

In the classical setting where only key material is input, we use exactly the specifications of SKINNY described previously. However, when some tweak material is to be used in the tweakey state, we dedicate $TK1$ for this purpose and XOR a bit set to "1" every round to the second bit of the top cell of the third column (i.e. the second bit of $IS_{0,2}$). In other words, when there is some tweak material, we add an extra "1" in the constant matrix from AddConstants). Besides, in situations where the user might use different tweak sizes, we recommend to dedicate some cells of $TK1$ to encode the size of the tweak material, in order to ensure proper separation. Note that these are only recommendations, thus not strictly part of the specifications of SKINNY.

3 Rationale of SKINNY

Several design choices of SKINNY have been borrowed from existing ciphers, but most of our components are new, optimized for our goal: a cipher well suited for most lightweight applications. When designing SKINNY, one of our main criteria was to only add components which are vital for the security of the primitive, removing any unnecessary operation (hence the name of our proposal). We end

---

[4] For simplicity we do not include here tweakey sizes that are not a multiple of s bits. However, such cases can be trivially handled by generalizing the tweakey schedule description to the bit level.

up with the sound property that removing any component or using weaker version of a component from SKINNY would lead to a much weaker (or actually insecure) cipher. Therefore, the construction of SKINNY has been done through several iterations, trying to reach the exact spot where good performance meets strong security arguments. We detail in this section how we tried to follow this direction for each layer of the cipher.

We note that one could have chosen a slightly smaller Sbox or a slightly sparser diffusion layer, but our preliminary implementations showed that these options represent worse tradeoff overall. For example, one could imagine a very simple cipher iterating thousands of rounds composed of only a single non-linear boolean operation, an XOR and some bit wiring. However, such a cipher will lead to terrible performance regarding throughput, latency or energy consumption.

When designing a lightweight encryption scheme, several use cases must be taken in account. While area optimized implementations are important for some very constrained applications, throughput or throughput-over-area optimized implementations are also very relevant. Actually, looking at recently introduced efficiency measurements [19], one can see that our designs choices are good for many types of implementations, which is exactly what makes a good general-purpose lightweight encryption scheme.

3.1   Estimating Area and Performances

In order to discuss the rationale of our design, we first quickly describe an estimation in Gate Equivalent (GE) of the ASIC area cost of several simple bit operations (for UMC 180 nm 1.8 V [35]): a NOR/NAND gate costs 1 GE, a OR/AND gate costs 1.33 GE, a XOR/XNOR gate costs 2.67 GE and a NOT gate costs 0.67 GE. Finally, one memory bit can be estimated to 6 GE (scan flip-flop). Of course, these numbers depend on the library used, but it will give us at least some rough and easy evaluation of the design choices we will make.

Besides, even though many tradeoffs exist, we distinguish between a serial implementation, a round-based implementation and a low-latency implementation. In the latter, the entire ciphering process is performed in a single clock cycle, but the area cost is then quite important as all rounds need to be directly implemented. For a round-based implementation, an entire round of the cipher is performed in a single clock cycle, thus ending with the entire ciphering process being done in r cycles and with a moderate area cost (this tradeoff is usually a good candidate for energy efficiency). Finally, in a serial implementation, one reduces the datapath and thus the area to the minimum (usually a few bits, like the Sbox bit size), but the throughput is greatly reduced. The ultimate goal of a good lightweight encryption primitive is to use lightweight components, but also to ensure that these components are compact and efficient for all these tradeoffs. This is what SIMON designers have managed to produce, but sacrificing a few security guarantees. SKINNY offers similar (sometimes even better) performances than SIMON, while providing much stronger security arguments with regard to classical differential or linear cryptanalysis.

3.2 General Design and Components Rationale

A first and important decision was to choose between a Substitution-Permutation Network (SPN), or a Feistel network. We started from a SPN construction as it is generally easier to provide stronger bounds on the number of active Sboxes. However, we note that there is a dual bit-sliced view of SKINNY that resembles some generalized Feistel network. Somehow, one can view the cipher as a primitive in between an SPN and an "AND-rotation-XOR" function like SIMON. We try to get the best of both worlds by benefiting the nice implementation tradeoffs of the latter, while organizing the state in an SPN view so that bounds on the number of active Sboxes can be easily obtained.

The absence of whitening key is justified by the reduction of the control logic: by always keeping the exact same round during the entire encryption process we avoid the control logic induced by having a last non-repeating layer at the end of the cipher. Besides, this simplifies the general description and implementation of the primitive. Obviously, having no whitening key means that a few operations of the cipher have no impact on the security. This is actually the case for both the beginning and the end of the ciphering process in SKINNY since the key addition is done in the middle of the round, with only half of the state being involved with this key addition every round.

A crucial feature of SKINNY is the easy generation of several block size or tweakey size versions, while keeping the general structure and most of the security analysis untouched. Going from the 64-bit block size versions to the 128-bit block size versions is simply done by using a 8-bit Sbox instead of a 4-bit Sbox, therefore keeping all the structural analysis identical. Using bigger tweakey material is done by following the STK construction [17], which allows automated analysis tools to still work even though the input space become very big (in short, the superposition trick makes the **TK2** and **TK3** analysis almost as time consuming as the normal and easy **TK1** case). Besides, unlike previous lightweight block ciphers, this complete analysis of the **TK2** and **TK3** cases allows us to dedicate a part of this tweakey material to be potentially some tweak input, therefore making SKINNY a flexible tweakable block cipher. Also, we directly obtain related-key security proofs using this general structure.

SubCells. The choice of the Sbox is obviously a crucial decision in an SPN cipher and we have spent a lot of efforts on looking for the best possible candidate. For the 4-bit case, we have designed a tool that searches for the most compact candidate that provides some minimal security guarantees. Namely, with the bit operations cost estimations given previously, for all possible combinations of operations (NAND/NOR/XOR/XNOR) up to a certain limit cost, our tool checks if certain security criterion of the tested Sbox are fulfilled. More precisely, we have forced the maximal differential transition probability of the Sbox to be 2^{-2} and the maximal absolute linear bias to be 2^{-2}. When both criteria are satisfied, we have filtered our search for Sbox with high algebraic degree.

Our results is that the Sbox used in the `PICCOLO` block cipher [31] is close to be the best one: our 4-bit Sbox candidate S_4 is essentially the `PICCOLO` Sbox with the last NOT gate at the end being removed (see Fig. 2). We believe this extra NOT gate was added by the `PICCOLO` designers to avoid fixed points (actually, if fixed points were to be removed at the Sbox level, the `PICCOLO` candidate would be the best choice), but in `SKINNY` the fixed points are handled with the use of constants to save some extra GE. Yet, omitting the last bit rotation layer removes already a lot of fixed points (the efficiency cost of this omission being null).

The Sbox S_4 can therefore be implemented with only 4 NOR gates and 4 XOR gates, the rest being only bit wiring (basically free in hardware). According to our previously explained estimations, this should cost 14.68 GE, but as remarked in [31], some libraries provide special gates that further save area. Namely, in our library the 4-input AND-NOR and 4-input OR-NAND gates with two inputs inverted cost 2 GE and they can be used to directly compute a XOR or an XNOR. Thus, S_4 can be implemented with only 12 GE. In comparison, the `PRESENT` Sbox [5] requires 3 AND, 1 OR and 11 XOR gates, which amounts to 27.32 GE (or 34.69 GE without the special 4-input gates).

All in all, our 4-bit Sbox S_4 has the following security properties: maximal differential transition probability of 2^{-2}, maximal absolute linear bias of 2^{-2}, branching number 2, algebraic degree 3 and one fixed point $S_4(\texttt{0xF}) = \texttt{0xF}$.

Regarding the 8-bit Sbox, the search space was too wide for our automated tool. Therefore, we instead considered a subclass of the entire search space: by reusing the general structure of S_4, we have tested all possible Sboxes built by iterating several times a NOR/XOR combination and a bit permutation. Our search found that the maximal differential transition probability and maximal absolute linear bias of the Sboxes are larger than 2^{-2} when we have less than 8 iterations of the NOR/XOR combination and bit permutation. With 8 iterations of the NOR/XOR combination and bit permutation, we found Sboxes with desired maximal differential transition probability of 2^{-2} and maximal absolute linear bias of 2^{-2} with algebraic degree 6. However, the algebraic degree of the inverse Sboxes of all these candidates is 5 rather than 6. In addition, having 8 iterations may result in higher latency when we consider a serial hardware implementation. Therefore, we considered having 2 NOR/XOR combinations in every iteration and reduce the number of iteration from 8 to 4. As a result, we found several Sboxes with the desired maximal differential probability and absolute linear bias, while reaching algebraic degree 6 for both the Sbox and its inverse (thus better than the 8 iterations case). Although such Sbox candidates have 3 fixed points when we omit the last bit permutation layer like the 4-bit case, we can easily reduce the number of fixed points by introducing a different bit permutation from the intermediate bit permutations to the last layer without any additional cost.

With 2 NOR/XOR combinations and a bit permutation iterated 4 times, S_8 can be implemented with only 8 NOR gates and 8 XOR gates (see Fig. 3), the rest being only bit wiring (basically free in hardware). The total area cost should be 24 GE according to our previously explained estimations and using special 4-input AND-NOR and 4-input OR-NAND gates. In comparison, while ensuring a maximal differential transition probability (resp. maximum absolute linear bias) of 2^{-6} (resp. 2^{-4}), the AES Sbox requires 32 AND/OR gates and 83 XOR gates to be implemented, which amounts to 198 GE. Even recent lightweight 8-bit Sbox proposal [10] requires 12 AND/OR gates and 26 XOR gates, which amounts to 64 GE, for a maximal differential transition probability (resp. maximum linear bias) of 2^{-5} (resp. 2^{-2}), but their optimization goal was different from ours.

All in all, we believe our 8-bit Sbox candidate $\mathcal{S}_8$ provides a good tradeoff between security and area cost. It has maximal differential transition probability of 2^{-2}, maximal absolute linear bias of 2^{-2}, branching number 2, algebraic degree 6 and a single fixed point $\mathcal{S}_8(\text{0xFF}) = \text{0xFF}$ (for the Sbox we have chosen, swapping two bits in the last bit permutation was probably the simplest method to achieve only a single fixed point).

Note that both our Sboxes $\mathcal{S}_4$ and $\mathcal{S}_8$ have the interesting feature that their inverse is computed almost identically to the forward direction (as they are based on a generalized Feistel structure) and with exactly the same number of operations. Thus, our design reasoning also holds when considering the decryption process.

AddConstants. The constants in SKINNY have several goals: differentiate the rounds (see Sect. 4.2), differentiate the columns and avoid symmetries, complicate subspace cryptanalysis (see Sect. 4.2) and attacks exploiting fixed points from the Sbox. In order to differentiate the rounds, we simply need a counter, and since the number of rounds of all SKINNY versions is smaller than 64, the most hardware friendly solution is to use a very cheap 6-bit affine LFSR (like in LED [15]) that requires only a single XNOR gate per update. The 6 bits are then dispatched to the two first rows of the first column (this will maximize the constants spread after the ShiftRows and MixColumns), which will already break the columns symmetry.

In order to avoid symmetries, fixed points and more generally subspaces to spread, we need to introduce different constants in several cells of the internal state. The round counter will already naturally have this goal, yet, in order to increase that effect, we have added a "1" bit to the third row, which is almost free in terms of implementation cost. This will ensure that symmetries and subspaces are broken even more quickly, and in particular independently of the round counter.

AddRoundTweakey. The tweakey schedule of SKINNY follows closely the STK construction from [17] (that allows to easily get bounds on the number of active Sboxes in the related-tweakey model). Yet, we have changed a few parts. Firstly, instead of using multiplications by 2 and 3 in a finite field, we have instead replaced these tweakey cells updates by cheap 4-bit or 8-bit LFSRs (depending on the size of the cell) to minimize the hardware cost. All our LFSRs require only a single XOR for the update, and we have checked that the differential cancellation behavior of these interconnected LFSRs is as required by the STK construction: for a given position, a single cancellation can only happen every 15 rounds for **TK2**, and same with two cancellations for **TK3**.

Another important generalization of the STK construction is the fact that every round we XOR only half of the internal cipher state with some subtweakey. The goal was clearly to optimize hardware performances of SKINNY, and it actually saves an important amount of XORs in a round-based implementation. The potential danger is that the bounds we obtain would dramatically drop because of this change. Yet, surprisingly, the bounds remained actually good and this was a good security/performance tradeoff to make. Another advantage is that we can now update the tweakey cells only before they are incorporated to the cipher internal state. Thus, half of tweakey cells only will be updated every round and the period of the cancellations naturally doubles: for a certain cell position, a single cancellation can only happen every 30 rounds for **TK2** and two cancellations can only happen every 30 rounds for **TK3**.

The tweakey permutation P_T has been chosen to maximize the bounds on the number of active Sboxes that we could obtain in the related-tweakey model (note that it has no impact in the single-key model). Besides, we have enforced for P_T the special property that all cells located in third and fourth rows are sent to the first and second rows, and vice-versa. Since only the first and second rows of the tweakey states are XORed to the internal state of the cipher, this ensures that both halves of the tweakey states will be equally mixed to the cipher internal state (otherwise, some tweakey bytes might be more involved in the ciphering process than others). Finally, the cells that will not be directly XORed to the cipher internal state can be left at the same relative position. On top of that, we only considered those variants of P_T that consist of a singe cycle.

We note that since the cells of the first tweakey word $TK1$ are never updated, they can be directly hardwired to save some area if the situation allows.

ShiftRows and MixColumns. Competing with SIMON's impressive hardware performance required choosing an extremely sparse diffusion layer for SKINNY, which was in direct contradiction with our original goal of obtaining good security bounds for our primitive. Note that since our Sboxes $\mathcal{S}_4$ and $\mathcal{S}_8$ have a branching number of two, we cannot use only a bit permutation layer as in the PRESENT block cipher: differential characteristics with only a single active Sbox per round would exist. After several design iterations, we came to the conclusion that binary matrices were the best choice. More surprisingly, while most block cipher designs are using very strong diffusion layers (like an MDS matrix), and even though a 4×4 binary matrices with branching number four exist, we preferred a much sparser candidate which we believe offers the best security/performance tradeoff (this can be measured in terms of Figure Of Adversarial Merit [19]).

Due to its strong sparseness, SKINNY binary diffusion matrix **M** has only a differential or linear branching number of two. This seems to be worrisome as it would again mean that differential characteristics with only a single active Sbox per round would exist (it would be the same for PRESENT block cipher if its Sbox did not have branching number three, which is the reason of the relatively high cost of the PRESENT Sbox). However, we designed $\mathbf{M}$ such that when a branching two differential transition occurs, the next round will likely lead to a much higher branching number. Looking at $\mathbf{M}$, the only way to meet branching two is to have an input difference in either the second or the fourth input only. This leads to an input difference in the first or third element for the next round, which then diffuses to many output elements. The differential characteristic with a single active Sbox per round is therefore impossible, and actually we will be able to prove at least 96 active Sboxes for 20 rounds. Thus, for the very cheap price of a differential branching two binary diffusion matrix, we are in fact getting a better security than expected when looking at the iteration of several rounds. The effect is the same with linear branching (for which we only need to look at the transpose of the inverse of $\mathbf{M}$, i.e. $(\mathbf{M}^{-1})^{\top}$).

We have considered all possibilites for $\mathbf{M}$ that can be implemented with at most three XOR operations and eventually kept the MixColumns matrices that, in combination with ShiftRows, guaranteed high diffusion and led to strong bounds on the minimal number of active Sboxes in the single-key model.

Note that another important criterion came into play regarding the choice of the diffusion layer of SKINNY: it is important that the key material impacts as fast as possible the cipher internal state. This is in particular a crucial point for SKINNY as only half of the state is mixed with some key material every round, and since there is no whitening keys. Besides, having a fast key diffusion will reduce the impact of meet-in-the-middle attacks. Once the two first rows of the state were arbitrarily chosen to receive the key material, given a certain subtweakey, we could check how many rounds were required (in both encryption and decryption directions) to ensure that the entire cipher state depends on this subtweakey. Our final choice of MixColumns is optimal: only a single round is required in both forward and backward directions to ensure this diffusion.

3.3   Comparing Differential Bounds

Our entire design has been crafted to allow good provable bounds on the minimal number of differential or linear active Sboxes, not only for the single-key model, but also in the related-key model (or more precisely the related-tweakey model in our case). We provide in Table 4 a comparison of our bounds with the best known proven bounds for other lightweight block ciphers at the same security level (all the ciphers in the table use 4-bit Sboxes with a maximal differential probability of 2^{-2}). We give in Sect. 4 more details on how the bounds of SKINNY were obtained.

First, we emphasize that most of the bounds we obtained for SKINNY are not tight, and we can hope for even higher minimal numbers of active Sboxes. This is not the case of LED or PRESENT for which the bounds are tight.

From the table, we can see that LED obtains better bounds for **SK**. Yet, the situation is inverted for **TK2**: due to a strong plateau effect in the **TK2** bounds of LED, it stays at 50 active Sboxes until Round 24, while SKINNY already

Table 4. Proved bounds on the minimal number of differential active Sboxes for SKINNY-64-128 and various lightweight 64-bit block 128-bit key ciphers. Model **SK** denotes the single-key scenario and model **TK2** denotes the related-tweakey scenario where differences can be inserted in both states $TK1$ and $TK2$.

Cipher	Model	Rounds															
		1	2	3	4	5	6	7	8	9	10	11	12	13	14	15	16
SKINNY	SK	1	2	5	8	12	16	26	36	41	46	51	55	58	61	66	75
(36 rounds)	TK2	0	0	0	0	1	2	3	6	9	12	16	21	25	31	35	40
LED	SK	1	5	9	25	26	30	34	50	51	55	59	75	76	80	84	100
(48 rounds)	TK2	0	0	0	0	0	0	0	0	1	5	9	25	26	30	34	50
PICCOLO	SK	0	5	9	14	18	27	32	36	41	45	50	54	59	63	68	72
(31 rounds)	TK2	0	0	0	0	0	0	0	5	9	14	18	18	23	27	27	32
MIDORI	SK	1	3	7	16	23	30	35	38	41	50	57	62	67	72	75	84
(16 rounds)	TK2	-	-	-	-	-	-	-	-	-	-	-	-	-	-	-	-
PRESENT	SK	-	-	-	-	10	-	-	-	-	20	-	-	-	-	30	-
(31 rounds)	TK2	-	-	-	-	-	-	-	-	-	-	-	-	-	-	-	-
TWINE	SK	0	1	2	3	4	6	8	11	14	18	22	24	27	30	32	-
(36 rounds)	TK2	-	-	-	-	-	-	-	-	-	-	-	-	-	-	-	-

reaches 72 active Sboxes at Round 24. Besides, LED performance will be quite bad compared to SKINNY, due to its strong MDS diffusion layer and strong Sbox.

Regarding PICCOLO, the bounds[5] are really similar to SKINNY for **SK** but worse for **TK2**. Yet, our round function is lighter (no use of a MDS layer), see Sect. 3.4.

No related-key bounds are known for MIDORI, PRESENT or TWINE. Besides, our **SK** bounds are better than PRESENT. Regarding MIDORI or TWINE in **SK**, while our bounds are slightly worse, we emphasize again that our round function is much lighter and thus will lead to much better performances.

Comparing differential bounds with SIMON is not as simple as with SPN ciphers. Yet, bounds on the best differential/linear characteristics for SIMON have been provided recently by [22][6].

Assuming (very) pessimistically for SKINNY that a maximum differential transition probability of 2^{-2} is always possible for each active Sbox in the differential paths with the smallest number of active Sboxes, we can directly obtain easy bounds on the best differential/linear characteristics for SKINNY. We provide in Table 5 a comparison between SIMON and SKINNY versions for the proportion of total number of rounds needed to provide a sufficiently good differential characteristic probability bound according to the cipher block size. One can see that

---

[5] We estimate the number of active Sboxes for PICCOLO to $\lceil 4.5 \cdot N_f \rceil$, where N_f is the number of active F-functions taken from [31].

[6] Their article initially contained results only for the smallest versions of SIMON, but the authors provided us updated results for all versions of SIMON.

SKINNY needs a much smaller proportion of its total number of rounds compared to SIMON to ensure enough confidence with regards to simple differential/linear attacks. Actually the related-key ratios of SKINNY are even smaller than single-key ratios of SIMON (no related-key bounds are known as of today for SIMON).

Table 5. Comparison between AES-128 and SIMON/SKINNY versions for the proportion of total number of rounds needed to provide a sufficiently good differential characteristic probability bound according to the cipher block size (i.e. $< 2^{-64}$ for 64-bit block size and $< 2^{-128}$ for 128-bit block size). Results for SIMON are updated results taken from [22].

Cipher	Model	
	Single-Key	Related-Key
SKINNY-64-128	$8/36 = \mathbf{0.22}$	$15/36 = \mathbf{0.42}$
SIMON-64-128	$19/44 = \mathbf{0.43}$	?
SKINNY-128-128	$15/40 = \mathbf{0.37}$	$19/40 = \mathbf{0.47}$
SIMON-128-128	$41/72 = \mathbf{0.57}$	?
AES-128	$4/10 = \mathbf{0.40}$	$6/10 = \mathbf{0.60}$

Finally, in terms of diffusion, all versions of SKINNY achieve full diffusion after only 6 rounds (forwards or backwards), while SIMON versions with 64-bit block size requires 9 rounds, and even 13 rounds for SIMON versions with 128-bit block size [22] (AES-128 reaches full diffusion after 2 of its 10 rounds). Again, the diffusion comparison according to the total number of rounds is at SKINNY's advantage.

3.4   Comparing Theoretical Performance

After some minimal security guarantee, the second design goal of SKINNY was to minimize the total number of operations. We provide in Table 6 a comparison of the total number of operations per bit for SKINNY and for other lightweight block ciphers, as well as some quality grade regarding its ASIC area in a round-based implementation. We explain in the full version of this article how these numbers have been computed.

One can see from the Table 6 that SIMON and SKINNY compare very favorably to other candidates, both in terms of number of operations and theoretical area grade for round-based implementations. This seems to confirm that when it comes to lightweight block ciphers, SIMON is probably the strongest competitor as of today. Besides, SKINNY has the best theoretical profile among all the candidates presented here, even better than SIMON for area. For speed efficiency, SKINNY outperforms SIMON when the key schedule is taken in account. This scenario is arguably the most important in practice: as remarked in [3], it is likely that lightweight devices will cipher very small messages and thus the back-end servers communicating with millions of devices will probably have to recompute the key schedule for every small message received.

In addition to its smaller key size, we note that KATAN-64-80 [9] theoretical area grade is slightly biased here as one round of this cipher is extremely light and such a round-based implementation would actually look more like a serial implementation and will have a very low throughput (KATAN-64-80 has 254 rounds in total).

While Table 6 is only a rough indication of the efficiency of the various designs, we observe that the ratio between the SIMON and SKINNY best software implementations, or the ratio between the smallest SIMON and SKINNY round-based hardware implementations actually match the results from the table (see full version of the paper).

4   Security Analysis

In this section, we provide a short summary of the in-depth analysis we conducted on the security of the SKINNY family of block ciphers. All details are provided in the full version of this article. We emphasize that we do not claim any security

Table 6. Total number of operations and theoretical performance of SKINNY and various lightweight block ciphers. N denotes a NOR gate, A denotes a AND gate, X denotes a XOR gate.

Cipher	nb. of rds	gate cost (per bit per round)			nb. of op. w/o key sch.	nb. of op. w/key sch.	round-based impl. area
		int. cipher	key sch.	total			
SKINNY -64-128	36	1 N 2.25 X	0.625 X	1 N 2.875 X	3.25×36 $= 117$	3.875×36 $= \mathbf{139.5}$	$1 + 2.67 \times 2.875$ $= \mathbf{8.68}$
SIMON -64/128	44	0.5 A 1.5 X	1.5 X	0.5 A 3.0 X	2×44 $= 88$	3.5×44 $= \mathbf{154}$	$0.67 + 2.67 \times 3$ $= \mathbf{8.68}$
PRESENT -128	31	1 A 3.75 X	0.125 A 0.344 X	1.125 A 4.094 X	4.75×31 $= 147.2$	5.22×31 $= \mathbf{161.8}$	$1.5 + 2.67 \times 4.094$ $= \mathbf{12.43}$
PICCOLO -128	31	1 N 4.25 X		1 N 4.25 X	5.25×31 $= 162.75$	5.25×31 $= \mathbf{162.75}$	$1 + 2.67 \times 4.25$ $= \mathbf{12.35}$
KATAN -64-80	254	0.047 N 0.094 X	3 X	0.047 N 3.094 X	0.141×254 $= 35.81$	3.141×254 $= \mathbf{797.8}$	$0.19 + 2.67 \times 3.094$ $= \mathbf{8.45}$
SKINNY -128-128	40	1 N 2.25 X		1 N 2.25 X	3.25×40 $= 130$	3.25×40 $= \mathbf{130}$	$1 + 2.67 \times 2.25$ $= \mathbf{7.01}$
SIMON -128/128	72	0.5 A 1.5 X	1 X	0.5 A 2.5 X	2×68 $= 136$	3×68 $= \mathbf{204}$	$0.67 + 2.67 \times 2.5$ $= \mathbf{7.34}$
NOEKEON -128	16	0.5 (A + N) 5.25 X	0.5 (A + N) 5.25 X	1 (A + N) 10.5 X	6.25×16 $= 100$	12.5×16 $= \mathbf{200}$	$2.33 + 2.67 \times 10.5$ $= \mathbf{30.36}$
AES -128	10	4.25 A 16 X	1.06 A 3.5 X	5.31 A 19.5 X	20.25×10 $= 202.5$	24.81×10 $= \mathbf{248.1}$	$7.06 + 2.67 \times 19.5$ $= \mathbf{59.12}$
SKINNY -128-256	48	1 N 2.25 X	0.56 X	1 N 2.81 X	3.25×48 $= 156$	3.81×48 $= \mathbf{183}$	$1 + 2.67 \times 2.81$ $= \mathbf{8.5}$
SIMON -128/256	72	0.5 A 1.5 X	1.5 X	0.5 A 3.0 X	2×72 $= 144$	3.5×72 $= \mathbf{252}$	$0.67 + 2.67 \times 3$ $= \mathbf{8.68}$
AES -256	14	4.25 A 16 X	2.12 A 7 X	6.37 A 23 X	20.25×14 $= 283.5$	29.37×14 $= \mathbf{411.2}$	$8.47 + 2.67 \times 23$ $= \mathbf{69.88}$

in the chosen-key or known-key model, but we *do* claim security in the related-key model. Moreover, we chose not to use any constant to differentiate between different block sizes or tweakey sizes versions of SKINNY, as we believe such a separation should be done at the protocol level, for example by deriving different keys (note that, if needed, this can easily be done by encoding these sizes and use them as fixed extra constant material every round).

4.1 Differential/Linear Cryptanalysis

In order to argue for the resistance of SKINNY against differential and linear attacks, we computed lower bounds on the minimal number of active Sboxes, both in the single-key and related-tweakey model. We recall that, in a differential (resp. linear) characteristic, an Sbox is called *active* if it contains a non-zero input difference (resp. input mask). In contrast to the single-key model, where the round tweakeys are constant and thus do not influence the activity pattern, an attacker is allowed to introduce differences (resp. masks) within the tweakey state in the related-tweakey model. For that, we considered the three cases of choosing input differences in **TK1** only, both **TK1** and **TK2**, and in all of the tweakey states **TK1**, **TK2** and **TK3**, respectively. Table 7 presents lower bounds on the number of differential active Sboxes for 16 up to 30 rounds. For computing these bounds, we generated a Mixed-Integer Linear Programming model following the approach explained in [25,32].

Table 7. Lowerbounds on the number of active Sboxes in SKINNY for large number of rounds. Note that the bounds on the number of linear active Sboxes in the single-key model are also valid in the related-tweakey model. In case the MILP optimization was too long, we provide upper bounds between parentheses.

Model	16	17	18	19	20	21	22	23	24	25	26	27	28	29	30
SK	75	82	88	92	96	102	108	(114)	(116)	(124)	(132)	(138)	(136)	(148)	(158)
TK1	54	59	62	66	70	75	79	83	85	88	95	102	(108)	(112)	(120)
TK2	40	43	47	52	57	59	64	67	72	75	82	85	88	92	96
TK3	27	31	35	43	45	48	51	55	58	60	65	72	77	81	85
SK Lin	70	76	80	85	90	96	102	107	(110)	(118)	(122)	(128)	(136)	(141)	(143)

For lower bounding the number of linear active Sboxes, we used the same approach. For that, we considered the inverse of the transposed linear transformation $\mathbf{M}^\top$. However, for the linear case, we only considered the single-key model. As it is described in [23], there is no cancellation of active Sboxes in linear characteristics. Thus, the bounds for **SK** give valid bounds also for the case where the attacker is allowed to not only control the message but also the tweakey input.

The above bounds are for single characteristic, thus it will be interesting to take a look at differentials and linear hulls. Being a rather complex task, we leave this as future work.

4.2   Further Cryptanalysis

Meet-in-the-Middle Attacks. Meet-in-the-middle attacks have been applied to block ciphers, e.g. [6,11]. From its application to the SPN structure [30], the number of attacked rounds can be evaluated by considering the maximum length of three features, partial-matching, initial structure and splice-and-cut. We conclude that meet-in-the-middle attack may work up to at most 22 rounds, so that 32+ rounds of SKINNY-64 provides a reasonable margin.

Remarks on Biclique Cryptanalysis. Biclique cryptanalysis improves the complexity of exhaustive search by computing only a part of encryption algorithm. The improved factor is often evaluated by the ratio of the number of Sboxes involved in the partial computation to all Sboxes in the cipher. The improved factor can be relatively big when the number of rounds in the cipher is small, which is not the case in SKINNY. We do not think improving exhaustive search by a small factor will turn into serious vulnerability in future. Therefore, SKINNY is not designed to resist biclique cryptanalysis with small improvement.

Impossible Differential Attacks. Impossible differential attack [4] finds two internal state differences Δ and Δ' such that Δ is never propagated to Δ'. The attacker then finds many plaintext/ciphertext pairs and tweakey values leading to (Δ, Δ'). Those tweakey values are wrong values, thus tweakey space can be reduced.

We found that the longest impossible differential characteristics reach 11 rounds and there are 16 such characteristics in total. While several rounds can be appended to turn this into a key-recovery attack, the number of rounds for SKINNY provide a sound security margin.

Integral Attacks. Integral attack [12,21] prepares a set of plaintexts so that particular cells can contain all the values in the set and the other cells are fixed to a constant value. Then properties of the multiset of internal state values after encrypting several rounds are considered. The integral distinguisher in the best attack we found covers 10 rounds that can be turned into a key-recovery attack on 14 rounds. The division property could be used to slightly extend those results. Again, given the number of rounds for SKINNY, integral attacks do not seem to be a threat for the security of the cipher.

Slide Attacks. In SKINNY, the distinction between the rounds of the cipher is ensured by the AddConstants operation and thus the straightforward slide attacks cannot be applied. We consider possible variants (in the full version of the paper), but we could not turn any of those into a valid attack.

Subspace Cryptanalysis. Invariant subspace cryptanalysis makes use of affine subspaces that are invariant under the round function. Given that SKINNY has a non-trivial key-scheduling, this technique does not seem well suited to launch an attack.

Algebraic Attacks. We detail in the full version of our paper why, not surprisingly, algebraic attacks do not threaten SKINNY.

5   Implementations, Performance and Comparison

We provide a complete study of SKINNY performance on various platforms (software, ASIC, FPGA, micro-controllers, ...) in the full version of the paper. Yet, we describe here our results regarding ASIC round-based implementations since it represents our top performance criterion.

We used Synopsys DesignCompiler version A-2007.12-SP1 to synthesize the designs considering UMCL18G212T3 [35] standard cell library, which is based on the UMC L180 $0.18\mu m$ 1P6M logic process with a typical voltage of 1.8 V. For the synthesis, we advised the compiler to keep the hierarchy and use a clock frequency of 100 KHz, which allows a fair comparison with the benchmark of other block ciphers reported in literature.

In a first step, we designed round-based implementations for all SKINNY variants providing a good trade-off between performance and area. All implementations compute a single round of SKINNY within a clock cycle. Besides, our designs take advantage of dedicated scan flip-flops rather than using simple flip-flops and additional multiplexers placed in front in order to hold round states and keys. Note that this approach leads to savings of 1 GE per bit to be stored. In order to allow a better and fairer comparison, we provide both throughput at a maximally achievable frequency and throughput at a frequency of 100 KHz.

Table 8 briefly summarizes the results of the round-based architectures of all SKINNY variants and compares it to other round-based implementations of lightweight ciphers taken from the literature. In particular, SKINNY-64-128 offers the smallest area footprint compared to other lightweight ciphers providing the same security level. Note, that even SIMON-64-128 implemented in a round-based fashion cannot compete with our design in terms of both area and throughput.

6   The Low-Latency Tweakable Block Cipher MANTIS

In this section, we present a tweakable block cipher design which is optimized for low-latency implementations.

The low-latency block cipher PRINCE already provides a very good starting point for a low-latency design. Its round function basically follows the AES structure, with the exception of using a MixColumns-like mapping with branch number 4 instead of 5. The main difference between PRINCE and AES (and actually all other ciphers) is that the design is symmetric around a linear layer in the middle. This allows to realize what was coined α-reflection: the decryption for a key K corresponds (basically) to encryption with a key $K \oplus \alpha$ where α is a

Table 8. Round-based implementations of SKINNY-64 and SKINNY-128.

	Area	Delay	Clock Cycles	Throughput @100 KHz	@maximum	Ref.
	GE	ns	#	KBit/s	MBit/s	
SKINNY-64-64	1223	1.77	32	200.00	1130.00	**New**
SKINNY-64-128	1696	1.87	36	177.78	951.11	**New**
SKINNY-64-192	2183	2.02	40	160.00	792.00	**New**
SKINNY-128-128	2391	2.89	40	320.00	1107.20	**New**
SKINNY-128-256	3312	2.89	48	266.67	922.67	**New**
SKINNY-128-384	4268	2.89	56	228.57	790.86	**New**
SIMON-64-128	1751	1.60	46	145.45	870.00	[2]
SIMON-128-128	2342	1.60	70	188.24	1145.00	[2]
SIMON-128-256	3419	1.60	74	177.78	1081.00	[2]
LED-64-64	2695	-	32	198.90	-	[15]
LED-64-128	3036	-	48	133.00	-	[15]
PRESENT-64-128	1884	-	32	200.00	-	[5]
PICCOLO-64-128	1773[a]	-	33	193.94	-	[31]

[a]This number includes 576 GE for key storage that is not considered in the original work.

fixed constant. Turning PRINCE into a tweakable block cipher is (conceptually) well understood when using e.g. the TWEAKEY framework [17]. First, define a tweakey-schedule and than simply increase the number of rounds until one can ensure that the cipher is secure against related-tweak attacks.

However, the problem is that the latency of a cipher is directly related to the number of rounds. Thus, it is crucial to find a design, i.e. a round function and a tweak-scheduling, that ensures security already with a minimal number of rounds. Here, components of the recently proposed block ciphers MIDORI [1] turn out to be very beneficial. In MIDORI, again an AES-like design, one of the key observations was that changing ShiftRows into a more general permutation allows to significantly improve upon the number of active Sboxes (in the single key model) while keeping a MixColumns-like layer with branch number 4 only. On top, the designers of MIDORI designed a 4-bit Sbox that was optimized with respect to circuit-depth. This directly leads to an improved version of PRINCE itself: replace the PRINCE round function by the MIDORI-round function while keeping the entire design symmetric around the middle to keep the α-reflection property. This simple change would result in a cipher with improved latency and improved security (i.e. number of active Sboxes) compared to PRINCE. It is actually exactly this PRINCE-like MIDORI that we use as a starting point for designing the low-latency block cipher MANTIS. The final step in the design of MANTIS was to find a suitable tweak-scheduling that would ensure a high number of active Sboxes not only in the single-key setting, but also in the setting where the attacker can control the difference in the tweak. Using, again, the MILP approach, we are able to demonstrate that a slight increase in the number of rounds (from 12 to 14) is already sufficient to ensure the resistance of MANTIS to differential (and linear) attacks in the related-tweak setting. Note that MANTIS is certainly not secure in the related-key model, as there always exist a probability one distinguisher caused by the α-reflection property.

MANTIS$_r$ has a 64-bit block length and works with a 128-bit key and 64-bit tweak. The parameter r specifies the number of rounds of one half of the cipher. The overall design is illustrated in Fig. 5.

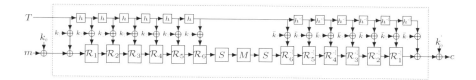

Fig. 5. Illustration of MANTIS$_6$.

We acknowledge the contribution of Roberto Avanzi to the design of MANTIS. He first suggested us to combine PRINCE with the TWEAKEY framework, and also to modify the latter by permuting the tweak independently from the key, in order to save on the Galois multiplications of the tweak cells. He then brainstormed with us on early versions of the design.

6.1 Description of the Cipher

MANTIS$_r$ is based on the FX-construction [20] and thus applies whitening keys before and after applying its core components. The 128-bit key is first split into $k = k_0 \parallel k_1$ with 64-bit subkeys k_0, k_1. Then, $(k_0 \parallel k_1)$ is extended to the 192 bit key

$$(k_0 \parallel k_0' \parallel k_1) := (k_0 \parallel (k_0 \ggg 1) \oplus (k_0 \gg 63) \parallel k_1),$$

and k_0, k_0' are used as whitening keys in an FX-construction. The subkey k_1 is used as the round key for all of the $2r$ rounds of MANTIS$_r$. We decided to stick with the FX construction for simplicity., even so other options as described in [8].

Initialization. The cipher receives a plaintext $m = m_0 \| m_1 \| \cdots \| m_{14} \| m_{15}$, where the m_i are 4-bit cells. The initialization of the cipher's internal state is performed by setting $IS_i = m_i$ for $0 \le i \le 15$.

The cipher also receives a tweak input $T = t_0 \| t_1 \| \cdots \| t_{15}$, where the t_i are 4-bit cells. The initialization of the cipher's tweak state is performed by setting

$T_i = t_i$ for $0 \leq i \leq 15$. Thus,

$$IS = \begin{bmatrix} m_0 & m_1 & m_2 & m_3 \\ m_4 & m_5 & m_6 & m_7 \\ m_8 & m_9 & m_{10} & m_{11} \\ m_{12} & m_{13} & m_{14} & m_{15} \end{bmatrix} \qquad T = \begin{bmatrix} t_0 & t_1 & t_2 & t_3 \\ t_4 & t_5 & t_6 & t_7 \\ t_8 & t_9 & t_{10} & t_{11} \\ t_{12} & t_{13} & t_{14} & t_{15} \end{bmatrix}$$

The Round Function. One round $\mathcal{R}_i(\cdot, tk)$ of MANTIS$_r$ operates on the cipher internal state depending on the round tweakey tk as

MixColumns $\circ$ PermuteCells $\circ$ AddTweakey$_{tk}$ $\circ$ AddConstant$_i$ $\circ$ SubCells.

In the following, we describe the components of the round function.

SubCells. The involutory MIDORI Sbox Sb$_0$ is applied to every cell of the internal state. Using the MIDORI Sbox is beneficial as this Sbox is especially optimized for small area and low circuit depth.

AddConstant. In the i-th round, the round constant RC_i is XORed to the internal state. The round constants are generated in a similar way as for PRINCE, that is we used the first digits of π to generate those constants (actually the very first digits correspond to α defined below). The round constants can be found in the full version of the paper. Note that, in contrast to PRINCE, the constants are added row-wise instead of column-wise.

AddRoundTweakey. In round $\mathcal{R}_i$, the (full) round tweakey state $h^i(T) \oplus k_1$ is XORed to the cipher internal state. In the i-th inverse round $\mathcal{R}_i^{-1}$, the tweakey state $h^i(T) \oplus \bar{k}_1 := h^i(T) \oplus k_1 \oplus \alpha$ with $\alpha =$ 0x243f6a8885a308d3 is XORed to the internal state. Note that this α, as the round constants, is chosen as the first digits of π. Thereby, it is $h(T) = t_{h(0)} \| t_{h(1)} \cdot \| t_{h(15)}$, where the tweak permutation h is defined as

$$h = [6, 5, 14, 15, 0, 1, 2, 3, 7, 12, 13, 4, 8, 9, 10, 11].$$

PermuteCells. The cells of the internal state are permuted according to the MIDORI permutation

$$P = [0, 11, 6, 13, 10, 1, 12, 7, 5, 14, 3, 8, 15, 4, 9, 2].$$

Note that the MIDORI permutation ensures a higher number of active Sboxes compared to the choice made in PRINCE.

MixColumns. Each column of the cipher internal state array is multiplied by the binary matrix used in MIDORI.

Encryption. In the following, we define H_r as the application of r rounds $\mathcal{R}_i$ and one additional SubCells layer. Similarly, we define H_r^{-1} as the application on one inverse SubCells layer plus r inverse rounds. Thus,

$$H_r(\cdot, T, k_1) = \text{SubCells} \circ \mathcal{R}_r(\cdot, h^r(T) \oplus k_1) \circ \cdots \circ \mathcal{R}_1(\cdot, h(T) \oplus k_1)$$
$$H_r^{-1}(\cdot, T, \bar{k}_1) = \mathcal{R}_1^{-1}(\cdot, h(T) \oplus \bar{k}_1) \circ \cdots \circ \mathcal{R}_r^{-1}(\cdot, h^r(T) \oplus \bar{k}_1) \circ \text{SubCells}.$$

With this notation, it is

$$\mathbf{Enc}_{(k_0, k_0', k_1)}(\cdot, T) = \texttt{AddTweakey}_{k_0' \oplus k_1 \oplus \alpha \oplus T} \circ H_r^{-1}(\cdot, T, k_1 \oplus \alpha)$$

$$\circ \texttt{MixColumns} \circ H_r(\cdot, T, k_1) \circ \texttt{AddTweakey}_{k_0 \oplus k_1 \oplus T}$$

Decryption. It is $\mathbf{Enc}^{-1}_{(k_0, k_0', k_1)}(\cdot, T) = \mathbf{Enc}_{(k_0', k_0, k_1 \oplus \alpha)}(\cdot, T)$ because of the α-reflection property.

6.2 Design Rationale

The goal was to design a cipher which is competitive to PRINCE in terms of latency with the advantage of being tweakable. In contrast to SKINNY, we distinguish between tweak and key input. In particular, we allow an attacker to control the tweak but not the key. Thus, similar to PRINCE, we do not claim related-key security. In order to reach this goal, again, several components are borrowed from already existing ciphers. In the following, we present the reasons for our design. Note that, as we aim for an efficient unrolled implementation, one is not restricted to a classical round-iterated design.

α-Reflection Property. $\texttt{MANTIS}_r$ is designed as a reflection cipher such that encryption under a key k equals decryption under a related key. This significantly reduces the implementation overhead for decryption. Therefore, the parameter r denotes only half the number of rounds, as the second half of the cipher is basically the inverse of the first half. It is advantageous that the diffusion matrix $\mathbf{M}$ is involutory since we need the middle part of the cipher to be an involution. Unlike in the description of PRINCE, we use the same round constant for the inverse $\mathcal{R}_i^{-1}$ of the i-th round and apply the addition of α to the round key k_1.

The Choice of the Diffusion Layer. To achieve low latency in an unrolled implementation, one is limited in the number rounds to be applied. Therefore, one has to achieve very fast diffusion while guaranteeing a high number of active Sboxes. To reach these requirements, we adopted the linear layer of MIDORI. It provides full diffusion only after three rounds and guarantees a high number of active Sboxes in the single-key setting. We refer to Table 4 for the bounds.

The Choice of the Sbox. For the Sbox in MANTIS we used the same Sbox as in MIDORI. The MIDORI Sbox has a significantly smaller latency than the PRINCE Sbox. The maximal linear bias is 2^{-2} and the best differential probability is 2^{-2} as well.

The Choice of the Tweakey Permutation h. Our aim was to choose a tweak permutation h such that five rounds (plus one additional SubCells layer) guarantee at least 16 active Sboxes in the related-tweak setting. This would guarantee at least 32 active Sboxes for $\texttt{MANTIS}_5$ which is enough to bound the differential probability (resp. linear bias) below $2^{-2.32}$. Since there are 16! possibilities for h, which is too much for an exhaustive search, we restricted ourself on a subclass of 8! tweak permutations. The restriction is that two complete rows (without changing the position of the cells in those rows) are permuted to different rows. In our case, the first and third row are permuted to the second and fourth row, respectively. The bounds were derived using the MILP tool. We tested several thousand choices for the permutation h and found out that 16 active Sboxes were the best possible to reach over H_5. Out of these optimal choices, we took the permutation that maximized the bound for MANTIS$_5$, and as a second step for MANTIS$_6$. We refer to Table 9 for the actual bounds.

Table 9. Lower bounds on the number of linear (and differential) active Sboxes in the single-key model and of differential active Sboxes in the related-tweak model.

	MANTIS$_2$	MANTIS$_3$	MANTIS$_4$	MANTIS$_5$	MANTIS$_6$	MANTIS$_7$	MANTIS$_8$
Linear	14	32	46	62	70	76	82
Rel. Tweak	6	12	20	34	44	50	56

Security Claim. For MANTIS$_7$, we claim that any adversary who in possession of 2^n chosen plain/ciphertext pairs which were obtained under chosen tweaks, but with a fixed unknown key, needs at least 2^{126-n} calls to the encryption function in order to recover the secret key. Thus, our security claims are the same as for PRINCE, except that we also claim related-tweak security. Moreover, already for MANTIS$_5$ we claim security against practical attacks, similar to what has been considered in the PRINCE challenge. More precisely, we claim that no related-tweak attack (better than the generic claim above) is possible against MANTIS$_5$ with less than 2^{30} chosen or 2^{40} known plaintext/ciphertext pairs. Note that because of the α-reflection, there exists a trivial related-key distinguisher with probability one. We especially encourage further cryptanalysis on the aggressive versions.

6.3 Security Analysis

As one round of MANTIS is almost identical to one round in MIDORI, most of the security analysis can simply be copied from the latter. This holds in particular for meet-in-the-middle attacks, integral attacks and slide attacks. We therefore only focus on the attacks where the changes in round constants and by adding the tweak actually result in different arguments.

Invariant Subspaces. The most successful attack against MIDORI-64 at the moment is an invariant subspace attack with a density of 2^{96} weak keys. The main observation here is that the round constants in MIDORI are too sparse and structured to avoid certain symmetries. More precisely, the round constants in MIDORI-64 only affect a single bit in each of the 16 4-bit cells. Together with a property of the Sbox this finally results in the mentioned attack. For MANTIS, the situation is very different as the round constants (in each half) are basically random values. This in particular ensures that the invariant subspace attack on MIDORI does not translate into an attack on MANTIS.

Differential and Linear Related-Tweak Attacks. Using the MILP approach, we are able to prove strong bounds against related-tweak linear and differential attacks. In particular, no related tweak linear or differential distinguisher based on a characteristics is possible for MANTIS$_5$, that is already for 12 layers of Sboxes. As MANTIS$_7$ has four more rounds, and additional key-whitening, we believe that is provides a small but sufficient security margin.

The results of unrolled implementations for MANTIS are listed in the full version of the paper.

Acknowledgements. The authors would like to thank the anonymous referees for their helpful comments. This work is partly supported by the Singapore National Research Foundation Fellowship 2012 (NRF-NRFF2012-06), the DFG Research Training Group GRK 1817 Ubicrypt and the BMBF Project UNIKOPS (01BY1040).

References

1. Banik, S., Bogdanov, A., Isobe, T., Shibutani, K., Hiwatari, H., Akishita, T., Regazzoni, F.: Midori: a block cipher for low energy. In: Iwata, T., Cheon, J.H. (eds.) ASIACRYPT 2015. LNCS, vol. 9453, pp. 411–436. Springer, Heidelberg (2015). doi:10.1007/978-3-662-48800-3_17
2. Beaulieu, R., Shors, D., Smith, J., Treatman-Clark, S., Weeks, B., Wingers, L.: Simon and speck: block ciphers for the internet of things. ePrint/2015/585 (2015)
3. Benadjila, R., Guo, J., Lomné, V., Peyrin, T.: Implementing lightweight block ciphers on x86 architectures. In: Lange, T., Lauter, K., Lisoněk, P. (eds.) SAC 2013. LNCS, vol. 8282, pp. 324–352. Springer, Heidelberg (2014)
4. Biham, E., Biryukov, A., Shamir, A.: Cryptanalysis of skipjack reduced to 31 rounds using impossible differentials. In: Stern, J. (ed.) EUROCRYPT 1999. LNCS, vol. 1592, pp. 12–23. Springer, Heidelberg (1999)
5. Bogdanov, A.A., Knudsen, L.R., Leander, G., Paar, C., Poschmann, A., Robshaw, M., Seurin, Y., Vikkelsoe, C.: PRESENT: an ultra-lightweight block cipher. In: Paillier, P., Verbauwhede, I. (eds.) CHES 2007. LNCS, vol. 4727, pp. 450–466. Springer, Heidelberg (2007)
6. Bogdanov, A., Rechberger, C.: A 3-subset meet-in-the-middle attack: cryptanalysis of the lightweight block cipher KTANTAN. In: Biryukov, A., Gong, G., Stinson, D.R. (eds.) SAC 2010. LNCS, vol. 6544, pp. 229–240. Springer, Heidelberg (2011)
7. Borghoff, J., Canteaut, A., Güneysu, T., Kavun, E.B., Knezevic, M., Knudsen, L.R., Leander, G., Nikov, V., Paar, C., Rechberger, C., Rombouts, P., Thomsen, S.S., Yalçın, T.: PRINCE – a low-latency block cipher for pervasive computing applications. In: Wang, X., Sako, K. (eds.) ASIACRYPT 2012. LNCS, vol. 7658, pp. 208–225. Springer, Heidelberg (2012)

8. Boura, C., Canteaut, A., Knudsen, L.R., Leander, G.: Reflection ciphers. In: Designs, Codes and Cryptography (2015)
9. De Cannière, C., Dunkelman, O., Knežević, M.: KATAN and KTANTAN — a family of small and efficient hardware-oriented block ciphers. In: Clavier, C., Gaj, K. (eds.) CHES 2009. LNCS, vol. 5747, pp. 272–288. Springer, Heidelberg (2009)
10. Canteaut, A., Duval, S., Leurent, G.: Construction of lightweight S-Boxes using Feistel and MISTY structures (Full Version). ePrint/2015/711 (2015)
11. Chaum, D., Evertse, J.-H.: Cryptanalysis of DES with a reduced number of rounds. In: Williams, H.C. (ed.) CRYPTO 1985. LNCS, vol. 218, pp. 192–211. Springer, Heidelberg (1986)
12. Daemen, J., Knudsen, L.R., Rijmen, V.: The block cipher SQUARE. In: Biham, E. (ed.) FSE 1997. LNCS, vol. 1267, pp. 149–165. Springer, Heidelberg (1997)
13. Daemen, J., Peeters, M., Assche, G.V., Rijmen, V.: Nessie proposal: the block cipher noekeon. Nessie submission (2000). http://gro.noekeon.org/
14. Gérard, B., Grosso, V., Naya-Plasencia, M., Standaert, F.-X.: Block ciphers that are easier to mask: how far can we go? In: Bertoni, G., Coron, J.-S. (eds.) CHES 2013. LNCS, vol. 8086, pp. 383–399. Springer, Heidelberg (2013)
15. Guo, J., Peyrin, T., Poschmann, A., Robshaw, M.J.B.: The LED block cipher. [29], pp. 326–341
16. Henson, M., Taylor, S.: Memory encryption: a survey of existing techniques. ACM Comput. Surv. 46(4), 1–53 (2013)
17. Jean, J., Nikolic, I., Peyrin, T.: Tweaks and keys for block ciphers: the TWEAKEY framework. In: Sarkar, P., Iwata, T. (eds.) ASIACRYPT 2014. LNCS, vol. 8874, pp. 274–288. Springer, Heidelberg (2014)
18. Jean, J., Nikolić, I., Peyrin, T.: Joltik v1.3 Submission to the CAESAR competition (2015). http://www1.spms.ntu.edu.sg/~syllab/Joltik
19. Khoo, K., Peyrin, T., Poschmann, A.Y., Yap, H.: FOAM: searching for hardware-optimal SPN structures and components with a fair comparison. In: Batina, L., Robshaw, M. (eds.) CHES 2014. LNCS, vol. 8731, pp. 433–450. Springer, Heidelberg (2014)
20. Kilian, J., Rogaway, P.: How to protect DES against exhaustive key search. In: Koblitz, N. (ed.) CRYPTO 1996. LNCS, vol. 1109, pp. 252–267. Springer, Heidelberg (1996)
21. Knudsen, L.R., Wagner, D.: Integral cryptanalysis. In: Daemen, J., Rijmen, V. (eds.) FSE 2002. LNCS, vol. 2365, pp. 112–127. Springer, Heidelberg (2002)
22. Kölbl, S., Leander, G., Tiessen, T.: Observations on the SIMON block cipher family. In: Gennaro, R., Robshaw, M.J.B. (eds.) CRYPTO 2015. LNCS, vol. 9215, pp. 161–185. Springer, Heidelberg (2015)
23. Kranz, T., Leander, G., Wiemer, F.: Linear cryptanalysis: on key schedules and tweakable block ciphers. Preprint (2016)
24. Moradi, A., Poschmann, A., Ling, S., Paar, C., Wang, H.: Pushing the limits: a very compact and a threshold implementation of AES. In: Paterson, K.G. (ed.) EUROCRYPT 2011. LNCS, vol. 6632, pp. 69–88. Springer, Heidelberg (2011)
25. Mouha, N., Wang, Q., Gu, D., Preneel, B.: Differential and linear cryptanalysis using mixed-integer linear programming. In: Wu, C.-K., Yung, M., Lin, D. (eds.) Inscrypt 2011. LNCS, vol. 7537, pp. 57–76. Springer, Heidelberg (2012)
26. National Institute of Standards and Technology: Recommendation for Key Management - NIST SP-800-57 Part 3 Revision 1. http://nvlpubs.nist.gov/nistpubs/SpecialPublications/NIST.SP.800-57Pt3r1.pdf
27. Peyrin, T., Seurin, Y.: Counter-in-Tweak: authenticated encryption modes for tweakable block ciphers. ePrint/2015/1049 (2015)

28. Piret, G., Roche, T., Carlet, C.: PICARO – a block cipher allowing efficient higher-order side-channel resistance. In: Bao, F., Samarati, P., Zhou, J. (eds.) ACNS 2012. LNCS, vol. 7341, pp. 311–328. Springer, Heidelberg (2012)

29. Preneel, B., Takagi, T. (eds.): CHES 2011. LNCS, vol. 6917. Springer, Heidelberg (2011)

30. Sasaki, Y.: Meet-in-the-Middle preimage attacks on AES hashing modes and an application to whirlpool. In: Joux, A. (ed.) FSE 2011. LNCS, vol. 6733, pp. 378–396. Springer, Heidelberg (2011)

31. Shibutani, K., Isobe, T., Hiwatari, H., Mitsuda, A., Akishita, T., Shirai, T.: Piccolo: an ultra-lightweight blockcipher. [29], pp. 342–357

32. Sun, S., Hu, L., Song, L., Xie, Y., Wang, P.: Automatic security evaluation of block ciphers with S-bP structures against related-key differential attacks. In: Lin, D., Xu, S., Yung, M. (eds.) Inscrypt 2013. LNCS, vol. 8567, pp. 39–51. Springer, Heidelberg (2014)

33. Suzaki, T., Minematsu, K., Morioka, S., Kobayashi, E.: TWINE: a lightweight block cipher for multiple platforms. In: Wu, H., Knudsen, L.R. (eds.) SAC 2012. LNCS, vol. 7707, pp. 339–354. Springer, Heidelberg (2013)

34. Grosso, V., Leurent, G., Standaert, F.-X., Varici, K., Journault, A., Durvaux, F., Gaspar, L., Kerckhof, S.: SCREAM v3 Submission to the CAESAR competition (2015)

35. Virtual Silicon Inc: 0.18 μm VIP Standard Cell Library Tape Out Ready, Part Number: UMCL18G212T3, Process: UMC Logic 0.18 μm Generic II Technology: 0.18μm, July 2004

36. Williams, P., Boivie, R.: CPU support for secure executables. In: McCune, J.M., Balacheff, B., Perrig, A., Sadeghi, A.-R., Sasse, A., Beres, Y. (eds.) Trust 2011. LNCS, vol. 6740, pp. 172–187. Springer, Heidelberg (2011)

Cryptanalytic Tools

Automatic Search of Meet-in-the-Middle and Impossible Differential Attacks

Patrick Derbez[1(✉)] and Pierre-Alain Fouque[1,2]

[1] Université Rennes 1 / IRISA, Rennes, France
{patrick.derbez,pierre-alain.fouque}@irisa.fr
[2] Institut Universitaire de France, Paris, France

Abstract. Tracking bits through block ciphers and optimizing attacks at hand is one of the tedious task symmetric cryptanalysts have to deal with. It would be nice if a program will automatically handle them at least for well-known attack techniques, so that cryptanalysts will only focus on finding new attacks. However, current automatic tools cannot be used as is, either because they are tailored for specific ciphers or because they only recover a specific part of the attacks and cryptographers are still needed to finalize the analysis.

In this paper we describe a generic algorithm exhausting the best meet-in-the-middle and impossible differential attacks on a very large class of block ciphers from byte to bit-oriented, SPN, Feistel and Lai-Massey block ciphers. Contrary to previous tools that target to find the best differential / linear paths in the cipher and leave the cryptanalysts to find the attack using these paths, we automatically find the best attacks by considering the cipher and the key schedule algorithms. The building blocks of our algorithm led to two algorithms designed to find the best simple meet-in-the-middle attacks and the best impossible truncated differential attacks respectively. We recover and improve many attacks on AES, mCRYPTON, SIMON, IDEA, KTANTAN, PRINCE and ZORRO. We show that this tool can be used by designers to improve their analysis.

Keywords: Automatic search · Meet-in-the-middle · Impossible truncated differential · Cryptanalysis

1 Introduction

To explore the exponential space of differentials or linears characteristics, cryptanalysts usually implement some algorithms. Many tools have been proposed for ciphers or hash functions [BDF11, DF13, FJP13, Leu12] but most of the time they are not publicly available. Moreover, they are not very convenient for block

Patrick Derbez was partially supported by the CORE ACRYPT project from the *Fond National de Recherche* (Luxembourg).

© International Association for Cryptologic Research 2016
M. Robshaw and J. Katz (Eds.): CRYPTO 2016, Part II, LNCS 9815, pp. 157–184, 2016.
DOI: 10.1007/978-3-662-53008-5_6

cipher designers and are rarely used for many reasons. On the one hand, some tools have been designed to explore precise ciphers and it is not easy to adapt them for new designs. The main reason is that we hope that taking into account some particularities of the cipher, would lead to more efficient attacks. Consequently, some details of the analyzed ciphers are hard-coded into the tool and it is not easy to make any changes. On the other hand, only cryptanalysts can used such tools which are more computational-aid than real tools. Indeed, some tools allow to find some differential paths, but more work has to be done by cryptanalysts to find the best attack. However, this last part is usually not completely trivial and it is not always the best differential paths, that would lead to the best attacks. For instance, the best differential attack on DES does not use the best and longest differential path [BS93] on 15 rounds, but a 13 rounds differential path is used with meet-in-the-middle technique to extend this path, leading to the so-called 3R attack. The meet-in-the-middle step is rarely considered in tools while it is computationally difficult to exhaust the most efficient combination of say a differential path with the number of guesses. Indeed, once a differential path is found, attackers have to guess some key bits in order to be able to check the differential part. Consequently, the overall complexity of the attack depends on the number of guesses and the probability of the differential. The best attack has a complexity that is the maximum of both stages. The last step is a meet-in-the-middle technique and it is well-known that it allows to find the most efficient attack since bad key guesses are quickly rejected. As a conclusion, if we want to automatically find the best attack, we need to be able to automatically solve each stage: find many good differential paths and for each of them evaluate the cost of the meet-in-the-middle part.

Automatic Tools. Automatic search of symmetric attacks boils down to solving a system of equations in many variables as Shannon described in his seminal work in 1949 on Communication Theory of Secrecy Systems: *Breaking a good cipher should require as much work as solving a system of simultaneous equations in a large number of unknowns of a complex type.* Algebraic cryptanalysis can be traced back to him and some attacks on stream ciphers have been very efficient [CM03]. However, solving these equations is not always easy and cryptanalysts have to take into account the structure of such systems if they want to efficiently solve them. Indeed, Gröbner basis algorithms have been used, but they never endanger the security of real block ciphers [BPW06a,BPW06b]. Cryptographers have to closely analyze the involved systems depending on the number of variables, their degree, some symmetries in the equations if they want to find some attacks. The other well-known tool consists in writing boolean equations and feed them to a SAT solver such as CryptoMiniSAT [SNC09]. Black box use of these two well-known solvers never lead to efficient attacks. They can be used either on a very small number of rounds [MS13] or when attacks are described in order to speed up the search [SKPI07,MZ06]. Since solving a polynomial system of equations in many variables is a NP-hard problem [GJ79], some cryptanalysts try to better take into account the structure of these systems. Since block ciphers are built iteratively in many rounds and each of them use a linear part (diffusion) and a non-linear part (Sbox essentially), one of the most interesting research directions consists in writing linear equations by adding new variables for each Sbox [BDF11,KBN09]. Consequently, we can write the system as a linear system in variables x and $S(x)$, where S is treated as an inert function. Such systems are not easy to solve because there is a relation between these two kinds of variables and classical gaussian technique does not work. In order to consider the system of equations, Bouillaguet *et al.* in [BDF11] have used well-known cryptographic techniques to solve such systems such as guess-and-determine and meet-in-the-middle. The consequence is that the tool is very versatile and solves any such systems by describing an algorithm to solve it with its average time/memory complexity. For instance, they were able to find attacks on MAC and stream-ciphers. However, it is not specific for block ciphers and it is not easy to search attacks involving many messages. This tool is nevertheless interesting since it is generic and for instance, Derbez and Fouque use it in [DF13] and Dinur and Jean in [DJ14].

Related Work. The original meet-in-the-middle attack [DS08] of Demirci and Selçuk against AES has been improved and generalized by many researchers and is nowadays well-understood. It relies on particular sets called δ-sets, which were first introduced by Daemen *et al.* against the block cipher SQUARE. At ASIACRYPT 2010, Dunkelman *et al.* [DKS10] presented several improvements for the attack including the *differential enumeration technique*, a clever and powerful memory/data trade-off that does not change the time. Then at EURO-CRYPT 2013, Derbez *et al.* [DFJ13] mainly showed that this technique leads to much better attacks than expected by Dunkelman *et al.*, and reached the best known attacks against 7-round AES-128 and 9-round AES-256 in the single-key model. Next, at FSE 2013, Derbez and Fouque [DF13] generalized the attack of Demirci and Selçuk by searching a match on some equation and not only on the byte state, and showed that approximately 2^{16} different attacks can be mounted against the AES. In order to find the best ones among them, they used the tool presented by Bouillaguet *et al.* [BDF11] at CRYPTO 2011 to take care of the key schedule relations between the subkey bytes involved in the attacks.

This kind of attacks is very efficient againsts round-reduced versions of the AES and actually it may also be efficient against non-SPN ciphers as showed by Li and Jia in [LJ14] where they successfully applied the technique against Camellia [AIK+00]. At ICISC 2013, Li *et al.* [LWWZ13] described an algorithm to find the best distinguishers one can use to mount a Demirci-Selçuk attack on a word-oriented block cipher. In particular, they showed that finding the distinguishers which have the least number of active cells can be turned into an integer linear program that they solved. As a result, they found new attacks against both Crypton-128 and mCrypton-96.

Our Contribution. Our first contribution is a new tool that allows to automatically find meet-in-the-middle and impossible differential attack. Contrary to other tools, this new one is publicly available and allows to recover differential paths and complete attacks by extending them using well-known meet-in-the-middle technique. One major contribution is that we determine specific problems that allow us to design a modular approach for our tool. Indeed, we will describe some building blocks that allow us to automatically find impossible differential attack, truncated differential path and meet-in-the-middle attacks when we combine them in a specific manner. Finally, we apply it on many block ciphers.

We show that our tool allows to discover automatically in a few seconds many of the best meet-in-the-middle and impossible differential attacks on some bit and byte oriented ciphers: CRYPTON, mCRYPTON, AES, SIMON, IDEA and XTEA. On SIMON, the tool allows to recover all the attacks found by hand by Boura *et al.* in [BNS14] and even improve them by one more round. Essentially, the tool was able to discover that we can save some guesses by guessing the xor of two key bits instead of each of the two bits. For IDEA, our results are noteworthy and we think it is a good example of bit-oriented cipher since it mixes various operations which prevent to use any larger field as in AES. This cipher has been unattacked during 10 years after its publications and in 2002, Biryukov and Demirci discovered a particular relation that allows them to break 2 rounds among the 8.5 rounds. About 10 years after, Biham, Dunkelman, Keller and Shamir use this relation to mount efficient meet-in-the-middle attacks [BDKS15]. In a few seconds, our tool was able to automatically recover the Biryukov-Demirci relation and to find all the attacks on 6 rounds [BDKS15]. On XTEA, the tool was also able to recover the best impossible differential path of [MHL+02] on 12 rounds. If we only want to recover differential path and not the complete attack, it is possible to ask it to the tool.

The main purpose of this tool is not only for cryptanalysis in order to find attacks, but also for designers in order to test their new ciphers. The ZORRO block cipher has been proposed by Gérard *et al.* at CHES 2013 [GGNS13] in order to be secure and efficient to mask. The main idea consists in using an easy to mask Sbox and to reduce the number of Sbox at each round since the overhead of masking comes from these two factors. The overall design is close to AES. However, many attacks have been discovered on this cipher including on the full number of rounds. Here, we exhaust using symmetries properties all the family of ZORRO ciphers and we show that some strategic positions of the Sboxes lead to stronger ciphers.

We describe a generic algorithm exhausting the best meet-in-the-middle and impossible differential attacks on a very large class of block ciphers. Unlike Li *et al.*'s algorithm, our is not restricted to word-oriented block ciphers and takes into account the key schedule relations to directly give as output the best attacks and their complexities. Actually, it is based on the tool of Bouillaguet *et al.* to estimate the complexity of the attacks. Thus our algorithm only requires as input a system of equations describing the targeted cipher and the type of each variable: plaintext, ciphertext, key or state. Incidentally, the building blocks of our algorithm led to two others algorithms designed to find the best simple meet-in-the-middle attacks and the best impossible truncated differential attacks respectively. Impossible differential cryptanalysis, which was simultaneously introduced by Knudsen [Knu98] and Biham *et al.* [BBS99], is a powerful technique against a large variety of block ciphers. While there are already algorithms designed to find impossible differential against various kind of block ciphers (for instance [WW12]), our is the first one which outputs the complexities of the best attacks. More precisely, our algorithm gives as output all the parameters required to compute the complexity according to the general formula given at ASIACRYPT 2014 by Boura *et al.* in [BNS14].

We implemented our algorithms in C++ and make them available at:

https://bitbucket.org/pderbez/crypto2016-tool/.

2   Preliminaries

First we present a generalization of the Demirci-Selçuk (GDS) meet-in-the-middle technique for iterated block ciphers. Then, we recall some definitions from Bouillaguet *et al.* [BDF11] about systems of AES-like equations.

The GDS attack is similar to the splice-and-cut technique [WRG+11] but works with differences rather than state values. Demirci-Selçuk attacks have been discovered for AES and first generalized by Derbez and Fouque in [DF13] to match on a byte relation involving many bytes rather than on one state byte. Here, we generalize it on iterated block ciphers.

2.1   Generalized Demirci-Selçuk (GDS) Attack

We illustrate GDS on an AES-like cipher and then we generalize it to other ciphers. The basic idea is the following and assume that we have a relation involving internal variables. It can be a linear relation between 5 active bytes in an AES computation around the MixColumn operation. On the second line of Fig. 1, we represent such a relation between two states. Once, the variables of this relation have been identified, we propagate them to the plaintext and ciphertext bits and we get, the bits that have to be guessed in the intermediate states from the ciphertext and plaintext. The main problem is that the number of bits that have to be guessed is very large as in the figure. The main trick to reduce them is to force some constraints on the differential path. They are described by the first line, where some conditions are proposed. We will search for plaintexts satisfying the differential path. It is classical in AES cryptanalysis to use the differential path with one transition from one byte to 4 active bytes after the MixColumn operation with probability one and we let it propagate to the plaintext and ciphertext part with probability one. Finally, we get a GDS attack on the third line that use the bytes in the intersection of the two sets used in each state.

More formally, the original attack of Demirci and Selçuk [DS08] mainly relies on two subcomponents: one truncated differential characteristic and one basic meet-in-the-middle attack. More precisely, let $E = E_3 \circ E_2 \circ E_1$ be an encryption function splitted into three parts. For the first step we pick a truncated difference Δ_X with b_i active bits, propagate it through E_1^{-1} (resp. $E_3 \circ E_2$) with probability 1 and denote the set of active bits by I_P (resp. I_C). Then, for the second step,

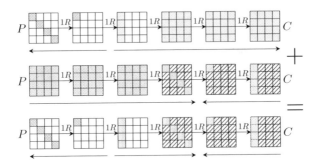

Fig. 1. Example of GDS attack (on 6-round AES). I_P is in blue, I_C in green, O_P in red and O_C in yellow. Hatched bytes play no roles and white bytes are constant. (Color figure online)

we mount a basic meet-in-the-middle attack against $E = E_3 \circ (E_2 \circ E_1)$: let Y be the output state of $E_2 \circ E_1$, we pick b_o bits of Y and denote by O_P (resp. O_C) the bits required to compute their difference in Y from the difference in the plaintexts (resp. ciphertexts).

To explain further the GDS attack we introduce the definition of a b-δ-set:

Definition 1 (b-δ-set). *A b-δ-set is a set of 2^b states such that b bits are active and take all the possible value while the others bits are constant. We also assume that the states of a b-δ-set are sorted according to differences (i.e. if $\{x^0, x^1, \ldots\}$ is a valid order then the valid orders are $\{x^i, x^{i+1}, \ldots\}$ for $0 \le i < 2^b$).*

The structure of the Generalized Demirci-Selçuk attack is the following:

- **Offline phase:**
  1. Consider the encryption of a b_i-δ-set $\{x^0, x^1, \ldots\}$ corresponding to the truncated difference Δ_X through E_2.
  2. Guess the value of $I_C \cap O_P$ for x^0.
  3. Deduce the differences in the b_o chosen bits of Y for the b_i-δ-set.
  4. Store them as a sequence of $2^{b_i} - 1$ b_o-bit values in a hash table.
- **Online phase:**
  1. Pick a plaintext P.
  2. Guess the value of I_P for P and identify a set $\{P, P^1, P^2, \ldots\}$ leading to a b_i-δ-set associated to Δ_X.
  3. Ask for the corresponding ciphertexts.
  4. Guess the value of O_C and partially decrypt the ciphertexts to compute the differences in the b_o chosen bits of Y.
  5. Check whether the sequence belongs to the hash table. If not, discard the guess.

The complexity of this procedure depends directly on how many values the sets I_P and $I_C \cap O_P$ can assume, $\mathcal{S}(I_C \cap O_P)$, and on how fast all the possible values of sets $I_P \cup O_C$ and $I_C \cap O_P$ can be enumerated, $\mathcal{T}(I_P \cup O_C)$:

- **Data:** $(2^{b_i} - 1) \cdot \mathcal{S}(I_P)$ adaptively chosen plaintexts,
- **Time (online):** $2^{b_i} \cdot \mathcal{T}(I_P \cup O_C)$ partial encryptions,
- **Memory:** $b_o \cdot (2^{b_i} - 1) \cdot \mathcal{S}(I_C \cap O_P)$ bits,
- **Time (offline):** $2^{b_i} \cdot \mathcal{T}(I_C \cap O_P)$ partial encryptions.

At the end of this attack we expect $\min(1, \mathcal{S}(I_C \cap O_P) \cdot 2^{-b_o(2^{b_i} - 1)}) \cdot \mathcal{S}(I_P \cup O_C)$ candidates to remain for $I_P \cup O_C$. Thus b_i and b_o have to be chosen such that they provide enough filtration, but expanding them also increases the size of the sets I_P, I_C, O_P and O_C which then may rise the complexity of the resulting attack.

Remarks:

- In the case where the truncated difference Δ_X does not fully active Δ_P, *i.e.* differences in some plaintext bits are null, the attack can be turned into a chosen-plaintext attack by starting by asking for a structure of plaintexts. Actually this is (almost) always better to do so since, in general, $(2^{b_i} - 1) \cdot \mathcal{S}(I_P)$ is higher than $2^{|\Delta_{P'}|}$.
- Some extra memory can be used to map each sequence to its corresponding value of $I_C \cap O_P$.
- Given two invertible matrices M_1 and M_2, we can rewrite the encryption function $E = (E_3 \circ M_2^{-1}) \circ (M_2 \circ E_2 \circ M_1^{-1}) \circ (M_1 \circ E_1)$. Hence the sentences "with b_i active bits" or "pick b_o bits of Y" should be understood as "with b_i active *linear combinations of* bits" or "pick b_o *linear combinations of* bits of Y".

2.2   Systems of AES-like Equations

In the sequel we recall some definitions of Bouillaguet *et al.* we will use in our algorithms. In particular, we detail the notion of linear variables that allows us to reduce variables. Indeed, in our system of equations that are linear in the variables x and $S(x)$, when all the equations only depend on $ax + bS(x)$ for specific value a and b, then we can replace the variable x by a new one in X that represents $ax + bS(x)$, so that if we recover X, we will be able to find x. Then, the second important notion is that for a system of equations describing the computation of the block cipher, the system is triangular from the plaintext and key variables to the ciphertext variables and so, from the ciphertext and key variables to the plaintext variables.

Given a finite field $\mathbb{F}_q$, where q is a power of a prime number, and a non-linear function $S : \mathbb{F}_q \longrightarrow \mathbb{F}_q$, an AES-like equation is defined as follows.

Definition 2 (AES-like equation). *An AES-like equation in variables* $\mathbf{X} = \{x_1, \ldots, x_n\}$ *is an equation of the form:*

$$\sum_{i=1}^{n} a_i x_i + \sum_{i=1}^{n} b_i S(x_i) + c = 0,$$

where $a_1, \ldots, a_n, b_1, \ldots, b_n, c \in \mathbb{F}_q$.

AES-like equations enjoy some very interesting properties. First the set of all the AES-like equations in variables $\mathbf{X} = \{x_1, \ldots, x_n\}$ is a vector space over $\mathbb{F}_q$. Indeed, this set 'is stable by the multiplication by a scalar and the sum of two AES-like equations is still an AES-like equation.

Definition 3 (AES-like system). *We denote by $\mathcal{V}(\mathbf{X})$ the vector space spanned by all the AES-like equations in variable $\mathbf{X}$. A system of AES-like equations in variables $\mathbf{X}$ is a subspace of $\mathcal{V}(\mathbf{X})$.*

Definition 4 (subsystem). *Let $\mathbb{E}$ be a system of AES-like equations in variables $\mathbf{X}$ and let $\mathbf{Y}$ be a subset of $\mathbf{X}$. We denote by $\mathbb{E}(\mathbf{Y})$ the subspace $\mathbb{E} \cap \mathcal{V}(\mathbf{Y})$. This subspace is the biggest subsystem of $\mathbb{E}$ composed of AES-like equations in variables $\mathbf{Y}$.*

Definition 5 (linear variable). *Let $\mathbb{E}$ be a system of AES-like equations in variables $\mathbf{X}$ and let be $x \in \mathbf{X}$. The variable x is a linear variable if and only if $\dim \mathbb{E} - \dim \mathbb{E}(\mathbf{X} - \{x\}) \leq 1$. The set of all the linear variables is denoted by $\textsc{Lin}(\mathbb{E})$.*

This definition may seem abstract and the following proposition clarifies it:

Property 1. Let $\mathbb{E}$ be a system of AES-like equations in variables $\mathbf{X}$ and let $x \in \textsc{Lin}(\mathbb{E})$. Then it exists $(a, b) \in \mathbb{F}_q^2$ such that each equation of $\mathbb{E}$ involving the variable x involves in fact a multiple of $ax + bS(x)$. In other words, if we replace $ax + bS(x)$ by X in the system of equations then x and $S(x)$ do not appear any more. In particular, $\textsc{Lin}(\mathbb{E}) \cap \mathbf{Y} \subseteq \textsc{Lin}(\mathbb{E}(\mathbf{Y}))$ for any subset $\mathbf{Y}$ of $\mathbf{X}$.

Linear variables are very important in the work of Bouillaguet *et al.*, in particular when the following assumption about the number of solutions of system of AES-like equations holds, we can estimate the complexity of our algorithms:

$$|\mathcal{S}ol\,(\mathbb{E}(\mathbf{Y}))| \approx q^{|\mathbf{Y}| - \dim \mathbb{E}(\mathbf{Y})}, \text{ for any subset } \mathbf{Y} \text{ of } \mathbf{X}.$$

Let us introduce a last definition related to linear variables:

Definition 6. *Let $\mathbb{E}$ be a system of AES-like equations in variables $\mathbf{X}$ and let $\mathbf{Y}$ be a subset of $\mathbf{X}$. Consider the following sequences:*

$$\mathbb{E}_0 := \mathbb{E}, \ \mathbb{E}_{i+1} := \mathbb{E}_i(\mathbf{X} - L_i), \ L_0 := \textsc{Lin}(\mathbb{E}) - \mathbf{Y}, \ L_{i+1} := L_i \cup (\textsc{Lin}(\mathbb{E}_{i+1}) - \mathbf{Y}).$$

The sequence $(\mathbb{E}_i)$ is decreasing and thus at some rank r it becomes constant. We denote by $\textsc{Lin}(\mathbb{E}, \mathbf{Y})$ the set of all variables occurring in the system $\mathbb{E}_r$.

3   New Set of Tools

In this section, we will first describe our generic GDS attack. Therefore, we need to explain how we can automatically find the useful relations (*minimal equations*) and how we automatically split the variables involved in these relations in order to perform an efficient meet-in-the-middle for instance in sets O_P and O_C. Then, we have to explain how we automatically find the truncated differential path and the sets I_P and I_C. Splitting variables in some sets appears to be quite obvious by hand when we consider the cipher round by round. However, using the system of equations, this task appears to be not easy. Moreover, we need to perform this split efficiently and without any redundancy since the number of splitting an equation involving n variables in two sets of k and $n - k$ variables becomes quickly very large. Finally, the intersections of the set of variables in some sets (I_P, I_C, O_P, O_C) define our attack and we use Bouillaguet *et al.* algorithm in order to find the best attack taking into account the key schedule equations.

It turns out that our tool is *modular* in the following sense. The algorithm used to find the sets O_P, O_C from the minimal equation is very similar to the one used to find the set I_P, I_C in the truncated differential path. Moreover, the algorithm used to find the impossible differential path used in fact two executions for the truncated differential path algorithms and by computing the intersection of both sets, we can automatically discover impossible differentials.

3.1   Generic Attack on Simple Block Cipher

Our idea is to build a tool finding the best GDS attacks on a block cipher, but where the block cipher is given as a system $\mathbb{E}$ of AES-like equations over $\mathbb{F}_q$. The only information assumed in our possession is the type of involved variables: plaintext ($\mathbf{P}$), ciphertext ($\mathbf{C}$), key ($\mathbf{K}$) or state ($\mathbf{X}$). To be a valid block cipher we impose three conditions on the system of equations:

$$|\mathbf{P}| = |\mathbf{C}|, \;\; \text{Lin} (\mathbb{E}, \mathbf{P} \cup \mathbf{K}) \cup \text{Lin} (\mathbb{E}, \mathbf{C} \cup \mathbf{K}) \subseteq \mathbf{K} \text{ and } \text{Lin} (\mathbb{E}(\mathbf{K}), \emptyset) = \emptyset.$$

These conditions are natural as they translate the fact that all variables can be computed step by step from $\mathbf{P}$ and $\mathbf{K}$ and also from $\mathbf{C}$ and $\mathbf{K}$, that all the key variables can be computed step by step from a master key and that the plaintext has the same size than the ciphertext (*i.e.* the blocksize).

For each non-key variable y we define 4 particular sets:

- $O_P(y) := \text{Lin} (\mathbb{E}, \mathbf{P} \cup \mathbf{K} \cup \{y\}) - \mathbf{K}$
- $O_C(y) := \text{Lin} (\mathbb{E}, \mathbf{C} \cup \mathbf{K} \cup \{y\}) - \mathbf{K}$
- $I_P(y) := \{x \in \mathbf{X} \cup \mathbf{P} \cup \mathbf{C} \mid y \in O_C(x)\}$
- $I_C(y) := \{x \in \mathbf{X} \cup \mathbf{P} \cup \mathbf{C} \mid y \in O_P(x)\}$

The set $O_P(y)$ (resp. $O_C(y)$) contains the state variables required to propagate the differences from the plaintexts (resp. ciphertexts) to both y and $S(y)$, *i.e.* the state variables required to compute y that go through an Sbox. In another hand, the set $I_P(y)$ (resp. $I_C(y)$) contains the state variables that are required to propagate a non-zero difference from y to the plaintext (resp. ciphertext). Those sets give us all the information we need about a block cipher. Interestingly, we distinguish two kinds of block ciphers: the SPNs for which $O_P(y) = I_P(y)$ and $O_C(y) = I_C(y)$ for all non-key variables y, and the other ones (Fig. 2).

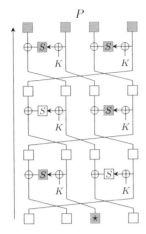
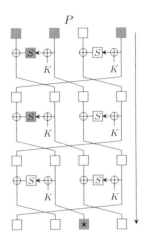

Fig. 2. Toy example. Variables of $I_P(\star)$ are in blue while variables of $O_P(\star)$ are in red. (Color figure online)

We can now give our algorithm finding the best GDS attacks which relies on four sub-algorithms. The aim of the first algorithm is to find a minimal equation involving a given variable. The two next ones are based on the guess-and-determine technique and are designed to exhaust the best building blocks of GDS attacks. Finally, the last sub-algorithm is just a merging procedure which also computes the complexities of the GDS attacks and sorts them.

Finding a Minimal Equation. In next algorithms we need to be able to find a minimal equation involving a particular variable y. Here *minimal* means that there is no equation involving y and a smaller subset (for the inclusion) of variables. For a system of AES-like equations it is rather simple as showed by Algorithm 1.

Algorithm 1. MinimalEquation

Data: A variable y and a system of equations $\mathbb{E}$ in variables $\mathbf{X} \supseteq \{y\}$
Result: A minimal equation involving y if any.

if y does not appear in $\mathbb{E}(\mathbf{X})$ **then return** $\{0\}$;
forall the $x \in \mathbf{X} - \{y\}$ **do**
$\quad F \leftarrow \mathbb{E}(\mathbf{X} - \{x\})$;
$\quad$ **if** y appears in F **then** $\mathbb{E} \leftarrow F$;
end
return *an equation of* $\mathbb{E}$ *involving* y

Truncated Differential Search. Given a value b, our goal is to exhaust all the minimal truncated differential characteristics that come from a truncated differential Δ_X of dimension b (at least), propagated in both way with probability 1. More precisely, we are interested by the corresponding couples (I_P, I_C) (defined in Sect. 2.1) that are minimal for the following partial order relation:

Algorithm 2. TruncatedDiffSearch

Data: A system of equations $\mathbb{E}$ representing a block cipher
Result: A list L containing all possible couples (I_P, I_C)

$L \leftarrow \emptyset$;
$S \leftarrow$ search state initialized such that each of the non-key variable can assume the 3 possible states;
forall the x *that may belong to* I_P *sorted according to* $|O_C(x)|$ **do**
$\quad S' \leftarrow S$;
$\quad$ Update S' with $x \in I_P^G$;
$\quad$ **if** S' is consistent **then** TruncatedDiffSearch$_{tmp}(\mathbb{E}, S', x, L)$;
$\quad$ Update S with $x \notin I_P$;
end
forall the x *that may belong to* I_C *sorted according to* $|O_P(x)|$ **do**
$\quad S' \leftarrow S$;
$\quad$ Update S' with $x \in I_C^G$;
$\quad$ **if** S' is consistent **then** TruncatedDiffSearch$_{tmp}(\mathbb{E}, S', x, L)$;
$\quad$ Set x to constant in S;
end
return L

$$(I_P, I_C) \preceq (I'_P, I'_C) \text{ if and only if } I_P \subseteq I'_P \text{ and } I_C \subseteq I'_C.$$

In other words, we would like to exhaust truncated differential characteristics for which the set of active bits is minimal for the inclusion.

To solve this problem we decided to use a guess-and-determine procedure. At the beginning each non-key variable has 3 possible states: it can belong to I_P, to I_C or be constant. Those states are exclusive, *i.e.* a variable can only be in one of them at the same time. Then the state search is easy to update thanks to the following rules:

- $x \in I_P \Rightarrow I_P(x) \subseteq I_P$ and $O_P(x) \cap I_C = \emptyset$.
- $x \in I_C \Rightarrow I_C(x) \subseteq I_C$ and $O_C(x) \cap I_P = \emptyset$.
- x constant $\Rightarrow O_P(x) \cap I_C = \emptyset$ and $O_C(x) \cap I_P = \emptyset$.

One could perform an exhaustive search using only those rules but this is not optimal. Instead we define two new subsets:

- $I_P^G := \{x \in I_P \mid \forall y \in I_P - \{x\}, \ x \notin I_P(y)\}$.
- $I_C^G := \{x \in I_C \mid \forall y \in I_C - \{x\}, \ x \notin I_C(y)\}$.

Those sets are somehow the generators of I_P and I_C respectively. Our idea is to begin by guessing one variable of I_P^G and then by flagging just enough variables to a non-constant state to ensure that the guessed one is *truly* non-constant. This is done by looking for minimal equations involving the guessed variable and only unset variables. Finally if the dimension of the zero differences is small enough then the couple (I_P, I_C) is stored. Otherwise, another variable of I_P^G or I_C^G is guessed and the procedure is repeated. Furthermore, the variables can be sorted such that at each step only two cases are possible: either the variable belongs to I_P^G or it does not belong to I_P. While being more generic, this is actually quite close than picking a round r, saying that variables of I_P belong to the first r rounds and then first guessing the state of variables of the r-th round. The whole procedure is described in an algorithmic manner in Algorithm 2 and 3.

Algorithm 3. TruncatedDiffSearch$_{tmp}$

Data: A system of equations $\mathbb{E}$, a search state $\mathcal{S}$, a variable y and a list L
Result: Fill the list L with all possible couples (I_P, I_C).

$s \leftarrow$ set of variables that may be or are constant;
$e \leftarrow$ MinimalEquation$(y, \mathbb{E}(s \cup \{y\}))$;
if $e \neq \{0\}$ **then**
> **forall the** x *involved in e that may be constant* **do**
>> $\mathcal{S}' \leftarrow \mathcal{S}$;
>> Update $\mathcal{S}'$ with $x \in I_P$;
>> **if** $\mathcal{S}'$ is consistent **then** TruncatedDiffSearch$_{tmp}(\mathbb{E}, \mathcal{S}', y, L)$;
>> $\mathcal{S}'' \leftarrow \mathcal{S}$;
>> Update $\mathcal{S}''$ with $x \in I_C$;
>> **if** $\mathcal{S}''$ is consistent **then** TruncatedDiffSearch$_{tmp}(\mathbb{E}, \mathcal{S}'', y, L)$;
>> Set x to constant in $\mathcal{S}$;
>
> **end**

else
> $d \leftarrow$ *dimension* of variables that may be or are constant;
> **if** $d > blocksize - b$ **then**
>> **forall the** x *that may belong to I_P sorted according to $|O_C(x)|$* **do**
>>> $\mathcal{S}' \leftarrow \mathcal{S}$;
>>> Update $\mathcal{S}'$ with $x \in I_P^G$;
>>> **if** $\mathcal{S}'$ is consistent **then** TruncatedDiffSearch$_{tmp}(\mathbb{E}, \mathcal{S}', x, L)$;
>>> Update $\mathcal{S}$ with $x \notin I_P$;
>>
>> **end**
>> **forall the** x *that may belong to I_C sorted according to $|O_P(x)|$* **do**
>>> $\mathcal{S}' \leftarrow \mathcal{S}$;
>>> Update $\mathcal{S}'$ with $x \in I_C^G$;
>>> **if** $\mathcal{S}'$ is consistent **then** TruncatedDiffSearch$_{tmp}(\mathbb{E}, \mathcal{S}', x, L)$;
>>> Set x to constant in $\mathcal{S}$;
>>
>> **end**
>
> **else**
>> $L \leftarrow L \cup \{(I_P, I_C)\}$;
>
> **end**

end

Basic Meet-in-the-middle Attack Search. Our algorithm to find the best couples (O_P, O_C) is quite similar to the previous one. Each non-key variable also has 3 possible states: it can belong to O_P, to O_C or be unused. The upgrade rules become:

- $x \in O_P \Rightarrow O_P(x) \subseteq O_P$ and $I_P(x) \cap O_C = \emptyset$.
- $x \in O_C \Rightarrow O_C(x) \subseteq O_C$ and $I_C(x) \cap O_P = \emptyset$.
- x unused $\Rightarrow I_P(x) \cap O_C = \emptyset$ and $I_C(x) \cap O_P = \emptyset$.

We also define two generators sets:

- $O_P^G := \{x \in O_P \mid \forall y \in O_P - \{x\}, \ x \notin O_P(y)\}$.
- $O_C^G := \{x \in O_C \mid \forall y \in O_C - \{x\}, \ x \notin O_C(y)\}$.

The search procedure begins by guessing a variable of O_P^G. Then we look for a minimal equation involving it, at least one unset variable and no variables flagged as unused. Next, two cases are possible: either we set to unused one of the involved variables and go back to the previous step, or we set to O_P or O_C all the involved variables. In the last case if (O_P, O_C) leads to enough equations we store it, otherwise we guess another variable of O_P^G or O_C^G and restart the procedure in order to increase the number of equations.

Merging Procedure. The merging procedure is quite simple and similar to the one used by Derbez *et al.* in [DF13]. In order to perform it we need to compute the number of values that the sets $I_P \cup O_C \cup \mathbf{P} \cup \mathbf{C}$ and $I_C \cap O_P$ can assume and the time required to enumerate them. Under the heuristic assumption of the number of solutions given in the previous section, the procedure described in Algorithm 4 can be used. This procedure takes as input a system of equations $\mathbb{E}$ and a set $\mathbf{Y}$ and gives as output a set $\mathbf{Z}$ containing $\mathbf{Y}$ such that the number of solutions of $\mathbb{E}(\mathbf{Z})$ is minimal. Furthermore, as we only consider systems $\mathbb{E}$ such that $\text{LIN}(\mathbb{E}, \emptyset) = \emptyset$ (*i.e.* systems that are *triangular*) then the time required to enumerate the solutions of a subsystem is equal to its number of solutions.

Algorithm 4. MinimalSolutions

Data: A system of equations $\mathbb{E}$ in variables $\mathbf{X}$ and $\mathbf{Y}$ a subset of $\mathbf{X}$
Result: A set $\mathbf{Z}$ such that $\mathbf{Y} \subseteq \mathbf{Z} \subseteq \mathbf{X}$ and $|Sol(\mathbb{E}(\mathbf{Z}))|$ is minimal

if $\text{LIN}(\mathbb{E}, \mathbf{Y}) - \mathbf{Y} = \emptyset$ then return $\mathbf{Y}$;
Pick $x \in \text{LIN}(\mathbb{E}, \mathbf{Y}) - \mathbf{Y}$;
$\mathbf{Z}_1 \leftarrow \text{MinimalSolution}(\mathbb{E}, \mathbf{X}, \mathbf{Y} \cup \{x\})$;
$\mathbf{Z}_2 \leftarrow \text{MinimalSolution}(\mathbb{E}(\mathbf{X} - \{x\}), \mathbf{X} - \{x\}, \mathbf{Y})$;
if $|Sol(\mathbb{E}(\mathbf{Z}_1))| < |Sol(\mathbb{E}(\mathbf{Z}_2))|$ then
  | return $\mathbf{Z}_1$
else
  | return $\mathbf{Z}_2$
end

3.2   Extension to a Larger Class of Block Ciphers

While interesting the previous algorithm can only handle a limited amount of block ciphers as many of them cannot be represented by a system of AES-like equations. So our idea is to make it work on systems of the following kind of equations:

$$\sum \alpha_i S_{i,j}(x_{\sigma(0)}, \ldots, x_{\sigma(j)}) + \sum \beta_j x_j + c = 0.$$

Indeed a very large variety of block ciphers can be written as systems of such equations and actually it is rather simple to extend our previous algorithms to handle them. The main difference is that instead of considering single variables we now have to consider set of variables. The notion of linear variable can be easily extended to set of variable as follows:

Definition 7 (linear set of variables). *Let $\mathbb{E}$ be a system of equations in variables $\mathbf{X}$ and let be $x_1, \ldots, x_n \in \mathbf{X}$. The set $\{x_1, \ldots, x_n\}$ is a linear set of variables if and only if $\dim \mathbb{E} - \dim \mathbb{E}(\mathbf{X} - \{x_1, \ldots, x_n\}) \leq n$.*

Obviously we do not consider all set of variables but only the ones which go through an Sbox. Also, two sets of variables can share some variables which may be a problem. To solve it we introduce new variables and equations as shown in the following example:

$$\left\{ S(x,y) + S(y,z) = 1 \right. \Rightarrow \left\{ \begin{array}{c} S(x,y) + S(t,z) = 1 \\ y - t = 0 \end{array} \right.$$

Finally, handling multi-variables S-boxes naturally leads to the particular case of AND and OR. While until now S-boxes were considered as black boxes, both those functions have a special property that we want to be properly handled. Indeed, the following equation holds for any variables x and y:

$$\mathrm{AND}(x,y) \oplus \mathrm{AND}(x \oplus \Delta x, y \oplus \Delta y) = \mathrm{AND}(x, \Delta y) \oplus \mathrm{AND}(\Delta x, y) \oplus \mathrm{AND}(\Delta x, \Delta y).$$

In particular, if $\Delta y = 0$ then $\mathrm{AND}(x,y) \oplus \mathrm{AND}(x \oplus \Delta x, y) = \mathrm{AND}(\Delta x, y)$, meaning that computing the difference after the AND requires Δx and y but not the actual value of x. This is also true for the OR operator since $\mathrm{OR}(x,y) = \mathrm{AND}(x,y) \oplus x \oplus y$. As a consequence, in the previous algorithms, we have to define new sets I'_P, I'_C, O'_P and O'_C containing the variables required to compute the differences in each variable of I_P, I_C, O_P and O_C respectively, and use them instead for the complexity computations.

3.3   Two Other Modes

The building blocks of the GDS search algorithm allow to make automatic search for two others kind of attacks.

Basic Meet-in-the-Middle Attack. It is actually one of the building block used in the GDS-attack search procedure and thus it can be used on itself to find very low data complexity attacks. Its application is quite marginal but it was successfully used during the PRINCE Challenge [Sem14] to win some of the contests and it automatically rediscovered the best attack on full KTANTAN [CDK09].

Impossible Differential Attack. Recently, Boura *et al.* [BNS14] proposed a generic vision of impossible differential attacks with the aim of simplifying and helping the construction and verification of this type of cryptanalysis. In particular, they provided a formula to compute the complexity of such an attack according to its parameters. To understand the formula we first briefly remain how an impossible differential attack is constructed. It starts by splitting the cipher in three parts: $E = E_3 \circ E_2 \circ E_1$ and by finding an impossible differential $(\Delta_X \nrightarrow \Delta_Y)$ through E_2. Then Δ_X (resp. Δ_Y) is propagated through E_1^{-1} (resp. E_3) with probability 1 to obtain Δ_{in} (resp. Δ_{out}). We denote by c_{in} and c_{out} the $\log_2$ of the probability of the transitions $\Delta_{in} \to \Delta_X$ and $\Delta_{out} \to \Delta_Y$ respectively. Finally we denote by k_{in} and k_{out} the key materials involved in those transitions. All in all the attack consists in discarding the keys k for which at least one pair follows the characteristic through E_1 and E_3 and in exhausting the remaining ones. The complexity of doing so is the following:

- **data:** C_{N_α}
- **memory:** N_α
- **time:** $C_{N_\alpha} + \left(1 + 2^{|k_{in} \cup k_{out}| - c_{in} - c_{out}}\right) N_\alpha C_{E'} + 2^{|k| - \alpha}$

where N_α is such that $(1 - 2^{-c_{in} - c_{out}})^{N_\alpha} < 2^{-\alpha}$, C_{N_α} is the number of chosen plaintexts required to generate N_α pairs satisfying $(\Delta_{in}, \Delta_{out})$, $|k|$ is the key size and $C_{E'}$ is the ratio of the cost of partial encryption to the full encryption.

As we already have an algorithm to find the kind of truncated differential characteristics used in impossible differential attack, making an automatic search for this kind of attacks is straightforward. The tool gives as output all the parameters used in the above formula.

3.4   Limitations and Usage

In this section we discuss the limitations of our tools and give some recommendations.

Generic VS Ad-Hoc. As our algorithms are very generic they are probably slower than an ad-hoc algorithm designed for a specific block cipher. In particular, we do not take into account the symmetries found in almost all modern ciphers. This could be a nice improvement of our algorithms and we are already thinking about such a feature.

ARX Ciphers. While in theory ARX ciphers are handled, in practice they are not. More precisely, fully describing all the modular additions to fit the expected representation leads to a lot of nested Sboxes and/or new variables which may make the search too slow. In such case, we recommend to describe them only for the 3-4 lower bits and to use black boxes for other ones as follows:

$$\{\, z = x + y \ [2^{32}] \,\} \Rightarrow \begin{cases} z_0 = x_0 \oplus y_0 \\ r_1 = \mathrm{AND}(x_0, y_0) \\ z_1 = x_1 \oplus y_1 \oplus r_1 \\ r_2 = \mathrm{AND}(x_1, y_1) \oplus \mathrm{AND}(x_1, r_1) \oplus \mathrm{AND}(y_1, r_1) \\ z_2 = x_2 \oplus y_2 \oplus r_2 \\ r_3 = \mathrm{AND}(x_2, y_2) \oplus \mathrm{AND}(x_2, r_2) \oplus \mathrm{AND}(y_2, r_2) \\ z_3 = x_3 \oplus y_3 \oplus r_3 \\ z_4 = S_4(x_3, \ldots, x_{31}, y_3, \ldots, y_{31}, r_3) \\ \qquad \cdots \\ z_{31} = S_{31}(x_3, \ldots, x_{31}, y_3, \ldots, y_{31}, r_3) \end{cases}$$

In our opinion the issue comes more from our implementation than from our algorithms and we are currently working on it.

Complex Key Schedule. Too complex key schedules may also make the search too slow. For instance, if it is very hard to retrieve a part of the master key without almost all the subkeys like for CLEFIA [SSA+07] or Camellia [AIK+00] then we recommend to remove the subkeys generation process from the system of equations. Our tools should see a key size larger than expected but the user can give bounds for data, time and memory complexities of attacks.

Exhaustive Search. Unfortunately, it is not always possible to fully perform the algorithms described in Sect. 3.1 in a reasonable time (say less than a month). In order to decrease the running time, one thing we considered was to slightly modify the partial order relation into the following one:

$$(I_P, I_C) \preceq (I'_P, I'_C) \text{ if and only if } |I_P| \le |I'_P| \text{ and } |I_C| \le |I'_C|.$$

While in theory we may miss some of the best attacks, we never encounter a block cipher for which the building blocks of best attacks were not minimal for this order relation because the complexity of attacks is highly (but not fully) related to the number of variables to enumerate.

Differential Enumeration Technique. In [DKS10], Dunkelman *et al.* introduced a sophisticated trade-off for GDS attacks which reduces the memory without increasing the time complexity. The main idea is to add restrictions on the parameters used to build the table such that those restrictions can be checked (at least partially) during the online phase. More precisely, they impose that sequences stored come from a δ-set containing a message m which belongs to a pair (m, m') that follows a well-chosen differential path. Then the attacker first focus on finding such pair before identifying a δ-set and finally building the sequence. This technique is very powerful and was used to reach the best attacks against the AES [DFJ13, DF13, LJW13]. We did not make an algorithm finding the best GDS attacks under this trade-off mainly because it may be complicated to compute the exact complexity of the resulting attack. However, we distinguish two cases:

– _SPN_: for an SPN block cipher the sets $I_P(y)$ and $O_P(y)$ (resp. $I_C(y)$ and $O_C(y)$) are equal for all non-key variable y. Thus any GDS attack defined by the four sets I_P, I_C, O_P and O_C leads to only one truncated differential characteristic (say Δ) such that active variables are exactly the variables of $I_P \cup (I_C \cap O_P) \cup O_C$. This is the natural candidate to use the differential enumeration technique. Then both the data and memory complexities are modified according to the probability of Δ and are easy to compute. However the time complexities of the online and the offline phases are more complicated to compute since in both cases we have to find the best algorithm to enumerate the solutions of a set of variables under the constraint of Δ (see [LJW13] for instance).

- *non-SPN*: for non-SPN block ciphers there is no natural truncated differential characteristic to use, making the search of best attacks much more complicated. Furthermore, the technique is less powerful than against SPN but can still provide efficient attacks as shown by Li and Jia in [LJ14].

4 Applications

Our tools handle a very large class of block ciphers and we applied them on AES [NIS01], ARIA [KKP+03], CLEFIA [SSA+07], KLEIN [GNL11], KTAN-TAN [CDK09], LBlock [WZ11], PICCOLO [SIH+11], PRINCE [BCG+12], SIMON [BSS+13], TWINE [SMMK12], ZORRO [GGNS13] and more. In this section we present many applications highlighting some of the possibilities offered by our set of tools.

4.1 MCrypton

mCrypton is a 64-bit lightweight block cipher introduced in 2006 by Lim *et al.*, which is a reduced version of Crypton. It is specifically designed for resource-constrained devices like RFID tags and sensors in wireless sensor networks. Like AES, mCrypton is also a SPN block cipher. According to key length, mCrypton has three versions namely mCrypton-64/96/128, which is in high accordance with AES-128/192/256. All the three versions have 12 rounds and each round consists of 4 transformations as follows:

- **Non-linear Substitution** γ. This transformation consists of nibble-wise substitutions using four 4-bit S-boxes S_0, S_1, S_2 and S_3.
- **Bit Permutation** π. The bit permutation transformation π has the same function than MixColumns transformation of AES: mixing each column of the state matrix. Operation π restricted to the i-th column is defined as follows:

$$b = \pi_i(a) \iff b[j] = \bigoplus_{k=0}^{3}(a[k] \mathbin{\&} m_{i+j+k \bmod 4}),$$

where $m_0 = 1110$, $m_1 = 1101$, $m_2 = 1011$, $m_3 = 0111$ and where $\&$ is the bitwise operation AND.
- **Column-To-Row Transposition** τ. This is simply the ordinary matrix transposition.

– **Key Addition** σ. It is a simple bit-wise XOR operation and resembles the AddRoundKey operation of AES.

mCrypton also adds a linear operation $\phi = \tau \circ \pi \circ \tau$ after the last round so that the whole encryption process is:

$$c = \phi \circ \sigma_{k_{12}} \circ \tau \circ \pi \circ \gamma \circ \ldots \circ \sigma_{k_1} \circ \tau \circ \pi \circ \gamma \circ \sigma_{k_0}(p).$$

All best known attacks against mCrypton are GDS attacks combined to the *differential enumeration technique*. Hence it was a good target to check whether our tool could find better attacks. As a result, we found attacks on more rounds for the three standardized key lengths. We also found an attack against 11 rounds of Crypton-256 while the full version is composed of 12 rounds. Complexities of attacks are reported in Table 1.

Table 1. Complexities of GDS attacks against mCrypton and Crypton.

Version	Rounds	Data	Time	Memory	Reference
64	7	2^{57}	2^{57}	2^{44}	[HBL14]
96	7	2^{57}	2^{57}	2^{44}	[HBL14]
	8	2^{48}	2^{65}	$2^{81.6}$	[KJS+13]
	9	2^{57}	2^{83}	2^{83}	ours
128	7	2^{57}	2^{57}	2^{44}	[HBL14]
	8	2^{48}	2^{65}	$2^{81.6}$	[KJS+13]
	8	2^{57}	2^{96}	2^{44}	[HBL14]
	9	2^{53}	2^{116}	2^{120}	[HBL14]
	10	2^{55}	2^{117}	2^{103}	ours

Attack against 10-round mCrypton-128. Let us describe our GDS attack against 10 rounds of the 128-bit version of mCrypton, depicted on Fig. 3.

First we introduce some notations: x_i for the state just before the $i - 1$-th γ operation, y_i for the state just after $i - 1$-th γ operation and z_i for the state just after the $i - 1$-th π operation. Given a state a, $a[i]$ denotes the i-th nibble of a and $a[i]_b$ the b-th bit of nibble $a[i]$.

For this attack we consider δ-sets of 2^6 messages such that nibbles $y_1[2, 3..7, 10, 12..15]$ and $z_1[2, 3..7, 10, 12..15]$ are constant, exploiting the fact that the branch number of the π operation is 4. Then, the meet-in-the-middle is performed on the 4 bit-equations between $\Delta y_6[1, 3, 9, 11]$ and $\Delta z_6[1, 3, 9, 11]$, exploiting again the same property of the π operation.

Given a δ-set $\{p^0, p^1, \ldots, p^{63}\}$, the ordered sequence

$$\left[y_6^1[1, 3, 9, 11] \oplus y_6^0[1, 3, 9, 11], \ldots, y_6^{63}[1, 3, 9, 11] \oplus y_6^0[1, 3, 9, 11] \right],$$

is fully determined by 42 nibbles, which can assume only 2^{159} thanks to the key schedule relations. Furthermore, if we restrict ourself to the case where the

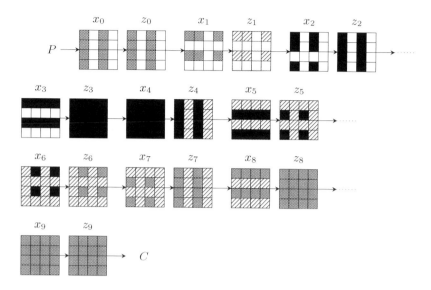

Fig. 3. Attack on 10-rounds mCrypton. Bytes of offline phase are in black. Bytes of online phase are in gray. Hatched bytes play no role. The differences are null in white squares

pair (p^0, p^1) follows the differential characteristic depicted on Fig. 3, the number of possible ordered sequences is reduced by a factor 2^{56}. Computing all these ordered sequences can be done using the same approach Derbez *et al.* used in [DFJ13]. On the other hand, the online phase requires to guess 42 state nibbles which can assume only 2^{117} values thanks to 51 key schedule equations.

Given a pair which may follow the differential characteristic, the 42 nibbles of the online phase can assume only $2^{117-32+6-64+4} = 2^{31}$ values. Enumerating those 2^{31} values in roughly 2^{31} is complicated but possible using a meet-in-the-middle procedure: the main idea is to compute all the possible values for involved nibbles of states x_0 and x_1 in one hand and all the possible values for involved nibbles of states x_7, x_8 and x_9 in an other hand, and then to match those sets according to key schedule equations.

All in all, we need 2^{23} structures of 2^{32} messages to get one pair following the differential characteristic. The probability for a wrong pair to pass the test is $2^{103-63*4} = 2^{-149}$ so we expect that only the right pair will pass it. Finally, the remaining key bits can be exhausted.

4.2   IDEA

IDEA was introduced by Lai and Massey in 1991 and became widely deployed due to its inclusion in the PGP package. It is a 64-bit, 8.5-round block cipher with 128-bit keys and it uses a composition of XOR operations, additions modulo 2^{16}, and multiplications over $GF(2^{16} + 1)$.

In order to apply our set of tools to IDEA, we chose to represent the multiplication by a black-box and to describe the modular addition only for the 4 least significant bits, other ones being handle by a black-box.

As a result, we automatically recovered the 6-round meet-in-the-middle attack described by Biham *et al.* in [BDKS15]. In particular, we retrieved the *keyless Biryukov-Demirci relation*, a linear equation involving the plaintext, the ciphertext, and several intermediate values computed during the IDEA encryption process. This equation is central in best known attacks against IDEA and was discovered only 15 years after IDEA was introduced.

4.3 XTEA

XTEA is an evolutionary improvement of TEA. XTEA makes essentially use of arithmetic and logic operations like TEA. New features of XTEA are to use two bits of δ_i and the shift of 11. This adjustments cause the indexes of round keys to be irregular. We can describe the output (Y_{i+1}, Z_{i+1}) of the i-th cycle of XTEA with the 64-bit input (Y_i, Z_i) as follows:

$$Y_{i+1} = Y_i \boxplus F(Z_i, K_{2i-1}, \delta_{i-1})$$
$$Z_{i+1} = Z_i \boxplus F(Y_i, K_{2i}, \delta_i)$$

where δ_i's are constants, K_i's round keys and where round function F is defined by:

$$F(X, K, \delta) = (((X \ll 4) \oplus (X \gg 5)) \boxplus X) \oplus (K \boxplus \delta).$$

Partially Described Modular Addition. In that case our tools can handle a large number of rounds. Unfortunately, resulting attacks were far from best known ones, in term of complexity and broken rounds, due to the information lost in the representation of the modular addition.

Fully Described Modular Addition. In that case our tools were not able to search for attacks on more than 10 rounds, in the sense that the search takes too much time. The main issue comes from the huge number of sets of variables for which the tools have to compute the number of possible values in order to compute complexities of resulting attacks. However this does not make our set of tools useless. Indeed, our idea was to run the tool searching for impossible differential attacks but with a bound equals to 0 on the time complexity of searched attacks. In that case, it becomes a simpler tool which only looks for truncated impossible differentials. As a result, we were able to recover the longest ones on XTEA in few minutes.

4.4 ZORRO

At CHES 2013, Gerard *et al.* presented the block cipher ZORRO [GGNS13]. It is an AES-like block cipher but with a partial non-linear layer. The 128-bit plaintext initializes the internal state viewed as a 4×4 matrix of bytes as values in the finite field $\mathbb{F}_{2^8}$. It has 24 rounds and the 128-bit master key is XORed with the internal state every four rounds. A round of ZORRO consists of 4 simple operations applied successively on the state matrix:

- **SubBytes** (SB^*) applies the same 8-bit to 8-bit invertible S-Box on each byte of the first row in parallel,
- **ShiftRows** (SR) shifts the i-th row left by i positions,
- **MixColumns** (MC) replaces each of the four column C of the state by $M \times C$ where M is a constant 4×4 maximum distance separable matrix over $\mathbb{F}_{2^8}$,
- **AddConstant** (AC) adds a constant on the first row.

Both the **MixColumns** and **ShiftRows** operations are the same than those used in the AES. The S-box however is different and was chosen in order to be easier to mask but in return has worse differential properties which were exploited by the differential attacks. In particular, ZORRO has been fully broken [GNPW13, BDD+14, RASA14] because of the existence of a high probability 4-round iterated differential (Fig. 4).

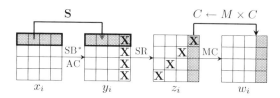

Fig. 4. A ZORRO-round applies $MC \circ SR \circ SB \circ AC$ to the state.

While ZORRO has been already broken, we will study it as a toy example to show how useful our tool can be to designers. Generalized Demirci-Selçuk attacks combined to the differential enumeration technique led to the best attacks on the three versions of the AES in the single-key setting and thus our idea was to study the resistance of ZORRO and its variants against such attacks. If the Sbox is applied on the same four bytes each round then there are 1820 variants of ZORRO. In order to not decrease the (already very low) resistance against differential cryptanalysis we considered only variants such that the S-box is applied on one byte per column and on one byte per diagonal, leading to 24 variants including the original one. Finally, and because of the symmetry in the structure of Zorro we focused on the 11 variants depicted on Fig. 5.

For each of those variants we wrote the corresponding system of equations and gave it to our tool. Interestingly, we found that the complexity and the number of rounds broken only depend on the number of rows having an S-box. More precisely, for all the variants with only one Sbox-free row we found that 16 rounds are secure against GDS attacks and 20 are fully secure against GDS attacks combined with the differential enumeration technique while 20 and 25

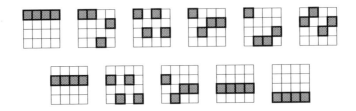

Fig. 5. Studied variants of Zorro.

rounds are required to provide the same security for the variants with two or three Sbox-free rows. As a consequence, the designers of ZORRO did not choose the most secure variant and the number of rounds chosen was too low. Actually, this enforces the results of Bar-On *et al.* [BDD+15], stating that the design behind Zorro may lead to both secure and easy to mask block ciphers as long as we take care of its specificities.

Note that here *fully secure against GDS attacks combined with the differential enumeration technique* means that there is no GDS attack with a time complexity strictly smaller than 2^k and a memory complexity strictly smaller than 2^{k+n}, where k is the keysize and n the blocksize, since, combined to the differential enumeration technique, such attack **may** (but not always) be turned into one with an overall complexity smaller than 2^k.

4.5   SIMON

SIMON [BSS+13] is family of lightweight block ciphers designed by the American National Security Agency (NSA) in 2013. It performs exceptionally well in both hardware and software, although SIMON is supposed to be more hardware-oriented. The SIMON family is based on a classical Feistel construction operating on two branches. The round function is composed of three simple operations: AND, XOR and rotations. More precisely, at each round the left branch is transformed using the following non-linear function:

$$F(L) := ((L \lll 8) \& (L \lll 1)) \oplus (L \lll 2).$$

Then, the output of F is XORed with the round key and the right branch to form the left branch of the next round. The SIMON family contains 10 ciphers and, in the sequel, we refer to them by SIMONn/k where n and k are respectively the blocksize and the keysize.

In [BNS14], Boura *et al.* described the best (in term of broken rounds) impossible differential attacks against all the versions of SIMON. However, after running our tool against SIMON we found that actually more rounds can be broken by using the exact same technique, highlighting how useful an automatic approach is.

20-Round Attack on SIMON32/64. To mount an impossible differential attack on 19-round SIMON32/64, Boura *et al.* used an impossible differential characteristic covering 11 rounds extended by 4 rounds in both directions such that $4 + 11 + 4 = 19$ rounds of the cipher were attacked. In our case we also use an 11-round impossible differential but our tool found one (see Table 2) that can be extended by $3 + 6$ rounds while still resulting in an attack faster than the exhaustive search according to the formula given Sect. 3.3.

Table 2. Impossible differential characteristic over 11 rounds of SIMON32/64. 0 denotes a bit with no difference, 1 a bit with a difference and $*$ a bit which may have a difference.

Round	Left branch L_r	Right branch R_r
3	0 0 0 0 0 0 0 0 0 0 0 0 0 0 0 0	1 0 0 0 0 0 0 0 0 0 0 0 0 0 0 0
4	1 0 0 0 0 0 0 0 0 0 0 0 0 0 0 0	0 0 0 0 0 0 0 0 0 0 0 0 0 0 0 0
5	0 0 0 0 0 0 0 0 * 0 0 0 0 0 1 *	1 0 0 0 0 0 0 0 0 0 0 0 0 0 0 0
6	* 0 0 0 0 0 * * 0 0 0 0 1 * * 0	0 0 0 0 0 0 0 0 * 0 0 0 0 0 1 *
7	0 0 0 0 * * * 0 * 0 1 * * * * *	* 0 0 0 0 0 * * 0 0 0 0 1 * * 0
8	* 0 * * * * * * 1 * * * * * * 0	0 0 0 0 * * * 0 * 0 1 * * * * *
9	* * * * * * * * * * * * * * * *	* 0 * * * * * * 1 * * * * * * 0
9	0 * 0 * * * * * * 0 0 0 0 * * *	* * * * * * * * 0 * 0 * * * * *
10	* 0 0 0 0 * * * 0 * 0 0 0 0 0 *	0 * 0 * * * * * * 0 0 0 0 * * *
11	0 * 0 0 0 0 0 * * 0 0 0 0 0 0 0	* 0 0 0 0 * * * 0 * 0 0 0 0 0 *
12	0 0 0 0 0 0 0 0 0 1 0 0 0 0 0 0	0 * 0 0 0 0 0 * * 0 0 0 0 0 0 0
13	0 0 0 0 0 0 0 0 0 0 0 0 0 0 0 0	0 0 0 0 0 0 0 0 0 1 0 0 0 0 0 0
14	0 0 0 0 0 0 0 0 0 1 0 0 0 0 0 0	0 0 0 0 0 0 0 0 0 0 0 0 0 0 0 0

The attack is depicted in Fig. 6. It can be seen that the difference in the plaintexts has to be zero in 16 bits and equals to 1 in 2 bits. Hence $c_{in} + c_{out} = 13 + 31 = 44$ and thus $N_\alpha \approx \alpha \cdot 2^{43.5}$. Given a structure of 2^{14} plaintexts such that bits $L_0[1..5, 8..11, 15]$ and $R_0[0..3, 7, 9]$ are constant and such that $L_0[12] = R_0[10]$, one can form $2^{14+13-1} = 2^{26}$ pairs lying in the right space and thus $C_{N_\alpha} = \alpha \cdot 2^{31.5}$. Finally, 70 subkey bits are involved in the attack (blue colored in Fig. 6) but they can assume only 2^{62} values thanks to the key schedule (see Appendix A). All in all, the complexity of our attack is $D = \alpha \cdot 2^{31.5}$, $M = \alpha \cdot 2^{43.5}$ and $T = \alpha \cdot 2^{31.5} + (1 + 2^{62-44}) \cdot \alpha \cdot 2^{43.5} C_{E'} + 2^{64-\alpha}$. As we estimate the ratio $C_{E'}$ to $70/(16 \cdot 20)$, the value of α minimizing the overall complexity is 4.17. However α has to be smaller than $2^{0.5}$ because of the data complexity and, for this particular value the complexity of our attack is:

$$D = 2^{32}, \ M = 2^{43.5} \text{ and } T = 2^{62.8},$$

which is similar to the complexity Boura et al. reached on 19 rounds ($D = 2^{32}$, $M = 2^{44}$ and $T = 2^{62.5}$).

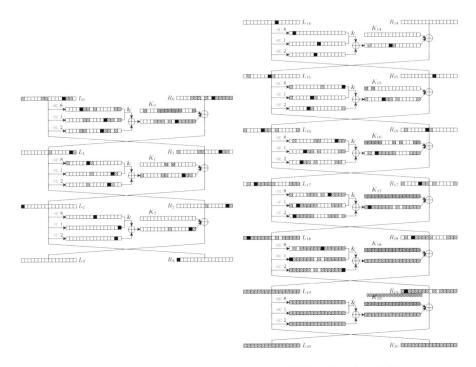

Fig. 6. Impossible differential attack against 20-round SIMON32/64. Difference equals to 0 in white bits, to 1 in black bits and unknown in red bits. Subkey material involved is in blue. (Color figure online)

Others Versions of SIMON. Running our tool against SIMON takes time (up to many days for the largest versions) so we did not exhaust all the best attacks yet. However, we found that SIMON32/64 is not the only version for which results of Boura *et al.* are suboptimal as, for instance, one more round can be broken for both SIMON48/64 and SIMON48/96.

5   Conclusion

In this paper we described powerful and versatile cryptanalysis tools handling a very large class of block ciphers. They are designed to find the best generalized Demirci-Selçk attacks, basic meet-in-the-middle attacks and impossible truncated differential attacks for a given target. They are publicly released, easy to use and their running time is reasonable (from few seconds for AES to many days for SIMON). Thus we believe they will be of great help for both designers and cryptanalysts. Furthermore, our approach is very generic and requires no *a priori* information about the targeted block cipher.

Future work will be to think about better algorithms/implementations, mainly in order to handle ARX ciphers faster. Including the last results concerning the differential enumeration technique would also be nice as well as handling systems from authenticated encryptions. Finally, currently we have to write code in order to generate the system of equations. It would be nice if we would be able to generate it from a C implementation.

A   Key Schedule Equations Used in the 20-Round Attack Against SIMON32/64

Actually, all the 70 key bits are not required to perform the impossible differential attack described Sect. 4.5 since bits $K_1[7] \oplus K_0[9]$, $K_1[9] \oplus K_0[11]$, $K_{15}[0] \oplus K_{16}[2]$ and $K_{15}[2] \oplus K_{16}[4]$ can be used instead of the 8 bits $K_0[9]$, $K_0[11]$, $K_1[7]$, $K_1[9]$, $K_{15}[0]$, $K_{15}[2]$, $K_{16}[2]$, $K_{16}[4]$. This already saves 4 bits.

Then, in SIMON32/64, the subkeys are related by the following equations:

$$K_{r+4} = K_r \oplus K_{r+1} \oplus (K_{r+1} \lll 1) \oplus (K_{r+3} \lll 3) \oplus (K_{r+3} \lll 4).$$

Switching K_r and K_{r+4} we can use this equation to express K_0 as a linear combination of K_{16}, K_{17}, K_{18} and K_{19}, and interestingly it has the following shape:

$$K_0 = K_{16} \oplus f(K_{17}, K_{18}, K_{19}),$$

where f is a linear function. Finally, thanks to this equation, we deduce bits 1, 8, 10 and 15 of K_0 from the same bits of K_{16} and the full subkeys K_{17}, K_{18} and K_{19}.

References

[AIK+00] Aoki, K., Ichikawa, T., Kanda, M., Matsui, M., Moriai, S., Nakajima, J., Tokita, T.: Camellia: A 128-bit block cipher suitable for multiple platforms - design and analysis. In: Stinson, D.R., Tavares, S. (eds.) SAC 2000. LNCS, vol. 2012, pp. 39–56. Springer, Heidelberg (2001)

[BBS99] Biham, E., Biryukov, A., Shamir, A.: Cryptanalysis of skipjack reduced to 31 rounds using impossible differentials. In: Stern, J. (ed.) EUROCRYPT 1999. LNCS, vol. 1592, pp. 12–23. Springer, Heidelberg (1999)

[BCG+12] Borghoff, J., Canteaut, A., Güneysu, T., Kavun, E.B., Knezevic, M., Knudsen, L.R., Leander, G., Nikov, V., Paar, C., Rechberger, C., Rombouts, P., Thomsen, S.S., Yalçın, T.: PRINCE – A low-latency block cipher for pervasive computing applications. In: Wang, X., Sako, K. (eds.) ASIACRYPT 2012. LNCS, vol. 7658, pp. 208–225. Springer, Heidelberg (2012)

[BDD+14] Bar-On, A., Dinur, I., Dunkelman, O., Lallemand, V., Tsaban, B.: Improved analysis of zorro-like ciphers. IACR Cryptology ePrint Archive **2014**, 228 (2014)

[BDD+15] Bar-On, A., Dinur, I., Dunkelman, O., Lallemand, V., Keller, N., Tsaban, B.: Cryptanalysis of SP networks with partial non-linear layers. In: Oswald, E., Fischlin, M. (eds.) EUROCRYPT 2015. LNCS, vol. 9056, pp. 315–342. Springer, Heidelberg (2015)

[BDF11]   Bouillaguet, C., Derbez, P., Fouque, P.-A.: Automatic search of attacks on round-reduced AES and applications. In: Rogaway, P. (ed.) CRYPTO 2011. LNCS, vol. 6841, pp. 169–187. Springer, Heidelberg (2011)

[BDKS15] Biham, E., Dunkelman, O., Keller, N., Shamir, A.: New attacks on IDEA with at least 6 rounds. J. Cryptol. **28**(2), 209–239 (2015)

[BNS14]   Boura, C., Naya-Plasencia, M., Suder, V.: Scrutinizing and improving impossible differential attacks: applications to CLEFIA, Camellia, LBlock and SIMON. In: Sarkar, P., Iwata, T. (eds.) ASIACRYPT 2014. LNCS, vol. 8873, pp. 179–199. Springer, Heidelberg (2014)

[BPW06a] Buchmann, J., Pyshkin, A., Weinmann, R.-P.: Block ciphers sensitive to gröbner basis attacks. In: Pointcheval, D. (ed.) CT-RSA 2006. LNCS, vol. 3860, pp. 313–331. Springer, Heidelberg (2006)

[BPW06b] Buchmann, J., Pyshkin, A., Weinmann, R.-P.: A zero-dimensional Gröbner basis for AES-128. In: Robshaw, M. (ed.) FSE 2006. LNCS, vol. 4047, pp. 78–88. Springer, Heidelberg (2006)

[BS93]   Biham, E., Shamir, A.: Differential Cryptanalysis of the Data Encryption Standard. Springer, New York (1993)

[BSS+13] Beaulieu, R., Shors, D., Smith, J., Treatman-Clark, S., Weeks, B., Wingers, L.: The simon and speck families of lightweight block ciphers. Cryptology ePrint Archive, Report 2013/404 (2013). http://eprint.iacr.org/

[CDK09]   De Cannière, C., Dunkelman, O., Knežević, M.: KATAN and KTANTAN — A family of small and efficient hardware-oriented block ciphers. In: Clavier, C., Gaj, K. (eds.) CHES 2009. LNCS, vol. 5747, pp. 272–288. Springer, Heidelberg (2009)

[CM03]   Courtois, N.T., Meier, W.: Algebraic attacks on stream ciphers with linear feedback. In: Biham, E. (ed.) EUROCRYPT 2003. LNCS, vol. 2656, pp. 345–359. Springer, Heidelberg (2003)

[DF13]   Derbez, P., Fouque, P.-A.: Exhausting Demirci-Selçuk meet-in-the-middle attacks against reduced-round AES. In: Moriai, S. (ed.) FSE 2013. LNCS, vol. 8424, pp. 541–560. Springer, Heidelberg (2014)

[DFJ13]   Derbez, P., Fouque, P.-A., Jean, J.: Improved key recovery attacks on reduced-round AES in the single-key setting. In: Johansson, T., Nguyen, P.Q. (eds.) EUROCRYPT 2013. LNCS, vol. 7881, pp. 371–387. Springer, Heidelberg (2013)

[DJ14]   Dinur, I., Jean, J.: Cryptanalysis of FIDES. In: Cid, C., Rechberger, C. (eds.) FSE 2014. LNCS, vol. 8540, pp. 224–240. Springer, Heidelberg (2015)

[DKS10]   Dunkelman, O., Keller, N., Shamir, A.: Improved single-key attacks on 8-Round AES-192 and AES-256. In: Abe, M. (ed.) ASIACRYPT 2010. LNCS, vol. 6477, pp. 158–176. Springer, Heidelberg (2010)

[DS08]   Demirci, H., Selçuk, A.A.: A meet-in-the-middle attack on 8-Round AES. In: Nyberg, K. (ed.) FSE 2008. LNCS, vol. 5086, pp. 116–126. Springer, Heidelberg (2008)

[FJP13]   Fouque, P.-A., Jean, J., Peyrin, T.: Structural evaluation of AES and chosen-key distinguisher of 9-round AES-128. In: Canetti, R., Garay, J.A. (eds.) CRYPTO 2013, Part I. LNCS, vol. 8042, pp. 183–203. Springer, Heidelberg (2013)

[GGNS13] Gérard, B., Grosso, V., Naya-Plasencia, M., Standaert, F.-X.: Block ciphers that are easier to mask: how far can we go? In: Bertoni, G., Coron, J.-S. (eds.) CHES 2013. LNCS, vol. 8086, pp. 383–399. Springer, Heidelberg (2013)

[GJ79]   Garey, M.R., Johnson, D.S.: Computers and Intractability: A Guide to the Theory of NP-Completeness. W.H. Freeman, New York (1979)

[GNL11] Gong, Z., Nikova, S., Law, Y.W.: KLEIN: A new family of lightweight block ciphers. In: Juels, A., Paar, C. (eds.) RFIDSec 2011. LNCS, vol. 7055, pp. 1–18. Springer, Heidelberg (2012)

[GNPW13] Guo, J., Nikolic, I., Peyrin, T., Wang, L.: Cryptanalysis of zorro. IACR Cryptology ePrint Archive 2013:713 (2013)

[HBL14] Hao, Y., Bai, D., Li, L.: A meet-in-the-middle attack on round-reduced mcrypton using the differential enumeration technique. In: Au, M.H., Carminati, B., Kuo, C.-C.J. (eds.) NSS 2014. LNCS, vol. 8792, pp. 166–183. Springer, Heidelberg (2014)

[KBN09] Khovratovich, D., Biryukov, A., Nikolic, I.: Speeding up collision search for byte-oriented hash functions. In: Fischlin, M. (ed.) CT-RSA 2009. LNCS, vol. 5473, pp. 164–181. Springer, Heidelberg (2009)

[KJS+13] Kang, J., Jeong, K., Sung, J., Hong, S., Lee, K.: Collision attacks on AES-192/256, crypton-192/256, mCrypton-96/128, anubis. J. Appl. Math. **2013**, 713673:1–713673:10 (2013). Observation of strains

[KKP+03] Kwon, D., et al.: New block cipher: ARIA. In: Lim, J.-I., Lee, D.-H. (eds.) ICISC 2003. LNCS, vol. 2971, pp. 432–445. Springer, Heidelberg (2004)

[Knu98] Knudsen, L.R.: Deal – a 128-bit block cipher. Technical Report Department of Informatics (1998)

[Leu12] Leurent, G.: Analysis of differential attacks in ARX constructions. In: Wang, X., Sako, K. (eds.) ASIACRYPT 2012. LNCS, vol. 7658, pp. 226–243. Springer, Heidelberg (2012)

[LJ14]   Li, L., Jia, K.: Improved meet-in-the-middle attacks on reduced-round camellia-192/256. Cryptology ePrint Archive, Report 2014/292 (2014)

[LJW13] Li, L., Jia, K., Wang, X.: Improved meet-in-the-middle attacks on aes-192 and prince. Cryptology ePrint Archive, Report 2013/573 (2013)

[LWWZ13] Lin, L., Wu, W., Wang, Y., Zhang, L.: General model of the single-key meet-in-the-middle distinguisher on the word-oriented block cipher. In: Lee, H.-S., Han, D.-G. (eds.) ICISC 2013. LNCS, vol. 8565, pp. 203–223. Springer, Heidelberg (2014)

[MHL+02] Moon, D., Hwang, K., Lee, W., Lee, S., Lim, J.: Impossible differential cryptanalysis of reduced round XTEA and TEA. In: Daemen, J., Rijmen, V. (eds.) FSE 2002. LNCS, vol. 2365, pp. 49–60. Springer, Heidelberg (2002)

[MS13]   Morawiecki, P., Srebrny, M.: A sat-based preimage analysis of reduced keccak hash functions. Inf. Process. Lett. **113**(10–11), 392–397 (2013)

[MZ06]   Mironov, I., Zhang, L.: Applications of SAT solvers to cryptanalysis of hash functions. In: Biere, A., Gomes, C.P. (eds.) SAT 2006. LNCS, vol. 4121, pp. 102–115. Springer, Heidelberg (2006)

[NIS01] NIST. Advanced Encryption Standard (AES), FIPS 197. Technical report, NIST, November 2001

[RASA14] Rasoolzadeh, S., Ahmadian, Z., Salmasizadeh, M., Aref, M.R.: Total break of zorro using linear and differential attacks. IACR Cryptology ePrint Archive 2014:220 (2014)

[Sem14] NXP Semiconductors. The PRINCE challenge (2014). https://www.emsec. rub.de/research/research_startseite/prince-challenge/

[SIH+11] Shibutani, K., Isobe, T., Hiwatari, H., Mitsuda, A., Akishita, T., Shirai, T.: *Piccolo*: An ultra-lightweight blockcipher. In: Preneel, B., Takagi, T. (eds.) CHES 2011. LNCS, vol. 6917, pp. 342–357. Springer, Heidelberg (2011)

[SKPI07] Sugita, M., Kawazoe, M., Perret, L., Imai, H.: Algebraic cryptanalysis of 58-Round SHA-1. In: Biryukov, A. (ed.) FSE 2007. LNCS, vol. 4593, pp. 349–365. Springer, Heidelberg (2007)

[SMMK12] Suzaki, T., Minematsu, K., Morioka, S., Kobayashi, E.: TWINE: A lightweight block cipher for multiple platforms. In: Knudsen, L.R., Wu, H. (eds.) SAC 2012. LNCS, vol. 7707, pp. 339–354. Springer, Heidelberg (2013)

[SNC09] Soos, M., Nohl, K., Castelluccia, C.: Extending SAT solvers to cryptographic problems. In: Kullmann, O. (ed.) SAT 2009. LNCS, vol. 5584, pp. 244–257. Springer, Heidelberg (2009)

[SSA+07] Shirai, T., Shibutani, K., Akishita, T., Moriai, S., Iwata, T.: The 128-bit blockcipher CLEFIA (Extended Abstract). In: Biryukov, A. (ed.) FSE 2007. LNCS, vol. 4593, pp. 181–195. Springer, Heidelberg (2007)

[WRG+11] Wei, L., Rechberger, C., Guo, J., Wu, H., Wang, H., Ling, S.: Improved meet-in-the-middle cryptanalysis of KTANTAN (Poster). In: Parampalli, U., Hawkes, P. (eds.) ACISP 2011. LNCS, vol. 6812, pp. 433–438. Springer, Heidelberg (2011)

[WW12] Wu, S., Wang, M.: Automatic search of truncated impossible differentials for word-oriented block ciphers. In: Galbraith, S., Nandi, M. (eds.) INDOCRYPT 2012. LNCS, vol. 7668, pp. 283–302. Springer, Heidelberg (2012)

[WZ11] Wu, W., Zhang, L.: LBlock: A lightweight block cipher. In: Lopez, J., Tsudik, G. (eds.) ACNS 2011. LNCS, vol. 6715, pp. 327–344. Springer, Heidelberg (2011)

Memory-Efficient Algorithms for Finding Needles in Haystacks

Itai Dinur[1]([✉]), Orr Dunkelman[2], Nathan Keller[3], and Adi Shamir[4]

[1] Computer Science Department, Ben-Gurion University, Beersheba, Israel
dinuri@cs.bgu.ac.il
[2] Computer Science Department, University of Haifa, Haifa, Israel
[3] Department of Mathematics, Bar-Ilan University, Ramat Gan, Israel
[4] Computer Science Department, The Weizmann Institute, Rehovot, Israel

Abstract. One of the most common tasks in cryptography and cryptanalysis is to find some interesting event (a needle) in an exponentially large collection (haystack) of $N = 2^n$ possible events, or to demonstrate that no such event is likely to exist. In particular, we are interested in finding needles which are defined as events that happen with an unusually high probability of $p \gg 1/N$ in a haystack which is an almost uniform distribution on N possible events. When the search algorithm can only sample values from this distribution, the best known time/memory tradeoff for finding such an event requires $O(1/Mp^2)$ time given $O(M)$ memory.

In this paper we develop much faster needle searching algorithms in the common cryptographic setting in which the distribution is defined by applying some deterministic function f to random inputs. Such a distribution can be modelled by a random directed graph with N vertices in which almost all the vertices have $O(1)$ predecessors while the vertex we are looking for has an unusually large number of $O(pN)$ predecessors. When we are given only a constant amount of memory, we propose a new search methodology which we call **NestedRho**. As p increases, such random graphs undergo several subtle phase transitions, and thus the log-log dependence of the time complexity T on p becomes a piecewise linear curve which bends four times. Our new algorithm is faster than the $O(1/p^2)$ time complexity of the best previous algorithm in the full range of $1/N < p < 1$, and in particular it improves the previous time complexity by a significant factor of $\sqrt{N}$ for any p in the range $N^{-0.75} < p < N^{-0.5}$. When we are given more memory, we show how to combine the **NestedRho** technique with the parallel collision search technique in order to further reduce its time complexity. Finally, we show how to apply our new search technique to more complicated distributions

O. Dunkelman was supported in part by the Israeli Science Foundation through grant No. 827/12 and by the Commission of the European Communities through the Horizon 2020 program under project number 645622 PQCRYPTO.

N. Keller was supported by the Alon Fellowship.

A. Shamir—Part of the work was done while the fourth author was visiting the Institute of Theoretical Studies. He would like to thank ITS, Dr. Max Rössler, the Walter Haefner Foundation and the ETH Foundation.

M. Robshaw and J. Katz (Eds.): CRYPTO 2016, Part II, LNCS 9815, pp. 185–206, 2016.
DOI: 10.1007/978-3-662-53008-5_7

with multiple peaks when we want to find all the peaks whose probabilities are higher than p.

Keywords: Cryptanalysis · Needles in haystacks · Mode detection · Rho algorithms · Parallel collision search

1   Introduction

Almost everything we do in the construction and analysis of cryptographic schemes can be viewed as searching for needles in haystacks: identifying the correct key among all the possible keys, finding preimages in hash functions, looking for biases in the outputs of stream ciphers, determining the best differential and linear properties of a block cipher, hunting for smooth numbers in factoring algorithms, etc. As cryptanalysts, our goal is to find such needles with the most efficient algorithm, and as designers our goal is to make sure that such needles either do not exist or are too difficult to find.

Needles can be defined in many different ways, depending on what distinguishes them from all the other elements in the haystack. One common theme which characterizes many types of needles in cryptography is that they are probabilistic events which have the highest probability p among all the $N = 2^n$ possible events in the haystack. Such an element is called the *mode of the distribution*, and for the sake of simplicity we will first consider the case in which the distribution is almost flat: a single peak has a probability of $p \gg 1/N$ and all the other events have a probability of about $1/N$ (as depicted in Fig. 1). Later on we will consider the more general case of distributions in which there are several peaks of varying heights, and we want to find all of them.

Our goal in this paper is to analyze the complexity of this probabilistic needle finding problem, assuming that the haystack distribution is given as a black box. By abstracting away the details of the task and concentrating on its essence, we make our techniques applicable to a wide variety of situations. On the other hand, in this general form we can not use specific optimization tricks that can make the search for particular types of needles more efficient.[1]

We will be interested in optimizing both the time complexity and the memory complexity of the search algorithm. Since random-access memory is usually much more expensive than time, we will concentrate primarily on memory-efficient algorithms: We will start by analyzing the best possible time complexity of algorithms which can use only a constant amount of memory, and then study how the time complexity can be reduced by using some additional memory.

The paper is organized as follows: Sect. 2 formalizes our computational model. Section 3 describes the best previously known folklore algorithms for solving the problem. We then show how to use standard collision detection algorithms to identify the mode when its probability p is sufficiently large in Sect. 4. We follow

---

[1] We leave to future work specific applications of our techniques to the concrete problems mentioned at the beginning of this Section.

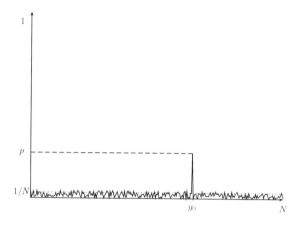

Fig. 1. An Example of the distributions that interest us

in Sect. 5 by introducing the new **2Rho** algorithm which uses a collision detection algorithm on the amplified mode probability obtained by running another collision detection algorithm on the original distribution. The algorithm is then extended to a general *i***Rho** by using even deeper nesting of the collision detection algorithm in Sect. 6. We consider time-memory tradeoffs in Sect. 7, and discuss the adaptations needed when the probability distribution has multiple peaks and we want to find all of them in Sect. 8. Finally, Sect. 9 concludes the paper.

2   Problem Statement and Model Description

The simplest conceptual model for our problem is one in which the sampling black box has a button, and each time we press it we are charged a unit of time and we get a freshly chosen event from the distribution. We can thus test whether a particular y is the mode y_0 by counting how many times this y was sampled from the distribution in $O(1/p)$ trials. Notice that when we have a single available counter, we have to run this algorithm separately for each candidate y. The simplest possible algorithm sequentially tries all the N possible candidates, but we can use the given distribution in order to make a better choice of the next candidate to test. Since the correct candidate is suggested with an enhanced probability of $1/p$, the time complexity is reduced from N/p to $1/p^2$. When we have M available counters, we can get a linear speed up by testing M candidate values simultaneously with the same number of samples, provided that $1/p \geq M$. This trivial approach yields the best known algorithms for finding the mode of a flat distribution with a single peak.

However, closer inspection of the problem shows that in most of our cryptographic applications, the distribution we want to analyze is actually generated by some deterministic function f whose input is randomly chosen from a uniform probability distribution. For example, when we look for biases in the first n

output bits of a stream cipher, we choose a random key, apply to it the deterministic bit generator, and define the (possibly non-uniform) output distribution by saying that a particular bit string has a probability of i/N if it occurs as a prefix of the output string for i out of the N possible keys. Similarly, when we look for a high probability differential property, we choose random pairs of plaintexts with a certain input difference, and deterministically encrypt them under some fixed key. This process generates a distribution by counting how many times we get each output difference, and the mode of this distribution suggests the best differential on the block cipher which uses the selected input difference.

In such situations, we replace the button in the black box by an input which can accept N possible values. The box itself becomes deterministic, and we sample the distribution by providing to the box a randomly chosen input value. The main difference between the two models is that when we repeatedly press the button we get unrelated samples, but when we repeatedly provide the same input value we always get the same output value. As we show in this paper, this small difference leads to surprising new kinds of mode-finding algorithms which have much better complexities than the trivial algorithm outlined above.

The mapping from inputs to outputs defined by the function f can be viewed as a random directed bipartite graph such as the one presented in Fig. 2, in which one of the vertices has a large in-degree. For the sake of simplicity, we assume that the function f has the same number N of possible inputs and outputs[2], and then we can merge input and output vertices which have the same name to get the standard model of a random single-successor graph on N vertices. When we iterate the application of the function f in this graph, we follow a Rho-shaped path which starts with a tail and then gets into a cycle. The graph consists of a small number of disjoint cycles, and all the other vertices in the graph are hanging in the form of trees around these cycles.

As we increase the probability p from $1/N$ to 1, one of the vertices y_0 becomes increasingly popular as a target, and the graph changes its properties. For example, it is easy to show that when p crosses the threshold of $O(1/\sqrt{N})$, there is a sudden phase transition in which y_0 is expected to move from a tree into one of the cycles (where it becomes much easier to locate), and the expected length of its cycle starts to shrink (whereas earlier it was always the same). As we show later in the paper, more subtle phase changes happen when p crosses several earlier thresholds, and thus the log-log complexity of our mode-searching algorithm becomes a piecewise linear function that bends several times at those thresholds, as depicted by the solid line in Fig. 5. Compared to the dotted line which depicts the best previous $1/p^2$ complexity, we get a significant improvement in the whole range of possible p values.

---

[2] If there is some discrepancy, we can use the same truncation trick that Hellman used in his time/memory tradeoff to deal with cryptosystems in which the key and ciphertext sizes are different.

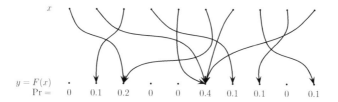

$y = F(x)$
Pr $=$ 0 0.1 0.2 0 0 0.4 0.1 0.1 0 0.1

Fig. 2. A graph representation of a biased function f

2.1 Notations and Conventions

Notation 1. *The set $\{1, 2, \ldots, N\}$ is denoted by $[N]$. Throughout the paper, all functions are from the set $[N]$ to itself. The* mode *of a function is the value in its range with the largest number of preimages.*

Problem Setup: The basic problem we study is the following. We are given a value $0 < p < 1$ and a function $f : [N] \to [N]$ which is generated by the following three-step process:

1. Choose $y_0 \in [N]$ uniformly at random.
2. Choose a subset $S \subset [N]$ uniformly at random amongst all subsets of size pN. Set $f(x) = y_0$ for all $x \in S$.
3. For each $x \notin S$, choose $f(x) \in [N] \setminus \{y_0\}$ uniformly at random.

By definition, the values of f on $[N] \setminus S$ can be simulated by a truly random oracle returning values in $[N] \setminus \{y_0\}$. Our initial goal is to detect y_0 with the fastest possible algorithm that uses only $O(1)$ memory cells. We can assume that the attacker knows p, since otherwise he can run a simple search algorithm with a geometrically decreasing sequence of probabilities (e.g., $1, 1/2, \ldots, 1/2^i, \ldots$) to find the highest value of p for which his attack succeeds (or stop when the attack becomes too expensive, which provides an upper bound on the probability of y_0, but does not identify it).

3 Trivial Memoryless Algorithms

In this section we formally present the simplest possible memoryless algorithms for detecting the mode for various values of p. They are based on sampling random points, and then checking whether they are indeed the required mode.

3.1 Memoryless Mode Verification Algorithm

We start the discussion by presenting a mode verification algorithm. The algorithm accepts a candidate y, and checks whether it is the mode y_0. The checking is done by choosing $O(1/p)$ random values, and verifying that sufficiently many

---
Algorithm 1. Mode Verification: Determining Whether a Given y is y_0

Initialize a counter $ctr \leftarrow 0$.
for $i = 1$ to c/p **do**
    Pick at random $x \in [N]$, and compute $y' = f(x)$.
    if $y' = y_0$ **then**
        Increment ctr.
    end if
end for
if $ctr \geq t$ **then**
    print y is y_0.
end if

---

of them are mapped to y under the function f. The algorithm is presented in Algorithm 1.

It is easy to see that Algorithm 1 makes c/p queries to $f(\cdot)$. Its success depends on the picked constants c and t. Assuming that indeed y is y_0 we expect that the number of times the chosen x leads to y is distributed according to a Poisson distribution with a mean value of c (otherwise, the distribution follows a Poisson distribution with a mean value of $c/Np \ll c$). Hence, for any desired success rate, one can easily choose c and the threshold t. For example, setting $c = 4$ and $t = 2$ offers a success rate of about 90.8 %.

3.2 Memoryless Sampling Algorithm

The sampling algorithm suggested in Algorithm 2 is based on picking at random a value x, computing a candidate $y = f(x)$ for the verification algorithm, and verifying whether y is indeed the correct y_0. It is easy to see that the algorithm is expected to probe $O(p^{-1})$ values of y, until y_0 is encountered, and that each verification takes $O(p^{-1})$, resulting in a running time of $O(p^{-2})$.

Algorithm 2. Finding y_0 by Sampling:

---
while y_0 was not found **do**
    Pick $x \in [N]$ at random.
    Compute $y = f(x)$.
    Call Algorithm 1 to check y.
end while

---

4   Using Rho-based Collision Detection Algorithms

We now present a different class of algorithms for detecting the mode, using collision detection algorithms combined with the trivial mode verification algorithm (Algorithm 1). These algorithms (such as Floyd's [7] or its variants [2,8]) start from some random point x, and iteratively apply f to it, i.e., produce the sequence $x, f(x), f^2(x) = f(f(x)), f^3(x), \ldots$, until a repetition is detected[3]. In the sequel, we call such algorithms "**Rho** algorithms". We denote the first repeated value in the sequence $x, f(x), f^2(x), \ldots$ by $f^\mu(x)$ and its second appearance by $f^{\mu+\lambda}(x)$, and call this common value the cycle's entry point.

Optimal Detection when $p \gg N^{-1/2}$. First, we show that when $p \gg N^{-1/2}$, the mode y_0 can be found in time $O(1/p)$. This complexity is clearly the best possible: if the number of queries to f is $o(1/p)$, then with overwhelming probability no preimage of y_0 is queried and so y_0 cannot be detected.

The idea is simple: we run a **Rho** algorithm, with an arbitrary random starting point x and an upper bound c/p on the length of the sequence for some small constant c. For such a length, we expect the mode y_0 to appear twice in the sequence with high probability, whereas due to the fact that $c/p < \sqrt{N}$ the collision found by **Rho** is not expected to be one of the other random values. We show the full analysis of the algorithm in Appendix A.

By using a memoryless **Rho** algorithm, we get a time complexity of $O(1/p)$ and a memory complexity of $O(1)$. As usual, the probability that y_0 is detected can be enhanced even further by repeating the algorithm with other starting points and checking each suggested point in time $O(1/p)$ using the trivial mode verification algorithm.

The RepeatedRho Algorithm: Detection in $O(p^{-3}N^{-1})$ for Arbitrary p. The above approach can be used for any value of p. However, when $p < 1/\sqrt{N}$, the probability that the output of **Rho** (i.e., the cycle's entry point) is indeed y_0 drops significantly. Specifically, we have the following lower bound, and one can easily show that the actual value is not significantly larger.

Proposition 1. *Assume that $p < 1/\sqrt{N}$ and thus **Rho** encounters $O(\sqrt{N})$ different output values until a collision is detected. Then the probability that **Rho** outputs y_0 is $\Omega(p^2 N)$.*

Proof. Since the probability of obtaining y_0 as the output is non-decreasing as a function of p, there is no loss of generality in assuming $p = cN^{-1/2}$ for a small c. In such a case, a lower bound on the probability of **Rho** producing y_0 is the probability that in the first $\sqrt{N}/2$ steps of the sequence $(x, f(x), f^2(x), \ldots)$, each value $y' \neq y_0$ appears at most once, while y_0 appears twice. Formally, let $L' = (x, f(x), \ldots, f^t(x))$, where $t = \min(\mu+\lambda, \sqrt{N}/2)$. Denote by $E_{y'}$ the event: "Each $y' \neq y_0$ appears at most once in L'", and by E_{y_0} the event: "y_0 appears twice in L'". Then

$$\Pr[Output(\mathbf{Rho}) = y_0] \geq \Pr[E_{y'} \wedge E_{y_0}] = \Pr[E_{y'}]\Pr[E_{y_0}|E_{y'}].$$

As we show in Appendix A, we have $\Pr[E_{y'}] \geq e^{-1/4} \approx 0.78$ and

$$\Pr[E_{y_0}|E_{y'}] = \Pr[X \geq 2|X \sim Poi(|L'|p)] \geq \Pr[X \geq 2|X \sim Poi(\sqrt{N}p/2)].$$

---

[3] Such a repetition must occur due to the fact that $f : [N] \rightarrow [N]$.

Finally, for any small λ we have

$$\Pr[X \geq 2|X \sim Poi(\lambda)] = 1 - e^{-\lambda}(1 + \lambda) \approx 1 - (1 - \lambda)(1 + \lambda) = \lambda^2,$$

and thus, combination of the above inequalities yields

$$\Pr[Output(\mathbf{Rho}) = y_0] \geq 0.78(\sqrt{N}p/2)^2 = 0.19p^2N,$$

as asserted. $\qquad\square$

This yields the **RepeatedRho** algorithm – an $O(p^{-3}N^{-1})$ algorithm for detecting the mode: run the **Rho** algorithm $O(1/p^2N)$ times, and check each output point in $O(1/p)$ time using the mode verification algorithm. With a constant probability, y_0 is suggested by at least one of the **Rho** invocations and is thus verified. As $p^{-3}N^{-1} < p^{-2}$ for all $p > N^{-1}$, this algorithm outperforms the sampling algorithm (Algorithm 2) whose running time is $O(p^{-2})$ for all p. See Fig. 3 for comparison of the algorithms for different values of p.

The analysis above implicitly assumes that all the invocations of **Rho** are independent. However, this is clearly not the case if we apply **Rho** to the same function f, while changing only the starting point x in each invocation. Indeed, since $p < 1/\sqrt{N}$, y_0 is not expected to be on a cycle of f, and thus no matter how many times we run the **Rho** algorithm using the same f but different starting points, we will never encounter y_0 as a cycle entry point.

In order to make the invocations of **Rho** essentially independent, we introduce the notion of *flavors* of f, like the flavors used in Hellman's classical time-memory tradeoff attack [3]. More specifically, we define the v's flavor of f as the function $f_v(x) = f(x + v)$ where the addition is computed modulo N. The different flavors of f share some local properties (e.g., they preserve the number of preimages of each y, and thus y_0 remains the mode of the function), but have different global properties (e.g., when iterated, their graphs have a completely different partition into trees and cycles). In particular, it is common to consider the various flavors of f as unrelated functions, even though this is not formally justified. We define the output of the v's invocation of **Rho** as the entry point into the cycle defined by f_v when we start from point v, and run the **RepeatedRho** algorithm by calling **Rho** multiple times with different randomly chosen flavors.

5   The 2Rho Algorithm

In this section we introduce the **2Rho** algorithm, and show that running a **Rho** algorithm over the results of a **Rho** algorithm outperforms all the previously suggested algorithms.

The main idea behind the new algorithm is that a single application of **Rho** can be viewed as a *bootstrapping* step that amplifies the probability of y_0 to be sampled. Indeed, by Proposition 1, The probability that **Rho** with a randomly chosen flavor will output y_0 is $\Omega(p^2N)$, and as long as $p \gg N^{-1}$, this is significantly larger than the probability p that y_0 will be sampled by a single invocation

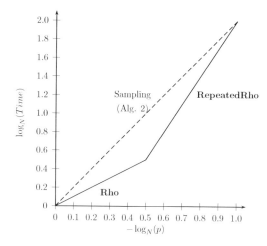

Fig. 3. Comparing the random sampling, **Rho**, and **RepeatedRho** algorithms

of f. Note that by symmetry the probabilities of all the other values of y to be returned by **Rho** with a random flavor remain uniformly low. We are thus facing exactly the same needle finding problem but with a magnified probability peak at exactly the same location y_0. In particular, if this new probability peak exceeds $N^{-0.5}$, we can find it by using a simple **Rho** algorithm. On the other hand, a single evaluation of **Rho** is now more time consuming than a single evaluation of f, and thus the bootstrapping yields a tradeoff between the total number of operations and the cost of each operation, so the parameters should be chosen properly in order to reduce the total complexity.

Our goal now is to formally define the new inner function $g(x)$ which will be used by the outer **Rho** algorithm. This g maps a flavor v to the cycle's entry point defined by running the **Rho** algorithm on the v's flavor of f (i.e., on f_v), starting with the initial value v. When we iterate g, we jump from a flavor to a cycle entry point, and then use the identity of the cycle's entry point to define the next flavor and starting point. This creates a large Rho structure over small Rho structures, as depicted in Fig. 4 in which the different colors indicate the different flavors of f we use in the algorithm. Each dotted line represents the first step we take when we switch to a new flavor, and is used only to visually separate the Rhos so that the figure will look more comprehensible. Note that the collision in the big cycle happens when we encounter the same cycle entry point a second time, but this does not imply that the two colliding small Rho's or their starting points are the same, since they typically use different flavors; it is only in the second and third times we meet the same cycle entry point that their corresponding Rho structures also becomes identical, and from then on we go through the same Rhos over and over.

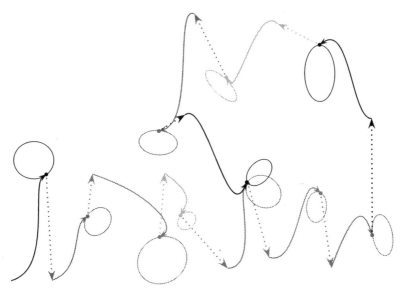

Different colors represent different flavors of f.

Fig. 4. The **2Rho** algorithm (Color figure online)

We now turn our attention to a specific range of probabilities p for which the **2Rho** algorithm (that runs an outer **Rho** algorithm over an inner **Rho** algorithm) offers a significant gain over the previously described algorithms.

5.1 Analysis of 2Rho in the Range $N^{-3/4} \ll p \leq N^{-1/2}$

Assume that $N^{-3/4} \ll p \leq N^{-1/2}$, and construct the function g as described above. Defining $p' = \Pr[g(x) = y_0]$, we have shown that $p' = \Omega(p^2 N) \gg N^{-1/2}$, and thus the mode of g can be found optimally in $O(1/p')$ evaluations of g using the **2Rho** algorithm.

In order to find the total complexity of **2Rho**, we have to compute the complexity of each evaluation of g, i.e., of an evaluation of **Rho** algorithm.

Note that for $c > 1$, the probability that all values $x, f(x), f^2(x), \ldots, f^{c\sqrt{N}}(x)$ are different is at most

$$\frac{(N-1)(N-2) \cdot \ldots \cdot (N - c\sqrt{N})}{(N-1)^{c\sqrt{N}}} \leq \frac{(N-1)^{\sqrt{N}}(N - \sqrt{N} - 1)^{(c-1)\sqrt{N}}}{(N-1)^{c\sqrt{N}}}$$

$$\leq \left(\frac{N - \sqrt{N}}{N}\right)^{(c-1)\sqrt{N}} \approx e^{-(c-1)}.$$

Hence, with an overwhelming probability, **Rho** finds a cycle in $O(\sqrt{N})$ operations. In order to avoid the rare cases where such algorithms take more time, we can slightly modify any **Rho** algorithm by stopping it after a predetermined number of f evaluations (e.g., $10\sqrt{N}$), in which case $g(x) = \textbf{Rho}(f, x+1)$.[4] In any case, the expected time complexity of an evaluation of g is $O(\sqrt{N})$ evaluations of f.

Therefore, the time complexity of **2Rho** is $O(\frac{1}{p'} \cdot \sqrt{N}) = O(p^{-2}N^{-1/2})$ operations. This is significantly faster than the Sampling algorithm, and also significantly faster than **RepeatedRho**, since $p^{-2}N^{-1/2} < p^{-3}N^{-1}$ for all $p < N^{-1/2}$.

Just like the **Rho** algorithm, the nested **2Rho** algorithm can be repeated when $p < N^{-3/4}$, to yield an algorithm for any p. Indeed, repeating **2Rho** until the mode is found (and verified by the verification algorithm), takes $O(p'^{-3}N^{-1}) = O((p^2N)^{-3}N^{-1}) = O(p^{-6}N^{-4})$ evaluations of g, or $O(p^{-6}N^{-3.5})$ evaluations of f. Hence, **Repeated2Rho** is better than the **RepeatedRho** algorithm for $p > N^{-5/6}$ and is worse for $p < N^{-5/6}$.

Table 1 describes our experimental verification of the **2Rho** algorithm for different values of p in the range $N^{-0.79} \le p \le N^{-0.5}$. We used a relatively small $N = 2^{28}$ (which makes the transition at $p = N^{-0.75}$ more gradual than we expect it to be for larger N), and repeated each experiment 100 times with different random functions f.

Table 1. Success rate of **2Rho** for $N = 2^{28}$ over 100 experiments

$p = \Pr[y_0]$		Success
Value	$\log_N(p)$	Rate
2^{-14}	-0.5	100%
2^{-15}	-0.54	100%
2^{-16}	-0.57	100%
2^{-17}	-0.61	97%
2^{-18}	-0.64	91%
2^{-19}	-0.68	71%
2^{-20}	-0.71	32%
2^{-21}	-0.75	8%
2^{-22}	-0.79	0%

6   Deeper Nesting of the Rho Algorithm

We now show how one can nest i**Rho** to obtain $(i+1)$**Rho**. We analyze the resulting complexities, and show that while for a small i, it yields better results,

---

[4] Of course, with a negligible probability, we may need to continue and define $g(x) = \textbf{Rho}(f, x+2)$, and so forth.

as i becomes larger it loses to simpler algorithms. In particular, it is advantageous to nest the **NestedRho** algorithm up to four times, but not a fifth time.

The 3Rho Algorithm for $N^{-7/8} \ll p \leq N^{-3/4}$. Assume that $N^{-7/8} \ll p \leq N^{-3/4}$, and define a new function $h(x)$ which maps an input flavor x into the cycle's entry point defined by the **2Rho** algorithm. As in the analysis of **2Rho** above, we define $p'' = \Pr[h(x) = y_0]$, and can show that $p'' = \Omega(p'^2 N) = \Omega(p^4 N^2 N) \gg N^{-1/2}$. Hence, the mode of h can be found optimally in $O(1/p'')$ evaluations of h using **Rho**. Since each evaluation of h requires $O(\sqrt{N})$ evaluations of g, and since each evaluation of g requires $O(\sqrt{N})$ evaluations of f, the overall complexity of the algorithm is $O(p^{-4}N^{-3}N) = O(p^{-4}N^{-2})$ evaluations of f. We call this algorithm **3Rho**, as it essentially performs yet another nesting layer of **2Rho**.

Algorithm 3. $(i+1)$**Rho** Algorithm for the Function $f(\cdot)$ (Based on i**Rho**)

Input: a random input $x \in [N]$.
Set $z \leftarrow i\mathbf{Rho}(f_x, x)$.   ▷ Note that in the recursion, the flavors of f add up.
while Repeated value of z is not encountered **do**
    Set $z \leftarrow i\mathbf{Rho}(f_z, z)$.
end while
Identify the repeated z value.[a]
return z.

---

[a]The identification can be done using Floyd's algorithm [7], or any of its variants.

The complexity of the **3Rho** algorithm is always better than that of **RepeatedRho** and is better than the $O(p^{-6}N^{-3.5})$ complexity of **Repeated2Rho** for all $p < N^{-3/4}$.

As in the case of **2Rho**, the **3Rho** algorithm can also be repeated as **Repeated3Rho** with mode verification to yield an algorithm for any p. The resulting complexity is $O(p''^{-3}N^{-1}) = O((p^4N^3)^{-3}N^{-1}) = O(p^{-12}N^{-10})$ evaluations of h, or $O(p^{-12}N^{-9})$ evaluations of f. This algorithm is better than the **RepeatedRho** for $p > N^{-8/9}$ and is worse for $p < N^{-8/9}$. However, it turns out that for $N^{-9/10} \ll p \leq N^{-7/8}$, we can do better by nesting **3Rho** yet another time.

The 4Rho Algorithm for $N^{-9/10} \ll p \leq N^{-7/8}$. Assume that $N^{-15/16} \ll p \leq N^{-7/8}$, and define a new mapping $\ell(x)$ which maps a flavor x into the cycle's entry point found by the **3Rho** algorithm. As in the above case of **3Rho**, we have $p''' = \Pr[\ell(x) = y_0] = \Omega(p''^2 N) = \Omega(p^8 N^6 N) \gg N^{-1/2}$. Hence, the mode of ℓ can be found optimally in $O(1/p''')$ evaluations of ℓ using **Rho**.

Since each evaluation of ℓ requires $O(N^{1.5})$ evaluations of f, the overall complexity of the algorithm is $O(p^{-8}N^{-7}N^{1.5}) = O(p^{-8}N^{-5.5})$ evaluations of f. We call this algorithm **4Rho**, as it performs a four-layer nesting of **Rho**.

Unlike the previous algorithms, **4Rho** is not better than all previous algorithms in the whole range $N^{-15/16} \ll p \leq N^{-7/8}$. Indeed, as $p \to N^{-15/16}$, the

Table 2. Summary of the best complexities of algorithms for detecting the mode

Probability range	Complexity formula	Complexity range	Algorithm
$p \geq N^{-0.5}$	$T = p^{-1}$	$T \leq N^{0.5}$	**Rho**
$N^{-0.75} \leq p \leq N^{-0.5}$	$T = p^{-2}N^{-0.5}$	$N^{0.5} \leq T \leq N$	**2Rho**
$N^{-0.875} \leq p \leq N^{-0.75}$	$T = p^{-4}N^{-2}$	$N \leq T \leq N^{1.5}$	**3Rho**
$N^{-0.9} \leq p \leq N^{-0.875}$	$T = p^{-8}N^{-5.5}$	$N^{1.5} \leq T \leq N^{1.7}$	**4Rho**
$N^{-1} \leq p \leq N^{-0.9}$	$T = p^{-3}N^{-1}$	$N^{1.7} \leq T \leq N^{2}$	**RepeatedRho**

complexity of **4Rho** approaches N^2, which is higher than even the straightforward Sampling Algorithm. In particular, **4Rho** is faster than **RepeatedRho** only as long as $p > N^{-0.9}$, which explains why the complexity curve reduces its slope at the top right corner of Fig. 5.

We note that the natural extension to **5Rho** is clearly inferior for any p since the complexity of each step of the outer **Rho** requires N^2 steps, which is already higher than the overall complexity of the Sampling algorithm.

The complexities of the best algorithms we were able to achieve (as a function of p) are presented in a mathematical form in Table 2 and in graphical form in Fig. 5.

7 Time-Memory Tradeoffs

In this section, we revisit the basic problem of detecting the mode, but assume that we have $O(M)$ memory cells available. Our goal is to detect the mode as efficiently as possible, where the complexity is formulated as a function of the parameters N, p and M.

Before starting, we note that we only deal with the case of $p < N^{-1/2}$, as we already have an optimal[5] memoryless algorithm for the case of $p \geq N^{-1/2}$ (as shown in Sect. 4).

We begin by describing the basic parallel collision search algorithm of [10]. We then describe a sequence of algorithms that extend the i**Rho** memoryless algorithms using parallel collision search.

7.1 Parallel Collision Search

The parallel collision search (**PCS**) algorithm presented by van Oorschot and Wiener [10] is a memory-efficient algorithm for finding multiple collisions at low amortized cost per collision in a function f that maps $[N]$ to $[N]$. Since its introduction, the algorithm has been extensively used in cryptanalysis (e.g., in [4–6,9]). Given M memory cells, the algorithm builds a structure of

---

[5] Given additional memory and/or CPUs allows parallelizing **Rho** algorithms. At the same time, the total computational complexity (which is the focus of this paper) remains the same, or (in some cases) may become worse.

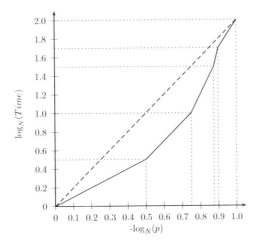

Fig. 5. Complexities of our best memoryless algorithms as a function of p

M chains which is similar to the one built in Hellman's time-memory tradeoff algorithm [3].

A chain in the structure starts at an arbitrary point x, and is evaluated by repeated applications of f (namely, $f^i(x) = f(f^{i-1}(x))$). The chains are terminated after about $\sqrt{N/M}$ evaluations of f, thus the structure contains a total of about $M \cdot \sqrt{N/M} = \sqrt{NM}$ points. Moreover, as $\sqrt{N/M} \cdot \sqrt{NM} = N$, according to the birthday paradox, each chain is expected to collide with another chain in the structure, and hence the chain structure contains $O(M)$ collisions. In order to find the $O(M)$ collisions efficiently, we define a set of *distinguished points* and terminate each chain once it reaches such a point. In our case, we define a set of $\sqrt{NM}$ distinguished points (e.g., the points whose $(\log_2(N) + \log_2(M))/2$ least significant bits are zero), and hence the expected chain size is $N/\sqrt{NM} = \sqrt{N/M}$ as required. The actual $O(M)$ collision points are recovered by sorting the M termination points of the chains (which are distinguished points), and restarting the chain computation for each colliding pair of chains. For sake of completeness we give in Appendix B the pseudo code for **PCS** (Algorithm 6). In total, the algorithm finds $O(M)$ collisions in $\sqrt{NM}$ time using $O(M)$ memory.

7.2 Mode Verification with Memory

The basic memoryless mode verification algorithm (Algorithm 1) can be extended to exploit memory, by checking multiple targets simultaneously. Namely, given M candidate y_i's, it is possible to check all of them at the same time for the cost of $O(1/p)$ queries to f, as suggested by Algorithm 4.

Using Algorithm 4, we can immediately improve the sampling algorithm (Algorithm 2). Instead of checking only one value at each call to the verification algorithm, we can now check M such values for the same complexity. Hence, Algorithm 5 picks each time M random values of y_i by random sampling, and calls Algorithm 4 to test which of them (if at all) is indeed y_0.

Algorithm 4. Mode Verification: Determining Whether y_0 is one of $y_1, y_2, \ldots y_M$

Initialize an array of counters $ctr[i] \leftarrow 0$ for $1 \leq i \leq M$.
for $j = 1$ to c/p **do**
    Pick at random $x \in [N]$, and compute $y' = f(x)$.
    if $y' = y_i$ for $1 \leq i \leq M$ **then**
        Increment $ctr[i]$.
    end if
end for
for $i = 1$ to N **do**
    if $ctr[i] \geq t$ **then**
        print y_i is y_0.
    end if
end for

Algorithm 5. Finding y_0 by Sampling (with Memory):

while y_0 was not found **do**
    for $i = 1$ to M **do**
        Pick $x_i \in [N]$ at random.
        Compute $y_i = f(x_i)$.
    end for
    Call Algorithm 4 to check $y_1, y_2, \ldots, y_M$.
end while

The probability that a single call to Algorithm 4 (testing M images) succeeds is about Mp (assuming[6] $M \leq p^{-1}$), and therefore we expect $O(M^{-1}p^{-1})$ calls to Algorithm 4. Each such call takes $O(p^{-1})$ evaluations of f, and hence the total time complexity of the algorithm is $O(M^{-1}p^{-2})$. Note that for $M = 1$ this algorithm reduces to Algorithm 2.

7.3 Mode Detection with Parallel Collision Search

This algorithm runs **PCS** with M chains and checks the $O(M)$ collision points found by running Algorithm 4. This process is repeated until it finds y_0, where each repetition is performed with a different flavor of f.

Since the M chains cover about $\sqrt{NM}$ distinct points, the probability that two distinct preimages of the mode y_0 (which are not expected to be distinguished points) are covered by the structure is about $(\sqrt{NM} \cdot p)^2 = NM \cdot p^2$ (assuming $\sqrt{NM} \cdot p < 1$, i.e., $p < (NM)^{-0.5}$). In this case, the algorithm will successfully recover the mode y_0 using the mode verification algorithm. Therefore, the algorithm is expected to execute **PCS** (and mode verification) about $N^{-1}M^{-1} \cdot p^{-2}$ times, where each execution requires $O(p^{-1})$ time (assuming $p < (NM)^{-1/2}$, mode verification dominates **PCS** in terms of time complexity). In total, the time complexity of the algorithm is $O(M^{-1}N^{-1} \cdot p^{-3})$. Note that for $M = 1$ we obtain **RepeatedRho**.

---

[6] We note that when $M > p^{-1}$, it is sufficient to fill $O(p^{-1})$ memory cells.

The formula above is only valid for $p < (NM)^{-0.5}$ or $M < p^{-2}N^{-1}$. Otherwise, we can utilize only $M = p^{-2}N^{-1}$ memory and obtain the essentially optimal time complexity of $O(p^{-1})$.

7.4   Mode Detection with Parallel Collision Search Over 2Rho

We now assume that $M < p^{-2}N^{-1}$ (otherwise, we use the previous **PCS** algorithm to detect the mode with optimal complexity) and extend the **2Rho** algorithm using **PCS**. This is done by defining a chain structure, computed by iterating the function g (as defined in Sect. 5) whose execution is computed by iterating a particular flavor of f until a collision point is found. Each chain starts with an arbitrary input to g (which defines a flavor of f) and is terminated at a distinguished point of g. Namely, the distinguished points are defined on the outputs of g (which are the collision points in f). Once again, we use Algorithm 4 to test the $O(M)$ collisions of g.

As calculated in Sect. 5 the probability that the mode y_0 will be the collision point in a single run of **Rho** (an iteration of g) is $p' = p^2 N$. Since the M chains of g cover about $\sqrt{NM}$ distinct collision points, the probability that two distinct preimages of the mode y_0 in g (which are not expected to be distinguished points) will be covered by the structure is about $(\sqrt{NM} \cdot p')^2 = NM \cdot p'^2 = NM \cdot p^4 N^2 = M \cdot N^3 p^4$ (assuming $\sqrt{NM} \cdot p' < 1$ or $p^2 N \cdot (NM)^{1/2} < 1$, namely $p^2 N^{3/2} M^{1/2} < 1$). As a result, we repeat the **PCS** algorithm (and the mode verification algorithm) $M^{-1} \cdot N^{-3} p^{-4}$ times (using distinct flavors of g). The **PCS** algorithm requires $(NM)^{1/2}$ invocations of g, each requiring $N^{1/2}$ time, namely, $N \cdot M^{1/2}$ time in total which dominates the complexity of the mode verification. Overall, the time complexity of the algorithm is $M^{-1} \cdot N^{-3} p^{-4} \cdot N \cdot M^{1/2} = M^{-1/2} \cdot N^{-2} p^{-4}$.

The formula above is only valid given that $p^2 N^{3/2} M^{1/2} < 1$ or $M < p^{-4} N^{-3}$. Otherwise, we can utilize only $M = p^{-4} N^{-3}$ (assuming[7] $p^{-4} N^{-3} \geq 1$) memory and obtain time complexity of $M^{-1/2} \cdot N^{-2} p^{-4} = p^2 N^{3/2} \cdot N^{-2} p^{-4} = p^{-2} N^{-1/2}$.

We now notice that it is possible to obtain more generic formulas that can be reused later. Essentially, the analysis of the algorithm depends on three parameters, as follows. The probability that the mode y_0 will be the collision point in a single run of **Rho** (an iteration of g) is $p' = p^2 N$, which we denote as $p^{x_1} N^{x_2}$ for $x_1 = 2, x_2 = 1$ in our case. In addition, each invocation of g requires $N^{1/2}$ time, which we denote as N^{x_3} for $x_3 = 1/2$ in this case. Based on these parameters, we can redo the analysis above symbolically and obtain that the time complexity of the algorithm is $M^{-1/2} \cdot N^{-2x_2 - 1/2 + x_3} p^{-2x_1}$.

This formula is only valid given that $M < p^{-2x_1} N^{-2x_2 - 1}$. Otherwise, we can utilize only $M = p^{-2x_1} N^{-2x_2 - 1}$ (assuming $p^{-2x_1} N^{-2x_2 - 1} \geq 1$) memory and obtain time complexity of $p^{-x_1} N^{-x_2 + x_3}$.

---

[7] When $p^{-4} N^{-3} < 1$, the algorithm is not applicable in its current form.

7.5 Mode Detection with Parallel Collision Search over 3Rho

We continue to analyze the sequence of algorithms that extend **3Rho** using **PCS**. The idea is essentially the same as in the extension of **2Rho**, where the difference is the function over which **PCS** is performed.

Here, **PCS** is executed over the function h (as defined in Sect. 6) while calling Algorithm 4 to test the $O(M)$ collisions points of h.

As calculated in Sect. 6 the probability that the mode y_0 will be the collision point in a single run of g is $p'' = p^4 N^3$, which we denote as $p^{x_1} N^{x_2}$ for $x_1 = 4, x_2 = 3$. In this case, each invocation of h requires N time, or N^{x_3} for $x_3 = 1$.

We now reuse the formulas obtained in Sect. 7.4 and consider our specific parameters $x_1 = 4, x_2 = 3, x_3 = 1$ for the case $M < p^{-2x_1} N^{-2x_2-1}$, or $M < p^{-8} N^{-7}$ assuming $p^{-8} N^{-7} \geq 1$ or $p \leq N^{-7/8}$. This gives time complexity of $M^{-1/2} \cdot N^{-2x_2-1/2+x_3} p^{-2x_1}$ or $M^{-1/2} \cdot N^{-5.5} p^{-8}$. Note that for $M = 1$ we obtain Algorithm **4Rho**.

For $M > p^{-8} N^{-7}$, we obtain time complexity of $p^{-x_1} N^{-x_2+x_3} = p^{-4} N^{-2}$. See Fig. 6 for comparison of different algorithms given $M = N^{1/4}$ memory.

Mode Detection with Parallel Collision Search over 2Rho. The extension of **PCS** over **4Rho** does not make sense since the function ℓ (defined in Sect. 6) used for **4Rho** is never iterated more than $N^{0.5}$ times in our algorithms. Hence, all its iterations can be covered by a single chain of **4Rho** and there is no benefit in using memory in this case.

7.6 Discussion

It is not intuitive to compare the algorithms described above, as their complexities are functions of both p and M. In order to get some intuition regarding their performance, we fix $M = N^{1/4}$ and summarize the complexity of the best algorithms for this case as a function of the single parameter p in Table 3. It is evident from the table that there is a range of p values for which we do not know how to efficiently exploit the memory. For example, consider $p = N^{-3/4}$, where our best algorithm is **PCS** over **2Rho**. However, it is actually a degenerate variant of **PCS** with $M = 1$ that coincides with the **2Rho** algorithm of Sect. 6.

8 Finding Multiple Peaks

We consider a generalization of our basic problem to the case that f is uniformly distributed except for k peaks. The peaks are denoted by $y_0, y_1, \ldots, y_{k-1}$, their associated probabilities are denoted by $p_0, p_1, \ldots, p_{k-1}$, and our goal is to find all of them.[8]

---

[8] In [1] a related problem is studied: Let f be a hash function. Assume that its range is smaller than its domain and that it is not balanced (i.e., not all outputs appear with the same probability). This work studies the effect of this irregularity on the complexity of the birthday collision search. In contrast, our work studies the algorithmic aspects of finding the collision (in a memory-efficient manner).

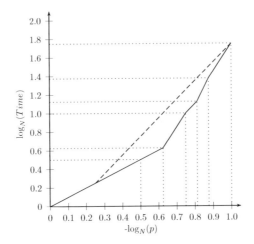

Fig. 6. Complexities of our best algorithms as a function of p given $M = N^{1/4}$ memory

Table 3. Summary of the best complexities of algorithms for detecting the mode with $M = N^{1/4}$

Probability range	Complexity formula	Complexity range	Algorithm
$p \geq N^{-0.5}$	$T = p^{-1}$	$T \leq N^{0.5}$	**Rho**
$N^{-5/8} \leq p \leq N^{-0.5}$	$T = p^{-1}$	$N^{0.5} \leq T \leq N^{5/8}$	**PCS**
$N^{-3/4} \leq p \leq N^{-5/8}$	$T = p^{-3}N^{-5/4}$	$N^{5/8} \leq T \leq N$	**PCS**
$N^{-13/16} \leq p \leq N^{-3/4}$	$T = p^{-2}N^{-1/2}$	$N \leq T \leq N^{9/8}$	**PCS over 2Rho**
$N^{-7/8} \leq p \leq N^{-13/16}$	$T = p^{-4}N^{-17/8}$	$N^{9/8} \leq T \leq N^{11/8}$	**PCS over 2Rho**
$N^{-1} < p \leq N^{-7/8}$	$T = p^{-3}N^{-5/4}$	$N^{11/8} \leq T \leq N^{7/4}$	**PCS**

The simplest case is one in which there are two peaks of equal height $p_0 = p_1$. By running the **NestedRho** algorithm several times with different flavors of f, we expect to find each one of y_0 and y_1 about half the time, and thus there is no need to modify anything.

The next case to consider is one in which there are only two peaks but $p_0 > p_1$. Due to the high power of p in our formulas, even moderate differences in the peak probabilities are amplified by the **NestedRho** algorithm to huge differences in the probability of finding the two peaks. For example, if p_0 is a thousand times bigger than p_1, and we run the algorithm multiple times, then we expect to find y_1 only in one in a million runs when we use **1Rho**, and only in one in a trillion runs when we use **2Rho**. Clearly, we have to reduce the attractiveness of y_0 before we have a realistic chance of noticing y_1.

The simplest way to neutralize the first peak we find (which is likely to be y_0), is to scatter its preimages so that they will point to different targets. Consider a modified function f' which is defined as f for any x for which $f(x) \neq y_0$, and as $f(x) + x$ for any x for which $f(x) = y_0$. In f', y_0 is no longer a peak, but y_1

remains at its original height. By applying **NestedRho** to f', we will find y_1 with high probability.

This can be easily generalized to a sequence of k peaks, provided that we have at least $O(k)$ memory to store all the peaks. Our algorithm is likely to discover them sequentially in decreasing order of probability, and we can decide to stop at any point when we run out of space or time.

The most general case is one in which we have a non-uniform distribution with no sharp peaks. In this case the output of the **NestedRho** algorithm has a preference to pick y values with higher probabilities, but may pick a lower probability y if there are many such values. In fact, the probability that our algorithm will pick a particular y is proportional to some power of its original probability, which depends on which nesting level we use (the detailed analysis is left for future work).

9 Conclusions and Open Problems

In this paper we introduced the generic problem of finding needles in haystacks, developed several novel techniques for its solution, and demonstrated the surprising complexity of its complexity function. Many problems remain open, such as:

1. Find non-trivial lower bounds on the time complexity of the problem.
2. Find better ways to exploit the available memory, beyond using **PCS**.
3. Extend the model to deal with other types of needles.
4. Find additional applications of the new **NestedRho** technique.

Acknowledgements. The authors thank Masha Gutman for her implementation of the experiments reported in Table 1.

A   Detailed Complexity Analysis of the Rho Approach for $p > 1/\sqrt{N}$

Consider the sequence $x, f(x), f^2(x), \ldots$. If we limit the length of the sequence by $4/p$ "random" steps, then with high enough probability, we expect to encounter y_0 twice. On the other hand, the probability that a "random" value is encountered twice is low, since $4/p$ is significantly smaller than the "birthday bound" $\sqrt{N}$. Hence, y_0 is expected to be the first repeated point, and hence, the output.

Formally, let A be a "truncated" **Rho** algorithm:[9]

1. Choose $x \in [N]$ uniformly at random.
2. Run **Rho** algorithm that computes the chain $x, f(x), \ldots, f^{4/p}(x)$ (or shorter chain if a collision is found before).

---

[9] The reader may think of the algorithm as Floyd's one, but the same analysis holds for any "reasonable" memoryless detection algorithm.

(a) If a collision is detected, denote its value by y, and run the verification algorithm on y.
(b) If no collision is detected, output "FAIL".

Proposition 2. *Assume that Algorithm A is run in the case $p \geq 16/\sqrt{N}$. Then $\Pr[Output(A) = y_0] \geq 0.69$.*

Proof. Throughout the proof we consider the sequence $L = (x, f(x), \ldots, f^{\mu+\lambda}(x))$ of values encountered by the algorithm until the first repetition (inclusive) or until the process terminates (if a repetition was not encountered). By the definition of f, this sequence is distributed like an *independent* sampling of $\mu + \lambda$ elements of the distribution of $range(f)$. Note that if a meeting point is detected at step t, this implies that the sequence $x, f(x), \ldots, f^{2t}(x)$ contains a repetition, and thus, $|L| \leq 8/p \leq \sqrt{N}/2$.

First, we bound from above $\Pr[\exists y' \neq y_0 : Output(A) = y']$, i.e., the probability that some $y' \neq y_0$ appears twice in L. Consider all values non-equal to y_0 that appear in L. Since $|L| \leq \sqrt{N}/2$, the probability that they are mutually different is at least

$$\frac{(N-1)(N-2)\cdots(N-|L|)}{(N-1)^{|L|}} \geq \left(\frac{N-|L|}{N}\right)^{|L|} \geq \left(\frac{N-\sqrt{N}/2}{N}\right)^{\sqrt{N}/2} \approx e^{-1/4}.$$

Hence, $\Pr[\exists y' \neq y_0 : Output(A) = y'] \leq 1 - e^{-1/4} \approx 0.22$.

Second, we bound from above $\Pr[Output(A) = FAIL]$, i.e., the probability that neither y_0 nor any other value appears twice in L. Note that in such a case, $|L| = 4/p$ since no repetition is encountered. By the definition of f, for any k, $\Pr[f^k(x) = y_0] = p$. Hence, the number of occurrences of y_0 in L is distributed like a $Bin(|L|, p) = Bin(4/p, p)$ random variable, that can be approximated by a $Poi(|L|p) = Poi(4)$ random variable. Hence,

$$\Pr[Output(A) = FAIL] \leq \Pr[Poi(4) \leq 1] = e^{-4} + 4e^{-4} \approx 0.09.$$

Combining the two bounds, we obtain

$$\Pr[Output(A) = y_0] = 1 - \Pr[\exists y' \neq y_0 : Output(A) = y'] - \Pr[Output(A) = FAIL]$$
$$\geq 1 - 0.22 - 0.09 = 0.69,$$

as asserted. $\qquad\qquad\square$

Algorithm 6. Par~~allel~~ ~~Collision~~ Search Algorithm

Initialize an empty table.
for $i = 1$ to M **do**
    Pick at random a point $x_i \in$ ~~...~~
    Set $tmp \leftarrow x_i$, $len \leftarrow 0$.
    while $f(tmp)$ is not a distinguished p~~oint~~
        $tmp \leftarrow f(tmp)$.
        Increment len.
    end while
    $tmp \leftarrow f(tmp)$.
    Increment len.
    Store in the table the pair (tmp, x_i, len).
end for
for All collisions $((p_i, x_i, len_i), (p_j, x_j, len_j))$ s.t. $p_i = p_j$ **do**
    Set $tmp_1 \leftarrow x_i$, $tmp_2 \leftarrow x_j$.
    if $len_1 > len_2$ **then**
        for $i = 1$ to $len_1 - len_2$ **do**
            $tmp_1 \leftarrow f(tmp_1)$
        end for
    end if
    if $len_2 > len_1$ **then**
        for $i = 1$ to $len_2 - len_1$ **do**
            $tmp_2 \leftarrow f(tmp_2)$
        end for
    end if
    while $f(tmp_1) \neq f(tmp_2)$ **do**
        $tmp_1 \leftarrow f(tmp_1)$, $tmp_2 \leftarrow f(tmp_2)$
    end while
    print tmp_1, tmp_2.
end for

References

1. Bellare, M., Kohno, T.: Hash function balance and its impact on birthday attacks. In: Cachin, C., Camenisch, J.L. (eds.) EUROCRYPT 2004. LNCS, vol. 3027, pp. 401–418. Springer, Heidelberg (2004)
2. Richard, R.P.: An improved monte carlo factorization algorithm. BIT Numer. Math. **20**(2), 176–184 (1980). doi:10.1007/BF01933190
3. Hellman, M.E.: A cryptanalytic time-memory trade-off. IEEE Trans. Inf. Theory **26**(4), 401–406 (1980)
4. Joux, A., Lucks, S.: Improved generic algorithms for 3-collisions. In: Matsui, M. (ed.) ASIACRYPT 2009. LNCS, vol. 5912, pp. 347–363. Springer, Heidelberg (2009)

206   I. Dinur et al.

...nostradamus attack. In: ...004, pp. 183–200. Springer,

5. Kelsey, J., Kohno, T.: Herdin... ...-bit hash functions for much less
Vaudenay, S. (ed.) EUROCR... CRYPT 2005. LNCS, vol. 3494, pp.
Heidelberg (2006)
6. Kelsey, J., Schneier, B.: ...ogramming: Seminumerical Algorithms, vol.
than 2^n work. In: Cra...
474–490. Springer, H... g a stack. Inf. Process. Lett. 90(3), 135–140 (2004)
7. Knuth, D.E.: The ...ppelbaum, J., Lenstra, A., Molnar, D., Osvik, D.A.,
II. Addison-Wesle...
8. Nivasch, G.: Cy... sen-prefix collisions for MD5 and the creation of a rogue
9. Stevens, M., ...alevi, S. (ed.) CRYPTO 2009. LNCS, vol. 5677, pp. 55–69.
de Weger, ...rg (2009)
CA certifi....C., Wiener, M.J.: Parallel collision search with cryptanalytic appli-
10. van O... ryptol. 12(1), 1–28 (1999)
Springer...
catio...

Breaking Symmetric Cryptosystems Using Quantum Period Finding

Marc Kaplan[1,2(✉)], Gaëtan Leurent[3], Anthony Leverrier[3],
and María Naya-Plasencia[3]

[1] LTCI, Télécom ParisTech, 23 avenue d'Italie, 75214 Paris CEDEX 13, France
kapmarc@gmail.com
[2] School of Informatics, University of Edinburgh,
10 Crichton Street, Edinburgh EH8 9AB, UK
[3] Inria Paris, Paris, France

Abstract. Due to Shor's algorithm, quantum computers are a severe threat for public key cryptography. This motivated the cryptographic community to search for quantum-safe solutions. On the other hand, the impact of quantum computing on secret key cryptography is much less understood. In this paper, we consider attacks where an adversary can query an oracle implementing a cryptographic primitive in a quantum superposition of different states. This model gives a lot of power to the adversary, but recent results show that it is nonetheless possible to build secure cryptosystems in it.

We study applications of a quantum procedure called *Simon's algorithm* (the simplest quantum period finding algorithm) in order to attack symmetric cryptosystems in this model. Following previous works in this direction, we show that several classical attacks based on finding collisions can be dramatically sped up using Simon's algorithm: finding a collision requires $\Omega(2^{n/2})$ queries in the classical setting, but when collisions happen with some hidden periodicity, they can be found with only $O(n)$ queries in the quantum model.

We obtain attacks with very strong implications. First, we show that the most widely used modes of operation for authentication and authenticated encryption (*e.g.* CBC-MAC, PMAC, GMAC, GCM, and OCB) are completely broken in this security model. Our attacks are also applicable to many CAESAR candidates: CLOC, AEZ, COPA, OTR, POET, OMD, and Minalpher. This is quite surprising compared to the situation with encryption modes: Anand *et al.* show that standard modes are secure with a quantum-secure PRF.

Second, we show that Simon's algorithm can also be applied to slide attacks, leading to an exponential speed-up of a classical symmetric cryptanalysis technique in the quantum model.

Keywords: Post-quantum cryptography · Symmetric cryptography · Quantum attacks · Block ciphers · Modes of operation · Slide attack

© International Association for Cryptologic Research 2016
M. Robshaw and J. Katz (Eds.): CRYPTO 2016, Part II, LNCS 9815, pp. 207–237, 2016.
DOI: 10.1007/978-3-662-53008-5_8

1 Introduction

The goal of post-quantum cryptography is to prepare cryptographic primitives to resist quantum adversaries, *i.e.* adversaries with access to a quantum computer. Indeed, cryptography would be particularly affected by the development of large-scale quantum computers. While currently used asymmetric cryptographic primitives would suffer from devastating attacks due to Shor's algorithm [43], the status of symmetric ones is not so clear: generic attacks, which define the security of ideal symmetric primitives, would get a quadratic speed-up thanks to Grover's algorithm [24], hinting that doubling the key length could restore an equivalent ideal security in the post-quantum world. Even though the community seems to consider the issue settled with this solution [6], only very little is known about real world attacks, that determine the real security of used primitives. Very recently, this direction has started to draw attention, and interesting results have been obtained. New theoretical frameworks to take into account quantum adversaries have been developed [2,11,12,15,20,23].

Simon's algorithm [44] is central in quantum algorithm theory. Historically, it was an important milestone in the discovery by Shor of his celebrated quantum algorithm to solve integer factorization in polynomial time [43]. Interestingly, Simon's algorithm has also been applied in the context of symmetric cryptography. It was first used to break the 3-round Feistel construction [31] and then to prove that the Even-Mansour construction [32] is insecure with superposition queries. While Simon's problem (which is the problem solved with Simon's algorithm) might seem artificial at first sight, it appears in certain constructions in symmetric cryptography, in which ciphers and modes typically involve a lot of structure.

These first results, although quite striking, are not sufficient for evaluating the security of actual ciphers. Indeed, the confidence we have on symmetric ciphers depends on the amount of cryptanalysis that was performed on the primitive. Only this effort allows researchers to define the security margin which measures how far the construction is from being broken. Thanks to the large and always updated cryptanalysis toolbox built over the years in the *classical* world, we have solid evaluations of the security of the primitives against classical adversaries. This is, however, no longer the case in the post-quantum world, *i.e.* when considering quantum adversaries.

We therefore need to build a complete cryptanalysis toolbox for quantum adversaries, similar to what has been done for the classical world. This is a fundamental step in order to correctly evaluate the post-quantum security of current ciphers and to design new secure ciphers for the post-quantum world.

Our Results. We make progresses in this direction, and open new surprising and important ranges of applications for Simon's algorithm in symmetric cryptography:

1. The original formulation of Simon's algorithm is for functions whose collisions happen only at some hidden period. We extend it to functions that have

more collisions. This leads to a better analysis of previous applications of Simon's algorithm in symmetric cryptography.

2. We then show an attack against the LRW construction, used to turn a block-cipher into a tweakable block cipher [33]. Like the results on 3-round Feistel and Even-Mansour, this is an example of construction with provable security in the classical setting that becomes insecure against a quantum adversary.

3. Next, we study block cipher modes of operation. We show that some of the most common modes for message authentication and authenticated encryption are completely broken in this setting. We describe forgery attacks against standardized modes (CBC-MAC, PMAC, GMAC, GCM, and OCB), and against several CAESAR candidates, with complexity only $O(n)$, where n is the size of the block. In particular, this partially answers an open question by Boneh and Zhandry [13]: "Do the CBC-MAC or NMAC constructions give quantum-secure PRFs?".

Those results are in stark contrast with a recent analysis of encryption modes in the same setting: Anand *et al.* show that some classical encryption modes are secure against a quantum adversary when using a quantum-secure PRF [3]. Our results imply that some authentication and authenticated encryption schemes remain insecure with *any* block cipher.

4. The last application is a quantization of slide attacks, a popular family of cryptanalysis that is independent of the number of rounds of the attacked cipher. Our result is the first exponential speed-up obtained directly by a quantization of a classical cryptanalysis technique, with complexity dropping from $O(2^{n/2})$ to $O(n)$, where n is the size of the block.

These results imply that for the symmetric primitives we analyze, doubling the key length is not sufficient to restore security against quantum adversaries. A significant effort on quantum cryptanalysis of symmetric primitives is thus crucial for our long-term trust in these cryptosystems.

The Attack Model. We consider attacks against classical cryptosystems using quantum resources. This general setting broadly defines the field of post-quantum cryptography. But attacking specific cryptosystems requires a more precise definition of the operations the adversary is allowed to perform. The simplest setting allows the adversary to perform local quantum computation. For instance, this can be modeled by the quantum random oracle model, in which the adversary can query the oracle in an arbitrary superposition of the inputs [11,14,45,49]. A more practical setting allows quantum queries to the hash function used to instantiate the oracle on a quantum computer.

We consider here a much stronger model in which, in addition to local quantum operations, an adversary is granted an access to a possibly remote cryptographic oracle in superposition of the inputs, and obtains the corresponding superposition of outputs. In more detail, if the encryption oracle is described by a classical function $\mathcal{O}_k : \{0,1\}^n \rightarrow \{0,1\}^n$, then the adversary can make standard quantum queries $|x\rangle|y\rangle \mapsto |x\rangle|\mathcal{O}_k(x) \oplus y\rangle$, where x and y are arbitrary

n-bit strings and $|x\rangle$, $|y\rangle$ are the corresponding n-qubit states expressed in the computational basis. A circuit representing the oracle is given in Fig. 1. Moreover, any superposition $\sum_{x,y} \lambda_{x,y} |x\rangle |y\rangle$ is a valid input to the quantum oracle, who then returns $\sum_{x,y} \lambda_{x,y} |x\rangle |y \oplus O_k(x)\rangle$. In previous works, these attacks have been called *superposition attacks* [20], *quantum chosen message attacks* [13] or *quantum security* [48].

Simon's algorithm requires the preparation of the uniform superposition of all n-bit strings, $\frac{1}{\sqrt{2^n}} \sum_x |x\rangle |0\rangle$[1]. For this input, the quantum encryption oracle returns $\frac{1}{\sqrt{2^n}} \sum_x |x\rangle |O_k(x)\rangle$, the superposition of all possible pairs of plaintext-ciphertext. It might seem at first that this model gives an overwhelming power to the adversary and is therefore uninteresting. Note, however, that the laws of quantum mechanics imply that the measurement of such a $2n$-qubit state can only reveal $2n$ bits of information, making this model nontrivial.

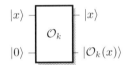

Fig. 1. The quantum cryptographic oracle.

The simplicity of this model, together with the fact that it encompasses any reasonable model of quantum attacks makes it very interesting. For instance, [12] gave constructions of message authenticated codes that remain secure against superposition attacks. A similar approach was initiated by [20], who showed how to construct secure multiparty protocols when an adversary can corrupt the parties in superposition. A protocol that is proven secure in this model may truthfully be used in a quantum world.

Our work shows that superposition attacks, although they are not trivial, allow new powerful strategies for the adversary. Modes of operation that are provably secure against classical attacks can then be broken. There exist a few options to prevent the attacks that we present here. A possibility is to forbid all kind of quantum access to a cryptographic oracle. In a world where quantum resources become available, this restriction requires a careful attention. This can be achieved for example by performing a quantum measurement of any incoming quantum query to the oracle. But this task involves meticulous engineering of quantum devices whose outcome remains uncertain. Even information theoretically secure quantum cryptography remains vulnerable to attacks on their implementations, as shown by attacks on quantum key distribution [35,46,50].

A more realistic approach is to develop a set of protocols that remains secure against superposition attacks. Another advantage of this approach is that it also covers more advanced scenarios, for example when an encryption device is given

---

[1] When there is no ambiguity, we write $|0\rangle$ for the state $|0 \ldots 0\rangle$ of appropriate length.

to the adversary as an obfuscated algorithm. Our work shows how important it is to develop protocols that remain secure against superposition attacks.

Regarding symmetric cryptanalysis, we have already mentioned the protocol of Boneh and Zhandry for MACs that remains secure against superposition attacks. In particular, we answer negatively to their question asking wether CBC-MAC is secure in their model. Generic quantum attacks against symmetric cryptosystems have also been considered. For instance, [28] studies the security of iterated block ciphers, and Anand et al. investigated the security of various modes of operations for encryption against superposition attacks [3]. They show that OFB and CTR remain secure, while CBC and CFB are not secure in general (with attacks involving Simon's algorithm), but are secure if the underlying PRF is quantum secure. Recently, [29] considers symmetric families of cryptanalysis, describing quantum versions of differential and linear attacks.

Cryptographic notions like indistinguishability or semantic security are well understood in a classical world. However, they become difficult to formalize when considering quantum adversaries. The quantum chosen message model is a good framework to study these [2,15,23].

In this paper, we consider forgery attacks: the goal of the attacker is to forge a tag for some arbitrary message, without the knowledge of the secret key. In a quantum setting, we follow the EUF-qCMA security definition that was given by Boneh and Zhandry [12]. A message authentication code is broken by a quantum existential forgery attack if after q queries to the cryptographic oracle, the adversary can generate at least $q+1$ valid messages with corresponding tags.

Organization. The paper is organized as follows. First, Sect. 2 introduces Simon's algorithm and explains how to modify it in order to handle functions that only approximately satisfy Simon's promise. This variant seems more appropriate for symmetric cryptography and may be of independent interest. Section 3 summarizes known quantum attacks against various constructions in symmetric cryptography. Section 4 presents the attack against the LRW constructions. In Sect. 5, we show how Simon's algorithm can be used to obtain devastating attacks on several widely used modes of operations: CBC-MAC, PMAC, GMAC, GCM, OCB, as well as several CAESAR candidates. Section 6 shows the application of the algorithm to slide attacks, providing an exponential speed-up. The paper ends in Sect. 7 with a conclusion, pointing out possible new directions and applications.

2   Simon's Algorithm and Attack Strategy

In this section, we present Simon's problem [44] and the quantum algorithm for efficiently solving it. The simplest version of our attacks directly exploits this algorithm in order to recover some secret value of the encryption algorithm. Previous works have already considered such attacks against 3-round Feistel schemes and the Even-Mansour construction (see Sect. 3 for details).

Unfortunately, it is not always possible to recast an attack in terms of Simon's problem. More precisely, Simon's problem is a promise problem, and in many cases, the relevant promise (that only a structured class of collisions can occur) is not satisfied, far from it in fact. We show in Theorem 1 below that, however, these additional collisions do not lead to a significant increase of the complexity of our attacks.

2.1   Simon's Problem and Algorithm

We first describe Simon's problem, and then the quantum algorithm for solving it. We refer the reader to the recent review by Montanaro and de Wolf on quantum property testing for various applications of this algorithm [38]. We assume here a basic knowledge of the quantum circuit model. We denote the addition and multiplication in a field with 2^n elements by "$\oplus$" and "$\cdot$", respectively.

We consider that the access to the input of Simon's problem, a function f, is made by querying it. A classical query oracle is a function $x \mapsto f(x)$. To run Simon's algorithm, it is required that the function f can be queried quantum-mechanically. More precisely, it is supposed that the algorithm can make arbitrary quantum superpositions of queries of the form $|x\rangle|0\rangle \mapsto |x\rangle|f(x)\rangle$.

Simon's problem is the following:

Simon's Problem: Given a Boolean function $f : \{0,1\}^n \to \{0,1\}^n$ and the promise that there exists $s \in \{0,1\}^n$ such that for any $(x,y) \in \{0,1\}^n$, $[f(x) = f(y)] \Leftrightarrow [x \oplus y \in \{0^n, s\}]$, the goal is to find s.

This problem can be solved classically by searching for collisions. The optimal time to solve it is therefore $\Theta(2^{n/2})$. On the other hand, Simon's algorithm solves this problem with quantum complexity $O(n)$. Recall that the Hadamard transform $H^{\otimes n}$ applied on an n-qubit state $|x\rangle$ for some $x \in \{0,1\}^n$ gives $H^{\otimes n}|x\rangle = \frac{1}{\sqrt{2^n}} \sum_{y \in \{0,1\}^n} (-1)^{x \cdot y}|y\rangle$, where $x \cdot y := x_1 y_1 \oplus \cdots \oplus x_n y_n$.

The algorithm repeats the following five quantum steps.

1. Starting with a $2n$-qubit state $|0\rangle|0\rangle$, one applies a Hadamard transform $H^{\otimes n}$ to the first register to obtain the quantum superposition

$$\frac{1}{\sqrt{2^n}} \sum_{x \in \{0,1\}^n} |x\rangle|0\rangle.$$

2. A quantum query to the function f maps this to the state

$$\frac{1}{\sqrt{2^n}} \sum_{x \in \{0,1\}^n} |x\rangle|f(x)\rangle.$$

3. Measuring the second register in the computational basis yields a value $f(z)$ and collapses the first register to the state:

$$\frac{1}{\sqrt{2}}(|z\rangle + |z \oplus s\rangle).$$

4. Applying again the Hadamard transform $H^{\otimes n}$ to the first register gives:

$$\frac{1}{\sqrt{2}}\frac{1}{\sqrt{2^n}}\sum_{y\in\{0,1\}^n}(-1)^{y\cdot z}\left(1+(-1)^{y\cdot s}\right)|y\rangle.$$

5. The vectors y such that $y\cdot s=1$ have amplitude 0. Therefore, measuring the state in the computational basis yields a random vector y such that $y\cdot s=0$.

By repeating this subroutine $O(n)$ times, one obtains $n-1$ independent vectors orthogonal to s with high probability, and s can be recovered using basic linear algebra. Theorem 1 gives the trade-off between the number of repetitions of the subroutine and the success probability of the algorithm.

2.2   Dealing with Unwanted Collisions

In our cryptanalysis scenario, it is not always the case that the promise of Simon's problem is perfectly satisfied. More precisely, by construction, there will always exist an s such that $f(x)=f(x\oplus s)$ for any input x, but there might be many more collisions than those of this form. If the number of such unwanted collisions is too large, one might not be able to obtain a full rank linear system of equations from Simon's subroutine after $O(n)$ queries. Theorem 1 rules this out provided that f does not have too many collisions of the form $f(x)=f(x\oplus t)$ for some $t\notin\{0,s\}$.

For $f:\{0,1\}^n\to\{0,1\}^n$ such that $f(x\oplus s)=f(x)$ for all x, consider

$$\varepsilon(f,s)=\max_{t\in\{0,1\}^n\setminus\{0,s\}}\Pr_x[f(x)=f(x\oplus t)]. \tag{1}$$

This parameter quantifies how far the function is from satisfying Simon's promise. For a random function, one expects $\varepsilon(f,s)=\Theta(n2^{-n})$, following the analysis of [19]. On the other hand, for a constant function, $\varepsilon(f,s)=1$ and it is impossible to recover s.

The following theorem, whose proof can be found in Appendix A, shows the effect of unwanted collisions on the success probability of Simon's algorithm.

Theorem 1 (Simon's algorithm with approximate promise). *If $\varepsilon(f,s)\le p_0<1$, then Simon's algorithm returns s with cn queries, with probability at least $1-\left(2\left(\frac{1+p_0}{2}\right)^c\right)^n$.*

In particular, choosing $c\ge 3/(1-p_0)$ ensures that the error decreases exponentially with n. To apply our results, it is therefore sufficient to prove that $\varepsilon(f,s)$ is bounded away from 1.

Finally, if we apply Simon's algorithm without any bound on $\varepsilon(f,s)$, we can not always recover s unambiguously. Still if we select a random value t orthogonal to all vectors u_i returned by each step of the algorithm, t satisfy $f(x\oplus t)=f(x)$ with high probability.

Theorem 2 (Simon's algorithm without promise). *After cn steps of Simon's algorithm, if t is orthogonal to all vectors u_i returned by each step of the algorithm, then $\Pr_x[f(x\oplus t)=f(t)]\ge p_0$ with probability at least $1-\left(2\left(\frac{1+p_0}{2}\right)^c\right)^n$.*

In particular, choosing $c \geq 3/(1 - p_0)$ ensures that the probability is exponentially close to 1.

2.3   Attack Strategy

The general strategy behind our attacks exploiting Simon's algorithm is to start with the encryption oracle $E_k : \{0,1\}^n \rightarrow \{0,1\}^n$ and exhibit a new function f that satisfies Simon's promise with two additional properties: the adversary should be able to query f in superposition if he has quantum oracle access to E_k, and the knowledge of the string s should be sufficient to break the cryptographic scheme. In the following, this function is called Simon's function.

In most cases, our attacks correspond to a classical collision attack. In particular, the value s will usually be the difference in the internal state after processing a fixed pair of messages (α_0, α_1), i.e. $s = E(\alpha_0) \oplus E(\alpha_1)$. The input of f will be inserted into the state with the difference s so that $f(x) = f(x \oplus s)$.

In our work, this function f is of the form:

$$f^1 : x \quad \mapsto P(\widetilde{E}(x) + \widetilde{E}(x \oplus s)) \quad \text{or,}$$

$$f^2 : b, x \quad \mapsto \begin{cases} \widetilde{E}(x) & \text{if } b = 0, \\ \widetilde{E}(x \oplus s) & \text{if } b = 1, \end{cases}$$

where $\widetilde{E}$ is a simple function obtained from E_k and P a permutation. It is immediate to see that f^1 and f^2 have periods s for f^1 or $1\|s$ for f^2.

In most applications, Simon's function satisfies $f(x) = f(y)$ for $y \oplus x \in \{0, s\}$, but also for additional inputs x, y. Theorem 1 extends Simon's algorithm precisely to this case. In particular, if the additional collisions of f are random, then Simon's algorithm is successful. When considering explicit constructions, we can not in general prove that the unwanted collisions *are* random, but rather that they *look random enough*. In practice, if the function $\varepsilon(f, s)$ is not bounded, then some of the primitives used in the construction have are far from ideal. We can show that this happens with low probability, and would imply an classical attack against the system. Applying Theorem 1 is not trivial, but it stretches the range of application of Simon's algorithm far beyond its original version.

Construction of Simon's Functions. To make our attacks as clear as possible, we provide the diagrams of circuits computing the function f. These circuits use a little number of basic building blocks represented in Fig. 2.

In our attacks, we often use a pair of arbitrary constants α_0 and α_1. The choice of the constant is indexed by a bit b. We denote by U_α the gate that maps b to α_b (See Fig. 2). For simplicity, we ignore here the additional qubits required in practice to make the transform reversible through padding.

Although it is well known that arbitrary quantum states cannot be cloned, we use the $CNOT$ gate to copy classical information. More precisely, a CNOT gate can copy states in the computational basis: $CNOT : |x\rangle|0\rangle \rightarrow |x\rangle|x\rangle$. This transform is represented in Fig. 2.

Finally, any unitary transform U can be controlled by a bit b. This operation, denoted U^b maps x to $U(x)$ if $b = 1$ and leaves x unchanged otherwise. In the quantum setting, the qubit $|b\rangle$ can be in a superposition of 0 and 1, resulting in a superposition of $|x\rangle$ and $|U(x)\rangle$. The attacks that we present in the following sections only make use of this procedure when the attacker knows a classical description of the unitary to be controlled. In particular, we do not apply it to the cryptographic oracle.

When computing Simon's function, *i.e.* the function f on which Simon's algorithm is applied, the registers containing the value of f must be unentangled with any other working register. Otherwise, these registers, which might hinder the periodicity of the function, have to be taken into account in Simon's algorithm and the whole procedure could fail.

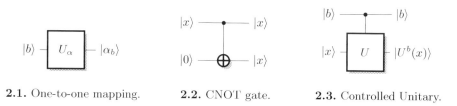

2.1. One-to-one mapping.   **2.2.** CNOT gate.   **2.3.** Controlled Unitary.

Fig. 2. Circuit representation of basic building blocks.

3   Previous Works

Previous works have used Simon's algorithm to break the security of classical constructions in symmetric cryptography: the Even-Mansour construction and the 3-round Feistel scheme. We now explain how these attacks work with our terminology and extend two of the results. First, we show that the attack on the Feistel scheme can be extended to work with random functions, where the original analysis held only for random permutations. Second, using our analysis Simon's algorithm with approximate promise, we make the number of queries required to attack the Even-Mansour construction more precise. These observations have been independently made by Santoli and Schaffner [41]. They use a slightly different approach, which consists in analyzing the run of Simon's algorithm for these specific cases.

3.1   Applications to a Three-Round Feistel Scheme

The Feistel scheme is a classical construction to build a random permutation out of random functions or random permutations. In a seminal work, Luby and Rackoff proved that a three-round Feistel scheme is a secure pseudo-random permutation [34].

A three-round Feistel scheme with input (x_L, x_R) and output $(y_L, y_R) = E(x_L, x_R)$ is built from three round functions R_1, R_2, R_3 as (see Fig. 3):

$$(u_0, v_0) = (x_L, x_R), \quad (u_i, v_i) = (v_{i-1} \oplus R_i(u_{i-1}), u_{i-1}), \quad (y_L, y_R) = (u_3, v_3).$$

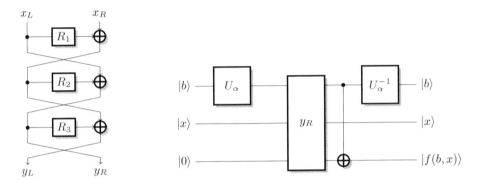

Fig. 3. Three-round Feistel scheme.

Fig. 4. Simon's function for Feistel.

In order to distinguish a Feistel scheme from a random permutation in a quantum setting, Kuwakado and Morii [31] consider the case were the R_i are permutations, and define the following function, with two arbitrary constants α_0 and α_1 such that $\alpha_0 \neq \alpha_1$:

$$f : \{0,1\} \times \{0,1\}^n \to \{0,1\}^n$$
$$b, x \quad \mapsto y_R \oplus \alpha_b, \quad \text{where } (y_R, y_L) = E(\alpha_b, x)$$
$$f(b, x) = R_2(x \oplus R_1(\alpha_b))$$

In particular, this f satisfies $f(b, x) = f(b \oplus 1, x \oplus R_1(\alpha_0) \oplus R_1(\alpha_1))$. Moreover,

$$f(b', x') = f(b, x) \Leftrightarrow x' \oplus R_1(\alpha_{b'}) = x \oplus R_1(\alpha_b)$$
$$\Leftrightarrow \begin{cases} x' \oplus x = 0 & \text{if } b' = b \\ x' \oplus x = R_1(\alpha_0) \oplus R_1(\alpha_1) & \text{if } b' \neq b \end{cases}$$

Therefore, the function satisfies Simon's promise with $s = 1 \parallel R_1(\alpha_0) \oplus R_1(\alpha_1)$, and we can recover $R_1(\alpha_0) \oplus R_1(\alpha_1)$ using Simon's algorithm. This gives a distinguisher, because Simon's algorithm applied to a random permutation returns zero with high probability. This can be seen from Theorem 2, using the fact that with overwhelming probability [19], there is no value $t \neq 0$ such that $\Pr_x[f(x \oplus t) = f(x)] > 1/2$ for a random permutation f (Fig. 4).

We can also verify that the value $R_1(\alpha_0) \oplus R_1(\alpha_1)$ is correct with two additional classical queries $(y_L, y_R) = E(\alpha_0, x)$ and $(y'_L, y'_R) = E(\alpha_1, x \oplus R_1(\alpha_0) \oplus R_1(\alpha_1))$ for a random x. If the value is correct, we have $y_R \oplus y'_R = \alpha_0 \oplus \alpha_1$.

Note that in their attack, Kuwakado and Morii implicitly assume that the adversary can query in superposition an oracle that returns solely the left part y_L of the encryption. If the adversary only has access to the complete encryption oracle E, then a query in superposition would return two *entangled* registers containing the left and right parts, respectively. In principle, Simon's algorithm requires the register containing the input value to be completely disentangled from the others.

Feistel Scheme with Random Functions. Kuwakado and Morii [31] analyze only the case where the round functions R_i are permutations. We now extend this analysis to *random functions* R_i. The function f defined above still satisfies $f(b, x) = f(b \oplus 1, x \oplus R_1(\alpha_0) \oplus R_1(\alpha_1))$, but it doesn't satisfy the exact promise of Simon's algorithm: there are additional collisions in f, between inputs with random differences. However, the previous distinguisher is still valid: at the end of Simon's algorithm, there exist at least one non-zero value orthogonal to all the values y measured at each step: s. This would not be the case with a random permutation.

Moreover, we can show that $\varepsilon(f, 1 \parallel s) < 1/2$ with overwhelming probability, so that Simon's algorithm still recovers $1 \parallel s$ following Theorem 1. If $\varepsilon(f, 1 \parallel s) > 1/2$, there exists (τ, t) with $(\tau, t) \notin \{(0, 0), (1, s)\}$ such that: $\Pr[f(b, x) = f(b \oplus \tau, x \oplus t)] > 1/2$. Assume first that $\tau = 0$, this implies:

$$\Pr[f(0, x) = f(0, x \oplus t)] > 1/2 \quad \text{or} \quad \Pr[f(1, x) = f(1, x \oplus t)] > 1/2.$$

Therefore, for some b, $\Pr[R_2(x \oplus R_1(\alpha_b)) = R_2(x \oplus t \oplus R_1(\alpha_b))] > 1/2$, *i.e.* $\Pr[R_2(x) = R_2(x \oplus t)] > 1/2$. Similarly, if $\tau = 1$, $\Pr[R_2(x \oplus R_1(\alpha_0)) = R_2(x \oplus t \oplus R_1(\alpha_1))] > 1/2$, *i.e.* $\Pr[R_2(x) = R_2(x \oplus t \oplus R_1(\alpha_0) \oplus R_1(\alpha_1))] > 1/2$.

To summarize, if $\varepsilon(f, 1 \parallel s) > 1/2$, there exists $u \neq 0$ such that $\Pr[R_2(x) = R_2(x \oplus u)] > 1/2$. This only happens with negligible probability for a random choice of R_2 as shown in [19].

3.2   Application to the Even-Mansour Construction

The Even-Mansour construction is a simple construction to build a block cipher from a public permutation [22]. For some permutation P, the cipher is:

$$E_{k_1, k_2}(x) = P(x \oplus k_1) \oplus k_2.$$

Even and Mansour have shown that this construction is secure in the random permutation model, up to $2^{n/2}$ queries, where n is the size of the input to P (Figs. 5 and 6).

However, Kuwakado and Morii [32] have shown that the security of this construction collapses if an adversary can query an encryption oracle with a superposition of states. More precisely, they define the following function:

$$f : \{0, 1\}^n \to \{0, 1\}^n$$
$$x \mapsto E_{k_1, k_2}(x) \oplus P(x) = P(x \oplus k_1) \oplus P(x) \oplus k_2.$$

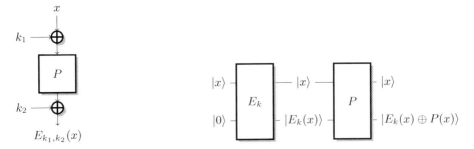

Fig. 5. Even-Mansour scheme.   **Fig. 6.** Simon's function for Even-Mansour.

In particular, f satisfies $f(x \oplus k_1) = f(x)$ (interestingly, the slide with a twist attack of Biryukov and Wagner [8] uses the same property). However, there are additional collisions in f between inputs with random differences. As in the attack against the Feistel scheme with random round functions, we use Theorem 1, to show that Simon's algorithm recovers k_1.[2]

We show that $\varepsilon(f, k_1) < 1/2$ with overwhelming probability for a random permutation P, and if $\varepsilon(f, k_1) > 1/2$, then there exists a classical attack against the Even-Mansour scheme. Assume that $\varepsilon(f, k_1) > 1/2$, that is, there exists t with $t \notin \{0, k_1\}$ such that $\Pr[f(x) = f(x \oplus t)] > 1/2$, *i.e.*,

$$p = \Pr[P(x) \oplus P(x \oplus k_1) \oplus P(x \oplus t) \oplus P(x \oplus t \oplus k_1) = 0] > 1/2.$$

This correspond to higher order differential for P with probability $1/2$, which only happens with negligible probability for a random choice of P. In addition, this would imply the existence of a simple classical attack against the scheme:

1. Query $y = E_{k_1,k_2}(x)$ and $y' = E_{k_1,k_2}(x \oplus t)$
2. Then $y \oplus y' = P(x) \oplus P(x \oplus t)$ with probability at least one half

Therefore, for any instantiation of the Even-Mansour scheme with a fixed P, either there exist a classical distinguishing attack (this only happens with negligible probability with a random P), or Simon's algorithm successfully recovers k_1. In the second case, the value of k_2 can then be recovered from an additional classical query: $k_2 = E(x) \oplus P(x \oplus k_1)$.

In the next sections, we give new applications of Simon's algorithm, to break various symmetric cryptography schemes.

4   Application to the LRW Construction

We now show a new application of Simon's algorithm to the LRW construction. The LRW construction, introduced by Liskov, Rivest and Wagner [33], turns

---

[2] Note that Kuwakado and Morii just assume that each step of Simon's algorithm gives a random vector orthogonal to k_1. Our analysis is more formal and captures the conditions on P required for the algorithm to be successful.

a block cipher into a tweakable block cipher, *i.e.* a family of unrelated block ciphers. The tweakable block cipher is a very useful primitive to build modes for encryption, authentication, or authenticated encryption. In particular, tweakable block ciphers and the LRW construction were inspired by the first version of OCB, and later versions of OCB use the tweakable block ciphers formalism. The LRW construction uses a (almost) universal hash function h (which is part of the key), and is defined as (see also Fig. 7):

$$\widetilde{E}_{t,k}(x) = E_k(x \oplus h(t)) \oplus h(t).$$

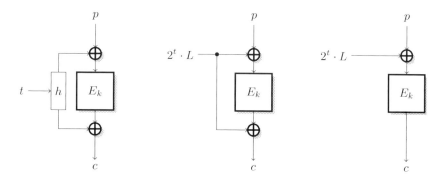

7.1. LRW construction.   **7.2.** XEX construction.   **7.3.** XE construction.

Fig. 7. The LRW construction, and efficient instantiations XEX (CCA secure) and XE (only CPsecure).

We now show that the LRW construction is not secure in a quantum setting. We fix two arbitrary tweaks t_0, t_1, with $t_0 \neq t_1$, and we define the following function:

$$f : \{0,1\}^n \rightarrow \{0,1\}^n$$
$$x \mapsto \widetilde{E}_{t_0,k}(x) \oplus \widetilde{E}_{t_1,k}(x)$$
$$f(x) = E_k\big(x \oplus h(t_0)\big) \oplus h(t_0) \oplus E_k\big(x \oplus h(t_1)\big) \oplus h(t_1).$$

Given a superposition access to an oracle for an LRW tweakable block cipher, we can build a circuit implementing this function, using the construction given in Fig. 8. In the circuit, the cryptographic oracle $\widetilde{E}_{t,k}$ takes two inputs: the block x to be encrypted and the tweak t. Since the tweak comes out of $\widetilde{E}_{t,k}$ unentangled with the other register, we do not represent this output in the diagram. In practice, the output is forgotten by the attacker.

It is easy to see that this function satisfies $f(x) = f(x \oplus s)$ with $s = h(t_0) \oplus h(t_1)$. Furthermore, the quantity $\varepsilon(f,s) = \max_{t \in \{0,1\}^n \setminus \{0,s\}} \Pr[f(x) = f(x \oplus t)]$ is bounded with overwhelming probability, assuming that E_k behaves as a random permutation. Indeed if $\varepsilon(f,s) > 1/2$, there exists some t with $t \notin \{0,s\}$ such that $\Pr[f(x) = f(x \oplus t)] > 1/2$, i.e.,

$$\Pr[E_k(x) \oplus E_k(x \oplus s) \oplus E_k(x \oplus t)) \oplus E_k(x \oplus t \oplus s) = 0] > 1/2$$

This correspond to higher order differential for E_k with probability $1/2$, which only happens with negligible probability for a random permutation. Therefore, if E is a pseudo-random permutation family, $\varepsilon(f,s) \leq 1/2$ with overwhelming probability, and running Simon's algorithm with the function f returns $h(t_0) \oplus h(t_1)$. The assumption that E behaves as a PRP family is required for the security proof of LRW, so it is reasonable to make the same assumption in an attack. More concretely, a block cipher with a higher order differential with probability $1/2$ as seen above would probably be broken by classical attacks. The attack is not immediate because the differential can depend on the key, but it would seem to indicate a structural weakness. In the following sections, some attacks can also be mounted using Theorem 2 without any assumptions on E.

In any case, there exist at least one non-zero value orthogonal to all the values y measured during Simon's algorithm: s. This would not be the case if f is a random function, which gives a distinguisher between the LRW construction and an ideal tweakable block cipher with $O(n)$ quantum queries to $\widetilde{E}$.

In practice, most instantiations of LRW use a finite field multiplication to define the universal hash function h, with a secret offset L (usually computed as $L = E_k(0)$). Two popular constructions are:

- $h(t) = \gamma(t) \cdot L$, used in OCB1 [40], OCB3 [30] and PMAC [10], with a Gray encoding γ of t,
- $h(t) = 2^t \cdot L$, the XEX construction, used in OCB2 [39].

In both cases, we can recover L from the value $h(t_0) \oplus h(t_1)$ given by the attack.

This attack is important, because many recent modes of operation are inspired by the LRW construction, and the XE and XEX instantiations, such as CAESAR candidates AEZ [25], COPA [4], OCB [30], OTR [37], Minalpher [42], OMD [18], and POET [1]. We will see in the next section that variants of this attack can be applied to each of these modes.

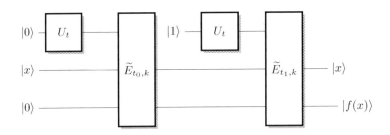

Fig. 8. Simon's function for LRW.

5   Application to Block Cipher Modes of Operations

We now give new applications of Simon's algorithm to the security of block cipher modes of operations. In particular, we show how to break the most popular and widely used block-cipher based MACs, and message authentication schemes: CBC-MAC (including variants such as XCBC [9], OMAC [26], and CMAC [21]), GMAC [36], PMAC [10], GCM [36] and OCB [30]. We also show attacks against several CAESAR candidates. In each case, the mode is proven secure up to $2^{n/2}$ in the classical setting, but we show how, by a reduction to Simon's problem, forgery attacks can be performed with superposition queries at a cost of $O(n)$.

Notations and Preliminaries. We consider a block cipher E_k, acting on blocks of length n, where the subscript k denotes the key. For simplicity, we only describe the modes with full-block messages, the attacks can trivially be extended to the more general modes with arbitrary inputs. In general, we consider a message M divided into ℓ n-bits block: $M = m_1 \parallel \dots \parallel m_\ell$. We also assume that the MAC is not truncated, $i.e.$ the output size is n bits. In most cases, the attacks can be adapted to truncated MACS.

5.1   Deterministic MACs: CBC-MAC and PMAC

We start with deterministic Message Authentication Codes, or MACs. A MAC is used to guarantee the authenticity of messages, and should be immune against forgery attacks. The standard security model is that it should be hard to forge a message with a valid tag, even given access to an oracle that computes the MAC of any chosen message (of course the forged message must not have been queried to the oracle).

To translate this security notion to the quantum setting, we assume that the adversary is given an oracle that takes a quantum superposition of messages as input, and computes the superposition of the corresponding MAC.

CBC-MAC. CBC-MAC is one of the first MAC constructions, inspired by the CBC encryption mode. Since the basic CBC-MAC is only secure when the queries are prefix-free, there are many variants of CBC-MAC to provide security for arbitrary messages. In the following we describe the Encrypted-CBC-MAC variant [5], using two keys k and k', but the attack can be easily adapted to other variants [9,21,26]. On a message $M = m_1 \parallel \dots \parallel m_\ell$, CBC-MAC is defined as (see Fig. 9):

$$x_0 = 0 \qquad x_i = E_k(x_{i-1} \oplus m_i) \qquad \text{CBC-MAC}(M) = E_{k'}(x_\ell)$$

CBC-MAC is standardized and widely used. It has been proved to be secure up to the birthday bound [5], assuming that the block cipher is indistinguishable from a random permutation.

$$x_0 = 0 \qquad x_i = E_k(x_{i-1} \oplus m_i) \qquad \text{CBC-MAC}(M) = E_{k'}(x_\ell)$$

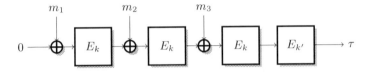

Fig. 9. Encrypt-last-block CBC-MAC.

Attack. We can build a powerful forgery attack on CBC-MAC with very low complexity using superposition queries. We fix two arbitrary message blocks α_0, α_1, with $\alpha_0 \neq \alpha_1$, and we define the following function:

$$f : \{0,1\} \times \{0,1\}^n \rightarrow \{0,1\}^n$$
$$b, x \quad \mapsto \text{CBC-MAC}(\alpha_b \parallel x) = E_{k'}\left(E_k\left(x \oplus E_k(\alpha_b)\right)\right).$$

The function f can be computed with a single call to the cryptographic oracle, and we can build a quantum circuit for f given a black box quantum circuit for CBC-MAC$_k$. Moreover, f satisfies the promise of Simon's problem with $s = 1 \parallel E_k(\alpha_0) \oplus E_k(\alpha_1)$:

$$f(0, x) = E_{k'}(E_k(x \oplus E_k(\alpha_1))),$$
$$f(1, x) = E_{k'}(E_k(x \oplus E_k(\alpha_0))),$$
$$f(b, x) = f(b \oplus 1, x \oplus E_k(\alpha_0) \oplus E_k(\alpha_1)).$$

More precisely:

$$f(b', x') = f(b, x) \Leftrightarrow x \oplus E_k(\alpha_b) = x' \oplus E_k(\alpha_{b'})$$
$$\Leftrightarrow \begin{cases} x' \oplus x = 0 & \text{if } b' = b \\ x' \oplus x = E_k(\alpha_0) \oplus E_k(\alpha_1) & \text{if } b' \neq b \end{cases}$$

Therefore, an application of Simon's algorithm returns $E_k(\alpha_0) \oplus E_k(\alpha_1)$. This allows to forge messages easily:

1. Query the tag of $\alpha_0 \parallel m_1$ for an arbitrary block m_1;
2. The same tag is valid for $\alpha_1 \parallel m_1 \oplus E_k(\alpha_0) \oplus E_k(\alpha_1)$.

In order to break the formal notion of EUF-qCMA security, we must produce $q + 1$ valid tags with only q queries to the oracle. Let $q' = O(n)$ denote the number of of quantum queries made to learn $E_k(\alpha_0) \oplus E_k(\alpha_1)$. The attacker will repeats the forgery step step $q' + 1$ times, in order to produce $2(q' + 1)$ messages with valid tags, after a total of $2q' + 1$ classical and quantum queries to the cryptographic oracle. Therefore, CBC-MAC is broken by a quantum existential forgery attack.

After some exchange at early stages of the work, an extension of this forgery attack has been found by Santoli and Schaffner [41]. Its main advantage is to handle oracles that accept input of fixed length, while our attack works for oracles accepting messages of variable length.

PMAC. PMAC is a parallelizable block-cipher based MAC designed by Rogway [39]. PMAC is based on the XE construction: the construction uses secret offsets Δ_i derived from the secret key to turn the block cipher into a tweakable block cipher. More precisely, the PMAC algorithm is defined as

$$c_i = E_k(m_i \oplus \Delta_i) \qquad\qquad \mathrm{PMAC}(M) = E_k^*\Big(m_\ell \oplus \sum c_i\Big)$$

where E^* is a tweaked variant of E. We omit the generation of the secret offsets because they are irrelevant to our attack.

First Attack. When PMAC is used with two-block messages, it has the same structure as CBC-MAC: $\mathrm{PMAC}(m_1 \parallel m_2) = E_k^*(m_2 \oplus E_k(m_1 \oplus \Delta_0))$. Therefore we can use the attack of the previous section to recover $E_k(\alpha_0) \oplus E_k(\alpha_1)$ for arbitrary values of α_0 and α_1. Again, this leads to a simple forgery attack. First, query the tag of $\alpha_0 \parallel m_1 \parallel m_2$ for arbitrary blocks m_1, m_2. The same tag is valid for $\alpha_1 \parallel m_1 \parallel m_2 \oplus E_k(\alpha_0) \oplus E_k(\alpha_1)$. As for CBC-MAC, these two steps can be repeated $t + 1$ times, where t is the number of quantum queries issued. The adversary then produces $2(t + 1)$ messages after only $2t + 1$ queries to the cryptographic oracle.

Second Attack. We can also build another forgery attack on PMAC where we recover the difference between two offsets Δ_i, following the attack against LRW given in Sect. 4. More precisely, we use the following function:

$$\begin{aligned} f : \{0,1\}^n &\to \{0,1\}^n \\ m &\mapsto \mathrm{PMAC}(m \parallel m \parallel 0^n) = E_k^*\left(E_k(m \oplus \Delta_0) \oplus E_k(m \oplus \Delta_1)\right). \end{aligned}$$

In particular, it satisfies $f(m \oplus s) = f(m)$ with $s = \Delta_0 \oplus \Delta_1$. Furthermore, we can show that $\varepsilon(f, s) \le 1/2$ when E is a good block cipher[3], and we can apply Simon's algorithm to recover $\Delta_0 \oplus \Delta_1$. This allows to create forgeries as follows:

1. Query the tag of $m_1 \parallel m_1$ for an arbitrary block m_1;
2. The same tag is valid for $m_1 \oplus \Delta_0 \oplus \Delta_1 \parallel m_1 \oplus \Delta_0 \oplus \Delta_1$.

As mentioned in Sect. 4, the offsets in PMAC are defined as $\Delta_i = \gamma(i) \cdot L$, with $L = E_k(0)$ and γ a Gray encoding. This allows to recover L from $\Delta_0 \oplus \Delta_1$, as $L = (\Delta_0 \oplus \Delta_1) \cdot (\gamma(0) \oplus \gamma(1))^{-1}$. Then we can compute all the values Δ_i, and forge arbitrary messages.

---

[3] Since this attack is just a special case of the LRW attack of Sect. 4, we don't repeat the detailed proof.

We can also mount an attack without any assumption on $\varepsilon(f, s)$, using Theorem 2. Indeed, with a proper choice of parameters, Simon's algorithm will return a value $t \neq 0$ that satisfies $\Pr_x[f(x \oplus t) = f(x)] \geq 1/2$. This value is not necessarily equal to s, but it can also be used to create forgeries in the same way, with success probability at least $1/2$.

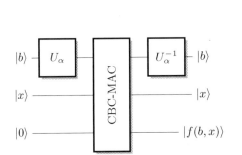
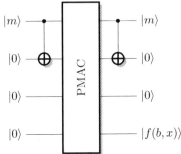

Fig. 10. Simon's function for CBC-MAC.

Fig. 11. Simon's function for the second attack against PMAC.

5.2   Randomized MAC: GMAC

GMAC is the underlying MAC of the widely used GCM standard, designed by McGrew and Viega [36], and standardized by NIST. GMAC follows the Carter-Wegman construction [16]: it is built from a universal hash function, using polynomial evaluation in a Galois field. As opposed to the constructions of the previous sections, GMAC is a randomized MAC; it requires a second input N, which must be non-repeating (a nonce). GMAC is essentially defined as (Figs. 10 and 11):

$$\mathrm{GMAC}(N, M) = \mathrm{GHASH}(M \parallel \mathrm{len}(M)) \oplus E_k(N \parallel 1)$$

$$\mathrm{GHASH}(M) = \sum_{i=1}^{\mathrm{len}(M)} m_i \cdot H^{\mathrm{len}(M)-i+1} \quad \text{with } H = E_k(0),$$

where $\mathrm{len}(M)$ is the length of M.

Attack. When the polynomial is evaluated with Horner's rule, the structure of GMAC is similar to that of CBC-MAC (see Fig. 12). For a two-block message, we have $\mathrm{GMAC}(m_1 \parallel m_2) = ((m_1 \cdot H) \oplus m_2) \cdot H \oplus E_k(N \parallel 1)$. Therefore, we us the same f as in the CBC-MAC attack, with fixed blocks α_0 and α_1:

$$f_N : \{0,1\} \times \{0,1\}^n \to \{0,1\}^n$$
$$b, x \quad \mapsto \mathrm{GMAC}(N, \alpha_b \parallel x) = \alpha_b \cdot H^2 \oplus x \cdot H \oplus E_k(N \parallel 1).$$

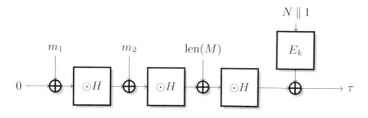

Fig. 12. GMAC

In particular, we have:

$$f(b', x') = f(b, x) \Leftrightarrow \alpha_b \cdot H^2 \oplus x \cdot H = \alpha_{b'} \cdot H^2 \oplus x' \cdot H$$

$$\Leftrightarrow \begin{cases} x' \oplus x = 0 & \text{if } b' = b \\ x' \oplus x = (\alpha_0 \oplus \alpha_1) \cdot H & \text{if } b' \neq b \end{cases}$$

Therefore f_N satisfies the promise of Simon's algorithm with $s = 1 \parallel (\alpha_0 \oplus \alpha_1) \cdot H$.

Role of the Nonce. There is an important caveat regarding the use of the nonce. In a classical setting, the nonce is chosen by the adversary under the constraint that it is non-repeating, *i.e.* the oracle computes $N, M \mapsto \mathrm{GMAC}(N, M)$. However, in the quantum setting, we don't have a clear definition of non-repeating if the nonce can be in superposition. To sidestep the issue, we use a weaker security notion where the nonce is chosen at random by the oracle, rather than by the adversary (following the IND-qCPA definition of [13]). The oracle is then $M \mapsto (r, \mathrm{GMAC}(r, M))$. If we can break the scheme in this model, the attack will also be valid with any reasonable CPA security definition.

In this setting we can access the function f_N only for a random value of N. In particular, we cannot apply Simon's algorithm as is, because this requires $O(n)$ queries to the *same* function f_N. However, a single step of Simon's algorithm requires a single query to the f_N function, and returns a vector orthogonal to s, for any random choice of N. Therefore, we can recover $(\alpha_0 \oplus \alpha_1) \cdot H$ after $O(n)$ steps, even if each step uses a different value of N. Then, we can recover H easily, and it is easy to generate forgeries when H is known:

1. Query the tag of $N, m_1 \parallel m_2$ for arbitrary blocks m_1, m_2 (under a random nonce N).
2. The same tag is valid for $m_1 \oplus 1 \parallel m_2 \oplus H$ (with the same nonce N).

As for CBC-MAC, repeating these two steps leads to an existential forgery attack.

5.3 Classical Authenticated Encryption Schemes: GCM and OCB

We now give applications of Simon's algorithm to break the security of standardized authenticated encryption modes. The attacks are similar to the attacks against authentication modes, but these authenticated encryption modes are nonce-based. Therefore we have to pay special attention to the nonce, as in the attack against GMAC. In the following, we assume that the nonce is randomly chosen by the MAC oracle, in order to avoid issues with the definition of non-repeating nonce in a quantum setting.

Extending MAC Attacks to Authenticated Encryption Schemes. We first present a generic way to apply MAC attacks in the context of an authenticated encryption scheme. More precisely, we assume that the tag of the authenticated encryption scheme is computed as $f(g(A), h(M, N))$, *i.e.* the authentication of the associated data A is independent of the nonce N. This is the case in many practical schemes (*e.g.* GCM, OCB) for efficiency reasons.

In this setting, we can use a technique similar to our attack against GMAC: we define a function $M \mapsto f_N(M)$ for a fixed nonce N, such that for any nonce N, $f_N(M) = f_N(M \oplus \Delta)$ for some secret value Δ. Next we use Simon's algorithm to recover Δ, where each step of Simon's algorithm is run with a random nonce, and returns a vector orthogonal to Δ. Finally, we can recover Δ, and if f_N was carefully built, the knowledge of Δ is sufficient for a forgery attack.

The CCM mode is a notable exception, where all the computations depend on the nonce. In particular, there is no obvious way to apply our attacks to CCM.

Extending GMAC Attack to GCM. GCM is one of the most widely used authenticated encryption modes, designed by McGrew and Viega [36]. GMAC is the composition of the counter mode for encryption with GMAC (computed over the associated data and the ciphertext) for authentication.

In particular, when the message is empty, GCM is just GMAC, and we can use the attack of the previous section to recover the hash key H. This immediately allows a forgery attack.

OCB. OCB is another popular authenticated encryption mode, with a very high efficiency, designed by Rogaway *et al.* [30, 39, 40]. Indeed, OCB requires only ℓ block cipher calls to process an ℓ-block message, while GCM requires ℓ block cipher calls, and ℓ finite field operations. OCB is build from the LRW construction discussed in Sect. 4. OCB takes as input a nonce N, a message $M = m_1 \| \ldots \| m_\ell$, and associated data $A = a_1 \| \ldots a_@$, and returns a ciphertext $C = c_1 \| \ldots \| c_\ell$ and a tag τ:

$$c_i = E_k(m_i \oplus \Delta_i^N) \oplus \Delta_i^N, \quad \tau = E_k\left(\Delta_\ell'^N \oplus \sum m_i\right) \oplus \sum b_i, \quad b_i = E_k(a_i \oplus \Delta_i).$$

Extending PMAC Attack to OCB. In particular, when the message is empty, OCB reduces to a randomized variant of PMAC:

$$\mathrm{OCB}_k(N, \varepsilon, A) = \phi_k(N) \oplus \sum b_i, \qquad b_i = E_k(a_i \oplus \Delta_i).$$

Note that the Δ_i values used for the associated data are independent of the nonce N. Therefore, we can apply the second PMAC attack previously given, using the following function:

$$f_N : \{0,1\}^n \to \{0,1\}^n$$
$$x \quad \mapsto \mathrm{OCB}_k(N, \varepsilon, x \parallel x)$$
$$f_N(x) = E_k(x \oplus \Delta_0) \oplus E_k(x \oplus \Delta_1) \oplus \phi_k(N)$$

Again, this is a special case of the LRW attack of Sect. 4. The family of functions satisfies $f_N(a \oplus \Delta_0 \oplus \Delta_1) = f_N(a)$, for any N, and $\varepsilon(f_N, \Delta_0 \oplus \Delta_1) \le 1/2$ with overwhelming probability if E is a PRP. Therefore we can use the variant of Simon's algorithm to recover $\Delta_0 \oplus \Delta_1$. Two messages with valid tags can then be generated by a single classical queries:

1. Query the authenticated encryption C, τ of $M, a \parallel a$ for an arbitrary message M, and an arbitrary block a (under a random nonce N).
2. C, τ is also a valid authenticated encryption of $M, a \oplus \Delta_0 \oplus \Delta_1 \parallel a \oplus \Delta_0 \oplus \Delta_1$, with the same nonce N.

Repeating these steps lead again to an existential forgery attack.

Alternative Attack Against OCB. For some versions of OCB, we can also mount a different attack targeting the encryption part rather than the authentication part. The goal of this attack is also to recover the secret offsets, but we target the Δ_i^N used for the encryption of the message. More precisely, we use the following function:

$$f_i : \{0,1\}^n \to \{0,1\}^n$$
$$m \quad \mapsto c_1 \oplus c_2, \text{where } (c_1, c_2, \tau) = \mathrm{OCB}_k(N, m \parallel m, \varepsilon)$$
$$f_i(m) = E_k(m \oplus \Delta_1^N) \oplus \Delta_1^N \oplus E_k(m \oplus \Delta_2^N) \oplus \Delta_2^N$$

This function satisfies $f_N(m \oplus \Delta_1^N \oplus \Delta_2^N) = f_N(m)$ and $\varepsilon(f_N, \Delta_0^N \oplus \Delta_1^N) \le 1/2$, with the same arguments as previously. Moreover, in OCB1 and OCB3, the offsets are derived as $\Delta_i^N = \Phi_k(N) \oplus \gamma(i) \cdot E_k(0)$ for some function Φ (based on the block cipher E_k). In particular, $\Delta_1^N \oplus \Delta_2^N$ is independent of N:

$$\Delta_1^N \oplus \Delta_2^N = (\gamma(1) \oplus \gamma(2)) \cdot E_k(0).$$

Therefore, we can apply Simon's algorithm to recover $\Delta_1^N \oplus \Delta_2^N$. Again, this leads to a forgery attack, by repeating the following two steps:

1. Query the authenticated encryption $c_1 \parallel c_2, \tau$ of $m \parallel m, A$ for an arbitrary block m, and arbitrary associated data A (under a random nonce N).
2. $c_2 \oplus \Delta_0^N \oplus \Delta_1^N \parallel c_1 \oplus \Delta_0^N \oplus \Delta_1^N, \tau$ is also a valid authenticated encryption of $m \oplus \Delta_0^N \oplus \Delta_1^N \parallel m \oplus \Delta_0^N \oplus \Delta_1^N, A$ with the same nonce N.

The forgery is valid because we swap the inputs of the first and second block ciphers. In addition, we have $\sum m_i = \sum m_i'$, so that the tag is still valid.

5.4   New Authenticated Encryption Schemes: CAESAR Candidates

In this section, we consider recent proposals for authenticated encryption, submitted to the ongoing CAESAR competition. Secret key cryptography has a long tradition of competitions: AES and SHA-3 for example, were chosen after the NIST competitions organized in 1997 and 2007, respectively. The CAESAR competition[4] aims at stimulating research on authenticated encryption schemes, and to define a portfolio of new authenticated encryption schemes. The competition is currently in the second round, with 29 remaining algorithms.

First, we point out that the attacks of the previous sections can be used to break several CAESAR candidates:

- CLOC [27] uses CBC-MAC to authenticate the message, and the associated data is processed independently of the nonce. Therefore, the CBC-MAC attack can be extended to CLOC[5].
- AEZ [25], COPA [4], OTR [37] and POET [1] use a variant of PMAC to authenticate the associated data. In both cases, the nonce is not used to process the associated data, so that we can extend the PMAC attack as we did against OCB[6].
- The authentication of associated data in OMD [18] and Minalpher [42] are also variants of PMAC (with a PRF that is not block cipher), and the attack can be applied.

In the next section, we show how to adapt the PMAC attack to Minalpher and OMD, since the primitives are different.

Minalpher. Minalpher [42] is a permutation-based CAESAR candidate, where the permutation is used to build a tweakable block-cipher using the tweakable Even-Mansour construction. When the message is empty (or fixed), the authentication part of Minalpher is very similar to PMAC. With associated data $A = a_1 \parallel \ldots a_@$, the tag is computed as:

$$b_i = P(a_i \oplus \Delta_i) \oplus \Delta_i \qquad\qquad \tau = \phi_k\left(N, M, a_@ \oplus \sum_{i=1}^{@-1} b_i\right)$$

$$\Delta_i = y^i \cdot L' \qquad\qquad\qquad L' = P(k \parallel 0) \oplus (k \parallel 0)$$

where ϕ_k is a permutation (we omit the description of ϕ_k because it is irrelevant for our attack). Since the tag is a function of $a_@ \oplus \sum_{i=1}^{@-1} b_i$, we can use the same attacks as against PMAC. For instance, we define the following function:

$$f_N : \{0,1\} \times \{0,1\}^n \to \{0,1\}^n$$
$$b, x \quad \mapsto \text{Minalpher}(N, \varepsilon, \alpha_b \parallel x) = \phi_k(N, \varepsilon, P(\alpha_b \oplus \Delta_1) \oplus \Delta_1 \oplus x).$$

---

[4] http://competitions.cr.yp.to/.
[5] This is not the case for the related mode SILC, because the nonce is processed before the data in CBC-MAC.
[6] Note that AEZ, COPA and POET also claim security when the nonce is misused, but our attacks are nonce-respecting.

In particular, we have:

$$f_N(b', x') = f_N(b, x) \Leftrightarrow P(\alpha_{b'} \oplus \Delta_1) \oplus x' = P(\alpha_b \oplus \Delta_1) \oplus x$$

$$\Leftrightarrow \begin{cases} x' \oplus x = 0 & \text{if } b' = b \\ x' \oplus x = P(\alpha_0 \oplus \Delta_1) \oplus P(\alpha_1 \oplus \Delta_1) & \text{if } b' \neq b \end{cases}$$

Since $s = P(\alpha_0 \oplus \Delta_1) \oplus P(\alpha_1 \oplus \Delta_1)$ is independent of N, we can easily apply Simon's algorithm to recover s, and generate forgeries.

OMD. OMD [18] is a compression-function-based CAESAR candidate. The internal primitive is a keyed compression function denoted F_k. Again, when the message is empty the authentication is very similar to PMAC. With associated data $A = a_1 \parallel \ldots a_@$, the tag is computed as:

$$b_i = F_k(a_i \oplus \Delta_i) \qquad\qquad \tau = \phi_k(N, M) \oplus \sum b_i$$

We note that the Δ_i used for the associated data do not depend on the nonce. Therefore we can use the second PMAC attack with the following function:

$$f_N : \{0, 1\}^n \to \{0, 1\}^n$$
$$x \mapsto \text{OMD}(N, \varepsilon, x \parallel x)$$
$$f_N(x) = \phi_k(N, \varepsilon) \oplus F_k(x \oplus \Delta_1) \oplus F_k(x \oplus \Delta_2)$$

This is the same form as seen when extending the PMAC attack to OCB, therefore we can apply the same attack to recover $s = \Delta_1 \oplus \Delta_2$ and generate forgeries.

6   Simon's Algorithm Applied to Slide Attacks

In this section we show how Simon's algorithm can be applied to a cryptanalysis family: slide attacks. In this case, the complexity of the attack drops again exponentially, from $O(2^{n/2})$ to $O(n)$ and therefore becomes much more dangerous. To the best of our knowledge this is the first symmetric cryptanalytic technique that has an exponential speed-up in the post-quantum world.

The Principle of Slide Attacks. In 1999, Wagner and Biryukov introduced the technique called *slide attack* [7]. It can be applied to block ciphers made of r applications of an identical round function R, each one parametrized by the same key K. The attack works independently of the number of rounds, r. Intuitively, for the attack to work, R has to be vulnerable to known plaintext attacks.

The attacker collects $2^{n/2}$ encryptions of plaintexts. Amongst these couples of plaintext-ciphertext, with large probability, he gets a "slid" pair, that is, a pair of couples (P_0, C_0) and (P_1, C_1) such that $R(P_0) = P_1$. This immediately implies that $R(C_0) = C_1$. For the attack to work, the function R needs to allow for an efficient recognition of such pairs, which in turns makes the key extraction from R easy. A trivial application of this attack is the key-alternate cipher with blocks of n bits, identical subkeys and no round constants. The complexity is then approximately $2^{n/2}$. The speed-up over exhaustive search given by this attack is then quadratic, similar to the quantum attack based on Grover's algorithm.

This attack is successful, for example, to break the TREYFER block cipher [47], with a data complexity of 2^{32} and a time complexity of $2^{32+12} = 2^{44}$ (where 2^{12} is the cost of identifying the slid pair by performing some key guesses). Comparatively, the cost for an exhaustive search of the key is 2^{64}.

Exponential Quantum Speed-Up of Slide Attacks. We consider the attack represented in Fig. 13. The unkeyed round function is denoted P and the whole encryption function E_k.

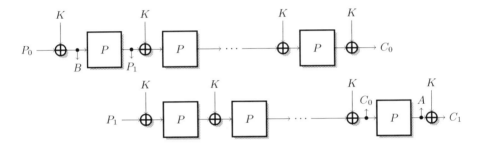

Fig. 13. Representation of a slid-pair used in a slide attack.

We define the following function:

$$f : \{0,1\} \times \{0,1\}^n \to \{0,1\}^n$$

$$b, x \mapsto \begin{cases} P(E_k(x)) \oplus x & \text{if } b = 0, \\ E_k(P(x)) \oplus x & \text{if } b = 1. \end{cases}$$

The slide property shows that all x satisfy $P(E_k(x)) \oplus k = E_k(P(x \oplus k))$. This implies that f satisfies the promise of Simon's problem with $s = 1 \| k$:

$$f(0, x) = P(E_k(x)) \oplus x = E_k(P(x \oplus k)) \oplus k \oplus x = f(1, x \oplus k).$$

In order to apply Theorem 1, we bound $\varepsilon(f, 1 \| k)$, assuming that both $E_k \circ P$ and $P \circ E_k$ are indistinguishable from random permutations. If $\varepsilon(f, 1 \| k) > 1/2$, there exists (τ, t) with $(\tau, t) \notin \{(0,0), (1,k)\}$ such that: $\Pr[f(b, x) = f(b \oplus \tau, x \oplus t)] > 1/2$. Let us assume $\tau = 0$. This implies

$$\Pr[f(0, x) = f(0, x \oplus t)] > 1/2 \quad \text{or} \quad \Pr[f(1, x) = f(1, x \oplus t)] > 1/2,$$

which is equivalent to

$$\Pr[P(E_k(x)) = P(E_k(x \oplus t)) \oplus t] > 1/2 \quad \text{or} \quad \Pr[E_k(P(x)) = E_k(P(x \oplus t)) \oplus t] > 1/2.$$

In particular, there is a differential in $P \circ E_k$ or $E_k \circ P$ with probability $1/2$. Otherwise, $\tau = 1$. This implies

$$\Pr[P(E_k(x)) \oplus x = E_k(P(x \oplus t)) \oplus x \oplus t] > 1/2$$
$$i.e. \quad \Pr[E_k(P(x \oplus k)) \oplus k = E_k(P(x \oplus t)) \oplus t] > 1/2.$$

Again, it means there is a differential in $E_k \circ P$ with probability $1/2$.

Finally we conclude that $\varepsilon(f, 1 \parallel k) \le 1/2$, unless $E_k \circ P$ or $P \circ E_k$ have differentials with probability $1/2$. If E_k behave as a random permutation, $E_k \circ P$ and $P \circ E_k$ also behave as random permutations, and these differential are only found with negligible probability. Therefore, we can apply Simon's algorithm, following Theorem 1, and recover k (Fig. 14).

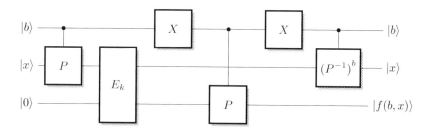

Fig. 14. Simon's function for slide attacks. The X gate is the quantum equivalent of the NOT gate that flips the qubit $|0\rangle$ and $|1\rangle$.

7   Conclusion

We have been able to show that symmetric cryptography is far from ready for the post quantum world. We have found exponential speed-ups on attacks on symmetric cryptosystems. In consequence, some cryptosystems that are believed to be safe in a classical world become vulnerable in a quantum world.

With the speed-up on slide attacks, we provided the first known exponential quantum speed-up of a classical attack. This attack now becomes very powerful. An interesting follow-up would be to seek other such speed-ups of generic techniques. For authenticated encryption, we have shown that many modes of operations that are believed to be solid and secure in the classical world, become completely broken in the post-quantum world. More constructions might be broken following the same ideas.

Acknowledgements. We would like to thank Thomas Santoli and Christian Schaffner for sharing an early stage manuscript of their work [41], Michele Mosca for discussions and LTCI for hospitality. This work was supported by the Commission of the European Communities through the Horizon 2020 program under project number 645622 PQCRYPTO. MK acknowledges funding through grants ANR-12-PDOC-0022-01 and ESPRC EP/N003829/1.

A   Proof of Theorem 1

The proof of Theorem 1 is based of the following lemma.

Lemma 1. *For $t \in \{0,1\}^n$, consider the function $g(x) := 2^{-n} \sum_{y \in t^\perp} (-1)^{x \cdot y}$, where $t^\perp = \{y \in \{0,1\}^n \ s.t. \ y \cdot t = 0\}$. for any x, it satisfies*

$$g(x) = \frac{1}{2}(\delta_{x,0} + \delta_{x,t}). \tag{2}$$

Proof. If $t = 0$ then $g(x) = \sum_{y \in \{0,1\}^n} (-1)^{x \cdot y} = \delta(x,0)$, which proves the claim. From now on, assume that $t \neq 0$. It is straightforward to check that $g(0) = g(t) = \frac{1}{2}$ because all the terms of the sum are equal to 1 and there are 2^{n-1} vectors y orthogonal to t. Since $\sum_{x \in \{0,1\}^n} g(x) = 1$, it is sufficient to prove that $g(x) \geq 0$ to establish the claim in the case $t \neq 0$. For this, decompose $g(x)$ into two terms:

$$g(x) = \sum_{y \in E_0} (-1)^{x \cdot y} - \sum_{y \in E_1} (-1)^{x \cdot y} = |E_0| - |E_1|,$$

where $E_i := \{y \in \{0,1\}^n \ s.t. \ y \cdot x = i \text{ and } y \cdot y = 0\}$ for $i = 0, 1$. Simple counting shows that:

$$|E_0| = \begin{cases} 2^{n-1} & \text{if } x = 0, \\ 2^{n-1} & \text{if } x = t, \\ 2^{n-2} & \text{otherwise.} \end{cases}$$

In particular, $|E_0| \geq |E_1|$ which implies that $g(x) \geq 0$.

We are now ready to prove Theorem 1. Each call to the main subroutine of Simon's algorithm will return a vector u_i. If cn calls are made, one obtains cn vectors $u_1, \ldots, u_{cn}$. By construction, f is such that $f(x) = f(x \oplus s)$ and consequently, the cn vectors $u_1, \ldots, u_{cn}$ are all orthogonal to s. The algorithm is successful provided one can recover the value of s unambiguously, which is the case if the cn vectors span the $(n-1)$-dimensional space orthogonal to s. (Let us note that if the space is $(n-d)$-dimensional for some constant d, one can still recover s efficiently by testing all the vectors orthogonal to the subspace.) In other words, the failure probability p_{fail} is

$$p_{\text{fail}} = \Pr[\dim(\text{Span}(u_1, \ldots, u_n)) \le n - 2]$$

$$\le \Pr[\exists t \in \{0,1\}^n \setminus \{0, s\} \text{ s.t. } u_1 \cdot t = u_2 \cdot t = \cdots = u_{cn} \cdot t = 0]$$

$$\le \sum_{t \in \{0,1\}^n \setminus \{0,s\}} \Pr[u_1 \cdot t = u_2 \cdot t = \cdots = u_{cn} \cdot t = 0]$$

$$\le \sum_{t \in \{0,1\}^n \setminus \{0,s\}} \left(\Pr[u_1 \cdot t = 0]\right)^{cn}$$

$$\le \max_{t \in \{0,1\}^n \setminus \{0,s\}} \left(2\Pr[u_1 \cdot t = 0]^c\right)^n$$

where the second inequality results from the union bound and the third inequality follows from the fact that the results of the cn subroutines are independent.

In order to establish the theorem, it is now sufficient to show that $\Pr[u \cdot t = 0]$ is bounded away from 1 for all t, where u is the vector corresponding to the output of Simon's subroutine. We will prove that for all $t \in \{0,1\}^n \setminus \{0, s\}$, the following inequality holds:

$$\Pr_u[u \cdot t = 0] = \frac{1}{2}\left(1 + \Pr_x[f(x) = f(x \oplus t)]\right) \le \frac{1}{2}(1 + \varepsilon(f, s)) \le \frac{1}{2}(1 + p_0). \quad (3)$$

In Simon's algorithm, one can wait until the last step before measuring both registers. The final state before measurement can be decomposed as:

$$2^{-n} \sum_{x \in \{0,1\}^n} \sum_{y \in \{0,1\}^n} (-1)^{x \cdot y} |y\rangle |f(x)\rangle = 2^{-n} \sum_{\substack{y \in \{0,1\}^n \\ \text{s.t. } y \cdot t = 0}} \sum_{x \in \{0,1\}^n} (-1)^{x \cdot y} |y\rangle |f(x)\rangle$$

$$+ 2^{-n} \sum_{\substack{y \in \{0,1\}^n \\ \text{s.t. } y \cdot t = 1}} \sum_{x \in \{0,1\}^n} (-1)^{x \cdot y} |y\rangle |f(x)\rangle.$$

The probability of obtaining u such that $u \cdot t = 0$ is given by

$$\Pr_u[u \cdot t = 0] = \left\| 2^{-n} \sum_{\substack{y \in \{0,1\}^n \\ \text{s.t. } y \cdot t = 0}} |y\rangle \sum_{x \in \{0,1\}^n} (-1)^{x \cdot y} |f(x)\rangle \right\|^2$$

$$= 2^{-2n} \sum_{\substack{y \in \{0,1\}^n \\ \text{s.t. } y \cdot t = 0}} \sum_{x,x' \in \{0,1\}^n} (-1)^{(x \oplus x') \cdot y} \langle f(x')|f(x)\rangle$$

$$= 2^{-2n} \sum_{x,x' \in \{0,1\}^n} \langle f(x')|f(x)\rangle \sum_{\substack{y \in \{0,1\}^n \\ \text{s.t. } y \cdot t = 0}} (-1)^{(x \oplus x') \cdot y}$$

$$= 2^{-2n} \sum_{x,x' \in \{0,1\}^n} \langle f(x')|f(x)\rangle 2^{n-1}(\delta_{x,x'} + \delta_{x',x \oplus t}) \quad (4)$$

$$= 2^{-(n+1)} \left[\sum_{x \in \{0,1\}^n} \langle f(x)|f(x)\rangle + \sum_{x \in \{0,1\}^n} \langle f(x \oplus t)|f(x)\rangle \right] \quad (5)$$

$$= \frac{1}{2} \left[1 + \Pr_x[f(x) = f(x \oplus t)] \right] \quad (6)$$

where we used Lemma 1 proven in the appendix in Eq. 4, and $\delta_{x,x'} = 1$ if $x = x'$ and 0 otherwise.

B   Proof of Theorem 2

Let t be a fixed value and $p_t = Pr_x[f(x \oplus t) = f(t)]$. Following the previous analysis, the probability that the cn vectors u_i are orthogonal to t can be written as $\Pr[u_1 \cdot t = u_2 \cdot t = \cdots = u_{cn} \cdot t = 0] = \left(\frac{1+p_t}{2}\right)^{cn}$.

In particular, we can bound the probability that Simon's algorithm returns a value t with $p_t < p_0$:

$$\Pr[p_t < p_0] = \sum_{t\,:\,p_t < p_0} \left(\frac{1+p_t}{2}\right)^{cn} \leq 2^n \times \left(\frac{1+p_0}{2}\right)^{cn}.$$

References

1. Abed, F., Fluhrer, S.R., Forler, C., List, E., Lucks, S., McGrew, D.A., Wenzel, J.: Pipelineable on-line encryption. In: Cid and Rechberger [17], pp. 205–223. http://dx.doi.org/10.1007/978-3-662-46706-0_11

2. Alagic, G., Broadbent, A., Fefferman, B., Gagliardoni, T., Schaffner, C., Jules, M.S.: Computational security of quantum encryption. arXiv preprint (2016). arXiv:1602.01441

3. Anand, M.V., Targhi, E.E., Tabia, G.N., Unruh, D.: Post-quantum security of the CBC, CFB, OFB, CTR, and XTS modes of operation. In: Takagi, T., et al. (eds.) PQCrypto 2016. LNCS, vol. 9606, pp. 44–63. Springer, Heidelberg (2016). http://dx.doi.org/10.1007/978-3-319-29360-8_4

4. Andreeva, E., Bogdanov, A., Luykx, A., Mennink, B., Tischhauser, E., Yasuda, K.: Parallelizable and authenticated online ciphers. In: Sako, K., Sarkar, P. (eds.) ASIACRYPT 2013, Part I. LNCS, vol. 8269, pp. 424–443. Springer, Heidelberg (2013). http://dx.doi.org/10.1007/978-3-642-42033-7_22

5. Bellare, M., Kilian, J., Rogaway, P.: The security of the cipher block chaining message authentication code. J. Comput. Syst. Sci. 61(3), 362–399 (2000). http://dx.doi.org/10.1006/jcss.1999.1694

6. Bernstei, D.J.: Introduction to post-quantum cryptography. In: Bernstein, D.J., Buchmann, J., Dahmen, E. (eds.) Post-Quantum Cryptography, pp. 1–14. Springer, Heidelberg (2009)

7. Biryukov, A., Wagner, D.: Slide attacks. In: Knudsen, L.R. (ed.) FSE 1999. LNCS, vol. 1636, pp. 245–259. Springer, Heidelberg (1999). http://dx.doi.org/10.1007/3-540-48519-8_18

8. Biryukov, A., Wagner, D.: Advanced slide attacks. In: Preneel, B. (ed.) EUROCRYPT 2000. LNCS, vol. 1807, pp. 589–606. Springer, Heidelberg (2000). http://dx.doi.org/10.1007/3-540-45539-6_41

9. Black, J.A., Rogaway, ... constructions. In: Bellare, ... Springer, Heidelberg (2000).

10. Black, J.A., Rogaway, P.: ...arbitrary-length messages: the three-key allelizable message authentica... 2000. LNCS, vol. 1880, pp. 197–215. CRYPT 2002. LNCS, vol. 2332, ... mode of operation for par- http://dx.doi.org/10.1007/3-540-46035-... /10.1007/3-540-44598-6_12

11. Boneh, D., Dagdelen, Ö., Fischlin, M., L.... ...udsen, L.R. (ed.) EURO- M.: Random oracles in a quantum world. ... Springer, Heidelberg (2002). ASIACRYPT 2011. LNCS, vol. 7073, pp. 41–6..., Schaffner, C., Zhandry, http://dx.doi.org/10.1007/978-3-642-25385-0_3 D.H., Wang, X. (eds.) ...er, Heidelberg (2011).

12. Boneh, D., Zhandry, M.: Quantum-secure message ...r, Heidelberg (2011). Johansson, T., Nguyen, P.Q. (eds.) EUROCRYPT 2013. LN... ...ation codes. In: 608. Springer, Heidelberg (2013). http://dx.doi.org/10.1007/978-7881, pp. 592–

13. Boneh, D., Zhandry, M.: Secure signatures and chosen ciphertex...38348-9_35 a quantum computing world. In: Canetti, R., Garay, J.A. (eds.) ...urity in 2013, Part II. LNCS, vol. 8043, pp. 361–379. Springer, Heidelberg (20..). http://dx.doi.org/10.1007/978-3-642-40084-1_21

14. Brassard, G., Høyer, P., Kalach, K., Kaplan, M., Laplante, S., Salvail, L.: Merkle puzzles in a quantum world. In: Rogaway, P. (ed.) CRYPTO 2011. LNCS, vol. 6841, pp. 391–410. Springer, Heidelberg (2011)

15. Broadbent, A., Jeffery, S.: Quantum homomorphic encryption for circuits of low t-gate complexity. In: Gennaro, R., Robshaw, M. (eds.) CRYPTO 2015. LNCS, vol. 9216, pp. 609–629. Springer, Heidelberg (2015)

16. Carter, L., Wegman, M.N.: Universal classes of hash functions (extended abstract). In: Hopcroft, J.E., Friedman, E.P., Harrison, M.A. (eds.) Proceedings of the 9th Annual ACM Symposium on Theory of Computing, Boulder, Colorado, USA, 4–6 May 1977, pp. 106–112. ACM (1977). http://doi.acm.org/10.1145/800105.803400

17. Cid, C., Rechberger, C. (eds.): FSE 2014. LNCS, vol. 8540. Springer, Heidelberg (2015). http://dx.doi.org/10.1007/978-3-662-46706-0

18. Cogliani, S., Maimuţ, D., Naccache, D., do Canto, R.P., Reyhanitabar, R., Vaudenay, S., Vizár, D.: OMD: a compression function mode of operation for authenticated encryption. In: Joux, A., Youssef, A. (eds.) SAC 2014. LNCS, vol. 8781, pp. 112–128. Springer, Heidelberg (2014). http://dx.doi.org/10.1007/978-3-319-13051-4_7

19. Daemen, J., Rijmen, V.: Probability distributions of correlation and differentials in block ciphers. J. Math. Crypt. **1**(3), 221–242 (2007). http://dx.doi.org/10.1515/JMC.2007.011

20. Damgård, I., Funder, J., Nielsen, J.B., Salvail, L.: Superposition attacks on cryptographic protocols. In: Padró, C. (ed.) ICITS 2013. LNCS, vol. 8317, pp. 142–161. Springer, Heidelberg (2014). http://dx.doi.org/10.1007/978-3-319-04268-8_9

21. Dworkin, M.: Recommendation for block cipher modes of operation: the CMAC mode for authentication. NIST Special Publication 800–38B, National Institute for Standards and Technology, May 2005

22. Even, S., Mansour, Y.: A construction of a cipher from a single pseudorandom permutation. J. Crypt. **10**(3), 151–162 (1997). http://dx.doi.org/10.1007/s001459900025

23. Gagliardoni, T., Hülsing, A., Schaffner, C.: Semantic security and indistinguishability in the quantum world. arXiv preprint (2015). arXiv:1504.05255

24. Grover, L.K.: A fast quantum ~rithm for database search. In: Miller,
    G.L. (ed.) Proceedings of ~ghth Annual ACM Symposium on the
    Theory of Computing, p~ay, ~ennsylvania, USA, 22–24 May 1996, pp.
    212–219. ACM (1996). h~ ~.org/10.1145/237814.237866
25. Hoang, V.T., Krovetz, ~s. In: Oswald, E., Fischlin, M. (eds.) EURO-
    and the problem th~ 9056, pp. 15–44. Springer, Heidelberg (2015).
    CRYPT 2015. LN~ ~78-3-662-46800-5_2
    http://dx.doi.org/1 K.: OMAC: one-key CBC MAC. In: Johansson, T.
26. Iwata, T., Kur~ ~vol. 2887, pp. 129–153. Springer, Heidelberg (2003).
    (ed.) FSE 200~ ~007/978-3-540-39887-5_11
    http://dx.d~ ~atsu, K., Guo, J., Morioka, S.: CLOC: authenticated encryption
27. Iwata, T, ~t. In: Cid and Rechberger [17] , pp. 149–167. http://dx.doi.org/10.
    for sho~3-662-46706-0_8
28. K~ 1007/ ~n, M.: Quantum attacks against iterated block ciphers. CoRR abs/1410.1434
    ~014). http://arxiv.org/abs/1410.1434
29. Kaplan, M., Leurent, G., Leverrier, A., Naya-Plasencia, M.: Quantum differential
    and linear cryptanalysis. CoRR abs/1510.05836 (2015). http://arxiv.org/abs/1510.
    05836
30. Krovetz, T., Rogaway, P.: The software performance of authenticated-encryption
    modes. In: Joux, A. (ed.) FSE 2011. LNCS, vol. 6733, pp. 306–327. Springer,
    Heidelberg (2011). http://dx.doi.org/10.1007/978-3-642-21702-9_18
31. Kuwakado, H., Morii, M.: Quantum distinguisher between the 3-round Feistel
    cipher and the random permutation. In: 2010 IEEE International Symposium on
    Information Theory Proceedings (ISIT), June 2010, pp. 2682–2685 (2010)
32. Kuwakado, H., Morii, M.: Security on the quantum-type Even-Mansour cipher.
    In: 2012 International Symposium on Information Theory and Its Applications
    (ISITA), October 2012, pp. 312–316 (2012)
33. Liskov, M., Rivest, R.L., Wagner, D.: Tweakable block ciphers. J. Crypt. **24**(3),
    588–613 (2011). http://dx.doi.org/10.1007/s00145-010-9073-y
34. Luby, M., Rackoff, C.: How to construct pseudorandom permutations
    from pseudorandom functions. SIAM J. Comput. **17**(2), 373–386 (1988).
    http://dx.doi.org/10.1137/0217022
35. Lydersen, L., Wiechers, C., Wittmann, C., Elser, D., Skaar, J., Makarov, V.: Hack-
    ing commercial quantum cryptography systems by tailored bright illumination.
    Nat. Photonics **4**(10), 686–689 (2010)
36. McGrew, D.A., Viega, J.: The security and performance of the Galois/Counter
    Mode (GCM) of operation. In: Canteaut, A., Viswanathan, K. (eds.)
    INDOCRYPT 2004. LNCS, vol. 3348, pp. 343–355. Springer, Heidelberg (2004).
    http://dx.doi.org/10.1007/978-3-540-30556-9_27
37. Minematsu, K.: Parallelizable Rate-1 authenticated encryption from
    pseudorandom functions. In: Nguyen, P.Q., Oswald, E. (eds.) EURO-
    CRYPT 2014. LNCS, vol. 8441, pp. 275–292. Springer, Heidelberg (2014).
    http://dx.doi.org/10.1007/978-3-642-55220-5_16
38. Montanaro, A., de Wolf, R.: A survey of quantum property testing. arXiv preprint
    (2013). arXiv:1310.2035
39. Rogaway, P.: Efficient instantiations of tweakable blockciphers and refinements
    to modes OCB and PMAC. In: Lee, P.J. (ed.) ASIACRYPT 2004. LNCS, vol.
    3329, pp. 16–31. Springer, Heidelberg (2004). http://dx.doi.org/10.1007/978-3-
    540-30539-2_2

40. Rogaway, P., Bellare, M., Black, J., Krovetz, T.: OCB: a block-cipher mode of operation for efficient authenticated encryption. In: Reiter, M.K., Samarati, P. (eds.) CCS 2001, Proceedings of the 8th ACM Conference on Computer and Communications Security, Philadelphia, Pennsylvania, USA, 6–8 November 2001, pp. 196–205. ACM (2001). http://doi.acm.org/10.1145/501983.502011
41. Santoli, T., Schaffner, C.: Using simon's algorithm to attack symmetric-key cryptographic primitives. arXiv preprint (2016). arXiv:1603.07856
42. Sasaki, Y., Todo, Y., Aoki, K., Naito, Y., Sugawara, T., Murakami, Y., Matsui, M., Hirose, S.: Minalpher v1.1. CAESAR submission, August 2015
43. Shor, P.W.: Polynomial-time algorithms for prime factorization and discrete logarithms on a quantum computer. SIAM J. Comput. 26(5), 1484–1509 (1997). http://dx.doi.org/10.1137/S0097539795293172
44. Simon, D.R.: On the power of quantum computation. SIAM J. Comput. 26(5), 1474–1483 (1997)
45. Unruh, D.: Non-interactive zero-knowledge proofs in the quantum random oracle model. In: Oswald, E., Fischlin, M. (eds.) EUROCRYPT 2015. LNCS, vol. 9057, pp. 755–784. Springer, Heidelberg (2015). Preprint on IACR ePrint 2014/587
46. Xu, F., Qi, B., Lo, H.K.: Experimental demonstration of phase-remapping attack in a practical quantum key distribution system. New J. Phys. 12(11), 113026 (2010)
47. Yuval, G.: Reinventing the travois: Encryption/MAC in 30 ROM bytes. In: Biham, E. (ed.) FSE 1997. LNCS, vol. 1267, pp. 205–209. Springer, Heidelberg (1997). http://dx.doi.org/10.1007/BFb0052347
48. Zhandry, M.: How to construct quantum random functions. In: 53rd Annual IEEE Symposium on Foundations of Computer Science, FOCS 2012, New Brunswick, NJ, USA, 20–23 October 2012, pp. 679–687. IEEE Computer Society (2012). http://dx.doi.org/10.1109/FOCS.2012.37
49. Zhandry, M.: Secure identity-based encryption in the quantum random oracle model. Int. J. Quan. Inf. 13(04), 1550014 (2015)
50. Zhao, Y., Fung, C.H.F., Qi, B., Chen, C., Lo, H.K.: Quantum hacking: experimental demonstration of time-shift attack against practical quantum-key-distribution systems. Phys. Rev. A 78(4), 042333 (2008)

Hardware-Oriented Cryptography

Efficiently Computing Data-Independent Memory-Hard Functions

Joël Alwen[1] and Jeremiah Blocki[2,3(✉)]

[1] IST Austria, Klosterneuburg, Austria
[2] Microsoft Research, Cambridge, USA
jblocki@microsoft.com
[3] Purdue, West Lafayette, USA

Abstract. A memory-hard function (MHF) f is equipped with a *space cost* σ and *time cost* τ parameter such that repeatedly computing $f_{\sigma,\tau}$ on an application specific integrated circuit (ASIC) is not economically advantageous relative to a general purpose computer. Technically we would like that any (generalized) circuit for evaluating an iMHF $f_{\sigma,\tau}$ has area × time (AT) complexity at $\Theta(\sigma^2 * \tau)$. A data-independent MHF (iMHF) has the added property that it can be computed with almost optimal memory and time complexity by an algorithm which accesses memory in a pattern independent of the input value. Such functions can be specified by fixing a directed acyclic graph (DAG) G on $n = \Theta(\sigma * \tau)$ nodes representing its computation graph.

In this work we develop new tools for analyzing iMHFs. First we define and motivate a new complexity measure capturing the amount of *energy* (i.e. electricity) required to compute a function. We argue that, in practice, this measure is at least as important as the more traditional AT-complexity. Next we describe an algorithm $\mathcal{A}$ for repeatedly evaluating an iMHF based on an arbitrary DAG G. We upperbound both its energy and AT complexities per instance evaluated in terms of a certain combinatorial property of G.

Next we instantiate our attack for several general classes of DAGs which include those underlying many of the most important iMHF candidates in the literature. In particular, we obtain the following results which hold for all choices of parameters σ and τ (and thread-count) such that $n = \sigma * \tau$.

- The Catena-Dragonfly function of [FLW13] has AT and energy complexities $O(n^{1.67})$.
- The Catena-Butterfly function of [FLW13] has complexities is $O(n^{1.67})$.
- The Double-Buffer and the Linear functions of [CGBS16] both have complexities in $O(n^{1.67})$.
- The Argon2i function of [BDK15] (winner of the Password Hashing Competition [PHC]) has complexities $O(n^{7/4}\log(n))$.
- The Single-Buffer function of [CGBS16] has complexities $O(n^{7/4}\log(n))$.
- *Any* iMHF can be computed by an algorithm with complexities $O(n^2/\log^{1-\epsilon}(n))$ for all $\epsilon > 0$. In particular when $\tau = 1$ this shows

© International Association for Cryptologic Research 2016
M. Robshaw and J. Katz (Eds.): CRYPTO 2016, Part II, LNCS 9815, pp. 241–271, 2016.
DOI: 10.1007/978-3-662-53008-5_9

that the goal of constructing an iMHF with AT-complexity $\Theta(\sigma^2 * \tau)$ is unachievable.

Along the way we prove a lemma upper-bounding the depth-robustness of any DAG which may prove to be of independent interest.

1 Introduction

Moderately hard to compute functions have proven to be useful security primitives. In this work we focus on "memory-hard functions" (MHF) introduced in [Per09]. These aim to serve as password hashing algorithms for storing passwords in a login system, as Key Derivation Functions (also called "key stretching" functions) for password-based cryptography and for building Proof-of-Effort protocols (in particular for use in cryptocurrencies such as Litecoin [Cha11, Bil13] and others). In each case the main security property we would like to achieve for the MHF is that brute-force attacks (i.e. evaluating the MHF on many inputs) using an application-specific integrated circuit (ASIC) should not be economically viable.

1.1 Memory-Hard Functions and Their Complexity

We interpret this intuitive goal in two ways. Either the cost of *building* the ASIC should be prohibitively expensive (in terms of say USD) or the cost of *running* the ASIC should be prohibitively expensive. In fact, given that the former is a one-time cost which can be amortized over the life-time of the device while the later is a recurring cost, it may often be the case that the later is the most interesting goal to achieve.

The cost of building a circuit is often approximated by its *AT-complexity* [Tho79, BL13, BK15, AS15]; that is the product of the area of the chip and the time it takes the chip to produce the output. In this work we consider MHFs built as modes of operation over an underlying compression function H. Thus, we measure time in units of *tocks*; namely the time it takes to evaluate one instance of H from start to finish[1]. We measure area in units of "memory-area" (MAr); namely the area required to store one output of H (called a *block*). Finally we parametrize our AT-complexity notion AT_R with the *core-memory area ratio* [BK15] $R > 0$, a positive real denoting the number of MAr required to implement one copy of H.[2]

To estimate the cost of running the chip we will use a new notion which we call the "energy-complexity" or *E-complexity* of the circuit. Intuitively, it approximates the energy (say in kilo-Watt-hours) used in an execution of the chip. More precisely the unit of measure for E-complexity is a "memory-Watt-tock" (MWt) – the number of kWh it takes to store one block for one tock. We also parametrize the complexity notion $E_{\bar{R}}$ with the *core-memory energy ratio*,

---

[1] I.e. without considering pipelining and other such amortized optimizations.

[2] This allows our analysis to be applied regardless of the particular VLSI technology employed and the particular implementation of H used when constructing the ASIC.

a positive real $\bar{R} > 0$ which is the number of MWt required to evaluate one instance of H.

To see why this is an interesting measure for achieving our stated security goal consider the case of password hashing. (The case for KDFs follows essentially the same reasoning.) Suppose an attacker manages to pilfer the credentials file from a login-server and now executes an off-line brute-force attack $\mathcal{A}$ implemented in an ASIC with core-memory energy ratio $\bar{R}$ using $E_{\bar{R}}(\mathcal{A})$ MWt per password guess. We model the monetary income from such an attack as being proportional to the number of password guesses made which we denote by $\#eval$.[3] Conversely, we model the running cost as being proportional to the electricity consumed by the ASIC while executing the attack, namely its $E_{\bar{R}}$-complexity times $\#eval$. The attacker can always increase income (i.e. increase $\#eval$) simply by adding more implementations of $\mathcal{A}$ to the ASIC or running the ASIC for more time. Therefore, the attack is profitable (in this model) if and only if the USD cost c of one MWt and the income i per password guess are such that $i > c * E_{\bar{R}}(\mathcal{A})$. Thus we can use $E_{\bar{R}}(\mathcal{A})$ as a key indicator for which ranges of (c, i) an attack is economically viable.

Quality of an Attack. A candidate MHF F is specified via an algorithm which evaluates it. (E.g. [Per09,FLW13,BDK15].) We refer to this algorithm as the *naïve* algorithm $\mathcal{N}$ for F and it is understood to be the algorithm used by the honest party. $\mathcal{N}$ is intended to be an algorithm that can be evaluated efficiently on *typical* (i.e. general purpose) computer architectures — where we may not be able to evaluate H multiple times in parallel. As usual, we are interested in what advantage an adversarial evaluation algorithm can have over the honest party. Therefore, one measure of the quality of a given algorithm $\mathcal{A}$ for evaluating (multiple instances of) an MHF F is to compare its complexity to that of $\mathcal{N}$. In particular for given core-memory ratios R and $\bar{R}$ the *AT-quality* and *energy-quality* of $\mathcal{A}$ are given by

$$\mathsf{AT\text{-}quality}_R(\mathcal{A}) = \frac{AT_R(\mathcal{N})}{AT_R(\mathcal{A})} \quad \text{and} \quad \mathsf{E\text{-}quality}_{\bar{R}}(\mathcal{A}) = \frac{E_{\bar{R}}(\mathcal{N})}{E_{\bar{R}}(\mathcal{A})}.$$

Here, $AT_R(\mathcal{A})$ (resp. $E_{\bar{R}}(\mathcal{A})$) measures the *amortized* AT complexity (resp. *amortized* energy complexity).[4] That is $AT_R(\mathcal{A})$ is smallest $AT_{\mathbb{R}}$ complexity of a chip implementing $\mathcal{A}$ divided by $\#inst(\mathcal{A})$ — the number of instances of F computed in an execution of $\mathcal{A}$. We consider $\mathcal{A}$ an "attack" if either one of these quality measures is greater than 1. (However we remark that all attacks in this work have both qualities *simultaneously* tending towards infinity as $\#inst$ grows.)

Data-Independent and Ideal MHFs. An *data-independent* memory-hard function (iMHF) is a function f for which the associated naïve algorithm $\mathcal{N}$, on

---

[3] Intuitively, the more passwords guesses made the higher the expected number of password (equivalents) recovered by the adversary which can then be monetized.

[4] Generally, unless explicitly specified otherwise, we are only interested in the *amortized* AT and energy complexities per instance of the MHF computed.

input x, computes $f(x)$ using a memory access pattern that is independent of x. These take on special importance in applications where the MHF is to be evaluated on secret input in an (at least somewhat) hostile environment. This is because (in contrast to their siblings data-*dependent* MHFs) it is much easier to implement an iMHF in such a way that it avoids information leakage via certain side-channel attacks such as Timing attacks. In these attacks, the variation in the time taken to perform certain operations is used to deduce information about the inputs upon which the MHF is being evaluated. Similar attacks have in the past been mounted by local adversarial processes [BM06], adversarial virtual machines in a cloud environment [RTSS09] or even completely remotely [Ber, ASK07]. Therefore, in the context of both KDFs and password hashing data-independence is a desirable property. All MHFs considered in this work are of this form.

In general, an iMHF f can be described via a fixed DAG G representing its computation graph. Each node represents an intermediary value, which is computed via some deterministic round function, using the values represented by the parent nodes in G (e.g. via a single call to H). The source node of G represents the input x while $f(x)$, the output of the computation, is the value represented by the sink node.

Let f be an iMHF given by some DAG G of size n with constant in-degree. There exists a trivial algorithm triv which can always compute f with AT and energy complexities $\Theta(n^2)$.[5] Given a constant $c > 1$ we consider f to be a *c-ideal* iMHF if, when we take the naïve algorithm to be triv, there exist no attack $\mathcal{A}$ on f with better quality than c (i.e. $\forall \mathcal{A}$ E-quality$_{\bar{R}}(\mathcal{A}) \leq c$ and AT-quality$_R(\mathcal{A}) \leq c$). A primary goal of research in this field is to find an ideal iMHF.[6]

1.2   MHF Candidates

Due to the growing interest in MHFs there are a number of candidate functions. For example in the recently completed Password Hashing Competition [PHC] most entrants claimed some form of memory-hardness. The goal of the PHC was to select a winning algorithm to act as a new standard for password hashing.

Catena. To the best of our knowledge the earliest candidate iMHF is the PHC finalist Catena [FLW13]. It received special recognition for its agile framework and its resistance to side-channel attacks. In [FLW13] the authors proposed two different DAGs giving rise two separate functions. The first, called Catena Bit Reversal, is based on an λ-layered graph BRG_λ^n with n nodes. The second is called Catena Double Butterfly and is based on a different $O(\lambda \log n)$-layered graph DBG_λ^n. The Catena designers recommended choosing $\lambda \in \{1, 2, 3, 4\}$ [FLW13].

---

[5] Simply compute each intermediary value in topological order, one value at a time, storing all results in memory until the computation is complete.

[6] Hopefully one permuting as simple as possible an explicit description and naïve implementation and as lightweight as possible round-function.

Argon2. One of the most important MHF candidates is Argon2 [BDK15]. Notably, it is the winner of the Password Hashing Competition [PHC]. Argon2 is equipped with a data-dependent mode of operation and an independent mode which is called Argon2i. Argon2i is recommended for password hashing.

Balloon Hashing. Most recently, three new candidate iMHFs are proposed in [CGBS16]. These are called the Single-Buffer (SB), Double-Buffer and Linear constructions respectively and are jointly referred to as the Balloon Hashing constructions. The authors provide strong evidence for the memory-hardness of all three candidates albeit assuming the absence of parallelism.

In general iMHF candidates are equipped with a *space-cost* parameter σ (in which the memory required per evaluation is intended to scale) and a *time-cost* parameter τ (in which the time required for an evaluation is intended to scale). Additionally, Argon2i, the Double-Buffer and the Linear functions are also equipped with a *parallelism* parameter ϕ the property that the naïve algorithm can make efficient use of (up to) ϕ concurrent threads. Viewing these functions as DAGs gives rise to a graph on $n = \sigma * \tau$ nodes with depth n/ϕ. The hope is that for all settings of (σ, τ, ϕ) the AT and energy complexity lie in $\Theta(\sigma^2 * \tau/\phi)$.

1.3   Our Contributions

In this work we introduce and motivate the notion of (amortized) energy complexity. Next we give a generic evaluation algorithm PGenPeb for data-independent iMHFs based on arbitrary DAG G. We analyze PGenPeb's energy and AT complexities in terms of a combinatorial property of G. In particular, we obtain an attack against any iMHF for which G is not *depth-robust*. Informally, a DAG G is not depth-robust if there is a relatively small set S of nodes such that after removing S from G the resulting graph (denoted $G - S$) has low depth (i.e. contains only short paths).

We instantiate the attack for various classes of DAGs. In particular, we exhibit a "depth-reducing" node set S for the Argon2i DAG, both types of Catena DAGs and all three Balloon Hashing DAGs. For example, for any parameters $(\sigma, \tau, \phi = 1)$ with $n = \sigma * \tau$ we obtain an attack on both Catena, the Double-Buffer and the Linear iMHFs with quality $\Omega(n^{1/3})$. Similarly we demonstrate an attack on Argon2i and the Single-Buffer iMHF with quality $\Omega(\frac{n^{1/4}}{\ln n})$.[7]

In fact we demonstrate that no DAG with constant indegree is sufficiently depth-robust to completely resist the attack. More precisely, we show that any iMHF is at best c-ideal for $c = \Omega(\log^{1-\epsilon} n)$ and any $\epsilon > 0$. In particular this means that ideal iMHFs, as described above, do not exist.

---

[7] For the cases when $\phi > 1$ PGenPeb maintains the same complexities but the resulting quality decreases somewhat as the complexity of the naïve algorithm improves for Argon2i, the Double-Buffer and the Linear functions. In other words quality decreases not because memory-hardness increases but because the honest algorithm becomes more efficient.

General Attack on Non-depth Robust DAGs. We first present in Sect. 3, a generic evaluation algorithm GenPeb which takes as inputs a node subset S. Because G is not depth-robust there exists a small set S of nodes such that $d = \mathsf{depth}(G-S)$ is relatively small. The basic idea behind our attack is to divide computation steps into two phases: balloon phases and light phases. Each light phase lasts roughly $g \gg d$ time steps. During light phases we discard most of the values that we have computed from memory keeping only values corresponding to nodes in S, the highest node i whose value has been computed and the parents of the nodes whose values we plan to compute in the next g time steps. As the name suggests, light phases are cheap. Our memory usage is low during these light phases and we will compute one instance of the round function (e.g. call to H) during each time step. During a Balloon Phase we quickly restore all of the discarded values to memory so that we can complete the next light phase. Unlike light phases, the balloon phases are more expensive because we are storing up to $O(n)$ values in memory and because we will often make multiple calls to the round function in parallel. However, the key observation is that we will not incur these higher cost in too many time steps. In particular, because the graph $G - S$ has small depth $d \ll g$ and we never discard values for nodes in S the Balloon Phase can be completed very quickly (i.e., in at most $d \ll g$ times steps) by making parallel calls to the round function.

While for any non-depth-robust graph the GenPeb algorithm has good energy complexity, obtaining an evaluation algorithm with low AT-complexity requires a bit more work. Notice that during a light phase most of the memory capacity and round function implementations needed for a balloon phase are no longer being used. Moreover light phases run for significantly more time than the balloon phases. These observations give rise to the low AT-complexity parallel algorithm PGenPeb which evaluates g/d instances of the iMHF concurrently such that at any given time only a single instance is in a balloon phase while all other instances are in light phases. Intuitively this results in more efficient use of available hardware while technically we get that the energy complexity of the algorithm is approximately equal to the AT complexity (Theorem 3).

Stacked Sandwich Graphs. In Sect. 4 we focus on two classes of DAGs called (strict) stacked sandwich graphs. Informally, a DAG G is a λ-stacked sandwich DAG if the nodes can be partitioned into $\lambda+1$ layers such that, with the possible exception of node i, all of the parents of node $i + 1$ are from previous layers. These classes include the DAGs implicit to both Catena iMHFs as well as the Double-Buffer and Linear iMHFs. We prove that no λ-stacked sandwich graph is depth-robust (Lemma 1). For any $t > 1$ there is a set S of n/t nodes such that $\mathsf{depth}(G - S) \leq (\lambda + 1)t$.

(n, δ, w)-Random Graphs. In Sect. 5 we turn to a class of random graphs called (n, δ, w)-random DAGs. We remark that the graphs implicit to Argon2i and the Single-Buffer iMHF (for a randomly chosen salt) fall into this category of random DAGs. We show (in Lemma 4) that, with high probability, by removing just a few nodes these graphs can be transformed into stacked sandwich graphs and are thus not depth-robust.

Attack on any iMHF. In Sect. 6 we prove that no DAG with constant indegree is sufficiently depth-robust to resist at least some form of attack (Theorem 8). In our proof, we rely on a result due to Valiant [Val77] which states that for any DAG G with m edges and depth d there is a set S of $m/\log d$ edges s.t. by deleting them we obtain a graph of depth at most $d/2$ (see Lemma 6). Given $\epsilon > 0$ we can repeatedly apply this result obtain a set S of $o\left(\frac{\delta n}{\log^{1-\epsilon} n}\right)$ nodes s.t $\mathsf{depth}(G - S) \leq \frac{n}{\log^2 n}$. Thus if we let the naïve algorithm be (any algorithm complexity comparable to) triv then we have a generic attack $\mathcal{A}$ with quality $\mathsf{AT\text{-}quality}_R(\mathcal{A}) = \Omega\left(\delta^{-1}\log^{1-\epsilon} n\right)$ and $E_R(\mathcal{A}) = \Omega\left(\delta^{-1}\log^{1-\epsilon} n\right)$.

Exact Security Analysis. Finally we present exact bounds for the energy and AT complexities of all of our attacks. Our analysis demonstrate that our attacks have high quality for practical values of n and $\bar{R}$ — not just as $n \to \infty$. For example setting $n = 2^{18}$ we already have an attack $\mathcal{A}$ against Argon2i with $\mathsf{AT\text{-}quality}_R(\mathcal{A})$, $\mathsf{E\text{-}quality}_{\bar{R}}(\mathcal{A}) > 1$ — using a realistic value $\bar{R} = 3,000$. In general, $\mathsf{E\text{-}quality}_{\bar{R}}(\mathcal{A})$ will increase as n increases or as $\bar{R}$ decreases.

1.4   Related Work

The intuitive goal of constructing functions for which VLSI implementations are prohibitively expensive was first laid out by Percival in [Per09]. This property was formalized by asking that evaluating such a function on a PRAM requires large ST-complexity. In particular evaluation algorithms with low amortized complexity such as those in this work were not considered. Percival also introduced the first, and currently most widely deployed, candidate MHF called scrypt. A full proof of security under a strong security definition remains a central open problem in the area. However recently significant progress has been made in this direction in [ACK+16]. It is interesting to note though that despite scrypt being data-*dependent* the (conditional) lower bound in [ACK+16] still does not exceed the upper-bound of Sect. 6 on the best possible quality of an iMHF.

Catena. In [FLW13] the authors of Catena restricted their analysis of its security to a sequential setting. That is they restrict an adversary to only being able to evaluate one instance of the underlying function H at a time. In this setting and for the case when $\lambda = 1$ the results of [LT82] show that, in a simplified computational model, BRG_1^n has ST-complexity $\Omega\left(n^2\right)$. Here ST-complexity denotes the product of the space and time required by any algorithm which evaluates Catena Bit Reversal. The intuition being that large ST-complexity implies large AT-complexity of any implementation in a custom chip.

Argon2. Argon2 [BDK15] was the winner of the Password Hashing Competition [PHC]. Argon2 is equipped with a data-dependent mode of operation and an independent mode which is called Argon2i. The authors recommend using Argon2i for password hashing due to its resistance to side channel attacks [BDK15]. Our attacks only apply to Argon2i, the data independent mode. Recently, Corrigan-Gibbs et al. [CGBS16] gave an attack on Argon2i which reduces the cost of computing Argon2i by a factor of 4.

Balloon Hashing. In [CGBS16] the authors also proposed three iMHFs which resist their attack on Argon2i. These are called Single-Buffer (SB), Double-Buffer and Linear and collectively referred to as the Balloon Hashing iMHFs.[8] Our attacks reduce the cost of computing both Argon2i and SB by a factor of $\tilde{\Omega}\left(n^{1/4}\right)$.

A Provably Secure MHF. Currently, the only candidate MHF equipped with a full proof of security is the one in [AS15]. There, the authors show an iMHF F for which the energy-complexity of the required storage alone (i.e. disregarding the cost of evaluating the round function) is within a polylogarithmic factor in n of the energy-complexity of the trivial algorithm triv. Moreover triv uses only a single instance of H (i.e. it is sequential) which implies that, roughly speaking, any evaluation algorithm for F can have E-quality $= O(\mathsf{polylog}(n))$. The results in Sect. 6 show that this is optimal for any iMHF up to the exponent in the polylogarithmic factor.

Attacking MHFs. The Catena Dragonfly iMHF has been attacked previously [BK15, AS15]. In particular, [AS15] demonstrated an attack on Catena Dragonfly $\mathsf{BRG}_{\lambda=1}^n$ which has energy quality E-quality $= O(\sqrt{n})$. The attack from [BK15] has slightly worse quality $O\left(n^{1/5}\right)$, but it applies even for Dragonfly variants in which $\lambda > 1$. At a high level the ideas behind both of these attacks is to divide memory into segments, store the leading block in each segment and then recompute the remaining blocks as needed. These attacks only work because the underlying Catena Dragonfly DAG BRG_{λ}^n allows for quick re-computation of the remaining blocks. In this work we observe that this key idea can be generalized to attack *any* non depth-robust iMHF. In particular, our techniques can be used to attack other iMHFs like Catena Butterfly, Argon2i [BDK15] and SB [CGBS16]. In fact, our attacks can be extended to *any* iMHF because no DAG is sufficiently depth-robust to resist at least some form of attack.

Memory-Bound Functions. An important precursor to memory-hard functions are memory-*bound* functions. First introduced in [ABMW05] here the complexity measure of interest is the number of cache misses required to evaluate the function. On the highest level the motivation is the same as that of memory-hard functions; namely to build moderately hard functions which are more equally hard across different computational devices (compared to the rather unbalanced notion of plain computational complexity). In particular it was observed that while computational speeds may vary greatly between different devices the same is not as true for memory latency speeds [DGN03]. In contrast memory-hard

---

[8] Corrigan-Gibbs et al. [CGBS16] use "Balloon Hashing" as a title for their iMHF however this similarity with the balloon phase in our evaluation algorithm is a slightly unfortunate coincidence.

functions aim to achieve egalitarian hardness by making the cost of custom hardware prohibitively large [Per09]. The first provably secure memory-bound function was (implicitly) given in [DGN03] where it was used to construct a protocol for fighting SPAM email. The construction was later improved in [DNW05] which was also the first result in cryptography to make use of a version of the pebbling model of computation; a technique later adapted in [AS15].

Password Storage. Recent high-profile security breaches (e.g., RockYou, Sony, LinkedIN, Ashley Madison[9]) highlight the importance of proper password storage practices like salting [Ale04] and key stretching [MT79][10]. However, hash iteration, the technique used by password hash functions like PBKDF2 [Kal00] and bcrypt [PM], is typically an insufficient defense against an adversary who could build customized hardware to evaluate the underlying hash function. In particular, the cost of computing a hash function H like SHA256 or MD5 on an ASIC is orders of magnitude smaller than the cost of computing H on traditional hardware [DGN03, NB+15]. By contrast, memory costs tend to be relatively stable across different architectures [DGN03], which motivates the use of memory-hard functions for password hashing [Per09].

Several orthogonal lines of research have explored defenses such as: distributing the storage and/or computation of a password hash across multiple servers (e.g., [BJKS03, CLN12]), storing fake password hashes on the server (e.g., [JR13, BBBB10]), the inclusion of secret salt values (e.g., "pepper") in password hashes [Man96, BD16] and the inclusion of the solution(s) to hard AI challenges in password hashes [CHS06, DC08, BBD13].

2 Preliminaries

We begin with some notation. Given a directed acyclic graph (DAG) $G = (V, E)$ of *size* $|V| = n$ and a subset $S \subseteq V$ we use $G - S$ to denote the resulting DAG after removing all nodes in S. We denote by $\mathsf{depth}(G)$ the length of the longest (directed) path in G and we denote by $\mathsf{indeg}(G)$ the maximum number of directed edges entering a single node. For integers $a \leq b$ we write $[a, b]$ as shorthand for the set $\{a, a + 1, \ldots, b\}$ and we write $[a]$ for the set $[1, a]$.

We use $H_\lambda = \sum_{i=1}^{\lambda} \frac{1}{i}$ to denote the λ'th harmonic number. In particular H_λ can be approximated by the natural logarithm $H_\lambda \approx \ln \lambda$.

---

[9] See http://www.privacyrights.org/data-breach/ (Retrieved 9/1/2015).

[10] Users routinely select lower entropy password [Bon12], which are especially vulnerable to an offline attacker when the underling password hash function is inexpensive to compute. Furthermore, stricter password restrictions (e.g., requiring a mix of numbers and upper/lower case letters) [SS09] have not been found to greatly improve the entropy of the resulting passwords [KSK+11, BKPS13]. In fact, sometime these policies reduced the entropy of user selected passwords [KSK+11]. These policies are often associated with high usability costs [FH10].

2.1   Complexity and Quality of Attacks

We consider algorithms in the parallel random oracle model (pROM) [AS15] of computation.[11] That is an algorithm is repeatedly invoked. At invocation $i \in \{1, 2, \ldots\}$ the algorithm is given the state (bit-string) σ_{i-1} it produced at the end of the previous invocation. Next $\mathcal{A}$ can make a batch of calls $\mathbf{q}_i = (q_{1,i}, q_{2,i}, \ldots)$ to the underlying round function H (modeled as a random oracle (RO)). Then it receives the response from H and can perform arbitrary computation before finally outputting an updated state σ_i. The initial state σ_0 contains the input to the computation which terminates once a special final state is produced by $\mathcal{A}$. Apart from the explicit states σ the algorithm may keep no other state between invocations. For a input x and coins r we denote by $\mathcal{A}(x; r; H)$ the corresponding (deterministic) execution of $\mathcal{A}$.

We define the runtime time($\mathcal{A}$) to be the maximum running time of $\mathcal{A}$ in any execution (over all choices of x, r and H). Then the *cumulative memory complexity* (CMC) and *cumulative RO complexity* are defined as

$$\mathsf{cmc}(\mathcal{A}) = \max_{x,r,H} \sum_{i \in [T-1]} |\sigma_i| \qquad \mathsf{crc}(\mathcal{A}) = \max_{x,r,H} \sum_{i \in [T]} |\mathbf{q}_i|$$

where $|\sigma|$ is the bit-length of state σ, $|\mathbf{q}|$ is the dimension of the vector $\mathbf{q}$ and $\max_{x,r,H}$ denotes the maximum over all possible executions of $\mathcal{A}$. Similarly the *absolute memory complexity* (AMC) and *absolute RO complexity* are defined to be (ARC)

$$\mathsf{amc}(\mathcal{A}) = \max_{x,r,H} \max_{i \in [T-1]} |\sigma_i| \qquad \mathsf{arc}(\mathcal{A}) = \max_{x,r,H} \max_{i \in [T]} |\mathbf{q}_i|.$$

We remark that these complexity measures are stricter then is common, especially with respect to maximizing over all random oracles H. However we use them to upper-bound the complexity of our attacks so this strictness can only serve to strengthen the results.

Using these tools we can now define the complexity of an algorithm as follows.

Definition 1 (AT and Energy Complexities). *Let $\mathcal{A}$ be a pROM algorithm which computes #inst($\mathcal{A}$) instances of an iMHF in parallel. Then for any core-memory area ratio $R > 0$ and any core-memory energy ratio $\bar{R} > 0$ the (amortized) AT-complexity and the (amortized) energy-complexity of $\mathcal{A}$ are defined to be*

$$AT_R(\mathcal{A}) = [\mathsf{amc}(\mathcal{A}) + R \cdot \mathsf{arc}(\mathcal{A})] \times \frac{\mathsf{time}(\mathcal{A})}{\#inst(\mathcal{A})} \qquad E_{\bar{R}}(\mathcal{A}) = \frac{\mathsf{cmc}(\mathcal{A}) + \bar{R} \cdot \mathsf{crc}(\mathcal{A})}{\#inst(\mathcal{A})}.$$

Finally we can define the quality of an attack in terms of how much (if at all) it improves on the naïve algorithm

---

[11] Alternatively the results in this work also apply to the random access machine model of computation.

Definition 2 (Attack Quality). *Let f be an MHF with naïve algorithm $\mathbb{N}$ and let $\mathcal{A}$ be a pROM algorithm for evaluating $\#inst(\mathcal{A})$ instance(s) of f. Then for any* core-memory area ratio $R > 0$ *and any* core-memory energy ratio $\bar{R} > 0$ *the* AT-quality *and* energy-quality *of $\mathcal{A}$ is defined to be*

$$\mathsf{AT\text{-}quality}_R(\mathcal{A}) = \frac{AT_R(\mathcal{N})}{AT_R(\mathcal{A})} \qquad \mathsf{E\text{-}quality}_{\bar{R}}(\mathcal{A}) = \frac{E_{\bar{R}}(\mathcal{N})}{E_{\bar{R}}(\mathcal{A})}.$$

In particular if either quantity is less than 1 then we call $\mathcal{A}$ an attack *on f.*

Let f be an iMHF based on some DAG G of size n with constant in-degree. Observe f can always be evaluated by computing one intermediate value at a time in topological order while never deleting a computed value. Clearly this always results in correctly computing f and it corresponds to a well defined pROM algorithm triv for evaluating f. Moreover $AT_R(\mathsf{triv}) = \Theta(n(n + R))$ and $E_{\bar{R}}(\mathsf{triv}) = \Theta(n(n+\bar{R}))$. Given a constant $c > 0$ we say that f is a c-ideal iMHF if, when $\mathsf{triv} = \mathcal{N}$ is the naïve, for any attack $\mathcal{A}$ we have $\mathsf{AT\text{-}quality}_R(\mathcal{A}) \geq c$ and $\mathsf{E\text{-}quality}_{\bar{R}}(\mathcal{A}) \geq c$. This is motivated by the observation that for any iMHF algorithm triv is always a possible way to evaluate it. An ideal iMHF captures the property that triv is (approximately) the best evaluation strategy possible.

Unfortunately we will later show that c-ideal iMHFs do not exist for any constant $c > 0$. As $n \to \infty$ we will have $\mathsf{AT\text{-}quality}_R(\mathcal{A}) = \omega(1)$ and $\mathsf{E\text{-}quality}_{\bar{R}}(\mathcal{A}) = \omega(1)$.

2.2 Pebbling and Graph Theory

We provide some shorthand for describing algorithms and give some useful graph theoretic definitions and lemmas.

Graph Pebbling. To simplify exposition, our attacks are often described in the language of parallel graph pebbling [AS15]. However, unlike in [AS15], we merely think of this as shorthand for describing an evaluation strategy of an iMHF rather then describing an algorithm in a distinct model of computation.

In particular any iMHF f which we consider is based on some fixed underlying DAG G with (a single source and sink node) which describes which values are used as inputs to which calls to the round function. To compute f on some input x each node of G is assigned a value (bit-string). The source receives the value x. The value of any other node v is defined to be the output of the round function applied to the values of the parent nodes of v. Finally $f(x)$ is defined to be the value of the sink node.[12]

With this in mind, each round of pebbling corresponds to one invocation in an execution. Placing a pebble on a node v in some round is shorthand for computing the value of v by computing the round function on the values of v's

---

[12] For concreteness, though not relevant to this work, in most cases the round function is simply the compression function H (with the exception of the Linear iMHF of [CGBS16]).

parents. Clearly this can only be done if $(x_1, \ldots, x_z)$ are stored in memory and so, if an algorithm places a pebble on a node whose parents do not all contain a pebble then we call such a move *illegal*. Thus we will always show that our pebbling strategies only produce legal pebblings in order to ensure that they correspond to a feasible pROM algorithm for evaluating iMHF. Finally having a pebble on a node at the end of a round corresponds to storing the value of that node in the state σ for that invocation.

Graph Theory. The key insight behind our attacks is that if a graph is not depth-robust enough then it can be efficiently pebbled.

Definition 3 (Depth Robust and Depth Reducible DAGs). *For* $e, d \in \mathbb{N}$ *a DAG* $G = (V, E)$ *is called* (e, d)-depth-robust *if*

$$\forall S \subseteq V : \quad |S| \leq e \Rightarrow \mathsf{depth}(G - S) \geq d.$$

If G *is not* (e, d)-depth-robust *then we say that* G *is* (e, d)-reducible.

In order to prove the generic attack on any iMHF we rely on a lemma, originally due to Valiant [Val77], to show that no graph is depth-robust enough not to permit at least some sort of attack.

3   Generic Attack

In this section we describe a general pebbling attack GenPeb against any (e, d)-reducible graph. GenPeb(G, S, g, d) takes as input a DAG $G = (V, D)$ and a set $S \subseteq V$ of size e such that $\mathsf{depth}(G - S) \leq d$ and a parameter $g \geq d$ which we will define below. In every round GenPeb makes progress (i.e., places a pebble on node i in the i'th round). Thus, $\mathsf{time}(\mathsf{GenPeb}) = n$ as the algorithm will place a pebble on the final node n in the n'th rounds. Intuitively, GenPeb is divided into two types of phases: Balloon Phases and a Light Phases. During light phases we throw out most of the pebbles on the graph keeping only pebbles on nodes in S, the highest pebbled node i and the parents of the nodes $[i, i + g]$ that we plan to pebble in the next g rounds. Every g rounds we execute a balloon phase to ensure that we will always have pebbles placed on the parents of the nodes that we plan to pebble in the next g rounds. Because we never remove pebbles on nodes in S and the DAG $G - S$ has depth $\leq d$ we will be able to accomplish this goal in at most d rounds. During light phases we keep at most $\delta g + e$ pebbles on the graph and we place at most one new pebble on G in every round. Thus the total cost during all light phases is at most $n(\delta g + e + \bar{R})$. While we may incur higher costs during a balloon phase we are only in the balloon phase for at most $\frac{dn}{g}$ rounds.

We analyze the energy complexity of GenPeb in terms of the depth reduction parameters e and d. These results are summarized in Theorem 2. While GenPeb will lead to attacks with good energy-quality $\mathsf{E\text{-}quality}_R$ the attack may not necessarily have good AT-quality $\mathsf{AT\text{-}quality}_R$. This is because GenPeb may still have high absolute memory and RO complexity due to the balloon phase. However, we can easily circumvent this problem by pebbling multiple copies of the DAG G in parallel, which corresponds to evaluating multiple independent instances of the iMHF. In particular, PGenPeb pebbles $\lfloor g/d \rfloor$ instances of G in parallel. We stagger evaluation of the different iMHF instances so that at most one of the $\lfloor g/d \rfloor$ pebbling instances is in a balloon phase at any point in time. To accomplish this PGenPeb simply waits $(i-1)d$ steps to begin pebbling the i'th instance of G. Thus, PGenPeb takes at most $n + \lfloor g/d \rfloor d \leq 2n$ steps to complete. The absolute memory and RO complexity of PGenPeb is essentially just the cost of the balloon phase for a single iMHF instance. Thus, PGenPeb leads to attacks with good AT-quality AT-quality$_R$ because the cost of the balloon phase can be amortizes among the $\lfloor g/d \rfloor$ iMHF instances we compute. Theorem 3 states both the energy and AT complexity of PGenPeb. The energy complexity of PGenPeb is roughly equivalent to the energy complexity of GenPeb, and the AT-complexity of PGenPeb is roughly twice the energy complexity of PGenPeb.

In the rest of the paper we will consider several specific families of DAGs like the underlying DAGs in the Catena and Argon2i iMHFs. For Catena, we can find a set $S \subseteq V$ of size $e = n/t$ such that depth$(G - S) = O(t)$ for every $t > 1$. For Argon2i we can find a set S with expected size $O\left(n/t + (n \ln \lambda)/\lambda\right)$ such that depth$(G - S) \leq t \cdot \lambda$. Combined with Theorem 3 we will obtain an attack on Catena with quality $\Omega\left(n^{1/3}\right)$ and an attack on Argon2i with quality $\Omega\left(n^{1/4}/\ln n\right)$.

GenPeb makes use of two subroutines need and keep. In our complexity analysis we omit the cost of computing these functions. However we stress that in all our attacks they are either trivial (constant) or very easy to compute. By "easy to compute" we mean that the sets returned by these subroutines will have a short description size (e.g., "all nodes" or $[i,j]$) and that it will be trivial to decide whether a given node v is in these sets.

We begin with some useful notation. Fix a DAG G of size n and number its nodes in (arbitrary) topological order from 1 to n. For $i \in [n]$ and $j \geq i$ we write parents(i,j) for the set of nodes v with an edge (v,u) for some $u \in [i, \min\{j, n\}]$. Next we fix the class of functions from which need and keep must be chosen in order to prove that GenPeb produces a legal pebbling (and thus defines a pROM evaluation algorithm).[13]

Definition 4 (Needed Pebbles). *Fix a subset of target nodes $T \subseteq V$ and a pebbling configuration $C \subseteq V$ of G.[14] Then a node $v \in V$ is needed for T within d' steps if there exists a completely unpebbled path P[15] of length $\geq d'$ from v to some node in T. We use $N_{C,T,d'}$ to denote the set of all such nodes. We use $K_{C,T}$ to denote the set of all nodes $v \in C$ such that $v \in T$ or v has a child v' such that $v' \in \bigcup_{i=0}^{n} N_{C,T,i}$.*

---

[13] Later on we instantiate need and keep in several ways but will always prove that they are valid for the inputs we use them for.

[14] That is fix a set C of nodes of V which currently have a pebble on them.

[15] That is $P \cap C = \emptyset$.

Algorithm 1. GenPeb (G, S, g, d)

Arguments : $G = (V, E)$, $S \subseteq V$, $g \in [\text{depth}(G - S), |V|]$, $d \geq \text{depth}(G - S)$
Local Variables: $n = |V|$

1 **for** $i = 1$ *to* n **do**
2 $\quad$ Pebble node i.
3 $\quad$ $l \leftarrow \lfloor i/g \rfloor * g + d + 1$
4 $\quad$ **if** $i \mod g \in [d]$ **then** $\hspace{4cm}$ // Balloon Phase
5 $\quad\quad$ $d' \leftarrow d - (i \mod g) + 1$
6 $\quad\quad$ $N \leftarrow \text{need}(l, l + g, d')$
7 $\quad\quad$ Pebble every $v \in N$ which has all parents pebbled.
8 $\quad\quad$ Remove pebble from any $v \notin K$ where $K \leftarrow S \cup \text{keep}(i, i + g) \cup \{n\}$.
9 $\quad$ **else** $\hspace{6cm}$ // Light Phase
10 $\quad\quad$ $K \leftarrow S \cup \text{parents}(i, i + g) \cup \{n\}$
11 $\quad\quad$ Remove pebbles from all $v \notin K$.
12 $\quad$ **end**
13 **end**

Definition 5 (Valid need and keep). *We say that the pair of functions* need *and* keep *is valid for* $\text{GenPeb}(G, S, d, g)$ *if we always have* $\text{need}(i, j, d') \supseteq N_{C,[i,j],d'}$ *and* $\text{keep}(i, j) \supseteq K_{C,[i,j]}$ *whenever* $\text{GenPeb}(G, S, \text{depth}(G - S), g)$ *queries* need *or* keep.

In our generic iMHF attack we use the trivial functions need and keep which always output V (e.g., during the balloon phase we pebble every node we can during each round and we never discard any pebbles). The following fact is easy to see:

Fact 1 (Generic Valid Subroutine). *Fix a DAG $G = (V, E)$ and let* need *and* keep *be the constant function returning V. Then the pair* need *and* keep *is valid for* $\text{GenPeb}(G, S, d, g)$ *for any set $S \subseteq V$ and any parameters $g \geq d \geq \text{depth}(G - S)$.*

While we would already obtain high quality attacks on Catena and Argon2i by using the generic need and keep subroutines, we show how our attacks can be optimized further by defining the subroutines need and keep more carefully.

We remark that by leaving need and keep undefined for now we leave some flexibility in the implementation of the balloon phase in $\text{GenPeb}(G, S, g, d)$. During each round of a balloon phase we may pebble any $v \in V$ which has all parents pebbled, but we are only required to add pebbles to these nodes once it becomes absolutely necessary to finish the balloon phase in time (e.g., there are only d' rounds left in the balloon phase and the vertex v is part of an completely unpebbled path to T of length $\geq d'$. Similarly, we are allowed to remove pebbles provided that they are no longer needed for the balloon phase (e.g., every path to T from that node has an intermediate pebble).

The easiest way to satisfy these conditions is to simply pebble every $v \in V$ which has all parents pebbled, and to never remove pebbles during the balloon phase (Fact 1). Indeed this is exactly what we do in our general attack on iMHFs. However, we demonstrate that further optimizations are possible against the Catena and Argon2i iMHFs (e.g., each of the new pebbles we add during a Catena balloon phase does not need to remain on the DAG very long). In each case the subroutines need and keep will have very simple instantiations — we will not need to perform complicated computations like breadth first search to find these sets.

Fix any G, S, g and d and let $\mathsf{M}(G, S, g, d)$ be the largest number of pebbles simultaneously on $G - S - \mathsf{parents}(i, i + g)$ during any round i which is in a Balloon phase of $\mathsf{GenPeb}(G, S, g, d)$[16]. Similarly let $\mathsf{C}(G, S, g, d)$ be the largest number of pebbles placed on G during any single round in a Balloon Phase. In the following we prove that GenPeb always produces a legal pebbling. Thus it describes a well formed pROM algorithm $\mathcal{A}$ for evaluating an iMHF based on G. We also show how to use $\mathsf{M}(G, S, g, d)$ and $\mathsf{C}(G, S, g, d)$ to upper-bound the energy-complexity of $\mathcal{A}$ with hardcoded inputs (G, S, g, d).

Theorem 2 (Energy Complexity of GenPeb). *Let $G = (V, E)$ be a DAG, with $\mathsf{indeg}(G) = \delta$. Further let $S \subseteq V$ with $|S| = e$ and $d \geq \mathsf{depth}(G - S)$ and let integer $g \in [d, n]$. Fix any valid pair of subroutines* need *and* keep *and let $\mathcal{A}$ be the pROM algorithm described by GenPeb with hardcoded inputs (G, S, g, d). Then $\mathcal{A}$ produces a valid pebbling and for any core-memory energy ratio $\bar{R}$ and $\mathsf{M} = \mathsf{M}(G, S, g, d)$ and $\mathsf{C} = \mathsf{C}(G, S, g, d)$ it holds that:*

$$\mathsf{cmc}(\mathcal{A}) \leq n \left(\frac{d \cdot \mathsf{M}}{g} + \delta g + e \right) \qquad \mathsf{crc}(\mathcal{A}) \leq n \left(\frac{\min\{d \cdot \mathsf{C}, n\}}{g} + 1 \right)$$

$$E_{\bar{R}}(\mathcal{A}) \leq n \left(\frac{d \cdot \mathsf{M} + \min\{d\mathsf{C}, n\} \cdot \bar{R}}{g} + \delta g + e + \bar{R} \right).$$

Proof. We first prove that $\mathsf{GenPeb}(G, S, g, d)$ produces a legal pebbling to ensure that $\mathcal{A}$ is a well defined algorithm. Then we upper-bound its energy complexity.

Recall that pebbles can be removed at will and by definition, in Step 7, GenPeb only places a pebble if it is legal to do so. Thus the only illegal move could come due to Step 2. Assume no illegal pebble has been placed up to node i. To show that i is then also pebbled legally it suffices to show that each of its parents $P \subseteq V$ are have a pebble at the beginning of round i. The most recent Balloon Phase to have completed before round i (if any) consisted of rounds $B = [i', i'+d]$ where $i-(i'+d) \leq g$. Consider the partition $P_1 = P \cap [i'+d+1, i-1]$, $P_2 = P \cap [i', i'+d]$ and $P_3 \cap [1, i'-1]$ of the set of parents P. By assumption all $v \in P_1$ were pebbled (legally) in the previous g rounds using Step 2 and so were not removed (by definition of K in Step 10). Moreover by assumption all nodes in P_2 where pebbled by Step 2 during B and so were not removed (by definition of K in Step 8 and the validity of the subroutine keep the pebble is not removed during the balloon phase $B = [i', i' + d]$ and by definition of K in

---

[16] Recall that there are n/g Balloon Phases and the jth Balloon Phase consists of rounds $\{jg + 1, \ldots, jg + d\}$.

Step 10 the pebble was not removed during rounds $[i' + d + 1, i - 1]$). Thus it suffices to prove that all $v \in P_3$ contained a pebble at some point during B since then by definition of K in Steps 2 and 10 they too will not be removed.

Let P_4 be the subset of P_3 which don't contain a pebble at the start of B. (If it is empty we are done.) Otherwise, for a given round j let $\mathbf{p}_j$ be all paths which end with a node in P_4 and are unpebbled at the beginning of the round. Let l_j be the length of the longest path in $\mathbf{p}_j$. We argue that $\forall j \in [i', i' + d - 1]$ then $l_j \leq d - (j - i')$. If this is the case then we are done. Entering the final round $i' + d - 1$ the length of the longest unpebbled path is $l_{i'+d-1} \leq 1$ so by the end of the final round of B all nodes P_4 – the end points of paths $\mathbf{p}_j$ – are pebbled.

We argue that $l_j \leq d - (j - i')$ by induction. Clearly, this is true when $j = i'$ as $l_j \leq \mathsf{depth}(G - S) \leq d$. Now assume that $l_j \leq d - (j - i')$ for some $j \in [i', i' + d - 1]$, let $p \in \mathbf{p}_j$ denote a longest path and let v denote the starting node of p. We first observe that either the starting node v of p has no parents or they are all pebbled.[17] Second, we observe that, because need is valid, in round j we must either have $v \in N$ or we must have $l_j < d - (j - i')$. In the latter case we have $l_{j+1} \leq l_j \leq d - (j + 1 - i')$ — because keep is valid we are not allowed to remove pebbles from any of the parents of v. In the former case we have $l_{j+1} \leq l_j - 1 \leq d - (j + 1 - i')$ because $v \in \mathsf{need}(i', i' + g, d' = d - (j + 1 - i'))$ by the validity of need. Thus in Step 7 of round j node v is pebbled. Finally it remains there till the end of the round since $v \in \mathsf{keep}(j, j + g)$ because there is a completely unpebbled path from v's children in p to $[i', i' + g + d]$. This completes the proof that GenPeb produces a legal pebbling.

Recall that the energy-complexity of $\mathcal{A}$ can be computed as $E_{\bar{R}}(\mathcal{A}) = \mathsf{cmc}(\mathcal{A}) + \bar{R} \cdot \mathsf{crc}(\mathcal{A})$. To upper-bound $\mathsf{cmc}(\mathcal{A})$ we can sum upper-bounds on the cmc of the Balloon phases and the cmc of the Light phases. To compute the Balloon phase term notice that GenPeb is in a Balloon Phase for nd/g steps and during each round i of a balloon phase there are, by definition, at most $\mathsf{M}(G, S, g, d)$ extra pebbles on $G - S - \mathsf{parents}(i, i + g)$. On the other hand, there are clearly at most n Light phase steps and at the start of each round i of a light phase there are no pebbles on $G - S - \mathsf{parents}(i, i + g)$. Finally, during each round i we pay cumulative memory cost at most e to keep pebbles on nodes in S and at most δg to keep pebbles on nodes in the set $\mathsf{parents}(i, i + g)$, which can be of size at most δg. Adding these three terms and factoring out an n term we get that $\mathsf{cmc}(\mathcal{A}) \leq n\left(\frac{d \cdot \mathsf{M}(G,S,g,d)}{g} + \delta g + e\right)$.

Placing a pebbled on G corresponds to making a call to H. To upper-bound $\mathsf{crc}(\mathcal{A})$ we observe that in any round of a Light phase only one pebble is ever placed on G (namely in Step 2). During each balloon phase we place at most $\mathsf{C}(G, S, g, d)$ pebbles on the graph in each rounds, and at most n pebbles on the graph in total. Thus we can write $\mathsf{crc}(\mathcal{A}) \leq n\left(\frac{\min\{n, d \cdot \mathsf{C}(G,S,g,d)\}}{g} + 1\right)$. Combing this with the bound on cmc and rearranging terms we obtain the theorem. □

---

[17] As otherwise it wouldn't be a longest path in $\mathbf{p}_j$.

Algorithm 2. PGenPeb (G, S, g, d, k)

Arguments   : G, $S \subseteq V$, $g \in [\mathsf{depth}(G - S), |V|]$ $d \geq \mathsf{depth}(G - S)$, $k \leq \lfloor \frac{g}{d} \rfloor$
Local Variables: $n = |V|$, copies $G_1, \ldots, G_k = G$, $S_1, \ldots, S_k = S$

---

1 **for** $t = 1$ *to* $n + kd$ **do**
2 | **Parallel for** $j = \max\{1, \frac{t-n}{d}\}$ *to* $\min\{k, \frac{t-1}{d}\}$ **do**
3 | | $i \leftarrow t - jd$
4 | | Pebble node i in G_j.
5 | | **if** $i = n$ **then**
6 | | | Remove pebbles from all $v \notin \{n\}$ in G_j
7 | | | **Break**
8 | | **end**
9 | | $l \leftarrow \lfloor i/g \rfloor * g + d + 1$
10 | | **if** $i \bmod g \in [d]$ **then**   // Balloon Phase
11 | | | $d' \leftarrow d - (i \bmod g) + 1$
12 | | | $N_j \leftarrow \mathsf{need}_j(l, l + g, d')$
13 | | | Pebble any $v \in N_j$ which has all parents pebbled.
14 | | | Remove pebble from any $v \notin K_j$ where
    | | | $K_j \leftarrow S_j \cup \mathsf{keep}_j(i, i + g) \cup \{n\}$.
15 | | **else**   // Light Phase
16 | | | $K_j \leftarrow S_j \cup \mathsf{parents}_j(i, i + g) \cup \{n\}$
17 | | | Remove pebbles from all $v \notin K_j$.
18 | | **end**
19 | **end**
20 **end**

The following Theorem 3 upper-bounds the complexity of PGenPeb. The proof in the full version [AB16] closely follows the analysis of GenPeb in Theorem 2. The key difference is that we evaluate multiple instances, and at that at most one of these instances is in a balloon phase at any point in time. Thus, we get a much tighter bound on AT-complexity because the worst-case memory usage M is approximately the same as the average memory usage of PGenPeb.

Theorem 3 (Complexity of PGenPeb). *Let* $G = (V, E)$ *be a DAG, with* $\mathsf{indeg}(G) = \delta$. *Further let* $S \subseteq V$ *with* $|S| = e$ *and* $d \geq \mathsf{depth}(G - S)$ *and let integer* $g \in [d, n]$. *Fix any valid pair of subroutines* need *and* keep *and let* $\mathcal{A}$ *be the pROM algorithm described by* PGenPeb *with hardcoded inputs* $(G, S, g, d, \lfloor \frac{g}{d} \rfloor)$. *Then for any core-memory area and energyratios* $R > 0$ *and* $\bar{R} > 0$ *and* $\mathsf{M} = \mathsf{M}(G, S, g, d)$ *and* $\mathsf{C} = \mathsf{C}(G, S, g, d)$ *it holds that:*

$$AT_R(\mathcal{A}) \leq 2n \left[\frac{d(\mathsf{M} + R\mathsf{C})}{g} + \delta g + e + R \right] \quad and$$

$$E_{\bar{R}}(\mathcal{A}) \leq n \left[\frac{d\mathsf{M} + \min\{d\bar{R}\mathsf{C}, n\bar{R}\})}{g} + \delta g + e + \bar{R} + 1 \right].$$

4   Sandwich Graph Attacks

In this section we focus on the two Catena hash functions [FLW13] as well as the second two Balloon Hashing constructions of [CGBS16]. The first Catena iMHF is given by the *Catena Bit Reversal Graph* (which we denote BRG_λ^n); an n node DAG which consists of a stack of $\lambda \in \mathbb{N}_{\geq 1}$ bit-reversal graphs [LT82]. Each node in a layer is associated with a $\log_2\left(\frac{n}{\lambda+1}\right)$ bit string and edges between layers correspond to the bit reversal operation[18]. The Catena designers recommended choosing $\lambda \in \{1, 2, 3, 4\}$ [FLW13]. The second Catena hash function is an iMHF based on the *Catena Double Butterfly Graph*, denoted DBG_λ^n. It is an n node DAG with $O(\lambda \log n)$ layers of nodes.

The "Double-Buffer" and "Linear" iMHFs of [CGBS16] consist of τ layers of σ nodes for a total of $n = \tau * \sigma$ nodes. Each layer is a path with its origin connected to the final node in the path of the previous layer. Moreover all nodes at layers $\tau \geq i \geq 1$ have 20 incoming edges from nodes selected uniformly and independently in the previous layer. In the "Double-Buffer" construction the hash of a node is given by hashing the concatenation of all parent node labels while in the "Linear" construction the parent node labels are first XORed together before being hashed for greater throughput. However this difference will not affect the results in this work.[19]

In this section we demonstrate that all of these DAGs can be computed with lower then hoped for energy and AT complexities (simultaneously) regardless of the random choices made when constructing the graphs. In particular, the iMHF corresponding to both BRG_λ^n and DBG_λ^n can be evaluated with amortized AT complexity $AT_R(\mathcal{A}) = O\left(n^{5/3} + Rn^{4/3}\right)$ and energy complexity $E_{\bar{R}}(\mathcal{A}) = O\left(n^{5/3} + Rn^{4/3}\right)$ for any value of λ. In fact, our attacks hold for a more general class of graphs characterized by Definition 6 below. Thus, to understand our attacks it is not critical to know the exact specification of these DAGs just that both DAGs are strict sandwich graphs. We refer an interested reader to the full version [AB16] of this paper for the actual definitions of the Catena DAGs BRG_λ^n and DBG_λ^n.

Definition 6 ((Strict) λ-Stacked Sandwich Graphs). *Let $n, \lambda \in \mathbb{N}_{\geq 1}$ be a integers such that $\lambda + 1$ divides n and let $k = n/(1 + \lambda)$ and let G be a DAG with n nodes. We say that G is a λ-stacked sandwich DAG if G contains a directed path of n nodes $(v_1, \ldots, v_n)$ with arbitrary additional edges connecting*

---

[18] The parameter λ in Catena is related to the parameter τ in Argon2i. The intended space complexity of Catena is $\sigma = 2n/(\lambda+1)$ and the intended computation time is n. Thus, the intended energy complexity is $2n^2/(\lambda+1)$.

[19] We remark that we have assumed that the thread count parameter $p = 1$. However we observe that for $p > 1$ the resulting DAG has an almost identical distribution except that every $s/p^t h$ edge along the path forming a layer is removed. This can only make the job easier of an evaluation algorithm. In particular the complexity of our attacks for the case $p = 1$ are an upper-bound on the complexities of these constructions for $p > 1$.

nodes from lower layers $L_j \doteq \{v_{jk+1}, \ldots, v_{jk+k}\}$ *with* $j \leq i$ *to the* $i + 1^{st}$ *layer* L_{i+1}. *If the DAG has no edges of the form* (u, v) *with* $u \in L_j$ *and* $v \in L_{j+2+i}$ *for* $i \geq 0$ *then we say it is a* strict λ-stacked sandwich DAG.

In particular, the Catena bit reversal graph BRG_λ^n is a strict λ-stacked sandwich DAG with n nodes and maximum indegree $\mathsf{indeg} = 2$. The Catena double butterfly graph DBG_λ^n is a strict $\big(\lambda(2x - 1) + 1\big)$-stacked sandwich DAG with n nodes, where $x \leq \log n$ is the integer such that $n = 2^x \cdot \big(\lambda(2x-1)+1\big)$ — see the full version [AB16] of this paper for additional details about the construction of BRG_λ^n and DBG_λ^n. Finally, for any parameters t and s, a randomly chosen DAG for the Double-Buffer and Linear iMHFs is a strict t-stacked sandwich graph on $n = ts$ nodes with probability 1.

Summary of the Results in this Section. Lemma 1 upper-bounds the depth-robustness of any λ-stacked sandwich DAG G — any λ-stacked sandwich DAG is $(n/t, \lambda t + t)$-reducible. Thus we can apply the generic attack (Theorem 3) to get an upper-bound on the energy and AT complexities of such graphs (Theorem 4) and so, in particular, also for the 4 constructions mentioned above. Theorem 4 states that there is an attack $\mathcal{A}$ with $AT_R(\mathcal{A})$ and energy complexity $E_{\bar{R}}(\mathcal{A}) = O\big((\lambda + \delta)n^{5/3} + \bar{R}n^{4/3}\big)$, where δ denotes the maximum indegree of the DAG.

While these results will also be useful in the next section focused on Argon2i for the 4 constructions above the results can be improved somewhat by observing that the constructions are actually based on *strict* stacked sandwich DAGs. In particular, we can further decrease the resulting complexity if we first define more targeted need and keep functions and prove that they are valid for $\mathsf{PGenPeb}$ when G is a *strict* λ-stacked sandwich DAG (Lemma 2). Theorem 5 says that there is an attack $\mathcal{A}$ with $AT_R(\mathcal{A})$ and energy complexity $E_{\bar{R}}(\mathcal{A}) = O\big(\delta n^{5/3} + \bar{R}n^{4/3}\big)$.

The following Lemma upper-bounds the depth-robustness of any λ-stacked sandwich DAG G. By combining this observation with the generic attack from the previous section we can obtain strong attacks on any λ-stacked sandwich DAG G.

Lemma 1 (Sandwich Graphs are Reducible). *Let G be a λ-stacked sandwich DAG then for any integer $t \geq 1$ G is $(n/t, \lambda t + t - \lambda - 1)$-reducible.*

Proof. Let $S = \big\{v_{it} \,\big|\, 1 \leq i \leq n/t\big\}$. We claim that $\mathsf{depth}(G - S) \leq \lambda t + t - \lambda - 1$. Consider any path P in $G - S$. For each layer L_j the path P can contain at most $t - 1$ nodes from layer L_j because any sequence of t consecutive nodes $v_i, v_{i+1} \ldots, v_{i+1}$ must contain at least one node in S. Thus,

$$|P| \leq \sum_{i=0}^{\lambda} \Big|P \bigcap L_j\Big| \leq (\lambda + 1)(t - 1). \qquad \square$$

The next lemma states that keep and need from Algorithm 3 and Algorithm 4 are valid for strict sandwich DAGs.

Algorithm 3. Function: need(x, y, d')

Arguments: x, $y \geq x$, $d' \geq 0$
Constants : Pebbling round i, g, t.
1 $j \leftarrow (i \mod g)$ // Current Layer is $L_{\lfloor j/t \rfloor}$
2 **Return** $L_{\lfloor j/t \rfloor} \cap \{it + j \mid i \leq \frac{n}{t}\}$

Algorithm 4. Function: keep(x, y)

Arguments: x, $y \geq x$
Constants : Pebbling round i, g, t.
1 $j \leftarrow (i \mod g)$
2 $\ell \leftarrow \lfloor j/t \rfloor$ // Current Layer
3 **Return** $L_{\geq \ell - 1}$

Lemma 2 (Valid need and keep for Strict Sandwich Graphs). *Let G be a strict λ-stacked sandwich DAG on n nodes, let $S = \left\{ it \,\middle|\, i \leq \frac{n}{\lambda+1} \right\}$, $d = (\lambda + 1)t$ and $g \geq d$ then the functions* need *and* keep *from Algorithms 3 and 4 are valid for* GenPeb(G, S, g, d).

If we modify keep to simply return the entire vertex set then we obtain a valid pair need and keep for general sandwich DAGs. The proofs of Lemmas 2 and 3 are in the full version [AB16].

Lemma 3 (Valid need and keep for Sandwich Graphs). *Let G be a λ-stacked sandwich DAG on n nodes, let $S = \left\{ it \,\middle|\, i \leq \frac{n}{\lambda+1} \right\}$, $d = (\lambda + 1)t$ and $g \geq d$. Further, let* keep *be the constant function that returns V and let* need *be the function from Algorithm 3. Then the pair* need *and* keep *are valid for* GenPeb(G, S, g, d).

Theorem 4 follows easily from Theorem 3, Lemma 3 and Lemma 1 by setting $g = n^{2/3}$ and $t = n^{1/3}$.

Theorem 4 (Complexity of Sandwich Graph). *Let F be an iMHF based on DAG G; a λ-stacked sandwich DAG on n nodes with $\lambda < n^{1/3}$ and maximum indegree* indeg$(G) = \delta$. *Then for any core-memory area and energy ratios R and $\bar{R}$ there exists an evaluation algorithm $\mathcal{A}$ with*

$$AT_R(\mathcal{A}) \leq 2n^{5/3} \left[(1 + \delta) + (\lambda + 1) + \frac{R}{n^{1/3}} + \frac{2R + (\lambda + 1)R}{n^{2/3}} \right] \quad and$$

$$E_{\bar{R}}(\mathcal{A}) \leq n^{5/3} \left[(\lambda + 1) + (\delta + 1) + \frac{3\bar{R} + 1}{n^{2/3}} + \frac{\bar{R}}{n^{1/3}} \right]$$

For *strict* λ-stacked sandwich DAGs Theorem 5 improves on the attack Theorem 4 by instantiating the keep function with Algorithm 4 instead of the constant function keep$(\cdot) = V$ that returns all vertices (permissible by Lemma 2).

A formal proof of Theorems 4 and 5 can be found in the full version of this paper [AB16].

Theorem 5 (Complexity of Strict Sandwich Graph). *Let G be a strict λ-stacked sandwich DAG on n nodes with $\lambda < n^{1/3}$ and maximum indegree* $\mathsf{indeg}(G) = \delta$ *then for any core-memory area and energy ratios R and $\bar{R}$ there exists an evaluation algorithm $\mathcal{A}$ for the corresponding iMHF with*

$$AT_R(\mathcal{A}) \leq 2n^{5/3} \times \left[3 + \delta + \frac{R}{n^{1/3}} + \frac{3R}{n^{2/3}} \right] \; and$$

$$E_{\bar{R}}(\mathcal{A}) \leq n^{5/3} \times \left[3 + \delta + \frac{\bar{R}}{n^{1/3}} + \frac{\bar{R}+1}{n^{2/3}} \right].$$

Theorem 5 follows easily from Theorem 3, Lemma 2 and Lemma 1 by setting $t = n^{1/3}$ and $g = n^{2/3}$. While Theorems 4 and 5 only hold for $\lambda < n^{1/3}$ we note there is an trivial pebbling algorithm for strict sandwich graphs with complexity $O(n^2/\lambda)$, or $O(n^{5/3})$ whenever $\lambda > n^{1/3}$ — see [AB16].

We remark that for special cases (e.g., $\lambda = 1$) an alternative instantiation of the functions need and keep in GenPeb allows us to immediately generalize a result of [AS15]. For any $\lambda = 1$-sandwich DAG G with maximum indeg $= 2$ there is an algorithm $\mathcal{A}$ with $\mathsf{amc}(\mathcal{A}) = O(\mathsf{indeg}\sqrt{n})$ and $\mathsf{cmc}(\mathcal{A}) = O(\mathsf{indeg} \times n^{1.5})$ — our result is slightly more general in that we do not require that G has maximum indeg $= 2$. Briefly, we can use need from Algorithm 3 and we can redefine keep to simply return the exact same set as need in each round of GenPeb (so that we don't immediately throw out pebbles in step 14 of the same round). It is easy to show that the pair keep and need is valid whenever G is a $\lambda = 1$-stacked sandwich DAGs. We refer an interested reader to the full version [AB16] of this paper for details.

5 (n, δ, w)-Random Graph Attacks

In this section we demonstrate how to extend our attacks to two recent iMHF proposals. The first is the Argon2i iMHF — the variant of Argon2 in which data access patterns are data independent [BDK15]. The authors recommended using Argon2i for password hashing applications to avoid potential side-channel leakage through data dependent memory access patterns [BDK15]. The basic Argon2i DAG is a (pseudo) randomly generated DAG with maximum indegree indeg $= 2$. Thus, we view the Argon2i DAG as a distribution over n node DAGs — see Definition 7. The second iMHF considered is the Single-Buffer (SB) construction of [CGBS16] (we considered the Double-Buffer and Linear iMHFs from [CGBS16] in Sect. 4).

We begin with the following definition which fixes a class of random graphs key to our analysis.

Definition 7 ((n, δ, σ)-random DAG). *Let $n \in \mathbb{N}$, $1 < \delta < n$, and $1 \leq \sigma \leq n$ such that σ divides n. An (n, δ, w)-random DAG is a randomly generated directed acyclic (multi)graph with n nodes $v_1, \ldots, v_n$ and with maximum in-degree δ for each vertex. The graph has directed edges (v_i, v_{i+1}) for $1 \leq i < n$ and random forward edges $(v_{r(i,1)}, v_i), \ldots, (v_{r(i,\delta-1)}, v_i)$ for each vertex v_i. Here, $r(i, j)$ is independently chosen uniformly at random from the set $[\max\{0, i - \sigma\}, i - 1]$.*

We observe that a τ-pass instance of Argon2i iMHF[20] is an $(n, 2, n/\tau)$-random DAG[21]. Similarly the τ-pass "Single-Buffer" construction of [CGBS16] is based on an $(n, 21, n/\tau)$-random DAG. Here, $\sigma = n/\tau$ denotes the size of the memory window used by the naive pebbling algorithm $\mathcal{N}$ (e.g., $\mathsf{amc}(\mathcal{N}) = \sigma$). The hope is that such graphs will have complexity $\theta(\sigma^2 \times \tau)$.

We cannot directly apply our results from the previous section because a (n, δ, σ)-random DAG will not be λ-stacked sandwich DAG with high probability. However, we can show that these graphs are 'close' to λ-stacked sandwich DAGs in the sense that, with high probability, there is a 'small' set S such that we can turn G into a λ-stacked sandwich DAGs just by removing edges incident to vertices in S. Definition 8 formalizes this intuition.

Definition 8 ((m, λ)-Layered Graph). *Let G be a DAG with $n = k\lambda$ nodes $v_1, \ldots, v_n$ with directed edges (v_i, v_{i+1}) for $1 \leq i < n$. Given a set S of nodes we use G_S to denote the resulting DAG if we removed all directed edges (v, v') that are incident to nodes in S (e.g., $v \in S$ or $v' \in S$) except for edges of the form (v_i, v_{i+1}). We say that G is a (m, λ)-layered DAG if we can find a set S with at most m nodes such that G_S is a λ-stacked sandwich DAG.*

Demonstrating that a graph is (m, λ)-layered is useful because Theorem 6 upper-bounds the AT and energy complexities of an iMHF based on such a graph. In particular, Theorem 6 relies on Lemma 4 which states that any layered graph is also depth reducible. In Lemma 5, for any given probability $\gamma > 0$ we upper-bound the size m when viewing an Argon2i graph as an (m, λ)-layered graph. Thus we get Theorem 7 which describes an evaluation algorithm for Argon2i and the Balloon Hashing algorithms and bound their AT and energy complexities. In particular it states both their expected values and upper-bounds holding with a given probability γ.

Lemma 4 (Layered Graphs are Reducible). *Let G be a (m, λ)-layered DAG then for any integer $t \geq 1$ G is $(n/t + m, \lambda t + t - \lambda - 1))$-reducible.*

---

[20] In the notation of [BDK15] the case when $\tau = 1$ corresponds to a single pass.

[21] In response to our attack and the attack of Gibbs et al. [CGBS16] the Argon2i team recently 'tweaked' their construction by adding additional edges of the form $(i, i + \sigma)$, where $\sigma = n/\tau$ is the size of the memory window. While this tweak does seem to eliminate the attack of Gibbs et al. [CGBS16], it is ineffective against our attack. In fact, as long as $\tau < n^{1/4}$ we can completely ignore these extra edges when constructing our depth-reducing set S in Lemma 5.

Proof By definition there is a set S_1 of m nodes such that G_{S_1} is a λ-stacked sandwich DAG. Now by Lemma 1 we can find a set $S_2 \subseteq V(G)$ of size n/t such that $G_{S_1} - S_2$ has depth at most $\mathsf{depth}(G_{S_1} - S_2) \leq \lambda t + t - \lambda - 1$. Now we set $S = S_1 \bigcup S_2$ we have $|S| \leq m + n/t$ and $\mathsf{depth}(G - S) \leq \mathsf{depth}(G_{S_1} - S_2) \leq \lambda t - \lambda - 1$.

Theorem 6, which upper bounds the complexity of a layered graph, now follows directly from Lemmas 3, 4 and Theorem 2. The proof is in the full version [AB16] of the paper.

Theorem 6 (Complexity of Layered Graph). *Let G be a (m, λ)-layered DAG on n nodes with $\lambda < n^{1/3}$ and maximum indegree $\mathsf{indeg}(G) = \delta$ and fix any $t > 0$, $g \geq t(\lambda + 1)$ then there exists an attack $\mathcal{A}$ on the corresponding iMHF such that for any core-memory ratio $R > 0$ and $\bar{R} > 0$ the energy and AT complexities of $\mathcal{A}$ are at most*

$$AT_R(\mathcal{A}) \leq 2n \left[\frac{n}{t} + m + \delta g + R + \frac{(\lambda + 1)t \cdot (n + 2R) + R \cdot n}{g} \right] \ and$$

$$E_{\bar{R}}(\mathcal{A}) \leq n \left[\frac{(\lambda + 1)t(n + 2\bar{R}) + \bar{R} \cdot n}{g} + \delta g + \frac{n}{t} + m + \bar{R} \right].$$

Lemma 5 is the key technical result in this section. It states that, in particular, for any $m \leq n$ and any constants τ and δ a $(n, \delta, \sigma = n/\tau)$-random DAG will be a $\left(O(m \log n), O\left(\frac{n}{m}\right)\right)$-layered DAG with high probability.

In a bit more detail we show how to construct a set S of (expected) size $O(m \log n)$. We note that our construction is computationally efficient. Intuitively, we partition the nodes of G into equal sized layers of consecutive nodes $L_0, \ldots, L_\lambda$. We add a node $v \in L_i$ to our set S if *any* of v's parents are also in layer L_i. In the single pass case $(w = n)$ we will add a vertex $v \in L_i$ to S with probability at most $(\delta - 1)/i$. Thus, in expectation we will add at most $(\delta - 1)|L_i|/i$ nodes from layer L_i to S. In total we add at most $\frac{n}{\lambda+1} \sum_{i=1}^{\lambda} \frac{1}{i} = \frac{nH_\lambda}{\lambda+1}$ nodes from layers $L \geq 1$ in expectation. Recall that by H_λ we denote the λth harmonic number.

Lemma 5 ($((n, \delta, \sigma)$-random DAGs are Layered). *Fix any $\lambda \geq 1$ and let $q = \lceil \frac{n}{\lambda+1} \rceil$. Then a $(n, \delta, \sigma = n/\tau)$-random DAG is a (m, λ)-layered DAG, where the random variable m has expected value $\mathbb{E}[m] = \tau(\delta - 1)q + q + \frac{(\delta-1)q \cdot H_\lambda}{2}$. Furthermore, for any $\gamma > e^{-3\left(\mathbb{E}[m]-q\right)/4}$ we have $m \leq \mathbb{E}[m] + \sqrt{3\left(\mathbb{E}[m] - q\right) \ln \gamma^{-1}}$ except with probability γ.*

Proof of Lemma 5. Let G be a $(n, \delta, \sigma = n/\tau)$-random DAG with nodes $v_1, \ldots, v_n$. For simplicity assume that $\lambda = n/q - 1$ is an integer so that we can divide G into λ layers $L_0, \ldots, L_\lambda$ of equal size q (If λ is not an integer then the first $\lceil \lambda \rceil - 1$ layers will contain $q + 1$ nodes each and the last layer will contain $\leq q$ nodes.). Layer L_i contains nodes $L_i = \{v_{iq+1}, \ldots, v_{(i+1)q}\}$. We construct $S \subseteq V(G)$ as follows:

$$S = \bigcup_i \{v_j \in L_i \mid \exists s \leq d - 1.v_{r(j,s)} \in L_i\}.$$

That is if any of v's parents are in the same layer as v we add v to S.

Given $v_j \in L_i$ we let x_j denote the indicator random variable that is 1 if and only if $v_{r(j,t)} \in L_i$ for some index $t \leq \delta - 1$. We note that by linearity of expectation we have

$$\mathbb{E}\left[|S|\right] \leq \sum_{t=0}^{\lambda} \sum_{j=1}^{q} \mathbb{E}\left[x_{iq+j}\right].$$

Let $\lambda' = \lfloor \frac{\sigma}{q} \rfloor - 1$. Suppose first that $i \leq \lambda'$ and $0 < j \leq q$ so that $iq+j \leq \sigma$ and $v_{iq+j} \in L_i$ ($i > 0$) then the probability that we add v_{iq+j} to S because one of its parents is also in layer L_i is at most $(\delta-1)(j-1)/(iq)$. Thus, $\mathbb{E}\left[x_{iq+j}\right] \leq (\delta-1)/i$ for each $v_j \in L_i$. Now suppose that $iq + j > \sigma$ then the probability that we add $v_{r(iq+j)}$ to S because one of its parents is in the same layer is at most $(\delta - 1)(j - 1)/\sigma = (\delta - 1)(j - 1)\tau/n$. Thus, $\mathbb{E}\left[x_{iq+j}\right] \leq (\delta-1)(q-1)\tau/n$. Thus, in expectation we have

$$\mathbb{E}\left[|S|\right] \leq q + (\delta - 1) \sum_{i=1}^{\lambda'} \frac{(q - 1)}{2i} + q(\delta - 1) \sum_{i=\lambda'+1}^{\lambda} \frac{(q - 1)\tau}{n}$$

$$\leq q + (\delta - 1) \sum_{i=1}^{\lambda'} \frac{(q - 1)}{2i} + q(\delta - 1)(\lambda - \lambda') \frac{(q - 1)\tau}{n}$$

$$\leq q + (\delta - 1)\frac{(q - 1)H_{\lambda'}}{2} + q(\delta - 1)\tau,$$

where $H_\lambda \approx \ln \lambda$ denotes the λ'th harmonic number and the last inequality follows because $(q - 1)\lambda \leq n$. Observe that the DAG G_S with all directed edges originating in S deleted (i.e., delete edges of the form (v, v') with $v \in S$) will be a λ-stacked sandwich graph because there are no forward edges within each layer apart from the chain (v_i, v_{i+1}). Let $X = \sum_{j=q+1}^{n} x_j$. Because the random variables $x_j \in \{0, 1\}$ are independent standard concentration bounds imply that $\Pr\left[X \geq \mathbb{E}\left[X\right] + \tau\right] \leq \exp\left(\frac{-\tau^2}{2\mathbf{Var}(X)+2\tau/3}\right)$. In our case we have

$$\mathbf{Var}(X) = \sum_{i=q+1}^{n} \sum_{j=q+1}^{n} \mathbb{E}\left[x_i x_j\right] - \mathbb{E}\left[x_i\right]\mathbb{E}\left[x_j\right] = \sum_{i=q+1}^{n} \mathbb{E}\left[x_i\right] - \mathbb{E}\left[x_i\right]^2$$

$$\leq \mathbb{E}\left[X\right] = (\delta - 1)\frac{(q - 1)H_{\lambda'}}{2} + q(\delta - 1)\tau.$$

Thus, for any $\gamma > \exp\left(-3\mathbb{E}\left[X\right]/4\right)$ we can set $\tau = \sqrt{3\mathbb{E}\left[X\right]\ln\gamma^{-1}}$ to obtain

$$\Pr\left[X \geq \mathbb{E}\left[X\right] + \tau\right] \leq \exp\left(\frac{-\tau^2}{\mathbb{E}\left[X\right] + 2\tau/3}\right) \leq \exp\left(\frac{3\mathbb{E}\left[X\right]\ln\gamma}{3\mathbb{E}\left[X\right]}\right) \leq \gamma.$$

Where the second inequality follows from the observation that $2\tau/3 < \mathbb{E}\left[X\right]$ whenever $\gamma > \exp\left(-3\mathbb{E}\left[X\right]/4\right)$. As $|S| = X + q$ we have

$$|S| \leq q + (\delta - 1)\frac{(q-1)H_{\chi'}}{2} + q(\delta-1)\tau + \sqrt{3\left((\delta-1)\frac{(q-1)H_{\chi'}}{2} + q(\delta-1)\tau\right)\ln\gamma^{-1}}$$

except with probability γ^{22}. □

As an immediate consequence of Lemma 5 we get an attack on the static mode of operation of the Password Hashing Competition winner Argon2. Specifically we can attack the Argon2i variant in which memory accesses patterns are not input dependent — a desirable property to prevent side-channel attacks based on cache timing. As another immediate consequence we also get an attack on the Single-Buffer construction of [CGBS16].

Theorem 7. *Let G be a $(n, \delta, n/\tau)$-random DAG on n nodes. There exists an evaluation algorithm $\mathcal{A}$ for the corresponding iMHF such that for any core-memory ratios $R > 0$ and $\bar{R} > 0$ the expected AT and energy complexities of $\mathcal{A}$ are at most $\mathbb{E}\left[AT_R(\mathcal{A})\right] \leq U_{AT}$ and $\mathbb{E}\left[E_{\bar{R}}(\mathcal{A})\right] \leq U_E$, where*

$$U_{AT} = 2n^{7/4}\left[1 + 2\delta + \frac{(\delta-1)H_{n^{1/4}/\tau}}{2} + \frac{1}{\tau} + \frac{n^{1/4}R + \sqrt{n} + R}{n^{3/4}}\right] \text{ and}$$

$$U_E = n^{7/4}\left[1 + 2\delta + \frac{(\delta-1)H_{n^{1/4}/\tau}}{2} + \frac{1}{\tau} + \frac{n^{1/4}\bar{R} + \sqrt{n} + \bar{R}}{n^{3/4}}\right].$$

Furthermore, except with probability $\gamma > e^{-n^{3/4}\left(\delta-1\right)\left(1+H_{n^{1/4}/\tau}/2\right)}$, we have

$$AT_R(\mathcal{A}) \leq U_{AT} + 2n^{7/4} \times \sqrt{\frac{3(\delta-1)\left(H_{n^{1/4}/\tau} + 2\right)\ln\gamma^{-1}}{2n^{3/4}}} \text{ and}$$

$$E_{\bar{R}}(\mathcal{A}) \leq U_E + n^{7/4} \times \sqrt{\frac{3(\delta-1)\left(H_{n^{1/4}/\tau} + 2\right)\ln\gamma^{-1}}{2n^{3/4}}}.$$

In the special case $\tau = 1$ (single-pass variants) Theorem 7 follows by setting $q = n^{3/4}$ in Lemma 5 and $\lambda = n^{1/4} - 1$, $g = n^{3/4}$ and $t = n^{1/4}$ in Theorem 6. For

---

[22] If $\gamma < \exp\left(-3\mathbb{E}\left[X\right]/4\right)$ then we can set $\tau = \sqrt{3\mathbb{E}\left[X\right]\ln\gamma^{-1}}$ to obtain a slightly weaker concentration bound. However, the term $\exp\left(-3\mathbb{E}\left[X\right]/4\right)$ is already negligibly small in all of our applications.

the general case (multi-pass variants) we can customize the function keep. The basic idea is simple: during a balloon phase we only need to keep about σ extra pebbles on the $(n, \delta, \sigma = n/\tau)$-random DAG because we can discard pebbles outside of the current memory window. The proof is included in the full version of this paper [AB16].

6   Ideal iMHFs Don't Exist

In this section we show that ideal iMHFs do not exist. More specifically we show that for every graph G there exists node set S and positive integer $g \geq$ depth $(G-S)$ such that the iMHF evaluation algorithm $\mathcal{A} = \mathsf{PGenPeb}(G, S, g, d, \lfloor g/d \rfloor)$ has AT and energy-complexity $o(n^2/\log^{1-\epsilon} n)$ for any constant $\epsilon > 0$. In particular, if we take the naïve algorithm to be $\mathcal{N} = \mathsf{triv}$ then $\mathcal{A}$ is an attack with energy-quality $\omega(\log^{1-\epsilon} n)$. We first prove (Lemma 7) that all DAGs are reducible provided that the maximum indegree δ is sufficiently small (e.g., $\delta \leq \log^{0.999} n$). The proof of Lemma 7 follows from a result of Valiant [Val77]. Once we have established that all DAGs are reducible we can use $\mathsf{PGenPeb}$ to obtain a high quality attack on any iMHF.

Lemma 6 ([Val77] **Extension**). *Given a DAG G with m edges and depth* $\mathsf{depth}(G) \leq d = 2^i$ *there is a set of m/i edges s.t. by deleting them we obtain a graph of depth at most $d/2$.*

Lemma 7 (All DAGs are Reducible). *Let $G = (V, E)$ be an arbitrary DAG of size $|V| = n = 2^k$ with $\mathsf{indeg}(G) = \delta$. Then for every integer $t \geq 1$ there is a set $S \subseteq V$ of size $|S| \leq \frac{t\delta n}{\log(n)-t}$ such that $\mathsf{depth}(G - S) \leq 2^{k-t}$. Furthermore, there is an efficient algorithm to find S.*

In a nutshell, Lemma 7 follows by invoking Lemma 6 t times. The detailed proof is in the full version [AB16]. Theorem 8 now follows from Lemma 7 and Theorem 3.

Theorem 8 (Complexities of any iMHF). *Let F be an iMHF based on arbitrary DAG $G = (V, E)$ of size $|V| = n$ with in-degree $\mathsf{indeg}(G) = \delta$. Then for every constant $\epsilon > 0$ and fixed ratios $R > 0$ and $\bar{R} > 0$ there exists an evaluation algorithm $\mathcal{A}$ such that*

$$E_{\bar{R}}(\mathcal{A}) = o\left(n\left(\frac{\delta n}{\log^{1-\epsilon}} + \hat{R}\right)\right) = AT_R(\mathcal{A})$$

where $\hat{R} = \max\{R, \bar{R}\}$.

In particular if we let the naïve algorithm for F be $\mathcal{N} = \mathsf{triv}$ then algorithm $\mathcal{A}$ is an attack with qualities

$$\mathsf{E\text{-}quality}(\mathcal{A}) = \Omega\left(\frac{\hat{R} + n}{\hat{R} + \delta n/\log^{1-\epsilon} n}\right) = \mathsf{AT\text{-}quality}(\mathcal{A})$$

or, for constant R and $\bar{R}$, simply $\Omega\left(\delta^{-1} \log^{1-\epsilon} n\right)$.

In a nutshell, in the proof we set $t = O(\log \log n)$ in Lemma 7 to obtain a set S s.t. $|S| \leq O(n \log \log(n)/\log(n))$ and $d = \mathsf{depth}(G - S) \leq n/\log^2(n)$. We then set $g = n/\log^{1+\epsilon}(n)$ in Theorem 3. See the full version of this paper for a detailed proof [AB16].

Remarks. We remark that for any constants δ and $\epsilon > 1$ Theorem 8 yields an attack with quality $\mathsf{E\text{-}quality}(\mathcal{A}) = \Omega(\log^{1-\epsilon} n) = \mathsf{AT\text{-}quality}(\mathcal{A})$. Furthermore, provided that $\delta \leq \log^{1-\epsilon'} n$ where $\epsilon' > 1$ Theorem 8 yields an attack with quality $\mathsf{E\text{-}quality}(\mathcal{A}) = \omega(1) = \mathsf{AT\text{-}quality}(\mathcal{A})$. If $\bar{R} \neq R$ then we can obtain separate attacks $\mathcal{A}_1$ and $\mathcal{A}_2$ optimizing E-quality and AT-quality respectively.

Erdos et al. [EGS75] constructed a graph G of any size n with $\mathsf{indeg}(G) = O(\log(n))$ which is $(\alpha n, \beta n)$-depth robust for some constants $0 < \beta < \alpha < 1$.[23] This would imply that our bounds in Theorem 8 and Lemma 7 are essentially tight. Alwen and Serbinenko [AS15] used the depth robust DAGs from [MMV13] as a building block to construct a family of DAGs with provably high pebbling complexity $\tilde{\Omega}(n^2)$. Thus, our general attack in Theorem 8 is optimal up to polylogarithmic factors.

7 Practical Considerations

In this section we demonstrate that our attacks have high quality for practical values of n. For example, we obtain positive attack quality against Argon2i, the winner of the Password Hashing Competition, when n is only 2^{18}. Figure 1a plots attack quality vs n for Argon2i and the Single-Buffer (SB) construction for various values of varies $\tau \in \{1, 3, 5\}$, the number of passes through memory. The full version [AB16] includes additional plots showing that we achieve an even greater attack quality against Catena Dragonfly and Butterfly variants. Figure 1b shows the results for our generic attack on any iMHF.

Parameter Optimization. We remark that we optimized the parameters of our attack for each specific value of n in our plots. For example, we showed that any λ-stacked sandwich DAG is $(n/t, t(\lambda + 1))$-reducible for any $t \geq 1$. For each different value of n we ran a script to find the optimal values of t and $g \geq t(\lambda + 1)$ which minimize the energy complexity (resp. AT-complexity) of $\mathsf{PGenPeb}(G, S, g, d = t(\lambda + 1), k = g/d)$. In our general iMHF attack we used a script to find the optimal value of t in Lemma 7 and the optimal value of g.

Naïve Algorithms. The naïve algorithm $\mathcal{N}$ for the Catena Butterfly iMHF has absolute memory complexity $\mathsf{amc}(\mathcal{N}) = n/(\lambda \log n)$ and energy complexity $E_{\bar{R}}(\mathcal{N}) = n(\mathsf{amc}(\mathcal{N}) + \bar{R})$. Similarly, the naïve algorithm $\mathcal{N}$ for Catena Dragonfly has absolute memory complexity $\mathsf{amc}(\mathcal{N}) = n/(\lambda + 1)$ and the naïve k-pass algorithm for Argon2i and SB has absolute memory complexity $\mathsf{amc}(\mathcal{N}) = n/k$.

---

[23] In [MMV13] the authors give an explicit construction of a DAG which has $\mathsf{indeg}(G) = \log^2 n$ which is $(\alpha n, \beta n)$-depth robust for any α and β arbitrarily close to 1.

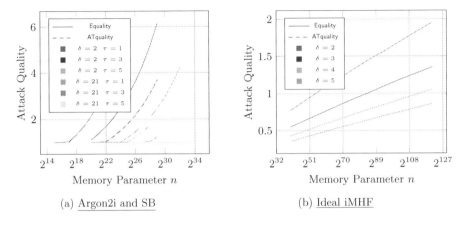

(a) <u>Argon2i and SB</u>   (b) <u>Ideal iMHF</u>

Fig. 1. Attack Quality ($R = \bar{R} = 3000$)

Thus, our attack quality decreases with λ or k. We stress that this is not because our attacks becomes less efficient as λ and k increases, but because the $\mathcal{N}$ algorithm requires less and less memory (thus, as λ, k increase the iMHFs become increasingly less ideal). By contrast, the naïve algorithm $\mathcal{N} = \mathsf{triv}$ for our general iMHF (and for Argon2i) has $E_{\bar{R}}(\mathcal{N}) = n(n + \bar{R})$.

Customized Attack Architecture. We have outlined efficient attacks on Catena, Argon2i and the Balloon Hashing iMHFs in the theoretical Parallel Random Oracle Machine (pROM) model of computation. Because pROM is a theoretical model of computation it is not obvious a priori that our attacks translate to practically efficient attacks that could be implemented in real hardware because it can be difficult to dynamically reallocate memory between processes in an ASIC (the amount of memory used during each round of a balloon phase is significantly greater than the amount of memory used during each round of a light phase). In the full version we argue that this architecture challenge would not be a fundamental barrier to an adversary. In particular, we outline an architecture for our algorithm PGenPeb using Argon2i as an example.

Briefly, we execute $n^{1/4}$ instances of the iMHF in parallel. Our architecture includes $n^{1/4}$ "light phase" chips and a single "Balloon Phase" chip which is responsible for executing all of the balloon phases in a round robin fashion. Each light phase chip only needs $O(n^{3/4} \ln n)$ memory and a single instance of the compression function H. The central balloon phase chip needs to have $O(n \ln n)$ memory and $\sqrt{n}$ instances of the compression functions H.

8   Conclusions

The results in this work show that (at the very least asymptotically speaking) most candidate iMHFs fall far short of their stated goals, and even of the weaker general upper-bound in Sect. 6. The notable exception is the construction of [AS15]. However, currently it can be viewed mainly as a theoretical result rather then a practical one due to the recursive nature of the construction and the high degree of the polylog factor complexity lower-bound (though this can partially be tightened with a slightly more fine grained security proof). Thus we are left with the central open problem of finding a practical construction of an iMHF which get as close as possible to the general upper-bound.

References

[AB16] Alwen, J., Blocki, J.: Efficiently computing data-independent memory-hard functions. Cryptology ePrint Archive, Report 2016/115 (2016). http://eprint.iacr.org/

[ABMW05] Abadi, M., Burrows, M., Manasse, M., Wobber, T.: Moderately hard, memory-bound functions. ACM Trans. Internet Technol. 5(2), 299–327 (2005)

[ACK+16] Alwen, J., Chen, B., Kamath, C., Kolmogorov, V., Pietrzak, K., Tessaro, S.: On the complexity of scrypt and proofs of space in the parallel random oracle model. Cryptology ePrint Archive, Report 2016/100 (2016). http://eprint.iacr.org/

[Ale04] Alexander, S.: Password protection for modern operating systems. login, June 2004

[AS15] Alwen, J., Serbinenko, V.: High parallel complexity graphs and memory-hard functions. In: Proceedings of the Eleventh Annual ACM Symposium on Theory of Computing, STOC 2015 (2015). http://eprint.iacr.org/2014/238

[ASK07] Acıiçmez, O., Schindler, W., Koç, Ç.K.: Cache based remote timing attack on the AES. In: Abe, M. (ed.) CT-RSA 2007. LNCS, vol. 4377, pp. 271–286. Springer, Heidelberg (2006)

[BBBB10] Bojinov, H., Bursztein, E., Boyen, X., Boneh, D.: Kamouflage: loss-resistant password management. In: Gritzalis, D., Preneel, B., Theoharidou, M. (eds.) ESORICS 2010. LNCS, vol. 6345, pp. 286–302. Springer, Heidelberg (2010)

[BBD13] Blocki, J., Blum, M., Datta, A.: Gotcha password hackers! In: Proceedings of the 2013 ACM Workshop on Artificial Intelligence and Security, pp. 25–34. ACM (2013)

[BD16] Blocki, J., Datta, A.: Cash: a cost asymmetric secure hash algorithm for optimal password protection. In: 29th IEEE Computer Security Foundations Symposium, CSF (2016, to appear)

[BDK15] Biryukov, A., Dinu, D., Khovratovich, D.: Fast and tradeoff-resilient memory-hard functions for cryptocurrencies and password hashing. Cryptology ePrint Archive, Report 2015/430 (2015). http://eprint.iacr.org/

[Ber] Bernstein, D.J.: Cache-Timing Attacks on AES

[Bil13] Markus, B.: Dogecoin (2013)

[BJKS03] Brainard, J.G., Juels, A., Kaliski, B., Szydlo, M.: A new two-server approach for authentication with short secrets. In: USENIX Security, vol. 3, pp. 201–214 (2003)

[BK15]   Biryukov, A., Khovratovich, D.: Tradeoff cryptanalysis of memory-hard functions. Cryptology ePrint Archive, Report 2015/227 (2015). http://eprint.iacr.org/

[BKPS13] Blocki, J., Komanduri, S., Procaccia, A., Sheffet, O.: Optimizing password composition policies. In: Proceedings of the Fourteenth ACM Conference on Electronic Commerce, pp. 105–122. ACM (2013)

[BL13]   Bernstein, D.J., Lange, T.: Non-uniform cracks in the concrete: the power of free precomputation. In: Sako, K., Sarkar, P. (eds.) ASIACRYPT 2013, Part II. LNCS, vol. 8270, pp. 321–340. Springer, Heidelberg (2013)

[BM06]   Bonneau, J., Mironov, I.: Cache-collision timing attacks against AES. In: Goubin, L., Matsui, M. (eds.) CHES 2006. LNCS, vol. 4249, pp. 201–215. Springer, Heidelberg (2006)

[Bon12]   Bonneau, J.: The science of guessing: analyzing an anonymized corpus of 70 million passwords. In: 2012 IEEE Symposium on Security and Privacy (SP), pp. 538–552. IEEE (2012)

[CGBS16] Corrigan-Gibbs, H., Boneh, D., Schechter, S.: Balloon hashing: provably space-hard hash functions with data-independent access patterns. Cryptology ePrint Archive, Report 2016/027 (2016). http://eprint.iacr.org/

[Cha11]   Lee, C.: Litecoin (2011)

[CHS06]   Canetti, R., Halevi, S., Steiner, M.: Mitigating dictionary attacks on password-protected local storage. In: Dwork, C. (ed.) CRYPTO 2006. LNCS, vol. 4117, pp. 160–179. Springer, Heidelberg (2006)

[CLN12]   Camenisch, J., Lysyanskaya, A., Neven, G.: Practical yet universally composable two-server password-authenticated secret sharing. In: Proceedings of the 2012 ACM Conference on Computer and Communications Security, pp. 525–536. ACM (2012)

[DC08]   Daher, W., Canetti, R.: Posh: a generalized captcha with security applications. In: Proceedings of the 1st ACM workshop on Workshop on AISec, pp. 1–10. ACM (2008)

[DGN03]   Dwork, C., Goldberg, A.V., Naor, M.: On memory-bound functions for fighting spam. In: Boneh, D. (ed.) CRYPTO 2003. LNCS, vol. 2729, pp. 426–444. Springer, Heidelberg (2003)

[DNW05]   Dwork, C., Naor, M., Wee, H.: Pebbling and proofs of work. In: Shoup, V. (ed.) CRYPTO 2005. LNCS, vol. 3621, pp. 37–54. Springer, Heidelberg (2005)

[EGS75]   Erdoes, P., Graham, R.L., Szemeredi, E.: On sparse graphs with dense long paths. Technical report, Stanford, CA, USA (1975)

[FH10]   Florêncio, D., Herley, C.: Where do security policies come from? In: Proceedings of SOUPS, p. 10 (2010)

[FLW13]   Forler, C., Lucks, S., Wenzel, J.: Catena: A memory-consuming password scrambler. IACR Cryptology ePrint Archive 2013:525 (2013)

[JR13]   Juels, A., Rivest, R.L.: Honeywords: making password-cracking detectable. In: Proceedings of the 2012 ACM Conference on Computer and Communications Security. ACM (2013)

[Kal00]   Kaliski, B.: Pkcs# 5: Password-based cryptography specification version 2.0 (2000)

[KSK+11] Komanduri, S., Shay, R., Kelley, P.G., Mazurek, M.L., Bauer, L., Christin, N., Cranor, L.F., Egelman, S.: Of passwords and people: measuring the effect of password-composition policies. In: Proceedings of the 2011 Annual Conference on Human Factors in Computing Systems, pp. 2595–2604. ACM (2011)

[LT82] Lengauer, T., Tarjan, R.E.: Asymptotically tight bounds on time-space trade-offs in a pebble game. J. ACM (JACM) **29**(4), 1087–1130 (1982)

[Man96] Manber, U.: A simple scheme to make passwords based on one-way functions much harder to crack. Comput. Secur. **15**(2), 171–176 (1996)

[MMV13] Mahmoody, M., Moran, T., Vadhan, S.P.: Publicly verifiable proofs of sequential work. In: Kleinberg, R.D. (ed.) Innovations in Theoretical Computer Science, ITCS 2013, Berkeley, CA, USA, 9–12 January 2013, pp. 373–388. ACM (2013)

[MT79] Morris, R., Thompson, K.: Password security: a case history. Commun. ACM **22**(11), 594–597 (1979)

[NB+15] Narayanan, A., Bonneau, J., Felten, E.W., Miller, A., Goldfeder, S.: Bitcoin and Cryptocurrency Technology (manuscript) (2015). Accessed 6 Aug 2015

[Per09] Percival, C.: Stronger key derivation via sequential memory-hard functions. In: BSDCan 2009 (2009)

[PHC] Password hashing competition. https://password-hashing.net/

[PM] Provos, N., Mazieres, D.: Bcrypt algorithm

[RTSS09] Ristenpart, T., Tromer, E., Shacham, H., Savage, S.: Hey, you, get off of my cloud: exploring information leakage in third-party compute clouds. In: Al-Shaer, E., Jha, S., Keromytis, A.D. (eds.) Proceedings of the 2009 ACM Conference on Computer and Communications Security, CCS 2009, pp. 199–212. ACM, New York (2009)

[SS09] Scarfone, K., Souppaya, M.: Nist special publication 800–118: Guide to enterprise password management (draft), April 2009

[Tho79] Thompson, C.D.: Area-time complexity for VLSI. In: Fischer, M.J., DeMillo, R.A., Lynch, N.A., Burkhard, W.A., Aho, A.V. (eds.) Proceedings of the 11h Annual ACM Symposium on Theory of Computing, 30 April - 2 May 1979, Atlanta, Georgia, USA, pp. 81–88. ACM, New York (1979)

[Val77] Valiant, L.G.: Graph-theoretic arguments in low-level complexity. In: Gruska, J. (ed.) MFCS 1977. LNCS, vol. 53, pp. 162–176. Springer, Heidelberg (1977)

Towards Sound Fresh Re-keying with Hard (Physical) Learning Problems

Stefan Dziembowski[1], Sebastian Faust[2(✉)], Gottfried Herold[2],
Anthony Journault[3], Daniel Masny[2], and François-Xavier Standaert[3]

[1] Institute of Informatics, University of Warsaw, Warsaw, Poland
[2] Fakultät für Mathematik, University of Bochum, Bochum, Germany
sebastian.faust@gmail.com
[3] ICTEAM – Crypto Group, Université catholique de Louvain,
Louvain-la-Neuve, Belgium

Abstract. Most leakage-resilient cryptographic constructions aim at limiting the information adversaries can obtain about secret keys. In the case of asymmetric algorithms, this is usually obtained by secret sharing (aka masking) the key, which is made easy by their algebraic properties. In the case of symmetric algorithms, it is rather key evolution that is exploited. While more efficient, the scope of this second solution is limited to stateful primitives that easily allow for key evolution such as stream ciphers. Unfortunately, it seems generally hard to avoid the need of (at least one) execution of a stateless primitive, both for encryption and authentication protocols. As a result, fresh re-keying has emerged as an alternative solution, in which a block cipher that is hard to protect against side-channel attacks is re-keyed with a stateless function that is easy to mask. While previous proposals in this direction were all based on heuristic arguments, we propose two new constructions that, for the first time, allow a more formal treatment of fresh re-keying. More precisely, we reduce the security of our re-keying schemes to two building blocks that can be of independent interest. The first one is an assumption of Learning Parity with Leakage, which leverages the noise that is available in side-channel measurements. The second one is based on the Learning With Rounding assumption, which can be seen as an alternative solution for low-noise implementations. Both constructions are efficient and easy to mask, since they are key homomorphic or almost key homomorphic.

1 Introduction

Side-channel attacks are an important concern for the security of cryptographic implementations. Since their apparition in the late 1990s, a large body of work has investigated solutions to prevent them efficiently, e.g. based on algorithmic and protocol ingredients. Masking (i.e., data randomization) and shuffling (i.e., operation randomization) are well studied representatives of the first category [41]. Leakage-resilient cryptography [29] is a popular representative of the second one. Interestingly, it has been shown recently that these approaches are

M. Robshaw and J. Katz (Eds.): CRYPTO 2016, Part II, LNCS 9815, pp. 272–301, 2016.
DOI: 10.1007/978-3-662-53008-5_10

complementary. Namely, leakage-resilient cryptography brings strong (concrete) security guarantees for stateful primitives such as stream ciphers (where key evolution prevents attacks taking advantage of multiple leakages per key). However, these stateful primitives generally require to be initialized with some fresh data, for example new session keys [52][1]. In practice, this initialization typically involves a stateless primitive such as a Pseudo Random Function (PRF), for which leakage-resilience is significantly less effective, since nothing prevents the adversary to repeat measurements for the same plaintext and key in this case [10]. Hence, the state-of-the-art in leakage-resilient symmetric cryptography is torn between two contradicting observations. On the one hand, leakage-resilient PRFs (and encryption schemes) such as [1,24,57,58] cannot be used for this initialization[2]. On the other hand, it seems that the execution of at least one stateless primitive (e.g., a PRF or a block cipher) is strictly needed for the deployment of leakage-resilient (symmetric) encryption and MACs [51]. This leaves the efficient protection of such stateless primitives with algorithmic countermeasures such as masking and shuffling as an important research goal.

In particular, masking appears as a promising solutions for this purpose, since it benefits from a good theoretical understanding [21,26,27,37,53]. Unfortunately, the secure masking of a block cipher like the AES also comes with significant drawbacks, especially when the number of shares increases (i.e., for so called higher-order masking schemes). First, it implies implementation overheads that are quadratic in the number of shares [33] (although some optimizations are possible for low number of shares, especially in hardware, e.g. [12]). Second, it has large randomness requirements (since the masked execution of non-linear operations at high-orders requires frequent refreshings of the shares). Third and probably most importantly, it assumes that the leakages of all these shares are independent, a condition that is frequently contradicted both in software (because of transition-based leakages [6,22]) and hardware implementations (because of so-called glitches [42,43]. Besides, standard algorithmic countermeasures able to deal with such independence issues usually imply additional implementation constraints, sometimes reflected by performance losses (e.g., the threshold implementations in [13,50] prevent glitches by increasing the number of shares).

Quite naturally, an extreme solution to this problem is to take advantage of asymmetric cryptographic primitives, for which algebraic properties usually make the masking countermeasure much easier to implement, as suggested for El Gamal encryption [38] and pairing-based MACs [44]. While these solutions may indeed lead to better efficiency vs. security tradeoffs than the direct protection of a block cipher with masking in the long term, they still imply significant performance overheads that may not be affordable for low-cost devices, and can only be amortized for quite high-order masking schemes.

---

[1] Note that using the key-evolution approach for the session key derivation (i.e. computing session key K_i as an "evolved" session key K_{i-1}) is often impractical, since it requires synchronization between the sender and the receiver.

[2] Excepted if combined with additional heuristic assumptions such as in [47].

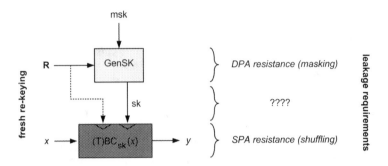

Fig. 1. Fresh re-keying and its leakage requirements.

Taking these challenges into account, an appealing intermediate path called *fresh re-keying* has been initiated by industrial and academic research [30, 46]. As illustrated in Fig. 1, its main idea is to exploit a good "separation of duties" between a re-keying function *GenSK* and a block cipher or tweakable block cipher [40][3]. That is, the function *GenSK*, which is used to generate the fresh session keys sk, needs to resist Differential Power Analysis attacks (DPA), i.e. attacks exploiting multiple measurements per key. By contrast, the (possibly tweakable) block cipher only needs to resist Simple Power Analysis (SPA) attacks, i.e. attacks exploiting a single measurement per key (or DPA attacks with limited trace count if the key refreshing is amortized). Quite naturally, this solution is useless in case *GenSK* is also a (tweakable) block cipher (since it would then be equally difficult to protect with masking). So previous fresh re-keying schemes additionally came with heuristic arguments justifying that this function does not need to be cryptographically strong, and only has to fulfill a limited set of properties (e.g., good diffusion). On top of this, they suggested to exploit key homomorphic *GenSK*'s, so that their masked implementation is (much) simplified. Taking advantage of key homomorphism indeed reduces the computational overheads and randomness requirements of masking to the minimum (i.e., the refreshing of the secret master key and the computation of the key homomorphic for each share). Besides, it also allows avoiding issues related to the independent leakage assumption, since we can then compute *GenSK* on each share independently. A polynomial multiplication in (e.g., a ring) was finally proposed as a possible instance for such functions [30, 46].

Yet, and while conceptually elegant, Fig. 1 also suggests the important caveat of existing fresh re-keying schemes. Namely, between the re-keying function *GenSK* that can be well protected against DPA thanks to masking, and the underlying (tweakable) block cipher that has to be secure against SPA (e.g., thanks to shuffling), one has to re-combine the shares to produce a fresh session

---

[3] As discussed in [23], using a tweakable block cipher allows obtaining beyond birthday security for this fresh re-keying scheme, while a standard block cipher only provides birthday security. Whether one or the other option is chosen will be essentially equivalent for the discussions in this paper.

key sk: an operation of which the leakage was essentially left out of the analysis so far. More precisely, the only (informal) guidelines were that it should be difficult to precisely extract hard information (e.g., bits) about sk, in order to avoid the algebraic attacks outlined in [45]. Recent results from Belaid et al. and Guo and Johansson then made it clear that a small noise may not be sufficient to secure *GenSK* against leakage on sk [9,11,35].

Our Contribution. Based on this state-of-the-art, we initiate the first formal study of fresh re-keying functions that are at the same time easy to mask (since key homomorphic or almost key homomorphic [18]) and cryptographically strong. To this end, we propose new security models using the ideal/real world paradigm and show that our instantiations described below can be proven secure under reasonable assumptions. Informally, our security guarantees state that even given continuous leakage (for instance, probing leakage or noisy leakage), the adversary will not be able to attack the re-keying function any better than an adversary that just obtains uniformly random session keys. We prove the security of two different instantiations of a re-keying function in this model (making different assumptions on the type of leakage as outlined below).

On the one hand, we start from the observation that in the context of re-keying, the function *GenSK*'s output is in fact never given to the adversary completely. Instead, the adversary learns only some *partial* leakage information about *GenSK*. Taking advantage of this observation, we first introduce a new assumption of Learning Parity with Leakage (LPL), of which the main difference with the standard Learning Parity with Noise (LPN) problem is that it relies on additive Gaussian (rather than Bernoulli) noise [14,15,32]. Note that we use the name Learning Parity with Leakage (and not with Gaussian noise) to reflect the fact that the amount of noise can be much larger than in the standard LPN assumption (since in a re-keying scheme, the authorized parties only deal with noise-free information). Then, we show that our new LPL assumption can be reduced to the standard LPN assumption. Finally, we instantiate a re-keying scheme based on LPL that is trivial to mask (since key homomorphic) and provide the actual noise values required to reach different security levels against adversaries targeting the re-combination step of the fresh key sk in Fig. 1. Conceptually, the main advantage of this construction is that it exploits the noise that is naturally available in side-channel leakages. However, our study also suggests that this physical noise may have to be increased by design to reach high security levels – see Sect. 6 for a discussion.

On the other hand, we consider the complementary context of small embedded devices with too limited noise for the previous LPL problem to be hard. In this case, we take advantage of the recently introduced Learning with Rounding (LWR) assumption [7], and describe a re-keying that is perfectly suitable for a low-noise environment. In order to make it most efficient, we instantiate it with computations in $\mathbb{Z}_q$ with $q = 2^b$, and a rounding function that can be simply implemented by dropping bits. This allows us to directly take advantage of standard arithmetic operators available in most computing platforms (e.g., recent ARM devices perform 32-bit multiplications in one cycle), without any additional hassle due to complex reductions. We then show that this re-keying function based on the LWR assumption can be efficiently masked thanks to an additional error correction step, which makes it almost key homomorphic. We finally provide parameters to instantiate it for various security levels, including very aggressive choice of parameters for which the security is not proven (or at least it is not based on the standard assumptions). Conceptually, the main advantage of this construction is that it ensures stronger cryptographic properties (i.e., computational indistinguishability from uniform) and therefore may be of interest beyond the re-keying scenario considered here.

As a result, we obtain two cryptographic constructions that can be used for fresh re-keying in both low-noise and high-noise contexts, for which masked implementations have minimum overheads and randomness requirements, and can easily fulfill the independent leakage assumption.

Besides the formal modeling and the security proofs of our constructions, we also present preliminary implementation results for our re-keying functions. Concretely, we report in Sect. 6 on implementations of our re-keying function on a 32-bit ARM and an 8-bit Atmel device. We give a comparison with masked AES implementations and show that for certain choices of the parameters (and under reasonable noise assumptions for the LPL-based construction), we can achieve improved efficiency.

Related Works. Our two re-keying constructions are naturally connected to previous cryptographic primitives based on LPN and LWR, such as LAPIN [36] and SPRING [8]. Interestingly, when it comes to their resistance against side-channel attacks, these new constructions also bring a neat solution to the main drawbacks of LAPIN and SPRING. Namely, for LAPIN, it remained that the generation and protection of the Bernoulli noise was challenging [31]. But when relying on the LPL assumption, we gain the advantage that this noise does not have to be generated (since it corresponds to the leakage noise that anyway has to be available on chip for the masking of F to be effective – see again [21,26,27, 37,53]). As for SPRING, the main challenge was to deal with the masking of the (non–linear) rounding operation [19]. But as described in Sect. 5, this masking is made easier with our re-keying function based on LWR.

A similar technique to our reduction from LPL to LPN was used in [11], who also analyze physical noise used as a countermeasure to leakage in the context of finite field multiplication and attack this by deriving LPN instances.

2   Preliminaries

Notations. We denote scalars u, v by small italic single characters. Vectors $\mathbf{u}, \mathbf{r}, \mathbf{k}$ are denoted by small bold letters. Matrices $\mathbf{R}, \mathbf{T}$ are denote by capital bold letters. We use capital letters R, U to denote scalars if we want to emphasize that we treat them as random variables and argue about probabilities.

Standard Assumptions. Our constructions will be based on the Learning Parity with Noise (LPN) assumption and the Offset Learning with Uniform Noise assumption. To analyse these, we recall some relevant standard assumptions:

Definition 1 (LPN). *Let $0 < \tau < \frac{1}{2}$ be fixed and $n \in \mathbb{N}$. For (unknown) $\mathbf{k} \in \mathbb{Z}_2^n$, the $\mathsf{LPN}_{n,\tau}$ sample distribution is is given by*

$$\mathcal{D}_{\mathsf{LPN},n,\tau} := (\mathbf{r}, \ell) \; for \; \mathbf{r} \in \mathbb{Z}_2^n \; uniform, \; e \leftarrow \mathcal{B}_\tau, \; \ell := \langle \mathbf{r}, \mathbf{k} \rangle + e \bmod 2,$$

where $\mathcal{B}_\tau$ denotes a Bernoulli distribution with $\Pr[e = 1] = \tau, \Pr[e = 0] = 1 - \tau$.

Given query access to $\mathcal{D}_{\mathsf{LPN},n,\tau}$ for uniformly random $\mathbf{k}$, the search $\mathsf{LPN}_{n,\tau}$-problem asks to find $\mathbf{k}$. The decision $\mathsf{LPN}_{n,\tau}$-problem asks to distinguish an oracle for $\mathcal{D}_{\mathsf{LPN},\tau}$ for uniformly random $\mathbf{k}$ from an oracle that outputs uniformly random values from $\mathbb{Z}_2^n \times \mathbb{Z}_2$. The search/decision $\mathsf{LPN}_{n,\tau}$-Assumption asserts that these problems are infeasible for PPT algorithms. $(n, \tau$ are given functions of the security parameter).

Definition 2 (LWE [54]). *Let Φ be some efficiently sampleable noise distribution on $\mathbb{Z}$, $n \in \mathbb{N}$ and $q > 0$ (often, but not necessarily prime). For (unknown) $\mathbf{k} \in \mathbb{Z}_q^n$, the $\mathsf{LWE}_{n,\Phi}$ sample distribution is given by*

$$\mathcal{D}_{\mathsf{LWE},n,q,\Phi} := (\mathbf{r}, \ell) \; for \; \mathbf{r} \in \mathbb{Z}_q^n \; uniform, \; e \leftarrow \Phi, \; \ell := \langle \mathbf{r}, \mathbf{k} \rangle + e \bmod q.$$

Given query access to $\mathcal{D}_{\mathsf{LWE},n,q,\Phi}$ for uniformly random $\mathbf{k} \in \mathbb{Z}_q^n$, the search $\mathsf{LWE}_{n,q,\Phi}$-problem asks to find $\mathbf{k}$. The decision $\mathsf{LWE}_{n,q,\Phi}$-problem asks to distinguish an oracle for $\mathcal{D}_{\mathsf{LWE},q,\Phi}$ for uniformly random $\mathbf{k}$ from an oracle that outputs uniformly random values from $\mathbb{Z}_q^n \times \mathbb{Z}_q$. The search/decision $\mathsf{LWE}_{n,q,\Phi}$-Assumption asserts that these problems are infeasible for PPT algorithms.

Usually, Φ is taken to be a discrete Gaussian, i.e. a probability distribution whose density $\Pr_{E \leftarrow \Phi}[E = x]$ is proportional to $\exp\left(-\frac{x^2}{2s}\right)$. In this case, one usually takes s as a parameter of the scheme rather than Φ. Note that LPN is an important special case of LWE with $q = 2$ and Φ a Bernoulli distribution.

Definition 3 (LWU [25,48]). *Another important special case is when the error distribution Φ is uniform from some interval, say $\{0, \ldots, B - 1\}$. In this paper, we call it the Learning with Uniform Noise distribution/problem/assumption $\mathsf{LWU}_{n,q,B}$. Note that only the length of the interval matters in this case, as the adversary can add a constant shift itself.*

Definition 4 (LWR [7]). *The Learning with Rounding (LWR) distribution/problem/assumption is often seen as a deterministic variant of LWE, where instead of adding some random noise $e \leftarrow \Phi$ to perturb $\langle \mathbf{r}, \mathbf{k} \rangle$, we round $\langle \mathbf{r}, \mathbf{k} \rangle$.*

More precisely, for appropriately chosen integers $p < q$, the rounding function $\lfloor \cdot \rceil_p : \mathbb{Z}_q \to \mathbb{Z}_p$ is $\lfloor x \rceil_p := \lfloor x \frac{p}{q} \rceil$, where $x \in \mathbb{Z}_q$ is represented as $x \in \{0, \ldots, q-1\}$. When applying $\lfloor \cdot \rceil_p$ to a vector in $\mathbb{Z}_q^n$, we apply it component-wise.

For (unknown) $\mathbf{k} \in \mathbb{Z}_q^n$, the $\mathsf{LWR}_{n,q,p}$ sample distribution is given by

$$\mathcal{D}_{\mathsf{LWR},n,q,p} := (\mathbf{r}, \ell) \; for \; \mathbf{r} \in \mathbb{Z}_q^n \; uniform \; and \; \ell = \lfloor \langle \mathbf{r}, \mathbf{k} \rangle \rceil_p.$$

Again, given query access to $\mathcal{D}_{\mathsf{LWR},n,q,p}$ for uniformly random $\mathbf{k} \in \mathbb{Z}_q^n$, the search $\mathsf{LWR}_{n,q,p}$-problem asks to find $\mathbf{k}$. The decision problem asks to distinguish

$\mathcal{D}_{\mathsf{LWR},n,q,p}$ *for uniformly random* $\mathbf{k}$ *from an oracle that outputs samples* $(\mathbf{r}, \lfloor u \rfloor_p)$ *for* $\mathbf{r} \in \mathbb{Z}_q^n, u \in \mathbb{Z}_q$ *uniform. Note that* $\lfloor u \rfloor_p$ *is not uniform in* $\mathbb{Z}_p$ *unless* $p \mid q$. *The search/decision* $\mathsf{LWR}_{n,q,p}$-*Assumption asserts that the problem is infeasible for PPT algorithms.*

3   General Framework

3.1   Re-keying Schemes

A re-keying scheme RK is a cryptographic primitive which generates session keys sk from a secret key msk and some public randomness $\mathbf{R}$. More precisely, $\mathsf{RK} = (Gen, GenSK, CorSK, \mathcal{D})$ consists of the following three PPT algorithms and an efficiently samplable distribution $\mathcal{D}$ from which the randomness is sampled:

$Gen(1^\lambda)$: Outputs a secret key msk and d shares thereof that we denote with $(\mathsf{msk})_d$.

$GenSK((\mathsf{msk})_d, \mathbf{R})$: Outputs a session key sk, new shares $(\mathsf{msk}')_d$ and potentially correction information v.

$CorSK(\mathsf{msk}, \mathbf{R}, v)$: Outputs a session key sk.

Concretely, $GenSK$ will be run by the chip to protect while $CorSK$ will be run by the other party. RK is called correct iff for $(\mathsf{msk}, (\mathsf{msk})_d) \leftarrow Gen(1^\lambda)$, $\mathbf{R} \leftarrow \mathcal{D}$, $(\mathsf{sk}, (\mathsf{msk}')_d, v) \leftarrow GenSK((\mathsf{msk})_d, \mathbf{R})$, we have that $CorSK(\mathsf{msk}, \mathbf{R}, v) = \mathsf{sk}$ holds with overwhelming probability. Further, we require that $(\mathsf{msk}')_d$ and $(\mathsf{msk})_d$ follow the same distribution, conditioned on msk.

One may think of $(\mathsf{msk}')_d$ and $(\mathsf{msk})_d$ as some form of encoding that protects against side-channel attacks[4]. The correction information v may be needed in some constructions since the session key when computed from $(\mathsf{msk})_d$ by $GenSK$ may be different when computed from msk by $CorSK$.

For the security definition of a re-keying scheme, we define three interactive PPT algorithms *Real*, *Ideal* and *Sim*. An adversary A will interact with them during a polynomially bounded amount of sessions (see Fig. 2).

We denote this process with $A^{Real((\mathsf{msk})_d)}(1^\lambda)$ when A interacts with *Real* and $A^{(Sim^{Ideal_c}, Ideal_c)}(1^\lambda)$ when A interacts with *Sim* and *Ideal*. In the latter case, $(Sim^{Ideal_c}, Ideal_c)$ is the concatenation of their outputs. During each session A receives the following output from *Real*, *Ideal* and *Sim*:

$Real((\mathsf{msk})_d)$: Takes $(\mathsf{msk})_d$ from the input and sample $\mathbf{R} \leftarrow \mathcal{D}$. Then, run $(\mathsf{sk}, (\mathsf{msk}')_d, v) \leftarrow GenSK((\mathsf{msk})_d, \mathbf{R})$. It outputs $(\mathbf{R}, \mathsf{sk}, v)$ to A. Further, it has additional, model specific inputs/outputs, e.g. probes which are leaked. It then overwrites $(\mathsf{msk})_d := (\mathsf{msk}')_d$ to be used in the next session.

---

[4] For the reader familiar with side-channel resistant implementations $(\mathsf{msk})_d$ denotes a masking of msk and $(\mathsf{msk}')_d$ denotes a refreshing of the shares of the masking. Indeed, in all our constructions, $(\mathsf{msk})_d$ will be d uniformly chosen values msk_i with $\sum_i \mathsf{msk}_i = \mathsf{msk}$. While we call $(\mathsf{msk})_d$ "shares" in the definition to match our later notation more closely, $(\mathsf{msk})_d$ could in principle be anything.

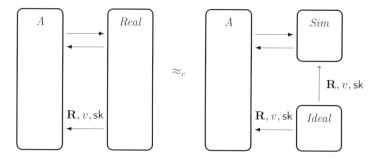

Fig. 2. An adversary A breaking the security of a re-keying scheme distinguishes the following cases: A interacts with *Real* or he interacts with *Sim* and *Ideal*. The session key is sk, the randomness for the session key generation $\mathbf{R}$ and some correction information v. Additionally A sees some model specific leakage.

$Ideal_c(1^\lambda)$: outputs in each session $(\mathbf{R}, v, \mathsf{sk})$ for uniform random sk, independent v and $\mathbf{R} \leftarrow \mathcal{D}$. Its random tape is c.

$Sim(1^\lambda)$: simulates model specific outputs while accessing the outputs of $Ideal_c$.

A re-keying scheme is called secure iff for any PPT A:

$$\left| \Pr\left[A^{Real((\mathsf{msk})_d)}(1^\lambda) = 1 \right] - \Pr\left[A^{(Sim^{Ideal_c}, Ideal_c)}(1^\lambda) = 1 \right] \right| \leq negl(\lambda),$$

where the probability is taken over the random tape of A, *Real*, *Sim*, c and $(\mathsf{msk}, (\mathsf{msk})_d) \leftarrow Gen(1^\lambda)$. It is easy to see that session keys sk need to be indistinguishable from uniform chosen keys to fulfill this security notion. This needs to hold even given the model specific leakage. The next two sections will describe two different leakage models that we consider in this work.

3.2 The Leakage Model for Re-Keying Schemes

For the two different instantiations that we present in Sect. 4 and Sect. 5 we propose two different leakage models. The leakage model specifies what additional information the adversary can obtain in the real world. In the ideal world the leakage then has to be simulated in a consistent way by the simulator *Sim*.

Re-keying Schemes in the t-probing Model. An important model to analyze the security of side-channel countermeasures is a security proof in the t-probing model [37]. In the t-probing model, the adversary is allowed to learn up to t-intermediate values of the computation of *GenSK*, i.e. of the generation of the session key. Notice that the definition of an intermediate value typically depends on the underlying scheme and its implementation. Our schemes are naturally described using group operations and internal values are group elements, notably elements from $\mathbb{Z}_p$ for p being a power of 2 (this includes the case of bits with $p = 2$). This means that the adversary A specifies a set of t probes $\mathcal{P}$, where $|\mathcal{P}| \leq t$ and the adversary obtains back from $Real((\mathsf{msk})_d)$ the intermediate values $V = \{v_{w_i}\}$, where $w_i \in \mathcal{P}$ and v_{w_i} is the value carried on the intermediate result labeled with w_i. Hence, if the computation was carried out over $\mathbb{Z}_p$, then the adversary obtains a set V with t elements, where each value in V corresponds to one of the intermediate values produced during the computation of *GenSK*. To show security of the re-keying function in the t-probing model, we need to construct an efficient simulator *Sim* that can simulate the replies of the adversary's probes V_i without probing access (i.e. from the values *Sim* obtains in the ideal world).

Re-keying Schemes in the t-noisy Probing Model. For our first re-keying scheme, we will show security in a weaker model than the standard t-probing model. We will assume that each of the t-probes is perturbed with additive Gaussian noise. This is a common assumption in works on side-channel analysis and can for instance be guaranteed using a physical noise generator [34]. For simplicity we assume that all intermediate values are in $\mathbb{Z}_2$ and for a probe on a wire that carries the bit $b \in \mathbb{Z}_2 = \{0, 1\}$ the adversary learns $b + e$, where e corresponds to noise from a continuous Gaussian distribution. It is important to note that the addition of the noise is a normal addition in the reals. The above can be generalized but for ease of exposition we stick to these simplifications in the rest of the paper. Using this terminology, in the t-noisy probing model the adversary A specifies a set of t probes $\mathcal{P}$, but instead of obtaining the exact values of the intermediate wires, he obtains a noisy version of them. That is, the set of replies V is $V = \{v_{w_i} + e_i\}$. Since the adversary only sees a noisy version of the intermediate values, the t-noisy probing model offers a weaker security guarantee than the t-probing model. Trivially, the leakage obtained in the t-noisy probing model can be simulated by t-probing leakage.

Security Against Continuous (Noisy) Probes. In the above two section we considered an adversary that can specify a set of probes and obtains the corresponding intermediate values. In the continuous probing model, this notion is extended by letting the adversary specify for *each* execution adaptively a new set of probes $\mathcal{P}$. More precisely, during the execution in the real world at the beginning of the i-th session, the adversary specifies adaptively a set of probes $\mathcal{P}^i$, and obtains the (noisy) intermediate values V^i that correspond to the wires specified in $\mathcal{P}^i$. Notice that the choice of $\mathcal{P}^i$ can depend on all information that the adversary A has seen previously, i.e. on $\{(V^j, \mathsf{sk}^j, \mathbf{R}^j, v^j)\}_{j \in [i-1]}$. Observe that the continuous probing model is significantly stronger than the one-shot (noisy) probing model as the adversary obtains significantly more information that he can exploit in breaking the scheme.

3.3 Masking Schemes

Masking schemes are a method to achieve security in the t-probing model (and hence also in the t-noisy-probing model since this model is weaker). In a masking scheme, each sensitive intermediate variable is split into d shares such that knowing only $(d-1)$ shares does not reveal information about the sensitive variable. Consider for instance the Boolean maskings scheme [37] where a sensitive variable k is represented by d random shares $k_1, \ldots, k_d$ such that $k = \sum_i k_i$. Clearly, knowing only $d-1$ arbitrary shares does not allow to recover the secret value k. The main difficulty in designing secure masking schemes is in computing with shared variables in a secure way. To this end one needs to design masked operations. In traditional masking schemes one typically designs masked algorithms for the basic operations of the underlying group (e.g., for addition and multiplication). While linear operations can be masked very efficiently, masking the non-linear operations, e.g., the multiplication, is significantly more costly. For instance, a masked multiplication in the Boolean masking scheme results into an overhead of $O(d^2)$ for each masked multiplication used in the computation.

We will apply masking schemes to protect the re-keying function against t-(noisy) probing attacks. To obtain better efficiency when executed in the masked domain, we design in the next sections cryptographically strong re-keying functions that are almost linear. Concretely, for our construction *GenSK* is divided into d sub-computations where each sub-computation only takes as input one share msk_i of $(\mathsf{msk})_d$ (this is the linear part of the re-keying function *GenSK*). Only at the very end of the computation the outputs of this linear part are recombined to obtain the final session key. We emphasize that by following this approach our construction also obtains strong glitch resistance – unlike normal masked implementations. Glitches can occur in hardware implementations due to synchronization problems. Since in our construction the sub-computations only depend on individual shares, glitches are prevented as we do not use operations that access multiple shares jointly.

Masking in the Continuous Leakage Model. To guarantee security of a masked implementation in the continuous leakage model, the secret shares used in the computation and in particular the shares of the key need to be refreshed frequently. Such a refreshing is typically done by a probabilistic Refresh algorithm. In our case, the refreshing is part of the *GenSK* algorithm and takes as input the shared master secret key $(\mathsf{msk})_d$ and produces a refreshed master secret key $(\mathsf{msk}')_d$. The correctness requirement of the refreshing algorithm says that both $(\mathsf{msk})_d$ and $(\mathsf{msk}')_d$ correspond to the same master secret key msk. If the underlying masking scheme is the Boolean masking scheme then this means that $\sum_i \mathsf{msk}_i = \sum_i \mathsf{msk}'_i$. Further, we require that the the distribution of (msk') is uniform among all possible such sharings of msk. Besides correctness, the refreshing algorithm also shall guarantee that side-channel information (i.e., the (noisy) probes) from different executions of the re-keying functions cannot be combined in an exploitable way. Informally, the refreshing schemes Refresh is said to be secure in the t-probing model against continuous attacks, if the leakage for probe sets $\mathcal{P}^j$ can be simulated without knowledge of the master secret key. Typically, the simulation is statistically indistinguishable from the continuous real execution of the Refresh algorithm.

To prove the security of our construction, we will need a secure instantiation of a refreshing algorithm. Several variants have been proposed in the literature with varying efficiency [4,26,37]. Since the focus of this work is not on designing secure refreshing schemes, we mainly ignore them for the rest of this paper. The following lemma can be proven about the Refresh from [26].

Lemma 1. *For any set of probes $\mathcal{P}$ with cardinality t, there exists a PPT simulator $\mathcal{S}$ and a set $\mathcal{I}$ of cardinality t such that for any $k \in \mathcal{K}$ and $k_1, \ldots, k_d$, $k'_1, \ldots, k'_d$ chosen uniformly at random subject to the constraint that $k := \sum_i k_i = \sum_i k'_i$ we have:*

$$(\mathcal{P}(\mathbf{k}' \leftarrow \mathsf{Refresh}(\mathbf{k})), \mathbf{k}_\mathcal{I}, \mathbf{k}'_\mathcal{I}) \equiv (\mathcal{S}(\mathcal{P}, \mathbf{k}_\mathcal{I}, \mathbf{k}'_\mathcal{I}), \mathbf{k}_\mathcal{I}, \mathbf{k}'_\mathcal{I}),$$

where in the above "$\equiv$" denotes the statistical equivalence and for a vector $\mathbf{k}$ and a set $\mathcal{I} \subseteq [d]$ the vector $\mathbf{k}_\mathcal{I}$ is the vector $\mathbf{k}$ restricted to the positions in $\mathcal{I}$.

The refreshing algorithm from Lemma 1 has complexity $O(d^2)$ (where the hidden constant in the O-notation are small). Recently, an improved refreshing algorithm based on expander graphs was proposed which has complexity $O(d)$ [4]. We notice that while asymptotically better, the hidden constants in $O(d)$ are much larger and hence are of less practical relevance. In our applications, the keys and all shares are vectors $k, k_i \in \mathbb{Z}_p^n$ for some n and the sharing of each of the n coordinates can be done independently. Consequently, we assume that if we probe only a total of t individual coordinates of some k_i's, then there exists a subset $J \subset \{1, \ldots, n\}$ of coordinates with $|J| \leq t$, such that the above Lemma 1 holds when restricted to coordinates from J.

4   Fresh Re-keying with Physical Noise

In this section, we instantiate the abstract re-keying scheme described above in an environment where sufficient physical noise is available. Our construction exploits the physical noise available in side-channel measurements in a constructive manner: the computation of the re-keying function is tailored in such a way that if the adversary obtains t-noisy probing leakage, he will not be able to break the re-keying scheme. While we believe that exploiting physical noise in a constructive way (i.e., for designing new cryptographic primitives) is an interesting conceptual contribution by itself, it also leads to potential efficiency improvements as we show in the implementation section (cfr. Sect. 6).

To show security of our re-keying scheme, we introduce a new learning assumption that we call the *Learning Parity with Leakage* (LPL) assumption. The LPL assumption says that inner product with physical noise cannot be distinguished from uniform samples. The main technical step is to show that the LPL assumption can be reduced to the classical LPN assumption (this is shown in Sect. 4.1). Notice that the most important difference between these two assumptions is that in LPL we add additive Gaussian noise (and no modular reduction is carried out), while in LPN the noise comes from the binomial distribution (and a modular reduction is carried out).

Of course, the requirement that the physical noise follows a Gaussian distribution is a strong assumption, and may not be perfectly fulfilled in practice:

physical noise indeed originates from a variety of sources (transistor noise, measurement noise, noise engines added by the cryptographic designers, ...). Yet, it has been observed in many practical settings that this assumption holds to a good extent [41]. More importantly, it is the starting point of most of the (e.g. template and regression-based) attacks that are usually considered in side-channel security evaluations [20,56]. Besides, it has been shown recently how to verify empirically that deviations from this Gaussian assumption do not significantly impact the security level of an implementation [28], which can therefore be done for our primitives as well[5].

Our Re-keying Function. We present our proposed LPN-based re-keying scheme $\Pi_{\text{noisy}} = (Gen, GenSK, CorSK, \mathcal{D})$, which will be proven secure when sufficient physical noise is available in the leakage measurements. Let n be the length of the master secret key, $m < n$ be the length of the session keys and d the number of shares used for the masking scheme. The distribution $\mathcal{D}$ from which the fresh randomness is sampled is defined as drawing uniformly at random $\mathbf{R} \leftarrow \mathbb{Z}_2^{m \times n}$. Let $H \colon \mathbb{Z}_2^m \to \mathbb{Z}_2^m$ be a hash function (modeled as a random oracle, see discussion below) and assume we have some secure refresh algorithm Refresh that satisfies the property of Lemma 1. Our re-keying function is then defined as follows:

$Gen(1^\lambda)$: Samples $\mathsf{msk} \leftarrow \mathbb{Z}_2^n$.
    It creates d shares $(\mathsf{msk})_d = \mathsf{msk}_1, \ldots, \mathsf{msk}_d$ such that $\sum \mathsf{msk}_i = \mathsf{msk}$.
$GenSK((\mathsf{msk})_d, \mathbf{R})$: A probabilistic algorithm working as follows:
    1. Compute $\mathbf{u}_i = \mathbf{R} \cdot \mathsf{msk}_i$
    2. Compute $\mathbf{u} = \sum_i \mathbf{u}_i$ iteratively as $((\ldots (\mathbf{u}_1 + \mathbf{u}_2) + \mathbf{u}_3) + \ldots) + \mathbf{u}_d$.
       Notice that other ways of computing this sum are possible, but will make the analysis more involved.
    3. The session key is computed as $\mathsf{sk} = H(\mathbf{u})$.
    4. Finally refresh the shares $(\mathsf{msk}')_d \leftarrow \mathsf{Refresh}((\mathsf{msk})_d)$.
    5. Output $(\mathbf{R}, \mathbf{u})$.
$CorSK(\mathsf{msk}, \mathbf{R})$: Output $H(\mathbf{R} \cdot \mathsf{msk})$.

From the above description it is clear that the additional value v (the correction term) is not used in this construction. It will be used in our construction from Sect. 5. Moreover, the reader may notice that the re-keying function is linear (except for the application of H at the end). The security against side-channel attacks comes from the fact that the adversary in the t-noisy probing model only obtains noisy intermediate values.

---

[5] Note that if significant deviations from the Gaussian assumptions were observed, it would not imply that our following constructions are directly broken – just that the parameters of our reductions below, and hence the parameters of our construction, will have to be changed, cfr. Remark 1 below.

On the Use of the Random Oracle. Our construction outputs $\mathsf{sk} = H(\mathbf{u})$ as the session key rather than $\mathbf{u} = \mathbf{R} \cdot \mathsf{msk}$. The use of the random oracle H is only for simplifying and unifying the security analysis. In particular, in case the preliminary session key $\mathbf{u}$ is used directly in an accompanied block cipher – as is the typical application of a re-keying scheme – then the additional hash function execution is not needed. We notice that in such a case the analysis would (at least) require that the block-cipher is secure against related key attacks. One way to enforce this (and obtain a security proof) is to model the block cipher as an ideal cipher in the analysis. Finally, we want to mention that of course the adversary can learn noisy probes of the preliminary session key $\mathbf{u}$ – however, in any applications of the re-keying function one has to make sure that these values are never seen directly by the adversary (i.e., without the noise).

Allowed Probes. In the t-(noisy) probing model, we allow the adversary to select probes from the following intermediate values: individual bits of msk_i or msk'_i, internal values of the computation of the $\mathbf{u}_i = \mathbf{R}\mathsf{msk}_i$, internal wires of Refresh or H (unless in the ROM) and individual bits of any $\sum_{i=1}^{k} \mathbf{u}_i$ (as specified above). The adversary is not allowed to obtain multiple noisy probes from the same value during a single session. This assumption is the same as made in all works on the noisy leakage model [21,26,53]. Finally, we assume that the adversary never probes $\mathbf{R}$ and sk as these values he obtains for free anyway.

4.1   Security of Our Construction Based on Physical Noise

In this section, we prove the security of our construction under the LPL (Learning Parity with Leakage) assumption. To this end, we will first formally define our new assumption, and show that it can be reduced to the classical LPN assumption. We then generalize LPL to an assumption that also models leakage from intermediate values from the computation of the session key and show that this change in the assumption does not affect the reduction by much. The LPL assumption with noisy probes then allow us to prove the security of our construction in the above specified model. So to summarize we show:

$$\mathsf{LPN} \text{ is hard} \implies \mathsf{LPL} \text{ is hard} \implies \Pi_{\mathrm{noisy}} \text{ is secure.}$$

Learning Parity with Leakage (LPL). We now give a formal definition of the LPL assumption, in which we model the physical noise distribution with a continuous Gaussian distribution Φ_s with density function $\Phi_s(x) := \frac{1}{\sqrt{2\pi}s} \exp(-\frac{x^2}{2s^2})$ and standard deviation s. First, the $\mathsf{LPL}_{n,s}$ sample distribution for secret $\mathbf{k} \in \mathbb{Z}_2^n$ is defined as

$$\mathcal{D}_{\mathsf{LPL}} := (\mathbf{r}, \ell = \langle \mathbf{r}, \mathbf{k} \rangle + e) \text{ for } \mathbf{r} \leftarrow \mathbb{Z}_2^n, e \leftarrow \Phi_s,$$

where $\langle \mathbf{r}, \mathbf{k} \rangle \in \{0,1\}$ is computed over $\mathbb{Z}_2$ and $\langle \mathbf{r}, \mathbf{k} \rangle + e$ is taken over the reals. Similarly, we define a distribution $\mathcal{D}_{\mathsf{UniformL}}$ that outputs $(\mathbf{r}, \ell)$, where $\mathbf{r} \leftarrow \mathbb{Z}_2^n$ and $\ell = u + e \in \mathbb{R}$ with $u \leftarrow \{0,1\}$ is a uniform bit and $e \leftarrow \Phi_s$.

The $\mathsf{LPL}_{n,s}$-search problem asks to find the secret, uniform $\mathbf{k}$, given query access to $\mathcal{D}_{\mathsf{LPL}}$[6]. The decision problem asks to distinguish $\mathsf{LPL}_{n,s}$ from $\mathcal{D}_{\mathsf{UniformL}}$. The search/decision LPL-Assumption is the assumption that these problems are hard for PPT adversaries.

Security Proof for LPL. We now show that LPL is at least as hard as LPN for appropriate choices of parameters. For this, we first show that LPL is actually equivalent to a variant of LPN, where the error probability τ is per-sample random and known. Formally, for a (sampleable) distribution Ψ on $[0, \frac{1}{2}]$, the $\mathsf{LPN}_{n,\Psi}$ sample distribution for uniform secret $\mathbf{k}$ is given by

$$\mathcal{D}_{\mathsf{LPN},\Psi} := (\mathbf{r}, \langle \mathbf{r}, \mathbf{k} \rangle + e, \tau) \text{ for } \mathbf{r} \leftarrow \mathbb{Z}_2^n, \tau \leftarrow \Psi, e \leftarrow \mathcal{B}_\tau.$$

Given query access to the sample distributions for fixed, random $\mathbf{k}$, the search $\mathsf{LPN}_{n,\Psi}$-problem ask to find $\mathbf{k}$ and the decision $\mathsf{LPN}_{n,\Psi}$-problem asks to distinguish $\mathcal{D}_{\mathsf{LPN},\Psi}$ from $\mathcal{D}_{\mathsf{Uniform},\Psi}$, where $\mathcal{D}_{\mathsf{Uniform},\Psi}$ outputs samples $(\mathbf{r}, \ell, \tau)$ with $\tau \leftarrow \Psi$ and $(\mathbf{r}, \ell) \leftarrow \mathbb{Z}_2^{n+1}$.

Lemma 2. *The search resp. decision $\mathsf{LPL}_{n,s}$-problem is equivalent to the search resp. decision $\mathsf{LPN}_{n,\Psi}$-problem via a tight, sample-preserving reduction. Here, the distribution for Ψ is given by sampling $\widetilde{U} \leftarrow \{0, 1\}, \widetilde{E} \leftarrow \Phi_s, \widetilde{L} = U + E$, $\widetilde{R}_{>1} := \exp\left(\frac{|L - \frac{1}{2}|}{s^2}\right)$ and outputting $\widetilde{\tau} = \left(\widetilde{R}_{>1} + 1\right)^{-1}$.*

A full proof is given in the extended (ePrint) version of this work. Here, we only give an intuition and explain the distribution of Ψ.

The key idea is to set Ψ in such a way that the amount of information learned about $\langle \mathbf{r}, \mathbf{k} \rangle$ from a single LPN sample is the same as the amount of information learned about $\langle \mathbf{r}, \mathbf{k} \rangle$ from a single LPL sample. To this end, we consider a quantity called R_{Bayes}, defined below, that measures exactly the amount of information learned about $\langle \mathbf{r}, \mathbf{k} \rangle$. We then compute this value for both the LPN case and for the LPL case. In the LPN case, R_{Bayes} is a function of τ. In the LPL-case, R_{Bayes} is a function of ℓ_{LPL}, where $(\mathbf{r}, \ell_{\mathsf{LPL}})$ is the output from LPL. Equating the values of R_{Bayes} will give us the involved definition of Ψ given above.

In fact, the proof given in the extended version of this work uses this value R_{Bayes} to transform LPL samples into LPN samples and vice versa via the correspondence $\ell_{\mathsf{LPL}} \leftrightarrow R_{\mathrm{Bayes}} \leftrightarrow \tau$.

Intuitively, giving an LPN sample $(\mathbf{r}, \ell) = (\mathbf{r}, u + e)$ for $u := \langle \mathbf{r}, \mathbf{k} \rangle, e \leftarrow \mathcal{B}_\tau$ with $\mathbf{r} \neq \mathbf{0}$ is (information-theoretically) equivalent to giving out $(\mathbf{r}, P_0, P_1)$, where $P_i = \Pr_\Omega[u = i \mid \ell \text{ is observed}]$. Since $P_0 + P_1 = 1$, we consider the fraction $R = \frac{P_0}{P_1}$ instead, which uniquely determines P_0, P_1. The probability space Ω for the definition of P_i takes $u \in \mathbb{Z}_2$ uniform for simplicity. For a general "prior" distribution $\Pr[u = 0]$ of u, Bayes' rule gives

$$\frac{\Pr[u = 0 \mid \ell \text{ is observed}]}{\Pr[u = 1 \mid \ell \text{ is observed}]} = R_{\mathrm{Bayes}} \cdot \frac{\Pr[u = 0]}{\Pr[u = 1]} \tag{1}$$

---

[6] We assume that for any value in $\mathbb{R}$, the adversary A receives an arbitrarily precise but polynomial representation. In particular A chooses how values in $\mathbb{R}$ are represented.

for the random variable R_{Bayes}, defined as a function of ℓ via

$$R_{\text{Bayes}} = \frac{Q_0}{Q_1} \in [0, \infty], \quad Q_i = \Pr_{E \leftarrow \mathcal{B}_\tau, L = E + u}[L = \ell \mid u = i].$$

We have $R = R_{\text{Bayes}}$ and the definition of R_{Bayes} does not depend on how u is chosen and completely captures what can be learned (in addition to any prior knowledge) from a given LPN sample about u via Eq. (1). For LPN, we easily compute $R_{\text{Bayes}} = \frac{1-\tau}{\tau}$ for $\ell = 0$ and $R_{\text{Bayes}} = \frac{\tau}{1-\tau}$ for $\ell = 1$. In the case $\ell = 0$, this means that $\tau = (R_{\text{Bayes}} + 1)^{-1}$. For $\ell = 1$ we have $1 - \tau = (R_{\text{Bayes}} + 1)^{-1}$. Similarly, an individual LPL sample $(\mathbf{r}, \ell_{\text{LPL}}) = (\mathbf{r}, u + e_{\text{LPL}})$ for continuous error $e_{\text{LPL}} \leftarrow \Phi_s$ provides statistical information about u and we can define R_{Bayes} analogously. A simple computation yields $R_{\text{Bayes}} = \frac{\Phi_s(\ell_{\text{LPL}})}{\Phi_s(\ell_{\text{LPL}} - 1)} = \exp\left(-\frac{\ell_{\text{LPL}} - \frac{1}{2}}{s^2}\right)$. The distribution of Ψ is constructed such that R_{Bayes} follows the same distribution in both the $\text{LPN}_{n,\Psi}$ and the $\text{LPL}_{n,s}$ case. Indeed, in the definition of Ψ, we mimic the distribution of R_{Bayes} by sampling $\widetilde{U}, \widetilde{E}$ and defining $\widetilde{R}_{\text{Bayes}, > 1}$ in a similar way to R_{Bayes}. Taking the absolute value in $\widetilde{R}_{\text{Bayes}}$ corresponds to normalizing τ into $1 - \tau$, if τ would otherwise be larger than $\frac{1}{2}$. The latter can be done by replacing ℓ_{LPL} by $1 - \ell_{\text{LPL}}$ in LPN.

Note that this information-theoretic argument does not show how to efficiently transform $\text{LPN}_{n,\Psi}$-samples into $\text{LPL}_{n,s}$-samples and vice versa. This is done in the full reductionist proof in the extended version of this work.

Next, we reduce standard LPN with fixed noise rate τ' to $\text{LPN}_{n,\Psi}$ with varying noise rate $\tau \leftarrow \Psi$. Clearly, if $\tau \geq \tau'$, this is very easy by just adding additional noise. If $\tau < \tau'$, the reduction fails, but any single sample with small noise rate can reveal at most 1 bit of information about $\mathbf{k}$. Hence, we need to bound the number of such outliers. Consequently, we have the following theorem:

Theorem 1. *Consider $s > 0$ and $0 < \tau' < \frac{1}{2}$. Then, provided s is sufficiently large, the $\text{LPL}_{n,s}$ problem is at least as hard as the $\text{LPN}_{n,\tau'}$ problem.*
More precisely, if $\text{LPN}_{n-X,\tau'}$ is (t, ε, Q)-secure, then $\text{LPL}_{n,s}$ is (t', ε', Q') secure with $Q = Q' - X, t \approx t', \varepsilon \approx \varepsilon'^7$. Here, X is a random variable measuring the loss of dimension. X follows a Bernoulli distribution on Q tries with success probability p, where $p = \Pr_{\tau \leftarrow \Psi}[\tau < \tau']$ and where Ψ is defined as in Lemma 2.
Let $0 < \Delta n$ be a (small) real number measuring the acceptable loss of dimension. Then by setting s such that

$$s \ln\left(\frac{1 - \tau'}{\tau'}\right) > \text{Fc}^{-1}\left(\frac{\Delta n}{2Q}\right) + \frac{1}{2s}, \tag{2}$$

where $\text{Fc}(x) = 1 - \frac{1}{\sqrt{2\pi}} \int_{-\infty}^{x} \exp(-\frac{t^2}{2}) dt$ is the complementary cdf of the normal distribution, we can ensure $\mathbb{E}[X] = pQ < \Delta n$.

Proof. See the extended version of this work.

---

[7] Note that here the (known) dimension of the LPN secret is random as well. If $X > n$, the problem is considered trivial.

To make the above theorem useful, we ask for $\Delta n \leq 1$. In this case, X is approximately Poisson distributed with parameter Δn. By making Δn negligibly small, we can have $X = 0$ with overwhelming probability. For setting parameters, we consider $\Delta n = 1$ acceptable, which gives $s = \mathcal{O}\left(\sqrt{\log Q}/\log\left(\frac{1-\tau}{\tau}\right)\right)$.

Remark 1. We observe that in Lemma 2, we showed that $\mathsf{LPL}_{n,s}$-samples $(\mathbf{r}, \ell)$ provide the more information about $\langle \mathbf{r}, \mathbf{k} \rangle$, the further away ℓ is from $\frac{1}{2}$. This is due to the superexponential decay of the Gaussian error function, which means that very large values of $\ell > 1$ are extremely more likely to have come from $\langle \mathbf{r}, \mathbf{k} \rangle = 1$ than to have come from $\langle \mathbf{r}, \mathbf{k} \rangle = 0$. Since we see a usually very large number Q of samples, it is the most extreme outliers for ℓ that determine the noise level τ' (and hence security) of LPN via the correspondence of Theorem 1. In particular, we care about outliers that appear with probability Q^{-1}. Note that this means that our reduction is sensitive to the behavior of the distribution tail of the physical noise, for which the assumption that this is Gaussian is much harder to verify. For example, a faster asymptotic decay than quadratic-exponential would hurt our reduction in terms of parameters, while a slower decay would lead to better parameters. Ideally, one would want a single-exponential decay rate. Also, since an adversary might choose to ignore all samples except for the outliers, there are actually attacks corresponding to the parameter loss of our reduction, provided the attacks do not use many samples.

LPL with Leakage of Intermediate Values. To adequately model the fact that the adversary may probe intermediate values, we consider a variant $\mathsf{LPL}_{n,s,d}$ of the LPL-problem, which is tailored to our particular application. Note that this variant models a situation where the adversary is able to probe *all* $\langle \mathbf{r}, \mathsf{msk}_i \rangle$ and also *all* partial sums $\sum_{i=1}^{k} \langle \mathbf{r}, \mathsf{msk}_i \rangle$ for $k \geq 2$, without being restricted to t probes. We do not model probes on msk_i's, internal values of $\mathbf{R} \cdot \mathsf{msk}_i$ or internal wires of Refresh here. The latter will be justified in Lemma 4, which shows that these probes do not help the adversary much, provided their number is restricted (the restriction on the number of probes only appears there). We now define the $\mathsf{LPL}_{n,s,d}$ distribution:

For secret $\mathbf{k} \in \mathbb{Z}_2^n$, $d \geq 2$, the $\mathsf{LPL}_{n,s,d}$ sample distribution is given as follows:

1. First, sample $\mathbf{r} \in \mathbb{Z}_2^n$.
2. Set $u = \langle \mathbf{r}, \mathbf{k} \rangle \bmod 2$ and share u into d uniform values $u_i \in \mathbb{Z}_2$ conditioned on $u = \sum_{i=1}^{d} u_i \bmod 2$.
3. For any $2 \leq k \leq d$, we define u'_k as the partial sum $u'_k = \sum_{i=1}^{k} u_i$.
4. Sample independent noise e_k for $1 \leq k \leq d$ and e'_k for $2 \leq k \leq d$, where each e_k, e'_k independently follows Φ_s.
5. Output $\mathbf{r}$ and all $u'_k + e'_k$ and all $u_k + e_k$.

Given query access to $\mathsf{LPL}_{n,s,d}$ for unknown, uniform $\mathbf{k} \in \mathbb{Z}_2^n$, the corresponding $\mathsf{LPL}_{n,s,d}$ search problem is to find $\mathbf{k}$. The decision problem ask to distinguish $\mathsf{LPL}_{n,s,d}$ for secret, uniform $\mathbf{k}$ from a distribution where u is chosen uniformly in the second step above. The decision/search $\mathsf{LPL}_{n,s,d}$-Assumption asserts that these problems are intractable for PPT algorithms.

Note that in the definition above, we split u into shares rather than $\mathbf{k}$ as in our scheme Π_{noisy}. Due due to linearity, this is equivalent, provided $\mathbf{R} \leftarrow \mathcal{D}$ selected in Π_{noisy} is full-rank.

Lemma 3. *Let $d \geq 2$. Then the $\mathsf{LPL}_{n,\sqrt{2}s,d}$-search problem is at least as hard as the search-$\mathsf{LPL}_{n,s}$ problem. More precisely, if search-$\mathsf{LPL}_{n,s}$ is (t, ε)-hard with q samples, then $\mathsf{LPL}_{n,\sqrt{2}s,d}$ is (t', ε')-hard with q samples, where $t \approx t', \varepsilon \approx \varepsilon'$.*

Proof. Note that if the adversary knows all shares u_i for $1 \leq i < d$ (which contain no information about $\mathbf{k}$) except for the last, then the only useful data are $u_d + e_d$ and $u'_d + e'_d$. With the given data, u_d can be computed from u'_d and vice versa. Having two independent noisy samples for the same value u_d is equivalent to reducing the noise by a factor $\sqrt{2}$. See the extended version of this work for details.

Remark 2. The above shows that we only lose at worst a factor of $\sqrt{2}$ in the noise rate due to the probes for intermediate values. We remark that this is an upper bound on the parameter loss and is not matched by real attacks: the reduction assumes that the u_i for $1 \leq i < d$ are known in clear (without the noise), which in reality is not the case. In fact, the more precise parameter loss is determined by the tail distribution of the R_{Bayes} value for $\mathsf{LPL}_{n,s,d}$, defined as in the proof of Lemma 2. Unfortunately, this distribution is difficult to compute.

We now give an intuition why the parameter loss of $\sqrt{2}$ is an exaggeration. The security level of $\mathsf{LPL}_{n,s}$ and $\mathsf{LPL}_{n,s,d}$ is essentially determined by outliers in those leakages (cfr. Remark 1). If we assume in favor of the adversary that we know all u_i but u_d, u_{d-1}, then we know all intermediate values except for u'_{d-1} and $u = u'_d$. We are interested in what can be learned about the latter.

The leakages with noise rate s for u_{d-1} and u'_{d-1} are as good a single leakage for u'_{d-1} with noise rate $s/\sqrt{2}$ by an argument similar to Lemma 3. For the leakage of u_d, we get a noise rate of s, which gives us some information about $u'_d = u_d + u'_{d-1}$. However, by taking the sum, the amount of information (measured by some $R'_{d,\mathrm{Bayes}}$ defined as in the proof of Lemma 2) that we learn about u'_d from the set of all leakages excluding that of u'_d is limited. Indeed, it can be at most as large as what we can learn about (the worse of) u_d and u'_{d-1}.

What we learn about u'_d from all leakages is then determined by R'_{Bayes} and what we learn from the leakage of u'_d directly. Since it is extremely more unlikely that the leakage of u'_{d-1} and u_d are *both* outliers than that the single measurement for u'_d is an outlier, the tail distribution for the information learned is mostly determined by the leakage of u'_d alone. Consequently, we expect to lose almost no security at all by revealing intermediate values. For that reason, we do not include the $\sqrt{2}$-factor in our concrete parameters.

The above definition only models probes on the computation of $\mathbf{u} = \sum_i \mathbf{u}_i$ in our re-keying scheme. It does not include probes for bits of shares $\mathsf{msk}_i, \mathsf{msk}'_i$ of the master key, probes for internal values of $\mathbf{R}\mathsf{msk}_i$'s or probes for interval wires of Refresh. The following lemma shows that this is indeed adequate, as these additional possibilities do not help the adversary anyway.

Lemma 4. *Consider our re-keying scheme Π_{noisy} with parameters n, d, m. Assume $n - m > \lambda + d$, where λ is the security parameter. Let $2t < d$. Model H as a random oracle. Assume Π_{noisy} is secure in the continuous t-noisy probing model with Gaussian noise s, where the adversary is only allowed to probe bits of the inner products $u_i = \mathbf{R} \cdot \mathsf{msk}_i$ or of partial sums $u'_k = \sum_{i=1}^{k} u_i$ thereof for $k \geq 2$. Then Π_{noisy} is secure in the continuous t-noisy probing model with Gaussian noise $\sqrt{t+1}s$, but without the latter restriction, i.e. when we also allow probes on bits of master key shares msk_i, msk'_i, on bits of internal values of the computation of $u_i = \mathbf{R}\mathsf{msk}_i$ or probes on internal wires of Refresh.*

Proof. By assumption, we are given some simulator Sim that simulates answers to bits of $\mathbf{u}_i = \mathbf{R} \cdot \mathsf{msk}_i$ and to bits of partial sums $\sum_{i=1}^{k} \mathbf{u}_i$ thereof. We need to show that we can extend Sim to a simulator Sim' that also simulates probes on bits of msk_i's, msk'_i's and on internal wires of Refresh and the computations of $\mathbf{u}_i = \mathbf{R}\mathsf{msk}_i$.

Our simulator Sim' will use Sim for the probes to $\mathbf{u}_i$ or $\sum_{i=0}^{k} \mathbf{u}_i$. To simulate other probes, Sim' will fix appropriate bits of some msk_i's or msk'_j's and use these to simulate the missing queries (conditioned on $\mathbf{u}_i$ and $\sum_{i=0}^{k} \mathbf{u}_i$). Let $M_{i,j}$ resp. $M'_{i,j}$ be the jth bit of msk_i resp. msk'_i. For a query to the jth bit of msk_i or msk'_i, we fix the value of $M_{i,j}$ resp. $M'_{i,j}$. For t' probes on internal wires of Refresh, we can fix some $t' \times t'$ submatrix of both M and M' by the properties of Refresh guaranteed by Lemma 1, which allows us to perfectly simulate the desired probes. Further, some bits of $M_{i,j}$ might already be fixed from probes (on the then-called M') from the previous session. Since the number of probes per session is limited by t, all the fixed bits are contained in $2t \times 2t$ submatrices $M_{I,J}, M'_{I,J}$ for $I \subset \{1, \ldots, d\}, J \subset \{1, \ldots, n\}$. Since $|I| < d$, there exists a share on which nothing is fixed. Consequently, the real value of the bits that we fixed to uniform are uniform, and independent from msk. Since $n - m > \lambda + d$, we have that the rows of R, together with the unit vectors corresponding to J are linearly independent with overwhelming probability. Due to that, the fixed bits are independent from both msk and the $\mathbf{u}_i$'s. It follows that the simulation of the probes on M, M' and Refresh can be done perfectly, independent from Sim.

What remains is the probes on intermediate values of the computation of $\mathbf{u}_i = \mathbf{R}\mathsf{msk}_i$. Assume that the individual output bits are computed independently as scalar products $\langle \mathbf{R}_j, \mathsf{msk}_i \rangle$ where $\mathbf{R}_j$ is the j-th row of $\mathbf{R}$. Then for a natural implementation, intermediate values correspond to inner products $\langle \mathbf{w}, \mathsf{msk}_i \rangle$, where $\mathbf{w}$ is obtained from some row of $\mathbf{R}$ by zeroing bits. Now, we fix the inner products $\langle \mathbf{w}, \mathsf{msk}_i \rangle$ to a uniformly random value and use that to simulate the probes. (They are completely analogous to coordinates of msk_i.) The only thing that may go wrong is that some $\mathbf{w}$ are linearly dependent to previously fixed coordinates of msk_i, where individually fixed bits of $M_{i,j}$ correspond to a unit vector for $\mathbf{w}$. In this case, we need to set the $\langle \mathbf{w}, \mathsf{msk}_i \rangle$ according to the linear combination. If this linear combination involves $\mathbf{w}$, we need to use $\mathbf{u}_i$, which we do not know. Essentially, in this situation, the adversary is probing the same unknown value c times with independent Gaussian noise, which is equivalent to probing once with a noise width reduced by a factor $\sqrt{c}$.

Formally, we show in the extended version of this work that we can still simulate the probe responses.   □

Theorem 2. *Consider our re-keying scheme Π_{noisy} with parameters n, d, m. Assume $n - m - d > \lambda, 2t < d$ and $\frac{n}{m} = \Theta(1)$. Model H as a random oracle. Then Π_{noisy} is secure in the continuous t-noisy probing model with Gaussian noise $\Phi_{\sqrt{t+1}s}$ under the Search-$\mathsf{LPL}_{n,s,d}$-Assumption.*

Clearly, using Lemma 3 together with Theorems 1 and 2 proves our re-keying scheme Π_{noisy} secure under LPN.

Proof. See the extended version of this work.

4.2   Concrete Parameters

In the previous section we proved our proposed scheme Π_{noisy} secure under the LPN assumption. We target our physical noise s such that the $\mathsf{LPL}_{n,s}$-assumption holds. Note that, as we argued in Remark 1, we expect the reduction in Theorem 1 relating $\mathsf{LPL}_{n,s}$ and $\mathsf{LPN}_{n,\tau'}$ to be matched by actual attacks, hence we really need these parameters. By contrast, the loss in the reduction of Theorem 2 is due to technical reasons and we do not expect there to be matching attacks. We argued in Remark 2 why the $\sqrt{2}$ loss factor for intermediate values that we obtained in Lemma 3 is far from tight. For the $\sqrt{t+1}$-factor from Lemma 4, one can actually show that it is not there if one uses a binary tree to carry out the computation of the sum in Π_{noisy}. Intuitively, it is better for the adversary to probe values as late in the computation as possible, as the best the adversary can hope is to learn some $\langle \mathbf{r}, \mathsf{msk} \rangle$, which is computed at the end. The $\sqrt{t+1}$-factor came from probes on internal values at the start of the computation. Unfortunately, we cannot prove Lemma 3 with our methods for a binary tree. The argument from Remark 2 becomes more complicated, but essentially still holds, which is why we ignore that $\sqrt{t+1}$ factor as well.

Hence, we believe that setting parameters such that the $\mathsf{LPL}_{n,s}$-Problem becomes hard is sufficient for our scheme to be secure.

We follow the proposal of Bogos et al. [17] of parameter choices (n, τ') for $\epsilon := 2^{-80}$-hard LPN. We can then use the relationship between the number of samples Q, the Gaussian width s and the Bernoulli noise τ' to determine s via. Theorem 1. Concretely, for $Q = 2^{80}$ and $\Delta n = 1$, we have $\mathrm{Fc}^{-1}\left(\frac{\Delta n}{2Q}\right) \approx 10.2846$, which allows us to relate s and τ'.

Note that in the context of side-channel analysis, a further reduction of the data complexity parameter could be considered. A choice of $Q = 2^{40}$ would already imply the capture of $\approx 2^{40}$ leakage traces, which corresponds to weeks of measurements with current acquisition hardware [49]. Since $\mathrm{Fc}^{-1}\left(\frac{\Delta n}{2Q}\right) \approx 7.1436$, this reduces the require noise level by a factor of approximately $\frac{10.28}{7.14} \approx 1.43$. Altogether, this approach leads to the choice of parameters given in Table 1.

Note also that the 80-bit security level of Table 1 corresponds to security against side-channel attacks. Of course, it remains that if no leakage is provided to the adversary, then the security of the re-keying scheme directly relates to the key size of the underlying (tweakable) block cipher.

Table 1. Standard deviations required for LPL with 80-bit hardness based on LPN with parameter τ. We assume a bound Q on the number of LPL-samples the adversary may see. To obtain these parameters, we numerically solved a (slightly) better Equation given in the extended version of this work rather than Eq. (2) from Theorem 1 with $\Delta n = 1$.

Dimension (n)	1280	640	512	448	384	256
LPN noise (τ')	0.05	0.125	0.25	0.325	0.4	0.45
LPL noise (s) for $Q = 2^{80}$	≈ 3.52	≈ 5.31	≈ 9.37	≈ 14.1	≈ 25.4	≈ 51.3
LPL noise (s) for $Q = 2^{40}$	≈ 2.46	≈ 3.70	≈ 6.52	≈ 9.79	≈ 17.6	≈ 35.6

5   Fresh Re-keying Without Physical Noise

For settings where no physical noise is given, or when it is not sufficient to achieve the desired security level, we now give an alternative solution for a fresh re-keying scheme, based on a variant of Learning with Rounding (LWR) assumption that we call Offset Learning with Rounding (OLWR). We will show in Theorem 3 below that for an unbounded amount of samples (OLWR) is at least as hard as Learning with uniform Errors (LWU). Before providing our OLWR-based re-keying scheme Π_{LWR}, we recall our rounding function and the OLWR assumption. For appropriately chosen integers $p < q$, the rounding function $\lfloor \cdot \rceil_p : \mathbb{Z}_q \to \mathbb{Z}_p$ is $\lfloor x \rceil_p := \lfloor x \frac{p}{q} \rceil$, where $x \in \mathbb{Z}_q$ is represented as $x \in \{0, \ldots, q-1\}$. When applying $\lfloor \cdot \rceil_p$ to a vector in $\mathbb{Z}_q^n$, we apply it component-wise. OLWR samples for dimension n and secret $\mathbf{k} \leftarrow \mathbb{Z}_q^n$ and an adversarialy chosen offset $\mathbf{o} \in \mathbb{Z}_q^n$, which is freshly (and adaptively) chosen for each sample, follow the distribution

$$\mathcal{D}_{\text{OLWR},n,q,p}(\mathbf{o}) := (\mathbf{r}, \lfloor \langle \mathbf{r}, \mathbf{k} + \mathbf{o} \rangle \rceil_p \mid \mathbf{r} \leftarrow \mathbb{Z}_q^n).$$

As usual, given query access to $\mathcal{D}_{\text{OLWR},n,q,p}$ for uniform $\mathbf{k}$, the search $\text{OLWR}_{n,q,p}$ problem asks to find $\mathbf{k}$. The search problem asks to distinguish this distribution from the uniform distribution $\mathcal{D}_{\text{Uniform}}$. Note that the uniform distribution does not depend on the input $\mathbf{o}$. The search/decision $\text{OLWR}_{n,q,p}$ Assumption asserts that this is infeasible for PPT algorithms.

We can define similar offset variants OLWU, OLWE etc. of LWU, LWE etc. where the adversary is allowed to add an (adaptively chosen) offset to $\mathbf{k}$.

For Learning with Errors (LWE) or Learning with Uniform Noise (LWU) it is easy to see that their offset variants are still as hard as LWE, LWU respectively. An adversary can simply compute samples with an arbitrary offset itself using the linearity of LWE and LWU. Since LWR is not linear this does not work for LWR. The reduction given in [7] from LWE to LWR also works from offset LWE to OLWR. Unfortunately it does not work for the parameters proposed in this work. Assuming the hardness of LWU for an unbounded amount of samples, OLWR will be also hard, as we show in the following theorem. A similar statement for their non-offset variants using similar techniques was shown by Bogdanov et al. [16].

Theorem 3 (Relationship Between LWU, OLWU *and* OLWR).

(a) For both the search and decision variants, $\mathsf{LWU}_{n,q,B}$ and $\mathsf{OLWU}_{n,q,B}$ are equivalent via a tight, sample-preserving reduction.

(b) Assume $p \mid q$. Then hardness of search resp. decision $\mathsf{LWU}_{n,q,B}$ implies hardness of search resp. decision $\mathsf{OLWR}_{n,q,p}$, where $B = \frac{q}{p}$.
More precisely, if we can solve the search resp. decision $\mathsf{OLWR}_{n,q,p}$-problem with advantage ε in time T, using Q samples, then we can solve the search resp. decision $\mathsf{LWU}_{n,q,B}$ problem with advantage $\varepsilon' = \varepsilon$ in expected time $T + \mathcal{O}(QB)$ using an expected $Q' = QB$ samples.

Proof. (a) follows immediately by linearity, as the offset $\mathbf{o}$ just adds $\langle \mathbf{R}, \mathbf{o} \rangle$, which is known. For (b), we show that we can transform $\mathsf{OLWU}_{n,q,B}$-samples for secret $\mathbf{k}$ into $\mathsf{OLWR}_{n,q,p}$ samples for the same unknown secret $\mathbf{k}$. By (a), this implies the claim.

To this end, suppose our reduction has to produce a (simulated) $\mathsf{OLWR}_{n,q,p}$ sample with offset $\mathbf{o}$. Then we repeatedly query $(\mathbf{r}, \ell_{\mathsf{OLWU}}) \leftarrow \mathsf{OLWU}_{n,q,B}(\mathbf{o})$ until $\ell_{\mathsf{OLWU}} + 1$ is divisble by B. We then output $(\mathbf{r}, \lfloor \ell_{\mathsf{OLWU}} \rfloor_p)$.

To analyse the output distribution, recall that $\ell_{\mathsf{OLWU}} = \langle \mathbf{r}, \mathbf{k} + \mathbf{o} \rangle + e$ for $0 \le e < B$. It follows that $\langle \mathbf{r}, \mathbf{k} + \mathbf{o} \rangle \in \{ \ell_{\mathsf{OLWU}}, \ell_{\mathsf{OLWU}} - 1, \ldots, \ell_{\mathsf{OLWU}} - B + 1 \}$. The condition that $\ell_{\mathsf{OLWU}} + 1$ is divisble by B is equivalent that *all* possible values for $\langle \mathbf{r}, \mathbf{k} + \mathbf{o} \rangle$ map to the same value under $\lfloor . \rfloor_p$.

Furthermore, note that the probability to reject an $\mathsf{OLWU}_{n,q,B}$ sample is always $\frac{1}{B}$, independent from $\mathbf{r}, \mathbf{k}, \mathbf{o}$, as it can be viewed as a condition on $e \leftarrow \{0, \ldots, B - 1\}$ alone. It follows that we output the correct distribution. $\square$

Unfortunately Theorem 3 does not work the ring version Ring-OLWR of OLWR. In this paper we choose parameter for Ring-OLWR such that ring LWR and ring LWU would be hard, even though there is no reduction to Ring-OLWR known for a polynomial modulus and their decisional variant which we will use.

5.1   Offset LWR-based Re-Keying

Our proposed re-keying scheme Π_{LWR} based on $\mathsf{OLWR}_{n,q,p}$ is defined as follows: The session keys are in $\mathbb{Z}_{p'}^m$ for $p' \mid p$ and the distribution $\mathcal{D}$ for $\mathbf{R}$ is the uniform distribution over $\mathbb{Z}_q^{m \times n}$.

$Gen(1^\lambda)$: Samples $\mathsf{msk} \leftarrow \mathbb{Z}_q^n$. Create d shares $(\mathsf{msk})_d := \mathsf{msk}_1, \ldots, \mathsf{msk}_d$ such that $\mathsf{msk} = \sum_{i=1}^d \mathsf{msk}_i$, uniformly among all possibilities. Output msk and $(\mathsf{msk})_d$.

$GenSK((\mathsf{msk})_d, \mathbf{R})$: For each of the shares msk_i, compute $\mathsf{sk}_i := \lfloor \mathbf{R} \cdot \mathsf{msk}_i \rfloor_p$. Then set $\mathsf{sk} := \lfloor \sum_{i=1}^d \mathsf{sk}_i \rfloor_{p'}$. Define the error correction information as $v := \sum_{i=1}^d \mathsf{sk}_i \bmod p/p'$. Finally, use a secure refresh operation $\mathsf{Refresh}$ as in Lemma 1 to refresh the shares $(\mathsf{msk}')_d \leftarrow \mathsf{Refresh}((\mathsf{msk})_d)$. Output sk, $(\mathsf{msk}')_d$ and v.

$CorSK(\mathsf{msk}, \mathbf{R}, v)$: Computes $y := \lfloor \mathbf{R} \cdot \mathsf{msk} \rfloor_p$ and $z := y + (v - y \bmod p/p') \bmod$
p. Output $\lfloor z \rfloor_{p'}$.

Theorem 4. *Let* $n, q, p, p', d \in \mathbb{N}$ *such that* $q > p > p'$, $p/p' > d$ *and* $p' \mid p$, $p \mid q$. *Then the* $\mathsf{OLWR}_{n,q,p}$*-based re-keying scheme is correct.*

Proof. First notice for $p/p' > d$ that the error term

$$e := \sum_{i=1}^{t} \lfloor \mathbf{R} \cdot \mathsf{msk}_i \rfloor_p - \lfloor \mathbf{R} \cdot \mathsf{msk} \rfloor_p$$

is bounded in each component by: $0 \leq |e| \leq p/p' - 1$. This follows directly from the fact that a value strictly smaller than 1 is rounded away per round operation, and $\mathsf{msk} = \sum_i \mathsf{msk}_i$, which implies $\mathbf{R} \cdot \mathsf{msk} = \sum_i \mathbf{R} \cdot \mathsf{msk}_i$. Next, $p' \mid p$ guarantees that any potential error component corresponds to a uniquely determined coset mod p/p'. Hence, we further have:

$$v - y \bmod p/p' := \left(\sum_i \lfloor \mathbf{R} \cdot \mathsf{msk}_i \rfloor_p \bmod p/p' \right) - \left(\lfloor \mathbf{R} \cdot \mathsf{msk} \rfloor_p \bmod p/p' \right) \bmod p/p'$$
$$= \left(\lfloor \langle \mathbf{R} \cdot \mathsf{msk} \rangle \rfloor_p + e \bmod p/p' \right) - \left(\lfloor \mathbf{R} \cdot \mathsf{msk} \rfloor_p \bmod p/p' \right) \bmod p/p'$$
$$= e.$$

Therefore, we have:

$$z' := y + (v - y \bmod p/p') \bmod p = \mathbf{R} \cdot \mathsf{msk} + e \bmod p$$
$$= \sum_i \lfloor \mathbf{R} \cdot \mathsf{msk}_i \rfloor_p \bmod p = \mathsf{sk}. \qquad \square$$

Theorem 5. *For moduli* $p, q \in \mathbb{N}$ *with* $p \mid q$ *and dimension* $n \in \mathbb{N}$, *the proposed re-keying scheme* Π_{LWR} *is secure under the* $\mathsf{OLWR}_{n,q,p}$ *assumption in the* $2t < d$ *probing model.*

Proof. We assume that $GenSK$ just outputs $\sum_{i=1}^{d} \lfloor \mathbf{R} \cdot \mathsf{msk}_i \rfloor_p$ which is equivalent to outputting $\mathsf{sk} = \left\lfloor \sum_{i=1}^{d} \lfloor \mathbf{R} \cdot \mathsf{msk}_i \rfloor_p \right\rfloor_{p'}$ and $v = \sum_{i=1}^{d} \lfloor \mathbf{R} \cdot \mathsf{msk}_i \rfloor_p$ mod p/p'. Therefore *Ideal* will simply output $\mathbf{R}, \mathsf{sk}'$ where $\mathsf{sk}' \leftarrow \mathbb{Z}_p^m$. *Sim* simulates t probes as follows:

– Let msk_i denote the shares at the beginning of the session for $i \in [d]$. *Sim* receives t probe requests and forwards the probes which affect the refresh procedure to the simulator for the refresh procedure. This simulator will w.l.o.g. respond with the $d - 1$ input shares msk_i of the refresh procedure and the probes which were targeted within the refreshing procedure. The output shares of the refresh procedure are not accessed by *Sim* during this session, but the next session after the probes were made.

– *Sim* computes for the $d-1$ shares $\mathsf{sk}_i := \lfloor \mathbf{R} \cdot \mathsf{msk}_i \rfloor_p$. Let msk_j be the share that is missing and that has not been directly targeted by any probe request. *Sim* defines $\mathsf{sk}_j = \mathsf{sk}' - \sum_{i \neq j} \mathsf{sk}_i$. Since all sk_i are known, *Sim* can answer any probe request on any the intermediate values which arise when computing $\sum_{i=1}^{d} \lfloor \mathbf{R} \cdot \mathsf{msk}_i \rfloor_p$.

Now we show the following statement using a reduction to OLWR: for the given simulator *Sim*, for any PPT A with

$$ | \Pr[A^{Real(\mathsf{msk})}(1^\lambda) = 1] - \Pr[A^{(Sim^{Ideal_c}, Ideal_c)}(1^\lambda) = 1] | = \varepsilon, $$

there is an algorithm D distinguishing OLWR with probability ε from uniform.

During each session, A requests t probes to D. It calls the simulator of the refresh scheme and receives $d-1$ shares msk_i, but not msk_j as it has been the case for *Sim*. Note that A has already sent all probe requests for the shares msk_i at the beginning of this and the previous session. Therefore D can identify a index j of a share msk_j which will never be requested by A. D requests a sample $(\mathbf{R}, \ell)$ for offset $-\sum_{i \neq j} \mathsf{msk}_i$ and has to decide in the end whether all of the samples $(\mathbf{R}, \ell)$ that D collects over the sessions are OLWR or uniformly distributed. D defines $\mathsf{sk}_j = \ell$ and computes $\mathsf{sk}_i = \lfloor \mathbf{R} \cdot \mathsf{msk}_i \rfloor_p$ for all $i \neq j$. Now D can similar to *Sim* respond to all the probe requests using sk_i. D computes the session key sk and v simply by computing $\sum_{i=1}^{d} \mathsf{sk}_i$. Afterwards he outputs $\left(\mathbf{R}, \sum_{i=1}^{d} \mathsf{sk}_i\right)$ to finish the current session. After finishing all sessions D outputs the output bit of A.

Let us assume that the samples $\mathbf{R}, \ell$ are OLWR distributed. The offset is $-\sum_{i \neq j} \mathsf{msk}_i$ such that $\ell := \lfloor \mathbf{R}(\mathsf{msk} - \sum_{i \neq j} \mathsf{msk}_i) \rfloor_p = \lfloor \mathbf{R} \cdot \mathsf{msk}_j \rfloor_p$. Hence D successfully simulates *Real*.

If that $(\mathbf{R}, \ell)$ is uniform, then D simulates the output of *Ideal* by outputting $\left(\mathbf{R}, \ell + \sum_{i \neq j} \mathsf{sk}_i\right)$. This is clearly uniformly random for uniform ℓ. Further $\mathsf{sk}_j := \ell$ used by D is the same as *Sim* would have computed: $\mathsf{sk}_j = \mathsf{sk}' - \sum_{i \neq j} \mathsf{sk}_i = \ell + \sum_{i \neq j} \mathsf{sk}_i - \sum_{i \neq j} \mathsf{sk}_i = \ell$. All other sk_i are derived from the $d-1$ outputs of the simulator of the key refreshing and hence have the same distribution. Therefore D simulates *Ideal* and *Sim* perfectly.   □

We emphasize that the presented results directly translate to the ring setting with Ring-OLWR. The reason for this is that the error correction of Π_{LWR}, the rounding $\lfloor \cdot \rfloor_p$, and the addition in the ring are carried out component-wise which is sufficient for both correctness and security.

5.2   Concrete Parameters

Historically, LWR has always been understood as a deterministic variant of LWE where the noise is rounded away applying the rounding function $\lfloor \cdot \rfloor_p$ to a LWE sample. This technique was used by Banerjee, Peikert and Rosen to reduce LWE to LWR [7]. Unfortunately their reduction only holds for a super-polynomial modulus q. This was improved by Alwen et al. [3] by using lossy pseudorandom samples. They achieve a reduction for a polynomial modulus q but with the drawback of a polynomially bounded amount of samples.

Because of these issues, the previous PRF construction by Banerjee et al. (called SPRING) [8] is based on parameters which are not obtained by choosing parameters for a hard instance of LWE and using one of the known reductions to LWR to get appropriate parameters for LWR. Interestingly, and despite their choice of parameters is not based on LWE, it seems that the best way to solve LWR for such a choice of parameters is still to exploit algorithms designed to solve LWE. We follow a similar approach, but using a different choice of parameters. Banerjee et al. propose two parameter choices for Ring-LWR, namely for $n = 128$ and $p = 2$ they choose either $q = 257$ or $q = 514$. For our application $p = 2$ is not sufficient, since we need $p/p' = \lceil \log(d) \rceil$ to correct errors in our rekeying scheme Π_{LWR} and $p > p'$. This means that for masking with up to 4 shares, we need at least $p \geq 4$.

Table 2 shows our choices of parameters for LWR, LWU and Ring-LWR, Ring-LWU which we will also use to instantiate OLWR and Ring-OLWR. The absolute noise level $\log q - \log p$ is large compared with modulus q. Furthermore, our secret msk is picked uniformly from $\mathbb{Z}_q^n$. Together, this will rule out the BKW algorithm and its variants [14,39]. The same holds for the algorithm of Arora and Ge [5]. Besides, for comparably small dimensions n, good lattice reductions exist, but since the noise is large, shortest vector sieving requires to find a vector of very small norm which seems to be hard. The size of n is compensated by the large noise to modulus ratio. Such a LWR or OLWR instance can also be seen as a hidden number problem.

Table 2. 128-bit security parameters for our re-keying scheme Π_{LWR}. For the ring version we use a irreducible polynomial $f \in \mathbb{Z}_q[X]/f$ with $\deg = n$. $\log q - \log p$ corresponds to the absolute noise level. The uniform noise for LWU is bounded by q/p. All our moduli are powers of two, which will guarantee that the output of the rounding function $\lfloor \cdot \rceil_p$ is uniform in $\mathbb{Z}_p$ for a uniform input in $\mathbb{Z}_q$.

Assumption	Dimension n	Modulus $\log q$	Modulus $\log p$
LWR,LWU	128	16	4
LWR,LWU	128	32	10
Ring-LWR, Ring-LWU	128	16	2
Ring-LWR, Ring-LWU	128	32	3

For other concrete parameters, Albrecht, Player and Scott survey how algorithms for solving LWE perform and give estimated running times [2]. These estimates on the running times affirm our choice of parameters. In the LWR setting, the LWE standard deviation corresponds roughly to $\frac{q}{p}$. As for Ring-LWR, we choose the parameters more conservatively, since for our ring over $\mathbb{Z}_{2^{16}}$ or $\mathbb{Z}_{2^{32}}$, it is likely that the ring product $\mathbf{R}$ msk of secret msk and public sample vector

$\mathbf{R}$ lies within an ideal of the ring. This might leak some information about the noise vector and the product might not depend on all the bits of msk.

Note finally that we chose a 128-bit security level which seem most relevant for general purpose applications. Reducing the security level to 80 could be acceptable for low-cost applications of Π_{LWR}. A simple (and conservative) solution for this purpose would be to reduce n from 128 to 80. By contrast, reducing the security parameters further (e.g., down to 40 bits) is not possible as in Sect. 4.2. Indeed, there is no guarantee that the attacks against our construction would require large data complexity (which is the only quantity that can be reasonably reduced in the context of side-channel attacks).

6   Implementation Results

In order to confirm the efficiency of our constructions, we implemented them on a 32-bit ARM7 device and on an 8-bit Atmel AVR device, for the standard parameters that we would select for concrete applications. For the LPL-based re-keying, we choose $n = 512$ and for the wPRF-based re-keying we choose $n = 128$, $q = 2^{32}$, $p = 2^{10}$ and p' ranging from 4 to 16 depending on the number of shares considered for masking. We then compared our implementation results with the ones obtained for the AES in [33], where higher-order masked implementations are evaluated on an Atmel AVR device. For illustration, we further extrapolated the cycle counts of the masked AES on the ARM7 device as 4 times lower than for the Atmel ones (since moving from an 8-bit to a 32-bit architecture). Note that in all cases, we refreshed the master key with a simple (linear) refreshing algorithm based on the addition of a vector of shares summing to zero, and we assumed a cost of 10 clock cycles to generate each byte of fresh randomness. This is consistent with the approach used in earlier re-keying papers [30,46]. More generally, and as for our selection of security parameters, such implementations correspond to the best tradeoff between security against state-of-the-art attacks and efficiency, that we suggest for further concrete investigation.

These performance evaluations are reported in Fig. 3 from which we can extract a number of interesting observations. First, and as expected, the cycle counts of our new constructions scale linearly in the number of shares, with small discontinuities for the wPRF-based re-keying (corresponding to the addition of one bit for error correction, each time $\lceil \log(d) \rceil$ increases). Second, the performances of the LPL-based re-keying and wPRF-based re-keying are similar on a 32-bit ARM device (where the multiplication is easy both in $\mathbb{Z}_2$ and $\mathbb{Z}_q$). They more significantly differ in the Atmel AVR case, because inner product operations in $\mathbb{Z}_2$ only involve simple (AND and XOR) operations, and directly lead to efficient implementations on this platform. By contrast, the wPRF-based re-keying on the Atmel device implies additional overheads for the 32-bit multiplication based on 8-bit operations (which takes approximately 40 clock cycles). Third, comparisons with the AES shows that (as expected as well), the interest of our new constructions increases with the number of shares (hence security level). In this respect, it is important to note that a masked software implementations of the AES protected with Boolean masking will generally suffer from independence issues, e.g. the recombination of the shares due to transition-based leakages [22]. Since our (almost) key homomorphic constructions do not suffer from this risk (because we can manipulate the shares independently), comparisons with this curve are overly conservative. As a quite optimistic comparison point, we can observe the cost of the "glitch-free" software implementations proposed in [55] which is already higher than the one of our primitives for 2 shares[8]. Alternatively, one can also use the simple reduction in [6] and double the number of shares of the masking scheme to obtain similar security, which also makes our masked re-keying schemes more efficient than the corresponding masked AES with 3 shares. Quite naturally, these comparisons are only informal. Yet, they illustrate the good implementation properties of fresh-rekeying. In this respect, the simplicity of the implementations, and limited constraints regarding the need of independent leakages, are certainly two important advantages.

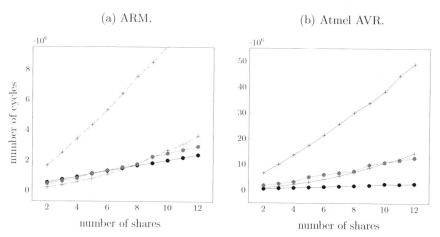

(a) ARM.   (b) Atmel AVR.

Fig. 3. Performance comparisons on ARM and Atmel AVR devices assuming 10 clock cycles per random byte. LPL-based re-keying ●, wPRF-based re-keying: ●, AES Boolean masking: −+−, AES glitch-free masking: −+−. Dashed ARM curves are extrapolated by scaling down the corresponding Atmel AVR performances by 4. (Color figure online)

Finally, a key difference between the LPL- and wPRF-based re-keying is that the latter one offers significantly stronger guarantees (since it is secure even in front of noise-free leakages), which explains its lower performances. By contrast, the LPL-based re-keying implementations have to include noise in the adversary's measurements. Hence, we end this section with a brief discussion of the noise levels required for secure LPL implementations, and how to generate them. For this purpose, a conservative estimate is to assume that this noise will be generated thanks to additive algorithmic noise. Typically, this could imply implementing a parasitic Linear Feedback Shift Register (LFSR) in parallel to

---

[8] Note that this curve is linear which corresponds to the amortized complexity of the best "packed secret sharing" at each order.

the inner product computations to which we have to add noise. Since the noise variance corresponding to 1 bit equals 0.25, we typically need an LFSR of size $N = \lceil 4 \times \sigma^2 \rceil$ to reach our estimated security levels[9]. For illustration, some numbers are given in Table 3, where we can see a tradeoff between the cost of computing the inner products and the cost of generating the noise. So already for these estimates, we see that the $n = 640$ and $n = 512$ instances should allow efficient implementations. Yet and importantly, in case more efficient noise engines are embedded on chip (based on supply noise, clock jitter, shuffling, ...), these figures can only become more positive for our re-keying, and the same holds if some parallelism is considered for the inner product computations (in which case the cost of noise generation will be amortized).

Table 3. Concrete parameters for noise generation.

Dimension (n)	1280	640	512	448	384	256
Bits of additive noise ($Q = 2^{80}$)	49	112	361	807	2601	10744
Bits of additive noise ($Q = 2^{40}$)	24	55	177	384	1272	5269

So overall, LPL-based re-keying is conceptually interesting since it leverages the intrinsic noise that is anyway present in side-channel measurements. But the noise levels that we require to reach high security levels are admittedly larger than these intrinsic noise levels. So it leads to interesting design challenges regarding the tradeoff between the cost of noise generation vs. the cost of inner product computations. By contrast, wPRF-based re-keying is more conservative, since based on a stronger cryptographic primitive that requires less assumptions, at the cost of reasonable performance overheads. The combination of these solutions therefore brings an interesting toolbox to cryptographic engineers, for secure and efficient cryptographic implementations in software and hardware.

Acknowledgements. Stefan Dziembowski is supported by the Foundation for Polish Science WELCOME/2010-4/2 grant founded within the framework of the EU Innovative Economy Operational Programme. Sebastian Faust is funded by the Emmy Noether Program FA 1320/1-1 of the German Research Foundation (DFG). Gottfried Herold is funded by the ERC grant 307952 (acronym FSC). Anthony Journault is funded by the INNOVIRIS project SCAUT. Daniel Masny is supported by the DFG Research Training Group GRK 1817/1. François-Xavier Standaert is a research associate of the Belgian Fund for Scientific Research (FNRS-F.R.S.). His work was funded in parts by the ERC project 280141 (acronym CRASH) and the ARC project NANOSEC.

---

[9] This assumes that the computation of every bit requires a similar amount of energy, which is usually observed in practice [56], and certainly holds to a good extent when considering blocks of bits as we do. The proposed values should anyway only be taken as an indication that generating the required amount of noise is feasible with existing hardware. Besides, note that for such an "algorithmic" noise generated by LFSR, the Gaussian distribution is ensured by design which avoids any risk related to faster decreasing tails.

References

1. Abdalla, M., Belaïd, S., Fouque, P.-A.: Leakage-resilient symmetric encryption via re-keying. In: Bertoni, G., Coron, J.-S. (eds.) CHES 2013. LNCS, vol. 8086, pp. 471–488. Springer, Heidelberg (2013)

2. Albrecht, M.R., Player, R., Scott, S.: On the concrete hardness of learning with errors. J. Math. Cryptol. $9(3)$, 169–203 (2015)

3. Alwen, J., Krenn, S., Pietrzak, K., Wichs, D.: Learning with rounding, revisited - new reduction, properties and applications. In: Canetti, R., Garay, J.A. (eds.) CRYPTO 2013, Part I. LNCS, vol. 8042, pp. 57–74. Springer, Heidelberg (2013)

4. Andrychowicz, M., Dziembowski, S., Faust, S.: Circuit compilers with $O(1 = \log(n))$ leakage rate. In: EUROCRYPT (2016)

5. Arora, S., Ge, R.: New algorithms for learning in presence of errors. In: ICALP (2011)

6. Balasch, J., Gierlichs, B., Grosso, V., Reparaz, O., Standaert, F.-X.: On the cost of lazy engineering for masked software implementations. In: Joye, M., Moradi, A. (eds.) CARDIS 2014. LNCS, vol. 8968, pp. 64–81. Springer, Heidelberg (2015)

7. Banerjee, A., Peikert, C., Rosen, A.: Pseudorandom functions, lattices. In: EUROCRYPT (2012)

8. Banerjee, A., Brenner, H., Leurent, G., Peikert, C., Rosen, A.: SPRING: fast pseudorandom functions from rounded ring products. In: Cid, C., Rechberger, C. (eds.) FSE 2014. LNCS, vol. 8540, pp. 38–57. Springer, Heidelberg (2015)

9. Belaïd, S., Fouque, P., Gérard, B.: Side-channel analysis of multiplications in GF(2128) - application to AES-GCM. In: ASIACRYPT (2014)

10. Belaïd, S., Grosso, V., Standaert, F.: Masking and leakage-resilientprimitives: one, the other(s) or both? Crypt. Commun. $7(1)$, 163–184 (2015)

11. Belaïd, S., Coron, J.-S., Fouque, P.-A., Gérard, B., Kammerer, J.-G., Prouff, E.: Improved side-channel analysis of finite-field multiplication. In: Güneysu, T., Handschuh, H. (eds.) CHES 2015. LNCS, vol. 9293, pp. 395–415. Springer, Heidelberg (2015)

12. Bilgin, B., Gierlichs, B., Nikova, S., Nikov, V., Rijmen, V.: A more efficient AES threshold implementation. In: AFRICACRYPT (2014)

13. Bilgin, B., Gierlichs, B., Nikova, S., Nikov, V., Rijmen, V.: Higher-order threshold implementations. In: Sarkar, P., Iwata, T. (eds.) ASIACRYPT 2014, Part II. LNCS, vol. 8874, pp. 326–343. Springer, Heidelberg (2014)

14. Blum, A., Kalai, A., Wasserman, H.: Noise-tolerant learning, the parity problem, and the statistical query model. In: ACM STOC (2000)

15. Blum, A., Furst, M.L., Kearns, M., Lipton, R.J.: Cryptographic primitives based on hard learning problems. In: Stinson, D.R. (ed.) CRYPTO 1993. LNCS, vol. 773, pp. 278–291. Springer, Heidelberg (1994)

16. Bogdanov, A., Guo, S., Masny, D., Richelson, S., Rosen, A.: On the hardness of learning with rounding over small modulus. In: Kushilevitz, E., Malkin, T. (eds.) TCC 2016-A. LNCS, vol. 9562, pp. 209–224. Springer, Heidelberg (2016). doi:10. 1007/978-3-662-49096-9_9

17. Bogos, S., Tramér, F., Vaudenay, S.: On solving LPN using BKW and variants. In: IACR Cryptology ePrint Archive (2015)

18. Boneh, D., Lewi, K., Montgomery, H.W., Raghunathan, A.: Key homomorphic PRFs, their applications. In: CRYPTO (2013)

19. Brenner, H., Gaspar, L., Leurent, G., Rosen, A., Standaert, F.-X.: FPGA implementations of SPRING. In: Batina, L., Robshaw, M. (eds.) CHES 2014. LNCS, vol. 8731, pp. 414–432. Springer, Heidelberg (2014)

20. Chari, S., Rao, J.R., Rohatgi, P.: Template attacks. In: CHES (2002)
21. Chari, S., Jutla, C.S., Rao, J.R., Rohatgi, P.: Towards sound approaches to counteract power-analysis attacks. In: CRYPTO (1999)
22. Coron, J.-S., Giraud, C., Prouff, E., Renner, S., Rivain, M., Vadnala, P.K.: Conversion of security proofs from one leakage model to another: a new issue. In: Schindler, W., Huss, S.A. (eds.) COSADE 2012. LNCS, vol. 7275, pp. 69–81. Springer, Heidelberg (2012)
23. Dobraunig, C., Koeune, F., Mangard, S., Mendel, F., Standaert, F.: Towards fresh, hybrid re-keying schemes with beyond birthday security. In: CARDIS (2015)
24. Dodis, Y., Pietrzak, K.: Leakage-resilient pseudorandom functions and side-channel attacks on feistel networks. In: CRYPTO (2010)
25. Döttling, N., Müller-Quade, J.: Lossy codes, a new variant of the learning-with-errors problem. In: EUROCRYPT (2013)
26. Duc, A., Dziembowski, S., Faust, S.: Unifying leakage models: from probing attacks to noisy leakage. In: Nguyen, P.Q., Oswald, E. (eds.) EUROCRYPT 2014. LNCS, vol. 8441, pp. 423–440. Springer, Heidelberg (2014)
27. Duc, A., Faust, S., Standaert, F.-X.: Making masking security proofs concrete. In: Oswald, E., Fischlin, M. (eds.) EUROCRYPT 2015. LNCS, vol. 9056, pp. 401–429. Springer, Heidelberg (2015)
28. Durvaux, F., Standaert, F.-X., Veyrat-Charvillon, N.: How to certify the leakage of a chip? In: Nguyen, P.Q., Oswald, E. (eds.) EUROCRYPT 2014. LNCS, vol. 8441, pp. 459–476. Springer, Heidelberg (2014)
29. Dziembowski, S., Pietrzak, K.: Leakage-resilient cryptography. In: IEEE FOCS (2008)
30. Gammel, B., Fischer, W., Mangard, S.: Generating a session key for authentication and secure data transfer. US Patent App. 14/074,279, November 2013
31. Gaspar, L., Leurent, G., Standaert, F.: Hardware implementation and side-channel analysis of Lapin. In: CT-RSA (2014)
32. Goldreich, O., Krawczyk, H., Luby, M.: On the existence of pseudorandom generators. SIAM J. Comput. **22**(6), 1163–1175 (1993)
33. Grosso, V., Standaert, F., Faust, S.: Masking vs. multiparty computation: how large is the gap for AES? J. Crypt. Eng. **4**(1), 47–57 (2014)
34. Güneysu, T., Moradi, A.: Generic side-channel countermeasures for reconfigurable devices. In: Preneel, B., Takagi, T. (eds.) CHES 2011. LNCS, vol. 6917, pp. 33–48. Springer, Heidelberg (2011)
35. Guo, Q., Johansson, T.: A new birthday-type algorithm for attacking the fresh re-keying countermeasure. Cryptology ePrint Archive, Report 2016/225 (2016)
36. Heyse, S., Kiltz, E., Lyubashevsky, V., Paar, C., Pietrzak, K.: Lapin: an efficient authentication protocol based on Ring-LPN. In: Canteaut, A. (ed.) FSE 2012. LNCS, vol. 7549, pp. 346–365. Springer, Heidelberg (2012)
37. Ishai, Y., Sahai, A., Wagner, D.: Private circuits: securing hardware against probing attacks. In: Boneh, D. (ed.) CRYPTO 2003. LNCS, vol. 2729, pp. 463–481. Springer, Heidelberg (2003)
38. Kiltz, E., Pietrzak, K.: Leakage resilient ElGamal encryption. In: Abe, M. (ed.) ASIACRYPT 2010. LNCS, vol. 6477, pp. 595–612. Springer, Heidelberg (2010)
39. Kirchner, P., Fouque, P.: An improved BKW algorithm for LWE with applications to cryptography and lattices. In: CRYPTO (2015)
40. Liskov, M., Rivest, R.L., Wagner, D.: Tweakable block ciphers. J. Crypt. **24**(3), 588–613 (2011)
41. Mangard, S., Oswald, E., Popp, T.: Power Analysis Attacks - Revealing the Secrets of Smart Cards. Springer, Heidelberg (2007)

42. Mangard, S., Popp, T., Gammel, B.M.: Side-channel leakage of masked CMOS gates. In: Menezes, A. (ed.) CT-RSA 2005. LNCS, vol. 3376, pp. 351–365. Springer, Heidelberg (2005)

43. Mangard, S., Pramstaller, N., Oswald, E.: Successfully attacking masked AES hardware implementations. In: CHES (2005)

44. Martin, D.P., Oswald, E., Stam, M., Wójcik, M.: A leakage resilient MAC. In: Groth, J. (ed.) IMACC 2015. LNCS, vol. 9496, pp. 295–310. Springer, Heidelberg (2015). doi:10.1007/978-3-319-27239-9_18

45. Medwed, M., Petit, C., Regazzoni, F., Renauld, M., Standaert, F.-X.: Fresh re-keying II: securing multiple parties against side-channel and fault attacks. In: Prouff, E. (ed.) CARDIS 2011. LNCS, vol. 7079, pp. 115–132. Springer, Heidelberg (2011)

46. Medwed, M., Standaert, F., Großschädl, J., Regazzoni, F.: Fresh rekeying: security against side-channel and fault attacks for low-cost devices. In: AFRICACRYPT (2010)

47. Medwed, M., Standaert, F.-X., Joux, A.: Towards super-exponential side-channel security with efficient leakage-resilient PRFs. In: Prouff, E., Schaumont, P. (eds.) CHES 2012. LNCS, vol. 7428, pp. 193–212. Springer, Heidelberg (2012)

48. Micciancio, D., Peikert, C.: Hardness of SIS and LWE with small parameters. In: CRYPTO (2013)

49. Moradi, A., Poschmann, A., Ling, S., Paar, C., Wang, H.: Pushing the limits: a very compact and a threshold implementation of AES. In: EUROCRYPT (2011)

50. Nikova, S., Rijmen, V., Schläffer, M.: Secure hardware implementation of nonlinear functions in the presence of glitches. J. Crypt. **24**(2), 292–321 (2011)

51. Pereira, O., Standaert, F., Vivek, S.: Leakage-resilient authentication and encryption from symmetric cryptographic primitives. In: ACM CCS (2015)

52. Petit, C., Standaert, F., Pereira, O., Malkin, T., Yung, M.: A block cipher based pseudo random number generator secure against side-channel key recovery. In: ASIACCS (2008)

53. Prouand, E., Rivain, M.: Masking against side-channel attacks: a formal security proof. In: EUROCRYPT (2013)

54. Regev, O.: On lattices, learning with errors, random linear codes, and cryptography. In: ACM STOC (2005)

55. Roche, T., Prou, E.: Higher-order glitch free implementation of the AES using secure multi-party computation protocols - extended version. J. Crypt. Eng. **2**(2), 111–127 (2012)

56. Schindler, W., Lemke, K., Paar, C.: A stochastic model for dierential side channel cryptanalysis. In: CHES (2005)

57. Standaert, F., Pereira, O., Yu, Y., Quisquater, J., Yung, M., Oswald, E.: Leakage resilient cryptography in practice. In: Towards Hardware-Intrinsic Security - Foundations and Practice (2010)

58. Yu, Y., Standaert, F.: Practical leakage-resilient pseudorandom objects with minimum public randomness. In: CT-RSA 2013 (2013)

ParTI – Towards Combined Hardware Countermeasures Against Side-Channel and Fault-Injection Attacks

Tobias Schneider[1](✉), Amir Moradi[1], and Tim Güneysu[2]

[1] Horst Görtz Institute for IT Security,
Ruhr-Universität Bochum, Bochum, Germany
{tobias.schneider-a7a,amir.moradi}@rub.de
[2] University of Bremen and DFKI, Bremen, Germany
tim.gueneysu@uni-bremen.de

Abstract. Side-channel analysis and fault-injection attacks are known as major threats to any cryptographic implementation. Hardening cryptographic implementations with appropriate countermeasures is thus essential before they are deployed in the wild. However, countermeasures for both threats are of completely different nature: Side-channel analysis is mitigated by techniques that hide or mask key-dependent information while resistance against fault-injection attacks can be achieved by redundancy in the computation for immediate error detection. Since already the integration of any single countermeasure in cryptographic hardware comes with significant costs in terms of performance and area, a combination of multiple countermeasures is expensive and often associated with undesired side effects.

In this work, we introduce a countermeasure for cryptographic hardware implementations that combines the concept of a provably-secure masking scheme (i.e., threshold implementation) with an error detecting approach against fault injection. As a case study, we apply our generic construction to the lightweight LED cipher. Our LED instance achieves first-order resistance against side-channel attacks combined with a fault detection capability that is superior to that of simple duplication for most error distributions at an increased area demand of 12 %.

1 Introduction

Over the last years, implementation attacks have seen a rise in popularity due to their ability to break cryptographic implementations which were believed to be cryptanalytically secure. Their power is based on vulnerabilities in the physical implementation instead of flaws in the cryptographic algorithm. The two most popular types of implementation attacks are side-channel analysis (SCA) and fault injection (FI) attacks.

SCA are passive attacks which exploit the information leakage related to cryptographic device internals through side channels, e.g., power consumption of a device [24]. Usually they involve a considerable number of measurements

© International Association for Cryptologic Research 2016
M. Robshaw and J. Katz (Eds.): CRYPTO 2016, Part II, LNCS 9815, pp. 302–332, 2016.
DOI: 10.1007/978-3-662-53008-5_11

and statistical tools to extract the sensitive information from the device. Over the years, various different types of attacks have been proposed with a large variety in capabilities and complexity. As a consequence, a wide range of countermeasures has been developed to thwart these attacks. Compared to other types of countermeasures, masking (as a form of secret sharing) has attracted the most interest inside the side-channel community. With its sound theoretical foundation, masking can be applied at different levels of abstraction to secure designs. Still the secure implementation of a masking scheme remains a major challenge since effects such as glitches in hardware circuits can completely invalidate the security assumptions of the schemes [26,27]. In response, Threshold Implementation (TI) [32], as a concept between Boolean masking and multiparty computation, has been specifically developed for hardware platforms to maintain security properties even in the presence of glitches. The TI concept has been applied to many algorithms including PRESENT [36], AES [5,16,29], KATAN [6,31], Keccak [4], arithmetic addition [43], Simon [44], PRINCE and Midori [30], and all 4-bit Sboxes [8].

Active FI attacks pose a further serious threat to instantiated cryptographic algorithms [3] by injecting a fault during its execution. The adversary then derives sensitive information from the erroneous output of the device. For more sophisticated attacks on symmetric schemes to work, multiple of these erroneous outputs need to be combined. Like for SCA, there are a wealth of attacks and possibilities to generate faults during the computation, e.g., by clock or power glitches or positioned photon injection using lasers. In terms of countermeasures, the majority of published concepts are based on the principle of concurrent error detection (CED). The main idea is to utilize redundancy in time or area to enable quasi-immediate detection of faults. Some CED schemes integrate the use of error detecting codes to enhance their level of protection. Over the years, various different codes have been studied to harden cryptographic implementations against FI attacks. Due to its simplicity, parity check codes are commonly used in this context [2,23]. Other schemes based on non-linear codes (e.g., [21,22]) were brought up to their beneficial fault coverage. Recently, the class of infective countermeasures have been put forward which do not require an explicit final check before returning the result. [17].

As previously discussed, it is mandatory for cryptographic devices to integrate dedicated SCA and FI countermeasures if they are operated in untrusted environments. Still the majority of proposed SCA and FI countermeasures have been solely evaluated separately, though both classes need to be integrated in a single device. For simple countermeasures (e.g., applying plain redundancy in area and time), a separate evaluation is justified since multiple executions of the same SCA-protected operation are admissible (with few exceptions). However, more sophisticated FI countermeasures are likely to affect the SCA countermeasure to a higher degree which can have a severe impact on the security and efficiency of the combined scheme. For example, if parity bits used by FI countermeasures are computed over *unmasked* intermediate values, it leads to a side-channel leakage even if the rest of the design in perfectly masked. Thus, a careless integration may easily lead to contradicting the assumptions of the underlying masking scheme, and hence failure of the masked design.

Related Work. In response, a few countermeasures have been proposed providing resistance against both kind of attacks. At the gate level, we refer to dual-rail logic styles (e.g., WDDL [46]) which – due to the additional presence of dual counterparts of the circuit – inherently offer a fault detection feature. However, the error detection rate is limited to the concept of simple duplication.

Furthermore, coding schemes have been used for combined countermeasures as well. Wiretap codes that have been applied as an SCA countermeasure [12,28] at algorithm level, can also provide a certain level of fault detection. Additionally, there are further examples [11] that use coding techniques for enhanced resistance against both types of attacks. However, most of the schemes are either designed for software implementations or provide only limited security at the expense of high overheads.

Besides combined countermeasures, there are also combined attacks which use a combination of fault injection and side-channel analysis to extract a secret. Several different attacks have been proposed against protected AES implementations where masking together with various fault countermeasures are integrated [13,15,39]. Our analyses consider this powerful threat as well.

Our Contribution. We propose a new combined physical protection scheme targeting hardware platforms. As mentioned before, the integration of CED schemes by simple *time* or *area redundancy* into masked designs is straightforward (see [48] for definitions). However, such constructions are not able to detect certain types of faults (e.g., identical faults which are injected in both instances of the design) and rather costly. Therefore, our target in this work is to merge more sophisticated *information redundancy* approaches (namely *error detecting codes*) with provably secure masked hardware designs. More precisely, we demonstrate how to integrate an error detecting code into first- (or higher-) order TI designs, while preserving all security requirements and features of the underlying TI concept. We formalize our methodology to allow various types of codes which provide the most flexibility in terms of protection and area requirement. We include a thorough analysis on the resistance of the combined countermeasure regarding the chosen order of TI and the parameters of the code. Note that the straightforward hardware duplication can be regarded as subtype of our combined countermeasure, but our generic concept enables to tweak the protection of the resulting design by the choice of code.

For practical evaluation we present a case-study on the cipher LED [19] that simplifies the explanation of the underlying concept due to its simple structure. We provide practical evaluations of our design implemented on an FPGA with respect to any detectable first-order leakage. Moreover, we evaluate the performance, area overhead as well as the fault coverage of the integrated *information redundancy* scheme. Note that the representations included in this work primarily discuss the case of a first-order TI design of LED with fault detection facility based on Hamming codes. But we like to emphasize that our generic construction can be similarly applied to any-order TI designs of other ciphers that are using different error detecting codes.

2   Background

2.1   Threshold Implementation

We briefly review the concept of Threshold Implementations (TI). For detailed information we refer the interested reader to the original articles [6,33].

For simplicity but without loss of generality, let us assume a 4-bit intermediate value of an arbitrary cipher with 4-bit S-Box, e.g., PRESENT [10] or LED [19]. We denote this corresponding 4-bit value as $\boldsymbol{x} = \langle x_1, \ldots, x_4 \rangle$. In TI, $\boldsymbol{x}$ is represented in $n-1$ order Boolean masked form $(\boldsymbol{x}^1, \ldots, \boldsymbol{x}^n)$, where $\boldsymbol{x} = \bigoplus\limits_{i=1}^{n} \boldsymbol{x}^i$ and each $\boldsymbol{x}^i$ similarly denotes a 4-bit vector $\langle x_1^i, \ldots, x_4^i \rangle$.

The linear functions, such as MixColumns of AES or LED, can be simply applied to the shares of $\boldsymbol{x}$ by $\mathsf{L}(\boldsymbol{x}) = \bigoplus\limits_{i=1}^{n} \mathsf{L}(\boldsymbol{x}^i)$. However, the realization of the S-Box over Boolean masking is not trivial. If the algebraic degree of the S-Box is denoted by t, the minimum number of shares to realize an S-Box protected against first-order attacks is $n = t+1$. To ensure *correctness* for the computation, this S-Box needs to provide the output $\boldsymbol{y} = \mathsf{S}(\boldsymbol{x})$ in shared form $(\boldsymbol{y}^1, \ldots, \boldsymbol{y}^m)$ with $\boldsymbol{y} = \bigoplus\limits_{i=1}^{m} \boldsymbol{y}^i$ and $m \geq n$ for bijective S-Boxes. In case of bijective S-Boxes (e.g., as for AES, PRESENT, and LED) the bit length of $\boldsymbol{x}$ and $\boldsymbol{y}$ (respectively of their shared forms) are identical.

Each output share $\boldsymbol{y}^{j \in \{1, \ldots, m\}}$ is given by a component function $\mathsf{f}^j(\cdot)$ over a subset of input shares. For first-order security, each component function $\mathsf{f}^{j \in \{1, \ldots, m\}}(\cdot)$ must be independent of at least one input share. This requirement on the independence from at least one share is defined as *non-completeness* property.

Since the security of masking schemes is based on the uniform distribution of the masks, the output of a TI S-Box must be also uniform since it is used as input in further parts of the implementation (e.g., the S-Box output of one cipher round which is given to the next S-Box after being processed by the linear diffusion layers). This property of *uniformity* requires for a bijective S-Box ($n = m$) each $(\boldsymbol{x}^1, \ldots, \boldsymbol{x}^n)$ to be mapped to a unique $(\boldsymbol{y}^1, \ldots, \boldsymbol{y}^n)$. In other words, it is sufficient in this case to check whether the TI S-Box forms a bijection with a $4 \cdot n$ input (and output) bit length.

As an example, take a function with an algebraic degree of $t = 3$. Hence, the number of input and output shares $n = m > 3$ directly affects the complexity of the circuit and its associated area overhead. Therefore, it is preferable to decompose the S-Box $\mathsf{S}(\cdot)$ into smaller bijections, e.g., $\mathsf{g} \circ \mathsf{f}(.)$, each with maximum algebraic degree of 2. The authors of [36] presented a decomposition of the PRESENT S-Box into two bijections g and f, each of which with an algebraic degree of 2. These parameters keep the number of shares for input and output at a minimum, i.e., $n = m = 3$.

2.2 Error Detecting Codes

Error detecting codes (EDC) are primarily used to transmit data over an unreliable communication channel. Those properties and notation of EDC that are also relevant for remainder of this work will be highlighted in the following [25].

Definition 1. *A linear code* **C** *of length n over $\mathbb{F}_q$ is a vector subspace over $\mathbb{F}_q^n$.*

We only consider binary codes (i.e., $q = 2$) in this work since they provide the best performance for our projected use-case in symmetric cryptography. A linear code **C** that maps messages of length k to codewords of length n is commonly denoted as an $[n, k]$-code.

Definition 2. *A generator matrix G of an $[n, k]$-code* **C** *comprises n basis vectors of* **C** *with length k.*

A generator matrix can be used to transform a given message $m \in \mathbb{F}_q^k$ to the corresponding code word $c \in$ **C** as $c = m \cdot G$.

Definition 3. *A matrix $H \in \mathbb{F}_q^{(n-k) \times n}$ with the property*

$$\mathbf{0} = H \cdot c^T, \ \forall c \in \mathbf{C} \tag{1}$$

is denoted as parity check matrix of the code **C**.

Such matrix can be used to easily check if a given c is a valid codeword of **C**.

Definition 4. *The minimum distance d of a linear code* **C** *is defined as*

$$d = min(\{wt\,(c_1 \oplus c_2)\,|c_1, c_2 \in \mathbf{C}, c_1 \neq c_2\}), \tag{2}$$

where $wt(x)$ returns the number of 1's in the vector x (known as Hamming weight). We denote a linear code **C** of length n, rank k and minimum distance d as an $[n, k, d]$-code. The minimum distance of a code determines its error detection and correction property.

Definition 5. *A code* **C** *with minimum distance d can be used to either detect $u = d - 1$ or correct $v = \lfloor \frac{d-1}{2} \rfloor$ errors. If d is even,* **C** *can simultaneously detect $u = \frac{d}{2}$ and correct $v = \frac{d-2}{2}$ errors.*

Given an erroneous codeword $c' = c \oplus e$, where e is known as the error vector, a u-error detecting code is able to detect that c' is faulty as long as $wt(e) \leq u$.

Definition 6. *The generator matrix G of a systematic code* **C** *is of the form $G = [I_k|P]$ where I_k denotes the identity matrix of size k.*

Each codeword c of a systematic code consist of the message itself which is padded by *check bits*, i.e., $c = [m|p]$. The check bits p[1] are generated by the rearward part of the generator matrix G represented by P. Note that all linear non-systematic codes can be transformed into a systematic code with the same minimum distance [9].

---

[1] Note that p can be also considered as a form of parity bits.

2.3 Concurrent Error Detection

Concurrent Error Detection (CED) systems are commonly used to detect arbitrary faults during the execution of an operation what makes them also an appropriate countermeasure against FI attacks [20]. Typically CED techniques rely on different types of redundancy to detect faulty computations. The most straightforward approach implements redundancy by multiple executions which results either in an increased area or in an increased time complexity. Certain intermediate values of different runs are compared with each other to detect errors.

As already indicated in the introduction, some CED schemes use error detecting codes following a structure similar to Fig. 1 to achieve a better fault coverage. In this basic example, CED is used to protect an OPERATION which is applied to a given INPUT. Initially, the CHECKBITS of the INPUT are generated by means of the GENERATOR matrix of the code. A PREDICTOR takes INPUT and CHECK-BITS and returns the predicted check bits of the output of OPERATION. These are compared with the actual CHECKBITS of the output. If a detectable error (depending on the type of the code) occurred during the execution, these two types of CHECKBITS will not be identical. Thus, a possible attack can be detected and averted. It should be noted that, depending on the target algorithm and the integrated code, the prediction functions can have an exalted level of complexity. Thus, the overhead of some CED schemes using EDC can be similar to a complete duplication of the operation.

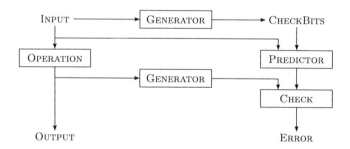

Fig. 1. A common structure of CED schemes using EDC.

Traditionally, the effectiveness of these fault detection countermeasures was examined in a uniform fault model. However, recent publications [20] have shown that this model does not closely resemble real-world attacks and that some of the presented countermeasures are in fact vulnerable to *biased fault attacks* [35].

3 Methodology

In this section, we introduce our methodology to develop a combined countermeasure against side-channel and fault injection attacks that is specifically tailored for hardware platforms. We first discuss the necessary considerations and restrictions of a combined scheme. This is followed by a detailed description of the attacker model and how to design a scheme to support arbitrary applications.

3.1 Design Considerations

First, our countermeasure is designed for hardware platforms. Thus, efficiency in software is not a concern in the design process. As hardware circuits are often used to achieve high performance, a primary design goal is to minimize the impact of the countermeasure on the performance.

Second, in terms of SCA countermeasure we aim at providing provable security (at least to a certain order). Therefore, hiding techniques are not applicable and we have to rely on masking. Given the first design goal, this leaves us with TI as it comes with a reasonable performance overhead compared to other masking schemes in hardware circuits [37].

Third, our scheme aims to be more secure against (realistic) FI attacks than simple duplication. Doubling the masked hardware circuit is a straightforward way to combine masking with some form of redundancy. However as mentioned before, simple duplication can be highly vulnerable to fault attacks if the fault model follows a different distribution than uniform. Therefore, we aim at building a scheme that is more robust against adversaries exploiting the effect of (reasonably) biased distributions. In this context, we choose EDC due to their sound theoretical foundation providing solid bounds on the number of detectable errors. Nevertheless, the balance between the error detection capability and runtime performance is essential to not severely impact our first design goal.

3.2 Attacker Model

Since our scheme aims to provide resistance against both SCA and FI attacks, we evaluate our methodology in a model that incorporates both types of threats. In the following, we assume an adversary that can observe the physical characteristics of the design during execution and further is able to inject faults in the circuit.

We assume a computationally bounded adversary that can observe the power consumption of our design during a finite number of executions. Note that security guarantees of TI also hold with respect to other side channels, e.g., electromagnetic emanations. Due to the computational restriction of the adversary, we can bound the number of possible observations. Given that the complexity of an attack increases with its order, we bound the adversary by the highest order of an attack he is able to mount. In other words, the adversary is able to observe a limited number of executions that is just enough to perform attacks of order d but not of order $d+1$. The actual order depends on the platform and the desired level of security.

Furthermore, the adversary is able to inject faults in the hardware circuit. In our model, we assume that injected faults only target the data path of the implementation and exclude the control flow. This is a different aspect of fault attacks which is not specific to our scheme. Given that the control flow usually does not need to be protected by masking, it is not considered in our combined countermeasure. Nevertheless, our combined countermeasure needs to be implemented together with a protected control flow to ensure complete security. There are various solutions to this problem. Even the EDC aspect from our combined countermeasure can be used to harden the control flow as described in [45]. Therefore, we model the injected faults as an error random variable E following a specific distribution $\mathcal{E}$. In our model, an error vector $e \in \mathbb{F}_q^n$ with probability $Pr[E = e]$ is sampled for each injected fault from the distribution and added (XORed) to the current *state* of the execution as *state'* = *state* $\oplus$ *e*. The execution continues the computation with the altered state *state'*. In the following, we consider two different types of the error distribution.

Since most existing works assume a uniform fault model, we also first examine our combined countermeasures against an adversary with a uniform distribution $\mathcal{E}_U$ so that $Pr[E = e_1] = Pr[E = e_2]$, $\forall e_1, e_2 \in E$.

Furthermore, we consider a biased distribution $\mathcal{E}_B$, where one specific set of error vectors $E_1 \subset E$ is significantly more probable than the set of remaining error vectors $E_2 \subset E$ with $Pr[E = e_1] \gg Pr[E = e_2]$, $\forall e_1 \in E_1, e_2 \in E_2$. The sets are determined by the type of faults that are considered in the model, e.g., $E_1 : \forall e, wt(e) \leq u$. In an extreme case, E_1 only contains one specific error vector e with $wt(e) = 1$. This scenario is akin to laser-based fault injections in which single bits can be targeted.

3.3   Code Selection

Obviously, the choice of the code strongly affects the efficiency and fault coverage of the resulting combined scheme. In this context, it is not possible to provide one specific code that exhaustively fits to all possible application scenarios. Instead, the code needs to be specifically chosen according to the target algorithm to yield optimal results. A poorly chosen code can cause a significant overhead while offering only little benefit in terms of fault coverage. In this subsection we discuss about necessary considerations made in the code selection process and give guidelines on the criteria how to pick a code.

Linear Codes. One important aspect in the design of a TI is the algebraic degree of the targeted functions. As explained in Sect. 2.1, the algebraic degree determines the minimum number of necessary shares. Given that the prediction functions are also part of the intended TI, it is crucial that they possess the same algebraic degree as the original function. Otherwise, the requirement of an additional share negatively affects the area complexity of the resulting design. This property is trivially fulfilled by linear codes. The encoding and decoding functions of linear codes are linear. Therefore, adding a decoding function before and an encoding function after the target function (cf. Fig. 3) to obtain the predictor guarantees that the emerging function has the same algebraic degree as the target function. For non-linear codes this property is not always satisfied. In addition, the encoding/decoding functions of linear codes can be implemented extremely efficiently which makes the necessary error check also very efficient. In the remainder of this work, we therefore only consider linear codes.

Systematic Codes. Systematic codes are advantageous to improve the efficiency. Due to their specially structured generator matrix (cf. Definition 6), the output of the targeted function does not need to be decoded to recover the correct result since the message is part of the codeword. This helps to eliminate one otherwise necessary step at the end of the design. Furthermore, the distinction between target function and predictor – as depicted in Fig. 1 – is otherwise not easily possible. Since one half of the design is nearly completely unaltered by the inclusion of fault countermeasure, it also allows the reuse of existing TI designs. This design decision does not limit the choice of codes since (as already mentioned in Sect. 2.2) every linear non-systematic code can be transformed into a systematic code with the same minimum distance.

Code Parameters. The choice of the three parameters of a linear code n, k, and d depends on the target algorithm. A good practice is to derive the code dimension k from the size of a single element that is used in most functions of the targeted algorithm, e.g., for an algorithm that performs most of its operations in $GF(2^8)$ it is advisable to set $k = 8$. This way unnecessary overhead due to the split or merge of check bits is avoided. Furthermore, the code length n also affects both the efficiency and fault coverage of the design. To achieve a desired error detecting level of $u = d - 1$, a certain minimal size of n is required. However, if n is chosen too large, the number of check bits increases resulting in a high area complexity. Therefore, it is important to find a good tradeoff between the length of the code n and its minimum distance d. In the following, we aim at a design in which the predictors work solely on the check bits. To achieve this, it is necessary that the message m can be fully recovered using the check bits p. Assuming that the message has full entropy (which is usually the case in symmetric cryptographic applications), it is advisable to set the rank to at least $n \geq 2k$.

3.4   Threshold Implementations with Error Detecting Codes

To achieve the desired level of security against SCA adversaries, it is necessary to implement all required functions according to the principles of TI. In particular, this includes the prediction functions as well. As it was already thoroughly discussed in [7,8] we omit the detailed explanation how to construct TI-compliant shared representations of arbitrary functions. Instead, we describe the specifics of including EDC in a TI design and how to easily find the TI of the predictors.

Notation. In the following, we assume a systematic linear code, which allows message recovery from the check bits. Further, we denote the input to the target algorithm by $\boldsymbol{m}_i$ with $n_m = |\boldsymbol{m}_i|$ as its bit length and the corresponding check bits as $\boldsymbol{p}_i$ with $n_p = |\boldsymbol{p}_i|$. The output of the target algorithm and its corresponding check bits are indicated by $\boldsymbol{m}_o$ and $\boldsymbol{p}_o$ respectively. Since the code does not change during the execution, the outputs have the same size as their corresponding inputs. Further, we assume that the TI of the target algorithm

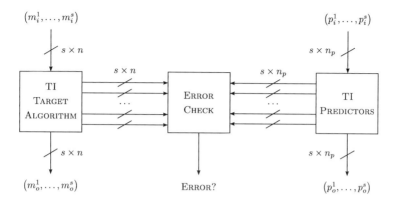

Fig. 2. The basic structure of our combined scheme.

requires a minimum of s shares to be secure. To this end, the messages and their corresponding check bits need to be masked accordingly as

$$\boldsymbol{m}_i = \bigoplus_{j=1}^{s} \boldsymbol{m}_i^j, \qquad \boldsymbol{p}_i = \bigoplus_{j=1}^{s} \boldsymbol{p}_i^j, \qquad \boldsymbol{m}_o = \bigoplus_{j=1}^{s} \boldsymbol{m}_o^j, \qquad \boldsymbol{p}_o = \bigoplus_{j=1}^{s} \boldsymbol{p}_o^j.$$

Basic Structure. Due to the special characteristics of the chosen code, it is possible to split up the computations of the underlying target algorithm and the predictors. The two output values $\boldsymbol{m}_o$ and $\boldsymbol{p}_o$ are calculated completely independent of each other. This leads to the basic structure as depicted in Fig. 2. There is an additional element (ERROR CHECK) which receives intermediate states of both circuits as input and checks if an error has occurred. The frequency for these checks is a variable in the specific design process, but it affects both the area and the fault coverage of the complete circuits. The higher the check frequency, the higher is the fault coverage but also the area requirements. In the most basic approach, only $\boldsymbol{m}_o$ and $\boldsymbol{p}_o$ are checked after a cipher run is complete.

For some TI designs a mask refresh during the execution is necessary to retain uniformity, e.g., for the AES S-Box [5]. Given that our proposed predictors are identical to the target function with an initial and final affine transformation, it is likely that they require a mask refresh depending on the shared function. Obviously, this can lead to a non-negligible overhead depending on the target algorithm. However, the separate computation paths (of the original and predictors) allow reusing the fresh randomness to some degree. Since both parts are completely independent and their respective intermediate values are never given to a joint function (except for the error check), it is possible to use the same random bits to refresh both sides. In the other case, where the predictors get inputs from both sides, this is not feasible without harming the uniformity property which would violate the security proofs of TI. For the error check, it is necessary to compute a function which takes inputs from both sides. However, in our scheme (and many others) this check can be implemented in a way that it only leaks the occurrence of a fault. For this to work, it is necessary that both sides use the same random masks which enables a separate error check on every share. This procedure would thwart most combined attacks. Alternatively without-out reused masks, the countermeasure against combined attacks from [15,39], which performs the check on masked values, could be easily modified to fit our scheme.

We illustrate this problem with an example. Let us assume a function F with two input bits a, b with $F(a,b) = ab$. The corresponding check bit is defined as $c = a + b$ with the predictor $F_p(a,c) = a + ac$. As noted in [8], there is no uniform sharing of F. Instead, a virtual share is added to achieve uniformity. The shared functions using one virtual share are

$$F_1 = a_2 b_2 + a_2 b_3 + a_3 b_2 + r \tag{3}$$
$$F_2 = a_3 b_3 + a_1 b_3 + a_3 b_1 + a_1 r + b_1 r \tag{4}$$
$$F_3 = a_1 b_1 + a_1 b_2 + a_2 b_1 + a_1 r + b_1 r + r, \tag{5}$$

where r is randomly drawn from a uniform distribution. Analogously, the predictor can be shared as

$$F_{p1} = a_2 + a_2 c_2 + a_2 c_3 + a_3 c_2 + r \tag{6}$$
$$F_{p2} = a_3 + a_3 c_3 + a_1 c_3 + a_3 c_1 + c_1 r \tag{7}$$
$$F_{p3} = a_1 + a_1 c_1 + a_1 c_2 + a_2 c_1 + c_1 r + r. \tag{8}$$

If both (F_1, F_2, F_3) and (F_{p1}, F_{p2}, F_{p3}) share the same r, the resulting six output bits would not be jointly uniform. Meaning that, they cannot be used as input to another joint function (i.e., another predictor) without violating the uniform input property of TI. To fix this, double amount of fresh randomness (i.e., one r bit for each part) is required.

Shared Predictors. Contrary to ordinary CED schemes, our predictors need to comply with the requirements of TI. In other words, the prediction functions work on masked check bits and fulfill the non-completeness, correctness, and uniformity properties. Finding functions with all these characteristics can be difficult for certain codes. However, in our presented scenario (i.e., a systematic linear code with a *sufficiently large* rank) it can be significantly simplified.

The general approach is shown in Fig. 3 with the example of affine and non-linear functions with three shares. $\pi : \mathbb{F}_2^{n_m} \to \mathbb{F}_2^{n_p}$ denotes the generation of the check bits using P, the right part the generator matrix. Respectively, $\pi^{-1} : \mathbb{F}_2^{n_p} \to \mathbb{F}_2^{n_m}$ is defined as its inverse, i.e., recovery of the message from the check bits. To derive the shared representation of the predictor from the target functions, each input share is first transformed using π^{-1}. Then the target function is applied and each resulting share is run through π to generate the corresponding check bits again. Of course, the steps do not have to be performed segregated. Instead, they are merged and subsequently optimized to achieve a better performance. The resulting functions trivially comply with the correctness property. Given that π and π^{-1} operate on single shares, non-completeness

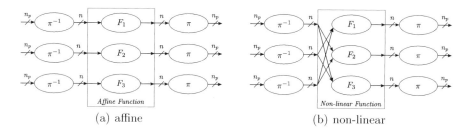

Fig. 3. Derivation of shared predictors for three shares.

is also maintained. Regarding the uniformity of the output shares, we need to differentiate between two cases. For $n_p = n_m$, the uniformity property is preserved from the target functions as noted in [41]. The encoding and decoding operations are only affine transformations which do not influence the uniformity in this setting. However, for $n_p > n_m$ this observation does not generally hold. If the steps are performed in an isolated manner, the reduction of the input shares to size n_m will come with a reduction in entropy. In result, the enlarged output shares are no longer uniform. A trivial solution would be the inclusion of a fresh random value to restore uniformity. However, this reduces the performance of the design and is therefore undesirable. A more efficient solution is to merge the three steps (i.e., π^{-1}, F, π) and eliminate the reduction of the input shares.

Depending on the operation, this optimization can be very effective. Especially if π and π^{-1} are linear over the F they can be partially canceled out. As mentioned before, functions with a high degree are often decomposed to reduce the number of shares. Usually there are multiple possibilities for decomposition with different efficiencies. Depending on the scenario, these decompositions do not need to be the same for the predictors. In these cases, the final result is still the same but not necessarily the intermediate values. This enables more efficient designs while leading to some limitations in the error detection, as discussed later on.

Error Detection. As noted before, the rate of error detection inside the algorithm affects the performance, area consumption and fault coverage of the design. Frequent error checking thwarts potential optimizations of the predictors what finally leads to larger circuits.

The error checking is performed similar to Fig. 1. However, in our basic scenario (without reusing randomness) the intermediate values are split up into multiple shares via Boolean masking. To still detect if an error has occurred, a two-step approach denoted as CHECK-AND-COMBINE is required. In the first step CHECK, the parity check matrix is multiplied with each share of the codeword. Thus, the resulting error check vectors v_{int}^j are computed as

$$v_{int}^j = H \cdot \left(c_{int}^j \right)^T = \pi(m_{int}^j) \oplus p_{int}^j, \; 1 \leq j \leq s. \tag{9}$$

If no error has occurred, these error vectors are a random sharing of the null vector. To check this, the error vectors are combined via XOR in the second

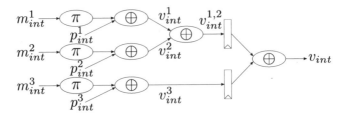

Fig. 4. Computation and unmasking of the error check vector for three shares in a first-order secure design.

step COMBINE. However, without any registers this procedure is equivalent to a function which has all shares of both parts of the circuit as input. This certainly violates the non-completeness property of TI. To this end, it is necessary to split up the second step COMBINE into multiple parts and include registers in between. In case of a first-order secure design, all but one of the shares are first combined. The result and the last share are then stored in a register and combined as

$$\boldsymbol{v}_{int} = \left(\bigoplus_{j=1}^{s-1} \boldsymbol{v}_{int}^{j} \right) \oplus \boldsymbol{v}_{int}^{s}. \tag{10}$$

If $\boldsymbol{v}_{int}$ is not the null vector, an error has been detected. The last XOR technically violates the non-completeness property as it unmasks $\boldsymbol{v}_{int}$ by merging all shares. However, $\boldsymbol{v}_{int}$ holds no information about the sensitive intermediate values of the circuit. Therefore, the SCA resistance of the design is not jeopardized by this.

An exemplary error check procedure with three shares is depicted in Fig. 4. It should be noted that the initial input values are indeed in compliance with the uniformity property of TI. The input values $v_{int}^{1,2}$ and v_{int}^{3} to the second part (right of the registers) are not jointly uniform given that if no error has occurred they are identical. Yet this does not affect the security of the resulting design since (as argued before) $\boldsymbol{v}_{int}$ does not hold any information related to sensitive intermediate values. This security guarantee still holds if the same randomness is used for masking both $\boldsymbol{m}_{int}$ and $\boldsymbol{p}_{int}$. Even though the input to the multiplication with H is not uniform, it does not pose a problem as it is applied to each share separately and the resulting $\boldsymbol{v}_{int}$ does not hold any information related to sensitive intermediate values.

The CHECK-AND-COMBINE procedure can be further simplified. To this end, it is necessary that all randomness is reused and the check bits are carefully generated and predicted during the cipher run. One possibility to generate the check bits assuming $n_p = n_m$ is

$$\boldsymbol{p}_i^j = \begin{cases} \pi\left(m_i\right) \oplus r^j, & \text{for } 1 \leq j < s \\ \pi\left(m_i\right) \oplus \left(\bigoplus_{j=1}^{s-1} r^j \right), & \text{for } j = s \end{cases} \tag{11}$$

where r^j denotes uniformly distributed fresh random masks with $r^j \in_R \mathbb{F}_2^k$, $1 \leq j < s$. Hence, the same masks are used for m_i and p_i. However, each share of the codeword $c_i^j = [m_i^j | p_i^j]$ is for itself not a valid codeword. The COMBINE-step is still necessary for error detection. To avoid this, the generation of the check bits need to be adjusted to

$$p_i^j = \pi(m_i^j), \ 1 \leq j \leq s. \tag{12}$$

Now each share of the codeword is valid and can be checked separately. In other words, each p_i^j can now be used to check its related m_i^j which makes the COMBINE-step unnecessary. Instead, if no error has occurred every v_i^j will be the null vector. To maintain this property, it is necessary that the predictors match exactly the main functions with additional encoding. Therefore, the aforementioned optimization technique regarding the decompositions of functions cannot be applied. Otherwise the p_i^j would lose this characteristic and an additional COMBINE-step becomes necessary. It should be noted that this only works given that $n_p = n_m$. Otherwise additional fresh randomness is required to achieve a uniform sharing of p_i making it impossible to check each share separately.

Overhead. The overhead of our scheme obviously depends on the chosen code and the underlying algorithm. Simple duplication, for example, is just an extreme case of our combined countermeasure in which P of the generator matrix is set to the identity matrix. However, if randomness is reused and $n_p \leq n$, the amount of fresh randomness is independent of the chosen code. In this case, our combined countermeasure uses the same amount of randomness as simple duplication. For other metrics it is not possible to give such a definite rule. Both area and performance can be worse or better than simple duplication, depending on how good the predictors can be optimized.

Combined Attacks. As mentioned in the introduction, there are combined attacks which can break AES implementations with certain combinations of countermeasures. Our proposed countermeasure can be also vulnerable to these kind of attacks depending on the underlying cipher and chosen code. However, most of these attacks focus on the error check and exploit that usually a combination of multiple shares is required. As described before our scheme can be instantiated without the necessity of a COMBINE-step which helps to prevent these attacks that rely on this as a point of attack. In this case, the leakage only contains information if an error has occurred but not more.

3.5   Security Analysis

We now discuss about the security properties of our combined countermeasure under the previously defined attacker model. Here, we distinguish between resistance against SCA attacks and FI attacks. In the latter case, our combined countermeasure is generally compared with a simple duplication of the TI.

SCA Resistance. As mentioned before, the security of a TI is derived from its order. A first-order TI is provable secure against first-order attacks [32]. Given that the adversary in our model can perform attacks up to order d, a TI of order d is accordingly required to protect our design. By following our proposed approach, the shared predictors are in compliance with the principle of a d-order TI. Therefore, they provide the same level of security as the d-order TI of the main circuit. Therefore, our proposed combined countermeasure has the exact level of SCA-security as a plain d-order TI without FI countermeasures. Furthermore, this level is independent of the chosen code meaning that simple duplication does not provide better or worse SCA-protection than a more complex EDC.

FI Resistance. The level of security against FI attacks depends on the parameters of the chosen code. In particular, the code distance d is important for the detection of certain types of errors. In this context, we can model the simple duplication countermeasure as a linear $[2k, k, 2]$-code $\mathbf{D}$ with $d = 2$. This is a comparably low distance given that such a distance can be achieved by a (in most cases) much shorter parity $[k + 1, k, 2]$-code.

The efficiency of a fault countermeasure can be assessed by its fault coverage rate which measures the proportion of undetectable faults. To simplify the analysis, we first assume that the fault is injected into an intermediate state of the execution which is used for error detection. That is, $\boldsymbol{p}_{int}$ are valid check bits for $\boldsymbol{m}_{int}$. As defined before, a fault is modeled as an error vector $e \neq \mathbf{0}$ that is added to the state $\boldsymbol{c}_{int} = [\boldsymbol{m}_{int} | \boldsymbol{p}_{int}]$. For a fault to be undetectable, e needs to be a valid codeword of the deployed code $\mathbf{C}$. This is rooted in the characteristic of linear codes in which every valid codeword can be written as the sum of two valid codewords as

$$c_3 = c_1 + c_2 = m_1 \cdot G + m_2 \cdot G = (m_1 + m_2) \cdot G,$$

with $c_1, c_2, c_3 \in \mathbf{C}$. Therefore, if e is not a valid codeword of $\mathbf{C}$ the erroneous result would also not be a valid codeword. Note that the aforementioned addition property of linear codes still holds for shared codewords. Meaning that if a valid codeword is added to one of the shares, it would result in a new shared codeword. With this, we can formally define the fault coverage of a code $\mathbf{C}$ as

$$Coverage_{\mathbf{C}}[E \sim \mathcal{E}] = 1 - Pr[e \in \mathbf{C} \wedge e \neq \mathbf{0}], \tag{13}$$

where the error variable E follows an error distribution $\mathcal{E}$.

Usually the rank of the code k is not chosen to be equal to the size of the whole input of the algorithm for efficiency reasons. Therefore, the intermediate state of the execution consists of multiple valid codewords. To further simplify the analysis we first assume that the adversary only injects one fault in one share of one codeword of the intermediate state. With $|C| = 2^k$ and $|E| = 2^n$ we can derive $Pr[e \in \mathbf{C} \wedge e \neq \mathbf{0}] = (2^k - 1)/2^n$ and define the fault coverage of the code $\mathbf{C}$ as

$$Coverage_{\mathbf{C}}[E \sim \mathcal{E}_U] = 1 - \frac{2^k - 1}{2^n}, \tag{14}$$

in the uniform fault model. Notably, the fault coverage in this model is independent of the code distance d. It means that it depends only on the rank k and the length n. Consequently, simple duplication provides the same fault coverage as any other code with the same k and n against this type of faults. For $\mathbf{D}$ the length is derived from the rank as $n = 2k$. The coverage can then be simplified to

$$Coverage_{\mathbf{D}}[E \sim \mathcal{E}_U] = 1 - \frac{1}{2^k} + \frac{1}{2^{2k}}. \tag{15}$$

As noted before, the uniform fault model is not a realistic assumption for all scenarios. Therefore, it is closer to reality to assume that the error distribution is biased to a certain degree [20]. For example, a clock glitch might cause similar errors in identical circuits which are close together (i.e., simple duplication). In this scenario, the fault coverage is severely reduced given that simple duplication cannot detect identical errors in both circuits. In the following, we assume that only a limited number of bits is affected by the fault. In the most extreme case, only one bit is affected which is related to laser fault injection[2]. We consider a biased distribution $\mathcal{E}_{B_b}$ with the corresponding subsets

$$E_1 = \{e \mid e \in E \wedge wt(e) \leq b\} \text{ with } Pr[e \in E_1] = 1, \tag{16}$$
$$E_2 = \{e \mid e \in E \wedge wt(e) > b\} \text{ with } Pr[e \in E_2] = 0. \tag{17}$$

We assume further that the error vectors in E_1 have per se identical probabilities. Depending on the method of fault injection, certain values of b are easier to achieve than others. Following this definition, $\mathcal{E}_{B_n}$ is equivalent to $\mathcal{E}_U$. In this fault model, it is possible to give specific bounds in which a complete fault coverage is achieved by our proposed countermeasure. It is trivial to see that an $[n, k, d]$-code which can detect $u = d - 1$ errors still achieves a complete fault coverage in the model following $\mathcal{E}_{B_u}$. However, it depends on the specific code how the fault coverage evolves for higher values of $b > u$. For simple duplication it can be easily calculated as

$$Coverage_{\mathbf{D}}[E \sim \mathcal{E}_{B_b}] = 1 - \sum_{j=1}^{\lfloor \frac{b}{2} \rfloor} \binom{k}{j} \Big/ \sum_{i=1}^{b} \binom{n}{i}. \tag{18}$$

It is notable that a simple duplication scheme achieves full fault coverage only for $b = 1$.

Depending on the scenario, there are other possible biased distributions. For example, if the attacker is only able to inject faults in one part of the design

---

[2] Note that bit flips which we assume in our attacker model might not be realistic for laser fault injection in certain scenarios [40]. However, we still use it in our model. The ability to set and reset bits instead of flipping enables trivial attacks in which the adversary tests each bit of the key to be zero or one. This attack cannot be directly prevented by our method without additional logic (e.g., allow only a certain number of faults). However, this is true for a majority of countermeasures and therefore not an issue unique to our methodology. The designer needs to include further countermeasures against this attack vector, e.g., splitting the key into multiple shares can increase the complexity of the attack.

(target algorithm or check bits) the full fault coverage is achieved for all codes with $d > 1$. Furthermore, the ability to inject symmetric errors in both parts strongly reduces the security of simple duplication. In the most extreme case, the adversary can pick bits to fault, e.g., by laser injection. In this case the error detecting capability is directly proportional to the attack complexity assuming that targeting more single bits by laser at different places increases the costs of the attack.

In reality, it might not be possible to only target one specific codeword, e.g., with round-based architectures. This affects the fault coverage since the error vector e needs to be valid codeword for every element of the state. Therefore, the estimation of the coverage can be adapted to include the number of state elements n_s

$$Coverage_{\mathbf{C}}[E \sim \mathcal{E}] = 1 - (Pr[e \in \mathbf{C} \wedge e \neq \mathbf{0}])^{n_s} . \tag{19}$$

We assume an error check in which each share is not checked separately. Therefore, the number of shares does not play any role in this estimation since it is enough to check whether the sum of all shares is a valid codeword as

$$(c^1 \oplus e^1) \oplus (c^2 \oplus e^2) \oplus (c^3 \oplus e^3) = c \oplus e. \tag{20}$$

If each share is checked separately, the fault coverage needs to include the number of shares in the calculation similar to n_s.

Up to now, we only considered faults that are injected at one point in time into an encoded state which is checked for errors. Depending on the power of the adversary, this scenario may be realistic. However, there are also other cases in which an attacker is more powerful and can inject more sophisticated types of errors.

One of these types are faults which are injected into a state between layers that is not directly checked. Instead, multiple operations are first performed on the erroneous state before it is checked. In this case, the fault coverage rate stays the same based on the fact that none of the operations change the validity of a codeword. In other words, if the error is detectable in one state, it should be also detectable in every following state.

Another important aspect for fault coverage is multiple faults at different points in time. Assume for example a linear transformation F of a codeword c in which an error e is injected is applied. This results in

$$F(c \oplus e) = F(c) \oplus F(e) \tag{21}$$

meaning that the output of F is combined with a transformed error $F(e)$. Given the structure of the functions and predictors, $F(.)$ cannot make a valid codeword $F(e)$ if e is not a valid codeword. Therefore, the fault coverage is not impaired for faults at a single point in time. However, $F(.)$ can increase the Hamming weight of e making it easier for an attacker to inject an additional error after the transformation. In the most extreme case, an attacker injects only two errors e_1, e_2 with $wt(e_1) = wt(e_2) = 1$ and can create an undetectable fault as

$$F(c \oplus e_1) \oplus e_2 = F(c) \oplus F(e_1) \oplus e_2, \tag{22}$$

with $wt(F(e_1)) = d - 1$. This approach works similarly for non-linear layers. However, in this case the output of the function is not the sum of the two transformed values. This attack vector can be prevented by introducing more error checks in the design. If every encoded state before a transformation is checked, this attack can be thwarted since the injection of the first error $F(c \oplus e_1)$ would be detected. Introducing more checks can obviously result in an increased area complexity.

As of now, all of the errors are added to an encoded state between layers. However, depending on the scenario it might be also possible to inject faults inside the combinatorial logic between these states. Since the logic usually consists of a cascade of multiple gates modeling, the fault as an addition of an error vectors is not trivial. However, depending on the abilities of the attacker this type of fault can be powerful. For example, an attacker can target one gate which derives multiple output bits. In this case, we have the same scenario as in the previous example that the injected fault e has $wt(e) = 1$ but it cannot be detected when the check is performed on the output of the combinatorial circuit where such e leads to $e' \in \mathbf{C}$. To completely avoid this type of attack it is necessary to isolate the logic for all output lines from each other. This way a faulty gate can only affect one of the output bits which prevents the aforementioned attack.

As illustrated by the previous example, it is important to realistically estimate the power of potential FI attackers. Choosing a code with a large distance and implementing the previously proposed countermeasures might lead to a highly secure system. However, each of these aspects can negatively influence the size of the design. As for many other systems, the balance between area and the level of security is an important aspect in the design process.

4   Case Study: LED

Up to now, our combined countermeasure has been only discussed from the theoretic perspective without targeting a specific algorithm. To better illustrate the rationales and parameters of the design process, we implement a block cipher according to our methodology. For the sake of comprehensibility, a relatively straightforward example is picked to explain the design choices in detail.

The most obvious target for this would be AES as it is the most widely deployed cipher. However, while the predictors for the linear layers of AES are comparably easy to implement, the TI of its non-linear layer poses still a challenge even without FI resistance [5]. In particular, it requires a significant amount of fresh randomness to achieve all the necessary TI properties. Another standardized cipher, for which an efficient TI exists, is PRESENT [10]. Its 4-bit S-Box can be efficiently implemented in various ways [7,41]. Contrary to AES, its permutation layer is very efficient in hardware, but its predictors are comparably inefficient.

A better example to demonstrate our combined countermeasure is LED. It combined the best aspects of AES and PRESENT by incorporating the

PRESENT S-Box and AES-like linear layers. Thus, an efficient TI and predictors can easily be achieved. In our case study, we present one way to implement LED with our methodology. Note that depending on the targeted attacker model, different choices are possible, e.g., higher-order TI or another code with a large distance. The SCA security of the final design is practically evaluated using an FPGA prototype, while the FI resistance is examined using the previously introduced attacker models.

4.1 Cipher Description

LED is a lightweight block cipher introduced in 2011 [19]. It has a 64-bit state and can be instantiated with different key sizes (primarily 64 or 128 bits). The basic structure of the cipher consists of addition of the round keys (ADDROUNDKEY) and so-called steps (STEP). In each step, four rounds of encryption are applied to the state. One round is made up of four layers ADDCONSTANTS, SUB-CELLS, SHIFTROWS and MIXCOLUMNSSERIAL. During ADDCONSTANTS constants which are derived from an LFSR are added to half of the state. The following three layers are similar to the layers of AES [34] and consist of a nibble-wise substitution and row/column-wise affine transformations. For 128-bit key (resp. 64-bit keys) LED-128 (resp. LED-64) performs 12 steps in total (resp. 8 steps) with key additions between them.

One important characteristic of LED is its very simple key schedule. Instead of using different round keys derived by a schedule function applied on a main key, the cipher directly uses 64 bits from the user-defined key for each round. This means that for the 64-bit version all round keys are the same, while in the 128-bit instantiation the key halves are used alternately.

4.2 Design and Implementation

We implement a design that is secure against first-order attacks. We decompose the S-Box that allows us to implement TI using three shares. In the following, we explain the selection of the code and the predictors for each layer of LED in detail.

Code Selection. Given that LED is a nibble-oriented cipher in which all operations work on either one or multiple nibbles of the state, we consider only codes with a rank of $k = 4$. This way, expensive merge or split of codewords can be minimized. Furthermore, we decided to set the length of the code to $n = 8 = 2 \cdot k$ to avoid additional fresh randomness. It would be beneficial to select a code over $GF(2^4)$, since most of the LED operations are in this field[3]. However, none of the 16 possible $[8, 4]$-codes has a distance larger than $d = 3$. Therefore, to achieve a higher level of protection against FI attacks, we choose a different code outside of $GF(2^4)$ but with a better error detection property.

---

[3] In this case, P is chosen in such a way that $\boldsymbol{p} = \pi(\boldsymbol{m}) = \boldsymbol{m} \cdot x$ with $x \in GF(2^4)$.

The extended Hamming code is a basic extension of the $[7, 4, 3]$-Hamming code. By adding an extra parity bit the code is transformed to a $[8, 4, 4]$-code, i.e., with $d = 4$. In our implementation we use the following generator and parity check matrices:

$$G = \begin{pmatrix} 1\,0\,0\,0 & 1\,1\,1\,0 \\ 0\,1\,0\,0 & 1\,1\,0\,1 \\ 0\,0\,1\,0 & 1\,0\,1\,1 \\ 0\,0\,0\,1 & 0\,1\,1\,1 \end{pmatrix}, \qquad H = \begin{pmatrix} 1\,1\,1\,0 & 1\,0\,0\,0 \\ 1\,1\,0\,1 & 0\,1\,0\,0 \\ 1\,0\,1\,1 & 0\,0\,1\,0 \\ 0\,1\,1\,1 & 0\,0\,0\,1 \end{pmatrix}. \tag{23}$$

Due to its simplicity, the code enables the use of efficient predictors while still achieving a high error detection capability with respect to its length.

Linear Layers. As described before, LED consists of four different linear layers. We discuss the application of the extended Hamming code to each layer without specifically considering TI, since every linear layer and corresponding predictor can be applied to each share separately as explained in Sect. 3.4. Note that the key and constants are not shared, following the same design strategy as in [5,29,36,41]. Therefore, in the two layers (ADDROUNDKEY, ADDCONSTANTS) where a value is added to the state, it is applied only to one share (of three).

ADDROUNDKEY. Since this layer only consists of a basic addition in $GF(2^4)$ of the round key to the state of the cipher, its predictor can be implemented very efficiently. It can be optimized to

$$\begin{aligned} \boldsymbol{p}_{int_2} &= \pi\left(\pi^{-1}\left(\boldsymbol{p}_{int_1}\right) \oplus key\right) \\ &= \boldsymbol{p}_{int_1} \oplus \pi\left(key\right), \end{aligned} \tag{24}$$

where $\boldsymbol{p}_{int_1}$ (resp. $\boldsymbol{p}_{int_2}$) denotes the input (resp. output) check bits to ADDROUNDKEY, and key a round key. Furthermore, LED does not include a key schedule. Thus, by computing $\pi\left(key\right)$ (of both key halves for LED-128) once at the start of the cipher, the predictor for the key addition can be easily realized without additional overhead.

ADDCONSTANTS. Two types of round constants are added to the state. One is derived from the key length and does not change over the course of the cipher. The bit size of the key length is stored in eight bits ($ks_7 ks_6 ks_5 ks_4\ ks_3 ks_2 ks_1 ks_0$). The lower and upper four bits of the bitstring are each considered as one encoded element. Since the key size does not change during the execution, this type of constant does not need to be updated. For LED-128 the specific bits are

$$(ks_7 ks_6 ks_5 ks_4\ ks_3 ks_2 ks_1 ks_0) = (1000\ 0000), \tag{25}$$
$$(ksp_7 ksp_6 ksp_5 ksp_4\ ksp_3 ksp_2 ksp_1 ksp_0) = (1110\ 0000)$$

where ksp_i denotes the corresponding check bits for this constant.

The other constant consists of six bits ($rc_5 rc_4\ rc_3 rc_2 rc_1 rc_0$) which are updated for every round by an LFSR. The update function can be represented by a matrix multiplication in $GF(2)$ as

$$
\begin{pmatrix} rc_0' \\ rc_1' \\ rc_2' \\ rc_3' \\ rc_4' \\ rc_5' \end{pmatrix} = \underbrace{\begin{pmatrix} 0\,0\,0\,0\,1\,1 \\ 1\,0\,0\,0\,0\,0 \\ 0\,1\,0\,0\,0\,0 \\ 0\,0\,1\,0\,0\,0 \\ 0\,0\,0\,1\,0\,0 \\ 0\,0\,0\,0\,1\,0 \end{pmatrix}}_{U} \cdot \begin{pmatrix} rc_0 \\ rc_1 \\ rc_2 \\ rc_3 \\ rc_4 \\ rc_5 \end{pmatrix} + \underbrace{\begin{pmatrix} 1 \\ 0 \\ 0 \\ 0 \\ 0 \\ 0 \end{pmatrix}}_{c},
\tag{26}
$$

where U denotes the update matrix. The related check bits defined as

$$
(rcp_3 rcp_2 rcp_1 rcp_0) = \pi(rc_3 rc_2 rc_1 rc_0) \tag{27}
$$
$$
(rcp_3 rcp_2 rcp_1 rcp_0) = \pi(00|rc_5 rc_4) \tag{28}
$$

need to be updated accordingly. To this end, the update matrix is first enlarged to incorporate the two padded zeros to

$$
U_L = \begin{pmatrix} 0\,0\,0\,0\,1\,1\,0\,0 \\ 1\,0\,0\,0\,0\,0\,0\,0 \\ 0\,1\,0\,0\,0\,0\,0\,0 \\ 0\,0\,1\,0\,0\,0\,0\,0 \\ 0\,0\,0\,1\,0\,0\,0\,0 \\ 0\,0\,0\,0\,1\,0\,0\,0 \\ 0\,0\,0\,0\,0\,0\,0\,0 \\ 0\,0\,0\,0\,0\,0\,0\,0 \end{pmatrix}.
\tag{29}
$$

The update matrix for the check bits $(U_{L_{check}})$ can be derived from by $\pi\left(U_L\left(\pi^{-1}(\cdot)\right)\right)$. Therefore, we can write (note that P is self-inverse):

$$
U_{L_{check}} = \begin{pmatrix} 1\,1\,1\,0\,0\,0\,0\,0 \\ 1\,1\,0\,1\,0\,0\,0\,0 \\ 1\,0\,1\,1\,0\,0\,0\,0 \\ 0\,1\,1\,1\,0\,0\,0\,0 \\ 0\,0\,0\,0\,1\,1\,1\,0 \\ 0\,0\,0\,0\,1\,1\,0\,1 \\ 0\,0\,0\,0\,1\,0\,1\,1 \\ 0\,0\,0\,0\,0\,1\,1\,1 \end{pmatrix} \cdot U_L \cdot \begin{pmatrix} 1\,1\,1\,0\,0\,0\,0\,0 \\ 1\,1\,0\,1\,0\,0\,0\,0 \\ 1\,0\,1\,1\,0\,0\,0\,0 \\ 0\,1\,1\,1\,0\,0\,0\,0 \\ 0\,0\,0\,0\,1\,1\,1\,0 \\ 0\,0\,0\,0\,1\,1\,0\,1 \\ 0\,0\,0\,0\,1\,0\,1\,1 \\ 0\,0\,0\,0\,0\,1\,1\,1 \end{pmatrix}.
\tag{30}
$$

The same procedure is applied to the constant factor of the update function (denoted as c in Eq. (26)). Overall, the check bits of the round constant can be updated as

$$
\begin{pmatrix} rcp_0' \\ rcp_1' \\ rcp_2' \\ rcp_3' \\ rcp_4' \\ rcp_5' \\ rcp_6' \\ rcp_7' \end{pmatrix} = \underbrace{\begin{pmatrix} 0\,0\,1\,1\,0\,0\,1\,1 \\ 0\,1\,0\,1\,0\,0\,1\,1 \\ 0\,1\,1\,0\,0\,0\,1\,1 \\ 1\,0\,0\,0\,0\,0\,0\,0 \\ 0\,1\,1\,1\,1\,1\,1\,0 \\ 0\,1\,1\,1\,1\,1\,1\,0 \\ 0\,1\,1\,1\,0\,0\,0\,1 \\ 0\,0\,0\,0\,1\,1\,1\,0 \end{pmatrix}}_{U_{L_{check}}} \cdot \begin{pmatrix} rcp_0 \\ rcp_1 \\ rcp_2 \\ rcp_3 \\ rcp_4 \\ rcp_5 \\ rcp_5 \\ rcp_5 \end{pmatrix} + \underbrace{\begin{pmatrix} 1 \\ 1 \\ 1 \\ 0 \\ 0 \\ 0 \\ 0 \\ 0 \end{pmatrix}}_{cp} . \tag{31}
$$

It is obvious that the update of the check bits requires additional resources. Still this overhead is negligible since the round constant update is only a small part of the cipher and not split up into multiple shares.

SHIFTROWS. This layer manipulates the state in a nibble-wise fashion. Since the codewords are not modified in any way, it is sufficient to apply the same permutation on the check bits.

MIXCOLUMNSERIAL. Four nibbles of the state are combined using a matrix A four consecutive times. The matrix multiplication is performed in $GF(2^4)$. Since addition is linear over $GF(2)$, we do not need to change the values of A for the check bits. Only the field multiplications with 2 and 4 need to be adapted to the predictor. The two multiplications with the reduction polynomial $X^4 + X + 1$ can be represented as a matrix multiplications in GF(2) as

$$
2 \cdot \begin{pmatrix} m_0 \\ m_1 \\ m_2 \\ m_3 \end{pmatrix} = \begin{pmatrix} 0\,0\,0\,1 \\ 1\,0\,0\,1 \\ 0\,1\,0\,0 \\ 0\,0\,1\,0 \end{pmatrix} \cdot \begin{pmatrix} m_0 \\ m_1 \\ m_2 \\ m_3 \end{pmatrix}, \quad 4 \cdot \begin{pmatrix} m_0 \\ m_1 \\ m_2 \\ m_3 \end{pmatrix} = \begin{pmatrix} 0\,0\,1\,0 \\ 0\,0\,1\,1 \\ 1\,0\,0\,1 \\ 0\,1\,0\,0 \end{pmatrix} \cdot \begin{pmatrix} m_0 \\ m_1 \\ m_2 \\ m_3 \end{pmatrix}. \tag{32}
$$

For the check bits, these matrices need to be adapted similar to Eq. (30) but with 4×4 matrices. The resulting matrices for the check bits are

$$
\pi \left(2 \cdot \pi^{-1} \begin{pmatrix} p_0 \\ p_1 \\ p_2 \\ p_3 \end{pmatrix} \right) = \begin{pmatrix} 0\,0\,1\,1 \\ 0\,1\,0\,1 \\ 0\,0\,0\,1 \\ 1\,1\,1\,1 \end{pmatrix} \cdot \begin{pmatrix} p_0 \\ p_1 \\ p_2 \\ p_3 \end{pmatrix}, \quad \pi \left(4 \cdot \pi^{-1} \begin{pmatrix} p_0 \\ p_1 \\ p_2 \\ p_3 \end{pmatrix} \right) = \begin{pmatrix} 1\,1\,1\,0 \\ 1\,0\,1\,0 \\ 1\,1\,1\,1 \\ 1\,0\,0\,0 \end{pmatrix} \cdot \begin{pmatrix} p_0 \\ p_1 \\ p_2 \\ p_3 \end{pmatrix}. \tag{33}
$$

This layer is also slightly more costly for the check bits. However, the overhead is not as significant as for PRESENT.

Non-linear Layer. Similar to [36], we decomposed the S-Box into two steps to reduce the number of required shares to three. The functions for the state and check bits are optimized independently of each other. As mentioned before, this procedure results in a more efficient implementation in terms of area with the penalty of not being able to check the correctness of each share individually. To find an area-efficient representation, we applied the same idea as in [7]. In particular, different affine transformations with different combinations of quadratic bijective classes (as defined in [8]) are tested and compared by their number of XOR and AND operations [36]. For the non-encoded TI we tested combinations of the form

$$S = A_3 \circ T_2 \circ A_2 \circ T_1 \circ A_1, \tag{34}$$

where A_1, A_2, A_3 are affine transformations and T_1, T_2 are quadratic bijections. We tested all possible valid combinations of Table 1 from [41] and decomposed the S-Box as $S(m) = F(G(x)), \forall m$ with

$$F = A_3 \circ T_2, \qquad\qquad G = A_2 \circ T_1 \circ A_1.$$

For the check bits Eq. (34) is slightly adjusted to

$$S_p = \pi \circ S \circ \pi^{-1} = \pi \circ A_3 \circ T_2 \circ \pi^{-1} \circ \pi \circ A_2 \circ T_1 \circ A_1 \circ \pi^{-1}, \tag{35}$$

and the S-Box for the check bits is split as $S_p(p) = Q(R(p)), \forall p$ with

$$Q = \pi \circ A_3 \circ T_2 \circ \pi^{-1}, \qquad\qquad R = \pi \circ A_2 \circ T_1 \circ A_1 \circ \pi^{-1}.$$

We found the most efficient decomposition for the classical TI using the quadratic class $\mathcal{Q}_{12}$ for both T_1, T_2 (see [7]). For the check bits the most efficient decomposition was obtained by the quadratic classes $\mathcal{Q}_{294}$ and $\mathcal{Q}_{299}$ for T_1 and T_2, respectively.

As a side note, since $R \neq \pi \circ G \circ \pi^{-1}$ (and likewise for Q and F), the error-checking procedure cannot be performed in-between the S-Box computation. Below we list the algebraic normal form (ANF) of the derived (and applied) functions (a and e as least significant bits).

$$G(d, c, b, a) = (h, g, f, e): \quad e = a + c + d + cb \qquad f = a \tag{36}$$
$$g = 1 + a + d + b + cb \quad h = 1 + a + bc + bd + cd$$

$$F(d, c, b, a) = (h, g, f, e): \quad e = a \qquad\qquad f = c + d + bd \tag{37}$$
$$g = 1 + a + b + c + cd \qquad h = c + bd$$

$$R(d, c, b, a) = (h, g, f, e): \quad e = a + b + db + dc \qquad f = b + c \tag{38}$$
$$g = c + ba + ca \qquad\qquad h = d + b + cb$$

$$Q(d, c, b, a) = (h, g, f, e):$$
$$e = 1 + a + b + c + db + dc \qquad f = 1 + a + b \tag{39}$$
$$g = a + d + db + dc \qquad\qquad h = c + ab + ac + ad + bc + bd$$

The uniform shared representations of the component functions (G_1, G_2, G_3), (F_1, F_2, F_3), (R_1, R_2, R_3), (Q_1, Q_2, Q_3) can be derived by direct sharing [8].

Basic Structure. We implemented the LED encryption with our countermeasure following a round-based architecture. The basic structure of our design is depicted in Fig. 5 with the predictors in the left half. As stated above, the S-Box and its corresponding function on check bits do not follow the same decomposition. Therefore, we perform the error check only at the first registered state $State_1^{i\in\{1,2,3\}}$. The ERROR CHECK module has been implemented following the concept of CHECK-AND-COMBINE, illustrated in Sect. 3.4. Both ADDROUND-KEY and ADDCONSTANTS are only applied to the first share since the key and the constants are not shared. An additional register stage is necessary inside SUBCELLS (between G and F as well as between R and Q) to avoid the propagation of glitches. The initial randomness is shared between both parts of the circuit and none of the layers requires additional fresh randomness to achieve uniformity. It should be noted that except for the initial loading (right half with shared plaintext, and left half with shared corresponding check bits) the two halves of the design do not interact with each other, and each one operates independently. At every clock cycle, the ERROR CHECK module examines the consistency of the state and its corresponding check bits.

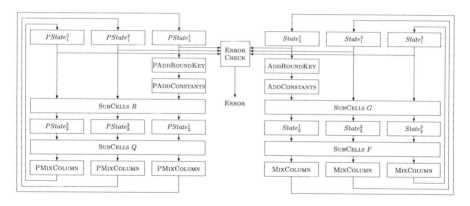

Fig. 5. The basic structure of our proposed LED design. Multiplexers for the plaintext and ADDROUNDKEY are omitted.

The proposed design can be easily extended to provide security against higher-order attacks by increasing the number of shares. As the linear functions are applied on each share separately, their basic structure does not change, while non-linear functions require further adjustment. A second-order TI of the PRESENT S-Box is given in [31]. However, mask refreshing might be necessary to ensure resistance against multivariate higher-order attacks as indicated in [38]. Note, however, that for higher-order TI the error check needs to be also adjusted accordingly to comply with the TI properties. In other words, extra registers should be integrated into COMBINE step of CHECK-AND-COMBINE module (see Fig. 4) and the COMBINE should be performed in several clock cycles to ensure that the desired higher-order resistance is not violated.

4.3   Area Comparison

We synthesized our implementations with the Synopsys Design Compiler using the UMCL18G212T3 [47] ASIC standard cell library (UMC 0.18 μm). The results are presented in Table 1.

As expected, the state registers constitute a significant portion of each circuit part (in the following referred to as *Original* and *Predictors*). Furthermore, the decomposed S-Box is in both cases the largest layer of the design. Since we make use of CHECK-AND-COMBINE, the error detection circuitry is relatively large due to the required additional registers of the COMBINE step. Overall, the predictors require around 27 % more area than the original TI. With the same error detection module, our design with the extended Hamming code is around 12 % bigger than simple duplication.

The synthesized circuit can operate at the maximum frequency of 148 MHz and requires 96 clock cycles for one encryption. The design forms a pipeline, where two plaintexts can be consecutively fed. This results in a maximum throughput of 197.3 Mbit/s. In comparison, the unprotected round-based implementation requires 46 clock cycles for one encryption and can operate at a maximum frequency of 131 MHz. This results in a throughput of 174.7 Mbit/s since the design does not allow a pipeline.

Table 1. Size of our design for an ASIC platform.

Module	Area [GE]			
	Original	Predictors	Error detection	Control
AddRoundKey	171	171	–	–
AddConstants	32	32	–	–
SubCells 1	1750	1584	–	–
SubCells 2	1051	2795	–	–
ShiftRows	0	0	–	–
MixColumnSerial	1532	2048	–	–
Total	7891	10028	2023	270
LED-ParTI	20212			

4.4   Resistance Against SCA

Given that all functions are compliant to the principles of TI, our design is provably secure against first-order attacks. Nevertheless, we also evaluated the security of our design experimentally using an FPGA and ported our design to the FPGA-based side-channel evaluation platform SAKURA-G [1] populated with a Xilinx Spartan-6 FPGA. The power traces obtained for our the design have been collected by means of a digital oscilloscope at sampling rate of 500 MS/s while the design was operating at a frequency of 3 MHz.

As an evaluation metric we used the non-specific t-test as proposed in [14,18] which has become a popular generic evaluation method in recent years [42]. In such a test, the leakages related to two sets of measurements are compared, one with a fixed input (plaintext) and the other one with randomly selected input. During the measurements, for both sets (which are also randomly interleaved) the 128-bit masks (used for initial sharing of the plaintexts as well as the check bits) are randomly selected with a uniform distribution.

While the test can examine the existence of detectable leakage at certain orders, we omit the details here. For further information, the interested reader is referred to the original articles [14,18,42]. Figure 6 depicts the results for univariate tests at first, second and third orders using 100 million measurements. The diagram show that our design is indeed first-order secure while – as expected – leakages for higher orders can be observed.

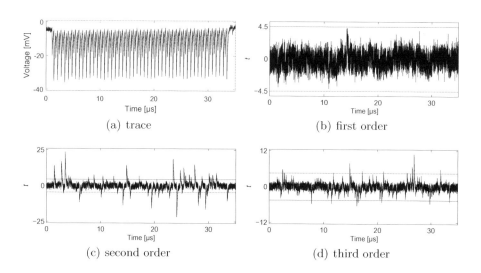

(a) trace

(b) first order

(c) second order

(d) third order

Fig. 6. A sample trace and the result of non-specific t-tests at orders one to three.

4.5   Resistance Against FI

We further examined the fault coverage of our scheme considering the previously introduced attacker model. Given that the extended Hamming $[8, 4, 4]$-code has a distance of $d = 4$, it can detect errors up to $wt(e) \le u = 3$. The coverage of this code is compared to the coverage that can be achieved with a simple $[8, 4, 2]$-duplication code with $u = 1$. We compute the fault coverage of both codes for the uniform distribution as well as for biased distributions $\mathcal{E}_{B_1}$ to $\mathcal{E}_{B_8}$. We consider both the best case (BC) and worst case (WC) for an attacker.

In the best case, the attacker is able to inject a fault into one share of a single codeword. In the worst case, he can inject faults into all shares of all codewords simultaneously. Given that the TI of LED operates on a 16-element state, the fault coverage is significantly increased in this case. Since we do not check each share separately, the number of shares does not influence the fault coverage rate of the worst case.

Table 2 represents the fault coverage rates for the examined cases for both codes. We already discussed how to compute the fault coverage for a duplication code in Sect. 3.5. In order to derive the corresponding fault coverage for the $[8,4,4]$-code, we look at the distribution of the Hamming weight of the codewords. Since $k = 4$, there exist 16 different codewords. 14 of them have Hamming weight $wt(c) = 4$, while there are two codewords with $wt(c) = 0$ and $wt(c) = 8$ respectively. Therefore, only some error vectors with Hamming weight of 4 or 8 (excluding the zero error vector) have the possibility to be undetectable by our scheme[4].

Table 2. Fault coverage for different distributions and codes.

		$\mathcal{E}_U$	$\mathcal{E}_{B_1}$	$\mathcal{E}_{B_2}$	$\mathcal{E}_{B_3}$	$\mathcal{E}_{B_4}$	$\mathcal{E}_{B_5}$	$\mathcal{E}_{B_6}$	$\mathcal{E}_{B_7}$
$[8,4,4]$	BC	0.94	1.00	1.00	1.00	0.91	0.93	0.94	0.94
	WC	$1 - 2^{-65}$	1.00	1.00	1.00	$1 - 2^{-56}$	$1 - 2^{-63}$	$1 - 2^{-66}$	$1 - 2^{-66}$
$[8,4,2]$	BC	0.94	1.00	0.89	0.95	0.93	0.95	0.94	0.94
	WC	$1 - 2^{-65}$	1.00	$1 - 2^{-51}$	$1 - 2^{-72}$	$1 - 2^{-64}$	$1 - 2^{-71}$	$1 - 2^{-66}$	$1 - 2^{-66}$

This observation is confirmed by the results in Table 2. The $[8,4,4]$-code provides full fault coverage in the biased model up to $\mathcal{E}_{B_3}$. Given that most of the valid codewords have a Hamming weight of 4, and $d = 4$, the biased distribution $\mathcal{E}_{B_4}$ leads to the lowest fault coverage. In short, $\mathcal{E}_{B_4}$ and $\mathcal{E}_{B_5}$ are the only distributions with which the simple duplication scheme is better than the extended Hamming code. For all other cases, the extended Hamming code outperforms (or is equal to) the simple duplication scheme. As expected, the worst case leads to very high fault coverage given that the probability to inject an error which results in a valid codeword for every element of the state is very low.

5   Conclusions

We presented an advanced hardware countermeasure which offers resistance both against SCA and FI attacks. In short, we proposed a construction to combine error detecting codes with the concept of threshold implementations. We have identified and discussed generic strategies to that additions for information redundancy do not contradict to the assumptions and requirements of the underlying masking scheme.

---

[4] Since $d = 4$, errors which flip 4 or 8 bits can turn a valid codeword into another valid codeword, and are hence undetectable.

From an general point of view, our combined countermeasure can be applied to arbitrary ciphers and supports different level of protections, i.e., first- or higher-order SCA resistance as well as various fault coverage settings. As an example, we have illustrated how to apply our methodology on the LED block cipher with the aim of maintaining first-order SCA protection while integrating an extended Hamming code to detect faults. Supported by our experimental validation, we have demonstrated how to realize an efficient design that satisfies the requirement to provide protection against SCA and FI.

Acknowledgment. The authors want to thank Falk Schellenberg for his helpful discussions and comments. The research in this work was supported in part by the DFG Research Training Group GRK 1817/1.

References

1. Side-channel attack user reference architecture. http://satoh.cs.uec.ac.jp/SAKURA/index.html
2. Bertoni, G., Breveglieri, L., Koren, I., Maistri, P., Piuri, V.: Error analysis and detection procedures for a hardware implementation of the advanced encryption standard. IEEE Trans. Comput. **52**(4), 492–505 (2003)
3. Biham, E., Shamir, A.: Differential fault analysis of secret key cryptosystems. In: Kaliski Jr., B.S. (ed.) CRYPTO 1997. LNCS, vol. 1294, pp. 513–525. Springer, Heidelberg (1997)
4. Bilgin, B., Daemen, J., Nikov, V., Nikova, S., Rijmen, V., Van Assche, G.: Efficient and first-order DPA resistant implementations of keccak. In: Francillon, A., Rohatgi, P. (eds.) CARDIS 2013. LNCS, vol. 8419, pp. 187–199. Springer, Heidelberg (2014)
5. Bilgin, B., Gierlichs, B., Nikova, S., Nikov, V., Rijmen, V.: A more efficient AES threshold implementation. In: Pointcheval, D., Vergnaud, D. (eds.) AFRICACRYPT 2014. LNCS, vol. 8469, pp. 267–284. Springer, Heidelberg (2014)
6. Bilgin, B., Gierlichs, B., Nikova, S., Nikov, V., Rijmen, V.: Higher-order threshold implementations. In: Sarkar, P., Iwata, T. (eds.) ASIACRYPT 2014, Part II. LNCS, vol. 8874, pp. 326–343. Springer, Heidelberg (2014)
7. Bilgin, B., Nikova, S., Nikov, V., Rijmen, V., Stütz, G.: Threshold implementations of all 3×3 and 4×4 S-boxes. In: Prouff, E., Schaumont, P. (eds.) CHES 2012. LNCS, vol. 7428, pp. 76–91. Springer, Heidelberg (2012)
8. Bilgin, B., Nikova, S., Nikov, V., Rijmen, V., Tokareva, N., Vitkup, V.: Threshold implementations of small S-boxes. Crypt. Commun. **7**(1), 3–33 (2015)
9. Blahut, R.E.: Algebraic Codes for Data Transmission. Cambridge University Press, Cambridge (2003)
10. Bogdanov, A.A., et al.: PRESENT: an ultra-lightweight block cipher. In: Paillier, P., Verbauwhede, I. (eds.) CHES 2007. LNCS, vol. 4727, pp. 450–466. Springer, Heidelberg (2007)
11. Bringer, J., Carlet, C., Chabanne, H., Guilley, S., Maghrebi, H.: Orthogonal direct sum masking. In: Naccache, D., Sauveron, D. (eds.) WISTP 2014. LNCS, vol. 8501, pp. 40–56. Springer, Heidelberg (2014)
12. Bringer, J., Chabanne, H., Le, T.: Protecting AES against side-channel analysis using wire-tap codes. J. Cryptographic Eng. **2**(2), 129–141 (2012)

13. Clavier, C., Feix, B., Gagnerot, G., Roussellet, M.: Passive and active combined attacks on AES—combining fault attacks and side channel analysis. In: FDTC, pp. 10–19. IEEE Computer Society (2010)

14. Cooper, J., Demulder, E., Goodwill, G., Jaffe, J., Kenworthy, G., Rohatgi, P.: Test Vector Leakage Assessment (TVLA) methodology in practice. In: International Cryptographic Module Conference (2013)

15. Dassance, F., Venelli, A.: Combined fault and side-channel attacks on the AES key schedule. In: FDTC, pp. 63–71. IEEE Computer Society (2012)

16. De Cnudde, T., Bilgin, B., Reparaz, O., Nikov, V., Nikova, S.: Higher-order threshold implementation of the AES S-box. In: CARDIS 2015 (2015)

17. Gierlichs, B., Schmidt, J.-M., Tunstall, M.: Infective computation and dummy rounds: fault protection for block ciphers without check-before-output. In: Hevia, A., Neven, G. (eds.) LatinCrypt 2012. LNCS, vol. 7533, pp. 305–321. Springer, Heidelberg (2012)

18. Goodwill, G., Jun, B., Jaffe, J., Rohatgi, P.: A testing methodology for side channel resistance validation. In: NIST Non-invasive Attack Testing Workshop (2011)

19. Guo, J., Peyrin, T., Poschmann, A., Robshaw, M.: The LED block cipher. In: Preneel, B., Takagi, T. (eds.) CHES 2011. LNCS, vol. 6917, pp. 326–341. Springer, Heidelberg (2011)

20. Guo, X., Mukhopadhyay, D., Jin, C., Karri, R.: Security analysis of concurrent error detection against differential fault analysis. J. Cryptographic Eng. $5(3)$, 153–169 (2015)

21. Karpovsky, M.G., Kulikowski, K.J., Taubin, A.: Differential fault analysis attack resistant architectures for the advanced encryption standard. In: Quisquater, J.-J., Paradinas, P., Deswarte, Y., El Kalam, A.A. (eds.) CARDIS. IFIP, vol. 153, pp. 177–192. Kluwer/Springer, USA (2004)

22. Karpovsky, M.G., Kulikowski, K.J., Taubin, A.: Robust protection against fault-injection attacks on smart cards implementing the advanced encryption standard. In: DSN, pp. 93–101. IEEE Computer Society (2004)

23. Karri, R., Kuznetsov, G., Gössel, M.: Parity-based concurrent error detection of substitution-permutation network block ciphers. In: Walter, C.D., Koç, Ç.K., Paar, C. (eds.) CHES 2003. LNCS, vol. 2779, pp. 113–124. Springer, Heidelberg (2003)

24. Kocher, P.C., Jaffe, J., Jun, B.: Differential power analysis. In: Wiener, M. (ed.) CRYPTO 1999. LNCS, vol. 1666, pp. 388–397. Springer, Heidelberg (1999)

25. MacWilliams, F.J., Sloane, N.: The Theory of Error Correcting Codes. North-Holland Mathematical Library. North-Holland Publishing Co., New York (1977). Includes index

26. Mangard, S., Popp, T., Gammel, B.M.: Side-channel leakage of masked CMOS gates. In: Menezes, A. (ed.) CT-RSA 2005. LNCS, vol. 3376, pp. 351–365. Springer, Heidelberg (2005)

27. Mangard, S., Pramstaller, N., Oswald, E.: Successfully attacking masked AES hardware implementations. In: Rao, J.R., Sunar, B. (eds.) CHES 2005. LNCS, vol. 3659, pp. 157–171. Springer, Heidelberg (2005)

28. Moradi, A.: Wire-tap codes as side-channel countermeasure — an FPGA-based experiment. In: Meier, W., Mukhopadhyay, D. (eds.) INDOCRYPT 2014. LNCS, vol. 8885, pp. 341–359. Springer, Switzerland (2014)

29. Moradi, A., Poschmann, A., Ling, S., Paar, C., Wang, H.: Pushing the limits: a very compact and a threshold implementation of AES. In: Paterson, K.G. (ed.) EUROCRYPT 2011. LNCS, vol. 6632, pp. 69–88. Springer, Heidelberg (2011)

30. Moradi, A., Schneider, T.: Side-channel analysis protection and low-latency in action - case study of PRINCE and Midori. Cryptology ePrint Archive, Report 2016/481 (2016). http://eprint.iacr.org/
31. Moradi, A., Wild, A.: Assessment of hiding the higher-order leakages in hardware. In: Güneysu, T., Handschuh, H. (eds.) CHES 2015. LNCS, vol. 9293, pp. 453–474. Springer, Heidelberg (2015)
32. Nikova, S., Rechberger, C., Rijmen, V.: Threshold implementations against side-channel attacks and glitches. In: Ning, P., Qing, S., Li, N. (eds.) ICICS 2006. LNCS, vol. 4307, pp. 529–545. Springer, Heidelberg (2006)
33. Nikova, S., Rijmen, V., Schläffer, M.: Secure hardware implementation of nonlinear functions in the presence of glitches. J. Cryptology **24**(2), 292–321 (2011)
34. NIST: FIPS PUB 197: advanced encryption standard, 14 June 2016. http://csrc.nist.gov/publications/fips/fips197/fips-197.pdf
35. Patranabis, S., Chakraborty, A., Nguyen, P.H., Mukhopadhyay, D.: A biased fault attack on the time redundancy countermeasure for AES. In: Mangard, S., Poschmann, A.Y. (eds.) COSADE 2015. LNCS, vol. 9064, pp. 189–203. Springer, Heidelberg (2015)
36. Poschmann, A., Moradi, A., Khoo, K., Lim, C., Wang, H., Ling, S.: Side-channel resistant crypto for less than 2,300 GE. J. Cryptology **24**(2), 322–345 (2011)
37. Prouff, E., Roche, T.: Higher-order glitches free implementation of the AES using secure multi-party computation protocols. In: Preneel, B., Takagi, T. (eds.) CHES 2011. LNCS, vol. 6917, pp. 63–78. Springer, Heidelberg (2011)
38. Reparaz, O., Bilgin, B., Nikova, S., Gierlichs, B., Verbauwhede, I.: Consolidating masking schemes. In: Gennaro, R., Robshaw, M. (eds.) CRYPTO 2015. LNCS, vol. 9215, pp. 764–783. Springer, Heidelberg (2015)
39. Roche, T., Lomné, V., Khalfallah, K.: Combined fault and side-channel attack on protected implementations of AES. In: Prouff, E. (ed.) CARDIS 2011. LNCS, vol. 7079, pp. 65–83. Springer, Heidelberg (2011)
40. Roscian, C., Sarafianos, A., Dutertre, J., Tria, A.: Fault model analysis of laser-induced faults in SRAM memory cells. In: FDTC, pp. 89–98. IEEE Computer Society (2013)
41. Sasdrich, P., Moradi, A., Güneysu, T.: Affine equivalence and its application to tightening threshold implementations. In: Dunkelman, O., et al. (eds.) SAC 2015. LNCS, vol. 9566, pp. 263–276. Springer, Heidelberg (2016). doi:10.1007/978-3-319-31301-6_16. http://eprint.iacr.org/2015/749
42. Schneider, T., Moradi, A.: Leakage assessment methodology. In: Güneysu, T., Handschuh, H. (eds.) CHES 2015. LNCS, vol. 9293, pp. 495–513. Springer, Heidelberg (2015)
43. Schneider, T., Moradi, A., Güneysu, T.: Arithmetic addition over boolean masking. In: Malkin, T., et al. (eds.) ACNS 2015. LNCS, vol. 9092, pp. 559–578. Springer, Heidelberg (2015). doi:10.1007/978-3-319-28166-7_27
44. Shahverdi, A., Taha, M., Eisenbarth, T.: Silent simon: a threshold implementation under 100 slices. In: HOST 2015, pp. 1–6. IEEE (2015)
45. Sunar, B., Gaubatz, G., Savas, E.: Sequential circuit design for embedded cryptographic applications resilient to adversarial faults. IEEE Trans. Comput. **57**(1), 126–138 (2008)
46. Tiri, K., Verbauwhede, I.: A logic level design methodology for a secure DPA resistant ASIC or FPGA implementation. In: DATE, pp. 246–251. IEEE Computer Society (2004)

47. Virtual Silicon Inc.: 0.18 µm VIP Standard cell library tape out ready, part number: UMCL18G212T3, process: UMC logic 0.18 µm Generic II technology: 0.18 µm, July 2004
48. Xiaofei Guo, D.M., Karri, R.: Provably secure concurrent error detection against differential fault analysis. Cryptology ePrint Archive, Report 2012/552 (2012). http://eprint.iacr.org/

Secure Computation and Protocols I

Network-Hiding Communication
and Applications to Multi-party Protocols

Martin Hirt[1], Ueli Maurer[1], Daniel Tschudi[1(✉)], and Vassilis Zikas[2]

[1] ETH Zurich, Zürich, Switzerland
{hirt,maurer,tschudid}@inf.ethz.ch
[2] RPI, Troy, USA
vzikas@cs.rpi.edu

Abstract. As distributed networks are heavily used in modern applications, new security challenges emerge. In a multi-party computation (in short, MPC) protocol over an incomplete network, such a challenge is to hide, to the extent possible, the topology of the underlying communication network. Such a topology-hiding (aka network hiding) property is in fact very relevant in applications where anonymity is needed.

To our knowledge, with the exception of two recent works by Chandran *et al.* [ITCS 2015] and by Moran *et al.* [TCC 2015], existing MPC protocols do not hide the topology of the underlying communication network. Moreover, the above two solutions are either not applicable to arbitrary networks (as is [ITCS 2015]) or, as in [TCC 2015], they make non-black-box and recursive use of cryptographic primitives resulting in an unrealistic communication and computation complexity even for simple, i.e., low degree and diameter, networks.

Our work suggests the first topology-hiding communication protocol for incomplete networks which makes black-box use of the underlying cryptographic assumption—in particular, a public-key encryption scheme—and tolerates any adversary who passively corrupts arbitrarily many network nodes. Our solutions are based on a new, enhanced variant of threshold homomorphic encryption, in short, TH-PKE, that requires no a-priori setup and allows to circulate an encrypted message over any (unknown) incomplete network and then decrypt it without revealing any network information to intermediate nodes. We show how to realize this enhanced TH-PKE from the DDH assumption. The black-box nature of our scheme, along with some optimization tricks that we employ, makes our communication protocol more efficient than existing solutions.

We then use our communication protocol to make any semi-honest secure MPC protocol topology-hiding with a reasonable—i.e., for simple

D. Tschudi—Research was supported by the Swiss National Science Foundation (SNF), project no. 200020-132794.

V. Zikas—Work done in part while the author was at ETH Zurich supported by the Swiss NSF Ambizione grant PZ00P2_142549, and while the author was visiting the Simons Institute for the Theory of Computing, supported by the Simons Foundation and by the DIMACS/Simons Collaboration in Cryptography through NSF grant #CNS-1523467.

M. Robshaw and J. Katz (Eds.): CRYPTO 2016, Part II, LNCS 9815, pp. 335–365, 2016.
DOI: 10.1007/978-3-662-53008-5_12

networks, polynomial with small constants—communication and computation overhead. We further show how to construct anonymous broadcast without using expensive MPCs to setup the original pseudonyms.

1   Introduction

Secure communication is perhaps the central goal of cryptography. It allows a sender, Alice, to securely transmit a message to a receiver, Bob so that even if some eavesdropper, Eve, is intercepting their communication she can not figure out anything about the transmitted message. When Alice and Bob share a physical (but potentially tappable) communication channel, this task can be easily carried out by use of standard public-key cryptography techniques, e.g., Bob sends Alice his public key who uses it to encrypt her message and send it over the physical communication channel to Bob. But this idealized scenario occurs rarely in modern networks, such as the Internet, where Alice and Bob would most likely not share a physical channel and would, instead, have to communicate over some (potentially incomplete) network of routers. Without further restrictions, the above modification marginally complicates the problem as it can be directly solved by means of a private flooding scheme. In such a scheme, Alice encrypts her message, as before, and sends it to all her immediate neighbors, i.e., network routers with which she shares physical links, who then forward it to their immediate neighbors, and so on, until it reaches Bob. Clearly, if Alice has a path to Bob and the forwarding step is repeated as many times as the length of this path, the message will reach Bob. And the fact that the intermediate routers only see encryptions of the transmitted message means that they do not learn anything about the message.

But modern distributed protocols often require much more than just privacy of the transmitted message. For example, ensuring anonymity in communication is a major goal of security as it, for example, protects against censorship or coercion. Similarly, as privacy awareness in social networks increases, users might not be willing to reveal information about the structure of their peer graph (i.e., their Facebook friends graph) to outsiders. Other applications might require to hide a communicating agent's location, as is the case in espionage or when using mobile agents to propagate information through some ad-hoc network, e.g., in vehicle-to-vehicle communication. All these applications require a routing scheme, that hides the topology of the underlying communication network. Evidently, using the simple private flooding strategy does not hide the topology of the underlying communication network as, for example, an eavesdropping router can easily determine its distance (and direction) to the sender by observing in which round (and from whom) it receives the first encryption.

1.1   Related Literature

The problem of routing through an incomplete network has received a lot of attention in communication networks with a vast amount of works aiming at optimizing communication complexity in various network types. In the following, however, we focus on the cryptographic literature which is more relevant to our goals—namely network hiding communication—and treatment.

Perhaps the main venue of work in which keeping the network hidden is a concern is the literature on anonymous communication, e.g., [Cha03, RR98, SGR97]. These works aim to hide the identity of the sender and receiver in a message transmission, in a way that protects these identities even against traffic analysis. In a different line of work initiated by Chaum [Cha81], so called *mix* servers are used as proxies which shuffle messages sent between various peers to disable an eavesdropper from following a message's path. This technique has been extensively studied and is the basis of several practical anonymization tools. An instance of the mix technique is the so called onion routing [SGR97, RR98], which is perhaps the most wide-spread anonymization technique. Roughly, it consists of the sender applying multiple encryptions in layers on his message, which are then "peeled-off" as the cipher-text travels through a network of onion routers towards its destination. An alternative anonymity technique by Chaum [Cha88] and implemented in various instances (e.g., [Bd90, GJ04, GGOR14]) is known as *Dining Cryptographers networks*, in short DC-nets. Here the parties themselves are responsible for ensuring anonymity. The question of hiding the communication network was also recently addressed in the context of secure multi-party computation by Chandran *et al.* [CCG+15]. This work aims to allow n parties to compute an arbitrary given function in the presence of an adaptive adversary, where each party communicates with a small (sublinear in the total number of parties) number of its neighbors. Towards this goal, [CCG+15] assumes that parties are secretly given a set of neighbors that they can communicate with. Because the adversary is adaptive, it is crucial in their protocol that the communication does not reveal much information about the network topology, as such information would allow the adversary to potentially discover the neighbors of some honest party, corrupt them, and isolate this party, thereby breaking its security[1]. Another work which considers such an adaptive corruption setting is the work of King and Saia [KS10], which is tailored to the Byzantine agreement problem. We note in passing that the result of [CCG+15, KS10] was preceded by several works which considered the problem of MPC over incomplete networks. However, these works do not aim to keep the network hidden as they either only consider a static adversary[2], e.g., [BGT13], and/or they only achieve so called *almost everywhere computation* [GO08, KSSV06a, KSSV06b, CGO15] where the adversary is allowed to isolate a small number of honest parties.

---

[1] In fact, by a factor $\sqrt{n}$ increase on the number of neighbors of each party, [CCG+15] can avoid the assumption of a trusted setup privately distributing the neighborhoods and achieve the same level of security while having the parties generate these neighborhoods themselves.

[2] A static adversary chooses all the parties to corrupt at the beginning of the protocol execution and therefore learning the network topology through the communication cannot help him isolate any honest party.

Most related to the goals of our work is the recent work of Moran et al. [MOR15], which considers the problem of *topology-hiding secure multi-party computation* over an incomplete network in the computational setting (i.e., assuming secure public-key encryption) tolerating a semi-honest (passive) and static adversary. At a very high level, [MOR15] uses public-key encryption and (semi-honest) multi-party computation to implement a proof-of-concept network-hiding communication protocol, which emulates a complete network of secure channels. This emulated network is then used to execute an arbitrary multi-party protocol in which parties communicate over a complete communication network, e.g., [GMW87,Pas04]. In fact, as noted in [MOR15], relying on a computational assumption seems inevitable, as in the information-theoretic setting the work of Hinkelmann and Jakoby [HJ07] excludes fully topology-hiding communication[3]. Due to the similarity to our goal we include a detailed comparison of our results with [MOR15] in Sect. 1.3.

1.2 Our Contributions

In this work we present the first network-hiding communication protocol which makes black-box use of public-key encryption and, for networks with moderate degree and diameter, has a moderate communication and computation complexity. Our protocol allows the parties to communicate over an incomplete network of point-to-point channels in a way which computationally hides both the transmitted message and the neighborhood of honest parties from an adversary *passively* corrupting arbitrary many parties. We remark that as pointed out in [CCG+15], when the communication graph is to be kept hidden, the adversary cannot be eavesdropping on communication channels, and in particular cannot be informed when a message is transmitted over some channel. We resolve this issue by assuming, along the lines of [MOR15], a special network functionality (cf. Sect. 2).

A bit more concretely, the high-level idea of our construction is to enhance the naïve private flooding-protocol by using homomorphic public-key encryption (in short, PKE). The starting point of our approach is the observation—underlying also the construction from [MOR15]—that the flooding protocol would be topology-hiding if the parties could not read intermediate messages. But instead of using, as in [MOR15], expensive nested MPCs for ensuring this fact (see below for a high-level description of [MOR15]) we use a version of threshold PKE with additional network hiding properties. We also show how to implement our enhanced threshold PKE definition assuming hardness of the Decisional Diffie-Hellmann (DDH) problem.

To demonstrate our ideas, imagine there was a world in which parties (corresponding to all intermediate routers) could encrypt with a homomorphic public-key encryption scheme where the private (decryption) key is known to nobody,

---

[3] To our understanding the result of [HJ07] does not apply to the case where a strong information-theoretic setup, e.g., sufficiently long correlated randomness, is available to the parties. Extending this results to that setting is an interesting open problem.

but instead parties have access to a decryption oracle. Provided that the associated PKE-scheme is semantically secure, parties can enhance the flooding protocol as follows: Alice encrypts its message and starts the flooding; in each step of the flooding protocol, the intermediate party—which, recall, is supposed to forward the received ciphertext—first re-randomizes the ciphertext and then forwards it. Once the message arrives to Bob, he invokes the decryption oracle to open its final ciphertext. We observe that in this case the adversary does no longer learn anything from intermediate messages, the protocol is thus topology-hiding.

There are two major challenges with the above approach. First, if intermediate parties are silent until a message reaches them during the flooding, then the adversary observing this fact can use it to deduce information about the network. E.g., if a neighbor p_i of a corrupted party has not sent anything by the second round of the flooding protocol, then the adversary can deduce that p_i is not a neighbor of Alice. Secondly, we need a way to implement the decryption oracle. Observe that using a off-the-shelf threshold decryption scheme and have decryption shares exchanged by means of flooding would trivially destroy the topology-hiding property; and the same is the case if we would use an MPC protocol for this purpose, unless the MPC were itself topology-hiding. In the following we discuss how we solve each of the protocols, separately.

The first issue—information leakage from silent parties—can be solved by having every party send messages in every round. As simple as this idea might seem, it has several difficulties. For starters, the messages that are injected by intermediary parties should be indistinguishable from encryptions, as otherwise adding this noise makes no difference. But now, there is a new issue that the intermediate parties cannot tell which of the indistinguishable messages they receive contains the initial message sent by Alice. The naive solution to this would be to have parties re-randomize everything they receive and add their own noise-message. But this would impose an exponential, in the graph diameter, factor both in the message and communication complexity. Our solution, instead, is to use the homomorphic properties of the encryption scheme and build an efficient process which allows every party to compute an encryption of the OR of the messages it receives from its neighbors. Thus, to transfer a bit b, Alice encrypts b and starts flooding, whereas every party encrypts a zero-bit and starts flooding simultaneously. In each following round of the flooding scheme, every party homomorphically computes the OR of the messages it receives and continues flooding with only this encryption. Bob keeps computing the OR of the encryptions he receives, and once sufficiently many rounds have passed, the decryption is invoked to have him obtain Alice's bit. Note that we only treat the case of semi-honest parties here, thus no party will input an encryption of a one-bit into this smart flooding scheme which would destroy its correctness.

To solve the second issue—i.e., implement the decryption oracle in a topology hiding manner—we introduce a new variant of threshold homomorphic public-key encryption (TH-PKE) with enhanced functionality, which we call *multi-homomorphic threshold encryption with reversible randomization.* Roughly speaking, our new TH-PKE assumes a strongly correlated setup, in which secret (sub)keys are nested in a way which is consistent with the network topology and which allows parties to decrypt messages in a topology hiding manner. We provide a security definition for the new primitive and describe a topology-hiding protocol for establishing the necessary setup using no setup-assumption whatsoever. And we also describe how to instantiate our schemes under the DDH assumptions. We believe that both the general definition of this augmented TH-PKE and the concrete instantiation could be of independent interest and can be used for anonymizing communication.

Applications. Building on our topology hiding network and utilizing the functionality of our topology hiding homomorphic OR protocol we present the following applications:

- Anonymous broadcast: We consider a variant of anonymous broadcast where parties can broadcast messages under a pseudonym. The presented protocol allows to realize anonymous broadcast directly from the topology hiding homomorphic OR protocol without using expensive MPC to setup the pseudonyms.
- Topology hiding MPC: Having a topology-hiding network, we can execute on top of it any MPC protocol from the literature that is designed for point-to-point channels which will render it topology hiding.

1.3 Comparison with [MOR15]

The work by Moran *et al.* [MOR15] provides the first, to the best of our knowledge, work that solves this problem for general graphs in the computational setting. Our goals are closely related to theirs. In fact, our security definition of topology-hiding communication and, more general, computation is a refinement of their simulation-based definition of topology-hiding MPC. But our techniques are very different. In light of this similarity in goals, in the following we include a more detailed comparison to our work.

More concretely, the solution of [MOR15] also follows the approach of enhancing the naïve flooding protocol to make it topology hiding. The key idea is to use nested MPCs, recursively, to protect sensitive information during the execution of the flooding protocol. Roughly, in the basic topology-hiding communication protocol of [MOR15], each party P_i is replaced by a virtual-party $\hat{P}_i$, which is emulated by its immediate neighbors by invoking locally (i.e., in the neighborhood) an off-the-shelf MPC protocol. The complete network of point-to-point channels required by the MPC protocol is emulated by use of a PKE-scheme over the star network centered around P_i, i.e., by naïve flooding where P_i is used as the routing node. The above ensures that P_i cannot analyze the messages that are routed through him, as they are actually handled by its corresponding virtual party $\hat{P}_i$. However, there is now a new problem to be solved, namely, how do virtual parties use the underlying (incomplete) communication network to flood messages in a topology hiding manner? This is solved as follows: To enable secure communication between adjacent virtual-parties a PKE-scheme is used (once more). Here each virtual-party generates a key-pair and sends the encryption key to the adjacent virtual-parties using real parties as intermediaries. This basic protocol is topology-hidingly secure as long as the adversary does not corrupt an entire neighborhood. But this is of course not enough for arbitrarily many corruptions to be tolerated. Thus, to ensure that the overall flooding protocol is also topology hiding, each virtual party is replaced, again by means of MPC, by a "doubly virtual" party $\hat{\hat{P}}$. This will ensure that only adversaries corrupting all the parties that emulate $\hat{\hat{P}}$ can break the topology hiding property. To extend the set of tolerable adversaries, the doubly virtual parties are again emulated, and this process is continued until we reach an emulated party that is emulated by all parties in the network. This requires in the worst case a number of nested MPCs in the order of the network diameter.

In the following we provide a comparison of the solution of [MOR15] with ours demonstrating the advantages of our solution both in terms of simplicity and efficiency. In all fairness, we should remark that the solution of [MOR15] was explicitly proposed as a proof-of-concept solution. The major advantage of our work over [MOR15] is that our communication protocol makes no use of generic MPC, and makes black-box use of the underlying PKE. This not only yields a substantial efficiency improvement, in terms of both communication and computation, but it also yields a more intuitive solution to the problem, as it uses the natural primitive to make communication private, namely encryption, instead of MPC.

More concretely, the player-virtualization protocol from [MOR15] makes non-black-box use of public-key encryption, i.e., the circuit which is computed via MPC is a public-key encryption/decryption circuit. This is typically a huge circuit which imposes an unrealistic slowdown both on the computation complexity and on the round and/or communication complexity[4]. And this is just at the first level of recursion; the computation of the second level, computes a circuit, which computes the circuit, which computes PK encryptions/decryptions, and so on. Due to the lack of concrete suggestions of instantiation of the PKE and MPC used in [MOR15] we were unable to compute exact estimates on the running time and communication complexity of the suggested protocols. Notwithstanding it should be clear that even for the simple case in which the network has constant degree and logarithmic diameter—for which their communication protocol in [MOR15] achieves a polynomial complexity—and even for the best MPC instantiation the actual constants are huge.

Instead, our solutions make black-box use of the underlying PKE scheme and are, therefore, not only more communication and computation efficient, but also easier to analyze. In fact, in our results we include concrete upper bounds on the communication complexity[5] of all our protocols. Indicatively, for a network with diameter D and maximum degree d our network-hiding broadcast protocol

---

[4] Of course the latter can be traded off by choosing to use either a communication heavy or a round heavy protocol.

[5] We note that the computation complexity of our protocols is similar to their communication complexity.

communicates at most $(d + 1)^D \cdot n \cdot \lambda$ bits within just $5 \cdot D$ rounds, where λ is linear (with small constant, less than 5)[6] in the security parameter κ of the underlying PKE scheme. We note that many natural network graphs, such as social networks or the internet have a small diameter[7].

1.4   Preliminaries and Notation

We consider an MPC-like setting where n parties $\mathcal{P} = \{P_1, \ldots, P_n\}$ wish to communicate in a synchronous manner over some incomplete network of secure channels. When the communication is intended to be from P_i, the *sender*, to P_j, the *receiver*, we will refer to the parties in $\mathcal{P} \setminus \{P_i, P_j\}$ as the *intermediate parties*. We will assume a passive and non-adaptive (aka static) computationally bounded adversary who corrupts an arbitrary subset $\overline{H} \subseteq \mathcal{P}$ of parties. Parties in $\overline{H}$ are called *dishonest* or *corrupted* while parties in $H = \mathcal{P} \setminus \overline{H}$ are called *honest*. We use simulation based security to prove our results. For simplicity our proofs are in Canetti's modular composition framework [Can98] but all our results translate immediately to the universal composition UC framework [Can00]. (Recall that we consider semi-honest static security.) In fact, to make this transition smoother, we describe our hybrids in the form of UC functionalities. For compactness, for any functionalities $\mathcal{F}$ and $\mathcal{G}$, we will denote by $\{\mathcal{F}, \mathcal{G}\}$ the composite functionality that gives parallel access to $\mathcal{F}$ and $\mathcal{G}$.

Throughout this work, we assume an, at times implicit, security parameter κ and write $\mathrm{neg}(\kappa)$ to refer to a negligible function of κ. (See [Gol01] for a formal definition of negligible functions.) For an algorithm A we write $(y_1, \ldots, y_k) \leftarrow A(x_1, \ldots, x_k)$ to denote that $(y_1, \ldots, y_k)$ are outputs of A given inputs $(x_1, \ldots, x_k)$. For a probabilistic algorithm B we write $(y_1, \ldots, y_k) \leftarrow B(x_1, \ldots, x_k; r)$ where r is the chosen randomness. If we write $(y_1, \ldots, y_k) \leftarrow B(x_1, \ldots, x_k)$ instead, we assume that the randomness has been chosen uniformly.

1.5   Organization of the Paper

The remainder of the paper is organized as follows. In Sect. 2 we give our definition of topology-hiding security. In Sect. 3 we present a construction which allows to realize topology-hiding communication. The construction is based on multi-homomorphic threshold encryption with reversible randomization (RR-MHT-PKE) which is introduced in Sect. 3.1. Next, in Sect. 3.2 we describe a topology-hiding threshold encryption protocol based on RR-MHT-PKE. This protocol is used in Sect. 3.3 to topology-hidingly realize the Boolean-OR functionality. This allows to give a topology-hiding construction of broadcast and secure channels in Sect. 3.4. Finally, in Sect. 4 we present topology-hiding MPC

---

[6] This can be contrasted with the complexity $O(d)^D \cdot n \cdot \lambda$ obtained by [MOR15].

[7] Backstrom *et al.* [UKBM11] showed that a sub-graph of the Facebook social network consisting of 99.6 % of all users had a diameter of 6. In this particular case the broadcast protocol would communicate at most $n^7 \cdot \lambda$ bits within 30 rounds.

and topology-hiding anonymous broadcast as applications of the protocols from the previous section.

2   Topology Hiding Security Definition

In this section we provide the formal simulation-based definition of topology-hiding computation. Our definition is an adaptation of the original simulation-based definition of Moran et al. [MOR15]. More concretely, the topology-hiding property requires that parties learn no information on the underlying communication network other than the description of their local neighborhood, i.e., the identities of their neighbors. To capture this property, we assume that the parties (in the real world) have access to a *network* functionality $\mathcal{N}$ which has knowledge of every party P_i's neighborhood (i.e., the set of point-to-point channels connected to P_i) and allows P_i to communicate (only) to its neighbors.

Clearly, a protocol execution over such a network $\mathcal{N}$ allows an adversary using it knowledge of the neighborhood of corrupted parties; thus the simulator needs to also be able to provide this information to its environment. To give this power to the simulator, [MOR15] augments the ideal functionality with an extra component which allows the simulator access to this information. In this work we use $\mathcal{N}$ itself in the ideal world to provide this information to the simulator. Note that this does not affect the security statements, as the trivial $\mathcal{N}$-dummy protocol $\phi^{\mathcal{N}}$ securely realizes $\mathcal{N}$[8].

A conceptual point in which our model of topology-hiding computation deviates from the formulation of Moran *et al.* has to do with respect to how the communication graph is chosen. At first thought, one might think that parameterizing the network functionality with the communication graph does the trick. This is, however, not the case because the parameters of hybrid-functionalities are known to the protocol which invokes them and are therefore also known to the adversary. The only information which is not known to the adversary are inputs of corrupted parties and internal randomness of the functionality; thus, as a second attempt, one might try to have the network functionality sample the communication graph from a given distribution[9]. Unfortunately this also fails to capture the topology-hiding property in full, as we would like to make sure that the adversary (or simulator) gets no information on any *given* (hidden) graph.

Motivated by the above, [MOR15] defines topology-hiding computation using the following trick: they assume an extra incorruptible party, whose only role is to provide the network graph as input to the network functionality. Because this network-choosing party is (by assumption) honest, the simulator cannot see its input and needs to work having only the knowledge that $\mathcal{N}$ allows him to obtain, i.e., the neighborhood of corrupted parties.

---

[8] In any case, our protocol will not output anything other than the output of the functionality, hence the simulator will only use $\mathcal{N}$ to learn the corrupted parties neighborhood.

[9] Intuitively, this would correspond to the hidden graph model of [CCG+15].

In this work we take a slightly different, but equivalent in its effect, approach to avoid the above hack of including a special purpose honest party. We assume that each party provides its desired neighborhood to $\mathcal{N}$ as (a special part of) its input. Since the inputs are explicitly chosen by the environment, we are effectively achieving the same topology-hiding property as [MOR15] but without the extra special-purpose honest party.

In the remainder of this section we provide a formal specification of our network functionality (also referred to as network resource) and our formal security definition of topology-hiding computation.

The Network. The network topology is captured by means of an undirected graph $G = (V, E)$ with vertex-set $V = \mathcal{P}$ and edge-set $E \subseteq \mathcal{P} \times \mathcal{P}$. An edge $(P_v, P_u) \in E$ indicates that P_u is in the neighborhood of P_v, which, intuitively, means that P_u and P_v can communicate over a bilateral secure channel. For a party P_v denote by $\mathbf{N}_G(v)$ its neighborhood in G. We will refer to $\mathbf{N}_G[v] = \{P_v\} \cup \mathbf{N}_G(v)$ as P_v's closed neighborhood. Furthermore let $\mathbf{N}_G[v]^k$ be all nodes in G which have distance k or less to P_v. (Clearly $P_v \in \mathbf{N}_G[v]^k$.)

The network functionality allows two types of access: (1) any party $P_v \in \mathcal{P}$ can submit its neighbors $\mathbf{N}_G(v)$, and (2) every party can submit a vector $\vec{m}$ of messages, one for each of its neighbors, which are then delivered in a batch form to their intended recipients. In order to be able to make statements for restricted classes of graphs, e.g., expanders, we parameterize the network functionality by a family $\mathcal{G}$ of setups and require that $\mathcal{N}_\mathcal{G}$ only allows (the environment on behalf of) the honest parties to chose their neighborhood from this class. Note, that the adversary is not bound to choose a neighborhood from a graph in $\mathcal{G}$, i.e., any valid neighborhood is accepted for corrupted parties. This is not an issue in the semi-honest setting considered in this work as a semi-honest adversary will submit whatever input the environment hands it. Thus, for the semi-honest case it suffices that the functionality becomes unavailable (halts) upon receiving an invalid neighborhood from the adversary (or from some honest party)[10]. In the full version of this paper [HMTZ16] we also describe a network functionality that adequately captures the guarantees needed to prevent a malicious adversary from using the check of whether or not the neighborhood he submits results in an invalid-graph message from $\mathcal{N}_\mathcal{G}$ to obtain information on the neighborhood of honest parties.

In the description of $\mathcal{N}_\mathcal{G}$ we use the following notation: For a graph G with vertex set V, and for any $V' \subseteq V$, we denote by $G|_{V'}$ the restriction of G to the vertices in V', i.e., the graph that results by removing from G all vertices in $V \setminus V'$ and their associated edges.

---

[10] Note that the environment knows/chooses all the inputs and therefore knows whether or not the submitted neighborhoods are allowed by the graph class.

Functionality $\mathcal{N}_{\mathcal{G}}$

The network initializes a topology graph $G = (V, E) := (\mathcal{P}, \emptyset)$.

Info Step:

1. Every party $P_i \in \mathcal{P}$ (and the adversary on behalf of corrupted parties) sends (input) $(\texttt{MyNeigborhood}, \mathbf{N}_G[i])$ to $\mathcal{N}$; if $\mathbf{N}_G[i]$ is a valid neighborhood for P_i, i.e., $\mathbf{N}_G[i] \subseteq \{(P_i, P_j) \mid P_j \in \mathcal{P}\}$, then $\mathcal{N}_{\mathcal{G}}$ updates $E := E \cup \mathbf{N}_G[i]$.
2. If there exist no $G' \in \mathcal{G}$ such that $G' = G$ then $\mathcal{N}_{\mathcal{G}}$ sets $E := \emptyset$ and halts. (Every future input is answered by outputting a special symbol $(\texttt{BadNetwork})$ to the sender of this input.)

Communication Step:

1. For each $P_i \in \mathcal{P}$ let $\mathbf{N}_G(i) = \{P_{i_1}, \ldots, P_{i_{\nu_i}}\}$.
2. Every $P_i \in \mathcal{P}$ sends $\mathcal{N}_{\mathcal{G}}$ input $(\texttt{send}, \vec{m}_i)$, where $\vec{m}_i = (m_{i,i_1}, \ldots, m_{i,i_{\nu_i}})$; if P_i does not submit a vector $\vec{m}_i$ of the right size or format, then $\mathcal{N}_{\mathcal{G}}$ adopts $\vec{m}_i = (\perp, \ldots, \perp)$.
3. Every P_i receives (output) $\vec{m}^i = (m_{i_1,i}, \ldots, m_{i_{\nu_i},i})$ from $\mathcal{N}_{\mathcal{G}}$.

An important feature of the above functionality is that the communication pattern (i.e., which parties send or receive messages) does not reveal to the adversary any information other than the neighborhood of corrupted parties. Thus, the simulator cannot use this functionality in the ideal world to extract information about the network. However, when using this network-functionality (in the real-world protocol) to emulate, e.g., a complete communication network, the adversary might use the messages exchanged in the protocol to extract information that the simulator cannot. In fact, the challenge of a topology-hiding protocol is exactly to ensure that the exchanged messages cannot be used by the adversary in such a way.

Definition 1. *Let $\mathcal{G}$ be a family of graphs with vertex set $\mathcal{P}$. Let also $\mathcal{F}$ be a functionality and $\mathcal{N}_{\mathcal{G}}$ denote the network functionality (as specified above) and π be a $\mathcal{N}_{\mathcal{G}}$-hybrid protocol. We say that $\pi^{\mathcal{N}_{\mathcal{G}}}$ securely realizes the functionality $\mathcal{F}$ in a topology-hiding manner with respect to network class $\mathcal{G}$ if and only if π securely realizes the composite functionality $\{\mathcal{F}, \mathcal{N}_{\mathcal{G}}\}$.*

3   Topology-Hiding Communication

In this section we present a construction which allows to securely and topology-hidingly realize different types of communication channels using black-box PKE. The section consists of the following four steps, each treated in a separate subsection.

RR-MHT-PKE: In Sect. 3.1 we introduce *multi-homomorphic threshold encryption with reversible randomization* (RR-MHT-PKE), which is a special type of threshold public-key encryption. In addition to the (common) homomorphic property of ciphertexts RR-MHT-PKE features homomorphic public-keys and decryption-shares. This allows for a decentralized generation of shared keys which enables parties to generate securely and topology-hidingly a public-key setup where the private-key is shared among all parties. Its reversible randomization property allows parties to transmit public-keys and/or ciphertexts through the network such that the adversary can not track them. We give a practical implementation of RR-MHT-PKE based on the DDH assumption in the full version [HMTZ16].

Topology-Hiding Encryption: In Sect. 3.2, we present a topology-hiding threshold encryption protocol based on *black-box* RR-MHT-PKE. More precisely, we provide (1) a distributed setup protocol, (2) an information-transmission protocol, and (3) a distributed decryption protocol.

Topology-hiding Boolean-OR: In Sect. 3.3 we present a protocol which, for networks with moderate degree and diameter, securely and topology-hidingly realizes the multiparty Boolean-OR functionality using the topology-hiding threshold encryption protocol from the previous section.

Topology-hiding Broadcast and Secure Channels: Finally, in Sect. 3.4 we use the Boolean-OR functionality to securely and topology-hidingly realize secure channels and broadcast. The main result of this section is the following theorem.

Theorem 1. *Given a network $\mathcal{N}_\mathcal{G}$ with diameter D and maximum degree d where $d^D = poly(\kappa)$ there exists a protocol which securely and topology-hidingly realizes broadcast using black-box RR-MHT-PKE. The protocol communicates at most $(d+1)^D \cdot n \cdot \lambda$ bits within $5 \cdot D$ rounds, where λ is linear (with small constant, less than 5) in κ.*

3.1 Multi-homomorphic Threshold Encryption with Reversible Randomization

In this section we introduce *multi-homomorphic threshold encryption with reversible randomization*, a special type of threshold public-key encryption, which will allow us to securely and topology-hidingly realize a distributed encryption scheme. We first start by recalling some standard definitions. A public-key encryption (PKE) scheme consists of three algorithms, `Keygen` for key generation, `Enc` for encryption and `Dec` for decryption. Since in this work we consider semi-honest adversaries, we will only need encryption satisfying the standard IND-CPA security definition. *Threshold* public-key encryption (T-PKE) is PKE in which the private key SK is distributed among l parties $p_1, \ldots, p_l$, such that each party p_i holds a share (aka sub-key) sk_i of SK with the property that any $l - 1$ sub-keys have no information on SK. Importantly, such a scheme allows for distributed decryption of any given ciphertext: any party p_i can locally compute, using its own sub-key sk_i of the private key SK, a decryption share x_i, so that if someone gets a hold of decryption shares (for the same c) from all parties (i.e., with each of the shares of the private key) he can combine them and recover the plaintext. *Homomorphic* (threshold) PKE allows to add up encrypted messages. Here, the message space $\langle \mathcal{M}, + \rangle$ and the ciphertext space $\langle \mathcal{C}, \cdot \rangle$ are groups such that $m_1 + m_2 = \mathsf{Dec}(\mathsf{SK}, \mathsf{Enc}(\mathsf{PK}, m_1; r_1) \cdot \mathsf{Enc}(\mathsf{PK}, m_2; r_2))$. for any key pair $(\mathsf{PK}, \mathsf{SK}) \leftarrow \mathsf{KeyGen}$ and any messages $m_1, m_2 \in \mathcal{M}$.

Multi-homomorphic Threshold Encryption. We first present *multi-homomorphic threshold encryption* which is in essence HT-PKE with two additional properties. The first property is a decentralized key-generation. The idea is that parties locally generate public/private-key pairs. By combining those local public keys they can then generate a public key with shared private-key where the local private keys act as key shares. More formally, its required that the public-key space $\langle \mathcal{PK}, \cdot \rangle$ and the private-key space $\langle \mathcal{SK}, + \rangle$ are groups. Moreover its is required (1) that there exists a *key-generation algorithm* KeyGen, which outputs a public/private-key pair $(\mathsf{pk}_i, \mathsf{sk}_i) \in \mathcal{PK} \times \mathcal{SK}$, and (2) that for any key pairs $(\mathsf{pk}_1, \mathsf{sk}_1), (\mathsf{pk}_2, \mathsf{sk}_2) \in \mathcal{PK} \times \mathcal{SK}$ it holds that $\mathsf{pk}_1 \cdot \mathsf{pk}_2$ is the public key corresponding to private key $\mathsf{sk}_1 + \mathsf{sk}_2$. In other words a multi-homomorphic threshold encryption scheme is homomorphic with respect to public/private keys. We point out this is not a standard property of threshold PKE schemes. For instance, the scheme of [Pai99], does not satisfy this property. Secondly, a versatile homomorphic threshold encryption scheme is required to be homomorphic with respect to decryption shares and private keys. That is, for any key pairs $(\mathsf{pk}_1, \mathsf{sk}_1), (\mathsf{pk}_2, \mathsf{sk}_2)$ and any ciphertext c it must hold that $\mathsf{ShareDecrypt}(\mathsf{sk}_1, c) \cdot \mathsf{ShareDecrypt}(\mathsf{sk}_2, c) = \mathsf{ShareDecrypt}(\mathsf{sk}_1 + \mathsf{sk}_2, c)$.

Definition 2. *A multi-homomorphic threshold encryption (MHT − PKE) scheme with security parameter κ consists of four spaces $\mathcal{M}$, $\mathcal{C}$, $\mathcal{SK}$, and $\mathcal{PK}$ and four algorithms* KeyGen, Enc, $\mathsf{ShareDecrypt}$, *and* $\mathsf{Combine}$ *which are parametrized by κ where:*

1. *The message space $\langle \mathcal{M}; + \rangle$, the public-key space $\langle \mathcal{PK}; \cdot \rangle$, the private-key space $\langle \mathcal{SK}; + \rangle$, the ciphertext space $\langle \mathcal{C}; \cdot \rangle$, and the decryption-share space $\langle \mathcal{DS}; \cdot \rangle$ are cyclic groups of prime order.*
2. *The (probabilistic) key-generation algorithm* KeyGen *outputs a public key $\mathsf{pk} \in \mathcal{PK}$ and a private key $\mathsf{sk} \in \mathcal{SK}$ where for any key pairs $(\mathsf{pk}_1, \mathsf{sk}_1)$, $(\mathsf{pk}_2, \mathsf{sk}_2) \in \mathcal{PK} \times \mathcal{SK}$ it holds that $\mathsf{pk}_1 \cdot \mathsf{pk}_2$ is the public key corresponding to private key $\mathsf{sk}_1 + \mathsf{sk}_2$.*
3. *The (probabilistic) encryption algorithm* Enc *takes a public key $\mathsf{pk} \in \mathcal{PK}$ and a message $m \in \mathcal{M}$ and outputs a ciphertext $c \leftarrow \mathsf{Enc}(\mathsf{PK}, m; r)$.*
4. *The decryption share algorithm* $\mathsf{ShareDecrypt}$ *takes a private key $\mathsf{sk}_i \in \mathcal{SK}$ and a ciphertext $c \in \mathcal{C}$ as inputs and outputs a decryption share $\mathsf{x}_i \leftarrow \mathsf{ShareDecrypt}(\mathsf{sk}_i, c)$. For any ciphertext $c \in \mathcal{C}$ and private keys $\mathsf{sk}_1, \mathsf{sk}_2 \in \mathcal{SK}$ where $\mathsf{x}_1 \leftarrow \mathsf{ShareDecrypt}(\mathsf{sk}_1, c)$ and $\mathsf{x}_2 \leftarrow \mathsf{ShareDecrypt}(\mathsf{sk}_2, c)$ it holds that $\mathsf{x}_1 \cdot \mathsf{x}_2 = \mathsf{ShareDecrypt}(\mathsf{sk}_1 + \mathsf{sk}_2, c)$.*

5. *The* combining algorithm Combine *takes a decryption share* $x \in \mathcal{DS}$ *and a ciphertext* $c \in \mathcal{C}$ *and outputs a message* $m \leftarrow$ Combine(x, c).

A MHT-PKE scheme satisfies the following correctness property: For any key pairs $(\mathsf{pk}_1, \mathsf{sk}_1), \ldots, (\mathsf{pk}_l, \mathsf{sk}_l) \leftarrow$ KeyGen *and any message* $m \in \mathcal{M}$ *it holds that* $m =$ Combine$(x_1 \cdot \ldots \cdot x_l, c)$ *where* $x_i =$ ShareDecrypt(sk_i, c), $c =$ Enc$(\mathsf{pk}, m; r)$ *and* $\mathsf{pk} = \mathsf{pk}_1 \cdot \ldots \cdot \mathsf{pk}_l$. *Moreover, given a message* m *and a ciphertext* c *one can efficiently invert* Combine, *i.e., compute a decryption share* x *with* $m =$ Combine(x, c).

We define the security of MHT-PKE with respect to a threshold variant of the IND-CPA security definition.

Definition 3. *A MHT-PKE scheme is* IND-TCPA *secure if the adversary's advantage in winning the following game is negligible in* κ.

1. *The game generates key pairs* $(\mathsf{pk}_1, \mathsf{sk}_1), \ldots (\mathsf{pk}_l, \mathsf{sk}_l) \leftarrow$ KeyGen *and chooses a random bit* b. *Then the adversary gets* $\mathsf{pk} = \mathsf{pk}_1 \cdot \ldots \cdot \mathsf{pk}_l$, $\mathsf{pk}_1, \ldots, \mathsf{pk}_l$ *and* $\mathsf{sk}_2, \ldots, \mathsf{sk}_l$. *This allows him to generate encryptions of arbitrary messages and to generate decryption shares for all key pairs except* $(\mathsf{pk}_1, \mathsf{sk}_1)$.
2. *The adversary specifies two messages* m_0 *and* m_1 *and the game returns* $c =$ Enc(PK, m_b).
3. *The adversary specifies a bit* b'. *If* $b = b'$ *the adversary has won the game.*

Furthermore for any chosen public-key $\mathsf{pk} \in \mathcal{PK}$, it should be hard to distinguish between $(\mathsf{pk}, \mathsf{pk} \cdot \mathsf{pk}_1)$ and $(\mathsf{pk}, \mathsf{pk}_2)$ where $\mathsf{pk}_1, \mathsf{pk}_2$ are distributed according to KeyGen. More formally, we require that the scheme has the *indistinguishability under chosen public-key attack* (IND-CKA) property.

Definition 4. *A MHT-PKE scheme is* IND-CKA *secure if the adversary's advantage in winning the following game is negligible in* κ.

1. *The adversary specifies a public key* $\mathsf{pk} \in \mathcal{PK}$.
2. *The game generates a key pair* $(\mathsf{pk}_1, \mathsf{sk}_1) \leftarrow$ KeyGen *and chooses a uniform random bit* b. *Then the adversary gets public key* pk_2 *where*

$$\mathsf{pk}_2 = \begin{cases} \mathsf{pk}_1 & \text{if } b = 0 \\ \mathsf{pk}_1 \cdot \mathsf{pk} & \text{if } b = 1 \end{cases}$$

3. *The adversary specifies a bit* b'. *If* $b = b'$ *the adversary has won the game.*

Reversible Randomization. Next, we introduce multi-homomorphic threshold encryption with *reversible randomization* which is MHT-PKE with additional randomization properties.

Randomization of Public Keys. The first property required is the randomization of public keys. More concretely, a MHT-PKE with reversible randomization allows a party P_i with public key pk_i to "randomize" pk_i, i.e., compute a new masked public-key $\widetilde{\mathsf{pk}}_i$ so that anyone seeing $\widetilde{\mathsf{pk}}_i$ is unable to tell whether it is a freshly generated public-key or a randomized version of pk_i. Importantly, we require the randomization algorithm to be reversible in the following sense. The randomization algorithm must provide P_i with information rk_i, the *de-randomizer*, which allows it to map any encryption with $\widetilde{\mathsf{pk}}_i$ back to an encryption with its original key pk_i. Looking ahead, the randomization of public-keys property will ensure that the adversary can not trace public keys while they travel the network. This allows us to build a topology-hiding information-transmission protocol.

Randomization of Ciphertexts. The second property required is the randomization of ciphertexts. More concretely, a MHT-PKE with reversible randomization allows a party P_i with ciphertext c_i to "randomize" c_i, i.e., compute a new masked ciphertext $\widehat{c}_i$ so that anyone seeing $\widehat{c}_i$ is unable to tell whether it is a freshly generated ciphertext (using an arbitrary public-key) or an randomized version of c_i. Importantly, we require the randomization algorithm to be reversible. This means it must provide P_i with information rk_i, the *de-randomizer*, which allows it to map any decryption share of $\widehat{c}_i$ and decryption key sk back to a decryption share of the original ciphertext c_i and sk. Looking ahead, the randomization of ciphertexts will ensure that the adversary can not trace ciphertexts and decryption-shares while they travel the network. This will allow us to build a topology-hiding decryption protocol. We remark that this property differs from the usual ciphertext re-randomization in homomorphic PKE schemes where one randomizes a ciphertext by adding up an encryption of 0.

MHT-PKE with Reversible Randomization. We can now give the formal definition of a MHT-PKE with reversible-randomization scheme.

Definition 5. *A* MHT-PKE *with reversible-randomization (RR-MHT-PKE) scheme is a MHT-PKE scheme with extra algorithms* RandKey, DerandCipher, RandCipher, DerandShare *where:*

1. *The (probabilistic) (key) randomization algorithm* RandKey *takes a public key* $\mathsf{pk} \in \mathcal{PK}$ *and outputs a new public key* $\widetilde{\mathsf{pk}} \in \mathcal{PK}$ *and a de-randomizer* $\mathsf{rk} \in \mathcal{RK}_P$.
2. *The* (ciphertext) de-randomization algorithm DerandCipher *takes a de-randomizer* $\mathsf{rk} \in \mathcal{RK}_P$ *and a ciphertext* $\widetilde{c} \in \mathcal{C}$ *and outputs a new ciphertext* $c \in \mathcal{C}$ *such that the following property holds. For any key pair* $(\mathsf{pk}, \mathsf{sk})$, $(\widetilde{\mathsf{pk}}, \mathsf{rk}) \leftarrow$ RandKey$(\mathsf{pk}; r')$, *any message* $m \in \mathcal{M}$, *and any ciphertext* $\widetilde{c} \leftarrow$ Enc$(\widetilde{\mathsf{pk}}, m; \widetilde{r})$ *there exists an* r *such that* Enc$(\mathsf{pk}, m; r) =$ DerandCipher$(\mathsf{rk}, \widetilde{c})$. *Moreover, given a ciphertext* c *and a de-randomizer* rk *one can efficiently invert* DerandCipher, *i.e., compute a ciphertext* $\widetilde{c}$ *such that* $c =$ DerandCipher$(\mathsf{rk}, \widetilde{c})$.

3. *The (probabilistic)* (ciphertext) randomization algorithm RandCipher *takes a ciphertext $c \in \mathcal{C}$ and outputs a new ciphertext $\hat{c} \in \mathcal{C}$ and a de-randomizer* rk $\in \mathcal{RK}_C$.
4. *The* (share) de-randomization algorithm DerandShare *takes a de-randomizer* rk $\in \mathcal{RK}_C$ *and a decryption share* $\hat{x} \in \mathcal{DS}$ *and outputs a decryption share* $x \in \mathcal{DS}$ *such that the following property holds. For any key pair* (pk, sk), *any ciphertext* $c \in \mathcal{C}$, $(\text{rk}, \hat{c}) \leftarrow$ RandCipher$(c; r)$, *and* $\hat{x} \leftarrow$ ShareDecrypt$(\text{sk}_i, \hat{c})$ *we have* DerandShare$(\text{rk}, \hat{x}) =$ ShareDecrypt(sk_i, c). *More over given a decryption share* x *and a de-randomizer* rk *one can efficiently invert* DerandShare, *i.e., compute a decryption shares* $\hat{x}$ *such that* $x =$ DerandShare$(\text{rk}, \hat{x})$.

For any public key pk it should be hard (for the adversary) to distinguish between $(\text{pk}, \text{RandKey}(\text{pk}))$ and (pk, pk') where pk' is freshly generated using KeyGen. Similar, for any ciphertext c it should be hard to distinguish between $(c, \text{RandCipher}(c))$ and (c, c') where c' is a randomly chosen ciphertext. More formally, the scheme should have the *indistinguishability under chosen public-key and chosen ciphertext attack* (IND-CKCA) property.

Definition 6. *A RR-MHT-PKE scheme is* IND-CKCA *secure if the adversary's advantage in winning the following game is negligible in κ.*

1. *The adversary specifies a public key* $\text{pk} \in \mathcal{PK}$ *and a ciphertext* $c \in \mathcal{C}$.
2. *The game generates key pairs* $(\text{pk}_1, \text{sk}_1), (\text{pk}_2, \text{sk}_2) \leftarrow$ KeyGen *and a uniform random message* $m \in \mathcal{M}$. *The game then chooses uniform random bits* b_1 *and* b_2. *The adversary gets public key* $\widetilde{\text{pk}}$ *and ciphertext* $\hat{c}$ *where*

$$\widetilde{\text{pk}} = \begin{cases} \text{RandKey}(\text{pk}) & \textit{if } b_1 = 0 \\ \text{pk}_1 & \textit{if } b_1 = 1 \end{cases}$$

and

$$\hat{c} = \begin{cases} \text{RandCipher}(c) & \textit{if } b_2 = 0 \\ \text{Enc}(\text{pk}_2, m) & \textit{if } b_2 = 1 \end{cases}.$$

3. *The adversary specifies bits* b_1' *and* b_2'. *If* $b_1 = b_1'$ *or* $b_2 = b_2'$ *the adversary has won the game.*

The security of a RR-MHT-PKE scheme is defined with respect to the above security properties.

Definition 7. *A RR-MHT-PKE scheme is* secure *if it is* IND-TCPA, *IND-CKA, and* IND-CKCA *secure.*

DDH Based RR-MHT-PKE. One can practically implement secure RR-MHT-PKE using an extended variant of the ElGamal cryptosystem [ElG84] over a group G of prime order $q(\kappa)$ where the DDH assumption holds. We refer to the full version [HMTZ16] for more details.

Lemma 1. *Given a DDH group one can securely implement RR-MHT-PKE.*

3.2 Topology-Hiding Threshold Encryption

In this section we build a topology-hiding threshold encryption protocol using a secure RR-MHT-PKE scheme. More precisely, we provide (1) a distributed setup protocol, (2) an information-transmission protocol, and (3) a distributed decryption protocol. Looking ahead, those protocols will allow us to topology-hidingly realize the Boolean-OR functionality.

The RR-MHT-PKE Scheme: We assume that the parties have access to a secure RR-MHT-PKE scheme with security parameter κ, where $n = poly(\kappa)$. In particular, each party has local (black-box) access to the algorithms of the RR-MHT-PKE scheme.

The Network Graph: A prerequisite for our protocols to work is that the network graph G of $\mathcal{N}_G$ is connected. Otherwise (global) information transmission is not possible. The parties also need to know upper bounds on the maximum degree and the diameter of the network graph. We therefore assume that the parties have access to an initialized network $\mathcal{N}_{\mathcal{G}}^{d,D}$ where the graphs in the family $\mathcal{G}$ are connected, have a maximum degree of $d \leq n$, and a diameter of at most $D \leq n$ where d and D are publicly known. For simplicity we restrict ourselves to present protocols for d-regular network graphs. We point out that one can extend the presented protocols to the general case where parties may have less than d neighbors. The idea is that a party which lacks d neighbors pretends to have d neighbors by emulating (messages from) virtual neighbors (cf. [MOR15]).

Setup Protocol. In this section we present a protocol which allows to topology-hidingly generate a threshold-setup where each party P_i holds a public key PK_i such that the corresponding private-key is shared among all parties. The high-level idea of our protocol is as follows. We first observe that the D-neighborhood of P_i consists of all parties. The setup thus provides party P_i with a public key where the corresponding private-key is shared among the parties in the D-neighborhood $\mathbf{N}_G[i]^D$ of P_i. This implies that one can generate the setup recursively. In order to generate a k-neighborhood public-key $\mathsf{PK}_i^{(k)}$, P_i asks each of its neighbors to generate a public key where the private key is shared in the neighbors $(k-1)$-neighborhood. It can then compute $\mathsf{PK}_i^{(k)}$ by combining the received public-keys.

Definition 8. *A setup for topology-hiding threshold encryption over a network $\mathcal{N}_{\mathcal{G}}^{d,D}$ consists of the following parts.*

Private-Key Shares: *Each party P_i holds a vector $(\overline{\mathsf{SK}}_i^{(0)}, \ldots, \overline{\mathsf{SK}}_i^{(D)})$ of $D+1$ private keys which we call its private-key shares. For any $0 \leq k \leq D$ we denote by $\overline{\mathsf{PK}}_i^{(k)}$ the public key corresponding to $\overline{\mathsf{SK}}_i^{(r)}$.*

Public-Keys: *Each party P_i holds a vector $(\mathsf{PK}_i^{(0)}, \ldots, \mathsf{PK}_i^{(D)})$ of $D+1$ public keys where $\mathsf{PK}_i^{(0)} = \overline{\mathsf{PK}}_i^{(0)}$ and $\mathsf{PK}_i^{(k)} = \overline{\mathsf{PK}}_i^{(k)} \cdot \prod_{P_j \in \mathbf{N}_G(i)} \mathsf{PK}_j^{(k-1)}$. We call $\mathsf{PK}_i^{(k)}$ the level-k public-key of P_i and denote by $\mathsf{SK}_i^{(r)}$ the corresponding (shared) private key. The public-key of P_i is $\mathsf{PK}_i := \mathsf{PK}_i^{(D)}$ and the shared private-key is $\mathsf{SK}_i := \mathsf{SK}_i^{(D)}$.*

Local Pseudonyms: *Each party P_i privately holds a injective random function $\nu_i(\cdot) : \mathbf{N}_G(i) \to \{1, \ldots, d\}$ which assigns each neighbor $P_j \in \mathbf{N}_G(i)$ a unique local identity $\nu_i(j) \in \{1, \ldots, d\}$. W.l.o.g. we will assume that $\nu_i(i) = 0$.*

We remark that the condition on the public-keys ensures that any $0 \leq k \leq D$ (and for reasonably large $\mathcal{PK}$) the private key $\mathsf{SK}_i^{(k)}$ is properly shared among the k neighborhood of P_i, i.e., each party in the k-neighborhood holds a non-trivial share.

Definition 9. *A protocol is a secure (topology-hiding) setup protocol over a network $\mathcal{N}_{\mathcal{G}}^{d,D}$ if it has the following properties.*

Correctness: *The protocol generates with overwhelming probability a setup for topology-hiding threshold encryption over the network $\mathcal{N}_{\mathcal{G}}^{d,D}$.*

Topology-Hiding Simulation: *The adversarial view in an actual protocol-execution can be simulated with overwhelming probability given the neighborhood of dishonest parties in $\mathcal{N}_{\mathcal{G}}^{d,D}$ and the output of dishonest parties, i.e., given the values*

$$\left\{ \mathbf{N}_G(i), \nu_i(\cdot), \overline{\mathsf{SK}}_i^{(0)}, \ldots, \overline{\mathsf{SK}}_i^{(D)}, \mathsf{PK}_i^{(0)}, \ldots, \mathsf{PK}_i^{(D)} \right\}_{P_i \in \overline{H}}$$

The simulation property ensures in particular that (a) the adversary does not learn more about the network topology and that (b) the adversary does not learn the private key corresponding to the public key PK_i of party P_i unless it corrupts the entire k-neighborhood of P_i.

Protocol GenerateSetup

Require: Parties have access to an initialized $\mathcal{N}_{\mathcal{G}}^{d,D}$.
1: Each P_i generates the local identities $\nu_i(\cdot)$ and sub-key pair $(\overline{\mathsf{PK}}_i^{(0)}, \overline{\mathsf{SK}}_i^{(0)}) \leftarrow \mathsf{KeyGen}$. Then it sets $\mathsf{PK}_i^{(0)} = \overline{\mathsf{PK}}_i^{(0)}$.
2: **for** $k = 1, \ldots, D$ **do**
3:     Each P_i sends $\mathsf{PK}_i^{(k-1)}$ to each $P_j \in \mathbf{N}_G(i)$ using $\mathcal{N}$.
4:     Each P_i generates sub-key pair $(\overline{\mathsf{PK}}_i^{(k)}, \overline{\mathsf{SK}}_i^{(k)}) \leftarrow \mathsf{KeyGen}$.
5:     Each P_i computes $\mathsf{PK}_i^{(k)} = \overline{\mathsf{PK}}_i^{(k)} \cdot \prod_{P_j \in \mathbf{N}_G(i)} \mathsf{PK}_j^{(k-1)}$.
6: **end for**
Output: P_i outputs $\nu_i(\cdot)$, $(\overline{\mathsf{SK}}_i^{(0)}, \ldots, \overline{\mathsf{SK}}_i^{(D)})$, and $(\mathsf{PK}_i^{(0)}, \ldots, \mathsf{PK}_i^{(D)})$.

Lemma 2. *Given a secure RR-MHT-PKE scheme the protocol GenerateSetup is a secure setup protocol. The protocol communicates $D \cdot d \cdot n \cdot \log |\mathcal{PK}|$ bits within D rounds.*

Proof. (sketch) **Correctness:** It follows directly from protocol inspection that the setup generated by `GenerateSetup` is valid for $\mathcal{N}_{\mathcal{G}}^{d,D}$. **Topology-Hiding Simulation:** The view of the adversary during an actual protocol execution is

$$\left\{ \mathbf{N}_G(i), \nu_i(\cdot), \left\{ \mathsf{PK}_i^{(k)}, \overline{\mathsf{PK}}_i^{(k)}, \overline{\mathsf{SK}}_i^{(k)} \right\}_{0 \leq r \leq D}, \left\{ \mathsf{PK}_j^{(k)} \right\}_{P_j \in \mathbf{N}_G(i), 0 \leq r \leq D-1} \right\}_{P_i \in \overline{H}}.$$

Now consider the view where the public keys $\left\{ \mathsf{PK}_j^{(k)} \right\}_{P_j \in \mathbf{N}_G(i) \cap H, 0 \leq r \leq D-1}$ are replaced by freshly generated public keys using `KeyGen`, i.e.,

$$\left\{ \mathbf{N}_G(i), \nu_i(\cdot), \left\{ \mathsf{PK}_i^{(k)}, \overline{\mathsf{PK}}_i^{(k)}, \overline{\mathsf{SK}}_i^{(k)} \right\}_{0 \leq r \leq D}, \left\{ \widetilde{\mathsf{PK}}_j^{(k)} \right\}_{P_j \in \mathbf{N}_G(i) \cap H, 0 \leq r \leq D-1} \right\}_{P_i \in \overline{H}}.$$

Note that the second view can be easily computed by a simulator given the outputs of dishonest parties. It remains to show that those views are computationally indistinguishable. Note that for any $P_j \in \mathbf{N}_G(\overline{H}) \cap H$ the public-key $\mathsf{PK}_j^{(k)}$ has the form $\mathsf{pk}_1 \cdot \mathsf{pk}$ where $\mathsf{pk}_1 = \overline{\mathsf{PK}}_j^{(k)}$ and $\mathsf{pk} = \prod_{P_i \in \mathbf{N}_G(j)} \mathsf{PK}_i^{(k-1)}$. The indistinguishability therefore follows from the IND-CKA security of the RR-MHT-PKE scheme. **Communication Complexity:** The protocol runs for D rounds and in each round $n \cdot d$ public-keys are sent.

Information-Transmission Protocol. In this section we present a topology-hiding information-transmission protocol. Here, each party has a message m_i and a public-key pk_i[11] as input. The output of party P_i is a ciphertext c_i under the public key pk_i. If all parties input the 0-message, c_i is an encryption of 0. Otherwise, c_i is an encryption of a random, non-zero message. The information-transmission protocol has a recursive structure and is thus parametrized by a level k. The protocol requires that parties have generated local pseudonyms. We therefore assume that the parties have access to a setup for topology-hiding threshold encryption over $\mathcal{N}_{\mathcal{G}}^{d,D}$.

Definition 10. *A protocol is a level-k (topology-hiding) secure information-transmission protocol over a network $\mathcal{N}_{\mathcal{G}}^{d,D}$ if it has the following properties.*

Setup, Inputs, and Outputs: *The parties initially hold a setup for topology-hiding threshold encryption over $\mathcal{N}_{\mathcal{G}}^{d,D}$ (cf. Definition 8). Each party holds as input a message $m_i \in \mathcal{M}$ and a public key $\mathsf{pk}_i \in \mathcal{PK}$ (not necessarily part of its setup).*
The output of each party P_i is a ciphertext $c_i \in \mathcal{C}$.
Correctness: *With overwhelming probability the output c_i is the encryption of message s_i under pk_i and randomness ρ_i (i.e. $c_i = \mathsf{Enc}(\mathsf{pk}_i, s_i; \rho_i)$) with*

$$s_i = \begin{cases} 0 & if \ m_j = 0 \ for \ all \ P_j \in \mathbf{N}_G[i]^k \\ x_i & if \ m_j \neq 0 \ for \ at \ least \ one \ P_j \in \mathbf{N}_G[i]^k \end{cases}$$

where $x_i \in \mathcal{M} \setminus \{0\}$ uniform at random.

---

[11] For notational simplicity we use uppercase letters for public-/private-keys which are part of the setup for $\mathcal{N}_{\mathcal{G}}^{d,D}$ and lowercase letters for arbitrary public-/private-keys.

Topology-Hiding Simulation: *The adversarial view in a real protocol-execution can be simulated with overwhelming probability given the following values*

$$\big\{ \mathbf{N}_G(i), m_i, \mathsf{pk}_i, c_i, \nu_i(\cdot) \big\}_{P_i \in \overline{H}} \cup \big\{ s_i, \rho_i \big\}_{\mathbf{N}_G[i]^k \subseteq \overline{H}}.$$

In other words the simulator gets the neighborhood of dishonest parties (in $\mathcal{N}_\mathcal{G}^{d,D}$), their protocol in- and outputs, and their local pseudonyms from the setup. For any party P_i where the whole k-neighborhood is dishonest the simulator is additionally given the content s_i and the randomness ρ_i of output c_i.

The simulation property ensures in particular that (a) the adversary does not learn more about the network topology and that (b) the adversary does not learn the content of ciphertext c_i of party P_i unless it corrupts the entire k-neighborhood of P_i.

Protocol InfoTransmisson $\big(k, (m_1, \mathsf{pk}_1), \ldots, (m_n, \mathsf{pk}_n) \big)$

Require: Parties have access to an initialized $\mathcal{N}_\mathcal{G}^{d,D}$ and have generated local pseudonyms.
Input: Each P_i inputs a message m_i and a public key pk_i.
1: **if** $k = 0$ **then**
2:      Each P_i computes $c_i = \mathsf{Enc}(\mathsf{pk}_i, 0)$ if $m_i = 0$ or $c_i = \mathsf{Enc}(\mathsf{pk}_i, x_i)$ if $m_i \neq 0$ where $x_i \in \mathcal{M} \setminus \{0\}$ uniform at random.
3: **else**
4:      Each P_i computes $(\widetilde{\mathsf{pk}}_i, \mathsf{rk}_i) \leftarrow \mathsf{RandKey}(\mathsf{pk}_i)$ and sends $\widetilde{\mathsf{pk}}_i$ to each $P_j \in \mathbf{N}_G[i]$ which denotes the received key by $\mathsf{pk}_{j,\nu_j(i)}$.
5:      **for** $l = 0, \ldots, d$ **do**
6:         The parties compute ciphertexts $(\widetilde{c}_{1,l}, \ldots, \widetilde{c}_{n,l})$ by invoking subprotocol InfoTransmisson $\big(k - 1, (m_1, \mathsf{pk}_{1,l}), \ldots, (m_n, \mathsf{pk}_{n,l}) \big)$.
7:      **end for**
8:      Each P_i sends $\widetilde{c}_{i,\nu_i(j)}$ to $P_j \in \mathbf{N}_G[i]$.
9:      Each P_i computes $c_i = \big(\prod_{P_j \in \mathbf{N}_G[i]} \mathsf{DerandCipher}(\mathsf{rk}_i, \widetilde{c}_{j,\nu_j(i)}) \big)^{r_i}$ for a uniform random $r_i \in \{1, \ldots, |\mathcal{M}| - 1\}$.
10: **end if**
Output: Each P_i outputs c_i.

Lemma 3. *Given a secure RR-MHT-PKE scheme and for any parameter $0 \leq k \leq D$ with $d^k = poly(\kappa)$, InfoTransmisson $(k, (m_1, \mathsf{pk}_1), \ldots, (m_n, \mathsf{pk}_n))$ is a secure level-k information-transmission protocol. The protocol communicates at most $(d + 1)^k \cdot n \cdot (\log |\mathcal{PK}| + \log |\mathcal{C}|)$ bits within $2k$ rounds.*

Proof. (sketch) **Correctness:** For $k = 0$ each party locally computes c_i as specified by the correctness property. The protocol thus achieves correctness perfectly. For $k > 0$ assume that the protocol achieves correctness for $(k - 1)$. More precisely, the output of a party P_j for parameter $(k - 1)$ is computed perfectly correct if all $(k - 1)$-neighbors have input 0. Otherwise, the output of P_j for parameter $(k - 1)$ is computed correctly except with error probability ε_{k-1}.

First, we consider the case where all parties in the k-neighborhood of P_i have input 0. The assumption for $(k-1)$ implies that all $\widetilde{c}_{i,\nu_i(j)}$ contain 0. The properties of the RR-MHT-PKE scheme imply that $s_i = r_i \cdot 0 = 0$. In the second case at least one party in the k-neighborhood of P_i has a non-zero input. This implies that at least one $\widetilde{c}_{i,\nu_i(j)}$ contains a uniform random, non-zero message (with error probability of at most ε_{k-1}). The properties of the RR-MHT-PKE thus ensure that c_i contains a uniform random, non-zero message (except with error probability $\varepsilon_k := \varepsilon_{k-1} + \frac{1}{|\mathcal{M}|}$). This implies an overall success probability of at least $1 - (\frac{k \cdot n}{|\mathcal{M}|})$. **Topology-Hiding Simulation:** To simulate the view of the adversary the simulator is given

$$\left\{ \mathbf{N}_G(i), m_i, \mathsf{pk}_i, c_i, \nu_i(\cdot) \right\}_{P_i \in \overline{H}} \cup \left\{ s_i, \rho_i \right\}_{\mathbf{N}_G[i]^k \subseteq \overline{H}}.$$

For $k = 0$ those values correspond exactly to the view of the adversary during an actual protocol execution. Simulation is thus easy. For the case $k > 0$ assume that the view of the adversary can be simulated for $k' < k$. The view of the adversary can now be simulated as follows. At the beginning, the simulator generates all public keys and de-randomizers seen by the adversary. For each dishonest P_i the simulator computes $\mathsf{rk}_i, \widetilde{\mathsf{pk}}_i$ using RandKey. For each honest P_j in the neighborhood of $\overline{H}$ the simulator sets $\widetilde{\mathsf{pk}}_j$ to a random public-key using KeyGen. Due to the IND-CKCA property of the RR-MHT-PKE scheme these public keys are indistinguishable from the corresponding public-keys seen by the adversary in an actual protocol-execution. The above values also determine all keys $\mathsf{pk}_{i,\nu_i(j)}$ for $P_i \in \overline{H}$ and $P_j \in \mathbf{N}_G(i)$. Now, we consider the ciphertexts seen by the adversary in the second part of the protocol. In essence the simulator must generate all $\widetilde{c}_{j,\nu_j(i)}$ where P_i and/or P_j are dishonest. If the whole $(k-1)$-neighborhood of P_j is dishonest the simulator must also provide the content and the randomness of $\widetilde{c}_{j,\nu_j(i)}$ which are required for the sub-simulation of the recursive protocol invocations. We recall that DerandCipher is efficiently invertible if the de-randomizer is known. First, simulator generates a random $r_i \in \{1, \ldots, |\mathcal{M}| - 1\}$ for each dishonest P_i. If the whole k-neighborhood of P_i is dishonest (i.e., $\mathbf{N}_G[i]^k \subseteq \overline{H}$) the simulator is additionally given s_i and ρ_i. This allows the simulator to compute for each neighbor $P_j \in \mathbf{N}_G[i]$ a valid $s_{j,\nu_j(i)}$, randomness $\rho_{j,\nu_j(i)}$, and an encryption $\widetilde{c}_{j,\nu_j(i)} = \mathsf{Enc}(\widetilde{\mathsf{pk}}_i, s_{j,\nu_j(i)}; \rho_{j,\nu_j(i)})$ such that $c_i = \left(\prod_{P_j \in \mathbf{N}_G[i]} \mathsf{DerandCipher}(\mathsf{rk}_i, \widetilde{c}_{j,\nu_j(i)}) \right)^{r_i}$. If there exists a honest party in the k-neighborhood of P_i, the simulator is not given s_i and ρ_i. However, in this case there is at least one P_j in $\mathbf{N}_G[i]$ such that $\mathbf{N}_G[j]^{k-1} \not\subseteq \overline{H}$. This allows the simulator to first generates all $s_{j,\nu_j(i)}, \rho_{j,\nu_j(i)}$ and $\widetilde{c}_{j,\nu_j(i)} = \mathsf{Enc}(\widetilde{\mathsf{pk}}_i, s_{j,\nu_j(i)}; \rho_{j,\nu_j(i)})$ where $\mathbf{N}_G[j]^{k-1} \subseteq \overline{H}$. Then it chooses the remaining $\widetilde{c}_{j,\nu_j(i)}$ randomly under the constraint that $c_i = \left(\prod_{P_j \in \mathbf{N}_G[i]} \mathsf{DerandCipher}(\mathsf{rk}_i, \widetilde{c}_{j,\nu_j(i)}) \right)^{r_i}$. In a final step the adversary generates for any honest $P_j \in \mathbf{N}_G[i]$ the values $s_{i,\nu_i(j)}, \rho_{i,\nu_i(j)}$ and $\widetilde{c}_{i,\nu_i(j)} = \mathsf{Enc}(\widetilde{\mathsf{pk}}_j, s_{i,\nu_i(j)}; \rho_{i,\nu_i(j)})$. The $IND - TCPA$ property of the RR-MHT-PKE scheme and the correctness property of the protocol ensure that the generated ciphertexts are indistinguishable from the ones seen by the adversary in an actual protocol execution. Now all values required for the simulation of the $d + 1$ invocations of InfoTransmisson with parameter $(k - 1)$ are given. The simulator can thus use the sub-simulator to generate the view of the adversary in the middle part of the protocol. **Communication Complexity:** Let $f(k)$ be the communication complexity of InfoTransmisson$(k, \ldots)$. Then we have $f(0) = 0$ and $f(k) = d \cdot n \cdot (\log |\mathcal{PK}| + \log |\mathcal{C}|) + (d + 1) \cdot f(k - 1)$. This results in a communication complexity of at most $(d+1)^k \cdot n \cdot (\log |\mathcal{PK}| + \log |\mathcal{C}|)$ bits. The round complexity follows from the observation that one can invoke the subprotocols InfoTransmisson$(k - 1, \ldots)$ in parallel.

Decryption Protocol. In this section we describe a distributed decryption protocol which allows each party P_i to decrypt a ciphertext c_i under its shared private-key SK_i which has been generated by the setup protocol. The decryption protocol consists of two parts. First the parties jointly compute for each ciphertext c_i a decryption-share x_i under the shared private-key of P_i. In a second phase each party P_i can locally decrypt c_i using the decryption share x_i. First, we present a subprotocol which allows to compute the required decryption shares. The key-idea is to use the homomorphic property of decryption-shares which allows a recursive computation. The subprotocol is therefore parametrized by k.

Definition 11. *A protocol is a secure level-k (topology-hiding) decryption-share protocol over a network $\mathcal{N}_{\mathcal{G}}^{d,D}$ if it has the following properties.*

Setup, Inputs, and Outputs: *The parties initially hold a setup for topology-hiding threshold encryption over $\mathcal{N}_{\mathcal{G}}^{d,D}$ (cf. Definition 8). Each party P_i inputs a ciphertext $c_i \in \mathcal{C}$. The output of party P_i is a decryption share $\mathsf{x}_i \in \mathcal{DS}$.*
Correctness: *With overwhelming probability $\mathsf{x}_i = \mathsf{ShareDecrypt}(\mathsf{SK}_i^{(k)}, c_i)$ for $\mathsf{SK}_i^{(k)}$ the level-k shared private-key of P_i from the setup.*
Topology-Hiding Simulation: *The adversarial view in a real protocol-execution can be simulated with overwhelming probability given the following values*
$$\left\{ \mathbf{N}_G(i), c_i, \mathsf{x}_i, \nu_i(\cdot), \overline{\mathsf{SK}}_i^{(0)}, \ldots, \overline{\mathsf{SK}}_i^{(k)} \right\}_{P_i \in \overline{H}}$$

In other words the simulator gets the neighborhood of dishonest parties (in $\mathcal{N}_{\mathcal{G}}^{d,D}$), their protocol in- and outputs, their local pseudonyms, and their private-key shares (up to level-k) of the assumed setup.

The simulation property ensures in particular that the adversary does not learn more about the network topology.

Protocol DecShares$(k, c_1, \ldots, c_n)$

Require: Parties have access to an initialized $\mathcal{N}_{\mathcal{G}}^{d,D}$ and have generated a setup for topology-hiding threshold encryption over $\mathcal{N}_{\mathcal{G}}^{d,D}$.

Input: Each P_i inputs a ciphertext c_i.

1: **if** $k = 0$ **then**
2:     Each P_i computes $\mathsf{x}_i = \mathtt{ShareDecrypt}(\overline{\mathsf{SK}}_i^{(0)}, c_i)$.
3: **else**
4:     Each P_i computes $(\mathsf{rk}_i, \widehat{c}_i) = \mathtt{RandCipher}(c_i)$ and sends $\widehat{c}_i$ to each $P_j \in \mathbf{N}_G(i)$ which denotes the received value by $c_{j,\nu_j(i)}$.
5:     **for** $l = 1, \ldots, d$ **do**
6:         Parties compute $(\mathsf{x}_{1,l}, \ldots, \mathsf{x}_{n,l}) = \mathtt{DecShares}(k-1, c_{1,l}, \ldots, c_{n,l})$.
7:     **end for**
8:     Each P_i sends $\mathsf{x}_{i,\nu_i(j)}$ to each $P_j \in \mathbf{N}_G(i)$.
9:     Each P_i computes first $\widehat{\mathsf{x}}_i = \prod_{P_j \in \mathbf{N}_G(i)} \mathsf{x}_{j,\nu_j(i)}$ and then computes $\mathsf{x}_i = \mathtt{DerandShare}(\mathsf{rk}_i, \widehat{\mathsf{x}}_i) \cdot \mathtt{ShareDecrypt}(\overline{\mathsf{SK}}_i^{(k)}, c_i)$.

10: **end if**

Output: Each P_i outputs x_i.

Lemma 4. *Given a secure RR-MHT-PKE scheme and for any parameter $0 \leq k \leq D$ with $d^k = poly(\kappa)$ the above protocol* DecShares$(k, c_1, \ldots, c_n)$ *is a secure level-k decryption-share protocol. The protocol communicates $d^k \cdot n \cdot (\log|\mathcal{DS}| + \log|\mathcal{C}|)$ bits within $2k$ rounds.*

Proof. (sketch) **Correctness:** The correctness essentially follows from the structure of the assumed setup and from the properties of the RVHT-PKE scheme. In the case $k = 0$ we have $\mathsf{SK}_i^{(0)} = \overline{\mathsf{SK}}_i^0$ which implies $\mathsf{x}_i = \mathtt{ShareDecrypt}(\mathsf{SK}_i^{(0)}, c_i)$. For $k > 0$ we have $\mathsf{SK}_i^{(k)} = \overline{\mathsf{SK}}_i^{(k)} + \sum_{P_j \in \mathbf{N}_G(i)} \mathsf{SK}_j^{(k-1)}$. The properties of the RVHT-PKE scheme thus imply that $\mathsf{x}_i = \mathtt{ShareDecrypt}(\mathsf{SK}_i^{(k)}, c_i)$ (c.f. protocol line 9). **Simulation:** In the case $k = 0$ the view of the adversary is directly determined by values given to the simulator. Simulation is therefore easy to achieve. In the case $k > 0$ the simulation of the adversarial view works similar as for the information-transmission protocol (we recall that $\mathtt{DerandShare}$ is efficiently invertible if the de-randomizer is known). The simulator essentially emulates the protocol run. The IND-CKCA property of the RVHT-PKE scheme allows the simulator to choose random ciphertexts for $c_{i,\nu_i(j)}$ of honest P_j. Moreover, the decryption shares $\mathsf{x}_{j,\nu_j(i)}$ for honest P_j can also be chosen randomly (where the distribution is conditioned on the outputs of dishonest parties). The view during the executions of $\mathtt{DecShares}$ with parameter $k - 1$ can be generated using the $(k - 1)$-subsimulator guaranteed by the induction hypothesis. **Communication Complexity:** Denote by $f(k)$ be the communication complexity of $\mathtt{DecShares}(k, \ldots)$. Then we have $f(0) = 0$ and $f(k) = n \cdot d \cdot (\log|\mathcal{DS}| + \log|\mathcal{C}|) + d \cdot f(k-1)$. This results in a communication complexity of $f(k) = d^k \cdot n \cdot (\log|\mathcal{DS}| + \log|\mathcal{C}|)$. The round complexity follows from the observation that one can invoke the subprotocols $\mathtt{DecShares}(k-1,\dots)$ in parallel.

Definition 12. *A protocol is a* secure (topology-hiding) threshold decryption protocol *for network* $\mathcal{N}_{\mathcal{G}}^{d,D}$ *if it has the following properties.*

Setup, Inputs and Outputs: *The parties initially hold a setup for topology-hiding threshold encryption over* $\mathcal{N}_{\mathcal{G}}^{d,D}$ *(cf. Definition 8). Each party* P_i *inputs a ciphertext* $c_i \in \mathcal{C}$. *The output of party* P_i *is a message* m_i.

Correctness: *With overwhelming probability it holds for each party* P_i *that* $m_i = \mathtt{Combine}(\mathtt{ShareDecrypt}(\mathsf{SK}_i, c_i))$ *where* SK_i *is the shared private-key of* P_i.

Topology-Hiding Simulation: *The adversarial view in a real-protocol execution can be simulated with overwhelming probability given the following values*

$$\left\{ \mathbf{N}_G(i), c_i, m_i, \nu_i(\cdot), \overline{\mathsf{SK}}_i^{(0)}, \dots, \overline{\mathsf{SK}}_i^{(D)} \right\}_{P_i \in \overline{H}}$$

In other words the simulator gets the neighborhood of dishonest parties (in $\mathcal{N}_{\mathcal{G}}^{d,D}$), *their protocol in- and outputs, their local pseudonyms, and their private-key shares of the assumed setup.*

Protocol $\mathtt{Decryption}(c_1, \dots, c_n)$

Require: Parties have access to an initialized $\mathcal{N}_{\mathcal{G}}^{d,D}$ and have generated a setup for topology-hiding threshold encryption over $\mathcal{N}_{\mathcal{G}}^{d,D}$.
Input: Each P_i inputs a ciphertext c_i.
  1: The parties compute $(\mathsf{x}_1, \dots, \mathsf{x}_n) = \mathtt{DecShares}(D, c_1, \dots, c_n)$.
Output: Each P_i outputs $\mathtt{Combine}(\mathsf{x}_i, c_i)$.

Lemma 5. *Given a secure RR-MHT-PKE scheme,* $\mathtt{Decryption}(k, c_1, \dots, c_n)$ *is a secure threshold decryption protocol. The protocol communicates* $d^D \cdot n \cdot (\log |\mathcal{DS}| + \log |\mathcal{C}|)$ *bits within* $2D$ *rounds.*

Proof. (sketch) The correctness follows directly from Lemma 4 and the properties of the RVHT-PKE scheme. The adversarial view in a real protocol execution can be simulated as follows (recall that $\mathtt{Combine}$ is efficiently invertible). First the simulator computes for each pair (c_i, m_i) a decryption share x_i such that $m_i = \mathtt{Combine}(\mathsf{x}_i, c_i)$. The rest of the view can then be generated using the sub-simulator for $\mathtt{DecShares}(D, \dots)$. The communication complexity and the number of rounds follows directly from the invocation of $\mathtt{DecShares}$ with parameter D.

3.3 Multi-party Boolean OR

In this section we present a protocol which securely and topology-hidingly realizes the multi-party Boolean-OR functionality $\mathcal{F}_{\mathsf{OR}}$ using the topology-hiding threshold encryption protocol from the previous section. The functionality $\mathcal{F}_{\mathsf{OR}}$

takes from each party P_i an input bit b_i and computes the OR of those bit, i.e., $b = b_1 \vee \cdots \vee b_n$.

Functionality $\mathcal{F}_{\mathrm{OR}}$

1. Every party P_i (and the adversary on behalf of corrupted parties) sends (input) bit b_i; if P_i does not submit a valid input, then $\mathcal{F}_{\mathrm{OR}}$ adopts $b_i = 0$.
2. Every party P_i receives (output) $b = b_1 \vee \cdots \vee b_n$.

Assumptions. We assume in the following that the parties have access to a secure RR-MHT-PKE scheme with security parameter κ, where $n = poly(\kappa)$. Moreover, parties are given the network $\mathcal{N}_{\mathcal{G}}^{d,D}$ where the graphs in the family $\mathcal{G}$ are connected, have a maximum degree of $d \leq n$, and a diameter of at most $D \leq n$ where d and D are publicly known.

Protocol Boolean-OR$(b_1, \ldots, b_n)$

Initialization:

1: Each party P_i inputs its neighborhood $\mathbf{N}_G[i]$ into $\mathcal{N}_{\mathcal{G}}^{d,D}$.
2: The parties generate a setup for topology-hiding threshold encryption over $\mathcal{N}_{\mathcal{G}}^{d,D}$ using GenerateSetup.

Computation:

Input: Each party P_i inputs a bit b_i.
1: Each party P_i sets $m_i = 0$ if $b_i = 0$. Otherwise, its sets m_i to an arbitrary message in $\mathcal{M} \setminus \{0\}$.
2: The parties compute
   $(c_1, \ldots, c_n) = \mathtt{InfoTransmisson}\big(D, (m_1, \mathsf{PK}_1), \ldots, (m_n, \mathsf{PK}_n)\big)$.
3: The parties compute $(m_1', \ldots, m_n') = \mathtt{Decryption}(c_1, \ldots, c_n)$.
Output: If $m_i' = 0$ P_i outputs 0. Otherwise it outputs 1.

Lemma 6. *Given a secure RR-MHT-PKE scheme and for d, D with $d^D = poly(\kappa)$ the protocol* Boolean $-$ OR$(b_1, \ldots, b_n)$ *securely and topology-hidingly realizes $\mathcal{F}_{\mathrm{OR}}$ (in the $\mathcal{N}_{\mathcal{G}}^{d,D}$-hybrid model). In the initialization phase the protocol* Boolean $-$ OR$(b_1, \ldots, b_n)$ *communicates $D \cdot d \cdot n \cdot \log |\mathcal{PK}|$ bits within D rounds. In the computation phase the protocol communicates at most $(d+1)^D \cdot n \cdot (\log |\mathcal{DS}| + \log |\mathcal{PK}| + 2 \log |\mathcal{C}|)$ bits within $4 \cdot D$ rounds.*

Proof. **Correctness:** We assume the condition $d^D = poly(\kappa)$. The correctness thus follows directly from the properties of Lemmas 2, 3, and 5 as the information-transmission protocol essentially allows to compute Boolean-ORs.

Topology-Hiding Simulation: Given the values $\big\{\mathbf{N}_G(i), b_i, b\big\}_{P_i \in \overline{H}}$. the view of the adversary can be simulated as follows. First the simulator generates a setup for $\mathcal{N}_{\mathcal{G}}^{d,D}$. Next, for each dishonest P_i the simulator computes the messages m_i and m_i'. It generates the corresponding ciphertext c_i (including the randomness). With those values the simulator now runs the the sub-simulators for GenerateSetup, InfoTransmisson($D, \dots$), and Decryption($\dots$). The properties of Lemmas 2, 3, and 5 ensure that the generated view is indistinguishable (for the adversary) from a real protocol execution.

Communication Complexity: The claimed communication complexity follows directly from the used subprotocols.

Remark 1. If the RR-MHT-PKE is instantiated using the DDH based construction, the computation complexity of the Boolean-OR protocol is similar to its communication complexity.

3.4 Topology-Hiding Broadcast and Secure Channels

In this section we describe a protocol which securely realizes the (bit) broadcast functionality $\mathcal{F}_{BC}^s$, while making-black box use of the $\mathcal{F}_{OR}$ functionality from the previous section. The functionality $\mathcal{F}_{BC}^s$ allows sender P_s to input a bit b_s which is output to all parties. This result directly implies that one can securely and topology-hidingly realize secure channels and broadcast using black-box RR-MHT-PKE.

Protocol Broadcast(P_s, b_s)

Require: The sender P_s inputs a bit b_s.
1: The parties compute $(b, \dots, b) = \mathcal{F}_{OR}(0, \dots, b_s, \dots, 0)$.
Output: Each party P_i outputs b.

Lemma 7. *The protocol* Broadcast(P_s, b_s) *securely realizes the $\mathcal{F}_{BC}^s$ functionality in the $\mathcal{F}_{OR}$-hybrid model.*

Proof. We have that $b = 0 \vee \cdots \vee b_s \vee \cdots \vee 0 = b_s$ which implies correctness. The view of the adversary in an actual protocol execution consists of inputs and outputs of dishonest parties and is therefore easy to simulate.

Corollary 1. *For d, D with $d^D = poly(\kappa)$ one can securely and topology-hidingly realize $\mathcal{F}_{BC}^s$ (in the $\mathcal{N}_{\mathcal{G}}^{d,D}$-hybrid model) using black-box RR-MHT-PKE while communicating at most $(d+1)^D \cdot n \cdot (\log |\mathcal{DS}| + \log |\mathcal{PK}| + 2 \log |\mathcal{C}|) + D \cdot d \cdot n \cdot \log |\mathcal{PK}|$ bits within $5 \cdot D$ rounds per invocation.*

Moreover, parties can simply realize secure channels given broadcast. First the receiver generates a key pair and broadcasts the public-key. The sender then broadcasts his message encrypted under this public-key.

Corollary 2. *For d, D with $d^D = poly(\kappa)$ one can securely and topology-hidingly realize secure channels (in the $\mathcal{N}_{\mathcal{G}}^{d,D}$-hybrid model) using black-box RR-MHT-PKE. The communication complexity is twice the one of the broadcast protocol.*

4   Applications

In this section we provide two applications of our network-hiding communication protocols. Namely, one can securely and topology-hidingly realize MPC and anonymous brodcast.

4.1   Topology-Hiding Secure Multi-Party Computation

The protocols from the previous section allow parties to topology-hidingly realize a complete network of secure channels (including broadcast channels). They can then use this network to execute a multi-party protocol of their choice, e.g., [GMW87,Pas04]. This easily proves the following result.

Theorem 2. *For d, D with $d^D = poly(\kappa)$ one can securely and topology-hidingly realize any given multiparty functionality (in the $\mathcal{N}_{\mathcal{G}}^{d,D}$-hybrid model) using black-box RR-MHT-PKE.*

4.2   Anonymous Broadcast

Theorem 2 implies that one can topology-hidingly realize anonymous channels given black-box access to a RR-MHT-PKE scheme. But using generic MPC to achieve an anonymous channel is expensive in terms of communication complexity. We therefore provide a protocol in the $\mathcal{F}_{OR}$-hybrid model which directly realizes *anonymous broadcast* $\mathcal{F}_{ABC}$.

The functionality $\mathcal{F}_{ABC}$ generates for each party a unique but random pseudonym. In the subsequent communication rounds each party can publish messages under its pseudonym. Message are linkable which means that parties can relate messages to pseudonyms. Parties can prevent this by generating fresh pseudonyms (e.g., after each communication round).

Functionality $\mathcal{F}_{ABC}$

Initialization:

1: The functionality generates a random permutation σ of n elements.
2: Each party P_i gets output $\sigma(i)$.

Communication Step:

Require: Each party P_i inputs a bit b_i.
Output: The parties get the vector $(o_1, \ldots, o_n)$ as output where $o_{\sigma(i)} = b_i$.

Anonymous Broadcast Protocol. The high-level idea of our construction is as follows. In a scheduling phase each party gets a random (but unique) communication slot $\sigma(i)$ assigned. In a communication round for each slot $\sigma(i)$ the $\mathcal{F}_{OR}$ functionality is invoked which allows P_i to broadcast its bit.

The major challenge is to compute the slot assignment. We solve this issue with a scheduling loop[12]. At the beginning each party selects a random slot. Then over several scheduling rounds the parties resolve colliding selections by computing a reservation matrix. The size of this matrix (parametrized by m) determines the collision detection probability. A larger m means a faster expected run time at the cost of increased communication costs per round.

Protocol AssignSlots(m)

1: Each party P_i chooses a random slot $s_i \in \{1, \ldots, n\}$.
2: **repeat**
3:     Each party P_i chooses a random token $r_i \in \{1, \ldots, m\}$ and computes the $n \times m$-matrix $A^{(i)} = (a_{x,y}^{(i)})$ where $a_{s_i, r_i}^{(i)} = 1$ and $a_{x,y}^{(i)} = 0$ otherwise.
4:     The parties compute the matrix $A = (a_{x,y})$ where $a_{x,y} = a_{x,y}^{(1)} \vee \cdots \vee a_{x,y}^{(n)}$ by invoking $\mathcal{F}_{\mathsf{OR}}$.
5:     If there exists an $r < r_i$ such that $a_{s_i, r} = 1$ party P_i chooses a new random slot $s_i \in \{1, \ldots, n\}$ such that s_i-th row of A contains only zeros.
6: **until** Each row of A contains exactly one 1.
Output: Each party P_i outputs s_i.

Lemma 8. *The protocol* AssignSlots(m) *for the* $\mathcal{F}_{\mathsf{OR}}$*-hybrid model securely computes a random permutation* σ *of* n *elements where each party* P_i *learns* $\sigma(i)$. *The expected number of rounds the protocol requires to compute the permutation is bounded by* $\frac{m}{m-1} \cdot n$ *where* $\mathcal{F}_{\mathsf{OR}}$ *is invoked* $n \cdot m$ *times per round.*

Proof. The protocol terminates if each row of A contains exactly one non-zero entry. Thus each slot in $\{1, \ldots, n\}$ has been chosen at least by one party. As there are n parties this also means that no slot was chosen twice. The output is therefore a valid permutation. Inspection of the protocol also reveals that the permutation is chosen uniform at random (we consider semi-honest security).

Next, we show that the protocol eventually terminates. Each slot is in one of three states. Either its empty, or its selected by multiple parties, or it is assigned to a single party. We observe that the state transition function for slots is monotone. A selected slot cannot become empty and an assigned slot stays assigned to the same party. In each round where a collision is detected at least one empty slot becomes assigned. After at most n such rounds there are no empty slots left. But this also means that each slot is selected by at least one party and the protocol terminates. This also leads to a crude upper bound on the number of expected rounds. We observe that a collision between two parties is detected with a probability of at least $p = (1 - \frac{1}{m})$. The expected number of rounds required to detect a collision is therefore at most $\frac{1}{p} = \frac{m}{m-1}$ (geometric distribution). The number of expected rounds is thus bounded by $\frac{m}{m-1} \cdot n$. It remains to consider the simulation of the adversarial view. We observe that the (current) slot selection of dishonest parties is enough to simulate the view of the

---

[12] A similar idea was used recently in [KNS15].

adversary in a scheduling round. The simulator can therefore essentially emulate the protocol (conditioned on the final slots of dishonest parties).

Protocol AnonymousBroadcast(m)

Initialization:

1: The parties compute $(\sigma(1), \ldots, \sigma(n)) = \texttt{AssignSlots}(m)$.

Communication Step:

Require: Each party P_i inputs a bit b_i.

1: **for** $s = 1, \ldots, n$ **do**
2:     The parties compute $(o_s, \ldots, o_s) = \texttt{Boolean-OR}(0, \ldots, b_{\sigma^{-1}(s)}, \ldots, 0)$.
3: **end for**
Output: Each party P_i outputs vector $(o_1, \ldots, o_n)$.

Lemma 9. *The protocol* AnonymousBroadcast(m) *securely realizes the functionality* $\mathcal{F}_{\mathsf{ABC}}$ *in the* $\mathcal{F}_{\mathsf{OR}}$*-hybrid model.*

Proof. The statement follows directly from Lemmas 7 and 8.

Corollary 3. *For d, D with $d^D = poly(\kappa)$ one can securely and topology-hidingly realize* $\mathcal{F}_{\mathsf{ABC}}$ *(in the* $\mathcal{N}_{\mathcal{G}}^{d,D}$*-hybrid model) using black-box RR-MHT-PKE.*

References

[Bd90] Bos, J.N.E., den Boer, B.: Detection of disrupters in the DC protocol. In: Quisquater, J.-J., Vandewalle, J. (eds.) EUROCRYPT 1989. LNCS, vol. 434, pp. 320–327. Springer, Heidelberg (1990)

[BGT13] Boyle, E., Goldwasser, S., Tessaro, S.: Communication locality in secure multi-party computation. In: Sahai, A. (ed.) TCC 2013. LNCS, vol. 7785, pp. 356–376. Springer, Heidelberg (2013)

[Can98] Canetti, R.: Security and composition of multi-party cryptographic protocols. Cryptology ePrint Archive, Report 1998/018 (1998). http://eprint.iacr.org/1998/018

[Can00] Canetti, R.: Universally composable security: a new paradigm for cryptographic protocols. Cryptology ePrint Archive, Report 2000/067 (2000). http://eprint.iacr.org/2000/067

[CCG+15] Chandran, N., Chongchitmate, W., Garay, J.A., Goldwasser, S., Ostrovsky, R., Zikas, V.: The hidden graph model: communication locality and optimal resiliency with adaptive faults. In: Roughgarden, T. (ed.) ITCS 2015, pp. 153–162. ACM, January 2015

[CGO15] Chandran, N., Garay, J.A., Ostrovsky, R.: Almost-everywhere secure computation with edge corruptions. J. Crypt. **28**(4), 745–768 (2015)

[Cha81] Chaum, D.L.: Untraceable electronic mail, return addresses, and digital pseudonyms. Commun. ACM **24**(2), 84–90 (1981)

[Cha88] Chaum, D.: The dining cryptographers problem: unconditional sender and recipient untraceability. J. Crypt. **1**(1), 65–75 (1988)

[Cha03] Chaum, D.: Untraceable electronic mail, return addresses and digital pseudonyms. In: Gritzalis, D. (ed.) Secure Electronic Voting. Advances in Information Security, pp. 211–219. Springer, Heidelberg (2003)

[ElG84] El Gamal, T.: A public key cryptosystem and a signature scheme based on discrete logarithms. In: Blakely, G.R., Chaum, D. (eds.) CRYPTO 1984. LNCS, vol. 196, pp. 10–18. Springer, Heidelberg (1985)

[GGOR14] Garay, J.A., Givens, C., Ostrovsky, R., Raykov, P.: Fast and unconditionally secure anonymous channel. In: Halldórsson, M.M., Dolev, S. (ed.) 33rd ACM PODC, pp. 313–321. ACM, July 2014

[GJ04] Golle, P., Juels, A.: Dining cryptographers revisited. In: Cachin, C., Camenisch, J.L. (eds.) EUROCRYPT 2004. LNCS, vol. 3027, pp. 456–473. Springer, Heidelberg (2004)

[GMW87] Goldreich, O., Micali, S., Wigderson, A.: How to play any mental game or a completeness theorem for protocols with honest majority. In: Aho, A. (ed.) 19th ACM STOC, pp. 218–229. ACM Press, May 1987

[GO08] Garay, J.A., Ostrovsky, R.: Almost-everywhere secure computation. In: Smart, N.P. (ed.) EUROCRYPT 2008. LNCS, vol. 4965, pp. 307–323. Springer, Heidelberg (2008)

[Gol01] Goldreich, O.: The Foundations of Cryptography - Basic Techniques, vol. 1. Cambridge University Press, Cambridge (2001)

[HJ07] Hinkelmann, M., Jakoby, A.: Communications in unknown networks: preserving the secret of topology. Theor. Comput. Sci. **384**(2–3), 184–200 (2007)

[HMTZ16] Hirt, M., Maurer, U., Tschudi, D., Zikas, V.: Network-hiding communication and applications to multi-party protocols. Cryptology ePrint Archive, Report 2016/556 (2016). http://eprint.iacr.org/

[KNS15] Krasnova, A., Neikes, M., Schwabe, P.: Footprint scheduling for dining-cryptographer networks. Cryptology ePrint Archive, Report 2015/1213 (2015). http://eprint.iacr.org/

[KS10] King, V., Saia, J.: Breaking the $o(n^2)$ bit barrier: scalable byzantine agreement with an adaptive adversary. In: Proceedings of the 29th Annual ACM Symposium on Principles of Distributed Computing, PODC 2010, Zurich, Switzerland, pp. 420–429, 25–28 July 2010

[KSSV06a] King, V., Saia, J., Sanwalani, V., Vee, E.: Scalable leader election. In: SODA, pp. 990–999 (2006)

[KSSV06b] King, V., Saia, J., Sanwalani, V., Vee, E.: Towards secure and scalable computation in peer-to-peer networks. In: FOCS, pp. 87–98 (2006)

[MOR15] Moran, T., Orlov, I., Richelson, S.: Topology-hiding computation. In: Dodis, Y., Nielsen, J.B. (eds.) TCC 2015, Part I. LNCS, vol. 9014, pp. 159–181. Springer, Heidelberg (2015)

[Pai99] Paillier, P.: Public-key cryptosystems based on composite degree residuosity classes. In: Stern, J. (ed.) EUROCRYPT 1999. LNCS, vol. 1592, pp. 223–238. Springer, Heidelberg (1999)

[Pas04] Pass, R.: Bounded-concurrent secure multi-party computation with a dishonest majority. In: Babai, L. (ed.) 36th ACM STOC, pp. 232–241. ACM Press, June 2004

Network Oblivious Transfer

ıjit Kumaresan, Srinivasan Raghuraman, and Adam Sealfon[✉]

MIT, Cambridge, USA
{ranjit,srirag,asealfon}@csail.mit.edu

Abstract. Motivated by the goal of improving the concrete efficiency of secure multiparty computation (MPC), we study the possibility of implementing an infrastructure for MPC. We propose an infrastructure based on oblivious transfer (OT), which would consist of OT channels between some pairs of parties in the network. We devise information-theoretically secure protocols that allow additional pairs of parties to establish secure OT correlations using the help of other parties in the network in the presence of a dishonest majority. Our main technical contribution is an upper bound that matches a lower bound of Harnik, Ishai, and Kushilevitz (Crypto 2007), who studied the number of OT channels necessary and sufficient for MPC. In particular, we characterize which n-party OT graphs G allow t-secure computation of OT correlations between all pairs of parties, showing that this is possible if and only if the complement of G does not contain the complete bipartite graph $K_{n-t,n-t}$ as a subgraph.

1 Introduction

Protocols for secure multiparty computation [8,16,31,66] allow a set of mutually distrusting parties to carry out a distributed computation without compromising the privacy of inputs or the correctness of the end result. As a research area, secure computation has witnessed several breakthroughs in the last decade [40,41,43,47,52–54,57,59,67]. However, despite a wide array of potential game-changing applications, there is nearly no practical adoption of secure computation today (with the notable exceptions of [11,12]). Computations wrapped in a secure computation protocol do not yet deliver results efficiently enough to be acceptable in many cloud-computing applications. For instance, state-of-the-art semihonest 2-party protocols incur a factor ≈ 100 slowdown even for simple computations.

In the absence of practical real-world protocols for secure computation which are secure in the presence of any number of dishonest parties, there is a need for

R. Kumaresan—Supported by Qatar Foundation, MIT Translational Fellowship Program, ONR N00014-11-1-0486 and NSF CNS1413920.

S. Raghuraman—Supported by the Irwin Mark Jacobs and Joan Klein Jacobs Presidential Fellowship, NSF CNS1413920, DARPA W911NF-15-C-0236 and the Simons Foundation.

A. Sealfon—Supported by DOE CSGF fellowship, NSF CNS1413920, DARPA W911NF-15-C-0236 and the Simons Foundation.

M. Robshaw and J. Katz (Eds.): CRYPTO 2016, Part II, LNCS 9815, pp. 366–396, 2016.
DOI: 10.1007/978-3-662-53008-5_13

relaxations that are meaningful and yet provide significant performance benefits. As an example, classic protocols for secure computation [8,16,63] (with subsequent improvements e.g., [4,9,19–21,23]) offer vastly better efficiency at the cost of tolerating only a small constant fraction of adversaries. The resilience offered is certainly acceptable when the number of participating parties is large, e.g., the setting of *large-scale* secure computation [13,14,25,68]. Although large-scale secure computation is well-suited for several interesting applications (such as voting, census, surveys), we posit that typical settings involve computations over data supplied by a few end users. In such cases, the overhead associated with interaction among a large number of *helper parties* is likely to render these protocols more expensive than a standard secure computation protocol among the end users. If the number of helper parties is small, security against a small fraction of corrupt parties may be a very weak guarantee, since a handful of corrupt parties could render the protocol insecure.

An orthogonal approach for reducing the online cost of secure computation protocols is the use of *preprocessing* [1,3,10,24]. This approach can dramatically reduce the cost of secure computation: for instance, given preprocessing [3], the ≈100 factor slowdown for simple computations no longer applies. Recent theoretical research has shown that many primitives can even be made *reusable* (e.g. [34]). Perhaps the most important drawback of this approach (other than the fact that the preprocessing phase is typically very expensive) is that the preprocessing is not *transferable*. Clearly, a pair of parties that want to perform a secure computation cannot benefit from this approach without performing the expensive preprocessing step. Moreover, this seems to hold even if each of the two parties have set up the preprocessing with multiple others. Typically, the cost of the preprocessing phase is quite high, presenting a barrier for the practical use of preprocessed protocols. This is especially true in settings where parties are unlikely to run many secure computations that would amortize the cost of preprocessing.

Motivated by the discussion above, we conclude that some directions that seem to offer efficiency benefits for secure computation are (1) highly resilient protocols that use only a small number of helper parties, and (2) a preprocessing procedure that allows a notion of transferability between users. Taken together, these two ideas have the potential to provide an *infrastructure* for efficient secure computation. Some sets of parties might run a preprocessing phase among themselves. These parties can then act as helper parties and "transfer" their preprocessing to help users who want to run a secure computation protocol. We informally describe some desiderata for such an infrastructure:

– *Reusability/Amortization.* Setting up an infrastructure component could be expensive, but using it and maintaining it should be inexpensive relative to setting up a new component.
– *Transferability/Routing.* It should be possible to combine different components of the infrastructure to deliver benefits to the end users.
– *Robustness/Fault-tolerance.* Failure or unavailability of some components of the infrastructure should not nullify the usefulness of the infrastructure.

It is not hard to see that the above criteria are fulfilled for infrastructures that we use in daily life, for e.g., the infrastructure for online communication (e-mail, instant messaging, etc.) consisting of transatlantic undersea cables, routers, wireless access points, etc. What cryptographic primitives would be good candidates for a *secure computation infrastructure*? In this work, we explore the possibility of using *oblivious transfer* [27,62] for this purpose.

1.1 Our Model: Network Oblivious Transfer

Oblivious transfer (OT) is a fundamental building block of secure computation [45,46]. As discussed in [45], some of the benefits of basing secure computation on OT include:

– *Preprocessing.* OT enables precomputation in an offline stage before the inputs or the function to be computed are known. The subsequent online phase is extremely efficient [3].
– *Amortization.* The cost of computing OTs can be accelerated using efficient OT extension techniques [2,43,45,59].
– *Security.* OTs can be realized under a wide variety of computational assumptions [18,27,58,60,62] or under physical assumptions.

In this work, we consider n parties connected by a synchronous network with secure point-to-point private communication channels between every pair of parties. In addition, some pairs of parties on the network have established *OT channels* between them providing them with the ability to perform arbitrarily many OT operations. We represent the OT channel network via an *OT graph* G. The vertices of G represent the n parties, and pairs of parties that have an established OT channel are connected by an edge in G. Since OT can be reversed unconditionally [64], we make no distinction between the sender and the receiver in an OT channel. This OT graph represents the infrastructure we begin with. The OT channels could either represent poly(λ) 1-out-of-2 OT correlations for a computational security parameter λ, or a physical channel (e.g., noisy channel) that realizes, say δ-Rabin OT [62].[1] We are interested in obtaining security against adaptive semihonest adversaries. We also discuss security against adaptive malicious adversaries under computational assumptions.

Two parties that are connected by an edge can use the corresponding existing OT channel to run a secure computation protocol between themselves. What about parties that are not connected by an edge? Clearly, they can establish an OT channel between themselves via an OT protocol [18,60] or perhaps using a physical channel. The latter option, if possible, is likely to be expensive and the costs of setting up a physical channel may be infeasible unless the two parties are likely to execute many secure computation protocols. The former option

---

[1] Recall that λ 1-out-of-2 OT correlations can be extended to poly(λ) 1-out-of-2 OT correlations via OT extension using just symmetric-key cryptography (e.g. one-way functions [2] or correlation-robust hash functions [43]).

is also expensive as it involves use of public-key cryptography which is somewhat necessary in the light of [42].[2] This motivates the question of whether additional parties can use an existing OT infrastructure to establish an OT channel between themselves unconditionally or relying only on the existence of symmetric-key cryptography. A positive result to this question would show that expensive cryptographic operations are not required to set up additional OT channels which could be used for efficient secure computation. In this work we construct OT protocols with information-theoretic security against a threshold adversary.

The Generality of an OT Infrastructure. Consider the following candidate for an infrastructure. Suppose there is a channel between a pair of parties that allows them to securely evaluate any function. Since OT is complete for secure computation, one can apply the results of [45,46] to use the OT channel to implement a secure evaluation channel. In the other direction, one can use a secure evaluation channel to trivially implement OT channels. Consequently, such a channel is equivalent to an OT channel. The same argument extends to channels that implement any 2-party primitive that is complete for secure computation [5,55]. Furthermore, the above argument also applies to the setting where a *set* of parties have a secure evaluation channel. Such a channel is equivalent to an OT graph where parties in the set have pairwise OT channels with everyone in the set.

Assuming a Full Network of Secure Channels. Secure channels between two parties can be implemented either via non-interactive key exchange and hybrid encryption or via a physical assumption. We emphasize that the one-time setup cost of emulating a secure channel (e.g. via Diffie-Hellman key exchange) is much lower than the one-time setup cost of emulating an OT channel that allows unbounded OT calls via an OT protocol even using OT extension. Furthermore, our assumption of secure channels is identical to the setting of [33,45,46], who show that secure computation reduces to OT under information-theoretic reductions.

1.2   Related Work and Our Contributions

Related Work. As mentioned previously, there is a large body of work on secure computation in the offline/online model (cf. [10,24,50,51,59,61] and references therein). These protocols exhibit an extremely fast online phase at the expense of a slow preprocessing phase (sometimes using MPC [51] or more typically, OT correlations [59] or a somewhat homomorphic encryption scheme [24]). To the best of our knowledge, the question of *transferability* of preprocessing has not been explicitly investigated in the literature with the notable exception of [36], which we will discuss in greater detail below. There is a large body of work

---

[2] As a rule of thumb, use of public-key cryptography is computationally around 4–6 orders of magnitude more expensive than using symmetric-key cryptography [7].

on secure computation against a threshold adversary (e.g. [8,16,31,63]). Popular regimes where secure computation against threshold adversaries have been investigated are for $t < n/3$, $t < n/2$, or $t = n - 1$. In this work we are interested in threshold adversaries for a dishonest majority, that is, adversaries which can corrupt t out of n parties for $n/2 \leq t < n$.[3] Such regimes were investigated in other contexts such as authenticated broadcast [29] and fairness in secure computation [6,39,44]. Infrastructures for *perfectly secure message transmission* (PSMT) were investigated in the seminal work of [26] (see also [28] and references therein). While the task of PSMT is similar to our question regarding OT channels, there are inherent differences. For example, our protocols can implement OT even between two parties that are isolated in the OT graph (i.e., not connected to any other party via an OT channel).[4] In PSMT, on the other hand, there is no hope of achieving secure communication with a node that is not connected by any secure channel.

Most relevant to our results is the work of Harnik et al. [36]. The main question in their work is an investigation of the number of OT channels sufficient to implement a n-party secure computation protocol. In a nutshell, they show against an adaptive t-threshold adversary for $t = (1 - \delta)n$, an explicit construction of an OT graph consisting of $(n + o(n))\binom{\lceil 1/\delta \rceil}{2}$ OT channels that suffices to implement secure computation among the n parties. They note further that against a static adversary, $\binom{\lceil s/\delta \rceil}{2}$ OT channels suffice, where s denotes a statistical security parameter. On the negative side, they show that a complete OT graph is necessary for secure computation when dealing with an adversary that can corrupt $t = n - 1$ parties. They derive this result by showing that in a 3-party OT graph with two OT channels, it is not possible to obtain OT correlations between the third pair of parties with security against two corruptions. Moreover they generalize their 3-party negative result to any OT graph whose complement contains the complete bipartite graph $K_{n-t,n-t}$ as a subgraph. In our paper we extend and generalize the results of [36], fully characterizing the networks for which it is possible to obtain OT correlations between a designated pair of parties. We now proceed to explain our contributions in more detail.

Our Contributions. We introduce our main result:

Theorem (informal). Let $G = (V, E)$ be an OT graph on n parties $P_1, \ldots P_n$, so that any pair of parties P_i, P_j which are connected by an edge may make an unbounded number of calls to an OT oracle. Let $\mathbb{A}$ be the class of semihonest t-threshold adversaries which may adaptively corrupt at most t parties.[5] Then two parties A and B in $\{P_1, \ldots, P_n\}$ can information-theoretically emulate an OT oracle while being secure against all adversaries $\mathcal{A} \in \mathbb{A}$ if and only if

---

[3] When $t < n/2$, there is no need to rely on an OT infrastructure [63].

[4] Recall that the model considered in this work, we assume a *full* network of secure private communication channels.

[5] Combining our work with results from [32,35], we can also obtain computational security against malicious adversaries in both the nonadaptive and adaptive settings.

1. (honest majority) it holds that $t < n/2$; or
2. (trivial) A and B are connected by an edge in G; or
3. (partition) there exists no partition V_1, V_2, V_3 of G such that all of the following conditions are satisfied: (a) $|V_1| = |V_2| = n - t$ and $|V_3| = 2t - n$; (b) $A \in V_1$ and $B \in V_2$; and (c) for every $A' \in V_1$ and $B' \in V_2$ it holds that $(A', B') \notin E$.

Our main theorem gives a complete characterization of networks for which a pair of parties can utilize the OT network infrastructure to execute a secure computation protocol. The first two conditions in our theorem are straightforward: (1) if $t < n/2$, then we are in the honest majority regime, and thus it is possible to implement secure computation (or emulate an OT oracle) using the honest majority information-theoretically secure protocols of [63]; (2) clearly if A and B are connected by an OT edge then by definition they can emulate an OT oracle.

Condition (3) applies when $t \geq n/2$ and when A and B do not have an OT edge between them. This condition is effectively the converse of the impossibility result of [36], which states that any n-party OT graph whose complement contains $K_{n-t,n-t}$ as a subgraph cannot allow a n-party secure computation that tolerates t semihonest corruptions. Condition (3) implies that any n-party OT graph whose complement does not contain $K_{n-t,n-t}$ as a subgraph can run n-party secure computations tolerating t semihonest corruptions.

Applying Our Main Theorem. We first compare our positive results to those of [36]. They investigate how to construct an OT graph with the minimum number of edges allowing n parties to execute a secure computation protocol. They show a construction for a graph with $(n + o(n))\binom{\lceil 1/\delta \rceil}{2}$ edges which they prove is sufficient for resilience against an adversary that corrupts $(1 - \delta)n$ parties. Our result provides a complete, simple characterization of which OT graphs on n vertices are sufficient to run a t-secure protocol generating OT correlations between all pairs of vertices for any $t \geq n/2$, which is sufficient to obtain a protocol for secure computation among the n parties [45, 46]. Our main theorem also implies that determining the minimum number of OT edges needed to execute a secure computation protocol for general $n, t \geq n/2$ is equivalent to an open problem in graph theory posed by Zarankiewicz in 1951 [48].

Our results immediately imply that for some values of t, extremely simple sparse OT graphs suffice for achieving secure multiparty computation. For n even and $t = n/2$, we have that the t-claw graph (cf. Fig. 4(a)) has t edges and suffices to achieve t-secure multiparty computation. For n odd and $t = (n+1)/2$, the $(t + 1)$-cycle has $t + 1$ edges and suffices to achieve t-secure multiparty computation. We show in the full version that these examples are the sparsest possible graphs which can achieve $\lfloor (n + 1)/2 \rfloor$-secure multiparty computation.

Next, our results are also well-suited to make use of an OT infrastructure for secure computation. Specifically, let G_I denote the OT graph consisting of existing OT edges between parties that are part of the infrastructure. Now suppose a pair of parties A, B not connected by an OT edge wish to execute a secure computation protocol. Then they can find a subgraph G of G_I with $A, B \in G$

and $|G| = n$ such that they agree that at most t out of the n parties can be corrupt and the partition condition in our main theorem holds for G. Since it is possible to handle a dishonest majority, parties do not have to settle for a lower threshold and can enjoy increased confidence in the security of their protocol by making use of the infrastructure. Surprisingly, it turns out the OT subgraph G need not even contain t OT edges to offer resilience against t corruptions (cf. Fig. 2(c) with $n = 4, t = 2$).

A pair of parties may use the OT correlations generated as the base OTs for an OT extension protocol and inexpensively generate many OT correlations that can be saved for future use or to add to the OT infrastructure. In any case, it should be clear that our protocols readily allow load-balancing across the OT infrastructure and are also abort-tolerant in the sense that if some subgraph G ends up not delivering the output, then one can readily use a different subgraph G'. Thus we believe that our results can be used to build a *scalable* infrastructure for secure computation that allows (1) amortization, (2) routing, and (3) is robust.

An Important Caveat Regarding Efficiency. In the special cases $t = n/2 + \mathcal{O}(1)$ and $t = n - \mathcal{O}(1)$, determining whether a graph satisfies the partition condition requires at most $\text{poly}(n)$ time. However, in general the problem is coNP-complete, since it can be restated in the graph complement as subgraph isomorphism of a complete bipartite graph [30]. Our protocols are efficient in n only for $t = n/2 + \mathcal{O}(1)$ and $t = n - \mathcal{O}(1)$.[6] In particular, our protocol is quite efficient for small values of n, a setting in which computing OT correlations in the presence of a dishonest majority may be especially useful in practice.

2   Preliminaries

2.1   Notation and Definitions

Let $\mathcal{X}, \mathcal{Y}$ be two probability distributions over some set S. Their *statistical distance* is

$$\mathbf{SD}\,(\mathcal{X},\mathcal{Y}) \overset{\text{def}}{=} \max_{T \subseteq S}\{\Pr\,[\mathcal{X} \in T] - \Pr\,[\mathcal{Y} \in T]\}$$

We say that $\mathcal{X}$ and $\mathcal{Y}$ are ϵ-close if $\mathbf{SD}\,(\mathcal{X},\mathcal{Y}) \leq \epsilon$ and this is denoted by $\mathcal{X} \approx_\epsilon \mathcal{Y}$. We say that $\mathcal{X}$ and $\mathcal{Y}$ are identical if $\mathbf{SD}\,(\mathcal{X},\mathcal{Y}) = 0$ and this is denoted by $\mathcal{X} \equiv \mathcal{Y}$.

All graphs addressed in this work are undirected. We denote a graph as $G = (V, E)$ where V is a set of vertices and E is a set of edges. We denote an edge e as $e = \{v_1, v_2\}$, where $v_1, v_2 \in V$.

---

[6] For $t = n/2 + \mathcal{O}(1)$, we achieve efficiency using computationally-secure OT extension (e.g. [2,43]). Our protocol with information-theoretic security is quasipolynomial-time for $t = n/2 + \mathcal{O}(1)$. We do, however, achieve information-theoretic security in polynomial time for $t = n - \mathcal{O}(1)$.

For $n \in \mathbb{N}$, let K_n denote the complete graph on n vertices. Let Λ_a^s denote the graph $G = (V, E)$ on $2a + s$ vertices with $V = V_A \dot{\cup} V_S \dot{\cup} V_B$, where $|V_A| = |V_B| = a$ and $|V_S| = s$, and

$$E = \{\{v_1, v_2\} : v_1 \notin V_A \vee v_2 \notin V_B\}$$

We will sometimes consider subgraphs of Λ_a^s which preserve labels of vertices. In this case we will always label the vertices so that vertex $A \in V_A$ and vertex $B \in V_B$.

For two graphs $G_1 = (V, E_1)$ and $G_2 = (V, E_2)$ with the same vertex set V, we say that G_1 and G_2 are $(v_1, \ldots, v_\ell)$-*isomorphic*, denoted by $G_1 \simeq_{v_1, \ldots, v_\ell} G_2$, if the two graphs are isomorphic to one another while fixing the labelings of vertices $v_1, \ldots, v_\ell \in V$, that is, there exists an isomorphism σ such that $\sigma(v_i) = v_i$ for all $i \in [\ell]$.

Similarly, given graphs $G_1 = (V_1, E_1)$ and $G_2 = (V_2, E_2)$ with $V_1 \subseteq V_2$ and $v_1, \ldots, v_\ell \in V_1$, we say that G_1 is a $(v_1, \ldots, v_\ell)$-*subgraph* of G_2, denoted $G_1 \subseteq_{v_1, \ldots, v_\ell} G_2$, if G_1 is $(v_1, \ldots, v_\ell)$-isomorphic to some subgraph of G_2.

In particular, in the special case that graph $G = (V, E)$ contains vertices $A, B \in V$, we say that G is an (A, B)-*subgraph* of Λ_a^s (or that $G \subseteq_{A,B} \Lambda_a^s$) if there is an isomorphism σ between G and a subgraph of Λ_a^s such that A is mapped into set V_A and B is mapped into set V_B (that is, $\sigma(A) \in V_A$ and $\sigma(B) \in V_B$).

Call an n-vertex graph $G = (V, E)$ k-*unsplittable* for $k \leq n/2$ if any two disjoint sets of k vertices have some edge between them. That is, G is k-unsplittable if for all partitions of the vertices V into three disjoint sets V_1, V_2, V_3 of sizes $|V_1| = |V_2| = k$ and $|V_3| = n - 2k$, there exists some edge $(u, v) \in E$ with $u \in V_1, v \in V_2$. It is immediate from this definition that G is k-unsplittable if and only if $G \nsubseteq \Lambda_k^{n-2k}$.

Similarly, call G (k, A, B)-*unsplittable* for $k \leq n/2$ and $A, B \in V$ if any two disjoint sets of k vertices containing A and B, respectively, have some edge between them. That is, G is (k, A, B)-unsplittable if for all partitions of the vertices of V into three disjoint sets V_1, V_2, V_3 of sizes $|V_1| = |V_2| = k$ and $|V_3| = n - 2k$ such that $A \in V_1$ and $B \in V_2$, there exists some edge $(u, v) \in E$ with $u \in V_1, v \in V_2$. From this definition we have immediately that G is (k, A, B)-unsplittable if and only if $G \nsubseteq_{A,B} \Lambda_k^{n-2k}$.

2.2   Secure Computation

Consider the scenario of n parties $P_1, \ldots, P_n$ with private inputs $x_1, \ldots, x_n \in \mathcal{D}$ computing a function $f : \mathcal{D}^n \to \mathcal{D}^n$. Let Π be a protocol computing f. We consider security against adaptive t-threshold adversaries, that is, adversaries that adaptively corrupt a set of at most t parties, where $0 \leq t < n$.[7] We assume the adversary to be semihonest (i.e. honest-but-curious). That is, the corrupted parties follow the prescribed protocol, but the adversary may try to infer additional information about the inputs of the honest parties. As noted in [36], in

---

[7] Note that when $t = n$, there is nothing to prove.

the computational setting, using zero-knowledge proofs, it is possible to generically compile a protocol which is secure against semihonest adversaries into another protocol which is secure against adaptive malicious adversaries [32].[8] This justifies our focus on the semihonest setting here.

For a PPT adversary $\mathcal{A}$, let random variable $\mathrm{REAL}_{\Pi,\mathcal{A}}^{x_1,\ldots,x_n}$ consist of the views of the corrupted parties when the protocol Π is run on parties $P_1,\ldots,P_n$ with inputs $x_1,\ldots,x_n$ respectively. In the ideal world, the honest parties are replaced with a simulator $\mathcal{S}$ that does not receive input values and knows only the output value of each corrupted party in an honest execution of the protocol. We define the random variable $\mathrm{IDEAL}_{\Pi,\mathcal{A},\mathcal{S}}^{x_1,\ldots,x_n}$ as the output of the adversary $\mathcal{A}$ in the ideal game with the simulator when the inputs to parties $P_1,\ldots,P_n$ are $x_1,\ldots,x_n$, respectively.

Definition 1. *A protocol Π is said to t-securely compute the function f if*

- *For all $x_1,\ldots,x_n \in \mathcal{D}^n$, party P_i receives y_i, where $(y_1,\ldots,y_n) = f(x_1,\ldots,x_n)$, at the end of the protocol.*
- *For all adaptive semihonest PPT t-threshold adversaries $\mathcal{A}$, there exists a PPT simulator $\mathcal{S}$ such that for all $x_1,\ldots,x_n \in \mathcal{D}^n$*

$$\left\{ \mathrm{REAL}_{\Pi,\mathcal{A}}^{x_1,\ldots,x_n} \right\} \equiv \left\{ \mathrm{IDEAL}_{\Pi,\mathcal{A},\mathcal{S}}^{x_1,\ldots,x_n} \right\}$$

This definition is for secure computation with perfect information-theoretic security and a nonadaptive adversary. By [15], in the semihonest setting with information-theoretic security, any protocol which is nonadaptively secure is also adaptively secure. Consequently, satisfying this definition suffices to achieve adaptive security.

In the discussion below, we will sometimes relax security to statistical or computational definitions. A protocol is statistically t-secure if the random variables $\mathrm{REAL}_{\Pi,\mathcal{A}}^{x_1,\ldots,x_n}$ and $\mathrm{IDEAL}_{\Pi,\mathcal{A},\mathcal{S}}^{x_1,\ldots,x_n}$ are statistically close, and computationally t-secure if they are computationally indistinguishable.

2.3   Oblivious Transfer

In this work OT refers to 1-out-of-2 oblivious transfer defined as follows.

Definition 2. *We define 1-out-of-2 oblivious transfer f_{OT} for a sender $A = P_1$ with inputs $x_0, x_1 \in \{0,1\}^m$, a receiver $B = P_2$ with input $b \in \{0,1\}$ and $n-2$ parties $P_3,\ldots,P_n$ with input $\perp$ as*

$$f_{\mathrm{OT}}((x_0,x_1), b, \perp, \ldots, \perp) = (\perp, x_b, \perp, \ldots, \perp)$$

Note that while OT is typically defined as a 2-party functionality, the definition above adapts it our setting and formulates OT as an n-party functionality where only two parties supply non-$\perp$ inputs.

---

[8] We note that in the computational setting, it is also possible to transform, in a *black-box* way, a protocol which is secure against semihonest adversaries into another protocol which is secure against static malicious adversaries [35].

Definition 3. *Let G be a network consisting of n parties $A = P_1, B = P_2, P_3, \ldots, P_n$. Then a t-secure OT protocol $\Pi_{A \to B}^{G,t}$ is a protocol that t-securely computes the function f_{OT} on the inputs of the parties with A as the sender and B as the receiver.*

We note that OT is symmetric, in the following sense.

Lemma 1 [64]. *If there exists a t-secure OT protocol $\Pi_{A \to B}^{G,t}$ for an n-party network G with n parties $A = P_1, B = P_2, P_3, \ldots, P_n$ with A as the sender and B as the receiver, then there exists a t-secure OT protocol $\widehat{\Pi}_{B \to A}^{G,t}$ for the same n parties with B as the sender and A as the receiver.*

We represent parties as nodes of a graph G where an edge $\{A, B\}$ indicates that parties A and B may run a 1-secure OT protocol with A as the sender and B as the receiver. By Lemma 1, the roles of the sender and receiver may be reversed, so it makes sense to define G as an undirected graph.

We note the following result regarding the completeness of OT for achieving arbitrary secure multiparty computation.

Lemma 2 [33,45,46]. *Consider the complete network $G \simeq K_n$ on n vertices. Then, for any function $f : \mathcal{D}^n \to \mathcal{R}^n$, there exists a protocol Π which $(n-1)$-securely computes f, where party i receives the ith input $x_i \in \mathcal{D}$ and produces the ith output $(f(x))_i \in \mathcal{R}$.*

3   Warm-Ups

Let $G = (V, E)$ be an n-vertex graph representing a network with n parties, where an edge $\{P_i, P_j\} \in E$ indicates that parties P_i and P_j may run a 1-secure 2-party OT protocol with P_i as the sender and P_j as the receiver. Let $t < n$ be an upper bound on the number of corruptions made by the adversary. The central question considered in this work is the following. For which graphs G and which pairs of parties $A, B \in V$ does there exist a t-secure OT protocol with A as the sender and B as the receiver?

We begin by discussing some simple special cases of small networks. These will provide useful intuition for our main results. For $t < n/2$, it is possible to obtain a t-secure OT protocol for any n-vertex graph $G = (V, E)$ between any $A, B \in V$, since we can perform secure multiparty computation without any pre-existing OT channels if there is an honest majority [63]. It remains to consider the setting where $t \geq n/2$.

A few small cases have been resolved in prior work. For $n = 2$, $t = 1$, a 1-secure OT protocol (with perfect security) between the vertices of the two-vertex graph G does not exist unless the parties were already connected by an OT channel [17,49]. This result is illustrated in Fig. 1(a).

For $n = 3$, $t = 2$, it is known that we can obtain a 2-secure OT protocol between a pair of vertices A, B only if those vertices are already connected by an OT channel, even if there are OT channels from both A and B to the third

Fig. 1. Known impossibility results. Securely computing f_{OT} between A' and B' is impossible for $t = 1$ in G_{CK} and is impossible for $t = 2$ in G_{HIK}.

vertex C as depicted in Fig. 1(b). More generally, for any $n \geq 2$ and $t = n - 1$, there exists a t-secure OT protocol with sender A and receiver B only if those vertices are already connected by an OT channel, even if all other $\binom{n}{2} - 1$ pairs of vertices are connected by OT channels [36]. This also resolves the question for $n = 4, t = 3$.

The remainder of this section is devoted to an exploration of the setting $n = 4, t = 2$. This is the smallest case not resolved by prior techniques, and will illustrate many of the tools used in subsequent sections to obtain our general protocols. The key cases for $n = 4, t = 2$ are shown in Fig. 2. As discussed below, these cases are sufficient to completely resolve the four-party setting.

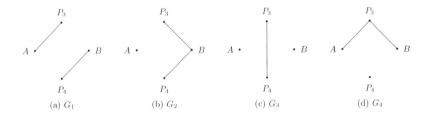

Fig. 2. Cases for $n = 4$ parties with $t = 2$ corruptions.

3.1 Case 1: Fig. 2(a)

We first show that if $G \simeq_{A,B} G_1$ then there does not exist a 2-secure OT protocol for G with A as the sender and B as the receiver.[9] This is a consequence of the impossibility result of [17,49]. An outline of the argument is as follows.

Consider components $\mathcal{C}_1 = \{A, P_3\}$ and $\mathcal{C}_2 = \{B, P_4\}$ of G, and let Π be a 2-secure protocol computing f_{OT} in G with A as the sender and B as the receiver. Then we can use Π to construct a 1-secure protocol Π' for the 2-party network G_{CK} in Fig. 1(a) with A' as the sender and B' as the receiver. In protocol Π', party A' runs Π for both parties of component $\mathcal{C}_1$ of G, and B' runs Π for both parties of component $\mathcal{C}_2$. OT channel invocations can be handled locally, since all OT channels in G are between parties in the same component. Since protocol Π is 2-secure, in particular it is secure against corruptions of parties in

---

[9] Recall that $H \simeq_{A,B} H'$ for two graphs H, H' if there exists an isomorphism between H and H' preserving the labels of vertices A and B.

C_1 or the parties in C_2. Consequently Π' is a 1-secure OT protocol for a network $G' \simeq_{A',B'} G_{CK}$ with A' as the sender and B' as the receiver. However, from [17,49], we know that no such protocol exists with perfect security. Consequently there is no 2-secure protocol Π for a network $G \simeq_{A,B} G_1$.

Note that this impossibility holds not only for $G \simeq_{A,B} G_1$ but for any (A, B)-subgraph of G_1. In particular, if $G = (V, E)$ is a four-vertex graph a single edge that is incident to vertex A or vertex B, then G cannot have a 2-secure protocol computing f_{OT} between A and B except in the trivial case when there is already an edge $\{A, B\} \in E$. This technique of reducing to the known impossiblity results of [17,36,49] to obtain lower bounds is described formally in Sect. 4.

3.2   Case 2: Fig. 2(b)

In this example we obtain a positive result, showing that there exists a 2-secure OT protocol with A as the sender and B as the receiver. Since B has degree 2 in G_2, we have that either B or one of its neighbors must be honest, and so one of the two OT channels must contain an honest party. This suggests the idea of using secret-sharing to ensure security against 2 corruptions.

Consider the following OT protocol where sender A has inputs $x_0, x_1 \in \{0, 1\}^m$ and receiver B has input $b \in \{0, 1\}$. A computes 2-out-of-2 shares (x_0^1, x_0^2) and (x_1^1, x_1^2) of its inputs x_0, x_1, respectively. A then sends shares x_0^1 and x_1^1 to party P_3 and x_0^2 and x_1^2 to party P_4. Parties P_3 and B invoke their secure OT channel with inputs (x_0^1, x_1^1) and b, and parties P_4 and B invoke their secure OT channel with inputs (x_0^2, x_1^2) and b respectively. B uses the obtained shares x_b^1, x_b^2 to reconstruct x_b.

We informally argue the 2-security of this protocol assuming that exactly one of A and B is corrupt.[10] Consider the case where A is corrupt and B is honest. The input of B is only used over secure OT channels, so by the 1-security of the OT channels with P_3 and P_4, the corrupt parties can learn nothing about B's input bit b. Now consider the case where B is corrupt and A is honest. Either P_3 or P_4 must be honest. If P_3 is honest then the security of OT channel $\{P_3, B\}$ implies that B learns nothing about share x_{1-b}^1, so the security of the secret sharing scheme implies that the corrupt parties do not use x_{1-b}. By symmetry, the same argument applies if P_4 is honest. This completes the argument.

Note that by Lemma 1, we can also obtain a 2-secure OT protocol from A to B whenever A has degree 2 in OT network. Furthermore, we can extend this idea to construct a t-secure OT protocol whenever either the sender or the receiver has degree at least t. We call this protocol the t-claw protocol and describe it in detail in Sect. 5.1.

---

[10] An additional step is needed to address the case in which P_3 and P_4 are corrupt and A and B are both honest. Then P_3 and P_4 can learn x_0 and x_1, the inputs of A, in the protocol just described. This can be handled with the technique of OT correction, using a one-time pad and the secure point-to-point channel between A and B. Equivalently, we could run the protocol on random inputs, and then use method of [3] to obtain 1-out-of-2 OT from random OT. If A and B are both corrupt then there is nothing to prove.

3.3   Case 3: Fig. 2(c)

Somewhat surprisingly, we can also show a positive result for graphs $G \simeq_{A,B} G_3$ even though the OT network has no edges involving either the sender A or the receiver B. The protocol is as follows. Since parties P_3 and P_4 have an OT channel between them, by Lemma 2, they can perform 1-secure MPC between them. P_3 and P_4 use MPC to compute 2-out-of-2 shares of OT correlations with uniformly random inputs and send corresponding shares to A and B, who can then reconstruct the correlations. More concretely, the MPC protocol computes 2-out-of-2 shares (r_0^1, r_0^2), (r_1^1, r_1^2) of two randomly sampled m-bit strings r_0, r_1, 2-out-of-2 shares (c^1, c^2) of a random bit $c \in \{0, 1\}$, and independent 2-out-of-2 shares (s^1, s^2) of the string r_c. Party P_3 receives the first share of each secret, and party P_4 receives the second share. Party P_3 then sends shares r_0^1, r_1^1 to A and s^1, c^1 to B, while P_4 sends shares r_0^2, r_1^2 to A and s^2, c^2 to B. A can then reconstruct r_0 and r_1, and B can reconstruct c and r_c. Parties A and B have now established a random OT correlation, which they can use to perform OT with their original inputs using OT correction [3].[11]

We now informally argue the 2-security of this protocol. If A and B are both honest, then the corrupt parties receive no information about their inputs, while if A and B are both corrupt then there is nothing to prove. Consequently we can assume that exactly one of A and B is corrupt and that either P_3 or P_4 is honest. If A is corrupt and P_3 or P_4 is honest, then the adversary learns nothing about c and r_c, since it only sees one of the two shares of each. The OT correction phase uses these strings as one-time pads for inputs which are unknown to the adversary, and consequently are information-theoretically hidden from the adversary. Consequently A learns nothing about B. The case where B is corrupt and P_3 or P_4 is honest follows by the same argument.

This construction can be extended to obtain a t-secure OT protocol whenever the OT graph contains a t-clique consisting of t parties which are not the OT sender or receiver. We call protocol the t-clique protocol and describe it in detail in Sect. 5.2.

3.4   Case 4: Fig. 2(d)

We also obtain a positive result for graphs $G \simeq_{A,B} G_4$. We introduce here a technique we call cascading. The idea is as follows. Using the protocol described in Sect. 3.2 for network G_2 of Fig. 2(b), we have 2-secure OT protocol with P_3 as the sender and P_4 as the receiver. This effectively gives us an OT channel between P_3 and P_4. Applying the protocol from Sect. 3.3 on the augmented network, we obtain a 2-secure OT protocol with A as the sender and B as the receiver. We describe this pictorially in Fig. 3.

The 2-security of the protocol follows from the 2-security of the underlying protocols of Sects. 3.2 and 3.3. The technique of cascading for combining t-secure protocols is described in detail in Sect. 5.3.

---

[11] This OT correction step can be performed as follows. Party B sends $b' = b \oplus c$ to A. A responds with $y_0 = x_0 \oplus r_{b'}$ and $y_1 = x_1 \oplus r_{1-b'}$. Finally, B computes $y_b \oplus r_c = x_b$.

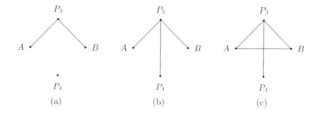

Fig. 3. Illustrating the cascading protocol for Case 4: Fig. 2(d); (a) → (b) → (c)

3.5   Cases 1–4 are Exhaustive

Note that a t-secure OT protocol with sender A and receiver B in an OT network G trivially yields a t-secure protocol for any network G' such that $G \subseteq_{A,B} G'$. From cases 1 and 3, we can securely compute f_{OT} in a network G containing at most a single edge if and only if the edge is $\{A, B\}$ or $\{P_3, P_4\}$. From cases 1, 2, and 4, we can compute f_{OT} in a network G containing two or more edges including neither of $\{A, B\}$ or $\{P_3, P_4\}$ if and only if there is some vertex with degree at least 2 in the OT graph. This completes the characterization of 4-party networks with 2 corruptions.

4   Lower Bound

We now describe a family of impossibility results using a generic reduction to the impossiblity result in [36], which we restate in our language below.

Lemma 3 [36]. *Consider any three party network G with $G \simeq_{A',B'} G_{\mathrm{HIK}}$, the graph in Fig. 1(b). Then any 2-secure OT protocol with A' as the sender and B' as the receiver can be used (as a black box) to obtain a 1-secure OT protocol for a network G' with $G' \simeq_{A',B'} G_{\mathrm{Kus}}$, the graph in Fig. 1(b), with A' as the sender and B' as the receiver.*

The theorem below describes an impossibility result over a family of networks. We note that this result was observed in [36]; we restate it our language and defer the formal proof to the full version.

Theorem 1. *Let $n \geq 2$ and $n/2 \leq t < n$, and let G be an n party network such that $G \subseteq \Lambda_{n-t}^{2t-n}$, with $P_1 \in V_A$ and $P_2 \in V_B$. Any t-secure OT protocol for G with P_1 as the sender and P_2 as the receiver can be used (as a black box) to obtain a 1-secure OT protocol for a network G' with $G' \simeq_{A,B} G_{\mathrm{CK}}$ with A' as the sender and B' as the receiver.*

5   Building Blocks

In this section, we describe a few key protocols and techniques that we use in the subsequent sections to prove our main theorem.

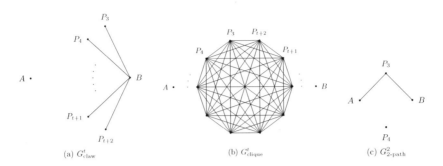

Fig. 4. Building block networks. (a) t-claw graph (b) t-clique graph (c) 2-path graph

5.1 The t-claw Protocol

The first protocol we describe is the t-claw protocol, where the graph G describing the network is such that $G \simeq_{A,B} G_{\text{claw}}^t$. The protocol is described in Protocol 1. The protocol is a straightforward generalization of the one described in Sect. 3.2. The idea is for A to compute t-out-of-t shares of its inputs and distribute them among the t parties connected to B. These t parties then perform OT with B so that B receives the shares to reconstruct his output.

Protocol 1: t-claw Protocol

Preliminaries: Let $A, B, P_3, \ldots, P_{t+2}$ be the $t + 2$ parties in a network $G \simeq_{A,B}$ G_{claw}^t. A has inputs $x_0, x_1 \in \{0, 1\}^m$ and B has input $b \in \{0, 1\}$.

Protocol:

1. B chooses a random bit $c \in \{0, 1\}$ and sends $b' = b \oplus c$ to A.
2. A chooses two random one-time pads $r_0, r_1 \in \{0, 1\}^m$ and sends $y_0 = x_0 \oplus r_{b'}$ and $y_1 = x_1 \oplus r_{1-b'}$ to B.
3. A then computes t-out-of-t shares $(r_0^1, \ldots, r_0^t)$ and $(r_1^1, \ldots, r_1^t)$ of r_0, r_1, respectively.
4. For each $i \geq 3$, A sends shares r_0^i and r_1^i to party P_i.
5. For each $i \geq 3$, parties P_i and B invoke the OT protocol $\Pi_{P_i \to B}^{G,1}$ with inputs (r_0^i, r_1^i) and c respectively.
6. B uses the obtained shares $r_c^1, \ldots, r_c^t$ to reconstruct r_c.
7. B finally computes $y_b \oplus r_c = x_b$.

Lemma 4. *Protocol 1 is an efficient t-secure OT protocol for a network $G \simeq_{A,B}$ G_{claw}^t with A as the sender and B as the receiver.*

Proof Intuition. The t-security of the protocol can be seen as follows. Steps 1, 2 and 7 perform OT correction, that is, they perform a transformation from random OT to 1-out-of-2 OT. This transformation protects against the case that the parties $P_3, \ldots, P_{t+2}$ (that is, all but A and B) are corrupt. Suppose A were corrupt and B were honest. Clearly, A colluding with any of the parties $P_3, \ldots, P_{t+2}$ provides A with no additional information since all they possess are shares sent by A. Next, if A were honest and B corrupt, at least one of the parties $P_3, \ldots, P_{t+2}$ must be honest. B has no information about those shares and hence does not learn anything. Finally, if both A and B were corrupt, there is nothing to prove.

5.2   The t-clique Protocol

The next protocol we describe is the t-clique protocol, where the graph G describing the network is such that $G \simeq_{A,B} G^t_{\text{clique}}$. The protocol is described in Protocol 2. The protocol is a straightforward generalization of the one described in Sect. 3.3. The idea is for the parties $P_3, \ldots, P_{t+2}$ to compute t-out-of-t shares of OT correlations and send them to A and B respectively. The parties have a complete network of OT channels, so this can be done via multiparty computation (Lemma 2). A and B then perform OT correction using their secure channel. We state the lemma, give a proof outline and defer the full proof to the full version.

Lemma 5. *Protocol 2 is an efficient t-secure OT protocol for a network $G \simeq_{A,B} G^t_{\text{clique}}$ with A as the sender and B as the receiver.*

Proof Intuition. The t-security of the protocol can be seen as follows. Steps 4, 5 and 6 perform OT correction, that is, they perform a transformation from random OT to 1-out-of-2 OT. This transformation protects against the case that all of parties $P_3, \ldots, P_{t+2}$ (that is, all but A and B) are corrupt. If one of A and B were corrupt, there exists at least one honest party among the parties $P_3, \ldots, P_{t+2}$. Hence, even by colluding, A or B would have no information about those shares and would not learn anything. Finally, if both A and B were corrupt, there is nothing to prove.

5.3   Cascading

The following building block is a generalization of the technique described in Sect. 3.4. The technique describes a general method of combining protocols iteratively. In our context, this can be thought of a tool for transforming a network described by a graph G to one described by a graph G', where $G \subseteq_V G'$ and G and G' are both graphs on the same vertex set V. In other words, it describes protocols as adding new edges indicating the establishment of OT correlations between new pairs of parties in the network. With this abstraction, it is easy to view the technique of cascading as one which combines protocols iteratively to transform the underlying network by adding new edges. This is described formally below.

Definition 4. *Let $G = (V, E)$ and $G' = (V, E')$ be two graphs on the same set of vertices, V, with $G \subseteq_V G'$. We say that a protocol Π t-transforms a network*

Protocol 2: t-clique Protocol

Preliminaries: Let $A, B, P_3, \ldots, P_{t+2}$ be the $t + 2$ parties in a network $G \simeq_{A,B} G^t_{\text{clique}}$. A has inputs $x_0, x_1 \in \{0, 1\}^m$ and B has input $b \in \{0, 1\}$.

Protocol:

1. Parties $P_3, \ldots, P_{t+2}$ use their pairwise OT channels to run t-secure MPC for the function f using the protocol from Lemma 2 for the function f described ahead. The function f is to securely compute t-out-of-t shares $(r_0^1, \ldots, r_0^t)$, $(r_1^1, \ldots, r_1^t)$ of two randomly sampled one-time pad keys r_0, r_1, $(c^1, \ldots, c^t)$ of a random bit $c \in \{0, 1\}$, and independent shares $(s^1, \ldots, s^t)$ of key r_c, so that party $i + 2$ receives only shares r_0^i, r_1^i, s^i, c^i for each i.
2. Each party P_{i+2} for $i \geq 1$ sends shares r_0^i, r_1^i to A and s^i, c^i to B.
3. A uses shares $(r_0^1, \ldots r_0^t)$ and $(r_1^1, \ldots, r_1^t)$ to reconstruct r_0 and r_1.
4. B uses shares $(c^1, \ldots, c^t)$ and $(s^1, \ldots, s^t)$ to reconstruct c and r_c and sends $b' = b \oplus c$ to A.
5. A computes $y_0 = x_0 \oplus r_{b'}$ and $y_1 = x_1 \oplus r_{1-b'}$ and sends both to B.
6. B computes $y_b \oplus r_c = x_b$.

G into the network G' if for each $\{P_i, P_j\} \in E' \setminus E$, Π is a t-secure OT protocol for a network G with P_i as the sender and P_j as the receiver.[12]

Lemma 6. *If Π_1 is a protocol that runs in time T_1 and t-transforms network G_1 into G_2, and Π_2 is a protocol that runs in time T_2 and t-transforms network G_2 into G_3, then there exists a protocol Π that runs in time $T_1 T_2$ and t-transforms G_1 into G_3.*

Proof. The protocol Π simply runs Π_2, running protocol Π_1 to obtain the necessary correlations whenever Π_2 invokes OT on an edge of $G_2 \setminus G_1$. Let $\mathcal{S}_1$ and $\mathcal{S}_2$ be the simulators associated with Π_1 and Π_2 respectively. The simulator for Π simply runs $\mathcal{S}_2$, invoking $\mathcal{S}_1$ for OT calls made on edges in $G_2 \setminus G_1$. $\square$

Using OT extension [2,43], we can also obtain a computationally secure version of cascading with improved efficiency.

Lemma 7. *Let λ be a computational security parameter. Assuming one-way functions or correlation-robust hash functions, if Π_1 is a protocol that runs in time T_1 and t-transforms network G_1 into G_2, and Π_2 is a protocol that runs in time T_2 and t-transforms network G_2 into G_3, then there exists a computationally secure protocol Π that runs in time $\lambda \cdot T_1 + T_2 \cdot \text{poly}(\lambda)$ and t-transforms G_1 into G_3.*

---

[12] Note that a single protocol Π may set up independent random OT correlations for several pairs of parties $\{P_i, P_j\} \in E' \setminus E$. These correlations can be used to run 1-out-of-2 OT using OT correction.

Proof. First, run protocol Π_1 λ times on random inputs to obtain λ independent OT correlations for each edge of $G_2 \setminus G_1$. Then run Protocol Π_2, using OT extension to obtain OT correlations for OT calls made on edges in $G_2 \setminus G_1$. □

5.4 The 2-path Graph

The protocol described in this section is a commonly used subroutine in several of the protocols which follow. It is a particular combination of the tools encountered in Sects. 5.1, 5.2 and 5.3. The subroutine, which we call 2-path, is the same as the one described in Sect. 3.4. It is used to obtain OT correlations between parties who have a common neighbor in a four-party network with at most two corruptions (see Fig. 4(c)). The following lemma is immediate from Lemma 6 and the 2-security of Protocols 1 and 2 for $t = 2$ (Lemmata 4 and 5).

Lemma 8. *Protocol 3 is an efficient 2-secure OT protocol for a network $G \simeq_{A,B} G_{2\text{-path}}^2$ with A as the sender and B as the receiver.*

Protocol 3: 2-path

Preliminaries: Let A, B, C, D be the parties, and let there exist OT channels (A, C) and (B, C). A has input (x_0, x_1), and B has input $b \in \{0, 1\}$.

Protocol:

1. Invoke Protocol 1 (2-claw) on parties (D, C, A, B) to obtain OT correlations on edge (D, C).
2. By Lemma 6, we have an OT channel between D and C.
3. Invoke Protocol 2 (2-clique) on parties (A, B, C, D).

5.5 Combiners

OT combiners aim to combine several insecure candidate protocols for establishing OT correlations between two parties into a single secure protocol. For a class of adversaries $\mathbb{A}$, it is possible to achieve this when the candidate protocols satisfy the property that a majority of them are secure against each adversary $\mathcal{A} \in \mathbb{A}$. The following lemma is due to [37,56], relying on prior work by [38,65] and based on a construction by [22].

Lemma 9 [37,56]. *Let $\mathbb{A}$ be an adversary class. Suppose there exist m protocols $\Pi_1, \ldots, \Pi_m$ for $f_{OT}(A, B, P_1, \ldots, P_n)$ such that for any adversary $\mathcal{A} \in \mathbb{A}$ a majority of the protocols are secure. Then, there exists a protocol $\Pi^*(\Pi_1, \ldots, \Pi_m)$ for $f_{OT}(A, B, P_1, \ldots, P_n)$ which is secure against all adversaries $\mathcal{A} \in \mathbb{A}$. Moreover, if each protocol Π_i is efficient and perfectly secure, then so is Π^*.*

6   The Case $t = n/2$

We now consider the specific case of $t = n/2$, that is, when at most half the parties are corrupt. We note that this is the smallest value of t for which the question is non-trivial. From the lower bounds proven in Theorem 1, we already have that for all n-party networks G containing A and B such that $G \subseteq_{A,B} \Lambda^0_{n/2}$, there exists no $n/2$-secure OT protocol with A as the sender and B as the receiver. Surprisingly Theorem 2 shows that these are the only networks for which $(n/2)$-secure OT between A and B is impossible. Below, we provide an explicit $n/2$-secure OT protocol between A and B whenever the network G is $(n/2, A, B)$-unsplittable.

Theorem 2. *Let G be an n-party network OT containing parties A and B. Then Protocol 5 is an $n/2$-secure OT protocol between A and B if and only if G is $(n/2, A, B)$-unsplittable.*

We analyze the efficiency of the protocol in Theorem 3 below. The protocol as stated runs in quasi-polynomial time. We can also obtain a computationally secure protocol which runs in polynomial time. The protocol we describe proceeds in two stages. In the first stage, the protocol transforms every connected component of the network into a clique. This transformation is very specific to the case of $t = n/2$, and in particular, for $t > n/2$ a connected component cannot in general function as a clique. This transformation is carried out by means of repeatedly calling Protocol 4, which obtains OT correlations between a pair of parties who have a common neighbour. This protocol uses the building block Protocol 3 from Sect. 5.4 along with machinery of OT combiners described in Sect. 5.5.

Lemma 10. *Let G be an n-vertex OT network with edges $\{A, C\}$ and $\{B, C\}$. Protocol 4 is an $n/2$-secure OT protocol for the network G with A as the sender and B as the receiver.*

Proof. We consider cases depending on the number of corrupted parties in the set $T = \{A, B, C\}$. If T contains at most one corrupted party, then each tuple (A, B, C, P_i) for $i \geq 4$ contains at most 2 corrupted parties, so each protocol Π_i in step 1 is secure. If T contains two corrupted parties, then there are at most $t - 2 = (n - 4)/2$ corrupted parties among $P_4, \ldots, P_n$, so a majority of these parties are honest. Consequently a majority of the protocols Π_i which are combined in step 1 are secure. Thus, in either case, by Lemma 9 the protocol is secure. Finally, if all three parties of T are corrupted, then all uncorrupted parties receive no input, so the simulator $\mathcal{S}$ can perfectly simulate the uncorrupted parties by running the honest protocol. Therefore Protocol 4 is $n/2$-secure.   □

We now complete the proof of Theorem 2.

Proof Intuition (Theorem 2): It is easy to see that by invoking Protocol 4 repeatedly, one can obtain OT correlations between any pair of parties in the same connected component. In other words, using cascading (Lemma 6), we can assume

---

Protocol 4: Completing Triangles

Preliminaries: Let $A, B, C, P_4, \ldots, P_n$ be the n parties, and let there exist OT channels (A, C) and (B, C). A has input (x_0, x_1), and B has input $b \in \{0, 1\}$.

Protocol:

1. Run a combined protocol $\Pi^*(\Pi_4, \ldots, \Pi_n)$ on the $n - 3$ protocols $\Pi_4, \ldots, \Pi_n$, where
   - For each $i \geq 4$, Π_i denotes an invocation of Protocol 3 (2-path) with the four parties A, B, C, P_i with A as the sender and B as the receiver.

---

that we are given a network which consists of disjoint cliques. This is done in step 1 of Protocol 5. Hence, if A and B were in the same connected component in G, this process would end up with correlations between A and B and we can terminate the protocol (step 2).

If A and B are in different components, then a natural next step is to run the clique protocol described in Sect. 5.2 with each of the cliques and parties A and B with the intent of setting up OT correlations between A and B. However, the number of corruptions t may be greater than the size of any clique, and so Protocol 2 may not be secure. However, for an invocation to be secure, we only require that the clique contains at least one honest party. A majority of parties must be in cliques containing at least one honest party, so if we invoke Protocol 2 for each of the parties on their respective cliques, for any adversary a majority of the invocations is secure. By Lemma 9 we can combine these candidate protocols to obtain a single secure protocol. This is performed in step 5 of Protocol 5. Finally, we note that steps 3, 4 and 6 perform OT correction, that is, they perform a transformation from random OT to 1-out-of-2 OT. This yields the $n/2$-security of Protocol 5.

Proof (Theorem 2). The "only if" part of theorem has been proven by virtue of the lower bound of Theorem 1 with $t = n/2$. We now prove the "if" part. We note that in the case where A and B are in the same connected component in the network G, by the $n/2$-security of Protocol 4 and Lemma 6, we note that Protocol 5 is an $n/2$-secure OT protocol with A as the sender and B as the receiver, thus proving the theorem.

We now proceed to the case where A and B are not in the same connected component in G. We must show that the protocol is secure against t-threshold adversaries as long as the vertices cannot be partitioned into two sets V_A, V_B each of size $t = n/2$ with $A \in V_A, B \in V_B$ such that there are no edges between V_A and V_B. Let $\mathcal{A}$ be a t-threshold adversary which corrupts parties T, $|T| \leq t$. We will construct a simulator $\mathcal{S}$ which plays the role of the uncorrupted parties.

Protocol 5: $n/2$ corruptions

Preliminaries: Let $P_1 = A, P_2 = B, P_3, \ldots, P_n$ be the n parties in a network $G = (V, E)$. A has input (x_0, x_1), and B has input $b \in \{0, 1\}$.

Protocol:

1. While there exist parties $P_i, P_j, P_k \in V$ such that $\{P_i, P_j\} \in E$, $\{P_j, P_k\} \in E$, but $\{P_i, P_k\} \notin E$:
   (a) Let S be the set of triples of distinct vertices $(X, Y, Z) \in V^3$ with $\{X, Y\} \in E, \{Y, Z\} \in E$, and $\{X, Z\} \notin E$.
   (b) For each triple $(X, Y, Z) \in S$, invoke Protocol 4 with independent random inputs $(r_0^{i,k}, r_1^{i,k})$ and $b^{i,k}$, to obtain OT correlations along edge $\{X, Z\}$.
   (c) Invoking cascading (Lemma 6), we can add $\{X, Z\}$ to the edge set E for all triples $(X, Y, Z) \in S$.
   The OT network G now consists of disjoint cliques $\mathcal{C}_1, \ldots, \mathcal{C}_\ell$.
2. If A and B are in the same clique, then halt.
3. B samples a random bit c and sends $b' = b \oplus c$ to A.
4. A chooses random one-time pads r_0, r_1 and sends $y_0 = x_0 \oplus r_{b'}$ and $y_1 = x_1 \oplus r_{1-b'}$ to B.
5. Let $\mathcal{C}_1$ be the clique containing A and $\mathcal{C}_2$ be the clique containing B. For each party P_i, $i \geq 3$, let $\mathcal{C}(i)$ denote the clique containing party i, and let $P_{j_1}, \ldots, P_{j_{|\mathcal{C}(i)|}}$ denote the parties in clique $\mathcal{C}(i)$.
   Run a combined protocol $\Pi^*(\Pi_1, \ldots, \Pi_n)$ on the n protocols $\Pi_1, \ldots, \Pi_n$, where
   – For each $i \in [n]$, Π_i denotes an invocation of Protocol 2 on the $|\mathcal{C}(i)| + 2$ parties $A, B, P_{j_1}, \ldots, P_{j_{|\mathcal{C}(i)|}}$ with inputs (r_0, r_1) and c.[a]
6. Finally, B computes $x_b = y_b \oplus r_c$.

---

[a] In the case $\mathcal{C}(i) = \mathcal{C}_1$, A is both the OT sender and a member of the clique. A similar condition holds for B in the case $\mathcal{C}(i) = \mathcal{C}_2$.

If $\{A, B\} \subset T$ then the uncorrupted parties receive no input, so the simulator can perfectly simulate the uncorrupted parties. If $\{A, B\} \cap T = \emptyset$ then $\mathcal{S}$ chooses arbitrary inputs x_0, x_1, b and runs the protocol. Since the only steps which depend on the input at all are on point-to-point channels between A and B, the view of the adversary in the real and ideal worlds is identical.

Otherwise, we have that the corrupted parties T include exactly one of A, B. If $A \in T$ but $B \notin T$, then $\mathcal{S}$ chooses an arbitrary bit b and runs the protocol, invoking the OT simulator for each invocation of Protocol 4. It follows that as long as the combined protocol Π^* in step 5 is secure against $\mathcal{A}$, Protocol 5 is secure against $\mathcal{A}$. It remains to show that a majority of the n protocols $\Pi_1, \ldots, \Pi_n$ are secure against $\mathcal{A}$. Since party B is honest, by Lemma 5, protocol Π_i is secure against $\mathcal{A}$ as long as at least one of the parties in clique $\mathcal{C}(i)$ is honest. In particular, if party P_i is honest then protocol Π_i is secure against $\mathcal{A}$. At most t of the parties $P_1, \ldots, P_n$ are corrupt, so the only protocols which may be insecure against $\mathcal{A}$ are the t protocols Π_i corresponding to the corrupted parties P_i. Assume that all t of these protocols are insecure against $\mathcal{A}$. Then the corrupted parties lie in completely corrupted cliques who sizes sum to $n/2$. This then gives a set $V_A = T$ of $n/2$ parties containing A but not B such that there are no edges from V_A to the remaining vertices $V_B = \overline{T}$. However, we know that G possesses no such partition. Hence, at most $t - 1 < n/2$ of the n protocols are insecure against $\mathcal{A}$ and hence by Lemma 9, the combined protocol Π^* in step 5 is secure and hence Protocol 5 is secure against $\mathcal{A}$.

The remaining case that $B \in T$ but $A \notin T$ is similar. Here, the simulator $\mathcal{S}$ is given the output value x_b. $\mathcal{S}$ runs the protocol with (x_b, x_b) as the input to A, again invoking the OT simulator for each invocation of Protocol 4. As above, as long as the combined protocol Π^* in step 5 is secure against $\mathcal{A}$, Protocol 5 is secure against $\mathcal{A}$. By the same argument, the only protocols Π_i which may be insecure against $\mathcal{A}$ are the t protocols corresponding to the corrupted parties P_i. If all t of these protocols are insecure against $\mathcal{A}$, we have a set $V_A = \overline{T}$ of $n/2$ parties containing A but not B such that there are no edges from V_A to the remaining vertices $V_B = T$. However, we know that G possesses no such partition, so at most $t - 1 < n/2$ of the n protocols are insecure against $\mathcal{A}$. By Lemma 9, the combined protocol Π^* in step 5 is secure and so Protocol 5 is secure against $\mathcal{A}$. $\qquad\square$

We now analyze the efficiency of Protocol 5.

Theorem 3. *Protocol 5 runs in quasi-polynomial time. Assuming one-way functions, we can obtain a computationally secure protocol which runs in polynomial time using computationally secure cascading (Lemma 7).*

Proof. Each iteration of step 1 decreases the length of a path between any pair of vertices from ℓ to $\lceil \ell + 1 \rceil / 2$. Consequently, after $O(\log n)$ iterations the graph will consist of a collection of disjoint cliques, and the protocol will move on to the next step. By Lemma 6 (Cascading), if each iteration can be performed in time at most T assuming the augmented graph, then the full cascaded protocol runs in time at most $T^{O(\log n)}$. Since $T = \text{poly}(n)$ and each other step of the protocol is efficient, this implies that Protocol 5 runs in quasi-polynomial time.

Replacing the cascading of step 1 with the more efficient but computationally secure cascading of Lemma 7, we have the cascaded protocol runs in time $O(T \text{poly}(\lambda) \cdot \log n)$. Since each other step of the protocol is efficient, this implies that assuming one-way functions, we have a computationally-secure version of Protocol 5 that runs in quasi-polynomial time. $\qquad\square$

7   The Case $t = n - 2$

On account of the lower bound proven in [36], we note that $t = n - 2$ is the largest value of t for which the question is non-trivial. In this section we present an improved computationally efficient OT protocol between A and B for the special case $t = n - 2$ for all $(2, A, B)$-unsplittable networks G.

Protocol 6: $n - 2$ corruptions

Preliminaries: Let $P_1 = A, P_2 = B, P_3, \ldots, P_n$ be the n parties, and let graph $G = (V, E)$ be the OT network among the parties. A has input (x_0, x_1), and B has input $b \in \{0, 1\}$.

Protocol:

1. For all pairs of parties $P_i, P_j \in V$ with $i, j \geq 3$ such that $\{P_i, P_j\} \notin E$:
   (a) Invoke Protocol 5 (or any 2-secure protocol for $n' = 4$) on the induced OT subgraph $G_{i,j} := G \cap \{P_i, P_j, A, B\}$ with independent random inputs $(r_0^{i,j}, r_1^{i,j})$ and $b^{i,j}$, to obtain OT correlations along edge $\{P_i, P_j\}$.
   (b) By virtue of cascading (Lemma 6), we can add edge $\{P_i, P_j\}$ to the graph G.[a]
   The OT network G now contains a $(n - 2)$-clique among vertices $P_i, \ldots, P_n$.
2. Invoke Protocol 2 (t-clique) with input (x_0, x_1) and b.

---

[a] We will only have OT security over this edge when at least two of the parties P_i, P_j, A, B are honest, but we obtain the functionality of the edge regardless. We address security of the overall protocol in the proof.

Theorem 4. *Let G be an n-party OT network containing parties A and B. Then Protocol 6 is an efficient $(n - 2)$-secure OT protocol between A and B if and only if G is $(2, A, B)$-unsplittable.*

The protocol is built upon the following structural aspect of the network G under consideration. Since G is $(2, A, B)$-unsplittable, for any two sets of vertices $V_A A$ and $V_B B$ such that $|V_A| = |V_B| = 2$, there exists an edge from a vertex of V_A to a vertex of V_B. In particular, this implies that for any two parties P_i, P_j where $i, j \geq 3$, the sub-network $G_{i,j}$ induced by parties A, B, P_i and P_j is $(2, A, B)$-unsplittable. Then for any i, j, we also have that the sub-network $G_{i,j}$ is $(2, P_i, P_j)$-unsplittable. Hence, we could try to obtain OT correlations between every pair of vertices P_i, P_j by running Protocol 5 on every $G_{i,j}$ for $n = 4$ parties. Notice that if these invocations were secure, then we would obtain an $(n - 2)$-clique in the network after which we can execute Protocol 2 in order to obtain OT correlations between A and B. This is described in Protocol 6. However, each of the execution of Protocol 5 is only guaranteed to be secure if at most two of the corresponding parties are corrupt. This need not be true in general, and so we cannot directly leverage the security of Protocol 5. Nonetheless, we will argue that Protocol 6 is secure against $t = n - 2$ corruptions.

Proof Intuition (Theorem 4): In order to analyze the $(n - 2)$-security of Protocol 6, we consider each invocation of Protocol 5 on a sub-network $G_{i,j}$. If at most two of the four parties in $G_{i,j}$ are corrupt, then that invocation of Protocol 5 is secure and yields secure OT correlations between parties P_i and P_j. Appealing to Lemma 6, we can augment G to include edge $\{P_i, P_j\}$.

Each $G_{i,j}$ must contain at least one honest party since either A or B must be honest (otherwise, there is nothing to prove). It remains to consider sub-networks $G_{i,j}$ in which three of the parties are corrupt. Since at least one of A or B is honest, this implies that both P_i and P_j are corrupt. Thus, there is nothing to prove regarding the security of the invocation of Protocol 5 on $G_{i,j}$ since we are establishing OT correlations between a pair of corrupt parties P_i and P_j. Combining these claims, we have that each of the invocations of Protocol 5 is secure and yields secure OT correlations between the pairs of parties P_i, P_j for all $i, j \geq 3$. By virtue of Lemma 6, we obtain an $(n-2)$-clique in the network and the $(n-2)$-security of Protocol 2 with $t = n - 2$ proves the $(n-2)$-security of Protocol 6.

The formal proof is deferred to the full version.

8   The General Case: $t \geq n/2$

In this section, we resolve the network OT question for general $t \geq n/2$. Note that from the protocols in Sects. 6 and 7 we already have tight answers for the special cases $t = n/2$ and $t = n - 2$. We address the general question from both ends of the spectrum, namely for t larger than $n/2$ and t smaller than $n - 2$. These analyses yield two distinct protocols which employ the protocols from Sects. 6 and 7 as their respective base cases. The two protocols we describe are efficient in different parameter regimes. Protocol 7 described in Sect. 8.1 is quasi-polynomially efficient[13] when $t = n/2 + \mathcal{O}(1)$, and Protocol 8 described in Sect. 8.2 is (polynomially) efficient when $t = n - \mathcal{O}(1)$. Putting these protocols together, we obtain a single protocol that is efficient under computational security when either $t = n/2 + \mathcal{O}(1)$ or $t = n - \mathcal{O}(1)$. We note that the problem of recognizing whether there exists a t-secure OT protocol is efficient in these cases, while the recognition problem for general n, t is coNP-complete.

8.1   General Protocol (Quasi-polynomial for $t = n/2 + \mathcal{O}(1)$)

We now describe a t-secure OT protocol between A and B for all $(n - t, A, B)$-unsplittable networks G. As a consequence of the lower bound described in Sect. 4, this result is tight.

Theorem 5. *Let G be an n-party OT network containing parties A and B, and let $t \geq n/2$. Then Protocol 7 is a t-secure OT protocol between A and B if and only if G is $(n - t, A, B)$-unsplittable. The protocol achieves perfect security and runs in quasi-polynomial time for $t = n/2 + \mathcal{O}(1)$. Assuming one-way functions, we can also obtain a protocol which achieves computational security and runs in polynomial time for $t = n/2 + \mathcal{O}(1)$.*

The protocol proceeds by recursion, reducing the problem of obtaining an OT protocol on an n-vertex graph with $t > n/2$ corrupted parties to a number

---

[13] Or polynomially efficient under computational security.

of instances of n'-vertex graphs, a majority of which have at most t' corrupted parties, for $n' = n - 1$ and $t' = t - 1$. As shown below, each n'-vertex subgraph G' has a structure similar to G in the sense that G' is $(n' - t', A, B)$-unsplittable whenever G is $(n - t, A, B)$-unsplittable. We can now recurse on these smaller problem instances, invoking an OT combiner to obtain the full protocol.

More precisely, the protocol constructs $n - 2$ subgraphs on $n - 1$ vertices, where each subgraph is obtained by deleting a single vertex other than A and B. We can recursively run a $(t - 1)$-secure OT protocol on each of the subgraphs. The final protocol invokes a combiner on these $n - 2$ candidate protocols. It remains to be shown that a majority of the subgraphs G' contain at most $t - 1$ corrupt parties.

Proof Intuition (Theorem 5): We may assume that at least one of A or B is honest. As described above, we wish to argue that a majority of the subgraphs G' contain at most $t - 1$ corrupt parties. Combining this with the claim that these subgraphs preserve an unsplittability property of G and invoking Lemma 9 completes the proof.

However, this claim follows from the following observation. Since $t > n/2$, if exactly t parties are corrupt then a majority of the subgraphs contain at most $t - 1$ corrupt parties since A and B are not both corrupt. If strictly fewer than t parties are corrupt then all of the sub-graphs contain at most $t - 1$ corrupt parties. In either case, for a majority of subgraphs, at most $t - 1$ of the parties are corrupt.

We first present and prove a structure lemma.

Lemma 11. *Given graph $G = (V, E)$ and a vertex i, let G_i be the induced graph on the $n - 1$ vertices $V \setminus \{i\}$. If G is $(n - t, A, B)$-unsplittable, then G_i is also $(n - t, A, B)$-unsplittable.*

Proof. We will prove the contrapositive. Suppose that $G_i \subseteq_{A,B} \Lambda_{n-t}^{2t-n-1}$. This means there exists a partition of the vertex set of G_i as $V \setminus \{i\} = V_A \dot\cup V_S \dot\cup V_B$ with no edges between V_A and V_B, where $A \in V_A$, $B \in V_B$, $|V_A| = |V_B| = n - t$ and $|V_S| = 2t - n - 1$. But then we can partition the vertex set of G as $V = V_A \dot\cup V_S' \dot\cup V_B$, where $V_S' = V_S \cup \{i\}$. We have that $|V_A| = |V_B| = n - t$ and $|V_S'| = 2t - n$, and there are no edges between V_A and V_B, so $G \subseteq_{A,B} \Lambda_{n-t}^{2t-n}$, which is a contradiction. $\square$

As an immediate consequence, the condition described in Theorem 5 is both necessary and sufficient in order to obtain a complete network of OT channels and perform secure multiparty computation among all parties in the network.

Corollary 1. *Let G be an n-party network. For $t \geq n/2$, we can t-securely generate OT correlations between all pairs of parties (thus, completing the OT network) if and only if the G is $(n - t)$-unsplittable.*

The formal proofs of Theorem 5 and Corollary 1 are deferred to the full version.

Protocol 7: General Protocol I

Preliminaries: Let $A, B, P_3, \ldots, P_n$ be the n parties in a network G and let $t \geq n/2$ be the maximum number of corruptions. A has input (x_0, x_1), and B has input $b \in \{0, 1\}$.

Protocol:

1. If $t = n/2$, then invoke Protocol 5 and halt.
2. Otherwise, run a combined protocol $\Pi^*(\Pi_3, \ldots, \Pi_n)$, where
   - For each $i \geq 3$, Π_i denotes the recursive invocation of this protocol on the $n - 1$ parties excluding party P_i with the induced sub-network $G \setminus \{P_i\}$ and $t' = t - 1$ corruptions.

8.2   General Protocol (Efficient for $t = n - \mathcal{O}(1)$)

We now describe another t-secure OT protocol for all networks G with A as the sender and B as the receiver whenever the network G is $(n-t, A, B)$-unsplittable. This protocol uses, in spirit, a reduction in the opposite sense than the one described in Sect. 8.1. The protocol is efficient whenever $t = n - \mathcal{O}(1)$.

Theorem 6. *Let G be an n-party OT network containing parties A and B, and let $t \geq n/2$. Protocol 8 is a t-secure OT protocol between A and B if and only if G is $(n - t, A, B)$-unsplittable. The protocol is efficient for $t = n - \mathcal{O}(1)$.*

The idea behind this protocol is the following. We increase the size of the network in order to obtain a large number N of well-connected additional simulated parties such that at least one them is guaranteed to be honest. We may assume that at least one of A and B is honest, as otherwise there is nothing to prove. Consequently there are at least two honest parties in the augmented network. We will now apply the protocol from Sect. 7. It remains to describe the construction of these simulated parties, to show that at least one of them is honest, and to prove a structural lemma that if the original network G is $(n - t, A, B)$-unsplittable then the augmented network G' is $(2, A, B)$-unsplittable.

Proof Intuition (Theorem 6): We first describe the new network generated by Protocol 8. The parties other than A and B in the newly constructed network consist of all subsets of size $n - t - 1$ of the parties in G containing neither A nor B. Lemma 12 below shows that this new network G' is $(2, A, B)$-unsplittable whenever G is $(n - t, A, B)$-unsplittable, where the edges of G' are as described in Protocol 8. A party X in G' will be considered honest if all constituent parties $P_i \in X$ from G are honest. Since one of A and B is honest and at most t parties are corrupt, at least $n - t$ parties are honest and in particular, at least $n - t - 1$ of the parties other than A and B must be honest. This means that one of the subsets is completely honest. Since A or B is also honest, G' is guaranteed to have at least two honest parties. Combining these facts and invoking Theorem 4 completes the argument.

Protocol 8: General protocol II

Preliminaries: Let $P_1 = A, P_2 = B, P_3, \ldots, P_n$ be the n parties in a network $G = (V, E)$. A has input (x_0, x_1), and B has input $b \in \{0, 1\}$. Let $k = n - t$.

Protocol:

1. Invoke Protocol 6 with $t' = n - 2$ on the n'-node network G' with inputs (x_0, x_1) and b, where $n' = \binom{n-2}{k-1} + 2$, and
   - S_{k-1} is the set of subsets of $\{P_3, \ldots, P_n\}$ of size $k - 1$.
   - The n' vertices of G' correspond to A, B, and the $\binom{n-2}{k-1}$ subsets of S_{k-1}.
   - The edges of G' are defined as follows. Two subsets $X, Y \in S_{k-1}$ will have an edge if either $X \cap Y \neq \emptyset$ or there exists a pair of parties $P_i \in X$ and $P_j \in Y$ with $\{P_i, P_j\} \in E$.
   - Invocation of OT over an edge $\{X, Y\}$ in G' with inputs (z_0, z_1) and c is performed as follows.
     - If $X \cap Y \neq \emptyset$, then choose some party $P_i \in X \cap Y$. $P_i \in X$ and hence knows (z_0, z_1); similarly, $P_i \in Y$ and knows c. Consequently P_i knows z_c, and sends it to the other members of set Y.
     - If $X \cap Y = \emptyset$, there is a pair of parties $P_i \in X, P_j \in Y$ such that $\{P_i, P_j\} \in E$. P_i knows (z_0, z_1) and P_j knows c, so they can invoke OT over the channel (P_i, P_j) in G, and P_j can then send the value z_c to the other members of set Y.

We will use the following structural lemma about the network G' constructed in Protocol 8. The formal proof of Theorem 6 is deferred to the full version.

Lemma 12. *If G is $(n-t, A, B)$-unsplittable, then G' is a $(2, A, B)$-unsplittable network on $n' = \binom{n-2}{n-t-1} + 2$ vertices, where G' is the network from Protocol 8.*

Proof. We prove the contrapositive. Assume that $G' \subseteq_{A,B} \Lambda_2^{n'-2}$. Let $k = n - t$, and for $i \in \mathbb{N}$, let S_i denote the set of subsets of $V \setminus \{A, B\} = \{P_3, \ldots, P_n\}$ of size i. Then there exist vertices $X, Y \in S_{k-1}$ such that there are no edges in G' between any of the parties in $\{A, X\}$ and any of the parties in $\{B, Y\}$. In particular, $X \cap Y = \emptyset$, since otherwise $\{X, Y\}$ would be an edge of G'. This implies that we have $2k = 2(n-t)$ parties $\{A, B\} \cup X \cup Y$ such that there are no edges in G from the $n - t$ parties $\{A\} \cup X$ to any of the $n - t$ parties $\{B\} \cup Y$. By definition, this means that $G \subseteq_{A,B} \Lambda_{n-t}^{2t-n}$, which is a contradiction. $\square$

References

1. Applebaum, B., Ishai, Y., Kushilevitz, E., Waters, B.: Encoding functions with constant online rate or how to compress garbled circuits keys. In: Canetti, R., Garay, J.A. (eds.) CRYPTO 2013, Part II. LNCS, vol. 8043, pp. 166–184. Springer, Heidelberg (2013)
2. Beaver, D.: Correlated pseudorandomness and the complexity of private computations. In: STOC, pp. 479–488 (1996)

3. Beaver, D.: Precomputing oblivious transfer. In: Coppersmith, D. (ed.) CRYPTO 1995. LNCS, vol. 963, pp. 97–109. Springer, Heidelberg (1995)

4. Beerliová-Trubíniová, Z., Hirt, M.: Perfectly-secure MPC with linear communication complexity. In: Canetti, R. (ed.) TCC 2008. LNCS, vol. 4948, pp. 213–230. Springer, Heidelberg (2008)

5. Beimel, A., Malkin, T., Micali, S.: The all-or-nothing nature of two-party secure computation. In: Wiener, M. (ed.) CRYPTO 1999. LNCS, vol. 1666, pp. 80–97. Springer, Heidelberg (1999)

6. Beimel, A., Omri, E., Orlov, I.: Protocols for multiparty coin toss with dishonest majority. In: Rabin, T. (ed.) CRYPTO 2010. LNCS, vol. 6223, pp. 538–557. Springer, Heidelberg (2010)

7. Bellare, M., Hoang, V.T., Keelveedhi, S., Rogaway, P.: Efficient garbling from a fixed-key blockcipher. In: IEEE Security and Privacy, pp. 478–492 (2013)

8. Ben-Or, M., Goldwasser, S., Wigderson, A.: Completeness theorems for noncryptographic fault-tolerant distributed computations. In: STOC, pp. 1–10 (1988)

9. Ben-Sasson, E., Fehr, S., Ostrovsky, R.: Near-linear unconditionally-secure multiparty computation with a dishonest minority. In: Safavi-Naini, R., Canetti, R. (eds.) CRYPTO 2012. LNCS, vol. 7417, pp. 663–680. Springer, Heidelberg (2012)

10. Bendlin, R., Damgård, I., Orlandi, C., Zakarias, S.: Semi-homomorphic encryption and multiparty computation. In: Paterson, K.G. (ed.) EUROCRYPT 2011. LNCS, vol. 6632, pp. 169–188. Springer, Heidelberg (2011)

11. Bogdanov, D., Laur, S., Willemson, J.: Sharemind: a framework for fast privacy-preserving computations. In: Jajodia, S., Lopez, J. (eds.) ESORICS 2008. LNCS, vol. 5283, pp. 192–206. Springer, Heidelberg (2008)

12. Bogetoft, P., et al.: Secure multiparty computation goes live. In: Dingledine, R., Golle, P. (eds.) FC 2009. LNCS, vol. 5628, pp. 325–343. Springer, Heidelberg (2009)

13. Boyle, E., Chung, K.-M., Pass, R.: Large-scale secure computation: multi-party computation for (parallel) RAM programs. In: Gennaro, R., Robshaw, M. (eds.) CRYPTO 2015. LNCS, vol. 9216, pp. 742–762. Springer, Heidelberg (2015)

14. Boyle, E., Goldwasser, S., Tessaro, S.: Communication locality in secure multiparty computation - how to run sublinear algorithms in a distributed setting. In: Sahai, A. (ed.) TCC 2013. LNCS, vol. 7785, pp. 356–376. Springer, Heidelberg (2013)

15. Canetti, R., Damgård, I., Dziembowski, S., Ishai, Y., Malkin, T.: On adaptive vs. non-adaptive security of multiparty protocols. In: Pfitzmann, B. (ed.) EUROCRYPT 2001. LNCS, vol. 2045, pp. 262–279. Springer, Heidelberg (2001)

16. Chaum, D., Crépeau, C., Damgård, I.: Multiparty unconditionally secure protocols. In: STOC, pp. 11–19 (1988)

17. Chor, B., Kushilevitz, E.: A zero-one law for boolean privacy (extended abstract). In: STOC, pp. 62–72 (1989)

18. Chou, T., Orlandi, C.: The simplest protocol for oblivious transfer. In: Lauter, K., Rodríguez-Henríquez, F. (eds.) LatinCrypt 2015. LNCS, vol. 9230, pp. 40–58. Springer, Heidelberg (2015)

19. Damgård, I., Ishai, Y.: Scalable secure multiparty computation. In: Dwork, C. (ed.) CRYPTO 2006. LNCS, vol. 4117, pp. 501–520. Springer, Heidelberg (2006)

20. Damgård, I., Ishai, Y., Krøigaard, M.: Perfectly secure multiparty computation and the computational overhead of cryptography. In: Gilbert, H. (ed.) EUROCRYPT 2010. LNCS, vol. 6110, pp. 445–465. Springer, Heidelberg (2010)

21. Damgård, I., Ishai, Y., Krøigaard, M., Nielsen, J.B., Smith, A.: Scalable multiparty computation with nearly optimal work and resilience. In: Wagner, D. (ed.) CRYPTO 2008. LNCS, vol. 5157, pp. 241–261. Springer, Heidelberg (2008)

22. Damgård, I., Kilian, J., Salvail, L.: On the (im)possibility of basing oblivious transfer and bit commitment on weakened security assumptions. In: Stern, J. (ed.) EUROCRYPT 1999. LNCS, vol. 1592, pp. 56–73. Springer, Heidelberg (1999)

23. Damgård, I., Nielsen, J.B.: Scalable and unconditionally secure multiparty computation. In: Menezes, A. (ed.) CRYPTO 2007. LNCS, vol. 4622, pp. 572–590. Springer, Heidelberg (2007)

24. Damgård, I., Pastro, V., Smart, N., Zakarias, S.: Multiparty computation from somewhat homomorphic encryption. In: Safavi-Naini, R., Canetti, R. (eds.) CRYPTO 2012. LNCS, vol. 7417, pp. 643–662. Springer, Heidelberg (2012)

25. Dani, V., King, V., Movahedi, M., Saia, J.: Brief announcement: breaking the o(nm) bit barrier, secure multiparty computation with a static adversary. In: PODC, pp. 227–228 (2012)

26. Dolev, D., Dwork, C., Waarts, O., Yung, M.: Perfectly secure message transmission. In: FOCS, pp. 36–45 (1990)

27. Even, S., Goldreich, O., Lempel, A.: A randomized protocol for signing contracts. In: Chaum, D., Rivest, R.L., Sherman, A.T. (eds.) Advances in Cryptology, pp. 205–210. Springer, New York (1983)

28. Fitzi, M., Franklin, M.K., Garay, J.A., Vardhan, S.H.: Towards optimal and efficient perfectly secure message transmission. In: Vadhan, S.P. (ed.) TCC 2007. LNCS, vol. 4392, pp. 311–322. Springer, Heidelberg (2007)

29. Garay, J.A., Katz, J., Koo, C.-Y., Ostrovsky, R.: Round complexity of authenticated broadcast with a dishonest majority. In: FOCS, pp. 658–668 (2007)

30. Garey, M., Johnson, D.: Computers and Intractability: A Guide to the Theory of NP-Completeness. W. H. Freeman, New York (1979)

31. Goldreich, O., Micali, S., Wigderson, A.: How to play any mental game, or a completeness theorem for protocols with honest majority. In: STOC, pp. 218–229 (1987)

32. Goldreich, O., Micali, S., Wigderson, A.: Proofs that yield nothing but their validity for all languages in NP have zero-knowledge proof systems. J. ACM 38(3), 691–729 (1991)

33. Goldreich, O., Vainish, R.: How to solve any protocol problem - an efficiency improvement. In: Pomerance, C. (ed.) CRYPTO 1987. LNCS, vol. 293, pp. 73–86. Springer, Heidelberg (1988)

34. Goldwasser, S., Kalai, Y., Popa, R., Vaikuntanathan, V., Zeldovich, N.: Reusable garbled circuits and succinct functional encryption. In: STOC, pp. 555–564 (2013)

35. Haitner, I.: Semi-honest to malicious oblivious transfer—the black-box way. In: Canetti, R. (ed.) TCC 2008. LNCS, vol. 4948, pp. 412–426. Springer, Heidelberg (2008)

36. Harnik, D., Ishai, Y., Kushilevitz, E.: How many oblivious transfers are needed for secure multiparty computation? In: Menezes, A. (ed.) CRYPTO 2007. LNCS, vol. 4622, pp. 284–302. Springer, Heidelberg (2007)

37. Harnik, D., Ishai, Y., Kushilevitz, E., Nielsen, J.B.: OT-combiners via secure computation. In: Canetti, R. (ed.) TCC 2008. LNCS, vol. 4948, pp. 393–411. Springer, Heidelberg (2008)

38. Harnik, D., Kilian, J., Naor, M., Reingold, O., Rosen, A.: On robust combiners for oblivious transfer and other primitives. In: Cramer, R. (ed.) EUROCRYPT 2005. LNCS, vol. 3494, pp. 96–113. Springer, Heidelberg (2005)

39. Hirt, M., Lucas, C., Maurer, U.: A dynamic tradeoff between active and passive corruptions in secure multi-party computation. In: Canetti, R., Garay, J.A. (eds.) CRYPTO 2013, Part II. LNCS, vol. 8043, pp. 203–219. Springer, Heidelberg (2013)

40. Huang, Y., Katz, J., Evans, D.: Efficient secure two-party computation using symmetric cut-and-choose. In: Canetti, R., Garay, J.A. (eds.) CRYPTO 2013, Part II. LNCS, vol. 8043, pp. 18–35. Springer, Heidelberg (2013)

41. Huang, Y., Katz, J., Kolesnikov, V., Kumaresan, R., Malozemoff, A.J.: Amortizing garbled circuits. In: Garay, J.A., Gennaro, R. (eds.) CRYPTO 2014, Part II. LNCS, vol. 8617, pp. 458–475. Springer, Heidelberg (2014)

42. Impagliazzo, R., Rudich, S.: Limits on the provable consequences of one-way permutations. In: STOC, pp. 44–61 (1989)

43. Ishai, Y., Kilian, J., Nissim, K., Petrank, E.: Extending oblivious transfers efficiently. In: Boneh, D. (ed.) CRYPTO 2003. LNCS, vol. 2729, pp. 145–161. Springer, Heidelberg (2003)

44. Ishai, Y., Kushilevitz, E., Lindell, Y., Petrank, E.: On combining privacy with guaranteed output delivery in secure multiparty computation. In: Dwork, C. (ed.) CRYPTO 2006. LNCS, vol. 4117, pp. 483–500. Springer, Heidelberg (2006)

45. Ishai, Y., Prabhakaran, M., Sahai, A.: Founding cryptography on oblivious transfer – efficiently. In: Wagner, D. (ed.) CRYPTO 2008. LNCS, vol. 5157, pp. 572–591. Springer, Heidelberg (2008)

46. Kilian, J.: Founding cryptography on oblivious transfer. In: STOC, pp. 20–31 (1988)

47. Kolesnikov, V., Schneider, T.: Improved garbled circuit: free XOR gates and applications. In: Aceto, L., Damgård, I., Goldberg, L.A., Halldórsson, M.M., Ingólfsdóttir, A., Walukiewicz, I. (eds.) ICALP 2008, Part II. LNCS, vol. 5126, pp. 486–498. Springer, Heidelberg (2008)

48. Kovári, T., Sós, V., Turán, P.: On a problem of K. Zarankiewicz. Colloquium Math. 3(1), 50–57 (1954)

49. Kushilevitz, E.: Privacy and communication complexity. In: FOCS, pp. 416–421 (1989)

50. Larraia, E., Orsini, E., Smart, N.P.: Dishonest majority multi-party computation for binary circuits. In: Garay, J.A., Gennaro, R. (eds.) CRYPTO 2014, Part II. LNCS, vol. 8617, pp. 495–512. Springer, Heidelberg (2014)

51. Lindell, Y., Pinkas, B., Smart, N.P., Yanai, A.: Efficient constant round multiparty computation combining BMR and SPDZ. In: Gennaro, R., Robshaw, M. (eds.) CRYPTO 2015. LNCS, vol. 9216, pp. 319–338. Springer, Heidelberg (2015)

52. Lindell, Y., Riva, B.: Cut-and-choose yao-based secure computation in the online/offline and batch settings. In: Garay, J.A., Gennaro, R. (eds.) CRYPTO 2014, Part II. LNCS, vol. 8617, pp. 476–494. Springer, Heidelberg (2014)

53. Lindell, Y.: Fast cut-and-choose based protocols for malicious and covert adversaries. In: Canetti, R., Garay, J.A. (eds.) CRYPTO 2013, Part II. LNCS, vol. 8043, pp. 1–17. Springer, Heidelberg (2013)

54. Lindell, Y., Pinkas, B.: An efficient protocol for secure two-party computation in the presence of malicious adversaries. In: Naor, M. (ed.) EUROCRYPT 2007. LNCS, vol. 4515, pp. 52–78. Springer, Heidelberg (2007)

55. Maji, H.K., Prabhakaran, M., Rosulek, M.: A zero-one law for cryptographic complexity with respect to computational UC security. In: Rabin, T. (ed.) CRYPTO 2010. LNCS, vol. 6223, pp. 595–612. Springer, Heidelberg (2010)

56. Meier, R., Przydatek, B., Wullschleger, J.: Robuster combiners for oblivious transfer. In: Vadhan, S.P. (ed.) TCC 2007. LNCS, vol. 4392, pp. 404–418. Springer, Heidelberg (2007)

57. Mohassel, P., Riva, B.: Garbled circuits checking garbled circuits: more efficient and secure two-party computation. In: Canetti, R., Garay, J.A. (eds.) CRYPTO 2013, Part II. LNCS, vol. 8043, pp. 36–53. Springer, Heidelberg (2013)

58. Naor, M., Pinkas, B.: Efficient oblivious transfer protocols. In: SODA, pp. 448–457 (2001)
59. Nielsen, J.B., Nordholt, P.S., Orlandi, C., Burra, S.S.: A new approach to practical active-secure two-party computation. In: Safavi-Naini, R., Canetti, R. (eds.) CRYPTO 2012. LNCS, vol. 7417, pp. 681–700. Springer, Heidelberg (2012)
60. Peikert, C., Vaikuntanathan, V., Waters, B.: A framework for efficient and composable oblivious transfer. In: Wagner, D. (ed.) CRYPTO 2008. LNCS, vol. 5157, pp. 554–571. Springer, Heidelberg (2008)
61. Prabhakaran, M., Prabhakaran, V.: On secure multiparty sampling for more than two parties. In: Information Theory Workshop (ITW) (2012)
62. Rabin, M.: How to exchange secrets by oblivious transfer (1981)
63. Rabin, T., Ben-Or, M.: Verifiable secret sharing and multiparty protocols with honest majority. In: STOC, pp. 73–85 (1989)
64. Wolf, S., Wullschleger, J.: Oblivious transfer is symmetric. In: Vaudenay, S. (ed.) EUROCRYPT 2006. LNCS, vol. 4004, pp. 222–232. Springer, Heidelberg (2006)
65. Wullschleger, J.: Oblivious-transfer amplification. In: Naor, M. (ed.) EUROCRYPT 2007. LNCS, vol. 4515, pp. 555–572. Springer, Heidelberg (2007)
66. Yao, A.C.-C.: How to generate and exchange secrets. In: FOCS, pp. 162–167 (1986)
67. Zahur, S., Rosulek, M., Evans, D.: Two halves make a whole. In: Oswald, E., Fischlin, M. (eds.) EUROCRYPT 2015. LNCS, vol. 9057, pp. 220–250. Springer, Heidelberg (2015)
68. Zamani, M., Movahedi, M., Saia, J.: Millions of millionaires: Multiparty computation in large networks. In: ePrint 2014/149

On the Power of Secure Two-Party Computation

Carmit Hazay[1]([⊠]) and Muthuramakrishnan Venkitasubramaniam[2]

[1] Bar-Ilan University, Ramat Gan, Israel
carmit.hazay@biu.ac.il
[2] University of Rochester, Rochester, NY 14611, USA
muthuv@cs.rochester.edu

Abstract. Ishai, Kushilevitz, Ostrovsky and Sahai (STOC 2007, SIAM JoC 2009) introduced the powerful "MPC-in-the-head" technique that provided a general transformation of information-theoretic MPC protocols secure against passive adversaries to a ZK proof in a "black-box" way. In this work, we extend this technique and provide a generic transformation of any semi-honest secure two-party computation (2PC) protocol (with mild adaptive security guarantees) in the so called *oblivious-transfer* hybrid model to an *adaptive* ZK proof for any NP-language, in a "black-box" way assuming only one-way functions. Our basic construction based on Goldreich-Micali-Wigderson's 2PC protocol yields an adaptive ZK proof with communication complexity proportional to quadratic in the size of the circuit implementing the NP relation. Previously such proofs relied on an expensive Karp reduction of the NP language to Graph Hamiltonicity (Lindell and Zarosim (TCC 2009, Journal of Cryptology 2011)). We also improve our basic construction to obtain the first linear-rate adaptive ZK proofs by relying on efficient maliciously secure 2PC protocols. Core to this construction is a new way of transforming 2PC protocols to efficient (adaptively secure) instance-dependent commitment schemes.

As our second contribution, we provide a general transformation to construct a randomized encoding of a function f from any 2PC protocol that securely computes a related functionality (in a black-box way). We show that if the 2PC protocol has mild adaptive security guarantees then the resulting randomized encoding (RE) can be decomposed to an offline/online encoding.

As an application of our techniques, we show how to improve the construction of Lapidot and Shamir (Crypto 1990) to obtain a four-round ZK proof with an "input-delayed" property. Namely, the honest prover's algorithm does not require the actual statement to be proved until the last round. We further generalize this to obtain a four-round "commit and prove" zero-knowledge with the same property where the prover

C. Hazay—Research was partially supported by the European Research Council under the ERC consolidators grant agreement n. 615172 (HIPS), and by the BIU Center for Research in Applied Cryptography and Cyber Security in conjunction with the Israel National Cyber Bureau in the Prime Minister's Office.
M. Venkitasubramaniam—Research supported by Google Faculty Research Grant and NSF Award CNS-1526377.

M. Robshaw and J. Katz (Eds.): CRYPTO 2016, Part II, LNCS 9815, pp. 397–429, 2016.
DOI: 10.1007/978-3-662-53008-5_14

commits to a witness w in the second message and proves a statement x regarding the witness w that is determined only in the fourth round.

Keywords: Adaptive zero-knowledge proofs · Secure two-party computation · Randomized encoding · Interactive hashing · Instance-dependent commitments

1 Introduction

In this work we establish new general connections between three fundamental tasks in cryptography: secure two-party computation, zero-knowledge proofs and randomized encoding. We begin with some relevant background regarding each of these tasks.

Secure Multiparty Computation. The problem of *secure multiparty computation* (MPC) [Yao86, CCD87, GMW87, BGW88] considers a set of parties with private inputs that wish to jointly compute some function of their inputs while preserving certain security properties. Two of these properties are *privacy*, meaning that the output is learned but nothing else, and *correctness*, meaning that no corrupted party or parties can cause the output to deviate from the specified function. Security is formalized using the simulation paradigm where for every adversary $\mathcal{A}$ attacking a real protocol, we require the existence of a simulator $\mathcal{S}$ that can cause the same damage in an ideal world, where an incorruptible trusted third party computes the function for the parties and provides them their output.

Honest vs. Dishonest Majority. Generally speaking, there are two distinct categories for MPC protocols: (1) one for which security is guaranteed only when a *majority* of the parties are honest, and (2) one for which security is guaranteed against an arbitrary number of corrupted parties. In the former category it is possible to construct "information-theoretic" secure protocols where security holds unconditionally,[1] whereas in the latter only computational security can be achieved while relying on cryptographic assumptions.[2] The former setting necessarily requires 3 or more parties while the latter can be constructed with just two parties. In this work, we will focus on the latter setting, considering secure two-party computation.

Semi-honest vs. Malicious Adversary. The adversary may be *semi-honest*, meaning that it follows the protocol specification but tries to learn more than allowed, or *malicious*, namely, arbitrarily deviating from the protocol specification in order to compromise the security of the other players in the protocol. Constructing semi-honestly secure protocols is a much easier task than achieving security against a malicious adversary.

---

[1] Namely, against computationally unbounded adversaries.

[2] If one is willing to provide ideal access to an oblivious-transfer functionality then one can achieve information-theoretic security even in the honest minority setting [GMW87, CvdGT95, IPS08].

Static vs. Adaptive Corruption. The initial model considered for secure computation was one of a *static adversary* where the adversary controls a subset of the parties (who are called *corrupted*) before the protocol begins, and this subset cannot change. A stronger corruption model allows the adversary to choose which parties to corrupt throughout the protocol execution, and as a function of its view; such an adversary is called *adaptive*. Adaptive corruptions model "hacking" attacks where an external attacker breaks into parties' machines in the midst of a protocol execution and are much harder to protect against. In particular, protocols that achieve adaptivity are more complex and the computational hardness assumptions needed seem stronger; see [CLOS02,KO04,CDD+04,IPS08]. Achieving efficiency seems also to be much harder.

Zero-Knowledge. Zero-knowledge (ZK) interactive protocols [GMR89] are paradoxical constructs that allow one party (denoted the prover) to convince another party (denoted the verifier) of the validity of a mathematical statement $x \in \mathcal{L}$, while providing *zero additional knowledge* to the verifier. Beyond being fascinating in their own right, ZK proofs have numerous cryptographic applications and are one of the most fundamental cryptographic building blocks. The zero-knowledge property is formalized using the *simulation paradigm*. That is, for every malicious verifier $\mathcal{V}^*$, we require the existence of a simulator $\mathcal{S}$ that reproduces a view of $\mathcal{V}^*$ that is indistinguishable from a view when interacting with the honest prover, given only the input x. Zero-knowledge protocols can be viewed as an instance of secure two-party computation where the function computed by the third-party simply verifies the validity of a witness held by the prover.

Static vs. Adaptive. Just as with general secure computation, the adversary in a zero-knowledge protocol can be either static or adaptive. Security in the presence of a statically corrupted prover implies that the protocol is sound, namely, a corrupted prover cannot convince a verifier of a false statement. Whereas security in the presence of a statically corrupted verifier implies that the protocol preserves zero-knowledge. Adaptive security on the other hand requires a simulator that can simulate adaptive corruptions of both parties.

Much progress has been made in constructing highly efficient ZK proofs in the static setting. In a recent breakthrough result, Ishai, Kushilevitz, Ostrovsky and Sahai [IKOS09] provided general constructions of ZK proofs for any NP relation $\mathcal{R}(x, \omega)$ which make a "black-box" use of an MPC protocol for a related multiparty functionality f, where by black-box we mean that f can be programmed to make only black-box (oracle) access to the relation $\mathcal{R}$. Leveraging the highly efficient MPC protocols in the literature [DI06] they obtained the first "constant-rate" ZK proof. More precisely, assuming one-way functions, they showed how to design a ZK proof for an arbitrary circuit C of size s and bounded fan-in, with communication complexity $O(s) + \mathsf{poly}(\kappa, \log s)$ where κ is the security parameter. Besides this, the work of [IKOS07,IKOS09] introduced the very powerful "MPC-in-the-head" technique that has found numerous applications in obtaining "black-box" approaches, such as unconditional two-party computation [IPS08], secure computation of arithmetic circuits [IPS09], non-malleable commitments

[GLOV12], zero-knowledge PCPs [IW14], resettably-sound ZK [OSV15] to name a few, as well as efficient protocols, such as oblivious-transfer based cryptography [HIKN08,IPS08,IPS09] and homomorphic UC commitments [CDD+15].

In contrast, in the adaptive setting, constructing adaptive zero-knowledge proofs is significantly harder and considerably less efficient. Beaver [Bea96] showed that unless the polynomial hierarchy collapses the ZK proof of [GMR89] is not secure in the presence of adaptive adversaries. Quite remarkably, Lindell and Zarosim showed in [LZ11] that adaptive zero-knowledge proofs for any NP language can be constructed assuming only one-way functions. However, it is based on reducing the statement that needs to be proved to an NP complete problem, and is rather inefficient. In fact, the communication complexity of the resulting zero knowledge is $O(s^4)$ where s is the size of the circuit. A first motivation for our work is the goal of finding alternative approaches of constructing (efficient) adaptive ZK proofs without relying on the expensive Karp-reduction step.

Randomized Encoding (RE). The third fundamental primitive considered in this work is *randomized encoding* (RE). Formalized in the works of [IK00,IK02, AIK06], randomized encoding explores to what extent the task of securely computing a function can be simplified by settling for computing an "encoding" of the output. Loosely speaking, a function $\widehat{f}(x,r)$ is said to be a randomized encoding of a function f if the output distribution depends only on $f(x)$. More formally, the two properties required of a randomized encoding are: (1) given the output of $\widehat{f}$ on (x,r), one can efficiently compute (decode) $f(x)$, and (2) given the value $f(x)$ one can efficiently sample from the distribution induced by $\widehat{f}(x,r)$ where r is uniformly sampled. One of the earliest constructions of a randomized encoding is that of "garbled circuits" and originates in the work of Yao [Yao86]. Additional variants have been considered in the literature in the early works of [Kil88,FKN94]. Since its introduction, randomized encoding has found numerous applications, especially in parallel cryptography where encodings with small parallel complexity yields highly efficient secure computation [IK00,IK02,AIK06]. (See also [GKR08,GGP10,AIK10,GIS+10,BHHI10,BHR12,App14] for other applications).

Statistical vs. Computational. Randomized encodings can be statistical or computational depending on how close the sampled distribution is to the real distribution of $\widehat{f}$. While statistical randomized encodings exist for functions computable by NC^1 circuits, only computational REs are known for general polynomial-time computable function. We refer the reader to [AIKP15] for a more detailed investigation on the class of languages that have statistical REs.

Online/Offline Complexity. In an online/offline setting [AIKW13], one considers an encoding $\widehat{f}(x,r)$ which can be split as an offline part $\widehat{f}_{\mathrm{OFF}}(r)$ which only depends on the function f, and an online part $\widehat{f}_{\mathrm{ON}}(x,r)$ that additionally depends on input x. This notion is useful in a scenario where a weak device is required to perform some costly operation f on sensitive information x: In an offline phase

$\widehat{f}_{\text{OFF}}(r)$ is published or transmitted to a cloud, and later in an online phase, the weak device upon observing the sample x, transmits the encoding $\widehat{f}_{\text{ON}}(x,r)$. The cloud then uses the offline and online parts to decode the value $f(x)$ and nothing else. The goal in such a setting is to minimize the online complexity, namely the number of bits in $\widehat{f}_{\text{ON}}(x,r)$. In the classic garbled circuit construction, the online complexity is proportional to $|x|\mathsf{poly}(\kappa)$ where κ is the security parameter. More recently, Applebaum, Ishai, Kushilevitz and Waters showed in [AIKW13] how to achieve constant online rate of $(1+o(1))|x|$ based on concrete number-theoretic assumptions.

A notoriously hard question here is to construct an *adaptively secure* RE where privacy is maintained even if the online input x is *adaptively* chosen based on the offline part. In fact, the standard constructions of garbled circuits (with short keys) do not satisfy this stronger property unless some form of "exponentially-hard" assumption is made [GKR08] or analyzed in the presence of the so-called *programmable random-oracle* model [AIKW13]. In fact, it was shown in [AIKW13] that any adaptively secure randomized encoding must have an online complexity proportional to the output length of the function. The work of Hemenway [HJO+15] provided the first constructions of adaptively-secure RE based on the minimal assumption of one-way functions.

While the connection between RE and secure computation has been explored only in one direction, where efficient RE yield efficient secure computation, we are not aware of any implication in the reverse direction. A second motivation of our work is to understand this direction while better understanding the complexity of constructing secure protocols by relying on the lower bounds established for the simpler RE primitive.

1.1   Our Contribution

In this work we present the following transformations:

1. A general construction of a *static* zero-knowledge proof system $\Pi_{\mathcal{R}}$ for any NP relation $\mathcal{R}(x,\omega)$ that makes a black-box use[3] of a two-party protocol Π_f^{OT}, carried out between parties P_1 and P_2, for a related functionality f in the oblivious-transfer (OT) hybrid model,[4] along with a (statically secure) bit commitment protocol,[5] that can be realized assuming only one-way functions. The requirement on our protocol Π_f^{OT} is: Perfect (UC) security against static corruptions by semi-honest adversaries. For example, the standard versions

---

[3] The functionality f can be efficiently defined by making only a black-box (oracle) access to the NP relation $\mathcal{R}$.

[4] Where all parties have access to an idealized primitive that implements the OT functionality, namely, the functionality upon receiving input (s_0, s_1) from the sender and a bit b from the receiver, returns s_b to the receiver and nothing the sender.

[5] We will be able to instantiate our commitment schemes using a statistically-binding commitment scheme for commitments made by the prover in the ZK protocol, and by a statistically-hiding commitment scheme for commitments made by the verifier.

of the known [GMW87] protocol (denoted by GMW) and [Yao86]'s protocol satisfy these requirements.

2. A general construction of an *adaptively secure* zero-knowledge proof system $\Pi_{\mathcal{R}}$ for any NP relation $\mathcal{R}(x, \omega)$ that makes a black-box use of a two-party protocol Π_f^{OT}, carried out between parties P_1 and P_2, for a related functionality f in the oblivious-transfer (OT) hybrid model, along with a (statically secure) bit commitment protocol, that can be realized assuming only one-way functions. The requirements on our protocol Π_f^{OT} are: (1) Perfect (UC) security against semi-honest parties admitting a static corruption of P_1 and an adaptive corruption of P_2, and (2) P_1 is the sender in all OT invocations. We remark that the semi-honest version of the GMW protocol satisfies these requirements. In fact, we will only require milder properties than perfect privacy (namely, robustness and invertible sampleability) and adaptive corruption (namely, one-sided semi-adaptive [GWZ09]) which will be satisfied by the standard Yao's protocol [Yao86] based on garbled circuits.

3. A general construction of a *randomized encoding* for any function f that makes a black-box use (a la [IKOS09]) of a two-party computation protocol Π_f^{OT}, carried out between parties P_1 and P_2, for a related functionality g in the OT-hybrid assuming only one-way functions. If we start with the same requirements as our first transformation (namely, only security against static adversaries) then we obtain a standard randomized encoding. However, if we start with a protocol as required in our second transformation with the additional requirement that it admits (full) adaptive corruption of P_2, we obtain an online/offline RE. Moreover, our construction makes a black-box use of a randomized encoding for the functionality f. Finally, we also show how to obtain an adaptive ZK proof for an NP relation $\mathcal{R}$ using a slightly stronger version of RE (that our second instantiation above will satisfy). An important corollary we obtain here is that starting from an RE that is additionally secure against adaptive chosen inputs we obtain the—so called—input-delayed ZK proof in the static setting.

A few remarks are in order.

Remark 1. In transformations 2 and 3 we require the underlying 2PC protocol to be one-sided semi-adaptive (where the sender is statically corrupted, and the receiver is adaptively corrupted). This security notion is a weak requirement and almost all known protocols that are secure in the static setting are also semi-adaptive secure. Namely, the 2PC protocols based on [Yao86, GMW87] are one-sided semi-adaptive secure in our sense. In most cases, the semi-adaptive simulation can be accomplished by honestly generating the simulation of one party and then upon adaptive corruption of the other party, simulation can be accomplished by relying on the semi-adaptive simulation of OT calls (which in turn can be achieved using only one-way functions).

Remark 2. Our online/offline RE based on (semi-adaptive) 2PC protocols is efficient only for certain protocols. Looking ahead, the offline complexity of the resulting RE is proportional to the honest algorithm of party P_1 and the online complexity is proportional to the semi-adaptive simulation of party P_2. In the case of [Yao86], applying our transformation yields the standard RE based on garbled circuits. We note that while we do not obtain any new constructions of RE, our transformation relates the semi-adaptive simulation complexity of a protocol to the efficiency of a corresponding RE.

Comparison with [IKOS09]. We remark that the approach of [IKOS09] that transforms general MPC protocols cannot be used "directly" to yield our first result concerning static ZK. This is because all constructions presented in their work require to instantiate the MPC protocol with at least three parties. In work subsequent to this, Ishai et. al [IKPY16] show how to extend the [IKOS09]-transformation to obtain our first result in a more communication efficient way. Our second and third tranformations, allows a strengthening of our first result to additionally achieves an input-delayed property. We obtain this stronger property by crucially relying on the semi-adaptive simulation. We remark that, both the approaches of [IKOS09,IKPY16] cannot yield such a protocol as the views of all parties are committed to by the prover in the first round and there is no mechanism to equivocate the views as required in the application. Another important distinction is that we only commit to the transcript of the interaction in the first round while [IKOS09] commits to each individual view. On the other hand, our approach cannot be applied to information theoretic protocols as the transcript of the interaction information theoretically binds the inputs and outputs of all parties.

1.2   Applications

We list a few of the applications of our techniques and leave it as future work to explore the other ramifications of our transformations.

COMMIT AND PROVE INPUT-DELAYED ZK PROOFS. In [LS90], a three-round witness-indistinguishable (WI) proof had been shown for Graph Hamiltonicity with a special "input-delayed" property: namely, the prover uses the statement to be proved only in the last round. Recently, in [CPS+15] it was shown how to obtain efficient input-delayed variants of the related "Sigma protocols" when used in a restricted setting of an OR-composition. We show that starting from a robust RE that is additionally secure against adaptive inputs, we can obtain general constructions of input-delayed zero-knowledge proofs that yield an efficient version of the protocol of [LS90] for arbitrary NP-relations. We remark that our work is stronger than [CPS+15] in that it achieves the stronger adaptive soundness property (which is satisfied by [LS90,FLS99]). The communication complexity in our protocol depends only linearly on the size of the circuit implementing the NP relation. As in our other transformation, this transformation will only depend on the relation in a black-box way. Finally, we show how to realize robust RE secure against adaptive inputs based on recent work of Hemenway et al. [HJO+15].

The "commit-and-prove" paradigm considers a prover that first commits to a witness w and then, in a second phase upon receiving a statement x asserts whether a particular relation $R(x, w) = 1$ without revealing the committed value,. This paradigm implicit in the work of [GMW87], later formalized in [CLOS02], is a powerful mechanism to strengthen semi-honest secure protocols to maliciously secure ones. The MPC-in-the-head approach of [IKOS09] shows how to obtain a commit and prove protocol in the commitment-hybrid model thereby providing a construction that relies on the underlying commitment (in turn the one-way function) in a black-box way. This has been used extensively in several works to close the gap between black-box and non-black-box constructions relying on one-way functions (cf. [GLOV12, GOSV14, OSV15] for a few examples). We show that our input-delayed ZK proof further supports the commit-and-prove paradigm. In fact, using our approach, we provide the first constructions of commit-and-prove protocol with this property that relies on the underlying commitment functionality in a black-box way. Instantiating the underlying non-interactive commitment scheme with one-way permutation, we obtain a black-box construction of a 4-round commit and prove protocol with the input-delayed property.

INSTANCE-DEPENDENT TRAPDOOR COMMITMENT SCHEMES. As a side result, we show that our constructions imply instance-dependent trapdoor commitment schemes, for which the witness ω serves as a trapdoor that allows to equivocate the commitment into any value. Specifically, this notion implies the same hiding/binding properties as any instance-dependent commitment scheme with the additional property that the witness allows to decommit a commitment into any message. To the best of our knowledge, our construction is the first trapdoor commitment for all NP. Prior constructions were known only for Σ-protocols [Dam10] and for Blum's Graph-Hamiltonicity [FS89].

1.3 Our Techniques

In this section, we provide an overview of our transformations and the techniques.

Static ZK via (semi-honest) 2PC or "2PC-in-the-head". We begin with a perfectly-correct 2PC protocol Π_f between parties P_1 and P_2 that securely implements the following functionality f: $f(x, \omega_1, \omega_2)$ outputs 1 if and only if $(x, \omega_1 \oplus \omega_2) \in \mathcal{R}$ where ω_1 and ω_2 are the private inputs of P_1 and P_2 in the two party protocol Π_f. We require that the 2PC protocol admits semi-honest UC security against static corruption of P_1 and P_2. Our first step in constructing a ZK proof involves the prover P simulating an honest execution between P_1 and P_2 by first sampling ω_1 and ω_2 at random such that $\omega_1 \oplus \omega_2 = \omega$, where ω is the witness to the statement x and then submitting the transcript of the interaction to the verifier V. The verifier responds with a bit b chosen at random. The prover then reveals the view of P_1 if $b = 0$ and the view of P_2 if $b = 1$, namely it just provides the input and randomness of the respective parties. Soundness follows from the perfect correctness of the protocol. Zero-knowledge, on the other hand, is achieved by invoking the simulation of parties P_1 and P_2 depending on the guess that the simulator makes for the verifier's bit b.

This general construction, however, will inherit the hardness assumptions required for the 2PC, which in the case of [Yao86, GMW87] protocols will require the existence of an oblivious-transfer protocol. We next show how to modify the construction to rely only on one-way functions. The high-level idea is that we encode the transcript of all oblivious-transfer invocations by using a "randomized encoding" of the oblivious-transfer functionality based on one-way functions as follows:

– For every OT call where P_1's input is (s_0, s_1) and P_2's input is t, we incorporate it in the transcript τ by generating a transcript containing the commitments c_0 and c_1 of s_0 and s_1 using a statistically binding commitment scheme , (which can be based on one-way functions), placing the decommitment information of c_t in P_2's random tape.[6]

This protocol results in an interactive commitment phase as we rely on a statistically-binding commitment scheme and the first message corresponding to all commitments needs to be provided by the receiver.

Compared to [IKOS09, IPS08], we remark that our ZK proof does not provide efficiency gains (using OT-preprocessing) as we require a commitment for every oblivious-transfer and in the case of compiling [GMW87] results in $O(s)$ commitments where s is the size of the circuit. Nevertheless, we believe that this compilation illustrates the simplicity of obtaining a ZK proof starting from a 2PC protocol.

Adaptive ZK via "2PC-in-the-head". First, we recall the work of Lindell and Zarosim [LZ11] that showed that constructing adaptively secure ZK proofs can be reduced to constructing *adaptive instance-dependent commitment* schemes [BMO90, IOS97, OV08, LZ11]. In fact, by simply instantiating the commitments from the prover in the (static) ZK proofs of [IKOS09] with instance-dependent commitments, we can obtain an adaptive ZK proof. Briefly, instance-dependent commitment schemes are defined with respect to a language $\mathcal{L} \in \mathsf{NP}$ such that for any statement x the following holds. If $x \in \mathcal{L}$ then the commitment associated with x is computationally hiding, whereas if $x \notin \mathcal{L}$ then the commitment associated with x is perfectly binding. An adaptively secure instance-dependent commitment scheme additionally requires that there be a "fake" commitment algorithm which can be produced using only the statement x, but later, given a witness ω such that $(x, \omega) \in \mathcal{R}$, be opened to both 0 and 1.

First, we describe an instance-dependent commitment scheme using a (perfectly-correct) 2PC protocol Π_f engaged between parties P_1 and P_2 that securely implements the following functionality f: $f(x, \omega_1, \omega_2)$ outputs 1 if and only if $(x, \omega_1 \oplus \omega_2) \in \mathcal{R}$ where ω_1 and ω_2 are the private inputs of P_1 and P_2 in the two party protocol Π_f. We will require that only P_2 receives an output and that Π_f is (UC) secure against the following adversaries: (1) A semi-honest adversary $\mathcal{A}_1$ that statically corrupts P_1, and (2) A semi-honest adversary $\mathcal{A}_2$ that statically corrupts P_2.

---

[6] Note that, in Naor's statistically binding commitment scheme [Nao91] the decommitment information is the inverse under a pseudorandom generator that is uniformly sampled, and hence can be placed in the random tape.

Given such a 2PC Π_f a commitment to the message 0 is obtained by committing to the view of party P_1 in an interaction using Π_f, using the simulator $\mathcal{S}_1$ for adversary $\mathcal{A}_1$ as follows. The commitment algorithm runs $\mathcal{S}_1$ on input a random string ω_1 that serves as the input of P_1. The output of the commitment on input 0 is τ where τ is the transcript of the interaction between P_1 and P_2 obtained from the view of P_1 generated by $\mathcal{S}_1$. A commitment to 1 is obtained by running the simulator $\mathcal{S}_2$ corresponding to $\mathcal{A}_2$ where the input of P_2 is set to a random string ω_2. The output of the commitment is transcript τ obtained from the view of P_2 output by $\mathcal{S}_2$. Decommitting to 0 simply requires producing input and output (ω_1, r_1) for P_1 such that the actions of P_1 on input ω_1 and random tape r_1 are consistent with the transcript τ. Decommitting to 1 requires producing input and randomness (ω_2, r_2) for P_2 consistent with τ and P_2 outputs 1 as the output of the computation. The hiding property of the commitment scheme follows from the fact that the transcript does not reveal any information regarding the computation (i.e. transcript can be simulated indistinguishably). The binding property for statements $x \notin \mathcal{L}$, on the other hand, relies on the perfect correctness of the protocol. More precisely, if a commitment phase τ is decommitted to both 0 and 1, then we can extract inputs and randomness for P_1 and P_2 such that the resulting interaction with honest behavior yields τ as the transcript of messages exchanged and P_2 outputting 1. Note that this is impossible since the protocol is perfectly correct and 1 is not in the image of f for $x \notin \mathcal{L}$.

Next, to obtain an *adaptively secure* instance-dependent commitment scheme we will additionally require that Π_f be secure against a semi-honest adversary $\mathcal{A}_3$ that first statically corrupts P_1 and then adaptively corrupts P_2 at the end of the execution. This adversary is referred to as a semi-adaptive adversary in the terminology of [GWZ09]. The fake commitment algorithm follows the same strategy as committing to 0 with the exception that it relies on the simulator $\mathcal{S}_3$ of $\mathcal{A}_3$. $\mathcal{S}_3$ is a simulator that first produces a view for P_1 and then post execution produces a view for P_2. More formally, the fake commitment algorithm sets P_1's input to a random string ω_1 and produces P_1's view using $\mathcal{S}_3$ and outputs τ where, τ is the transcript of the interaction. Decommitting to 0 follows using the same strategy as the honest decommitment. Decommitting to 1, on the other hand, requires producing input and randomness for P_2. This can be achieved by continuing the simulation by $\mathcal{S}_3$ post execution. However, to run $\mathcal{S}_3$ it needs to produce an input for party P_2 such that it outputs 1. This is possible as the decommitting algorithm additionally receives the real witness ω for x, using which it sets P_2's input as $\omega_2 = \omega \oplus \omega_1$.

In fact, we will only require adversaries $\mathcal{A}_2$ and $\mathcal{A}_3$, as the honest commitment to 0 can rely on $\mathcal{S}_3$. Indistinguishability of the simulation will then follow by comparing the simulations by $\mathcal{S}_2$ and $\mathcal{S}_3$ with a real-world experiment with adversaries $\mathcal{A}_2, \mathcal{A}_3$ where the parties inputs are chosen at random subject to the condition that they add up to ω and using the fact that the adversaries are semi-honest.

We will follow an approach similar to our previous transformation to address calls to the OT functionality. We will additionally require that P_1 plays the sender's role in all OT invocations. We note that our encoding accommodates an adaptive corruption of P_2, as it enables us to equivocate the random tape of P_2 depending on its input t.

To instantiate our scheme, we can rely on [Yao86] or [GMW87] to obtain an adaptive instance-dependent commitment scheme. Both commitments results in a communication complexity of $O(s \cdot \mathsf{poly}(\kappa))$ where s is the size of the circuit implementing the relation $\mathcal{R}$ and κ is the security parameter. Achieving adaptive zero-knowledge is then carried out by plugging in our commitment scheme into the prover's commitments in the [IKOS09] zero-knowledge (ZK) construction, where it commits to the views of the underlying MPC protocol. The resulting protocol will have a complexity of $O(s^2 \cdot \mathsf{poly}(\kappa))$ and a negligible soundness error. We remark that this construction already improves the previous construction of Lindell and Zarosim that requires the expensive Karp reduction to Graph Hamiltonicity. Our main technical contribution is showing how we can further improve our basic construction to achieve a complexity of $O(s \cdot \mathsf{poly}(\kappa))$ and therefore obtaining a "linear"-rate *adaptive* ZK proof.

RE from (semi-honest) 2PC. To construct a RE for a function f, we consider an arbitrary 2PC protocol that securely realizes the related function g that is specified as follows: $g(a_1, a_2) = f(a_1 \oplus a_2)$ where a_1 and a_2 are the private inputs of P_1 and P_2 in the two party protocol Π_g. We will make the same requirements on our 2PC as in the previous case, namely, security with respect to adversaries $\mathcal{A}_1$ and $\mathcal{A}_2$. The offline part of our encoding function $\widehat{f}_{\mathrm{OFF}}(r)$ is defined using the simulator $\mathcal{S}_3$ for adversary $\mathcal{A}_3$ that proceeds as follows. Upon corrupting P_1, $\mathcal{S}_3$ is provided with a random input string a_1, where the simulation is carried out till the end of the execution and temporarily stalled. The output of $\widehat{f}_{\mathrm{OFF}}(r)$ is defined to be the simulated transcript of the interaction between parties P_1 and P_2. Next, upon receiving the input x, the online part $\widehat{f}_{\mathrm{ON}}(x, r)$ continues the simulation by $\mathcal{S}_1$ which corrupts P_2 post execution (at the end of the protocol execution), where P_2's input is set as $a_2 = x \oplus a_1$ and its output is set as $f(x)$. Finally, the output of $\widehat{f}_{\mathrm{ON}}(x, r)$ is defined by the input and random tape of P_2. In essence, $\widehat{f}(x, r) = (\widehat{f}_{\mathrm{OFF}}(r), \widehat{f}_{\mathrm{ON}}(x, r))$ constitutes the complete view of P_2 in an execution using Π_g. The decoder simply follows P_2's computation in the view and outputs P_2's output, which should be $f(x)$ by the correctness of the algorithm. The simulation for our randomized encoding $\mathcal{S}$ relies on the simulator for the adversary $\mathcal{A}_2$, denoted by $\mathcal{S}_2$. Namely, upon receiving $f(x)$, $\mathcal{S}$ simply executes $\mathcal{S}_2$. Recalling that $\mathcal{S}_2$ corrupts P_2, $\mathcal{S}$ simply provides a random string a_2 as its input and $f(x)$ as the output. Finally, the offline and online parts are simply extracted from P_2's view accordingly. Privacy will follow analogously as in our previous case.

Note that the offline complexity of our construction is equal to the communication complexity of the underlying 2PC protocol Π_g, whereas the online complexity amounts to the input plus the randomness complexity of P_2. The efficiency of our randomized encoding ties the offline part with the static simulation of party P_1 and the online part with the semi-adative simulation of P_2. Moreover, this protocol can be instantiated by the [Yao86, GMW87] protocols, where the OT sub-protocols are implemented using one-way functions as specified before. We remark that the protocol of [Yao86] does not, in general, admit adaptive corruptions, yet it is secure in the presence of a semi-adaptive adversary that adaptively corrupts P_2 after statically corrupting P_1. The [Yao86] based protocol will result in an offline complexity of $O(s \cdot \mathsf{poly}(\kappa))$ and an online complexity of $O(n \cdot \mathsf{poly}(\kappa))$ where s is the size of the circuit implementing f and n is the input length.[7] Whereas the [GMW87] protocol will result in an offline and online complexities of $O(s \cdot \mathsf{poly}(\kappa))$. While this might not be useful in the "delegation of computation" application of randomized encoding as the online encoding is not efficient, it can be used to construct an instance-dependent commitment scheme where we are interested only in the total complexity of the encoding. Finally, we remark that if we are not interested in an offline/online setting and just require a standard randomized encoding we will requite Π_f to be secure only against a static corruption of P_2 by $\mathcal{A}_2$ and the honest encoding can be carried out by emulating the real world experiment (as opposed to relying on the simulation by $\mathcal{S}_3$).

Next, we provide a construction of instance-dependent commitments based on online/offline RE. Standard RE will not be sufficient for this and we introduce a stronger notion of *robustness* for RE and show that the preceeding construction already satisfies this. Then based on a robust RE we show how to get an instant-dependent commitment scheme. In fact, we can get an adaptive instance-dependent commitment scheme if the underlying RE has a corresponding adaptive property. Since adaptive instance-dependent comitment schemes are sufficient to realize adaptive ZK, this provides a transformation from RE to adaptive ZK.

"Linear"-Rate Adaptive ZK Proof from Malicious 2PC. The main drawback in our first construction of adaptive ZK proofs was in the equivocation parameter of our instance-dependent commitment. Namely, to equivocate one bit, we incurred a communication complexity of $O(s \cdot \mathsf{poly}(\kappa))$. To improve the communication complexity one needs to directly construct an instance-dependent commitment scheme for a larger message space $\{0,1\}^\ell$. We show how to construct a scheme where the communication complexity depends only *additively* on the equivocation parameter, implying $O((s+\ell)\mathsf{poly}(\kappa))$ overhead. Combining such a scheme with the [IKOS09] ZK proof results in a protocol with communication complexity of $O(n \cdot s \cdot \mathsf{poly}(\kappa) + \sum_{i=1}^{n} \ell_i \cdot \mathsf{poly}(\kappa))$ where ℓ_i is the length of the ith commitment made by the prover. Setting $n = \omega(\log k)$ results and using $\sum_i \ell_i = s \cdot \mathsf{poly}(\kappa)$ in an adaptive ZK proof with negligible soundness error and complexity $O(s \cdot \mathsf{poly}(\kappa))$. We remark here that by linear rate, we mean we obtain a protocol whose communication complexity that depends linearly on the circuit size. This stands in contrast of the previous approach by Lindell and

---

[7] We note that the online complexity can be improved by relying on the work of [AIKW13].

Zarosim [LZ11] that depends at least cubic in the circuit size. In comparison, for the static case, [IKOS09] provide a "constant" rate Static ZK proof, i.e. a ZK proof whose communication complexity is $O(s + \mathsf{poly}(k))$.

Our approach to construct an instance-dependent commitment scheme for larger message spaces is to rely on a maliciously secure two-party computation. Specifically, suppose that for a polynomial-time computable Boolean function $f(x,y)$ we have a 2PC protocol Π_f with parties P_1 and P_2, where P_2 receives the output of the computation and satisfies all the conditions required in our original transformation. In addition we require it to satisfy statistical security against a malicious P_1 (in the OT-hybrid). In fact, it suffices for the protocol to satisfy the following "soundness" condition: If there exists no pair of inputs x, y such that $f(x,y) = 1$ then for any malicious P_1^*, the probability that an honest P_2 outputs 1 is at most 2^{-t}, where the probability is taken over the randomness of party P_2. Then, using such a protocol, we can provide a framework to construct an instance-dependent commitment scheme where the soundness translates to the equivocation parameter, namely, it will be $O(t)$ for soundness 2^{-t}.

Concretely, given an input statement x we consider a protocol Π_f that realizes function f defined by: $f(\omega_1, \omega_2) = 1$ iff $(x, \omega_1 \oplus \omega_2) \in \mathcal{R}$. We first describe an (incorrect) algorithm as a stepping stone towards explaining the details of the final construction. The commitment algorithm on input a message m, (just as in our transformation to RE) invokes the simulator $\mathcal{S}_2$ that corresponds to the adversary $\mathcal{A}_2$, which statically corrupts P_2 with an input set to a random string ω_2 and output 1. Upon completing the simulation, the committer submits to the receiver the transcript of the interaction and $\mathsf{Ext}(r_2) \oplus m$ where r_2 is the randomness of P_2 output by the simulation and $\mathsf{Ext}(\cdot)$ is a randomness extractor that extracts $R - \Omega(t)$ bits where R is the length of P_2's random tape. A decommitment simply produces m along with P_2's input and randomness corresponding to the transcript output in the commitment phase. Intuitively, binding follows directly from the soundness condition as no adversarial committer can produce two different random strings for P_2, as the entropy of all "accessible" random tapes for P_2 is "extracted" out by Ext.[8] The fake commitment, on the other hand, relies as above on a simulator corresponding to $\mathcal{A}_1$ that statically corrupts P_1 and adaptively corrupts P_2, where instead of $\mathsf{Ext}(r_2) \oplus m$ it simply sends a random string. Equivocation, on the other hand, is achievable if the simulation can additionally access the entire space of consistent random tapes of P_2 and invert Ext. Several problems arise when materializing this framework.

The first issue is that we cannot rely on an extractor as the adversary can adaptively decide on r_2 given the description of Ext. Now, since extractors are only statistically secure, this implies that for certain (very small) set of values for r_2 there could be multiple pre-images with respect to Ext. Instead, we rely on an *interactive hashing* protocol [NOVY98, DHRS04, HR07] that guarantees binding against computationally unbounded adversaries. More precisely, an interactive hashing protocol ensures that if the set of random tapes accessible to the

---

[8] This is not entirely accurate and is presented just for intuition. More details are presented in next paragraph.

adversary is at most $2^{R-\Omega(t)}$ then except with negligible probability it cannot obtain two random tapes that are consistent with the transcript of the hashing protocol. This protocol will additionally require to satisfy an invertible sampleability property where given an interaction it is possible to compute efficiently a random input consistent with the transcript. We will not be able to rely on the efficient 4-message protocol of [DHRS04] but will rely on the original protocol of [NOVY98] that proceeds in a linear number of rounds (linear in the message length) where inverting simply requires solving a system of linear equations in a finite field.

Another major issue is that the space of consistent random tapes might not be "nice" to be invertible. Namely, to adaptively decommit a fake commitment to an arbitrary message we require that the space of consistent random tapes for P_2, i.e. consistent with the transcript τ of the protocol and the transcript of the interactive-hashing protocol in the commitment phase, to be "uniform" over a nice domain. We thus consider a variant of the protocol in [IPS08] so that the space of consistent random tapes will be uniform over the bits of a specified length. While this modification solves the problem of "nice" random tapes, it requires re-establishing a certain "soundness" condition in the compilation of [IPS08].

As mentioned before we combine our adaptive instance-dependent commitment scheme with the ZK protocol of [IKOS09]. We will rely on a variant where the MPC protocol in their construction will be instantiated with the classic [BGW88] protocol, as opposed to highly-efficient protocol of [DI06]. The reason is that we will additionally require a reconstructability property[9] of the MPC protocol that can be shown to be satisfied by [BGW88]. Secondly, relying on this efficient variant anyway does not improve the asymptotic complexity to beyond a linear-rate. As an independent contribution we also provide a simple adaptive ZK protocol based on garbled circuits that satisfies reconstructability but will only achieve soundness error $1/2$ (see Sect. 6).

1.4   Perspective

Our work is similar in spirit to the work of [IKOS09, IPS08] that demonstrated the power information-theoretic MPC protocols in constructing statically-secure protocols. Here, we show the power of (adaptively-secure) 2PC protocols in the OT-hybrid helps in constructing adaptively-secure protocols and randomized encodings. Instantiating our 2PC with the standard protocols of [Yao86, GMW87] yields simple constructions of adaptive ZK proofs and randomized encodings. While ZK can be viewed as a special instance of a two-party computation protocol, the resulting instantiation requires stronger assumptions (such as enhanced trapdoor permutations). On the other hand, our transformation requires only one-way functions. As mentioned earlier, we not only provide adaptive ZK proofs, but we obtain two new simple static ZK proofs from our instance-based commitments.

---

[9] Informally, reconstructability requires that given the views of t out of n players in an instance of the protocol, and the inputs of all parties, it is possible to reconstruct the views of the remaining parties consistent with views of the t parties.

A second contribution of our construction shows a useful class of applications for which 2PC protocols can be used to reduce the round complexity of black-box constructions. The well known and powerful "MPC-in-the-head" technique has found extensive applications in obtaining black-box construction of protocols that previously depended on generic Karp reductions. In many cases their approach was used to close the gap between black-box and non-black-box constructions. In particular, their approach provided the first mechanism to obtain a commit-and-prove protocol that depended on the underlying commitment in a black-box way. We believe that our technique yields an analogous "2PC-in-the-head" technique which in addition to admitting similar commit-and-prove protocols can improve the round complexity as demonstrated for the case of non-malleable commitments. This is because of the input-delayed property that is achievable for our commit-and-prove protocols.

In addition, we believe it will be useful in applications that rely on certain special properties of the Blum's Graph-Hamiltonicity ZK proof (BH). Concretely, we improve the [LZ11] adaptive ZK proof and the input-delayed protocol from [LS90] both of which relied on BH ZK proof. More precisely, by relying on our ZK proof based on our instance-dependent commitment schemes that, in turn, depends on the NP relation in a black-box way, we save the cost of the expensive Karp reduction to Graph Hamiltonicity. We leave it as future work to determine if other applications that rely on the BH ZK proof can be improved (e.g., NIZK).

2   Preliminaries

We denote the security parameter by κ. We say that a function $\mu : \mathbb{N} \to \mathbb{N}$ is *negligible* if for every positive polynomial $p(\cdot)$ and all sufficiently large κ's it holds that $\mu(\kappa) < \frac{1}{p(\kappa)}$. We use the abbreviation PPT to denote probabilistic polynomial-time. For an NP relation $\mathcal{R}$, we denote by $\mathcal{R}_x$ the set of witnesses of x and by $\mathcal{L}_{\mathcal{R}}$ its associated language. That is, $\mathcal{R}_x = \{\omega \mid (x, \omega) \in \mathcal{R}\}$ and $\mathcal{L}_{\mathcal{R}} = \{x \mid \exists\, \omega\ s.t.\ (x, \omega) \in \mathcal{R}\}$.

2.1   Adaptive Instance-Dependent Commitment Schemes [LZ11]

We extend the instance-dependent commitment scheme definition of [LZ11], originally introduced for the binary message space, to an arbitrary message space $\mathcal{M}$.

Syntax. Let $\mathcal{R}$ be an NP relation and $\mathcal{L}$ be the language associated with $\mathcal{R}$. A (non-interactive) adaptive instance dependent commitment scheme (AIDCS) for $\mathcal{L}$ is a tuple of probabilistic polynomial-time algorithms (Com, Com', Adapt), where:

- Com is the commitment algorithm: For a message $m \in \mathcal{M}_n$, an instance $x \in \{0, 1\}^*$, $|x| = n$ and a random string $r \in \{0, 1\}^{p(|x|)}$ (where $p(\cdot)$ is a polynomial), $\mathsf{Com}(x, m; r)$ returns a commitment value c.
- Com' is a "fake" commitment algorithm: For an instance $x \in \{0, 1\}^*$ and a random string $r \in \{0, 1\}^{p(|x|)}$, $\mathsf{Com}'(x; r)$ returns a commitment value c.

– Adapt is an adaptive opening algorithm: Let $x \in \mathcal{L}$ and $\omega \in \mathcal{R}_x$. For all c and $r \in \{0,1\}^{p(|x|)}$ such that $\mathsf{Com}'(x;r) = c$, and for all $m \in \mathcal{M}_n$, $\mathsf{Adapt}(x, \omega, c, m, r)$ returns a pair (m, r') such that $c = \mathsf{Com}(x, m; r')$. (In other words, Adapt receives a "fake" commitment c and a message m, and provides an explanation for c as a commitment to the message m.)

Security. We now define the notion of security for our commitment scheme.

Definition 21 (AIDCS). *Let $\mathcal{R}$ be an NP relation and $\mathcal{L} = \mathcal{L}_{\mathcal{R}}$. We say that (Com, Com', Adapt) is a secure AIDCS for $\mathcal{L}$ if the following holds:*

1. *Computational hiding: The ensembles $\{\mathsf{Com}(x, m)\}_{x \in \mathcal{L}, m\{0,1\}^{|x|}}$, and $\{\mathsf{Com}'(x)\}_{x \in \mathcal{L}}$ are computationally indistinguishable.*
2. *Adaptivity: The distributions $\{\mathsf{Com}(x, m; U_{p(|x|)}), m, U_{p(|x|)}\}_{x \in \mathcal{L}, \omega \in \mathcal{R}_{\mathcal{L}}, m \in \{0,1\}^{|x|}}$ and*
   $\{\mathsf{Com}'(x; U_{p(|x|)}), m, \mathsf{Adapt}(x, \omega, \mathsf{Com}'(x; U_{p(|x|)}), m)\}_{x \in \mathcal{L}, \omega \in \mathcal{R}_{\mathcal{L}}, m \in \{0,1\}^{|x|}}$ are computationally indistinguishable (that is, the random coins that are generated by Adapt are indistinguishable from real random coins used by the committing algorithm Com).
3. *Statistical binding: For all $x \notin \mathcal{L}$, $m, m' \in \mathcal{M}_{|x|}$, and a commitment c, the probability that there exist r, r' for which $c = \mathsf{Com}(x, m; r)$ and $c = \mathsf{Com}(x, m'; r')$ is negligible in κ.*

2.2   Zero-Knowledge Proofs

Definition 22 (Interactive proof system). *A pair of PPT interactive machines $(\mathcal{P}, \mathcal{V})$ is called an interactive proof system for a language $\mathcal{L}$ if there exists a negligible function negl such that the following two conditions hold:*

1. COMPLETENESS: *For every $x \in \mathcal{L}$,*

$$\Pr[\langle \mathcal{P}, \mathcal{V} \rangle(x) = 1] \geq 1 - \mathsf{negl}(|x|).$$

2. SOUNDNESS: *For every $x \notin L$ and every interactive PPT machine B,*

$$\Pr[\langle B, \mathcal{V} \rangle(x) = 1] \leq \mathsf{negl}(|x|).$$

Definition 23 (Zero-knowledge). *Let $(\mathcal{P}, \mathcal{V})$ be an interactive proof system for some language $\mathcal{L}$. We say that $(\mathcal{P}, \mathcal{V})$ is computational zero-knowledge if for every PPT interactive machine $\mathcal{V}^*$ there exists a PPT algorithm $\mathcal{S}$ such that*

$$\{\langle \mathcal{P}, \mathcal{V}^* \rangle(x)\}_{x \in \mathcal{L}} \overset{c}{\approx} \{\langle \mathcal{S} \rangle(x)\}_{x \in \mathcal{L}}$$

where the left term denote the output of $\mathcal{V}^*$ after it interacts with $\mathcal{P}$ on common input x whereas, the right term denote the output of $\mathcal{S}$ on x.

Input-Delayed Zero-Knowledge Proofs. We will construct zero-knowledge proofs with an "input-delayed" property. Roughly speaking, this property allows an honest prover to generate all messages except from the last one, without knowledge of the statement. In such a situation, the soundness and zero-knowledge properties can additionally be required to be *adaptively* secure. Namely, soundness is required to hold even if the cheating prover adaptively chooses the statement (before the last message). Zero-knowledge, in the other hand, is required to hold even if the malicious verifier chooses a (true) statement before the last round.

Adaptive Zero-Knowledge. This notion considers the case for which the prover is adaptively corrupted. Loosely speaking, the simulator obtains a statement $x \in \mathcal{L}$. Moreover, at any point of the execution, the adaptive adversary is allowed to corrupt the prover. It is then required that zero-knowledge holds even in the presence of an adaptive adversary.

2.3 Garbled Circuits

Our notion of garbled circuits includes an additional algorithm of oblivious generation of a garbled circuit. Namely, given the randomness used to produce a garbled circuit $\widetilde{C}$ of some circuit C, the algorithm generates new randomness that explains $\widetilde{C}$ as the outcome of the simulated algorithm. We note that this modified notion of garbled circuits can be realized based on one-way functions, e.g., the construction from [LP09], for instance when the underlying symmetric key encryption used for garbling has an additional property of oblivious ciphertext generation (where a ciphertext can be sampled without the knowledge of the plaintext). Then the simulated garbling of a gate produces a garbled table using three obliviously generated ciphertexts and one ciphertext that encrypts the output label. We note that the ability to switch from a standard garbled circuit to a simulated one will be exploited in our constructions below in order to equivocate a commitment to 0 into a commitment to 1. Towards introducing our definition of garbled circuits we denote vectors by bold lower-case letters and use the parameter n to denote the input and output length for the Boolean circuit C.

Definition 24 (Garbling scheme). *A garbling scheme* Garb = (Grb, Enc, Eval, Dec) *consists of four polynomial-time algorithms that work as follows:*

- $(\widetilde{C}, \mathbf{dk}, \mathsf{sk}) \leftarrow \mathsf{Grb}(1^\kappa, C; r_{\mathsf{Grb}})$: *is a probabilistic algorithm with randomness* r_{Grb} *that takes as input a circuit* C *with* $2n$ *input wires and* n *output wires and returns a garbled circuit* $\widetilde{C}$, *a set of decoding keys* $\mathbf{dk} = (\mathrm{dk}_1, \ldots, \mathrm{dk}_n)$ *and a secret key* sk.
- $\widetilde{\mathbf{x}} := \mathsf{Enc}(\mathsf{sk}, \mathbf{x})$ *is a deterministic algorithm that takes an input a secret key* sk, *an input* $\mathbf{x}$ *and returns an encoded input* $\widetilde{\mathbf{x}}$. *We denote this algorithm by* $\widetilde{\mathbf{x}} := \mathsf{Enc}(\mathsf{sk}, \widetilde{\mathbf{x}})$. *In this work we consider decomposable garbled schemes. Namely, the algorithm takes multiple input bits* $\mathbf{x} = (x_1, \ldots, x_n)$, *runs* $\mathsf{Enc}(\mathsf{sk}, \cdot)$ *on each* x_i *and returns the garbled inputs* $\widetilde{x}_1$ *through* $\widetilde{x}_n$, *denoted by input labels.*

- $\widetilde{\mathbf{y}} := \mathsf{Eval}(\widetilde{C}, \widetilde{\mathbf{x}})$: *is a deterministic algorithm that takes as input a garbled circuit $\widetilde{C}$ and encoded inputs $\widetilde{\mathbf{x}}$ and returns encoded outputs $\widetilde{\mathbf{y}}$.*
- $\{\bot, y_i\} := \mathsf{Dec}(\mathrm{dk}_i, \widetilde{y}_i)$: *is a deterministic algorithm that takes as input a decoding key dk_i and an encoded output $\widetilde{y}_i$ and returns either the failure symbol $\bot$ or an output y_i. We write $\{\bot, \mathbf{y}\} := \mathsf{Dec}(\mathbf{dk}, \widetilde{\mathbf{y}})$ to denote the algorithm that takes multiple garbled outputs $\widetilde{\mathbf{y}} = (\widetilde{y}_1 \ldots \widetilde{y}_n)$, runs $\mathsf{Dec}(\mathrm{dk}_i, \cdot)$ on each $\widetilde{y}_i$ and returns the outputs y_1 through y_n.*

Correctness. We say that Garb is correct if for all $n \in \mathbb{N}$, for any polynomial-size circuit C, for all inputs $\mathbf{x}$ in the domain of C, for all $(\widetilde{C}, \mathbf{dk}, \mathsf{sk})$ output by $\mathsf{Grb}(1^\kappa, C)$, for $\widetilde{\mathbf{x}} := \mathsf{Enc}(\mathsf{sk}, \mathbf{x})$ and $\widetilde{\mathbf{y}} := \mathsf{Eval}(\widetilde{C}, \widetilde{\mathbf{x}})$ and for all $i \in [n]$, $y_i := \mathsf{Dec}(\mathrm{dk}_i, \widetilde{y}_i)$, where $(y_1, \ldots, y_n) = C(\mathbf{x})$.

Security. We say that a garbling scheme Garb is secure if there exists a PPT algorithm SimGC such that for any polynomial-size circuit C, for all inputs $\mathbf{x}$ in the domain of C, for all $(\widetilde{C}, \mathbf{dk}, \mathsf{sk})$ output by $\mathsf{Grb}(1^\kappa, C)$ and $\widetilde{\mathbf{x}} := \mathsf{Enc}(\mathsf{sk}, \mathbf{x})$ it holds that,

$$(\widetilde{C}, \widetilde{\mathbf{x}}, \mathbf{dk}) \overset{c}{\approx} \mathsf{SimGC}\,(1^\kappa, C, \mathbf{y})\,, where \;\; \mathbf{y} = C(\mathbf{x}).$$

Oblivious Sampling. There exists a PPT algorithm OGrb such that for any polynomial-time circuit C and for all input/output pairs $(\mathbf{x}, \mathbf{y})$ such that $C(\mathbf{x}) = \mathbf{y}$ it holds that,

$$\{r'_{\mathsf{Grb}}, \mathsf{SimGC}\,(1^\kappa, C, \mathbf{y}; r'_{\mathsf{Grb}})\}_{r'_{\mathsf{Grb}} \leftarrow \{0,1\}^*} \overset{c}{\approx} \{\hat{r}_{\mathsf{Grb}}, \widetilde{C}, \widetilde{x}, \mathbf{dk}\}_{(\hat{r}_{\mathsf{Grb}}, \widetilde{x}) \leftarrow \mathsf{OGrb}(1^\kappa, C, \mathbf{x}, r_{\mathsf{Grb}})}$$

where $(\widetilde{C}, \mathbf{dk}, \mathsf{sk}) \leftarrow \mathsf{Grb}(1^\kappa, C; r_{\mathsf{Grb}})$.

Note that correctness is perfect by our definition, which implies that a garbled circuit must be evaluated to the correct output. We further note that this notion is achieved by employing the point-and-permute optimization [PSSW09] to the garbling construction, as the evaluator of an honestly generated circuit always decrypts a single ciphertext for each gate which leads to the correct output. Furthermore, we assume that giving the secret key it is possible to verify that the garbled circuit was honestly generated. Again, this holds with respect to existing garbling schemes, as the secret key includes the encoding of all input labels which allows to recompute the entire garbling and verifying the correctness of each gate.

2.4   Randomized Encoding

We review the definition of randomized encoding from [IK00, AIK04].

Definition 25 (Randomized Encoding). *Let $f : \{0,1\}^n \to \{0,1\}^\ell$ be a function. Then a function $\widehat{f} : \{0,1\}^n \times \{0,1\}^m \to \{0,1\}^s$ is said to be a randomized encoding of f, if:*

Correctness: *There exists a decoder algorithm B such that for any input $x \in \{0,1\}^n$, except with negligible probability over the randomness of the encoding and the random coins of B, it holds that $B(\widehat{f}(x, U_m)) = f(x)$.*

Computational (Statistical) Privacy: *There exists a PPT simulator $\mathcal{S}$, such that for any input $x \in \{0,1\}^n$ the following distributions are computationally (statistically) indistinguishable over $n \in \mathbb{N}$:*

- $\{\widehat{f}(x, U_m)\}_{n\in\mathbb{N}, x\in\{0,1\}^n}$,
- $\{\mathcal{S}(f(x))\}_{n\in\mathbb{N}, x\in\{0,1\}^n}$.

We require our randomized encoding to satisfy some additional properties:

1. **Robustness:** Applebaum et al. introduced in [AIKW13] the measures of *offline* and *online* complexities of an encoding, where the offline complexity refers to the number of bits in the output of $\widehat{f}(x,r)$ that solely depend on r and the online complexity refers to the number of bits that depend on both x and r. The motivation in their work was to construct *online efficient* randomized encoding, where the online complexity is close to the input size of the function. In our construction, we are not concerned specifically with the online complexity, but we require that there exists an offline part of the randomized encoding that additionally satisfies a robustness property. We present the definition of robustness for boolean functions f as it suffices for our construction.

   We say that $\widehat{f}$ is a *robust encoding* of f if there exist functions $\widehat{f}_{\mathrm{OFF}}$ and $\widehat{f}_{\mathrm{ON}}$ such that $\widehat{f}(x,r) = (\widehat{f}_{\mathrm{OFF}}(r), \widehat{f}_{\mathrm{ON}}(x,r))$ and, in addition, it holds that: if there exists no x such that $f(x) = 1$, then for any r, there exists no z such that $B(\widehat{f}_{\mathrm{OFF}}(r), z)$ outputs 1. Intuitively, robustness ensures that if the offline part was honestly computed using $\widehat{f}_{\mathrm{OFF}}$ then there cannot exist any online part that can make the decoder output an element not in the range of the function f. We remark that it is possible to rewrite any randomized encoding as $(\widehat{f}_{\mathrm{OFF}}(r), \widehat{f}_{\mathrm{ON}}(x,r))$ for some functions $\widehat{f}_{\mathrm{OFF}}$ and $\widehat{f}_{\mathrm{ON}}$ (for instance, by setting $\widehat{f}_{\mathrm{OFF}}$ to be the function that outputs the empty string and $\widehat{f}_{\mathrm{ON}} = \widehat{f}$). Nevertheless, in order for the encoding to be robust there must exist a way to split the output bits of $\widehat{f}(x,r)$ into an offline part $\widehat{f}_{\mathrm{OFF}}(r)$ and online part $\widehat{f}_{\mathrm{ON}}(x,r)$ such that they additionally satisfy the robustness property. As mentioned before, it will not always be important for us to minimize the online complexity, where instead we require that the encoding is robust while minimizing the total (online+offline) complexity. We note that our definition is in the spirit of the authenticity definition with respect to garbled schemes from [BHR12].

2. **Oblivious sampling:** We require an additional oblivious property, as for the definition of garbling schemes, (that, looking ahead, will enable equivocation in our instance-dependence commitment schemes where a randomized encoding of function f can be explained as a simulated encoding). We denote this algorithm by ORE and define this new security property as follows.

   For any function f as above and for all input/output pairs (x, y) such that $f(x) = y$ it holds that, $\{r', \mathcal{S}(y; r')\}_{r' \leftarrow \{0,1\}^*} \overset{c}{\approx} \{r', \widehat{f}_{\mathrm{OFF}}(r), \widehat{f}_{\mathrm{ON}}(x, r)\}_{r' \leftarrow \mathrm{ORE}(x,r)}$ where r is the randomness for generating $\widehat{f}$.

In Sect. 5, we show how to realize a robust randomized encoding scheme based on any two-party computation protocol (that meets certain requirements), which, in particular is satisfied by the [Yao86, GMW87] protocols. While this construction does not achieve any "non-trivial" online complexity, it will be sufficient for our application, as the total complexity will be $O(s\kappa)$. We note that garbling schemes meet our definition of robust randomized encoding. Therefore, we have the following theorem:

Theorem 26. *Assuming the existence of one-way functions. Then, for any polynomial time computable boolean function $f : \{0,1\}^n \rightarrow \{0,1\}$, there exists a robust randomized encoding scheme $(\widehat{f}_{\mathrm{OFF}}, \widehat{f}_{\mathrm{ON}}, \mathcal{S})$ such that the offline complexity is $O(s \cdot \mathsf{poly}(\kappa))$ and online complexity is $O(n \cdot \mathsf{poly}(\kappa))$ where s is the size of the circuit computing f, n is the size of the input to f and κ is the security parameter.*

3   Warmup: Static Zero-Knowledge Proofs from 2PC

Our technique also imply static ZK proofs from any two-party protocol that provides perfect correctness. Intuitively speaking, consider a two-party protocol that is secure in the presence of static adversaries with perfect correctness. Then, the prover generates the transcript of an execution where the parties' inputs are secret shares of the witness ω. That is, the parties' inputs are ω_1 and ω_2, respectively, such that $\omega = \omega_1 \oplus \omega_2$. Upon receiving a challenge bit from the verifier, the prover sends either the input and randomness of P_1 or P_2, for which the verifier checks for consistency with respect to the transcript, and that P_2 outputs 1. From the correctness of the underlying two-party protocol it holds that a malicious prover will not be able to answer both challenges, as that requires generating a complete accepting view. On the other hand, zero-knowledge is implied by the privacy of the two-party protocol. We now proceed with the formal description of our zero-knowledge proof. Let x denote a statement in an NP language $\mathcal{L}$, associated with relation $\mathcal{R}$, let C be a circuit that outputs 1 on input (x, ω) only if $(x, \omega) \in \mathcal{R}$, and let $\Pi_g^{\mathrm{OT}} = \langle \pi_1, \pi_2 \rangle$ denote a two-party protocol that privately realizes C with perfect correctness; see Sect. 5 for the complete details of protocol Π_g^{OT} when embedded with our OT encoding. Our protocol is specified in Fig. 1. We note that our protocol implies the *first static zero-knowledge proof* based on (the two-party variant of) [GMW87, Yao86]. In Sect. 5 we discuss how to rely solely on one-way functions. In [HV16] we prove the following claim,

Theorem 31. *Assume the existence of one-way functions. Then, the protocol presented in Fig. 1 is a static honest verifier zero-knowledge proof for any language in NP.*

Static Zero-Knowledge Proof for any Language $\mathcal{L} \in$ NP

Inputs: A circuit C that computes the function $f(x, \omega) = \mathcal{R}(x, \omega)$ and a public statement $x \in \mathcal{L}$ for both. A witness ω for the validity of x for the prover $\mathcal{P}$.

The protocol:

1. $\mathcal{P} \rightarrow \mathcal{V}$: $\mathcal{P}$ invokes Π_g^{OT} and emulates the roles of P_1 and P_2 on random shares ω_1, ω_2 of ω, and randomness r_1, r_2. Let τ be the transcript of messages exchanged between these parties. $\mathcal{P}$ sends τ to the verifier.
2. $\mathcal{V} \rightarrow \mathcal{P}$: The verifier sends a random challenge bit $b \leftarrow \{0, 1\}$.
3. $\mathcal{P} \rightarrow \mathcal{V}$: Upon receiving the bit b the prover continues as follows,
   - If $b = 0$ then the prover sends (r_1, ω_1).
   - Else, if $b = 1$ then the prover sends (r_2, ω_2).
4. The verifier checks that the randomness and input are consistent with τ by emulating the corresponding party. In case of emulating P_2, the verifier checks that it further outputs 1.

Fig. 1. Static zero-knowledge proof for any language $\mathcal{L} \in$ NP

4 Instance-Dependent Commitments from Garbled Schemes

As a warmup, we present our first adaptive instance-dependent commitment scheme based on our garbled circuits notion as formally defined in Sect. 2.3 which, in turn, implies a construction for the binary message space $\{0, 1\}$ based on one-way functions (see more detailed discussion in Sect. 2.3). Let x denote a statement in an NP language $\mathcal{L}$, associated with relation $\mathcal{R}$, and let C be a circuit that outputs 1 on input (x, ω) only if $(x, \omega) \in \mathcal{R}$.[10] Intuitively speaking, our construction is described as follows.

A commitment to the bit 0 is defined by a garbling of circuit C , i.e., $\mathsf{Grb}(C)$, and a commitment to the secret key whereas a commitment to the bit 1 is defined by a simulated garbling of the circuit C with output set to 1, i.e., the garbled circuit output by $\mathsf{SimGC}(C, 1)$, and a commitment the input encoding $\tilde{z}$ that is output by $\mathsf{SimGC}(C, 1)$. The decommitment to the bit 0 requires revealing the secret key (all input labels) with which the receiver checks that $\mathsf{Grb}(C)$ is indeed a garbling of C. On the other hand, the decommitment to the bit 1 requires decommitting to $\tilde{z}$ with which the receiver checks that the simulated garbled circuit evaluates to 1. Importantly, if the committer knows a witness ω for the validity of x in $\mathcal{L}$, then it can always honestly commit to a garbling of circuit C and later decommit to both 0 and 1. For statements $x \in \mathcal{L}$, the hiding property of the commitment scheme follows directly from the indistinguishability of the

---

[10] More explicitly, we assume that the common statement x is embedded inside the circuit and only ω is given as its input.

simulated garbled circuit and the hiding property of the underlying commitment scheme. Whereas, for $x \notin \mathcal{L}$, the commitment is perfectly binding as even an unbounded committer cannot provide a honestly generated garbled circuit, and at the same time provide an encoding of some input that evaluates the garbled circuit to 1 (as there exists no witness ω for x). Finally, considering garbling constructions from the literature, such as the [LP09] scheme, we note that the communication complexity of our construction for committing a single bit equals $O(s \cdot \mathsf{poly}(\kappa))$ where s is the circuit's size and κ is the security parameter. In [HV16], a formal proof of the following theorem is provided.

Theorem 41. *Assume the existence of one-way functions. Then, there exists a secure adaptive instance-dependent commitment scheme for any language in* NP.

5   Randomized Encoding from Two-Party Computation

In this section, we show how to construct a randomized encoding for any function f, given a two-party computation in the oblivious transfer (OT)-hybrid. This is opposed to prior works that have established the usefulness of randomized encoding in constructing efficient multiparty computation [IK00, AIK04, DI06].

Let $f : \{0,1\}^n \to \{0,1\}$ be an arbitrary polynomial-time computable function. We define $g(a,b) = f(a \oplus b)$ and view g as a two-party functionality. Then let $\Pi_g^{\mathrm{OT}} = \langle \pi_1, \pi_2 \rangle$ be a two-party protocol which realizes g with the following guarantees:

1. It guarantees UC security against semi-honest adversaries in the OT-hybrid that can statically corrupt either P_1 or P_2 and adaptively corrupt P_2. Looking ahead, we consider two different adversaries: (1) adversary $\mathcal{A}_1$ that corrupts P_1 at the beginning of the execution and adaptively corrupts P_2 post-execution (further denoted as a semi-adaptive adversary [GWZ09]) and (2) adversary $\mathcal{A}_2$ that corrupts P_2 at the beginning of the execution. We denote the corresponding simulators by $\mathcal{S}_1$ and $\mathcal{S}_2$.
2. Finally, we require that P_1 is the (designated) sender for all OT instances and that the output of the computation is obtained only by P_2.

We remark that both the classic Yao's garbled circuit construction [Yao86] and the [GMW87] protocol satisfy these conditions in the OT-hybrid. We further stress that while garbled circuit constructions do not (in general) admit adaptive corruptions, we show that the specific corruption by adversary $\mathcal{A}_1$ can be simulated in the OT-hybrid. In [HV16] we discuss these two realizations in more details. We next demonstrate how to transform any two-party computation protocol that satisfies the properties listed above to a randomized encoding. Our first construction will rely on trapdoor permutations to realize the OT functionality. We then relax this requirement and show how to rely on one-way functions.

Given any protocol Π_g^{OT} we consider a protocol $\widetilde{\Pi}$ that is obtained from Π_g^{OT} by replacing every OT call with the enhanced trapdoor permutation based OT protocol of [EGL85]. Let $\{f_{\mathrm{TDP}} : \{0,1\}^n \to \{0,1\}^n\}$ be a family of trapdoor permutations and h be the corresponding hard-core predicate. More precisely,

– For every OT call where P_1's input is (s_0, s_1) and P_2's input is t, we require P_1 to send the index of a trapdoor permutation f_{TDP} to P_2. Next, P_2 samples v_{1-t} and u_t uniformly at random from $\{0, 1\}^n$ and sets $v_t = f_{\mathrm{TDP}}(u_t)$. P_2 sends (v_0, v_1) to P_1, that is followed by the message (c_0, c_1) from P_1 to P_2 where $c_0 = h(u_0) \oplus s_0$ and $c_1 = h(u_1) \oplus s_1$ and $u_0 = f_{\mathrm{TDP}}^{-1}(v_0), u_1 = f_{\mathrm{TDP}}^{-1}(v_1)$.

We need to verify that $\widetilde{\Pi}$ satisfies all the required properties.

1. It follows from the fact that if Π_g^{OT} implements g with UC security against semi-honest adversaries $\mathcal{A}_1$ and $\mathcal{A}_2$, then $\widetilde{\Pi}$ achieves the same against corresponding adversaries that corrupt the same parties and finally output the view of P_2. In more details, recall that $\mathcal{A}_1$ corrupts P_1 at the beginning and P_2 post execution (adaptively). Now, since Π_g^{OT} admits simulation of $\mathcal{A}_1$ in the OT-hybrid, for the same property to hold for $\widetilde{\Pi}$, it suffices to achieve simulation of the OT protocol where the sender is corrupted at the beginning and the receiver is corrupted post execution. It is easy to see that the [EGL85] protocol satisfies this requirement since the receiver is equivocable. Next, to see that $\mathcal{A}_2$ can be simulated we rely on the fact that the OT protocol described above admits (semi-honest) receiver's simulation. Therefore, $\widetilde{\Pi}$ satisfies all the required properties.
2. This property directly holds as we rely on the same instructions to determine the sender and receiver of the OT calls.

Our Randomized Encoding. We now proceed with the description of our robust randomized encoding of f as formalized in Definition 25 by specifying the functions $\widehat{f}_{\mathrm{OFF}}, \widehat{f}_{\mathrm{ON}}$ and the simulation $\mathcal{S}$. Towards describing our algorithms, we consider a real world experiment carried out between parties P_1 and P_2 that engage in an execution of $\widetilde{\Pi}$ with environment $\mathcal{Z}$. Let $\mathbf{REAL}_{\widetilde{\Pi},\mathcal{A},\mathcal{Z}}(\kappa, x, \mathbf{r})$ denote the output of $\mathcal{Z}$ on input x, random tape $r_{\mathcal{Z}}$ and a security parameter κ upon interacting with $\mathcal{A}$ with random tape $r_{\mathcal{A}}$ and parties P_1, P_2 with random tapes r_1, r_2, respectively, that engage in protocol $\widetilde{\Pi}$ where the inputs are determined by $\mathcal{Z}$ and $\mathbf{r} = (r_{\mathcal{Z}}, r_{\mathcal{A}}, r_1, r_2)$. Let $\mathbf{REAL}_{\widetilde{\Pi},\mathcal{A},\mathcal{Z}}(\kappa, x)$ denote a random variable describing $\mathbf{REAL}_{\widetilde{\Pi},\mathcal{A},\mathcal{Z}}(\kappa, x, \mathbf{r})$ where the random tapes are chosen uniformly. We denote by $\mathbf{IDEAL}_{g,\mathcal{S},\mathcal{Z}}(\kappa, x, \mathbf{r})$ the output of $\mathcal{Z}$ on input x, random tape $r_{\mathcal{Z}}$ and security parameter κ upon interacting with $\mathcal{S}$ and parties P_1, P_2, running an ideal process with random tape $r_{\mathcal{S}}$, where $\mathbf{r} = (r_{\mathcal{Z}}, r_{\mathcal{S}})$. Let $\mathbf{IDEAL}_{g,\mathcal{S},\mathcal{Z}}(\kappa, x)$ denote a random variable describing $\mathbf{IDEAL}_{g,\mathcal{S},\mathcal{Z}}(\kappa, x, \mathbf{r})$ when the random tapes $r_{\mathcal{Z}}$ and $r_{\mathcal{S}}$ are chosen uniformly.

Encoding: Consider a (semi-honest) adversary $\mathcal{A}_1$ that corrupts P_1 at the beginning of the execution. At the end of the execution, $\mathcal{A}_1$ first sends τ to $\mathcal{Z}$ where τ is the transcript of messages exchanged between P_1 and P_2. Next it (adaptively) corrupts P_2 and sends (a_2, r_2) to $\mathcal{Z}$ where a_2 and r_2 are the respective input and randomness used by party P_2. Let $\mathcal{S}_1$ be the corresponding simulator as guaranteed by the properties of $\widetilde{\Pi}$.

1. $\widehat{f}_{\mathrm{OFF}}(r)$: The offline encoding is obtained by running $\mathcal{S}_1$ with randomness $r_{\mathcal{S}_1}$ until it sends the first message to the environment. Recall that $\mathcal{S}_1$ statically corrupts P_1, where upon completing the execution, $\mathcal{S}_1$ sends the transcript of the messages to the environment. We define the output of $\widehat{f}_{\mathrm{OFF}}(r)$ to be this output where the input a_1 of party P_1 is sampled uniformly at random. Notice that the offline part of the encoding does not depend on the input x as required.

2. $\widehat{f}_{\mathrm{ON}}(x,r)$: To obtain the online part, we continue the execution of $\mathcal{S}_1$ in the execution corresponding to the transcript τ generated by $\widehat{f}_{\mathrm{OFF}}(r)$. Recall that after sending τ, $\mathcal{S}_1$ adaptively corrupts P_2 and sends the input and random tape of P_2 to the environment. $\widehat{f}_{\mathrm{ON}}(x,r)$ continues the emulation of $\mathcal{S}_1$, where upon corrupting party P_2 it feeds $\mathcal{S}_1$ with the input of P_2 as $a_2 = x \oplus a_1$ and $f(x)$ as the output. The simulation returns the view of P_2 and $\widehat{f}_{\mathrm{ON}}(x,r)$ is set to (a_2, r_2) where r_2 is the random tape of P_2 output by $\mathcal{S}_1$.

Decoder: The decoder B on input $(z_{\mathrm{OFF}}, z_{\mathrm{ON}})$ recomputes the view of P_2 from the messages sent by P_1 to P_2 in z_{OFF} and the input and randomness of P_2 in z_{ON}. It checks if the messages sent from P_2 to P_1 are consistent with what is in z_{OFF} and finally outputs what P_2 outputs in the execution.

Simulation: Consider the (semi-honest) adversary $\mathcal{A}_2$ that statically corrupts P_2. At the end of the execution $\mathcal{A}_2$ sends $(\tau, (a_2, r_2))$ to $\mathcal{Z}$ where τ is the transcript of messages exchanged between P_1 and P_2 and a_2 and r_2 are the respective input and randomness used by party P_2. Let $\mathcal{S}_2$ be the corresponding simulator. Then the simulation algorithm of the randomized encoding $\mathcal{S}$ is defined as follows. Upon receiving $y = f(x)$, $\mathcal{S}$ invokes $\mathcal{S}_2$ where P_2's input is set to a uniformly chosen random string a_2 and its output is set to y. Recall that $\mathcal{S}_2$ outputs $(\tau, (a_2, r_2))$ at the end of the execution. Then the output of $\mathcal{S}$ is defined by $(s_{\mathrm{OFF}}, s_{\mathrm{ON}})$ where $s_{\mathrm{OFF}} = \tau$ and $s_{\mathrm{ON}} = (a_2, r_2)$.

Theorem 51. *Let $(\widehat{f}(x,r), \mathcal{S}, B)$ be as above. Then $\widehat{f}(x,r)$ is a randomized encoding of f with computational privacy. Assuming the existence of enhanced trapdoor permutations, we obtain an encoding with offline complexity $C_{\Pi} + \rho_{\Pi}\kappa$ and online complexity $|x| + r_{\Pi} + \rho_{\Pi}\kappa$ where C_{Π} is the communication complexity of Π_g^{OT} in the OT-hybrid, ρ_{Π} in the number of OT invocations made by P_2, r_{Π} is the randomness complexity of P_2 in Π_g^{OT} and κ is the security parameter. If we instead rely on one-way functions we achieve an encoding with offline and online complexities $C_{\Pi} + \rho_{\Pi}\mathsf{poly}(\kappa)$ and $|x| + r_{\Pi} + \rho_{\Pi}\mathsf{poly}(\kappa)$, respectively.*

In [HV16] we discuss the relaxation to one-way functions and the proof.

Complexity. Finally, we measure the complexity of our encoding. Note first that for each OT call the offline encoding is a pair of image elements of the one-way permutation incurring $O(\kappa)$ overhead, while the online complexity is a preimage of length κ. Then the offline encoding of the overall construction is the communication complexity of $\widetilde{\Pi}$ which equals to the communication of Π_g^{OT}, denoted by C_{Π}, together with the number of OT calls, denoted by ρ_{Π}, which overall yields $C_\Pi + \rho_\Pi O(\kappa)$. Moreover, the online encoding includes P_2's input a_2 and randomness r_2 where the latter includes the randomness complexity of Π_g^{OT} and the complexity of the receiver's randomness for the OT invocations which is $|x| + r_\Pi + \rho_\Pi \kappa$. If we rely on one-way functions then the OT calls are incorporated as commitments and incur $\mathsf{poly}(\kappa)$ per invocation for the commitment as well as the decommitment algorithms.

5.1   Corollaries and Applications

Below, we demonstrate the power of the proceeding transformation by proving lower bounds and providing additional applications. We discuss instance-dependent commitment schemes in [HV16] as well as realizations for our RE.

Input-Delayed Zero-Knowledge Proofs. In this section, we extend the basic construction of instance-dependent commitment schemes from our previous construction to additionally allow constructing input-delayed zero-knowledge proofs. We show how randomized-encoding that is secure against adaptive chosen inputs can be used to realize input-delayed zero-knowledge proofs. Then relying on the recent construction of such a randomized encoding [HJO+15] we obtain a constant-rate input-delayed zero-knowledge proof, namely whose communication complexity is $O(s) + \mathsf{poly}(\kappa)$ where s is the size of the circuit realizing the NP-relation and κ is the security parameter. We achieve this in two steps. First, we extend our notion of instance-dependent commitment scheme to one where the actual commitment scheme do not require the input statement. Then using such an instance-dependent commitment scheme we will show how to realize an input-delayed zero-knowledge proofs. We provide next definitions for the above primitives.

Our first notion is that of input-delayed instant-dependent commitment scheme. On a high-level, this primitive is a variant of the plain instant-dependent commitment scheme where the real and fake commitment algorithms do not require the knowledge of the input statement in the commit phase. The statement can be adaptively chosen based on the commit phase and will be required only in the decommit phase. Second, we will not require an Adapt algorithm that can explain a fake commitment as an honest commitment of any message by generating random coins for an honest committer that would have produced the same commitment. Instead, we will only require the slightly weaker property of the fake commitment being equivocable. Towards this, we will introduce a decommitment algorithm for the honest commitment that additionally takes as input the statement x and produces a decommitment to the corresponding message m. The receiver then verifies the decommitment with respect to the statement x. Corresponding to the fake commitment algorithm, we now require an algorithm that, given the statement and the witness can reveal a commitment (i.e. produce decommitments) to any message m.

Definition 52. (Input-delayed IDCS). *Let $\mathcal{R}$ be an NP relation and $\mathcal{L}$ be the language associated with $\mathcal{R}$. A (non-interactive) instance dependent commitment scheme (IDCS) for $\mathcal{L}$ is a tuple of probabilistic polynomial-time algorithms $(\widetilde{\mathsf{Com}}, \widetilde{\mathsf{Decom}}, \widetilde{\mathsf{Ver}}, \widetilde{\mathsf{Com}}', \mathsf{Equiv})$, where:*

- $\widetilde{\mathsf{Com}}$ *is the commitment algorithm: For a message $m \in \mathcal{M}_n$, and a random string $r \in \{0,1\}^{p(n)}$, $\widetilde{\mathsf{Com}}(1^n, m; r)$ returns a commitment value c where n is the length of the input-instance and $p(\cdot)$ is a polynomial.*
- $\widetilde{\mathsf{Decom}}$ *is the decommitment algorithm that on input a statement x, commitment c, mesage m and randomness r outputs a decommitment d.*
- $\widetilde{\mathsf{Ver}}$ *is the verification algorithm that on input x, m, c, d outputs accept or reject.*
- $\widetilde{\mathsf{Com}}'$ *is a "fake" commitment algorithm: For a random string $r \in \{0,1\}^{q(n)}$, $\widetilde{\mathsf{Com}}'(1^n, r)$ returns a commitment value c where n is the length of the input instance and $q(\cdot)$ is a polynomial.*
- Equiv *is an equivocation algorithm: Let $x \in \mathcal{L}$ and $\omega \in \mathcal{R}_x$. For all c and $r \in \{0,1\}^{q(|x|)}$ such that $\mathsf{Com}'(r) = c$, and for all $m \in \mathcal{M}_n$, $\mathsf{Equiv}(x, \omega, c, m, r)$ outputs d such that $\widetilde{\mathsf{Ver}}(x, m, c, d)$ outputs accept.*

The hiding property now requires that for any message m, an honest commitment and decommitment to m be indistinguishable from a fake commitment and decommitment to m even when the input statement is adaptively chosen after the commitment phase. The binding property on the other hand will require that for any commitment c and a false statement $x \notin \mathcal{L}$, there exists no values m, d and m', d' such that $\widetilde{\mathsf{Ver}}(x, m, c, d) = \widetilde{\mathsf{Ver}}(x, m', c, d') =$ accept. Finally, in Fig. 2 we describe our input-delayed zero-knowledge proof.

Theorem 53. *Assume the existence of one-way functions. Then, the protocol presented in Fig. 2 is an input-delayed zero-knowledge proof with soundness $1/2$ for any language in NP.*

See [HV16] for the proof. Finally, we need to show how our input-delayed IDCS can be constructed from a robust randomized encoding that is secure against an adaptive chosen input. We begin with a randomized encoding for the following function f: $f(x, \omega) = (\mathcal{R}(x, \omega), x)$. Since the randomized encoding is secure against adaptive choice of inputs, the simulation algorithm of the RE is decomposed into two algorithms, namely the offline part s_{OFF} and online part s_{ON}. Now, we can define our commitment algorithm as follows: A commitment to 0 returns the offline part of the encoding $\widehat{f}_{\mathrm{OFF}}(r)$ whereas a commitment to 1 returns the offline part of the simulation $s_{\mathrm{OFF}}(r')$ where r and r' are the randomness used for the algorithms. A decommitment to 0 requires revealing randomness showing that the commitment was generated honestly using $\widehat{f}_{\mathrm{OFF}}(r)$ and a decommitment to 1 requires providing the online part s_{ON} that along with the commitment decodes to $(1, x)$ where x is the statement. Finally, the fake commitment algorithm is defined as a commitment to 0. Observe that both the honest and fake commitment algorithms do not depend on the input statement. This is enabled by the adaptive input security of the randomized encoding. The

---

Input-Delayed Zero-Knowledge Proof for any Language $\mathcal{L} \in$ NP

Building block: Input delayed IDCS $(\widetilde{\mathsf{Com}}, \widetilde{\mathsf{Decom}}, \widetilde{\mathsf{Ver}}, \widetilde{\mathsf{Com}}', \mathsf{Equiv})$ for $\mathcal{L}$.

Inputs: A circuit C that computes the function $f(x, \omega) = (\mathcal{R}(x, \omega), x)$.

The protocol:

1. $\mathcal{P} \to \mathcal{V}$: $\mathcal{P}$ invokes $\mathsf{com}_0 \leftarrow \widetilde{\mathsf{Com}}'(1^\kappa; r)$ and $\mathsf{com}_1 \leftarrow \widetilde{\mathsf{Com}}'(1^\kappa; r)$ and sends $(\mathsf{com}_0, \mathsf{com}_1)$ to the verifier.
2. $\mathcal{V} \to \mathcal{P}$: The verifier sends a random challenge $b \leftarrow \{01, 10\}$.
3. $\mathcal{P} \to \mathcal{V}$: Upon receiving the input statement x and witness ω,
   - If $b = 01$ then the prover sends the decommitments to $(\mathsf{com}_0, \mathsf{com}_1)$ by computing $\mathsf{Equiv}(x, \omega, \mathsf{com}_0, 0, r)$ and $\mathsf{Equiv}(x, \omega, \mathsf{com}_1, 1, r)$.
   - If $b = 10$ then the prover sends the decommitments to $(\mathsf{com}_0, \mathsf{com}_1)$ by computing $\mathsf{Equiv}(x, \omega, \mathsf{com}_0, 1, r)$ and $\mathsf{Equiv}(x, \omega, \mathsf{com}_1, 0, r)$.
4. The verifier checks that the decommitments are valid with respect to x.

Fig. 2. Input-delayed zero-knowledge proof for any language $\mathcal{L} \in$ NP

hiding property of the commitment for bit 0 holds directly, whereas the hiding property for the bit 1 follows from the simulation property of the randomized encoding. Binding on the other hand follows directly from the robustness property of the randomized encoding. The complete description is given in [HV16]. We note that the work of Hemenway et al. [HJO+15] shows how to obtain a randomized encoding that is secure against adaptively chosen inputs. We show in [HV16] how to extend it to achieve the stronger robustness property. Combining their work with our construction, we have the following corollary.

Corollary 54. *Assuming the existence of one-way functions. Then for any NP-relation $\mathcal{R}$, there exists an input-delayed ZK proof with communication complexity $O(s \cdot \mathsf{poly}(k))$ where s is the size of the circuit computing the NP relation.*

5.2   Commit-and-Prove Zero-Knowledge Proofs

In the "commit-and-prove" paradigm, the prover first commits to its witness and then proves that the statement, along with the decommitment value maintains the underlying NP relation. This paradigm has turned useful for constructing maliciously secure protocols [GMW87, CLOS02]. In this section we show how to design such an *input-delayed* proof, namely, where the statement is determined only at the last round and the underlying commitment scheme (in turn the one-way function) is used in a black-box way. Specifically, in this input-delaying flavour the witness is known ahead of time but not the statement, and hence not the NP relation.

As above, we employ a robust randomized encoding that is secure in the presence of adaptive choice of inputs, where the simulation algorithm is split into an offline and online phases, that computes the function $f_{\omega_0}(x, \omega_1) = (\mathcal{R}(x, \omega_0 \oplus \omega_1), x, \omega_1)$ where ω_0 is hardwired into the circuit that computes this functionality. The reason we need to hardwire it is because the offline phase must be associated with this share. Whereas the other share ω_1 is output by the circuit in order to enforce the usage of the right share.

Achieving Negligible Soundness. In order to improve the soundness parameter of our ZK proof we need to repeat the basic proof sufficiently many times in parallel, using fresh witness shares each time. This, however, does not immediately work as the dishonest prover may use different shares for each proof instance. In order to overcome this problem we use the [IKOS09] approach in order to add a mechanism that verifies the consistency of the shares. Namely, suppose we wish to repeat the basic construction in parallel $N = O(t)$ times where $t = O(\kappa)$ and κ is the security parameter. Formally,

- The verifier picks a random t-subset I of $[N]$. It also picks t random challenge bit $\{ch_i\}_{i \in I}$ and commits to them.
- The prover then continues as follows:
    1. It first generates N independent XOR sharings of w, say $\{(w_{i,0}, w_{i,1})\}_{i \in [N]}$.
    2. It generates the views of $2N$ parties $P_{i,0}$ and $P_{i,1}$ for $i \in [N]$ executing a t-robust t-private MPC protocol, where $P_{i,j}$ has input $w_{i,j}$, that realizes the functionality that checks if $w_{i,0} \oplus w_{i,1}$ are equal for all i. Let $V_{i,j}$ be view of party $P_{i,j}$.
    3. Next, it computes N offline encodings of the following set of functions:

    $$f_{w_{i,0}, V_{i,0}}(x, w_{i,1}, V_{i,1}) = (b, x, w_{i,1}, V_{i,1})$$

    for $i \in [N]$, where $b = 1$ if and only if $\mathcal{R}(x, w_{i,0} \oplus w_{i,1})$ holds and the views $V_{i,0}$ and $V_{i,1}$ are consistent with each other.
    4. Finally, the prover sends:

    $$\left\{ f^{\mathrm{OFF}}_{w_{i,0}}(r_i), , (r_i), , (w_{i,0}), , (w_{i,1}), , (V_{i,0}), , (V_{i,1}) \right\}_{i \in [N]}.$$

- The verifier decommits to all its challenges.
- For every index i in the t subset the prover replies as follows:
    - If $ch_i = 0$ then it decommits to r_i, $w_{i,0}$ and $V_{i,0}$. The verifier then checks if the offline part was constructed correctly (as in our basic proof).
    - If $ch_i = 1$ then i sends $f^{\mathrm{ON}}_{w_{i,0}}(r_i, x, w_{i,1})$ and decommits $w_{i,1}$. The verifier then runs the decoder and checks if it obtains $(1, x, w_{i,0})$.
    Furthermore, for every index i, the prover decommits the views V_{i,ch_i} for which the verifier checks if the MPC-in-the-head protocol was executed correctly.

Theorem 55. *Assume the existence of one-way functions. Then, the above protocol is a commit-and-prove input-delayed zero-knowledge proof with negligible soundness for any language in* NP.

6   Constructing Adaptive Zero-Knowledge Proofs

We describe next how to construct adaptive zero-knowledge proofs for all NP languages based on our instance-dependent commitment schemes from Sects. 4 and 5.

Let x denote a statement to be proven by the prover relative to some language $\mathcal{L}$ associated with relation $\mathcal{R}$. Then the prover generates a garbled circuit C that takes (x, ω) and outputs 1 only if $(x, \omega) \in \mathcal{R}$, and commits to this garbling and the secret key sk using the commitment scheme from Sect. 4. Next, upon receiving a challenge bit b from the verifier, the prover continues as follow. If $b = 0$ then the prover decommits to the commitment of the secret key and the garbled circuit for which the verifier verifies the correctness of garbling. Else, if $b = 1$ then the prover decommits a "path" in the garbled circuit and provides an encoding for ω that evaluates the path to 1. Namely, we consider the concrete garbling construction by [Yao86, LP09] for which each evaluation induces a path of computation, where each gate evaluation requires the decryption of a single ciphertext out of four ciphertexts, where this ciphertext can be part of the decommitted information handed to the verifier when $b = 1$. The verifier then evaluates the garbling on this path and checks that the outcome if 1. We note that it is not clear how to generalize this property (where only part of the garbled circuit is decommitted) nor the following reconstructability property for the notion of randomized encodings.

Let Garb = (Grb, Enc, Eval, Dec) denote a garbling scheme as in Sect. 2.3. Then, we will require one more property that Garb should satisfy:

Reconstructability: Given any path of computation in the garbled circuit it is possible to reconstruct the rest of the garbled circuit as being honestly generated by Grb.

We note that the [LP09] garbling scheme meets this notion. The description of our protocol can be found in Fig. 3 and the proof of the following theorem in [HV16].

Theorem 61. *Assume the existence of one-way functions. Then, the protocol presented in Fig. 3 is an adaptively secure honest verifier zero-knowledge proof for any language in* NP *with soundness error* $1/2$.

We note that the communication complexity of our protocol is $O(\kappa s^2)$ where κ is the security parameter and s is the size of C. In the full version we extend this construction to achieve a linear-rate adaptive ZK proof and obtain the following theorem.

Theorem 62. *Assume the existence of one-way functions. Then, for any* NP *relation* $\mathcal{R}$ *that can be verified by a circuit of size* s *(using bounded fan-in gates), there exists an adaptive zero-knowledge proof with communication complexity* $O(s) \cdot \mathsf{poly}(\kappa, \log s)$ *where* κ *is the security parameter.*

Adaptive Zero-Knowledge Proof for Any Language $\mathcal{L} \in$ NP

Building block: Instance-dependent commitment scheme Com for language $\mathcal{L}$.

Inputs: A circuit C as above and a public statement $x \in \mathcal{L}$ for both. A witness ω for the validity of x for the prover $\mathcal{P}$.

The protocol:

1. $\mathcal{P} \rightarrow \mathcal{V}$: $\mathcal{P}$ generates $(\widetilde{C}, \mathbf{dk}, \mathsf{sk}) \leftarrow \mathsf{Grb}(1^\kappa, C)$ and sends $\mathsf{Com}(\widetilde{C}, \mathbf{dk})$ and $\mathsf{Com}(\mathsf{sk})$ to the verifier (where the commitments are computed using the real commitment algorithm).
2. $\mathcal{V} \rightarrow \mathcal{P}$: The verifier sends a random challenge bit $b \leftarrow \{0, 1\}$.
3. $\mathcal{P} \rightarrow \mathcal{V}$:
   - If $b = 0$ then the prover decommits to $\widetilde{C}, \mathbf{dk}$ and sk. The verifier accepts if the decommitments are valid and that the garbling was honestly generated.
   - If $b = 1$ then the prover decommits to $\mathbf{dk}$ and further provides the decommitment for the encoding of ω and the path of computation in the commitment to $\widetilde{C}$ that is evaluated during the computation of $\mathsf{Eval}(\widetilde{C}, \widetilde{\omega})$. Namely, the prover invokes $\widetilde{\omega} := \mathsf{Enc}(\mathsf{sk}, \omega)$ and then decommits to the encoding of $\widetilde{\omega}$ within the commitment of sk (recall that this is possible due to the decomposability of the garbled scheme), as well as the path of computation. The verifier then invokes $\widetilde{y} := \mathsf{Eval}(\widetilde{C}, \widetilde{\omega})$ and accepts if $\mathsf{Dec}(\mathbf{dk}, \widetilde{\omega})$ equals 1.

Fig. 3. Adaptive zero-knowledge proof for any language $\mathcal{L} \in$ NP

References

[AIK04] Applebaum, B., Ishai, Y., Kushilevitz, E.: Cryptography in NC^0. In: FOCS, pp. 166–175 (2004)

[AIK06] Applebaum, B., Ishai, Y., Kushilevitz, E.: Cryptography in NC^0. SIAM J. Comput. **36**(4), 845–888 (2006)

[AIK10] Applebaum, B., Ishai, Y., Kushilevitz, E.: From secrecy to soundness: efficient verification via secure computation. In: Abramsky, S., Gavoille, C., Kirchner, C., Meyer auf der Heide, F., Spirakis, P.G. (eds.) ICALP 2010. LNCS, vol. 6198, pp. 152–163. Springer, Heidelberg (2010)

[AIKP15] Agrawal, S., Ishai, Y., Khurana, D., Paskin-Cherniavsky, A.: Statistical randomized encodings: a complexity theoretic view. In: Halldórsson, M.M., Iwama, K., Kobayashi, N., Speckmann, B. (eds.) ICALP 2015. LNCS, vol. 9134, pp. 1–13. Springer, Heidelberg (2015)

[AIKW13] Applebaum, B., Ishai, Y., Kushilevitz, E., Waters, B.: Encoding functions with constant online rate or how to compress garbled circuits keys. In: Canetti, R., Garay, J.A. (eds.) CRYPTO 2013, Part II. LNCS, vol. 8043, pp. 166–184. Springer, Heidelberg (2013)

[App14] Applebaum, B.: Key-dependent message security: generic amplification and completeness. J. Cryptol. **27**(3), 429–451 (2014)

[Bea96] Beaver, D.: Correlated pseudorandomness and the complexity of private computations. In: STOC, pp. 479–488 (1996)

[BGW88] Ben-Or, M., Goldwasser, S., Wigderson, A.: Completeness theorems for non-cryptographic fault-tolerant distributed computation (extended abstract). In: STOC, pp. 1–10 (1988)

[BHHI10] Barak, B., Haitner, I., Hofheinz, D., Ishai, Y.: Bounded key-dependent message security. In: Gilbert, H. (ed.) EUROCRYPT 2010. LNCS, vol. 6110, pp. 423–444. Springer, Heidelberg (2010)

[BHR12] Bellare, M., Hoang, V.T., Rogaway, P.: Foundations of garbled circuits. In: CCS, pp. 784–796 (2012)

[BMO90] Bellare, M., Micali, S., Ostrovsky, R.: Perfect zero-knowledge in constant rounds. In: STOC, pp. 482–493 (1990)

[CCD87] Chaum, D., Crépeau, C., Damgård, I.B.: Multiparty unconditionally secure protocols. In: Pomerance, C. (ed.) CRYPTO 1987. LNCS, vol. 293, pp. 462–462. Springer, Heidelberg (1988)

[CDD+04] Canetti, R., Damgård, I., Dziembowski, S., Ishai, Y., Malkin, T.: Adaptive versus non-adaptive security of multi-party protocols. J. Cryptol. 17(3), 153–207 (2004)

[CDD+15] Cascudo, I., Damgård, I., David, B., Giacomelli, I., Nielsen, J.B., Trifiletti, R.: Additively homomorphic UC commitments with optimal amortized overhead. In: Katz, J. (ed.) PKC 2015. LNCS, vol. 9020, pp. 495–515. Springer, Heidelberg (2015)

[CLOS02] Canetti, R., Lindell, Y., Ostrovsky, R., Sahai, A.: Universally composable two-party and multi-party secure computation. In: STOC, pp. 494–503 (2002)

[CPS+15] Ciampi, M., Persiano, G., Scafuro, A., Siniscalchi, L., Visconti, I.: Improved OR composition of sigma-protocols. IACR Cryptology ePrint Archive, 2015:810 (2015)

[CvdGT95] Crépeau, C., van de Graaf, J., Tapp, A.: Committed oblivious transfer and private multi-party computation. In: Coppersmith, D. (ed.) CRYPTO 1995. LNCS, vol. 963, pp. 110–123. Springer, Heidelberg (1995)

[Dam10] Damgård, I.: On Σ-protocols (2010). http://www.cs.au.dk/ivan/Sigma.pdf

[DHRS04] Ding, Y.Z., Harnik, D., Rosen, A., Shaltiel, R.: Constant-round oblivious transfer in the bounded storage model. In: Naor, M. (ed.) TCC 2004. LNCS, vol. 2951, pp. 446–472. Springer, Heidelberg (2004)

[DI06] Damgård, I.B., Ishai, Y.: Scalable secure multiparty computation. In: Dwork, C. (ed.) CRYPTO 2006. LNCS, vol. 4117, pp. 501–520. Springer, Heidelberg (2006)

[EGL85] Even, S., Goldreich, O., Lempel, A.: A randomized protocol for signing contracts. Commun. ACM 28(6), 637–647 (1985)

[FKN94] Feige, U., Kilian, J., Naor, M.: A minimal model for secure computation (extended abstract). In: STOC, pp. 554–563 (1994)

[FLS99] Feige, U., Lapidot, D., Shamir, A.: Multiple noninteractive zero knowledge proofs under general assumptions. SIAM J. Comput. 29(1), 1–28 (1999)

[FS89] Feige, U., Shamir, A.: Zero knowledge proofs of knowledge in two rounds. In: Brassard, G. (ed.) CRYPTO 1989. LNCS, vol. 435, pp. 526–544. Springer, Heidelberg (1990)

[GGP10] Gennaro, R., Gentry, C., Parno, B.: Non-interactive verifiable computing: outsourcing computation to untrusted workers. In: Rabin, T. (ed.) CRYPTO 2010. LNCS, vol. 6223, pp. 465–482. Springer, Heidelberg (2010)

[GIS+10] Goyal, V., Ishai, Y., Sahai, A., Venkatesan, R., Wadia, A.: Founding cryptography on tamper-proof hardware tokens. In: Micciancio, D. (ed.) TCC 2010. LNCS, vol. 5978, pp. 308–326. Springer, Heidelberg (2010)

[GKR08] Goldwasser, S., Kalai, Y.T., Rothblum, G.N.: One-time programs. In: Wagner, D. (ed.) CRYPTO 2008. LNCS, vol. 5157, pp. 39–56. Springer, Heidelberg (2008)

[GLOV12] Goyal, V., Lee, C.-K., Ostrovsky, R., Visconti, I.: Constructing non-malleable commitments: a black-box approach. In: FOCS, pp. 51–60 (2012)

[GMR89] Goldwasser, S., Micali, S., Rackoff, C.: The knowledge complexity of interactive proof systems. SIAM J. Comput. **18**(1), 186–208 (1989)

[GMW87] Goldreich, O., Micali, S., Wigderson, A.: How to play any mental game or a completeness theorem for protocols with honest majority. In: STOC, pp. 218–229 (1987)

[GOSV14] Goyal, V., Ostrovsky, R., Scafuro, A., Visconti, I.: Black-box non-black-box zero knowledge. In: Symposium on Theory of Computing, STOC 2014, New York, NY, USA, 31 May – 3 June 2014, pp. 515–524 (2014)

[GWZ09] Garay, J.A., Wichs, D., Zhou, H.-S.: Somewhat non-committing encryption and efficient adaptively secure oblivious transfer. In: Halevi, S. (ed.) CRYPTO 2009. LNCS, vol. 5677, pp. 505–523. Springer, Heidelberg (2009)

[HIKN08] Harnik, D., Ishai, Y., Kushilevitz, E., Nielsen, J.B.: OT-Combiners via secure computation. In: Canetti, R. (ed.) TCC 2008. LNCS, vol. 4948, pp. 393–411. Springer, Heidelberg (2008)

[HJO+15] Hemenway, B., Jafargholi, Z., Ostrovsky, R., Scafuro, A., Wichs, D.: Adaptively secure garbled circuits from one-way functions. IACR Cryptology ePrint Archive, 2015: 1250 (2015)

[HR07] Haitner, I., Reingold, O.: A new interactive hashing theorem. In: CCC, pp. 319–332 (2007)

[HV16] Hazay, C., Venkitasubramaniam, M.: On the power of secure two-party computation. IACR Cryptology ePrint Archive, 2016: 74 (2016)

[IK00] Ishai, Y., Kushilevitz, E.: Randomizing polynomials: a new representation with applications to round-efficient secure computation. In: FOCS, pp. 294–304 (2000)

[IK02] Ishai, Y., Kushilevitz, E.: Perfect constant-round secure computation via perfect randomizing polynomials. In: Widmayer, P., Triguero, F., Morales, R., Hennessy, M., Eidenbenz, S., Conejo, R. (eds.) ICALP 2002. LNCS, vol. 2380, pp. 244–256. Springer, Heidelberg (2002)

[IKOS07] Ishai, Y., Kushilevitz, E., Ostrovsky, R., Sahai, A.: Zero-knowledge from secure multiparty computation. In: Proceedings of the 39th Annual ACM Symposium on Theory of Computing, San Diego, California, USA, 11–13 June 2007, pp. 21–30 (2007)

[IKOS09] Ishai, Y., Kushilevitz, E., Ostrovsky, R., Sahai, A.: Zero-knowledge proofs from secure multiparty computation. SIAM J. Comput. **39**(3), 1121–1152 (2009)

[IKPY16] Ishai, Y., Kushilevitz, E., Prabhakaran, M., Sahai, A., Yu, C.H.: Secure protocol transformations. In: Robshaw, M., Katz, J. (eds.) CRYPTO 2016. LNCS, vol. 9815, pp. 430–458. Springer, Heidelberg (2016)

[IOS97] Itoh, T., Ohta, Y., Shizuya, H.: A language-dependent cryptographic primitive. J. Cryptol. **10**(1), 37–50 (1997)

[IPS08] Ishai, Y., Prabhakaran, M., Sahai, A.: Founding cryptography on oblivious transfer – efficiently. In: Wagner, D. (ed.) CRYPTO 2008. LNCS, vol. 5157, pp. 572–591. Springer, Heidelberg (2008)

[IPS09] Ishai, Y., Prabhakaran, M., Sahai, A.: Secure arithmetic computation with no honest majority. In: Reingold, O. (ed.) TCC 2009. LNCS, vol. 5444, pp. 294–314. Springer, Heidelberg (2009)

[IW14] Ishai, Y., Weiss, M.: Probabilistically checkable proofs of proximity with zero-knowledge. In: Lindell, Y. (ed.) TCC 2014. LNCS, vol. 8349, pp. 121–145. Springer, Heidelberg (2014)

[Kil88] Kilian, J.: Founding cryptography on oblivious transfer. In: STOC, pp. 20–31 (1988)

[KO04] Katz, J., Ostrovsky, R.: Round-optimal secure two-party computation. In: Franklin, M. (ed.) CRYPTO 2004. LNCS, vol. 3152, pp. 335–354. Springer, Heidelberg (2004)

[LP09] Lindell, Y., Pinkas, B.: A proof of security of Yao's protocol for two-party computation. J. Cryptol. **22**(2), 161–188 (2009)

[LS90] Lapidot, D., Shamir, A.: Publicly verifiable non-interactive zero-knowledge proofs. In: Menezes, A., Vanstone, S.A. (eds.) CRYPTO 1990. LNCS, vol. 537, pp. 353–365. Springer, Heidelberg (1991)

[LZ11] Lindell, Y., Zarosim, H.: Adaptive zero-knowledge proofs and adaptively secure oblivious transfer. J. Cryptol. **24**(4), 761–799 (2011)

[Nao91] Naor, M.: Bit commitment using pseudorandomness. J. Cryptol. **4**(2), 151–158 (1991)

[NOVY98] Naor, M., Ostrovsky, R., Venkatesan, R., Yung, M.: Perfect zero-knowledge arguments for NP using any one-way permutation. J. Cryptol. **11**(2), 87–108 (1998)

[OSV15] Ostrovsky, R., Scafuro, A., Venkitasubramanian, M.: Resettably sound zero-knowledge arguments from OWFs - the (semi) black-box way. In: Dodis, Y., Nielsen, J.B. (eds.) TCC 2015, Part I. LNCS, vol. 9014, pp. 345–374. Springer, Heidelberg (2015)

[OV08] Ong, S.J., Vadhan, S.P.: An equivalence between zero knowledge and commitments. In: Canetti, R. (ed.) TCC 2008. LNCS, vol. 4948, pp. 482–500. Springer, Heidelberg (2008)

[PSSW09] Pinkas, B., Schneider, T., Smart, N.P., Williams, S.C.: Secure two-party computation is practical. In: Matsui, M. (ed.) ASIACRYPT 2009. LNCS, vol. 5912, pp. 250–267. Springer, Heidelberg (2009)

[Yao86] Yao, A.C.C: How to generate and exchange secrets (extended abstract). In: FOCS, pp. 162–167 (1986)

Secure Protocol Transformations

Yuval Ishai[1,3], Eyal Kushilevitz[1,3], Manoj Prabhakaran[2(✉)], Amit Sahai[3], and Ching-Hua Yu[2]

[1] Technion, Haifa, Israel
{yuvali,eyalk}@cs.technion.il
[2] University of Illinois, Urbana-Champaign, USA
{mmp,cyu17}@cs.illinois.edu
[3] University of California, Los Angeles, USA
sahai@cs.ucla.edu

Abstract. In the rich literature of secure multi-party computation (MPC), several important results rely on "protocol transformations," whereby protocols from one model of MPC are transformed to protocols from another model. Motivated by the goal of simplifying and unifying results in the area of MPC, we formalize a general notion of black-box protocol transformations that captures previous transformations from the literature as special cases, and present several new transformations. We motivate our study of protocol transformations by presenting the following applications.

– Simplifying feasibility results:
  - Easily rederive a result in Goldreich's book (2004), on MPC with full security in the presence of an honest majority, from an earlier result in the book, on MPC that offers "security with abort."
  - Rederive the classical result of Rabin and Ben-Or (1989) by applying a transformation to the simpler protocols of Ben-Or et al. or Chaum et al. (1988).
– Efficiency improvements:
  - The first "constant-rate" MPC protocol for a constant number of parties that offers full information-theoretic security with an optimal threshold, improving over the protocol of Rabin and Ben-Or;
  - A fully secure MPC protocol with optimal threshold that improves over a previous protocol of Ben-Sasson et al. (2012) in the case of "deep and narrow" computations;
  - A fully secure MPC protocol with near-optimal threshold that improves over a previous protocol of Damgård et al. (2010) by improving the dependence on the security parameter from linear to polylogarithmic;
  - An efficient new transformation from passive-secure two-party computation in the OT-hybrid and OLE-hybrid model to zero-knowledge proofs, improving over a recent similar transformation of Hazay and Venkitasubramaniam (2016) for the case of static zero-knowledge, which is restricted to the OT-hybrid model and requires a large number of commitments.

Finally, we prove the *impossibility* of two simple types of black-box protocol transformations, including an unconditional variant of a previous negative result of Rosulek (2012) that relied on the existence of one-way functions.

© International Association for Cryptologic Research 2016
M. Robshaw and J. Katz (Eds.): CRYPTO 2016, Part II, LNCS 9815, pp. 430–458, 2016.
DOI: 10.1007/978-3-662-53008-5_15

1   Introduction

Secure multi-party computation (MPC) is one of the central topics around which modern cryptography has been shaped. Research in MPC has led to major innovations in cryptography, including effective definitional approaches (e.g., simulation-based security [16,17]), powerful and vastly applicable algorithmic techniques (starting with secret-sharing [29] and garbling schemes [31]), sharp impossibility results (e.g., [9]) and even several cryptographic concepts ahead of their time (like fully-homomorphic encryption [30]). Significantly, in recent years, some of these results have started moving from theory to practice, spurring significant further theoretical and engineering effort to optimize their performance and usability.

Over 35 years of active research, MPC has grown into a rich and complex topic, with many incomparable flavors and numerous protocols and techniques. Indeed, just cataloguing the state of the art results is a non-trivial research project in itself, as exemplified by the recent work of Perry et al. [26], which proposes classifying the existing protocols using 22 dimensions.

This diversity of models and questions forms a wide spectrum of possible tradeoffs between functionality, security, and efficiency, which partially explains the massive amount of research in the area. But this diversity also poses the risk of misdirected research efforts. For instance, if a new technique is introduced in order to obtain an efficiency improvement in one model, it is not clear a-priori to which other models the same technique may apply; and even when the same technique directly applies to other models, one typically needs to manually modify protocols and their analysis to ensure it.

While developing and maintaining a systematic database like the one in [26] is certainly helpful, we propose a complementary approach to taming the complex landscape of MPC protocols. Our approach is to relate the various flavors of MPC problems to each other by means of general *protocol transformations*. More concretely, our work studies the following high level question:

> *To what extent can results in one MPC model be "automatically" transformed to other models?*

This question is motivated by the following goals.

- *Simplicity.* The current proofs of the main feasibility results in the area of MPC are quite involved, and results for different models share few common ingredients. We would like to obtain a simpler and more modular *joint* derivation of different feasibility results from the 1980s [3,7,16,25,27,32], which were originally proved using very different techniques.
- *Efficiency.* Despite a lot of progress on the efficiency of MPC, there are still significant gaps between the efficiency of the best known protocols in different models. For instance, viewing the number of parties n as a constant, n-party protocols that offer full-security (with guaranteed output delivery) against $t < n/2$ malicious parties [10,27] are asymptotically less efficient compared to similar protocols with security against $t < n/3$ parties [3], or even to protocols that offer "security with abort" against $t < n$ malicious parties [22].

A classical example of a general protocol transformation is the well known "GMW compiler," [16], which transforms any MPC protocol that offers security against passive corruptions into one that offers security against active corruptions, with the help of zero-knowledge proofs. Considering that this transformation has been behind several subsequent feasibility results, one may legitimately consider that *the GMW transformation is as important as – if not more important than – the GMW protocol itself is, as an object of study.* More recent examples include the IKOS transformation using "MPC-in-the-head" [20] and the IPS transformation that combines player-virtualization with "watchlists" [22]. Common to all these techniques is the idea that they generically transform any set of protocols that are secure for some ("easier") flavors of MPC into a protocol that is secure for another ("harder") flavor.

While these previous results demonstrate the plausibility of general MPC protocol transformations in some interesting cases, they are still far from covering the space of all desirable transformations between different MPC models and leave open several natural questions.

In this work, we initiate a systematic study of such MPC protocol transformations. We define a framework to formalize these transformations, and present a few positive and negative results. We are interested in obtaining conceptually simpler alternative proofs for known feasibility results by means of new transformations, as well as in obtaining new results. We now discuss the goals of this research in more detail.

The main theoretical motivation for studying protocol transformations is that they highlight the *essential new challenges* presented in a harder flavor of MPC compared to an easier flavor. For instance, the GMW-transformation distilled out verifying claims in zero-knowledge as the essential challenge in moving from semi-honest security to security against active corruption. As another example, in this work, we present a new transformation, that can recover the classical feasibility result of Rabin and Ben-Or [27] regarding security with guaranteed output delivery with an honest majority, from two simpler feasibility results (both of which were solved in [3,7]): (i) security against passive corruption with an honest majority and (ii) security with guaranteed output delivery but only with an arbitrarily large fraction of honest parties. We identify achieving an intermediate security notion – security with partially identifiable abort – as the key challenge in this transformation.

As noted above, another important motivation behind studying protocol transformations is the possibility of *efficiency improvements*. On the face of it, protocol transformations are not ideal for obtaining *efficient* protocols, as one can hope to obtain extra efficiency by engineering fine details of the protocols as applicable to the specific flavor of MPC. While that may indeed be true, a protocol transformation can leverage advances in one flavor of MPC to obtain efficiency improvements in another flavor. As it turns out, this lets us obtain several *new asymptotic efficiency results* based on a single new transformation. Considering that efficiency of MPC is a well-studied area, obtaining several new result at once illustrates the power of such transformations.

There are other practical and theoretical motivations that led to this work, which we mention below.

– From a pragmatic point of view, understanding the connections across flavors of MPC will help in *modular implementations* of protocols. Indeed, the implementation of a transformation from one flavor to another would tend to be significantly simpler than an entire protocol in the latter flavor, specified and implemented from scratch.

– Roles of important techniques can often be *encapsulated as transformations* among appropriate intermediate security notions (e.g., "player elimination" can be encapsulated as implementing a transformation from "identifiable-abort-security" to full-security). In the absence of such abstraction, these techniques remain enmeshed within more complex protocols, and may not benefit from research focus that a transformation can attract.

– More generally, transformations are important in *reducing duplicated research effort*. For instance, if a new technique is introduced in order to obtain an efficiency improvement in one model, it is not clear *a priori* to which other models the same technique may apply; and even when the same technique directly applies to other models, one typically needs to manually modify protocols and their analysis to ensure it. On the other hand, if generic transformations are available across models, techniques can be easily adapted across models.

– Finally, a theoretical framework is necessary to understand the *limitations of protocol transformations*, via formal impossibility theorems. Indeed, without a rigorous notion of "black-box" transformations, it is not clear how to rule out the possibility of a "transformation" which simply discards the protocol it is given and builds one from scratch. This is especially the case for unconditional security, where the standard notions of black-box use of computational assumptions are not helpful in differentiating a legitimate transformation from one which builds its own (unconditionally secure) protocol from scratch.

A Motivating Example. As an illustration of the use of protocol transformations in simplifying the landscape of MPC protocols, we consider two protocol schemes from Goldreich's book [15, Chapter 7]. The first one obtains (stand-alone) security-with-abort against arbitrary number of corruptions by an active, probabilistic polynomial time (PPT) adversary[1] (under standard cryptographic assumptions), for general function evaluation, in a model with broadcast channels only. The second one obtains full-security (i.e., guaranteed output delivery) in the same setting, but restricting the adversary to corrupt less than half the parties. Both these protocol schemes are obtained using the GMW transformation. However, *the latter feasibility result does not take advantage of the former*, but instead uses verifiable secret-sharing (VSS) and several other techniques to achieve full-security, while retaining certain elements from the previous construction.

We point out that in fact, one could avoid the duplicated effort by giving a protocol transformation from the former flavor to the latter flavor of MPC. For this, we abstract out a slightly stronger security guarantee provided by the first

---

[1] One may consider static or adaptive corruption here. By default, we shall consider adaptive adversaries in all constructions in this paper.

protocol: while it allows an adversary to abort the protocol after learning its own input, aborting always leads to identification of at least one party that is corrupted by the adversary. This notion of security is often referred to as security with identifiable-abort [21]. In Sect. 4.1, we show that one can easily transform such a protocol into a protocol with full-security.

Security Augmentation and Efficiency Leveraging. Typically, an MPC protocol transformation falls into one of two broad (informally defined) classes: *security augmentation* and *efficiency leveraging*. Security augmentation refers to building MPC protocols with strong security guarantees by transforming MPC protocols with weaker security guarantees. The IPS compiler [22] is an instance of security augmentation. Efficiency leveraging, on the other hand, aims to improve the efficiency of MPC protocols, without necessarily increasing their security guarantee. In such a transformation, the original (inefficient) protocol will typically be used on a "small" sub-computation task, in combination with other cheaper (but less secure) protocols applied to the original "large" computation task. The goal of the sub-computation task is usually to ensure that the strong attacks on the final protocol has the effect of weak attacks on an execution of the cheaper, less secure protocol. An instance of efficiency leveraging is given by Bracha's transformation [5], in which the strength of the security guarantee corresponds to the corruption-threshold (i.e., what fraction of parties are corrupted) that can be tolerated.

1.1   Our Contributions

Framework. Firstly, we formalize the notion of a Black-Box Transformation (BBT) from protocol schemes satisfying some security (or efficiency) requirements to a protocol scheme satisfying some other requirements.[2] Towards this, we formalize notions like protocol schemes (which map functionalities to protocols) and security definitions (which are just sets of pairs of functionalities and protocols), all in a fairly abstract fashion. A BBT itself is modeled using a circuit that describes a protocol's structure as a program built from various components.

The framework is general enough to cast all of the above mentioned transformation (GMW, Bracha, IKOS and IPS) as instances of BBT.

We remark that we treat security notions highly abstractly, and do not impose any conditions on how security is proven. However, in all our positive results and examples, security definitions use a simulation paradigm, and one could define a "fully" blackbox transformation by requiring that the simulator of the protcol resulting from the transformation be constructed in a black-box manner from the simulators of the given protocols. For the sake of simplicity, and to keep the focus on the structure of the constructions rather than on the proofs of security,

---

[2] The term "Black-Box" refers to the fact that (the next-message function of) the resulting protocol uses (the next-message function of) all the constituent protocols and the functionality itself as oracles; however, note that the constituent protocols themselves may depend on their functionalities in a non-black-box manner.

we do not formally include this restriction in our definition of BBT. We also point out that this strengthens our impossibility results.

New Transformations and Consequences. We present a new transformation which can be used to obtain known and new results about (information-theoretically) secure MPC for general function evaluation, with guaranteed output delivery, given an honest-majority and a broadcast channel. Our transformation yields such an MPC scheme starting from two protocol schemes – one achieving full-security, but for a lower threshold (βn corruption threshold, for some $\beta > 0$) and one achieving semi-honest security under honest-majority (Corollary 1). (See the next section for an overview of the transformation, and the various intermediate transformations that lead to it.) From this transformation we obtain the following results:

1. We readily obtain the result of Rabin and Ben-Or [27] as a consequence of the earlier work of Ben-Or et al. and Chaum et al. [3,7], via the above transformation.
2. We obtain the first "constant-rate" MPC protocol scheme with guaranteed output delivery against corruption of less than $n/2$ parties, provided the number of parties is constant (Corollary 2). That is, the total communication in this protocol is at most $c_n|C|$, where C is the circuit representation of the function, and c_n is a constant independent of the security parameter and C but dependent only on the number of parties. This result is obtained – following the lead of [22][3] – by applying our transformation to the scheme of [12] (combined with a secret-sharing scheme due to [8]) and the semi-honest secure scheme of [3].
3. Next, we present an *efficiency leveraging* transformation, which is designed to improve the efficiency of a protocol scheme with full-security, by combining it with a (cheaper) protocol which achieves security-with-abort (Theorem 8). By applying this transformation to the above protocol with full-security and an efficient protocol with security-with-abort from [14], we obtain a "scalable" MPC protocol with full-security and optimal corruption-threshold – i.e., tolerating corruption of less than $n/2$ parties (Corollary 3).[4] For an arguably natural class of functions (namely, sequential computations, where the size of a circuit implementing the function is comparable to its depth), this is the first scalable protocol with full-security and optimal threshold (complementing a result of [4], which obtains similar efficiency for circuits which are of relatively low depth).
4. We present an efficient new transformation from two-party protocols in the OT-hybrid or OLE-hybrid model that offer security against passive corruptions to zero-knowledge proofs in the commitment-hybrid model, improving

---

[3] In [22], these two protocol schemes were combined to obtain a similar constant-rate protocol, but in the oblivious-transfer (OT) hybrid model and with security-with-abort.

[4] Here the term "scalable" denotes that for evaluating large circuits C, the *communication complexity per party* scales as $\tilde{O}(|C|)$ (up to polylog multiplicative factors and polynomial additive terms of the security parameter and the number of partiesh).

over a recent similar transformation of Hazay and Venkitasubramaniam [18] for the case of static zero-knowledge. (We note that the IKOS transformation for protocols in such hybrid models requires at least 3 parties.) The transformation from [18] cannot be applied in the OLE-hybrid model, and when applied to natural protocols in the OT-hybrid model such as the GMW protocol, it requires several separate commitments for each gate in the circuit. Our transformation for the OLE-hybrid model can be applied towards efficient zero-knowledge proofs for *arithmetic* circuits and in both hybrids our transformation requires just a constant number of commitments overall (for a constant soundness error). This transformation may have relevance to the recent line of work on practical zero-knowledge proofs initiated in [24]. In contrast to [18], we do not consider here the goal of adaptive zero-knowledge in the plain model.

5. Our final application considers the problem of relaxing the corruption threshold from the optimal $n/2$ to $n(1/2 - \epsilon)$, for any constant $\epsilon > 0$. In this case, we obtain a *highly scalable protocol* in which the *total* communication for evaluating a circuit C is $\tilde{O}(|C|)$, ignoring additive terms that depend on the number of parties, but not the size of the circuit (Corollary 4). This improves over a result of [13].[5]

   For this, we apply Bracha's transformation [5] to one of the above protocols. Specifically, we use Bracha's transformation to combine an outer protocol that has a relatively low corruption threshold but is highly scalable with respect to communication and computation (in our case the one from [13]), and an inner protocol with optimal threshold (in our case, the one from item 2 above), to obtain a protocol with a near-optimal threshold.

Impossibility Results. One may ask if security against active corruption can solely be based on security against semi-honest adversaries. Such questions can be formalized as questions about the existence of a BBT. We present two impossibility results:

1. We consider the question of functionally-black-box protocol schemes, introduced by Rosulek [28]. (This is a special case of protocol transformations where no protocol scheme is provided to the transformation.) Rosulek demonstrated a two-party functionality family for which there is no functionally black-box protocol, *assuming the existence of one-way functions*. We present an unconditional version of this result (Theorem 1).
2. We show a functionality family – namely, zero-knowledge proof functionalities – for which there is no BBT from semi-honest security to security (with abort) against active adversaries (Theorem 2).

We remark that the proof of our second result breaks down if we expanded the family of functionalities from ZK functionalities to all efficient functionalities.

---

[5] In [13], in the absence of broadcast channels, the near-optimal threshold of $n(\frac{1}{3} - \epsilon)$ was considered. We can extend our result to this setting by implementing broadcast channels among a constant number of parties, with a constant factor blow-up in communication.

We leave it as an important open problem to prove broader impossibility results for *general* computation (in which the family considered is the family of all functionalities).

1.2   Technical Overview

Black-Box Transformations. We make precise a notion of a black-box transformation among protocol schemes. Given a functionality f, a black-box transformation can define new functionalities (which are syntactically just programs) that access f in a black-box manner. Then, it can invoke a given protocol scheme on any such functionality, to obtain a protocol (which is, again, a program). The transformation can repeat these steps of defining new functionalities in terms of programs it already has, and of invoking given protocol schemes on such functionalities any number of times. At the end, it outputs one of the programs as its protocol.

We point out that the "protocol step" (invoking a protocol scheme on a functionality) is *not* limited to using the functionality as a black-box. However, it is a black-box step in the sense that the transformation can be instantiated with *any* protocol scheme with the requisite security guarantees.

Example: IPS Transformation. An example of a black-box transformation (that we shall build on later) is the IPS transformation [22]. We shall graphically represent a transformation using a circuit diagram like the one in Fig. 1.

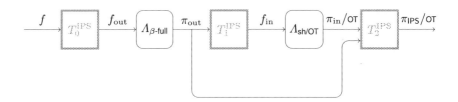

Fig. 1. Black-Box Transformation in the IPS compiler

Here, each rectangular node (labeled T_0^{IPS}, T_1^{IPS} and T_2^{IPS}) outputs a program which makes black-box access to one or more programs input to that node. T_0^{IPS} converts an n-party functionality f into a functionality f_{out} involving n "clients" and N "servers". T_1^{IPS} defines f_{in} to be an n-party functionality in which the trusted party carries out the program of a server in the protocol π_{out}. The bulk of the compiler is part of the transformation T_2^{IPS}, which combines the programs of two protocols π_{out} and π_{in} in a black-box way to define the final protocol.

The diagram also shows two other nodes, labeled $\Lambda_{\beta\text{-full}}$ and $\Lambda_{\mathrm{sh/OT}}$, each of which take as input a functionality (f_{out} and f_{in} resp.) and produces a protocol (π_{out} and π_{in} resp.). The labels on the nodes indicate the security guarantees required of these protocols (security against active corruption of strictly less than a $\beta > 0$ fraction of the parties, and security against semi-honest corruption, in the OT hybrid model resp.). [22] show that irrespective of what protocol schemes are used to define the protocols produced by these nodes, as long as those schemes meet the required security conditions, the resulting protocol will be a protocol for f with security against active corruption of any number of parties.

New Transformations. We present several new transformations, some of which are summarized in Table 1. In particular, we show how to transform a low-threshold fully-secure protocol scheme and a high/optimal-threshold semi-honest secure protocol scheme to a high/optimal-threshold protocol with full-security (presented as Corollary 1). The main step is to achieve a weaker notion of security (called "security with partially-identifiable-abort") against the same high fraction of corruption. Then, we show how a protocol with partially-identifiable-abort security can be transformed to one with full-security.

The second of these two transformations turns out to be easy, using "Error-Correcting Secret-Sharing" or ECSS (also known as robust secret-sharing) [6], which can be realized easily using ordinary Secret-Sharing and one-time message authentication codes (MAC) (see the full version). Partially-identifiable-abort-security allows us to perform, in case of an abort, a *player elimination* process, so that an honest majority is maintained. By carrying this out not on the original function, but on a function which accepts ECSS-shared inputs and produces ECSS-shared outputs, we show how to obtain full-security. The more challenging transformations is obtaining partially-identifiable-abort-security in the first place, as discussed below.

Obtaining Partially-Identifiable-Abort Security. This transformation is based on the IPS transformation [22] which, however, was not designed for the setting with an honest majority. Hence, it relied on an OT-hybrid model, and could obtain only "security with abort." We modify this transformation in a couple of ways to obtain partially-identifiable-abort security in the honest-majority setting, in the plain model (with a broadcast channel). There are two major modifications we introduce, summarized below.

Watchlist Channels in the Plain Model. An important aspect of the IPS transformation is a collection of "watchlist channels" used by each party to monitor secretly chosen instances of a semi-honest secure inner protocol. In the IPS transformation, Rabin OT is used to implement the watchlist channel. Instead, we rely on a weaker variant, $\widetilde{\mathsf{OT}}$, which we can directly implement in the honest-majority setting (without even broadcast channels), using Shamir's secret-sharing. $\widetilde{\mathsf{OT}}$ allows an adversary to selectively cause aborts when there is no erasure. The reason this suffices for building a watchlist channel is that this functionality will be applied to random inputs, and when an abort occurs, we can safely identify a pair of inconsistent parties – at least one of which is corrupt – by having all parties reveal their views in the protocol (over a broadcast channel).[6]

---

[6] When no abort occurs, the adversary can indeed learn some information (i.e., that an erasure occurred), but this can happen only in a small number of instances before an abort occurs.

Table 1. A summary of the main black-box transformations in this paper. The first column lists the type of the protocol scheme(s) given, and the second column lists the type of protocol scheme obtained. t stands for the number of parties that can be corrupted. id_α-security denotes partially-identifiable-abort security, in which, in the event of an abort, a set of parties, at least α fraction of which are corrupt, is identified by all honest parties. sh-security stands for security against semi-honest corruption, abort and full-security stand for security against active corruption, with the latter having guaranteed output delivery.

From	To	Theorem	Notes
id_α-security, $t < \alpha n$	full, $t < \alpha n$	Theorem 3, Theorem 4	Using player-elimination. Theorem 4 relies on a non-blackbox decomposition of the function, and yields efficiency close to the non-abort-case efficiency of the given protocol
(sh-security, $t < \alpha n$) and (full-security, $t < \beta n$)	id_α, $t < \alpha n$	Theorem 5, Theorem 6	An honest-majority version of the IPS transformation. Any $\beta > 0$ suffices. Theorem 6 saves a factor of n using an expander graph-based watchlist scheme
(sh-security, $t < \alpha n$) and (full-security, $t < \beta n$)	full, $t < \alpha n$	Corollary 1	Combining the above two
(abort-secure π_1, $t < \alpha n$) and (id_α-secure π_2, $t < \alpha n$)	id_α, $t < \alpha n$	Theorem 7	Efficiency Leveraging: resulting protocol almost as efficient as π_1 when there is no abort[a]
(abort-secure π_1, $t < \alpha n$) and (full-secure π_2, $t < \alpha n$)	full, $t < \alpha n$	Theorem 8	Efficiency Leveraging: resulting protocol is almost as efficient as π_1. From Theorem 7 and Theorem 4. Relies on a non-blackbox decomposition of the function

[a]Note that a naïve protocol which runs π_1 first and in the event of an abort, runs π_2 for the same functionality does not work. If π_1 aborting is considered as an abort event, then it gives the same efficiency guarantee, but is not an id_α-secure scheme, because if π_2 completes without an abort, the protocol fails to identify an α-corrupt set. If π_1 aborting is not considered an abort event, the protocol fails to meet the efficiency guarantee.

Obtaining Partially-Identifiable Abort Instead of Abort. In the original IPS transformation, even if the outer protocol has security with guaranteed output delivery, the final protocol offers only security with abort (without any identification of the corrupt parties). This is due to the fact that when a party detects an inconsistency, it simply aborts the protocol. In the setting with honest majority, we show how to modify the IPS transformation, so as to obtain partially-identifiable abort, such that a set of two parties can be identified of which at least one is guaranteed to be corrupt.

Consider when P_i detects an inconsistency in the messages reported over a watchlist channel that it has access to, in an inner protocol session. In this case, P_i cannot exactly identify the source of inconsistency, but only localize it to a pair of parties P_{i_1}, P_{i_2}, one of which is corrupt. However, since P_i itself could be a corrupt party, at this point the honest parties can agree on one of (P_i, P_{i_1}, P_{i_2}) being corrupt. But being able to identify a set in which only $1/3$ fraction is guaranteed to be corrupt falls below our required guarantee of 1 out of 2 being corrupt.

To further localize corruption, we require all the parties to broadcast their views in the inner-protocol session in which an inconsistency was detected, as they had earlier communicated over the watchlist channel to P_i. If an inconsistency is detected among the broadcast views, then all parties can identify a pair (P_{i_1}, P_{i_2}) which are inconsistent with each other. On the other hand, if all the views that are broadcast are consistent with each other, then, if P_i had indeed observed an inconsistency earlier, it can point out one party P_{i_1} which reported a view over the watchlist channel different from the one it reported over the broadcast channel. Then P_i is required to broadcast this party's identity, and all parties agree on the pair (P_i, P_{i_1}).

To see that this transformation retains security, note that by causing an abort, the adversary can cause at most one server's computation to be revealed over the broadcast channel. This corresponds to the adversary corrupting one extra server in the outer protocol. Since the choice of parameters in the IPS compiler leaves a comfortable margin for the number of server corruptions, this does not affect the overall security.

Efficiency Improvements. When considering a non-constant number of parties, there are a couple of major sources of inefficiency in the transformation above, which we can address.

Firstly, in the transformation from partially-identifiable-abort security to full security, the protocol could be restarted $\Theta(n)$ times. To avoid this overhead, we require the function to be given in the form of a composition of $\Theta(n)$ functions (for instance, a layered circuit with $\Theta(n)$ layers), each one of approximately the same size complexity. Then, one can restrict the duplicated effort for each restart to correspond to a single component, and can ensure that overall $O(n)$ restarts can only about double the cost.

Secondly, in the IPS compiler, every party can potentially watch every inner protocol session. This requires that all the communication in each inner-protocol session is sent out (encrypted with one-time pads) to all the n parties. To avoid this overhead, we can use an expander graph to define which parties may watch the execution of which servers. Specifically, we can use an expander graph between the set of parties and the set of servers in the outer protocol, in which *the degree of each server is a constant*, but any subset of $n/2$ parties has in its neighborhood (i.e., will potentially watch) almost all of the servers. Thus, the communication in each inner-protocol session (corresponding to the servers in the outer protocol) is sent out to only a constant number of parties.

Efficiency Leveraging: Transformations for Improving Efficiency. We present a new instance of efficiency leveraging, in which an MPC protocol scheme with full-security is "extended" by leveraging the efficiency of cheaper MPC protocols which only offer security with abort. Specifically, we show how to combine a protocol which guarantees only security with abort given an honest majority (e.g., from [14]) and a protocol with full-security given honest majority (like the one we constructed above) to obtain one which approaches the efficiency of the former protocol while enjoying full-security like the latter.

The basic idea is simple. We can obtain a protocol with $1/2$-identifiable-abort security as follows: given a functionality, we will run a protocol with security-with-abort to compute it; if the protocol terminates without aborting (as confirmed with the help of broadcast messages), then our protocol terminates successfully. If it aborts, then we run an (inefficient) MPC protocol with full-security for a functionality which accepts the views in the first protocol and detects a pair of parties with conflicting views, at least one of which is corrupt (if no conflict is detected, then a party who aborted in the first place can be identified as a corrupt party, since, as part of the security guarantees, we shall require zero probability for abort if all parties run honestly). To make this idea work, we need to ensure that the inefficient MPC is called only on a small piece of computation. With appropriate parameters for decomposition of the function, this indeed gives new asymptotic results (for relatively "narrow" circuits).

Negative Results. We prove two negative results. Firstly, we show that there is a function family $\mathcal{F}$ such that there is no "functionally blackbox" protocol scheme [28] for $\mathcal{F}$ (even for semi-honest security). The family $\mathcal{F}$ consists of boolean functions of the form f_α, where $\alpha \in \{0,1\}^k$ and $f_\alpha(x,y) = 1$ if and only if $x \oplus y = \alpha$.

Our second negative result shows a function family $\mathcal{G}$ such that semi-honest secure protocol schemes for $\mathcal{G}$ cannot be converted in a blackbox manner to protocols with active security (with abort). We choose $\mathcal{G}$ to be the family of zero-knowledge proofs for a class of relations. Then, there is a semi-honest secure protocol for $\mathcal{G}$ which only accesses the given functionality $f \in \mathcal{G}$ in a blackbox manner. Hence, a blackbox transformation from semi-honest secure protocol schemes to schemes with active security translates to a functionally blackbox protocol scheme for $\mathcal{G}$ with active security.

To complete the proof, we show how to define $\mathcal{G}$ (assuming the existence of a pseudorandom function) such that there is no active secure, functionally blackbox protocol scheme for $\mathcal{G}$.

1.3   Organization of the Paper

The rest of the paper is organized as follows (with some of the details deferred to the full version). Section 2 includes several basic definitions of the framework, and Sect. 3 defines the notion of a blackbox transformation. In Sect. 4, we give some simple transformations, including a new transformation that improves on a recent result by [18]. Section 5 presents two impossibility results regarding black-box transformations. Section 6 through Sect. 8 present several transformations, which are summarized in Table 1. Section 9 presents the results we obtain by applying these transformations to protocol schemes in the literature.

2   Preliminaries

The basic objects in our framework are *protocols*. Technically, a protocol is specified by a single program (say, Turing Machine) for the "next-message function" of all the parties in the protocol (formally defined in the full version). We shall write Π to denote the set of all protocols.

A *functionality* is technically just a special instance of a protocol, involving a trusted party. We often abuse our notation and refer to the trusted party's program as the functionality. We shall often refer to a *functionality family $\mathcal{F}$*, which is simply a set of functionalities, i.e., $\mathcal{F} \subseteq \Pi$. We denote the family of all probabilistic polynomial time computable secure function evaluation functionalities by $\mathcal{F}^*$ (represented by circuits).

We use a *synchronous model* of communication (with rushing adversaries), so that all parties in a protocol proceed in a round-by-round fashion. Note that this is applicable to ideal functionalities too. However, typically we are not interested in the exact number of rounds in the ideal functionality, as long as it finishes within a polynomial number of rounds.

2.1   Security Definitions

Technically, a *security definition* for a functionality family $\mathcal{F}$ is formalized as a relation $\Lambda \subseteq \mathcal{F} \times \Pi$. The intention is that $(f, \pi) \in \Lambda$ iff π is a secure protocol for f. For a security notion named secure, the corresponding relation will typically be written as Λ_{secure}.

In Table 2 we name some of the main security definitions considered in our results. For instance, $\Lambda^{\mathcal{F}}_{\alpha\text{-full/BC}}$ includes all pairs (f, π) such that f is a functionality in the family $\mathcal{F}$, and π is a UC-secure protocol with guaranteed output delivery (within a polynomial number of rounds), against computationally unbounded adversaries who may adaptively corrupt strictly less than α fraction of the parties, and BC means that the protocol uses a broadcast channel. In all our security notions, for simplicity of our transformations, we require that an honest party aborts the protocol only if there is no possible honest execution of the protocol that is consistent with its view. We also define a security notion generalizing the notion of security with identifiable abort:

Table 2. Terminology used for guarantees from protocols.

$\Lambda_{\text{secure}}^{\mathcal{F}}$	(f,π) s.t. $f \in \mathcal{F}$ and π meets the definition secure (for a polynomial-round version of f). If $\mathcal{F} = \mathcal{F}^*$, the family of all probabilistic polynomial time function evaluation functionalities, we simply write Λ_{secure}		
α-secure	secure, restricted to corruption of strictly less than α fraction of the parties	secure/F	protocol is in the F-hybrid model. e.g., secure/BC denotes protocols using broadcast channels
sa	standalone security (default is UC security)	ppt	adversary is PPT (default is unbounded adversary)
sh	semi-honest adversary	full	active adversary (with guaranteed output delivery)
abort	adversary may learn its output and then decide which honest parties get their outputs and which do not	id_θ	same as abort, but on abort, honest parties agree on a non-empty set of parties, at least a θ fraction of which is corrupt. We shall abbreviate α-id_α as α-id

Security with θ-Identifiable Abort. Given a functionality f, we define a functionality $f^{\langle \text{id}_\theta \rangle}$ to formalize the notion of security with θ-identifiable abort. As defined in the full version, we require the functionalities to be in a normal form, involving a computation phase and an output delivery phase. ——————— $f^{\langle \text{id}_\theta \rangle}$ internally runs f and interacts with Adv as follows.

1. Accept the inputs from all parties (including honest parties and parties corrupted by Adv) and forward to f. (If there is no input from P_i, substitute it with a dummy input.) Set the output vector as set by f.
2. If Adv sends getoutput, then send the corrupted parties' outputs to Adv.
3. If Adv sends (corrupt, T) s.t. T is a subset of parties in which at least a θ fraction are corrupt, then change the output of all honest parties to be (corrupt, T).
4. **Output phase:** Deliver the (current) output to all parties.

2.2   Protocol Schemes

A *protocol scheme* maps a functionality to a protocol (with a desired security property).

Definition 1 (Λ-scheme). $\mathcal{P} : \mathcal{F} \to \Pi$ *is said to be a Λ-scheme if $\mathcal{F}$ is a functionality family such that $\Lambda \subseteq \mathcal{F}^* \times \Pi$, and for every $f \in \mathcal{F}$, $(f, \mathcal{P}(f)) \in \Lambda$.*

For example, the semi-honest BGW-protocol scheme is a $\Lambda_{\alpha\text{-sh}}^{\mathcal{F}}$-scheme where $\mathcal{F}$ is the family of all circuit-evaluation functionalities and $\alpha = \frac{1}{2}$. Typical protocol schemes are *uniform*, in that there is a Turing Machine which, on input a standardized description of f, for $f \in \mathcal{F}$, outputs the code of $\mathcal{P}(f)$.

Complexity Notation. To discuss asymptotic efficiency guarantees of protocol schemes, we augment the notation for security definitions to include protocols' communication (and sometimes, computational) cost. Typically, a protocol's complexity is measured as a function of some complexity measure of the functionality f that it is realizing, as well as the number of parties n and the security parameter k of the protocol execution. For each functionality family, we shall require a cost measure size : $\mathcal{F} \to \mathbb{Z}^+$, that maps $f \in \mathcal{F}$ to a positive integer. We stress that a functionality f denotes a specific implementation (of a trusted party in a protocol), and so there can be different $f \in \mathcal{F}$ which are all functionally equivalent, but with differing values of size(f).

To capture the typical efficiency guarantees in the literature, we define a p-$\Lambda_{\text{secure}}^{\mathcal{F}}$ scheme as a $\Lambda_{\text{secure}}^{\mathcal{F}}$ scheme $\mathcal{P}$ such that for any $f \in \mathcal{F}$, $\mathcal{P}(f)$ is a protocol whose communication cost (for n parties, and security parameter k) is

$$O(p(n, k) \cdot \text{size}(f) + \text{poly}(n, k)). \tag{1}$$

For typical functionality families $\mathcal{F}$, a functionality $f \in \mathcal{F}$ is represented as a circuit C_f, and size(f) is the size of C_f. The function $p(n, k)$ reflects the multiplicative overhead of secure computation, on top of the size of the (insecure) computation.

Often, protocol schemes which offer a smaller value for $p(n, k)$ incur additive costs. To denote protocol schemes with such complexities, we use a more detailed notation: $(p, q, r; \mathsf{D})$-$\Lambda_{\text{secure}}^{\mathcal{F}}$ schemes are $\Lambda_{\text{secure}}^{\mathcal{F}}$ schemes $\mathcal{P}$ such that for all $f \in \mathcal{F}$, the communication cost of $\mathcal{P}(f)$ is $O(p(n, k) \cdot \text{size}(f) + \text{poly}(n, k) \cdot \mathsf{D}(f))$, its *computation cost* is $O(q(n, k) \cdot \text{size}(f) + \text{poly}(n, k) \cdot \mathsf{D}(f))$, and its *randomness cost* is $O(r(n, k) \cdot \text{size}(f) + \text{poly}(n, k) \cdot \mathsf{D}(f))$. Here D is a secondary cost measure – typically the depth of the circuit C_f – which is often much smaller than size(f). We omit D to indicate that $\mathsf{D}(f)$ is a constant and omit q and/or r to leave them as unspecified poly(n, k) functions. We omit $\mathcal{F}$ if it equals $\mathcal{F}^*$, the family of all probabilistic polynomial time function evaluation functionalities.

For functionality families using circuit representation, a traditional choice for D is depth: depth(f) denotes the depth of the circuit C_f representing f. We shall find it useful to define another function width, defined as follows. For any topological sorting of the gates in the circuit, define a sorted-cut as a partition of the gates into two sets so that all the gates in one part appear before any gate in the other part, in the topologically sorted order; the max-sorted-cut for a sort order is the maximum number of wires crossing a sorted-cut. width(f) is the value of the max-sorted-cut of C_f minimized over all topological sorts of C_f. (Alternately, we could require the topological sort to be part of the circuit specification. In this case, an appropriate model of computation would be a *linear bijection straight-line program* [2], and width would correspond to the number of "registers" in the program.)

For protocol schemes providing partially-identifiable security, like α-id-schemes, we sometimes want to distinguish the cost of an execution without an abort event and that with an abort event (and identification): a $\langle \gamma, \delta \rangle$-$\Lambda_{\alpha\text{-id}}$ scheme denotes a $\Lambda_{\alpha\text{-id}}$ scheme $\mathcal{P}$ such that the communication cost of $\mathcal{P}(f)$ is

$O(\gamma(n,k) \cdot \mathsf{size}(f) + \mathrm{poly}(n,k))$ without abort events and $O(\delta(n,k) \cdot \mathsf{size}(f) + \mathrm{poly}(n,k))$ with abort.

Finally, we write $(p,q,r;\mathsf{D}) \sim \Lambda_{\mathsf{secure}}^{\mathcal{F}}$ instead of $(p,q,r;\mathsf{D}) \text{-} \Lambda_{\mathsf{secure}}^{\mathcal{F}}$ and so on, if we intend to use $\widetilde{O}(\cdot)$ instead of $O(\cdot)$ in the above costs.[7] The notation is summarized in Table 3.

Table 3. Additional notation for protocol schemes (for n parties, and security parameter k).

$(p,q,r;\mathsf{D}) \text{-} \Lambda_{\mathsf{secure}}$	$\Lambda_{\mathsf{secure}}$ scheme $\mathcal{P}$ s.t. the communication cost of $\mathcal{P}(f)$ is $O(p(n,k) \cdot \mathsf{size}(f) + \mathrm{poly}(n,k) \cdot \mathsf{D}(f))$, the computation cost is $O(q(n,k) \cdot \mathsf{size}(f) + \mathrm{poly}(n,k) \cdot \mathsf{D}(f))$ and randomness cost is $O(r(n,k) \cdot \mathsf{size}(f) + \mathrm{poly}(n,k) \cdot \mathsf{D}(f))$
$(p,q;D) \text{-} \Lambda_{\mathsf{secure}}$	$(p,q,r;\mathsf{D}) \text{-} \Lambda_{\mathsf{secure}}$, where $r(n,k)$ is $\mathrm{poly}(n,k)$
$(p,q) \text{-} \Lambda_{\mathsf{secure}}$	$(p,q;\mathsf{D}) \text{-} \Lambda_{\mathsf{secure}}$, where $D(f)$ is a constant
$(p;D) \text{-} \Lambda_{\mathsf{secure}}$	$(p,q;\mathsf{D}) \text{-} \Lambda_{\mathsf{secure}}$, where $q(f)$ is $\mathrm{poly}(n,k)$
$p \text{-} \Lambda_{\mathsf{secure}}$	$(p,q;\mathsf{D}) \text{-} \Lambda_{\mathsf{secure}}$, where $D(f)$ is a constant and $q(f)$ is $\mathrm{poly}(n,k)$
$\langle \gamma, \delta \rangle \text{-} \Lambda_{\alpha\text{-id}}$	$\Lambda_{\mathsf{secure}}$ scheme $\mathcal{P}$ s.t. the communication cost of $\mathcal{P}(f)$ is $O(\gamma(n,k) \cdot \mathsf{size}(f) + \mathrm{poly}(n,k))$ without abort events and $O(\delta(n,k) \cdot \mathsf{size}(f) + \mathrm{poly}(n,k))$ with abort
$(\mathrm{params}) \sim \Lambda_{\mathsf{secure}}$	Similar to $(\mathrm{params}) \text{-} \Lambda_{\mathsf{secure}}$ scheme, but with $\widetilde{O}(\cdot)$ instead of $O(\cdot)$

2.3   Error-Correcting Secret-Sharing

Some of our transformations rely on a simple variant of secret-sharing that has been referred to as robust secret-sharing or as honest-dealer VSS [6,11,27]. To clarify the nature of this primitive, we shall call it *Error-Correcting Secret-Sharing (ECSS)*, and define it formally below.

Definition 2 (Error-Correcting Secret Sharing). *A pair of algorithms* (share, reconstruct) *is said to be an* (n,t)-*Error-Correcting Secret Sharing (ECSS) scheme over a message space* $\mathcal{M}$ *if the following hold:*

1. **Secrecy:** *For all* $s \in \mathcal{M}$ *and* $N_c \subseteq [n], |N_c| < t$, *the distribution of* $\{\sigma_i\}_{i \in N_c}$ *is independent of* s, *where* $(\sigma_1, ..., \sigma_n) \leftarrow \mathsf{share}(s)$.
2. **Reconstruction from upto** t **Erroneous Shares:** *For all* $s \in \mathcal{M}$, *and all* $(\sigma_1, ..., \sigma_n)$ *and* $(\sigma'_1, ..., \sigma'_n)$ *such that* $\Pr[(\sigma_1, ..., \sigma_n) \leftarrow \mathsf{share}(s)] > 0$ *and* $|\{i \mid \sigma'_i = \sigma_i\}| \geq n - t$, *it holds that* $\mathsf{reconstruct}(\sigma'_1, ..., \sigma'_n) = s$.

---

[7] $\widetilde{O}(h)$ denotes $O(h \cdot \mathrm{polylog} h)$.

3   Defining Black-Box Transformations

In this section, we present our framework of black-box transformations, which operates on protocol schemes (Definition 1). More specifically, a black-box transformation defines a Λ-scheme in terms of Λ'-schemes, for one or more other security notions Λ'. We present our definition in two parts – first the syntax of a transformation, followed by its security requirements.

Definition 3 (Black-Box Transformation (BBT): Syntax). *A BBT for a functionality family $\mathcal{F}$ is defined as a circuit C with*

- *a single input wire taking a functionality $f \in \mathcal{F}$,*
- *a single output wire outputting a protocol $\pi \in \Pi$,*
- *one or more black-box nodes labeled with oracle TMs $T_1, \cdots, T_s$,*
- *one or more protocol nodes labeled with relations $\Lambda_1, \cdots, \Lambda_t$ where $\Lambda_i \subseteq \mathcal{F}_i \times \Pi$ for some functionality family $\mathcal{F}_i$.*

For a black-box node labeled with T_i we require that the number of oracles accessed by T_i is equal to the number of input wires to that node. For a protocol node, we require that there is only one input wire.

Given such a circuit C and protocol schemes $\mathcal{P}_1, \cdots, \mathcal{P}_t$ such that each $\mathcal{P}_i$ is a Λ_i-scheme, we define $C^{\mathcal{P}_1,\dots,\mathcal{P}_t}(f) \in \Pi$ as follows. We shall set the value on each wire in C to be a protocol in Π (possibly a functionality), starting with the input wire and ending with the output wire, which is taken as the value $C^{\mathcal{P}_1,\dots,\mathcal{P}_t}(f)$. First, set the value on the input wire to be f. Then, for any black-box node with all its input wires' values already set to values $\pi_1, \cdots, \pi_d$, set its output wire's value to $T_i^{\pi_1,\dots,\pi_d}$, where T_i is the label on the node. For any protocol node with its input wire's value set to π, set its output wire's value to $\mathcal{P}_i(\pi)$, where i is the index of the protocol node in C (if $\mathcal{P}_i(\pi)$ is undefined, then $C^{\mathcal{P}_1,\dots,\mathcal{P}_t}(f)$ is undefined).

Definition 4 (Black-Box Transformation (BBT)). *We say that a BBT C, for a functionality family $\mathcal{F}$, is a BBT from $\{\Lambda_1, \cdots, \Lambda_t\}$ to Λ, if C has t protocol nodes labeled with $(\Lambda_1, \cdots, \Lambda_t)$ and, for all $f \in \mathcal{F}$ and all $(\mathcal{P}_1, \cdots, \mathcal{P}_t)$ such that each $\mathcal{P}_i$ is a Λ_i-scheme, we have $(f, C^{\mathcal{P}_1,\dots,\mathcal{P}_t}(f)) \in \Lambda$.*

4   Examples of Black-Box Transformations

In the full version, we illustrate how several important constructions from the literature are in fact BBTs from simpler security notions or simpler function families, to more demanding ones. This list includes Bracha's compiler [5] (from high-threshold (and low-efficiency) security and low-threshold (and high-efficiency) security to a high-threshold (and high-efficiency) security), the IKOS compiler [20] (from semi-honest secure MPC and and honest-majority secure MPC to active security for Zero-Knowledge proofs) and the IPS compiler [22] (as above, but for arbitrary MPC). The GMW compiler [16] could also be viewed as a BBT (from semi-honest security and active security specialized to zero-knowledge functionality, to active security).

It is helpful to visualize these transformations using "circuit diagrams." An example of the IPS transformation was given in Fig. 1. Similar diagrams for the other examples mentioned above are given in the full version.

Below we discuss two new simple BBTs, which yield much simpler alternatives to more complex constructions in the literature.

Improving Over [18]. Very recently, Hazay and Venkitasubramaniam [18], presented an IKOS-like transformation that starts from any (semi-honest) two-party protocol *in the OT-hybrid model* and gives a zero-knowledge proof system in the commitment-hybrid model. We present a different transformation that has several advantages over [18]: our transformation may start with a two-party protocol in the OLE-hybrid model,[8] whereas the one from [18] seems inherently restricted to the OT-hybrid model. Perhaps more importantly, to achieve a constant level of soundness our transformation uses only a constant number of commitments (to long strings), compared to the protocol in [18] that uses as many commitments as the number of OT calls. For the simplest case of the GMW protocol applied to a boolean circuit of size s, our protocol requires only 6 commitments whose total length is $O(|C|)$ whereas the protocol from [18] requires $O(|C|)$ separate bit-commitments. These features of our transformation make it appealing for the design of practical ZK protocols based on OT-hybrid and OLE-hybrid protocols such as GMW.

Our transformation, as well as the IKOS transformation on which it is based, are presented in the full version. At a high-level, we give a simple BBT from a 2-party semi-honest MPC protocol scheme in the OLE-hybrid model to a 3-party 1-private MPC protocol scheme in the plain model; this transformation is then readily composed with the IKOS transformation (which can be applied to a 1-private protocol) to obtain our full transformation.

4.1   A Pedagogical Application

One of the results from Goldreich's textbook [15] can be simplified using a BBT. In [15], two separate protocols for $\Lambda_{\text{abort-ppt-sa-id}}$ (i.e., security-with-identifiable-abort) and $\Lambda_{1/2-\text{full-ppt-sa}}$ (i.e., security with guaranteed output delivery, with an honest majority) are presented, with the latter relying on VSS. Below, we give a BBT from $\Lambda_{\text{abort-ppt-sa-id}}$ to $\Lambda_{1/2-\text{full-ppt-sa}}$, that uses ECSS (see Sect. 2.3) instead of VSS.

To evaluate an n-party function f, each party shares its input using an $\lceil n/2 \rceil$-out-of-n error-correcting secret-sharing (ECSS) scheme (see Sect. 2.3), and sends the resulting shares to the n parties. We remark that an ECSS is much simpler

---

[8] OLE stands for Oblivious Linear function Evaluation. It is a generalization of Oblivious Transfer where a sender has (a, b) in a field $\mathbb{F}$ and the receiver has $x \in \mathbb{F}$. At the end of the protocol, the receiver will learn $ax + b$ while the sender learns nothing. OLE-based protocols are useful for arithmetic computation. Such protocols are obtained in [23] by generalizing the OT-based GMW protocol [16].

than, say, a VSS protocol, and can be constructed readily by adding message authentication code (MAC) tags to the shares of any threshold secret sharing scheme (such as Shamir's scheme). Then, the parties use a protocol π from the protocol scheme with security-with-identifiable-abort to evaluate a function f', which takes shares as its inputs, reconstructs them to get inputs for f, evaluates f and reshares the outputs among all parties, again using ECSS. If the shares given as inputs have fewer than $n/2$ errors, f' can error-correct and recover the original input being shared; otherwise it defines the reconstructed value to be a default value (this corresponds to the shares not being generated correctly in the first place). If the protocol π for f' does not abort, then all the parties are expected to redistribute the shares they received from π, so that each party gets all the shares of its output; due to the error-correcting property, and since the adversary can corrupt less than $n/2$ of the shares received by each honest party, every honest party will be able to correctly recover its output. On the other hand, if the protocol π aborts, due to the identifiable-abort security guarantee, all honest parties will agree on the identity of one corrupt party. Note that at this point, even though the adversary may learn its outputs from π (i.e., outputs of f'), these carry no information and can be efficiently simulated (by a simulator running the protocol with arbitrary inputs for the honest parties). Hence, the parties can simply eliminate the identified party (and still retain honest majority), and restart the entire protocol on a smaller functionality in which the eliminated party's input is replaced by a default value. This process must eventually terminate, after at most $\lceil n/2 \rceil$ attempts, guaranteeing output for all honest parties.

An ad-hoc use of the above "player elimination" technique was made in several previous MPC protocols (see, e.g., [19] and references therein). In contrast, our use of this technique yields a *completely general transformation* from a weaker flavor of MPC to a stronger one.

5   Impossibility of Black-Box Transformations

In this section, we present some impossibility results for BBT. Before proceeding, we emphasize that in the definition of BBT, we *do not* require the security proofs to be black-box in any form. In particular, the simulators used to define security can arbitrarily depend on the functionality in a non-black-box manner. As such, the impossibility results on BBT are of a rather strong nature.

Our first impossibility results relates to an interesting special case of a BBT, namely, BBT from $\emptyset$ to Λ. This corresponds to the notion of a *functionally-black-box* protocol introduced by Rosulek [28], wherein there is an oracle TM such that for all $f \in \mathcal{F}$, T^f is a secure protocol (according to Λ) for f. Rosulek demontrated a two-party functionality family for which there is no functionally black-box protocol, *assuming the existence of one-way functions.* We present an unconditional version of this result.

Theorem 1. *There exists a two-party functionality family $\mathcal{F}$ such that there is no BBT from $\emptyset$ to $\Lambda_{\mathsf{sh}}^{\mathcal{F}}$. In particular, there is no BBT from $\emptyset$ to $\Lambda_{\mathsf{sh}}^{\mathcal{F}^*}$.*

The detailed proof is given in the full version. Here we sketch the main ideas of the proof.

Proof Sketch. The family $\mathcal{F}$ we shall use to prove the theorem consists of boolean functions of the form f_α, $\alpha \in \{0,1\}^k$, where $f_\alpha(x,y) = 1$ if and only if $x \oplus y = \alpha$. To show that there can be no secure protocol for f_α, in which the two parties access the function only in a blackbox manner, we consider the following experiment. Pick x, y, α uniformly and independently at random, and run the protocol for f_α with inputs x, y. Then we argue that the probability for both of the following events should be negligible:

(A) Either party queries their oracle with (p,q) such that $p \oplus q = \alpha$.
(B) Either party queries their oracle with (p,q) such that $p \oplus q = x \oplus y$.

The probability of event A is negligible since α is chosen uniformly at random, and the parties make only a polynomial number of queries. The reason for the probability of event B being negligible is the security of the protocol: in an ideal world, since $x \oplus y \neq \alpha$, a corrupt party (simulator), even given α, can learn only a negligible amount of information about the other party's input. Now, we consider a "coupled" experiment in which instead of α, we pick $\alpha^* = x \oplus y$, and run the same protocol but now for f_{α^*}. It can be argued that for the random tapes in the protocol for which events (A) and (B) does not occur in the first case, they will not occur in the second run too. Thus with high probability, both the executions produce the same output, violating the correctness of the protocol. $\qquad\square$

Also, we consider the question of showing impossibility of BBT from semi-honest security to active security. We present such a result conditioned on the existence of one-way functions.

Theorem 2. *Assuming the existence of one-way functions, there exists a two-party functionality family $\mathcal{G}$ such that there is no BBT from $\{\Lambda_{\mathsf{sh}}^{\mathcal{G}}\}$ to $\Lambda_{\mathsf{abort}}^{\mathcal{G}}$.*

We present the intuition behind the proof below, and defer the detailed proof to the full version.

Proof Sketch. We will let $\mathcal{G}$ to be the family of zero-knowledge proofs for a class of relations. Then, there is a semi-honest secure protcol for $\mathcal{G}$ which only access the given functionality $f \in \mathcal{G}$ in a blackbox manner. Hence, a black-box transform from semi-honest secure protocol schemes to schemes with active security translates to a functionally blackbox protocol scheme for $\mathcal{G}$ with active security. To show that this does not exist, we assume the existence of a pseudorandom function F and define $\mathcal{G}$ as follows. The relations associated with $\mathcal{G}$ are $R_s = \{(x,w) \mid F_s(w) = x\}$, where F_s denotes F with seed s.

To show that there can be no ZK protocol for this relation in which the parties only have blackbox access to an oracle for the relation R_s (but the simulator may depend on s), we consider a cheating prover as follows. When given (x, w) and access to R_s, it uses a wrapper around R_s to turn it into relation which accepts (x, w) (and does not accept (x', w) for $x' \neq x$), but otherwise behaves like R_s. Then the cheating prover runs the honest prover with access to the modified oracle. Using the ZK property we can argue that an honest verifier, when given a random x, cannot detect the difference between interacting with the real prover and the cheating prover. Thus, if the protocol is complete, the cheating prover will be able to break soundness.   □

6   A BBT from Partially-Identifiable-Abort to Full Security

We present a simple black-box transformation from *partially-identifiable abort security* (formalized using $\Lambda_{\alpha\text{-id}}$ below) to full security. This will be an important ingredient in our applications in Sect. 9. First, we present a simple but general version of this transformation (which suffices for feasibility results); in Theorem 4, we shall present a more efficient variant.

Theorem 3. *For any $0 \leq \alpha \leq 1/2$, there exists a BBT from $\Lambda_{\alpha-\text{id}/\text{BC}}$ to $\Lambda_{\alpha\text{-full}/\text{BC}}$. Specifically, there is a BBT from $p\text{-}\Lambda_{\alpha-\text{id}/\text{BC}}$ to $(np; \mathsf{D})\text{-}\Lambda_{\alpha\text{-full}/\text{BC}}$, where $\mathsf{D}(f)$ is the input plus output size of f.*

Our tools behind this construction are relatively simple. In particular, we do not use verifiable secret-sharing (VSS), but instead use the much simpler primitive Error-Correcting Secret-Sharing (ECSS) (see Sect. 2.3), which can be realized easily using ordinary Secret-Sharing and one-time message authentication codes (MAC).

Here we give a high level overview of the construction, with a complete description defered to the full version. The idea behind this BBT is that if we have a protocol which either completes the computation or identifies a set of parties such that at least α fraction of which are corrupt, then, in the event of an abort, we can remove the identified set of parties from active computation and restart the computation. Note that this preserves the corruption threshold of α (i.e., strictly less than α fraction remains corrupt) among the set of "active" parties.

For this idea to work, we need to keep the outputs secret-shared (so that by aborting, the adversary does not learn any useful information, even though it receives its outputs from the computation), and after the computation finishes, guarantee reconstruction. Further, we need to use secret-sharing to let all the parties deliver their inputs to the set of active parties. All this will be achieved using ECSS in a straightforward manner, for $\alpha \leq 1/2$.

A More Efficient Variant. In the above BBT, we restarted the entire computation in the event of an abort. To avoid this, we rely on having access to a "layered representation" of the function. Formally, consider a parametrized functionality $\hat{f}$, parametrized by an index $i \in \{1, \cdots, d\}$, such that $f = \hat{f}[d] \circ ... \circ \hat{f}[1]$, such that $\mathsf{size}(\hat{f}[i]) = O(\mathsf{size}(f)/d)$, for all i. We define $\mathsf{width}_d(f)$ to be the smallest number w such that there exists a decomposition of f into d layers, each of size $O(\mathsf{size}(f)/d)$, such that the number of output wires from any layer is at most w. We shall typically take d to be a polynomial $d(n,k)$. Note that $\mathsf{width}(f)$ defined in Sect. 2.2 is an upper-bound on $\mathsf{width}_d(f)$ for all d.

Since decomposing f into $\hat{f}$ is not a black-box operation, we require a "protocol scheme" that carries out this decomposition. For this we define a $\Lambda_{\mathsf{layer}[d]}$ scheme to be one which maps f to a parametrized function $\hat{f}$ such that

$$f = \hat{f}[d] \circ \cdots \circ \hat{f}[1],$$

and $\forall i \in [d]$, $\mathsf{size}(\hat{f}[i]) = O(\mathsf{size}(f)/d)$ and the number of bits output by $\hat{f}[i] \leq \mathsf{width}_d(f)$.

Then, as shown in the full version, we obtain the following efficiency improvement over Theorem 3.

Theorem 4. *For any* $0 < \alpha \leq 1/2$, *there exists a BBT from* $\{\Lambda_{\mathsf{layer}[d]}, \langle \gamma, \delta \rangle \text{-} \Lambda_{\alpha\text{-id}}\}$ *to* $(\gamma; \mathsf{D}) \text{-} \Lambda_{\alpha\text{-full}}$, *where* $d(n,k) = n \cdot \frac{\delta(n,k)}{\gamma(n,k)}$ *and* $\mathsf{D}(f) = \mathsf{width}_d(f)$.

7 A BBT from $\{\Lambda_{\alpha\text{-sh}}, \Lambda_{\beta\text{-full}}\}$ to $\Lambda_{\alpha\text{-id}}$

Our goal in this section is to obtain a BBT that increases the corruption threshold of a fully secure protocol, by combining it with a semi-honest protocol which has the higher threshold. Given Theorem 3, it suffices to obtain a protocol with partially-identifiable-abort against the higher corruption threshold. Formally, we shall prove the following theorem, which is interesting when $\beta < \alpha$.

Theorem 5. *For any* $0 < \alpha, \beta \leq 1/2$, *there exists a BBT from* $\{\Lambda_{\alpha\text{-sh}}, \Lambda_{\beta\text{-full}}\}$ *to* $\Lambda_{\alpha-\mathsf{id}/\mathsf{BC}}$.

This BBT (detailed in the full version) resembles the IPS compiler, but achieves $1/2$-identification in case of abort, and also avoids the use of OT in watchlists. For this, it replaces T_2^{IPS} in IPS (see Fig. 1) with a black-box transformation T_2. Figure 2 compares T_2^{IPS} and T_2. T_2^{IPS} consists of a "core" compiler $\mathrm{IPS}_{\mathrm{core}}$, which produces a protocol in a "watchlist-channel hybrid" model (also using OT if it is needed by the inner protocol). Separately, a watchlist-channel functionality $\mathcal{W}$ was realized using a protocol w_{IPS} in the OT-hybrid model. Finally, the former was composed with the latter to obtain a protocol in the OT-hybrid model.

In T_2, firstly the OT used in the watchlist protocol is replaced with a functionality $\widetilde{\mathsf{OT}}$, which is then implemented by a protocol $\pi_{\widetilde{\mathsf{OT}}}$ in the honest-majority setting; further this watchlist protocol is modified in a simple manner to achieve $1/2$-identification. The functionality of the resulting protocol is captured by $\mathcal{W}^*$. Next, the protocol generated by $\mathrm{IPS}_{\mathrm{core}}$ is modified to facilitate $1/2$-identification (even if given the watchlist functionality $\mathcal{W}^*$ instead of $\mathcal{W}$), following the outline sketched in Sect. 1.2 (see paragraph *Obtaining Partially-Identifiable-Abort Security*). The final protocol is obtained by composing this protocol with the watchlist protocol for $\mathcal{W}^*$.

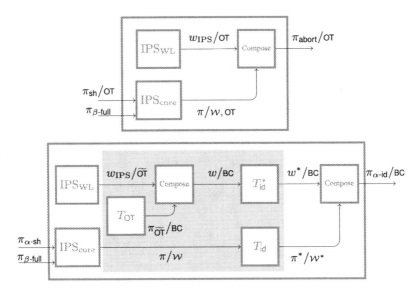

Fig. 2. T_2^{IPS} and T_2. The shaded region shows the new components in T_2. Note that T_2 retains $\mathrm{IPS}_{\mathrm{core}}$ and $\mathrm{IPS}_{\mathrm{WL}}$ from T_2^{IPS} as it is.

7.1   Using a Sparse Watchlist

The BBT in Theorem 5 is in fact a BBT from $\{(p_{\mathrm{in}}, q_{\mathrm{in}}, r_{\mathrm{in}}) \text{-} \Lambda_{\alpha\text{-sh}}, (p_{\mathrm{out}}, q_{\mathrm{out}}) \text{-} \Lambda_{\beta\text{-full}}\}$ to $p \text{-} \Lambda_{\alpha\text{-id/BC}}$, where $p = n^2 \cdot (p_{\mathrm{in}} + r_{\mathrm{in}}) \cdot (q_{\mathrm{out}} + n \cdot p_{\mathrm{out}})$. But by exploiting the honest majority guarantee which was absent in the setting of [22], we can state the following version.

Theorem 6. *For any $0 < \alpha, \beta \le 1/2$, and polynomials $p_{\mathrm{in}}, q_{\mathrm{in}}, r_{\mathrm{in}}, p_{\mathrm{out}}, q_{\mathrm{out}}$, there exists a BBT from $\{(p_{\mathrm{in}}, q_{\mathrm{in}}, r_{\mathrm{in}}) \text{-} \Lambda_{\alpha\text{-sh}}, (p_{\mathrm{out}}, q_{\mathrm{out}}) \text{-} \Lambda_{\beta\text{-full}}\}$ to $p \text{-} \Lambda_{\alpha\text{-id/BC}}$, where $p = n \cdot (p_{\mathrm{in}} + r_{\mathrm{in}}) \cdot (q_{\mathrm{out}} + n \cdot p_{\mathrm{out}})$.*

The above result saves a factor of n compared to the previous transformation. The efficiency improvement comes from a sparser watchlist mechanism (using an expander graph to define which parties may watch the execution of which servers) in the BBT from $(\Lambda_{\beta\text{-full}}, \Lambda_{\alpha\text{-sh}})$ to $\Lambda_{\alpha-\mathrm{id/BC}}$. We present the details in the full version.

8   Efficiency Leveraging

Bracha's transformation is a classical example of efficiency leveraging. It was originally proposed in the context of byzantine agreement [5], and later applied to MPC protocols (see, e.g., [13]). Below, we record a version of this result that is sufficient for our applications.

Proposition 1 (Bracha's Transformation [5]). *Let $0 < \epsilon, \beta \le \alpha \le 1/2$, and let $p'(n, k) = c_n$ be independent of k. Then, for each* secure $\in \{\mathsf{sh}, \mathsf{abort}, \mathsf{full}\}$ *and* *any function* D, *there exists a BBT from* $\{(p,q;\mathsf{D}) \text{-} \Lambda^{\mathcal{F}}_{\beta-\text{secure}}, \ p' \text{-} \Lambda_{\alpha-\text{secure}}\}$ *to* $(p'';\mathsf{D}) \text{-} \Lambda^{\mathcal{F}}_{(\alpha-\epsilon)-\text{secure}}$, *where* $p''(n,k) = p(n,k) + q(n,k)$.

In this section, we present a new instance of efficiency leveraging for full-security: a simple BBT from $\{\Lambda_{\alpha\text{-abort}}, \Lambda_{\alpha\text{-full}}\}$ to $\Lambda_{\alpha\text{-full}}$, in which the resulting protocol's efficiency is comparable to that of the protocol in $\Lambda_{\alpha\text{-abort}}$.

First we present a efficiency leveraging transformation for $\Lambda_{\alpha\text{-id}}$ which can then be combined with Theorem 4 to obtain efficiency leveraging for $\Lambda_{\alpha\text{-full}}$. In our efficiency leveraging transformation for $\Lambda_{\alpha\text{-id}}$ the efficiency of the resulting protocol, *when there is no abort event*, is comparable to that of a cheaper $\Lambda_{\alpha\text{-abort}}$ protocol. Formally, we have the following theorem.

Theorem 7. *For any* $0 \leq \alpha \leq 1/2$, *and functions* $p, q, p' \in \text{poly}(n,k)$, *there exists a BBT from* $\{(p,q) \text{-} \Lambda_{\alpha\text{-abort}}, p' \text{-} \Lambda_{\alpha\text{-id}}\}$ *to* $\langle \gamma, \delta \rangle \text{-} \Lambda_{\alpha\text{-id}}$, *where* $\gamma = p$ *and* $\delta = p' \cdot (p + q)$.

The protocol scheme claimed in Theorem 7 is shown in Fig. 3. The first node is a protocol node of $p \text{-} \Lambda_{\alpha\text{-abort}}$, which converts a functionality f into a protocol π_{abort}.

The second node is a black-box node T_1, which converts the protocol π_{abort} to an (n-party) functionality f^*, in which the trusted party takes the view of each party in an execution of π_{abort} as the input, carries out the execution of π_{abort}, and identifies a set of two parties which have inconsistent views, if it exists.[9] When there is none, it outputs $\emptyset$. The third node $\Lambda_{\alpha\text{-id}}$ compiles f^* into a protocol π_{id}.

Finally, a black-box node T_2 combines π_{abort} and π_{id} together and transforms them into a protocol π, which works as follows: initially the parties execute π_{abort} on the given input, and on finishing this execution successfully, each party broadcasts "done." If all parties broadcast "done," then each party outputs the output from the execution of π_{abort} and terminates. If not, they execute π_{id} with their views in the execution of π_{abort} as input. If this latter execution itself aborts, π_{id} identifies a set of parties S at least an α fraction of which is corrupt (where $\alpha \leq 1/2$). otherwise (i.e., if π_{id} finishes without an abort event), then all parties agree on the output of f^*, namely a set S of two parties at least one of which is corrupt, or the emptyset $\emptyset$; if the output is $\emptyset$, the parties set S to be

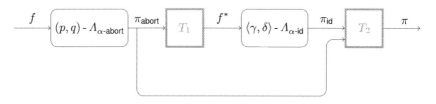

Fig. 3. Black-Box Transformation from $\{(p,q) \text{-} \Lambda_{\alpha\text{-abort}}, p' \text{-} \Lambda_{\alpha\text{-id}}\}$ to $\langle \gamma, \delta \rangle \text{-} \Lambda_{\alpha\text{-id}}$, where $\gamma = p$ and $\delta = p' \cdot (p + q)$.

---

[9] Recall that the view of a party involves its initial input, the randomness, and all the received messages.

the singleton set consisting of the lexicographically smallest party who did not broadcast "done" after the execution of π_{abort}. In all cases, if π_{abort} resulted in an abort, the honest parties agree on a set of parties S of which at least an α fraction is corrupt.

We verify that the complexity of π is as claimed in the theorem. When there is no abort event, the communication cost is essentially the same as that of π_{abort}, namely $p(n, k)$; otherwise, there is an additional the cost from π_{id}, which is $\widetilde{O}(p(n, k) + p'(n, k) \cdot \text{size}(f^*))$, where $\text{size}(f^*) = \widetilde{O}((p(n, k) + q(n, k)) \cdot \text{size}(f))$. Hence the whole scheme is in $\langle \gamma, \delta \rangle$ - $\Lambda_{\alpha\text{-id}}$ with $\gamma = p$ and $\delta' = p' \cdot (p + q)$.

Combining Theorem 7 with Theorem 4 we get the following result. Here we state it as efficiency leveraging for full-security; however, the result holds as a BBT from $\{\Lambda_{\text{layer}[d]}, (p, q)\text{-}\Lambda_{\alpha\text{-abort}}, p'\text{-}\Lambda_{\alpha\text{-id}}\}$ as well.

Theorem 8. *For all $0 \leq \alpha \leq 1/2$, and for all functions $p, q, p' \in \text{poly}(n, k)$, there exists a BBT from $\{\Lambda_{\text{layer}[d]}, (p, q)\text{-}\Lambda_{\alpha\text{-abort}}, p'\text{-}\Lambda_{\alpha\text{-full}}\}$ to $(p; \mathsf{D})\text{-}\Lambda_{\alpha\text{-full}}$, where $d = \frac{n \cdot p' \cdot (p+q)}{p}$ and $\mathsf{D}(f) = \text{width}_d(f)$.*

9   Applications

In Sect. 4.1, we already saw a pedagogical application of BBT, in simplifying the exposition of security with guaranteed output delivery (with computationally bounded adversaries). In this section, we give several interesting examples regarding how to use the BBTs in the previous sections for deriving both feasibility and efficiency results.

○ **Rabin and Ben-Or without Honest-Majority VSS.** As our first example, we reproduce the classic feasibility result of Rabin and Ben-Or [27] for fully secure MPC for corruption against $t < n/2$ parties. The core new tool developed in this paper (and used in subsequent results in this regime of corruption) was Verifiable Secret-Sharing (VSS) that is secure against corruption of $t < n/2$ parties. Interestingly, our construction by-passes the need for an explicit VSS protocol for this corruption regime, instead showing that one can directly use fully secure MPC from prior work [3,7]. Our construction is based on the following direct corollary of Theorems 3 and 5.

Corollary 1. *For any $0 < \alpha, \beta \leq 1/2$, there exists a BBT from $\{\Lambda_{\alpha\text{-sh}}, \Lambda_{\beta\text{-full}}\}$ to $\Lambda_{\alpha\text{-full}/\text{BC}}$.*

To obtain the result of [27] we simply apply Corollary 1 to the protocols in [3,7].

○ **Constant-Rate MPC with Full-Security for Small Number of Parties.** Our first quantitative result is a "constant-rate" honest-majority MPC protocol with guaranteed output delivery, *when the number of parties involved is constant.* That is, as the size of the function grows, the communication complexity of the protocol grows linearly at a rate that is independent of the security parameter. For MPC of large circuits, against the optimal corruption threshold $n/2$, this gives an amortized complexity of $O(1)$ per gate, compared to $O(k)$ per gate in the previously best result from [4].

Corollary 2. *There exists a* p-$\Lambda_{1/2\text{-full}/\mathsf{BC}}$*-scheme, where* $p(n,k) = c_n$ *is independent of* k.

This result is obtained as a corollary of Theorems 3 and 6[10] First we obtain a p-$\Lambda_{1/2-\mathsf{id}/\mathsf{BC}}$ scheme by applying the BBT from Theorem 6 to the $\Lambda_{1/2\text{-sh}}$-scheme from [3] and the constant rate $\Lambda_{\beta\text{-full}}$-scheme (for some $\beta > 0$) that is obtained by instantiating the protocol scheme from [12] using the constant-rate ramp scheme of [8]. (The same "outer protocol" was used in [22] to obtain a constant-rate $\Lambda_{\mathsf{abort}/\mathsf{OT}}$-scheme.) Then by further applying the BBT from Theorem 3, we obtain the p-$\Lambda_{1/2\text{-full}/\mathsf{BC}}$ protocol as claimed.

○ **Scalable MPC with Full-Security, Optimal Threshold.** Our next result is a "scalable" honest-majority MPC protocol with guaranteed output delivery. We define the function class $\mathcal{F}_{\mathsf{arith}}$ of functions represented as arithmetic circuits over a field $\mathbb{F}$ such that $\log|\mathbb{F}| > k$. For $f \in \mathcal{F}_{\mathsf{arith}}$, $\mathsf{size}(f)$ refers to $\log|\mathbb{F}| \cdot |C_f|$, where $|C_f|$ is the number of gates in the circuit C_f representing f. Equivalently, $\mathsf{size}(f)$ measures the number of binary wires in the circuit C_f; similarly $\mathsf{width}(f)$ measures the width of C_f in bits.

Corollary 3. *There exists a* $(p; \mathsf{D})$-$\Lambda^{\mathcal{F}_{\mathsf{arith}}}_{\frac{1}{2}\text{-full}/\mathsf{BC}}$*-scheme, where* $p(n,k) = n$ *and* $\mathsf{D} = \mathsf{width}(f)$.

That is, for MPC of large arithmetic circuits over a large field, with security against the optimal corruption threshold $n/2$, we get an amortized communication cost of $O(n)$ bits per binary wire in the circuit. This result is obtained as a corollary of Theorems 4 and 7, by applying the BBTs to the $\Lambda^{\mathcal{F}_{\mathsf{arith}}}_{1/2\text{-abort}}$-scheme from [14] and the p-$\Lambda_{1/2\text{-id}}$-scheme from Corollary 2. Note that we have used $\mathsf{width}(f)$ as an upper-bound on $\mathsf{width}_d(f)$ over all d.

Our result complements a similar result of Ben-Sasson et al. [4] in which the secondary complexity measure is depth, instead of width. We remark that a natural regime for scalable MPC involves long sequential computations (carried out by a small or moderate number of parties), so that a circuit for the computation would be deep and narrow. In such a regime, the above result, which yields a cost of $O(n \cdot \mathsf{size}(f) + \mathsf{poly}(n,k))$, compares favorably to the protocols of [4] which yield a cost of $\widetilde{\Omega}(n \cdot \mathsf{size}(f) + n^2 \cdot \mathsf{depth}(f) + \mathsf{poly}(n,k))$.

○ **Highly Scalable MPC with Full-Security, Near Optimal Threshold.** Our final application considers the problem of relaxing the corruption threshold from the optimal $\alpha = 1/2$ to $\alpha = 1/2 - \epsilon$, for any constant ϵ.

Corollary 4. *For every* $\epsilon > 0$, *there exists a* $(p_\epsilon; \mathsf{D})$-$\Lambda_{(\frac{1}{2}-\epsilon)\text{-full}/\mathsf{BC}}$*-scheme, where* $p_\epsilon(n,k) = c_\epsilon$ *is independent of* n *and* k *and* $\mathsf{D}(f) = \mathsf{depth}(f)$.

---

[10] The construction leading to Theorem 5 also suffices here. We point to Theorem 6 only because it makes the parameters explicit; the optimization in Sect. 7.1 is not important for this result.

This generalizes a result in [13], which obtained a similar result (without using a broadcast channel) for the threshold $\frac{1}{3} - \epsilon$. We obtain this result by applying Proposition 1 (Bracha's efficiency leveraging transformation) to our c_n - $\Lambda_{\frac{1}{2}\text{-full}/\text{BC}}$ scheme from Corollary 2 and the $(c_1, c_2; \mathsf{depth})$ - $\Lambda_{\beta\text{-full}}$ scheme from [13] (for, say, $\beta = 1/6$ and c_1, c_2 being constants), with $\alpha = 1/2$.

Acknowledgments. This research was done in part while the authors were visiting the Simons Institute for the Theory of Computing, supported by the Simons Foundation and by the DIMACS/Simons Collaboration in Cryptography through NSF grant #CNS-15-23467. The individual authors were supported during this work by the following grants: ISF grant 1709/14, BSF grant 2012378, ERC starting grant 259426, a DARPA/ARL SAFEWARE award, NSF Frontier Award 14-13955, NSF grants 12-28856, 12-28984, 11-36174, 11-18096, and 10-65276, a Xerox Faculty Research Award, a Google Faculty Research Award, an equipment grant from Intel, and an Okawa Foundation Research Grant. This material is in part based upon work supported by the Defense Advanced Research Projects Agency through the ARL under Contract W911NF-15-C-0205. The views expressed are those of the authors and do not reflect the official policy or position of the Department of Defense, the National Science Foundation, or the U.S. Government.

References

1. Proceedings of the 20th STOC. ACM (1988)
2. Ben-Or, M., Cleve, R.: Computing algebraic formulas using a constant number of registers. SIAM J. Comput. **21**(1), 54–58 (1992)
3. Ben-Or, M., Goldwasser, S., Wigderson, A.: Completeness theorems for non-cryptographic fault-tolerant distributed computation. In: Proceedings of the 20th STOC, pp. 1–10. ACM (1988)
4. Ben-Sasson, E., Fehr, S., Ostrovsky, R.: Near-linear unconditionally-secure multiparty computation with a dishonest minority. In: Safavi-Naini, R., Canetti, R. (eds.) CRYPTO 2012. LNCS, vol. 7417, pp. 663–680. Springer, Heidelberg (2012)
5. Bracha, G.: An o(log n) expected rounds randomized byzantine generals protocol. J. ACM **34**(4), 910–920 (1987)
6. Cevallos, A., Fehr, S., Ostrovsky, R., Rabani, Y.: Unconditionally-secure robust secret sharing with compact shares. In: Pointcheval, D., Johansson, T. (eds.) EUROCRYPT 2012. LNCS, vol. 7237, pp. 195–208. Springer, Heidelberg (2012)
7. Chaum, D., Crépeau, C., Damgård, I.: Multiparty unconditionally secure protocols. In: Proceedings of 20th STOC, pp. 11–19. ACM (1988)
8. Chen, H., Cramer, R.: Algebraic geometric secret sharing schemes and secure multiparty computations over small fields. In: Dwork, C. (ed.) CRYPTO 2006. LNCS, vol. 4117, pp. 521–536. Springer, Heidelberg (2006)
9. Cleve, R.: Limits on the security of coin flips when half the processors are faulty (extended abstract). In: STOC, pp. 364–369. ACM (1986)
10. Cramer, R., Damgård, I.B., Dziembowski, S., Hirt, M., Rabin, T.: Efficient multiparty computations secure against an adaptive adversary. In: Stern, J. (ed.) EUROCRYPT 1999. LNCS, vol. 1592, pp. 311–326. Springer, Heidelberg (1999)
11. Cramer, R., Damgård, I.B., Fehr, S.: On the cost of reconstructing a secret, or VSS with optimal reconstruction phase. In: Kilian, J. (ed.) CRYPTO 2001. LNCS, vol. 2139, pp. 503–523. Springer, Heidelberg (2001)

12. Damgård, I.B., Ishai, Y.: Scalable secure multiparty computation. In: Dwork, C. (ed.) CRYPTO 2006. LNCS, vol. 4117, pp. 501–520. Springer, Heidelberg (2006)

13. Damgård, I., Ishai, Y., Krøigaard, M.: Perfectly secure multiparty computation and the computational overhead of cryptography. In: Gilbert, H. (ed.) EUROCRYPT 2010. LNCS, vol. 6110, pp. 445–465. Springer, Heidelberg (2010)

14. Genkin, D., Ishai, Y., Prabhakaran, M.M., Sahai, A., Tromer, E.: Circuits resilient to additive attacks with applications to secure multiparty computation. In: The Proceedings of the 46th Annual Symposium on the Theory of Computing (STOC) (2014)

15. Goldreich, O.: Foundations of Cryptography: Basic Applications, vol. 2. Cambridge University Press, New York (2004). ISBN:0521830842

16. Goldreich, O., Micali, S., Wigderson, A.: How to play ANY mental game. In: ACM, ed. Proceedings of 19th STOC, pp. 218–229. ACM (1987). See [14, Chap. 7] for more details

17. Goldwasser, S., Micali, S., Rackoff, C.: The knowledge complexity of interactive proof-systems. In: Proceedings of 17th STOC, pp. 291–304. ACM (1985)

18. Hazay, C., Venkitasubramaniam, M.: On the power of secure two-party computation. Cryptology ePrint Archive, Report 2016/074. http://eprint.iacr.org/2016/074. (2016 to appear in Proceedings of Crypto 2016)

19. Hirt, M., Nielsen, J.B.: Upper bounds on the communication complexity of optimally resilient cryptographic multiparty computation. In: Roy, B. (ed.) ASIACRYPT 2005. LNCS, vol. 3788, pp. 79–99. Springer, Heidelberg (2005)

20. Ishai, Y., Kushilevitz, E., Ostrovsky, R., Sahai, A.: Zero-knowledge from secure multiparty computation. In: STOC, pp. 21–30. ACM (2007)

21. Ishai, Y., Ostrovsky, R., Zikas, V.: Secure multi-party computation with identifiable abort. In: Garay, J.A., Gennaro, R. (eds.) CRYPTO 2014, Part II. LNCS, vol. 8617, pp. 369–386. Springer, Heidelberg (2014)

22. Ishai, Y., Prabhakaran, M., Sahai, A.: Founding cryptography on oblivious transfer – efficiently. In: Wagner, D. (ed.) CRYPTO 2008. LNCS, vol. 5157, pp. 572–591. Springer, Heidelberg (2008)

23. Ishai, Y., Prabhakaran, M., Sahai, A.: Secure arithmetic computation with no honest majority. In: Reingold, O. (ed.) TCC 2009. LNCS, vol. 5444, pp. 294–314. Springer, Heidelberg (2009)

24. Jawurek, M., Kerschbaum, F., Orlandi, C.: Zero-knowledge using garbled circuits: how to prove non-algebraic statements efficiently. In: 2013 ACM SIGSAC Conference on Computer and Communications Security, CCS 2013, Berlin, Germany, 4–8 November 2013, pp. 955–966 (2013)

25. Kilian, J.: Founding cryptography on oblivious transfer. In: STOC, pp. 20–31. ACM (1988)

26. Perry, J., Gupta, D., Feigenbaum, J., Wright, R.N.: Systematizing secure computation for research and decision support. In: Abdalla, M., De Prisco, R. (eds.) SCN 2014. LNCS, vol. 8642, pp. 380–397. Springer, Heidelberg (2014)

27. Rabin, T., Ben-Or, M.: Verifiable secret sharing and multiparty protocols with honest majority. In: Proceedings of 21st STOC, pp. 73–85. ACM (1989)

28. Rosulek, M.: Must you know the code of f to securely compute f? In: Safavi-Naini, R., Canetti, R. (eds.) CRYPTO 2012. LNCS, vol. 7417, pp. 87–104. Springer, Heidelberg (2012)

29. Shamir, A.: How to share a secret. Commun. ACM **22**(11), 612–613 (1979)

30. Shamir, A., Rivest, R.L., Adleman, L.M.: Mental poker. Technical report LCS/TR-125, Massachusetts Institute of Technology, April 1979

31. Yao, A.C.: Protocols for secure computation. In: Proceedings of 23rd FOCS, pp. 160–164. IEEE (1982)
32. Yao, A.C.: How to generate and exchange secrets. In: Proceedings of 27th FOCS, pp. 162–167. IEEE (1986)

On the Communication Required
for Unconditionally Secure Multiplication

Ivan Damgård$^{(\boxtimes)}$, Jesper Buus Nielsen, Antigoni Polychroniadou,
and Michael Raskin

Department of Computer Science, Aarhus University, Aarhus, Denmark
{ivan,jbn,antigoni,raskin}@cs.au.dk

Abstract. Many information-theoretic secure protocols are known for general secure multi-party computation, in the honest majority setting, and in the dishonest majority setting with preprocessing. All known protocols that are efficient in the circuit size of the evaluated function follow the same "gate-by-gate" design pattern: we work through an arithmetic (boolean) circuit on secret-shared inputs, such that after we process a gate, the output of the gate is represented as a random secret sharing among the players. This approach usually allows non-interactive processing of addition gates but requires communication for every multiplication gate. Thus, while information-theoretic secure protocols are very efficient in terms of computational work, they (seem to) require more communication and more rounds than computationally secure protocols. Whether this is inherent is an open and probably very hard problem. However, in this work we show that it is indeed inherent for protocols that follow the "gate-by-gate" design pattern. We present the following results:

- In the honest majority setting, as well as for dishonest majority with preprocessing, any gate-by-gate protocol must communicate $\Omega(n)$ bits for every multiplication gate, where n is the number of players.
- In the honest majority setting, we show that one cannot obtain a bound that also grows with the field size. Moreover, for a constant number of players, amortizing over several multiplication gates does not allow us to save on the computational work, and – in a restricted setting – we show that this also holds for communication.

All our lower bounds are met up to a constant factor by known protocols that follow the typical gate-by-gate paradigm. Our results imply that a fundamentally new approach must be found in order to improve the communication complexity of known protocols, such as BGW, GMW, SPDZ etc.

The authors of this work were supported by the Danish National Research Foundation and the National Science Foundation of China (under the grant 61061130540) for the Sino-Danish Center for the Theory of Interactive Computation, within which part of this work was performed; by the CFEM research center (supported by the Danish Strategic Research Council). Ivan Damgård was also supported by the Advanced ERC grant MPCPRO. Jesper Buus Nielsen is supported by European Research Council Starting Grant 279447. This work was done in [part] while Antigoni Polychroniadou was visiting the Simons Institute for the Theory of Computing, supported by the Simons Foundation and by the DIMACS/Simons Collaboration in Cryptography through NSF grant CNS-1523467.

M. Robshaw and J. Katz (Eds.): CRYPTO 2016, Part II, LNCS 9815, pp. 459–488, 2016.
DOI: 10.1007/978-3-662-53008-5_16

1   Introduction

Secure Multi-Party Computation (MPC) allows n players to compute an agreed function on privately held inputs, such that the desired result is correctly computed and is the only new information released. This should hold, even if t out of n players have been actively or passively corrupted by an adversary.

If point-to-point secure channels between players are assumed, any function can be computed with unconditional (perfect) security, against a passive adversary if $n \geq 2t+1$ and against an active adversary if $n \geq 3t+1$ [BOGW88, CCD88]. If we assume a broadcast channel and accept a small error probability, $n \geq 2t+1$ is sufficient to get active security [RBO89].

The protocols behind these results require a number of communication rounds that is proportional to the depth of an (arithmetic) circuit computing the function. Moreover, the communication complexity is proportional to the size of the circuit. Whether we can have constant round protocols and/or communication complexity much smaller than the size of the circuit and still be efficient (polynomial-time) in the circuit size of the function is a long-standing open problem. Note that this is indeed possible if one makes computational assumptions. Note also that if we give up on being efficient in the circuit size, then there are unconditionally secure and constant round protocols for any function [IK00] (which will, however, be very inefficient in general with respect to the computation). Moreover, there are works that apply to special classes of circuits (e.g., constant-depth circuits [BI05]) or protocols that require exponential amount of computation [BFKR90, NN01] and exponential storage complexity [IKM+13].

The above issues are not only of theoretical interest: the methods we typically use in information-theoretic secure protocols tend to be computationally much more efficient than the cryptographic machinery we need for computational security. So unconditionally secure protocols are very attractive from a practical point of view, except for the fact that they seem to require a lot of interaction.

The Gate-by-gate Design Pattern. The fact that existing information-theoretic secure protocols (which are efficient in the circuit size of the function) have large round and communication complexity is a natural consequence of the fact that all such protocols follow the same *typical* "gate-by-gate" design pattern: Initially all inputs are secret-shared among the players. Then, for each gate in the circuit, where both its inputs have been secret-shared, we execute a subprotocol that produces the output from the gate in a secret-shared form. The protocol maintains as an invariant that for all gates that have been processed so far, the secret-sharing of the output value is of the same form used for the inputs (so we can continue processing gates) and is appropriately randomised such that one could open this sharing while revealing only that output value. As a result, it is secure to reveal/open the final outputs from the circuit.

For all known constructions which are efficient in the circuit size of the function, it is the case that multiplication gates require communication to be processed (while addition/linear gates usually do not). The number of rounds is at least the (multiplicative) depth of the circuit, and the communication complexity is $\Omega(ns)$ for a circuit of size s (the size being measured as the number of multiplication gates) in the worst case for $t < n/3$ and $t < n/2$ see the results of [DN07, BTH08] and [BSFO12, GIP+14, GIP15], respectively. Note that protocols that tolerate a sub-optimal number of corrupted parties (e.g., $t < 0.49n$) and are based on packed secret-sharing techniques can reduce the amortised cost of multiplications if they can be parallelised [DIK+08, IPS09, DIK10, GIP15]. These techniques do not apply to all circuits, in particular not to "tall and skinny" circuits whose multiplicative depth is comparable to their size. In addition, they can at best save an $\mathcal{O}(n)$ factor in communication and computational work.

The situation is essentially the same for recent protocols that are designed for dishonest majority in the preprocessing model [DPSZ12, NNOB12] (except that amortization based on packed secret-sharing does not apply here due to the dishonest majority setting).

1.1   Contributions

In this paper, we ask a very natural question for unconditionally secure protocols which, to the best of our knowledge, has not been studied in detail before:

Is it really inherent that the typical *gate-by-gate approach to secure computation requires communication for each multiplication operation?*

Our Model. To avoid misunderstandings, let us be more precise about the model we assume: we consider synchronous protocols that are semi-honest and statistically secure against static corruption of at most t of the n players. We assume that point-to-point secure channels are available, and protocols are allowed to have dynamic communication patterns (in a certain sense we make precise later), i.e., it is not fixed a priori whether a protocol sends a message in a given time slot. Moreover, there is no bound on the computational complexity of protocols, in particular arbitrary secret sharing schemes are allowed. A *gate-by-gate* protocol is a protocol that evaluates an arithmetic circuit and for every multiplication gate, it calls a certain type of subprotocol we call a *Multiplication Gate Protocol* (MGP). We define MGPs precisely later, but they basically take as input random *shares* of two values a, b from a field and output random *shares* of $c = ab$. Neither the MGP nor the involved secret sharing schemes have to be the same for all gates. We do not even assume that the same secret sharing scheme is used for the inputs and outputs of an MGP, we only require that the *reconstruction threshold* for the output sharing is at most $2t$ for honest majority and at most n for dishonest majority.

An *ordered* gate-by-gate protocol must call the MGP's in an order corresponding to the order in which one would visit the gates when evaluating the circuit, whereas this is not required in general. Thus the gate-by-gate notion is somewhat more general than what one might intuitively expect and certainly includes much more than, say the standard BGW protocol – which, of course, makes our negative results stronger.

Note that if multiplications did not require communication, it would immediately follow (for semi-honest security) that we would have an unconditionally secure two-round protocol for computing any function. But as mentioned above this is not *a priori* impossible: it follows, for instance, from [IK00, IKM+13], that if less than a third of the players are corrupted, there is indeed such a two-round protocol (which, however, requires super polynomial computational work in general).

Honest Majority Setting. For honest majority protocols it is relatively easy to show that multiplications do require communication: we argue in the paper that any MGP secure against t corruptions requires that at least $2t + 1$ players communicate. For protocols with dynamic communication pattern this bound holds in expectation. It turns out that a protocol beating this bound would imply an unconditionally secure two-party protocol computing a multiplication, which is well known to be impossible. This implies that the communication complexity of any gate-by-gate protocol for honest majority must be proportional to $n \cdot s$ where s is the circuit size and that the round complexity of an ordered gate-by-gate protocol must be at least proportional to the multiplicative depth of the circuit. This matches the best protocols we know for general Boolean circuits up to a constant factor. For arithmetic circuits over large fields one might wonder whether the communication must grow with the field size. However, this cannot be shown via a general bound on MGPs: we give an example secret sharing scheme allowing for an MGP with communication complexity independent of the field size.

A gate-by-gate protocol is not allowed to amortise over several multiplications that can be done in parallel. This is anyway not possible in general, for instance if we evaluate a "tall and skinny" circuit forcing us to do multiplications sequentially. But for more benign circuits, amortization is indeed an option. However, we show that in a restricted setting, MGPs doing k multiplication gates in parallel must have communication that grows linearly with k. We also show (in full generality) that amortization can save at most an $\mathcal{O}(n)$ factor in the computational work, matching what we can get from known techniques based on packed secret-sharing. This proof technique for this bound is quite interesting: We base it on a lower bound by Winkler and Wullschleger [WW10] on the amount of preprocessed data one needs for (statistically) secure two-party computation of certain functions. We find it somewhat surprising that an information theoretic bound on the size of data translates to a bound on local computation.

Dishonest Majority Setting with Preproccesing. The argument used for the honest majority case breaks down if we consider protocols in the preprocessing model (where correlated randomness is considered): here it is indeed possible to compute multiplications with unconditional security, even if $t = n - 1$ of the n players are corrupt. Nevertheless, we show similar results for this setting: here, any MGP secure against $t = n - 1$ corruptions must have all n players communicate. This implies that, also in this setting, any gate-by-gate protocol has communication complexity $\Omega(n \cdot s)$. Note that existing constructions [DPSZ12] meet the resulting bound for gate-by-gate protocols up to a constant factor.

To obtain the result, we exploit again the lower bound by Winkler and Wullschleger, but in a different way. In a nutshell, we show that constructions beating our bound would imply a protocol that is too good to be true according to [WW10].

The result holds exactly as stated above assuming that the target secret-sharing scheme that the protocol outputs shares in is of a certain type that includes the simple additive secret-sharing scheme (which is also used in [DPSZ12, NNOB12]). If we put no restrictions on the target scheme, the results get a bit more complicated. Essentially what we show is the following: suppose we replace the multiplication gate by a more general gate that does some computation on a fixed number of inputs, such as the inner product of two vectors. Then we show that once the computation done by the gate gets large enough (in a certain sense we define in the paper), again a protocol handling such a gate must communicate a lot. It is the target secret-sharing scheme that determines how "large" the gate needs to be, see more details within.

Comparison to Related Work. There is a lot of prior work on lower bounding communication in interactive protocols, see for instance [Kus92, FY92, CK93, FKN94, KM97, KR94, BSPV99, GR03] (see [DPP14] for an overview of these results). They typically provide lower bounds for very specific functions such as modular addition, and are not applicable to our situation. Probably the most relevant previous work is [DPP14]. Their model does not match ours, as they consider three parties where only two have input and only the third party gets output. Hence we cannot use their results directly, but it is instructive to consider their techniques as it shows why our problem is more tricky than it may seem at first. One important idea used in [DPP14] is to make a "cut", i.e., one considers a (small) subset $\mathcal{C}$ of the parties and then argue that either the communication between $\mathcal{C}$ and the rest of the world must be large enough to determine their inputs, since otherwise other players could not compute the output; or that $\mathcal{C}$ must receive information of sufficient size to be able to compute its own outputs.

It turns out that these ideas are not sufficient for us: recall that we start from a situation where players already have shares of the input values a, b. Now, if $\mathcal{C}$ is large enough to be qualified in the input secret sharing scheme, then $\mathcal{C}$ already has information enough to determine a, b (and for some secret sharing schemes even the shares of all players). So $\mathcal{C}$ can in principle compute correct shares of $c = ab$ by itself without communicating with anyone. On the other hand, if $\mathcal{C}$ is unqualified, then the complement of $\mathcal{C}$ is typically qualified, and therefore does not need information from $\mathcal{C}$ to compute output. But one might think that $\mathcal{C}$ needs to *receive* information to determine its output, in particular, the output shares must be properly coordinated to form a consistent sharing of c. Remember, however, that players already have properly coordinated shares of *the inputs*, and they might be able to use those to form a correct output sharing while communicating less. Indeed, this is what happens for addition gates, where there is no communication, players just add their shares locally.

It follows that the idea of a cut is not enough, one must exploit in some non-trivial way that we are handling a *multiplication* gate, which is exactly what we do. It is possible that one could use the fact that we do multiplication together with the concept of residual information which was also used in [DPP14], to get better bounds than we achieve here, but this remains a speculation.

Note that our model does not count communication needed to construct the shares that are input, nor does it count any communication needed to reconstruct results from the output shares. This does count in the standard model and makes lower bounds easier to prove. For instance, in [DNOR15] lower bounds were recently proved on the message complexity of computing a large class of functions securely, primarily by showing that a significant number of messages must be sent before the input are uniquely determined. In fact, if we included a secret sharing phase before the multiplication protocol and a reconstruction phase after it, these would entail so much communication that the bounds obtained from existing results would leave nothing to explain why the *privacy preserving* multiplication step is communication intensive.

It is also easy to see that one cannot get bounds in our model based only on correctness, for instance by methods from communication complexity. If parties have shares in a and b, no communication is needed to produce some set of *correct* shares in ab: one can simply consider the shares in a and b together as a (redundant) sharing of ab. Indeed this satisfies all our demands to a multiplication gate protocol except privacy: the output threshold is the same and we can correctly reconstruct ab, but privacy is of course violated because reconstruction would tell us more than ab. So, our bounds arguably require privacy.

2   Preliminaries

Notation. We say that a function ε is negligible if $\forall c \; \exists \; \sigma_c \in \mathbb{N}$ such that if $\sigma \geq \sigma_c$ then $\varepsilon(\sigma) < \sigma^{-c}$. We write $[n]$ to denote the set $\{1, 2, ..., n\}$. Moreover, calligraphic letters denote sets. The complement of a set $\mathcal{A}$ is denoted by $\overline{\mathcal{A}}$. The distribution of a random variable X over $\mathcal{X}$ is denoted by P_X. Given the distribution P_{XY} over $\mathcal{X} \times \mathcal{Y}$, the marginal distribution is denoted by $P_X(x) := \sum_{y \in \mathcal{Y}} P_{XY}(x, y)$. A conditional distribution $P_{X|Y}(x, y)$ over $\mathcal{X} \times \mathcal{Y}$ defines for every $y \in \mathcal{Y}$ a distribution $P_{X|Y=y}$. The *statistical distance* between two distributions P_X and P'_X over the domain $\mathcal{X}$ is defined as the maximum, over all (inefficient) distinguishers $D : \mathcal{X} \to \{0, 1\}$, of the distinguishing advantage $\text{SD}(P_X, P'_X) = \big| Pr[D(X) = 1] - Pr[D(X') = 1] \big|$. The *conditional Shannon entropy* of X given Y is defined as $H(X|Y) := -\sum_{x,y} P_{XY}(x, y) \log P_{X|Y}(x, y)$ where all logarithms are binary and the *mutual information* of X and Y as $I(X; Y) = H(X) - H(X|Y)$. We also use $h(p) = -p \log p - (1 - p) \log(1 - p)$ for the binary entropy function. Furthermore, we denote by Π_f an n-party protocol for a function f and by $\Pi_f^{A,B}$ a two-party protocol between parties A and B.

Protocols. We consider protocols involving n parties, denoted by the set $\mathcal{P} = \{\mathsf{P}_1, \ldots, \mathsf{P}_n\}$. The parties communicate over synchronous, point-to-point secure channels. We consider non-reactive secure computation tasks, defined by a deterministic or randomized functionality $f : \mathcal{X}_1 \times \ldots \times \mathcal{X}_n \to \mathcal{Z}_1 \times \ldots \times \mathcal{Z}_n$. The functionality specifies a mapping from n inputs to n outputs the parties want to compute. The functionality can be fully specified by a conditional probability distribution $P_{Z_1 \cdots Z_n | X_1 \cdots X_n}$, where X_i is a random variable over $\mathcal{X}_i$, Z_i is a random variable over $\mathcal{Z}_i$, and for all inputs $(x_1, \ldots, x_n)$ we have a probability function $P_{Z_1 \cdots Z_n | X_1 \cdots X_n = (x_1, \ldots, x_n)}$ and $P_{Z_1 \cdots Z_n | X_1 \cdots X_n = (x_1, \ldots, x_n)}(z_1, \ldots, z_n)$ is the probability that the output is $(z_1, \ldots, z_n)$ when the input is $(x_1, \ldots, x_n)$. Vice versa, we can consider any conditional probability distribution $P_{Z_1 \cdots Z_n | X_1 \cdots X_n}$ as a specification of a probabilistic functionality. In the following we will freely switch between the terminology of probabilistic functionalities and conditional probability distributions.

We consider stand-alone security as well as static and passive corruptions of t out of n parties for some $t \leq n$. This means that a set of t parties are announced to be corrupted before the protocol is executed, and the corrupted parties still follow the protocol but might pool their views of the protocol to learn more than they should. We consider statistical correctness and statistical security. We allow simulators to be inefficient. Except that we do not consider computational security, the above model choices are the possible weakest ones, which just makes our impossibility proofs stronger.

The Security Parameter. The security is measured in a security parameter σ and we require that the "insecurity" goes to 0 as σ grows. We do not allow n to grow with σ, i.e., we require that the protocol can be made arbitrarily secure when run among a fixed set of parties by just increasing σ. The literature sometimes consider protocol which only become secure when run among a sufficiently large number of parties. We do not cover such protocols.

Communication Model. We assume that each pair of parties are connected by a secure communication channel, which only leaks to the adversary the length of each message sent[1]. We consider protocols proceeding in synchronous rounds. Following [DPP14] we assume that in each round each pair of parties $(\mathsf{P}_i, \mathsf{P}_j)$ will specify a prefix free code $M_{i,j} \subset \{0,1\}^*$ and then P_i will send a message $m \in M_{i,j}$. The codes might be dynamically chosen, but we require that the parties agree on the codes. If the length of a sent message does not match the length specified by the receiver, the receiver will terminate with an error symbol $\perp$ as output, which will make it count as a violation of correctness.

Let ϵ denote the empty string and let $E = \{\epsilon\}$. If $M_{i,j} = E$, then we say that P_i sends no message to P_j in that round, i.e., we use the empty string to denote the lack of a message. Notice that if $M_{i,j} \neq E$, then $\epsilon \notin M_{i,j}$ as $M_{i,j}$ must be prefix free. Therefore, at the point where P_j specifies the code $M_{i,j}$ for a given round, P_j already knows whether or not P_i will send a message in that round.

---

[1] This is a standard way to model secure communication by an ideal functionality since any implementation using crypto would leak the message length.

We in particular say that P_j anticipates a message from P_i when $M_{i,j} \neq E$. We will only be interested in counting the number of messages sent, not their size. When the protocol is correct, the number of messages sent is obviously equal to the number of messages anticipated.

Definition 1 (Anticipated message complexity). *We say that the expected message complexity of a party is the expected number of times a non-empty message is sent or anticipated by the party. The expected message complexity of a protocol is simply the sum of the expected message complexity of the parties, divided by 2. We divide by 2 to avoid counting a transmitted message twice. The expectation is taken over the randomness of the players and maximised over all inputs.*

The reason for insisting on a prefix free code for this slightly technical notion is to avoid a problem we would have if we allowed the communication pattern to vary arbitrarily: consider a setting where P_j wants to send a bit b to P_i. If $b = 0$ it sends no message to P_i or say the empty string. If $b = 1$ it sends 0 to P_i. If b is uniformly random, then in half the cases P_j sends a message of length 0 and in half the cases it sends a message of length 1. This means that a more liberal way of counting the communication complexity would say that the expected communication complexity is $\frac{1}{2}$. This would allow to exchange 1 bit of information with an expected $\frac{1}{2}$ bits of communication. This does not seem quite reasonable. The prefix-free model avoids this while still allowing the protocol to have a dynamic communication pattern. Note that since we want to prove impossibility it is stronger to allow protocols with dynamic rather than fixed communication patterns.

Protocols with Preprocessing. We will also consider protocols for the preprocessing model. In the preprocessing model, the specification of a protocol also includes a joint distribution $P_{R_1 \cdots R_n}$ over $\mathcal{R}_1 \times \ldots \times \mathcal{R}_n$, where the $\mathcal{R}_i$'s are finite randomness domains. This distribution is used for sampling correlated random inputs $(r_1, \ldots, r_n) \leftarrow P_{R_1 \cdots R_n}$ received by the parties before the execution of the protocol. Therefore, the preprocessing is independent of the inputs. The actions of a party P_i in a given round may in this case depend on the private random input r_i received by P_i from the distribution $P_{R_1 \cdots R_n}$ and on its input x_i and the messages received in previous rounds. In addition, the action might depend on the statistical security parameter σ which is given as input to all parties along with x_i and r_i. Using the standard terminology of secure computation, the preprocessing model can be thought of as a hybrid model where the parties have one-time access to an ideal randomized functionality P (with no inputs) providing them with correlated, private random inputs r_i.

Security Definition. A protocol securely implements an ideal functionality with an error of ε, if the entire view of each corrupted player can be simulated with an error of at most ε in an ideal setting, where the players only have black-box access to the ideal functionality. Formally, consider Definition 2 below.

Definition 2. *Let Π be a protocol for the $P_{R_1 \cdots R_n}$-preprocessing model. Let $P_{Z_1 \cdots Z_n | X_1 \cdots X_n}$ be an n-party functionality. Let* Adv *be a randomized algorithm,* *which chooses to corrupt a set $\mathcal{A} \subseteq \{1, \ldots, n\}$ of at most $t \in \mathbb{N}$ parties. Let $\boldsymbol{x} = (x_1, \ldots, x_n) \in \mathcal{X}_1 \times \ldots \times \mathcal{X}_n$ be an input. Let $\mathsf{Pattern}^{\Pi}(\sigma, \boldsymbol{x})$ denote the communication pattern in a random run of the protocol Π, i.e., the list of the length of the messages exchanged between all pairs of parties in all rounds, on input $\boldsymbol{x}$ and with security parameter σ. Define $\mathsf{View}_{\mathsf{Adv}}^{\Pi}(\sigma, \boldsymbol{x})$ to be the $\mathsf{Pattern}^{\Pi}(\sigma, \boldsymbol{x})$ concatenated with the view of the parties P_i for $i \in \mathcal{A}$ in the same random run of the protocol Π. Let $\mathsf{Output}_{\mathcal{A}}^{\Pi}(\sigma, \boldsymbol{x})$ be just the inputs and outputs of the honest parties P_i for $i \notin \mathcal{A}$ in the same random run of the protocol Π. Let*

$$\mathsf{Exec}_{\mathsf{Adv}}^{\Pi}(\sigma, \boldsymbol{x}) = (\mathsf{View}_{\mathsf{Adv}}^{\Pi}(\sigma, \boldsymbol{x}), \mathsf{Output}_{\mathcal{A}}^{\Pi}(\sigma, \boldsymbol{x})).$$

Let S be a randomized function called the simulator. *Sample $\boldsymbol{z}$ according to $P_{Z_1 \cdots Z_n | X_1 \cdots X_n}(\boldsymbol{x})$. Give input $\{(x_i, z_i)\}_{i \in \mathcal{A}}$ to S. Let $\mathsf{S}(\{(x_i, z_i)\}_{i \in \mathcal{A}})$ denote the random variable describing the output of S. Let*

$$\mathsf{Sim}_{\mathsf{S}}(\sigma, \boldsymbol{x}) = \left(\mathsf{S}(\{(x_i, z_i)\}_{i \in \mathcal{A}}), \{(x_i, z_i)\}_{i \notin \mathcal{A}} \right).$$

The protocol is ε-semi-honest secure with threshold t if there exist S such that for all $\boldsymbol{x}$ and all $\mathcal{A}$ with $|\mathcal{A}| \leq t$ it holds that

$$\mathsf{SD}(\mathsf{Exec}_{\mathsf{Adv}}^{\Pi}(\sigma, \boldsymbol{x}), \mathsf{Sim}_{\mathsf{S}}(\sigma, \boldsymbol{x})) \leq \varepsilon(\sigma).$$

The protocol is statistically *semi-honest secure with threshold t if it is ε-semi-honest secure for a negligible ε.*

Secret-Sharing. A $(t+1)$-out-of-n secret-sharing scheme takes as input a secret s from some input domain and outputs n shares, with the property that it is possible to efficiently reconstruct s from every subset of $t + 1$ shares, but every subset of at most t shares reveals nothing about the secret s. The value t is called the privacy *threshold* of the scheme.

A secret-sharing scheme consists of two algorithms: the first algorithm, called the *sharing algorithm* Share, takes as input the secret s and the parameters t and n, and outputs n shares. The second algorithm, called the *recovery algorithm* $\mathsf{Recover}$, takes as input $t + 1$ shares and outputs a value s. It is required that the reconstruction of shares generated from a value s produces the same value s. Formally, consider the above definition.

Definition 3 (Secret-sharing). *Let $\mathbb{F}$ be a finite field and let $n, t \in \mathbb{N}$. A pair of algorithms $\mathcal{S}_t^n = (\mathsf{Share}, \mathsf{Recover})$ where Share is randomized and $\mathsf{Recover}$ is deterministic are said to be a secret-sharing scheme if for every $n, t \in \mathbb{N}$, the following conditions hold.*

Reconstruction: *For any set $\mathcal{T} \subseteq \{1, \ldots, n\}$ such that $|\mathcal{T}| > t$ and for any $s \in \mathbb{F}$ it holds that*

$$\Pr[\mathsf{Recover}(\mathsf{Share}_{\mathcal{T}}(s, n, t)) = s] = 1$$

where $\mathsf{Share}_{\mathcal{T}}$ is the restriction of the outputs of Share to the elements in $\mathcal{T}$.

Privacy: *For any set $\mathcal{T} \subseteq \{1, \ldots, n\}$ such that $|\mathcal{T}| \leq t$ and for any $s, s' \in \mathbb{F}$ it holds that*

$$\mathsf{Share}_{\mathcal{T}}(s, n, t) \equiv \mathsf{Share}_{\mathcal{T}}(s', n, t)$$

where we use $\equiv$ to denote that two random variables have the same distribution.

Additive Secret-Sharing. In an additive secret-sharing scheme, n parties hold shares the sum of which yields the desired secret. By setting all but a single share to be a random field element, we ensure that any subset of $n - 1$ parties cannot recover the initial secret.

Definition 4 (Additive secret-sharing). *Let $\mathbb{F}$ be a finite field and let $n \in \mathbb{N}$. Consider the secret-sharing scheme $\mathcal{A}^n = (\mathsf{Share}, \mathsf{Recover})$ defined below.*

- *The algorithm Share on input (s, n) performs the following:*
  1. *Generate $(s_1, \ldots, s_{n-1})$ uniformly at random from $\mathbb{F}$ and define $s_n = s - \sum_{i=1}^{n-1} s_i$.*
  2. *Output $(s_1, \ldots, s_n)$ where s_i is the share of the i-th party.*
- *The recovery algorithm $\mathsf{Recover}$ on input $(s_1, \cdots, s_n)$, outputs $\sum_{i=1}^{n} s_i$.*

It is easy to show that the distribution of any $n - 1$ of the shares is the uniform one on $\mathbb{F}^{n-1}$ and hence independent of s.

Secret-sharing Notation. In the sequel for a value $s \in \mathbb{F}$ we denote by $[s]^{\mathcal{S}_t^n}$ a random sharing of s for the secret-sharing scheme $\mathcal{S}_t^n$. That is, $[s]^{\mathcal{S}_t^n} \leftarrow \mathsf{Share}(s, n, t)$ where $[s]^{\mathcal{S}_t^n} = (s_1, \ldots, s_n)$. Similarly, we denote by $[s]^{\mathcal{A}^n}$ a random additive sharing of s secret shared among n parties.

Primitives. In the sequel we consider the following two-party functionalities which naturally extend to the multi-party setting.

Definition 5 (Multiplication MULT functionality). *Let $\mathbb{F}$ be a finite field. Consider two parties A and B. We define the two-party functionality $\mathrm{MULT}(a, b)$ which on input $a \in \mathbb{F}$ from party A and $b \in \mathbb{F}$ from party B outputs $\mathrm{MULT}(a, b) = a \cdot b$ to both parties.*

Definition 6 (Inner Product IP_κ functionality). *Let $\mathbb{F}$ be a finite field and let $\kappa \geq 1$. Consider two parties A and B. We define the two-party functionality $\mathrm{IP}_\kappa(a, b)$ which on input $a \in \mathbb{F}^\kappa$ from party A and $b \in \mathbb{F}^\kappa$ from party B outputs $\mathrm{IP}_\kappa(a, b) = \sum_{i=1}^{\kappa} a_i b_i$ to both parties.*

3   Secure Computation in the Plain Model

We first investigate the honest majority scenario. As explained in the introduction, we will consider protocols that compute arithmetic circuits over some field

securely using secret-sharing. All known protocols of this type handle multiplication gates by running a subprotocol that takes as input shares in the two
inputs a and b to the gate and output shares of the product ab, such that the
output shares contain only information about ab (and no side information on
a nor b). Accordingly, we define below a *multiplication gate protocol* (MGP) to
be an interactive protocol for n players that does exactly this, and then show a
lower bound on the communication required for such a protocol.

Definition 7 (Multiplication Gate Protocol $\Pi_{\mathbf{MULT}}$). *Let $\mathbb{F}$ be a finite
field and let $n \in \mathbb{N}$. Let $\mathcal{S}_t^n$ and $\hat{\mathcal{S}}_{t'}^n$ be two secret-sharing schemes as per Definition 3. A protocol Π_{MULT} is an n-party Multiplication Gate Protocol (MGP)
with thresholds t, t', input sharing-scheme $\mathcal{S}_t^n$ and output sharing-scheme $\hat{\mathcal{S}}_{t'}^n$ if
it satisfies the following properties:*

Correctness: *In the interactive protocol Π_{MULT}, players start from sets of
shares $[a]^{\mathcal{S}_t^n} \leftarrow \mathsf{Share}(a, n, t)$ and $[b]^{\mathcal{S}_t^n} \leftarrow \mathsf{Share}(b, n, t)$. Each player outputs a share such that these together form a set of shares $[ab]^{\hat{\mathcal{S}}_{t'}^n}$. Moreover,
$t' < 2t$.*

t-privacy: *If the protocol is run on randomly sampled shares $[a]^{\mathcal{S}_t^n}$ and $[b]^{\mathcal{S}_t^n}$,
then the only new information the output shares can reveal to the adversary
is ab. We capture this by requiring that for any adversary corrupting a player
subset $\mathcal{A}$ of size at most t, there exists a simulator $\mathsf{S}_{\mathcal{A}}$ which when given the
input shares of the parties in $\mathcal{A}$ (denoted by $[a]_{\mathcal{A}}^{\mathcal{S}_t^n}$, $[b]_{\mathcal{A}}^{\mathcal{S}_t^n}$) and the product ab,
will simulate the honest parties' output shares (denoted by $[ab]_{\overline{\mathcal{A}}}^{\hat{\mathcal{S}}_{t'}^n}$) and the
view of the parties in $\mathcal{A}$ with statistically indistinguishable distribution.
Formally, for any adversary ADV corrupting a player set $\mathcal{A}$ with $|\mathcal{A}| \le t$
there exist $\mathsf{S}_{\mathcal{A}}$ such that for randomly sampled shares $[a]^{\mathcal{S}_t^n} \leftarrow \mathsf{Share}(a, n, t)$
and $[b]^{\mathcal{S}_t^n} \leftarrow \mathsf{Share}(b, n, t)$, it holds that*

$$\mathrm{SD}\left(\left(\mathsf{View}_{\mathsf{ADV}}^{\Pi_{\mathrm{MULT}}}(\sigma, [a]^{\mathcal{S}_t^n}, [b]^{\mathcal{S}_t^n})), [ab]_{\overline{\mathcal{A}}}^{\hat{\mathcal{S}}_{t'}^n}\right), \mathsf{S}_{\mathcal{A}}(\sigma, [a]_{\mathcal{A}}^{\mathcal{S}_t^n}, [b]_{\mathcal{A}}^{\mathcal{S}_t^n}, ab)\right) \le \varepsilon(\sigma), \quad (1)$$

*where σ is a security parameter and where, in the underlying random experi-
ment, probabilities are taken over the choice of input shares as well as random
coins of the protocol and simulator.*

Note that we do not require the input and output sharing schemes to be the
same, we only require that the output threshold is not too large ($t' < 2t$).
Known MPG's actually have $t' = t$ to allow continued computation, we want to
be more generous to make our lower bound stronger. Note also that we do not
require the simulators to be efficient.

Recall that we use the term *gate-by-gate* protocol to refer to any protocol
that computes an arithmetic circuit securely by invoking an MGP for each
multiplication gate in the circuit such that the sets of shares that are input are
randomly chosen. We leave unspecified what happens with addition gates as this
is irrelevant for the bounds we show. An *ordered* gate-by-gate protocol invokes

MGP's for multiplication gates in an order corresponding to the order in which one would visit the gates when evaluating the circuit.

In the following we show that any MGP in a gate-by-gate protocol must communicate for every multiplication gate in the honest majority setting even if only semi-honest security is required. The technique of our proof is as follows. We build an information-theoretic two-party computation protocol utilizing an n-party MGP by emulating multiple parties (in the head) and then use the impossibility result on the existence of an information-theoretic two-party computation protocol to show a contradiction.

Theorem 1. *There exists no MGP Π_{MULT} as per Definition 7 with thresholds t, t', and with expected anticipated message complexity $\leq 2t$.*

Proof. Suppose for contradiction that there exists an MGP Π_{MULT} with expected anticipated communication complexity at most $2t$. We first show a proof in the simpler case where the communication pattern is fixed. This means that at most $2t$ parties are communicating, i.e., they send or receive messages and the set of parties that communicate is known and fixed. For simplicity of exposition, suppose that these parties are $P_1, \ldots, P_{2t}$. We are going to use Π_{MULT} to construct a two-party unconditionally secure protocol $\Pi_{\text{MULT}}^{A,B}$ which securely computes the MULT function between parties A, B as per Definition 5.

In particular, given two parties A and B, with inputs $a, b \in \mathbb{F}$, respectively, involved in the $\Pi_{\text{MULT}}^{A,B}$ protocol, we are going to let A emulate the first t parties that communicate and B emulate the other t parties, say $P_{t+1}, \ldots, P_{2t}$. The protocol $\Pi_{\text{MULT}}^{A,B}$ proceeds as follows:

Protocol $\Pi_{\text{MULT}}^{A,B}(\sigma, a, b)$
    Input Phase:
        1. Parties A, B secret share their inputs a, b using the secret-sharing scheme $\mathcal{S}_t^n$. More specifically, A computes $[a]^{\mathcal{S}_t^n} \leftarrow \text{Share}(a, n, t)$ and B computes $[b]^{\mathcal{S}_t^n} \leftarrow \text{Share}(b, n, t)$.
        2. Party A sends the input shares $(a_{t+1}, \ldots, a_{2t})$ to party B and Party B sends the input shares $(b_1, \ldots, b_t)$ to party A.
    Evaluation Phase:
        1. Parties A, B invoke the protocol $\Pi_{\text{MULT}}(\sigma, a_1, \ldots a_n, b_1, \ldots b_n)$. The emulation of Π_{MULT} yields a set of shares $[c]^{\mathcal{S}_{t'}^n}$ and outputs $(c_1, \ldots, c_t)$ to party A and $(c_{t+1}, \ldots, c_{2t})$ to party B.
    Output Phase:
        2. Party A sends the output shares $(c_1, \ldots, c_t)$ to party B and Party B sends the output shares $(c_{t+1}, \ldots, c_{2t})$ to party A.
        3. Each party given $2t > t'$ shares of c recovers the output $c = a \cdot b$.

We now show that the above protocol is correct and secure. Correctness follows immediately from $t' < 2t$ - as then $2t$ shares are enough to reconstruct. The protocol is secure (private) due to the t-privacy property of Π_{MULT}. More precisely, if party A is corrupted, we need to simulate his view of the protocol given a and the product ab. We do this as follows: Let $\mathcal{A}$ be the set of parties A

emulates in the MGP. We now compute $[a]^{S_t^n} \leftarrow \mathsf{Share}(a, n, t)$ and sample $[b]_{\mathcal{A}}^{S_t^n}$ which can be done by the privacy property of S_t^n. We then run the simulator $\mathsf{S}_{\mathcal{A}}$ guaranteed by the t-privacy property to get $\mathsf{S}_{\mathcal{A}}(\sigma, [a]_{\mathcal{A}}^{S_t^n}, [b]_{\mathcal{A}}^{S_t^n}, ab)$. Note that this output includes $\mathcal{A}$'s view of the MGP as well as all output shares.

The simulator now outputs $[a]^{S_t^n}$, $[b]_{\mathcal{A}}^{S_t^n}$ and $\mathsf{S}_{\mathcal{A}}(\sigma, [a]_{\mathcal{A}}^{S_t^n}, [b]_{\mathcal{A}}^{S_t^n}, ab)$. This is statistically indistinguishable from A's view of $\Pi_{\mathrm{MULT}}^{A,B}(\sigma, a, b)$ by the privacy property of S_t^n and Eq. (1). A similar simulator for B's view is easy to construct.

However, the above leads to a contradiction since it is well known [BGW88, CCD88] that it is impossible to realize passively secure two-party multiplication (such as the $\Pi_{\mathrm{MULT}}^{A,B}$ protocol) in the information theoretic setting (even if inefficient simulators are allowed). Therefore, the theorem follows.

We now address the case where the communication pattern might be dynamic. We say that a party communicated if it sent a non-empty message or if it anticipated a non-empty message. So by definition, the expected number of communicating parties is $\leq 2t$. Since the observed value is an integer, there is some non-zero, constant probability p such that the observed value of the number of communication parties is at most $2t$. We can therefore pick a subset $\mathcal{C}$ of the parties of size $2t$ such that it happens with probability at least $p/\binom{n}{2t}$ that only the parties in $\mathcal{C}$ communicate. Since we can increase the security parameter σ independently of n, the number $p/\binom{n}{2t}$ is a positive constant (in σ). We can then modify $\Pi_{\mathrm{MULT}}^{A,B}(a, b)$ such that B runs t parties in $\mathcal{C}$ and A runs the other t parties. The protocol runs as $\Pi_{\mathrm{MULT}}^{A,B}(a, b)$ except that if it A or B observe that a party in $\mathcal{C}$ anticipates a non-empty message from a party outside $\mathcal{C}$, then the execution is terminated. In case the protocol terminates, the two parties just try again. Since $p/\binom{n}{2t}$ is a positive constant this succeeds in an expected constant number of tries. Notice that when the protocol succeeds, all parties in $\mathcal{C}$ received all the messages they would have received in a run of $\Pi_{\mathrm{MULT}}^{A,B}(a, b)$ where all the parties were active, as parties only receive the messages they anticipate. Hence the parties in $\mathcal{C}$ have correct outputs (except with negligible probability). For the same reason the output of the parties simulated by A and B will be correct. Hence A and B can reconstruct the output from the $2t$ shares. We can also argue that the protocol is private: We will simulate A's (or B)'s view by running the simulator $\mathsf{S}_{\mathcal{A}}$ (where again $\mathcal{A}$ is the set of parties emulated by A) repeatedly until a view is produced where no party in $\mathcal{C}$ anticipates a message from outside of $\mathcal{C}$. Note that $\mathsf{S}_{\mathcal{A}}$ simulates the view of an adversary corrupting $\mathcal{A}$, and this view includes the communication pattern from which it is evident who anticipates messages. □

The above theorem immediately implies:

Corollary 1. *Any gate-by-gate protocol that is secure against $t = \Theta(n)$ corruptions must communicate $\Omega(n \cdot |C|)$ bits where $|C|$ is the size of the circuit C to compute, and moreover, an ordered gate-by-gate protocol must have a number of rounds that is proportional to the (multiplicative) depth of C.*

Jumping ahead, we note that the arguments for this conclusion break down completely when we consider secure computation in the preprocessing model with dishonest majority since in such a model it is no longer true that two-party unconditionally secure multiplication is impossible: just a single preprocessed multiplication triple will be enough to compute a multiplication. We return to this issue in the next section.

A bound that grows with the field size? It is natural to ask if we can get a lower bound on the complexity of an MPG that grows with the field size? after all, existing MGPs do need to send more bits for larger fields. However, the answer is no, as the following example shows: for $a \in \mathbb{F}$, define z_a to be 0 if $a = 0$ and 1 otherwise. Then we represent an element $a \in \mathbb{F}$ as a pair (z_a, ℓ_a) where ℓ_a is randomly chosen if $a = 0$ and otherwise $\ell_a = \log_g(a)$, where g is a fixed generator of the multiplicative group $\mathbb{F}^*$. Let $u = |\mathbb{F}^*|$. Observe that now we have $(z_{ab}, \ell_{ab}) = (z_a \cdot z_b, (\ell_a + \ell_b) \bmod u)$.

We now construct a secret sharing scheme: given a secret $a \in \mathbb{F}$, we first compute (z_a, ℓ_a) and then share z_a using, e.g., Shamir's scheme and share ℓ_a additively modulo u. An MGP for this scheme can use a standard protocol to compute shares in $z_a \cdot z_b$ and local addition to get shares in $(\ell_a + \ell_b) \bmod u$. Clearly, the communication complexity of this MGP does not depend on $|\mathbb{F}|$.

Of course, the secret sharing scheme we defined is not efficient (at least not in all fields) because one needs to take discrete logs. This is not formally a problem since we did not make any assumptions on the efficiency of secret sharing schemes. But we can in fact get a more satisfactory solution by replacing the additive sharing of the discrete log with black-box sharing directly over the group $\mathbb{F}^*$ [CF02]. This is doable in polynomial time, will cost a factor that is logarithmic in the number of players, but since black-box secret-sharing is homomorphic over the group operation, the resulting MGP still has communication independent of $|\mathbb{F}|$.

Amortized Multiplication Gate Protocols. There is one clear possibility for circumventing the bounds we just argued for gate-by-gate protocols, namely: what if the circuit structure allows us to do, say k multiplications in parallel? Perhaps this can be done more efficiently than k separate multiplications? Of course, this will not help for a worst case circuit whose depth is comparable to its size. But in fact, for "nicer" circuits, we know that such optimizations are possible, based on so-called packed secret-sharing. The catch, however, is that apart from loosing in resilience this only works if there is a gap of size $\Theta(k)$ between the privacy and reconstruction thresholds of the secret-sharing scheme used, so the number of players must grow with k.

One may ask if this is inherent, i.e., can we save on the *communication* needed for many multiplication gates in parallel, only by increasing the number of players? While we believe this is true, we were not able to show it in full generality. But we were able to do so for *computational* complexity, as detailed below. Furthermore, for a restricted setting we explain below and a fixed number of players, we could show that the communication must grow linearly with k.

First, we can trivially extend Definition 3 to cover schemes in which the secret is a vector $\boldsymbol{a} = (a_1, \ldots, a_k)$ of field elements instead of a single value. A further extension covers *ramp* schemes in which there are two thresholds: the privacy threshold t which is defined as in Definition 3 and a reconstruction threshold $r > t$, where any set of size at least r can reconstruct the secret. Such a scheme is denoted by $S_{t,r}^n$. Note that the shares in this case may be shorter than the secret, perhaps even a single field element per player. We can now define a simple extension of the multiplication gate protocol concept:

Definition 8 (*k-Multiplication Gate Protocol Π_{MULT^k}*)*. Let* $\mathbb{F}$ *be a finite field and let* $n \in \mathbb{N}$. *Let* $S_{t,r}^n$ *and* $\hat{S}_{t,r}^n$ *be two ramp sharing schemes defined over* $\mathbb{F}$, *for sharing vectors in* $\mathbb{F}^k$. Π_{MULT^k} *is said to be a k-Multiplication Gate Protocol* (k-MGP) *with thresholds* t, r, *input sharing scheme* $S_{t,r}^n$ *and output sharing scheme* $\hat{S}_{t,r}^n$ *if it satisfies the following properties:*

Correctness: *In the interactive protocol* Π_{MULT^k}, *players start from sets of shares* $[\boldsymbol{a}]^{S_{t,r}^n}$ *and* $[\boldsymbol{b}]^{S_{t,r}^n}$. *Each player outputs a share such that these together form a set of shares* $[\boldsymbol{a} * \boldsymbol{b}]^{\hat{S}_{t,r}^n}$, *where* $\boldsymbol{a} * \boldsymbol{b}$ *is the coordinatewise product of* $\boldsymbol{a}$ *and* $\boldsymbol{b}$.

t-privacy: *If the protocol is run on randomly sampled shares* $[\boldsymbol{a}]^{S_t^n}$ *and* $[\boldsymbol{b}]^{S_t^n}$, *then the only new information the output shares can reveal to the adversary is* $\boldsymbol{a} * \boldsymbol{b}$. *We capture this by requiring that for any adversary corrupting player subset* $\mathcal{A}$ *of size at most* t, *there exists a simulator* $S_{\mathcal{A}}$ *which when given the input shares of the parties in* $\mathcal{A}$ *(denoted by* $[\boldsymbol{a}]_{\mathcal{A}}^{S_t^n}$, $[\boldsymbol{b}]_{\mathcal{A}}^{S_t^n}$ *) and the product* ab, *will simulate the honest parties' output shares (denoted by* $[\boldsymbol{a} * \boldsymbol{b}]_{\overline{\mathcal{A}}}^{\hat{S}_{t'}^n}$ *) and the view of the parties in* $\mathcal{A}$ *with statistically indistinguishable distribution.*

Formally, for any adversary ADV *corrupting player set* $\mathcal{A}$ *with* $|\mathcal{A}| \leq t$ *there exist* $S_{\mathcal{A}}$ *such that for randomly sampled shares* $[\boldsymbol{a}]^{S_t^n} \leftarrow \mathsf{Share}(\boldsymbol{a}, n, t)$ *and* $[\boldsymbol{b}]^{S_t^n} \leftarrow \mathsf{Share}(\boldsymbol{b}, n, t)$, *it holds that*

$$\mathrm{SD}\left(\left(\mathsf{View}_{\mathrm{ADV}}^{\Pi_{\mathrm{MULT}}^k}(\sigma, [\boldsymbol{a}]^{S_t^n}, [\boldsymbol{b}]^{S_t^n})), [\boldsymbol{a} * \boldsymbol{b}]_{\overline{\mathcal{A}}}^{\hat{S}_{t'}^n}\right), \; S_{\mathcal{A}}(\sigma, [\boldsymbol{a}]_{\mathcal{A}}^{S_t^n}, [\boldsymbol{b}]_{\mathcal{A}}^{S_t^n}, \boldsymbol{a} * \boldsymbol{b})\right) \leq \varepsilon(\sigma), \quad (2)$$

where σ *is a security parameter and where, in the underlying random experiment, probabilities are taken over the choice of input shares as well as random coins of the protocol and simulator.*

Before giving our result on k-MGPs we note that for any interactive protocol, it is always possible to represent the total computation done by the players as an arithmetic circuit over a finite field (arithmetic circuits can emulate Boolean circuit which can in turn emulate Turing machines). We can encode messages as field elements and represent sending of messages by wires between the parts of the circuit representing sender and receiver. For a protocol Π, we refer to an algorithm outputting such a circuit as *an arithmetic representation of Π*. Note that such a representation is not in general unique, but once we have chosen one, it makes sense to talk about, e.g., the number of multiplications done by a player in Π.

Theorem 2. *Let $t < r \leq n \in \mathbb{N}$. Also let $\mathcal{P} = \{P_1, \ldots, P_n\}$ be a set of parties. Assume that the* k-MGP Π_{MULT^k} *defined over $\mathbb{F}$ has thresholds t, r. Then for any arithmetic representation of Π_{MULT^k} (over any finite field) and for each subset $\mathcal{S} \subset \mathcal{P}$ of size $n - 2t$, the total number of multiplications done by players in $\mathcal{S}$ is $\Omega(k)$.*

Proof. Suppose for contradiction that there exists a k-MGP Π_{MULT^k} in which the total number of multiplications done by players in $\mathcal{S}$ is $o(k)$. Assume for notational convenience that $\mathcal{S} = \{P_{2t+1}, \ldots, P_n\}$. We are going to use it to construct a two-party unconditionally secure protocol $\Pi_{\mathrm{MULT}}^{A,B}$ in the preprocessing model which securely computes k multiplications as follows. We let $u \leftarrow P_U$ denote the correlated randomness we will use in $\Pi_{\mathrm{MULT}}^{A,B}$. Given two parties A and B involved in the $\Pi_{\mathrm{MULT}}^{A,B}$ protocol, the idea is to use the assumed k-MGP where A emulates t players and B emulates another t players. In addition, parties A, B together emulate the rest of the parties in $\mathcal{S}$. This can be done using the preprocessed data u: we consider the parties in $\mathcal{S}$ as a reactive functionality $f_{\mathcal{S}}$ which can be implemented using an existing protocol in the preprocessing model. One example of such a protocol is the SPDZ protocol [DPSZ12] denoted by $\Pi_{f_{\mathcal{S}}}^{SPDZ}$[2] which uses additive-secret sharing. Therefore, protocol $\Pi_{\mathrm{MULT}}^{A,B}$ proceeds as follows:

Protocol $\Pi_{\mathrm{MULT}}^{A,B}(\{a_i\}_{i \in [k]}, \{b_i\}_{i \in [k]}, u)$:

  Input Phase:

  1. $\forall i \in [k]$, parties A, B secret share their inputs a_i, b_i using the ramp sharing scheme $\mathcal{S}_{t,r}^n$. So A computes $[\boldsymbol{a}]^{\mathcal{S}_{t,r}^n} \leftarrow \mathsf{Share}((\boldsymbol{a}), n, t)$ and B computes $[\boldsymbol{b}]^{\mathcal{S}_{t,r}^n} \leftarrow \mathsf{Share}((\boldsymbol{b}), n, t)$. For simplicity of exposition, we denote by $(\bar{a}_1, \ldots, \bar{a}_n), (\bar{b}_1, \ldots, \bar{b}_n)$ the shares of $[\boldsymbol{a}]^{\mathcal{S}_{t,r}^n}$ and $[\boldsymbol{b}]^{\mathcal{S}_{t,r}^n}$, respectively.
  2. Party A sends the input shares $(\bar{a}_1, \ldots, \bar{a}_t)$ to party B and Party B sends the input shares $(\bar{b}_t, \ldots, \bar{b}_{2t})$ to party A.
  3. Additively secret share the inputs $(\bar{a}_{2t+1}, \ldots, \bar{a}_n, \bar{b}_{2t+1}, \ldots, \bar{b}_n)$ of the parties in $\mathcal{S}$ between A and B using the additive secret-sharing $\mathcal{A}^2$ and obtain the shares $([\bar{a}_{2t+1}]^{\mathcal{A}^2}, \ldots, [\bar{a}_n]^{\mathcal{A}^2}, [\bar{b}_{2t+1}]^{\mathcal{A}^2}, \ldots, [\bar{b}_n]^{\mathcal{A}^2})$. For the following phase, as we mentioned above, we will think of the computation done by the parties in $\mathcal{S}$ as a reactive functionality $f_{\mathcal{S}}$ which is implemented using the protocol $\Pi_{f_{\mathcal{S}}}^{SPDZ}$ in the preprocessing model.

  Evaluation Phase:

  Parties A, B invoke the protocol $\Pi_{\mathrm{MULT}^k}([\boldsymbol{a}]^{\mathcal{S}_{t,r}^n}, [\boldsymbol{b}]^{\mathcal{S}_{t,r}^n})$ in which A, B emulate t parties each, and they together emulate the rest, $n - 2t$ players, using the preprocessed data u invoking protocol $\Pi_{f_{\mathcal{S}}}^{SPDZ}$. To this end, note that $\Pi_{f_{\mathcal{S}}}^{SPDZ}$ represents data by additive secret-sharing. Values $(\bar{a}_{2t+1}, \ldots, \bar{a}_n, \bar{b}_{2t+1}, \ldots, \bar{b}_n)$ of the parties in $\mathcal{S}$ were already additively shared, so they can be used directly as input to $\Pi_{f_{\mathcal{S}}}^{SPDZ}$.

  Now, the emulation of Π_{MULT^k} is augmented with the protocol $\Pi_{f_{\mathcal{S}}}^{SPDZ}$

---

[2] We do passive security here, so a simpler variant of SPDZ will suffice, without authentication codes on the shared values.

as follows: when a party in S would do a local operation, we do the same operation in $\Pi_{f_S}^{SPDZ}$. When a party outside S sends a message to a party in S an additive secret-sharing of that message is formed between A and B. When a party in S sends a message to a party outside S the corresponding additive secret-sharing is reconstructed towards A or B, depending on who emulates the receiver. In the end, we will obtain additive sharings between A and B of the outputs of parties in S, namely $([\bar{c}_{2t+1}]^{A^2}, \ldots, [\bar{c}_n]^{A^2})$.

Output Phase:

1. A sends the output shares $(\bar{c}_1, \ldots, \bar{c}_t)$ to B, B sends the output shares $(\bar{c}_{t+1}, \ldots, \bar{c}_{2t})$ to A computed by Π_{MULT^k}, and A and B exchange their additive shares $([\bar{c}_{2t+1}]^{A^2}, \ldots, [\bar{c}_n]^{A^2})$ in order to recover $(\bar{c}_{2t+1}, \ldots, \bar{c}_n)$.

2. Now both A and B have $n \geq r$ shares of the output and can recover the result $a * b$.

We now show that the above protocol is correct and secure. Correctness follows immediately from the correctness of Π_{MULT^k} and $\Pi_{f_S}^{SPDZ}$. We argue that the protocol is secure (private) due to the security of $\Pi_{f_S}^{SPDZ}$ and the t-privacy property of the MGP Π_{MULT^k} (see Eq. (1)). For the case where A is corrupted, we first observe that by using the simulator for the $\Pi_{f_S}^{SPDZ}$ protocol, we can argue that the view of A in the real protocol is statistically close to the one obtained by replacing players in S by the ideal functionality f_S.

We can then make a simulator for corrupt A in the f_S-hybrid model, as follows: The shares received by A in the input phase can be simulated by the privacy property of the input sharing scheme, and the rest of the view can be simulated by invoking the simulator S_A of the protocol Π_{MULT^k} guaranteed by Definition 7, on input $[a]_A^{S_t^n}, [b]_A^{S_t^n}, a * b$. Note that S_A is in charge of simulating f_S. It can therefore define the responses of f_S such that they are consistent with the view generated by S_A.[3]

We therefore conclude from Eq. (2) that S_A generates a view that is statistically indistinguishable from the real view of an adversary corrupting A. A similar argument holds for B.

Now note that the preprocessed data required by the protocol $\Pi_{f_P}^{SPDZ}$ amount to a constant number of field elements for each multiplication done. This means that our 2-party protocol needs $o(k)$ preprocessed field elements by assumption on Π_{MULT^k}. However, this leads to a contradiction since by results in [WW10], it is impossible for two parties to compute k multiplications with statistical security using preprocessed data of size $o(k)$ field elements.   □

What this theorem shows is, for instance, that if we want each player to do only a constant number of local multiplications in a k-MGP, then n needs to be $\Omega(k)$. Since this is precisely what protocols based on packed sharing can achieve (see, e.g., [DIK+08]), the bound in the theorem is in this sense tight. What the

---

[3] Note that $\Pi_{f_S}^{SPDZ}$ reveals the structure of the circuit for f_S. This is secure as we assume that the parties in S are represented as known arithmetic circuits.

theorem also says is that *every* subset of size $n - 2t$ needs to work hard, so in the case where we tolerate a maximal number of corruptions, i.e., $n = 2t + 1$, we see that a gate by gate protocol in this case must have computational complexity $\Omega(n|C|)$, for *any* circuit of size $|C|$, not only for "tall and skinny" circuits as we had before.

A restricted result on communication complexity. Our final result on honest majority concerns k-MGPs that are *regular* by which we mean, first that the output shares they produce follow the same distribution that is also produced by the Share algorithm of the output secret sharing scheme. This is a rather natural condition that is satisfied by all known k-MGP's. Second, we will assume that the input and output schemes are the same, have $r = t + 1$ (thus excluding packed sharing), and is ideal, i.e., each share is a single field element. This is satisfied by Shamir's scheme, for instance. The ideal assumption can be replaced by much weaker conditions requiring various symmetry properties, but we stick with the simpler case for brevity.

Theorem 3. *A regular* k-MGP *has (expected)* $\Omega(k)$ *communication complexity.*

Proof. The message pattern and -lengths in the k-MGP must not depend on the inputs $a_1, ..., a_k, b_1, ..., b_k$, so we are done if we show the bound for some fixed distribution of inputs. We choose the uniform distribution, and we write A_1, A_2 etc. for the corresponding random variables. Now consider any subset A of t players and another player P. Let U denote the joint view of players in A before we do the k-MGP and U' denote the view after. V, V' denote the corresponding views of P. Finally, C_A, C_P denote the messages sent and received by A and P during the k-MGP. Note that $I(U; V) = 0$: We have not sent any messages yet, and furthermore, even given the t shares of A in some a_i, since a_i is uniform in $\mathbb{F}$ and the scheme is ideal, the share of P is uniform in $\mathbb{F}$ as well. Also, without loss of generality, we can set $U' = (U, C_A)$ and $V' = (V, C_P)$.

However, since the $C_i = A_i B_i$ are not uniform, we do have common information after the protocol. We see this as follows: since 0 times any value is 0, the value 0 is more likely for c_i than others. So $H(C_i) \leq \log q - \epsilon$ for some constant $\epsilon > 0$ that depends on $|\mathbb{F}| = q$. Furthermore, given the vector of shares $\boldsymbol{S}_A$ of A in C_i, there is a 1-1 correspondence between possible values of C_i and values of the share S_p of P. This means that $H(S_p | \boldsymbol{S}_A) \leq \log |\mathbb{F}| - \epsilon$. On the other hand, $H(S_p) = \log q$ because the scheme is ideal, so therefore $I(\boldsymbol{S}_A; S_P) \geq \epsilon$. This applies to every C_i, so we have that $I(U'; V') \geq k\epsilon$. We can now compute as follows, using a standard chain rule for mutual information:

$$\begin{aligned}
\epsilon k &\leq I((U, C_A); (V, C_P)) \\
&= I(U; (V, C_P)) + I(C_A; (V, C_P)|U) \\
&\leq I(U; (V, C_P)) + H(C_A) \\
&= I(U; V) + I(U; C_P|V) + H(C_A) \\
&\leq I(U; V) + H(C_P) + H(C_A) \\
&= H(C_P) + H(C_A).
\end{aligned}$$

So indeed, the expected size of the communication grows linearly with k.

4   Secure Computation in the Preprocessing Model

It is well known that all functions can be computed with unconditional security in the setting where $n-1$ of the n players may be corrupted, and where the players are given correlated randomness, also known as preprocessed data, that does not have to depend on the function to be computed, nor on the inputs. Winkler and Wullschleger [WW10] proved lower bounds on the the amount of preprocessed data needed to compute certain functions with statistical security where the bound depends on certain combinatorial properties of the target function.

All existing protocols in the preprocessing model that are efficient in the circuit size of the function, work according to the gate-by-gate approach we encountered in the previous section. We can define (ordered) gate-by-gate protocols and MGPs exactly as for the honest majority setting, with two exceptions: MGPs are allowed to consume preprocessed data, and the output threshold t' must equal the input threshold t. This is because we typically have $t = n - 1$ in this setting, and then it does not make sense to consider $t' > t$, then even all players cannot reconstruct the output,

As before, we want to show that multiplication gate protocols require a certain amount of communication, but as mentioned before, we can no longer base ourselves on impossibility of unconditionally secure multiplication for two parties, since this is in fact possible in the preprocessing model. Instead, the contradiction will come from the known lower bounds on the size of the preprocessed data needed to compute certain functions.

4.1   Protocols Based on Additive Secret-Sharing

We start by showing that any gate-by-gate protocol must communicate for every multiplication gate when the underlying secret sharing scheme is the additive one. We show that an MGP that does not communicate enough implied a protocol that contradicts the lower bound by Winkler and Wullschleger [WW10] on the the amount of preprocessed data needed to compute certain functions with statistical security.

Theorem 4. *Consider the preprocessing model where $n-1$ of the n players may be passively corrupted. In this setting, there exists no MGP Π_{MULT} with expected anticipated communication complexity $\leq n - 1$ and with additive secret-sharing $\mathcal{A}^n$ as output sharing scheme.*

Proof. Suppose for contradiction that there exists an MGP Π_{MULT} (with preprocessed data $u \leftarrow P_U$) which contradicts the claim of the theorem. Similar to Theorem 1 we will first assume a fixed communication pattern. Assume for notational convenience that only the parties $P_1, \ldots, P_{n-1}$ communicate. Given two parties A and B, we are going to construct a two-party protocol $\Pi_{\mathrm{MULT}}^{A,B}$ which on input $a, b \in \mathbb{F}$ from A, B, respectively, securely computes ab. The idea is for A to emulate the $n - 1$ players who communicate in Π_{MULT} while B emulates the last player. In particular, protocol $\Pi_{\mathrm{MULT}}^{A,B}$ proceeds as follows:

Protocol $\Pi_{\mathrm{MULT}}^{A,B}$

Input Phase:
1. Parties A, B secret share their inputs a, b using the input secret-sharing scheme $\mathcal{A}^n$ of Π_{MULT}. More specifically, A computes $[a]^{\mathcal{A}^n} \leftarrow$ Share$(a, n, n-1)$ and B computes $[b]^{\mathcal{A}^n} \leftarrow$ Share$(b, n, n-1)$.
2. Party A sends the input share a_n to party B and Party B sends the input shares $(b_1, .., b_{n-1})$ to party A.

Evaluation Phase:
1. Parties A, B invoke the MGP Π_{MULT} as per Definition 7 in the preprocessing model where A emulates the $n-1$ players who communicate, and we assume these are the first $n-1$ players. This means that this phase involves no communication between A and B, but it may consume some preprocessed data u. The execution of Π_{MULT} yields a sharing of $[c]^{\mathcal{A}^n}$ and outputs $(c_1, ..., c_{n-1})$ to party A and c_n to party B.

Output Phase:
1. A sends $\sum_{i=1}^{n-1} c_i$ to B and B sends c_n to A. The parties add the received values to recover the output $c = a \cdot b$.

Correctness of this protocol follows immediately. The protocol can be argued to be secure(private). In particular, the simulator S for $\Pi_{\mathrm{MULT}}^{A,B}$ proceeds as follows. The preprocessing data to be used by the corrupted party can be simulated with the correct distribution without any knowledge of the inputs. In the input phase, the corrupted party receives only an unqualified set of shares whose distribution can be simulated perfectly. There is no communication to be simulated in the evaluation phase. In the output phase, it is the case that whenever the protocol computes the correct result, then the share received from the honest party is trivial to simulate because it is determined from the corrupted party's own share and the result ab. Hence, the only source of error is the negligible probability that the output is wrong in the real execution, so it follows that

$$\mathsf{SD}(\mathsf{Exec}_{\mathsf{Adv}}^{\Pi_{\mathrm{MULT}}^{A,B}}(\sigma, (a, b)), \mathsf{Sim}_{\mathsf{S}}(\sigma, (a, b))) \leq \epsilon(\sigma).$$

However, we can say even more: Let $u \leftarrow P_U$ be the preprocessed data that is consumed during the protocol (Π_{MULT} uses preprocessed data). We now define a new protocol $\Pi_{\mathrm{MULT}^k}^{A,B}$ that will compute k *independent* multiplications (do not confuse this protocol with the amortized and honest majority protocol in Definition 8). It does this by running k instances of $\Pi_{\mathrm{MULT}}^{A,B}$, *using the same preprocessed data u for all instances.*

Normally, it is of course not secure to reuse preprocessed data, but in this particular case it works because the communication in $\Pi_{\mathrm{MULT}}^{A,B}$ is independent of u, and so is the simulation. More precisely, $\Pi_{\mathrm{MULT}^k}^{A,B}$ is clearly correct because each instance of $\Pi_{\mathrm{MULT}}^{A,B}$ runs with correctly distributed preprocessed data. It is also private: we can simulate by first simulating the corrupted party's part of u and then running k instances of the rest of S's code. Again, the only source of error is the case where the real protocol computes an incorrect result, but the probability of this happening for any of the k instances is at most a factor k larger than for a single instance, by a union bound, and so is still negligible.

However, this leads to a contradiction with the result of [WW10]: they showed that the amount of preprocessed data needed for a secure multiplication is at least some non-zero number of bits w. It also follows from [WW10] that if we want k multiplications on independently chosen inputs this requires kw bits. So if we consider a k large enough that kw is larger than the size of u, we have a contradiction and the theorem follows.

We now generalise to dynamic communication patterns. As in the proof of Theorem 1 we can find a party P_i such that with some constant positive probability p the party P_i does not send a message and no party anticipates a message from P_i. Assume without loss of generality that this is party P_n. Assume first that p is negligibly close to 1. In that case the parties can apply the above protocol unmodified. Consider then the case where p is not negligibly close to 1. We also have that p is not negligibly close to 0. Hence there is a non-negligible probability that P_n sends a message and a non-negligible probability that P_n does not send a message. The decision of P_n to communicate or not can depend only on four values:

- Its share a_n of a.
- Its share b_n of b.
- Its share u_n of the correlated randomness.
- Its private randomness, call it r_n.

This means that there exist a function $\varrho(a_n, b_n, u_n, r_n) \in \{0, 1\}$ such that P_n communicates iff $\varrho(a_n, b_n, u_n, r_n) = 1$. Observe that the decision can in fact not depend more than negligibly on a_n and b_n. If it did, this would leak information on these shares to the parties $P_1, \ldots, P_{n-1}$ which already know all the other shares. This would in turn leak information on a or b to the parties $P_1, \ldots, P_{n-1}$, which would contradict the simulatability property of the protocol. We can therefore without loss of generality assume that there exist a function $\varrho(u_n, r_n) \in \{0, 1\}$ such that P_n communicates iff $\varrho(u_n, r_n) = 1$.

Assume that with non-negligible probability over the choice of the u_n received by P_n it happens that the function $\varrho(u_n, r_n)$ depends non-negligibly on r_n, i.e., for a uniform r_n it happens with non-negligible probability that $\varrho(u_n, r_n) = 0$ and it also happens with non-negligible probability that $\varrho(u_n, r_n) = 1$. Since r_n is independent of the view of the parties $P_1, \ldots, P_{n-1}$, as it is the private randomness of P_n, it follows that the probability that one of the other parties anticipate a message from P_n is independent of whether $\varrho(u_n, r_n) = 0$ or $\varrho(u_n, r_n) = 1$. Hence it either happens with non-negligible probability that $\varrho(u_n, r_n) = 0$ and yet one of the other parties anticipate a message from P_n or it happens with non-negligible probability that $\varrho(u_n, r_n) = 1$ and yet none of the other parties anticipate a message from P_n. Both events contradict the correctness of the protocol. We can therefore without loss of generality assume that there exist a function $\varrho(u_n) \in \{0, 1\}$ such that P_n communicates iff $\varrho(u_n) = 1$. By assumption we have that p is non-zero, so there exist some u_n such that $\varrho(u_n) = 0$. We can therefore condition the execution on the event $\varrho(u_n) = 0$. Let P_U be the distribution from which u is sampled. Consider then the random variable $P_{U'}$ which is distributed as P_U under the condition that $\varrho(u_n) = 0$. We claim that if we run Π_{MULT} with $P_{U'}$ instead of P_U then the protocol is still secure. Assuming that this claim is true, A and B can apply the above protocol, but simply use $(\Pi_{\mathrm{MULT}}, P_{U'})$ instead of $(\Pi_{\mathrm{MULT}}, P_U)$.

What remains is therefore only to argue that $(\Pi_{\mathrm{MULT}}, P_{U'})$ is secure. To simulate the protocol, run the simulator S'_A for $(\Pi_{\mathrm{MULT}}, P_U)$ until it outputs a simulated execution where P_n did not communicate. Let E be the event that P_n does not communicate. Since it can be checked from just inspecting the view of the real execution of $(\Pi_{\mathrm{MULT}}, P_U)$ (or the simulation) whether E occurred, it follows that E occurs with the same probability in the real execution and the simulation (or at least probabilities which are negligible close) or we could use the occurrence of E to distinguish. Since E happens with a positive constant probability it then also follows that the real execution conditioned on E and the simulation condition on E are indistinguishable, or we could apply a distinguisher for the conditioned distributions when E occurs and otherwise make a random guess to distinguish the real execution of $(\Pi_{\mathrm{MULT}}, P_U)$ from its simulation. This shows that S'_A simulates $(\Pi_{\mathrm{MULT}}, P_{U'})$. $\square$

A generalisation. We note that Theorem 3 easily extends to any output secret sharing scheme with the following property: Given shares $c_1, ..., c_n$ of c, there is a function ϕ such that one can reconstruct c from $c_1, ..., c_{n-1}, \phi(c_n)$ and given c and $c_1, ..., c_{n-1}$ one can simulate $\phi(c_n)$ with statistically close distribution. The proof is the same as above except that in the output phase, B sends $\phi(c_n)$ to A, who computes c and sends it to B.

Theorem 3 shows, for instance, that the SPDZ protocol [DPSZ12] has optimal communication for the class of gate-by-gate protocols using additive secret-sharing: it sends $O(n)$ messages for each multiplication gate, and of course one needs to send $\Omega(n)$ messages if all n players are to communicate, as mandated in the theorem. Note also that in the dishonest majority setting, the privacy threshold of the secret-sharing scheme used has to be $n - 1$, so we cannot have a gap between the reconstruction and privacy thresholds, and so amortisation tricks based on packed secret-sharing cannot be applied. We therefore do not consider any lower bounds for amortised MGP's.

4.2   Protocols Based on Any Secret-Sharing Scheme

Note that if we consider an MGP whose output sharing scheme is not the additive scheme, the protocol $\Pi_{\mathrm{MULT}}^{A,B}$ in the proof of Theorem 4 may not work. This is because it is no longer clear that given your own share of the product and the result, the other party's share is determined. In particular, the distribution of the other share may depend on the preprocessed data we consume and so if we just send that share in the clear, it is not obvious that we can reuse the preprocessing.

The solution is to not send shares in the clear, but have the parties securely compute the output from their shares. This can be done using an existing general protocol for secure computation in the preprocessing model. This will mean that we can indeed reuse preprocessed data consumed by the MGP protocol itself. However, we now consume new preprocessed data for every instance of the reconstruction protocol since this protocol requires communication. It turns out that if we use a variant of the MGP that computes, not just one product, but an inner product of long enough vectors, we can still obtain a contradiction. This works because we can show that computing the inner product of long vectors requires lots of preprocessed data. On the other hand, the inner product itself is just one field element, therefore the cost of reconstructing such a small result is not significant.

In order to obtain the above result and give more details, we proceed by proving some auxiliary results with lower bounds on the amount of preprocessed data needed for a secure evaluation of a function f.

Lower Bounds for Secure Function Evaluation in the Preprocessing Model. In this section we will give lower bounds for secure implementations of functions $f : \mathcal{X} \times \mathcal{Y} \to \mathcal{Z}$ in the P_U, P_V-preprocessing model, which for simplicity of exposition we refer to as P_{U_f, V_f}, that outputs correlated randomness for the semi-honest setting. In particular, we are in the setting where the parties A, B have access to a functionality that gives a random variable U_f to A and V_f to B with some guaranteed joint distribution P_{U_f, V_f} of U_f, V_f. Given this, the parties compute securely a function $f : \mathcal{X} \times \mathcal{Y} \to \mathcal{Z}$ where A holds $x \in \mathcal{X}$, and B holds $y \in \mathcal{Y}$. This function should have no redundant inputs for party A[4]:

$$\forall x, x' \in \mathcal{X}(x \neq x' \to \exists y \in \mathcal{Y} : f(x, y) \neq f(x', y)) \tag{3}$$

The authors of [WW10] obtained Theorem 5 that gives a lower bound on the conditional entropy of P_{U_f, V_f}. Their bound applies for input distributions X and Y which are independent and uniformly distributed. This implies worst case communication complexity. Our bound in Theorem 6 also applies to independent and uniform distributions.

Theorem 5. *Let* $f : \mathcal{X} \times \mathcal{Y} \to \mathcal{Z}$ *be a function that satisfies property (3). Assume there exists a protocol having access to* P_{U_f, V_f} *which is an* ε*-secure implementation of* f *in the semi-honest model with* $t = 1$ *corruptions. Then*

$$H(U_f | V_f) \geq \max_y H(X | f(X, y)) - (3|\mathcal{Y}| - 2)(\varepsilon \log |\mathcal{Z}| + h(\varepsilon)) - \varepsilon \log |\mathcal{X}| - h(\varepsilon).$$

Our general result will only apply to functions where the output lives in a ring $\mathcal{Z}$. As it will become apparent, for the next theorem we require the following property for a function $f : \mathcal{X} \times \mathcal{Y} \to \mathcal{Z}$:

$$\forall x, x' \in \mathcal{X}(x \neq x' \to \exists y_1, y_2 \in \mathcal{Y} : f(x, y_1) - f(x, y_2) \neq f(x', y_1) - f(x', y_2)) \tag{4}$$

---

[4] Party A must enter all the information about X into the protocol. An example of a function that satisfies this property is the inner product IP.

Note that the bound in Theorem 5 still applies for functions f that satisfy properties (3) and (4).

In the following we explore the lower bounds on the amount of preprocessed data with respect to composition of functions. In Theorem 6 we prove a lower bound on the conditional entropy of P_{U_h, V_h} for a function h which is a linear combination of two functions f and g. Our bound also applies to compositions of k functions where k is an arbitrary number. Basically, we show that the amount of preprocessed data you need to compute the sum of f and g is the sum of what you need to compute f and g separately, as long as f and g are applied to distinct and independent inputs. We clearly need this assumption, as otherwise the theorem is clearly false, just think of applying $f = g$ on the same inputs.

Theorem 6. *Let $f : \mathcal{X} \times \mathcal{Y} \to \mathcal{Z}_f$, $g : \mathcal{Z} \times \mathcal{W} \to \mathcal{Z}_g$ be functions that satisfy properties (3) and (4). Assume that $\mathcal{Z}_f = \mathcal{Z}_g$. Let h be a linear combination of f and g, namely: $\forall x \in \mathcal{X}, y \in \mathcal{Y}, z \in \mathcal{Z}, w \in \mathcal{W}, h(x, z, y, w) := \alpha f(x, y) + \beta g(z, w)$ for some $\alpha, \beta \neq 0$. If there exists a protocol that securely implements the function h with access to P_{U_h, V_h}, then it holds that*

$$H(U_h | V_h) \geq \max_y H(X | f(X, y)) + \max_w H(Z | g(Z, w)).$$

Furthermore, the function h will have the following property:

$$\forall x \neq x' \in \mathcal{X}, z \neq z' \in \mathcal{Z} \; \exists y_1, y_2 \in \mathcal{Y}, w_1, w_2 \in \mathcal{W} :$$
$$h(x, z, y_1, w_1) - h(x, z, y_2, w_2) \neq h(x', z', y_1, w_1) - h(x', z', y_2, w_2) \, (5)$$

Proof. We start by proving that the function h has this property:

$$\forall x, x' \in \mathcal{X}, z, z' \in \mathcal{Z}((x, z) \neq (x', z')) \to$$
$$\exists y \in \mathcal{Y}, w \in \mathcal{W} : h(x, z, y, w) \neq h(x', z', y, w) \, (6)$$

By assumption we consider the following two properties on the function g:

$$\forall z \neq z' \in \mathcal{Z} \; \exists w \in \mathcal{W} : g(z, w) \neq g(z', w) \tag{7}$$

$$\forall z \neq z' \in \mathcal{Z} \; \exists w_1, w_2 \in \mathcal{W} : g(z, w_1) - g(z, w_2) \neq g(z', w_1) - g(z', w_2) \tag{8}$$

and properties (3) and (4).

In order to prove properties (6) and (5) for the function h we proceed as follows:

Case 1. $x = x', z \neq z'$:
   Suppose that $\exists y$ such that $f(x', y) = f(x', y)$. By assumption $\exists w \in \mathcal{W} : g(z, w) \neq g(z', w)$. Therefore, it follows that $f(x', y) - f(x, y) \neq g(z, w) - g(z', w)$ and property (6) holds.
Case 2. $x \neq x', z = z'$:
   Suppose that $\exists w$ such that $g(z', w) = g(z', w)$. By assumption $\exists y \in \mathcal{Y} : f(x, y) \neq g(x', y)$. It follows that $f(x', y) - f(x, y) \neq g(z, w) - g(z', w)$ and property (6) holds.

Case 3. $x \neq x', z \neq z'$:

      Let $c = f(x', y) - f(x, y)$ for some $y \in \mathcal{Y}$. By assumption $\exists w_1, w_2 \in \mathcal{W}$ such that $c_1 = g(z, w_1) - g(z', w_1)$ and $c_2 = g(z, w_2) - g(z', w_2)$ such that $c_1 \neq c_2$. Without loss of generality, assume that $c \neq c_1$ then $f(x', y) - f(x, y) \neq g(z, w_1) - g(z', w_1)$ and property (5) follows.

Since the function h satisfy property (6) it also has property (3) and hence we get from Theorem 5 that

$$H(U_h|V_h) \geq \max_{y,w} H(X, Z|h(X, Z, y, w)).$$

We then get that:

$$H(U_h|V_h) \geq \max_{y,w} H(X, Z|\alpha f(X, y) + \beta g(Z, w)) \tag{9}$$

$$\geq \max_{y,w} H(X, Z|f(X, y), g(Z, w)) \tag{10}$$

$$\geq \max_{y} H(X|f(X, y)) + \max_{w} H(Z|g(Z, w)) \tag{11}$$

Inequality (11) follows from the independence of X, Z. This proves the theorem. $\square$

Remark 1. The above theorem also applies to multiplicative relations ruling out the cases where $g(z, w) = 0$ and $f(x, y) = 0$.

Exploiting Theorem 6 we prove a lower bound for the inner product function IP_k as per Definition 6.

Lemma 1. *Let $\kappa \geq 1$ and let $f : \mathcal{X} \times \mathcal{Y} \to \mathcal{Z}$ be a multiplication function as per Definition 5. If there exist a protocol Π_{IP_k} which securely implements the inner product function IP_k with error probability ε in the semi-honest model and having access to $P_{U_{\text{IP}_k} V_{\text{IP}_k}}$ then*

$$H(U_{\text{IP}_k}|V_{\text{IP}_k}) \geq k \cdot \max_{y} H(X|f(X, y)) \tag{12}$$

Proof. Since the function f satisfies properties (3) and (4), a straightforward application of Theorem 6 for $k = 2$ yields $H(U_{\text{IP}_2}|V_{\text{IP}_2}) \geq 2 \cdot \max_y H(X|f(X, y))$. However it is easy to see that the proof of Theorem 6 extends to addition of k functions for any k, so the lemma follows in the same way from this more general result. $\square$

Utilising Theorem 6 in the following we prove that any function whose "preprocessing complexity" is large enough requires lots of communication. What "large enough" means here is determined by the output secret-sharing scheme used in the protocol, in a sense we make precise below. In the following, when f is a function with two inputs and one output, we will speak about *a protocol for computing shares of an f-output*, denoted by $\Pi_{f-output}$. This is essentially the same as an MGP except that we replace multiplication by f. So the protocol takes as input shares of x_1 and x_2 and computes shares of $f(x_1, x_2)$ as output.

Note that the inputs x_1, x_2 may be vectors of field elements, whereas we will by default assume that the output is a single field element.

In the sequel, for simplicity of exposition let L_f denote a lower bound on the amount of preprocessed data needed for a secure implementation of f in the preprocessing model and let U_f denote an upper bound.

Reconstruction Protocol Π_{rec}. Let $\mathcal{S}_t^n$ be the secret-sharing scheme as per Definition 3 and let $f'_{\mathcal{S}_t^n}$ be the reconstruction function of $\mathcal{S}_t^n$. Then, we can securely implement the function $f'_{\mathcal{S}_t^n}$ in the preprocessing model via the protocol Π_{SPDZ} yielding the protocol Π_{rec}[5]. It follows that Π_{rec} demands communication and that its complexity depends only on the underlying secret-sharing scheme $\mathcal{S}_t^n$. In this case we obtain an upper bound U_{rec} on the amount of preprocessed data consumed by Π_{rec}.

Theorem 7. *Consider the preprocessing model where t of the n players may be passively corrupted. Let Π_{rec} be a secure output reconstruction protocol with access to $P_{\mathsf{U}_{rec}, \mathsf{V}_{rec}}$ for the secret-sharing scheme $\hat{\mathcal{S}}_t^n$. Let f be a function with two inputs and one field element as output such that $\mathsf{U}_{rec} < \mathsf{L}_f$. There exists no passively secure n-player protocol $\Pi_{f-output}$ with expected anticipated communication complexity $\leq t$ for computing shares of an f-output with $\hat{\mathcal{S}}_t^n$ as output secret-sharing scheme.*

Proof. We start by assuming a fixed communication pattern. Suppose for contradiction that there exists a protocol Π_f where at most t players communicate. Assume that it is the t first parties. Given two parties A and B, we are going to construct a two-party protocol $\Pi_f^{A,B}$ which on input a, b from A, B, respectively, securely computes $f(a, b)$. The idea is to execute the $\Pi_{f-output}$ protocol in which A emulates the t players who communicate while B emulates the rest of the parties but we are interest just for one additional party, say P_{t+1}. In particular, protocol $\Pi_f^{A,B}(a, b)$ proceeds as follows:

Protocol $\Pi_f^{A,B}(a, b)$:
  Input Phase:
  1. Parties A, B secret share their inputs a, b using the secret-sharing scheme $\mathcal{S}_t^n$. More specifically, A computes $[a]^{\mathcal{S}_t^n} \leftarrow \mathsf{Share}(a, n, t)$ and B computes $[b]^{\mathcal{S}_t^n} \leftarrow \mathsf{Share}(b, n, t)$.
  2. Party A sends the input share $(a_{t+1}, \ldots, a_n)$ to party B and Party B sends the input shares $(b_1, \ldots, b_t)$ to party A.
  Evaluation Phase:
  1. Parties A, B invoke the protocol $\Pi_{f-output}$ where A emulates the t players who communicate, and we assume these are the first t players. This means that this phase involves no communication between A and B, but it may consume some preprocessed data. The execution of $\Pi_{f-output}$ yields a sharing of $[c]^{\mathcal{S}_t^n}$ and outputs $(c_1, ..., c_t)$ to party A and $(c_{t+1}, \ldots, c_n)$ to party B.

---

[5] Note that any protocol in the preprocessing model can be used.

Output Phase:
1. Both parties locally invoke protocol Π_{Rec} with access to $P_{U_{rec}, V_{rec}}$ which on input $[c]^{\hat{S}_t^n}$ outputs the result $f(a, b)$.

Correctness of the protocol follows immediately from the correctness of $\Pi_{f-output}$ and Π_{Rec}. The protocol can be argued to be secure(private). More specifically, the simulator $S_{\mathcal{A}}$ of $\Pi_f^{A,B}$ proceeds as follows. In the input phase, the parties receive only an unqualified set of shares whose distribution can be simulated perfectly. There is no communication to be simulated in the evaluation phase. In the output phase, simulation is guaranteed by the invocations of the sub-simulator of the secure protocol Π_{Rec}. Hence, it follows that

$$\mathsf{SD}(\mathsf{Exec}_{\mathsf{Adv}}^{\Pi_f^{A,B}}(\sigma, (a, b)), \mathsf{Sims}_{\mathsf{S}_{\mathcal{A}}}(\sigma, (a, b))) \leq \varepsilon(\sigma).$$

We can claim the following: Note that the communication in $\Pi_f^{A,B}$ is actually independent of the preprocessed data needed in order to securely compute f. Therefore, while reusing the same preprocessed data for each invocation of $\Pi_{f-output}$, we could have executed ℓ instances of $\Pi_f^{A,B}$ on independent inputs without affecting correctness since the simulation is independent of the preprocessed data. However, since protocol Π_{Rec} is interactive its preprocessed data must be refreshed for each of the ℓ executions of Π_{Rec}. This means that the amount of preprocessed data needed in order to compute ℓ instances of f is $U_f + \ell \cdot U_{rec}$. So if we consider an ℓ large enough such that $\ell \cdot L_f > U_f + \ell \cdot U_{rec}$, we have a contradiction and the theorem follows.

The generalization to dynamic communication patterns follows along the lines of the proof of Theorem 4: there we split the players in a maximal unqualified set ($n-1$ players) and the rest (1 player). Here we do the same except that the maximal unqualified set has t players and $n-t$ remain. We then argue exactly as in the proof of Theorem 4 that decisions to send/receive cannot depend on private randomness or shares, and therefore we can build a new protocol that can be used in our construction of a 2-party protocol.   □

Given a function f with one output and a non-zero lower bound, we can add it to itself on distinct inputs a sufficient number of times in order to satisfy the condition in the above theorem. An example of a function f is the inner product function IP_k which is the composition of k MULT functions. In Lemma 1 we obtained a lower bound L_{IP^k} on the amount of preprocessed data consumed by a protocol that securely implements the function IP^k. Now, if k is large enough to satisfy the condition $U_{rec} < L_{IP_k}$, then it holds that $\ell \cdot U_{rec} + L_{MULT} < \ell \cdot L_{IP_k}$ for large enough ℓ leading to a contradiction with Theorem 7.

5   Conclusions

We have shown that any information-theoretic secure protocol that follows the *typical* gate-by-gate design pattern must communicate for every multiplication gate, even if only semi-honest security is required, for both honest majority and dishonest majority with preprocessing where the target secret sharing scheme is an additive one. We have also shown similar results for any target secret sharing scheme in the dishonest majority setting. This highlights a reason why, even with preprocessing, all known protocols which are efficient in the circuit size $|C|$ of the evaluated function require $\Omega(n|C|)$ communication and $\Omega(d_C)$ rounds where d_C is the depth of C. Our result implies that a fundamental new approach must be found in order to construct protocols with reduced communication complexity that beat the complexities of BGW, GMW, SPDZ etc. Of course, it is also possible that our bounds hold for *any* protocol efficient in the circuit size of the function, and this is the main problem we leave open. Another open problem is to find unrestricted bounds on MGPs for parallel multiplications.

References

[BFKR90] Beaver, D., Feigenbaum, J., Kilian, J., Rogaway, P.: Security with low communication overhead. In: Menezes, A., Vanstone, S.A. (eds.) CRYPTO 1990. LNCS, vol. 537, pp. 62–76. Springer, Heidelberg (1991)

[BGW88] Ben-Or, M., Goldwasser, S., Wigderson, A.: Completeness theorems for non-cryptographic fault-tolerant distributed computation (extended abstract). In: 20th Annual ACM Symposium on Theory of Computing, pp. 1–10. ACM Press, May 1988

[BI05] Barkol, O., Ishai, Y.: Secure computation of constant-depth circuits with applications to database search problems. In: Shoup, V. (ed.) CRYPTO 2005. LNCS, vol. 3621, pp. 395–411. Springer, Heidelberg (2005)

[BOGW88] Ben-Or, M., Goldwasser, S., Wigderson, A.: Completeness theorems for non-cryptographic fault-tolerant distributed computation. In: Proceedings of the Twentieth Annual ACM Symposium on Theory of Computing, STOC 1988, pp. 1–10. ACM, New York (1988)

[BSFO12] Ben-Sasson, E., Fehr, S., Ostrovsky, R.: Near-linear unconditionally-secure multiparty computation with a dishonest minority. In: Safavi-Naini, R., Canetti, R. (eds.) CRYPTO 2012. LNCS, vol. 7417, pp. 663–680. Springer, Heidelberg (2012)

[BSPV99] Blundo, C., De Santis, A., Persiano, G., Vaccaro, U.: Randomness complexity of private computation. Comput. Complex. **8**(2), 145–168 (1999)

[BTH08] Beerliová-Trubíniová, Z., Hirt, M.: Perfectly-Secure MPC with linear communication complexity. In: Canetti, R. (ed.) TCC 2008. LNCS, vol. 4948, pp. 213–230. Springer, Heidelberg (2008)

[CCD88] Chaum, D., Crépeau, C., Damgård, I.: Multiparty unconditionally secure protocols (extended abstract). In: 20th Annual ACM Symposium on Theory of Computing, pp. 11–19. ACM Press, May 1988

[CF02] Cramer, R., Fehr, S.: Optimal black-box secret sharing over arbitrary abelian groups. In: Yung, M. (ed.) CRYPTO 2002. LNCS, vol. 2442, pp. 272–287. Springer, Heidelberg (2002)

[CK93] Chor, B., Kushilevitz, E.: A communication-privacy tradeoff for modular addition. Inf. Process. Lett. **45**(4), 205–210 (1993)

[DIK+08] Damgård, I., Ishai, Y., Krøigaard, M., Nielsen, J.B., Smith, A.: Scal-
          able multiparty computation with nearly optimal work and resilience. In:
          Wagner, D. (ed.) CRYPTO 2008. LNCS, vol. 5157, pp. 241–261. Springer,
          Heidelberg (2008)

[DIK10]   Damgård, I., Ishai, Y., Krøigaard, M.: Perfectly secure multiparty com-
          putation and the computational overhead of cryptography. In: Gilbert,
          H. (ed.) EUROCRYPT 2010. LNCS, vol. 6110, pp. 445–465. Springer,
          Heidelberg (2010)

[DN07]   Damgård, I., Nielsen, J.B.: Scalable and unconditionally secure multiparty
          computation. In: Menezes, A. (ed.) CRYPTO 2007. LNCS, vol. 4622, pp.
          572–590. Springer, Heidelberg (2007)

[DNOR15] Damgård, I., Nielsen, J.B., Ostovsky, R., Rosen, A.:Unconditionally secure
          computation with reduced interaction. Cryptology ePrint Archive, Report
          2015/630 (2015). http://eprint.iacr.org/

[DPP14]   Data, D., Prabhakaran, M.M., Prabhakaran, V.M.: On the communi-
          cation complexity of secure computation. In: Garay, J.A., Gennaro, R.
          (eds.) CRYPTO 2014, Part II. LNCS, vol. 8617, pp. 199–216. Springer,
          Heidelberg (2014)

[DPSZ12] Damgård, I., Pastro, V., Smart, N., Zakarias, S.: Multiparty computation
          from somewhat homomorphic encryption. In: Safavi-Naini, R., Canetti, R.
          (eds.) CRYPTO 2012. LNCS, vol. 7417, pp. 643–662. Springer, Heidelberg
          (2012)

[FKN94]   Feige, U., Kilian, J., Naor, M.: A minimal model for secure computation
          (extended abstract). In: 26th Annual ACM Symposium on Theory of Com-
          puting, pp. 554–563. ACM Press, May 1994

[FY92]   Franklin, M.K., Yung, M.: Communication complexity of secure computa-
          tion (extended abstract). In: 24th Annual ACM Symposium on Theory of
          Computing, pp. 699–710. ACM Press, May 1992

[GIP+14] Genkin, D., Ishai, Y., Prabhakaran, M., Sahai, A., Tromer, E.: Circuits
          resilient to additive attacks with applications to secure computation. In:
          Shmoys, D.B. (ed.) 46th Annual ACM Symposium on Theory of Comput-
          ing, pp. 495–504. ACM Press, May / June 2014

[GIP15]   Genkin, D., Ishai, Y., Polychroniadou, A.: Efficient multi-party computa-
          tion: from passive to active security via secure SIMD circuits. In: Gennaro,
          R., Robshaw, M. (eds.) CRYPTO 2015. LNCS, vol. 9216, pp. 721–741.
          Springer, Heidelberg (2015)

[GR03]   Gál, A., Rosén, A.: Lower bounds on the amount of randomness in private
          computation. In: 35th Annual ACM Symposium on Theory of Computing,
          pp. 659–666. ACM Press, June 2003

[IK00]   Ishai, Y., Kushilevitz, E.: Randomizing polynomials: A new representation
          with applications to round-efficient secure computation. In: 41st Annual
          Symposium on Foundations of Computer Science, pp. 294–304. IEEE Com-
          puter Society Press, November 2000

[IKM+13] Ishai, Y., Kushilevitz, E., Meldgaard, S., Orlandi, C., Paskin-Cherniavsky,
          A.: On the power of correlated randomness in secure computation. In:
          Sahai, A. (ed.) TCC 2013. LNCS, vol. 7785, pp. 600–620. Springer,
          Heidelberg (2013)

[IPS09]   Ishai, Y., Prabhakaran, M., Sahai, A.: Secure arithmetic computation with
          no honest majority. In: Reingold, O. (ed.) TCC 2009. LNCS, vol. 5444, pp.
          294–314. Springer, Heidelberg (2009)

[KM97]   Kushilevitz, E., Mansour, Y.: Randomness in private computations. SIAM J. Discrete Math. **10**(4), 647–661 (1997)

[KR94]   Kushilevitz, E., Rosén, A.: A randomness-rounds tradeoff in private computation. In: Desmedt, Y.G. (ed.) CRYPTO 1994. LNCS, vol. 839, pp. 397–410. Springer, Heidelberg (1994)

[Kus92]   Kushilevitz, E.: Privacy and communication complexity. SIAM J. Discrete Math. **5**(2), 273–284 (1992)

[NN01]   Naor, M., Nissim, K.: Communication preserving protocols for secure function evaluation. In: 33rd Annual ACM Symposium on Theory of Computing, pp. 590–599. ACM Press, July 2001

[NNOB12]   Nielsen, J.B., Nordholt, P.S., Orlandi, C., Burra, S.S.: A new approach to practical active-secure two-party computation. In: Safavi-Naini, R., Canetti, R. (eds.) CRYPTO 2012. LNCS, vol. 7417, pp. 681–700. Springer, Heidelberg (2012)

[RBO89]   Rabin, T., Ben-Or, M.: Verifiable secret sharing and multiparty protocols with honest majority (extended abstract). In: 21st Annual ACM Symposium on Theory of Computing, pp. 73–85. ACM Press, May 1989

[WW10]   Winkler, S., Wullschleger, J.: On the efficiency of classical and quantum oblivious transfer reductions. In: Rabin, T. (ed.) CRYPTO 2010. LNCS, vol. 6223, pp. 707–723. Springer, Heidelberg (2010)

Obfuscation

Universal Constructions and Robust Combiners for Indistinguishability Obfuscation and Witness Encryption

Prabhanjan Ananth[1(✉)], Aayush Jain[1], Moni Naor[2], Amit Sahai[1], and Eylon Yogev[2]

[1] Center for Encrypted Functionalities and Department of Computer Science, UCLA, Los Angeles, USA
{prabhanjan,aayush,sahai}@cs.ucla.edu
[2] Department of Computer Science, Weizmann Institute of Science, Rehovot, Israel
{moni.naor,eylon.yogev}@weizmann.ac.il

Abstract. Over the last few years a new breed of cryptographic primitives has arisen: on one hand they have previously unimagined utility and on the other hand they are not based on simple to state and tried out assumptions. With the on-going study of these primitives, we are left with several different candidate constructions each based on a different, not easy to express, mathematical assumptions, where some even turn out to be insecure.

A *combiner* for a cryptographic primitive takes several candidate constructions of the primitive and outputs one construction that is as good as any of the input constructions. Furthermore, this combiner must be efficient: the resulting construction should remain polynomial-time even when combining polynomially many candidate. Combiners are especially important for a primitive where there are several competing constructions whose security is hard to evaluate, as is the case for indistinguishability obfuscation (IO) and witness encryption (WE).

One place where the need for combiners appears is in design of a *universal construction*, where one wishes to find "one construction to rule

P. Ananth—Partially supported by grant #360584 from the Simons Foundation. Partially supported by grants under Amit Sahai.

A. Jain—Supported by grants under Amit Sahai.

M. Naor—Research supported in part by grants from the Israel Science Foundation grant no. 1255/12, BSF and from the I-CORE Program of the Planning and Budgeting Committee and the Israel Science Foundation (grant no. 4/11). Moni Naor is the incumbent of the Judith Kleeman Professorial Chair.

A. Sahai—University of California Los Angeles and Center for Encrypted Functionalities. Research supported in part from a DARPA/ARL SAFEWARE award, NSF Frontier Award 1413955, NSF grants 1228984, 1136174, 1118096, and 1065276, a Xerox Faculty Research Award, a Google Faculty Research Award, an equipment grant from Intel, and an Okawa Foundation Research Grant. This material is based upon work supported by the Defense Advanced Research Projects Agency through the ARL under Contract W911NF-15-C-0205. The views expressed are those of the author and do not reflect the official policy or position of the Department of Defense, the National Science Foundation, or the U.S. Government.

M. Robshaw and J. Katz (Eds.): CRYPTO 2016, Part II, LNCS 9815, pp. 491–520, 2016.
DOI: 10.1007/978-3-662-53008-5_17

them all": an explicit construction that is secure if *any* construction of the primitive exists.

In a recent paper, Goldwasser and Kalai posed as a challenge finding universal constructions for indistinguishability obfuscation and witness encryption. In this work we resolve this issue: we construct universal schemes for IO, and for witness encryption, and also resolve the existence of combiners for these primitives along the way. For IO, our universal construction and combiners can be built based on *either* assuming DDH, or assuming LWE, with security against subexponential adversaries. For witness encryption, we need only one-way functions secure against polynomial time adversaries.

1   Introduction

We live in a golden, but dangerous, age for cryptography. New primitives are proposed along with candidate constructions that achieve things that were previously in the realm of science fiction. Two such notable examples are *indistinguishability obfuscation*[1] (IO), and *witness encryption*[2] (WE). However, at the same time, we are seeing a steady stream of new attacks on assumptions that are underlie, or at least are closely related to, these new candidates. With this proliferation of constructions and assumptions comes the question: how do we evaluate these various assumptions, which constructions do we choose and how do we actually use them?

What is better: one candidate construction of indistinguishability obfuscation (IO) or two such candidate constructions? What about a polynomial-sized family of candidates? The usual approach should be "the more the merrier", but how do we use these several candidates to actually obfuscate? The relevant notion is that of a *combiner*: it takes several candidates for a primitive and produces one instance of the primitive so that if any of the original ones is a secure construction then the result is a secure primitive. Furthermore, this combiner must be efficient: the resulting construction should remain polynomial-time. Another issue is what do we assume about the insecure constructions. Are they at least correct, i.e. do they maintain the functionality, or can they be arbitrarily faulty? We are interested in a combiner that adds very little complexity to the basic underlying schemes and assumes as little as possible regarding the insecure schemes, i.e. they may be completely dysfunctional. Furthermore, we would like the assumptions underlying our combiner to be as minimal and standard as possible.

One Candidate to Rule Them All (Theoretically Speaking). In fact, we can even go further: A closely related issue to the existence of combiners is that

---

[1] Indistinguishability obfuscation is the ability to scramble a program so that it is not possible to decide what was the source code out of two semantically equivalent options.

[2] Witness encryption is a method for encrypting a message relative to a string x and language L so that anyone with a witness w that $x \in L$ can decrypt but if $x \notin L$ then no information about the message is leaked.

of a *universal construction* of a primitive: a concrete construction of the primitive that is secure if *any* secure construction exists. In the context of candidate constructions, a universal IO candidate would change the game considerably between attacker and defender: Currently, each IO candidate is based on specific mathematical techniques, and a cryptanalysis of each candidate can be done by finding specific weaknesses in the underlying mathematics. With a universal IO candidate, the only way to give a cryptanalysis of this candidate would be to prove that no secure IO scheme exists. To the best of our knowledge, no plausible approaches have been proposed for obtaining such a proof. Thus, a universal IO scheme would vastly raise the bar on what an attacker must do.

Furthermore, intriguingly, we note that IO exists if **P=NP**. In contrast to other objects in cryptography, IO by itself does not imply hardness. This raises the possibility of a future non-constructive existence proof for IO, even without needing to resolve **P** vs **NP**. If we have a universal IO scheme, then any such non-constructive proof would be made explicit: the universal IO scheme would be guaranteed to be secure.

Indeed, in a recent opinion paper regarding assumptions Goldwasser and Kalai [19] wrote:

> *We pose the open problem of finding a universal instantiations for other generic assumptions, in particular for IO obfuscation, witness encryption, or 2-message delegation for NP.*

In this work we resolve two out of those three primitives, namely IO and witness encryption, for security against subexponential adversaries for IO, and polynomial adversaries for witness encryption. Our universal constructions also resolve the existence of combiners for these primitives along the way. For IO, our universal construction and combiners can be built based on *either* assuming DDH, or assuming LWE, with security against subexponential adversaries. For witness encryption, we need only one-way functions secure against polynomial adversaries.

Robust IO Combiners. We construct both (standard) combiners and robust combiners. A (standard) combiner handles only security: the promise is that all given candidates are correct, but only one is promised to be secure. These combiners are useful when different schemes are based on different hardness assumptions, but they all have a proof of correctness. The resulting combined scheme will be correct and as secure as all the underlying assumptions.

A robust combiner handles the case where security *and* correctness are both promised only for a single candidate. We only know of constructing universal schemes from *robust* combiners and in particular, (standard) combiners does not suffice.

The Status of IO Schemes or – Are We Dead Yet? The state of the art of IO is in flux. There is a steady stream of proposals for constructions and a similar stream of attacks on various aspects of the constructions. In order to clarify the state of the art in the full version [2] we provide a detailed explanation of the constructions, the attacks and what implications they have (a summary is provided in Fig. 13 of the full version). As of now (June 2016) there is no argument or attack known that implies that all iO schemes or primitives used by them are broken.

Brief History of Combiners and Universal Cryptographic Primitives. The notion of a combiner and its connection to universal construction were formalized by Harnik [21] (see also Herzberg [22,23]). An early instance of a combiner for encryption is that of Asmuth and Blakely [8]. A famous example of a universal construction (and the source of the name) is that of one-functions due to Levin [27] (for details see Goldreich [17, Sect. 2.4.1]).

Related Work. Concurrent to our work, Fischlin et al. [13], building upon [24], also studied the notion of robust obfuscation combiners. The security notions considered in their work also deal with virtual black box obfuscation and virtual gray box obfuscation, that are not dealt with in our work. However, they achieve a much weaker result: they can only combine a *constant* number of candidates and furthermore, they assume that a majority of the candidates are correct. Thus, their combiners are not useful to obtaining any implication to universal indistinguishability obfuscation.

1.1   Our Results

Our first result is a construction of an IO combiner. We give two separate constructions, one using LWE, and other using DDH. Thus, we can build IO combiners from two quite different assumptions.

Theorem 1 (Informal). *Under the hardness of Learning with Errors (LWE) and IO secure against sub-exponential adversaries, there exist an IO combiner.*

Theorem 2 (Informal). *Under the hardness of Decisional Diffie-Hellman (DDH) and IO secure against sub-exponential adversaries, there exist an IO combiner.*

We show how to adapt the LWE-based IO combiner to obtain a universal IO scheme.

Theorem 3 (Informal). *Under the hardness of Learning with Errors (LWE) against sub-exponential adversaries and the existence of IO secure against sub-exponential adversaries, there exists a universal IO scheme.*

For witness encryption, we have similar results, under assumptions widely believed to be weaker. We prove the following theorem.

Theorem 4 (Informal). *If one-way functions exist, then there exist a secure witness encryption combiner.*

Again, we extend this and get a universal witness encryption scheme.

Theorem 5 (Informal). *If one-way functions and witness encryption exist, then there is a universal witness encryption scheme.*

Theorem 5 assumes the existence of one-way functions. Notice that if P = NP then WE exist, however, one-way functions do not. Thus, in most cryptographic application one-way functions are used as an additional assumption. Nevertheless, we can make a stronger statement: If there exist any hard-on-average language in NP then there is a universal WE scheme. In [25] it was shown that the existence of witness encryption and a hard-on-average language in NP implies the existence of one-way functions. By combing this with Levin's universal one-way function [27] we obtain our result.

In full version, we present the constructions of universal secret sharing for NP and universal witness PRFs. Both these constructions assume only one-way functions.

2   Techniques

We present the technical challenges and describe how we overcome them.

2.1   Universal Obfuscation

A natural starting point is to revisit the construction of universal one-way functions [27] – constructions of other known universal cryptographic primitives [21] have the same flavor. An explicit function f is said to be a universal one-way function if the mere existence of any one-way function implies that f is one-way.

The universal one-way function f_{univ} on input $x = y_1||\ldots||y_\ell$, where $|x| = \ell^2$, executes as follows[3]:

1. Interpret the integer $i \in \{1, \ldots, \ell\}$ as a Turing machine M_i. This interpretation is quite standard in the computational complexity literature[4].
2. Output $M_1(y_1)||\cdots||M_\ell(y_\ell)$.

To argue security, we exploit the fact that there exists a secure one-way function represented by Turing machine M_{owf}. Let ℓ_0 be an integer that can be interpreted as M_{owf}. We argue that it is hard to invert $M_{owf}(x)$, where x has length at least ℓ_0^2 and is drawn uniformly at random. To see why, notice that in Step 1, M_{owf} will be included in the enumeration. From the security of M_{owf} it follows that it is hard to invert $M_{owf}(y_{\ell_0})$, where y_{ℓ_0} is the ℓ_0^{th} block of x. This translates to the un-invertibility of $f_{univ}(x)$. This proves that f_{univ} is one-way[5].

Let us try to emulate the same approach to obtain universal indistinguishability obfuscation. On input circuit C, first enumerate the Turing machines $M_1, \ldots, M_\ell$, where ℓ here is the size of the circuit C. We interpret M_i's as indistinguishability obfuscators. It is not clear how to implement the second step in

---

[3] If x can not be expressed of this form then suitably truncate x till it is of this form.

[4] This fact was used to prove the famous Gödel's incompleteness theorem [16].

[5] Note that the definition of one-way function only requires un-invertibility to hold for sufficiently long inputs. This requirement is satisfied by f_{univ} as its un-invertibility holds for inputs of lengths greater than ℓ_0^2.

the context of obfuscation – unlike one-way functions we cannot naïvely break the circuit into blocks and individually obfuscate each block. We need a mechanism to jointly obfuscate a circuit using multiple obfuscators $M_1, \ldots, M_\ell$ such that the security of the joint obfuscation is guaranteed as long as one of the obfuscators is secure. This is where indistinguishability obfuscation combiners come in. Designing combiners for indistinguishability obfuscation involves a whole new set of challenges and we deal with them in a separate section (Sect. 2.2). For now, we assume we have such combiners at our disposal.

Warmup Attempt. Using combiners for IO, we propose the following approach to achieve universal obfuscation. The universal obfuscator IO_{univ} on input circuit C executes the following steps:

1. Interpret the integer $i \in \{1, \ldots, \ell\}$ as a Turing machine M_i.
2. Obfuscate C by applying the IO combiner on the machines $M_1, \ldots, M_\ell$. Output the result $\overline{C}$ of the IO combiner.

Unlike the case of one-way functions, in addition to security we need to argue correctness of the above scheme. An obfuscator M_i is said to be correct if the obfuscated circuit $M_i(C)$ is equivalent to C (or agrees on most inputs) and this should be true *for every circuit* C. This in turn depends on the correctness of obfuscators $M_1, \ldots, M_\ell$. But we don't have any guarantee on the correctness of $M_1, \ldots, M_\ell$.

Test-and-Discard. We handle this by first checking for every i whether the obfuscator M_i is correct. This is infeasible in general. However, we test the correctness of M_i only on the particular circuit obfuscated by M_i during the execution of the universal obfuscation. In more detail, suppose we execute IO_{univ} on circuit C and during the execution of the IO combiner, let $[\mathbf{C}]_i$ (derived from C) be the circuit that we obfuscate using machine M_i. Then we test whether $M_i([\mathbf{C}]_i)$ agrees with M_i on significant fraction of inputs. This can be done by picking inputs at random and testing whether both circuits (obfuscated and unobfuscated) agree on these inputs. If M_i fails the test, it is discarded. If it passes the test, then M_i cannot be used directly since $M_i([\mathbf{C}]_i)$ could agree with $[\mathbf{C}]_i$ on $(1 - 1/\text{poly})$-fraction of inputs and yet it could pass the test with non-negligible probability. So we need to reduce the error probability of $M_i([\mathbf{C}]_i)$ to negligible before it is ready to be used.

Correctness Amplification. A first thought would be to use the recent work that shows an elegant correctness amplification for IO by Bitansky and Vaikuntanathan [6]. In particular, they show how to transform an obfuscator that is correct on at least $(1/2 + 1/\text{poly})$-fraction of inputs into one that is correct on all inputs. At first glance this seems to be "just what the doctor ordered", there is, however, one catch here: their transformation is guaranteed to work if the obfuscator is correct *for every circuit* C on at least $(1/2 + 1/\text{poly})$-fraction of inputs. However, we are only ensured that it is approximately correct on *only one* circuit! Nonetheless we show how to realize correctness amplification with respect to a single circuit and ensure that $M_i([\mathbf{C}]_i)$ does not agree with $[\mathbf{C}]_i$ on only negligible fraction of inputs. Once we perform the error amplification, the obfuscator M_i will be used in the IO combiner. In the end, the result of the IO combiner will be an obfuscated circuit $\overline{C}$; the correctness guarantees of $M_i([\mathbf{C}]_i)$, for every i, translate to the corresponding correctness guarantee of $\overline{C}$.

Handling Selective Abort Obfuscators. We now move on to security. For two equivalent circuits C_0, C_1, we need to argue that their obfuscations are computationally indistinguishable. To do this, we need to rely on the security of IO combiner. The security of IO combiner requires that as long as one of the machines M_i is a secure obfuscator[6] then the joint obfuscation of C_0 using $M_1, \ldots, M_\ell$ is indistinguishable from the joint obfuscation of C_1 using the same candidates. The fact that same candidates are used is crucial here since the final obfuscated circuit could potentially reveal the description of the obfuscators combined.

However, there is no such guarantee offered in our case! Recall that we have a 'test-and-discard' phase where we potentially throw out some obfuscators. It might be the case that a particular candidate M_{mal} is correct only on circuits derived from C_0 but fails on circuits derived from C_1. We call such obfuscators *selective abort obfuscators*. Clearly, selective abort obfuscators can lead to a complete break of security. In fact, if there are ℓ obfuscators used then potentially $\ell - 1$ of them could be of selective abort type. To protect against these adversarial obfuscators we ensure that the distribution of the ℓ derived circuits is computationally independent from the circuit to obfuscate.

Issue of Runtime. While the above ideas ensure correctness and security, we haven't yet shown that our scheme is efficient. In fact it could potentially be the case our scheme never halts on some inputs[7]. This could happen since we have no a priori knowledge on the runtime of the obfuscators considered. We propose a naïve solution to this problem: we assume the knowledge of an upper bound on the runtime of the actually secure obfuscator. In some sense, the assumption of time bound might be inherent – without this we are required to predict a bound on the runtime of a Turing machine and we know in general this is an undecidable problem.

2.2   Combiners for Indistinguishability Obfuscation

We now focus our attention on constructing an IO combiner. Recall, in the setting of IO combiner we are given multiple IO candidates[8] with all of them satisfying correctness but with only one of them being secure. We then need to combine all of them to produce a joint obfuscator that is secure.

---

[6] Just as in the case of one-way functions, for sufficiently large circuits C, one of the enumerated machines will be a secure obfuscator.

[7] This is not a problem for the case of one-way functions because of a well established result that given any one-way function that runs in arbitrary polynomial time we can transform it into a different one-way function that takes quadratic time.

[8] IO candidates are just indistinguishability obfuscation schemes. The scheme of [3] is an example of an IO candidate, scheme of [30] is another example and so on.

This scenario is reminiscent of a concept we are quite familiar with: Secure Multi-Party Computation (MPC). In the secure multi-party computation setting, there are multiple parties with individual inputs and the goal of all these parties is to jointly compute a functionality. The privacy requirement states that the inputs of the honest parties are hidden other than what can be leaked by the output.

Indeed, MPC provides a natural template to solve the problem of building an IO combiner: Let $\Pi_1, \ldots, \Pi_n$ be the IO candidates and let C be the circuit to be obfuscated.

- Secret share the circuit C into n shares $s_1, \ldots, s_n$.
- Take any n-party MPC protocol for the functionality $\mathcal{F}$ that can tolerate all-but-one malicious adversaries [18]. The n-input functionality $\mathcal{F}$ takes as input $((s_1, x_1), (s_2, x_2), \ldots, (s_n, x_n))$; reconstructs C from the shares and outputs $C(x)$ only if $x = x_1 = \cdots = x_n$.
- Obfuscate the "code" (or algorithmic description) of the ith party using Π_i.
- The joint obfuscation of all the parties is the final obfuscated circuit!

To evaluate on an input x, perform the MPC protocol on the obfuscated parties with (s_i, x) being the input of the ith party.

Could the above approach lead to a secure IO combiner? The hope is that the security of MPC can be used to argue that one of the shares (corresponding to the honest party) is hidden which then translates to the hiding of C.

However, we face some fundamental challenges in our attempt to realize the above template, and in particular we will not be able to just invoke general solutions like [18], and we will need to leverage more specialized cryptographic objects.

Challenge #1: Single-Input versus Multi-Inputs Security. Recall that in the context of MPC, we argue the security only for a *particular set of inputs* (one for every party) in one session. In particular, a fresh session needs to be executed to compute the functionality on a different set of inputs. However, obfuscation is *re-usable* – it enables multiple evaluations of the obfuscated circuit. The obfuscated circuit should hide the original circuit independent of the number of times the obfuscated circuit is evaluated. On the other hand, take the classical Yao's garbled circuits [32], used in two party secure computation, for example. Suppose we are provided with the ability to evaluate the garbled circuit on two different inputs then the security completely breaks down.

Challenge #2: Power of the Adversary. Suppose we start with an arbitrary multi-round MPC protocol. In the world of IO combiners, this corresponds to executing a candidate multiple times during the evaluation of a single input. While the party in the MPC protocol can maintain state in between executions, a candidate does not have the same luxury since it is *stateless*. This enables the adversarial evaluator to launch so called *resetting attacks*: during the evaluation of the IO combiner on a single input x, a secure candidate could first be executed on transcripts consistent with x and later executed on transcripts consistent with a different input x'. Since, the secure candidate cannot maintain state, it is possible that it cannot recognize such a malicious execution. We need to devise additional mechanisms to prevent such attacks.

Challenge #3: Virtual Black Box Obfuscation versus IO. The above two challenges exist even if we had started off with virtual black box (VBB) obfuscation. Dealing with indistinguishability obfuscation as opposed to VBB presents us with fresh challenges. Indeed, in MPC, we take for granted that an honest party hides its input from the adversary. However, if we obfuscate the parties using IO, it is not clear whether the relevant input – the share of C – is hidden at all. Arguing this requires importing IO-friendly tools (for instance, [31]) studied in the recent literature and making it compatible with the tools of MPC that we want to use.

We will see next how to address the above challenges.

Our Approach. We present two different approaches to construct IO combiners. The first solution, in addition to existence of IO, assumes the hardness of Decisional Diffie Hellman. The second solution assumes additionally the hardness of learning with errors. Common to both these solutions is a technique of [9] that we'll call the *partition-programming technique*. We give a brief overview of this technique below.

Partition-Programming Technique: Consider a randomized algorithm $P(\cdot, \cdot)$ that takes as input secret sk, public instance $x \in \{0,1\}^\lambda$ and produces a distribution $\mathcal{D}_x$. Suppose there exists a simulator Sim that on input x outputs a distribution $\mathcal{D}_x^*$ such that the distributions $\mathcal{D}_x$ and $\mathcal{D}_x^*$ are statistically close.

Let us say we are given obfuscation of $P(sk, \cdot)$ (sk is hardwired in the program), we show how to use the partition-programming technique to *remove the secret* sk. We proceed in 2^λ hybrids: In the ith hybrid, we have a hybrid obfuscated program that on input x, executes $P(sk, x)$ if $x \leq i$ but otherwise it executes $\mathsf{Sim}(x)$. Now, the indistinguishability of ith hybrid and $(i + 1)^{th}$ hybrid can be argued directly from the security of IO: here we are using the fact that the simulated distribution and the real distribution are statistically close. In the $(2^{\lambda+1})^{th}$ hybrid, we have a program that only uses Sim, on every input, to generate the output distribution. Thus, we have removed the secret sk from the program.

This technique will come in handy when we address Challenge #1. We will see below how this technique will be used in both the solutions.

DDH-Based Solution. We begin by tackling Challenge #2. We noted that using interactive MPC solutions are bound to result in resetting attacks. Hence, we restrict our attention to non-interactive solutions. We need to determine our communication pattern between the candidates. In particular, we consider the *"line"* communication pattern: Suppose there are n candidates $\Pi_1, \ldots, \Pi_n$ and let C be the circuit to be obfuscated. For this discussion, we use the same notation Π_i to also refer to the circuit obfuscated by the candidate Π_i. The first obfuscated circuit Π_1 produces an output that will be input to Π_2 and so on.

In the end, Π_n will receive the input from Π_{n-1} and the output of Π_n will determine the final output.

Lets examine how to achieve a solution in the above communication model, by first considering a naïve approach: Π_1 has hardwired into it an encryption $\mathsf{Enc}(pk, C)$ of circuit C to be obfuscated. It receives an input x, it performs a part of the computation and sends the result to the next candidate Π_2 who performs another part of the computation, sends it to Π_3 and so on. In the end, the last candidate Π_n has the secret key sk to decrypt the output. This is clearly insecure because if both Π_1 and Π_n are broken then using sk and $\mathsf{Enc}(pk, C)$ we can recover the circuit C. This suggests the use of a re-encryption scheme. A re-encryption scheme is associated with public keys $pk_1, \ldots, pk_{n+1}$ and corresponding re-encryption keys $rk_{1\rightarrow 2}, \ldots, rk_{n\rightarrow(n+1)}$. The first candidate Π_1 will have hardwired into it $\mathsf{Enc}(pk_1, C)$ and the ith candidate has hardwired into it the re-encryption key $rk_{i\rightarrow i+1}$. Thus, the ith candidate performs part of the computation, re-encrypts with respect to pk_{i+1} using its re-encryption key $rk_{i\rightarrow i+1}$. We provide the secret key sk_{n+1}, corresponding to public key pk_{n+1}, as part of the obfuscated circuit. Using this, the evaluator can decrypt the output and produce the answer. Intuitively, as long as one candidate hides one secret key, the circuit C should be safe.

The natural next step is to figure out how to implement the "computation" itself: one direction would be to consider re-encryption schemes that are homomorphic with respect to arbitrary computations. However, we currently do not know of the existence of such schemes based on DDH (for LWE-based solutions, see below). We note that [1] faced similar hurdles while designing DDH-based multi-server delegation schemes. They employed the use of re-randomizable garbled circuits to implement the "computation" aspect of the above approach. A re-randomizable garbling scheme is a garbling scheme which is accompanied by a re-randomization algorithm that takes as input garbled circuit-input wire keys pair (GC, w_x) and outputs (GC^r, w_x^r).

Following along the lines of the approach of [1], we propose the following solution template:

1. First we compute the garbled circuit-wire keys pair (GC^1, w^1) of circuit C corresponding to the re-randomizable garbled circuits scheme. Here, w^1 comprises of keys associated to bits 0 and 1 with respect to every position. Π_1 has hardwired into it, $\mathsf{Enc}(pk_1, (\mathsf{GC}^1, w^1))$.
2. Π_1 takes as input x and produces $\mathsf{Enc}(pk_2, (\mathsf{GC}^2, w_x^2))$, where (GC^2, w_x^2) is obtained by first re-randomizing (GC^1, w^1) and then choosing the wire keys corresponding to x. This process is enabled using the re-encryption key $rk_{1\rightarrow 2}$. In addition, we require that the re-encryption process allows for homomorphic operations – in particular, it should allow for homomorphism of re-randomization operation of the garbling schemes.
3. The ith candidate takes as input $\mathsf{Enc}(pk_i, (\mathsf{GC}^i, w_x^i))$; homomorphically re-randomizes the garbled circuit while simultaneously re-encrypting the ciphertext to obtain $\mathsf{Enc}(pk_{i+1}, (\mathsf{GC}^{i+1}, w_x^{i+1}))$.

4. In the end, the nth candidate Π_n outputs $\mathsf{Enc}(pk_{n+1}, (\mathsf{GC}^{n+1}, w_x^{n+1}))$. Using the secret key sk_n, we can decrypt the output $(\mathsf{GC}^{n+1}, w_x^{n+1})$. We then evaluate the garbled circuit GC^{n+1} using the wire keys w_x^{n+1} to recover the output.

We employ a specific re-randomizable garbled circuits by [15] and homomorphic re-encryption scheme by [7], where both these primitives can be based on DDH. The above template does not immediately work since an adversarial evaluator could feed in incorrect inputs to the secure candidate. While [1] used non-interactive zero knowledge proofs (NIZKs) to resolve this issue, we need to employ "IO-friendly" proofs such as statistically-sound NIZKs [5,31]. Refer to full version [2] for the formal construction.

Security: To argue security, we need to rely on the security of re-encryption schemes in addition to the security guarantees of the other schemes. The security property of a re-encryption scheme states that given re-encryption keys $\{rk_{i\to i+1}\}_{i\in[n]} \setminus \{rk_{i\to i+1}\}$ and a secret key sk_{n+1}, it is computationally hard to distinguish $\mathsf{Enc}(pk_1, m_0)$ from $\mathsf{Enc}(pk_1, m_1)$.

To argue the security of universal obfuscator, we have to get rid of the re-encryption key corresponding to the secure candidate – indeed, in the case of [1] the re-encryption key corresponding to the honest party is removed in the security proof. In our scenario, however, this can only be implemented if we hardwire all possible outputs inside the code of the secure candidate. Clearly, this is not possible since there are exponentially many outputs. This is where we will use the *partition-programming technique* to remove the re-encryption key. To apply the technique, we argue that the re-encrypted ciphertexts are statistically close to freshly generated ciphertexts (which will be our simulated distribution) and this property holds for the particular instantiation of [7] we are considering.

LWE-Based Solution. We give an alternate construction based on the learning with errors (LWE) assumption. One potential approach is to take the above solution and replace the DDH-based primitives with LWE-based primitives. Namely, we replace re-randomizable garbled circuits and re-encryption schemes with fully homomorphic encryption schemes. While we believe this is a viable approach, it turns out we can give an arguably more elegant solution by using the notion of *multi-key fully homomorphic encryption* [10,28,29]. A multi-key FHE allows for generating individual public key-secret key pairs $\{pk_i, sk_i\}$ such that they can be later combined to obtain a joint public key **pk**. To be more precise, given a ciphertext with respect to pk_i, there is an "Expand" operation that transforms it into a ciphertext with respect to a joint public key **pk**. Once this done, the resulting ciphertext can be homomorphically evaluated just like any FHE scheme. The resulting ciphertexts can then be individually decrypted using sk_i's to obtain partial decryptions. Finally, there is a mechanism to combine the partial decryptions to obtain the final output.

Before we outline the solution below, we first fix the communication model. We consider a *"star"* interaction network: suppose there are n candidates $\Pi_1, \ldots, \Pi_n$. Each candidate Π_i is executed on the same input x. The joint outputs of all these candidates are then combined to obtain the final output. We propose the solution template based on multi-key FHE below.

1. We first secret share C into different shares $s_1, \ldots, s_n$.
2. Generate public key-secret key pairs $\{pk_i, sk_i\}$ for all $i \in [n]$. Encrypt s_i with respect to pk_i to obtain the ciphertext CT_i.
3. "Expand" every ciphertext CT_i into another ciphertext $\widehat{\mathsf{CT}}_i$ with respect to the joint public key $\mathbf{pk}$ which is a function of $(pk_1, \ldots, pk_n)$.
4. Every candidate Π_i has hardwired into it the secret key sk_i and ciphertext $\widehat{\mathsf{CT}}_i$. It takes as input x and first homomorphically evaluates the universal circuit U_x on $\widehat{\mathsf{CT}}_i$ to obtain an encryption of $C(x)$, namely $\widehat{\mathsf{CT}}_i^{C(x)}$, with respect to $\mathbf{pk}$. Finally, using sk_i it outputs the partial decryption of $\widehat{\mathsf{CT}}_i^{C(x)}$.
5. The different partial decryptions output by the candidates are later combined to obtain the final output.

Security: We rely on the semantic security of the MFHE scheme to argue the security of the obfuscator. The security notion of multi-key FHE intuitively guarantees that the semantic security on ciphertext CT_i can be argued as long as the adversary never gets the secret key sk_i for some $i \in [n]$. A naïve approach is to remove the secret key sk_i from the secure candidate Π_i. A similar issue that we encountered in the case of DDH-based solution arises here as well – we need to hardwire exponentially many outputs. Here comes partition-programming technique to the rescue! We show how to use this technique to remove sk_i after which we can argue the semantic security of MFHE, and thus the security of the obfuscator. To apply this technique, we need an alternate simulated distribution that simulates the partial decryption keys. We use the scheme of [29] who define such a simulatability property where the simulated distribution is statistically close to the real distribution. Refer to Sect. 4 for the formal construction.

The above LWE-based construction, unlike the DDH-based construction, satisfies some additional properties that are used to design a special type of IO combiner (we call this decomposable IO combiner in Sect. 3.1) which will then be used to construct universal indistinguishability obfuscation.

Robust IO Combiners. The description above details how to construct a (standard) IO combiner, that is, one that assumes all candidates are correct. The construction on a robust combiner is similar to the construction of the universal IO scheme. We discard candidates that are not approximately correct and boost the correctness of those that are. The difference between a universal scheme and a robust combiner is that in a robust combiner we are given n arbitrary candidates whereas in a universal scheme we construct the candidates by enumerating over TMs in a lexicographic order.

2.3   Universal Witness Encryption

We have discussed the construction of an IO combiner, and how to use the combiner to achieve a universal construction of IO. We describe our construction of a universal witness encryption (WE) scheme. We show that a universal WE scheme exists on the sole assumption of the existence of a one-way function.

First, we construct a WE combiner. This is achieved similarly to combiners for public-key encryption [21], using secret sharing. To encrypt a message m one secret shares the message to n shares such that all of them are needed to recover the message. Then, he encrypts each share using a different candidate. If at least one of the candidate schemes is secure then at least one share is unrecoverable and the message remains hidden.

The main challenge constructing a universal WE scheme is handling correctness. In the universal IO construction we had two main steps. The first was to test whether a candidate is approximately correct. This step was accomplished easily by sampling the obfuscated circuit on random inputs and verifying its correctness. Notice that although we cannot verify that the candidate is approximately correct for *all* circuits, we can verify that it is correct for the circuit in hand. The second step was to boost the correctness to achieve (almost) perfect security. This was obtained by suitably adapting the transformation described by Bitansky and Vaikuntanathan [6] to work in our setting where we only have a correctness guarantee for a single circuit.

The techniques used for the universal IO scheme seem not to apply for WE. Consider a language L with a relation R and a candidate scheme Π. To test correctness on an input x and a message m, one needs to encrypt the message and decrypt the resulting ciphertext. However, decryption requires a valid witness for x, where it might be NP-hard to find one! Testing, therefore, is limited to instances where it is easy to find a witness, a regime where witness encryption is trivial. Moreover, even given an approximate candidate, the boosting techniques used for the universal IO scheme do not apply for witness encryption.

Witness Injection. We describe a transformation that modifies any WE candidate scheme to be "testable" and also show how to boost the correctness of such testable schemes. Our first technique is to inject a "fake" witness for any x such that it will be easy to find this witness, for a party which has a trapdoor and computationally hard without the trapdoor (this is as in Feige and Shamir [12]). Moreover, this transformation will be indistinguishable for the (computationally bounded) candidate scheme.

Denote $(x, w) \in R$ for an instance x with a valid witness w. Let PRG be a length doubling pseudorandom generator. For any string z, we augment the language L and define L_z with the relation R_z such that

$$(x, w) \in R_z \iff (x, w) \in R \vee \mathsf{PRG}(w) = z.$$

Notice that if we choose $z = \mathsf{PRG}(s)$ for a random seed s, then L_z is the trivial language of all strings. Whereas, if z is chosen uniformly at random then with high probability L_z is equivalent to L, and these two cases are indistinguishable for anyone not holding the seed. This step enables us to test a candidate for some specific instance x: We choose $z \leftarrow \mathsf{PRG}(s)$, encrypt relatively to L_z, decrypt using the "fake" witness s and verify the output. After testing, we replace z with a random string (outside the range of the PRG) to get back the original language L. The problem is that this guarantees correctness only on our specific witness. The decryption algorithm, however, might refuse to cooperate for any other witness the user chooses to use.

Witness Protection Program. The next step is to apply what we call a *witness-worst-case* transformation. That is, a scheme that works on all witnesses with the same probability. Our main tool is a non-interactive zero knowledge (NIZK) proof system with statistical soundness. Suppose (P, V) is such a NIZK scheme with a common random string σ. Then we further augment the language L_z to $L_{z,\sigma}$ with relation $R_{z,\sigma}$ such that:

$$(x, \pi) \in R_{z,\sigma} \iff V(\sigma, x, \pi) = 1.$$

If $(x, w) \in R$ is a valid instance witness pair for L, then the corresponding witness for $L_{z,\sigma}$ will be $\pi \leftarrow P(\sigma, x, w)$. That is, executing the transformed scheme on x, w relative to the language L translate to executing the original scheme on x, π relative to the language $L_{z,\sigma}$ for a randomly chosen z. Finally, to boost the success probability we apply a standard "BPP amplification"; encrypt many times and take the majority.

The result is roughly the following algorithm. We take any scheme and apply our witness-worst-case transformation for $z \leftarrow \mathsf{PRG}(s)$. Afterwards, we can test it on a fake witness while we are assured that it will work the same for any other witness. Then, if the scheme passes all tests, we replace z with a random string, and boost the correctness such that it will work for any witness with all but negligible probability. Finally, we apply the WE combiner to get a universal scheme. For the exact details see Sect. 2.3.

Relying on One-Way Functions. The description above of a universal witness encryption scheme used NIZK proof system as a building block, where we promised using only one-way functions. These proofs are not known to be implied by one-way functions and moreover no universal NIZK scheme is known (and this is an interesting open problem!). However, standard interactive zero knowledge can be constructed for any language in NP for one-way functions and moreover there exist a universal one-way function [27]. Of course, we cannot use an interactive protocol, but, taking a closer look we observe that we can simulate a protocol between a verifier and a prover before the actual witness is given. That is, we can simulate a zero-knowledge protocol that might have many rounds, however, only the final round depends on the witness itself. Such protocols are known as *pre-process non-interactive zero-knowledge protocols* and where studied in [11,26] where they proved how to construct them based on way-one functions.

For the final scheme, we will run the pre-process protocol to get two private states σ_V and σ_P for the verifier and the prover respectively, just before the final round. The modified language will be L_{z,σ_V} with relation R_{z,σ_V}, where

$$(x, \pi) \in R_{z,\sigma_V} \iff V(\sigma_V, x, \pi) = 1.$$

We will publish σ_P as part of the encryption so that a user, given witness w can produce the corresponding final round of the proof $\pi \leftarrow P(\sigma_P, x, w)$. Notice that the given the state of the prover, σ_P, the proof π is *not* zero-knowledge. However, since the decryption algorithm of the scheme does not get the state of the prover (only the state of the verifier) then from his perspective it is zero-knowledge.

3   Indistinguishability Obfuscation (IO) Combiners

Suppose we have many indistinguishability obfuscation (IO) schemes, also referred to as *IO candidates*). We are additionally guaranteed that one of the candidates is secure. No guarantee is placed on the rest of the candidates and they could all be potentially broken. Indistinguishability obfuscation combiners provides a mechanism of combining all these candidates into a single monolithic IO scheme *that is secure*. We emphasize that the only guarantee we are provided is that one of the candidates is secure and in particular, it is unknown exactly which of the candidates is secure.

We give a thorough formal treatment of the concept of IO combiners next. We start by providing the syntax of an obfuscation scheme and then present the definitions of a (secure) IO candidate. Then, in Sect. 3.1 we finally present the definition of IO combiner.

Syntax of Obfuscation Scheme. An obfuscation scheme associated to a class of circuits $\mathcal{C} = \{\mathcal{C}_\lambda\}_{\lambda \in \mathbb{N}}$ consists of two PPT algorithms (Obf, Eval) defined below.

- **Obfuscate,** $\overline{C} \leftarrow \mathsf{Obf}(1^\lambda, C)$: It takes as input security parameter λ, a circuit $C \in \mathcal{C}_\lambda$ and outputs an obfuscation of C, $\overline{C}$.
- **Evaluation,** $y \leftarrow \mathsf{Eval}(\overline{C}, x)$: This is a deterministic algorithm. It takes as input an obfuscation $\overline{C}$, input $x \in \{0,1\}^\lambda$ and outputs y.

Throughout this work, we will only be concerned with *uniform* Obf algorithms. That is, Obf and Eval are represented as Turing machines (or equivalently uniform circuits).

μ-Correct IO Candidate. We define the notion of an IO candidate below. The following definition of obfuscation scheme incorporates *only* the correctness and polynomial slowdown properties of an indistinguishability obfuscation scheme [4,14,20].

Definition 1 (μ-Correct IO Candidate). *An obfuscation scheme $\Pi = (\mathsf{Obf}, \mathsf{Eval})$ is an μ-correct IO candidate for a class of circuits $\mathcal{C} = \{\mathcal{C}_\lambda\}_{\lambda \in \mathbb{N}}$, with every $C \in \mathcal{C}_\lambda$ has size $\mathrm{poly}(\lambda)$, if it satisfies the following properties:*

- **Correctness**: *For every $C : \{0,1\}^\lambda \to \{0,1\} \in \mathcal{C}_\lambda, x \in \{0,1\}^\lambda$ it holds that:*

$$\Pr\left[\mathsf{Eval}\left(\mathsf{Obf}(1^\lambda, C), x\right) = C(x)\right] \geq \mu(\lambda),$$

   over the random coins of Obf.
- **Polynomial Slowdown**: *For every $C : \{0,1\}^\lambda \to \{0,1\} \in \mathcal{C}_\lambda$, we have the running time of* Obf *on input $(1^\lambda, C)$ to be $\mathrm{poly}(|C|, \lambda)$. Similarly, we have the running time of* Eval *on input $(\overline{C}, x)$ is $\mathrm{poly}(|\overline{C}|, \lambda)$.*

Note that an identity function I is a valid IO candidate. We make use of this fact later on.

Remark 1. We say that Π is an IO candidate if it is a μ-correct IO candidate with $\mu = 1$.

μ-Correctϵ-Secure IO Candidate. If any IO candidate additionally satisfies the following (informal) security property then we define it to be a *secure* IO candidate: for every pair of circuits C_0 and C_1 that are equivalent we have obfuscations of C_0 and C_1 to be indistinguishable by any PPT adversary.

Definition 2 (μ-Correct ϵ-Secure IO Candidate). *An obfuscation scheme $\Pi = (\mathsf{Obf}, \mathsf{Eval})$ for a class of circuits $\mathcal{C} = \{\mathcal{C}_\lambda\}_{\lambda \in \mathbb{N}}$ is a μ-**correct** ϵ-**secure IO candidate** if it satisfies the following conditions:*

- *Π is a μ-correct IO candidate with respect to $\mathcal{C}$,*
- **Security.** *For every PPT adversary $\mathcal{A}$, for every sufficiently large $\lambda \in \mathbb{N}$, for every $C_0, C_1 \in \mathcal{C}_\lambda$ with $C_0(x) = C_1(x)$ for every $x \in \{0,1\}^\lambda$ and $|C_0| = |C_1|$, we have:*

$$\left| \Pr\left[0 \leftarrow \mathcal{A}\left(\mathsf{Obf}(1^\lambda, C_0), C_0, C_1 \right) \right] - \Pr\left[0 \leftarrow \mathcal{A}\left(\mathsf{Obf}(1^\lambda, C_1), C_0, C_1 \right) \right] \right| \leq \epsilon(\lambda)$$

We remarked earlier that identity function is an IO candidate. However, note that the identity function is *not* a secure IO candidate.

Remark 2. We say that Π is a secure IO candidate if it is a μ-correct ϵ-secure IO candidate with $\mu = 1$ and $\epsilon(\lambda) = \mathsf{negl}(\lambda)$, for some negligible function negl.

In the literature [14,31], a secure IO candidate is simply referred to as an indistinguishability obfuscation scheme.

We have the necessary ingredients to define an IO combiner.

3.1 Definition of IO Combiner

We present the formal definition of IO combiner below. First, we provide the syntax of the IO combiner. Later we present the properties associated with an IO combiner.

There are two PPT algorithms associated with an IO combiner, namely, CombObf and CombEval. Procedure CombObf takes as input circuit C along with the description of multiple IO candidates[9] and outputs an obfuscation of C. Procedure CombEval takes as input the obfuscated circuit, input x, the description of the candidates and outputs the evaluation of the obfuscated circuit on input x.

Syntax of IO Combiner. We define an IO combiner $\Pi_{\mathsf{comb}} = (\mathsf{CombObf}, \mathsf{CombEval})$ for a class of circuits $\mathcal{C} = \{\mathcal{C}_\lambda\}_{\lambda \in \mathbb{N}}$.

- **Combiner of Obfuscate algorithms, $\overline{C} \leftarrow \mathsf{CombObf}(1^\lambda, C, \Pi_1, \ldots, \Pi_n)$:** It takes as input security parameter λ, a circuit $C \in \mathcal{C}$, description of IO candidates $\{\Pi_i\}_{i \in [n]}$ and outputs an obfuscated circuit $\overline{C}$.
- **Combiner of Evaluation algorithms, $y \leftarrow \mathsf{CombEval}(\overline{C}, x, \Pi_1, \ldots, \Pi_n)$:** It takes as input obfuscated circuit $\overline{C}$, input x, descriptions of IO candidates $\{\Pi_i\}_{i \in [n]}$ and outputs y.

---

[9] The description of an IO candidate includes the description of the obfuscation and the evaluation algorithms.

We define the properties associated to any IO combiner. There are three main properties – correctness, polynomial slowdown, and security. The correctness and the polynomial slowdown properties are defined on the same lines as the corresponding properties of the IO candidates.

The intuitive security notion of IO combiner says the following: suppose one of the candidates is a secure IO candidate then the output of obfuscator (CombObf) of the IO combiner on C_0 is computationally indistinguishable from the output of the obfuscator on C_1, where C_0 and C_1 are equivalent circuits.

Definition 3 $((\mu', \mu)$-Correct (ϵ', ϵ)-Secure IO Combiner). *Consider a circuit class $\mathcal{C} = \{\mathcal{C}_\lambda\}_{\lambda \in \mathbb{N}}$. We say that $\Pi_{\text{comb}} = (\text{CombObf}, \text{CombEval})$ is a (μ', μ)-correct (ϵ', ϵ)-secure IO combiner if the following conditions are satisfied: Let $\Pi_1, \ldots, \Pi_n$ be n μ-correct IO candidates for P/poly, where μ is a function of μ' and ϵ is a function of ϵ'.*

- **Correctness.** *Let $C \in \mathcal{C}_{\lambda \in \mathbb{N}}$ and $x \in \{0,1\}^\lambda$. Consider the following process: (a) $\overline{C} \leftarrow \text{CombObf}(1^\lambda, C, \Pi_1, \ldots, \Pi_n)$, (b) $y \leftarrow \text{CombEval}(\overline{C}, x, \Pi_1, \ldots, \Pi_n)$. Then, $\Pr[y = C(x)] \geq \mu'(\lambda)$ over the randomness of CombObf.*
- **Polynomial Slowdown.** *For every $C : \{0,1\}^\lambda \to \{0,1\} \in \mathcal{C}_\lambda$, we have the running time of CombObf on input $(1^\lambda, C, \Pi_1, \ldots, \Pi_n)$ to be at most $\text{poly}(|C| + n + \lambda)$. Similarly, we have the running time of CombEval on input $(\overline{C}, x, \Pi_1, \ldots, \Pi_n)$ to be at most $\text{poly}(|\overline{C}| + n + \lambda)$.*
- **Security.** *Let Π_i be ϵ-secure for some $i \in [n]$. For every PPT adversary $\mathcal{A}$, for every sufficiently large $\lambda \in \mathbb{N}$, for every $C_0, C_1 \in \mathcal{C}_\lambda$ with $C_0(x) = C_1(x)$ for every $x \in \{0,1\}^\lambda$ and $|C_0| = |C_1|$, we have:*

$$\left| \Pr\left[0 \leftarrow \mathcal{A}\left(\overline{C_0}, C_0, C_1, \Pi_1, \ldots, \Pi_n\right)\right] - \Pr\left[0 \leftarrow \mathcal{A}\left(\overline{C_1}, C_0, C_1, \Pi_1, \ldots, \Pi_n\right)\right] \right|$$
$$\leq \epsilon'(\lambda),$$

where $\overline{C_b} \leftarrow \text{CombObf}(1^\lambda, C_b, \Pi_1, \ldots, \Pi_n)$ for $b \in \{0,1\}$.

Some remarks are in order.

Remark 3. We say that Π_{comb} is an IO combiner if it is a (μ', μ)-correct (ϵ', ϵ)-secure IO combiner, where (a) $\mu' = 1$, (b) $\mu = 1$, (c) $\epsilon' = \text{negl}'$ and, (d) $\epsilon = \text{negl}$ with negl and negl' being negligible functions.

4 Constructions of IO Combiners

We propose constructions of combiners for indistinguishability obfuscation. Here we present a construction based on the learning with errors assumption. In full version [2], we also present a construction based on the decisional Diffie Hellman assumption. We present the formal construction below. For an informal explanation of the construction, we refer the reader to Introduction.

Construction. Consider a circuit class $\mathcal{C}$. We use a threshold multi-key FHE scheme TMFHE = (Setup, KeyGen, Enc, Expand, FHEEval, Dec, PartDec, FinDec). We additionally use a puncturable PRF family $\mathcal{F}$.

We construct an IO combiner $\Pi_{\text{comb}} = \Pi_{\text{comb}}[\Pi_1, \dots, \Pi_n]$ for $\mathcal{C}$ below.

$\underline{\text{CombObf}(1^\lambda, C, \Pi_1, \dots, \Pi_n)}$: It takes as input security parameter λ, circuit $C \in \mathcal{C}_\lambda$, description of candidates $\{\Pi_i = (\Pi_i.\text{Obf}, \Pi_i.\text{Eval})\}_{i \in [n]}$ and does the following.

1. **Initialization of TMFHE Parameters:**
   - Execute the setup of the threshold multi-key FHE scheme, params $\leftarrow$ Setup$(1^\lambda, 1^d)$, where $d = \text{poly}(\lambda, |C|)^{10}$. Execute $\{(sk_i, pk_i) \leftarrow \text{KeyGen}$ (params)$\}_{i \in [n]}$.
   - Sample n random strings $\{S_i\}_{i \in [n]}$ of size $|C|$ such that $\bigoplus_{i \in [n]} S_i = C$.
   - For all $i \in [n]$, encrypt the string S_i using pk_i, $\text{CT}_i \leftarrow \text{Enc}(pk_i, S_i)$.
   - For every $i \in [n]$, generate the expanded ciphertext under pk_i by executing $\widehat{\text{CT}}_i \leftarrow \text{Expand}((pk_1, \dots, pk_n), i, \text{CT}_i)$.
2. **Obfuscating Circuits using IO Candidates:**
   - For every $i \in [n]$, sample puncturable PRF keys $K^i \xleftarrow{\$} \{0, 1\}^\lambda$.
   - For every $i \in [n]$, construct circuit $G_i = G_i\left[K^i, sk_i, \{pk_i\}_{i \in [n]}, \{\widehat{\text{CT}}_i\}_{i \in [n]}\right] \in \mathcal{C}^i$ as described in Fig. 1.
   - Generate $\overline{G_i} \leftarrow \Pi_i.\text{Obf}(1^\lambda, G_i)$.

Output the obfuscation $\overline{C} = (\overline{G_1}, \dots, \overline{G_n})$.

$\underline{\text{CombEval}(\overline{C}, x, \Pi_1, \dots, \Pi_n)}$: On input an obfuscation $\overline{C}$, an input x, descriptions of candidates $\{\Pi_i\}_{i \in [n]}$ evaluate the obfuscations on input x to obtain $p_i \leftarrow \Pi_i.\text{Eval}(\overline{G_i}, x)$ for all $i \in [n]$. Execute the final decryption algorithm, $y \leftarrow \text{FinDec}(p_1, \dots, p_n)$. Output y.

5   Universal Obfuscation

We introduce the notion of universal obfuscation. We define a pair of Turing machines $\Pi_{\text{univ}}.\text{Obf}$ and $\Pi_{\text{univ}}.\text{Eval}$ to be a universal obfuscation if the existence of a secure IO candidate implies that $(\Pi_{\text{univ}}.\text{Obf}, \Pi_{\text{univ}}.\text{Eval})$ is also a secure IO candidate. Constructing a universal obfuscation scheme means that we can turn the mere existence of a secure IO candidate into an explicit construction. Formally, we have the following definition:

Definition 4 $((T, \epsilon)$-Universal Obfuscation). *We say that a pair of Turing machines $\Pi_{\text{univ}} = (\Pi_{\text{univ}}.\text{Obf}, \Pi_{\text{univ}}.\text{Eval})$ is a **universal obfuscation**, parameterized by T and ϵ, if there exists an ϵ-secure indistinguishability obfuscator for $P/poly$ with time function T then Π_{univ} is an indistinguishability obfuscator for $P/poly$ with time function $\text{poly}(T)$.*

---

[10] Looking ahead, we set d to be the size of C as against its depth so that a PPT adversary will not be able to distinguish obfuscations of two functionally equivalent circuits C_0 and C_1 with the same size but potentially different depths by just measuring the size of params.

$$G_i \left[K^i, sk_i, \{pk_i\}_{i\in[n]}, \{\widehat{\mathsf{CT}_i}\}_{i\in[n]} \right]$$

Hardwired values: PRF key K^i, TMFHE partial decryption key sk_i, TMFHE public keys $\{pk_i\}_{i\in[n]}$, TMFHE expanded ciphertext $\{\widehat{\mathsf{CT}_i}\}_{i\in[n]}$.
Input: $x \in \{0,1\}^\lambda$.

  - Perform $\widehat{\mathsf{CT}_{\mathsf{out}}} \leftarrow \mathsf{FHEEval}\left(\mathsf{params}, U_x(\cdot), \widehat{\mathsf{CT}_1}, \ldots, \widehat{\mathsf{CT}_n}\right)$, where $U_x(\cdot)$ is a universal circuit that takes as input n strings $S_1, .., S_n$ and first computes $\bigoplus_{j\in[n]} S_j = C$ where $C \in \mathcal{C}_\lambda$ and outputs $C(x)$.
  - Generate randomness $r_i \leftarrow PRF_{K^i}(x)$.
  - Execute the partial decryption algorithm, $p_i \leftarrow \mathsf{PartDec}\left(\widehat{\mathsf{CT}_{\mathsf{out}}}, pk_1, \ldots, pk_n, i, sk_i; r_i\right)$
  - Output p_i.

Fig. 1. Circuit G_i

5.1 Construction of (T, ϵ)-Universal Obfuscation

We proceed to construct a (T, ϵ)-universal obfuscation. The core building block in our construction is a *decomposable* IO combiner – this is a specific type of IO combiner that satisfies additional properties (explained below).

Main Ingredient: Decomposable IO Combiner. A decomposable IO combiner is a type of IO combiner, where the obfuscate algorithm has a specific structure. In particular, the obfuscate algorithm takes as input circuit C to be obfuscated, the description of the candidates $\Pi_1, \ldots, \Pi_n$ and executes in two main steps. In the first step, circuit C is preprocessed into n circuits $[\mathbf{C}]_1, \ldots, [\mathbf{C}]_n$. In the second step, each individual circuit $[\mathbf{C}]_i$ is obfuscated using the candidate Π_i. The concatenation of the resulting obfuscated circuits is the final output.

In addition to the standard properties of IO combiner, we require that the decomposable IO combiner satisfies two more properties: *Circuit-Specific Correctness* and *Decomposable Security*. The formal description is given below.

Definition 5 (Decomposable IO Combiner). *A (ϵ', ϵ)-secure IO combiner $\Pi_{\mathsf{comb}} = (\Pi_{\mathsf{comb}}.\mathsf{Obf}, \Pi_{\mathsf{comb}}.\mathsf{Eval})$ of $(\Pi_1, \ldots, \Pi_n)$ for a class of circuits $\mathcal{C} = \{\mathcal{C}_\lambda\}$ is said to be (ϵ', ϵ)-secure (η', η)-decomposable IO combiner if there exists a PPT algorithm $\mathsf{Preproc}$ such that the following holds: $\Pi_{\mathsf{comb}}.\mathsf{Obf}$ on input $(1^\lambda, C \in \mathcal{C}_\lambda, \Pi_1, \ldots, \Pi_n)$ executes the steps:*

(a) (Preprocessing step) $\overline{C} = ([\mathbf{C}]_1, \ldots, [\mathbf{C}]_n, aux) \leftarrow \mathsf{Preproc}(1^\lambda, 1^n, C)$,
(b) (Candidate Obfuscation step) for all $i \in [n]$, $\overline{[\mathbf{C}]_i} \leftarrow \Pi_i.\mathsf{Obf}(1^\lambda, [\mathbf{C}]_i)$,
(c) Outputs $\overline{C} = \left(\overline{[\mathbf{C}]_1}, \ldots, \overline{[\mathbf{C}]_n}, aux\right)$.

Additionally, we require the following properties to hold:

– (η', η)-**Circuit-Specific Correctness.** *Consider a circuit* $C \in \mathcal{C}_\lambda$. *Let* $([\mathbf{C}]_1, \ldots, [\mathbf{C}]_n, aux) \leftarrow \mathsf{Preproc}(1^\lambda, 1^n, C)$. *Let for all* $i \in [n]$, $\overline{[\mathbf{C}]}_i \leftarrow \Pi_i.\mathsf{Obf}(1^\lambda, [\mathbf{C}]_i)$. *Denote* $\overline{C} = (\overline{[\mathbf{C}]}_1, \ldots, \overline{[\mathbf{C}]}_n)$.
If for all $i \in [n]$, $\Pr_{x \xleftarrow{\$} \{0,1\}^\lambda} \left[\overline{[\mathbf{C}]}_i(x) = [\mathbf{C}]_i(x) \right] \geq \eta(\lambda)$ *then*
$\Pr_{x \xleftarrow{\$} \{0,1\}^\lambda} \left[\overline{C}(x) = C(x) \right] \geq \eta'(\lambda)$.

– **Decomposable Security**: *For every* $C_0, C_1 \in \mathcal{C}_\lambda$ *such that* $|C_0| = |C_1|$, *for every* $\mathbf{i} \in [n]$, *we have:*

$$\left\{ \left\{ [\mathbf{C}]_i^0 \right\}_{\substack{i \neq \mathbf{i}, \\ i \in [n]}} \right\} \approx_c \left\{ \left\{ [\mathbf{C}]_i^1 \right\}_{\substack{i \neq \mathbf{i}, \\ i \in [n]}} \right\},$$

where $[\mathbf{C}]_i^b \leftarrow \mathsf{Preproc}(1^\lambda, 1^n, C_b \in \mathcal{C}_\lambda)$ *for* $b \in \{0, 1\}$.

We claim that the construction of IO Combiner in Sect. 4 is already a decomposable IO combiner. To show this, we first note that the obfuscator $\Pi_{\mathsf{univ}}.\mathsf{Obf}$ in the construction in Sect. 4 can be decomposed in a preprocessing step and candidate obfuscation step: the preprocessing step comprises of all the steps till the generation of circuits $\{G_i\}_{i \in [n]}$ (Fig. 1). The output of the preprocessing step is $(G_1, \ldots, G_n)$.

Furthermore, the circuit-specific correctness property was already proved in [2]. More specifically, we showed the aforementioned construction satisfies $(1 - n\mu, 1 - \mu)$-circuit specific correctness property. All is remaining is to show that the construction satisfies decomposable security. We prove the following theorem. The proof can be found in [2].

Theorem 6. *The construction presented in Sect. 4 is a* $(\mathsf{negl}, \epsilon)$-*secure* $(1 - \frac{1}{\lambda}, 1 - \frac{1}{\lambda^2})$-*decomposable IO combiner, where the number of candidates is* λ.

Step I: Construction of Approx. Correct(T, ϵ)-Universal Obfuscation.

We construct a universal obfuscation scheme $\Pi_{\mathsf{univ}} = (\mathsf{Obf}, \mathsf{Eval})$ for a class of circuits $\mathcal{C}$ below. Our scheme will be approximately correct. The main ingredient is a decomposable IO combiner (Definition 5) $\Pi_{\mathsf{comb}} = (\Pi_{\mathsf{comb}}.\mathsf{Obf}, \Pi_{\mathsf{comb}}.\mathsf{Eval})$ for $\mathcal{C}$. But first, we establish some notation.

Notation. Let $\mathcal{S}$ be the class of all possible Turing machines. It is well known result [16] that there is a one-to-one correspondence between $\mathcal{S}^2$ and $\mathbb{N}$ given by $\phi : \mathbb{N} \to \mathcal{S}^2$. Furthermore, there is a fixed polynomial f such that the time to compute $\phi(j)$ is at most $\leq f(j)$, for every $j \in \mathbb{N}$.

$\Pi_{\mathsf{univ}}.\mathsf{Obf}(1^\lambda, C)$: It takes as input security parameter λ, circuit $C \in \mathcal{C}_\lambda$ and executes the following steps:

1. Let $\phi(i) = (\Pi_i.\mathsf{Obf}, \Pi_i.\mathsf{Eval})$, for $i \in \{1, \ldots, \lambda\}$. Denote $\Pi_i = (\Pi_i.\mathsf{Obf}, \Pi_i.\mathsf{Eval})$.

2. **Preprocessing phase of Decomposable IO Combiner.** First compute the preprocessing step, $([\mathbf{C}]_1, \ldots, [\mathbf{C}]_n, aux) \leftarrow \mathsf{Preproc}(1^\lambda, 1^n, C)$ $(n = \lambda)$.

3. **Eliminating Candidates with Large Runtimes.** For all $i \in [\lambda]$, execute $\Pi_i.\mathsf{Obf}(1^\lambda, [\mathbf{C}]_i)$ for at most $t = T(\lambda, |[\mathbf{C}]_i|)$ number of steps. For every $i \in [\lambda]$, if the computation of $\Pi_i.\mathsf{Obf}(1^\lambda, [\mathbf{C}]_i)$ does not abort within t number of time steps re-assign $\Pi_i.\mathsf{Obf} = I$ and $\Pi_i.\mathsf{Eval} = UTM$, where I is an identity TM[11] and UTM is a universal TM[12].

   At the end of this step, the execution of $\Pi_i.\mathsf{Obf}(1^\lambda, [\mathbf{C}]_i)$ takes time at most $T(\lambda, |[\mathbf{C}]_i|)$.

4. **Eliminates Candidates with Imperfect Correctness.** For all $i \in [\lambda]$, execute $\Pi_i.\mathsf{Obf}(1^\lambda, [\mathbf{C}]_i)$ for at most $t = T(\lambda, |[\mathbf{C}]_i|)$ number of steps. Denote $\overline{[\mathbf{C}]_i}$ to be the result of computation. Denote ℓ to be the input length of $[\mathbf{C}]_i$. For every $i \in [n]$, sample λ^3 points $x_{1,i}, \ldots, x_{\lambda^3,i} \xleftarrow{\$} \{0,1\}^\ell$. Check if the following condition holds:

$$\bigwedge_{j=1}^{\lambda^3} \left([\mathbf{C}]_i(x_{j,i}) = \Pi_i.\mathsf{Eval}\left(\overline{[\mathbf{C}]_i}, x_{j,i}\right) \right) = 1 \tag{1}$$

   If for any $i \in [\lambda]$ the above condition does not hold, re-assign $\Pi_i.\mathsf{Obf} = I$ and $\Pi_i.\mathsf{Eval} = UTM$. At the end of this step, every candidate satisfies the above condition.

5. **Candidate Obfuscation Phase of Decomposable IO Combiner.** For all $i \in [\lambda]$, execute $\Pi_i.\mathsf{Obf}(1^\lambda, [\mathbf{C}]_i)$ for at most $t = T(\lambda, |[\mathbf{C}]_i|)$ number of steps. Denote $\overline{[\mathbf{C}]_i}$ to be the result of computation.

6. Output $\overline{C} = \left((\Pi_1, \ldots, \Pi_\lambda), (\overline{[\mathbf{C}]_1}, \ldots, \overline{[\mathbf{C}]_\lambda}, aux)\right)$.

$\underline{\Pi_{\mathsf{univ}}.\mathsf{Eval}(\overline{C}, x)}$: On input the obfuscated circuit $\overline{C}$ and input x, do the following. First parse $\overline{C}$ as $\left((\Pi_1, \ldots, \Pi_\lambda), \overline{C_{comb}} = (\overline{[\mathbf{C}]_1}, \ldots, \overline{[\mathbf{C}]_\lambda}, aux)\right)$. Compute $y \leftarrow \Pi_{\mathsf{comb}}.\mathsf{Eval}\left(\overline{C_{comb}}, x, \Pi_1, \ldots, \Pi_\lambda\right)$. Output y.

Theorem 7. *Assuming that Π_{comb} is a $(\mathsf{negl}, \epsilon)$-secure $\left(1 - \frac{1}{\lambda}, 1 - \frac{1}{\lambda^2}\right)$-decomposable IO combiner, the above scheme Π_{univ} is a (T, ϵ)-universal obfuscation that is $\left(1 - \frac{1}{\lambda}\right)$-correct.*

Proof. We first remark about the running time of the obfuscator and the evaluator algorithms. First, we consider $\Pi_{\mathsf{univ}}.\mathsf{Obf}$. The running time of first step (Bullet 1) is $\lambda f(\lambda) = \mathsf{poly}(\lambda)$ (where f was defined earlier in the proof). The running time of each of the rest of the steps is $\mathsf{poly}(\lambda, t, |C|)$. Plugging in the fact that $t = T(\lambda, \mathsf{poly}(\lambda, |C|))$, we have that the total running time of all the steps to be $\mathsf{poly}(T(\lambda, |C|))$[13]. We move on to $\Pi_{\mathsf{univ}}.\mathsf{Eval}$. Here, the running time is governed by the running time of the $\Pi_{\mathsf{comb}}.\mathsf{Eval}$ algorithm which is $\mathsf{poly}(T(\lambda, |C|))$. And hence, the running time of $\Pi_{\mathsf{univ}}.\mathsf{Eval}$ is again $\mathsf{poly}(T(\lambda, |C|))$.

---

[11] An identity TM on input C outputs C.

[12] A universal TM on input circuit-input pair (C, x) outputs $C(x)$.

[13] Observe that here we used two facts of the time function: (a) $T(\lambda, |C|) \geq |C| + \lambda$ and, (b) $T(\lambda, \mathsf{poly}(|C|)) = \mathsf{poly}'(T(\lambda, |C|))$.

Correctness. Consider the following lemma.

Lemma 1. Π_{univ} *is a* $\left(1 - \frac{1}{\lambda}\right)$*-correct IO candidate.*

Proof. Consider a circuit $C \in \mathcal{C}_\lambda$. We prove the following claim. For all $i \in [n]$, let $([\mathbf{C}]_1, \ldots, [\mathbf{C}]_n, aux) \leftarrow \text{Preproc}(1^\lambda, 1^n, C)$ with $n = \lambda$. Also, let $\{\Pi_i\}_{i \in [n]}$ be the description of the candidates at the end of Bullet 3. Note that some of the candidates could be re-assigned in Bullets 2 and 3. Let $\overline{[\mathbf{C}]}_i \leftarrow \Pi_i.\text{Obf}(1^\lambda, [\mathbf{C}]_i)$. We prove the following claim in full version [2].

Claim 8. *Let $i \in [n]$ be such that*

$$\Pr_{x \xleftarrow{\$} \{0,1\}^\lambda} \left[[\mathbf{C}]_i(x) = \Pi_i.\text{Eval}(\overline{[\mathbf{C}]}_i, x) \right] \leq 1 - \frac{1}{\lambda^2}$$

*Then, the ith candidate Π_i satisfies Condition (1) (Bullet 4) with **negligible** probability (over the random coins of $x_{j,i}$).*

The above claim proves that at the end of Bullet 4, with overwhelming probability the following holds for every $i \in [n]$:

$$\Pr_{x \xleftarrow{\$} \{0,1\}^\lambda} \left[[\mathbf{C}]_i(x) = \Pi_i.\text{Eval}(\overline{[\mathbf{C}]}_i, x) \right] \geq 1 - \frac{1}{\lambda^2}$$

We now apply the circuit-specific completeness property of the $(1 - \frac{1}{\lambda}, 1 - \frac{1}{\lambda^2})$-decomposable IO combiner Π_{comb} which ensures that the following holds:

$$\Pr_{x \xleftarrow{\$} \{0,1\}^\lambda} \left[C(x) = \Pi_{\text{comb}}.\text{Eval}(\overline{C}, x) \right] \geq 1 - \frac{1}{\lambda}$$

where $\overline{C} = (\overline{[\mathbf{C}]}_1, \ldots, \overline{[\mathbf{C}]}_n, aux)$. Note that $\overline{C}$ is the output of $\Pi_{\text{univ}}.\text{Obf}$. Also, the output of $\Pi_{\text{univ}}.\text{Eval}$ on input $(\overline{C}, x)$ is dictated by the result of $\Pi_{\text{comb}}.\text{Eval}(\overline{C}, x)$.

Thus, we have

$$\Pr_{x \xleftarrow{\$} \{0,1\}^\lambda} \left[C(x) = \Pi_{\text{univ}}.\text{Eval}(\overline{C}, x) \right] \geq 1 - \frac{1}{\lambda},$$

where $\overline{C} \leftarrow \Pi_{\text{univ}}.\text{Obf}(1^\lambda, C)$.

Security. We prove the following lemma.

Lemma 2. Π_{univ} *is a* (negl)*-secure IO candidate.*

Proof. Recall that the universal obfuscator proceeds in two phases. In the first phase, it chooses the "correct" candidates and then in the second phase, it combines all these candidates to produce the obfuscated circuit. At first glance, it should seem that as long as we ensure that one of the "correct" candidates is secure then the security of IO combiner should hold, and thus the security of universal obfuscator will follow. To make this more precise, lets say C_0 and C_1 are two equivalent circuits. Let $\overrightarrow{\Pi_0} = \Pi_1^0, \ldots, \Pi_{n_0}^0$ and $\overrightarrow{\Pi_1} = \Pi_1^1, \ldots, \Pi_{n_1}^1$ be the "correct" candidates chosen with respect to C_0 and C_1 respectively. Now, assuming that $\overrightarrow{\Pi_0}$ and $\overrightarrow{\Pi_1}$ have at least one secure candidate; the hope is that we can then invoke the security of IO combiner to argue computational indistinguishability of obfuscation of C_0 and C_1. This does not work because the security of IO combiner dictates that $\overrightarrow{\Pi_0} = \overrightarrow{\Pi_1}$. Indeed obfuscation of C_0 (resp., C_1) could potentially reveal $\overrightarrow{\Pi_0}$ (resp., $\overrightarrow{\Pi_1}$) at which point no security holds. While we cannot argue that $\overrightarrow{\Pi_0} = \overrightarrow{\Pi_1}$, because of the *selective abort obfuscators* described in Introduction, we can still show that $\overrightarrow{\Pi_0} \approx_c \overrightarrow{\Pi_1}$. Arguing the indistinguishability of the candidates then helps us invoke the security of IO combiner and then the proof of the theorem follows. Arguing the indistinguishability of candidates is performed by invoking the decomposable security property of the underlying IO combiner. We present the key lemmas here. Completed proof can be found in the full version [2].

Formal Details. We first introduce some notation. Consider a circuit $C \in \mathcal{C}_\lambda$. Let $((\Pi_1, \ldots, \Pi_\lambda), ([\mathbf{C}]_1, \ldots, [\mathbf{C}]_\lambda), aux)$ be the output of $\Pi_{\mathsf{univ}}.\mathsf{Obf}(1^\lambda, C)$. Note that many of the candidates $(\Pi_1, \ldots, \Pi_\lambda)$ could potentially be re-assigned during the execution of $\Pi_{\mathsf{univ}}.\mathsf{Obf}$. This re-assignment is a function of the circuit C that is obfuscated and the random coins of the algorithm. Hence, we can define a distribution $\mathsf{Dist}_{C,\lambda,\mathbf{i}}$, parameterized by C, λ, $\mathbf{i} \in [n]$, on $\{0,1\}^\lambda$ such that $x \xleftarrow{\$} \mathsf{Dist}_{C,\lambda,\mathbf{i}}$ defines which of the candidates gets re-assigned. That is, the ith bit $x_i = 1$ indicates that Π_i will remain unchanged and $x_i = 0$ indicates that Π_i is re-assigned. Furthermore, $x_{\mathbf{i}}$ is always 1.

In more detail, we define the sampling algorithm of distribution $\mathsf{Dist}_{\lambda,C,\mathbf{i}}$ as follows: denote by $\Pi_1', \ldots, \Pi_\lambda'$ the set of candidates enumerated in Bullet 1 and let $\Pi_{\mathbf{i}}$ be an IO candidate that is always correct. Note that the description of these candidates are independent of the circuit C and they only depend on the security parameter λ. At the end of Bullet 4, denote the candidates to be $(\Pi_1, \ldots, \Pi_\lambda)$. We then assign x to be such that the ith bit of x, namely, $x_i = 1$ if $\Pi_i' = \Pi_i$ else $x_i = 0$ if $\Pi_i' \neq \Pi_i$. Output x. Note that $x_{\mathbf{i}} = 1$ since $\Pi_{\mathbf{i}}$ is always correct.

The formal description of the sampling algorithm of $\mathsf{Dist}_{\lambda,C,\mathbf{i}}$ is given next.

Sampler of $\mathsf{Dist}_{\lambda,C,\mathbf{i}}$:

- Let $\phi(i) = (\Pi_i.\mathsf{Obf}, \Pi_i.\mathsf{Eval})$, for $i \in \{1, \ldots, \lambda\}$. Denote $\Pi_i = (\Pi_i.\mathsf{Obf}, \Pi_i.\mathsf{Eval})$.
- First compute the preprocessing step, $([\mathbf{C}]_1, \ldots, [\mathbf{C}]_n, aux) \leftarrow \mathsf{Preproc}(1^\lambda, 1^n, C)$. Here, $n = \lambda$. Maintain another copy of the set of candidates - for every $i \in [\lambda]$, set $\Pi_i' = \Pi_i$.
- For all $i \in [\lambda]$, execute $\Pi_i.\mathsf{Obf}(1^\lambda, [\mathbf{C}]_i)$ for at most $t = T(\lambda, |[\mathbf{C}]_i|)$ number of steps. For every $i \in [\lambda]$, if the computation of $\Pi_i.\mathsf{Obf}(1^\lambda, [\mathbf{C}]_i)$ does not abort

within t number of time steps re-assign $\Pi_i.\mathsf{Obf} = I$ and $\Pi_i.\mathsf{Eval} = UTM$, where I is an identity TM and UTM is a universal TM.

- For all $i \in [\lambda]$, execute $\Pi_i.\mathsf{Obf}(1^\lambda, [\mathbf{C}]_i)$ for at most $t = T(\lambda, |[\mathbf{C}]_i|)$ number of steps. Denote $\overline{[\mathbf{C}]}_i$ to be the result of computation. Denote ℓ to be the input length of $[\mathbf{C}]_i$. For every $i \in [n]$, sample λ^3 points $x_{1,i}, \ldots, x_{\lambda^3,i} \xleftarrow{\$} \{0,1\}^\ell$. Check if the following condition holds:

$$\bigwedge_{j=1}^{\lambda^3} \left([\mathbf{C}]_i(x_{j,i}) = \Pi_i.\mathsf{Eval}\left(\overline{[\mathbf{C}]}_i, x_{j,i}\right) \right) = 1 \tag{2}$$

If for any $i \in [\lambda]$ the above condition does not hold, re-assign $\Pi_i.\mathsf{Obf} = I$ and $\Pi_i.\mathsf{Eval} = UTM$.

- Construct a string $x \in \{0,1\}^\lambda$ such that the ith bit x_i is generated as:

$$x_i = \begin{cases} 1, & \text{if} \quad \Pi_i = \Pi_i' \\ 0, \text{otherwise} \end{cases}$$

- Output x.

Remark 4. For every x in the support of $\mathsf{Dist}_{\lambda,C,\mathbf{i}}$ we have $x_{\mathbf{i}} = 1$ ($\mathbf{i}^{th}$ bit of x) since the $\mathbf{i}^{th}$ candidate is always correct.

We prove the following useful sub-lemma. For every two circuits C_0, C_1 we claim that the outputs of the corresponding distributions $\mathsf{Dist}_{\lambda,C_0,\mathbf{i}}$ and $\mathsf{Dist}_{\lambda,C_1,\mathbf{i}}$ are computationally indistinguishable. Here, $\mathbf{i}$ corresponds to the candidate that is always correct. The proof can be found in [2].

SubLemma 1 (Candidate Indistinguishability Lemma). *For large enough security parameter λ, any two circuits $C_0, C_1 \in \mathcal{C}_\lambda$, $\mathbf{i} \in [n]$ we have $\{x \xleftarrow{\$} \mathsf{Dist}_{\lambda,C_0,\mathbf{i}}\} \approx_c \{x \xleftarrow{\$} \mathsf{Dist}_{\lambda,C_1,\mathbf{i}}\}$, where $\mathbf{i}^{th}$ candidate (respresented by $\phi(\mathbf{i})$) is always correct, assuming that Π_{comb} satisfies decomposable security property.*

We now proceed to prove the main lemma. Recall that we are assured the existence of a secure IO candidate that is always correct. Let $\mathbf{i} \in \mathbb{Z}_{>0}$ be such that $\phi(\mathbf{i})$ represents the secure candidate. Let $\lambda \geq \mathbf{i}$. Consider two equivalent circuits $C_0, C_1 \in \mathcal{C}_\lambda$. That is, $|C_0| = |C_1|$ and for every $x \in \{0,1\}^\lambda$ we have $C_0(x) = C_1(x)$. Our goal is to show that $\Pi_{\mathsf{univ}}.\mathsf{Obf}(1^\lambda, C_0) \approx_c \Pi_{\mathsf{univ}}.\mathsf{Obf}(1^\lambda, C_1)$.

We define the following experiment. The following experiment, parameterized by (C_0, C_1), is same as $\Pi_{\mathsf{univ}}.\mathsf{Obf}(1^\lambda, C_0)$ except that the decision to choose which of the candidates to obfuscate the derived circuits $\{[\mathbf{C}]_i\}$ is made solely based on the circuit C_1.

$\mathsf{ExptObf}(1^\lambda, C_0, C_1, \mathbf{i})$:

- Let $\phi(i) = (\Pi_i.\mathsf{Obf}, \Pi_i.\mathsf{Eval})$, for $i \in \{1, \ldots, \lambda\}$. Denote $\Pi_i = (\Pi_i.\mathsf{Obf}, \Pi_i.\mathsf{Eval})$.
- Compute the preprocessing step, $([\mathbf{C}]_1, \ldots, [\mathbf{C}]_n, aux) \leftarrow \mathsf{Preproc}(1^\lambda, 1^n, C_0)$ with $n = \lambda$.
- Sample x from $\mathsf{Dist}_{\lambda, C_1, \mathbf{i}}$, where $\mathbf{i} \in [\lambda]$. That is, x is sampled from the distribution Dist parameterized by $(\lambda, C_1, \mathbf{i})$.
- For every $i \in [\lambda]$ and $x_i = 0$, re-assign $\Pi_i.\mathsf{Obf} = I$ and $\Pi_i.\mathsf{Eval} = UTM$.
- Execute $\overline{[\mathbf{C}]}_i \leftarrow \Pi_i.\mathsf{Obf}(1^\lambda, [\mathbf{C}]_i)$ for at most $T(\lambda, |[\mathbf{C}]_i|)$ number of steps.
- Output $\overline{C} = (\overline{[\mathbf{C}]}_1, \ldots, \overline{[\mathbf{C}]}_\lambda)$.

Consider the following claims.

Claim 9. *The distributions* $D_0 = \{\mathsf{ExptObf}\left(1^\lambda, C_b, C_b, \mathbf{i}\right)\}$ *and* $D_1 = \{\Pi_{\mathsf{univ}}.\mathsf{Obf}\left(1^\lambda, C_b\right)\}$ *are identical, for* $b \in \{0, 1\}$.

The proof of the above claim follows directly from the description of $\mathsf{Dist}_{\lambda, C, \mathbf{i}}$.

Claim 10. *The distributions* $D_0 = \{\mathsf{ExptObf}\left(1^\lambda, C_b, C_0, \mathbf{i}\right)\}$ *and* $D_1 = \{\mathsf{ExptObf}(1^\lambda, C_b, C_1, \mathbf{i})\}$ *are computationally indistinguishable for* $b \in \{0, 1\}$.

The proof of the above claim follows from the Candidate Indistinguishability Lemma (Lemma 1).

Claim 11. *The distributions* $D_0 = \{\mathsf{ExptObf}\left(1^\lambda, C_0, C_b, \mathbf{i}\right)\}$ *and* $D_1 = \{\mathsf{ExptObf}(1^\lambda, C_1, C_b, \mathbf{i})\}$ *are computationally indistinguishable for* $b \in \{0, 1\}$.

We rely on the security (third bullet in Definition 3) of decomposable IO combiner to prove this claim. That is, the output of the IO combiner on two equivalent circuits are computationally indistinguishable. The proof can be found in the full version [2].

From Claims 9, 10 and 11, it follows that $\Pi_{\mathsf{univ}}.\mathsf{Obf}(1^\lambda, C_0) \approx_c \Pi_{\mathsf{univ}}.\mathsf{Obf}(1^\lambda, C_1)$. In more detail,

$$\Pi_{\mathsf{univ}}(1^\lambda, C_0) \equiv \mathsf{ExptObf}\left(1^\lambda, C_0, C_0, \mathbf{i}\right) \text{ (from Claim 9)}$$
$$\approx_c \mathsf{ExptObf}\left(1^\lambda, C_0, C_1, \mathbf{i}\right) \text{ (from Claim 10)}$$
$$\approx_c \mathsf{ExptObf}\left(1^\lambda, C_1, C_1, \mathbf{i}\right) \text{ (from Claim 11)}$$
$$\equiv \Pi_{\mathsf{univ}}(1^\lambda, C_1)\text{ (from Claim 9)}$$

We have demonstrated that Π_{univ} satisfies both the correctness and security properties. This proves the theorem.

Step II: Approx. Correct to Exact(T, ϵ)-Universal Obfuscation. In Step I, we showed how to construct a universal obfuscator that is $\left(1 - \frac{1}{\lambda}\right)$ correct. That is, for sufficiently large security parameter $\lambda \in \mathbb{N}$, every circuit $C \in \mathcal{C}_\lambda$, it holds that:

$$\Pr_{x \xleftarrow{\$} \{0,1\}^\lambda} \left[\Pi_{\mathsf{univ}}.C(x) = \mathsf{Eval}(\overline{C}, x) \ : \ \overline{C} \leftarrow \Pi_{\mathsf{univ}}.\mathsf{Obf}(1^\lambda, C)\right] \geq 1 - \frac{1}{\lambda}$$

We now apply the transformation of BV [6] to obtain a universal obfuscator that is exact (with overwhelming probability). In particular, we apply their transformation that is based on sub-exponential LWE assumption.

That is, for every $C \in \mathcal{C}_\lambda, x \in \{0,1\}^\lambda$, with high probability it holds that:

$$\Pr\left[\,\Pi_{\mathsf{univ}}.C(x) = \mathsf{Eval}(\overline{C}, x)\ :\ \overline{C} \leftarrow \Pi_{\mathsf{univ}}.\mathsf{Obf}(1^\lambda, C)\,\right] = 1$$

We state the formal theorem below.

Theorem 12. *Assuming learning with errors secure against adversaries running in time $2^{n^{\epsilon'}}$ and $(1 - 1/\lambda)$-correct (T, ϵ)-universal obfuscation, we have a (T, ϵ) -universal obfuscation that is exact (with overwhelming probability).*

Combining Step I and II $\Longrightarrow$ Main Result. Combining both the above steps and instantiating the decomposable IO combiner (Theorem 6) we get the following result:

Theorem 13. *Assuming LWE secure against adversaries running in time $2^{n^{\epsilon'}}$, there exists a (T, ϵ)-Universal Obfuscation with ϵ' being a function of ϵ.*

6   Witness Encryption Combiners

6.1   Definition of WE Combiner

We present the formal definition of a WE combiner below. The definition is similar to the definition of IO combiners. The task of the WE combiner is to take n candidates that are correct (in terms of encryption and decryption), and yield a scheme which is as secure as any one of the candidate schemes.

For a scheme Π we say that it is a *correct WE candidate* if it satisfies that correctness requirement and we say that a candidate is secure if it satisfies the security requirement (definitions found in the full version). We say that it is correct and secure if it satisfies both the requirements.

There are two PPT algorithms associated with an WE combiner, namely, CombEnc and CombDec. Procedure CombEnc takes as input an instance x, a message m along with the description of multiple WE candidates and outputs a ciphertext. Procedure CombDec takes as input the ciphertext, a witness w, the description of the candidates and outputs the original message. Since the execution times of the candidates could potentially differ, we require the algorithms CombEnc and CombDec in addition to their usual inputs also take a time function T as input. T dictates an upper bound on the time required to execute all the candidates.

Syntax of WE Combiner. We define an WE combiner $\Pi_{\mathsf{comb}} = ($CombEnc, CombDec$)$ for a language L.

- **Combiner of encryption algorithms, CT $\leftarrow$ CombEnc$(1^\lambda, x, m, \Pi_1, \ldots,$ $\Pi_n, T)$:** It takes as input security parameter λ, an instance x, a message m, description of WE candidates $\{\Pi_i\}_{i \in [n]}$, time function T and outputs a ciphertext.

– **Combiner of decryption algorithms,** $y \leftarrow$ CombDec(CT, w, $\Pi_1, \ldots, \Pi_n$, T): It takes as input a ciphertext CT, a witness for the instance x, descriptions of WE candidates $\{\Pi_i\}_{i \in [n]}$, time function T and outputs y.

We define the properties associated with a WE combiner scheme. There are two properties – correctness and security. We only consider the scenario where all the candidate WE schemes are (almost) perfectly correct but only one of them is secure.

Definition 6 (Secure WE Combiner). *Let $\Pi_1, \ldots, \Pi_n$ be n (almost) perfectly correct WE candidates for* NP *(that is all the schemes are correct, however all of them need not be secure). We say that $\Pi_{\mathsf{comb}} = (\mathsf{CombEnc}, \mathsf{CombDec})$ is a* **secure WE combiner** *if the following conditions are satisfied:*

– **Correctness.** *Consider the following process: (a)* CT $\leftarrow$ CombEnc($1^\lambda, x, m$, $\Pi_1, \ldots, \Pi_n, T$), *(b)* $y \leftarrow$ CombDec(CT, w, $\Pi_1, \ldots, \Pi_n, T$). *Then,* $\Pr[y = m] \geq 1 - \mathsf{negl}(\lambda)$ *over the randomness of* CombEnc.
– **Security:** *If for some $i \in [n]$ candidate Π_i is secure then, for any PPT adversary A and any polynomial $p(\cdot)$, there exists a negligible function $\mathsf{negl}(\cdot)$, such that for any $\lambda \in \mathbb{N}$, any $x \notin L$ and any two equal-length messages m_1 and m_2 such that $|x|, |m_1| \leq p(\lambda)$, we have that*

$$| \Pr[A(\mathsf{CombEnc}(1^\lambda, x, m_1, \Pi_1, \ldots, \Pi_n, T) = 1] -$$
$$\Pr[A(\mathsf{CombEnc}(1^\lambda, x, m_2, \Pi_1, \ldots, \Pi_n, T)) = 1]| \leq \mathsf{negl}(\lambda).$$

Henceforth, we set the time function to be an a priori fixed polynomial. In our constructions presented next, we drop the parameter T which is input to the above algorithms.

6.2   Construction of WE Combiner

We give a construction of a WE combiner. Formally, we prove the following theorem.

Theorem 14. *If one-way functions exist, then there exists a secure WE combiner.*

The construction is given below. As described in Sect. 2.3, the main ingredient of the construction is a (perfectly) secure secret sharing scheme.

CombEnc($1^\lambda, x, m, \Pi_1, \ldots, \Pi_n$) : It takes as input security parameter λ, instance x, message m, description of candidates $\{\Pi_i = (\Pi_i.\mathsf{Enc}, \Pi_i.\mathsf{Dec})\}_{i \in [n]}$ and does the following.

1. **Secret share the message.** Choose n random strings $r_1, \ldots, r_n \in \{0, 1\}^{|m|}$ such that $r_1 \oplus \ldots \oplus r_n = m$.
2. **Encrypt shares using candidates.** For $i \in [n]$, encrypt r_i using candidate Π_i: $y_i \leftarrow \Pi_i.\mathsf{Enc}(x, r_i)$.

3. Output $(y_1, \ldots, y_n)$.

$\underline{\mathsf{CombDec}(1^\lambda, \boldsymbol{y}, w, \Pi_1, \ldots, \Pi_n)}$: On input $\boldsymbol{y} = (y_1, \ldots, y_n)$, an input x with witness w, descriptions of candidates $\{\Pi_i\}_{i \in [n]}$ run the decryption candidates to obtain $r_i \leftarrow \Pi_i.\mathsf{Dec}(1^\lambda, y_i, w)$ for all $i \in [n]$. Compute $m \leftarrow r_1 \oplus \ldots \oplus r_n$ and output m.

Correctness: The correctness follows immediately from the scheme. For any $x \in L$ using the witness w we will get all r_i for $i \in [n]$ and from them we compute the correct message $m = r_1 \oplus \ldots \oplus r_n$.

Security: To prove security, assume that $x \notin L$ and let $i^* \in [n]$ be such that candidate Π_{i^*} is secure. Let m_0, m_1 be any two messages. Consider the following sequence of hybrids. Let H_0, parameterized by $(r_1, \ldots, r_n)$, be a distribution on the encryptions of m_0. That is, H_0 is a distribution over $(y_1, \ldots, y_n)$ where $y_i \leftarrow \Pi_i.\mathsf{Enc}(x, r_i)$ where r_i are random strings such that $r_1 \oplus \ldots \oplus r_n = m_0$. Then we define H_1, again parameterized by $(r_1, \ldots, r_n)$, to be a distribution on encryptions of the message $m_0 \oplus m_1 \oplus r_i$. That is, H_0 is a distribution over $y_{i^*} \leftarrow \Pi_{i^*}.\mathsf{Enc}(1^\lambda, x, r')$ where $r' = m_0 \oplus m_1 \oplus r_i$. From the security of Π_{i^*} we have that $\mathsf{H}_0 \approx \mathsf{H}_1$. Notice that

$$r_1 \oplus \ldots \oplus r_{i^*-1} \oplus r' \oplus r_{i^*+1} \ldots \oplus r_n = m_0 \oplus m_1 \oplus m_0 = m_1.$$

Moreover, the distribution of $r_1, \ldots, r_{i^*-1}, r', r_{i^*+1}, \ldots, r_n$ and the distribution $r_1, \ldots, r_n$ such that $r_1 \oplus \ldots \oplus r_n = m_1$ are identical. Therefore, if we define H_2 to be the distribution on the honest encryptions of the message m_1 (i.e., performed according to the scheme), we get that $\mathsf{H}_1 \equiv \mathsf{H}_2$. Thus we have that $\mathsf{H}_0 \approx \mathsf{H}_2$ which proves the security of the above scheme.

Acknowledgements. We thank Yuval Ishai for helpful discussions and for bringing to our notice the problem of universal obfuscation. We additionally thank Abhishek Jain and Ilan Komargodsky for useful discussions.

References

1. Ananth, P., Chandran, N., Goyal, V., Kanukurthi, B., Ostrovsky, R.: Achieving privacy in verifiable computation with multiple servers-without fhe and without pre-processing. In: PKC (2014)
2. Ananth, P., Jain, A., Naor, M., Sahai, A., Yogev, E.: Universal obfuscation and witness encryption: Boosting correctness and combining security. IACR Cryptology ePrint Archive (2016)
3. Barak, B., Garg, S., Kalai, Y.T., Paneth, O., Sahai, A.: Protecting obfuscation against algebraic attacks. In: Nguyen, P.Q., Oswald, E. (eds.) EUROCRYPT 2014. LNCS, vol. 8441, pp. 221–238. Springer, Heidelberg (2014)
4. Barak, B., Goldreich, O., Impagliazzo, R., Rudich, S., Sahai, A., Vadhan, S.P., Yang, K.: On the (im)possibility of obfuscating programs. In: Kilian, J. (ed.) CRYPTO 2001. LNCS, vol. 2139, p. 1. Springer, Heidelberg (2001)

5. Bitansky, N., Paneth, O.: ZAPs and non-interactive witness indistinguishability from indistinguishability obfuscation. In: Dodis, Y., Nielsen, J.B. (eds.) TCC 2015, Part II. LNCS, vol. 9015, pp. 401–427. Springer, Heidelberg (2015)

6. Bitansky, N., Vaikuntanathan, V.: Indistinguishability obfuscation: from approximate to exact. In: Kushilevitz, E., et al. (eds.) TCC 2016-A. LNCS, vol. 9562, pp. 67–95. Springer, Heidelberg (2016). doi:10.1007/978-3-662-49096-9_4

7. Blaze, M., Bleumer, G., Strauss, M.J.: Divertible protocols and atomic proxy cryptography. In: Nyberg, K. (ed.) EUROCRYPT 1998. LNCS, vol. 1403, pp. 127–144. Springer, Heidelberg (1998)

8. Asmuth, C.A., Blakley, G.R.: An efficient algorithm for constructing a cryptosystem which is harder to break than two other cryptosystems. Comput. Math. Appl. **7**(6), 447–450 (1981). doi:10.1016/0898-1221(81)90029-8. http://www.sciencedirect.com/science/article/pii/0898122181900298. ISSN: 0898-1221

9. Canetti, R., Lin, H., Tessaro, S., Vaikuntanathan, V.: Obfuscation of probabilistic circuits and applications. In: Dodis, Y., Nielsen, J.B. (eds.) TCC 2015, Part II. LNCS, vol. 9015, pp. 468–497. Springer, Heidelberg (2015)

10. Clear, M., McGoldrick, C.: Multi-identity and multi-key leveled FHE from learning with errors. In: Gennaro, R., Robshaw, M. (eds.) CRYPTO 2015. LNCS, vol. 9216, pp. 630–656. Springer, Heidelberg (2015)

11. De Santis, A., Micali, S., Persiano, G.: Non-interactive zero-knowledge with preprocessing. In: Goldwasser, S. (ed.) CRYPTO 1988. LNCS, vol. 403, pp. 269–282. Springer, Heidelberg (1990)

12. Feige, U., Shamir, A.: Witness indistinguishable and witness hiding protocols. In: STOC (1990)

13. Fischlin, M., Herzberg, A., Noon, H.B., Shulman, H.: Obfuscation combiners (2016)

14. Garg, S., Gentry, C., Halevi, S., Raykova, M., Sahai, A., Waters, B.: Candidate indistinguishability obfuscation and functional encryption for all circuits. In: FOCS (2013)

15. Gentry, C., Halevi, S., Vaikuntanathan, V.: i-Hop homomorphic encryption and rerandomizable yao circuits. In: Rabin, T. (ed.) CRYPTO 2010. LNCS, vol. 6223, pp. 155–172. Springer, Heidelberg (2010)

16. Gödel, K.: Über formal unentscheidbare sätze der principia mathematica und verwandter systeme i. Monatshefte für mathematik und physik (1931)

17. Goldreich, O.: The Foundations of Cryptography. Basic Techniques, vol. 1. Cambridge University Press, Cambridge (2001)

18. Goldreich, O., Micali, S., Wigderson, A.: How to play any mental game. In: STOC (1987)

19. Goldwasser, S., Tauman Kalai, Y.: Cryptographic assumptions: a position paper. In: Kushilevitz, E., et al. (eds.) TCC 2016-A. LNCS, vol. 9562, pp. 505–522. Springer, Heidelberg (2016). doi:10.1007/978-3-662-49096-9_21

20. Goldwasser, S., Rothblum, G.N.: On best-possible obfuscation. In: Vadhan, S.P. (ed.) TCC 2007. LNCS, vol. 4392, pp. 194–213. Springer, Heidelberg (2007)

21. Harnik, D., Kilian, J., Naor, M., Reingold, O., Rosen, A.: On robust combiners for oblivious transfer and other primitives. In: Cramer, R. (ed.) EUROCRYPT 2005. LNCS, vol. 3494, pp. 96–113. Springer, Heidelberg (2005)

22. Herzberg, A.: On tolerant cryptographic constructions. In: Menezes, A. (ed.) CT-RSA 2005. LNCS, vol. 3376, pp. 172–190. Springer, Heidelberg (2005)

23. Herzberg, A.: Folklore, practice and theory of robust combiners. J. Comput. Secur. **17**(2), 159–189 (2009). doi:10.3233/JCS-2009-0336

24. Herzberg, A., Shulman, H.: Robust combiners for software hardening. In: Acquisti, A., Smith, S.W., Sadeghi, A.-R. (eds.) TRUST 2010. LNCS, vol. 6101, pp. 282–289. Springer, Heidelberg (2010)

25. Komargodski, I., Moran, T., Naor, M., Pass, R., Rosen, A., Yogev, E.: One-way functions and (im)perfect obfuscation. In: FOCS (2014)

26. Lapidot, D., Shamir, A.: Publicly verifiable non-interactive zero-knowledge proofs. In: Menezes, A., Vanstone, S.A. (eds.) CRYPTO 1990. LNCS, vol. 537, pp. 353–365. Springer, Heidelberg (1991)

27. Levin, L.A.: One-way functions and pseudorandom generators. Combinatorica **7**(4), 357–363 (1987). doi:10.1007/BF02579323

28. López-Alt, A., Tromer, E., Vaikuntanathan, V.: On-the-fly multiparty computation on the cloud via multikey fully homomorphic encryption. In: STOC (2012)

29. Mukherjee, P., Wichs, D.: Two round multiparty computation via multi-key FHE. In: Fischlin, M., Coron, J.-S. (eds.) EUROCRYPT 2016. LNCS, vol. 9666, pp. 735–763. Springer, Heidelberg (2016). doi:10.1007/978-3-662-49896-5_26

30. Pass, R., Seth, K., Telang, S.: Indistinguishability obfuscation from semantically-secure multilinear encodings. In: Garay, J.A., Gennaro, R. (eds.) CRYPTO 2014, Part I. LNCS, vol. 8616, pp. 500–517. Springer, Heidelberg (2014)

31. Sahai, A., Waters, B.: How to use indistinguishability obfuscation: deniable encryption, and more. In: STOC (2014)

32. Yao, A.C.C.: How to generate and exchange secrets (extended abstract). In: FOCS (1986)

Obfuscation Combiners

Marc Fischlin[1]([✉]), Amir Herzberg[2], Hod Bin-Noon[2], and Haya Shulman[3]

[1] Technische Universität Darmstadt, Darmstadt, Germany
marc.fischlin@cryptoplexity.de
[2] Bar Ilan University, Ramat Gan, Israel
[3] Fraunhofer SIT, Darmstadt, Germany

Abstract. Obfuscation is challenging; we currently have practical candidates with rather vague security guarantees on the one side, and theoretical constructions which have recently experienced jeopardizing attacks against the underlying cryptographic assumptions on the other side. This motivates us to study and present *robust combiners for obfuscators*, which integrate several candidate obfuscators into a single obfuscator which is secure as long as a quorum of the candidates is indeed secure.

We give several results about building obfuscation combiners, with matching upper and lower bounds for the precise quorum of secure candidates. Namely, we show that one can build 3-out-of-4 obfuscation combiners where at least three of the four combiners are secure, whereas 2-out-of-3 structural combiners (which combine the obfuscator candidates in a black-box sense) with only two secure candidates, are impossible. Our results generalize to $(2\gamma + 1)$-out-of-$(3\gamma + 1)$ combiners for the positive result, and to 2γ-out-of-3γ results for the negative result, for any integer γ.

To reduce overhead, we define *detecting combiners*, where the combined obfuscator may sometimes produce an error-indication instead of the desired output, indicating that some of the component obfuscators is faulty. We present a $(\gamma + 1)$-out-of-$(2\gamma + 1)$ detecting combiner for any integer γ, bypassing the previous lower bound. We further show that γ-out-of-2γ structural detecting combiners are again impossible.

Since our approach can be used for practical obfuscators, as well as for obfuscators proven secure (based on assumptions), we also briefly report on implementation results for some applied obfuscator programs.

1 Introduction

Software obfuscation has a long tradition in aiming at protecting against reverse engineering. For example, the first International Obfuscated C Code Contest (www.ioccc.org) has been organized in 1984 and experienced the 23rd event in this series in 2014. There are obfuscators for all popular programming languages today. For example, for Java, there are several open-source projects like Pro-Guard, ClassEncrypt, or JavaGuard, and an even larger number of commercial products. These approaches are usually based on heuristics and best practices,

© International Association for Cryptologic Research 2016
M. Robshaw and J. Katz (Eds.): CRYPTO 2016, Part II, LNCS 9815, pp. 521–550, 2016.
DOI: 10.1007/978-3-662-53008-5_18

ranging from simple renaming of function and variables names, to elaborate schemes, e.g. [19]. However, these practical obfuscators do not provide verifiable, *proven* security guarantees.

Provably secure obfuscation, in the sense that it is based on some reasonable cryptographic assumption, has long been a highly desirable yet hard-to-reach goal. Even worse, there have been devastating impossibility results for the natural notion of virtual black-box obfuscation [8] and only limited positive results for special cases like point functions [16]. A significant breakthrough came with the work by Garg et al. [28], indicating that the relaxed yet useful notion of indistinguishability obfuscation may be achievable for general circuits. This notion basically says that one cannot distinguish the obfuscated codes of two functionally equivalent circuit programs.

It is fair to say that the underlying cryptographic assumption, on which is security of the construction of Garg et al. [28] is based upon, is non-standard and not well analyzed (yet). This is also true for the alternative approach to build indistinguishability obfuscators proposed by Pass et al. [48]. This is complemented by yet other proposals of Gentry et al. [32] based on a more standard-like computational assumption about multilinear maps, and of Ananth and Jain [3] based on compact functional encryption. At the same time, recent attacks [18,21–23,31] on multilinear maps, albeit currently not known to break the aforementioned obfuscation candidates, testify that constructions may suddenly turn out to lack the desired security guarantees. New suggestions and attacks keep on appearing at high frequency [6,30,41,46].

The above leaves us with multiple choices of candidates for building obfuscators, both in practice as well as in theory, and it is currently difficult to determine the best choice in terms of security. For the heuristic, practical obfuscators, it may be even harder to distinguish sound constructions from weak approaches, since the design strategies may be vague. A straightforward idea to boost confidence in obfuscator candidates, both in theory as well as in practice, is to interlock multiple solutions and approaches. This idea of failure-tolerant cryptographic designs has traditionally been subsumed under the notion of robust combiners.

1.1   Robust Combiners for Obfuscation

The notion of robust combiners has been introduced by Harnik et al. [34] based on the idea of tolerant cryptographic designs by Herzberg [35–37]. Such combiners take several candidates for a cryptographic task and provide a secure solution if a quorum of the candidates is indeed secure. The idea has been successfully applied to several cryptographic primitives, including hash functions [14,25,27,45,47,49,50], encryption [24,34], commitments [34–36], and oblivious transfer [34,43,44].

A robust combiner for obfuscation would take as input a program (abstractly in form of a circuit or a Turing machine[1]) and create an obfuscated version with

---

[1] In our presentation of our formal results we focus on circuits instead of Turing machines, since our approach applies equally well to both settings but the state of the art of solutions is much more advanced in the circuit setting.

the help of the candidate obfuscators $\mathcal{O}_1, \mathcal{O}_2, \ldots, \mathcal{O}_N$. As long as a sufficient number of candidate obfuscators is indeed secure, the combiner should also provide a secure obfuscator. In order to make formal claims about the robustness of the combiner, due to the lack of rigorous security properties for practical obfuscators, one inevitably needs to base the notion of security for the combiner on the various models in the cryptographic literature, such as virtual black-box obfuscation or indistinguishability obfuscation.[2]

What distinguishes the idea of combiners for obfuscation from the previous scenarios is that obfuscation combiners are higher-order combiners which are closely linked to the functionality of their inputs. Consider for instance the case of hash function combiners where it usually suffices that the combiner preserves the security property only, enabling solutions like the concatenation combiner $\mathsf{Comb}^{H_1, H_2}(x) = H_1(x) \| H_2(x)$ with longer output for collision resistance. Devising hash combiners with equal output size as H_1, H_2, retaining this mild functional property, is conceivably hard [14,49,50]. An obfuscation combiner, in contrast, must provide a circuit which computes the same function as the input circuit; it cannot implement a different function with a larger output. Indeed, note that functional preservation and input hiding are conflicting requirements for obfuscation and one is easy to achieve without the other.

As a concrete example consider combiners in the context of virtual black-box obfuscation. Herzberg and Shulman [38] show that the cascading construction $\mathsf{Comb}^{\mathcal{O}_1, \mathcal{O}_2}(\cdot) = \mathcal{O}_2(\mathcal{O}_1(\cdot))$ of two candidate obfuscators $\mathcal{O}_1, \mathcal{O}_2$ is robust for this notion *as long as functional correctness of the candidates is guaranteed*. If this is not granted and the inner obfuscator is corrupt then $\mathcal{O}_1$ may implement an arbitrary function, such that the combiner neither preserves functional correctness nor necessarily input hiding. The latter holds as one usually does not have any security guarantees for input circuits with diverging functionalities, even if obfuscator $\mathcal{O}_2$ is sound. Analogously, if the outer obfuscator is corrupt then the resulting cascade may no longer sustain functionality.

While functional correctness of an obfuscator is usually not based on unproven cryptographic assumptions, unlike the obfuscation property, there are two reasons why certifying functional correctness may still be hard. First, software implementations are error prone, and the complexity of previous theoretical proposals for obfuscation [28,32,48] seems to be inimical in this regard. Secondly, one may have little control over, or insights into, the actual obfuscation program. This is clearly true for commercial obfuscation programs; in fact, the programs of such obfuscators are often themselves obfuscated. The creation of a corrupt obfuscator, which intentionally leaks some information, is easy; to demonstrate this, we implemented demos of different types of corrupted-obfuscators, including obfuscators which leak information even when used in cascade.

The concern about corrupt-obfuscators may also emerge in theoretical solutions. As an example for the latter, in the universal-parameter generation setting [39] a *trusted* party publishes an obfuscated program which parties can use to

---

[2] There are approaches to define metrics for practical obfuscators [5,20], but mainly in terms of software complexity. We discuss them in Sect. 7.

generate common parameters. What if we now prefer to use several potentially *untrusted* authorities and combine their obfuscated programs?

1.2   Our Results

Our goal is to provide a general combiner for obfuscation. It should satisfy the formal requirements in order to allow for sound solutions both in theory and in practice. Ideally, the combiner should tolerate a large number of corrupt obfuscators, be very efficient, and ensure various notions of obfuscation simultaneously. Note that, while virtual black-box obfuscation may be impossible in general, for some functions and attack models [7,15] the notion may still be achievable, such that our combiner should also comply with this notion.

On the positive side we present a 3-out-of-4 combiner which can tolerate a single corrupt combiner out of four candidates. It is depicted in Fig. 1 and consists of two layers. In the first layer we insert the input circuit C into three combinations of three of the obfuscators each; in each combination, we output a circuit that produces the majority of the three obfuscated circuits. We only require three of the four combinations of picking three of four obfuscators. Each unit ensures that if at most one candidate is corrupt then functional correctness is still preserved. In the next layer we then run each of the first-layer majority circuits through the complementary fourth obfuscation candidate and again take the majority to ensure correctness. Obfuscation follows as either all three candidates on the first layer are sound and thus hide the input circuit, or the fourth candidate on the second layer ensures this.

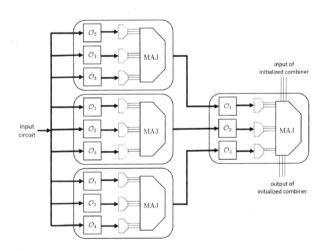

Fig. 1. Our 3-out-of-4 combiner. The MAJ circuit has three hardwired circuits C_1, C_2 and C_3 with equal input and output sizes, which also correspond to the input and output size of the MAJ circuit. For input x the MAJ circuit evaluates each of the three circuits for x and returns the bit-wise majority of the circuit's outputs.

Our combiner indeed works for different notions of obfuscation such as virtual black-box and grey box obfuscation,[3] indistinguishability obfuscation, and differing-input obfuscation. In total it requires twelve calls to obfuscators and has depth 2. The latter is important as obfuscation may cause a polynomial blow-up in size. Remarkably, while most theoretical solutions currently induce a significant size expansion, with a few exceptions [4,13], obfuscators in practice only display a mild increase in code size. Note that devising combiners of depth 1 with a structure as above is impossible as the corrupt obfuscator may then leak information about the input circuit via the output.

We then show an impossibility result for 2-out-of-3 combiners. There are, of course, trivial combiners in this case, such as the combiner which simply uses the sound candidate only, and the (inefficient) combiner for indistinguishability obfuscation that evaluates the input circuit and then outputs the lexicographic smallest equivalent circuit. We thus focus on *structural* combiners that use a fixed pattern, independently of the status of the candidates, and do not semantically interpret the input circuit. Our 3-out-of-4 combiner is structural in this regard. We show that no 2-out-of-3 structural combiner may ensure both functional correctness and obfuscation. This holds for the weaker notion of indistinguishability obfuscation and therefore also for the stronger notions of black-box and grey-box obfuscation. Note that this also applies to any 1-out-of-2 combiner.[4]

We extend the positive result as well as the negative result to the case of $(2\gamma+1)$-out-of-$(3\gamma+1)$ resp. 2γ-out-of-3γ combiners. That is, we give a construction which can be seen as a less efficient generalization of our basic solution if one can corrupt at most γ out of $3\gamma+1$ obfuscators. We then argue that one cannot have structural combiners if γ out of the 3γ obfuscators can be corrupt. For both settings we can draw on the ideas and techniques from the basic cases.

The combiners above are correcting in the sense that they guarantee functional correctness if a quorum of input obfuscator candidates is secure. One can also envision a weaker notion of combiners, which output circuits that either compute the correct output of the input circuit, but may also output, instead, a special error indicator $\perp$. This error indicator should be output only when one of the component obfuscator is faulty; if all obfuscators are sound then the combiner must never output $\perp$. In particular, such combiners cannot output false answers. We call them *detecting combiners* in analogy to coding theory.

For detecting combiners we achieve slightly different bounds. That is, we show that one can have $(\gamma+1)$-out-of-$(2\gamma+1)$ combiners for any γ. For the case $\gamma=1$ and 2-out-of-3 combiners we can again provide an optimized version similar to our original 3-out-of-4 combiner. Concerning lower bounds, we can apply the ideas of the other combiners to show that there cannot exist structural γ-out-of-2γ detecting combiners for any γ. The reduced overhead of detecting combiners may make them attractive option for practical implementations, where once detection ability exists, the attack-vector of providing faulty obfuscator appears unlikely.

---

[3] With respect to dependent auxiliary inputs [33].

[4] Every 1-out-of-2 combiner is also a 2-out-of-3 combiner if it ignores the third obfuscator.

While our main results follow the common approach in provably secure obfuscation, we stress that we view our approach to be equally well suited for practice. In Sect. 7 we therefore evaluate performance of our combiner when applied to practical obfuscators, and discuss the implications of our findings in this domain.

Concurrent Work. Independently of our work, Ananth et al. [2] also discuss the idea of obfuscation combiners. Their approach is fundamentally different from ours, and results in (non-structural) obfuscation combiners which are secure as long as a single candidate is secure. However, this comes at the cost of a significant overhead, and also requires additional cryptographic assumptions such as LWE or DDH, and indistinguishability obfuscation against sub-exponential adversaries.

2   Preliminaries

We exclusively treat circuits here; the approach can be transfered to the case of Turing machines straightforwardly. When speaking of circuits C from some class $\mathcal{C} = (\mathcal{C}_\lambda)_{\lambda \in \mathbb{N}}$ we usually mean some arbitrary (but efficiently computable) description of the circuit. When considering specific encodings with dedicated properties, as required for our lower bounds, we usually write $\langle C \rangle$ for the encoding of the circuit under scheme $\langle \cdot \rangle$. If, on the other hand, we consider the function implemented by the circuit we usually write $C(\cdot)$ instead, and $C(x)$ for the output of circuit C on input x. When writing $C(\cdot) = C'(\cdot)$ or $C \equiv C'$ we refer to functional equality of circuits C and C', comprising input and output length, whereas $C = C'$ or $\langle C \rangle = \langle C' \rangle$ means equal descriptions (under the encoding in question).

2.1   Obfuscators

Barak et al. [8] defined several notions of obfuscators, with *virtual black-box* (VBB) obfuscators being the strongest one. This notion says that the adversary cannot learn anything from an obfuscated circuit beyond the circuit's outputs for chosen inputs. While they also showed that this notion is in general unachievable, for specific cases such as point functions one may be able to attain this level of obfuscation. Below we mainly consider obfuscation of circuits, and we also consider the possibility that the obfuscator itself may be non-uniform and work specifically for different values of λ. The latter allows corrupt (also called malicious) obfuscators to match the algorithm class of adversaries and distinguishers. All obfuscators here, sound and corrupt ones, are nonetheless considered to be stateless.

Definition 1 (Virtual Black-Box Obfuscation). *A (possibly non-uniform) PPT algorithm $\mathcal{O}$ is a virtual black-box obfuscator for circuit class $\mathcal{C} = (\mathcal{C}_\lambda)_{\lambda \in \mathbb{N}}$ if the following holds:*

Functional Correctness: *For any $\lambda \in \mathbb{N}$, any circuit $C \in \mathcal{C}_\lambda$, any obfuscated version $O \leftarrow \mathcal{O}(1^\lambda, C)$ we have $C \equiv O$.*

VBB Obfuscation: *For any (possibly non-uniform) PPT algorithm $\mathcal{A}$ there exists a (possibly non-uniform) algorithm PPT $\mathcal{S}$ and a negligible function $\epsilon(\lambda)$ such that for all circuits $C \in \mathcal{C}_\lambda$ we have*

$$|Prob\left[\mathcal{A}(1^\lambda, \mathcal{O}(1^\lambda, C)) = 1\right] - Prob\left[\mathcal{S}^C(1^\lambda) = 1\right]| \leq \epsilon(\lambda),$$

where the probabilities are over the randomness of $\mathcal{O}$ and $\mathcal{A}$ resp. $\mathcal{S}$.

Virtual grey-box (VGB) obfuscation [10] is defined analogously, only that the simulator above is computationally unbounded but can make at most a polynomial number of queries to its oracle circuit. Clearly, VGB obfuscation implies VGB obfuscation. A stronger notion is based on the extension to *(dependent) auxiliary inputs* [33] where both the adversary and the simulator receive a random sample aux as additional input, where aux may depend on any circuit $C' \in \mathcal{C}_\lambda$.[5] We will use this version for proving the security of our combiners for VBB and VGB obfuscation.

Another meaningful relaxation, implied by both notions above in the non-uniform setting, is *indistinguishability obfuscation* [8] which basically says that the obfuscations of two functional equivalent circuits are indistinguishable:

Definition 2 (Indistinguishability Obfuscator). *A (possibly non-uniform) PPT algorithm iO is called an indistinguishability obfuscator for a circuit class $\mathcal{C} = (\mathcal{C}_\lambda)_{\lambda \in \mathbb{N}}$ if the following conditions hold:*

Functional Correctness: *For any $\lambda \in \mathbb{N}$, any circuit $C \in \mathcal{C}_\lambda$, any obfuscated version $O \leftarrow iO(1^\lambda, C)$ we have $C \equiv O$.*

Indistinguishability: *For any (possibly non-uniform) PPT distinguisher $\mathcal{D}$, there exists a negligible function $\epsilon(\lambda)$ such that for all circuits $C_0, C_1 \in \mathcal{C}_\lambda$ with $C_0 \equiv C_1$ we have*

$$|Prob\left[\mathcal{D}(1^\lambda, C_0, C_1, iO(1^\lambda, C_0)) = 1\right]$$
$$- Prob\left[\mathcal{D}(1^\lambda, C_0, C_1, iO(1^\lambda, C_1)) = 1\right]| \leq \epsilon(\lambda),$$

where the probabilities are over the randomness of iO and $\mathcal{D}$.

There are several variations of the above definitions. For one, we can allow for a negligible error in the functional correctness (over the random choices of the obfuscator). Both our positive and our negative result are robust with respect to such a change. That is, our 3-out-of-4-combiners uses a constant number of obfuscator calls such that the error would remain negligible; obfuscation would still hold, because the leakage due to incorrect obfuscator outputs has negligible probability. Similarly, our impossibility result about 2-out-of-3 combiners would still hold, even if the starting combiners would have perfect functional correctness, but the (fixed-size structural) combiner could have a negligible error. Alternatively, one may use the recent approach in [12] to eliminate the error first.

---

[5] This slightly strengthens the original auxiliary input setting [33] where only $C' = C$ is allowed.

Finally, yet another version of obfuscation, called *differing-inputs* obfuscation [8], demands indistinguishability of two obfuscated circuits, but only if the input circuits C_0, C_1 can be sampled such that finding inputs where C_0 and C_1 differ, is infeasible. More formally, we assume that there is a PPT algorithm Sampler associated to the circuit family C such that for any PPT algorithm A there exists a negligible function $\epsilon(\lambda)$ such that the probability that $C_0(x) \neq C_1(x)$, where $(C_0, C_1, \mathsf{aux}) \leftarrow \mathsf{Sampler}(1^\lambda)$ and $x \leftarrow A(1^\lambda, C_0, C_1, \mathsf{aux})$, is at most $\epsilon(\lambda)$. Note that we assume that $\mathsf{Sampler}(1^\lambda)$ only outputs circuits $C_0, C_1 \in \mathcal{C}_\lambda$.

A differing-inputs obfuscator diO for C and Sampler is now defined analogously to an indistinguishability obfuscator, only that it is infeasible to distinguish outputs $\mathsf{diO}(1^\lambda, C_0)$ and $\mathsf{diO}(1^\lambda, C_1)$ for $(C_0, C_1, \mathsf{aux}) \leftarrow \mathsf{Sampler}(1^\lambda)$, even if given aux as additional input. While the notion is also quite useful for the design of protocols [1], Garg et al. [29] argue that the notion may be hard to achieve.

2.2   Combiners for Obfuscators

Roughly, a combiner for obfuscators is a procedure which uses a set of obfuscators $\mathcal{O}_1, \mathcal{O}_2, \ldots$ to turn an input circuit C into an obfuscated one, with the guarantee that if an (unspecified) quorum of the underlying obfuscators is secure, then so is the combiner. In the definition below we abstractly speak of o-obfuscators, leaving open which obfuscation category $o \in \{\text{VBB, VGB, indistinguishability, differing-inputs}\}$ we refer to.

For combiners of primitives with multiple properties, such as functional correctness and obfuscation here, there are varying levels of combiners, called weak, mild, and strong [26,27]. A strong combiner preserves security "property-wise", i.e., for each property individually if sufficiently many candidates have this property then so does the combiner. A weak combiner only preserves all properties if there are enough candidates which are secure and thus have all properties simultaneously. The mild notion is in between where the candidates must somehow cover all properties but for each property possibly by different candidates. In [26,27] it has been discussed that strong robustness implies mild robustness which in turn implies weak robustness, and that the implications are strict in case of hash functions for some properties.

Definition 3 (Robust Combiner for o-Obfuscation). *Let Comb be a PPT oracle algorithm and let $\mathcal{O}_1, \ldots, \mathcal{O}_N$ be o-obfuscators candidates. Then Comb is called a*

- *strongly robust t-out-of-N combiner if for each of functional correctness and o-obfuscation, if at least t of the N candidates have this property, then so does the combiner $\mathsf{Comb}^{\mathcal{O}_1, \ldots, \mathcal{O}_N}$;*
- *mildly robust t-out-of-N combiner if, whenever functional correctness and o-obfuscation are each satisfied by at least t of the N candidates, then the combiner too has both properties;*

– *weakly robust t-out-of-N combiner if the combiner is a functional correct o-obfuscator if there are at least t out of N candidates which are simultaneously functionally correct and o-obfuscators.*

The definition assumes that the obfuscators and combiner all work for the same class $\mathcal{C}$ of obfuscatable circuits. This neglects an important aspect, though: If the combiner calls obfuscators recursively then the candidates need to be able to handle obfuscated circuits, too. We assume that this is indeed the case — and discuss it more explicitly for our structural combiners below— making the implicit assumption that the candidates also allow for a superclass $\mathcal{C}^{\mathsf{Comb}}$ of circuits which is rich enough tor capture intermediate circuits created by the specific combiner. Still, the task for the combiner is to obfuscate the "core" class $\mathcal{C}$ of circuits.

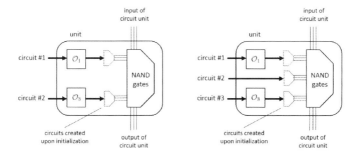

Fig. 2. Example of a unit (with pass-through version on the right-hand side).

As usual for combiners in general, there is always a secure obfuscation combiner, namely, the one which "obliviously" uses the secure obfuscator $\mathcal{O}_i$ and ignores the other ones in order to obfuscate the input circuit. However, this only provides an existential proof and says nothing about how to design an actual solution. Even worse, for indistinguishability obfuscation there is a trivial (non-efficient) combiner for obfuscators which can be described effectively [8]. The combiner takes as input (the description of) a circuit C and finds the (lexicographically) minimal circuit $C_{\min}$ which computes the same functionality as C and outputs this circuit $C_{\min}$. Then any two circuits C, C' with the same functionality yield the same obfuscated circuit $C_{\min}$. This combiner ignores the candidate obfuscators and already constitutes an unconditionally secure obfuscator itself. It is even efficient relative to a Σ_2^p oracle. Hence, any lower bound for combiners would need to bypass this result and therefore need to implicitly show that $\Sigma_2^p \neq P$.

One option to circumvent the first problem is to require to have an effective mean to turn attacks against the combiner into attacks for the candidate obfuscators. This option of so-called *black-box combiners* has been used for other lower bounds such as for hash function combiners [14,49,50]. Still, in our setting such black-box combiners would have to deal with the problem of the inefficient combiner.

An alternative path, which we also take here, is therefore to restrict the way how the combiner works. Whereas the above unconditional combiner approaches the circuit *semantically* by plotting its behavior, we look into what we call *structural* combiners here. Basically, these are combiners which have a prescribed structure with place-holder gates for the obfuscators, and they merely plug in the input circuit and derive the output circuit according to this fixed structure, without evaluating the circuits. It turns out that our 3-out-of-4 combiner is in fact structural.

2.3   Structural Combiners

A structural combiner for obfuscators is a circuit consisting of NAND gates and of obfuscator gates, where each one of the latter is labeled with one of the obfuscators $\mathcal{O}_i$. The layout is independent of the actual obfuscators and should thus work with any concrete obfuscator candidates, i.e., be black-box. The combiner is structured in so-called units. A *unit* is a sub circuit which takes as input the descriptions of circuits and itself describes a circuit. The unit first inserts the input circuits into some of the obfuscators, where we allow multiple appearances of obfuscators in a unit, and then processes the output circuits by a circuit consisting of NAND gates only. An example is given in the left part of Fig. 2. If the input circuit is given to obfuscators $i_1, i_2, \ldots$ then we call this an $\{i_1, i_2, \ldots\}$-unit for the multiset $\{i_1, i_2, \ldots\}$. The example in Fig. 2 describes a $\{1, 3\}$-unit. Furthermore, we can even let some input circuit be passed to the NAND-circuit completely, saying that the unit is *pass-through* in this case. Since it is irrelevant for our lower bound which circuit is passed through, we do not need to specify the identifier. The right hand side of Fig. 2 shows a pass-through version of a $\{1, 3\}$-unit.

The output of a unit can itself serve again as the input for another unit. We can therefore nest units in a tree-like structure as in Fig. 3. In particular, we can analogously to the notion of depths of circuits define the depth of a unit, starting with level-1 units, as well as paths from the input circuit to the final unit. We call the path of units form level-1 units to the final unit a full path. A unit which is level-1 always receive the combiner's input circuit as inputs, but potentially also other unit circuits if it is simultaneously a higher level unit. Every unit has at least one input circuit, and a unit can of course serve as multiple inputs to other units.

To complete the description of a structural combiner we need to specify the output of our combiner for some input circuit C, once the obfuscator candidates are determined. We call this the *initialization* of the combiner with C. Basically the output is again a circuit and it is derived by stepwise replacing the obfuscator gates in units (starting with level-1 units which receive C as input) with samples of the output of the corresponding obfuscator. Note that the structure of the combiner circuit remains, only the obfuscator gates are now filled in with concrete circuits. In case of pass-through units we additionally place the code of the unit's input circuit inside the new circuit at the corresponding position. Once a unit has been initialized we can use it as input to a higher-level unit and initialize that unit, till we have eventually initialized the final unit. Instructively, the reader may think of this as a left-to-right pass in Fig. 3 to compute the final output circuit, denoted as $\mathsf{Comb}^{\mathcal{O}_1,\mathcal{O}_2,\cdots}(C)$. Note that this is a random variable, depending on the randomness of the obfuscators. A sample of this random variable can then be fed with inputs x to produce some output y.

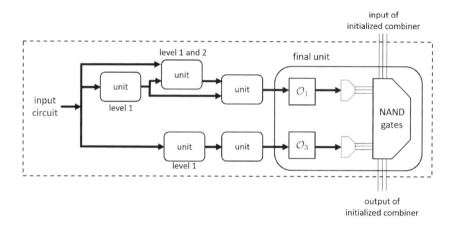

Fig. 3. Combiner circuit consisting of units.

The above assumes that the class of obfuscatable circuits for structural circuits is closed under recursive constructions of units. We note that for concrete constructions such as our 3-out-of-4 combiner in the next section it suffices that we can also obfuscate level-1 units of the original input circuits. Given a circuit class $\mathcal{C} = (\mathcal{C}_\lambda)_{\lambda \in \mathbb{N}}$, some fixed structural combiner Comb, and fixed obfuscators $\mathcal{O}_1, \mathcal{O}_2, \ldots$ we denote by $\mathcal{C}^{\mathsf{Comb}} = (\mathcal{C}_\lambda^{\mathsf{Comb}})_{\lambda \in \mathbb{N}}$ the class of circuits which, besides all circuits $C \in \mathcal{C}_\lambda$, for any C also includes all possible initializations of all units of the combiner (except for the final unit) for the given obfuscators. It is understood that, when considering a specific combiner Comb, all candidate obfuscators $\mathcal{O}_1, \mathcal{O}_2, \ldots$ must be able to handle the class $\mathcal{C}^{\mathsf{Comb}}$, whereas the combiner only works for the "inner" class $\mathcal{C}$. Instructively, one may think of $\mathcal{C}$ as the class one would like to obfuscate, although the candidate obfuscators allow for broader classes.

3   Robust 3-out-of-4 Combiner for Obfuscators

In this section we present a 3-out-of-4 (structural) combiner for obfuscation, depicted in Fig. 1 on Page 524.

3.1   Construction

The idea is to first obfuscate the input circuit C by all combinations of 3 out of the 4 given obfuscators $\mathcal{O}_1, \ldots, \mathcal{O}_4$ and for each combination taking the majority of the output of the three obfuscated circuits. Note that since at least 2 of the 3 obfuscators in such a combination work properly, the majority decision provides a functionally correct output. Formally, for the majority circuit MAJ combining three input circuits by evaluating each one for a given input x and taking the bit-wise majority of the outputs, we thus build the circuits

$$O_{i_1, i_2, i_3} \leftarrow \mathrm{MAJ}(\mathcal{O}_{i_1}(C), \mathcal{O}_{i_2}(C), \mathcal{O}_{i_3}(C)), \quad 1 \leq i_1 < i_2 < i_3 \leq 4$$

for all possible 4 combinations of i_1, i_2, i_3. Since we merely need an arbitrary 3 of these 4 circuits for the next stage, we take the combinations leaving out obfuscators $1, 2$ and 3 (in this order).

Of course, a corrupt obfuscator among $\mathcal{O}_{i_1}, \mathcal{O}_{i_2}, \mathcal{O}_{i_3}$ in the majority combination could still reveal information about the input circuit C. We hence add another layer where we now combine three of the majority combinations as before, by running each combination O_{i_1, i_2, i_3} through the complementary obfuscator $\mathcal{O}_{i_4}$ and taking the majority of these circuits again. Put differently, we now build the circuit

$$\mathrm{MAJ}(\mathcal{O}_1(O_{2,3,4}), \mathcal{O}_2(O_{1,3,4}), \mathcal{O}_3(O_{1,2,4})).$$

Functional correctness of our combiner is guaranteed because each of the input circuits $O_{2,3,4}, O_{1,3,4}, O_{1,2,4}$ computes the correct function and at least two of the level-2 obfuscators $\mathcal{O}_1, \mathcal{O}_2, \mathcal{O}_3$ are correct. The obfuscation property holds because if one of the level-2 obfuscators, say, $\mathcal{O}_1^*$, is malicious, then the level-1 obfuscators generating $O_{2,3,4}$ already hide the input circuit. Furthermore, the malicious obfuscator $\mathcal{O}_1^*$ cannot bias the functional correctness of the circuits $O_{1,3,4}$ and $O_{1,2,4}$ in the other branches, such that the sound second-layer obfuscators $\mathcal{O}_2, \mathcal{O}_3$ also hide $O_{1,3,4}$ and $O_{1,2,4}$ and thus the input circuit C, even if $\mathcal{O}_1^*$ on the first level reveals information about C.

3.2   Security

We start by showing that the combiner is (strongly) robust for indistinguishability obfuscation. Recall that strong robustness refers to the fact that each property, functional correctness and obfuscation, is preserved individually. Note that for our combiner (and also the security proof) it suffices that the parties merely have black-box access to all obfuscators.

Theorem 1. *The combiner in Fig. 1 is a strongly robust 3-out-of-4 combiner for indistinguishability obfuscation.*

Proof. Functional correctness is straightforward, given that for each unit at least two obfuscators are functionally correct and since we apply the majority of the outputs.

We next show indistinguishability. Take an arbitrary distinguisher $\mathcal{D}$ against our combiner. We need to show that there exists a negligible function ϵ such that for an arbitrary pair $C_0, C_1 \in \mathcal{C}_\lambda$ of circuits, the distinguishing advantage of $\mathcal{D}$ is smaller than $\epsilon(\lambda)$. The idea is to show that one can gradually replace the input circuits C_0 to the obfuscators in the combiner by circuit C_1, taking some care with the single corrupt obfuscator.

For the gradual replacement fix the order of the nine level-1 obfuscators $\mathcal{O}_2, \mathcal{O}_3, \mathcal{O}_4, \ldots, \mathcal{O}_1, \mathcal{O}_2, \mathcal{O}_4$ according to their appearance in Fig. 1 from top to down, with one exception: for a parameter $k \in \{1, 2, 3, 4\}$, a reminiscent for the index of the corrupt obfuscator $\mathcal{O}_k^*$, we move all occurrences of this obfuscator to the very end of the list. For instance, for $k = 2$ we would have the order $\mathcal{O}_3, \mathcal{O}_4, \mathcal{O}_1, \ldots, \mathcal{O}_4, \mathcal{O}_2^*, \mathcal{O}_2^*$. Let $K = K(k) \in \{7, 8\}$ be the first index of $\mathcal{O}_k^*$ in that list. Define now the random hybrid variables $H_i^k(C_0, C_1)$ for $i = 0, 1, \ldots, 9$ as the output of our combiner if we pass circuit C_0 for the first i obfuscators (according to our order) and C_1 for the remaining $9 - i$ ones. Then, clearly $H_9^k(C_0, C_1)$ corresponds to the distribution of our combiner for input C_0, and $H_0^k(C_0, C_1)$ to the one of our combiner for C_1. It hence suffices to show for any i that $\mathcal{D}$'s probability of distinguishing adjacent H_{i-1}^k, H_i^k is negligible.

To bound the advantage of $\mathcal{D}$ for each pair (H_{i-1}^k, H_i^k) we will wrap the algorithm into a sequence of distinguishers $\mathcal{D}_i^k$ for $i = 1, 2, \ldots, 9$. The distinguisher $\mathcal{D}_i^k$ works in two modes, depending on the status of the i-th obfuscator in our sequence:

– If i is such that the i-th obfuscator is not corrupt, i.e., $i < K(k)$, then $\mathcal{D}_i^k$ expects as input a pair C_0, C_1 and an obfuscated circuit O' generated by the i-th obfuscator in our order for C_b, $b \in \{0, 1\}$. Algorithm $\mathcal{D}_i^k$ computes the output of our combiner (with the given obfuscators) but inserts C_1 as input in the first $i - 1$ level-1 obfuscators, O' as the *output* of the i-th level-1 obfuscator, and C_0 as input in the final $9 - i$ slots. It completes the output O of the combiner for these data and lets $\mathcal{D}$ run on C_0, C_1 and O. Algorithm $\mathcal{D}_i^k$ returns whatever $\mathcal{D}$ outputs.

– If the i-th obfuscator is corrupt, i.e., $i \geq K(k)$, then $\mathcal{D}_i^k$ expects as extra auxiliary input a pair C_0, C_1 and a sample O' of one of the sound obfuscator candidates. Here the obfuscator $\mathcal{O}_j$ producing O' is determined by looking at the level-1 unit u in which the i-th (corrupt) obfuscator $\mathcal{O}_k^*$ appears. For this unit, and its three obfuscators, there exists the fourth, complementing obfuscator $\mathcal{O}_j$ to which the unit's output is fed to on the level-2 unit. For instance, if $k = 2$, $K = 8$, and $i = 8$, then the corresponding level-1 unit u is the top one in Fig. 1, and the complementing obfuscator is $\mathcal{O}_1$.

The input to the complementing obfuscator $\mathcal{O}_j$ for deriving O' is either a sample of the level-1 unit where all honest obfuscators are initialized with C_0 and the corrupt one with C_1, or all of them are initialized with C_0. By assumption, both samples are in the class $\mathcal{C}_\lambda^{\mathsf{Comb}}$ such that the sample can be passed to $\mathcal{O}_j$. Algorithm $\mathcal{D}_i^k$ now evaluates our combiner, by replacing inputs to obfuscators up to index i by C_1, for subsequent indices giving input C_0, and replacing the output of the complementing obfuscator $\mathcal{O}_j$ in unit u when evaluating our combiner by O'. Return $\mathcal{D}$'s output bit on input C_0, C_1 and the combiner's output O.

Assume i is such that the i-th obfuscator in order is still different from $\mathcal{O}_k^*$, i.e., $i < K$. Then if O' is the obfuscation of C_1, then $\mathcal{D}_i^k$ runs $\mathcal{D}$ exactly on the distribution of the hybrid variable $H_{i-1}^k(C_0, C_1)$. In particular, for $i = 1$ algorithm $\mathcal{D}_i^k$ runs $\mathcal{D}$ on a sample of our combiner's output for C_1. Analogously, for $i = 9$ and O' stemming from the complementing obfuscator for a sample of the level-1 unit with all C_0 inputs, the input to $\mathcal{D}$ is distributed like a sample of our combiner for C_0 (and thus of $H_9^k(C_0, C_1)$).

Assume that the k-th obfuscator is indeed corrupt. For $i < K$ it follows from the indistinguishability obfuscation of the sound obfuscators that there exist negligible functions $\epsilon_i(\lambda)$ such that for any C_0, C_1, the advantage of $\mathcal{D}_i^k$ in distinguishing the two input cases is at most $\epsilon_i(\lambda)$. For $i \geq K$ this follows as the input circuits to the two sound obfuscators in unit u are already C_0, such that the majority computation of the unit ensures that in both cases the unit circuit computes the function $C_0(\cdot)$. It follows that both input circuits to the complementing obfuscator $\mathcal{O}_j$ compute the same function and we can again conclude from the security of the obfuscator that the advantage must be bounded by some function $\epsilon_i(\lambda)$. Note that here we take advantage of the fact that indistinguishability holds for all circuits and therefore in paticular also for our partly combiner samples.

It therefore also holds for any i that the advantage of $\mathcal{D}$ in distinguishing $H_{i-1}(C_0, C_1)$ and $H_i(C_0, C_1)$ for any C_0, C_1 is at most $\epsilon_i(\lambda)$, too. Hence, the overall advantage of $\mathcal{D}$ is at most $\epsilon(\lambda) := \sum_{i=1}^{9} \epsilon_i(\lambda)$ and thus negligible. it suffices that the proof provides an existential result.[6]

The claim carries over to the case of differing-inputs obfuscation. Recall that the main difference to indistinguishability obfuscation is that, for the differing-inputs case, the circuits in question are generated by an algorithm Sampler such that the circuits may compute different functions, but Sampler ensures that finding differing inputs is infeasible. We can basically apply the same hybrid argument in this case as above. However, for the step $i \geq K$, when using the obfuscation of our level-1 unit, we need to specify sampler Sampler$_k'$ with oracle access to $\mathcal{O}_1, \ldots, \mathcal{O}_4$ to generate the input circuit for the complementing obfuscator. Algorithm Sampler$_k'$ first runs Sampler to get (C_0, C_1, aux), then generates two samples of the level-1 unit (one time using C_0 for the honest obfuscators and C_1 for $\mathcal{O}_k^*$, and the other time using C_0 everywhere), and finally outputs these two samples and aux$' = (C_0, C_1, \text{aux})$ as auxiliary data. Note that finding an input x where the two level-1 unit samples differ is impossible, as both implement the same function.

We next show that the claim remains true with respect to virtual black-box and grey-box obfuscation. For this we assume that the adversary and the simulator receive some circuit-dependent auxiliary input aux as additional input, as explained in Sect. 2.

---

[6] Note that we do not need to know the index k of the obfuscator; unlike the construction it suffices that the proof provides an existential result.

Proposition 1. *The combiner in Fig. 1 is a strongly robust 3-out-of-4 combiner for virtual black-box and grey-box obfuscation with respect to dependent auxiliary input.*

Proof. Functional correctness follows as in the case of indistinguishability obfuscation. We only discuss the VBB property here; the VGB property follows analogously.

Consider an adversary $\mathcal{A}_0$ against VBB obfuscation. This adversary receives an output sample O' of our combiner as input and some auxiliary input $\mathsf{aux}[0] = \mathsf{aux}[0](C)$. Let k be again the index of the malicious obfuscator and this time define $L = L(k) \in \{3, 5\}$ as follows. For $k = 4$ we would have the malicious obfuscator $\mathcal{O}_4^*$ only on first-level units and we only need to look at the $L = 3$ second-level obfuscators. For $k \in \{1, 2, 3\}$, on the other hand, the malicious combiner appears in a second-level unit and we thus consider the $L = 5$ sound obfuscators, consisting of the 3 obfuscators leading to the second-level appearance of $\mathcal{O}_k^*$ and the remaining 2 honest level-two obfuscators.

Assume now that we change the auxiliary input to include the obfuscator results of our combiner for all L sound obfuscators defined above. Denote these intermediate results, ordered according to the obfuscator application, by $O[1..L] = (O_{i_1}, O_{i_2}, O_{i_3}, \ldots, O_{i_L})$, and let $O[1..i]$ denote the first i entries in $O[1..L]$. Let $\mathsf{aux}[0..i]$ denote the sample given by a sample of first i obfuscator outputs, together with the (independent) sample $\mathsf{aux}[0]$ of $\mathcal{A}_0$.

Instead of considering $\mathcal{A}_0(1^\lambda, O', \mathsf{aux}[0])$ we construct an algorithm $\mathcal{A}_1$ which receives 1^λ and $\mathsf{aux}[0..L]$ as input, assembles a combiner output O' from $\mathsf{aux}[1..L]$ by possibly evaluating the (level-2) malicious obfuscator, and runs adversary $\mathcal{A}_0(1^\lambda, O', \mathsf{aux}[0])$. Then, clearly, the output distribution of both algorithms are identical. We can now view $\mathcal{A}_1$ as an algorithm which receives $\mathsf{aux}[0..L - 1]$ as auxiliary input, and the obfuscated circuit $\mathsf{aux}[L]$ together with 1^λ as regular input.[7] For this algorithm $\mathcal{A}_1$, by assumption about the security of $\mathcal{O}_{i_L}$ producing O_{i_L}, there exists a simulator $\mathcal{S}_1^C(1^\lambda, \mathsf{aux}[0..L - 1])$ with negligibly close output distribution.

Given $\mathcal{S}_1$ we construct an adversary $\mathcal{A}_2$ which receives auxiliary input $\mathsf{aux}[0..L - 2]$, and 1^λ and $\mathsf{aux}[L - 1]$ as regular input. It runs $\mathcal{S}_1(1^\lambda, \mathsf{aux}[0..L - 2])$ and uses $\mathsf{aux}[L - 1]$ to answer oracle calls. Note that, by the functional correctness of $\mathcal{O}_{i_{L-1}}$, using $\mathsf{aux}[L - 1]$ to simulate the oracle C of $\mathcal{S}_1$ is sound as both circuits compute the same function. We can set this argument forth to eventually obtain a simulator $\mathcal{S}_L^C(1^\lambda, \mathsf{aux}[0])$, producing some output distribution which is negligibly close to the one of our initial adversary $\mathcal{A}_0(1^\lambda, O', \mathsf{aux}[0])$. This shows VBB obfuscation.

---

[7] By construction, if we shift the input of a level-2 obfuscator then this is a sample of a level-1 unit, whereas the other auxiliary inputs are based on the original and functional equivalent circuit C.

4   Lower Bounds for Combiners

To illustrate how we use the two required security properties, function preservation and indistinguishability, against each other to derive our general result, it is useful to demonstrate our technique for some toy examples. In the examples we use an unspecified notion of indistinguishability of the obfuscators as we merely highlight the issues; the reader may think for sake of concreteness of the notion of indistinguishability obfuscation.

4.1   Simple Attempts that Fail

The first attempt to build a secure combiner consists of a single unit and is given in the left hand part of Fig. 4. It uses three obfuscators $\mathcal{O}_1, \mathcal{O}_2, \mathcal{O}_3$ and runs the input circuit through each of them. Then it combines the three obfuscated circuits by a majority circuit. Note that this means that this part takes the circuits and has some input wires for the input x, and it evaluates each circuit on x and outputs y as the bit-wise majority of the answers. While we cannot break the functional correctness of the combiner with a single corrupt obfuscator $\mathcal{O}_i^*$, we can easily break the indistinguishability property. To this end we take control of obfuscator $\mathcal{O}_1^*$ and let it simply output the input circuit in clear. Note that this means that the unit, after having been initialized, reveals the input circuit in clear as well, and this easy to distinguish.

In our second example we have two obfuscators $\mathcal{O}_1, \mathcal{O}_2$, let the output of them be combined arbitrarily, and then input the derived circuit into obfuscator $\mathcal{O}_3$. Note that in this case it is unclear how to break the indistinguishability property by corrupting a single obfuscator only. If it is $\mathcal{O}_3$ then the obfuscators of the first unit already hide the input circuit; if we corrupt one of the obfuscators $\mathcal{O}_1, \mathcal{O}_2$ then the final obfuscation hides the actual circuit.

We can, nonetheless, in the second example break the functional preservation property. Namely, assume that both $\mathcal{O}_1$ and $\mathcal{O}_2$ are secure and that there are two potential input circuits D_0, D_1 computing *different* functions for the same input and output length. If we control $\mathcal{O}_1$, then we let it on any input circuit rather obfuscate D_1. Vice versa, if we control $\mathcal{O}_2$ then we let it always obfuscate D_0, independently of the actual input. It follows that the initialized combiners

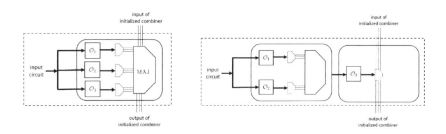

Fig. 4. Examples of (insecure) structural combiners

for D_0 (with our malicious $\mathcal{O}_1^*$ and with genuine $\mathcal{O}_2$) and for D_1 (with genuine $\mathcal{O}_1$ and our malicious $\mathcal{O}_2^*$) have the same distribution. For at least one of the two cases the computed function must then be incorrect, as the initialization samples for both input circuits D_0, D_1 have the same distributions in both cases.

4.2   The General Case of 2-out-of-3 Combiners

The attacks in the simple case show the path for our general impossibility result for 2-out-of-3 structural combiners. If one of the three obfuscators appears in all units on the path from some level-1 unit to the final unit, then it can pass on information about the input circuit C to the final unit. This is done by forwarding some information about the input circuit in the output of the obfuscator. This would clearly violate the indistinguishability property. Hence, on all paths there must be a unit which only uses (at most) the same two obfuscators. But then we can "confuse" the combiner as we did in the second example above. The argument, however, requires some care to deal with the fact that we have many paths. Note also that the confusion strategy fails in our 3-out-of-4 construction because the majority of the three combiners yields the correct function.

For sake of concreteness we use the notion of indistinguishability obfuscation for the obfuscators and the combiners. Recall that this means that for any functionally equivalent circuits C_0, C_1 from class $\mathcal{C}$ the combiners initialization with these two circuits must be computationally indistinguishable. To avoid trivial cases we assume that the class $\mathcal{C}$ contains at least two distinct but functionally equivalent circuits E_0, E_1, and that it also contains two circuits D_0, D_1 computing different functions. We call such classes *non-trivial*.

Since we pass on circuits as inputs we need to fix some encoding. Let $\langle \cdot \rangle$ denote such a function mapping circuits from the class $\mathcal{C}$ to strings. We assume that the encoding is such that given an encoding of a unit (after initialization) one can reconstruct the circuits output by the obfuscators. That is, there exists an efficient algorithm reconstruct such that given any initialization $V \leftarrow U(C_1, C_2, C_3, \dots)$ of a unit U, including obfuscated circuits $O_i \leftarrow \mathcal{O}_{i_j}(C_i)$ and possibly pass-through circuits $O_i = C_i$, we have reconstruct($\langle V \rangle$) = $(\langle O_1 \rangle, \langle O_2 \rangle, \langle O_3 \rangle, \dots)$. Furthermore, we assume that the encoding of no unit coincides with the encoding of our equivalent circuits E_0 or E_1 such that it is clear if each $\langle O_i \rangle$ is the result of a unit initialization or rather one of the circuits E_0 or E_1. We call such encodings $\langle \cdot \rangle$ *admissible*. Note that this means that the combiner itself cannot apply any obfuscation techniques beyond the ones provided by the obfuscators placed inside the unit.

We first show that on any full path (from level-1 units to the final unit) each of the three combiners does not appear in a unit (and that a unit in which it does not appear is not pass-through). This holds for any structural combiner, independently of the total number of obfuscators and the number of malicious ones:

Lemma 1. *Let* $\mathsf{Comb}^{\mathcal{O}_1, \mathcal{O}_2, \dots, \mathcal{O}_N}$ *be a structural combiner for a non-trivial circuit class* $\mathcal{C}$ *with admissible encoding* $\langle \cdot \rangle$. *Then for any full path of units of the* *combiner and for any* $i \in \{1, 2, \ldots, N\}$ *there must be a unit which is not pass-through and which is not an* $\{i, \ldots\}$*-unit, or else the combiner cannot be an indistinguishable obfuscator.*

Proof. Assume that there exists a full path of units and an $i \in \{1, 2, \ldots, N\}$ such that each unit on the path is an $\{i, \ldots\}$-unit or that it is pass-through (or both). Then we show how to break indistinguishability obfuscation as follows. Let E_0, E_1 be some functional equivalent circuits in the class with distinct encodings under $\langle \cdot \rangle$. We corrupt obfuscator $\mathcal{O}_i$ and for each input circuit let it, for each call, simply output the input circuit in clear, by duplicating the input description.

Since each unit on the pass includes the i-th obfuscator (or is pass through) a distinguisher can distinguish between a combiner obfuscation of E_0 and E_1 as follows. The distinguisher receives as input the initialization of the final unit U, and runs reconstruct($\langle U \rangle$) to recover all (obfuscated or pass-through) input circuits $O_1, O_2, \ldots$. Since the distinguisher knows the layout of the combiner it can recursively apply the reconstruction algorithm to outputs of the i-th combiner resp. to passed circuits; both are initialized units. Following the full path in question, the distinguisher eventually obtains either E_0 or E_1 as the input circuit, and can thus distinguish the two cases easily.   □

We next show that, given that each full path contains a unit in which, say, obfuscator $\mathcal{O}_3$ does not appear, we can confuse the combiner. This time, the claim only holds for 2-out-of-3 combiners:

Lemma 2. *Let* $\mathsf{Comb}^{\mathcal{O}_1, \mathcal{O}_2, \mathcal{O}_3}$ *be a structural combiner for a non-trivial circuit class* $\mathcal{C}$ *with admissible encoding* $\langle \cdot \rangle$. *Then the combiner cannot be perfectly correct.*

In particular, if u denotes the number of units in the structural combiner and m the maximal number of obfuscator gates in a unit, then with probability at least $2^{-u}(mu)^{-mu}$ (over the random choices of the obfuscators) the combiner's function is different from the one of the input circuit. If u and m are constant, for instance, this means a constant error in functional preservation.

Proof. By Lemma 1 for each path from level-1 units to the final unit there exists a unit which does not contain, say, the obfuscator $\mathcal{O}_3$ and which is neither pass-through. Put differently, such a unit contains (at most) the obfuscators $\mathcal{O}_1, \mathcal{O}_2$, each one possibly multiple times Let $U_1, U_2, \ldots$ be the corresponding units which we call *confusion units*. In the example in Fig. 5 the confusion units on the three paths are marked by dotted lines.

We consider two cases, one time corrupting obfuscator $\mathcal{O}_1$, the other time corrupting obfuscator $\mathcal{O}_2$. Let us first consider the case that we corrupt obfuscator $\mathcal{O}_1$. Our version $\mathcal{O}_1^*$ of the obfuscator will internally hold, and formally attributed to the non-uniformity, an initialization sample of $\mathsf{Comb}^{\mathcal{O}_1, \mathcal{O}_2, \mathcal{O}_3}(D_1)$ with the genuine obfuscators for input circuit D_1. In particular, for each confusion unit U_i it will include the $j \leq m$ circuit codes of $O_i^j[D_1]$ which the original obfuscator $\mathcal{O}_1$ output in unit U_i in this sample. In order to make our obfuscator state-free we will guess the right insertion positions and injected circuits.

That is, for each call (about some input circuit) our malicious obfuscator $\mathcal{O}_1^*$ tosses a coin. If it comes out as head, then the obfuscator proceeds as the genuine obfuscator would. If it is tail, then it picks one of the at most mu circuits $O_i^j[D_1]$ at random, and returns this circuit. An example of a run with good guesses is given in the left part of Fig. 5.

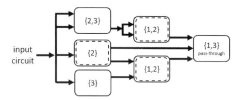

Fig. 5. Confusion units in this example are marked by dotted lines.

For the other case we corrupt $\mathcal{O}_2$ and include a sample of $\mathsf{Comb}^{\mathcal{O}_1,\mathcal{O}_2,\mathcal{O}_3}(D_0)$ of the genuine obfuscators, this time for input circuit D_0. Analogously to the other case denote the output of $\mathcal{O}_2$ in the confusion unit U_i by $O_i^j[D_0]$. When called, the malicious obfuscator $\mathcal{O}_2^*$ also generates an honest answer with probability $\frac{1}{2}$, and inserts one of the pre-sampled circuits $O_i^j[D_0]$, the choice made at random, in the other case.

For the analysis we start with the case of a malicious obfuscator $\mathcal{O}_1^*$. Note that, if we let u denote the number of units in the combiner, then with probability 2^{-u} we overwrite the obfuscator's behavior exactly for the confusion units, since we predict the status of each unit (confusion or not) exactly with probability $\frac{1}{2}$. If so, then we also inject the hardwired circuits $O_i^j[D_1]$ "correctly" in confusion unit U_i with probability at least $(mu)^{-mu}$, since we have at most u units with at most m obfuscator gates and need to guess for each unit correctly among the at most mu possibilities among all $O_i^j[D_1]$'s. If this happens, and the combiner receives

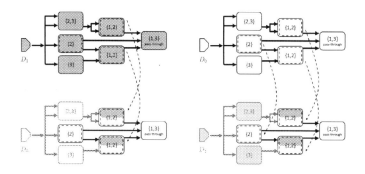

Fig. 6. Confusion strategy with malicious obfuscator $\mathcal{O}_1^*$ injecting parts of the upper D_1 initialization sample into the lower D_0 initialization (left), and malicious obfuscator $\mathcal{O}_2^*$ injecting parts of the upper D_0 initialization sample into the lower D_1 initialization (right).

circuit D_0 as input, then in the confusion units we have consistent samples for $\mathcal{O}_1$ gates (if present), as if the combiners input had been D_1. See the left part of Fig. 6 for an example. Simultaneously, in the same unit, we have consistent samples for $\mathcal{O}_2$ gates (if present), as if the overall input had been circuit D_0.

By symmetry, the same is true if we control obfuscator $\mathcal{O}_2^*$ and the combiner's input is D_1. Hence, with probability at least $2^{-u}(mu)^{-mu}$ either case creates the same output distribution. If this happens, then on each path to the final unit the corresponding confusion unit produces the same distribution upon the single initialization in both cases. It follows that the combiner must implement an incorrect function in one of the cases, showing that functional preservation is not satisfied. It follows that the combiner cannot be perfectly correct.

Noting that the combiner cannot work even if 2 of the 3 obfuscators both have both properties simultaneously, the previous lemmas imply that there are not even weakly robust 2-out-of-3 combiners for indistinguishability obfuscation. It follows that there cannot exist stronger forms of structural combiners either, such as 1-out-of-2 combiners, strong combiner, or virtual grey-box combiners.

Theorem 2. *For any $o \in \{VBB, VGB, indistinguishability, differing\text{-}inputs\}$ there is no structural weakly robust 2-out-of-3 o-obfuscation combiner* $\mathsf{Comb}^{\mathcal{O}_1, \mathcal{O}_2, \mathcal{O}_3}$ *for non-trivial circuit classes $\mathcal{C}$ with admissible encoding $\langle \cdot \rangle$.*

5 The General Case of $(2\gamma[+1])$-out-of-$(3\gamma[+1])$ Combiners

In this section we present a generalization of our 3-out-of-4 combiner to the case of $(2\gamma+1)$-out-of-$(3\gamma+1)$ combiners for any fixed integer γ. In fact, our combiner for $\gamma = 1$ in Sect. 3 can be seen as a special parallelized version of the general approach here. We then discuss that our lower bound for 2-out-of-3 structural combiners also carries over to the more general case of 2γ-out-of-3γ combiners, showing that our general combiner here is optimal in this regard.

5.1 Robust $(2\gamma + 1)$-out-of-$(3\gamma + 1)$ Combiners

Consider all sets I of subsets of $\{1, 2, \ldots, 3\gamma+1\}$ of size $2\gamma+1$. For each such set I form the unit which, similar to our 3-out-of-4 case, first in parallel obfuscates the input circuit with each obfuscator $\mathcal{O}_i$ for $i \in I$, and then compute the majority circuit over all these $2\gamma + 1$ obfuscated circuits. We write

$$O_I(\cdot) = \mathrm{MAJ}\{\mathcal{O}_i(\cdot) \mid i \in I\}$$

for this unit. To obfuscate a circuit C compose each of these units for all the I's sequentially, in arbitrary order. Let us denote this process by

$$\left(\prod_I O_I\right)(\cdot) = O_{I_\ell}(\cdots O_{I_3}(O_{I_2}(O_{I_1}(\cdot))) \cdots)$$

for constant $\ell = \binom{3\gamma+1}{2\gamma+1}$. Call this the *sequential-subset combiner* for γ.

Intuitively, the sequential-subset combiner guarantees robustness as there exists a subset I such that this subset only uses the $2\gamma + 1$ uncorrupt obfuscators. At the same time each circuit O_I computes the correct function as the majority of the $2\gamma + 1$ obfuscators faithfully computes the correct function.

Theorem 3. *For any constant γ the sequential-subset combiner is a strongly robust $(2\gamma + 1)$-out-of-$(3\gamma + 1)$ combiner for indistinguishability obfuscation, for differing-inputs obfuscation, for virtual black-box obfuscation, and for grey-box obfuscation, the latter ones for dependent auxiliary inputs.*

The proof is similar to our 3-out-of-4 combiner. Functionally correctness follows from the fact that the majority computation in each O_{I_i} ensures that the at most γ corrupt obfuscators cannot bias the outcome. Obfuscation follows as before because there must exist one set I which exclusively contains non-malicious obfuscators.

5.2   Impossibility for 2γ-out-of-3γ Combiners

In this section we discuss that our lower bound for 2-out-of-3 structural combiners carries over to the more general case of 2γ-out-of-3γ combiners.

Theorem 4. *There is no structural weakly robust 2γ-out-of-3γ obfuscation combiner $\mathsf{Comb}^{\mathcal{O}_1, \mathcal{O}_2, \ldots, \mathcal{O}_{3\gamma}}$ for non-trivial circuit classes $\mathcal{C}$ with admissible encoding $\langle \cdot \rangle$.*

Proof. Recall the proof for the 2-out-of-3 case. There, in the first step we have shown that on each path from a level-1 unit to the output unit there must be a unit in which obfuscator $\mathcal{O}_3$ does not appear and which is not pass-through. We called these units confusion units.

The same argument now applies here as well for the γ obfuscators with indices $2\gamma + 1, \ldots, 3\gamma$. Else, if there was a path in which one of these obfuscators appears in each unit (or if the unit is pass-through), then we could easily corrupt these obfuscators and forward information about the input circuit through the admissible encoding $\langle \cdot \rangle$. Hence, in the case here there must be a confusion unit on each path, which only uses circuits with indices $1, 2, \ldots, 2\gamma$ and which are not pass-through.

In the second step of the proof for the 2-out-of-3 case we then show that in the confusion units with obfuscators $\mathcal{O}_1$ and $\mathcal{O}_2$ we can confuse the combiner. One time we corrupt $\mathcal{O}_1$ and let it insert samples of circuit D_1, and the other time we corrupt $\mathcal{O}_2$ and insert samples for D_0, where D_0, D_1 compute different functions. Then the combiner's view when run on input D_0 in the first case, and on D_1 in the second case, has the same distribution and the combiner cannot provide functional correctness.

We apply the same argument here, one time corrupting the first γ obfuscators with indices $1, \ldots, \gamma$ and inserting a sample for D_1, and the other time corrupting obfuscators with indices between $\gamma + 1, \ldots, 2\gamma$ and using a sample for D_0. Then the combiner's views in both cases (for input circuit D_0 in the first case, and for D_1 in the second case) are identical again such that it cannot provide a correct combiner.

As in the 2-out-of-3 case the malicious obfuscators above insert the confusion samples at random positions, such that it only achieves confusion with the same bound as in the previous case. Note also that we took advantage of the fact that corrupt combiners are coordinated centrally by the adversary.

6   Detecting Combiners

The combiners in the previous section were correcting in the sense that they guaranteed functionality correctness if a quorum if obfuscator candidates is secure. Here we consider combiners which should create circuits which either output the correct value, but may give some error output $\perp$. We call them detecting combiners.

For detecting combiners we require a weaker correctness property, namely that for any circuit $C \in \mathcal{C}$, for any $O \leftarrow \mathsf{Comb}^{\mathcal{O}_1, \mathcal{O}_2, \cdots}(C)$ we have that $O(x) \in \{C(x), \perp\}$ for all $x \in \{0, 1\}^*$ in the domain of C. This means that the combiner may sometimes fail to compute the correct function value but then it signals this by outputting a special symbol $\perp$. To prevent trivial solutions like the combiner which outputs the circuit that always returns $\perp$ we assume that $C \equiv O$ if all obfuscators are secure. Note that our assumption about the obfuscators $\mathcal{O}_1, \mathcal{O}_2, \ldots$ being able to deal with (intermediate) combiner outputs in $\mathcal{C}^{\mathsf{Comb}}$ implies that the obfuscators may now also receive circuits which occasionally output $\perp$.

6.1   Robust $(\gamma + 1)$-out-of-$(2\gamma + 1)$ Detecting Combiners

To build our $(\gamma + 1)$-out-of-$(2\gamma + 1)$ combiner we follow the approach of our sequential-subset combiner. We can also straightforwardly give the optimized version for the case of a 1-out-of-3 combiner, akin to our 1-out-of-4 combiner, but omit this step here. To build the sequential-subset combiner consider here all sets I of subsets of $\{1, 2, \ldots, 2\gamma + 1\}$ of size $\gamma + 1$. For each such set I we first obfuscate the input circuit with each obfuscator $\mathcal{O}_i$ for $i \in I$. But now instead of completing the computation by adding a majority sub circuit, we now use the detecting version which (a) either outputs the string on which all circuits agree upon as output (even if it is $\perp$), or (b) returns $\perp$ is there is no such unanimous decision. Let

$$O_I(\cdot) = \mathsf{UNAN}\{\mathcal{O}_i(\cdot) \mid i \in I\}$$

denote this unit with the unanimity circuit at the end. For obfuscation of C now compute the sequential-subset combiner

$$\left(\prod_I O_I\right)(\cdot) = O_{I_\ell}(\cdots O_{I_3}(O_{I_2}(O_{I_1}(\cdot))) \cdots)$$

as before for constant $\ell = \binom{2\gamma + 1}{\gamma + 1}$.

Theorem 5. *For any constant γ the sequential-subset combiner is a strongly robust $(\gamma+1)$-out-of-$(2\gamma+1)$ detecting combiner for indistinguishability obfuscation, for differing-inputs obfuscation, for virtual black-box obfuscation, and for grey-box obfuscation, the latter ones for dependent auxiliary inputs.*

The proof is similar to the case of correcting combiners, except that we only guarantee the weaker functional correctness. This property is given since in each unit for index set I there is at least one honest obfuscator among the $\gamma + 1$ ones, the unanimity circuit either outputs the function value computed by the honest obfuscator (if all other circuits agree), which may either be the correct function value for some x or $\bot$, or it returns the error message $\bot$. It follows that the overall output of the combiner circuit can only comply with the circuit's output, or returns $\bot$. The obfuscation properties follow as before noting that the obfuscators are able to handle input circuits with output $\bot$, and that there must exist an index set I which only contains good obfuscators.

6.2 Impossibility of γ-out-of-2γ Detecting Combiners

The idea for the lower bound for correcting combiners carries over to detecting combiners, as follows.

Theorem 6. *There is no structural weakly robust γ-out-of-2γ obfuscation combiner $\mathsf{Comb}^{\mathcal{O}_1,\mathcal{O}_2,\ldots,\mathcal{O}_{2\gamma}}$ for non-trivial circuit classes $\mathcal{C}$ with admissible encoding $\langle\cdot\rangle$.*

Proof. As in the case of 2-out-of-3 combiners and 2γ-out-of-3γ combiners, here, there must be also (non-pass-through) confusion units in each path from input units to the final unit, where none of the obfuscators with indices $\gamma+1,\ldots,2\gamma$ appears. Assume now that we corrupt the obfuscators with indices $1,\ldots,\gamma$ and let these obfuscators insert intermediate samples of a circuit D_0 of the combiner's obfuscation in the confusion units, independently of the input. If the insertions happen at the right position with significant probability, then the combiner must output an obfuscated circuit *as if the combiner has been run on D_0 for honest obfuscators*. In particular, the combiner's circuit must then compute the function D_0 on every input. This holds even if the original input circuit was D_1, computing a different function than D_0, i.e., $D_0(z) \neq D_1(z)$ for some string z. But then the combiner's circuit produces a false output $D_0(z) \neq \bot$ for input z and cannot be detecting.

7 Implementation and Evaluation

Our formal results have been stated in terms of the common notion of circuit obfuscation. In practice, however, programs are usually considered to be better modeled for Turing machines. We stress that our results, especially for the majority-based combiner, hold for such Turing machine programs as well.

Namely, our 3-out-of-4 combiner would then output the program implementing the nested majority implementations.

Concerning provably secure instantiations for Turing machine obfuscation, we note that if the running time and the input length of the Turing machine are bounded then one can in principle transform such machines into corresponding circuits, albeit at the cost of increasing the complexity significantly. A more efficient solution is to use obfuscation techniques for Turing machines directly. Given the current state of constructions this is possible if the input length can be bounded [40] and, for other constructions, if the space is also bounded beforehand [11,17].

To evaluate the suggested combiners for typical obfuscation programs in practice we implemented the PyObf python package [9] that can be used to wrap existing obfuscators and to implement new combiners. Even though the conceptual construction of a combiner is not related to the concrete implementation of the underlying obfuscators, implementations for different programming languages might differ, since some constructions introduce new run-time parts (e.g., as the MAJ circuit in our case) to the program. We chose to use JavaScript as the implementation programming language because of the relatively high number of available obfuscators.

7.1   Performance Evaluation

For performance evaluation we used Yahoo!'s YUICompressor v2.4.8[8] as $\mathcal{O}_1$, a slightly randomized version of it as $\mathcal{O}_2$, Google's Closure Compiler v20151015[9] as $\mathcal{O}_3$, and jsPacker.pl v1.00b[10] as $\mathcal{O}_4$. Note that there is no essential difference between $\mathcal{O}_1$ and $\mathcal{O}_2$, especially in terms of obfuscation overhead, since the latter only uses different and randomized symbol selection routine. Security of combiners usually relies on somewhat independent components but since we are mainly interested in performance evaluation here we opted for using the related choice. The evaluated combiners are:

$$C_1(.) = \mathcal{O}_4(\mathcal{O}_3(\mathcal{O}_2(\mathcal{O}_1(.))))$$
$$C_2(.) = \mathcal{O}_2(\mathcal{O}_1(\mathcal{O}_4(\mathcal{O}_3(.))))$$
$$C_3(.) = \mathrm{MAJ}(\mathcal{O}_1(O_{2,3,4}), \mathcal{O}_2(O_{1,3,4}), \mathcal{O}_3(O_{1,2,4}))$$
$$C_4(.) = \mathrm{MAJ}(\mathcal{O}_3(O_{4,1,2}), \mathcal{O}_4(O_{3,1,2}), \mathcal{O}_1(O_{3,4,2}))$$

The evaluated programs (with varying input size, ranging from a few thousand bytes to a roughly million bytes) are Cookies.js v1.2.2[11] (6,637 bytes), Highlight.js v9.0.0[12] (22,604 bytes), jCarousel v0.3.4[13] (46,007 bytes), Backbone.js

---

[8] https://github.com/yui/yuicompressor.
[9] https://github.com/google/closure-compiler.
[10] http://dean.edwards.name/download/.
[11] https://github.com/ScottHamper/Cookies.
[12] https://highlightjs.org.
[13] http://sorgalla.com/jcarousel/.

v1.2.3[14] (71,415 bytes), Chart.js v1.0.2[15] (109,612 bytes), Epoch v0.8.4[16] (115,940 bytes), Swig v1.4.2[17] (143,975 bytes), PhysicsJS v0.7.0[18] (171,847 bytes), jQuery v1.6.4[19] (238,166 bytes), Raphaël v2.1.4[20] (304,254 bytes), Dojo v1.10.4[21] (629,481 bytes), Video.js v5.4.4[22] (675,527 bytes) and AngularJS v1.4.5[23] (1,052,336 bytes).

Note that the above circuit model describes a program as a function with input and output, in contrast to the common software design of JavaScript libraries that heavily depends on the JavaScript context (e.g., the window object). But this difference is irrelevant to performance evaluation.

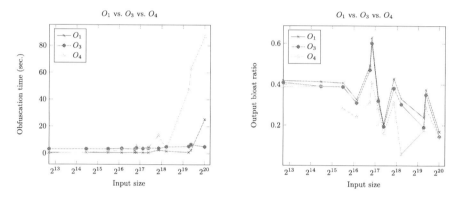

Fig. 7. Overhead of individual obfuscators (time and output bloat ratio) for the various programs (in relation to their input sizes for the obfuscator). Note that we do not display obfuscator $\mathcal{O}_2$ here as its performance is essentially identical to the one of $\mathcal{O}_1$.

Figure 7 gives the effectiveness of the obfuscators in terms of obfuscation time and of output-bloat ratio. Here we show the figures in relation of the sizes of the various programs (from 6,637 bytes to 1,052,336 bytes) given as input to the obfuscators. Note that the results may depend heavily on the specific input programs such that we cannot expect perfectly monotonic behavior in the graphs. Also, as mentioned before, many practical obfuscators come with techniques for code size reduction such that the output bloat ratio can be—and often is—smaller than 1. Next, we compare these figures to the results of the

[14] http://backbonejs.org/.
[15] http://www.chartjs.org/.
[16] http://epochjs.github.io/epoch/.
[17] http://paularmstrong.github.io/swig/.
[18] http://wellcaffeinated.net/PhysicsJS/.
[19] https://jquery.com/.
[20] https://github.com/DmitryBaranovskiy/raphael.
[21] https://dojotoolkit.org/.
[22] http://videojs.com/.
[23] https://angularjs.org/.

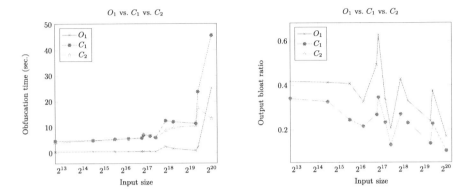

Fig. 8. Overhead of cascaded combiner (time and output bloat ration).

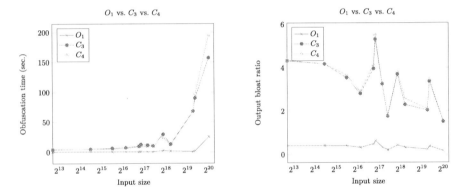

Fig. 9. Overhead of 3-out-of-4 combiner (time and output bloat ration).

suggested combiners, first to the cascade combiners in Fig. 8 and then to the 3-out-of-4 combiners in Fig. 9.

In summary, the proposed obfuscation combiners do not add significant run time overhead compared to a single obfuscator. The factor is roughly proportional to the number of invoked instances, with some gains presumably due to the intermediate code optimization. Due to the advanced compression techniques the code size of our cascaded combiners is in the same order as the individual obfuscators. For the 3-out-of-4 combiner we of course get an increased output size because of the tripling for each majority step, potentially also hampering some code reductions.

7.2 Security Evaluation

Due to the unclear situation about security properties of practical obfuscators we have proven robustness of our combiners with respect to the common theoretical notions of obfuscation in the literature. There are approaches to define metrics for practical obfuscators, though. A first approach is by Collberg et al. [20] who define notions for *potency* (the incomprehensibility of the transformed program for humans), *resilience* (the hardness of undoing the transformation through the joint effort of engineers and deobfuscation techniques), and *cost* (the overhead caused by the obfuscator). The measure of quality of an obfuscator is then given by a vector of these three metrics.

While the notion of cost in [20] even distinguishes the full range between exponential and constant overhead in execution resources, the metrics for potency and resilience in [20] are less rigorous. They are accompanied by suitable software complexity measures such as program length or cyclomatic complexity [42]. Anckaert et al. [5] later used software complexity measures, too, for establishing a benchmarking system for obfuscators for binary executables.

While it is beyond the scope of our work here, it may be interesting to benchmark our obfuscation combiners according to the metrics in [5]. Note that their metrics focus on resilience and somewhat neglect the overhead. Since our combiners in principle increase the software complexity at the cost of incurring additional steps, one should expect that combinations of benchmarked obfuscators yield better values in this regard.

8   Conclusion

Our positive results about combiners, and also our lower bounds, indicate how to proceed both in theory and practice. If you only have two available candidates then the best solution appears to be the sequential composition $\mathcal{O}_2(\mathcal{O}_1(\cdot))$, *if one can somehow guarantee that the inner obfuscator provides functional correctness.* For three candidates (out of which at least two are sound) then our 2-out-of-3 *detecting* combiner should be the primary choice. To ensure correct output, our 3-out-of-4 combiner provides a secure solution.

Acknowledgements. We are grateful to Christian Collberg for his feedback and encouragement. Marc Fischlin is supported by the Heisenberg grant Fi 940/3-2 and the SPP 1736 grant Fi 940/5-1 of the German Research Foundation (DFG). Amir Herzberg is support by the Israeli Ministry of Science and Technology.

References

1. Ananth, P., Boneh, D., Garg, S., Sahai, A., Zhandry, M.: Differing-inputs obfuscation and applications. Cryptology ePrint Archive, Report 2013/689 (2013). http://eprint.iacr.org/2013/689

2. Ananth, P., Jain, A., Naor, M., Sahai, A., Yogev, E.: Universal obfuscation and witness encryption: boosting correctness and combining security (2016). http://eprint.iacr.org/2016/281

3. Ananth, P., Jain, A.: Indistinguishability obfuscation from compact functional encryption. In: Gennaro, R., Robshaw, M. (eds.) CRYPTO 2015. LNCS, vol. 9216, pp. 308–326. Springer, Heidelberg (2015)

4. Ananth, P., Jain, A., Sahai, A.: Indistinguishability obfuscation with constant size overhead. IACR Cryptology ePrint Archive, Report 2015/1023 (2015). http://eprint.iacr.org/2015/1023

5. Anckaert, B., Madou, M., Sutter, B.D., Bus, B.D., Bosschere, K.D., Preneel, B.: Program obfuscation: a quantitative approach. In: Proceedings of the 3th ACM Workshop on Quality of Protection, QoP 2007, Alexandria, VA, USA, 29 October 2007, pp. 15–20. ACM (2007)

6. Badrinarayanan, S., Miles, E., Sahai, A., Zhandry, M.: Post-zeroizing obfuscation: new mathematical tools, and the case of evasive circuits. In: Fischlin, M., Coron, J.-S. (eds.) EUROCRYPT 2016. LNCS, vol. 9666, pp. 764–791. Springer, Heidelberg (2016). doi:10.1007/978-3-662-49896-5_27

7. Barak, B., Garg, S., Kalai, Y.T., Paneth, O., Sahai, A.: Protecting obfuscation against algebraic attacks. In: Nguyen, P.Q., Oswald, E. (eds.) EUROCRYPT 2014. LNCS, vol. 8441, pp. 221–238. Springer, Heidelberg (2014)

8. Barak, B., Goldreich, O., Impagliazzo, R., Rudich, S., Sahai, A., Vadhan, S.P., Yang, K.: On the (im)possibility of obfuscating programs. J. ACM **59**(2), 6 (2012)

9. Bin Noon, H.: Pyobf (2016). https://github.com/hodbn/pyobf

10. Bitansky, N., Canetti, R.: On strong simulation and composable point obfuscation. In: Rabin, T. (ed.) CRYPTO 2010. LNCS, vol. 6223, pp. 520–537. Springer, Heidelberg (2010)

11. Bitansky, N., Garg, S., Lin, H., Pass, R., Telang, S.: Succinct randomized encodings and their applications. In: Servedio, R.A., Rubinfeld, R. (eds.) 47th ACM STOC, pp. 439–448. ACM Press, New York (2015)

12. Bitansky, N., Vaikuntanathan, V.: Indistinguishability obfuscation: from approximate to exact. http://eprint.iacr.org/2015/704

13. Bitansky, N., Vaikuntanathan, V.: Indistinguishability obfuscation from functional encryption. In: IEEE 56th Annual Symposium on Foundations of Computer Science, FOCS 2015, Berkeley, CA, USA, 17–20 October 2015, pp. 171–190. IEEE Computer Society (2015)

14. Boneh, D., Boyen, X.: On the impossibility of efficiently combining collision resistant hash functions. In: Dwork, C. (ed.) CRYPTO 2006. LNCS, vol. 4117, pp. 570–583. Springer, Heidelberg (2006)

15. Brakerski, Z., Rothblum, G.N.: Virtual black-box obfuscation for all circuits via generic graded encoding. In: Lindell, Y. (ed.) TCC 2014. LNCS, vol. 8349, pp. 1–25. Springer, Heidelberg (2014)

16. Canetti, R.: Towards realizing random oracles: hash functions that hide all partial information. In: Kaliski Jr., B.S. (ed.) CRYPTO 1997. LNCS, vol. 1294, pp. 455–469. Springer, Heidelberg (1997)

17. Canetti, R., Holmgren, J., Jain, A., Vaikuntanathan, V.: Succinct garbling and indistinguishability obfuscation for RAM programs. In: Servedio, R.A., Rubinfeld, R. (eds.) 47th ACM STOC, pp. 429–437. ACM Press, New York (2015)

18. Cheon, J.H., Han, K., Lee, C., Ryu, H., Stehlé, D.: Cryptanalysis of the multilinear map over the integers. In: Oswald, E., Fischlin, M. (eds.) EUROCRYPT 2015. LNCS, vol. 9056, pp. 3–12. Springer, Heidelberg (2015)

19. Collberg, C., Thomborson, C.: Watermarking, tamper-proofing, and obfuscation-tools for software protection. IEEETSE. IEEE Trans. Softw. Eng. **28**, 735–746 (2002)

20. Collberg, C., Thomborson, C., Low, D.: A taxonomy of obfuscating transformations. Technical report #148, Department of Computer Science, The University of Auckland, New Zealand (1997)

21. Coron, J.S., et al.: Zeroizing without low-level zeroes: new MMAP attacks and their limitations. In: Gennaro, R., Robshaw, M. (eds.) CRYPTO 2015. LNCS, vol. 9216, pp. 247–266. Springer, Heidelberg (2015)

22. Coron, J.S., Lee, M., Lepoint, T., Tibouchi, M.: Cryptanalysis of GGH15 multilinear maps. Cryptology ePrint Archive, Report 2015/1037 (2015). http://eprint.iacr.org/2015/1037

23. Coron, J.S., Lepoint, T., Tibouchi, M.: Cryptanalysis of two candidate fixes of multilinear maps over the integers. Cryptology ePrint Archive, Report 2014/975 (2014). http://eprint.iacr.org/2014/975

24. Dodis, Y., Katz, J.: Chosen-ciphertext security of multiple encryption. In: Kilian, J. (ed.) TCC 2005. LNCS, vol. 3378, pp. 188–209. Springer, Heidelberg (2005)

25. Fischlin, M., Lehmann, A.: Security-amplifying combiners for collision-resistant hash functions. In: Menezes, A. (ed.) CRYPTO 2007. LNCS, vol. 4622, pp. 224–243. Springer, Heidelberg (2007)

26. Fischlin, M., Lehmann, A.: Multi-property preserving combiners for hash functions. In: Canetti, R. (ed.) TCC 2008. LNCS, vol. 4948, pp. 375–392. Springer, Heidelberg (2008)

27. Fischlin, M., Lehmann, A., Pietrzak, K.: Robust multi-property combiners for hash functions. J. Crypt. **27**(3), 397–428 (2014)

28. Garg, S., Gentry, C., Halevi, S., Raykova, M., Sahai, A., Waters, B.: Candidate indistinguishability obfuscation and functional encryption for all circuits. In: 54th FOCS, October 2013, pp. 40–49. IEEE Computer Society Press (2013)

29. Garg, S., Gentry, C., Halevi, S., Wichs, D.: On the implausibility of differing-inputs obfuscation and extractable witness encryption with auxiliary input. In: Garay, J.A., Gennaro, R. (eds.) CRYPTO 2014, Part I. LNCS, vol. 8616, pp. 518–535. Springer, Heidelberg (2014)

30. Garg, S., Mukherjee, P., Srinivasan, A.: Obfuscation without the vulnerabilities of multilinear maps. Cryptology ePrint Archive, Report 2016/390 (2016). http://eprint.iacr.org/2016/390

31. Gentry, C., Halevi, S., Maji, H.K., Sahai, A.: Zeroizing without zeroes: cryptanalyzing multilinear maps without encodings of zero. Cryptology ePrint Archive, Report 2014/929 (2014). http://eprint.iacr.org/2014/929

32. Gentry, C., Lewko, A., Sahai, A., Waters, B.: Indistinguishability obfuscation from the multilinear subgroup elimination assumption (2015)

33. Goldwasser, S., Kalai, Y.T.: On the impossibility of obfuscation with auxiliary input. In: 46th FOCS, October 2005, pp. 553–562. IEEE Computer Society Press (2005)

34. Harnik, D., Kilian, J., Naor, M., Reingold, O., Rosen, A.: On robust combiners for oblivious transfer and other primitives. In: Cramer, R. (ed.) EUROCRYPT 2005. LNCS, vol. 3494, pp. 96–113. Springer, Heidelberg (2005)

35. Herzberg, A.: Folklore, practice and theory of robust combiners. Cryptology ePrint Archive, Report 2002/135 (2002). http://eprint.iacr.org/2002/135

36. Herzberg, A.: On tolerant cryptographic constructions. In: Menezes, A. (ed.) CT-RSA 2005. LNCS, vol. 3376, pp. 172–190. Springer, Heidelberg (2005)

37. Herzberg, A.: Folklore, practice and theory of robust combiners. J. Comput. Secur. **17**(2), 159–189 (2009)

38. Herzberg, A., Shulman, H.: Robust combiners for software hardening. In: Acquisti, A., Smith, S.W., Sadeghi, A.-R. (eds.) TRUST 2010. LNCS, vol. 6101, pp. 282–289. Springer, Heidelberg (2010)

39. Hofheinz, D., Jager, T., Khurana, D., Sahai, A., Waters, B., Zhandry, M.: How to generate and use universal samplers. Cryptology ePrint Archive, Report 2014/507 (2014). http://eprint.iacr.org/2014/507

40. Koppula, V., Lewko, A.B., Waters, B.: Indistinguishability obfuscation for turing machines with unbounded memory. In: Servedio, R.A., Rubinfeld, R. (eds.) 47th ACM STOC, pp. 419–428. ACM Press, New York (2015)

41. Lin, H.: Indistinguishability obfuscation from constant-degree graded encoding schemes. In: Fischlin, M., Coron, J.-S. (eds.) EUROCRYPT 2016. LNCS, vol. 9665, pp. 28–57. Springer, Heidelberg (2016). doi:10.1007/978-3-662-49890-3_2

42. McCabe, T.J.: A complexity measure. IEEE Trans. Softw. Eng. **2**(4), 308–320 (1976)

43. Meier, R., Przydatek, B.: On robust combiners for private information retrieval and other primitives. In: Dwork, C. (ed.) CRYPTO 2006. LNCS, vol. 4117, pp. 555–569. Springer, Heidelberg (2006)

44. Meier, R., Przydatek, B., Wullschleger, J.: Robuster combiners for oblivious transfer. In: Vadhan, S.P. (ed.) TCC 2007. LNCS, vol. 4392, pp. 404–418. Springer, Heidelberg (2007)

45. Mennink, B., Preneel, B.: Breaking and fixing cryptophia's short combiner. In: Gritzalis, D., Kiayias, A., Askoxylakis, I. (eds.) CANS 2014. LNCS, vol. 8813, pp. 50–63. Springer, Heidelberg (2014)

46. Miles, E., Sahai, A., Zhandry, M.: Annihilation attacks for multilinear maps: cryptanalysis of indistinguishability obfuscation over GGH13. Cryptology ePrint Archive, Report 2016/147. http://eprint.iacr.org/2016/147

47. Mittelbach, A.: Cryptophia's short combiner for collision-resistant hash functions. In: Jacobson, M., Locasto, M., Mohassel, P., Safavi-Naini, R. (eds.) ACNS 2013. LNCS, vol. 7954, pp. 136–153. Springer, Heidelberg (2013)

48. Pass, R., Seth, K., Telang, S.: Indistinguishability obfuscation from semantically-secure multilinear encodings. In: Garay, J.A., Gennaro, R. (eds.) CRYPTO 2014, Part I. LNCS, vol. 8616, pp. 500–517. Springer, Heidelberg (2014)

49. Pietrzak, K.: Non-trivial black-box combiners for collision-resistant hash-functions don't exist. In: Naor, M. (ed.) EUROCRYPT 2007. LNCS, vol. 4515, pp. 23–33. Springer, Heidelberg (2007)

50. Pietrzak, K.: Compression from collisions, or why crhf combiners have a long output. In: Wagner, D. (ed.) CRYPTO 2008. LNCS, vol. 5157, pp. 413–432. Springer, Heidelberg (2008)

On Statistically Secure Obfuscation
with Approximate Correctness

Zvika Brakerski[1], Christina Brzuska[2], and Nils Fleischhacker[3(✉)]

[1] Weizmann Institute of Science, Rehovot, Israel
[2] Technical University of Hamburg, Hamburg, Germany
[3] CISPA, Saarland University, Saarbrücken, Germany
fleischhacker@cs.uni-saarland.de

Abstract. Goldwasser and Rothblum (TCC '07) prove that statistical indistinguishability obfuscation (iO) cannot exist if the obfuscator must maintain perfect correctness (under a widely believed complexity theoretic assumption: $\mathcal{NP} \not\subseteq \mathcal{SZK} \subseteq \mathcal{AM} \cap \mathbf{co}\mathcal{AM}$). However, for many applications of iO, such as constructing public-key encryption from one-way functions (one of the main open problems in theoretical cryptography), *approximate* correctness is sufficient. It had been unknown thus far whether statistical approximate iO (saiO) can exist.

We show that saiO does not exist, even for a minimal correctness requirement, if $\mathcal{NP} \not\subseteq \mathcal{AM} \cap \mathbf{co}\mathcal{AM}$, and if one-way functions exist. A simple complementary observation shows that if one-way functions do not exist, then average-case saiO exists. Technically, previous approaches utilized the behavior of the obfuscator on *evasive* functions, for which saiO always exists. We overcome this barrier by using a PRF as a "baseline" for the obfuscated program.

We broaden our study and consider relaxed notions of *security* for iO. We introduce the notion of *correlation obfuscation*, where the obfuscations of equivalent circuits only need to be mildly correlated (rather than statistically indistinguishable). Perhaps surprisingly, we show that correlation obfuscators exist via a trivial construction for some parameter regimes, whereas our impossibility result extends to other regimes. Interestingly, within the gap between the parameters regimes that we show possible and impossible, there is a small fraction of parameters that still allow to build public-key encryption from one-way functions and thus deserve further investigation.

Z. Brakerski—Supported by the Israel Science Foundation (Grant No. 468/14), the Alon Young Faculty Fellowship, Binational Science Foundation (Grant No. 712307) and Google Faculty Research Award.
Christina Brzuska is grateful to NXP for supporting her chair for IT Security Analysis.
C. Brzuska and N. Fleischhacker—Part of this work was done while Christina Brzuska and Nils Fleischhacker were working for Microsoft Research, Cambridge.
N. Fleischhacker—Supported by the German Federal Ministry of Education and Research (BMBF) through funding for the Center for IT-Security, Privacy and Accountability (CISPA – www.cispa-security.org) and the German research foundation (DFG) through funding for the collaborative research center 1223.

M. Robshaw and J. Katz (Eds.): CRYPTO 2016, Part II, LNCS 9815, pp. 551–578, 2016.
DOI: 10.1007/978-3-662-53008-5_19

1   Introduction

Constructing public-key cryptography (e.g. public-key encryption) from private-key cryptography (such as one-way functions) is one of the most fundamental questions in theoretical cryptography, going back to the seminal paper of Diffie and Hellman [9]. Diffie and Hellman suggested that *program obfuscators* with sufficiently strong security properties would allow to realize this transformation. A program obfuscator is a compiler that takes as input a program, and outputs another program with equivalent functionality, but which is harder to reverse engineer. Diffie and Hellman suggested to obfuscate the encryption circuit of a symmetric-key encryption scheme, and use the obfuscated program as a public key so as to obtain a public-key encryption scheme. An additional hint that obfuscation may be instrumental in solving this riddle was provided by Impagliazzo and Rudich [20,21], who proved that a transformation from symmetric to public-key must make *non black-box* use of the underlying symmetric primitive. Indeed, program obfuscation is one of very few non black-box techniques known in cryptography.

Modern research showed that the Diffie-Hellman transformation requires obfuscators with security guarantees that do not exist in general [1,2,16]. However, recent years have seen incredibly prolific study of weak notions of obfuscation, following the introduction of a candidate *indistinguishability obfuscator* (iO) by Garg et al. [10]. The security guarantee of iO is that the obfuscation of two functionally equivalent circuits should result in indistinguishable output distributions. That is, that reverse engineering could not detect which of two equivalent implementations had been the source of the obfuscated program. Sahai and Waters [30] showed that even this seemingly weak notion suffices for private-key to public-key transformation (via a clever construction that does not resemble the Diffie-Hellman suggestion).

One would have hoped that a weak notion such as iO may be realizable with *statistical* security, i.e. that reverse engineering (to the limited extent required by iO) will not be possible even to an attacker with unlimited computational power. The existence of such *statistical indistinguishability obfuscator* (siO) would resolve the question of constructing public key cryptography from one-way functions, as well as would allow to construct one-way functions based on the hardness of $\mathcal{NP}$ [23]. Alas, Goldwasser and Rothblum [14,15] proved that siO cannot exist unless the polynomial hierarchy collapses (in particular that it implies $\mathcal{NP} \subseteq \mathcal{SZK}$, and it is known that $\mathcal{SZK} \subseteq \mathcal{AM} \cap \mathbf{co}\mathcal{AM}$), which is considered quite unlikely in computational complexity, and at any rate way beyond the current understanding of complexity theory. This seems to put a damper on our hopes to achieve statistically secure obfuscation.

However, the [14,15] negative result crucially relies on the *correctness* of the obfuscator. That is, it only rules out such obfuscators that perfectly preserve the functionality of the underlying primitive (at least with high probability over the coins of the obfuscator). In contrast, the symmetric to public key transformation can be made to work with only *approximate* correctness, i.e. a non-negligible correlation between the functionality of the input circuit and that of the output circuit (where the probability is taken over the randomness of the obfuscator and the input domain). The question of whether statistical approximate iO (saiO) exists was therefore the new destination in the quest for understanding obfuscation. Interestingly, it turns out that ruling out *computational* notions of iO in some idealized models also boils down to the question of whether saiO exists (see Sect. 1.2 below). The study of this notion is the objective of this paper.

Our Results. We show that statistical approximate iO (saiO) does not exist if one-way functions exist (under the assumption that $\mathcal{NP} \not\subseteq \mathcal{AM} \cap \mathbf{co}\mathcal{AM}$). Thus, in particular, that saiO cannot be used for the transformation from symmetric to public-key cryptography. We show that if one-way functions exist, then any non-negligible correlation between the output of the obfuscator and the input program would imply an $\mathcal{SZK}$ algorithm for unique SAT (USAT). As SAT reduces to USAT via a randomized reduction [32], a result of Mahmoody and Xiao [27] shows that this implies that SAT is in $\mathcal{AM} \cap \mathbf{co}\mathcal{AM}$.

To complement our result, we observe that if one-way functions do not exist, then an average-case notion of saiO exists for any distribution. Specifically, for any efficiently samplable distribution over circuits, there exists an saiO obfuscator whose correctness holds with high probability over the circuits in that distribution (inverting the order of quantifiers would imply a worst-case saiO).

A Study of Correlation Obfuscation. Our impossibility results extend beyond the case of saiO. In fact, the result applies even when the *security* of the obfuscator is approximate. Namely, when we are only guaranteed that the obfuscation of functionally equivalent circuits results in distributions that have mild statistical distance (as opposed to negligible). This motivated us to explore the properties of this new kind of obfuscators, that as far as we know have not been studied in the literature before.

We consider statistical approximate *correlation* obfuscation sacO. A sacO obfuscator is characterized by two parameters $\epsilon \in [0, 1/2)$ and $\delta \in [0, 1)$. The requirement is that correctness holds with probability $1 - \epsilon$ (with respect to the randomness of the obfuscator and a random choice of input), and that obfuscating two functionally equivalent circuits results in distributions with statistical distance δ. The case of negligible δ is exactly saiO, discussed above, and the case of $\epsilon = 0$ corresponds to perfect correctness.

We observe that our impossibility result degrades gracefully and holds so long as $2\epsilon + 3\delta < 1$. We found this state of affairs unsatisfactory, and tried to extend the result to hold for the entire parameter range. However, it turns out that sacO exists via an almost trivial construction whenever $2\epsilon + \delta > 1$ (e.g. $\epsilon = \delta = 0.4$). We do not know if sacO exists in the intermediate parameter regime.

Lastly, we conduct a study of whether sacO is sufficient to construct public-key encryption from one-way functions. We present an amplified version of the Sahai-Waters construction using an amplification technique due to Holenstein. Interestingly, it appears that there is a region in the parameter domain that would allow to construct public-key encryption from one-way functions, but is

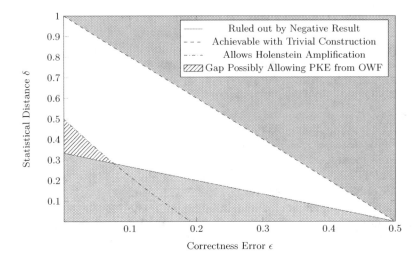

Fig. 1. The graph gives an overview over the possible range of parameters for sacO. In the upper right are parameter regimes that can be achieved using the construction described in Appendix A. In the lower left are the strong parameter regimes ruled out by our negative result in Sect. 3. The graph shows nicely the gap between the parameters that can be ruled out and those that can be used to construct public key encryption using the construction of Sahai and Waters as well as the amplification technique of Holenstein.

not ruled out by our current technique. See Fig. 1 for the landscape of sacO parameters. We leave it as an intriguing open problem to close the gap between the various parameter regimes.

1.1 Our Techniques

Our starting point is the Goldwasser-Rothblum impossibility result. Consider a statistical iO obfuscator such that for any pair of functionally equivalent circuits, the obfuscator generates statistically indistinguishable distributions, and in addition the output circuit of the obfuscator is always *functionally equivalent* to the input circuit (this can be relaxed to hold only with high probability over the random coins of the obfuscator). Goldwasser-Rothblum observe that an unsatisfiable SAT formula Ψ is functionally equivalent to the all-zero function $\mathbf{0}$ and therefore the distributions produced by a siO obfuscator in both cases should be statistically indistinguishable. Slightly more formally, let $X[C]$ denote the distribution output by the obfuscator on input circuit C, then we get that $X[\Psi] \equiv X[\mathbf{0}]$, where $\equiv$ denotes statistical indistinguishability. On the contrary, if Ψ is a satisfiable formula, then it has a different functionality than $\mathbf{0}$ and therefore the support of $X[\Psi]$ and $X[\mathbf{0}]$ will be disjoint (and thus obviously not statistically indistinguishable). It follows that in order to solve SAT, it suffices to tell whether $X[\Psi]$ is close to $X[\mathbf{0}]$. As we know due to Sahai and

Vadhan [29], there is an $\mathcal{SZK}$ protocol that takes two polynomial-time samplers, and decides whether they sample from distributions that are ϵ_1-statistically close or ϵ_2-statistically far, so long as $(\epsilon_2 - \epsilon_1)$ is a noticeable function. The conclusion is that an siO obfuscator implies an $\mathcal{SZK}$ protocol for SAT which in turn implies that $\mathcal{NP} \subseteq \mathcal{SZK}$.

To sum up the core argument, to show that an siO obfuscator does not exists unless $\mathcal{NP} \subseteq \mathcal{SZK}$, Goldwasser-Rothblum built the formula-indexed distribution $X[\Psi]$ that samples an siO obfuscation of Ψ and has the properties that it is (i) efficiently sampleable, (ii) if Ψ is not satisfiable, then $X[\Psi]$ and $X[\mathbf{0}]$ are close, while (iii) if Ψ not satisfiable, then $X[\Psi]$ and $X[\mathbf{0}]$ are far.

Allowing the obfuscator to have approximate correctness thwarts this approach completely. Hard SAT instances are obviously ones where the density of accepting inputs is sub-polynomial, since otherwise random sampling would yield a satisfying assignment with non-negligible probability. Therefore a satisfiable and unsatisfiable SAT formulae will have almost identical functionality. One could consider an saiO obfuscator that on any SAT formula that is not trivially satisfiable, would just produce an obfuscation of $\mathbf{0}$. This means that $X[\Psi]$ will have the same distribution whether Ψ is satisfiable or not and thus, property (iii) is not satisfied anymore.

In order to overcome this issue, we construct a different distribution on formula-indexed circuits $C_X[k, \Psi]$ (where k is some uniformly random key k) such that if Ψ is not satisfiable, then $C_X[k, \Psi]$ and $C_X[k, \mathbf{0}]$ have the same functionality, and if Ψ is satisfiable, then $C_X[k, \Psi]$ and $C_X[k, \mathbf{0}]$ differ on a single point. Then, assuming one-way functions exist, we show that, although these two circuits differ on a single point only, the obfuscator saiO of $C_X[k, \Psi]$ has to produce a distribution that is statistically far from saiO of $C_X[k, \mathbf{0}]$. To do this, we rely on the fact that the obfuscator itself is computationally efficient, and therefore it cannot break the hardness of one-way functions and derived cryptographic objects such as pseudorandom functions (PRFs) or *puncturable* PRFs (see below). This way, we construct a new formula-indexed distribution $X[\Psi]$ that satisfies properties (i), (ii) and (iii) as discussed above.

Puncturable PRFs were introduced simultaneously in [6,7,22] and were utilized as an essential building block for indistinguishability obfuscation in [30]. A standard PRF is a function that can be efficiently computable using a key k, but is indistinguishable from a random function via oracle access. A puncturable PRF is a PRF where one can generate a *punctured key* $k\{x_0\}$ which allows to compute the PRF at all points except x_0, but the value at x_0 is still indistinguishable from uniform, even given the punctured key. Punctured PRFs can be constructed from any one-way function.

Based on a puncurable PRF and an saiO obfuscator O, we now construct a distribution on pairs of circuits (for now not indexed by a formula) such that the two circuits differ on a single point only and yet, an saiO obfuscator will produce distributions that are far. Let k be a key for a puncturable PRF, let x_0 be a random point in the domain, let $k\{x_0\}$ be a key punctured at x_0 and consider the function $f_{k\{x_0\},y}$ that outputs $\mathsf{PRF}(k\{x_0\}, x) = \mathsf{PRF}(k, x)$ for all

$x \neq x_0$, and outputs y on input x_0. Then by definition $f_{k\{x_0\},y}$ for a random y and $f_{k\{x_0\},y_0} = \mathsf{PRF}(k,\cdot)$ for $y_0 = \mathsf{PRF}(k,x_0)$ are identical in functionality except maybe at point x_0. However, using puncturing, we can guarantee that the distributions $\mathsf{O}(f_{k\{x_0\},y})$ and $\mathsf{O}(f_{k\{x_0\},y_0})$, where k, x_0, y are chosen uniformly at random are *statistically far*. To see this, it is enough to show that $\mathsf{O}(f_{k\{x_0\},y})$ and $\mathsf{O}(\mathsf{PRF}(k,\cdot))$ are statistically far since $f_{k\{x_0\},y_0} = \mathsf{PRF}(k,\cdot)$ and thus $\mathsf{O}(f_{k\{x_0\},y_0}) \equiv \mathsf{O}(\mathsf{PRF}(k,\cdot))$. Consider the predicate that checks whether $\mathsf{O}(\mathsf{PRF}(k,\cdot))(x_0) = \mathsf{PRF}(k,x_0)$. This predicate must have non-negligible bias towards holding true, and is efficiently checkable, which also implies that $\mathsf{O}(f_{k\{x_0\},y})(x_0) = f_{k\{x_0\},y}(x_0)$ holds true with noticeable bias, since otherwise we will have an efficient distinguisher from $f_{k\{x_0\},y_0} = \mathsf{PRF}(k,\cdot)$ in contradiction to the puncturable PRF security. Finally, since $y \neq y_0$ with high probability (assume for simplicity that the PRF and the obfuscator have long outputs and keys of half the size), this implies that $\mathsf{O}(f_{k\{x_0\},y})$ and $\mathsf{O}(f_{k\{x_0\},y_0})$ have noticeable statistical distance, since they will have noticeable probability mass on circuits that respect the functionality on x_0. Note that we used a *computational* argument, the security of punctured PRFs, to derive a *statistical* statement about the output distribution of the obfuscator.

We would like to use the aforementioned distributions to distinguish between satisfiable and unsatisfiable formulae. Let us restrict our attention to Unique-SAT formulae that are either unsatisfiable or have only one satisfying assignment. Unique-SAT is known to be $\mathcal{NP}$-Hard via a randomized reduction [32], and a result of Mahmoody and Xiao [27] shows that if Unique-SAT is in $\mathcal{SZK} \subseteq \mathcal{AM} \cap \mathbf{co}\mathcal{AM}$, then SAT is in $\mathcal{AM} \cap \mathbf{co}\mathcal{AM}$ (See Sect. 2.1).

Let Ψ be a formula that has a unique satisfying assignment, then one can randomize the satisfying assignment (if it exists) to be uniformly distributed over the input space (e.g. by XORing all variables with a random string). Now, consider the function $f_{k,y,\Psi}$ defined s.t. $f_{k,y,\Psi}(x) = \mathsf{PRF}(k,x)$ if x does not satisfy Ψ, and $f_{k,y,\Psi}(x) = y$ otherwise. By definition, if Ψ is unsatisfiable then $f_{k,y,\Psi} = \mathsf{PRF}(k,\cdot)$ and if Ψ is satisfiable by some x_0 (which is uniformly distributed) then $f_{k,y,\Psi} = f_{k\{x_0\},y}$. Therefore $\mathsf{O}(f_{k,y,\Psi})$ is guaranteed to have a noticeable statistical distance in the case where Ψ is unsatisfiable (in which case it is close to $\mathsf{O}(f_{k,y,0})$) and in the case where it is uniquely satisfiable (in which case it is far from $\mathsf{O}(f_{k,y,0})$). This will allow us to produce an $\mathcal{SZK}$ protocol to distinguish the two possibilities.

In a World without OWFs. We recall that if OWFs do not exist then for any efficiently computable function f and with overwhelming probability over a y sampled from the output distribution of f, it is possible to efficiently sample (almost) uniformly (up to negligible error) from the set $f^{-1}(y) = \{x : f(x) = y\}$ [19]. Given an efficiently sampleable distribution over circuits, we can construct an average-case obfuscator for this family as follows. Let sampC be a sampler for this distribution of circuits and consider the function $f(r, x_1, \ldots, x_m)$ for a large polynomial m such that $f(r, x_1, \ldots, x_m) = (x_1, \ldots, x_m, C(x_1), \ldots, C(x_m))$, for $C = \mathsf{sampC}(r)$.

Now, to obfuscate a circuit C, sample $x_1, \ldots, x_m$ and compute $y_i = C(x_i)$. Then sample $(r, x_1, \ldots, x_m)$ from $f^{-1}(x_1, \ldots, x_m, y_1, \ldots, y_m)$ and finally output $C' = \mathsf{sampC}(r)$. This is clearly a perfect indistinguishability obfuscator (i.e. two circuits with the same functionality will produce identical distributions). It is also approximately correct on the average, because on average, if two circuits agree on a randomly chosen set of points, then they will have a large agreement altogether.

We note that a similar and even simpler argument shows that if all efficiently computable functions are PAC learnable [31], even allowing membership queries, then saiO with perfect indistinguishability exists. This follows immediately by definition by giving the learner (black-box) access to C, and outputting its hypothesis C' as the output of the obfuscator. In such case OWFs trivially do not exist.

The Landscape of Correlation Obfuscation. Extending our techniques to rule out sacO with $2\epsilon + 3\delta < 1$ follows from carefully analyzing the parameters in the proof outlined above (one can get $2\epsilon + 4\delta < 1$ by straightforward analysis, and the slight improvement comes from properly defining the random variables in the problem). We can show a trivial sacO obfuscator for $2\epsilon + \delta > 1$ as follows. Given an input circuit C, use random sampling to find the majority value of the truth table of C (if C is approximately balanced, then any value works). Then output the constant function taking the majority value with probability 2ϵ, and output C itself with probability $1 - 2\epsilon$. Correctness will hold with probability $1 - \epsilon$, since if C is output then correctness is perfect, and if the constant function is output then correctness is approximately $1/2$. The correlation between two functionally equivalent circuits is at least 2ϵ since the calculation of the majority value only depends on the truth table. We provide a more formal analysis in Appendix A. It seems that such a trivial obfuscator cannot imply any non-trivial results.

We notice that a sacO obfuscator can be plugged into the Sahai-Waters construction, and would imply weak notions of security and correctness for the resulting public-key encryption scheme. Holenstein [18] shows that, for some parameters, this weak notion can be amplified to standard security and correctness. Plugging in our parameters, we get that roughly when $\frac{1}{2} - 3\epsilon + 2\epsilon^2 > \delta$, sacO would imply symmetric to public key transformation using this method. This leaves a small region of parameters where sacO is not known to be impossible, and if it is possible it will imply highly non-trivial results. It is not clear whether other parameter regimes can also be useful, or whether our impossibility can be extended to rule out the entire useful regime. We refer to Fig. 1 again for a visual characterization of the parameter regimes.

1.2 Consequences of Our Result

Our result strengthens previous negative results for proving the existence of iO in several ideal models. Previous works show that a construction of statistically secure (perfectly correct) iO in any of those ideal models implies the existence of saiO in the standard model. Actually, one can generalize these results to also hold for saiO. Combined with our result, we now yield that a construction of iO or saiO in these ideal models implies that $\mathcal{NP} \subseteq \mathcal{AM} \cap \mathbf{co}\mathcal{AM}$ or the non-existence of one-way functions.

This line of research was initiated by Canetti et al. [8] who show that given a VBB obfuscator in the random oracle model, one can remove the random oracle at the cost of relaxing the correctness of the obfuscator. Pass and Shelat [28] show an analogous result for VBB obfuscators in the ideal constant-degree encoding model, and Mahmoody et al. [25] show analogous results for the generic group model and the generic trapdoor permutation model. All these results transform a VBB obfuscator in an oracle world into an approximately correct VBB obfuscator in the standard model. They yield an impossibility result for VBB obfuscation in the ideal models, as approximately correct VBB is known not to exist, assuming trapdoor permutations, see [3,8]. The crucial insight of Mahmoody et al. [26] is that all these oracle removal procedures are actually oblivious to the exact notion of obfuscation. The reason is that all proofs proceed by showing that the oracle-free obfuscation is as secure as the oracle-based obfuscation, i.e., the oracle-free obfuscated circuit can be simulated by an adversary in the oracle world, given the oracle-based obfuscated circuit. Therefore, if one has an iO obfuscator in any of the ideal models, via the oracle removal procedures, one obtains an saiO obfuscator in the standard model. Mahmoody et al. [26] conclude that, as an saiO obfuscator in the standard model allows to resolve the long-standing open problem of building public-key encryption from symmetric-key encryption, it seems very hard to construct such an object. In other words, their result rules out saiO assuming that building public-key encryption from symmetric-key encryption is impossible. Our result strengthens[1] their result by ruling out saiO based on the accepted complexity postulate that $\mathcal{NP} \nsubseteq \mathcal{AM} \cap \mathbf{co}\mathcal{AM}$ and the fundamental assumption of cryptography that one-way functions exist. Therefore, based on the same assumptions, iO in all aforementioned idealized models cannot exist.

1.3   Open Problems

The main question that we leave open is the set of parameters for sacO that are useful and that are (im)possible. Note that it is desirable to have more positive results not only for sacO, but also for acO, the *computational* variant of sacO, in the spirit of Bitansky-Vaikuntanathan [4] who give an assumption-based transformations from aiO to standard iO. Even if sacO for useful parameters turns out to be impossible, it might still be easier to build acO for useful parameters and then use amplification rather than to build fully secure fully correct iO directly.

In particular, note that for a certain parameter range of sacO, we do not know of any impossibility results of building sacO in ideal models.

---

[1] Note that our result is only a "stronger" result in a moral sense, but not in a formal sense. While the non-existence of one-way function would allow us to build a reduction from public-key encryption to symmetric-key encryption (as in this case, both do not exist), it is not known that $\mathcal{NP} \subseteq \mathcal{AM} \cap \mathbf{co}\mathcal{AM}$ implies that we can build a public-key encryption scheme from a one-way function.

The oracle removal procedures that we discuss in Sect. 1.2 maintain security and only weaken correctness. Therefore, a variant of the oracle removal procedures can also be proven for sacO (losing some amount of correctness). As not all useful parameters for sacO are ruled out by our results, one might aim for building sacO in an ideal model for these parameters. Note that one can use our result as a sanity check for any potential oracle construction: If the construction would also work for parameters that we rule out, then it is probably better to pursue a different approach.

Another direction for building useful statistical variants of iO is to relax the computational efficiency of the obfuscator in which case the distributions $X[\Psi]$ that we considered before are not efficiently sampleable anymore (condition (i)) and thus, the $\mathcal{SZK}$ argument fails. Interestingly, Lin et al. [24] recently showed that such a notion of iO that they call XiO has indeed useful applications to transformations on functional encryption.

2   Preliminaries

We first introduce some general notation. By $n \in \mathbb{N}$, we denote the security parameter that we give to all algorithms implicitly in unary representation 1^n. By $\{0,1\}^\ell$ we denote the set of all bit-strings of length ℓ. For a finite set S, we denote the action of sampling x uniformly at random from S by $x \leftarrow_\$ S$, and denote the cardinality of S by $|S|$. Algorithms are assumed to be randomized, unless otherwise stated. We call an algorithm efficient or PPT if it runs in time polynomial in the security parameter. If $\mathcal{A}$ is randomized then by $y \leftarrow \mathcal{A}(x; r)$ we denote that $\mathcal{A}$ is run on input x and with random coins r and produced output y. If no randomness is specified, then we assume that $\mathcal{A}$ is run with freshly sampled uniform random coins, and write this as $y \leftarrow_\$ \mathcal{A}(x; \mathcal{U})$ or in shorthand $y \leftarrow_\$ \mathcal{A}(x)$. For a circuit C we denote by $|C|$ the size of the circuit. We say a function $\mathsf{negl}(n)$ is negligible if for any positive polynomial $\mathsf{poly}(n)$, there exists an $N \in \mathbb{N}$, such that for all $n > N$, $\mathsf{negl}(n) \leq \frac{1}{\mathsf{poly}(n)}$. To define statistically secure variants of obfuscation we will use the following definition of statistical distance.

Definition 1 (Statistical Distance). *For two probability distributions X, Y we define the statistical distance $\mathsf{SD}(X, Y)$ as*

$$\mathsf{SD}(X, Y) = \max_{\mathcal{A}}(\Pr_{x \leftarrow_\$ X}[\mathcal{A}(x) = 1] - \Pr_{y \leftarrow_\$ Y}[\mathcal{A}(y) = 1])$$

where $\mathcal{A}$ ranges over all probabilistic algorithms including inefficient ones.

2.1   Complexity Theory

We refer the reader to Goldreich's book [11] for a detailed exposition of complexity theory. We now discuss a few object that are most relevant to our proof. We let SAT denote the set of all satisfiable CNF formulae, we let USAT denote the set of CNF formulae that have exactly one satisfying assignment, and UNSAT denote the set of CNF formulae that have no satisfying assignment. Given a formula Ψ, deciding whether $\Psi \in$ SAT is an $\mathcal{NP}$-Complete problem. We recall that a *promise problem* $\Pi = (\Pi_{\mathsf{Yes}}, \Pi_{\mathsf{No}})$ is a pair of disjoint subsets of $\{0, 1\}^*$. Of particular interest to us is the *unique SAT* (promise) problem UniqueSAT $=$ (USAT, UNSAT). Total problems (a.k.a languages) are a special case of promise problems, e.g. (SAT, UNSAT) is exactly the SAT problem. In such a case, it suffices to specify Π_{Yes} in order to completely define the problem.

We consider the notion of *randomized polynomial time Turing reductions* between problems. A *promise oracle* to a problem $\Pi = (\Pi_{\mathsf{Yes}}, \Pi_{\mathsf{No}})$, is one that always answers 1 on inputs in Π_{Yes} and always answers 0 on inputs in Π_{No}, but otherwise can answer arbitrarily, and even inconsistently between calls. We define the class $\mathcal{BPP}^{\Pi}$ as the class of problems solvable using a probabilistic polynomial time algorithm with access to a Π oracle. In other words, $\mathcal{BPP}^{\Pi}$ is the class of problems that are *reducible* to Π. One can verify that this class indeed composes, i.e. if $\widetilde{\Pi} \in \mathcal{BPP}^{\Pi}$ then $\mathcal{BPP}^{\widetilde{\Pi}} \subseteq \mathcal{BPP}^{\Pi}$. Valiant and Vazirani [32] showed that SAT is reducible to unique SAT.

Theorem 1 (Valiant-Vazirani). SAT $\in \mathcal{BPP}^{\mathsf{UniqueSAT}}$.

An additional promise problem which will be of interest to us is the GapSD problem, defined by Sahai and Vadhan [29]. This problem essentially captures the hardness of distinguishing between efficient samplers for statistically close distributions and ones for statistically far distributions. We recall that for a circuit C (which we regard as a sampler from a distribution), $C(\mathcal{U})$ denotes the distribution generated by running C on a random input.

Definition 2 (GapSD Problem). *The problem* GapSD $=$ (GapSD$_{\mathsf{Yes}}$, GapSD$_{\mathsf{No}}$) *is defined as follows. Consider tuples of the form* $(C_0, C_1, \nu, 1^\ell)$, *where* C_0, C_1 *are circuits,* ν *is a threshold value and* 1^ℓ *is a unary encoding of a probability gap. Define*

$$\mathsf{GapSD}_{\mathsf{Yes}} = \{(C_0, C_1, \nu, 1^\ell) : \mathsf{SD}(C_0(\mathcal{U}), C_1(\mathcal{U})) < \nu\},$$

and

$$\mathsf{GapSD}_{\mathsf{No}} = \{(C_0, C_1, \nu, 1^\ell) : \mathsf{SD}(C_0(\mathcal{U}), C_1(\mathcal{U})) > \nu + 1/\ell\}.$$

Combining results by Mahmoody and Xiao [27] and by Bogdanov and Lee [5] as follows implies that $\mathcal{BPP}^{\mathsf{GapSD}}$ is contained in $\mathcal{AM} \cap \mathbf{co}\mathcal{AM}$.[2]

Theorem 2. $\mathcal{BPP}^{\mathsf{GapSD}} \subseteq \mathcal{AM} \cap \mathbf{co}\mathcal{AM}$.

Proof. It follows from [5, Theorem 9] that GapSD $\in \mathcal{AM} \cap \mathbf{co}\mathcal{AM}$. This means that both (GapSD$_{\mathsf{Yes}}$, GapSD$_{\mathsf{No}}$) and its complement (GapSD$_{\mathsf{No}}$, GapSD$_{\mathsf{Yes}}$) have $\mathcal{AM}$ protocols, say with completeness $9/10$ and soundness $1/10$. Consider the

---

[2] In fact, by applying [27] we get that $\mathcal{BPP}^{\mathcal{SZK}} \in \mathcal{AM} \cap \mathbf{co}\mathcal{AM}$, which is almost what we need. However, it is only known that GapSD $\in \mathcal{SZK}$ under a somewhat weaker definition of the GapSD problem.

protocol that takes $(C_0, C_1, \nu, 1^\ell)$ and does the following. First, execute the $\mathcal{AM}$ protocol for $(\mathsf{GapSD}_{\mathsf{Yes}}, \mathsf{GapSD}_{\mathsf{No}})$ on input $x_1 = (C_0, C_1, \nu + 1/(4\ell), 1^{(4\ell)})$. Then, execute the $\mathcal{AM}$ protocol for $(\mathsf{GapSD}_{\mathsf{No}}, \mathsf{GapSD}_{\mathsf{Yes}})$ (note the reverse order) on $x_2 = (C_0, C_1, \nu - 1/(2\ell), 1^{(4\ell)})$. Accept only if the two executions accepted. Now, assume that $\nu = \mathsf{SD}(C_0, C_1)$. Then it holds that $x_1 \in \mathsf{GapSD}_{\mathsf{Yes}}$ and $x_2 \in \mathsf{GapSD}_{\mathsf{No}}$ and therefore our new protocol accepts with probability at least $8/10$. However, if $|\nu - \mathsf{SD}(C_0, C_1)| > 1/\ell$ then either $x_1 \in \mathsf{GapSD}_{\mathsf{No}}$ or $x_2 \in \mathsf{GapSD}_{\mathsf{Yes}}$ and therefore our new protocol accepts with probability at most $2/10$. This means that our protocol is an $\mathcal{AM}$ protocol that, for any ϵ, can decide given (C_0, C_1), $1^{\lceil 1/\epsilon \rceil}$ and ν whether $\nu = \mathsf{SD}(C_0(\mathcal{U}), C_1(\mathcal{U}))$ or whether $|\nu - \mathsf{SD}(C_0(\mathcal{U}), C_1(\mathcal{U}))| > \epsilon$.

Consider the class $\mathbb{R}$-**TFAM** as defined in [27, Definition 3.1] and consider the real valued function $f_{\mathsf{SD}} : \{0,1\}^* \to \mathbb{R}$ defined as $f_{\mathsf{SD}}(C_0, C_1, 1^k) = \mathsf{SD}(C_0(\mathcal{U}), C_1(\mathcal{U}))$ (note that the third parameter is ignored and is used only for padding purposes). Our protocol above implies, by definition, that $f_{\mathsf{SD}} \in \mathbb{R}$-**TFAM**.

Furthermore, it holds that $\mathcal{BPP}^{\mathsf{GapSD}} \subseteq \mathcal{BPP}^{\mathcal{O}_{f_{\mathsf{SD}}}}$, for any oracle $\mathcal{O}_{f_{\mathsf{SD}}}$ that on input $x \in \{0,1\}^n$ outputs a value y such that $|y - f_{\mathsf{SD}}(x)| \leq 1/n$. To see this, we notice that we can answer GapSD queries of the form $(C_0, C_1, \nu, 1^\ell)$ as follows: First compute $y = \mathcal{O}_{f_{\mathsf{SD}}}(C_0, C_1, 1^{2\ell})$, then if $y < \nu + 1/(2\ell)$ return Yes, otherwise return No. This implies that $\mathcal{BPP}^{\mathsf{GapSD}} \subseteq \mathcal{BPP}^{\mathbb{R}\text{-}\mathbf{TFAM}}$ by [27, Definition 3.2] (when choosing $\epsilon(n) = 1/n$).

Finally, [27, Theorem 1.1] states that $\mathcal{BPP}^{\mathbb{R}\text{-}\mathbf{TFAM}} \subseteq \mathcal{AM} \cap \mathbf{co}\mathcal{AM}$, which implies that $\mathcal{BPP}^{\mathsf{GapSD}} \subseteq \mathcal{AM} \cap \mathbf{co}\mathcal{AM}$ as desired.

We now state an important corollary of Theorem 2 which shows that there would be unlikely consequences if $\mathsf{UniqueSAT} \in \mathcal{BPP}^{\mathsf{GapSD}}$.

Corollary 3. *If* $\mathsf{UniqueSAT} \in \mathcal{BPP}^{\mathsf{GapSD}}$, *then* $\mathcal{NP} \subseteq \mathcal{AM} \cap \mathbf{co}\mathcal{AM}$.

Proof. By definition it holds that $\mathcal{NP} \subseteq \mathcal{BPP}^{\mathsf{SAT}}$. Theorem 1 implies that $\mathcal{BPP}^{\mathsf{SAT}} \subseteq \mathcal{BPP}^{\mathsf{UniqueSAT}}$. If $\mathsf{UniqueSAT} \in \mathcal{BPP}^{\mathsf{GapSD}}$ then $\mathcal{BPP}^{\mathsf{UniqueSAT}} \subseteq \mathcal{BPP}^{\mathsf{GapSD}}$. Together with $\mathcal{BPP}^{\mathsf{GapSD}} \subseteq \mathcal{AM} \cap \mathbf{co}\mathcal{AM}$ from Theorem 2, we get

$$\mathcal{NP} \subseteq \mathcal{BPP}^{\mathsf{SAT}} \subseteq \mathcal{BPP}^{\mathsf{UniqueSAT}} \subseteq \mathcal{BPP}^{\mathsf{GapSD}} \subseteq \mathcal{AM} \cap \mathbf{co}\mathcal{AM},$$

and the corollary follows.

2.2 Obfuscation

In this subsection, we define the statistically secure variant of approximately correct indistinguishability obfuscation (saiO) and its generalization that we call statistically secure *Approximately Correct Correlation Obfuscation* (sacO). We start with the generalized variant sacO first and then define saiO as a special case. The notion of correlation obfuscation, in contrast to standard indistinguishability obfuscation, does not require that the output of the obfuscator is *indistinguishable* for functionally equivalent circuits. Rather, it only requires that there is a noticeable correlation between the outputs.

Definition 3 (Approximately Correct Correlation Obfuscation). *Let* O *be a* PPT *algorithm that takes boolean circuits (with a single output bit) as inputs and produces boolean circuits as output. For a circuit* C, *we let* $O(C; r)$ *denote the output of running* O *on* C *with randomness* r, *and we let* $O(C)$ *denote the distribution* $O(C; r)$ *with uniform* r.

We say that O *is a* $(1-\epsilon)$-*approximately correct and* $(1-\delta)$-*secure correlation obfuscator sacO if the following conditions hold:*

Approximate Correctness. *For any circuit* C *it holds that*

$$\Pr_{r,x}[O(C; r)(x) = C(x)] \geq 1 - \epsilon(|C|, n).$$

Correlation. *For any pair of circuits* C_1, C_2 *which compute the same function and such that* $|C_1| = |C_2|$ *it holds that* $\mathsf{SD}(O(C_1), O(C_2)) \leq \delta(|C_1|, n)$.

The definition of statistically secure approximately correct indistinguishability obfuscation (saiO) follows by requiring negligible statistical distance δ.

Definition 4 (Approximately Correct Indistinguishability Obfuscation). *Let* O *be a* $(1 - \epsilon)$-*approximately correct and* $(1 - \delta)$-*secure correlation obfuscator. We say that* O *is also a* $(1 - \epsilon)$-*approximately correct statistically secure indistinguishability obfuscator (saiO) if there exists a negligible function* $\mathsf{negl}(|C|, n)$ *such that for all circuits* C *it holds that* $\delta(|C|, n) \leq \mathsf{negl}(|C|, n)$.

2.3   Puncturable Pseudorandom Functions

We use a weak notion of puncturable pseudorandom function. This notion suffices for our results and follows trivially from the stronger standard definition.

Definition 5 (Puncturable Pseudorandom Functions). *A pair of* PPT *algorithms* (PRF, Puncture) *is a puncturable pseudorandom function with one-bit output if, on input a key* $k \in \{0, 1\}^n$ *or a punctured key* k^* *and an input value* $x \in \{0, 1\}^n$, PRF *deterministically outputs a bit* b *and on input a key* $k \in \{0, 1\}^n$ *and an input value* x_0, Puncture *outputs a punctured key* k^* *such that the following two properties are satisfied.*

Functionality Preserved Under Puncturing. For all keys k, *all input values* x_0, *all punctured keys* $k^* \leftarrow_\$ \mathsf{Puncture}(k, x_0)$, *and all input values* $x \neq x_0$, *it holds that*

$$\mathsf{PRF}(k^*, x) = \mathsf{PRF}(k, x).$$

Security. For every PPT *adversary* $(\mathcal{A}_1, \mathcal{A}_2)$ *such that* $\mathcal{A}_1(1^n; r_1)$ *outputs an input value* x_0 *and state* st, *consider an experiment where* $\mathsf{k} \leftarrow_\$ \{0, 1\}^n$, $\mathsf{k}^* = \mathsf{Puncture}(\mathsf{k}, x_0; t)$, *and* $b \leftarrow_\$ \{0, 1\}$. *Then we have*

$$| \Pr_{k, r_1, t, r_2}[\mathcal{A}_2(\mathsf{st}, \mathsf{k}^*, x_0, \mathsf{PRF}(k, x_0); r_2) = 1]$$
$$- \Pr_{k, b, r_1, t, r_2}[\mathcal{A}_2(\mathsf{st}, \mathsf{k}^*, x_0, b; r_2) = 1] \leq \mathsf{negl}(n).$$

As observed by [6,7,22] puncturable PRFs can, for example, be constructed from pseudorandom generators (and thereby one-way functions [17]) via the GGM tree-based construction [12,13].

3   Negative Results for sacO and saiO

We now prove our main theorem that sacO for a large class of parameters, in particular the saiO parameters, is impossible assuming one-way functions and $\mathcal{NP} \not\subseteq \mathcal{AM} \cap \mathbf{co}\mathcal{AM}$.

Theorem 4 (Impossibility of sacO). *If $(1-\epsilon)$-approximately correct, $(1-\delta)$-secure sacO for $\mathcal{P}$ exists, and there exists some polynomial $\mathsf{poly}(|C|,n)$ such that $\delta(|C|,n) \leq \frac{1}{3} - \frac{2}{3}\epsilon(|C|,n) - \frac{1}{\mathsf{poly}(|C|,n)}$, then one-way functions do not exist or $\mathcal{NP} \subseteq \mathbf{co}\mathcal{AM} \cap \mathcal{AM}$.*

By setting δ to be some negligible function, impossibility of saiO follows immediately as a corollary.

Corollary 5 (Impossibility of saiO). *If $(1 - \epsilon)$-approximately correct, saiO for $\mathcal{P}$ exists, and there exists some polynomial $\mathsf{poly}(|C|,n)$ such that $\epsilon(|C|,n) \leq \frac{1}{2} - \frac{1}{\mathsf{poly}(|C|,n)}$, then one-way functions do not exist or $\mathcal{NP} \subseteq \mathbf{co}\mathcal{AM} \cap \mathcal{AM}$.*

Proof (Theorem 4). We define an efficiently samplable distribution $X[\Psi]$ that is parametrized by a formula Ψ, and we define a reference distribution Y that should be parametrized by the size of Ψ and the number of variables in Ψ, but we omit the dependency on Ψ for readability. We note that in the introduction, we discussed to use $Y = X[\mathbf{0}]$, where $\mathbf{0}$ is a canonical representation of an unsatisfiable formula of the same size as Ψ. It is intuitive to think of Y as being indeed equal to $X[\mathbf{0}]$. However, for the sake of tightness, jumping ahead, we will use a slightly different distribution and note that this allows us to gain an additive term of δ in Claim 11.

As in the proof by Goldwasser and Rothblum [14,15] that we sketched in the introduction, we want to define $X[\Psi]$ (and Y) in a way such that properties (1), (2) and (3) are satisfied, assuming one-way functions and sacO. If we manage to do so, then we suceed in showing that these assumptions imply the collapse of the polynomial hierarchy.

Our proof will rely on the promise problem (USAT, UNSAT) rather than the language SAT (See Subsect. 2.1) and therefore, instead of using the gap statistical distance problem GapSD directly as Goldwasser-Rothblum, we will consider $\mathcal{BPP}^{\mathsf{GapSD}}$ to be able to accommodate the randomized reduction from SAT to USAT (See Theorem 1).

Our proof does not rely on complexity-theoretic techniques, except for proving the following claim and showing that the theorem follows from it.

Claim 6. *Assume that there is a formula-indexed distribution $X[\Psi]$, a reference distribution Y, a function ν, and a polynomial $\mathsf{poly}(n)$ such that the following three conditions are satisfied.*

(1) There is a uniform polynomial-time algorithm $\mathcal{A}$, that on input Ψ, constructs two polynomial-size randomized circuits that sample from $X[\Psi]$ and Y respectively.

(2) If Ψ is in UNSAT, then $X[\Psi]$ is has statistical distance at most $\nu(n)$ from Y.

(3) If Ψ is in USAT, then $X[\Psi]$ has statistically distance at least $\nu(n) + \frac{1}{\text{poly}(n)}$ from Y.

Then USAT *is in* $\mathcal{BPP}^{\mathsf{GapSD}} \subseteq \mathcal{AM} \cap \mathbf{co}\mathcal{AM}$.

Proof. Given that conditions (1), (2) and (3) are satisfied, we construct an algorithm $\mathcal{B}$ such that for all GapSD oracles and all formulae Ψ, $\mathcal{B}^{\mathsf{GapSD}}(\Psi)$ outputs 1 with probability 1 if $\Psi \in$ USAT and 0 with probability 1 if $\Psi \in$ UNSAT. On input Ψ, the algorithm $\mathcal{B}$ runs $\mathcal{A}$ to get circuits for $X[\Psi]$ and Y and queries $(X[\Psi], Y, \nu(n), 1^{\text{poly}(n)})$ to the GapSD oracle. $\mathcal{B}$ returns whatever the oracle returns. By properties (1), (2) and (3), the query that $\mathcal{B}$ makes is in $\mathsf{GapSD}_{\mathsf{Yes}}$ if $\Psi \in$ USAT and in $\mathsf{GapSD}_{\mathsf{No}}$ if $\Psi \in$ UNSAT. Hence, $\mathcal{B}$ is correct and USAT is in $\mathcal{BPP}^{\mathsf{GapSD}}$. Moreover, due to Theorem 2 by Mahmoody and Xiao, $\mathcal{BPP}^{\mathsf{GapSD}} \subseteq \mathcal{AM} \cap \mathbf{co}\mathcal{AM}$.

To obtain the main theorem, we need to show that USAT is in $\mathcal{BPP}^{\mathsf{GapSD}}$ implies that $\mathcal{NP}$ is in $\mathcal{AM} \cap \mathbf{co}\mathcal{AM}$ which directly follows from Corollary 3 of Theorem 2 by Mahmoody and Xiao. Thus, if we can show that a distributions as described in conditions (1), (2) and (3) exist, then the theorem follows.

We now define $X[\Psi]$ and Y and then show that they satisfy (1), (2) and (3) assuming the existence of one-way functions and sacO with suitable correctness and security.

Definition 6 (Distribution). *Let $\ell(n)$ be a sufficiently large polynomial designating the size to which all circuits are padded before being obfuscated. Let Ψ be a formula, let $(\mathsf{PRF}, \mathsf{Puncture})$ be a puncturable pseudorandom function, and let O be a $(1 - \epsilon)$-correct, statistically $(1 - \delta)$-secure approximate correlation obfuscator, where $\delta(|C|, n) \leq \frac{1}{3} - \frac{2}{3}\epsilon(|C|, n) - \frac{1}{\text{poly}(|C|, n)}$. We now define the distribution $X[\Psi]$ and Y, where the circuits $\mathsf{C}_X[k, b, s, \Psi]$ and $\mathsf{C}_{prf}[k]$ are defined to the right of the distributions.*

$X[\Psi](1^n)$	$\mathsf{C}_X[k, s, \Psi](x)$	$Y(1^n)$	$\mathsf{C}_{prf}[k](x)$
$k \leftarrow_\$ \{0,1\}^n$	**if** $\Psi(x \oplus s) = 1$	$k \leftarrow_\$ \{0,1\}^n$	**return** $\mathsf{PRF}(k, x)$
$s \leftarrow_\$ \{0,1\}^n$	$\quad$ **return** $\mathsf{PRF}(k, x) \oplus 1$	$s \leftarrow_\$ \{0,1\}^n$	
$C := \mathsf{C}_X[k, s, \Psi]$	**else**	$C := \mathsf{C}_{prf}[k]$	
$C' \leftarrow_\$ \mathsf{O}(C)$	$\quad$ **return** $\mathsf{PRF}(k, x)$	$C' \leftarrow_\$ \mathsf{O}(C)$	
return (k, s, C')		**return** (k, s, C')	

Claim 7 (Distribution). *The distributions defined in Definition 6 satisfy the conditions demanded in Claim 6. I.e., there exists a function ν and a polynomial $\text{poly}(n)$ such that they satisfy the following:*

(1) There is a uniform polynomial-time algorithm $\mathcal{A}$, that on input Ψ, constructs two polynomial-size randomized circuits that sample from $X[\Psi]$ and Y respectively.

(2) If Ψ is in UNSAT, then $X[\Psi]$ is has statistical distance at most $\nu(n)$ from Y.

(3) If Ψ is in USAT, *then $X[\Psi]$ has statistically distance at least $\nu(n) + \frac{1}{\mathsf{poly}(n)}$ from Y.*

We will first state two claims and a lemma that will allow us to prove Claim 7. We will then prove Claim 7 and afterwards prove the claims and the lemma.

Claim 8 (Efficient Sampling). *There is a uniform polynomial-time algorithm $\mathcal{A}$, that on input Ψ, constructs two polynomial-size randomized circuits that sample from $X[\Psi]$ and Y respectively.*

Claim 9 (Statistical Proximity). *For all formulae $\Psi \in$ UNSAT, $X[\Psi]$ has statistical distance at most $\delta(\ell(n), n)$ from Y.*

Lemma 10 (Statistical Distance). *There exists a negligible function $\mathsf{negl}(n)$, such that for all formulae $\Psi \in$ USAT, $X[\Psi]$ has statistical distance at least $1 - 2\epsilon(\ell(n), n) - 2\delta(\ell(n), n) - \mathsf{negl}(n)$ from Y.*

Proof (Claim 7). Condition (1) follows immediately from Claim 8. Condition (2) follows from Claim 9 for a function $\nu(n) = \delta(\ell(n), n)$. From Lemma 10, it follows that, if Ψ is in USAT, then $X[\Psi]$ has statistically distance at least $1 - 2\epsilon(\ell(n), n) - 2\delta(\ell(n), n) - \mathsf{negl}(n)$ from Y. Combining this with the $\nu(n)$ obtained from Claim 9 we get that condition (3) holds, if there exists a polynomial $\mathsf{poly}(n)$, such that

$$\delta(\ell(n), n) + \frac{1}{\mathsf{poly}(n)} \leq 1 - 2\epsilon(\ell(n), n) - 2\delta(\ell(n), n) - \mathsf{negl}(n)$$

$$\Leftrightarrow \quad 3\delta(\ell(n), n) \leq 1 - 2\epsilon(\ell(n), n) - \frac{1}{\mathsf{poly}(n)} - \mathsf{negl}(n)$$

$$\Leftrightarrow \quad \delta(\ell(n), n) \leq \frac{1}{3} - \frac{2}{3}\epsilon(\ell(n), n) - \frac{1}{\mathsf{poly}(n)} - \mathsf{negl}(n). \qquad (1)$$

And, since $\mathsf{negl}(n)$ is dominated by an inverse polynomial, Eq. 1 is already ensured by Definition 6, condition (3) holds, and the claim follows.

Proof (Claim 8). Sampling k and s is efficient and so is constructing $\mathsf{C}_X[k, s, \Psi]$ and $\mathsf{C}_{\mathsf{prf}}[k]$. Finally, from the efficiency of the obfuscator, it follows that $X[\Psi]$ and Y are efficiently samplable by polynomial-size randomized circuits.

Proof (Claim 9). For all unsatisfiable formulae Ψ, the circuits $\mathsf{C}_X[k, s, \Psi]$ and $\mathsf{C}_{\mathsf{prf}}[k]$ are functionally equivalent and of same size $\ell(n)$. Hence, by statistical security of the obfuscator, the distributions $(k, s, \mathsf{O}(\mathsf{C}_X[k, s, \Psi]))$ and $(k, s, \mathsf{O}(\mathsf{C}_{\mathsf{prf}}[k]))$ have statistical distance at most $\delta(\ell(n), n)$.

We now turn to the most involved part of the proof, which is to show that Lemma 10 holds. In order to show that for all formulae $\Psi \in$ USAT, $X[\Psi]$ is statistically far from Y, we show that, if $\Psi \in$ USAT, then the distribution $X[\Psi]$ has a property that Y does not have. We state the property in two claims.

Claim 11. *For all x_0, it holds that*

$$\Pr_{(k,s,C') \leftarrow\$ Y(1^n)}[C'(x_0 \oplus s) \neq \mathsf{PRF}(k, x_0 \oplus s)] \leq \epsilon(\ell(n), n).$$

Claim 12. *If $\Psi \in \mathsf{USAT}$, then there exists x_Ψ, such that*

$$\Pr_{(k,s,C') \leftarrow\$\, X[\Psi](1^n)} [C'(x_\Psi \oplus s) \neq \mathsf{PRF}(k, x_\Psi \oplus s)]$$
$$\geq 1 - \epsilon(\ell(n), n) - 2\delta(\ell(n), n) - 2\mathsf{negl}(n).$$

Proof (Lemma 10). Lemma 10 follows directly from Claims 11 and 12, because the stated properties are statistical properties, i.e., we can give an inefficient distinguisher as follows: The distinguisher determines x_Ψ through exhaustive search and then, given a sample (k, s, C') from either $X[\Psi]$ or Y, checks whether $\mathsf{PRF}(k, \cdot)$ and C' differ on input $x_\Psi \oplus s$. If the sample is from $X[\Psi]$, they will differ with probability greater than $1 - \epsilon(\ell(n), n) - 2\delta(\ell(n), n) - \mathsf{negl}(n)$. If on the other hand the sample is from Y, then they will differ only with probability less than $\epsilon(\ell(n), n)$. This concludes the proof of Lemma 10, subject to proving the claims.

It now remains to prove Claims 11 and 12. The proof of the first property is relatively straightforward, while the proof of the second property contains the technical key arguments that we discussed above.

Proof (Claim 11). To prove the claim, we will argue that the following equalities hold:

$$\Pr_{(k,s,C') \leftarrow\$\, Y(1^n)} [C'(x_0 \oplus s) \neq \mathsf{PRF}(k, x_0 \oplus s)] \tag{2}$$
$$= \Pr_{k,s \leftarrow\$\, \{0,1\}^n, C' \leftarrow\$\, \mathsf{O}(\mathsf{C}_{\mathsf{prf}[k]})} [C'(x_0 \oplus s) \neq \mathsf{PRF}(k, x_0 \oplus s)] \tag{3}$$
$$= \Pr_{k,s \leftarrow\$\, \{0,1\}^n, C' \leftarrow\$\, \mathsf{O}(\mathsf{C}_{\mathsf{prf}[k]})} [C'(s) \neq \mathsf{PRF}(k, s)] \tag{4}$$
$$\leq \epsilon(\ell(n), n) \tag{5}$$

Equation 3 is simply a restatement of the claim. Given that s is uniformly and independently distributed, s and $x_0 \oplus s$ are distributed identically and therefore, also Eq. 4 holds. Finally, Eq. 4 simply checks whether an obfuscated circuit does not agree with the original circuit on a uniformly chosen input. This happens by definition of correctness with probability at most $\epsilon(\ell(n), n)$, yielding Eq. 5 and concluding the proof.

Proof (Claim 12). Let x_Ψ denote the accepting assignment of Ψ. We first define the following game

$$\mathsf{Game}_1(n)$$

$$(k, s, C') \leftarrow\$\, X[\Psi]$$
$$x_0 := x_\Psi \oplus s$$
$$b := \mathsf{PRF}(k, x_0) \oplus 1$$
$$\textbf{return } (C'(x_0) \overset{?}{=} b)$$

and observe that

$$\Pr_{(k,s,C') \leftarrow\$\, X[\Psi](1^n)} [C'(x_\Psi \oplus s) \neq \mathsf{PRF}(k, x_\Psi \oplus s)] = \Pr[\mathsf{Game}_1(n) = 1].$$

We will now bound this probability using a series of game hops. To specify the game hops, we need to specify an additional circuit $C_{\mathsf{punct}}[k^*, x_0, b](x)$, that is parametrized by a punctured PRF key k^*, an input x_0, and a bit b.

$$\underline{C_{\mathsf{punct}}[k^*, x_0, b](x)}$$

if $x = x_0$

    return b

else

    return $\mathsf{PRF}(k^*, x)$

Note that Game_2 is a re-write of Game_1 by making $X[\Psi]$ explicit.

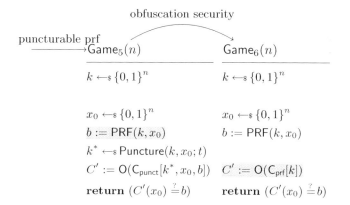

We will first bound the differences between each pair of consecutive games and then prove a bound for $\Pr[\mathsf{Game}_6(n) = 1]$.

Hop from Game_1 *to* Game_2. The changes between the two games are purely syntactic. I.e., the definition of the sampling process from $X[\Psi]$ is explicitely written down in Game_2. Therefore, the two games are perfectly equivalent, and it holds that

$$\Pr[\mathsf{Game}_1(n) = 1] = \Pr[\mathsf{Game}_2(n) = 1]. \tag{6}$$

Hop from Game$_2$ *to* Game$_3$. Here it is critical to observe that $C_X[k, s, \Psi]$ and $C_{\text{punct}}[k^*, x_0, b]$ are functionally equivalent. Even though the key is punctured on $x_0 = x_\Psi \oplus s$ in C_{punct}, this makes no difference, since PRF is never invoked on x_0 in the circuit. Instead the circuit outputs the hardcoded value $b = \text{PRF}(k, x_0) \oplus 1$ on input x_0, which is the same value output by $C_X[k, s, \Psi]$. Therefore, the two circuits are functionally equivalent and it follows from the statistical security of the obfuscator that the statistical difference between the distributions of C' in the two games is at most $\delta(\ell(n), n)$. It follows, that also the distribution of the outputs of Game$_2$ and Game$_3$ have a statistical distance of at most $\delta(\ell(n), n)$. I.e.,

$$|\Pr[\text{Game}_3(n) = 1] - \Pr[\text{Game}_2(n) = 1]| \leq \delta(\ell(n), n). \tag{7}$$

Hop from Game$_3$ *to* Game$_4$. Since s is no longer known to the obfuscator in Game$_3$, $x_0 := x_\Psi \oplus s$ is simply a uniformly distributed value. Thus, x_0 is distributed identically in Game$_3$ and Game$_4$ and it follows that

$$\Pr[\text{Game}_3(n) = 1] = \Pr[\text{Game}_4(n) = 1]. \tag{8}$$

Hop from Game$_4$ *to* Game$_5$. Note that x_Ψ is no longer required to evaluate Game$_4$ and Game$_5$. Therefore, the two games can be evaluated efficiently. This allows us to bound the difference between the two games by the security of the puncturable pseudorandom function. To bound the difference between games Game$_4(n)$ and Game$_5(n)$, we construct a distinguisher $(\mathcal{A}_1, \mathcal{A}_2)$ with advantage

$$\tfrac{1}{2} \cdot |\Pr[\text{Game}_4(n) = 1] - \Pr[\text{Game}_5(n) = 1]|$$

against the puncturable PRF as follows:

$\mathcal{A}_1(1^n; r_1)$	$\mathcal{A}_2(\text{st}, k^*, x_0, b; r_2)$
$x_0 \leftarrow_\$ \{0, 1\}^n$	$C' := \mathsf{O}(C_{\text{punct}}[k^*, x_0, b])$
return $(\bot, x_0)$	**return** $(C'(x_0) \overset{?}{=} b)$

Observe, that in the case where $\mathcal{A}_2$ receives the PRF value, it holds that

$$\Pr_{k, r_1, t, r_2}[\mathcal{A}_2(\text{st}, k^*, x_0, \text{PRF}(k, x_0); r_2) = 1] = \Pr[\text{Game}_5(n) = 1]. \tag{9}$$

If on the other hand, $\mathcal{A}_2$ receives a b chosen uniformly at random, then b is equal to $\text{PRF}(k, x_0)$ and $\text{PRF}(k, x_0) \oplus 1$ with probability $\frac{1}{2}$ respectively, and it holds that

$$\Pr_{k, b, r_1, t, r_2}[\mathcal{A}_2(\text{st}, k^*, x_0, b; r_2) = 1] = \frac{1}{2}\Pr[\text{Game}_4(n) = 1] + \frac{1}{2}\Pr[\text{Game}_5(n) = 1] \tag{10}$$

By security of the puncturable PRF, it must hold that

$$|\Pr_{k, r_1, t, r_2}[\mathcal{A}_2(\text{st}, k^*, x_0, \text{PRF}(k, x_0)); r_2) = 1]$$
$$- \Pr_{k, b, r_1, t, r_2}[\mathcal{A}_2(\text{st}, k^*, x_0, b; r_2) = 1]| \leq \text{negl}(n)$$

Combining this with Eqs. 9 and 10 yields

$$\left| \Pr[\mathsf{Game}_5(n) = 1] - \frac{1}{2}\Pr[\mathsf{Game}_4(n) = 1] - \frac{1}{2}\Pr[\mathsf{Game}_5(n) = 1] \right| \leq \mathsf{negl}(n)$$

$$\implies \frac{1}{2}|\Pr[\mathsf{Game}_5(n) = 1] - \Pr[\mathsf{Game}_4(n) = 1]| \leq \mathsf{negl}(n)$$

$$\implies |\Pr[\mathsf{Game}_5(n) = 1] - \Pr[\mathsf{Game}_4(n) = 1]| \leq 2\mathsf{negl}(n). \tag{11}$$

Hop from Game_5 *to* Game_6. Here it is critical to observe that $\mathsf{C}_{\mathsf{punct}}[k^*, x_0, b]$ and $\mathsf{C}_{\mathsf{prf}}[k]$ are functionally equivalent. Even though the key is punctured on x_0 in $\mathsf{C}_{\mathsf{punct}}$, this makes no difference, since PRF is never invoked on x_0 in the circuit. Instead the circuit outputs the hardcoded value $b = \mathsf{PRF}(k, x_0)$ on input x_0. Therefore, the two circuits are functionally equivalent and it follows from the statistical security of the obfuscator that the statistical difference between the distributions of C' in the two games is at most $\delta(\ell(n), n)$. It follows, that also the distribution of the outputs of Game_5 and Game_6 have a statistical distance of at most $\delta(\ell(n), n)$. I.e.,

$$|\Pr[\mathsf{Game}_5(n) = 1] - \Pr[\mathsf{Game}_6(n) = 1]| \leq \delta(\ell(n), n). \tag{12}$$

It remains to bound the probability $\Pr[\mathsf{Game}_6(n) = 1]$. Observe, that x_0 is a uniformly chosen input unknown to the obfuscator. Further, the $\mathsf{Game}_6(n)$ simply checks whether the output of circuit C' is the correct output value of the obfuscated circuit. Therefore, the correctness of the obfuscator implies that

$$\Pr[\mathsf{Game}_6(n) = 1] \geq 1 - \epsilon(\ell(n), n). \tag{13}$$

Finally, combining Eq. 13 with Eqs. 6 through 12, we get

$$\Pr[\mathsf{Game}_1(n) = 1]$$
$$\geq \Pr[\mathsf{Game}_6(n) = 1] - |\Pr[\mathsf{Game}_1(n) = 1] - \Pr[\mathsf{Game}_6(n) = 1]|$$
$$\geq 1 - \epsilon(\ell(n), n) - 2\delta(\ell(n), n) - 2\mathsf{negl}(n)$$

thus concluding the proof of Claim 12 and Theorem 4.

Acknowledgment. We are grateful to Andrej Bogdanov, Kai-Min Chung, Siyao Guo, Markulf Kohlweiss, Arno Mittelbach and Vinod Vaikuntanathan for helpful discussions. In particular, Andrej and Vinod pointed out that PAC-learneability implies approximate obfuscation and that thus, CNF formulae are PAC-learneable, which implies that impossibility results for saiO need to obfuscate more complex functions than CNF formulae. The discussions with Vinod at the Mathematisches Forschungsinstitut Oberwolfach (MFO) inspired the idea of embedding a formula into a PRF. Vinod also suggested that in the absence of one-way functions, there exists a perfectly secure variant of obfuscation where the correctness is on average over the circuit distribution, the input and the obfuscator.

A   A Positive Result for Correlation Obfuscation

In this appendix, we instantiate approximately correct correlation obfuscation for a large class of weak parameters. The idea of the construction is fairly simple and is based on two observations. For circuits with only a single bit output, we can efficiently estimate the majority of the outputs by using random sampling. This estimation depends only on the function computed by a circuit and not on the circuit itself. Therefore an obfuscator that simply outputs the estimated majority is fully secure but only correct with probability about $1/2$. An obfuscator, that simply outputs the circuit itself, on the other hand, is not secure at all (statistical distance is 1), but is fully correct.

By combining these two obfuscators and outputting the majority with probability 2ϵ and the circuit itself with probability $1 - 2\epsilon$ we can construct a roughly $(1 - \epsilon)$ approximatly correct and $(1 - 2\epsilon)$ secure obfuscator $O_{\epsilon,\mu}$ as detailed below. The parameter μ is some inverse polynomial function that describes the amount of approximation error that we allow (and that affects the correctness of $O_{\epsilon,\mu}$) when the obfuscators samples repeatedly from the output distribution of the circuit to see whether the circuit is closer to the constant 1 or constant 0 function.

For any circuit C, $\text{in}(C)$ denotes the number of input wires. For $b \in \{0,1\}$, Const_b^i is a canonical circuit with input length i and constant output b. The Bernoulli distribution of a parameter $p \in [0,1]$ is defined by Ber_p, i.e., it holds that $\text{Pr}_{b \leftarrow\$ \text{Ber}_p}[b = 1] = p$ and $\text{Pr}_{b \leftarrow\$ \text{Ber}_p}[b = 0] = 1 - p$. Depending on the desired error parameter μ, the obfuscation proceeds as follows.

$O_{\epsilon,\mu}(C, 1^n)$	$\text{EstMaj}(C, \mu, 1^n)$
$b \leftarrow\$ \text{Ber}_{2\epsilon}$	**for** $i := 1, \ldots, \lceil \frac{4n}{\mu^2} \rceil$
if $b = 1$:	$\quad x_i \leftarrow\$ \{0,1\}^{\text{in}(C)}$
$\quad m := \text{EstMaj}(C, \mu, 1^n)$	$\quad y_i := C(x_i)$
$\quad C' := \text{Const}_m^{\text{in}(C)}$	$\quad$ **return** $\text{maj}(y_1, \ldots, y_{\lceil \frac{4n}{\mu^2} \rceil})$
else	
$\quad C' := C$	
return C'	

Claim 13. *On input $(C, 1^n)$, the obfuscator $O_{\epsilon,\mu}$ runs in time linear in $\frac{4n}{\mu^2}|C|$ plus the time needed to sample from $\text{Ber}_{2\epsilon}$ and is an $(1 - (\epsilon + \mu))$ approximately correct and $(1 - 2\epsilon)$ secure correlation obfuscator for circuits with single bit output.*

Proof. Efficiency follow by construction and so does security, because EstMaj only uses the input-output behaviour of the circuit which is the same for two functionally identical circuits. If the function induced by the circuit C is less than $\frac{\mu}{4}$ from being balanced (i.e., 1 with probability $\frac{1}{2}$ on a uniformly random input), then the correctness error is at most $\frac{\mu}{2}$, if $b = 1$, and 0, if $b = 0$ and hence, the overall correctness error is upper bounded by $(1 - 2\epsilon) \cdot 0 + 2\epsilon \cdot \frac{\mu}{2} = \epsilon\mu \leq \epsilon + \mu$. If the function induced by the circuit C outputs a fixed value, w.l.o.g. 1, with probability at least $\frac{1}{2} + \frac{\mu}{4}$, then via a Chernoff bound, the probability that $\mathsf{EstMaj}(C, \mu, 1^n)$ outputs 1 is at least $1 - \mathsf{negl}(n)$ and in that case, the correctness error is at most $\frac{1}{2} - \frac{\mu}{4}$ and else, the correctness error is at most 1. Hence, for the case that $b = 1$, we obtain an upper bound on the correctness error of $(\frac{1}{2} - \frac{\mu}{4}) \cdot (1 - \mathsf{negl}(n)) + 1 \cdot \mathsf{negl}(n) = \frac{1}{2} - \frac{\mu}{4} + \mathsf{negl}(n)$. As before, when $b = 0$, the correctness error is 0 and hence, we obtain as upper bound on the correctness error $(1 - 2\epsilon) \cdot 0 + 2\epsilon \cdot (\frac{1}{2} - \frac{\mu}{4} + \mathsf{negl}(n)) = \epsilon - \frac{\mu}{2} \leq \epsilon + \mu$.

B   Correctness and Security Parameters for sacO to Build a Public-Key Encryption Scheme from a One-Way Function

By inspecting the Sahai-Waters [30] construction to transform a one-way function into a public-key encryption scheme (PKE) by using obfuscation, Bitansky and Vaikuntanathan [4] and Mahmoody et al. [26] observe that approximately correct iO suffices for this transformation. Both papers consider approximately correct variants of iO with "full" security, i.e., where the adversary has only negligible advantage in distinguishing obfuscations of two functionally equivalent circuits. As discussed in previous sections, approximately correct correlation obfuscation (sacO) with weaker security might still be useful. We therefore work out the exact correctness and security parameters required of a sacO for the Sahai-Waters transformation to work. Jumping ahead, we note that part of the bounds that we obtain here are ruled out by our impossibility result, but not all of them.

For much weaker parameters, we earlier gave a trivial construction of sacO. We do not deem this construction to be useful. As expected, there is a gap between the parameters that we can construct trivially and the parameters that we can rule out (else, we would have a proof that one-way functions imply the collapse of the polynomial hierarchy). Also, as expected, the trivial bounds do not suffice to instantiate the Sahai-Waters construction (according to our analysis that we have reasons to believe is tight).

On the other hand, our impossibility result does not rule out all useful bounds for sacO. It is an interesting question to (1) show that also for the parameters in this small gap, sacO cannot exist, or (2) show a construction for these parameters, and/or (3) improve the parameters that are needed for meaningful applications. Note that even if it turns out that sacO for these parameters cannot exist, (3) could still be a fruitful research direction, because it might be helpful to weaken the parameters also on variants of acO with *computational* security in order to obtain constructions from weaker assumptions.

We will consider sacO with $(1 - \delta)$-security and $(1 - \epsilon)$ correctness, and we will also yield a PKE that does not achieve full correctness and that does not achieve full security. In some cases, as observed by Holenstein [18], via amplification, it is possible to achieve full security and correctness with overwhelming probability. However, as we discuss now, amplification is not always possible.

B.1   Amplification

We define $(1 - \epsilon_{\mathsf{PKE}})$-correct and $(\frac{1}{2} - \delta_{\mathsf{PKE}})$-secure PKE as follows.

Definition 7 (Approximate Public Key Encryption). *Let* $\mathsf{PKE} = (\mathsf{KGen}, \mathsf{Enc}, \mathsf{Dec})$ *be a public key encryption scheme.*

Correctness *We say that* PKE *is* $(1 - \epsilon_{\mathsf{PKE}})$-*correct, if it holds that*

$$\Pr_{b, \mathsf{KGen}, \mathsf{Enc}}\left[\mathsf{Dec}(\mathsf{sk}, \mathsf{Enc}(b, pk)) = b, (\mathsf{pk}, \mathsf{sk}) \leftarrow_\$ \mathsf{PKE.KGen}(1^n)\right] \geq 1 - \epsilon_{\mathsf{PKE}}(n).$$

Security. *We say that* PKE *is* $(\frac{1}{2} - \delta_{\mathsf{PKE}})$-*secure, if for all efficient adversaries* $\mathcal{A}$, *there exists a negligible function* $\mathsf{negl}(n)$ *such that*

$$\Pr_{b \leftarrow_\$ \{0,1\}, \mathsf{KGen}, \mathsf{Enc}}\left[\mathcal{A}(\mathsf{pk}, \mathsf{Enc}(b, \mathsf{pk})) = b, (\mathsf{pk}, \mathsf{sk}) \leftarrow_\$ \mathsf{KGen}(1^n)\right]$$
$$\leq \frac{1}{2} + \delta_{\mathsf{PKE}}(n) + \mathsf{negl}(n).$$

We would like to amplify such a scheme into "standard" PKE, where ϵ_{PKE} and δ_{PKE} are negligible. We now discuss via a counterexample why such an amplification is not generally possible. Take a bit encryption scheme that outputs the message bit with probability α and a random bit with probability $1 - \alpha$ and where decryption is the identity function. This PKE scheme is $(\frac{1}{2} - \frac{\alpha}{2})$-secure and $(\frac{1}{2} + \frac{\alpha}{2})$-correct. Correctness parameters are thus only meaningful if ϵ_{PKE} and δ_{PKE} are bounded away from $\frac{1}{2}$ and if, moreover, there is a meaningful relationship between the security and the correctness parameter. Holenstein [18] shows (and we use the presentation of Mahmoody et al. [26] here) that amplification is possible if there exists a polynomial $\mathsf{poly}(n)$ such that

$$(1 - 2\epsilon_{\mathsf{PKE}}(n))^2 > 2\delta_{\mathsf{PKE}}(n) + \frac{1}{\mathsf{poly}(n)}.$$

Note that Holenstein also shows a tightness result for his amplification technique with respect to restricted black-box reductions.

B.2   The Sahai-Waters Construction

We now present the Sahai-Waters [30] construction of a public-key encryption scheme from a one-way function. We recall that by Håstad et al. [17], Goldreich et al. [13], and several independent proofs [6,7,22] that the GGM construction is a puncturable PRF, puncturable PRFs and OWFs are existentially equivalent. The key generation of the Sahai-Waters construction draws a key k for a puncturable PRF as the secret key sk and then outputs an obfuscation of the following circuit $\mathsf{C_{SW}}[k]$ as a public key pk:

$$\underline{\mathsf{C_{SW}}[k](m, r)}$$
$$r' := \mathsf{PRG}(r)$$
$$c := m \oplus \mathsf{PRF}(k, r')$$
$$\textbf{return } (r', c)$$

The encryption algorithm $\mathsf{Enc}(\mathsf{pk}, m, r)$ interprets the public key pk as a circuit, runs it on (m, r) and returns the result as a ciphertext. Finally, for decryption of a pair (r', c), the decryption algorithm $\mathsf{Dec}(\mathsf{sk}, (r', c))$ outputs $m :=$ $c \oplus \mathsf{PRF}(\mathsf{sk}, r')$.

Claim 14 (Sahai-Waters). *The Sahai-Waters construction instantiated with sacO with correctness $1 - \epsilon$ and security $1 - \delta$ yields a public-key encryption scheme with correctness error $\epsilon_{\mathsf{PKE}}(n) = \epsilon(|C|, n)$ and a distinguishing advantage of $\delta_{\mathsf{PKE}}(n) = \delta(|C|, n) + \epsilon(|C|, n)$.*

Before we prove this claim, we will first illustrate what this implies for the bounds on parameters allowing for Holenstein amplification. Combining the bound for Holenstein amplification with Claim 14, we get that

$$2\delta_{\mathsf{PKE}}(n) + \frac{1}{\mathsf{poly}(n)} < (1 - 2\epsilon_{\mathsf{PKE}}(n))^2 \qquad (14)$$

$$\implies \quad 2\delta(|C|, n) + 2\epsilon(|C|, n) + \frac{1}{\mathsf{poly}(n)} < (1 - 2\epsilon(|C|, n))^2 \qquad (15)$$

$$\implies \quad \delta(|C|, n) < \frac{1}{2} - 3\epsilon(|C|, n) + 2\epsilon(|C|, n)^2 - \frac{1}{2\mathsf{poly}(n)}. \qquad (16)$$

We thus get the following corollary.

Corollary 15. *Any $(1-\epsilon)$ correct and $(1-\delta)$ secure sacO implies a construction of public key encryption from one-way functions, if there exists some polynomial $\mathsf{poly}(|C|, n)$ such that*

$$\delta(|C|, n) < \frac{1}{2} - 3\epsilon(|C|, n) + 2\epsilon(|C|, n)^2 - \frac{1}{\mathsf{poly}(n)}.$$

Proof (Proof of Claim 14). Note that correctness of the encryption scheme is over a random message, the randomness of the key generation and the randomness of the encryption algorithm. The obfuscated circuit is therefore invoked on a uniformly random input and the probability that it does not output the correct ciphertext can thus be bounded by the correctness error of the obfuscator. Since the decryption of the scheme is perfectly correct, we thus get that $\epsilon_{\mathsf{PKE}}(n) = \epsilon(|C|, n)$.

To prove security, we first define the following game

$$\mathsf{Game}_1(n)$$

$k \leftarrow_\$ \{0, 1\}^n$

$r \leftarrow_\$ \{0, 1\}^{n/2}$

$\mathsf{pk} \leftarrow_\$ \mathsf{O}(C_{\mathsf{SW}}[k])$

$b \leftarrow_\$ \{0, 1\}$

$c := \mathsf{pk}(b, r)$

$b' \leftarrow_\$ \mathcal{A}(\mathsf{pk}, c)$

return $(b' \stackrel{?}{=} b)$

and observe that

$$\Pr_{b \leftarrow_{\$} \{0,1\},\mathsf{KGen},\mathsf{Enc}}\left[\mathcal{A}(\mathsf{pk}, \mathsf{Enc}(b, \mathsf{pk})) = b,\ (\mathsf{pk}, \mathsf{sk}) \leftarrow_{\$} \mathsf{KGen}(1^n)\right]$$
$$= \Pr\left[\mathsf{Game}_1(n) = 1\right].$$

We will now bound this probability using a series of game hops.

PRG security: $\mathsf{Game}_2(n) \rightarrow \mathsf{Game}_3(n)$; obfuscation security: $\mathsf{Game}_3(n) \rightarrow \mathsf{Game}_4(n)$

$\mathsf{Game}_2(n)$	$\mathsf{Game}_3(n)$
$k \leftarrow_{\$} \{0,1\}^n$	$k \leftarrow_{\$} \{0,1\}^n$
$r \leftarrow_{\$} \{0,1\}^{n/2}$	
$r' := \mathsf{PRG}(r)$	$r' \leftarrow_{\$} \{0,1\}^n$
$\mathsf{pk} \leftarrow_{\$} \mathsf{O}(\mathsf{C_{SW}}[k])$	$\mathsf{pk} \leftarrow_{\$} \mathsf{O}(\mathsf{C_{SW}}[k])$
$b \leftarrow_{\$} \{0,1\}$	$b \leftarrow_{\$} \{0,1\}$
$c := (b \oplus \mathsf{PRF}(k, r'), r')$	$c := (b \oplus \mathsf{PRF}(k, r'), r')$
$b' \leftarrow_{\$} \mathcal{A}(\mathsf{pk}, c)$	$b' \leftarrow_{\$} \mathcal{A}(\mathsf{pk}, c)$
return $(b' \stackrel{?}{=} b)$	**return** $(b' \stackrel{?}{=} b)$

PRF security: $\mathsf{Game}_4(n) \rightarrow \mathsf{Game}_5(n)$

$\mathsf{Game}_4(n)$	$\mathsf{Game}_5(n)$
$k \leftarrow_{\$} \{0,1\}^n$	$k \leftarrow_{\$} \{0,1\}^n$
$r' \leftarrow_{\$} \{0,1\}^n$	$r' \leftarrow_{\$} \{0,1\}^n$
$k^* \leftarrow_{\$} \mathsf{Puncture}(k, r'; t)$	$k^* \leftarrow_{\$} \mathsf{Puncture}(k, r'; t)$
$\mathsf{pk} \leftarrow_{\$} \mathsf{O}(\mathsf{C_{SW}}[k^*])$	$\mathsf{pk} \leftarrow_{\$} \mathsf{O}(\mathsf{C_{SW}}[k^*])$
$b \leftarrow_{\$} \{0,1\}$	$b \leftarrow_{\$} \{0,1\}$
	$s \leftarrow_{\$} \{0,1\}$
$c := (b \oplus \mathsf{PRF}(k, r'), r')$	$c := (b \oplus s, r')$
$b' \leftarrow_{\$} \mathcal{A}(\mathsf{pk}, c)$	$b' \leftarrow_{\$} \mathcal{A}(\mathsf{pk}, c)$
return $(b' \stackrel{?}{=} b)$	**return** $(b' \stackrel{?}{=} b)$

We will first bound the differences between each pair of consecutive games and then argue a bound for $\Pr[\mathsf{Game}_5(n) = 1]$.

Hop from Game_1 *to* Game_2. The change between the two games is that the ciphertext is now no longer computed using the obfuscated circuit. Instead, it is computed as specified in the unobfuscated circuit $\mathsf{C_{SW}}[k]$. Since the input to the circuit is uniformly and independently distributed, we can bound the probability that the two computations differ by the correctness of the sacO. I.e. it holds that

$$|\Pr[\mathsf{Game}_1(n) = 1] - \Pr[\mathsf{Game}_2(n) = 1]| \leq \epsilon(|C|, n). \qquad (17)$$

Hop from Game_2 *to* Game_3. The change between the two games is that the bit-string r' is no longer the output of a PRG and instead a uniformly chosen random string. We can thus bound the difference between the two games using the security of the pseudorandom generator. I.e., we can construct a distinguisher $\mathcal{D}$ with advantage $|\Pr[\mathsf{Game}_2(n) = 1] - \Pr[\mathsf{Game}_3(n) = 1]|$ as follows

$$\underline{\mathcal{D}(r')}$$

$k \leftarrow_\$ \{0, 1\}^n$

$\mathsf{pk} \leftarrow_\$ \mathsf{O}(C_{\mathsf{SW}}[k])$

$b \leftarrow_\$ \{0, 1\}$

$c := (b \oplus \mathsf{PRF}(k, r'), r')$

$b' \leftarrow_\$ \mathcal{A}(\mathsf{pk}, c)$

return $(b' \overset{?}{=} b)$

Observe, that in the case where $\mathcal{D}$ receives the output of the PRG, it holds that

$$\Pr_{r,\mathcal{D}}[\mathcal{D}(\mathsf{PRG}(r)) = 1] = \Pr[\mathsf{Game}_2(n) = 1]. \qquad (18)$$

If on the other hand, $\mathcal{D}$ receives an r' chosen uniformly at random, then it holds that

$$\Pr_{r',\mathcal{D}}[\mathcal{D}(r') = 1] = \Pr[\mathsf{Game}_3(n) = 1]. \qquad (19)$$

By definition of a secure PRG, there further exists a negligible function $\mathsf{negl}(n)$, such that

$$|\Pr_{r,\mathcal{D}}[\mathcal{D}(\mathsf{PRG}(r)) = 1] - \Pr_{r',\mathcal{D}}[\mathcal{D}(r') = 1]| \leq \mathsf{negl}(n).$$

Combining this with Eqs. 18 and 19, we get

$$|\Pr[\mathsf{Game}_2(n) = 1] - \Pr[\mathsf{Game}_3(n) = 1]| \leq \mathsf{negl}(n)(n). \qquad (20)$$

Hop from Game_3 *to* Game_4. In this hop, the obfuscated circuit is replaced. It is critical to observe, that if r' is *not in the range* of PRG, then the two circuits are functionally equivalent, since the PRF will never be invoked on the point the key is punctured on. In this case, the distance between the two games can therefore be bounded by the security of the sacO. If r' *is in the range* of PRG, then we have no guarantee, but this only occurs with probabilty $2^{-n/2}$. Thus it follows that

$$|\Pr[\mathsf{Game}_3(n) = 1] - \Pr[\mathsf{Game}_4(n) = 1]| \leq \delta(|C|, n) + 2^{-n/2}. \qquad (21)$$

Hop from Game_4 *to* Game_5. Note that in Game_5, the PRF value is replaced with a uniformly chosen random value. This allows us to bound the difference between the two games by the security of the puncturable pseudorandom function. To bound the difference between games Game_4 and Game_5, we construct a distinguisher $(\mathcal{D}_1, \mathcal{D}_2)$ with advantage

$$|\Pr[\mathsf{Game}_4(n) = 1] - \Pr[\mathsf{Game}_5(n) = 1]|$$

against the puncturable PRF as follows:

$\mathcal{D}_1(1^n; r_1)$	$\mathcal{D}_2(\mathsf{st}, k^*, r', s; r_2)$
$r' \leftarrow_\$ \{0,1\}^n$	$\mathsf{pk} \leftarrow_\$ O(C_{\mathsf{SW}}[k^*])$
$\mathbf{return}\ (\bot, r')$	$b \leftarrow_\$ \{0,1\}$
	$c := (b \oplus s)$
	$b' \leftarrow_\$ \mathcal{A}(\mathsf{pk}, c)$
	$\mathbf{return}\ (C'(x_0) \overset{?}{=} b)$

Observe, that in the case where $\mathcal{A}_2$ receives the PRF value, it holds that

$$\Pr_{k,r_1,t,r_2}[\mathcal{D}_2(\mathsf{st}, k^*, r', \mathsf{PRF}(k, r'); r_2) = 1] = \Pr[\mathsf{Game}_4(n) = 1]. \qquad (22)$$

If on the other hand, $\mathcal{D}_2$ receives an s chosen uniformly at random, it holds that

$$\Pr_{k,s,r_1,t,r_2}[\mathcal{D}_2(\mathsf{st}, k^*, r', s; r_2) = 1] = \Pr[\mathsf{Game}_5(n) = 1] \qquad (23)$$

By security of the puncturable PRF, it must hold that there exists a negligible function $\mathsf{negl}(n)$ such that

$$|\Pr_{k,r_1,t,r_2}[\mathcal{D}_2(\mathsf{st}, k^*, r', \mathsf{PRF}(k, r'); r_2) = 1]$$
$$- \Pr_{k,s,r_1,t,r_2}[\mathcal{D}_2(\mathsf{st}, k^*, r', s; r_2) = 1]| \leq \mathsf{negl}(n)$$

Combining this with Eqs. 22 and 23 yields

$$|\Pr[\mathsf{Game}_5(n) = 1] - \Pr[\mathsf{Game}_4(n) = 1]| \leq \mathsf{negl}(n) \qquad (24)$$

It remains to bound the probability $\Pr[\mathsf{Game}_5(n) = 1]$. However, the ciphertext in Game_5 is simply a uniformly distributed random value that does not reveal any information about b. Therefore, it is easy to see that $\Pr[\mathsf{Game}_5(n) = 1] = \frac{1}{2}$. Combining this with Eqs. 17, 20, 21, and 24, we can conclude that

$$\Pr[\mathsf{Game}_1(n) = 1] \leq \frac{1}{2} + \delta(|C|, n) + \epsilon(|C|, n),$$

thus concluding the proof.

References

1. Barak, B., Goldreich, O., Impagliazzo, R., Rudich, S., Sahai, A., Vadhan, S.P., Yang, K.: On the (im)possibility of obfuscating programs. In: Kilian, J. (ed.) CRYPTO 2001. LNCS, vol. 2139, pp. 1–18. Springer, Heidelberg (2001)
2. Barak, B., Goldreich, O., Impagliazzo, R., Rudich, S., Sahai, A., Vadhan, S.P., Yang, K.: On the (im)possibility of obfuscating programs. J. ACM **59**(2), 6 (2012)

3. Bitansky, N., Paneth, O.: On the impossibility of approximate obfuscation and applications to resettable cryptography. In: Boneh, D., Roughgarden, T., Feigenbaum, J. (eds.) 45th Annual ACM Symposium on Theory of Computing, Palo Alto, CA, USA, 1–4 June 2013, pp. 241–250. ACM Press (2013)
4. Bitansky, N., Vaikuntanathan, V.: Indistinguishability obfuscation: from approximate to exact. In: Kushilevitz, E., Malkin, T. (eds.) TCC 2016-A. LNCS, vol. 9562, pp. 67–95. Springer, Heidelberg (2016). doi:10.1007/978-3-662-49096-9_4
5. Bogdanov, A., Lee, C.H.: Limits of provable security for homomorphic encryption. In: Canetti, R., Garay, J.A. (eds.) CRYPTO 2013, Part I. LNCS, vol. 8042, pp. 111–128. Springer, Heidelberg (2013)
6. Boneh, D., Waters, B.: Constrained pseudorandom functions and their applications. In: Sako, K., Sarkar, P. (eds.) ASIACRYPT 2013, Part II. LNCS, vol. 8270, pp. 280–300. Springer, Heidelberg (2013)
7. Boyle, E., Goldwasser, S., Ivan, I.: Functional signatures and pseudorandom functions. In: Krawczyk, H. (ed.) PKC 2014. LNCS, vol. 8383, pp. 501–519. Springer, Heidelberg (2014)
8. Canetti, R., Kalai, Y.T., Paneth, O.: On Obfuscation with random oracles. In: Dodis, Y., Nielsen, J.B. (eds.) TCC 2015, Part II. LNCS, vol. 9015, pp. 456–467. Springer, Heidelberg (2015)
9. Diffie, W., Hellman, M.E.: Multiuser cryptographic techniques. In: American Federation of Information Processing Societies, 1976 National Computer Conference. AFIPS Conference Proceedings, New York, NY, USA, 7–10 June 1976, vol. 45, pp. 109–112. AFIPS Press (1976)
10. Garg, S., Gentry, C., Halevi, S., Raykova, M., Sahai, A., Waters, B.: Candidate indistinguishability obfuscation and functional encryption for all circuits. In: 54th Annual Symposium on Foundations of Computer Science, Berkeley, CA, USA, 26–29 October 2013, pp. 40–49. IEEE Computer Society Press (2013)
11. Goldreich, O.: Computational Complexity - A Conceptual Perspective. Cambridge University Press, Cambridge (2008)
12. Goldreich, O., Goldwasser, S., Micali, S.: How to construct random functions (extended abstract). In: 25th Annual Symposium on Foundations of Computer Science, Singer Island, Florida, 24–26 October 1984, pp. 464–479. IEEE Computer Society Press (1984)
13. Goldreich, O., Goldwasser, S., Micali, S.: How to construct random functions. J. ACM **33**(4), 792–807 (1986)
14. Goldwasser, S., Rothblum, G.N.: On best-possible obfuscation. In: Vadhan, S.P. (ed.) TCC 2007. LNCS, vol. 4392, pp. 194–213. Springer, Heidelberg (2007)
15. Goldwasser, S., Rothblum, G.N.: On best-possible obfuscation. J. Cryptology **27**(3), 480–505 (2014)
16. Hada, S., Sakurai, K.: A note on the (im)possibility of using obfuscators to transform private-key encryption into public-key encryption. In: Miyaji, A., Kikuchi, H., Rannenberg, K. (eds.) IWSEC 2007. LNCS, vol. 4752, pp. 1–12. Springer, Heidelberg (2007)
17. Håstad, J., Impagliazzo, R., Levin, L.A., Luby, M.: A pseudorandom generator from any one-way function. SIAM J. Comput. **28**(4), 1364–1396 (1999)
18. Holenstein, T.: Strengthening Key Agreement Using Hard-Core Sets. Ph.D. thesis, ETH Zurich (2006)
19. Impagliazzo, R., Luby, M.: One-way functions are essential for complexity based cryptography (extended abstract). In: 30th Annual Symposium on Foundations of Computer Science, Research Triangle Park, North Carolina, 30 October - 1 November 1989, pp. 230–235. IEEE Computer Society Press (1989)

20. Impagliazzo, R., Rudich, S.: Limits on the provable consequences of one-way permutations. In: 21st Annual ACM Symposium on Theory of Computing, Seattle, Washington, USA, 15–17 May 1989, pp. 44–61. ACM Press (1989)

21. Impagliazzo, R., Rudich, S.: Limits on the provable consequences of one-way permutations. In: Goldwasser, S. (ed.) CRYPTO 1988. LNCS, vol. 403, pp. 8–26. Springer, Heidelberg (1990)

22. Kiayias, A., Papadopoulos, S., Triandopoulos, N., Zacharias, T.: Delegatable pseudorandom functions and applications. In: Sadeghi, A.-R., Gligor, V.D., Yung, M. (eds.) ACM CCS 13, 20th Conference on Computer and Communications Security, Berlin, Germany, 4–8 November 2013, pp. 669–684. ACM Press (2013)

23. Komargodski, I., Moran, T., Naor, M., Pass, R., Rosen, A., Yogev, E.: One-way functions and (im)perfect obfuscation. In: 55th Annual Symposium on Foundations of Computer Science, Philadelphia, PA, USA, 18–21 October 2014, pp. 374–383. IEEE Computer Society Press (2014)

24. Lin, H., Pass, R., Seth, K., Telang, S.: Output-compressing randomized encodings and applications. Cryptology ePrint Archive, Report 2015/720 (2015). http://eprint.iacr.org/2015/720

25. Mahmoody, M., Mohammed, A., Nematihaji, S.: On the impossibility of virtual black-box obfuscation in idealized models. In: Kushilevitz, E., Malkin, T. (eds.) TCC 2016-A. LNCS, vol. 9562, pp. 18–48. Springer, Heidelberg (2016). doi:10.1007/978-3-662-49096-9_2

26. Mahmoody, M., Mohammed, A., Nematihaji, S., Pass, R., Shelat, A.: Lower bounds on assumptions behind indistinguishability obfuscation. In: Kushilevitz, E., Malkin, T. (eds.) TCC 2016-A. LNCS, vol. 9562, pp. 49–66. Springer, Heidelberg (2016). doi:10.1007/978-3-662-49096-9_3

27. Mahmoody, M., Xiao, D.: On the power of randomized reductions and the checkability of SAT. In: Proceedings of the 25th Annual IEEE Conference on Computational Complexity, CCC 2010, Cambridge, Massachusetts, 9–12 June 2010, pp. 64–75. IEEE Computer Society (2010)

28. Pass, R., Shelat, A.: Impossibility of VBB obfuscation with ideal constant-degree graded encodings. In: Kushilevitz, E., Malkin, T. (eds.) TCC 2016-A. LNCS, vol. 9562, pp. 3–17. Springer, Heidelberg (2016). doi:10.1007/978-3-662-49096-9_1

29. Sahai, A., Vadhan, S.P.: A complete promise problem for statistical zero-knowledge. In: 38th Annual Symposium on Foundations of Computer Science, Miami Beach, Florida, 19–22 October 1997, pp. 448–457. IEEE Computer Society Press (1997)

30. Sahai, A., Waters, B.: How to use indistinguishability obfuscation: deniable encryption, and more. In: Shmoys, D.B. (ed.) 46th Annual ACM Symposium on Theory of Computing, New York, NY, USA, 31 May - 3 June 2014, pp. 475–484. ACM Press (2014)

31. Leslie, G.: Valiant.: a theory of the learnable. Commun. ACM **27**(11), 1134–1142 (1984)

32. Valiant, L.G., Vazirani, V.V.: NP is as easy as detecting unique solutions. In: Sedgewick, R. (ed.) 17th Annual ACM Symposium on Theory of Computing, Providence, Rhode Island, USA, 6–8 May 1985, pp. 458–463. ACM Press (1985)

Revisiting the Cryptographic Hardness of Finding a Nash Equilibrium

Sanjam Garg[1], Omkant Pandey[2], and Akshayaram Srinivasan[1(✉)]

[1] University of California, Berkeley, USA
{sanjamg,akshayaram}@berkeley.edu
[2] Stony Brook University, Brookhaven, USA
omkant@gmail.com

Abstract. The exact hardness of computing a Nash equilibrium is a fundamental open question in algorithmic game theory. This problem is complete for the complexity class PPAD. It is well known that problems in PPAD cannot be NP-complete unless NP = coNP. Therefore, a natural direction is to reduce the hardness of PPAD to the hardness of problems used in cryptography.

Bitansky, Paneth, and Rosen [FOCS 2015] prove the hardness of PPAD assuming the existence of quasi-polynomially hard indistinguishability obfuscation and sub-exponentially hard one-way functions. This leaves open the possibility of basing PPAD hardness on simpler, polynomially hard, computational assumptions.

We make further progress in this direction and reduce PPAD hardness directly to polynomially hard assumptions. Our first result proves hardness of PPAD assuming the existence of *polynomially hard* indistinguishability obfuscation ($i\mathcal{O}$) and one-way permutations. While this improves upon Bitansky et al.'s work, it does not give us a reduction to simpler, polynomially hard computational assumption because constructions of $i\mathcal{O}$ inherently seems to require assumptions with sub-exponential hardness. In contrast, *public key functional encryption* is a much simpler primitive and does not suffer from this drawback. Our second result shows that PPAD hardness can be based on *polynomially hard* compact public key functional encryption and one-way permutations. Our results further demonstrate the power of polynomially hard compact public key functional encryption which is believed to be weaker than indistinguishability obfuscation. Our techniques are general and we expect them to have various applications.

1 Introduction

The problem of computing a *Nash equilibrium* is fundamental to algorithmic game theory. The hardness of this problem has attracted significant attention. Since a mixed Nash equilibrium is guaranteed to exist for every game [Nas51], the problem belongs to the complexity class TFNP [MP91]. In a series of works, originating from Papadimitriou [Pap94], the problem was established to be complete for the complexity class PPAD [DGP09, CDT09]. PPAD is a subclass of

M. Robshaw and J. Katz (Eds.): CRYPTO 2016, Part II, LNCS 9815, pp. 579–604, 2016.
DOI: 10.1007/978-3-662-53008-5_20

TFNP containing problems that reduce (in polynomial time) to a special problem called as END-OF-LINE (or EOL in short). Informally, EOL instance includes a "succinct" description of an exponential sized directed graph where each node has in-degree and out-degree at most 1 and a source node having in-degree 0 and out-degree 1. The goal is to find another source or a sink (having in-degree 1 and out-degree 0). It is easy to observe that such a node is guaranteed to exist by a simple parity argument.

The exact hardness of this problem, however, is still not fully understood. Since the class PPAD is *total*, it is unlikely to contain NP-complete problems unless polynomial hierarchy collapses to the first level [MP91, Pap94]. This is similar to the status of hardness assumptions in cryptography which are not believed to be NP-complete, but nevertheless, hard. Due to this similarity, cryptographic problems were suggested as natural candidates in [Pap94] for studying the hardness of PPAD. Indeed, the hardness of some total super-classes of PPAD, such as PPA and PPP, can already be reduced to "standard" cryptographic problems like factoring and collision-resistant hashing [Jer12]. However, such a reduction is not known for PPAD.

A natural extension of this idea is to consider cryptographic problems with a richer and more powerful structure. One of the richest cryptographic structure is *program obfuscation* as formulated by Barak et al. [BGI+12]. It is a compiler to transform any computer program into an "unintelligible one" while preserving its functionality. Ideally, the obfuscation of a program should be a "virtual black-box" (VBB), i.e., access to the obfuscated program should be no better than access to a black-box implementing the program [BGI+12]. Abbot et al. [AKV04] show that PPAD-hardness can be based on VBB obfuscation of a natural pseudo random function. Unfortunately, VBB obfuscation is impossible in general [BGI+12], and there are strong limitations to obfuscating pseudorandom functions [GK05, BCC+14], including the one in [AKV04].

A natural relaxation of VBB obfuscation is *indistinguishability obfuscation* ($i\mathcal{O}$) [BGI+12]. Informally, $i\mathcal{O}$ guarantees that the obfuscation of a circuit looks indistinguishable from the obfuscation of any other, functionally equivalent, circuit of same size. Starting from the work of Garg et al. [GGH+13b], several candidate constructions [BR14, BGK+14, PST14, GLSW15, Zim15, AB15, GMS16] for $i\mathcal{O}$ have been suggested based on various assumptions on multilinear maps [GGH13a] and public key functional encryption [AJ15, BV15a, AJS15].

Motivated by the progress on obfuscation, Bitansky et al. [BPR15] revisit the hardness of PPAD and provide an elegant reduction to the hardness of $i\mathcal{O}$. This is the first reduction of its kind which reduces PPAD-hardness to the security of a concrete and plausible cryptographic primitive. This, together with the progress on $i\mathcal{O}$, gives hope to the possibility of basing PPAD-hardness on simpler, more standard cryptographic primitives.

1.1   Our Contribution

In this work, we revisit the problem of reducing PPAD-hardness to rich and expressive cryptographic systems. We build upon the work of [BPR15] with two specific goals:

- **Rely on polynomial-hardness of** $i\mathcal{O}$**:** One drawback of the BPR reduction is that it requires $i\mathcal{O}$ schemes with at least quasi-polynomial security. It is not clear if such a large loss in the reduction is necessary. Our first goal is to obtain an improved, polynomial time reduction.
- **Rely on simpler, polynomially hard, assumptions:** While tremendous progress has been made on justifying the security of current $i\mathcal{O}$ schemes, ultimately the security of the resulting constructions still either relies on an exponential number of assumptions (basically, one per pair of circuits), or a polynomial set of assumptions with exponential loss in the reduction. Our second goal is thus to completely get rid of $i\mathcal{O}$ or any other component with non-polynomial time flavor, and reduce PPAD-hardness to simpler, polynomially hard, assumptions.

With respect to our first goal, we prove the following theorem:

Theorem 1. *Assuming the existence of polynomially hard one-way permutations and indistinguishability obfuscation for* P/poly, *the* END-OF-LINE *problem is hard for polynomial-time algorithms.*

This polynomially reduces the hardness of PPAD to $i\mathcal{O}$ since PPAD is the class of problems that are reducible to the END-OF-LINE problem.
With respect to our second goal, we show that PPAD-hardness can be reduced to the security of *compact* public-key *functional encryption* ($\mathcal{FE}$) in polynomial time. We note that polynomially hard public key functional encryption is a polynomially falsifiable assumption [Nao03].
A public key functional encryption ($\mathcal{FE}$) scheme for general circuits [BSW11, O'N10] is similar to an ordinary (public-key) encryption scheme with the crucial difference that there are many decryption keys, each of which has an associated function f; when an encryption of a message m is decrypted with a key for function f, it decrypts to the value $f(m)$. The intuitive security guarantee is that given the secret key corresponding to f and a ciphertext encrypting m, an adversary would not be able to get any information about m except $f(m)$. Our second result proves the following theorem:

Theorem 2. *Assuming the existence of polynomially-hard one-way permutations and compact public key functional encryption for general circuits, the* END-OF-LINE *problem is hard for polynomial-time algorithms.*

Compact functional encryption, as demonstrated by the recent results of Bitansky and Vaikuntanathan [BV15b] and Ananth et al. [AJS15], can be generically constructed from the so called "collusion-resistant function encryption with collusion-succinct ciphertexts", which in turn can be constructed from simpler polynomial hardness assumptions over multi-linear maps, as shown by Garg et al. [GGHZ16]. This is in sharp contrast to $i\mathcal{O}$ where all constructions still inherently seem to require exponential loss in the security reduction[1]. Combined with the results of [GGHZ16,BV15b,AJS15], Theorem 2 bases PPAD-hardness on simpler polynomial hardness assumptions. It is interesting to note that compact public key functional encryption implies indistinguishability obfuscators [AJ15,BV15a] but with sub-exponential security loss.

1.2   Our Techniques

We now present a technical overview of our approach. Building upon the work of [BPR15], it suffices to show a sampling procedure that samples hard instances of SINK-OF-VERIFIABLE-LINE problem. We will first show how to generate such instances using polynomially-hard $i\mathcal{O}$ and then discuss how to do the same using polynomially-hard $\mathcal{FE}$.

PPAD Hardness from Indistinguishability Obfuscation. Let us start by recalling the definition of PPAD. The class PPAD is defined to be the set of all *total* search problems that are polynomial time reducible to the END-OF-LINE (EOL) problem. Intuitively, an EOL instance includes a *succinct* description of an exponential sized directed graph with each node having in-degree and out-degree at most 1. Given a source node (which has in-degree 0 and out-degree 1), the goal is to find another source or a sink (which has in-degree 1 and out-degree 0). By a simple parity argument one can observe that such a node is guaranteed to exist.

The hardness of PPAD was proven in [BPR15] by considering a different problem, proposed in [AKV04], called SINK-OF-VERIFIABLE-LINE problem (SVL) in [BPR15]. It was shown that SVL reduces to the EOL problem [AKV04,BPR15], and therefore hardness of SVL implies hardness of EOL and PPAD.

An instance of the SVL problem is specified by a tuple $(x_s, \mathsf{Succ}, \mathsf{Ver}, T)$ where x_s is called the source node, Succ and Ver are called successor and verification circuits respectively, and T is a target index. Succ succinctly defines an (exponential sized) directed *line graph* starting from the source node x_s. That is, a node x is connected to a node y in the graph through an outgoing edge if and only if $y = \mathsf{Succ}(x)$. Ver is used to verify whether a given node is the ith node (starting from the source node x_s) on the path defined by Succ. To be more precise, $\mathsf{Ver}(x, i) = 1$ if and only if $x = \mathsf{Succ}^{i-1}(x_s)$. The goal, given the instance, is to find the T-th node (Target) on the path. We want to construct an efficiently samplable distribution over instances of SVL for which no polynomial time algorithm can find the T-th node with non-negligible probability.

BPR Approach. Bitansky et al., building upon [AKV04], consider a line graph where the i-th node is defined by the output of pseudorandom function (PRF) on i, i.e., the i-th node is (i, σ) such that $\sigma = \mathsf{PRF}_S(i)$ for a randomly chosen

---

[1] An informal explanation of this observation appears in [GLSW15].

key S. Intuitively, σ is a signature on i. The successor circuit of the hard SVL instance, Succ, is then defined by *obfuscating* a "verify and sign" circuit, VS_S, using general purpose $i\mathcal{O}$; VS_S simply outputs the next point $(i+1, \mathsf{PRF}_S(i+1))$ if the input is a valid point (i, σ) and rejects otherwise. The verification circuit Ver simply tests that a given input will not be rejected by the successor circuit. The source node is given by $(1, \mathsf{PRF}_S(1))$ and the target index T is set to a super-polynomial value in the security parameter.

Intuitively, the hardness of the above instance relies on the fact that it is impossible to obtain a signature on a node before obtaining the signature on the previous node in the path. Since T is super-polynomial in the security parameter, it follows that no polynomial time algorithm can obtain a signature on T. While the underlying idea of this reduction is intuitive, reducing its hardness to $i\mathcal{O}$ is more involved. This is shown by first changing the obfuscated circuit Succ so that it does not behave correctly on a randomly chosen point u, and simply outputs $\perp$. One can think of the Succ circuit being "punctured" at point u. This would also imply that the "punctured" circuit does not output a signature on $u+1$ unlike the original circuit. The next step uses *this fact* to "puncture" the circuit at the point $u+1$. This step is realized through the "punctured" programming approach of Sahai and Waters [SW14]. At a high level, this process is then repeated for the next point $u+2$, and then for $u+3$, and so on, until the circuit does not have the ability to sign on any point in the interval $[u, T]$. Once the circuit is "punctured" at T, it can be observed that no algorithm can find the Tth node with non-zero probability. Performing these changes however, requires more care since the number of points in $[u, T]$ is not polynomial. In hindsight, the primary reason for sub-exponential loss in this approach is because it is not possible to "puncture" a larger interval in a "single shot." In particular, to be able to use the security of $i\mathcal{O}$, this approach must increase the "punctured" interval by one point at a time.

Our Approach: Many Chains of Varying Length. Our main idea is to introduce a richer structure to the nodes in the graph, that avoids the need to increase the "punctured" interval by one point at a time. Instead, we want to make longer "jumps", sometimes of exponential length, in the proof strategy. Specifically, we aim to make only polynomially many jumps in total to travel from u to T.

In particular, instead of considering one signature per node, we consider κ signatures for every node where 2^κ is the total number of nodes on the line. That is, a node in our graph is of the form $(i, \sigma_1, \ldots, \sigma_\kappa)$ where σ_j is a signature on the first j bits of i computed using a key S_j (different for each index) for every $j \in [\kappa]$. The successor circuit is obfuscation of a program which simply checks each signature on appropriate prefixes of i, and if so, it signs all κ prefixes of $i+1$ using appropriate keys. The verification circuit is as before, the source node is simply the signatures on the first node, i.e., $(0^\kappa, \mathsf{PRF}_{S_1}(0), \ldots, \mathsf{PRF}_{S_\kappa}(0^\kappa))$, and $T = 2^\kappa - 1$. Observe that the BPR reduction is equivalent to having only σ_κ.

We now explain how this structure on the nodes helps us in achieving a polynomial loss in the reduction. As before, we start by "puncturing" the successor circuit on a random point u. To illustrate the main idea, let us assume that the binary representation of u has k trailing 1s, i.e., u is of the form: $u_1 \cdots u_{\kappa-k-1} \| 01^k$ where $1 \le k \le \kappa$. Then, $u + 1 = u_1 \cdots u_{\kappa-k-1} \| 10^k$, i.e., it has k trailing 0s. Observe that:

1. The first $\kappa - k$ prefix bits of $u + 1$ are *identical* to the first $\kappa - k$ prefix bits of *all points* in the interval $[u + 1, u + 2^k]$.
2. Signature $\sigma_{\kappa-k}$ (corresponding to the prefix of length $\kappa - k$) for the node $u + 1$ is not needed (for checking and signing) anywhere else on the line graph except for nodes in the interval $[u + 1, u + 2^k]$.

As before, suppose that we have punctured the successor circuit at a random node u. Then, the fact that the punctured circuit does not output *any* signature on $u + 1$ means that it does not output the signature $\sigma_{\kappa-k}$ on the first $\kappa - k$ bits of $u + 1$; consequently, and most importantly, this means that it does not output this signature on the first $\kappa - k$ bits of *any point in the interval* $[u + 1, u + 2^k]$. This allows us to increase the interval from $[u + 1, u + 2^k]$ by considering only a constant number of hybrids. We then repeat this process by considering $u + 2^k$ as our next point and iterate until we reach T.

Metaphorically, the signatures can be thought of as "virtual chains" emanating from each node and connecting to other nodes. The first chain coming out of a node i is connected to i's immediate neighbor which is $i + 1$. The second chain is connected to a node two hops away from i and the j-th chain is connected to a node 2^j hops away from i and so on. The number of chains coming out from a node i is one more than the number of trailing ones in the binary representation of i. Equivalently, the number of chains coming out of i is the number of bits that change from i to $i + 1$. Puncturing the circuit is viewed as cutting chains of appropriate lengths between points. While BPR strategy always cuts a chain of length 1, our proof strategy cuts the longest possible chain it can and then iterates the process again until it reaches the target T. See Fig. 1 for an illustration.

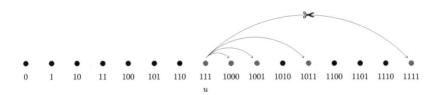

Fig. 1. Illustration of cutting a chain for $u = 0111$

While implementing the above idea we face the difficulty that for a random u the number of chains coming out of u could be very small (as small as 1). We get over this difficulty by initially cutting "smaller" length chains until we have the ability to cut "larger" length chains. Intuitively, this is made possible since the number of trailing 1 s in $u + 2^k$ is strictly larger than the number of trailing 1s (given by k) in u. We show that we need to cut no more than a linear (in the security parameter κ) number of chains to reach T and hence our reduction suffers only a polynomial (in fact linear) loss in the security parameter.

PPAD Hardness from Functional Encryption. We now give a technical overview of our hardness result for PPAD from *compact* functional encryption with *polynomial loss*. As noted earlier, although $i\mathcal{O}$ can be reduced to compact $\mathcal{FE}$ [AJ15,BV15a], we cannot directly rely on this reduction since it suffers subexponential security loss. Instead, we try to directly reduce PPAD-hardness to compact $\mathcal{FE}$.

To directly reduce PPAD-hardness to $\mathcal{FE}$, we follow the same approach as before, and generate hard on average instances of SVL using functional encryption. To demonstrate the technical challenges while proving the result from $\mathcal{FE}$ we will be considering a single PRF key, as in BPR [BPR15], instead of our idea of using κ keys to implement "multiple chains of varying length". The scenario with a single PRF key already captures the main technical challenges while keeping the exposition simple. Later, we will explain how to combine the two ideas together to obtain a direct polynomial reduction to $\mathcal{FE}$.

The line graph implicitly defined by this successor circuit will be similar to the BPR reduction as before. The successor circuit encodes a pseudo random function $\mathsf{PRF}_S : \{0,1\}^\kappa \to \{0,1\}^\kappa$ in its description. The source node is given by $(0^\kappa, \mathsf{PRF}_S(0^\kappa))$. A node (x,σ) is present on the line graph if and only if $\sigma = \mathsf{PRF}_S(x)$. The successor circuit takes as input (x,σ), checks the validity of the node and if the node is valid outputs $(x+1, \mathsf{PRF}_S(x+1))$. The target index is given by $2^\kappa - 1$.

Our goal is to produce an "obfuscated" (or encrypted) version of this successor circuit using $\mathcal{FE}$. To do this, we will rely on the "binary tree construction" idea of [AJ15,BV15a] for constructing $i\mathcal{O}$ from $\mathcal{FE}$. Note that though this reduction suffers from sub-exponential loss and we tailor the construction of our successor circuit so that it suffers only from a polynomial loss.

Binary Tree Based Evaluation [AJ15,BV15a]. Let us first recall the main ideas of [AJ15,BV15a] for constructing $i\mathcal{O}$ from $\mathcal{FE}$. We present an "over-simplified" version of their construction which is actually sufficient for our purposes but is not sufficient for achieving $i\mathcal{O}$ security.

An "obfuscation" for a circuit $C : \{0,1\}^\kappa \to \{0,1\}^*$ is a sequence of $\kappa + 1$ functional keys $\mathsf{FSK}_1, \cdots, \mathsf{FSK}_{\kappa+1}$ generated using independently sampled master secret keys $MSK_1, \cdots, MSK_{\kappa+1}$ along with a ciphertext c_ϕ encrypting the empty string under public-key PK_1 (corresponding to MSK_1). The first κ function keys implement the "bit-extension" functionality. That is, the ith function key corresponds to a function that takes in an $(i-1)$-bit string $y \in \{0,1\}^{i-1}$ and outputs functional encryptions of $y\|0$ and $y\|1$ under PK_{i+1}[2]. The function key $\mathsf{FSK}_{\kappa+1}$ corresponds to the circuit C.

To evaluate the obfuscated circuit on an input $x \in \{0,1\}^\kappa$, one does the following: decrypt c_ϕ under FSK_1 to obtain encryptions of 0 and 1. Depending

---

[2] The randomness needed for generating the encryptions is obtained using a PRF.

on the bit x_1, choose either the left or right encryption and decrypt it using FSK_2 and so on. Thus, in κ steps one can obtain an encryption of x under $PK_{\kappa+1}$ which can be used to compute $C(x)$ using $\mathsf{FSK}_{\kappa+1}$. One can think of the construction as having a binary tree structure where evaluating the circuit on an input x corresponds to traversing along the path labeled x.

Sub-exponential Loss. An intuitive reason for why this construction requires sub-exponential loss to achieve $i\mathcal{O}$ is that the behavior of the obfuscated circuit should be changed on all κ-bit inputs which are 2^κ in number. The key insight in our reduction is that we can achieve our goals by changing the behavior of the obfuscated circuit at only polynomial many inputs and thus incurring only a polynomial security loss.

Our Construction. We will motivate our construction through a series of attempts and fixes.

First Attempt. Our first attempt was to mimic the construction of [AJ15, BV15a]. We generate $2\kappa + 1$ functional keys $\mathsf{FSK}_1, \cdots, \mathsf{FSK}_{2\kappa+1}$ where the first 2κ of them correspond to the bit-extension function used for encrypting (x, σ) under $PK_{2\kappa+1}$ and $\mathsf{FSK}_{2\kappa+1}$ corresponds to the circuit Next that checks the validity of the node (x, σ) and outputs the next node in the graph if (x, σ) is valid. The main question with this approach is: How does the circuit Next check the validity of the input node and output the next node in the path? The circuit Next must somehow have access to the PRF key S but this access should not be "visible" to the outside world.

We definitely cannot hardwire the PRF key S in the circuit as the current constructions of public key functional encryption schemes do not provide any meaningful notions of "function-privacy". One possible approach is to "propagate" the key S along the entire tree. That is, encrypt the key S in the ciphertext c_ϕ and the bit extension functions output encryptions that also includes S. Though this approach sounds promising, we are unable to use the "punctured" programming techniques of Sahai and Waters that were crucial in the reduction of PPAD hardness to $i\mathcal{O}$. In particular, to puncture the key S at a point x we need to puncture the key along every path thus incurring a sub-exponential loss that we wanted to avoid. To fix this issue, we develop "fine-grained" puncturing techniques.

Second Attempt: "Prefix Puncturing." To solve the problem explained earlier, we develop techniques to "surgically" puncture the PRF key S along a path x without affecting the distribution on rest of the paths. We now explain the details.

Every string $y \in \{0, 1\}^{\leq \kappa}$ has a natural association with a node in the binary tree where the root is associated with the empty string ϕ. At a high level, we want the set of keys K_y appearing in node y to have the following properties:

– The keys derived from K_y can be used for checking the validity of every node in the subtree rooted at y. This translates to be able to compute the PRF value at x for every (x, σ) that appears in the subtree rooted at y. We denote this property as *prefix puncturability*.

– The keys derived from K_y can be used for computing the next node for every node in the subtree rooted at y. This would translate to the ability to compute the PRF value at $x + 1$ for every (x, σ) appearing at the subtree rooted at y.

A pseudorandom function that has a natural binary tree structure and has the prefix-puncturable property is the construction due to Goldreich et al. [GGM86]. We exploit this property in the GGM construction to propagate the "prefix-punctured" keys along the binary tree.

At every node $y \in \{0, 1\}^{\leq \kappa}$, we propagate two keys S_y, S_{y+1} where S_y denotes the key S prefix-punctured at string y. Intuitively, S_y is the key used for checking the input node is valid and S_{y+1} is used for generating the next node on the path[3]. The bit extension function generates $S_{y\|0}, S_{y\|0+1}$ and $S_{y\|1}, S_{y\|1+1}$ from S_y, S_{y+1} and propagates these values along with $y\|0$ and $y\|1$ respectively. The circuit Next receives S_x, S_{x+1} where $x \in \{0, 1\}^{\kappa}$ and checks the validity of the input signature using S_x and generates the next node in the path if the input is valid using S_{x+1}.

Note that the puncturing of the keys does not happen after the level κ as by this time we have parsed the x which completely determines the key S_x, S_{x+1}. Therefore, we need to propagate S_x, S_{x+1} along the entire subtree rooted at x where we parse σ. This creates the following problem: consider a scenario where the successor circuit already outputs $\bot$ on the point x and we are trying to extend the interval to include $x + 1$. Recall that the crucial idea behind the ability to increase the interval is that S_{x+1} does not occur anywhere else in the computation of the circuit. We observe that S_{x+1} gets propagated along the entire subtree (of exponential size) rooted at x where the input σ is parsed. Hence, to "remove all traces" of S_{x+1} along the subtree rooted at x, we need to incur a sub-exponential loss.

Final Construction: "Encrypt the Next Signature." We solve the above problem by "implicitly" checking whether the given node is valid. This implicit checking is facilitated by encrypting the signature on the next node by using the signature on the current node. Intuitively, an evaluator can obtain the signature on the next node if and only if he holds a valid signature on the current node.

Instead of propagating the keys S_x, S_{x+1} in clear in the subtree parsing σ, we "cut-short" the tree at level where x is parsed. Once x is parsed (and hence we have the values S_x and S_{x+1}), we apply a length doubling injective pseudo random generator PRG on the signature S_x to obtain two halves $\mathsf{PRG}_0(S_x)$ and $\mathsf{PRG}_1(S_x)$. We encrypt S_{x+1} under $\mathsf{PRG}_1(S_x)$ and output the encryption along with $\mathsf{PRG}_0(S_x)$. The Next circuit takes $\sigma, \mathsf{PRG}_0(S_x)$ and the encrypted version

---

[3] Note that instead of S_{y+1} it is enough to propagate $S_{y+1\|0^{\kappa - |y|}}$. It is in fact crucial for our reduction that we propagate $S_{y+1\|0^{\kappa - |y|}}$ instead of S_{y+1}. But we will use S_{y+1} for ease of notation and exposition.

of S_{x+1} and checks whether $\mathsf{PRG}_0(\sigma) = \mathsf{PRG}_0(S_x)$[4] and if yes it decrypts using $\mathsf{PRG}_1(\sigma)$ to obtain S_{x+1}. Notice that now we don't run into the same problem while trying to increase the interval to include S_{x+1}. This is because we can first change S_x to a random string by relying on pseudo randomness at punctured point property of GGM PRF and then relying on semantic security of secret key encryption we can change the encryption under $\mathsf{PRG}_1(S_x)$ to some junk value. Implementing these two steps is non-trivial and we rely on "hidden trapdoor" technique of Ananth et al. [ABSV15] while generating the function keys to achieve this.

Note that we still haven't explained how the successor circuit is "punctured" at a random point in the first place. To this end, we "artificially" change the honest execution of the circuit to have a hardwired random value v and the circuit checks if $\mathsf{PRG}(x) = v$ and if so outputs $\bot$. The honest execution does not output $\bot$ for any input x with overwhelming probability since PRG has sparse images. We then change this random v to $\mathsf{PRG}(u)$ for a random u relying on the security of the PRG. A consequence of this fix is that even our honest evaluation of the successor circuit looks somewhat "artificial". This seems necessary to circumvent the sub-exponential loss incurred while constructing obfuscation from functional encryption.

Putting it All Together. To show hardness of PPAD from $\mathcal{FE}$ by incurring polynomial loss in the security reduction we need to combine the above ideas with that of "multiple-chains of varying length". As explained in the chain-cutting technique we generate κ GGM keys $S_1, \cdots, S_\kappa$. We propagate the "prefix-punctured" keys corresponding to every index $i \in [\kappa]$ along every node in the binary tree. A careful reader might have noticed that though it is necessary to check the validity of the input signatures for every prefix, it is actually sufficient to generate signatures on the next node on the path only for those bit positions that change when incrementing by 1. This is because for the rest of the bit positions that share the same prefix with the input node and we can just output those input signatures along with those newly computed ones, provided the input is valid. This observation is in fact crucial to prove the security of our construction. We need to ensure that the Next circuit must have the ability to check the validity of every signature but it has access only to those prefix punctured keys corresponding to the bit positions that change when incrementing by 1.

We satisfy these two "conflicting" properties by decoupling the process of checking the input signatures and the process of generating the next node on the path. In order to check the input signatures we propagate $\mathsf{PRG}_0(S_{i,x})$ for every $i \in [\kappa]$ and to generate the signatures on the next node on the path we propagate an encrypted version of $S_{j,x+1}$ under $\mathsf{PRG}_1(S_{j,x})$ only for those bits j that change when incrementing x.

---

[4] We need this explicit check for the verification circuit to decide if a particular node is an ith node or not. Also, we need a stronger property on pseudo random generator called as left half injectivity for this check to be correct always.

1.3   Subsequent Work

Garg et al. in [GPSZ16] extended our techniques to base Trapdoor Permutations on polynomial hardness of compact Functional Encryption. In the same work, they also showed how to base Non-Interactive Key Exchange (NIKE) for unbounded parties from polynomially hard compact Functional Encryption. Recently, Garg and Srinivasan [GS16] extended our techniques to construct adaptively secure Functional Encryption against unbounded collusions from single-key, selectively secure Functional encryption with weakly compact ciphertexts.

Rosen et al. [RSS16] investigated the possibility of basing average-case PPAD hardness on standard cryptographic assumptions. They showed that average-case PPAD hardness does not imply one-way functions in a black-box manner and average-case SVL hardness cannot be based on injective trapdoor functions in a black-box manner. An implication of this work is that it might be possible to base PPAD hardness on one-way functions but such a result has to use techniques that significantly deviate from Bitansky et al. [BPR15] and our work.

Hubáček and Yogev [HY16] extended our result to base hardness of a complexity class CLS on compact Functional Encryption. CLS is a sub-class of PPAD and captures Continuous Local Search problems. They showed a reduction between the SVL problem and a problem called as END-OF-METERED-LINE which is contained in CLS. This allowed them to base hardness of CLS on polynomially hard compact Functional Encryption.

2   PPAD

A large part of this section is taken verbatim from [BPR15]. A search problem is given by a tuple (I, R). I defines the set of instances and R is an NP relation. Given $x \in I$, the goal is to find a witness w (if it exists) such that $R(x, w) = 1$. We say that a search problem (I_1, R_1) polynomial time reduces to another search problem (I_2, R_2) if there exists polynomial time algorithms P, Q such that for every $x_1 \in I_1$, $P(x_1) \in I_2$ and given w_2 such that $(P(x_1), w_2) \in R_2$, $R_1(x_1, Q(w_2)) = 1$.

A search problem is said to be *total* if for any $x \in \{0, 1\}^*$, there exists a polynomial time procedure to test whether $x \in I$ and for all $x \in I$, the set of witnesses w such that $R(x, w) = 1$ is non-empty. The class of total search problems is denoted by TFNP. PPAD [Pap94] is a subset of TFNP and is defined by its complete problem called as END-OF-LINE (abbreviated as EOL).

Definition 1 [Pap94]. $\mathsf{EOL} = \{I_{\mathsf{EOL}}, R_{\mathsf{EOL}}\}$ *where* $I_{\mathsf{EOL}} = \{(x_s, \mathsf{Succ}, \mathsf{Pred}) : \mathsf{Succ}(x_s) \neq x_s = \mathsf{Pred}(x_s)\}$ *and* $R_{\mathsf{EOL}}((x_s, \mathsf{Succ}, \mathsf{Pred}), w) = 1$ *iff* $(\mathsf{Pred}(\mathsf{Succ}(w)) \neq w) \vee (\mathsf{Succ}(\mathsf{Pred}(w)) \neq w \wedge w \neq x_s)$.

Definition 2 [Pap94]. *The complexity class* PPAD *is the set of all search problems* (I, R) *such that* $(I, R) \in$ TFNP *and* (I, R) *polynomial time reduces to* EOL.

A related problem to EOL is the SINK-OF-VERIFIABLE-LINE (abbreviated as SVL) which is defined as follows:

Definition 3 [AKV04,BPR15]. $\mathsf{SVL} = \{I_{\mathsf{SVL}}, R_{\mathsf{SVL}}\}$ *where* $I_{\mathsf{SVL}} = \{(x_s, \mathsf{Succ},$ $\mathsf{Ver}, T)\}$ *and* $R_{\mathsf{SVL}}((x_s, \mathsf{Succ}, \mathsf{Ver}, T), w) = 1$ *iff* $\big(\mathsf{Ver}(w, T) = 1\big)$.

SVL instance defines a single directed path with the source being x_s. Succ is the successor circuit and there is a directed edge between u and v if and only if $\mathsf{Succ}(u) = v$. Ver is the *verification* circuit and is used to test whether a given node is the ith node from x_s. That is, $\mathsf{Ver}(x, i) = 1$ iff $x = \mathsf{Succ}^{i-1}(x_s)$. The goal is to find the Tth node in the path. It is easy to observe that for every *valid* SVL instance the set of witness w is not empty. But SVL may not be total since there is no known efficient procedure to test whether the instance is valid or not. But it was shown in [AKV04,BPR15] that SVL polynomial time reduces to EOL.

Lemma 1 [AKV04,BPR15]. SVL *polynomial time reduces to* EOL.

3   Preliminaries

κ denotes the security parameter. A function $\mu(\cdot) : \mathbb{N} \to \mathbb{R}^+$ is said to be negligible if for all polynomials $\mathsf{poly}(\cdot)$, $\mu(\kappa) < \frac{1}{\mathsf{poly}(\kappa)}$ for large enough κ. For a probabilistic algorithm $\mathcal{A}$, we denote by $\mathcal{A}(x; r)$ the output of $\mathcal{A}$ on input x with the content of the random tape being r. We will omit r when it is implicit from the context. We denote $y \leftarrow \mathcal{A}(x)$ as the process of sampling y from the output distribution of $\mathcal{A}(x)$ with a uniform random tape. For a finite set S, we denote $x \xleftarrow{\$} S$ as the process of sampling x uniformly from the set S. We model non-uniform adversaries $\mathcal{A} = \{\mathcal{A}_\kappa\}$ as circuits such that for all κ, $\mathcal{A}_\kappa$ is of size $p(\kappa)$ where $p(\cdot)$ is a polynomial. We will drop the subscript κ from the adversary's description when it is clear from the context. We will also assume that all algorithms are given the unary representation of security parameter 1^κ as input and will not mention this explicitly when it is clear from the context. We will use PPT to denote Probabilistic Polynomial Time algorithm. We denote $[\kappa]$ to be the set $\{1, \cdots, k\}$. We will use $\mathsf{negl}(\cdot)$ to denote an unspecified negligible function and $\mathsf{poly}(\cdot)$ to denote an unspecified polynomial.

A binary string $x \in \{0, 1\}^\kappa$ is represented as $x_1 \cdots x_\kappa$. x_1 is the most significant (or the highest order bit) and x_κ is the least significant (or the lowest order bit). The i-bit prefix $x_1 \cdots x_i$ of the binary string x is denoted by $x_{[i]}$. We use $x \| y$ to denote concatenation of binary strings x and y. We say that a binary string y is a prefix of x if and only if there exists a string $z \in \{0, 1\}^*$ such that $x = y \| z$.

Injective Pseudo Random Generator. We give the definition of an injective Pseudo Random Generator PRG.

Definition 4. *An injective pseudo random generator* PRG *is a deterministic polynomial time algorithm with the following properties:*

- **Expansion:** *There exists a polynomial* $\ell(\cdot)$ *(called as the expansion factor) such that for all* κ *and* $x \in \{0, 1\}^\kappa$, $|\mathsf{PRG}(x)| = \ell(\kappa)$.

- **Pseudo randomness:** *For all κ and for all poly sized adversaries $\mathcal{A}$,*

$$|\Pr[\mathcal{A}(\mathsf{PRG}(U_\kappa)) = 1] - \Pr[\mathcal{A}(U_{\ell(\kappa)}) = 1]| \leq \mathsf{negl}(\kappa)$$

where U_i denotes the uniform distribution on $\{0,1\}^i$.
- **Injectivity:** *For every κ and for all $x, x' \in \{0,1\}^\kappa$ such that $x \neq x'$, $\mathsf{PRG}(x) \neq \mathsf{PRG}(x')$.*

We in fact need an additional property from an injective PRG. Let us consider PRG where the expansion factor (or the output length) is given by $2 \cdot \ell(\cdot)$. Let us denote the first $\ell(\cdot)$ bits of the output of the PRG by the function PRG_0 and the next $\ell(\cdot)$ bits of the output of the PRG by PRG_1.

Definition 5. *A pseudo random generator PRG is said to be left half injective if for every κ and for all $x, x' \in \{0,1\}^\kappa$ such that $x \neq x'$. $\mathsf{PRG}_0(x) \neq \mathsf{PRG}_0(x')$.*

Note that left half injective PRG is also an injective PRG. We note that the standard construction of pseudo random generator for arbitrary polynomial stretch from one-way permutations is left half injective. For completeness, we state the construction:

Lemma 2. *Assuming the existence of one-way permutations, there exists a pseudo random generator that is left half injective.*

Proof. Let $f : \{0,1\}^\kappa \rightarrow \{0,1\}^\kappa$ be a one-way permutation with hardcore predicate $B : \{0,1\}^\kappa \rightarrow \{0,1\}$ [GL89]. Let G be an algorithm defined as follows: On input $x \in \{0,1\}^\kappa$, $G(x) = f^n(x)\|B(x)\|B(f(x))\cdots B(f^{n-1}(x))$ where $n = 2\ell(\kappa) - \kappa$. Clearly, $|G(x)| = 2\ell(\kappa)$. The pseudo randomness property of $G(\cdot)$ follows from the security of hardcore bit. The left half injectivity property follows from the observation that f^n is a permutation.

Puncturable Pseudo Random Function. We recall the notion of puncturable pseudo random function from [SW14]. The construction of pseudo random function given in [GGM86] satisfies the following definition [BW13, KPTZ13, BGI14].

Definition 6. *A puncturable pseudo random function $\mathcal{PRF}$ is a tuple of PPT algorithms $(\mathsf{KeyGen}_{\mathcal{PRF}}, \mathsf{PRF}, \mathsf{Punc})$ with the following properties:*

- **Efficiently Computable:** *For all κ and for all $S \leftarrow \mathsf{KeyGen}_{\mathcal{PRF}}(1^\kappa)$, $\mathsf{PRF}_S : \{0,1\}^{\mathsf{poly}(\kappa)} \rightarrow \{0,1\}^\kappa$ is polynomial time computable.*
- **Functionality is preserved under puncturing:** *For all κ, for all $y \in \{0,1\}^\kappa$ and $\forall x \neq y$,*

$$\Pr[\mathsf{PRF}_{S\{y\}}(x) = \mathsf{PRF}_S(x)] = 1$$

where $S \leftarrow \mathsf{KeyGen}_{\mathcal{PRF}}(1^\kappa)$ and $S\{y\} \leftarrow \mathsf{Punc}(S, y)$.
- **Pseudo randomness at punctured points:** *For all κ, for all $y \in \{0,1\}^\kappa$, and for all poly sized adversaries $\mathcal{A}$*

$$|\Pr[\mathcal{A}(\mathsf{PRF}_S(y), S\{y\}) = 1] - \Pr[\mathcal{A}(U_\kappa, S\{y\}) = 1]| \leq \mathsf{negl}(\kappa)$$

where $S \leftarrow \mathsf{KeyGen}_{\mathcal{PRF}}(1^\kappa)$, $S\{y\} \leftarrow \mathsf{Punc}(S, y)$ and U_κ denotes the uniform distribution over $\{0,1\}^\kappa$.

Indistinguishability Obfuscator. We now define Indistinguishability obfuscator from [BGI+12, GGH+13b].

Definition 7. *A PPT algorithm $i\mathcal{O}$ is an indistinguishability obfuscator for a family of circuits $\{C_\kappa\}_\kappa$ that satisfies the following properties:*

- **Correctness:** *For all κ and for all $C \in C_\kappa$ and for all x,*

$$\Pr[i\mathcal{O}(C)(x) = C(x)] = 1$$

  where the probability is over the random choices of $i\mathcal{O}$.
- **Security:** *For all $C_0, C_1 \in C_\kappa$ such that for all x, $C_0(x) = C_1(x)$ and for all poly sized adversaries $\mathcal{A}$,*

$$|\Pr[\mathcal{A}(i\mathcal{O}(C_0)) = 1] - \Pr[\mathcal{A}(i\mathcal{O}(C_1)) = 1]| \leq \mathsf{negl}(\kappa)$$

Functional Encryption. We recall the notion of functional encryption with selective indistinguishability based security [BSW11, O'N10].

A functional encryption $\mathcal{FE}$ is a tuple of PPT algorithms (FE.Setup, FE.Enc, FE.KeyGen, FE.Dec) with the message space $\{0,1\}^*$ having the following syntax:

- FE.Setup(1^κ) : Takes as input the unary encoding of the security parameter κ and outputs a public key PK and a master secret key MSK.
- FE.Enc$_{PK}(m)$: Takes as input a message $m \in \{0,1\}^*$ and outputs an encryption C of m under the public key PK.
- FE.KeyGen(MSK, f) : Takes as input the master secret key MSK and a function f (given as a circuit) as input and outputs the function key FSK_f.
- FE.Dec(FSK_f, C): Takes as input the function key FSK_f and the ciphertext C and outputs a string y.

Definition 8 (Correctness). *The functional encryption scheme $\mathcal{FE}$ is correct if for all κ and for all messages $m \in \{0,1\}^*$,*

$$\Pr\left[y = f(m) \,\middle|\, \begin{array}{l} (PK, MSK) \leftarrow \mathsf{FE.Setup}(1^\kappa) \\ C \leftarrow \mathsf{FE.Enc}_{PK}(m) \\ \mathsf{FSK}_f \leftarrow \mathsf{FE.KeyGen}(MSK, f) \\ y \leftarrow \mathsf{FE.Dec}(\mathsf{FSK}_f, C) \end{array} \right] = 1 \tag{1}$$

Definition 9 (Selective Security). *For all κ and for all poly sized adversaries $\mathcal{A}$,*

$$\left|\Pr[\mathsf{Expt}_{1^\kappa, 0, \mathcal{A}} = 1] - \Pr[\mathsf{Expt}_{1^\kappa, 1, \mathcal{A}} = 1]\right| \leq \mathsf{negl}(\kappa)$$

where $\mathsf{Expt}_{1^\kappa, b, \mathcal{A}}$ is defined below:

- **Challenge Message Queries:** *The adversary $\mathcal{A}$ outputs two messages m_0, m_1 such that $|m_0| = |m_1|$ to the challenger.*
- *The challenger samples $(PK, MSK) \leftarrow \mathsf{FE.Setup}(1^\kappa)$ and generates the challenge ciphertext $C \leftarrow \mathsf{FE.Enc}_{PK}(m_b)$. It then sends (PK, C) to $\mathcal{A}$.*

- **Function Queries**: $\mathcal{A}$ submits function queries f to the challenger. The challenger responds with $\mathsf{FSK}_f \leftarrow \mathsf{FE.KeyGen}(MSK, f)$.
- If $\mathcal{A}$ makes a query f to functional key generation oracle such that $f(m_0) \neq f(m_1)$, output of the experiment is $\bot$. Otherwise, the output is b' which is the output of $\mathcal{A}$.

Remark 1. We say that the functional encryption scheme $\mathcal{FE}$ is **single-key, selectively secure** if the adversary $\mathcal{A}$ in $\mathsf{Expt}_{1^\kappa, b, \mathcal{A}}$ is allowed to query the functional key generation oracle $\mathsf{FE.KeyGen}(MSK, \cdot)$ on a single function f.

Definition 10. (Compactness, *[AJS15, BV15a, AJ15]*)**.** *The functional encryption scheme $\mathcal{FE}$ is said to be compact if for all $\kappa \in \mathbb{N}$ and for all $m \in \{0, 1\}^*$ the running time of the encryption algorithm $\mathsf{FE.Enc}$ is $\mathsf{poly}(\kappa, |m|)$.*

Prefix Puncturable Pseudo Random Functions. We now define the notion of prefix puncturable pseudo random function PPRF which is satisfied by the construction of the pseudo random function in [GGM86].

Definition 11. *A prefix puncturable pseudo random function $\mathcal{PPRF}$ is a tuple of PPT algorithms* $(\mathsf{KeyGen}_{\mathcal{PPRF}}, \mathsf{PrefixPunc})$ *satisfying the following properties:*

- **Functionality is preserved under** repeated *puncturing: For all κ, for all $y \in \cup_{k=0}^{\mathsf{poly}(\kappa)} \{0, 1\}^k$ and for all $x \in \{0, 1\}^{\mathsf{poly}(\kappa)}$ such that there exists a $z \in \{0, 1\}^*$ s.t. $x = y \| z$,*

$$\Pr[\mathsf{PrefixPunc}(\mathsf{PrefixPunc}(S, y), z) = \mathsf{PrefixPunc}(S, x)] = 1$$

where $S \leftarrow \mathsf{KeyGen}_{\mathcal{PPRF}}(1^\kappa)$.
- **Pseudorandomness at punctured prefix:** *For all κ, for all $x \in \{0, 1\}^{\mathsf{poly}(\kappa)}$, and for all poly sized adversaries $\mathcal{A}$*

$$|\Pr[\mathcal{A}(\mathsf{PrefixPunc}(S, x), \mathsf{Keys}) = 1] - \Pr[\mathcal{A}(U_\kappa, \mathsf{Keys}) = 1]| \leq \mathsf{negl}(\kappa)$$

where $S \leftarrow \mathsf{KeyGen}_{\mathcal{PRF}}(1^\kappa)$ and $\mathsf{Keys} = \{\mathsf{PrefixPunc}(S, x_{[i-1]} \| (1 - x_i))\}_{i \in [\mathsf{poly}(\kappa)]}$.

4 Hardness from Indistinguishability Obfuscation

In this section, we prove that SVL is hard on average assuming polynomial hardness of indistinguishability obfuscation, injective PRGs and puncturable pseudo random functions. Coupled with the fact that SVL reduces to EOL (Lemma 1) we have the following theorem.

Theorem 3. *Assume the existence of one-way permutations and indistinguishability obfuscation against polynomial time adversaries then we have that EOL problem is hard for polynomial time algorithms.*

4.1   Hard on Average SVL Instances

In this section, we describe an efficient sampler that provides hard on average instances $(x_s, \mathsf{Succ}, \mathsf{Ver}, 1^\kappa)$ of SVL. Here x_s is the source node and Succ is the successor circuit. We define a directed edge between u and v if and only if $\mathsf{Succ}(u) = v$. Ver is the verification circuit and is used to test whether a given node is the kth node from x_s. That is, $\mathsf{Ver}(x, k) = 1$ iff $x = \mathsf{Succ}^{k-1}(x_s)$. For the generated instances, we argue that it is hard to find the 1^κ node in the path.

The formal description of hard on average SVL instance sampler is provided in Fig. 3. Internally this sampler generates an obfuscation of the Next circuit provided in Fig. 2. Next we describe the SVL instances which we consider informally.

The instance we generate defines a line graph. The nodes in the graph are of the form: $(x, \sigma_1, \cdots, \sigma_\kappa)$ where $x \in \{0,1\}^\kappa$. The nodes satisfy the following relation: for all $i \in [\kappa]$, $\mathsf{PRF}_{S_i}(x_{[i]}) = \sigma_i$ and in that case we say that $(x, \sigma_1, \cdots, \sigma_\kappa)$ is *valid*. The node $(x, \sigma_1, \cdots, \sigma_\kappa)$ is connected to $(x+1, \sigma_1', \cdots, \sigma_\kappa')$ through an outgoing edge and is connected to $(x-1, \sigma_1'', \cdots, \sigma_\kappa'')$ through an incoming edge where $\sigma_1', \cdots, \sigma_\kappa'$ and $\sigma_1'', \cdots, \sigma_\kappa''$ satisfy the above described PRF relationship. The source node is given by $(0^\kappa, \mathsf{PRF}_{S_1}(0), \cdots, \mathsf{PRF}_{S_\kappa}(0^\kappa))$.

At a very high level successor circuit of our SVL instances provides a method for moving forward from one node to the next. The successor circuit in our instances corresponds to an obfuscation of the Next circuit. This circuit on input a node of the form $(x, \sigma_1, \cdots, \sigma_\kappa)$ checks for the validity of the input. If it is valid, it outputs the next node $(x+1, \sigma_1' \cdots \sigma_\kappa')$ where $\sigma_i' = \mathsf{PRF}_{S_i}((x+1)_{[i]})$ in the path. On an invalid input, it outputs $\perp$.

Input: $(x, \sigma_1, \cdots, \sigma_\kappa)$
Hardcoded Parameters: $S_1, \cdots, S_\kappa$

1. For any $i \in [\kappa]$, if $\sigma_i \neq \mathsf{PRF}_{S_i}(x_{[i]})$ then output $\perp$.
2. If $x = 1^\kappa$, then output SOLVED.
3. Else output $(x+1, \sigma_1', \cdots, \sigma_\kappa')$, where for all $i \in [\kappa]$ compute $\sigma_i' = \mathsf{PRF}_{S_j}((x+1)_{[i]})$.

Padding: This circuit is padded so that total size of the circuit is $p(\kappa)$, for some polynomial $p(\cdot)$ specified later.

Fig. 2. $\mathsf{Next}_{S_1, \cdots, S_\kappa}$

For the hard SVL instances we additionally need to provide a verification circuit. The verification circuit just uses the successor circuit in a very natural manner. The verification circuit on input $(x, \sigma_1, \cdots, \sigma_\kappa, j)$ outputs 1 if and only if $x = j - 1$ and $\mathsf{Next}_{S_1, \cdots, S_\kappa}(x, \sigma_1, \cdots, \sigma_\kappa) \neq \perp$.

Due to space constraints we defer the proof of hardness to full version of this paper [GPS15].

- **Sampled Ingredients:** Sample $\{S_i\}_{i \in [\kappa]} \leftarrow \mathsf{KeyGen}_{\mathcal{PRF}}(1^\kappa)$. For all $i \in [\kappa]$, S_i is a seed for a PRF mapping i bits to κ bits. That is, $\mathsf{PRF}_{S_i} : \{0,1\}^i \rightarrow \{0,1\}^\kappa$.
- **Source Node:** The source node $x_s = (0^\kappa, \mathsf{PRF}_{S_1}(0), \cdots, \mathsf{PRF}_{S_\kappa}(0^\kappa))$.
- **Successor Circuit:** The successor circuit is given by $i\mathcal{O}(\mathsf{Next}_{S_1, \cdots, S_\kappa})$ where the circuit $\mathsf{Next}_{S_1, \cdots, S_\kappa}$ is described in Figure 2.
- **Verification Circuit:** The verification circuit, given by Ver, on input $((x, \sigma_1 \cdots \sigma_\kappa), j)$ checks if $x = j - 1$ and $i\mathcal{O}(\mathsf{Next}_{S_1, \cdots, S_\kappa})((x, \sigma_1 \cdots \sigma_\kappa)) \neq \perp$.

Fig. 3. Sampler for hard on average instances of SVL based on hardness of $i\mathcal{O}$

5   Hardness Result Based on Functional Encryption

In this section we show that SVL is hard on average assuming polynomially hard functional encryption and one-way permutations. Coupled with the fact that SVL reduces to EOL (Lemma 1) we have the following theorem.

Theorem 4. *Assume the existence of one-way permutations and functional encryption against polynomial time adversaries then we have that* EOL *problem is hard for polynomial time algorithms.*

Recall that hard SVL instance based on $i\mathcal{O}$ (Sect. 4), required κ puncturable PRF keys. Basing hardness on polynomially hard functional encryption requires us to still maintain κ keys. However, now we need to use prefix-puncturing (see Definition 11) which is more delicate and needs to be handled carefully. Consequently the construction ends up being complicated. However, the special mechanism of prefix-puncturing that we use is crucial to understanding our construction. So towards simplifying exposition, we start by abstracting out the details of this puncturing and present a special tree structure and some properties about it next.

5.1   Special Tree Key Structure

Let $x_{[i]}$ denote the first i (higher order) bits of x i.e. $x_1 \cdots x_i$. Now note that any $y \in \{0,1\}^i$ can be identified with a node in a binary tree for which nodes at depth i correspond to strings $\{0,1\}^i$. Note that the root of the tree corresponds to the empty string ϕ. As previously mentioned our construction needs κ PPRF keys, namely $S_1, \ldots S_\kappa$. The key S_i works on inputs of length i. We use $S_{i,x}$ to denote the key S_i prefix punctured at a string $x \in \{0,1\}^{\leq i}$.

Looking ahead, in our hard-on-average instances of SVL each $x \in \{0,1\}^\kappa$ will be attached with *associated signature* values $\sigma_1, \ldots, \sigma_\kappa$ where for each $i \in [\kappa]$ we have that $\sigma_i = \mathsf{PrefixPunc}(S_i, x_{[i]})$. Furthermore in our construction given x and the associated signature values, we will need to verify these values and provide the associated signature values for $x + 1$, but this has to be done in a circuitous manner because of several security reasons. We do not delve into the security arguments right away, but focus on describing the prefix-puncturing that we need to perform.

We next describe the set V_x^i where $x \in \{0,1\}^{\leq i}$, which contains suitable prefix-puncturings of the key S_i. Intuitively, we want this set to contain all keys that will allow us to perform the task of *checking the validity* of the ith associated signature on any input of the form $x\|y$ where $y \in \{0,1\}^{\kappa-|x|}$ as well as *computing* the ith associated signature for $(x\|y) + 1$. Furthermore, it should suffice to generate $V_{x\|y}^i$ for all y. For any node $x \in \{0,1\}^{\leq i}$, this very naturally translates to the keys $S_{i,x}$ and $S_{i,x+1}$. A careful reader might have noticed that instead of $S_{i,x+1}$, it in fact suffices to just have $S_{i,(x+1)\|0^{i-|x|}}$. As it turns out we must only include $S_{i,(x+1)\|0^{i-|x|}}$. Including $S_{i,x+1}$ prevents the Derivability Lemma (Lemma 4) from going through.

Recall that the key S_i corresponds to a PPRF key for inputs of length i. Therefore, for $x\|y$ such that $|x| = i$, the key S_i can be prefix-punctured only for the prefix $x = (x\|y)_{[i]}$. This raises the following question. Should we include $S_{i,x}$ and $S_{i,x+1}$ in all $V_{x\|y}^i$? As we will see later, in our construction, we carefully decouple the checking of associated signatures from the generation of new associated signatures. An important consequence, relevant here is that, even though the checks need to be performed for all $x\|y$, a new i^{th} associated signature needs to be generated for only one choice of y, namely $1^{\kappa-|x|}$ (the all 1 string of length $\kappa - |x|$). This design choice (which is crucial for polynomial security loss) also allows us to set $V_{x\|y}^i$ for all other choices of y to be $\emptyset$. In terms of the binary tree structure one can think of this as V_x^i getting passed only along the rightmost path in the subtree rooted at x. At a very high level, this allows us to argue that the key S_i (proved formally in Lemma 4) can be punctured at a special point by removing keys fron V_x^i for only a polynomial number of choices of x and i. This is crucial for ensuring that our proof of security has only a polynomial number of hybrids.

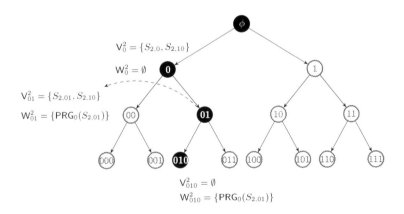

Fig. 4. Example of values contained in V_x^2 for $x \in \{0,1\}^{\leq 3}$.

Next note that dropping keys from $V^i_{x\|y}$ (such that $|x| = i$) hinders the checking of associated signatures provided along with inputs $x\|y$ where $y \neq 1^{\kappa-i}$. We tackle this issue by introducing a vestigial set $W^i_{x\|y}$ corresponding to each $V^i_{x\|y}$. This vestigial set contains remnants of the keys that were dropped from V^i_x. We craft these remnants to be such that they suffice for performing the necessary checks. In particular, we set these remnants to be the left half of an left half injective PRG evaluation on the dropped key.

More formally, V^i_x and V_x are defined as follows. In the following, for any $i \in [\kappa]$ we treat $1^i + 1$ as 1^i, and $\phi + 1$ as ϕ. Here 1^i is a string of i 1s and ϕ is the empty string.

$$V_x = \bigcup_{i\in[\kappa]} V^i_x \qquad V^i_x = \begin{cases} \{S_{i,x_{[i]}}, S_{i,x_{[i]}+1}\} & \text{if } |x| > i \text{ and } x = x_{[i]}\|1^{|x|-i} \\ \{S_{i,x}, S_{i,(x+1)\|0^{i-|x|}}\} & \text{if } |x| \leq i \\ \emptyset & \text{otherwise} \end{cases}$$

$$W_x = \bigcup_{i\in[\kappa]} W^i_x \qquad W^i_x = \begin{cases} \{\mathsf{PRG}_0(S_{i,x_{[i]}})\} & \text{if } |x| \geq i \\ \emptyset & \text{otherwise} \end{cases}$$

For the empty string $x = \phi$, these sets can be initialized as follows.

$$V_\phi = \bigcup_{i\in[\kappa]} V^i_\phi \qquad V^i_\phi = \{S_i\}$$

$$W_\phi = \bigcup_{i\in[\kappa]} W^i_\phi \qquad W^i_\phi = \emptyset$$

Illustration with an Example. Finally we explain what sets V^2_x, W^2_x contain when x is a prefix of 010 in Fig. 4. At the root node we have $V^2_\phi = \{S_2\}$ and $W_\phi = \emptyset$. The set V^2_0 contains $S_{2,0}$ and $S_{2,10}$ and the set W^2_0 is still empty. Next note that V^2_{01} contains $S_{2,01}, S_{2,10}$ and W^2_{01} contains $\mathsf{PRG}_0(S_{2,01})$. Finally set $V^2_{010} = \emptyset$ and W^2_{010} continues to contain $\mathsf{PRG}_0(S_{2,01})$.

Properties of the Special Tree Key Structure. We now prove several properties about the special tree key structure. Intuitively speaking the crux of the lemmas is the claim V-set for can a node can be used to derive its children. Furthermore each element in V-set for any node can only be derived from the V-set of nodes in exactly two different paths.

Lemma 3 (Computability Lemma). *There exists an explicit efficient procedure that given V_x, W_x computes $V_{x\|0}, W_{x\|0}$ and $V_{x\|1}, W_{x\|1}$.*

Proof. We start by noting that it suffices to show that for each i, given V^i_x, W^i_x one can compute $V^i_{x\|0}, W^i_{x\|0}$ and $V^i_{x\|1}, W^i_{x\|1}$. We argue this next. Observe that two cases arise either $|x| < i$ or $|x| \geq i$. We deal with the two cases:

- $|x| < i$: In this case V_x^i is $\{S_{i,x}, S_{i,(x+1)\|0^{i-|x|}}\}$ and these values can be used to compute $S_{i,x\|0}$, $S_{i,x\|1}$, $S_{i,(x\|0)+1} = S_{i,x\|1}$ and $S_{i,((x\|1)+1)\|0^{i-|x|-1}} = S_{i,(x+1)\|0\|0^{i-|x|-1}} = S_{i,(x+1)\|0^{i-|x|}}$. Observe by case by case inspection that these values are sufficient for computing $V_{x\|0}^i, W_{x\|0}^i$ and $V_{x\|1}^i, W_{x\|1}^i$ in all cases.

- $|x| \geq i$: Note that according to the constraints placed on x by the definition, if $V_x^i = \emptyset$ then both $V_{x\|0}^i$ and $V_{x\|1}^i$ must be $\emptyset$ as well. On the other hand if $V_x^i \neq \emptyset$ then $V_{x\|0}^i$ is still $\emptyset$ while $V_{x\|1}^i = V_x^i$. Additionally, $W_{x\|0}^i = W_{x\|1}^i = W_x^i$.

This concludes the proof.

Lemma 4 (Derivability Lemma). *For every $i \in [\kappa], x \in \{0,1\}^i$ and $x \neq 1^i$ we have that, $S_{i,x+1}$ can be derived from keys in V_y^i if and only if y is a prefix of $x\|1^{\kappa-i}$ or $(x+1)\|1^{\kappa-i}$. Additionally, $S_{i,0^i}$ can be derived from keys in V_y if and only if y is a prefix of $0^i\|1^{\kappa-i}$ (Fig. 5).*

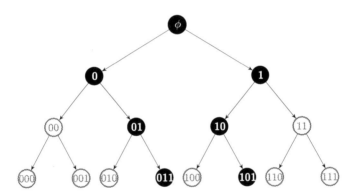

Fig. 5. Black nodes represent the choices of $x \in \{0,1\}^{\leq 3}$ such that V_x^2 can be used to derive $S_{2,10}$.

Proof. We start by noting that for any $y \in \{0,1\}^{>i} \cap \{0,1\}^{\leq \kappa}$, by definition of V-sets we have that $V_y^i = V_{y[i]}^i$ or $V_y^i = \emptyset$. Hence it suffices to prove the above lemma for $y \in \{0,1\}^{\leq i}$.

We first prove that if y is a prefix of x or $(x+1)$ then we can derive $S_{i,x+1}$ from V_y^i. Two cases arise:

- Observe that if y is a prefix of x then we must have that either y is a prefix of $x+1$ or $x+1 = (y+1)\|0^{i-|y|}$. Next note that by definition of V-sets we have that $V_y^i = \{S_{i,y}, S_{i,(y+1)\|0^{i-|y|}}\}$, and one of these values can be used to compute $S_{i,x+1}$.

- On the other hand if y is a prefix of $x+1$ then again by definition of V-sets we have that $V_y^i = \{S_{i,y}, S_{i,(y+1)\|0^{i-|y|}}\}$, and $S_{i,y}$ can be used to compute $S_{i,x+1}$.

Next we show that no other $y \in \{0,1\}^{\leq i}$ allows for such a derivation. Note that by definition of V-sets we have that $V_y^i = \{S_{i,y}, S_{i,(y+1)\|0^{i-|y|}}\}$. We will argue that neither $S_{i,y}$ nor $S_{i,(y+1)\|0^{i-|y|}}$ can be used to derive $S_{i,x+1}$.

- We are given that y is not a prefix of $x + 1$. This implies that $S_{i,y}$ cannot be used to derive $S_{i,x+1}$.
- Now we need to argue that $S_{i,(y+1)\|0^{i-|y|}}$ cannot be used to compute $S_{i,x+1}$. For this, it suffices to argue that $x+1 \neq (y+1)\|0^{i-|y|}$. If $x+1 = (y+1)\|0^{i-|y|}$ then y must be prefix of x. However, we are given that this is not the case. This proves our claim.

The argument for the value $S_{i,0^i}$ follows analogously. This concludes the proof.

5.2 Hard on Average SVL Instances

In this section, we describe our construction for hard on average instance of SVL. In particular, we describe our sampler that samples hard on average instances $(x_s, \mathsf{Succ}, \mathsf{Ver}, 1^\kappa)$. Here x_s is the source node and Succ is the successor circuit. We define a directed edge between u and v if and only if $\mathsf{Succ}(u) = v$. Ver is the verification circuit and is used to test whether a given node is the kth node from x_s. That is, $\mathsf{Ver}(x, k) = 1$ iff $x = \mathsf{Succ}^{k-1}(x_s)$. For the generated instances, we argue that it is hard to find the 1^κ node in the path.

In our construction we use a selectively secure functional encryption scheme ($\mathsf{FE.Setup}, \mathsf{FE.KeyGen}, \mathsf{FE.Enc}, \mathsf{FE.Dec}$), a prefix-puncturable PRF (Definition 11), a semantically secure symmetric key encryption ($\mathsf{SK.KeyGen}, \mathsf{SK.Enc}, \mathsf{SK.Dec}$) and injective PRGs having the left half injectivity property Definition 5. PRG_0 and PRG_1 denote the left and the right part of the output of this PRG.

The formal description of hard on average SVL instance sampler is provided in Fig. 6. Internally this sampler generates the successor circuit to include functional encryption secret keys for circuits provided in Fig. 7. Next we informally describe the SVL instances considered.

A sampled instance implicitly defines a line graph where each node in the graph is of the form $(x, \sigma_1, \cdots, \sigma_\kappa)$ where $\sigma_i = \mathsf{PrefixPunc}(S_i, x_{[i]})$ for all $i \in [\kappa]$. We say a node is *valid* if the above condition holds. The node $(x, \sigma_1, \cdots, \sigma_\kappa)$ is connected to $(x+1, \sigma_1', \cdots, \sigma_\kappa')$ by an outgoing edge and to $(x-1, \sigma_1'', \cdots, \sigma_\kappa'')$ by an incoming edge. The successor circuit on input $(x, \sigma_1, \cdots, \sigma_\kappa)$ checks for the validity of the node and if the node is valid it outputs $(x+1, \sigma_1', \cdots, \sigma_\kappa')$. The verification circuit on input $(x, \sigma_1, \cdots, \sigma_\kappa, j)$ outputs if and only if $x = j - 1$ and $(x, \sigma_1, \cdots, \sigma_\kappa)$ is valid.

We now explain how the successor circuit works. The successor circuit is described by a sequence of $\kappa + 1$ secret keys $\mathsf{FSK}_1, \cdots, \mathsf{FSK}_{\kappa+1}$ for appropriate functions. There keys are generated corresponding to independent instances of functional encryption. Along with the keys the successor circuit also contains a ciphertext c_ϕ that encrypts the empty string, ϕ, under PK_1 along with the key values V_ϕ and W_ϕ. Intuitively, the function key FSK_i corresponds to a function

- **Sampled Ingredients**:
    1. Sample $\{S_i\}_{i\in[\kappa]}$ and K_ϕ from $\mathsf{KeyGen}_{\mathcal{PPRF}}(1^\kappa)$. Here S_i's is a key that works for i bit inputs, namely $\mathsf{PPRF}_{S_i} : \{0,1\}^i \to \{0,1\}^\kappa$ for all $i \in [\kappa]$. Similarly, K_ϕ works on inputs of length $\mathsf{rand}(\kappa)$ where $\mathsf{rand}(\cdot)$ would be specified later. Initialize $\mathsf{V}_\phi^i = S_i$, $\mathsf{V}_\phi = \bigcup_{i\in[\kappa]} \mathsf{V}_\phi^i$ and $\mathsf{W}_\phi = \emptyset$.
    2. Sample $(PK_i, MSK_i) \leftarrow \mathsf{FE.Setup}(1^\kappa)$ for all $1 \le i \le \kappa + 1$.
    3. Sample $sk \leftarrow \mathsf{SK.KeyGen}(1^\kappa)$ and let $\Pi \leftarrow \mathsf{SK.Enc}_{sk}(\pi)$ and $\Lambda \leftarrow \mathsf{SK.Enc}_{sk}(\lambda)$ where $\pi = 0^{\ell(\kappa)}$ and $\lambda = 0^{\ell'(\kappa)}$. Here $\ell(\cdot)$ and $\ell'(\cdot)$ are appropriate length functions specified later.
    4. Sample $v \leftarrow \{0,1\}^{2\kappa}$.
- **Functional encryption ciphertext and keys to simulate obfuscation**:
    1. For each $i \in [\kappa]$ generate $\mathsf{FSK}_i \leftarrow \mathsf{FE.KeyGen}(MSK_i, F_{i,PK_{i+1},\Pi})$ and $\mathsf{FSK}_{\kappa+1} \leftarrow \mathsf{FE.KeyGen}(MSK_{\kappa+1}, G_{v,\Lambda})$, where $F_{i,PK_{i+1},\Pi}$ and $G_{v,\Lambda}$ are circuits described in Figure 7.
    2. Let $c_\phi = \mathsf{FE.Enc}_{PK_1}(\phi, \mathsf{V}_\phi, \mathsf{W}_\phi, 0^\kappa, 0)$
- **Source node**: The source node x_s is given by $(0^\kappa, \sigma_1, \cdots, \sigma_\kappa)$ where $\sigma_i = \mathsf{PPRF}_{S_i}(0^i)$ for all $i \in [\kappa]$.
- **Successor Circuit**: The successor circuit Succ in our setting takes as input $x, \sigma_1, \ldots, \sigma_\kappa$ and outputs $x + 1, \sigma_1', \ldots, \sigma_\kappa'$ if the associated signatures $\sigma_1, \cdots, \sigma_\kappa$ are valid. It proceeds as follows:
    1. For $i \in [\kappa]$ compute $c_{x_{[i-1]}\|0}, c_{x_{[i-1]}\|1} := \mathsf{FE.Dec}(\mathsf{FSK}_i, c_{x_{[i-1]}})$.
    2. Obtain $d_x = ((\alpha_1, \ldots, \alpha_\kappa), (\beta_j, \ldots, \beta_\kappa))$ as output of $\mathsf{FE.Dec}(\mathsf{FSK}_{\kappa+1}, c_x)$. Here $j = f(x)$ where $f(x)$ is the smallest j such that $x = x_{[j]}\|1^{\kappa-j}$.
    3. Output $\perp$ if $\mathsf{PRG}_0(\sigma_i) \ne \alpha_i$ for any $i \in [\kappa]$ or if $d_x = \perp$.
    4. If $x = 1^\kappa$, output SOLVED.
    5. For each $i \in [j-1]$ set $\sigma_i' = \sigma_i$.
    6. For each $i \in \{j, \ldots, \kappa\}$ set $\gamma_i = \mathsf{PRG}_1(\sigma_i)$ and σ_i' as $\mathsf{SK.Dec}_{\gamma_j, \ldots, \gamma_\kappa}(\beta_i)$, decrypting β_i encrypted under $\gamma_j, \ldots \gamma_\kappa$.
    7. Output $(x + 1, \sigma_1', \cdots, \sigma_\kappa')$.
- **Verification Circuit**: The verification circuit Ver on input $x, \sigma_1, \ldots, \sigma_\kappa, j$ outputs 1 if Succ on input $x, \sigma_1, \ldots, \sigma_\kappa$ doesn't output $\perp$ and $x = j - 1$ and 0 otherwise.

Fig. 6. Hard on average instance for SVL based on hardness of FE.

F_i that takes as input a binary string x of length i and outputs an encryption of $x\|0$ and $x\|1$ under PK_{i+1}. Additionally these ciphertexts, in addition to $x\|0$ and $x\|1$, also contain key values $\mathsf{V}_{x\|0}, \mathsf{W}_{x\|0}$ and $\mathsf{V}_{x\|1}, \mathsf{W}_{x\|1}$ respectively. Recall from Sect. 5.1 that the keys in these sets are used to test validity of signatures provides as input and to generate the new ones.

The successor circuit on an input of the form $(x, \sigma_1, \cdots, \sigma_\kappa)$ does the following. It first obtains an encryption of x along with key values V_x and W_x under the public key $PK_{\kappa+1}$. This is done as follows. Start with c_ϕ and decrypt it using key FSK_1 to obtain encryptions of 0 and 1. Choose one of them based on which one is a prefix of x and continue the process. Repeating this process

$$F_{i,PK_{i+1},\Pi}$$

Hardcoded Values: i, PK_{i+1}, Π.
Input: $(x \in \{0,1\}^{i-1}, \mathsf{V}_x, \mathsf{W}_x, K_x, sk, \mathsf{mode})$

1. If $(\mathsf{mode} = 0)$ then output $\mathsf{FE.Enc}_{PK_{i+1}}(x\|0, \mathsf{V}_{x\|0}, \mathsf{W}_{x\|0}, K_{x\|0}, sk, \mathsf{mode}; K'_{x\|0})$ and $\mathsf{FE.Enc}_{PK_{i+1}}(x\|1, \mathsf{V}_{x\|1}, \mathsf{W}_{x\|1}, K_{x\|1}, sk, \mathsf{mode}; K'_{x\|1})$, where for $b \in \{0,1\}$, $K_{x\|b} = \mathsf{PrefixPunc}(K_x, b\|0)$ and $K'_{x\|b} = \mathsf{PrefixPunc}(K_x, b\|1)$ and $(\mathsf{V}_{x\|0}, \mathsf{W}_{x\|0})$, $(\mathsf{V}_{x\|1}, \mathsf{W}_{x\|1})$ are computed using the efficient procedure from the Computability Lemma (Lemma 3).
2. Else recover $(x\|0, c_{x\|0})$ and $(x\|1, c_{x\|1})$ from $\mathsf{SK.Dec}_{sk}(\Pi)$ and output $c_{x\|0}$ and $c_{x\|1}$.

$$G_{v,\Lambda}$$

Hardcoded Values: v, Λ
Input: $x \in \{0,1\}^\kappa, \mathsf{V}_x, \mathsf{W}_x, K_x, sk, \mathsf{mode}$

1. If $(\mathsf{PRG}(x) = v)$ then output $\bot$.
2. If $\mathsf{mode} = 0$, (Below $j = f(x)$ where $f(x)$ is the largest j such that $x = x_{[j]}\|1^{\kappa-j}$.)
   (a) For each $i \in [\kappa]$, set $\alpha_i = \mathsf{PRG}_0(\sigma_i)$ (obtained from W_x^i for $i \leq j$ and from V_x^i for $i > j$).
   (b) For each $i \in \{j, \ldots, \kappa\}$ set $\gamma_i = \mathsf{PRG}_1(\sigma_i)$ and $\beta_i = \mathsf{SK.Enc}_{\gamma_j, \cdots, \gamma_\kappa}(S_{i,x_{[i]}+1})$, encrypting $S_{i,x_{[i]}+1}$ under $\gamma_j, \ldots \gamma_\kappa$. (Using randomness obtained by expanding K_x sufficiently.)
   (c) Output $((\alpha_1, \ldots, \alpha_\kappa), (\beta_j, \ldots, \beta_\kappa))$
3. Else recover (x, d_x) from $\mathsf{SK.Dec}_{sk}(\Lambda)$ and output d_x.

Fig. 7. Circuits for which functional encryption secret keys are given out.

κ times results in the desired ciphertext. Next decrypt the obtained ciphertext using $\mathsf{FSK}_{\kappa+1}$ and it provides some information essential for checking validity of provided input signatures and additional information to generate the signatures for the next node. More details are provided in Figs. 6 and 7.

Setting $\mathsf{rand}(\cdot)$ We set $\mathsf{rand}(\kappa) = 2\kappa + r(\kappa)$ where $r(\kappa)$ is the maximum number of random bits used for generating encryptions of $S_{i,x_{[i]}+1}$ under $\gamma_j, \cdots, \gamma_\kappa$ for every $i \in [j, \kappa]$.

Due to space constraints, we defer the proof of hardness of the sampled SVL instance to the full version of the paper [GPS15].

Acknowledgements. The first author would like to thank Sidharth Telang for useful discussions on related topics. Research supported in part from DARPA Safeware Award W911NF15C0210, AFOSR Award FA9550-15-1-0274, and NSF CRII Award 1464397. The views expressed are those of the author and do not reflect the official policy or position of the Department of Defense, the National Science Foundation, or the U.S. Government.

References

[AB15] Applebaum, B., Brakerski, Z.: Obfuscating circuits via composite-order graded encoding. In: Dodis, Y., Nielsen, J.B. (eds.) TCC 2015, Part II. LNCS, vol. 9015, pp. 528–556. Springer, Heidelberg (2015)

[ABSV15] Ananth, P., Brakerski, Z., Segev, G., Vaikuntanathan, V.: From selective to adaptive security in functional encryption. In: Gennaro, R., Robshaw, M. (eds.) CRYPTO 2015. LNCS, vol. 9216, pp. 657–677. Springer, Heidelberg (2015)

[AJ15] Ananth, P., Jain, A.: Indistinguishability obfuscation from compact functional encryption. In: Gennaro, R., Robshaw, M.J.B. (eds.) CRYPTO 2015. LNCS, vol. 9215, pp. 308–326. Springer, Heidelberg (2015)

[AJS15] Ananth, P., Jain, A., Sahai, A.: Achieving compactness generically: Indistinguishability obfuscation from non-compact functional encryption. IACR Cryptology ePrint Archive, 2015:730 (2015)

[AKV04] Abbot, T., Kane, D., Valiant, P.: On Algorithms for Nash Equilibria (2004). http://web.mit.edu/tabbott/Public/final.pdf

[BCC+14] Bitansky, N., Canetti, R., Cohn, H., Goldwasser, S., Kalai, Y.T., Paneth, O., Rosen, A.: The impossibility of obfuscation with auxiliary input or a universal simulator. In: Garay, J.A., Gennaro, R. (eds.) CRYPTO 2014, Part II. LNCS, vol. 8617, pp. 71–89. Springer, Heidelberg (2014)

[BGI+12] Barak, B., Goldreich, O., Impagliazzo, R., Rudich, S., Sahai, A., Vadhan, S.P., Yang, K.: On the (im)possibility of obfuscating programs. J. ACM **59**(2), 6 (2012)

[BGI14] Boyle, E., Goldwasser, S., Ivan, I.: Functional signatures and pseudorandom functions. In: Krawczyk, H. (ed.) PKC 2014. LNCS, vol. 8383, pp. 501–519. Springer, Heidelberg (2014)

[BGK+14] Barak, B., Garg, S., Kalai, Y.T., Paneth, O., Sahai, A.: Protecting obfuscation against algebraic attacks. In: Nguyen, P.Q., Oswald, E. (eds.) EUROCRYPT 2014. LNCS, vol. 8441, pp. 221–238. Springer, Heidelberg (2014)

[BPR15] Bitansky, N., Paneth, O., Rosen, A.: On the cryptographic hardness of finding a nash equilibrium. In: FOCS (2015)

[BR14] Brakerski, Z., Rothblum, G.N.: Virtual black-box obfuscation for all circuits via generic graded encoding. In: Lindell, Y. (ed.) TCC 2014. LNCS, vol. 8349, pp. 1–25. Springer, Heidelberg (2014)

[BSW11] Boneh, D., Sahai, A., Waters, B.: Functional encryption: definitions and challenges. In: Ishai, Y. (ed.) TCC 2011. LNCS, vol. 6597, pp. 253–273. Springer, Heidelberg (2011)

[BV15a] Bitansky, N., Vaikuntanathan, V.: Indistinguishability obfuscation from functional encryption. In: 56th FOCS, pp. 171–190. IEEE Computer Society Press (2015)

[BV15b] Bitansky, N., Vaikuntanathan, V.: Indistinguishability obfuscation from functional encryption. IACR Cryptology ePrint Archive, 2015:163 (2015)

[BW13] Boneh, D., Waters, B.: Constrained pseudorandom functions and their applications. In: Sako, K., Sarkar, P. (eds.) ASIACRYPT 2013, Part II. LNCS, vol. 8270, pp. 280–300. Springer, Heidelberg (2013)

[CDT09] Chen, X., Deng, X., Teng, S.-H.: Settling the complexity of computing two-player nash equilibria. J. ACM 56(3), 1–57 (2009)

[DGP09] Daskalakis, C., Goldberg, P.W., Papadimitriou, C.H.: The complexity of computing a nash equilibrium. Commun. ACM 52(2), 89–97 (2009)

[GGH13a] Garg, S., Gentry, C., Halevi, S.: Candidate multilinear maps from ideal lattices. In: Johansson, T., Nguyen, P.Q. (eds.) EUROCRYPT 2013. LNCS, vol. 7881, pp. 1–17. Springer, Heidelberg (2013)

[GGH+13b] Garg, S., Gentry, C., Halevi, S., Raykova, M., Sahai, A., Waters, B.: Candidate indistinguishability obfuscation and functional encryption for all circuits. In: 54th FOCS, pp. 40–49, Berkeley, CA, USA. IEEE Computer Society Press, 26–29 October 2013

[GGHZ16] Garg, S., Gentry, C., Halevi, S., Zhandry, M.: Fully secure functional encryption from multilinear maps. In: TCC (2016)

[GGM86] Goldreich, O., Goldwasser, S., Micali, S.: How to construct random functions. J. ACM 33(4), 792–807 (1986)

[GK05] Goldwasser, S., Kalai, Y.T.: On the impossibility of obfuscation with auxiliary input. In: FOCS, pp. 553–562 (2005)

[GL89] Goldreich, O., Levin, L.A.: A hard-core predicate for all one-way functions. In: Proceedings of the 21st Annual ACM Symposium on Theory of Computing, Seattle, Washigton, USA, pp. 25–32, 14–17 May 1989

[GLSW15] Gentry, C., Lewko, A.B., Sahai, A., Waters, B.: Indistinguishability obfuscation from the multilinear subgroup elimination assumption. In: 56th FOCS, pp. 151–170. IEEE Computer Society Press (2015)

[GMS16] Garg, S., Mukherjee, P., Srinivasan, A.: Obfuscation without the vulnerabilities of multilinear maps. IACR Cryptology ePrint Archive, 2016:390 (2016)

[GPS15] Garg, S., Pandey, O., Srinivasan, A.: On the exact cryptographic hardness of finding a nash equilibrium. Cryptology ePrint Archive, Report 2015/1078 (2015). http://eprint.iacr.org/2015/1078

[GPSZ16] Garg, S., Pandey, O., Srinivasan, A., Zhandry, M.: Breaking the subexponential barrier in obfustopia. Cryptology ePrint Archive, Report 2016/102 (2016). http://eprint.iacr.org/2016/102

[GS16] Garg, S., Srinivasan, A.: Unifying security notions of functional encryption. Cryptology ePrint Archive, Report 2016/524 (2016). http://eprint.iacr.org/

[HY16] Hubácek, P., Yogev, E.: Hardness of continuous local search: query complexity and cryptographic lower bounds. Electron. Colloquium Comput. Complex. (ECCC) 23, 63 (2016)

[Jer12] Emil Jerábek. Integer factoring and modular square roots. CoRR abs/1207.5220 (2012)

[KPTZ13] Kiayias, A., Papadopoulos, S., Triandopoulos, N., Zacharias, T.: Delegatable pseudorandom functions and applications. In: 2013 ACM SIGSAC Conference on Computer and Communications Security, CCS 2013, Berlin, Germany, pp. 669–684, 4–8 November 2013

[MP91] Megiddo, N., Papadimitriou, C.H.: On total functions, existence theorems and computational complexity. Theor. Comput. Sci. **81**(2), 317–324 (1991)

[Nao03] Naor, M.: On cryptographic assumptions and challenges. In: Boneh, D. (ed.) CRYPTO 2003. LNCS, vol. 2729, pp. 96–109. Springer, Heidelberg (2003)

[Nas51] Nash, J.: Non-cooperative games. Ann. Math. **54**(2), 286–295 (1951)

[O'N10] O'Neill, A.: Definitional issues in functional encryption. IACR Cryptology ePrint Archive, 2010:556 (2010)

[Pap94] Papadimitriou, C.H.: On the complexity of the parity argument and other inefficient proofs of existence. J. Comput. Syst. Sci. **48**(3), 498–532 (1994)

[PST14] Pass, R., Seth, K., Telang, S.: Indistinguishability obfuscation from semantically-secure multilinear encodings. In: Garay, J.A., Gennaro, R. (eds.) CRYPTO 2014, Part I. LNCS, vol. 8616, pp. 500–517. Springer, Heidelberg (2014)

[RSS16] Rosen, A., Segev, G., Shahaf, I.: Can PPAD hardness be based on standard cryptographic assumptions? Electron. Colloquium Comput. Complex. (ECCC) **23**, 59 (2016)

[SW14] Sahai, A., Waters, B.: How to use indistinguishability obfuscation: deniable encryption, and more. In: Symposium on Theory of Computing, STOC 2014, New York, NY, USA, 31 May–03 June 2014, pp. 475–484 (2014)

[Zim15] Zimmerman, J.: How to obfuscate programs directly. In: Oswald, E., Fischlin, M. (eds.) EUROCRYPT 2015. LNCS, vol. 9057, pp. 439–467. Springer, Heidelberg (2015)

Asymmetric Cryptography
and Cryptanalysis II

Cryptanalysis of GGH15 Multilinear Maps

Jean-Sébastien Coron[1]($\boxtimes$), Moon Sung Lee[1], Tancrède Lepoint[2], and Mehdi Tibouchi[3]

[1] University of Luxembourg, Luxembourg City, Luxembourg
jean-sebastien.coron@uni.lu
[2] CryptoExperts, Paris, France
[3] NTT Secure Platform Laboratories, Tokyo, Japan

Abstract. We describe a cryptanalysis of the GGH15 multilinear maps. Our attack breaks the multipartite key-agreement protocol in polynomial time by generating an equivalent user private key; it also applies to GGH15 with safeguards. We also describe attacks against variants of the GGH13 multilinear maps proposed by Halevi (ePrint 2015/866) aiming at supporting graph-induced constraints, as in GGH15.

1 Introduction

Multilinear Maps. For the past couple of years, cryptographic multilinear maps have found numerous applications in the design of cryptographic protocols, the most salient example of which is probably the construction of indistinguishability obfuscation (iO) [GGH+13b]. The first multilinear maps candidate (GGH13) was described by Garg, Gentry and Halevi [GGH13a] from ideal lattices. It was then followed by another candidate (aka, CLT13) due to Coron, Lepoint and Tibouchi [CLT13] using the same techniques but over the integers, and later by a third candidate (GGH15) by Gentry, Gorbunov and Halevi [GGH15], related to the homomorphic encryption scheme from [GSW13].

Unfortunately, these candidates do not rely on well-established hardness assumptions, and recent months have witnessed a number of attacks (including [CHL+15,CGH+15,HJ16,BGH+15,PS15,CFL+16]) showing that they fail to meet a number of desirable security requirements, and that they cannot be used to securely instantiate such and such protocols. Some attempts to protect against these attacks have also known a similar fate [CLT15,BGH+15]. The security of the constructions based on these multilinear maps is currently unclear to the community [Hal15]. While two recent works [CGH+15,MSZ16] have shown polynomial-time attacks against some obfuscation candidates, many iO candidates remain unaffected by the attacks proposed so far. The same cannot be said for the more immediate application of multilinear maps that is one-round multipartite key agreement.

One-Round Multipartite Key-Agreement Protocol. Since its discovery in 1976, the Diffie–Hellman protocol [DH76] is one of the most widely used

© International Association for Cryptologic Research 2016
M. Robshaw and J. Katz (Eds.): CRYPTO 2016, Part II, LNCS 9815, pp. 607–628, 2016.
DOI: 10.1007/978-3-662-53008-5_21

cryptographic protocol to create a common secret between two parties. A generalization of this one-round protocol to three parties was proposed in 2000 by Joux [Jou00] using cryptographic *bilinear* maps; it was later extended to $k \geq 4$ parties assuming the existence of a cryptographic $(k-1)$-linear map by Boneh and Silverberg [BS02]. In a nutshell, the protocol works as follows: assuming some public parameters are shared by all the parties, each party broadcasts some data and keeps some data secret, and then by combining their secret data with the other parties' published values using the multilinear map, they can derive a shared common secret key.

The first candidates for a k-partite Diffie–Hellman key-agreement protocol for arbitrary k were described in [GGH13a, CLT13] using respectively the GGH13 and CLT13 multilinear maps candidates. Unfortunately, the protocols were later shown to be insecure in [HJ16, CHL+15]: using the public parameters and the broadcast data, an eavesdropper can recover the shared common secret key in polynomial time.

The GGH15 Key-Agreement Protocol. Since the third proposed multilinear maps scheme, GGH15, does not fit the same graded encoding framework as the earlier candidates, one needs new constructions to use it to instantiate cryptographic protocols. And the first such application was again a Diffie–Hellman key-agreement protocol [GGH15, Sect. 5.1]. To avoid similar attacks as the one that targeted GGH13 and CLT13, based on encodings of zero, the protocol was designed in such a way that the adversary is never given encodings of the same element that could be subtracted without doing the full key-agreement computation. Namely, each party i has a directed path of matrices $\boldsymbol{A}_{i,1}, \ldots, \boldsymbol{A}_{i,k+1}$ all sharing the same end-point $\boldsymbol{A}_{i,k+1} = \boldsymbol{A}_0$, and has a secret value s_i. She can then publish encodings of s_i on the chains of the other parties in a "round robin" fashion, *i.e.* s_i is encoded on the j-th edge of the chain of the party $i' = j - i + 1$, with index arithmetic modulo k. The graph for 3 parties is illustrated in Fig. 1.

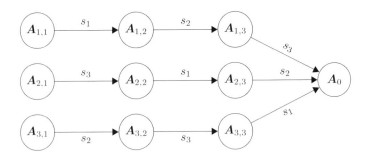

Fig. 1. Graph for a 3-partite key-agreement protocol with GGH15 multilinear maps.

On the i-th chain, Party i will then be able to multiply these encodings (the one he kept secret and the ones published by the other parties) to get an encoding of $\prod_j s_j$ relative to the path $\boldsymbol{A}_{i,1} \rightsquigarrow \boldsymbol{A}_0$. Now, since the encodings of s_i cannot be mixed before the end-point $\boldsymbol{A}_0$, it seems difficult to obtain an encoding of 0 on an edge in the middle of the graph to mount "zeroizing attacks" [GGH15].

Halevi's Candidate Key-Agreement Protocols. As no attack was known on GGH15 multilinear maps and in an attempt to reinstate a key-agreement protocol for GGH13, Halevi recently proposed, on the Cryptology ePrint Archive, two variants of GGH13 supporting a similar key-agreement protocol [Hal15].[1] The first variant uses the "asymmetric" GGH13 scheme to handle the graph structure [Hal15, Sect. 7]. Namely, in basic GGH13 each encoding is multiplicatively masked by a power z^i of a secret mask z; in asymmetric GGH13, the encodings can be masked by powers of multiple z_j's. Therefore, in this new key-agreement protocol candidate, the public encodings are now associated with independent masks $z_{i,j}$'s such that their product yields the same value Z, i.e. $\prod_j z_{i,j} = Z$ for all i (so that the final encoding shall extract to the same shared key). The graph for 3 parties is illustrated in Fig. 2.

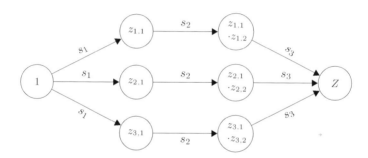

Fig. 2. Multipartite key agreement from asymmetric GGH13, with 3 parties, from [Hal15, Sect. 7].

Once again, the fact that the encodings of the same value s_i are multiplied with different masks gives hope that no encoding of 0 multiplied by a value other than Z can be obtained, and therefore that zeroizing attacks are impossible [GGH13a, CGH+15].

A second variant of GGH13, which we refer to as Graph-GGH13, mimics the structure of GGH15 encodings more closely and is described in [Hal15, Sect. 6]. An encoding $c \in \alpha + gR$ relative to a path $u \rightsquigarrow v$ is now a matrix $\tilde{\boldsymbol{C}} = \boldsymbol{P}_u^{-1} \cdot \boldsymbol{C} \cdot \boldsymbol{P}_v$, where $\boldsymbol{C} \in \mathbb{Z}_q^{n \times n}$ is the multiply-by-c matrix, and the $\boldsymbol{P}_w$'s are secret random matrices. In the key-agreement protocol, each party i has a directed path of matrices $\boldsymbol{P}_{i,1}, \ldots, \boldsymbol{P}_{i,k+1}$ all sharing the same end-point $\boldsymbol{P}_{i,k+1} = \boldsymbol{P}_0$

---

[1] As mentioned in the last remark of the paper, although the key-agreement protocol can be described also based on CLT13, the attacks from [CGH+15] can be used to break it.

and the same start-point $\boldsymbol{P}_{i,1} = \boldsymbol{P}_1$, and has a secret value s_i. She can then publish encodings of s_i on the chains of the other parties in a "round robin" fashion. The graph for 3 parties is illustrated in Fig. 3.

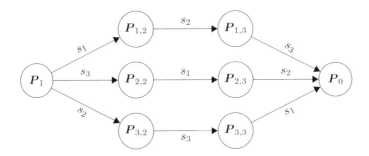

Fig. 3. Multipartite key agreement from GGH13 with graph constraints, with 3 parties, from [Hal15, Sect. 6].

And here again, the fact that the encodings corresponding to the same s_i are multiplied on the left and on the right by completely random matrices $\boldsymbol{P}_{i,j}$ makes it difficult to cancel them out and obtain an encoding of 0 without evaluating the full "chains" (that is, the operations of the key agreement itself).

Finally, in order to capture the intuition of what it means for an attacker to break the scheme, Halevi defined, for both schemes, the "core computational task" of an adversary as recovering any basis of the (hidden) plaintext space [Hal15, Sect. 2.2].

Our Contributions. Our main contribution is to describe a cryptanalysis of the Diffie–Hellman key-agreement protocol when instantiated with GGH15 multilinear maps. Our attack makes it possible to generate an equivalent user private key in polynomial time, which in turn allows to recover the shared session key. Our attack proceeds in two steps: in the first step, we express the secret exponent of one user as a linear combination of some other secret exponents corresponding to public encodings, using a variant of the Cheon *et al.* attack [CHL+15]. This does not immediately break the protocol because the coefficients of the linear combination can be large. In the second step, we use the previous linear combination to derive an encoding equivalent to the user private encoding, by correcting the error resulting from the large coefficients of the linear combination. Our attack also applies to GGH15 with safeguards; we extend the basic attack by using another linear relation to estimate the error incurred from the large coefficients, thus enabling to recover the shared session key.

In the full version of this paper [CLLT15], we also describe attacks that break both variants of GGH13 proposed by Halevi in [Hal15]. Our attacks apply some variant of the Cheon *et al.* attack [CHL+15] to recover a basis of the secret plaintext space R/gR in polynomial time. This was considered as the "core computational task of an attacker" in [Hal15].

Source Code. A proof-of-concept implementation of our cryptanalysis of GGH15, using the Sage [Dev16] mathematics software system, is available at: http://pastebin.com/7kZHnTXY

2   The GGH15 Multilinear Map Scheme

We briefly recall the GGH15 multilinear map scheme; we refer to [GGH15] for a full description. In the following we only consider the commutative variant from [GGH15, Sect. 3.2], as only that commutative variant can be used in the multipartite key-agreement protocol from [GGH15, Sect. 5.1].

2.1   GGH15 Multilinear Maps

The construction works over polynomial rings $R = \mathbb{Z}[x]/(f(x))$ and $R_q = R/qR$ for some degree n irreducible integer polynomial $f(x) \in \mathbb{Z}[x]$ and an integer q. The construction is parametrized by a directed acyclic graph $G = (V, E)$. To each node $u \in V$ a random row vector $\boldsymbol{A}_u \in R_q^m$ is assigned, where m is a parameter. An encoding of a small plaintext element $s \in R$ relative to path $u \rightsquigarrow v$ is a matrix with small coefficients $\boldsymbol{D} \in R^{m \times m}$ such that:

$$\boldsymbol{A}_u \cdot \boldsymbol{D} = s \cdot \boldsymbol{A}_v + \boldsymbol{E} \pmod{q}$$

where $\boldsymbol{E}$ is a small error vector of dimension m with components in R; we refer to [GGH15] for how such encoding $\boldsymbol{D}$ can be generated, based on a trapdoor sampling procedure from [MP12]. Only small plaintext elements $s \in R$ are encoded. As in [Hal15] we use the row vector notation for $\boldsymbol{A}_u$, rather than the column vector notation used in [GGH15].[2] It is easy to see that two encodings $\boldsymbol{D}_1$ and $\boldsymbol{D}_2$ relative to the same path $u \rightsquigarrow v$ can be added; namely from:

$$\begin{aligned} \boldsymbol{A}_u \cdot \boldsymbol{D}_1 &= s_1 \cdot \boldsymbol{A}_v + \boldsymbol{E}_1 \pmod{q} \\ \boldsymbol{A}_u \cdot \boldsymbol{D}_2 &= s_2 \cdot \boldsymbol{A}_v + \boldsymbol{E}_2 \pmod{q} \end{aligned}$$

we obtain:

$$\boldsymbol{A}_u \cdot (\boldsymbol{D}_1 + \boldsymbol{D}_2) = (s_1 + s_2) \cdot \boldsymbol{A}_v + \boldsymbol{E}_1 + \boldsymbol{E}_2 \pmod{q}.$$

Moreover two encodings $\boldsymbol{D}_1$ and $\boldsymbol{D}_2$ relative to path $u \rightsquigarrow v$ and $v \rightsquigarrow w$ can be multiplied to get an encoding relative to path $u \rightsquigarrow w$. Namely given:

$$\begin{aligned} \boldsymbol{A}_u \cdot \boldsymbol{D}_1 &= s_1 \cdot \boldsymbol{A}_v + \boldsymbol{E}_1 \pmod{q} \\ \boldsymbol{A}_v \cdot \boldsymbol{D}_2 &= s_2 \cdot \boldsymbol{A}_w + \boldsymbol{E}_2 \pmod{q} \end{aligned}$$

---

[2] With the column vector notation, the corresponding equation in [GGH15] is $\boldsymbol{D} \cdot \boldsymbol{A}_u = s \cdot \boldsymbol{A}_v + \boldsymbol{E} \pmod{q}$.

we obtain by multiplying the matrix encodings $\boldsymbol{D}_1$ and $\boldsymbol{D}_2$:

$$\begin{aligned} \boldsymbol{A}_u \cdot \boldsymbol{D}_1 \cdot \boldsymbol{D}_2 &= (s_1 \cdot \boldsymbol{A}_v + \boldsymbol{E}_1) \cdot \boldsymbol{D}_2 \pmod{q} \\ &= s_1 \cdot s_2 \cdot \boldsymbol{A}_w + s_1 \cdot \boldsymbol{E}_2 + \boldsymbol{E}_1 \cdot \boldsymbol{D}_2 \pmod{q} \\ &= s_1 \cdot s_2 \cdot \boldsymbol{A}_w + \boldsymbol{E}' \pmod{q} \end{aligned}$$

for some new error vector $\boldsymbol{E}'$. Since s_1, $\boldsymbol{E}_1$, $\boldsymbol{E}_2$ and $\boldsymbol{D}_2$ have small coefficients, $\boldsymbol{E}'$ still has small coefficients (compared to q), and therefore the product $\boldsymbol{D}_1 \cdot \boldsymbol{D}_2$ is an encoding of $s_1 \cdot s_2$ for the path $u \rightsquigarrow w$.

Finally, given an encoding $\boldsymbol{D}$ relative to path $u \rightsquigarrow w$ and the vector $\boldsymbol{A}_u$, extraction works by computing the high-order bits of $\boldsymbol{A}_u \cdot \boldsymbol{D}$. Namely we have:

$$\boldsymbol{A}_u \cdot \boldsymbol{D} = s \cdot \boldsymbol{A}_w + \boldsymbol{E} \pmod{q}$$

for some small $\boldsymbol{E}$, and therefore the high-order bits of $\boldsymbol{A}_u \cdot \boldsymbol{D}$ only depend on the secret exponent s.

Remark 1. As emphasized in [GGH15], only the plaintext space of the s_i's is commutative, not the space of the encoding matrices $\boldsymbol{D}_i$. The ability to multiply the plaintext elements s_i in arbitrary order will be used in the multipartite key-agreement protocol below.

2.2 The GGH15 Multipartite Key-Agreement Protocol

We briefly recall the multipartite key-agreement protocol from [GGH15, Sect. 5.1]. We consider the protocol with k users. As illustrated in Fig. 4 for $k = 3$ users, each user i for $1 \le i \le k$ has a directed path of vectors $\boldsymbol{A}_{i,1}, \ldots, \boldsymbol{A}_{i,k+1}$, all sharing the same end-point $\boldsymbol{A}_0 = \boldsymbol{A}_{i,k+1}$. The i-th user will use the resulting chain to extract the session key. Each user i has a secret exponent s_i. Each secret exponent s_i will be encoded in each of the k chains; the encoding of s_i on the j-th chain for $j \ne i$ will be published, while the encoding of s_i on the i-th chain will be kept private by user i. Therefore on the i-th chain only user i will be able to compute the session key. The exponents s_i are encoded in a "round robin" fashion; namely the i-th secret s_i is encoded on the chain of user j at edge $\ell = i + j - 1$, with index arithmetic modulo k. Only the vectors $\boldsymbol{A}_{i,1}$ for $1 \le i \le k$ are made public to enable extraction of the session-key; the others are kept private. We recall the formal description of the protocol in the full version of this paper [CLLT15].

We illustrate the protocol for $k = 3$ users. For the chain corresponding to User 1, we have the following encodings:

$$\begin{aligned} \boldsymbol{A}_{1,1} \cdot \boldsymbol{D}_{1,1} &= s_1 \cdot \boldsymbol{A}_{1,2} + \boldsymbol{F}_{1,1} \pmod{q} \\ \boldsymbol{A}_{1,2} \cdot \boldsymbol{D}_{1,2} &= s_2 \cdot \boldsymbol{A}_{1,3} + \boldsymbol{F}_{1,2} \pmod{q} \\ \boldsymbol{A}_{1,3} \cdot \boldsymbol{D}_{1,3} &= s_3 \cdot \boldsymbol{A}_0 + \boldsymbol{F}_{1,3} \pmod{q} \end{aligned}$$

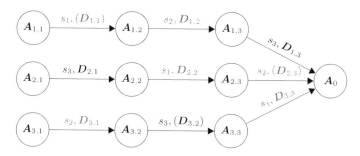

Fig. 4. Graph of a key agreement between 3 parties for GGH15. The vertices contain random vectors $\boldsymbol{A}_{ij}$, and encodings are represented on the edges. Each party is represented by a different color, keeps the encoding in parenthesis secret and publishes the two other encodings.

where $\boldsymbol{D}_{1,2}$ and $\boldsymbol{D}_{1,3}$ are public while $\boldsymbol{D}_{1,1}$ is kept private by User 1. Therefore User 1 can compute modulo q:

$$\boldsymbol{A}_{1,1} \cdot \boldsymbol{D}_{1,1} \cdot \boldsymbol{D}_{1,2} \cdot \boldsymbol{D}_{1,3} = (s_1 \cdot \boldsymbol{A}_{1,2} + \boldsymbol{F}_{1,1}) \cdot \boldsymbol{D}_{1,2} \cdot \boldsymbol{D}_{1,3} \pmod{q}$$
$$= (s_1 \cdot s_2 \cdot \boldsymbol{A}_{1,3} + s_1 \cdot \boldsymbol{F}_{1,2}$$
$$+ \boldsymbol{F}_{1,1} \cdot \boldsymbol{D}_{1,2}) \cdot \boldsymbol{D}_{1,3} \pmod{q}.$$

Letting $\hat{\boldsymbol{F}}_{1,2} := s_1 \cdot \boldsymbol{F}_{1,2} + \boldsymbol{F}_{1,1} \cdot \boldsymbol{D}_{1,2}$, we obtain:

$$\boldsymbol{A}_{1,1} \cdot \boldsymbol{D}_{1,1} \cdot \boldsymbol{D}_{1,2} \cdot \boldsymbol{D}_{1,3} = \left(s_1 \cdot s_2 \cdot \boldsymbol{A}_{1,3} + \hat{\boldsymbol{F}}_{1,2}\right) \cdot \boldsymbol{D}_{1,3} \pmod{q}$$
$$= s_1 \cdot s_2 \cdot s_3 \cdot \boldsymbol{A}_0 + s_1 \cdot s_2 \cdot \boldsymbol{F}_{1,3} + \hat{\boldsymbol{F}}_{1,2} \cdot \boldsymbol{D}_{1,3} \pmod{q}.$$

Since s_1, s_2 and s_3 are small and $\boldsymbol{F}_{1,3}$, $\hat{\boldsymbol{F}}_{1,2}$ and $\boldsymbol{D}_{1,3}$ have small components, User 1 can extract the most significant bits corresponding to $s_1 \cdot s_2 \cdot s_3 \cdot \boldsymbol{A}_0$. Similarly User 2 will compute the session key using the following chain, where $\boldsymbol{D}_{2,1}$ and $\boldsymbol{D}_{2,2}$ are public while $\boldsymbol{D}_{2,3}$ is private to User 2:

$$\boldsymbol{A}_{2,1} \cdot \boldsymbol{D}_{2,1} = s_3 \cdot \boldsymbol{A}_{2,2} + \boldsymbol{F}_{2,1} \pmod{q}$$
$$\boldsymbol{A}_{2,2} \cdot \boldsymbol{D}_{2,2} = s_1 \cdot \boldsymbol{A}_{2,3} + \boldsymbol{F}_{2,2} \pmod{q}$$
$$\boldsymbol{A}_{2,3} \cdot \boldsymbol{D}_{2,3} = s_2 \cdot \boldsymbol{A}_0 + \boldsymbol{F}_{2,3} \pmod{q}.$$

Namely User 2 can compute:

$$\boldsymbol{A}_{2,1} \cdot \boldsymbol{D}_{2,1} \cdot \boldsymbol{D}_{2,2} \cdot \boldsymbol{D}_{2,3} = (s_3 \cdot s_1 \cdot \boldsymbol{A}_{2,3} + s_3 \cdot \boldsymbol{F}_{2,2}$$
$$+ \boldsymbol{F}_{2,1} \cdot \boldsymbol{D}_{2,2}) \cdot \boldsymbol{D}_{2,3} \pmod{q}$$
$$= s_3 \cdot s_1 \cdot s_2 \cdot \boldsymbol{A}_0 + \boldsymbol{F} \pmod{q}$$

for some small vector $\boldsymbol{F}$, and extract the same most significant bits corresponding to $s_1 \cdot s_2 \cdot s_3 \cdot \boldsymbol{A}_0$; the same holds for User 3.

The previous encodings are generated by random linear combination of public encodings, corresponding to secret exponents $t_{i,\ell}$ for $1 \leq \ell \leq N$, for large enough N. More precisely, for each $1 \leq i \leq k$ one generates random small plaintext elements $t_{i,\ell}$ for $1 \leq \ell \leq N$, which are then encoded on all chains j at edge $i' = i + j - 1$ (with index modulo k), by $C_{j,i',\ell}$. This means that for $k = 3$ users, we have the following encodings corresponding to User 1:

$$A_{1,1} \cdot C_{1,1,\ell} = t_{1,\ell} \cdot A_{1,2} + E_{1,1,\ell} \pmod{q}$$
$$A_{2,2} \cdot C_{2,2,\ell} = t_{1,\ell} \cdot A_{2,3} + E_{2,2,\ell} \pmod{q}$$
$$A_{3,3} \cdot C_{3,3,\ell} = t_{1,\ell} \cdot A_0 + E_{3,3,\ell} \pmod{q}$$

and the tuple $(D_{1,1}, D_{2,2}, D_{3,3})$ is generated by linear combination of the tuple $(C_{1,1,\ell}, C_{2,2,\ell}, C_{3,3,\ell})$, so that the matrices $D_{1,1}$, $D_{2,2}$ and $D_{3,3}$ encode the same secret exponent s_1; the same holds for users 2 and 3. We refer to the full version of this paper [CLLT15] for the formal description of the protocol.

3   Cryptanalysis of GGH15 Without Safeguards

In the following we describe a cryptanalysis of the multipartite key-agreement protocol based on GGH15 multilinear maps recalled in the previous section. Heuristically our attack recovers the session-key from public element in polynomial-time. Our attack proceeds in two steps.

1. In the first step, we are able to express one secret exponent s_1 as a linear combination of the other secret exponents $t_{1,\ell}$, using a variant of the Cheon et al. attack [CHL+15]. However this does not immediately break the protocol, because the coefficients are not small.

2. In the second step, we compute an equivalent of the private encoding of User 1 from the previous linear combination, by correcting the error due to the large coefficients. This breaks the key-exchange protocol.

3.1   Description with 3 Users

For simplicity we first consider the protocol with only 3 users; the extension to $k \geq 3$ users is relatively straightforward and described in the full version of this paper [CLLT15]. Therefore we consider the following 3 rows corresponding to the 3 users:

$$A_{1,1} \cdot D_{1,1} = s_1 \cdot A_{1,2} + F_{1,1} \pmod{q} \qquad A_{1,1} \cdot C_{1,1,\ell} = t_{1,\ell} \cdot A_{1,2} + E_{1,1,\ell} \pmod{q}$$
$$A_{1,2} \cdot D_{1,2} = s_2 \cdot A_{1,3} + F_{1,2} \pmod{q} \qquad A_{1,2} \cdot C_{1,2,\ell} = t_{2,\ell} \cdot A_{1,3} + E_{1,2,\ell} \pmod{q}$$
$$A_{1,3} \cdot D_{1,3} = s_3 \cdot A_0 + F_{1,3} \pmod{q} \qquad A_{1,3} \cdot C_{1,3,\ell} = t_{3,\ell} \cdot A_0 + E_{1,3,\ell} \pmod{q}$$
$$A_{2,1} \cdot D_{2,1} = s_3 \cdot A_{2,2} + F_{2,1} \pmod{q} \qquad A_{2,1} \cdot C_{2,1,\ell} = t_{3,\ell} \cdot A_{2,2} + E_{2,1,\ell} \pmod{q}$$
$$A_{2,2} \cdot D_{2,2} = s_1 \cdot A_{2,3} + F_{2,2} \pmod{q} \qquad A_{2,2} \cdot C_{2,2,\ell} = t_{1,\ell} \cdot A_{2,3} + E_{2,2,\ell} \pmod{q}$$
$$A_{2,3} \cdot D_{2,3} = s_2 \cdot A_0 + F_{2,3} \pmod{q} \qquad A_{2,3} \cdot C_{2,3,\ell} = t_{2,\ell} \cdot A_0 + E_{2,3,\ell} \pmod{q}$$
$$A_{3,1} \cdot D_{3,1} = s_2 \cdot A_{3,2} + F_{3,1} \pmod{q} \qquad A_{3,1} \cdot C_{3,1,\ell} = t_{2,\ell} \cdot A_{3,2} + E_{3,1,\ell} \pmod{q}$$
$$A_{3,2} \cdot D_{3,2} = s_3 \cdot A_{3,3} + F_{3,2} \pmod{q} \qquad A_{3,2} \cdot C_{3,2,\ell} = t_{3,\ell} \cdot A_{3,3} + E_{3,2,\ell} \pmod{q}$$
$$A_{3,3} \cdot D_{3,3} = s_1 \cdot A_0 + F_{3,3} \pmod{q} \qquad A_{3,3} \cdot C_{3,3,\ell} = t_{1,\ell} \cdot A_0 + E_{3,3,\ell} \pmod{q}$$

where all encodings $C_{i,j,\ell}$ and $D_{i,j}$ are public, except $D_{1,1}$ which is private on Row 1, $D_{2,3}$ is private on Row 2, and $D_{3,2}$ is private on Row 3. The corresponding graph is illustrated in Fig. 4. Note that on each row we have used the same index ℓ for $t_{1,\ell}$, $t_{2,\ell}$ and $t_{3,\ell}$, but on a given row one can obviously compute product of encodings for different indices.

First Step: Linear Relations. In the first step of the attack, we show that we can express s_1 as a linear combinations of the $t_{1,\ell}$'s. For this we consider the rows 2 and 3, for which the encodings $D_{2,2}$ and $D_{3,3}$ corresponding to s_1 are public. In the remaining of the attack, we always consider a fixed index $\ell = 1$ for the encodings corresponding to $t_{3,\ell}$, and for simplicity we write $t_3 := t_{3,1}$, $C_{1,3} := C_{1,3,1}$, $C_{2,1} := C_{2,1,1}$ and $C_{3,2} := C_{3,2,1}$.

Since we always work with the same t_3, on Row 2 we define the product encodings $\hat{C}_{2,2,\ell} := C_{2,1} \cdot C_{2,2,\ell}$, and on Row 3 we define the product encodings $\hat{C}_{3,2,\ell} := C_{3,1,\ell} \cdot C_{3,2}$; recall that we use a fixed index for t_3. Therefore we can write:

$$A_{2,1} \cdot \hat{C}_{2,2,\ell} = t_{1,\ell} \cdot t_3 \cdot A_{2,3} + \hat{E}_{2,2,\ell} \pmod q \qquad (1)$$
$$A_{2,3} \cdot C_{2,3,\ell} = t_{2,\ell} \cdot A_0 + E_{2,3,\ell} \pmod q$$
$$A_{3,1} \cdot \hat{C}_{3,2,\ell} = t_{2,\ell} \cdot t_3 \cdot A_{3,3} + \hat{E}_{3,2,\ell} \pmod q$$
$$A_{3,3} \cdot C_{3,3,\ell} = t_{1,\ell} \cdot A_0 + E_{3,3,\ell} \pmod q$$

for some small error vectors $\hat{E}_{2,2,\ell}$ and $\hat{E}_{3,2,\ell}$.

For simplicity of notations, we first consider a fixed index i for the encodings corresponding to $t_{1,i}$, and we write $t_1 := t_{1,i}$, $\hat{C}_{2,2} := \hat{C}_{2,2,i}$ and $C_{3,3} := C_{3,3,i}$. Similarly we consider a fixed index j for the encodings corresponding to $t_{2,j}$ and we write $t_2 := t_{2,j}$, $C_{2,3} := C_{2,3,j}$ and $\hat{C}_{3,2} := \hat{C}_{3,2,j}$. We use similar notations for the corresponding error vectors.

All previous equations hold modulo q only. To get a result over R instead of only modulo q, we compute the difference between two rows, for the same product of secret exponents. More precisely, we compute:

$$\omega = A_{2,1} \cdot \hat{C}_{2,2} \cdot C_{2,3} - A_{3,1} \cdot \hat{C}_{3,2} \cdot C_{3,3} \qquad (2)$$
$$= t_1 \cdot t_3 \cdot t_2 \cdot A_0 + t_1 \cdot t_3 \cdot E_{2,3} + \hat{E}_{2,2} \cdot C_{2,3}$$
$$- t_2 \cdot t_3 \cdot t_1 \cdot A_0 - t_2 \cdot t_3 \cdot E_{3,3} - \hat{E}_{3,2} \cdot C_{3,3}$$

$$= t_1 \cdot t_3 \cdot E_{2,3} + \hat{E}_{2,2} \cdot C_{2,3} - t_2 \cdot t_3 \cdot E_{3,3} - \hat{E}_{3,2} \cdot C_{3,3}. \qquad (3)$$

Namely the latter equation holds over R (and not only modulo q) because all the terms in (3) have small coefficients; namely the only term $t_1 \cdot t_2 \cdot t_3 \cdot A_0$ with large coefficients modulo q is canceled when doing the subtraction.

We have that ω is a vector of dimension m. Now an important step is to restrict ourselves to the first component of ω. Namely in order to apply the same technique as in the Cheon *et al.* attack, we would like to express ω as the product of two vectors, where the left vector corresponds to User 1 and the right vector corresponds to User 2. However due to the "round-robin" fashion of exponent encodings, for this we would need to swap the product $\hat{E}_{3,2} \cdot C_{3,3}$ appearing in (3), since $\hat{E}_{3,2}$ corresponds to User 2 while $C_{3,3}$ corresponds to User 1; this cannot be done if we consider the full vector ω. By restricting ourselves to the first component of ω, the product $\hat{E}_{3,2} \cdot C_{3,3}$ becomes a simple scalar product that can be swapped; namely the scalar product of $\hat{E}_{3,2}$ by the first column vector $C'_{3,3}$ of the matrix $C_{3,3}$. We obtain the scalar:

$$\omega = t_1 \cdot t_3 \cdot E_{2,3} + \hat{E}_{2,2} \cdot C'_{2,3} - t_2 \cdot t_3 \cdot E_{3,3} - C'_{3,3} \cdot \hat{E}_{3,2}$$

where $C'_{2,3}$ and $C'_{3,3}$ are the first column vectors of $C_{2,3}$ and $C_{3,3}$ respectively, both of dimension m; similarly $E_{2,3}$ and $E_{3,3}$ are the first components of $E_{2,3}$ and $E_{3,3}$ respectively.

We can now write ω as the scalar product of 2 vectors, the left one corresponding only to User 1, and the right one corresponding only to User 2:

$$\omega = \begin{bmatrix} t_1 & \hat{E}_{2,2} & E_{3,3} & C'_{3,3} \end{bmatrix} \cdot \begin{bmatrix} t_3 \cdot E_{2,3} \\ C'_{2,3} \\ -t_2 \cdot t_3 \\ -\hat{E}_{3,2} \end{bmatrix}.$$

Note that the two vectors in the product have dimension $2m + 2$.

As in the Cheon *et al.* attack [CHL+15], we can now extend ω to a matrix by considering many left row vectors and many right column vectors. However instead of a square matrix as in the Cheon *et al.* attack, we consider a rectangular matrix with $2m + 3$ rows and $2m + 2$ columns. In Eq. (2), this is done by considering $2m + 3$ public encodings $\hat{C}_{2,2,i}$ and $C_{3,3,i}$ corresponding to User 1, and similarly $2m + 2$ encodings $C_{2,3,j}$ and $\hat{C}_{3,2,j}$ corresponding to User 2, for $1 \le i \le 2m+3$ and $1 \le j \le 2m+2$. More precisely we compute as previously over R the following matrix elements, restricting ourselves to the first component:

$$(W)_{ij} = A_{2,1} \cdot \hat{C}_{2,2,i} \cdot C'_{2,3,j} - A_{3,1} \cdot \hat{C}_{3,2,j} \cdot C'_{3,3,i} \tag{4}$$

and as previously we can write:

$$(W)_{ij} = \begin{bmatrix} t_{1,i} & \hat{E}_{2,2,i} & E_{3,3,i} & C'_{3,3,i} \end{bmatrix} \cdot \begin{bmatrix} t_3 \cdot E_{2,3,j} \\ C'_{2,3,j} \\ -t_{2,j} \cdot t_3 \\ -\hat{E}_{3,2,j} \end{bmatrix}.$$

We obtain a $(2m + 3) \times (2m + 2)$ matrix W with:

$$W = \underbrace{\begin{bmatrix} & & \cdots & \\ t_{1,i} & \hat{E}_{2,2,i} & E_{3,3,i} & C'_{3,3,i} \\ & & \cdots & \end{bmatrix}}_{A} \cdot \underbrace{\begin{bmatrix} & t_3 \cdot E_{2,3,j} & \\ & C'_{2,3,j} & \\ \vdots & -t_{2,j} \cdot t_3 & \vdots \\ & -\hat{E}_{3,2,j} & \end{bmatrix}}_{B}$$

where the matrix $\boldsymbol{A}$ has $2m+3$ rows vectors, each of dimension $2m+2$, and the matrix $\boldsymbol{B}$ has $2m+2$ column vectors, each of dimension $2m+2$; hence $\boldsymbol{B}$ is a square matrix.

By doing linear algebra, we can find a vector $\boldsymbol{u}$ over R of dimension $2m+3$ such that $\boldsymbol{u} \cdot \boldsymbol{W} = 0$, which gives:

$$(\boldsymbol{u} \cdot \boldsymbol{A}) \cdot \boldsymbol{B} = 0.$$

Heuristically with good probability the matrix $\boldsymbol{B}$ is invertible, which implies:

$$\boldsymbol{u} \cdot \boldsymbol{A} = 0.$$

Since the first column of the matrix $\boldsymbol{A}$ is the column vector given by the $t_{1,i}$'s, such vector $\boldsymbol{u}$ gives a linear relation among the secret exponents $t_{1,i}$.

Moreover, since the encodings $\boldsymbol{D}_{2,2}$ and $\boldsymbol{D}_{3,3}$ corresponding to s_1 are public, we can express s_1 as a linear combination of the $t_{1,i}$'s, over R. Namely we can define as previously the product encoding $\hat{\boldsymbol{D}}_{2,2} := \boldsymbol{C}_{2,1} \cdot \boldsymbol{D}_{2,2}$, with:

$$\boldsymbol{A}_{2,1} \cdot \hat{\boldsymbol{D}}_{2,2} = s_1 \cdot t_3 \cdot \boldsymbol{A}_{2,3} + \hat{\boldsymbol{F}}_{2,2} \quad (\bmod\ q)$$

for some small error vector $\hat{\boldsymbol{F}}_{2,2}$, and we can now compute the same $(\boldsymbol{W})_{ij}$ as in (4) but with $\hat{\boldsymbol{D}}_{2,2}$ and $\boldsymbol{D}'_{3,3}$ instead of $\hat{\boldsymbol{C}}_{2,2,i}$ and $\boldsymbol{C}'_{3,3,i}$, where $\boldsymbol{D}'_{3,3}$ is the first column of $\boldsymbol{D}_{3,3}$. More precisely, we compute for all $1 \leq j \leq 2m+2$:

$$\omega_j = \boldsymbol{A}_{2,1} \cdot \hat{\boldsymbol{D}}_{2,2} \cdot \boldsymbol{C}'_{2,3,j} - \boldsymbol{A}_{3,1} \cdot \hat{\boldsymbol{C}}_{3,2,j} \cdot \boldsymbol{D}'_{3,3}$$

which gives as previously:

$$\omega_j = \begin{bmatrix} s_1 & \hat{\boldsymbol{F}}_{2,2} & F_{3,3} & \boldsymbol{D}'_{3,3} \end{bmatrix} \cdot \begin{bmatrix} t_3 \cdot E_{2,3,j} \\ \boldsymbol{C}'_{2,3,j} \\ -t_{2,j} \cdot t_3 \\ -\hat{\boldsymbol{E}}_{3,2,j} \end{bmatrix}.$$

This implies that we can replace any row vector $[t_{1,i}\ \hat{\boldsymbol{E}}_{2,2,i}\ E_{3,3,i}\ \boldsymbol{C}'_{3,3,i}]$ in the matrix $\boldsymbol{A}$ by the row vector:

$$[s_1\ \hat{\boldsymbol{F}}_{2,2}\ F_{3,3}\ \boldsymbol{D}'_{3,3}] \tag{5}$$

where $\boldsymbol{D}'_{3,3}$ is the first column of $\boldsymbol{D}_{3,3}$, and $F_{3,3}$ is the first component of $\boldsymbol{F}_{3,3}$. Using the previous technique, we can therefore obtain a linear relation between s_1 and the $t_{1,i}$'s over R. More precisely, with overwhelming probability, such a relation can be put in the form:

$$\mu \cdot s_1 = \sum_{i=1}^{2m+2} \lambda_i \cdot t_{1,i} \tag{6}$$

with $\mu \in \mathbb{Z}$ and $\lambda_1, \ldots, \lambda_{2m+2} \in R$. Indeed, we obtain such a relation by computing the kernel of the matrix analogous to $\boldsymbol{W}$ above in echelon form over the fraction field of R, which gives the kernel of the corresponding matrix $\boldsymbol{A}$ (assuming that $\boldsymbol{B}$ is invertible). Unless a minor of that matrix vanishes, which happens with only negligible probability, this gives a relation where the coefficient of s_1 is 1 and the other coefficients are in the fraction field $R \otimes_{\mathbb{Z}} \mathbb{Q}$ of R. By clearing denominators, we get an expression of the form (6).

Then, by considering exactly one additional $t_{1,i}$ (say $t_{1,2m+3}$) and carrying out the same computations with indices $i = 2, \ldots, 2m + 3$ instead of $i = 1, \ldots, 2m + 2$, we get a second relation:

$$\nu \cdot s_1 = \sum_{i=2}^{2m+3} \lambda'_i \cdot t_{1,i}.$$

If the integers μ and ν are relatively prime, which happens with significant probability[3], we can apply Bézout's identity to obtain a linear relation in R where the coefficient of s_1 is 1:

$$s_1 = \sum_{i=1}^{2m+3} \alpha_i \cdot t_{1,i}. \tag{7}$$

Note that we have the same linear relations for the other components of the vector (5) corresponding to s_1, namely:

$$\hat{\boldsymbol{F}}_{2,2} = \sum_{i=1}^{2m+3} \alpha_i \cdot \hat{\boldsymbol{E}}_{2,2,i}, \quad F_{3,3} = \sum_{i=1}^{2m+3} \alpha_i \cdot E_{3,3,i}, \quad \boldsymbol{D}'_{3,3} = \sum_{i=1}^{2m+3} \alpha_i \cdot \boldsymbol{C}'_{3,3,i}. \tag{8}$$

Second Step: Equivalent Private-Key. In this second step, we show how to publicly compute an encoding equivalent to $\boldsymbol{D}_{1,1}$, which is private to User 1; this will break the key-agreement protocol. In the first step, we had considered rows 2 and 3 to derive the linear relations (7) and (8); we now consider Row 1. On Row 1, the encodings $\boldsymbol{D}_{1,2}$ and $\boldsymbol{D}_{1,3}$ are public, so we can define as previously the product encoding $\hat{\boldsymbol{D}}_{1,3} = \boldsymbol{D}_{1,2} \cdot \boldsymbol{D}_{1,3}$, which gives:

$$\boldsymbol{A}_{1,2} \cdot \hat{\boldsymbol{D}}_{1,3} = s_2 \cdot s_3 \cdot \boldsymbol{A}_0 + \hat{\boldsymbol{F}}_{1,3} \pmod{q}$$

for some small error vector $\hat{\boldsymbol{F}}_{1,3}$. Recall that the encoding $\boldsymbol{D}_{1,1}$ is private to User 1, with:

$$\boldsymbol{A}_{1,1} \cdot \boldsymbol{D}_{1,1} = s_1 \cdot \boldsymbol{A}_{1,2} + \boldsymbol{F}_{1,1} \pmod{q}. \tag{9}$$

Therefore only User 1 can privately compute:

$$\boldsymbol{A}_{1,1} \cdot \boldsymbol{D}_{1,1} \cdot \hat{\boldsymbol{D}}_{1,3} = s_1 \cdot s_2 \cdot s_3 \cdot \boldsymbol{A}_0 + s_1 \cdot \hat{\boldsymbol{F}}_{1,3} + \boldsymbol{F}_{1,1} \cdot \hat{\boldsymbol{D}}_{1,3} \pmod{q} \tag{10}$$

and extract the high order bits of $s_1 \cdot s_2 \cdot s_3 \cdot \boldsymbol{A}_0 \bmod q$ to generate the session key.

---

[3] Heuristically, it is the probability that two random elements of R have coprime norms, since the rational integer denominator of an element of the fraction field has the same prime factors as its norm. For $R = \mathbb{Z}[x]/(x^{2^n} + 1)$, that probability is close to 3/4: see the full version of this paper [CLLT15].

We cannot compute the previous equation since $\boldsymbol{D}_{1,1}$ is private. However since we know a linear relation (7) between s_1 and the $t_{1,i}$'s, and the encodings $\boldsymbol{C}_{1,1,i}$ corresponding to $t_{1,i}$ are public, with:

$$\boldsymbol{A}_{1,1} \cdot \boldsymbol{C}_{1,1,i} = t_{1,i} \cdot \boldsymbol{A}_{1,2} + \boldsymbol{E}_{1,1,i} \quad (\text{mod } q)$$

it is then natural to compute:

$$\tilde{\boldsymbol{D}}_{1,1} = \sum_{i=1}^{2m+3} \alpha_i \cdot \boldsymbol{C}_{1,1,i},$$

which gives:

$$\boldsymbol{A}_{1,1} \cdot \tilde{\boldsymbol{D}}_{1,1} = s_1 \cdot \boldsymbol{A}_{1,2} + \sum_{i=1}^{2m+3} \alpha_i \cdot \boldsymbol{E}_{1,1,i} \quad (\text{mod } q). \tag{11}$$

The difference with (9) is that the error term $\sum_{i=1}^{2m+3} \alpha_i \cdot \boldsymbol{E}_{1,1,i}$ is not necessarily small since the coefficients α_i can be large. Therefore if we compute:

$$\boldsymbol{A}_{1,1} \cdot \tilde{\boldsymbol{D}}_{1,1} \cdot \hat{\boldsymbol{D}}_{1,3} = s_1 \cdot s_2 \cdot s_3 \cdot \boldsymbol{A}_0 + s_1 \cdot \hat{\boldsymbol{F}}_{1,3} + \left(\sum_{i=1}^{2m+3} \alpha_i \cdot \boldsymbol{E}_{1,1,i} \right) \cdot \hat{\boldsymbol{D}}_{1,3} \quad (\text{mod } q) \tag{12}$$

then as opposed to (10) this does not reveal the high-order bits of $s_1 \cdot s_2 \cdot s_3 \cdot \boldsymbol{A}_0 \bmod q$. In the following, we show how to derive an approximation of $\sum_{i=1}^{2m+3} \alpha_i \cdot \boldsymbol{E}_{1,1,i}$ over R, in order to correct the error in (11) and break the protocol. This is the second part of our attack.

As in the first step of the attack, to get equations over R and not only modulo q, we consider the difference between two rows, this time the difference between rows 1 and 3 (instead of rows 2 and 3). We have the public encodings:

$$\boldsymbol{A}_{1,1} \cdot \boldsymbol{C}_{1,1,\ell} = t_{1,\ell} \cdot \boldsymbol{A}_{1,2} + \boldsymbol{E}_{1,1,\ell} \quad (\text{mod } q)$$
$$\boldsymbol{A}_{1,2} \cdot \hat{\boldsymbol{C}}_{1,3,\ell} = t_{2,\ell} \cdot t_3 \cdot \boldsymbol{A}_0 + \hat{\boldsymbol{E}}_{1,3,\ell} \quad (\text{mod } q)$$
$$\boldsymbol{A}_{3,1} \cdot \hat{\boldsymbol{C}}_{3,2,\ell} = t_{2,\ell} \cdot t_3 \cdot \boldsymbol{A}_{3,3} + \hat{\boldsymbol{E}}_{3,2,\ell} \quad (\text{mod } q)$$
$$\boldsymbol{A}_{3,3} \cdot \boldsymbol{C}_{3,3,\ell} = t_{1,\ell} \cdot \boldsymbol{A}_0 + \boldsymbol{E}_{3,3,\ell} \quad (\text{mod } q)$$

where we let $\hat{\boldsymbol{C}}_{1,3,\ell} := \boldsymbol{C}_{1,2,\ell} \cdot \boldsymbol{C}_{1,3}$, for some small error vector $\hat{\boldsymbol{E}}_{1,3,\ell}$. As previously we can compute over R, restricting ourselves to the first component, where $\hat{\boldsymbol{C}}'_{1,3,j}$ and $\boldsymbol{C}'_{3,3,i}$ are the first columns of $\hat{\boldsymbol{C}}_{1,3,j}$ and $\boldsymbol{C}_{3,3,i}$ respectively:

$$\omega_{ij} = \boldsymbol{A}_{1,1} \cdot \boldsymbol{C}_{1,1,i} \cdot \hat{\boldsymbol{C}}'_{1,3,j} - \boldsymbol{A}_{3,1} \cdot \hat{\boldsymbol{C}}_{3,2,j} \cdot \boldsymbol{C}'_{3,3,i}$$

$$= t_{1,i} \cdot \hat{\boldsymbol{E}}_{1,3,j} + \boldsymbol{E}_{1,1,i} \cdot \hat{\boldsymbol{C}}'_{1,3,j} - t_{2,j} \cdot t_3 \cdot \boldsymbol{E}_{3,3,i} - \hat{\boldsymbol{E}}_{3,2,j} \cdot \boldsymbol{C}'_{3,3,i}.$$

We can therefore compute over R, using the coefficients α_i from the linear relation (7):

$$\Omega_j = \sum_{i=1}^{2m+3} \alpha_i \cdot \left(\boldsymbol{A}_{1,1} \cdot \boldsymbol{C}_{1,1,i} \cdot \hat{\boldsymbol{C}}'_{1,3,j} - \boldsymbol{A}_{3,1} \cdot \hat{\boldsymbol{C}}_{3,2,j} \cdot \boldsymbol{C}'_{3,3,i} \right) \tag{13}$$

$$= \sum_{i=1}^{2m+3} \alpha_i \cdot \left(t_{1,i} \cdot \hat{E}_{1,3,j} + \boldsymbol{E}_{1,1,i} \cdot \hat{\boldsymbol{C}}'_{1,3,j} - t_{2,j} \cdot t_3 \cdot E_{3,3,i} - \hat{\boldsymbol{E}}_{3,2,j} \cdot \boldsymbol{C}'_{3,3,i} \right).$$

Using the linear relations (7) and (8), we obtain:

$$\Omega_j = s_1 \cdot \hat{E}_{1,3,j} - t_{2,j} \cdot t_3 \cdot F_{3,3} - \hat{E}_{3,2,j} \cdot \boldsymbol{D}'_{3,3} + \left(\sum_{i=1}^{2m+3} \alpha_i \cdot \boldsymbol{E}_{1,1,i} \right) \cdot \hat{\boldsymbol{C}}'_{1,3,j}$$

which gives:

$$\Omega_j = u_j + \left(\sum_{i=1}^{2m+3} \alpha_i \cdot \boldsymbol{E}_{1,1,i} \right) \cdot \hat{\boldsymbol{C}}'_{1,3,j} \tag{14}$$

for some small u_j in R. In summary we obtain a large scalar Ω_j because the coefficients α_i in (13) are large, but eventually what makes Ω_j large is only the contribution from $(\sum_{i=1}^{2m+3} \alpha_i \cdot \boldsymbol{E}_{1,1,i}) \cdot \hat{\boldsymbol{C}}'_{1,3,j}$; namely because of the linear relations (7) and (8) the other terms remain small.

We can now write (14) in vectorial form, where we let $\hat{\boldsymbol{C}}''_{1,3}$ be the square matrix whose columns are the column vectors $\hat{\boldsymbol{C}}'_{1,3,j}$ for $1 \leq j \leq m$; recall that the $\hat{\boldsymbol{C}}'_{1,3,j}$ are the first column vectors of the matrix encodings $\hat{\boldsymbol{C}}_{1,3,j}$. We obtain a row vector $\boldsymbol{\Omega}$ of dimension m, where:

$$\boldsymbol{\Omega} = \boldsymbol{u} + \left(\sum_{i=1}^{2m+3} \alpha_i \cdot \boldsymbol{E}_{1,1,i} \right) \cdot \hat{\boldsymbol{C}}''_{1,3} \tag{15}$$

where $\hat{\boldsymbol{C}}''_{1,3}$ is a public square matrix of dimension m.

Now the crucial observation is that because the vector $\boldsymbol{u}$ has small components, we can get an approximation of the vector $\sum_{i=1}^{2m+3} \alpha_i \cdot \boldsymbol{E}_{1,1,i}$ by reducing the vector $\boldsymbol{\Omega}$ modulo the matrix $\hat{\boldsymbol{C}}''_{1,3}$, assuming that $\hat{\boldsymbol{C}}''_{1,3}$ is an invertible matrix, which heuristically holds with good probability. This can be done by solving over the fraction field of R the linear system $\boldsymbol{\Omega} = \boldsymbol{y} \cdot \hat{\boldsymbol{C}}''_{1,3}$ and then rounding to R the coefficients of $\boldsymbol{y}$. Heuristically the vector $\boldsymbol{E} = \lfloor \boldsymbol{y} \rceil$ should be a good approximation of $\sum_{i=1}^{2m+3} \alpha_i \cdot \boldsymbol{E}_{1,1,i}$; namely letting:

$$\boldsymbol{E}' = \sum_{i=1}^{2m+3} \alpha_i \cdot \boldsymbol{E}_{1,1,i} - \boldsymbol{E} \tag{16}$$

we get using $y = \Omega \cdot \hat{C}_{1,3}^{''-1}$:

$$E' = (\Omega - u) \cdot \hat{C}_{1,3}^{''-1} - E$$

$$= y - E - u \cdot \hat{C}_{1,3}^{''-1}$$

and therefore since $y - E$ and u are small, the difference vector E' should be small if the norm of the transpose of the matrix $\hat{C}_{1,3}^{''-1}$ remains small. We know that such a bound holds with probability close to 1 if we model $\hat{C}_{1,3}^{''}$ as a random matrix (e.g. Rudelson [Rud08] provides a bound of the form $O(m^{3/2})$), and so we expect E' to be small (compared to q) for randomly generated encodings, since in the GGH15 parameter selection one takes $m = \Theta(\log q)$.

Combining (11) and (16), we get:

$$A_{1,1} \cdot \tilde{D}_{1,1} - E = s_1 \cdot A_{1,2} + E' \pmod{q}$$

for a small vector E'. Note that the previous equation is very similar to the original equation for the private encoding $D_{1,1}$:

$$A_{1,1} \cdot D_{1,1} = s_1 \cdot A_{1,2} + F_{1,1} \pmod{q}$$

the only difference being the publicly computed correction vector E. Therefore the pair $(\tilde{D}_{1,1}, E)$ gives us an equivalent of the private encoding $D_{1,1}$, which breaks the protocol. More precisely we can eventually compute from public parameters:

$$\left(A_{1,1} \cdot \tilde{D}_{1,1} - E\right) \cdot D_{1,2} \cdot D_{1,3} = \left(s_1 \cdot A_{1,2} + E'\right) \cdot \hat{D}_{1,3} \pmod{q}$$

$$= s_1 \cdot s_2 \cdot s_3 \cdot A_0$$

$$+ s_1 \cdot \hat{F}_{1,3} + E' \cdot \hat{D}_{1,3} \pmod{q}.$$

Since all the error terms are small, this enables to extract the high-order bits of $s_1 \cdot s_2 \cdot s_3 \cdot A_0 \bmod q$, and breaks the protocol.

3.2 Extension to $k \geq 3$ Users

The extension of our attack to $k \geq 3$ users is relatively straightforward and described in the full version of this paper [CLLT15].

4 Cryptanalysis of GGH15 with Safeguards

In [GGH15, Sect. 5.1] two safeguards for multipartite key agreement based on GGH15 multilinear maps are described:

1. Kilian-style randomization of the encodings, where C is replaced by $\bar{C} := R^{-1} \cdot C \cdot R'$ using the randomizer matrices R, R' belonging to two adjacent nodes.

2. Choosing the first encoding matrix in each chain to have large entries.

In the following, we show how to extend our previous attack when those two safeguards are used.

4.1   First Safeguard: Kilian-Style Randomization of the Encodings

The following safeguard for GGH15 multilinear maps is described in [GGH15], using Kilian-type randomization [Kil88]. For each internal node v in the graph one can choose a random invertible $m \times m$ matrix $\boldsymbol{R}_v$ modulo q, and for the sinks and sources we set $\boldsymbol{R}_v = \boldsymbol{I}$. Then each encoding $\boldsymbol{C}$ relative to path $u \leadsto v$ is replaced by a masked encoding $\bar{\boldsymbol{C}} := \boldsymbol{R}_u^{-1} \cdot \boldsymbol{C} \cdot \boldsymbol{R}_v$. Concretely, in the GGH15 key-agreement protocol, instead of publishing encodings $\boldsymbol{C}_{i,j}$ with:

$$\boldsymbol{A}_{i,j} \cdot \boldsymbol{C}_{i,j,\ell} = t_{1+(j-i \bmod k),\ell} \cdot \boldsymbol{A}_{i,j+1} + \boldsymbol{E}_{i,j,\ell} \pmod{q}$$

one would only publish the masked encodings modulo q:

$$\bar{\boldsymbol{C}}_{i,j,\ell} := \boldsymbol{R}_{i,j}^{-1} \cdot \boldsymbol{C}_{i,j,\ell} \cdot \boldsymbol{R}_{i,j+1} \tag{17}$$

with $\boldsymbol{R}_{i,1} = \boldsymbol{R}_{i,k+1} = \boldsymbol{I}$ for all i; the same masking is applied to the encodings $\boldsymbol{D}_{i,j}$. Since the product of encoding on any source-to-sink path remains the same, the same value is eventually extracted. Namely for all i we have:

$$\prod_{j=1}^{k} \bar{\boldsymbol{C}}_{i,j} = \prod_{j=1}^{k} \boldsymbol{C}_{i,j}$$

and therefore exactly the same session-key as before is computed by all users.

4.2   Second Safeguard: First Encodings with Large Entries

The second safeguard described in [GGH15, Sect. 5.1] consists in choosing the first encodings $\boldsymbol{C}_{i,1}$ in each chain to have large entries modulo q, instead of small entries. Namely the first encoding $\boldsymbol{C}_{i,1}$ does not contribute in the error term when computing the session-key, so it can have large entries.

4.3   Cryptanalysis of GGH15 with both Safeguards

In this section we show how to extend our attack from Sect. 3 when both safeguards are used. Note the first step of our attack still applies, since in the first step we are only using product of encodings from source to sink. Namely in Eq. (4) exactly the same value $(\boldsymbol{W})_{ij}$ is obtained when using masked encodings. Therefore we can still derive the same linear relation between secret exponents as in (7) and (8).

However the second step of our attack does not apply directly, since our second step requires the knowledge of the matrix $\hat{\boldsymbol{C}}_{1,3}''$ in (15), which is obtained from the first columns of the encodings $\hat{\boldsymbol{C}}_{1,3,j} = \boldsymbol{C}_{1,2,j} \cdot \boldsymbol{C}_{1,3}$. Since these are partial products only, such partial products would be masked by the unknown randomization matrix $\boldsymbol{R}_{1,2}^{-1}$ modulo q, hence the matrix $\hat{\boldsymbol{C}}_{1,3}''$ is unknown.

We can however adapt our second step as follows. For simplicity we keep the same notations as previously, that is we describe our extended attack in term of the original encodings $C_{i,j,\ell}$, instead of the masked encodings $\bar{C}_{i,j,\ell}$ from (17); in that case we are only allowed to use products of encodings from source to sink. We first start with a slightly different equation from (15):

$$\Omega = u + \left(\sum_{i=1}^{2m+3} \alpha_i \cdot E_{1,1,i} \right) \cdot \hat{G}''_{1,3} \qquad (18)$$

where $\hat{G}''_{1,3}$ is a matrix whose columns are the first column vectors of $D_{1,2} \cdot C_{1,3,j}$ for $1 \leq j \leq 2m + 2$. Note that in (12) the error term that we must estimate to recover the session key is:

$$E = \left(\sum_{i=1}^{2m+3} \alpha_i \cdot E_{1,1,i} \right) \cdot \hat{D}_{1,3} \qquad (19)$$

Using a similar approach as in the attack first step, our approach consists in finding a vector x with coefficients in the fraction field $R \otimes_{\mathbb{Z}} \mathbb{Q}$ of R such that:

$$\hat{D}'_{1,3} = \hat{G}''_{1,3} \cdot x$$

where $\hat{D}'_{1,3}$ is the first column vector of $\hat{D}_{1,3}$. Applying the vector x on (18) and rounding in R, we obtain:

$$\lfloor \Omega \cdot x \rceil = \lfloor u \cdot x \rceil + \left(\sum_{i=1}^{2m+3} \alpha_i \cdot E_{1,1,i} \right) \cdot \hat{D}'_{1,3}$$

Since the components of u (over R) are small, and moreover the coefficients of x (over $R \otimes_{\mathbb{Z}} \mathbb{Q}$) are heuristically also small, the scalar $\lfloor u \cdot x \rceil$ in R is small compared to q, and therefore we obtain a good estimate of the first component of the error vector E from (19), which enables to recover the first component of the session key and breaks the scheme.[4]

4.4 Detailed Description

First Step: Linear Relations in R. The first step of our attack is exactly the same as previously. Namely as mentioned previously the first step of our previous attack still applies, since in the first step we are only using product of encodings from source to sink. More precisely in Eq. (4) exactly the same value $(W)_{ij}$ is obtained when using masked encodings, and therefore we can still derive the same linear relations as in (7) and (8):

$$s_1 = \sum_{i=1}^{2m+3} \alpha_i \cdot t_{1,i}, \quad \hat{F}_{2,2} = \sum_{i=1}^{2m+3} \alpha_i \cdot \hat{E}_{2,2,i},$$

$$F_{3,3} = \sum_{i=1}^{2m+3} \alpha_i \cdot E_{3,3,i}, \quad D'_{3,3} = \sum_{i=1}^{2m+3} \alpha_i \cdot C'_{3,3,i}. \qquad (20)$$

---

[4] Other components of the session key can be also obtained analogously.

Note that as opposed to Sect. 3 we don't know the value of the encodings $D'_{3,3}$ and $C'_{3,3,i}$, since they are masked by the R_{ij} matrices; we only recover the coefficients α_i in R.

Second Step: Another Linear Relation. In the second step, our goal is to find a vector x with coefficients in the fraction field $R \otimes_{\mathbb{Z}} \mathbb{Q}$ of R such that:

$$D'_{1,3} = \sum_{i=1}^{2m+2} x_i \cdot C'_{1,3,i}$$

where $D'_{1,3}$ and $C'_{1,3,i}$ are the first column vectors of $D_{1,3}$ and $C_{1,3,i}$ respectively. We show that this can be done using the same approach as in the attack first step.

Namely letting $\hat{C}_{1,2,\ell} := C_{1,1,\ell} \cdot C_{1,2}$ where we let $C_{1,2} := C_{1,2,1}$ corresponding to $t_2 := t_{2,1}$, we obtain:

$$A_{1,1} \cdot \hat{C}_{1,2,\ell} = t_{1,\ell} \cdot t_2 \cdot A_{1,3} + \hat{E}_{1,2,\ell} \pmod{q}$$
$$A_{1,3} \cdot C_{1,3,\ell} = t_{3,\ell} \cdot A_0 + E_{1,3,\ell} \pmod{q}$$

Similarly letting $\hat{C}_{2,3,\ell} := C_{2,2,\ell} \cdot C_{2,3}$ where $C_{2,3} := C_{2,3,1}$, we get:

$$A_{2,1} \cdot C_{2,1,\ell} = t_{3,\ell} \cdot A_{2,2} + E_{2,1,\ell} \pmod{q}$$
$$A_{2,2} \cdot \hat{C}_{2,3,\ell} = t_1 \cdot t_{2,\ell} \cdot A_0 + \hat{E}_{2,3,\ell} \pmod{q}$$

We can therefore compute the following matrix elements in R, restricting ourselves as previously to the first component of the vectors:

$$(W)_{ij} = A_{1,1} \cdot \hat{C}_{1,2,i} \cdot C'_{1,3,j} - A_{2,1} \cdot C_{2,1,j} \cdot \hat{C}'_{2,3,i}$$
$$= t_{1,i} \cdot t_2 \cdot E_{1,3,j} + \hat{E}_{1,2,i} \cdot C'_{1,3,j} - t_{3,j} \cdot \hat{E}_{2,3,i} - E_{2,1,j} \cdot \hat{C}'_{2,3,i}$$

for all $1 \leq i \leq 2m+2$ and $1 \leq j \leq 2m+2$, where $C'_{1,3,j}$ and $\hat{C}'_{2,3,i}$ are the first column vectors of $C_{1,3,j}$ and $\hat{C}_{2,3,i}$ respectively. This gives:

$$(W)_{ij} = \begin{bmatrix} t_{1,i} t_2 & \hat{E}_{1,2,i} & \hat{E}_{2,3,i} & \hat{C}'_{2,3,i} \end{bmatrix} \cdot \begin{bmatrix} E_{1,3,j} \\ C'_{1,3,j} \\ -t_{3,j} \\ -E_{2,1,j} \end{bmatrix} .$$

Moreover, since the encodings $D_{1,3}$ and $D_{2,1}$ corresponding to s_3 on rows 1 and 2 are public, we can additionally compute the corresponding vector:

$$(V)_i = A_{1,1} \cdot \hat{C}_{1,2,i} \cdot D'_{1,3} - A_{2,1} \cdot D_{2,1} \cdot \hat{C}'_{2,3,i}$$
$$= \begin{bmatrix} t_{1,i} t_2 & \hat{E}_{1,2,i} & \hat{E}_{2,3,i} & \hat{C}'_{2,3,i} \end{bmatrix} \cdot \begin{bmatrix} F_{1,3} \\ D'_{1,3} \\ -s_3 \\ -F_{2,1} \end{bmatrix} .$$

where $\boldsymbol{D}'_{1,3}$ is the first column vector of $\boldsymbol{D}_{1,3}$. Therefore assuming that the matrix $\boldsymbol{W}$ is invertible, we can find $\boldsymbol{x}$ in $R \otimes_{\mathbb{Z}} \mathbb{Q}$ such that:

$$\boldsymbol{W} \cdot \boldsymbol{x} = \boldsymbol{V}$$

which gives as required:

$$\boldsymbol{D}'_{1,3} = \sum_{i=1}^{2m+2} x_i \cdot \boldsymbol{C}'_{1,3,i} \tag{21}$$

Note that the only difference with the linear relations from Step 1 is that we don't require the x_i's to be in R, only in the fraction field $R \otimes_{\mathbb{Z}} \mathbb{Q}$ of R; this implies that heuristically such coefficients should remain small in absolute value.

Third Step: Estimating the Error Term. In the third step our goal is to estimate the error term when computing the session-key, as in the second step of the basic attack. We first start with a slightly different equation from (15):

$$\Omega = \boldsymbol{u} + \left(\sum_{i=1}^{2m+3} \alpha_i \cdot \boldsymbol{E}_{1,1,i} \right) \cdot \hat{\boldsymbol{G}}''_{1,3} \tag{22}$$

where $\hat{\boldsymbol{G}}''_{1,3}$ is a matrix whose columns are the first column vectors of $\boldsymbol{D}_{1,2} \cdot \boldsymbol{C}_{1,3,j}$ for $1 \leq j \leq 2m + 2$. Therefore the only difference with (15) is that we use the matrix $\hat{\boldsymbol{G}}''_{1,3}$ instead of $\hat{\boldsymbol{C}}''_{1,3}$.

To obtain (22) we proceed as follows. Instead of letting $\hat{\boldsymbol{C}}_{1,3,\ell} = \boldsymbol{C}_{1,2,\ell} \cdot \boldsymbol{C}_{1,3}$ as in the basic attack, we let $\hat{\boldsymbol{C}}_{1,3,\ell} = \boldsymbol{D}_{1,2} \cdot \boldsymbol{C}_{1,3,\ell}$. Similarly we let $\hat{\boldsymbol{C}}_{3,2,\ell} := \boldsymbol{D}_{3,1} \cdot \boldsymbol{C}_{3,2,\ell}$. This is possible because on rows 1 and 3 the encodings $\boldsymbol{D}_{1,2}$ and $\boldsymbol{D}_{3,1}$ corresponding to s_2 are public. We obtain:

$$\boldsymbol{A}_{1,1} \cdot \boldsymbol{C}_{1,1,\ell} = t_{1,\ell} \cdot \boldsymbol{A}_{1,2} + \boldsymbol{E}_{1,1,\ell} \pmod{q}$$

$$\boldsymbol{A}_{1,2} \cdot \hat{\boldsymbol{C}}_{1,3,\ell} = s_2 \cdot t_{3,\ell} \cdot \boldsymbol{A}_0 + \hat{\boldsymbol{E}}_{1,3,\ell} \pmod{q}$$

$$\boldsymbol{A}_{3,1} \cdot \hat{\boldsymbol{C}}_{3,2,\ell} = s_2 \cdot t_{3,\ell} \cdot \boldsymbol{A}_{3,3} + \hat{\boldsymbol{E}}_{3,2,\ell} \pmod{q}$$

$$\boldsymbol{A}_{3,3} \cdot \boldsymbol{C}_{3,3,\ell} = t_{1,\ell} \cdot \boldsymbol{A}_0 + \boldsymbol{E}_{3,3,\ell} \pmod{q}$$

As previously we can compute over R, restricting ourselves to the first component, where $\hat{\boldsymbol{C}}'_{1,3,j}$ and $\boldsymbol{C}'_{3,3,i}$ are the first columns of $\hat{\boldsymbol{C}}_{1,3,j}$ and $\boldsymbol{C}_{3,3,i}$ respectively:

$$\omega_{ij} = \boldsymbol{A}_{1,1} \cdot \boldsymbol{C}_{1,1,i} \cdot \hat{\boldsymbol{C}}'_{1,3,j} - \boldsymbol{A}_{3,1} \cdot \hat{\boldsymbol{C}}_{3,2,j} \cdot \boldsymbol{C}'_{3,3,i}$$

$$= t_{1,i} \cdot \hat{\boldsymbol{E}}_{1,3,j} + \boldsymbol{E}_{1,1,i} \cdot \hat{\boldsymbol{C}}'_{1,3,j} - s_2 \cdot t_{3,j} \cdot \boldsymbol{E}_{3,3,i} - \hat{\boldsymbol{E}}_{3,2,j} \cdot \boldsymbol{C}'_{3,3,i}.$$

We can therefore compute over R, using the coefficients α_i from the linear relations (20):

$$\Omega_j = \sum_{i=1}^{2m+3} \alpha_i \cdot \omega_{ij}$$

$$= \sum_{i=1}^{2m+3} \alpha_i \cdot \left(t_{1,i} \cdot \hat{\boldsymbol{E}}_{1,3,j} + \boldsymbol{E}_{1,1,i} \cdot \hat{\boldsymbol{C}}'_{1,3,j} - s_2 \cdot t_{3,j} \cdot \boldsymbol{E}_{3,3,i} - \hat{\boldsymbol{E}}_{3,2,j} \cdot \boldsymbol{C}'_{3,3,i} \right)$$

Using the linear relations in (20), we obtain:

$$\Omega_j = s_1 \cdot \hat{E}_{1,3,j} - s_2 \cdot t_{3,j} \cdot F_{3,3} - \hat{E}_{3,2,j} \cdot D'_{3,3} + \left(\sum_{i=1}^{2m+3} \alpha_i \cdot E_{1,1,i} \right) \cdot \hat{C}'_{1,3,j}$$

where $D'_{3,3}$ is the first column vector of $D_{3,3}$. This gives:

$$\Omega_j = u_j + \left(\sum_{i=1}^{2m+3} \alpha_i \cdot E_{1,1,i} \right) \cdot \hat{C}'_{1,3,j}$$

for some small u_j in R. Since we have let $\hat{C}_{1,3,j} = D_{1,2} \cdot C_{1,3,j}$ for $1 \le j \le 2m+2$, in vectorial form we obtain (22) as required, where $\hat{G}''_{1,3}$ is the matrix whose columns are the first column vectors of $D_{1,2} \cdot C_{1,3,j}$ for $1 \le j \le 2m + 2$.

Recall that in (12) the error term that we must estimate to recover the session key is:

$$E = \left(\sum_{i=1}^{2m+3} \alpha_i \cdot E_{1,1,i} \right) \cdot \hat{D}_{1,3} \tag{23}$$

where $\hat{D}_{1,3} = D_{1,2} \cdot D_{1,3}$. In the following we will only estimate the first component, so we let $\hat{D}'_{1,3} = D_{1,2} \cdot D'_{1,3}$, where $\hat{D}'_{1,3}$ and $D'_{1,3}$ are the first column vectors of $\hat{D}_{1,3}$ and $D_{1,3}$ respectively.

We now use the vector x computed in the second step. In matrix notation, Eq. (21) gives:

$$D'_{1,3} = C''_{1,3} \cdot x$$

where $C''_{1,3}$ is the matrix whose columns are the first column vectors of $C_{1,3,i}$ for $1 \le i \le 2m + 2$. Using $\hat{G}''_{1,3} = D_{1,2} \cdot C''_{1,3}$, this gives:

$$\hat{D}'_{1,3} = D_{1,2} \cdot D'_{1,3} = D_{1,2} \cdot C''_{1,3} \cdot x = \hat{G}''_{1,3} \cdot x$$

where $\hat{D}'_{1,3}$ is the first column vector of $\hat{D}_{1,3}$. Applying the vector x on (22), we therefore get:

$$\Omega \cdot x = u \cdot x + \left(\sum_{i=1}^{2m+3} \alpha_i \cdot E_{1,1,i} \right) \cdot \hat{D}'_{1,3}$$

We claim that this provides a good estimate of the first component of the error vector E from (23). Recall that the components of x are in $R \otimes_{\mathbb{Z}} \mathbb{Q}$, so by rounding to the nearest integer we can get the following value in R:

$$E' = \lfloor \Omega \cdot x \rceil = \lfloor u \cdot x \rceil + \left(\sum_{i=1}^{2m+3} \alpha_i \cdot E_{1,1,i} \right) \cdot \hat{D}'_{1,3} \tag{24}$$

Since the components of u (over R) are small, and moreover the coefficients of x (over $R \otimes_{\mathbb{Z}} \mathbb{Q}$) are also small (heuristically), the scalar $\lfloor u \cdot x \rceil$ in R is small.

Finally, letting as previously:

$$\tilde{\boldsymbol{D}}_{1,1} = \sum_{i=1}^{2m+3} \alpha_i \cdot \boldsymbol{C}_{1,1,i},$$

we obtain:

$$\boldsymbol{A}_{1,1} \cdot \tilde{\boldsymbol{D}}_{1,1} = s_1 \cdot \boldsymbol{A}_{1,2} + \sum_{i=1}^{2m+3} \alpha_i \cdot \boldsymbol{E}_{1,1,i} \quad (\text{mod } q).$$

which gives as previously:

$$\boldsymbol{A}_{1,1} \cdot \tilde{\boldsymbol{D}}_{1,1} \cdot \hat{\boldsymbol{D}}'_{1,3} = s_1 \cdot s_2 \cdot s_3 \cdot A_0 + s_1 \cdot \hat{F}_{1,3} + \left(\sum_{i=1}^{2m+3} \alpha_i \cdot \boldsymbol{E}_{1,1,i} \right) \cdot \hat{\boldsymbol{D}}'_{1,3} \quad (\text{mod } q)$$

Therefore combining with (24) we can compute from public parameters:

$$\boldsymbol{A}_{1,1} \cdot \tilde{\boldsymbol{D}}_{1,1} \cdot \hat{\boldsymbol{D}}'_{1,3} - E' = s_1 \cdot s_2 \cdot s_3 \cdot A_0 + s_1 \cdot \hat{F}'_{1,3} - \lfloor \boldsymbol{u} \cdot \boldsymbol{x} \rceil \quad (\text{mod } q)$$

Since the terms $s_1 \cdot \hat{F}'_{1,3}$ and $\lfloor \boldsymbol{u} \cdot \boldsymbol{x} \rceil$ are small, this reveals the first component of the secret vector $s_1 \cdot s_2 \cdot s_3 \cdot \boldsymbol{A}_0$, which breaks the scheme.

Acknowledgements. This work has been supported in part by the European Union's H2020 Programme under grant agreement number ICT-644209.

References

[BGH+15] Brakerski, Z., Gentry, C., Halevi, S., Lepoint, T., Sahai, A., Tibouchi, M.: Cryptanalysis of the quadratic zero-testing of GGH. Cryptology ePrint Archive, Report 2015/845 (2015). https://eprint.iacr.org/2015/845

[BS02] Boneh, D.: Silverberg, Alice: Applications of multilinear forms to cryptography. Contemp. Math. **324**, 71–90 (2002)

[CFL+16] Cheon, J.H., Fouque, P.-A., Lee, C., Minaud, B., Ryu, H.: Cryptanalysis of the new CLT multilinear map over the integers. In: Fischlin, M., Coron, J.-S. (eds.) EUROCRYPT 2016. LNCS, vol. 9665, pp. 509–536. Springer, Heidelberg (2016). doi:10.1007/978-3-662-49890-3_20

[CGH+15] Coron, J.-S., Gentry, C., Halevi, S., Lepoint, T., Maji, H.K., Miles, E., Raykova, M., Sahai, A., Tibouchi, M.: Zeroizing without low-level zeroes: new MMAP attacks and their limitations. In: Gennaro, R., Robshaw, M. (eds.) CRYPTO 2015, Part I. LNCS, vol. 9215, pp. 247–266. Springer, Heidelberg (2015)

[CHL+15] Cheon, J.H., Han, K., Lee, C., Ryu, H., Stehlé, D.: Cryptanalysis of the multilinear map over the integers. In: Oswald, E., Fischlin, M. (eds.) EUROCRYPT 2015. LNCS, vol. 9056, pp. 3–12. Springer, Heidelberg (2015)

[CLLT15] Coron, J.-S., Lee, M.S., Lepoint, T., Tibouchi, M.: Cryptanalysis of GGH15 multilinear maps. Cryptology ePrint Archive, Report 2015/1037 (2015). http://eprint.iacr.org/

[CLT13] Coron, J.-S., Lepoint, T., Tibouchi, M.: Practical multilinear maps over the integers. In: Canetti, R., Garay, J.A. (eds.) CRYPTO 2013, Part I. LNCS, vol. 8042, pp. 476–493. Springer, Heidelberg (2013)

[CLT15] Coron, J.-S., Lepoint, T., Tibouchi, M.: New multilinear maps over the integers. In: Gennaro, R., Robshaw, M. (eds.) CRYPTO 2015, Part I. LNCS, vol. 9215, pp. 267–286. Springer, Heidelberg (2015)

[Dev16] The Sage Developers. Sage Mathematics Software (Version 7.0) (2016). http://www.sagemath.org

[DH76] Diffie, W., Hellman, M.E.: New directions in cryptography. IEEE Trans. Inf. Theory **22**(6), 644–654 (1976)

[GGH13a] Garg, S., Gentry, C., Halevi, S.: Candidate multilinear maps from ideal lattices. In: Johansson, T., Nguyen, P.Q. (eds.) EUROCRYPT 2013. LNCS, vol. 7881, pp. 1–17. Springer, Heidelberg (2013)

[GGH+13b] Garg, S., Gentry, C., Halevi, S., Raykova, M., Sahai, A., Waters, B.: Candidate indistinguishability obfuscation and functional encryption for all circuits. In: Reingold, O. (ed.) FOCS 2013, pp. 40–49. IEEE Computer Society, USA (2013)

[GGH15] Gentry, C., Gorbunov, S., Halevi, S.: Graph-induced multilinear maps from lattices. In: Dodis, Y., Nielsen, J.B. (eds.) TCC 2015, Part II. LNCS, vol. 9015, pp. 498–527. Springer, Heidelberg (2015)

[GSW13] Gentry, C., Sahai, A., Waters, B.: Homomorphic encryption from learning with errors: conceptually-simpler, asymptotically-faster, attribute-based. In: Canetti, R., Garay, J.A. (eds.) CRYPTO 2013, Part I. LNCS, vol. 8042, pp. 75–92. Springer, Heidelberg (2013)

[Hal15] Halevi, S.: Graded encoding, variations on a scheme. Cryptology ePrint Archive, Report 2015/866 (2015). https://eprint.iacr.org/2015/866

[HJ16] Hu, Y., Jia, H.: Cryptanalysis of GGH Map. In: Fischlin, M., Coron, J.-S. (eds.) EUROCRYPT 2016. LNCS, vol. 9665, pp. 537–565. Springer, Heidelberg (2016). doi:10.1007/978-3-662-49890-3_21

[Jou00] Joux, A.: A one round protocol for tripartite Diffie-Hellman. In: Bosma, W. (ed.) ANTS 2000. LNCS, vol. 1838, pp. 385–394. Springer, Heidelberg (2000)

[Kil88] Kilian, J.: Founding cryptography on oblivious transfer. In: Simon, J. (ed.) STOC 1988, pp. 20–31. ACM (1988)

[MP12] Micciancio, D., Peikert, C.: Trapdoors for lattices: simpler, tighter, faster, smaller. In: Pointcheval, D., Johansson, T. (eds.) EUROCRYPT 2012. LNCS, vol. 7237, pp. 700–718. Springer, Heidelberg (2012)

[MSZ16] Miles, E., Sahai, A., Zhandry, M.: Annihilation attacks for multilinear maps: cryptanalysis of indistinguishability obfuscation over GGH13. Cryptology ePrint Archive, Report 2016/147 (2016). https://eprint.iacr.org/2016/147

[PS15] Pellet-Mary, A., Damien Stehlé, D.: Cryptanalysis of Gu's ideal multilinear map. Cryptology ePrint Archive, Report 2015/759 (2015). https://eprint.iacr.org/2015/759

[Rud08] Rudelson, M.: Invertibility of random matrices: norm of the inverse. Ann. Math. **168**(2), 575–600 (2008)

Annihilation Attacks for Multilinear Maps: Cryptanalysis of Indistinguishability Obfuscation over GGH13

Eric Miles[1]([⊠]), Amit Sahai[1], and Mark Zhandry[2,3]

[1] Center for Encrypted Functionalities, UCLA, Los Angeles, USA
{enmiles,sahai}@cs.ucla.edu
[2] MIT, Cambridge, USA
mzhandry@gmail.com
[3] Princeton University, Princeton, USA

Abstract. In this work, we present a new class of polynomial-time attacks on the original multilinear maps of Garg, Gentry, and Halevi (2013). Previous polynomial-time attacks on GGH13 were "zeroizing" attacks that generally required the availability of low-level encodings of zero. Most significantly, such zeroizing attacks were not applicable to candidate indistinguishability obfuscation (iO) schemes. iO has been the subject of intense study.

To address this gap, we introduce *annihilation attacks*, which attack multilinear maps using non-linear polynomials. Annihilation attacks can work in situations where there are no low-level encodings of zero. Using annihilation attacks, we give the first polynomial-time cryptanalysis of candidate iO schemes over GGH13. More specifically, we exhibit two simple programs that are functionally equivalent, and show how to efficiently distinguish between the obfuscations of these two programs.

Given the enormous applicability of iO, it is important to devise iO schemes that can avoid attack. We discuss some initial directions for safeguarding against annihilating attacks.

1 Introduction

In this work, we present a new class of polynomial-time attacks on the original multilinear maps of Garg, Gentry, and Halevi [GGH13a]. Previous attacks on GGH13

A. Sahai—Research supported in part from a DARPA/ARL SAFEWARE award, NSF Frontier Award 1413955, NSF grants 1228984, 1136174, 1118096, and 1065276, a Xerox Faculty Research Award, a Google Faculty Research Award, an equipment grant from Intel, and an Okawa Foundation Research Grant. This material is based upon work supported by the Defense Advanced Research Projects Agency through the ARL under Contract W911NF-15-C-0205. The views expressed are those of the author and do not reflect the official policy or position of the Department of Defense, the National Science Foundation, or the U.S. Government.

M. Zhandry—Supported in part by the Defense Advanced Research Projects Agency (DARPA) and the U.S. Army Research Office under contract number W911NF-15-C-0226.

© International Association for Cryptologic Research 2016
M. Robshaw and J. Katz (Eds.): CRYPTO 2016, Part II, LNCS 9815, pp. 629–658, 2016.
DOI: 10.1007/978-3-662-53008-5_22

were not applicable to many important applications of multilinear maps, most notably candidate indistinguishability obfuscation (iO) schemes over GGH13 [GGH+13b, BR14, BGK+14, PST14, AGIS14, MSW14, BMSZ16]. Indeed, previous attacks on GGH13 can be classified into two categories:

- Works presenting polynomial-time attacks that either explicitly required the availability of low-level encodings of zero [GGH13a, HJ16], or required a differently represented low-level encoding of zero, in the form of an encoded matrix with a zero eigenvalue [CGH+15]. As a result, such "zeroizing" attacks do not apply to any iO candidates.
- Works that yield subexponential or quantum attacks [CDPR16, ABD16, CJL16]. This includes the works of [ABD16, CJL16] that were announced concurrently with the initial publication of our work. We note that the attacks of [ABD16, CJL16] on GGH13 mmaps, for example, require exponential running time if $n = \lambda \log^2 q$.

iO has been the subject of intense study. Thus, understanding the security of candidate iO schemes is of high importance. To do so, we need to develop new polynomial-time attacks that do not require, explicitly or implicitly, low-level encodings of zero.

Annihilation Attacks. To address this gap, we introduce *annihilation attacks*, which attack multilinear maps in a new way, using non-linear polynomials. Annihilation attacks can work in situations where there are no low-level encodings of zero. Using annihilation attacks, we give the first polynomial-time cryptanalysis of several candidate iO schemes over GGH13 from the literature. More specifically, we exhibit two simple programs that are functionally equivalent, and show how to efficiently distinguish between the obfuscations of these two programs. We also show how to extend our attacks to more complex candidate obfuscation schemes over GGH13, namely ones that incorporate the "dual-input" approach of [BGK+14]. (Note that, even without the dual-input structure, [BGK+14, AGIS14, MSW14, BMSZ16] were candidates for achieving iO security when implemented with [GGH13a].) Additionally, we give the first polynomial-time cryptanalysis of the candidate order revealing encryption scheme due to Boneh et al. [BLR+15] when instantiated over GGH3.

We now give an overview of our attack. The overview will introduce the main conceptual ideas and challenges in mounting our attack. After the overview, we will discuss potential defenses that may thwart our attack and generalizations of it.

1.1   Overview of the Attack

We begin with a simplified description of the GGH13 scheme, adapted from text in [CGH+15].

The GGH13 Scheme. For GGH13 [GGH13a] with k levels of multilinearity, the plaintext space is a quotient ring $R_g = R/gR$ where R is the ring of integers in a number field and $g \in R$ is a "small element" in that ring. The space of encodings is $R_q = R/qR$ where q is a "big integer". An instance of the scheme relies on two secret elements, the generator g itself and a uniformly random denominator $z \in R_q$. A small plaintext element α is encoded "at level one" as $u = [e/z]_q$ where e is a "small element" in the coset of α, that is $e = \alpha + gr$ for some small $r \in R$.

Addition/subtraction of encodings at the same level is just addition in R_q, and it results in an encoding of the sum at the same level, so long as the numerators do not wrap around modulo q. Similarly multiplication of elements at levels i, i' is a multiplication in R_q, and as long as the numerators do not wrap around modulo q the result is an encoding of the product at level $i + i'$.

The scheme also includes a "zero-test parameter" in order to enable testing for zero at level k. Noting that a level-k encoding of zero is of the form $u = [gr/z^k]_q$, the zero-test parameter is an element of the form $\mathbf{p}_{\mathrm{zt}} = [hz^k/g]_q$ for a "somewhat small element" $h \in R$. This lets us eliminate the z^k in the denominator and the g in the numerator by computing $[\mathbf{p}_{\mathrm{zt}} \cdot u]_q = h \cdot r$, which is much smaller than q because both h, r are small. If u is an encoding of a non-zero α, however, then multiplying by $\mathbf{p}_{\mathrm{zt}}$ leaves a term of $[h\alpha/g]_q$ which is not small. Testing for zero therefore consists of multiplying by the zero-test parameter modulo q and checking if the result is much smaller than q.

Note that above we describe the "symmetric" setting for multilinear maps where there is only one z, and its powers occur in the denominators of encodings. More generally, we will equally well be able to deal with the "asymmetric" setting where there are multiple z_i. However, we omit this generalization here as our attack is agnostic to such choices. Our attack is also agnostic to other basic parameters of the GGH13, including the specific choice of polynomial defining the ring R.

Setting of Our Attack. Recall that in our setting, we – as the attacker – will not have access to any low-level encodings of zero. Thus, in general, we are given as input a vector $\mathbf{u}$ of ℓ encodings, corresponding to a vector $\boldsymbol{\alpha}$ of ℓ values being encoded, and with respect to a vector $\mathbf{r}$ of ℓ random small elements. Thus, for each $i \in [\ell]$, there exists some value $j_i < k$ such that

$$u_i = \left[\frac{\alpha_i + gr_i}{z^{j_i}} \right]_q \qquad \alpha_i \neq 0$$

What a Distinguishing Attack Entails. In general, we consider a situation where there are two distributions over vectors of values: $\boldsymbol{\alpha}^{(0)}$ and $\boldsymbol{\alpha}^{(1)}$. Rather than directly viewing these as distributions over values, we can think of them as distinct vectors of multivariate polynomials over some underlying random variables, which then induce distributions over values via the distributions on the underlying random variables. Thus, from this viewpoint, $\boldsymbol{\alpha}^{(0)}$ and $\boldsymbol{\alpha}^{(1)}$ are just two distinct vectors of polynomials, that are known to us in our role as attacker.

Then a challenger chooses a random bit $b \in \{0, 1\}$, and we set $\boldsymbol{\alpha} = \boldsymbol{\alpha}^{(b)}$. Then we are given encodings $\boldsymbol{u}$ of the values $\boldsymbol{\alpha}$ using fresh randomness $\boldsymbol{r}$, and our goal in mounting an attack is to determine the challenger's bit b.

Note that to make this question interesting, it should be the case that all efficiently computable methods of computing top-level encodings of zero (resp. non-zero) using encodings of $\boldsymbol{\alpha}^{(0)}$ should also yield top-level encodings of zero (resp. non-zero) using encodings of $\boldsymbol{\alpha}^{(1)}$. Otherwise, an adversary can distinguish the encodings simply by zero testing.

Using Annihilating Polynomials. Our attack first needs to move to the polynomial ring R. In order to do so, the attack will need to build top-level encodings of zero, and then multiply by the zero-testing element $\mathbf{p}_{zt}$. Because we are in a setting where there are no low-level encodings of zero, top-level encodings of zero can only be created through algebraic manipulations of low-level encodings of nonzero values that lead to cancellation. Indeed, a full characterization of exactly how top-level encodings of zero can be created for candidate iO schemes over GGH13 was recently given by [BMSZ16]. In general, our attack will need to have access to a collection of valid algebraic manipulations that yield top-level encodings of zero, starting with the encodings $\boldsymbol{u}$.

Generally, then, a top-level encoding of zero e produced in this way would be stratified into levels corresponding to different powers of g, as follows:

$$e = \frac{g\gamma_1 + g^2\gamma_2 + \cdots g^k\gamma_k}{z^k}$$

and thus

$$f := [e \cdot \mathbf{p}_{zt}]_q = h \cdot (\gamma_1 + g\gamma_2 + \cdots g^{k-1}\gamma_k)$$

Above, each γ_i is a polynomial in the entries of $\boldsymbol{\alpha}$ and $\boldsymbol{r}$. As suggested by the stratification above, our main idea is to focus on just one level of the stratification. In particular, let us focus on the first level of the stratification, corresponding to the polynomial γ_1.

A Simple Illustrative Example. Suppose that we had three ways of generating top-level encodings of zero, e, e', and e'', which yield products f, f', and f'' in the ring R. Suppose further that e, e', and e'' contained polynomials $\gamma_1 = xr$; $\gamma_1' = xr^2$; and $\gamma_1'' = x$, where x is a random variable underlying $\boldsymbol{\alpha}^{(0)}$. Then we observe that there is an efficiently computable *annihilating polynomial,* $Q(a, b, c) := a^2 - bc$, such that $Q(\gamma_1, \gamma_1', \gamma_1'')$ is the zero polynomial. Further, because Q is homogeneous, $Q(h \cdot \gamma_1, h \cdot \gamma_1', h \cdot \gamma_1'')$ is also the zero polynomial. (We will always ensure that our annihilating polynomials are homogeneous, which essentially comes for free due to the homogeneity of the γ_1 polynomials in the iO setting; see Lemma 1.)

Thus, if we compute $Q(f, f', f'')$, we obtain an element in the ring R that is contained in the ideal $\langle hg \rangle$.

However, consider the top-level encodings of zero e, e', and e'' that arise from $\boldsymbol{\alpha}^{(1)}$, which is a different vector of polynomials over x than $\boldsymbol{\alpha}^{(0)}$. Suppose that in this case, the encodings e, e', and e'' contain polynomials $\gamma_1 = x^3r$; $\gamma_1' = xr$; and $\gamma_1'' = x$. In this scenario, the polynomial Q is no longer annihilating, and instead yields $Q(\gamma_1, \gamma_1', \gamma_1'') = x^6r^2 - x^2r$. Thus, what we have is that if the challenge bit $b = 0$, then $Q(f, f', f'')$ is contained in the ideal $\langle hg \rangle$, but if the challenge bit $b = 1$, then $Q(f, f', f'')$ is not contained in the ideal $\langle hg \rangle$.

Obtaining this distinction in outcomes is the main new idea behind our attack.

Central Challenge: How to Compute Annihilating Polynomials? While it was easy to devise an annihilating polynomial for the polynomials contained in the simple example above, in general annihilating polynomials can be hard to compute. Every set of $n + 1$ or more polynomials over n variables is algebraically dependent and hence must admit an annihilating polynomial. Indeed, therefore, if we do not worry about how to compute annihilating polynomials, our high-level attack idea as described above would apply to *every* published iO scheme that can be instantiated with GGH13 maps that we are aware of, and it would work for every pair of equivalent programs that output zero sufficiently often. This is simply because every published iO candidate can be written as an algebraic expression using only a polynomial number of underlying random variables, whereas the obfuscated program can be evaluated on an exponential number of inputs.

However, unless the polynomial hierarchy collapses (specifically, unless **Co-NP $\subseteq$ AM**), there are sets of (cubic) polynomials over n variables for which the annihilating polynomial cannot be represented by any polynomial-size arithmetic circuit [Kay09]. As a result, for our attack idea to be meaningful, we must show that the annihilating polynomials we seek are efficiently representable by arithmetic circuits and that such representations are efficiently computable. In particular, we seek to do this in the context of (quite complex) candidates for indistinguishability obfuscation.

We begin by looking deeper at the structure of the polynomials γ_1 that we need to annihilate. In particular, let's examine what these polynomials look like as a consequence of the stratification by powers of g. We see that by the structure of encodings in GGH13, each polynomial γ_1 will be linear in the entries of r and potentially non-linear in the entries of α. This is already useful, since the r variables are totally unstructured and unique to each encoding given out, and therefore present an obstacle to the kind of analysis that will enable us to find an annihilating polynomial.

To attack iO, we will first design two simple branching programs that are functionally equivalent but distinct as branching programs. To this end, we consider two branching programs that both compute the *always zero* functionality. The simplest such program is one where every matrix is simply the identity matrix, and this will certainly compute the constant zero functionality. To design another such program, we observe that the anti-identity matrix

$$B = \begin{pmatrix} 0 & 1 \\ 1 & 0 \end{pmatrix}$$

can be useful, because it has the property that $BB = I$. Thus, to make another branching program that computes the always zero functionality, we can create a two-pass branching program, where the (two) matrices corresponding to $x_1 = 0$ are both set to B, and all other matrices are set to I.

With these branching programs in mind, we analyze the γ_1 polynomials that arise. The main method that we use to prune the search space for annihilating polynomials is to find changes of variables that can group variables together in order to minimize the number of active variables. We use a number of methods, including inclusion-exclusion formulas, to do this. By changing variables, we are able to reduce the problem to finding the annihilating polynomial for a set of polynomials over only a *constant* number of variables. When only a constant number of variables are present, exhaustive methods for finding annihilating polynomials are efficient. For further details, refer to Sect. 5.

Moving to More Complex iO Candidates. The above discussion covers the main ideas for finding annihilating polynomials, and by generalizing our methods, we show that they extend to more challenging settings. Most notably, we can extend our methods to work for the *dual-input* technique of [BGK+14], which has been used in several follow-up works [AGIS14, MSW14, BMSZ16]. Previously, no cryptanalysis techniques were know to apply to this setting. For further details, see Sects. 4 and 5.

An Abstract Attack Model. We first describe our attacks within a new abstract attack model, which is closely related to a model proposed in [CGH+15, Appendix A]. The new model is roughly the same as existing generic graded encoding models, except that a successful zero test returns an *algebraic element* rather than a bit $b \in \{0, 1\}$. These algebraic elements can then be manipulated, say, by evaluating an annihilating polynomial over them. This model captures the fact that, in the GGH13 candidate graded encoding scheme [GGH13a], the zero test actually does return an algebraic element in a polynomial ring that can be manipulated.

We describe our attacks in this abstract model to (1) highlight the main new ideas for our attack, and (2) to demonstrate the robustness of our attack to simple "fixes" for multilinear maps that have been proposed.

Theorem 1. *Let $\mathcal{O}$ denote the single-input variant of the iO candidates in [BGK+14, PST14, AGIS14, MSW14, BMSZ16] (over GGH13 [GGH13a] maps).*

There exist two functionally-equivalent branching programs $\mathbf{A}, \mathbf{A}'$ such that $\mathcal{O}(\mathbf{A})$ and $\mathcal{O}(\mathbf{A}')$ can be efficiently distinguished in the abstract attack model described in Sect. 2.

Note that in the single input case, the [BGK+14, AGIS14, BMSZ16] obfuscators over GGH13 [GGH13a] maps were shown to achieve iO security in the standard generic graded encoding model. This theorem shows that such security does not extend to our more refined model.

The attack in Theorem 1 works by executing the obfuscated program honestly on several inputs, which produces several zero-tested top-level 0-encodings. Recall that in our model, each successful zero-test returns an algebraic element. We then give an explicit polynomial that annihilates these algebraic elements in the case of one branching program, but fails to annihilate in the other. Thus by evaluating this polynomial on the algebraic elements obtained and testing for zero, it is possible to distinguish the two cases.

Beyond the Abstract Attack Model. Our abstract attack does not immediately yield an attack on actual graded encoding instances. For example, when the graded encoding is instantiated with [GGH13a], the result of an annihilating polynomial is an element in the ideal $\langle hg \rangle$, whereas if the polynomial does not annihilate, then the element is not in this ideal. However, this ideal is not explicitly known, so it is not a priori obvious how to distinguish the two cases.

We observe that by evaluating the annihilating polynomial many times on different sets of values, we get many different vectors in $\langle hg \rangle$. With enough vectors, we (heuristically) can compute a spanning set of vectors for $\langle hg \rangle$. This is the only heuristic portion of our attack analysis, and it is similar in spirit to previous heuristic analysis given in other attacks of multilinear maps (see, e.g., [CGH+15]). With such a spanning set, we can then test to see if another "test" vector is in this ideal or not. This is the foundation for our attack on obfuscation built from the specific [GGH13a] candidate.

Dual-Input Obfuscation and Beyond. Moving on to the dual-input setting, we do not know an explicit annihilating polynomial for the set of algebraic elements returned by our model. However, we are able to show both that such a polynomial must exist, and furthermore that it must be efficiently computable because it has constant size. Thus we demonstrate that there *exists* an efficient distinguishing adversary in the abstract attack model. As before, we can turn this into a heuristic attack on obfuscation built from [GGH13a] graded encodings. We also show that modifying the branching programs to read $d > 2$ bits at each level does not thwart the attack for constant d, because the annihilating polynomial still has constant size (albeit a larger constant).

Theorem 2. *Let $\mathcal{O}$ denote the dual-input variant of the iO candidates found in [BGK+14, PST14, AGIS14, MSW14, BMSZ16] (over GGH13 [GGH13a] maps).*

There exist two functionally-equivalent branching programs $\mathbf{A}, \mathbf{A}'$ such that $\mathcal{O}(\mathbf{A})$ and $\mathcal{O}(\mathbf{A}')$ can be efficiently distinguished in the abstract attack model described in Sect. 2.

We do not currently know how to extend these attacks to the multilinear maps of Coron, Lepoint, and Tibouchi [CLT13].

1.2 Attacking Candidate Order-Revealing Encryption

We show how to apply annihilating polynomials to the candidate order-revealing encryption (ORE) scheme of Boneh et al. [BLR+15]. ORE is a symmetric key encryption scheme where it is possible to learn the order of plaintexts without knowing the secret key, but nothing else is revealed by the ciphertexts. Such a scheme would allow, for example, making range queries on an encrypted database without the secret key. The ORE of [BLR+15] is one of the few implementable applications of multilinear maps. We demonstrate a polynomial such that whether or not the polynomial annihilates depends on more than just the order of the plaintexts. We therefore get an attack in our refined abstract model:

Theorem 3. *Let $\mathcal{E}$ denote the ORE scheme of [BLR+15] (over GGH13 [GGH13a] maps). There exist two sequences of plaintexts $m_1^0 < \cdots < m_\ell^{(0)}$ and $m_1^{(1)} < \cdots < m_\ell^{(1)}$ such that $\mathcal{E}(m_1^0), \cdots, \mathcal{E}(m_\ell^0)$ and $\mathcal{E}(m_1^1), \cdots, \mathcal{E}(m_\ell^1)$ can be efficiently distinguished in the abstract attack model described in Sect. 2.*

We also show how to extend our attack to obtain an explicit attack when instantiated over GGH13 maps. This attack has an analogous heuristic component as in our attack on obfuscation.

1.3 Defenses

The investigation of the mathematics needed to build iO remains in its infancy. In particular, our work initiates the study of annihilation attacks, and significant future study is needed to see how such attacks can be avoided. We begin that process here with a brief discussion of some promising directions. (Following the initial publication of our work, Garg, Mukherjee, and Srinivasan [GMS16] gave a new candidate iO construction, based on a new variant of the [GGH13a] multi-linear maps, that provably resists all known polynomial-time attacks, including ours, assuming an explicit PRF in NC^1. Following that, we [MSZ16] gave a simpler candidate iO construction, using the original [GGH13a] multilinear maps, that provably resists all known polynomial-time attacks under a more general assumption.)

As noted above, the primary obstacle to mounting annihilation attacks is finding an efficiently representable annihilating polynomial. Indeed, at present we only know how to mount our attack for a small class of matrix branching programs. Many iO candidates work by first transforming a general program into one of a very specific class of branching programs, that does not include any of the matrix branching programs that we know how to attack using annihilation attacks. However, exactly which iO candidates can be attacked using annihilation attacks, and how annihilation attacks can be prevented, remains unclear.

Is it possible that more complex algebraic constructions of iO candidates can avoid the existence of such annihilating polynomials? For example, we do not know how to extend our attack to the original iO candidate of [GGH+13b], and it is still a possibility that their candidate is secure. The difficulty of extending our attack to their scheme stems from the extra defenses they apply, namely appending random elements to the diagonal of the branching program. This randomization of the branching program means our polynomials do not annihilate, and has so far prevented us from pruning the search space of polynomials to find new annihilating polynomials.

On the other hand, there may still be efficiently computable annihilating polynomials for [GGH+13b], or for that matter any other candidate iO scheme. Given any candidate, how would we argue that no such annihilating polynomials exist? As one approach, we propose exploring ideas from the proof of [Kay09] that shows the existence of sets of polynomials for which no efficient annihilating polynomial can be found, unless **Co-NP** $\subseteq$ **AM**. Perhaps these ideas can be combined with ideas from [BR14, MSW14] to identify a candidate iO scheme where finding relevant annihilating polynomials will be provably as hard as inverting a one-way function.

Going further, we propose exploring *non-algebraic* methods for randomizing the matrix branching programs being obfuscated, in such a way that this randomization destroys all algebraic descriptions of the α values that are being given out in encoded form. For example, suppose a matrix branching program can be randomized (while preserving functionality) using matrices drawn randomly from discrete matrix subgroups, resulting in matrices whose entries cannot be written as $\mathbb{Z}$-linear (or $\mathbb{Z}_p$-linear) polynomials. Then the usual algebraic notion of annihilating polynomials may no longer be capable of yielding an attack.

2 Model Description

We now describe an abstract model for attacks on current multilinear map candidates. There are "hidden" variables $X_1, \ldots, X_n$ for some integer n, $Z_1, \ldots, Z_m$ for another integer m, and g. Then there are "public" variable $Y_1, \ldots, Y_m$, which are set to $Y_i = q_i(\{X_j\}) + gZ_i$ for some polynomials q_i. All variables are defined over a field $\mathbb{F}$.

The adversary is allowed to make two types of queries:

- In a **Type 1** query, the adversary submits a "valid" polynomial p_k on the Y_i. Here "valid" polynomials come from some restricted set of polynomials. These restrictions are those that are enforceable using graded encodings. Next, we consider p as a polynomial of the formal variables X_j, Z_i, g. Write $p_k = p_k^{(0)}(\{X_j\}, \{Z_i\}) + gp_k^{(1)}(\{X_j\}, \{Z_i\}) + g^2 \ldots$. If p_k is identically 0, then the adversary receives $\perp$ in return. If $p_k^{(0)}$ is *not* identically 0, then the adversary receive $\perp$ in return. If p_k is *not* 0 but $p_k^{(0)}$ *is* identically 0, then the adversary receives a handle to a new variable W_k, which is set to be $p_k/g = p_k^{(1)}(\{X_j\}, \{Z_i\}) + gp_k^{(2)}(\{X_j\}, \{Z_i\}) + \ldots$.
- In a **Type 2** query, the adversary is allowed to submit *arbitrary* polynomials r with small algebraic circuits on the W_k that it has seen so far. Consider $r(\{W_k\})$ as a polynomial of the variables X_j, Z_i, g, and write $r = r^{(0)}(\{X_j\}, \{Z_i\}) + gr^{(1)}((\{X_j\}, \{Z_i\})) + g^2 \ldots$. If $r^{(0)}$ is identically zero, then the model responds with 0. Otherwise the model responds with 1.

In current graded encoding schemes, the set of "valid" polynomials is determined by the restrictions placed by the underlying set structure of the graded encoding. Here we consider a more abstract setting where the set of "valid" polynomials is arbitrary.

In the standard abstract model for graded encodings, **Type 1** queries output a bit as opposed to an algebraic element, and there are no **Type 2** queries. However, this model has been shown to improperly characterize the information received from **Type 1** queries in current candidate graded encoding schemes. The more refined model above more accurately captures the types of attacks that can be carried out on current graded encodings.

2.1   Obfuscation in the Abstract Model

We now describe an abstract obfuscation scheme that encompass the schemes of [AGIS14,BMSZ16], and can also be easily extended to incorporate the scheme of [BGK+14]. The obfuscator takes as input a branching program of length ℓ, input length n, and arity d. The branching program contains an input function $\mathsf{inp} : [\ell] \to 2^{[n]}$ such that $|\mathsf{inp}(i)| = d$ for all $i \in [\ell]$. Moreover, the branching program contains $2^d \ell + 2$ matrices $A_0, \{A_{i,S_i}\}_{i \in [\ell]}, A_{\ell+1}$ where S_i ranges over subsets of $\mathsf{inp}(i)$, and $A_0 A_{\ell+1}$ are the "bookend" vectors. To evaluate a branching program on input x, we associate x with the set $T \subseteq [n]$ where $i \in T$ if and only if $x_i = 1$. To evaluate the branching program on input x (set T) compute the following product.

$$\mathbf{A}(T) = A_0 \times \prod_{i=1}^{\ell} A_{i, T \cap \mathsf{inp}(i)} \times A_{\ell+1}$$

The output of the branching program is 0 if and only if $\mathbf{A}(T) = 0$.

The obfuscator first generates random matrices $\{R_i\}_{i \in [\ell+1]}$ and random scalars $\{\alpha_{i,S_i}\}_{i \in [\ell], S_i \subseteq \mathsf{inp}(i)}$. Then it computes the randomized branching program consisting of the matrices $\widetilde{A_{i,S_i}} = \alpha_{i,S_i}(R_i \cdot A_{i,S_i} \cdot R_{i+1}^{adj})$ and bookend vectors $\widetilde{A_0} = A_0 \cdot R_1^{adj}$ and $\widetilde{A_{\ell+1}} = R_{\ell+1} \cdot A_{\ell+1}$. Here R_i^{adj} denotes the adjugate matrix of R_i that satisfies $R_i^{adj} \times R_i = \det(R_i) \cdot I$. It is easy to see that this program computes the same function as the original branching program.

Finally, the obfuscator sets the "hidden" variables in the model to the $\widetilde{A}$ matrices. Denote the "public" variables as $Y_{i,S} = \widetilde{A_{i,S}} + g Z_{i,S} = \alpha_{i,S} R_i \cdot A_{i,S} \cdot R_{i+1}^{adj} + g Z_{i,S}$ (and define $Y_0, Y_{\ell+1}$ analogously). The set of valid **Type 1** polynomials is set up so that honest evaluations of the branching program are considered valid. That is, the polynomials

$$p_T = Y_0 \times \prod_{i=1}^{\ell} Y_{i, T \cap \mathsf{inp}(i)} \times Y_{\ell+1}$$

are explicitly allowed. Notice that the g^0 coefficient of p_T is exactly $\widetilde{\mathbf{A}}(\mathbf{T}) \equiv \mathbf{A}(T)$, so the evaluator can run the program by querying on p_T, and checking if the result is $\perp$.

In the case $d = 1$, this obfuscator corresponds to the basic single-input branching program obfuscator of [AGIS14,BMSZ16]. In the more restricted model where there are no **Type 2** queries, it was shown how to set the underlying graded encodings so that the only valid **Type 1** queries are linear combinations of arbitrarily-many p_T polynomials. This is sufficient for indistinguishability obfuscation. When $d = 2$ this corresponds to the dual-input version of these obfuscators, in which it was shown how to set up the underlying graded encodings so that the the linear combination has polynomial size. This is sufficient for virtual black box obfuscation (again in the more restricted model).

In any case, since for functionality the set of allowed queries *must* include honest executions of the program, we always allow queries on the p_T polynomials themselves. As such, our attacks will work by only making **Type 1** queries on honest evaluations of p_T. Thus with *any* restrictions in our abstract model that allow for such honest evaluations of p_T, we will demonstrate how to to break indistinguishability security.

An Equivalent Formulation. When the obfuscator described above is concretely implemented, the final step is to encode each element of each Y matrix in the GGH13 candidate multilinear map scheme [GGH13a]. Recall that for this, an element $a \in \mathbb{Z}_p$ is mapped to a polynomial $a + gr \in \mathbb{Z}[x]/(x^n + 1)$ (here we omit the level of the encodings, which is without loss of generality since we only compute honest evaluations of the branching program). Then, when evaluating on an input whose output is 0, the g^0-coefficient in p_T will be 0 in the GGH13 ring, namely it will be 0 *modulo the ideal* $\langle g \rangle$. However, we would like this coefficient to be identically 0.

To this end, we note that the obfuscation procedure can be viewed in a slightly different way that will guarantee this. Namely, we *first* encode the A matrices in the GGH13 ring, and *then* we perform the randomization steps over this ring. We note that, crucially, the necessary adjoint matrices R_i^{adj} can be computed over this ring. Then by the properties of the adjoint, we are guaranteed that the off-diagonal entries of each $(R_i^{adj} \times R_i)$ are identically 0, and this ensures that the g^0-coefficient of p_T is as well.

3   Abstract Attack

Here we describe an abstract attack on obfuscation in our generic model. For simplicity, we describe the attack for *single input* branching programs, which proves Theorem 1. We extend to dual-input and more generally d-input branching programs in Sect. 5, which will prove Theorem 2.

3.1   The Branching Programs

The first branching program **A** is defined as follows. It has $2n + 2$ layers, where the first and last layers consist of the row vector $A_0 := (0\ 1)$ and the column vector $A_{2n+1} := (1\ 0)^T$ respectively. The middle $2n$ layers scan through the input bits twice, once forward and once in reverse, with input selection function $\mathsf{inp}(i) := \min(i, 2n + 1 - i)$ (so x_1 is read in layers 1 and $2n$, x_2 is read in layers 2 and $2n - 1$, etc.)[1]. In each of these layers, both matrices are the identity, i.e. we have

$$A_{i,0} = A_{i,1} = \begin{pmatrix} 1 & 0 \\ 0 & 1 \end{pmatrix}$$

for $i \in [2n]$. Here, we adopt the more standard notation for branching programs where the matrix $A_{i,b}$ is selected if $x_{\mathsf{inp}(i)} = b$.

The branching program $\mathbf{A} = \{\mathsf{inp}, A_0, A_{2n+1}, A_{i,b} \mid i \in [2n], b \in \{0,1\}\}$ is evaluated in the usual way:

$$\mathbf{A}(x) := A_0 \times \prod_{i=1}^{2n} A_{i, x_{\mathsf{inp}(i)}} \times A_{2n+1}.$$

Clearly this satisfies $\mathbf{A}(x) = 0$ for all x.

The second branching program $\mathbf{A}' = \{\mathsf{inp}', A_0', A_{2n+1}', A_{i,b}' \mid i \in [2n], b \in \{0,1\}\}$ is defined almost identically. The sole difference is that, in the layers reading bits any of the bits $x_1, \ldots, x_k$ for some integer $k \leq n$, the matrices corresponding to "$x_i = 0$" are changed to be anti-diagonal. Namely, we have

$$A_{i,0}' = A_{2n+1-i,0}' = \begin{pmatrix} 0 & 1 \\ 1 & 0 \end{pmatrix} \text{ for } i \in [k]$$

and all other components remain the same (i.e. $\mathsf{inp}' = \mathsf{inp}$, $A_0' = A_0$, $A_{2n+1}' = A_{2n+1}$, and $A_{i,b}' = A_{i,b}$ for all (i, b) where $b = 1$ or $i \in [k+1, 2n-k]$). We again have $\mathbf{A}'(x) = 0$ for all x, because the anti-diagonal matrix above is its own inverse and all the matrices commute.

3.2   The Distinguishing Attack

We now specialize the abstract obfuscation scheme from Sect. 2 to the single-input case. We choose invertible matrices $\{R_i \in \mathbb{Z}_p^{2\times 2}\}_{i\in[2n+1]}$ and non-zero scalars $\{\alpha_{i,x} \in \mathbb{Z}_p\}_{i\in[2n], b\in\{0,1\}}$ uniformly at random. Next, we define

$$\tilde{A}_0 := A_0 \cdot R_1^{adj} \qquad \tilde{A}_{2n+1} := R_{2n+1} \cdot A_{2n+1} \qquad \tilde{A}_{i,b} := \alpha_{i,b} R_i \cdot A_{i,b} \cdot R_{i+1}^{adj}$$

for $i \in [2n]$, $b \in \{0,1\}$, where R_i^{adj} is the adjugate matrix of R_i. Finally, each of the entries of the various $\tilde{A}$ are what are actually encoded, meaning the "public" variables consist of

$$Y_{i,b} = \alpha_{i,b} R_i \cdot A_{i,b} \cdot R_{i+1}^{adj} + g Z_{i,b}$$

---

[1] Recall that in the single-input case, the set outputted by $\mathsf{inp}(i)$ is just a singleton set.

Next, by performing a change of variables on the $Z_{i,b}$, we can actually write

$$Y_{i,b} = \alpha_{i,b} R_i \cdot (A_{i,b} + g Z_{i,b}) \cdot R_{i+1}^{adj}$$

The underlying graded encodings guarantee some restrictions on the types of **Type 1** encodings allowed — however, the restrictions *must* allow evaluation of the branching program on various inputs. In particular, the query

$$p_x := Y_0 \times \prod_{i=1}^{2n} Y_{i,x_{\mathsf{inp}(i)}} \times Y_{2n+1}$$

is allowed. Now, the coefficient of g^0 in p_x is given by

$$p_x^{(0)} := \widetilde{A_0} \times \prod_{i=1}^{2n} \widetilde{A_{i,x_{\mathsf{inp}(i)}}} \times \widetilde{A_{2n+1}} = \rho \prod_i \alpha_{i,x_{\mathsf{inp}(i)}} A_0 \times \prod_{i=1}^{2n} A_{i,x_{\mathsf{inp}(i)}} \times A_{2n+1}$$

which evaluates to 0 by our choice of branching programs. (Note that by the discussion at the end of Sect. 2, we can take this coeffficient to be identically 0, and not merely divisible by g.) Here $\rho := \prod_i \det(R_i)$ satisfies $\rho I = \prod_i R_i R_i^{adj}$, and we abuse notation by letting $Y_{0,x_{\mathsf{inp}(0)}}$ denote Y_0 (and similarly for the other matrices).

Thus, the model, on **Type 1** query p_x, will return a handle to the variable

$$p_x^{(1)} := \rho \prod_i \alpha_{i,x_{\mathsf{inp}(i)}} \sum_{i=0}^{2n+1} \left(A_{0,x_{\mathsf{inp}(0)}} \cdots A_{i-1,x_{\mathsf{inp}(i-1)}} \cdot Z_{i,x_{\mathsf{inp}(i)}} \cdot A_{i+1,x_{\mathsf{inp}(i+1)}} \cdots A_{2n+1,x_{\mathsf{inp}(2n+1)}} \right)$$

As in Sect. 2, we will associate $x \in \{0,1\}^n$ with sets $T \subset [n]$ where $i \in T$ if and only if $x_i = 1$. For $i \in [2, n]$, write $\alpha_{i,b}' = \alpha_{i,b} \alpha_{2n+1-i,b}$. Also set $\alpha_{1,b}' = \rho \alpha_{1,b} \alpha_{2n,b}$. Thus $\rho \prod_{i=1}^{2n} \alpha_{i,x_{\mathsf{inp}(i)}} = \prod_{i=1}^{n} \alpha_{i,x_i}'$. Define this quantity as $U_x = U_T$. It is straightforward to show that the U_T satisfy the following equation[2] when $|T| \geq 2$.

$$U_T = U_\emptyset^{-(|T|-1)} \cdot \prod_{j \in T} U_{\{j\}}$$

Moreover, any equation satisfied by the U_T is generated by these equations.

For the other part of $p_x^{(1)} = p_T^{(1)}$, there are two cases:

- The branching program is all-identity, with bookends $(1 \ 0)$ and $(0 \ 1)^T$. Then $A_{i,0} = A_{i,1} =: A_i$. Here, we write $\beta_{i,b} = A_0 \cdots A_{i-1} \cdot Z_{i,b} \cdot A_{i+1} \cdots A_{2n+1}$. Notice that the $\beta_{i,b}$ are all independent. For $0 \leq i \leq n$, let $\beta_{i,b}' = \beta_{i,b} + \beta_{2n+1-i,b}$. Thus,

$$\sum_{i=0}^{2n+1} \left(A_{0,x_{\mathsf{inp}(0)}} \cdots A_{i-1,x_{\mathsf{inp}(i-1)}} \cdot Z_{i,x_{\mathsf{inp}(i)}} \cdot A_{i+1,x_{\mathsf{inp}(i+1)}} \cdots A_{2n+1,x_{\mathsf{inp}(2n+1)}} \right) = \sum_{i=0}^{n} \beta_{i,x_i}'$$

---

[2] See Theorem 4 for a proof of a more general identity.

Define this quantity as $V_x = V_T$. It is similarly straightforward to show that the V_T satisfy the following equation when $|T| \geq 2$.

$$V_T = -(|T| - 1)V_\emptyset + \sum_{j \in T} V_{\{j\}}$$

Moreover, any equation satisfied by the V_T is generated by these equations. Piecing together, we have that $p_T^{(1)} = U_T V_T$, where U_T, V_T satisfy the equations above.

– The branching program is as above, except that it has reverse diagonals for $b = 0$, $i \leq k$. Consider a term $\cdots A_{i-1,x_{\mathsf{inp}(i-1)}} \cdot Z_{i,x_{\mathsf{inp}(i)}} \cdot A_{i+1,x_{\mathsf{inp}(i+1)}} \cdots$. Suppose for the moment that $i \leq k + 1$. Since each $A_{i,b}$ is either diagonal or anti-diagonal, we have that $\cdots A_{i-1,x_{\mathsf{inp}(i-1)}} = \cdots A_{i-1,x_{i-1}}$ is equal to the row vector $(0 \ 1)$ if the parity of $x_{[1,i-1]}$ is zero, and is equal to $(1 \ 0)$ if the parity is 1. Similarly, $A_{i+1,x_{\mathsf{inp}(i+1)}} \cdots$ is equal to the column vector $(1 \ 0)^T$ if the parity of $x_{[1,i-1]}$ is zero, and $(0 \ 1)^T$ otherwise[3]. Therefore, $\cdots A_{i-1,x_{\mathsf{inp}(i-1)}} \cdot Z_{i,x_{\mathsf{inp}(i)}} \cdot A_{i+1,x_{\mathsf{inp}(i+1)}} \cdots$ is equal to $\left(Z_{i,x_{\mathsf{inp}(i)}}\right)_{1,2}$ or $\left(Z_{i,x_{\mathsf{inp}(i)}}\right)_{2,1}$, depending on the parity of $x_{[1,i-1]}$. Therefore, define $\gamma_{i,b,p}$ to be the result of the product when $x_i = b$ and the parity of $x_{[1,i-1]}$ is p. For $i \in [2n+k, 2n]$, the same holds, so we can absorb the product for this i into $\gamma_{i,b,p}$. For $i \in [k+2, 2n-k-1]$, the same holds true, except that it is only the parity of the bits $x_{[1,k]}$ that matter. Therefore, we can write the product as $\gamma_{i,b,p}$ where $x_i = b$ and the parity of $x_{[1,k]}$ is p. Notice that each of the $\gamma_{i,b,p}$ are independent. Define

$$W_T = W_x = \sum_{i=1}^{n} \gamma_{i,x_i,\mathsf{parity}(x_{[1,\min(i-1,k)]})}$$

Then we have that $p_T^{(1)} = U_T W_T$.

The W_T must satisfy *some* linear relationships, since the number of W is 2^n, but the number of γ is $4n$. We have not derived a general equation, but instead we will focus on two cases. If the bits $x_1, \ldots, x_k$ are fixed (say to 0), then the parity for these bits is always the same (0). Therefore, W_T for these T satisfy the same equations as the V_T. Thus, any equation satisfied by the $p_T^{(1)}$ for these T in the all-identity case will also be satisfied in the anti-diagonal case. In the other case, take $T \subseteq \{1, 2, 3\}$, and suppose $k = 1$. In this simple case, it is straightforward to show that the following are the only linear relationships among these W:

$$W_{1,2,3} + W_1 = W_{1,2} + W_{1,3}$$
$$W_{2,3} + W_\emptyset = W_2 + W_3$$

These are different, and fewer, than the equations satisfied by the V_T. This will be the basis for our distinguishing attack.

---

[3] The rest of the bits of x do not matter, since both matrices for each of these bits occur in the product $A_{i+1,x_{\mathsf{inp}(i+1)}} \cdots$, and therefore cancel out.

To distinguish the two branching programs, it suffices to find a polynomial Q that annihilates the $p_T^{(1)}$ for $T \subseteq \{1,2,3\}$ in the all-identity case, but does not annihilate in the anti-identity case. Here is such a polynomial; we note that, though it does not matter for our attack, this is in fact the minimal annihilating polynomial for $\{p_T^{(1)}\}_{T \subseteq \{1,2,3\}}$.

$$
\begin{aligned}
Q_{1,2,3} = {} & \left(p_\emptyset^{(1)} p_{1,2,3}^{(1)}\right)^2 + \left(p_1^{(1)} p_{2,3}^{(1)}\right)^2 + \left(p_2^{(1)} p_{1,3}^{(1)}\right)^2 + \left(p_3^{(1)} p_{1,2}^{(1)}\right)^2 \\
& - 2 \left(p_\emptyset^{(1)} p_{1,2,3}^{(1)} p_1^{(1)} p_{2,3}^{(1)} + p_\emptyset^{(1)} p_{1,2,3}^{(1)} p_2^{(1)} p_{1,3}^{(1)} + p_\emptyset^{(1)} p_{1,2,3}^{(1)} p_3^{(1)} p_{1,2}^{(1)} \right. \\
& \left. + p_1^{(1)} p_{2,3}^{(1)} p_2^{(1)} p_{1,3}^{(1)} + p_1^{(1)} p_{2,3}^{(1)} p_3^{(1)} p_{1,2}^{(1)} + p_2^{(1)} p_{1,3}^{(1)} p_3^{(1)} p_{1,2}^{(1)}\right) \\
& + 4(p_\emptyset^{(1)} p_{1,2}^{(1)} p_{1,3}^{(1)} p_{2,3}^{(1)} + p_{1,2,3}^{(1)} p_1^{(1)} p_2^{(1)} p_3^{(1)})
\end{aligned}
$$

The fact that $Q_{1,2,3}$ annihilates in the all-identity case can be verified by tedious computation. The fact that it does *not* annihilate in the anti-diagonal case can also be verified by tedious computation as follows. Consider a generic degree 4 polynomial Q in the $p_T^{(1)}$ for $T \subseteq \{1,2,3\}$. The condition "Q annihilates the $p_T^{(1)}$" can be expressed as a linear equation in the coefficients of Q. Since Q has degree 4 in 8 variables, the number of coefficients is bounded by a constant, so the linear constraints can be solved. The result of this computation is that $Q = 0$ is the only solution.

By Schwartz-Zippel, if Q does not annihilate, then with overwhelming probability over the randomness of the obfuscation, the result of applying Q is non-zero.

The attack thus works as follows. First query on inputs x which are zero in every location except the first three bits. Since the branching program always evaluates to zero, the model will return a handle to the element $p_T^{(1)}$, where $T \subseteq \{1,2,3\}$ is the set of bits where x is 1. Then, evaluate the polynomial $Q_{1,2,3}$ on the elements obtained. If the result is 0, then guess that we are in the all-identity case. If the result is non-zero, then guess that we are in the anti-diagonal case. As we have shown, this attack distinguishes the two cases with overwhelming probability.

We make one final observation that will be relevant for attacking the specific [GGH13a] candidate. We note that, for either branching program, the following is true. Let T_0 be some subset of $[k+1, n]$ of size 3, and write $T_0 = i_1, i_2, i_3$. Let T_1 some subset of $[1, n] \setminus T_0$. Then for any subset $T \subseteq [3]$, write $\hat{p}_T^{(1)} := p_{T'}^{(1)}$, where $T' = \{i : i \in T_1 \text{ or } i = i_j \text{ for some } j \in T\}$. If we then evaluate the above polynomial $Q_{1,2,3}$ over the $\hat{p}_T^{(1)}$, we see that it annihilates. This is because the corresponding $p_{T'}^{(1)}$ satisfy the same equations as above.

3.3   Extensions

Here we consider the extension of our attack to other settings.

More General Branching Programs. First, a straightforward extension of our analysis above shows that $Q_{1,2,3}$ will successfully annihilate for *any* "trivial" branching program where, for each layer i, the matrices $A_{i,0}$ and $A_{i,1}$ are the same. In other words, the evaluation of the branching program is completely independent of the input bits. In contrast, above we showed a very simple branching program which does not satisfy this property for which $Q_{1,2,3}$ does not annihilate. More generally, it appears that for more complicated branching programs, $Q_{1,2,3}$ will typically annihilate. Therefore, our attack generalizes to distinguish "trivial" branching programs from many complicated branching programs.

Padded-[BMSZ16]. Next, we observe that our attack does not require the branching programs to compute the all-0s function, and that essentially any desired functionality can be used.

Assume that we are given BPs A, A' that both compute the same function $f : \{0,1\}^n \to \{0,1\}$. We augment them to obtain new BPs B, B' by adding 6 extra "padding" layers anywhere in the program; the first two of these input layers read (new) input bit x_{n+1}, the next two read x_{n+2}, and the final two read x_{n+3}. For B we put the identity matrix everywhere in these layers, while for B' we put the anti-identity matrix when the bit read is 0.

The augmented BPs compute essentially the same function as before, namely $f' : \{0,1\}^{n+3} \to \{0,1\}$ where $f'(x) = f(x|_{1\cdots n})$. So, provided it is easy to find an input x for which the original function $f(x) = 0$, we can obtain the outputs $B(x \circ y)$ and $B'(x \circ y)$ for every $y \in \{0,1\}^3$ and evaluate the annihilating polynomial $Q_{1,2,3}$ on them. By the same analysis, this will distinguish the two BPs in our attack model.

[BGK+14] *and* [BR14]. Our attack also extends to the candidate obfuscator from [BGK+14]. This obfuscator differs slightly from the one described in Sect. 2, as we describe now. Assume that we start with a BP consisting solely of $w \times w$ matrices (i.e. without bookends), such that the product matrix = identity iff the function evaluates to 0. The [BGK+14] obfuscator first chooses random vectors $s, t \in \mathbb{Z}_p^w$, and adds these as the bookends. Then, in addition to giving out the encoded matrices $Y_{i,b}$ and bookends s and t, the obfuscator gives out the encoded value of the inner product $\langle s, t \rangle$, as well an encoding of each $\alpha_{i,b}$. Finally, evaluation of the obfuscated program on input x is given by

$$ s \times \prod_i Y_{i,x_{\mathsf{inp}(i)}} \times t \; - \; \langle s, t \rangle \cdot \prod_i \alpha_{i,x_{\mathsf{inp}(i)}} $$

which is an encoding of 0 iff the product of the original BP matrices = identity.

Note that the first term in the subtraction matches the polynomial p_x that was analyzed above, so to extend our attack to the [BGK+14] obfuscator we must account for the g^1-coefficient of the second term $\langle s, t \rangle \cdot \prod_i \alpha_{i,x_{\mathsf{inp}(i)}}$. Denoting $\alpha_{0,x_{\mathsf{inp}(0)}} := \langle s, t \rangle$, the polynomial we need to analyze becomes $\prod_i (\alpha_{i,x_{\mathsf{inp}(i)}} + g \cdot z_{i,x_{\mathsf{inp}(i)}})$, where as above the z variables represent the GGH13 randomness. The g^1-coefficient of this polynomial is $\sum_i \left(z_{i,x_{\mathsf{inp}(i)}} \prod_{j \neq i} \alpha_{j,x_{\mathsf{inp}(j)}} \right)$, which we can rewrite as

$$\prod_i \alpha_{i,x_{\mathsf{inp}(i)}} \sum_i \tilde{z}_{i,x_{\mathsf{inp}}(i)} \tag{1}$$

via the change of variables $\tilde{z}_{i,x_{\mathsf{inp}}(i)} = z_{i,x_{\mathsf{inp}}(i)}/\alpha_{i,x_{\mathsf{inp}}(i)}$. Finally, observe that expression (1) can be easily absorbed into the previous decomposition

$$p_x^{(1)} = \rho \prod_i \alpha_{i,x_{\mathsf{inp}(i)}} \sum_i \beta'_{i,x_i}$$

and indeed the same annihilating polynomial $Q_{1,2,3}$ works for the [BGK+14] obfuscator as well.

We also believe that our attack extends to the candidate obfuscator in [BR14], because evaluating the program in that setting corresponds to a subtraction similar to the one just analyzed. However, we have not completely verified this due to the complexity of the [BR14] construction.

[PST14]. The main difference between the candidate obfuscator in [PST14] and the one in Sect. 2 is that the initial BP matrices are first padded with extra 1 s along the diagonal, i.e. they transform

$$A_{i,b} \longmapsto \begin{pmatrix} A_{i,b} & \\ & I \end{pmatrix}$$

However, since this preserves the property of a layer having the same matrix for both bits, our analysis can be applied to attack this candidate as well.

[GGH+13b]. The only candidate branching program obfuscator to which we do not know how to apply our attack is the original candidate due to [GGH+13b]. In this candidate, the initial BP matrices are first padded with extra *random* elements on the diagonal, i.e. they transform

$$A_{i,b} \longmapsto \begin{pmatrix} A_{i,b} & \\ & D_{i,b} \end{pmatrix}$$

where $D_{i,b}$ is a random diagonal matrix of dimension d (and $A_{i,b}$ is assumed to have dimension 5). Then, bookend vectors s and t are chosen

$$s = (s_1, \dots, s_5, \underbrace{0, \dots, 0}_{d/2}, \underbrace{\$, \dots, \$}_{d/2}) \qquad t = (t_1, \dots, t_5, \underbrace{\$, \dots, \$}_{d/2}, \underbrace{0, \dots, 0}_{d/2})$$

subject to $\sum_{i\leq 5} s_i t_i = 0$. (This is a slight simplification of [GGH+13b], but it illustrates the core technical problem in applying our attack.)

Now consider the evaluation on input x:

$$p_x = s \times \prod_i \begin{pmatrix} A_{i,x_{\mathsf{inp}(i)}} & \\ & D_{i,x_{\mathsf{inp}(i)}} \end{pmatrix} \times t = s \times \begin{pmatrix} \prod_i A_{i,x_{\mathsf{inp}(i)}} & \\ & \prod_i D_{i,x_{\mathsf{inp}(i)}} \end{pmatrix} \times t$$

While this product indeed encodes value $\sum_{i\leq 5} s_i t_i = 0$ when $\prod_i A_{i,x_{\mathsf{inp}(i)}} = I$, the g^1-coefficient becomes quite complicated. This is due to the uniform entries in s and t, which select some of the GGH randomization variables from the "southwest" quadrant of the product matrix. As a result, we do not know how to extend our attack to this setting.

4   Attack on GGH13 Encodings

In this section we explain how the abstract attack above extends to actual obfuscation schemes [BGK+14, AGIS14, BMSZ16] when implemented with [GGH13a] multilinear maps. At a high level, this is done by implementing **Type 1** and **Type 2** queries ourselves, without the help of the abstract model's oracle.

Implementing **Type 1** queries is straightforward: for any honestly executed 0-output of the program, namely an encoding

$$p_x = \left[\left(p_x^{(0)}(\{X_j\}, \{Z_i\}) + gp_x^{(1)}(\{X_j\}, \{Z_i\}) + g^2...\right)/z^k\right]_q$$

with $p_x^{(0)}(\{X_j\}, \{Z_i\}) = 0$, we can multiply by the zero-testing parameter $\mathbf{p}_{zt} = [hz^k/g]_q$ to obtain

$$W_x := [p_x \cdot \mathbf{p}_{zt}]_q = h \cdot \left(p_x^{(1)}(\{X_j\}, \{Z_i\}) + gp_x^{(2)}(\{X_j\}, \{Z_i\}) + g^2...\right) \qquad (2)$$

This differs from what is returned in the abstract attack because of the factor h. To handle this, we ensure that our annihilating polynomial Q is *homogeneous*, and thus $Q(\{h \cdot p_x^{(1)}(\{X_j\}, \{Z_i\})\}_x) = 0$ whenever $Q(\{p_x^{(1)}(\{X_j\}, \{Z_i\})\}_x) = 0$. (Lemma 1 in fact shows we can assume Q is homogeneous without loss of generality, because the $p_x^{(1)}$ are all homogeneous and of the same degree.)

To implement **Type 2** queries, we must check whether a given polynomial Q over $\{W_x\}_{x \in S}$ (for some $S \subseteq \{0,1\}^n$) is an annihilating polynomial, i.e. whether $Q\left(\{h \cdot p_x^{(1)}(\{X_j\}, \{Z_i\})\}_{x \in S}\right) = 0$. To do this we observe that, for any such Q, $Q(\{W_x\}_{x \in S})$ produces a ring element in the ideal $\langle hg \rangle$. So, we compute many such elements $v_i = Q_i(\{W_x\}_{x \in S_i})$, where Q_i is the (homogeneous) polynomial that annihilates $\{p_x^{(1)}(\{X_j\}, \{Z_i\})\}_{x \in S_i}$ when the encodings were formed by obfuscating the all-identity branching program. More specifically, we compute enough v_i to (heuristically) form a basis of $\langle hg \rangle$. Then, we compute one more element v^* which is either in $\langle hg \rangle$ or not depending on which branching program was obfuscated, and finally we use the $\langle hg \rangle$-basis to test this.

4.1   The Attack

We use essentially the same pair of branching programs $\mathbf{A}, \mathbf{A}'$ that were used in the abstract attack (see Sect. 3.1): $\mathbf{A}$ consists of all identity matrices, while in $\mathbf{A}'$ the two matrices corresponding to $x_1 = 0$ are changed to be anti-diagonal.

Let $\mathcal{O}$ denote the obfuscator described in Sect. 2.1. This obfuscator is exactly the one from [BMSZ16], with two exceptions. First, it operates on a branching program reading only one bit per layer, while in [BMSZ16] the branching programs read two bits per layer. In Sect. 5, we show that our abstract attack, and thus also the concrete attack described here, extends to the dual-input setting. (In fact, we show that it extends to arity-d branching programs for any constant d.) Second, Eq. (2) (and the presence of z^k in $\mathbf{p}_{zt}$) assumes that all encodings output by $\mathcal{O}$ are level-1 GGH encodings, while in [BMSZ16] a more complicated level structure is used (following [BGK+14, MSW14]). However, since our attack only uses these encodings to honestly execute the obfuscated program, (2) holds even for this level structure.

Here is our attack:

- Let $m = n^{O(1)}$ be the dimension of the underlying encodings (this is a parameter of the [GGH13a] scheme). Note that any m linearly independent elements of $\langle hg \rangle$ form a basis for $\langle hg \rangle$. Let $m' \gg m$ be an integer.
- Repeat the following for $t = 1, \ldots, m'$:
  - Choose a random size-3 subset $T_0 = \{i_1, i_2, i_3\} \subseteq [n]$ that *does not* contain 1. T_0 will correspond to the set of input bits that we vary.
  - Choose a random subset $T_1 \subseteq ([n] \setminus T_0)$. T_1 will correspond to a fixing of the bits outside T_0.
  - For each $T \subseteq [3]$,
    * let $x_T \in \{0,1\}^n$ be the string such that $x_i = 1$ if and only if either $i \in T_1$, or $i = i_j$ for some $j \in T$ (recall that $T_0 = \{i_1, i_2, i_3\}$).
    * Run the obfuscated program on input x, until the zero test query. Let $p_T^{(1)}$ be the vector obtained from zero testing.
  - Evaluate the polynomial $Q_{1,2,3}$ in Sect. 3 on the $p_T^{(1)}$. Let the output be defined as v_t. That is, we let x_T vary over the the 8 possible values obtained by fixing all the input bits outside of T_0, run the obfuscated program on each of the x_T, and then evaluate the polynomial $Q_{1,2,3}$ on the results to get v_t.
- Find a linearly independent subset V of the v_t.
- Choose a random size-3 subset $T_0^* = \{i_1, i_2, i_3\} \subseteq [n]$ that *does* contain 1. For each $T \subseteq [3]$, compute $p_T^{(1)}$ as above. Then evaluate the polynomial $Q_{1,2,3}$ on the $p_T^{(1)}$ to obtain a vector v^*.
- Finally, test if v^* is in the span of V. If it is, output 1. Otherwise, output 0.

Analysis of Our Attack. As in Sect. 3, let $T_0 \subseteq [n]$, and choose an arbitrary fixing of the remaining bits. Suppose we evaluate the branching program on the 8 different inputs corresponding to varying the bits in T_0, and then run the polynomial $Q_{1,2,3}$ on the results. Then $Q_{1,2,3}$ annihilates annihilates in either of the following cases:

- T_0 does not contain 1.
- The branching program is the all-identity program, even if T_0 contains 1.

Therefore, we see that $Q_{1,2,3}$ annihilates for each $t = 1, \ldots, m'$. In the case of [GGH13a], $Q_{1,2,3}$ annihilating mans that the resulting vector v is an element of the ideal $\langle hg \rangle$.

Thus, each of the v_t are elements in the ideal, regardless of the branching program. We will heuristically assume that the v_t span the entire ideal. This is plausible since the number m' of v_t is much larger than the dimension of the ideal. Increasing m' relative to m should increase the likelihood of the heuristic being true.

For v^*, however, things are different. v^* is in the ideal if the branching program is the all-identity, but outside the ideal (with high probability) if the branching program has anti-diagonals, since in this case $Q_{1,2,3}$ does not annihilate. Therefore, our test for v^* being linearly independent from v will determine which branching program we were given.

5   Beyond Single-Input Branching Programs

In this section, we show an abstract attack on dual-input branching programs, proving Theorem 2. More generally, we show that generalizing to d-input branching programs for any constant d will not prevent the attack.

We first recall our semantics of branching programs in the general d-ary setting. Fix integers d, ℓ and n which respectively correspond to the number of bits read by each layer of the branching program, the length of the branching program, and the input length. Let inp $: [\ell] \to 2^{[n]}$ be any function such that $|\mathsf{inp}(i)| = d$ for all $i \in [\ell]$. A branching program of length ℓ then consists of $2^d \ell + 2$ matrices $A_0, \{A_{i,S_i}\}_{i \in [\ell]}, A_{\ell+1}$ where S_i ranges over subsets of inp(i), and $A_0 A_{\ell+1}$ are the "bookend" vectors.

We associate an input x with the subset $T \in 2^{[n]}$ of indices where x is 1. To evaluate the branching program on input x (set T) compute the product

$$\mathbf{A}(T) = A_0 \times \prod_{i=1}^{\ell} A_{i,T\cap\mathsf{inp}(i)} \times A_{\ell+1}$$

Consider the obfuscation of the branching program. Let R_i be the Kilian randomizing matrices. Let $\alpha_{i,S}$ be the extra randomization terms. Then the encoded values seen by the adversary are the matrices $Y_{i,S} = \alpha_{i,S} R_i . A_{i,S} . R_{i+1}^{adj} + g Z_{i,S}$

By performing a change of variables on the $Z_{i,S}$, we can actually write $Y_{i,S} = \alpha_{i,S} R_i \cdot (A_{i,S} + g Z_{i,S}) \cdot R_{i+1}^{adj}$

The encodings will guarantee some restrictions on the **Type 1** queries allowed — however they must allow evaluation of the branching program. Thus we assume that the following query is allowed for every $T \subseteq [n]$.

$$p_T = Y_0 \times \prod_{i=1}^{\ell} Y_{i,T\cap\mathsf{inp}(i)} \times Y_{\ell+1}$$

Now we will assume a trivial branching program where (1) within each layer, all matrices are the same ($A_{i,S_i} = A_{i,S_i'}$ for any $S_i, S_i' \in \mathsf{inp}(i)$), so in particular the program is constant, and (2), the branching program evaluates to 0 on all inputs. Therefore, the g^0 coefficient in p_T will evaluate to zero everywhere. Thus, a **Type 1** query will output a handle to the variable

$$p_T^{(1)} = \rho \left(\prod_i \alpha_{i,S\cap\mathsf{inp}(i)} \right) \sum_i \left(\cdots A_{i,T\cap\mathsf{inp}(i-1)} \cdot Z_{i,T\cap\mathsf{inp}(i)} \cdot A_{i+1,T\cap\mathsf{inp}(i+1)} \cdots \right)$$

For any sets $S' \subseteq S \subseteq [n]$ with $|S| = d$, define

$$\alpha_{S,S'} := \prod_{i:\mathsf{inp}(i)=S} \alpha_{i,S'} \qquad \beta_{S,S'} := \sum_{i:\mathsf{inp}(i)=S} \beta_{i,S'}$$

and for any set $T \subseteq [n]$, define

$$U_T := \prod_{S:|S|=d} \alpha_{S,T\cap S} \qquad V_T := \sum_{S:|S|=d} \beta_{S,T\cap S}$$

Then we have that $p_T^{(1)} = U_T V_T$.

The following theorem shows that, for $|T| > d$, U_T and V_T can each be written as rational polynomials in the variables $U_{T'}, V_{T'}$ for $|T'| \leq d$.

Theorem 4. *Let $T \subseteq [n]$ with $|T| > d$. Then,*

$$U_T = \prod_{T' \subseteq T:|T'| \leq d} U_{T'}^{(-1)^{d-|T'|} \cdot \binom{|T|-|T'|-1}{d-|T'|}}$$

and

$$V_T = \sum_{T' \subseteq T:|T'| \leq d} (-1)^{d-|T'|} \cdot \binom{|T|-|T'|-1}{d-|T'|} \cdot V_{T'}.$$

Proof. We prove this equation for V_T, the proof for U_T is analogous. Consider expanding the left and right sides of the equation in terms of the $\beta_{S,Z}$ and equating the coefficients of $\beta_{S,Z}$ on both sides, we see that the following claim suffices to prove the theorem:

Claim. For any sets T, S, Z,

$$\sum_{T' \subseteq T:|T'| \leq d, T' \cap S=Z} \binom{|T|-|T'|-1}{d-|T'|}(-1)^{d-|T'|} = \begin{cases} 1 & \text{if } T \cap S = Z \\ 0 & \text{if } T \cap S \neq Z \end{cases}$$

The left hand side (resp. right hand side) of the above equation corresponds to the coefficient of $\beta_{S,Z}$ in the right hand side (resp. left hand side) of the V equation in Theorem 4. Hence the theorem follows from the claim.

We now prove the claim. First, suppose $Z \not\subseteq T \cap S$. Then the sum on the right is empty, so the result is zero, as desired. Next, suppose $Z \subseteq T \cap S$. Then for any T' in the sum, we can write $T' = Z \cup T''$ where $T'' \subseteq T \setminus (S \cup Z)$ and $|T''| \leq d - |Z|$. Therefore, we can think of the sum as being over T''. The number of T'' of size i is $\binom{|T \setminus (S \cup Z)|}{i}$. Therefore, the sum on the left is equal to

$$\sum_{i=0}^{d-|Z|} \binom{|T \setminus (S \cup Z)|}{i}\binom{|T|-|Z|-i-1}{d-|Z|-i}(-1)^{d-i-|Z|}$$

Let $e = d - |Z|$, $t = |T| - |Z| = |T \setminus Z|$ (since $Z \subseteq (T \cap S) \subseteq T$), and $k = |T \setminus (S \cup Z)|$. Notice that $k \leq t$, and that $k = t$ if and only $Z = T \cap S$. Thus, we need to show that

$$\sum_{i=0}^{e} \binom{k}{i} \binom{t-i-1}{e-i} (-1)^{e-i} = \begin{cases} 1 & \text{if } k = t \\ 0 & \text{if } k < t \end{cases}$$

First, we use the identity $(-1)^s \binom{s-r-1}{s} = \binom{r}{s}$ with $s = e - i$ and $r = e - t$ to replace $\binom{t-i-1}{e-i}(-1)^{e-i}$ with $\binom{e-t}{e-i}$ (note that the binomial coefficients are defined for negative integers such as $e - t$).

Then we have that the left hand side becomes $\sum_{i=0}^{e} \binom{k}{i} \binom{e-t}{e-i}$. The Chu-Vandermonde identity shows that this is equal to $\binom{k+(e-t)}{e} = \binom{e-(t-k)}{e}$. Notice that if $t = k$, the result is 1. Moreover, if $k < t$, then the upper index of the binomial is less than the bottom index, so the result is 0. This proves the claim and hence the theorem.

Annihilating Polynomial for $p_T^{(1)}$. We now describe our abstract attack using annihilating polynomials. The first step is to argue that it is possible to efficiently devise a non-zero polynomial Q on several of the $p_T^{(1)}$ such that Q is identically zero when the $p_T^{(1)}$ come from the obfuscation. In particular, we need Q to be identically zero as a polynomial over the α's and β's. Using Theorem 4, it suffices to find Q that is identically zero as a rational function over the U_T, V_T for $|T| \leq d$.

We will first consider the values $p_T^{(1)}$ as polynomials in the $V_T, U_T, |T| \leq d$ over the rationals. Let $k = 2d + 2$, and consider all $p_T^{(1)}$ for $T \subseteq [k]$. Then each $p_T^{(1)}$ is a rational function of the U_T, V_T for $T \subseteq [k], |T| \leq d$. There are $\sum_{i=0}^{d} \binom{k}{i} < 2^{2d+1}$ such T, and therefore fewer than 2^{2d+2} such U_T, V_T. Yet there are 2^{2d+2} different $p_T^{(1)}$ for $T \subseteq [k]$ of arbitrary size. Thus, there must be some algebraic dependence among the $p_T^{(1)}$. Notice moreover that the expression for $p_T^{(1)}, T \subseteq [k]$ in terms of the $U_{T'}, V_{T'}, T' \subseteq [k], |T'| \leq d$ are fixed rational functions with integer coefficients, independent of the branching program, n, or ℓ; the only dependence is on d. Recall that we are taking d to be a constant, so the number of $p_T^{(1)}, V_{T'}, U_{T'}$ and the coefficients in the relation between them are all constants. Therefore, there is a fixed polynomial Q_d in the $p_T^{(1)}$ over the rationals such that Q_d is identically zero when the $p_T^{(1)}$ come from obfuscation.

We note that by a more tedious argument, it is actually possible to show there must be an algebraic dependence among the $p_T^{(1)}$, and hence an annihilating polynomial for them, when T varies over the subsets of $[k]$ for $k = 2d + 1$ (as opposed to $2d + 2$).

By multiplying by the LCM of the denominators of the rational coefficients, we can assume without loss of generality that Q_d has integer coefficients. Therefore, there is a fixed integer polynomial Q_d such that $Q_d(p_T^{(1)})$ is identically 0. Since the coefficients are integers, this polynomial actually also applies in any field or ring; we just need to verify that it is not identically zero in the field/ring.

This will be true as long as the characteristic of the ring is larger than the largest of the coefficients. Since in our case, the ring characteristic grows (exponentially) with the security parameter, for high enough security parameter, the polynomial Q_d will be non-zero over the ring.

Computing the Annihilating Polynomial Q_d. In Sect. 3, we gave an annihilating polynomial for the case $d = 1$. For more general d, we do not know a more general expression. However, we still argue that such a Q_d can be efficiently found for any d:

- The polynomial Q_d is just a fixed polynomial over the integers; in particular is has a constant-sized description for constant d. Thus, we can assume that Q_d is simply given to the adversary.
- If we want to actually compute Q_d, this is possible using linear algebra. Using degree bounds for the annihilating polynomial due to [Kay09], we can determine an upper bound t on the degree of Q_d. Then, the statement "Q_d annihilates the $p_T^{(1)}$" can be expressed as a system of linear equations in the coefficients of Q_d, where the equations themselves are determined by expressions for $p_T^{(1)}$ in terms of the $U_{T'}, V_{T'}$. By solving this system of linear equations, it is possible to obtain a polynomial Q_d. We note that, for constant d, t will be constant, the system of linear constraints will be constant, and hence it will take constant time to compute Q_d. In terms of d, the running time is necessarily exponential (since the number of variables $p_T^{(1)}$ is exponential).

The following lemma shows that we can take Q to be a homogeneous polynomial, which will be necessary for obtaining an attack over [GGH13a].

Lemma 1. *Let* $p_1, \ldots, p_k$ *be homogeneous polynomials each of the same degree* d. *Let* Q *be any polynomial that annihilates* $\{p_i\}_i$, *and let* Q_r *denote the homogeneous degree-r part of* Q. *Then* Q_r *annihilates* $\{p_i\}_i$ *for each* $r \leq \deg(Q)$.

Proof. If $Q_r(\{p_i\}_i) \neq 0$ for some $r \leq \deg(Q)$, then Q_r contains some degree-dr monomial m. Then because $\sum_{r=0}^{\deg(Q)} Q_r(\{p_i\}_i) = Q(\{p_i\}_i) = 0$, some $Q^{(r')}$ for $r' \neq r$ must contain the monomial $-m$. However, since $Q^{(r')}$ is homogeneous of degree $dr' \neq dr$, this is a contradiction.

Completing the Attack. Using the annihilating polynomial above, we immediately get an attack on the abstract model of obfuscation. The attack distinguishes the trivial branching program where all matrices across each layer are the same, from a more general all-zeros branching program that always outputs zero, but has a non-trivial branching program structure.

The attack proceeds as follows: query the model on **Type 1** queries for all p_T as T ranges over the subsets of $[k]$. Since the branching program always outputs 0, the model will return a handle to the $p_T^{(1)}$ polynomials. Then evaluate the annihilating polynomial Q_d above on the obtained $p_T^{(1)}$. If the result is non-zero (as will be the case for many non-trivial branching programs), then we know the branching program was *not* the trivial branching program. In contrast, if the result *is* zero, then we can safely guess that we are in the trivial branching program case. Hence, we breach the indistinguishability security of the obfuscator.

6 Attacking Order Revealing Encryption

In this section, we describe how to attack the order revealing encryption (ORE) scheme of Boneh et al. [BLR+15], proving Theorem 3.

Theorem 3. *Let $\mathcal{E}$ denote the ORE scheme of [BLR+15] (over GGH13 [GGH13a] maps). There exist two sequences of plaintexts $m_1^0 < \cdots < m_\ell^{(0)}$ and $m_1^{(1)} < \cdots < m_\ell^{(1)}$ such that $\mathcal{E}(m_1^0), \cdots, \mathcal{E}(m_\ell^0)$ and $\mathcal{E}(m_1^1), \cdots, \mathcal{E}(m_\ell^1)$ can be efficiently distinguished in the abstract attack model described in Sect. 2.*

We first recall the definition of an order revealing encryption scheme.

Definition 1. An order revealing encryption scheme consists of four algorithms (Gen, Enc, Dec, Comp) such that:

- Gen takes as input the security parameter, and outputs public parameters PP and a secret key sk.
- Enc(sk, m) is a secret key encryption algorithm that outputs a ciphertext c.
- Dec(sk, c) is a decryption algorithm that outputs a plaintext.
- Comp(PP, c_0, c_1) is a public key comparison procedure that takes as input two ciphertexts, and outputs a bit b.
- **Correct Decryption.** This is the standard correctness requirement for secret key encryption. For any m, with overwhelming probability over the choice of (PP, sk) and the random coins of Enc, we have that Dec(sk, Enc(sk, m)) outputs m.
- **Correct Comparison.** For any messages m_0, m_1, $m_0 < m_1$, we have that with overwhelming probability over the choice of (PP, sk) and the random coins of Enc, Comp(PP, Enc(sk, m_0), Enc(sk, m_1)) $= 0$ and Comp(PP, Enc(sk, m_1), Enc(sk, m_0)) $= 1$.
- **Security.** For any two polynomial-length sequences of ordered messages $m_0^{(0)} < m_1^{(0)} < \cdots < m_\ell^{(0)}$ and $m_0^{(1)} < m_1^{(1)} < \cdots < m_\ell^{(1)}$ of the same length ℓ, we have that the following two distributions are computationally indistinguishable: PP, Enc(sk, $m_0^{(0)}$), Enc(sk, $m_1^{(0)}$), ..., Enc(sk, $m_\ell^{(0)}$) and PP, Enc(sk, $m_0^{(1)}$), Enc(sk, $m_1^{(1)}$), ..., Enc(sk, $m_\ell^{(1)}$).

We note that the security definition is much weaker than that defined in [BLR+15], which allowed for adaptive message queries. Nonetheless, we will give an attack on their scheme even for our weaker definition

6.1   Description of [BLR+15] in Abstract Model

We now given an abstract description of the [BLR+15] order revealing encryption scheme in our model for graded encodings. We will actually describe a simplified variant for which, for any ciphertext, that ciphertext can be inserted into either the first or second ciphertext slot of Comp, but not both. That is, Enc now takes as input an additional bit b, and if $b = 0$, and Comp(PP, c_0, c_1) is only required to be correct where c_0 is encrypted using bit 0, and c_1 is encrypted using bit 1. This is how the Boneh et al. [BLR+15] protocol works; to obtain the usual notion of order revealing encryption, the encryption procedure simply encrypts twice, once to each input.

The starting point for the construction is a branching program $A_{1,0}, A_{1,1}, B_{1,0}, B_{1,1}, \ldots, A_{n,0}, A_{n,1}, B_{n,0}, B_{n,1}$ such that:

– For any two n-bit integers x, y, $\prod_{i=1}^{n} A_{x_i} \cdot B_{y_i} = 0$ if and only if $x < y$. Note that it is trivial to extend our attacks to work in the case where $<$ is replaced with $\leq, >$, or $\geq$.
– For any $j \in [n]$, the products $\prod_{i=1}^{j} A_{x_i} \cdot B_{y_i}$ and $\prod_{i=n-j+1}^{n} A_{x_i} \cdot B_{y_i}$, which will be vectors of some dimension, only depend on the result of comparing the first or last, respectively, j bits of x and y. That is, $\prod_{i=1}^{j} A_{x_i} \cdot B_{y_i}$ takes on one of 3 possible values, depending on the three possible results of comparing $x_{[1,i]}, y_{[1,i]}, <, >$, or $=$.

Note that here we describe a branching program without bookends, but where the matrices are shaped so that the output is a scalar. It is straightforward to obtain a branching program in this form by multiplying the branching program by the bookend vectors.

The secret key sk for the ORE scheme consists of $2n - 1$ random matrices R_i, as well as the necessary information to compute encodings in the graded encoding. The public key will be the description of the graded encoding scheme, which allows for **Type 1** and **Type 2** queries, with the class of valid **Type 1** queries to be specified later.

Encryption. To encrypt integer x into the left input to Comp, choose random $\alpha_{x,i}$ for $i \in [n]$, and compute $\widetilde{A_{x,i}} = \alpha_{x,i} R_{2i-2} \cdot A_{i,x_i} \cdot R_{2i-1}^{adj}$. Here, R_0 is just the integer 1. Then the $\widetilde{A_{x,i}}$ are encoded, meaning the public values seen by the adversary are

$$X_{x,i} = \alpha_{x,i} R_{2i-2} \cdot (A_{i,x_i} + gZ_{x,i}) \cdot R_{2i-1}^{adj}$$

for random $Z_{x,i}$. Here, we use the re-labeling of the Z variables used in Sect. 3.

Encryption in the right input to Comp is analogous. Choose random $\beta_{x,i}$ for $i \in [n]$, and compute $\widetilde{B_{x,i}} = \beta_{x,i} R_{2i-1} \cdot B_{i,x_i} \cdot R_{2i}^{adj}$. Here, R_{2n} is just the integer 1. Then the $\widetilde{B_{x,i}}$ are encoded, meaning the public values seen by the adversary are

$$Y_{x,i} = \beta_{x,i} R_{2i-1} \cdot (B_{i,x_i} + gW_{x,i}) \cdot R_{2i}^{adj}$$

for random $W_{x,i}$.

Comparison. To compare two ciphertexts c_0, c_1 consisting of $X_{x,i}$ and $Y_{y,i}$ for integers x, y, perform a **Type 1** query on the product

$$\prod_{i=1}^{n} X_{x,i} \cdot Y_{y,i} = \prod_{i=1}^{n} \alpha_{x,i}\beta_{y,i}\,(A_{i,x_i} + gZ_{x,i}) \cdot (B_{i,y_i} + gW_{y,i})$$

Notice that the g^0 term is exactly equal to $\prod_{i=1}^{n} A_{x_i} \cdot B_{y_i}$, up to scaling by the $\alpha_{x,i}, \beta_{y,i}$. Therefore, the result is zero if and only if $x < y$. Thus, it is possible to determine the order of the two plaintexts.

Note that these **Type 1** queries must be explicitly allowed for correctness. [BLR+15] analyze the types of queries that are allowed in the standard generic model for graded encodings; however, for our attack, we do not require any other **Type 1** queries.

6.2   Our Attack

Suppose Comp gives 0 on encryptions of x and y. We denote the coefficient of g^1 by $V(x, y)$, which is equal to the following expression.

$$\left(\prod_{i=1}^{n} \alpha_{x,i}\beta_{y,i}\right) \sum_{i=1}^{n} \left(\prod_{j=1}^{i-1} A_{j,x_j} \cdot B_{j,y_j}\right) \cdot (A_{i,x_i} \cdot W_{y,i} + Z_{x,i} \cdot B_{i,y_i}) \cdot \left(\prod_{j=i+1}^{n} A_{j,x_j} \cdot B_{j,y_j}\right)$$

Define $\alpha_x = \prod_{i=1}^{n} \alpha_{x,i}$ and $\beta_y = \prod_{i=1}^{n} \beta_{y,i}$. Recall that $\prod_{j=1}^{i-1} A_{j,x_j} \cdot B_{j,y_j}$ only depends on the result of comparing the first $i-1$ bits, and that $\prod_{j=i+1}^{n} A_{j,x_j} \cdot B_{j,y_j}$ only depends on the result of comparing the last $n - i$ bits. Therefore, we can re-write the g^1 coefficient as:

$$V(x, y) = \alpha_x \beta_y \sum_{i=1}^{n} \Big(Z_{x,i,\mathrm{Comp}(x_{[1,i-1]}, y_{[1,i]}), y_i, \mathrm{Comp}(x_{[i+1,n]}, y_{[i+1,n]})}$$

$$+ W_{y,i,\mathrm{Comp}(x_{[1,i-1]}, y_{[1,i-1]}), x_i, \mathrm{Comp}(x_{[i+1,n]}, y_{[i+1,n]})} \Big)$$

For variables $Z_{x,i,a,b,c}, W_{y,i,a,b,c}$ where $a, c \in \{<, =, >\}$ and $b \in \{0, 1\}$.

Choosing the Query Points. We now describe how we choose our query points. Let k be a positive integer, and $n = 2k + 4$. Let X_0, Y_0, X_1, Y_1 be sets of n-bit integers that have the form:

- X_b: $x = 0\,\hat{x}\,00\,0^k\,0$ for a k-bit integer $\hat{x}$. In particular $X_0 = X_1$.
- Y_b: $y = b\,1^k\,11\,\hat{y}\,b$ for a k-bit integer $\hat{y}$.

Then X_b, Y_b satisfy the following:

- For any $x \in X_b, y \in Y_b$, $x < y$.

- For any $x \in X_1, y \in Y_1$, $x_{[1,i]} < y_{[1,i]}$ for all $i \in [n]$. That is, the result of comparing the first i bits for any i is always $<$.
- For any $x \in X_0, y \in Y_0$, $x_{[1,i]} < y_{[1,i]}$, *unless*
  - $i = 1$
  - $i \in [2, k+1]$ and $\hat{x}_{[1,i-1]} = 1^{i-1}$.
- For any $x \in X_1, y \in Y_1$, $x_{[i,n]} < y_{[i,n]}$ for any $i \in [n]$. That is, the result of comparing the last $n - i + 1$ bits for any i is always $<$.
- For any $x \in X_0, y \in Y_0$, $x_{[i,n]} < y_{[i,n]}$, *unless*
  - $i = n$
  - $i \in [k+4, 2k+3]$ and $\hat{y}_{[i-k-3,k]} = 1^{i-k-3}$.

We first consider X_1, Y_1. For these x, y, $\mathsf{Comp}(x_{[1,i-1]}, y_{[1,i-1]})$ and $\mathsf{Comp}(x_{[i+1,n]}, y_{[i+1,n]})$ will always be the independent of the choice of $x \in X_1$ and $y \in Y_1$, namely $<$.[4] Moreover, for $i = 1$ or $i \in [k+2, n]$, x_i is independent of x. Therefore, the $Z_{x,i,\mathsf{Comp}(x_{[1,i-1]},y_{[1,i]}),y_i,\mathsf{Comp}(x_{[i+1,n]},y_{[i+1,n]})}$ for these i are independent of y. They can thus be absorbed into the other Z_x's. Similar statements hold for the y_i's for $i \in [1, k+3]$ or $i = n$.

This lets us write

$$V(x,y) = \alpha_x \beta_y \sum_{i=1}^{k} (Z_{x,i,\hat{y}_i} + W_{y,i,\hat{x}_i})$$

for $x \in X_1$, $y \in Y_1$, and variables $Z_{x,i,b}, W_{y,i,b}$. We write this as the sum of two inner products: $V(x,y) = Z_x \cdot \Gamma_y + \Delta_x \cdot W_y$ where

$$
\begin{aligned}
Z_x &= \alpha_x (Z_{x,1,0} \ Z_{x,1,1} \ Z_{x,2,0} \ Z_{x,2,1} \ \cdots \ Z_{x,n,0} \ Z_{x,n,1}) \\
W_y &= \beta_y (W_{y,1,0} \ W_{y,1,1} \ W_{y,2,0} \ W_{y,2,1} \ \cdots \ W_{y,n,0} \ W_{y,n,1}) \\
\Delta_x &= \alpha_x (\ (1-\hat{x}_1) \ \hat{x}_1 \ (1-\hat{x}_2) \ \hat{x}_2 \ \cdots \ (1-\hat{x}_n) \ \hat{x}_n \ ) \\
\Gamma_y &= \beta_y (\ (1-\hat{y}_1) \ \hat{y}_1 \ (1-\hat{y}_2) \ \hat{y}_2 \ \cdots \ (1-\hat{y}_n) \ \hat{y}_n \ )
\end{aligned}
$$

Let V be the matrix of $V(x,y)$ values as x, y vary over X_1, Y_1, respectively. Let Z, Δ be the matrices containing the vectors Z_x, Δ_x (respectively) as rows, and let W, Γ be the matrices containing the vectors W_y, Γ_y (respectively) as columns. Then we can write

$$V = Z \cdot \Gamma + \Delta \cdot W = (Z \ \Gamma) \cdot \begin{pmatrix} \Delta \\ W \end{pmatrix}$$

Now, the smallest dimension of the matrices Γ, Δ is $2k$, so their rank is clearly at most $2k$. We now argue that the rank is in fact at most $k + 1$ for each. To see this, note that the columns of Δ are spanned by the following $k + 1$ column vectors: $v_x^{(0)} = \alpha_x$, and $v_x^{(i)} = \hat{x}_i$ for $i \in [k]$. Thus the rank of Δ is at most

---

[4] Technically, in the case $i = 1$, $\mathsf{Comp}(x_{[1,i-1]}, y_{[1,i-1]})$ will give $=$. However, this is still independent of the choice of x and y, so all of the following arguments are still valid.

$k+1$. Moreover, it is straightforward to argue that any $k' \leq k+1$ rows of Δ are linearly independent. Similar arguments hold for Γ.

Since Z, W are full rank with overwhelming probability, $Z \cdot \Gamma$ and $\Delta \cdot W$ each have rank $\min(k+1, 2^k)$. Therefore, their sum V has rank at most $2k+2$. Moreover, since Z, W are random matrices, the ranks will add with overwhelming probability, so the total rank is $\min(2k+2, 2^k)$.

We now consider X_0, Y_0. Performing a similar treatment as we did in the case of X_1, Y_1, for any $x \in X_0, y \in Y_0$, we can therefore write

$$V(x,y) = \alpha_x \beta_x \left(\left(\sum_{i=1}^{k-1} Z_{x,i,\hat{y}_i,\delta_0(\hat{y}_{[i+1.k]})} \right) + \left(\sum_{i=2}^{k} W_{y,i,\hat{x}_i,\delta_1(\hat{x}_{[1.i-1]})} \right) \right)$$

where $\delta_b(z)$ is 1 if and only if all the bits of z are equal to b, and 0 otherwise. Note that one might expect there to be a $Z_{x,0,\delta_0(\hat{y})}$ term. However, $\delta_0(\hat{y})$ is determined by $\hat{y}_1$ and $\delta_0(\hat{y}_{[2,k]})$, and hence $Z_{x,0,\delta_0(\hat{y})}$ can be absorbed into $Z_{x,1,\hat{y}_1,\delta_0(\hat{y}_{[2.k]})}$. Similar statements hold for $W_{y,k+1,\delta_1(\hat{x})}$, $Z_{x,k,\hat{y}_k}$, and $W_{y,1,\hat{x}_1}$.

Through a similar analysis as in the X_1, Y_1 case, the matrix V whose entries are $V(x,y)$ can be written as $Z \cdot \Gamma + \Delta \cdot W$ for $2^k \times (4k-4)$ matrices Z, Δ and $(4k-4) \times 2^k$ matrices Γ, W, where Z, W contain the variables $Z_{x,i,b,c}, W_{y,i,b,c}$ and Δ, Γ are matrices that depend on the bits of x, y. Note that these matrices will be different than those computed in the X_1, Y_1 case. The matrices Δ, Γ can each be shown to have rank $\min(2k, 2^k)$. Then V has rank $\min(4k, 2^k)$.

Hence, the rank of V will depend on whether we consider X_0, Y_0 or X_1, Y_1. This will be the basis for our attack.

The Attack. We now describe our distinguishing attack. Set $k = 4$, and let X_0, Y_0, X_1, Y_1 be the sets of $2^k = 16$ integers each, as above.

- Query on the sequences (X_0, Y_0) and (X_1, Y_1), obtaining 32 ciphertexts corresponding to encryptions of (X_b, Y_b). Let D be the ciphertexts encrypting X_b, and E be the ciphertexts encrypting Y_b. Note that we only need the ciphertexts in D to be valid *left* inputs to Comp, and the ciphertexts in E to be valid *right* inputs.
- For each $d \in D, e \in E$, make a **Type 1** query on the polynomial corresponding to runing the comparison procedure on d, e. Since $x < y$ for each $x \in X_b, y \in Y_b$, the polynomial will evaluate to 0, and hence result of the query will be an algebraic element $V_{d,e}$.
- Assemble the $2^k = 16 \times 2^k = 16$ matrix V of the $V_{d,e}$ components.
- Compute the determinant of V. If the result is zero, output 1. Otherwise, output 0.

In the case $b = 1$, V will have rank $2k+2 = 10 < 16$. Hence the determinant gives 0. In the vase $b = 0$, V will have rank $4k = 16$. Hence the determinant will be non-zero with overwhelming probability. Thus, our attack successfully determines which set of encryptions it received.

Attack Over GGH13. We now describe how to turn this into an actual attack on ORE built on GGH13 multilinear maps. Let ℓ be some integer. Let $X^{(a)}$ be X_1, except with the ℓ-bit integer a prepended to each of the elements in X_1. Similarly define $Y^{(a)}$. Define X_b^*, Y_b^* as X_b, Y_b, except with 0^ℓ prepended to each element.

Let $m = n^{O(1)}$ be the dimension of the underlying encodings, and $m' \gg m$ be an integer. Let $S \subseteq \{0,1\}^\ell$ be a set of size m' that does not contain zero. We will attack ORE instantiated with $(12 + \ell)$-bit integers.

The attack works as follows:

- Query on sequences $(X_0^*, Y_0^*), \{(X^{(s)}, Y^{(s)})\}_{s \in S}$ and $(X_1^*, Y_1^*), \{(X^{(s)}, Y^{(s)})\}_{s \in S}$, obtaining ciphertexts $(D^*, E^*), \{(D^{(s)}, E^{(s)})\}_{s \in S}$. Note that we only need the D ciphertexts to be valid *left* inputs, and the E ciphertexts to be valid *right* inputs to Comp.
- For each $s \in S$, do the following:
  - Construct the matrix $V^{(s)}$, which consists of all the results of comparing d, e for $d \in D^{(s)}, e \in E^{(s)}$.
  - Compute the determinant polynomial on $V^{(s)}$, obtaining v_s.
- Find a linearly independent subset U of the v_s.
- Construct the matrix V^*, which consists of all the results of comparing d, e for $d \in D^*, e \in E^*$.
- Compute the determinant polynomial on V^*, obtaining v^*.
- Test if v^* is in the span of U. If it is, output 1, otherwise output 0.

From our prior analysis, $V^{(s)}$ is not full rank, so the determinant annihilates each of the $V^{(s)}$, giving a vector v_s in the idea $\langle hg \rangle^5$. We will heuristically assume that the v_s span the entire ideal, which is plausible since the number of s, namely m', is much larger relative to the dimension of the ideal. Meanwhile, the determinant only annihilates V^* in the case $b = 1$. Thus v^* will be in $\langle hg \rangle$ if $b = 1$, but not if $b = 0$. Our linear independence test therefore distinguishes the two cases.

References

[ABD16] Albrecht, M., Bai, S., Ducas, L.: A subfield lattice attack on overstretched NTRU assumptions. In: Advances in Cryptology, CRYPTO (2016)

[AGIS14] Ananth, P., Gupta, D., Ishai, Y., Sahai, A.: Optimizing obfuscation: avoiding Barrington's theorem. In: Proceedings of the 2014 ACM SIGSAC Conference on Computer and Communications Security, pp. 646–658 (2014)

[BGK+14] Barak, B., Garg, S., Kalai, Y.T., Paneth, O., Sahai, A.: Protecting obfuscation against algebraic attacks. In: Nguyen, P.Q., Oswald, E. (eds.) EUROCRYPT 2014. LNCS, vol. 8441, pp. 221–238. Springer, Heidelberg (2014)

---

[5] This follows from the discussion in Sect. 4 and the fact that the determinant is homogeneous.

[BLR+15] Boneh, D., Lewi, K., Raykova, M., Sahai, A., Zhandry, M., Zimmerman, J.: Semantically secure order revealing encryption: multi-input functional encryption without obfuscation. In: Proceedings of EuroCrypt (2015)

[BMSZ16] Badrinarayanan, S., Miles, E., Sahai, A., Zhandry, M.: Post-zeroizing obfuscation: new mathematical tools, and the case of evasive circuits. In: Fischlin, M., Coron, J.-S. (eds.) EUROCRYPT 2016. LNCS, vol. 9666, pp. 764–791. Springer, Heidelberg (2016). doi:10.1007/978-3-662-49896-5_27

[BR14] Brakerski, Z., Rothblum, G.N.: Virtual black-box obfuscation for all circuits via generic graded encoding. In: Lindell, Y. (ed.) TCC 2014. LNCS, vol. 8349, pp. 1–25. Springer, Heidelberg (2014)

[CDPR16] Cramer, R., Ducas, L., Peikert, C., Regev, O.: Recovering short generators of principal ideals in cyclotomic rings. In: Fischlin, M., Coron, J.-S. (eds.) EUROCRYPT 2016. LNCS, vol. 9666, pp. 559–585. Springer, Heidelberg (2016). doi:10.1007/978-3-662-49896-5_20

[CGH+15] Coron, J.-S., et al.: Zeroizing without low-level zeroes: new MMAP attacks and their limitations. In: Gennaro, R., Robshaw, M. (eds.) CRYPTO 2015. LNCS, vol. 9215, pp. 247–266. Springer, Heidelberg (2015)

[CJL16] Cheon, J.H., Jeong, J., Lee, C.: An algorithm for CSPR problems and cryptanalysis of the GGH multilinear map without an encoding of zero. Technical report, Cryptology ePrint Archive, report 2016/139 (2016)

[CLT13] Coron, J.-S., Lepoint, T., Tibouchi, M.: Practical multilinear maps over the integers. In: Canetti, R., Garay, J.A. (eds.) CRYPTO 2013, Part I. LNCS, vol. 8042, pp. 476–493. Springer, Heidelberg (2013)

[GGH13a] Garg, S., Gentry, C., Halevi, S.: Candidate multilinear maps from ideal lattices. In: Johansson, T., Nguyen, P.Q. (eds.) EUROCRYPT 2013. LNCS, vol. 7881, pp. 1–17. Springer, Heidelberg (2013)

[GGH+13b] Garg, S., Gentry, C., Halevi, S., Raykova, M., Sahai, A., Waters, B.: Candidate indistinguishability obfuscation and functional encryption for all circuits. In: Proceedings of FOCS (2013)

[GMS16] Garg, S., Mukherjee, P., Srinivasan, A.: Obfuscation without the vulnerabilities of multilinear maps. Cryptology ePrint Archive, Report 2016/390 (2016). http://eprint.iacr.org/

[HJ16] Hu, Y., Jia, H.: Cryptanalysis of GGH map. In: Fischlin, M., Coron, J.-S. (eds.) EUROCRYPT 2016. LNCS, vol. 9665, pp. 537–565. Springer, Heidelberg (2016). doi:10.1007/978-3-662-49890-3_21

[Kay09] Kayal, N.: The complexity of the annihilating polynomial. In: Proceedings of the 24th Annual IEEE Conference on Computational Complexity, CCC 2009, Paris, France, pp. 184–193, 15–18 July 2009

[MSW14] Miles, E., Sahai, A., Weiss, M.: Protecting obfuscation against arithmetic attacks. IACR Cryptology ePrint Archive 2014, p. 878 (2014)

[MSZ16] Miles, E., Sahai, A., Zhandry, M.: Secure obfuscation in a weak multilinear map model: a simplified construction secure against all known attacks. Cryptology ePrint Archive (2016). http://eprint.iacr.org/

[PST14] Pass, R., Seth, K., Telang, S.: Indistinguishability obfuscation from semantically-secure multilinear encodings. In: Garay, J.A., Gennaro, R. (eds.) CRYPTO 2014, Part I. LNCS, vol. 8616, pp. 500–517. Springer, Heidelberg (2014)

Three's Compromised Too: Circular Insecurity for Any Cycle Length from (Ring-)LWE

Navid Alamati and Chris Peikert[(✉)]

University of Michigan, Ann Arbor, USA
cpeikert@alum.mit.edu

Abstract. A public-key encryption scheme is *k-circular secure* if a cycle of k encrypted secret keys $(\mathsf{Enc}_{pk_1}(sk_2), \mathsf{Enc}_{pk_2}(sk_3), \ldots, \mathsf{Enc}_{pk_k}(sk_1))$ is indistinguishable from encryptions of zeros. Circular security has applications in a wide variety of settings, ranging from security of symbolic protocols to fully homomorphic encryption. A fundamental question is whether standard security notions like IND-CPA/CCA imply k-circular security.

For the case $k = 2$, several works over the past years have constructed counterexamples—i.e., schemes that are CPA or even CCA secure but not 2-circular secure—under a variety of well-studied assumptions (SXDH, decision linear, and LWE). However, for $k > 2$ the only known counterexamples are based on strong general-purpose obfuscation assumptions.

In this work we construct k-circular security counterexamples for any $k \geq 2$ based on (ring-)LWE. Specifically:

- for any constant $k = O(1)$, we construct a counterexample based on n-dimensional (plain) LWE for $\text{poly}(n)$ approximation factors;
- for any $k = \text{poly}(\lambda)$, we construct one based on degree-n ring-LWE for at most subexponential $\exp(n^\varepsilon)$ factors.

Moreover, both schemes are k'-circular insecure for $2 \leq k' \leq k$.

Notably, our ring-LWE construction does not immediately translate to an LWE-based one, because matrix multiplication is not commutative. To overcome this, we introduce a new "tensored" variant of LWE which provides the desired commutativity, and which we prove is actually equivalent to plain LWE.

1 Introduction

Classical security definitions for encryption, like semantic security [19], only consider messages that the *attacker itself* can generate. In certain contexts, however, a system must encrypt *secret keys*, which are unknown to the attacker, under corresponding public keys. Prominent examples of this include the anonymous credential scheme of Camenisch and Lysyanskaya [13], methods for proving

C. Peikert—This material is based upon work supported by the National Science Foundation under CAREER Award CCF-1054495 and CNS-1606362, and by the Alfred P. Sloan Foundation. The views expressed are those of the authors and do not necessarily reflect the official policy or position of the National Science Foundation or the Sloan Foundation.

M. Robshaw and J. Katz (Eds.): CRYPTO 2016, Part II, LNCS 9815, pp. 659–680, 2016.
DOI: 10.1007/978-3-662-53008-5_23

the computational soundness of symbolic protocols [2], password managers and disk encryption utilities, and Gentry's "bootstrapping" technique for obtaining (unbounded) fully homomorphic encryption [16,17].

For these reasons, the notions of *circular* and, more generally, *key-dependent message (KDM)* security have attracted much attention in recent years. Informally, a public-key cryptosystem is k-circular secure if an *encryption cycle* $(\mathsf{Enc}_{pk_1}(sk_2), \mathsf{Enc}_{pk_2}(sk_3), \ldots, \mathsf{Enc}_{pk_k}(sk_1))$ is indistinguishable from encryptions of "junk" messages. KDM security considers a broader setting in which (adversarially specified) functions of the secret keys may be encrypted under any of the public keys.

Early positive results on circular/KDM security go back to Black *et al.* [8,13], who proposed KDM-secure schemes in the random oracle model. Several years later, Boneh *et al.* [9] were the first to give a cryptosystem in the standard model with a proof of KDM-security (for affine functions) under a well-studied assumption, namely, Decision Diffie-Hellman (DDH). This was soon followed by constructions based on the learning with errors (LWE) [5] and quadratic residuosity [10] assumptions; constructions for richer notions like identity-based encryption [3]; and "KDM amplification" transforms that extended the class of functions far beyond affine ones [4,6,11,24].

Despite all this progress, a very basic yet still unresolved question about circular/KDM security—especially in light of the fact that almost all the systems cited above are *specially* designed to obtain it—is:

> *Do classical security notions like IND-CPA or IND-CCA imply k-circular security?*

For $k = 1$ there are trivial counterexamples, but for $k \geq 2$ the question is much more interesting, and has been studied extensively in recent years. To date there is a significant gap between what is known for the cases $k = 2$ and $k > 2$.

The case $k = 2$. In this setting there are several negative results based on well-studied assumptions. The first counterexamples were presented by Acar *et al.* [1] and Cash *et al.* [14], who respectively gave schemes that are CPA secure but *not* 2-circular secure, and schemes that are CPA/CCA secure but not even *weakly* two-circular secure. (Weak circular security refers to the secrecy of other encrypted messages in the presence of an encryption cycle.) In both works, CPA/CCA security was under the SXDH assumption for groups with asymmetric bilinear pairings.

Most recently, Bishop *et al.* [7] gave additional counterexamples for $k = 2$, based on the decision linear and LWE assumptions. In addition, they introduced the useful notion of a *cycle tester*, which simplifies and modularizes the construction of counterexamples. For example, they showed how to combine a k-cycle tester with any CPA/CCA-secure cryptosystem to obtain CPA/CCA-secure schemes that are not k-circular secure. (However, all their *concrete* cycle testers were for $k = 2$.)

The case $k > 2$. For larger values of k, the relationship between CPA/CCA and circular security remained open for many years. Intuitively, constructing a counterexample for this case is more difficult because encryption must set up a relation among k ciphertexts that can be efficiently detected; bilinear maps make this possible for $k = 2$, but seem less useful for $k > 2$. Indeed, the only negative results are two recent concurrent and independent works of Koppula *et al.* [20] and Marcedone and Orlandi [25], which used strong *obfuscation* assumptions to construct, for any k, encryption schemes that are CPA secure but k-circular inse-cure. More specifically, the counterexample in [20] is based on indistinguishability obfuscation (iO) for arbitrary circuits (e.g., the candidate construction proposed in [15]), whereas [25] used the even stronger assumption of virtual black box (VBB) obfuscation for a certain large enough class of functions. (Later, follow-ing [20], the authors of [25] refined their scheme to rely only on iO.) Separately, Koppula *et al.* also showed that any k-circular security counterexample can be generically transformed into one that is not even *weakly* circular secure, because an encryption cycle implicitly reveals all the secret keys.

In summary, for $k = 2$ we have circular-security counterexamples under a rea-sonably wide variety of well-studied assumptions, whereas for $k > 2$ the available evidence is weaker, since it is based on the more speculative assumption that secure iO exists. In particular, up to this point we do not have a candidate iO scheme with a proof of security under simple, plausible, and concrete assump-tions. This stands in contrast to well-studied problems like those relating to bilinear pairings or (ring-)LWE, the latter of which are provably hard assuming the *worst-case* hardness of certain lattice problems [12, 22, 27, 28].

1.1 Contributions

Our main contributions are k-circular security counterexamples, for *any* $k \geq 2$, based on the LWE [28] and ring-LWE [22] assumptions. We stress that these are the first circular security counterexamples for $k > 2$ that do not rely on general-purpose obfuscation assumptions. More specifically, we prove the following two main theorems (in what follows, λ denotes the security parameter):

Informal Theorem 1. *For any* $\mathrm{poly}(\lambda)$*-bounded* $k \geq 2$*, there exists (in the common random string model) a* k*-cycle tester based on* ring-LWE *in degree-*n *rings for* $\tilde{O}(nk)^{O(k)}$ *approximation factors. Moreover, it is also a* k'*-cycle tester for* $2 \leq k' \leq k$*.*

As example parameterizations, for any constant $k = O(1)$ we obtain a k-cycle tester based on $\mathrm{poly}(n)$ approximation factors, which are conjectured to offer $2^{\tilde{\Omega}(n)}$ hardness. For arbitrary $k = \mathrm{poly}(\lambda)$, we can obtain a k-cycle tester based on subexponential $2^{n^{\varepsilon}}$ factors for any desired constant $\varepsilon > 0$, by letting $n = \tilde{\Omega}(\lambda^{c/\varepsilon})$ be a sufficiently large polynomial in λ. For such factors, ring-LWE is conjectured to offer $2^{\tilde{\Omega}(n^{1-\varepsilon})} \geq 2^{\Omega(\lambda)}$ hardness.

Informal Theorem 2. *For any* constant $k \geq 2$*, there exists (in the common random string model) a* k*-cycle tester based on* plain LWE *in* n *dimensions for* $n^{O(k^2)}$ *approximation factors. Moreover, it is also a* k'*-cycle tester for* $2 \leq k' \leq k$*.*

We emphasize that unlike many lattice-based cryptographic schemes, the ring-LWE-based cycle tester from our first theorem does *not* appear to "mechanically" translate to plain LWE, so additional ideas are needed to prove our second theorem. In brief, this is because the ring-LWE problem is usually defined over a *commutative* ring, whereas in the plain LWE setting, the corresponding ring of n-by-n matrices is not commutative (see Sect. 1.2 below for further details). To overcome this obstacle, we introduce a new variant of LWE that we call *tensored LWE*, and prove that it is equivalent to plain LWE for corresponding parameters. We note, however, that this technique limits the solution to *constant* (but arbitrary) $k = O(1)$, because it induces key sizes that are exponential in k.

Finally, by combining our cycle testers with appropriate (ring-)LWE-based CPA/CCA-secure encryption schemes [18,26,28] using the generic transformations given in [7,20], we immediately obtain CPA/CCA-secure cryptosystems that are k-circular insecure, and (in the CPA-secure case) for which an encryption cycle even reveals all the encrypted secret keys.

Recent Related Work. In a concurrent and independent work, Koppula and Waters [21] also constructed a k-cycle tester for arbitrary (a priori bounded) k based on plain LWE; it can be easily adapted to ring-LWE using standard transformations. Like ours, their construction uses "telescoping products," but the exact way in which these are used to detect cycles differs significantly—in particular, their construction does not need secret keys to commute under multiplication (see Sect. 1.2 below for further details). This yields different simplicity and efficiency profiles for the schemes. Specifically, our *ring-LWE* scheme has public keys, secret keys, and ciphertexts that are all an $\Omega(n)$ factor smaller than in the ring-LWE version of their scheme, and is arguably technically simpler and more direct. However, their *plain-LWE* construction can handle any *polynomial* cycle length $k = \text{poly}(\lambda)$, whereas our plain-LWE construction is restricted to any constant $k = O(1)$ due to an n^k factor in our key and ciphertext lengths, which arises from our "tensored" form of plain LWE that yields commuting secrets. In addition, their scheme does not use a common random string, whereas ours does.

1.2   Techniques

Here we give an overview of our constructions and proof techniques. To start, we give a brief exposition of the LWE-based two-cycle tester from [7]. We recall that a k-cycle tester is a relaxed form of encryption scheme that does not require a decryption algorithm; it only requires an efficient algorithm that reliably detects when a k-tuple of ciphertexts forms an encryption cycle.

In the two-cycle tester from [7], a secret key is the randomness used to generate a uniformly random matrix $\mathbf{S} \in \mathbb{Z}_q^{n \times m}$ along with a "trapdoor" $T_{\mathbf{S}}$, using the GenTrap algorithm from, e.g., [26]. The matrix $\mathbf{S}$ is interpreted as a matrix of *LWE secrets*, and the public key is the LWE instance $(\mathbf{A}, \mathbf{B} \approx \mathbf{S}^t \mathbf{A})$ for a uniformly random $\mathbf{A} \in \mathbb{Z}_q^{n \times m}$.

To encrypt under a public key $(\mathbf{A}, \mathbf{B})$, we interpret the message as randomness for GenTrap, thereby generating some $\hat{\mathbf{S}}$ with trapdoor $T_{\hat{\mathbf{S}}}$. We then choose a random short integer vector $\mathbf{r}$, let $\mathbf{v} = \mathbf{Ar}$, and output the two-component ciphertext

$$\left(\mathbf{x} \leftarrow \hat{\mathbf{S}}^{-1}[\mathbf{v}] \,,\, \mathbf{u} = \mathbf{Br} \approx \mathbf{S}^t \mathbf{Ar} = \mathbf{S}^t \mathbf{v} \right) \in \mathbb{Z}^m \times \mathbb{Z}_q^m.$$

Here $\mathbf{x} \leftarrow \hat{\mathbf{S}}^{-1}[\mathbf{v}]$ denotes using the trapdoor $T_{\hat{\mathbf{S}}}$ to randomly sample a short solution to $\hat{\mathbf{S}}\mathbf{x} = \mathbf{v}$ without revealing any information about $T_{\hat{\mathbf{S}}}$, e.g., using a discrete Gaussian distribution [18]. (This is used in the proof of IND-CPA security.) Notice that $\mathbf{x}$ is a short integer vector, whereas $\mathbf{u}$ is "large."

Now consider an encryption cycle for two keys, which consists of ciphertexts

$$\left(\mathbf{x}_i = \mathbf{S}_{1-i}^{-1}[\mathbf{v}_i] \,,\, \mathbf{u}_i \approx \mathbf{S}_i^t \mathbf{v}_i \right)$$

for $i \in \{0, 1\}$, where $\mathbf{S}_i$ is the (secret) matrix produced by GenTrap using the ith secret key as randomness. Because the $\mathbf{x}_i$ are short, we have

$$\langle \mathbf{u}_0, \mathbf{x}_1 \rangle = \mathbf{u}_0^t \cdot \mathbf{x}_1 \approx \mathbf{v}_0^t \mathbf{S}_0 \cdot \mathbf{S}_0^{-1}[\mathbf{v}_1] = \mathbf{v}_0^t \cdot \mathbf{v}_1 = \langle \mathbf{v}_0, \mathbf{v}_1 \rangle$$
$$\langle \mathbf{u}_1, \mathbf{x}_0 \rangle = \mathbf{u}_1^t \cdot \mathbf{x}_0 \approx \mathbf{v}_1^t \mathbf{S}_1 \cdot \mathbf{S}_1^{-1}[\mathbf{v}_0] = \mathbf{v}_1^t \cdot \mathbf{v}_0 = \langle \mathbf{v}_1, \mathbf{v}_0 \rangle.$$

Because the inner product is commutative, testing whether $\langle \mathbf{u}_0, \mathbf{x}_1 \rangle \approx \langle \mathbf{u}_1, \mathbf{x}_0 \rangle$ (mod q) will therefore detect a two-cycle. (For ordinary ciphertexts, the approximation is unlikely to hold, because the inner products are essentially uniform and independent.)

Challenges Beyond Two-Cycles. Generalizing the above construction to work for cycle lengths larger than two comes with several technical challenges. One is that there does not appear to be an appropriate generalization of the inner product $\langle \cdot, \cdot \rangle$ to three or more vectors. However, a promising idea is to replace $\mathbf{v}$ with a *matrix* $\mathbf{V}$ of many columns, and likewise replace $\mathbf{x}$ with $\mathbf{X} \leftarrow \hat{\mathbf{S}}^{-1}[\mathbf{V}]$, so that $\hat{\mathbf{S}} \cdot \mathbf{X} = \mathbf{V}$. Then for, say, a 3-cycle, if we could somehow arrange for $\mathbf{V}_i = \mathbf{Z}_i \cdot \mathbf{S}_i$ for some $\mathbf{Z}_i$, we would have the "telescoping product"

$$\begin{aligned}
\mathbf{U}_0^t \cdot \mathbf{X}_1 \cdot \mathbf{X}_2 &= \mathbf{V}_0^t \cdot \mathbf{S}_0 \cdot \mathbf{S}_0^{-1}[\mathbf{V}_1] \cdot \mathbf{X}_2 \\
&= \mathbf{S}_0^t \cdot \mathbf{Z}_0^t \cdot \mathbf{Z}_1 \cdot \mathbf{S}_1 \cdot \mathbf{S}_1^{-1}[\mathbf{V}_2] \\
&= \mathbf{S}_0^t \cdot \mathbf{Z}_0^t \cdot \mathbf{Z}_1 \cdot \mathbf{Z}_2 \cdot \mathbf{S}_2,
\end{aligned}$$

and similarly for $\mathbf{U}_1 \cdot \mathbf{X}_2 \cdot \mathbf{X}_0$. Unfortunately, we do not see any way to generate $\mathbf{V}_i = \mathbf{Z}_i \cdot \mathbf{S}_i$ in the encryption algorithm, because $\mathbf{S}_i$ is *secret* (it can only be obtained from the ith secret key). Alternatively, we might try to obtain a more "LWE-like" *approximation* $\mathbf{V}_i \approx \mathbf{Z}_i \cdot \mathbf{S}_i$ using the public key, but then the above equations *do not even hold approximately*, because $\mathbf{V}_0$ is "large" and hence amplifies the errors too much.

Our Solution. With the above attempt in mind, we take a different and arguably simpler approach to LWE-based cycle testers, which resolves both of the difficulties identified above. Our approach is easiest to understand in the ring setting first. For concreteness, define $R = \mathbb{Z}[X]/(X^n + 1)$ for n a power of two, and define $R_q = R/qR = \mathbb{Z}_q[X]/(X^n + 1)$ for a suitably large modulus q.

As in [7], a secret key in our system is the randomness used by (a ring variant of) GenTrap to produce a row vector $\mathbf{a} \in R_q^m$ with a trapdoor $T_{\mathbf{a}}$. However, here we simply take $\mathbf{a}$ to be the *public* key, rather than using it as a vector of ring-LWE secrets.

To encrypt under public key $\mathbf{a}$, as in [7] we interpret the message as randomness for GenTrap to obtain an $\hat{\mathbf{a}} \in R_q^m$ and trapdoor $T_{\hat{\mathbf{a}}}$. We then choose an $s \in R$ from the ring-LWE error distribution, let $\mathbf{b} \approx s \cdot \mathbf{a} \in R_q^m$ (where the approximation hides ring-LWE errors), and output the ciphertext

$$\mathbf{C} \leftarrow \hat{\mathbf{a}}^{-1}[\mathbf{b}] \in R^{m \times m},$$

where $\hat{\mathbf{a}}^{-1}[\mathbf{b}]$ uses $T_{\hat{\mathbf{a}}}$ to randomly sample a short matrix $\mathbf{C}$ over R such that $\hat{\mathbf{a}} \cdot \mathbf{C} = \mathbf{b}$. Notice that in contrast with [7], the ciphertext is just one short matrix—it does not contain any "large" components, which will be important for cycle testing.

Consider now an encryption cycle of, say, three secret keys, which consists of ciphertexts

$$\mathbf{C}_i \leftarrow \mathbf{a}_{i-1}^{-1}[\mathbf{b}_i], \quad \mathbf{b}_i \approx s_i \cdot \mathbf{a}_i$$

for each $i \in \mathbb{Z}_3$ (where the subscript arithmetic is modulo three). We then have the telescoping product

$$\begin{aligned}
\mathbf{a}_2 \cdot \mathbf{C}_0 \cdot \mathbf{C}_1 \cdot \mathbf{C}_2 &= \mathbf{a}_2 \cdot \mathbf{a}_2^{-1}[\mathbf{b}_0] \cdot \mathbf{C}_1 \cdot \mathbf{C}_2 \\
&\approx s_0 \cdot \mathbf{a}_0 \cdot \mathbf{a}_0^{-1}[\mathbf{b}_1] \cdot \mathbf{C}_2 \\
&\approx s_0 \cdot s_1 \cdot \mathbf{a}_1 \cdot \mathbf{a}_1^{-1}[\mathbf{b}_2] \\
&\approx s_0 \cdot s_1 \cdot s_2 \cdot \mathbf{a}_2,
\end{aligned}$$

where the approximations hold because all the s_i and $\mathbf{C}_i$ are short. Similarly,

$$\mathbf{a}_0 \cdot \mathbf{C}_1 \cdot \mathbf{C}_2 \cdot \mathbf{C}_0 \approx s_1 \cdot s_2 \cdot s_0 \cdot \mathbf{a}_0.$$

Now because the ring R is commutative, the above right-hand sides are almost identical, except for the different public keys $\mathbf{a}_0, \mathbf{a}_2$. But this issue is easily addressed: the GenTrap algorithm comes in a version that takes a vector over R_q as a public parameter, and outputs an $\mathbf{a}$ having that vector as its *prefix*. Therefore, our cycle tester just checks whether the first entries of the above products (corresponding to the common prefix of $\mathbf{a}_0, \mathbf{a}_2$) are approximately equal. More precisely, the difference should be smaller than some bound that depends on the maximum cycle length k we want to be able to detect; this induces our choice of the modulus q. Finally, notice that the tester also works equally well for cycles of length k' for $2 \leq k' \leq k$.

Adapting to Plain LWE. There is a standard mechanical translation of cryptosystems from ring-LWE to plain LWE, which replaces every uniformly random $a \in R_q$ with a uniformly random matrix $\mathbf{A} \in \mathbb{Z}_q^{n \times n}$, and every error term $s \in R$ with a matrix $\mathbf{S} \in \mathbb{Z}^{n \times n}$ whose entries are drawn independently from the LWE error distribution. However, when this translation is applied to the above scheme, it is easy to see that the cycle tester does not work, because the error matrices $\mathbf{S}_i$ are unlikely to commute with each other under multiplication.

We resolve this difficulty by introducing a new *tensoring* technique that guarantees commutativity. (We believe that the technique will find additional applications.) The central fact we use is that the tensor product of square n-dimensional matrices obeys the following special case of the *mixed-product property*:

$$\mathbf{S}_1 \otimes \mathbf{S}_2 = (\mathbf{S}_1 \otimes \mathbf{I}_n) \cdot (\mathbf{I}_n \otimes \mathbf{S}_2) = (\mathbf{I}_n \otimes \mathbf{S}_2) \cdot (\mathbf{S}_1 \otimes \mathbf{I}_n) \in \mathbb{Z}^{n^2 \times n^2}.$$

In particular, the matrices $\mathbf{S}_1 \otimes \mathbf{I}_n$ and $\mathbf{I}_n \otimes \mathbf{S}_2$ commute under multiplication. (Naturally, the above equations generalize to the tensor product of any $k > 2$ matrices.)

We apply the above facts in our plain-LWE cycle tester as follows. When encrypting to the ith public key, we use an LWE secret matrix

$$\mathbf{S}'_i = \underbrace{\mathbf{I}_n \otimes \cdots \otimes \mathbf{I}_n}_{i \text{ terms}} \otimes \mathbf{S}_i \otimes \underbrace{\mathbf{I}_n \otimes \cdots \otimes \mathbf{I}_n}_{k-i-1 \text{ terms}} \in \mathbb{Z}^{n^k \times n^k},$$

where $\mathbf{S}_i \in \mathbb{Z}^{n \times n}$ has entries drawn from the error distribution. By the above, these $\mathbf{S}_i$ all commute with each other under multiplication, allowing us to conclude that (certain entries of) the telescoping products are approximately equal. Also notice that it is not necessary for all the $\mathbf{S}_i$ to appear in the final product, so the same cycle tester also detects k'-cycles for $2 \le k' \le k$.

In order for all this to work, the public key matrices $\mathbf{A}_i$ must have n^k rows, which is why our construction is limited to constant $k = O(1)$. Of course, it is not immediately obvious whether LWE is actually hard for such highly structured secret matrices $\mathbf{S}'_i$. Fortunately, we prove that this form of the problem is *equivalent* to n-dimensional LWE with the same error distribution, up to a polynomial factor in the number of samples given to the attacker. Known worst-case hardness theorems for LWE are essentially agnostic to the number of samples, so the reduction's lossiness in this respect is of little concern.

2 Preliminaries

For a positive integer t we let $[t] = \{0, \ldots, t-1\}$. The primary security parameter is denoted λ.

Tensor Products. The *tensor* (or *Kronecker*) product $\mathbf{A} \otimes \mathbf{B}$ of an m_1-by-n_1 matrix $\mathbf{A}$ with an m_2-by-n_2 matrix $\mathbf{B}$, both over a common ring $\mathcal{R}$, is the $m_1 m_2$-by-$n_1 n_2$ block matrix consisting of m_2-by-n_2 blocks, whose (i, j)th block is $a_{i,j} \cdot \mathbf{B}$, where $a_{i,j}$ denotes the (i, j)th entry of $\mathbf{A}$. Equivalently, we can view $\mathbf{A} \otimes \mathbf{B}$

as having rows indexed by $[m_1] \times [m_2]$ and columns indexed by $[n_1] \times [n_2]$, where the $((i_1, i_2), (j_1, j_2))$th entry is $a_{i_1, j_1} \cdot b_{i_2, j_2}$. This corresponds to the previous definition by "flattening" the row and column index sets using the bijection that maps $(k_1, k_2) \in [\ell_1] \times [\ell_2]$ to $k_1 \cdot \ell_2 + k_2 \in [\ell_1 \ell_2]$.

We extensively use the *mixed-product property* of tensor products, which says that

$$(\mathbf{A} \otimes \mathbf{B}) \cdot (\mathbf{C} \otimes \mathbf{D}) = (\mathbf{AC}) \otimes (\mathbf{BD})$$

for any matrices $\mathbf{A}, \mathbf{B}, \mathbf{C}, \mathbf{D}$ of compatible dimensions. In particular,

$$(\mathbf{A} \otimes \mathbf{B}) = (\mathbf{A} \otimes \mathbf{I}_{\text{height}(\mathbf{B})}) \cdot (\mathbf{I}_{\text{width}(\mathbf{A})} \otimes \mathbf{B}) = (\mathbf{I}_{\text{height}}(\mathbf{A}) \otimes \mathbf{B}) \cdot (\mathbf{A} \otimes \mathbf{I}_{\text{width}}(\mathbf{B})).$$

Subgaussians. For analyzing error growth in our schemes it will be convenient to use the notion of *subgaussian* random variables and matrices. We say that a real random variable X (or its distribution) is subgaussian with parameter s if for all $t \in \mathbb{R}$, the (scaled) moment-generating function satisfies[1]

$$\mathbb{E}[\exp(2\pi t X)] \leq (1 + \text{negl}(\lambda)) \cdot \exp(\pi s^2 t^2).$$

More generally, we say that a random matrix (over vector) $\mathbf{X}$ is subgaussian with parameter s if $\mathbf{u}^t \mathbf{X} \mathbf{v}$ is subgaussian with parameter s for all unit vectors $\mathbf{u}, \mathbf{v}$. It follows immediately from the definitions that a $\text{poly}(\lambda)$-dimensional matrix made up of independent subgaussian entries, or of independent subgaussian rows or columns, with common parameter s is itself subgaussian with parameter s.

The largest singular value, also known as *spectral norm*, of a matrix $\mathbf{X}$ is defined as $s_1(\mathbf{X}) := \max_{\mathbf{u} \neq \mathbf{0}} \|\mathbf{X}\mathbf{u}\| / \|\mathbf{u}\|$. It is clear that the spectral norm is sub-additive and sub-multiplicative: $s_1(\mathbf{X} + \mathbf{Y}) \leq s_1(\mathbf{X}) + s_1(\mathbf{Y})$ and $s_1(\mathbf{XY}) \leq s_1(\mathbf{X}) \cdot s_1(\mathbf{Y})$. We use the following standard fact about subgaussian matrices; see [29] for a proof.

Proposition 1. *For a subgaussian matrix $\mathbf{X} \in \mathbb{R}^{m \times n}$ with parameter s, we have $s_1(\mathbf{X}) \leq s \cdot O(\sqrt{m} + \sqrt{n})$ except with probability at most $2^{-\Omega(m+n)}$.*

2.1 Cryptographic Definitions

Here we present some cryptographic definitions. The definition of k-cycle tester is from [7].

Definition 1. *Let $\Pi = (\text{Setup}, \text{Gen}, \text{Enc})$ be a public-key encryption scheme (omitting the decryption algorithm) for message space $\mathcal{M} = \mathcal{M}_\lambda$. We say that Π is IND-CPA secure if every efficient adversary $\mathcal{A}$ has negligible (in λ) advantage in distinguishing the following two games for $b \in \{0, 1\}$:*

---

[1] We remark that the $1 + \text{negl}(\lambda)$ factor makes this a slight relaxation of the standard definition of subgaussian; it coincides with the notion of $\text{negl}(\lambda)$-subgaussian from [26].

1. *Generate $pp \leftarrow \mathsf{Setup}(1^\lambda)$ and $(pk, sk) \leftarrow \mathsf{Gen}(pp)$.*
2. *Given (pp, pk) to $\mathcal{A}$, which outputs a pair of messages $(m_0, m_1) \in \mathcal{M}^2$.*
3. *Generate $c \leftarrow \mathsf{Enc}(pk, m_b)$ and give c to the adversary.*

Definition 2. *Let $\Pi = (\mathsf{Setup}, \mathsf{Gen}, \mathsf{Enc})$ be a public-key encryption scheme (omitting the decryption algorithm) for message space $\mathcal{M} = \mathcal{M}_\lambda \supseteq \mathcal{S}_\lambda$, where $\mathcal{S}_\lambda$ denotes the secret-key space for security parameter λ. We say that Π is IND-CIRC-CPA^k secure if the following two games are computationally indistinguishable.*

1. *Generate $pp \leftarrow \mathsf{Setup}(1^\lambda)$ and $(pk_i, sk_i) \leftarrow \mathsf{Gen}(pp)$ for every $i \in \mathbb{Z}_k$.*
2. *In Game 0, let $c_i \leftarrow \mathsf{Enc}(pk_i, sk_{i-1})$ for $i \in \mathbb{Z}_k$ (where arithmetic in the subscripts is modulo k).*
   In Game 1, let $c_i \leftarrow \mathsf{Enc}(pk_i, 0)$ for $i \in \mathbb{Z}_k$ (where $0 \in \mathcal{M}$ denotes some arbitrary fixed message).
3. *Output $(pp, (pk_i)_{i \in \mathbb{Z}_k}, (c_i)_{i \in \mathbb{Z}_k})$.*

Definition 3 (Cycle Tester [7]). *Let $\Gamma = (\mathsf{Setup}, \mathsf{Gen}, \mathsf{Enc}, \mathsf{Test})$ be a tuple of randomized algorithms for which:*

- $\Pi = (\mathsf{Setup}, \mathsf{Gen}, \mathsf{Enc})$ *is a public-key encryption scheme for message space $\mathcal{M} = \mathcal{M}_\lambda \supseteq \mathcal{S}_\lambda$;*
- $\mathsf{Test}((pk_i, c_i)_{i \in \mathbb{Z}_k})$, *given a tuple of public keys pk_i and corresponding ciphertexts c_i, outputs a bit $b \in \{0, 1\}$.*

We say that Γ is a k-cycle tester if Π is IND-CPA secure, and if Test has non-negligible advantage in the IND-CIRC-CPA^k game against Π.

2.2   Learning with Errors

Definition 4. *For positive integer dimensions n, m, modulus q, and error distribution χ over $\mathbb{Z}$, the decision-LWE$_{n,q,\chi,m}$ problem is to distinguish, with non-negligible advantage, between $(\mathbf{A}; \mathbf{b}^t = \mathbf{s}^t \mathbf{A} + \mathbf{e}^t)$ where $\mathbf{A} \leftarrow \mathbb{Z}_q^{n \times m}$, $\mathbf{s} \leftarrow \chi^n$, $\mathbf{e} \leftarrow \chi^m$, and uniformly random $(\mathbf{A}; \mathbf{b}^t)$ of the same dimensions.[2]*

A standard instantiation of LWE is to let χ be a *discrete Gaussian* distribution (over $\mathbb{Z}$) with parameter $r = 2\sqrt{n}$, which is known to be subgaussian with parameter r (see [26]). For this parameterization, and for any polynomially bounded m, it is known that LWE is at least as hard as *quantumly* approximating certain "short vector" problems on n-dimensional lattices, in the worst case, to within $\tilde{O}(q\sqrt{n})$ factors [28]. Classical reductions are also known for different parameterizations [12,27].

---

[2] Notice that the coordinates of $\mathbf{s}$ are drawn from the error distribution χ; as shown in [5], this form of the problem is equivalent (up to a small difference in m) to the one where $\mathbf{s} \leftarrow \mathbb{Z}_q^n$ is drawn uniformly at random.

A standard hybrid argument shows that the *multi-secret* form of LWE—which is to distinguish

$$\begin{pmatrix} \mathbf{A} \\ \mathbf{B} = \mathbf{SA} + \mathbf{E} \end{pmatrix} \in \mathbb{Z}_q^{(n+t)\times m}$$

from uniform, where $\mathbf{A} \leftarrow \mathbb{Z}_q^{n\times m}$, $\mathbf{S} \leftarrow \chi^{t\times n}$, and $\mathbf{E} \leftarrow \chi^{t\times m}$ for some desired $t, m = \text{poly}(n)$—is equivalent to the above single-secret version, up to a t factor loss in the distinguishing advantage.

Tensored Form. In this work we rely on another equivalent form of LWE, which we call the *tensored* form. Let $m, t = \text{poly}(n)$ be as above, and additionally let $l, r = \text{poly}(n)$ be arbitrary. The problem is to distinguish

$$\begin{pmatrix} \mathbf{A} \\ \mathbf{B} = (\mathbf{I}_l \otimes \mathbf{S} \otimes \mathbf{I}_r) \cdot \mathbf{A} + \mathbf{E} \end{pmatrix} \in \mathbb{Z}_q^{l(n+t)r\times m}$$

from uniform, where $\mathbf{A} \leftarrow \mathbb{Z}_q^{lnr\times m}$, $\mathbf{S} \leftarrow \chi^{t\times n}$, and $\mathbf{E} \leftarrow \chi^{ltr\times m}$.

Lemma 1. *The tensored form of LWE for parameters n, t, m, l, r is equivalent to the multi-secret form for the same n, t and $M = mlr$ samples.*

Proof. The equivalence follows simply by an appropriate (efficient and reversible) reindexing. Specifically, given a multi-secret instance $(\mathbf{A}; \mathbf{B}) \in \mathbb{Z}_q^{(n+t)\times M}$, we transform it to a tensored instance $(\mathbf{A}'; \mathbf{B}') \in \mathbb{Z}_q^{l(n+t)r\times m}$ as follows. For convenience, we construct $\mathbf{A}'$ by indexing its rows by $[l] \times [n] \times [r]$ in the standard way, and similarly for $\mathbf{B}'$. We partition $\mathbf{A}$ into m blocks, each consisting of lr columns of dimension n. We arbitrarily index these columns by $[l] \times [r]$, and arrange them into a single column indexed by $[l] \times [n] \times [r]$ in the obvious way; the matrix $\mathbf{A}'$ is made up of these m columns. Similarly, we construct $\mathbf{B}'$ from $\mathbf{B}$ by grouping each block of lr columns of dimension t into a single column vector indexed $[l]\times[t]\times[r]$. It is easy to see that if $(\mathbf{A}; \mathbf{B})$ is uniformly random, then so is $(\mathbf{A}'; \mathbf{B}')$. Furthermore, by construction and by definition of matrix multiplication it can be verified that if $\mathbf{B} = \mathbf{SA} + \mathbf{E}$ for some $\mathbf{S}, \mathbf{E}$, then $\mathbf{B}' = (\mathbf{I}_l \otimes \mathbf{S} \otimes \mathbf{I}_r) \cdot \mathbf{A}' + \mathbf{E}'$, where $\mathbf{E}'$ is obtained from $\mathbf{E}$ in exactly the same way that $\mathbf{B}'$ is obtained from $\mathbf{B}$. Therefore, the transformation is a tight reduction from the multi-secret to the tensored form. Moreover, the transformation is efficiently reversible, which gives a reduction in the opposite direction.

2.3   Lattice Trapdoors

We recall some standard facts about trapdoors and preimage sampling for cryptographic lattices; for full details, see [18,26]. There exist efficient randomized algorithms GenTrap, SampleDom, and SamplePre having the following properties. For any positive integers n, q, there exist suitable $\bar{m} < m = O(n \log q)$ for which the following hold (the parameters $n, q, \bar{m}, m$ are implicit inputs to all the algorithms):

- GenTrap($\bar{\mathbf{A}}; \mathbf{R}$) takes some $\bar{\mathbf{A}} \in \mathbb{Z}_q^{n \times \bar{m}}$ and random coins $\mathbf{R} \in \mathcal{R}$ from a certain space $\mathcal{R}$, and outputs a matrix $\mathbf{A} \in \mathbb{Z}_q^{n \times m}$ whose first $\bar{m}$ columns are $\bar{\mathbf{A}}$, and for which $\mathbf{R}$ serves as a "trapdoor."
- SampleDom() outputs a random $\mathbf{x} \in \mathbb{Z}^m$, drawn from a certain distribution D. For brevity we usually write $\mathbf{x} \leftarrow D$.
- For any $\bar{\mathbf{A}} \in \mathbb{Z}_q^{n \times \bar{m}}$ and $\mathbf{R} \in \mathcal{R}$ defining $\mathbf{A}$ as above, and any $\mathbf{u} \in \mathbb{Z}_q^n$, SamplePre($\bar{\mathbf{A}}, \mathbf{R}, \mathbf{u}$) outputs a random $\mathbf{x} \in \mathbb{Z}^m$ (drawn from a certain distribution) such that $\mathbf{A}\mathbf{x} = \mathbf{u}$.

  When $\mathbf{A}$ and $\mathbf{R}$ are clear from context, we usually write $\mathbf{A}^{-1}[\mathbf{u}]$ for the sake of brevity, and because it satisfies the identity $\mathbf{A} \cdot \mathbf{A}^{-1}[\mathbf{u}] = \mathbf{u}$. (We stress that $\mathbf{A}^{-1}[\cdot]$ denotes a *randomized algorithm*, not a formal matrix inverse.)

We extend the above notation column-wise to matrices, i.e., D^ℓ is the distribution over $\mathbb{Z}^{m \times \ell}$ in which the columns are drawn independently from D, and $\mathbf{A}^{-1}[\mathbf{B}] \in \mathbb{Z}^{m \times \ell}$ for $\mathbf{B} \in \mathbb{Z}_q^{n \times \ell}$ applies $\mathbf{A}^{-1}$ independently to each column of $\mathbf{B}$.

Proposition 2. *The above algorithms satisfy the following statistical properties:*

1. *For uniformly random $\mathbf{A} \leftarrow \mathbb{Z}_q^{n \times m}$ and $\mathbf{x} \leftarrow D$, the distribution of $(\mathbf{A}, \mathbf{A}\mathbf{x})$ is within negligible statistical distance of uniform.*
2. *For uniformly random $\bar{\mathbf{A}}$ and $\mathbf{R} \leftarrow \mathcal{R}$, the distribution of $\mathbf{A} = \mathsf{GenTrap}(\bar{\mathbf{A}}, \mathbf{R})$ is within negligible statistical distance of uniform.*
3. *For any $\bar{\mathbf{A}}$ and any $\mathbf{R} \in \mathcal{R}$ defining $\mathbf{A} = \mathsf{GenTrap}(\bar{\mathbf{A}}; \mathbf{R})$, the following experiments are within negligible statistical distance:*
   (a) choose $\mathbf{x} \leftarrow D$ and output $(\mathbf{x}, \mathbf{u} = \mathbf{A}\mathbf{x})$;
   (b) choose uniformly random $\mathbf{u} \leftarrow \mathbb{Z}_q^n$, let $\mathbf{x} \leftarrow \mathbf{A}^{-1}[\mathbf{u}]$, and output $(\mathbf{x}, \mathbf{u})$.
4. *For any $\mathbf{A}$ output by $\mathsf{GenTrap}$ (on randomness $\mathbf{R}$), and any $\mathbf{u} \in \mathbb{Z}_q^n$, the distribution $\mathbf{A}^{-1}[\mathbf{u}]$ is subgaussian with parameter $\tilde{O}(m) = \tilde{O}(n \log q)$.*

Remark 1. We emphasize that Item 3 of Proposition 2 applies for *any* (possibly adversarial) choice of the trapdoor $\mathbf{R}$, which is needed in our application because the trapdoor will indeed be provided by the adversary. Fortunately, the GenTrap and SamplePre algorithms described in [26] can easily be instantiated to satisfy this property. In brief, this is GenTrap produces a short random matrix $\mathbf{R} \in \mathcal{R}$ as the trapdoor, and SamplePre works for any Gaussian parameter exceeding a certain $\tilde{O}(s_1(\mathbf{R}))$ bound. By defining $\mathcal{R}$ to be, say, the set of all binary matrices of appropriate dimensions, we ensure that $s_1(\mathbf{R}) \leq m$ for *every* $\mathbf{R} \in \mathcal{R}$, while also satisfying Item 2 via the leftover hash lemma.

2.4 The Ring Setting

Here we provide some background on rings, their geometry, and ring-LWE; then we recall analogous facts about trapdoors in the ring setting. For more details see [22,26]. (This material is only used for our ring-LWE construction in Sect. 4, and may be safely skipped.)

For simplicity, we work in the $2n$th cyclotomic ring $R := \mathbb{Z}[X]/(X^n + 1)$ for n a power of two. (However, all of our results can be adapted to arbitrary cyclotomics using the techniques from [23].) The *canonical embedding* $\sigma \colon R \to \mathbb{C}^n$ maps $r \in R$ to $(\sigma_i(r))_{i \in \mathbb{Z}_{2n}^*}$, where $\sigma_i(r) = r(\omega^i)$ and $\omega = \exp(\pi\sqrt{-1}/n) \in \mathbb{C}$ is the principal complex $2n$th root of unity. (Notice that this definition is agnostic to the choice of $\mathbb{Z}[X]$-representative of $r \in \mathbb{Z}[X]/(X^n + 1)$, which makes it "canonical.")

We use the canonical embedding to endow R with a geometry. Specifically, for a ring element $r \in R$ we define $\|r\| := \|\sigma(r)\|$ and $\|r\|_\infty := \|\sigma(r)\|_\infty$. We extend the norm notation to vectors and matrices by defining $\|\mathbf{x}\| = (\sum_i \|x_i\|^2)^{1/2}$ for any vector $\mathbf{x}$ over R, and $\|\mathbf{X}\|_\infty = \max\|x_{i,j}\|_\infty$ for any vector or matrix $\mathbf{X}$ over R. Finally, we define the *spectral norm* of $\mathbf{X}$ as

$$s_1(\mathbf{X}) := \sup_{\mathbf{u} \neq \mathbf{0}} \|\mathbf{X}\mathbf{u}\|/\|\mathbf{u}\|,$$

where the supremum is taken over all nonzero vectors (of appropriate dimension) over R. Clearly, the spectral norm is sub-additive and sub-multiplicative: $s_1(\mathbf{X} + \mathbf{Y}) \leq s_1(\mathbf{X}) + s_1(\mathbf{Y})$ and $s_1(\mathbf{X}\mathbf{Y}) \leq s_1(\mathbf{X}) \cdot s_1(\mathbf{Y})$. The following standard fact relates the spectral and ℓ_∞ norms.

Proposition 3. *For any matrix* $\mathbf{E} \in R^{l \times k}$ *we have* $s_1(\mathbf{E}) \leq \sqrt{lk} \cdot \|\mathbf{E}\|_\infty$.

The following standard fact bounds the coefficients of a ring element $r \in R$ by its ℓ_∞ norm.

Proposition 4. *For a ring element* $r \in R$, *let* $r = \sum_{j=0}^{n-1} r_j \cdot X^j \in \mathbb{Z}[X]$ *for* $r_j \in \mathbb{Z}$ *denote its canonical representative (with respect to the standard power basis of R). Then* $r_j \leq \|r\|_\infty$ *for every j.*

Ring-LWE. For an integer q, define $R_q := R/qR = \mathbb{Z}_q[X]/(X^n + 1)$.

Definition 5. *Let* χ *be an error distribution over* R. *The decision-RLWE$_{R,q,\chi,m}$ problem is to distinguish, with non-negligible advantage, between* $(\mathbf{a}; \mathbf{b} = s \cdot \mathbf{a} + \mathbf{e}) \in R_q^m \times R_q^m$ *where* $\mathbf{a} \leftarrow R_q^m$, $s \leftarrow \chi$, $\mathbf{e} \leftarrow \chi^m$, *and uniformly random* $(\mathbf{a}; \mathbf{b})$ *of the same dimensions.*

For appropriate parameters, decision-RLWE problem is (quantumly) at least as hard as the $(q \cdot \mathrm{poly}(n,m))$-approximate shortest vector problem on *any* ideal lattice in R, i.e., in the worst case [22]. The standard error distribution for which this theorem applies is a sufficiently wide discrete Gaussian distribution χ over R, for which

$$\Pr_{e \leftarrow \chi}[\|e\|_\infty > n^c] = \mathrm{negl}(n) \tag{1}$$

for some universal constant $c > 1$.

Trapdoors. Similarly to the plain setting, there are efficient randomized algorithms GenTrap, SampleDom, and SamplePre having the following properties. For any modulus q, there exist suitable $\bar{m} < m = \tilde{O}(\log q)$ for which the following hold (the parameters $R, q, \bar{m}, m$ are implicit inputs to all the algorithms):

- GenTrap($\bar{\mathbf{a}}; \mathbf{R}$) takes some $\bar{\mathbf{a}} \in R_q^{\bar{m}}$ and random $\mathbf{R} \in \mathcal{R}$ from a certain space $\mathcal{R}$, and outputs a vector $\mathbf{a} \in R_q^m$ whose first $\bar{m}$ components are $\bar{\mathbf{a}}$, and for which $\mathbf{R}$ serves as a "trapdoor."
- SampleDom() outputs a random column vector $\mathbf{x}^t \in R^m$, drawn from a certain distribution D. For brevity we usually write $\mathbf{x}^t \leftarrow D$.
- For any $\bar{\mathbf{a}} \in R_q^{\bar{m}}$ and $\mathbf{R} \in \mathcal{R}$ defining $\mathbf{a}$ as above, and any $u \in R_q$, SamplePre($\bar{\mathbf{a}}, \mathbf{R}, u$) outputs a random column vector $\mathbf{x}^t \in R^m$ (drawn from a certain distribution) such that $\mathbf{a} \cdot \mathbf{x}^t = u$.

  We usually write $\mathbf{a}^{-1}[u]$ for the sake of brevity, and because it satisfies the identity $\mathbf{a} \cdot \mathbf{a}^{-1}[u] = u$. Moreover, D^l is the distribution over $R^{m \times l}$ in which the columns are drawn independently from D. The notation $\mathbf{a}^{-1}[\mathbf{v}] \in R_q^{m \times l}$, where $\mathbf{v} \in R_q^l$, applies $\mathbf{a}^{-1}$ to each component of $\mathbf{v}$ independently.

The following proposition follows by a standard adaptation of "plain" trapdoor constructions (e.g., [26]) to the ring setting, and by the regularity lemma for rings given in [23].

Proposition 5. *The above algorithms satisfy the following statistical properties:*

1. *For uniformly random $\mathbf{a} \leftarrow R_q^m$ and $\mathbf{x}^t \leftarrow D$, the distribution of $(\mathbf{a}, \mathbf{a} \cdot \mathbf{x}^t) \in R^{m+1}$ is within negligible statistical distance of uniform.*
2. *For uniformly random $\bar{\mathbf{a}}$ and $\mathbf{R} \leftarrow \mathcal{R}$, the distribution of $\mathbf{a} = \mathsf{GenTrap}(\bar{\mathbf{a}}, \mathbf{R})$ is within negligible statistical distance of uniform.*
3. *For any $\bar{\mathbf{a}}$ and any $\mathbf{R} \in \mathcal{R}$ defining $\mathbf{a} = \mathsf{GenTrap}(\bar{\mathbf{a}}; \mathbf{R})$, the following experiments are within negligible statistical distance:*
   (a) *choose $\mathbf{x}^t \leftarrow D$ and output $(\mathbf{x}, u = \mathbf{a} \cdot \mathbf{x}^t)$;*
   (b) *choose uniformly random $u \leftarrow R_q$, let $\mathbf{x}^t \leftarrow \mathbf{a}^{-1}[u]$, and output $(\mathbf{x}, u)$.*
4. *There exists a universal constant $c > 1$ such that, for any $\mathbf{a}$ output by $\mathsf{GenTrap}$ (on randomness $\mathbf{R}$), and for any $u \in R_q$,*

$$\Pr[\|\mathbf{a}^{-1}[u]\|_\infty > n^c] = \mathrm{negl}(n).$$

3   LWE-Based Construction

In this section we construct, for any constant $k \geq 2$, a k-cycle tester that is IND-CPA secure based on the conjectured hardness of (plain) LWE, appropriately parameterized. The scheme involves the following parameters:

- $N := n^k$ for a positive integer n, an integer modulus q, and an error distribution χ over $\mathbb{Z}$, where n, q, χ are the parameters of the underlying LWE problem. For concreteness, we use the standard LWE error distribution χ, which is subgaussian with parameter $O(\sqrt{n})$.
- $\bar{M} < M = O(N \log q)$, where $\bar{M}, M$ are the dimensions associated with GenTrap for N, q.
- The secret-key and message spaces are both the randomness/trapdoor space $\mathcal{R}$ of GenTrap when given an N-by-$\bar{M}$ input.

Finally, each key is uniquely and arbitrarily identified with some $i \in \mathbb{Z}_k = \{0, \ldots, k-1\}$, which is provided to the key-generation algorithm. The tester is defined as follows.

- Setup(): output a uniformly random $\bar{\mathbf{A}} \leftarrow \mathbb{Z}_q^{N \times \bar{M}}$.
- Gen$(i, \bar{\mathbf{A}})$: let $\mathbf{A}_i = \mathsf{GenTrap}(\bar{\mathbf{A}}; \mathbf{R}_i)$ for $\mathbf{R}_i \leftarrow \mathcal{R}$, and output $(i, \mathbf{A}_i)$ as the public key and the trapdoor $\mathbf{R}_i$ as the secret key.
  Recall from Proposition 2 that the first $\bar{M}$ columns of $\mathbf{A}_i$ are $\bar{\mathbf{A}}$, and that $\mathbf{A}_i$ is negligibly far from uniform over the random choice of $\bar{\mathbf{A}}$ and $\mathbf{R}_i$.
- Enc$((i, \mathbf{A}_i), \mathbf{R} \in \mathcal{R})$: let $\mathbf{A} = \mathsf{GenTrap}(\bar{\mathbf{A}}; \mathbf{R})$, so that $\mathbf{R}$ is a trapdoor for $\mathbf{A}$. Choose an LWE secret matrix $\mathbf{S}_i \leftarrow \chi^{n \times n}$ and an error matrix $\mathbf{E}_i \leftarrow \chi^{N \times M}$, and output the ciphertext matrix

$$\mathbf{C} \leftarrow \mathbf{A}^{-1}[\mathbf{S}_i' \cdot \mathbf{A}_i + \mathbf{E}_i] \in \mathbb{Z}^{M \times M},$$
$$\text{where } \mathbf{S}_i' = (\mathbf{I}_{n^i} \otimes \mathbf{S}_i \otimes \mathbf{I}_{n^{k-i-1}}) \in \mathbb{Z}^{N \times N}.$$

  (The $\mathbf{A}^{-1}$ operation is performed using trapdoor $\mathbf{R}$.)
- Test$((\mathbf{A}_i, \mathbf{C}_i)_{i \in \mathbb{Z}_k})$: given public key matrices $\mathbf{A}_i$ and ciphertexts $\mathbf{C}_i$, check whether

$$(\mathbf{A}_{k-1} \cdot \mathbf{C}_0 \cdot \mathbf{C}_1 \cdots \mathbf{C}_{k-1} - \mathbf{A}_0 \cdot \mathbf{C}_1 \cdots \mathbf{C}_{k-1} \cdot \mathbf{C}_0) \cdot \bar{\mathbf{I}} \in (-q/4, q/4)^{N \times \bar{M}} \pmod{q}, \tag{2}$$

  where $\bar{\mathbf{I}} = \begin{pmatrix} \mathbf{I}_{\bar{M}} \\ \mathbf{0} \end{pmatrix} \in \mathbb{Z}^{M \times \bar{M}}$ for every i, which we use in the analysis below.)

Remark 2. In Eq. (2), the choice of products appearing in the difference is not special; the difference between any two products $\mathbf{A}_i \cdot \mathbf{C}_{i+1} \cdot \mathbf{C}_{i+2} \cdots \mathbf{C}_i$ for distinct values of $i \in \mathbb{Z}_k$ would work equally well.

Remark 3. The number and order of ciphertexts in an encryption cycle is also not too important. The Test algorithm naturally generalizes to work on any k' public keys and ciphertexts indexed by an ordered set $S \subseteq \mathbb{Z}_k$, for $2 \leq k' \leq k$. We simply take the difference of two products $\mathbf{A}_i \cdot \prod_{j \in S} \mathbf{C}_j$ for two distinct i, where the order of indices j cyclically follows the order of S and ends with $j = i$.

In the remainder of this section we prove the following theorem:

Theorem 1. *For any constant $k \geq 2$ and a sufficiently large $q = \tilde{O}(n^{3(k^2-1)/2})$, the above scheme is a k'-cycle tester for $2 \leq k' \leq k$, assuming the hardness of decision-LWE$_{n,q,\chi,M \cdot n^{k-1}}$.*

Recall that the LWE instantiation from Theorem 1 is at least as hard as (quantumly) approximating certain lattice problems on n-dimensional lattices, in the worst case, to within $\tilde{O}(n^{3k^2/2-1}) = \text{poly}(n)$ factors, which is conjectured to be exponentially hard in n.

In Sect. 3.1 below we prove IND-CPA security, in Sect. 3.2 we show that Test almost always accepts on an encryption cycle, and in Sect. 3.3 we show that Test almost never accepts on a non-cycle. Together these prove Theorem 1.

3.1 Security

Lemma 2. *The tuple* (Setup, Gen, Enc) *is IND-CPA secure under the LWE assumption from Theorem 1.*

Proof. We consider the following sequence of hybrid experiments, showing that adjacent hybrids are indistinguishable (either computationally or statistically), and that the last one does not depend on the adversary's choice of challenge message, which proves the claim. For simplicity, assume that the adversary names some target identity $i \in \mathbb{Z}_k$ at the start of the IND-CPA game. (The proof easily adapts to the case where the adversary adaptively chooses i after seeing all the public keys.)

Hybrid 1: Here the matrix $\mathbf{A}_i \in \mathbb{Z}_q^{N \times M}$ in the public key is generated uniformly at random, instead of by GenTrap. By Item 2 of Proposition 2, this experiment is statistically indistinguishable from the real IND-CPA game.

Hybrid 2: Here the matrix $\mathbf{B}_i \in \mathbb{Z}_q^{N \times M}$ given as input to the $\mathbf{A}^{-1}$ operation is chosen uniformly at random, rather than as $\mathbf{B}_i = \mathbf{S}_i' \cdot \mathbf{A}_i + \mathbf{E}_i$ (as in the previous hybrid).

Using the tensored form of LWE, which by Lemma 1 is equivalent to the one appearing in the theorem statement, a straightforward reduction shows that this experiment is computationally indistinguishable from the previous one. Specifically, given an instance $(\mathbf{A}'; \mathbf{B}')$ of the tensored form of LWE, the reduction sets $\mathbf{A}_i = \mathbf{A}'$, $\mathbf{B}_i = \mathbf{B}'$, and finally lets $\mathbf{C}_i \leftarrow \mathbf{A}^{-1}[\mathbf{B}_i]$, using the adversary's challenge message to define $\mathbf{A}$ and compute the $\mathbf{A}^{-1}[\cdot]$ operation (using the SamplePre algorithm) in the usual way.

Hybrid 3: Here the matrix $\mathbf{C}_i$ is drawn from D^M, i.e., each column is independently drawn from D, instead of by invoking $\mathbf{A}^{-1}[\mathbf{B}_i]$ for a matrix $\mathbf{A}$ defined by the adversary's challenge message.

We claim that for any choice of $\bar{\mathbf{A}}$ and challenge message, this experiment is within negligible statistical distance of the previous one. This follows immediately by Item 3 of Proposition 2, applied across each pair of corresponding columns of $\mathbf{U}_i$ and $\mathbf{C}_i$.

Clearly, the final hybrid experiment does not depend on the adversary's choice of challenge message, so the proof is complete.

3.2 Testing an Encryption Cycle

Lemma 3. *For a sufficiently large* $q = \tilde{O}(n^{3(k^2-1)/2})$, *the* Test *algorithm accepts with all but negligible probability when given an encryption k-cycle, i.e., in Game 0 of Definition 2.*

Remark 4. The lemma and its proof easily adapt to the case where Test is given a k'-cycle for $2 \le k' \le k$, as described in Remark 3. This is because the matrices $\mathbf{S}_i'$ commute with each other under multiplication, and the error terms are no larger in size and number.

Proof. We have $((i, \mathbf{A}_i), \mathbf{R}_i) \leftarrow \mathsf{Gen}(i, \bar{\mathbf{A}})$ and $\mathbf{C}_i \leftarrow \mathsf{Enc}((i, \mathbf{A}_i), \mathbf{R}_{i-1})$ for each $i \in \mathbb{Z}_k$, where all arithmetic in the subscripts is modulo k. Notice that when encrypting secret key $\mathbf{R}_{i-1}$ to produce $\mathbf{C}_i$, the encryption algorithm performs the $\mathbf{A}^{-1}$ operation for $\mathbf{A} = \mathbf{A}_{i-1}$. We therefore have

$$\mathbf{C}_i \leftarrow \mathbf{A}_{i-1}^{-1}[\mathbf{S}_i' \cdot \mathbf{A}_i + \mathbf{E}_i] \in \mathbb{Z}^{M \times M}$$
$$\text{where} \quad \mathbf{S}_i' = (\mathbf{I}_{n^i} \otimes \mathbf{S}_i \otimes \mathbf{I}_{n^{k-i-1}})$$
$$= (\underbrace{\mathbf{I}_n \otimes \cdots \otimes \mathbf{I}_n}_{i \text{ terms}} \otimes \mathbf{S}_i \otimes \underbrace{\mathbf{I}_n \otimes \cdots \otimes \mathbf{I}_n}_{k-i-1 \text{ terms}}) \in \mathbb{Z}^{N \times N}$$

for some error matrices $\mathbf{S}_i, \mathbf{E}_i$. Notice that because each $\mathbf{S}_i$ appears in a different position in its tensor product, the mixed-product property implies that the matrices $\mathbf{S}_i'$ commute with each other under multiplication, i.e.,

$$\mathbf{S}_i' \cdot \mathbf{S}_j' = \mathbf{S}_j' \cdot \mathbf{S}_i'.$$

Now observe that in Eq. (2), the minuend (left-hand term) of the difference expands as

$$\mathbf{L} := \mathbf{A}_{k-1} \cdot \mathbf{A}_{k-1}^{-1}[\mathbf{S}_0' \cdot \mathbf{A}_0 + \mathbf{E}_0] \cdot \mathbf{C}_1 \cdots \mathbf{C}_{k-1}$$
$$\approx \mathbf{S}_0' \cdot \mathbf{A}_0 \cdot \mathbf{A}_0^{-1}[\mathbf{S}_1' \cdot \mathbf{A}_1 + \mathbf{E}_1] \cdot \mathbf{C}_2 \cdots \mathbf{C}_{k-1} \qquad (\text{error } \mathbf{E}_0 \cdot \mathbf{C}_1 \cdots \mathbf{C}_{k-1})$$
$$\approx \mathbf{S}_0' \cdot \mathbf{S}_1' \cdot \mathbf{A}_1 \cdot \mathbf{A}_1^{-1}[\mathbf{S}_2' \cdot \mathbf{A}_2 + \mathbf{E}_2] \cdot \mathbf{C}_3 \cdots \mathbf{C}_{k-1} \quad (\text{error } \mathbf{S}_0' \cdot \mathbf{E}_1 \cdot \mathbf{C}_2 \cdots \mathbf{C}_{k-1})$$
$$\cdots$$
$$\approx \mathbf{S}_0' \cdots \mathbf{S}_{k-1}' \cdot \mathbf{A}_{k-1}. \qquad (\text{error } \mathbf{S}_0' \cdots \mathbf{S}_{k-2}' \cdot \mathbf{E}_{k-1})$$

(We analyze the error terms below.) Similarly, the subtrahend (right-hand term) of the difference expands in the same way as

$$\mathbf{R} := \mathbf{A}_0 \cdot \mathbf{C}_1 \cdots \mathbf{C}_{k-1} \cdot \mathbf{C}_0 \approx \mathbf{S}_1' \cdots \mathbf{S}_{k-1}' \cdot \mathbf{S}_0' \cdot \mathbf{A}_0$$
$$= \mathbf{S}_0' \cdot \mathbf{S}_1' \cdots \mathbf{S}_{k-1}' \cdot \mathbf{A}_0,$$

with error terms as in the previous expansion, but with all the subscripts incremented (modulo k). Finally, observe that

$$(\mathbf{L} - \mathbf{R}) \cdot \bar{\mathbf{I}} \approx \mathbf{S}_0' \cdots \mathbf{S}_{k-1}' \cdot (\mathbf{A}_{k-1} - \mathbf{A}_0) \cdot \bar{\mathbf{I}} = \mathbf{0},$$

where the approximation includes the errors (times $\bar{\mathbf{I}}$) from both of the above expansions.

It remains analyze the error terms from the above expansions. Recall that each $\mathbf{E}_i$ and $\mathbf{S}_i$ is made up of independent entries drawn from χ, which is subgaussian with parameter $O(\sqrt{n})$. Similarly, by Item 4 of Proposition 2, every $\mathbf{C}_i$ has independent subgaussian columns with parameter $\tilde{O}(M)$. Therefore, by Proposition 1,

$$s_1(\mathbf{E}_i) = O(\sqrt{nM}), \quad s_1(\mathbf{S}_i') = s_1(\mathbf{S}_i) = O(n), \quad s_1(\mathbf{C}_i) = \tilde{O}(M^{3/2})$$

except with negligible probability. It follows that in the analysis of $\mathbf{L}, \mathbf{R}$ above, the spectral norm of each error matrix—and thereby the magnitude of every entry—is bounded by $\tilde{O}(n^{1/2} \cdot M^{3k/2-1})$. Taking a sufficiently large $q = \tilde{O}(n^{3(k^2-1)/2})$ ensures that every entry in the sum of the error matrices has magnitude less than $q/4$, so the tester accepts.

3.3   Testing a Non-cycle

Lemma 4. *Under the LWE assumption from Theorem 1, the* Test *algorithm accepts with only negligible probability when given ciphertexts that all encrypt zero, i.e., in Game 1 of Definition 2.*

Proof. We consider the following sequence of hybrid experiments for generating the tester's input. We show that successive hybrids are indistinguishable (either computationally or statistically), which implies that the tester's acceptance probability differs by only a negligible amount in successive hybrids. Moreover, we show that its acceptance probability in the final hybrid is exponentially small, which proves the claim.

Hybrid 1: Here the public keys $\mathbf{A}_i$ are uniformly random and independent (modulo their common prefix $\bar{\mathbf{A}}$), and each ciphertext $\mathbf{C}_i$ is independently sampled from D^M.

   Following the proof of Lemma 2, this experiment is computationally indistinguishable from the real one (under the LWE assumption), and hence the tester's acceptance probability is only negligibly different in the two experiments.

Hybrids 2, 3, $\ldots$, $k+1$: In hybrid 2, in the cycle-test algorithm (Eq. (2)) we replace $\mathbf{A}_{k-1} \cdot \mathbf{C}_0$ with a uniformly random $\mathbf{A}_0'$, and similarly replace $\mathbf{A}_0 \cdot \mathbf{C}_1$ with a uniformly random $\mathbf{A}_1'$ (both independent of everything else). Hybrids 3 through $k+1$ are defined similarly, so that the final cycle-test algorithm simply tests whether $(\mathbf{A}_{k-1}' - \mathbf{A}_0') \cdot \bar{\mathbf{I}} \in (-q/4, q/4) \pmod{q}$ for uniformly random and independent $\mathbf{A}_{k-1}', \mathbf{A}_0'$. Clearly, this test accepts with probability bounded by the negligible quantity $2^{-N \cdot \bar{M}} \leq 2^{-n}$.

   We claim that each of these hybrids is within negligible statistical distance of the previous one. For Hybrid 2 this follows by Item 1 of Proposition 2: because $\mathbf{A}_{k-1}, \mathbf{A}_0$ are uniformly random, and $\mathbf{C}_0, \mathbf{C}_1$ are independent, $\mathbf{A}_{k-1} \cdot \mathbf{C}_0$ and $\mathbf{A}_0 \cdot \mathbf{C}_1$ are negligibly far from uniformly random and independent. (This is where we use the fact that $k \geq 2$.) The same argument applies for subsequent hybrids. This completes the proof.

4   Ring-LWE Construction

In this section we present a k-cycle tester that is IND-CPA secure assuming the hardness of ring-LWE (RLWE), appropriately parameterized. The construction works very similarly to the plain LWE one from Sect. 4. However, it is not limited to constant $k = O(1)$, but can be instantiated for any $k = \text{poly}(\lambda)$, because it does not use the tensoring technique. The scheme involves the following parameters:

- the ring $R = \mathbb{Z}[X]/(X^n + 1)$ for power-of-two n, the standard ring-LWE error distribution χ over R, and an integer modulus q (which we instantiate below);
- $\bar{m} < m = \tilde{O}(\log q)$, where $\bar{m}, m$ are the dimensions associated with the ring-based GenTrap for parameters R, q;
- The secret-key and message spaces are both $\mathcal{R}$, the randomness/trapdoor space of the ring-based GenTrap.

The construction is as follows.

- Setup(): output a uniformly random $\bar{\mathbf{a}} \in R_q^{\bar{m}}$.
- Gen($\bar{\mathbf{a}}$): let $\mathbf{a} \leftarrow \text{GenTrap}(\bar{\mathbf{a}}; \mathbf{R})$ for $\mathbf{R} \leftarrow \mathcal{R}$. Output $\mathbf{a}$ as the public key and the trapdoor $\mathbf{R}$ as the secret key.
- Enc($\mathbf{a}, \mathbf{R} \in \mathcal{R}$): let $\mathbf{v} \leftarrow \text{GenTrap}(\bar{\mathbf{a}}; \mathbf{R})$ where $\mathbf{v} \in R_q^m$. Choose $s \leftarrow \chi$ and $\mathbf{e} \leftarrow \chi^m$. Output the ciphertext

$$\mathbf{C} \leftarrow \mathbf{v}^{-1}[s \cdot \mathbf{a} + \mathbf{e}] \in R^{m \times m},$$

where the $\mathbf{v}^{-1}$ operation is performed using the trapdoor $\mathbf{R}$.
- Test($(\mathbf{a}_i, \mathbf{C}_i)_{i \in \mathbb{Z}_k}$): Given public keys $\mathbf{a}_i$ and ciphertexts $\mathbf{C}_i$, check whether

$$(\mathbf{a}_{k-1} \cdot \mathbf{C}_0 \cdot \mathbf{C}_1 \cdots \mathbf{C}_{k-1} - \mathbf{a}_0 \cdot \mathbf{C}_1 \cdots \mathbf{C}_{k-1} \cdot \mathbf{C}_0) \cdot \bar{\mathbf{I}} \in \mathcal{Q}^{\bar{m}} \pmod q, \quad (3)$$

where $\bar{\mathbf{I}} = \left(\begin{smallmatrix} \mathbf{I}_{\bar{m}} \\ 0 \end{smallmatrix} \right) \in R^{m \times \bar{m}}$, and $\mathcal{Q} \subseteq R$ is the set of ring elements whose coefficients (with respect to the standard power basis) all are in $(-q/4, q/4)$.

In the remainder of this section we prove the following theorem:

Theorem 2. *For any $k = \text{poly}(\lambda)$ and a sufficiently large $q = \tilde{O}(nk)^{O(k)}$, the above scheme is a k'-cycle tester for $2 \leq k' \leq k$, assuming the hardness of decision-RLWE$_{R,q,\chi,m}$.*

Recall that the Ring-LWE instantiation from Theorem 2 is at least as hard as (quantumly) approximating certain lattice problems on ideal lattices in R, in the worst case, to within $\tilde{O}(nk)^{O(k)}$ factors.

Lemma 5 below establishes IND-CPA security. In Sect. 4.1 we show that Test almost always accepts on an encryption cycle, and in Sect. 4.2 we show that Test almost never accepts on a non-cycle. Together these prove Theorem 2.

Lemma 5. *The tuple (Setup, Gen, Enc) is IND-CPA secure under the RLWE assumption from Theorem 2.*

Due to space restrictions, we omit the proof, which proceeds very similarly to the proof of Lemma 2.

4.1   Testing an Encryption Cycle

Lemma 6. *For a sufficiently large* $q = \tilde{O}(nk)^{O(k)}$, *the* Test *algorithm accepts with all but negligible probability when given an encryption k-cycle, i.e., in Game 0 of Definition 2.*

Proof. For input $(\mathbf{a}_i, \mathbf{C}_i)_{i \in \mathbb{Z}_k}$, we have

$$\mathbf{C}_i \leftarrow \mathbf{a}_{i-1}^{-1}[s_i \cdot \mathbf{a}_i + \mathbf{e}_i]$$

for some $s_i \leftarrow \chi$ and $\mathbf{e}_i \leftarrow \chi^m$. Moreover, by commutativity of R we have $s_i s_j = s_j s_i$ for any $i, j \in \mathbb{Z}_k$. Now for the left-hand term of Eq. (3) we have

$$
\begin{aligned}
\mathbf{l} := \mathbf{a}_{k-1} \cdot \mathbf{a}_{k-1}^{-1}[s_0 \cdot \mathbf{a}_0 + \mathbf{e}_0] \cdot \mathbf{C}_1 \cdots \mathbf{C}_{k-1} & \\
\approx s_0 \cdot \mathbf{a}_0 \cdot \mathbf{a}_0^{-1}[s_1 \cdot \mathbf{a}_1 + \mathbf{e}_1] \cdot \mathbf{C}_2 \cdots \mathbf{C}_{k-1} & \quad (\text{error } \mathbf{e}_0 \cdot \mathbf{C}_1 \cdots \mathbf{C}_{k-1}) \\
\approx s_0 \cdot s_1 \cdot \mathbf{a}_1 \cdot \mathbf{a}_1^{-1}[s_2 \cdot \mathbf{a}_2 + \mathbf{e}_2] \cdot \mathbf{C}_3 \cdots \mathbf{C}_{k-1} & \quad (\text{error } s_0 \cdot \mathbf{e}_1 \cdot \mathbf{C}_2 \cdots \mathbf{C}_{k-1}) \\
\cdots & \\
\approx s_0 \cdots s_{k-1} \cdot \mathbf{a}_{k-1}. & \quad (\text{error } s_0 \cdots s_{k-2} \cdot \mathbf{e}_{k-1})
\end{aligned}
$$

(We analyze the error terms below.) Similarly, for the right-hand term of Eq. (3), we have

$$\mathbf{r} := \mathbf{a}_0 \cdot \mathbf{C}_1 \cdots \mathbf{C}_{k-1} \cdot \mathbf{C}_0 \approx s_1 \cdots s_{k-1} \cdot s_0 \cdot \mathbf{a}_0$$
$$= s_0 \cdots s_{k-1} \cdot \mathbf{a}_0,$$

with error terms as in the previous expansion, but with all the subscripts incremented (modulo k). Therefore,

$$(\mathbf{l} - \mathbf{r}) \cdot \bar{\mathbf{I}} \approx s_0 \cdots s_{k-1} \cdot (\mathbf{a}_{k-1} - \mathbf{a}_0) \cdot \bar{\mathbf{I}} = \mathbf{0},$$

where the approximation includes the errors from the expansions of both $\mathbf{l}$ and $\mathbf{r}$, and where we use the fact that $\mathbf{a}_i \cdot \bar{\mathbf{I}} = \bar{\mathbf{a}}$ for every $i \in \mathbb{Z}_k$.

It remains to analyze the error terms. Recall that each $\mathbf{e}_i$ is made up of independent entries from χ. Also, each secret s_i comes from χ. Lastly, each ciphertext $\mathbf{C}_i \in R^{m \times m}$ is drawn as some $\mathbf{a}^{-1}[\cdot]$. Then by Eq. (1), Proposition 3, and Item 4 of Proposition 5, we have (except with negligible probability)

$$s_1(s_i) \le n^c, \quad s_1(\mathbf{e}_i) \le \sqrt{m} \cdot n^c, \quad s_1(\mathbf{C}_i) \le m \cdot n^c$$

for some universal constant $c > 1$. Let $\mathbf{e}$ denote the sum of all the error terms in the above approximations for $\mathbf{l}, \mathbf{r}$. We have

$$\|\mathbf{e}\|_\infty \le s_1(\mathbf{e}) \le 2k \cdot m^{k-1} \cdot n^{ck}.$$

Because $m = \tilde{O}(\log q)$, for a sufficiently large $q = \tilde{O}(nk)^{O(k)}$, Proposition 4 guarantees that every coefficient of every entry of $\mathbf{e}$ has the magnitude less than $q/4$, and therefore $\mathbf{e} \in \mathcal{Q}^m$ and Test accepts, as desired.

4.2   Testing a Non-cycle

Lemma 7. *Under the same RLWE assumption from Theorem 2, for $k \geq 2$ the* Test *algorithm accepts with only negligible probability on ciphertexts that all encrypt zero, i.e., in Game 1 of Definition 2.*

Proof. We consider the following sequence of hybrids. We show that adjacent hybrids are indistinguishable, either computationally or statistically. Hence, the tester's acceptance probability differs by only a negligible amount in successive hybrids.

Hybrid 1: In this hybrid, the public keys are uniformly random and independent (modulo their common prefix $\bar{\mathbf{a}}$), and each ciphertext is sampled independently from D^m. Following the proof of Lemma 5, this hybrid is computationally indistinguishable from real game.

Hybrids 2, 3, ..., $k + 1$: In the second hybrid, in Eq. (3) we replace $\mathbf{a}_{k-1} \cdot \mathbf{C}_0$ with a uniformly random $\mathbf{a}'_0$ and replace $\mathbf{a}_0 \cdot \mathbf{C}_1$ with a uniformly random $\mathbf{a}'_1$. We define hybrids 3 through $k + 1$ similarly. Hence, the final algorithm tests whether $(\mathbf{a}'_{k-1} - \mathbf{a}'_0) \cdot \bar{\mathbf{I}} \in \mathcal{Q}^{\bar{m}}$, where both terms in the difference are uniformly random and independent. The acceptance probability is therefore bounded by 2^{-n}.

Statistical indistinguishability of each of these hybrids from the previous one follows by Item 1 of Proposition 5. Therefore, the algorithm rejects on non-cycles with high probability, and proof is complete.

References

1. Acar, T., Belenkiy, M., Bellare, M., Cash, D.: Cryptographic agility and its relation to circular encryption. In: Gilbert, H. (ed.) EUROCRYPT 2010. LNCS, vol. 6110, pp. 403–422. Springer, Heidelberg (2010)
2. Adão, P., Bana, G., Herzog, J.C., Scedrov, A.: Soundness of formal encryption in the presence of key-cycles. In: di Vimercati, S.C., Syverson, P.F., Gollmann, D. (eds.) ESORICS 2005. LNCS, vol. 3679, pp. 374–396. Springer, Heidelberg (2005)
3. Alperin-Sheriff, J., Peikert, C.: Circular and KDM security for identity-based encryption. In: Fischlin, M., Buchmann, J., Manulis, M. (eds.) PKC 2012. LNCS, vol. 7293, pp. 334–352. Springer, Heidelberg (2012)
4. Applebaum, B.: Key-dependent message security: generic amplification and completeness. In: Paterson, K.G. (ed.) EUROCRYPT 2011. LNCS, vol. 6632, pp. 527–546. Springer, Heidelberg (2011)
5. Applebaum, B., Cash, D., Peikert, C., Sahai, A.: Fast cryptographic primitives and circular-secure encryption based on hard learning problems. In: Halevi, S. (ed.) CRYPTO 2009. LNCS, vol. 5677, pp. 595–618. Springer, Heidelberg (2009)
6. Barak, B., Haitner, I., Hofheinz, D., Ishai, Y.: Bounded key-dependent message security. In: Gilbert, H. (ed.) EUROCRYPT 2010. LNCS, vol. 6110, pp. 423–444. Springer, Heidelberg (2010)
7. Bishop, A., Hohenberger, S., Waters, B.: New circular security counterexamples from decision linear and learning with errors. In: Iwata, T., Cheon, J.H. (eds.) Advances in Cryptology – ASIACRYPT 2015. LNCS, vol. 9453, pp. 776–800. Springer, Heidelberg (2015)

8. Black, J., Rogaway, P., Shrimpton, T.: Encryption-scheme security in the presence of key-dependent messages. In: Nyberg, K., Heys, H.M. (eds.) SAC 2002. LNCS, vol. 2595, pp. 62–75. Springer, Heidelberg (2003)

9. Boneh, D., Halevi, S., Hamburg, M., Ostrovsky, R.: Circular-secure encryption from decision Diffie-Hellman. In: Wagner, D. (ed.) CRYPTO 2008. LNCS, vol. 5157, pp. 108–125. Springer, Heidelberg (2008)

10. Brakerski, Z., Goldwasser, S.: Circular and leakage resilient public-key encryption under subgroup indistinguishability - (or: quadratic residuosity strikes back). In: Rabin, T. (ed.) CRYPTO 2010. LNCS, vol. 6223, pp. 1–20. Springer, Heidelberg (2010)

11. Brakerski, Z., Goldwasser, S., Kalai, Y.T.: Black-box circular-secure encryption beyond affine functions. In: Ishai, Y. (ed.) TCC 2011. LNCS, vol. 6597, pp. 201–218. Springer, Heidelberg (2011)

12. Brakerski, Z., Langlois, A., Peikert, C., Regev, O., Stehlé, D.: Classical hardness of learning with errors. In: STOC, pp. 575–584 (2013)

13. Camenisch, J.L., Lysyanskaya, A.: An efficient system for non-transferable anonymous credentials with optional anonymity revocation. In: Pfitzmann, B. (ed.) EUROCRYPT 2001. LNCS, vol. 2045, pp. 93–118. Springer, Heidelberg (2001)

14. Cash, D., Green, M., Hohenberger, S.: New definitions and separations for circular security. In: Fischlin, M., Buchmann, J., Manulis, M. (eds.) PKC 2012. LNCS, vol. 7293, pp. 540–557. Springer, Heidelberg (2012)

15. Garg, S., Gentry, C., Halevi, S., Raykova, M., Sahai, A., Waters, B.: Candidate indistinguishability obfuscation and functional encryption for all circuits. In: FOCS, pp. 40–49 (2013)

16. Gentry, C.: A fully homomorphic encryption scheme. Ph.D. thesis, Stanford University (2009). http://crypto.stanford.edu/craig

17. Gentry, C.: Fully homomorphic encryption using ideal lattices. In: STOC, pp. 169–178 (2009)

18. Gentry, C., Peikert, C., Vaikuntanathan, V.: Trapdoors for hard lattices and new cryptographic constructions. In: STOC, pp. 197–206 (2008)

19. Goldwasser, S., Micali, S.: Probabilistic encryption. J. Comput. Syst. Sci. **28**(2), 270–299 (1984). Preliminary version in STOC 1982

20. Koppula, V., Ramchen, K., Waters, B.: Separations in circular security for arbitrary length key cycles. In: Dodis, Y., Nielsen, J.B. (eds.) TCC 2015, Part II. LNCS, vol. 9015, pp. 378–400. Springer, Heidelberg (2015)

21. Koppula, V., Waters, B.: Circular security separations for arbitrary length cycles from LWE. In: CRYPTO. (to appear 2016)

22. Lyubashevsky, V., Peikert, C., Regev, O.: On ideal lattices and learning with errors over rings. J. ACM **60**(6), 43:1–43:35 (2010). Preliminary version in Eurocrypt 2010

23. Lyubashevsky, V., Peikert, C., Regev, O.: A toolkit for Ring-LWE cryptography. In: Johansson, T., Nguyen, P.Q. (eds.) EUROCRYPT 2013. LNCS, vol. 7881, pp. 35–54. Springer, Heidelberg (2013)

24. Malkin, T., Teranishi, I., Yung, M.: Efficient circuit-size independent public key encryption with KDM security. In: Paterson, K.G. (ed.) EUROCRYPT 2011. LNCS, vol. 6632, pp. 507–526. Springer, Heidelberg (2011)

25. Marcedone, A., Orlandi, C.: Obfuscation ($\rightarrow$) (IND-CPA security ! $\rightarrow$ circular security). In: Abdalla, M., De Prisco, R. (eds.) SCN 2014. LNCS, vol. 8642, pp. 77–90. Springer, Heidelberg (2014)

26. Micciancio, D., Peikert, C.: Trapdoors for lattices: simpler, tighter, faster, smaller. In: Pointcheval, D., Johansson, T. (eds.) EUROCRYPT 2012. LNCS, vol. 7237, pp. 700–718. Springer, Heidelberg (2012)

27. Peikert, C.: Public-key cryptosystems from the worst-case shortest vector problem. In: STOC, pp. 333–342 (2009)
28. Regev, O.: On lattices, learning with errors, random linear codes, cryptography. J. ACM **56**(6), 1–40 (2009). Preliminary version in STOC 2005
29. Vershynin, R.: Compressed sensing, theory and applications, pp. 210–268. Cambridge University Press (2012). http://www-personal.umich.edu/ romanv/papers/ non-asymptotic-rmt-plain.pdf. chapter 5

Circular Security Separations for Arbitrary Length Cycles from LWE

Venkata Koppula$^{(\boxtimes)}$ and Brent Waters

University of Texas at Austin, Austin, USA
{kvenkata,bwaters}@cs.utexas.edu

Abstract. We describe a public key encryption that is IND-CPA secure under the Learning with Errors (LWE) assumption, but that is not circular secure for arbitrary length cycles. Previous separation results for cycle length greater than 2 require the use of indistinguishability obfuscation, which is not currently realizable under standard assumptions.

1 Introduction

The notion of key dependent message security departs from standard encryption security in that it allows the attacker to access ciphertexts where the messages are functions of the secret key. One prototypical example is k-circular security. An encryption scheme is said to be k-circular secure, if an adversary is unable to distinguish $\mathsf{Enc}(\mathsf{pk}_1, \mathsf{sk}_k), \mathsf{Enc}(\mathsf{pk}_2, \mathsf{sk}_1), \ldots, \mathsf{Enc}(\mathsf{pk}_k, \mathsf{sk}_{k-1})$ from k encryptions of the all 0 message.

The demand for encryption schemes that provide circular security has arisen in multiple applications. Camenisch and Lysyanskaya [13] applied circular secure encryption to anonymous credential systems, while Laud [22] and Adão et al. [2] use circular security to prove the soundness of symbolic protocols. Most notably Gentry's [19] bootstrapping technique shows how to achieve fully homomorphic encryption (FHE) for circuits of any depth chosen at evaluation time (i.e. not fixed at setup) from those of shallower depth if the FHE scheme is circular secure. There have been multiple constructions of circular secure schemes or more generally key-dependent message security, some proven in the random oracle model [8,13] and others in the standard model from particular assumptions [3–6,9–11].

One interesting question is whether k-circular security can come "for free". Is there some k such that *any* IND-CPA secure encryption scheme is guaranteed to be k-circular secure? If true, this would give an immediate path to applying Gentry's FHE bootstrapping technique among other applications.

A trivial folklore argument provides a separation for the case of $k = 1$. The first non-trivial example for $k = 2$ was given by Acar et al. [1] and extended by Cash et al. [14] using the Decisional Diffie-Hellman assumption over asymmetric

B. Waters—Supported by CNS-1228599 and CNS-1414082. DARPA through the U.S. Office of Naval Research under Contract N00014-11-1-0382, Google Faculty Research award, the Alfred P. Sloan Fellowship, Microsoft Faculty Fellowship, and Packard Foundation Fellowship.

© International Association for Cryptologic Research 2016
M. Robshaw and J. Katz (Eds.): CRYPTO 2016, Part II, LNCS 9815, pp. 681–700, 2016.
DOI: 10.1007/978-3-662-53008-5_24

bilinear groups. Subsequently, Bishop et al. [7] extended the result to include symmetric groups under the decision linear assumption as well as moving to the lattice setting with a counterexample under the Learning with Errors (LWE) assumption. However, they leave open the possibility of getting "free" circular security by simply extending the key cycle lengths to be greater than two.

The more general case of k-length cycles for arbitrary size k was considered by Koppula et al. [21] who showed that under the assumption of indistinguishability obfuscation (for polynomial sized circuits), for any k there exists schemes that are IND-CPA secure, but that are not k-circular secure. Marcedone and Orlandi [23] independently gave a similar result, but under the assumption of a virtual black box secure obfuscator for a certain functionality.

While these works cast doubts on the ability to get free circular security for larger cycle lengths, they do so by invoking a quite strong primitive of obfuscation. Notably, the only current candidates for obfuscation rely on the multilinear encodings for which the first candidate was proposed in 2013 [17]. In addition, to being relatively untested there have subsequently been several attacks discovered [15,16] on various multilinear encoding proposals.

Separation Without Obfuscation. This brings up to the central question of this paper.

Can we separate IND-CPA and circular security for arbitrary length cycles using standard assumptions (i.e. without invoking obfuscation or multilinear maps)?

Such a result would provide a firmer understanding of circular security. In addition, the introduction of the first general purpose obfuscation candidate [18] has lead to the realization of many cryptographic primitives that to this point were not realizable (e.g., deniable encryption, functional encryption, etc.). However, very few of these newly realized primitives have since been adapted to a standard assumption — one not involving obfuscation or multilinear maps. We believe that attacking this problem for one primitive can begin to crack the ice and hopefully begin to lead to insights for others.

A Separation Example from Learning with Errors. The main result of our paper is the introduction of a family of encryption systems that are IND-CPA secure under the LWE assumption, but which are made to not be k-circular secure for arbitrary k.

We first illustrate the challenges of building such a scheme by looking at the recent Bishop, Hohenberger and Waters construction [7], which gave a separation from LWE for $k = 2$. In this work, Bishop et al. first proposed a general framework for constructing circular security separations. This framework, called the *k-cycle tester framework*, consists of algorithms for setup, key generation, encryption and testing cycles. Note that unlike an encryption scheme, there is no decryption algorithm here. The setup algorithm outputs public parameters, which are used by the key generation algorithm to choose the public key and secret key. The encryption algorithm takes a public key and a message, and outputs its encryption. The cycle tester algorithm takes k public keys and k encryptions, and outputs 1 if the k public keys/ciphertexts form a key cycle (that is, $\mathsf{ct}_i \leftarrow \mathsf{Enc}(\mathsf{pk}_i, \mathsf{sk}_{i-1})$), else it outputs 0 with all but negligible probability. For security, encryptions of distinct messages must be computationally indistinguishable. Bishop et al. showed how to use such a k-cycle tester, together with an IND-CPA encryption scheme, to construct an IND-CPA encryption scheme that is not k-circular secure. They also showed several constructions of a 2-cycle tester from various assumptions, including one from LWE.

The BHW 2-cycle Tester from LWE: Unlike most existing LWE based encryption schemes where the message is part of a large norm vector, Bishop et al. used a novel approach for encrypting the message: via lattice trapdoors. A lattice trapdoor generation algorithm outputs a matrix $\mathbf{A}$ together with a trapdoor $T_\mathbf{A}$. The matrix looks uniformly random, while the trapdoor can be used to compute, for any matrix $\mathbf{U}$, a low norm matrix $\mathbf{S} = \mathbf{A}^{-1}(\mathbf{U})$ such that $\mathbf{A} \cdot \mathbf{S} = \mathbf{U}$.[1] Moreover, if $\mathbf{U}$ is chosen uniformly at random, then $\mathbf{S}$ reveals no information about the matrix $\mathbf{A}$, or the randomness used to sample $\mathbf{A}, T_\mathbf{A}$. Bishop et al. used the message vector as randomness for the lattice trapdoor generation algorithm.

   Their construction (with some modifications) can be described as follows. The setup algorithm simply outputs the LWE parameters. The key generation algorithm first samples a matrix $\mathbf{A}$ along with its lattice trapdoor $T_\mathbf{A}$. The secret key is the randomness used to compute $\mathbf{A}, T_\mathbf{A}$. To compute the public key, the algorithm chooses a matrix $\mathbf{C}$, computes $\mathbf{D} = \mathbf{C} \cdot \mathbf{A} + \mathsf{noise}$ and outputs $(\mathbf{C}, \mathbf{D})$ as the public key. The encryption algorithm uses the message msg as randomness for the trapdoor generation algorithm, computing a matrix $\mathbf{Z}$ and its trapdoor $T_\mathbf{Z}$. Next, it chooses a uniformly random $\{-1, 1\}$ vector $\mathbf{r}$ and computes $\mathbf{u} = \mathbf{C}^\top \cdot \mathbf{r}$ and $\mathbf{v} = \mathbf{D}^\top \cdot \mathbf{r} \approx \mathbf{A}^\top \cdot \mathbf{C}^\top \cdot \mathbf{r}$. The final ciphertext consists of a short vector $\mathbf{s} = \mathbf{Z}^{-1}(\mathbf{u})$ that contains the message, and a large vector $\mathbf{v}$ that is used for cycle testing. For IND-CPA security, one can use the LWE assumption and the Leftover Hash Lemma to argue that $\mathbf{C}^\top \cdot \mathbf{r}$ is indistinguishable from a uniformly random vector, and therefore $\mathbf{Z}^{-1}(\mathbf{C}^\top \cdot \mathbf{r})$ reveals no information about msg.

   The cycle testing algorithm takes as input two ciphertexts $(\mathbf{v}_1, \mathbf{s}_1), (\mathbf{v}_2, \mathbf{s}_2)$ and checks if $\mathbf{v}_1^\top \cdot \mathbf{s}_2$ is close to $\mathbf{v}_2^\top \cdot \mathbf{s}_1$. To see why this works, let us consider the case when the two ciphertexts form a key cycle; that is, $\mathbf{s}_1 = \mathbf{B}_2^{-1}(\mathbf{C}_1^\top \cdot \mathbf{r}_1)$, $\mathbf{v}_1^\top = \mathbf{r}_1^\top \cdot \mathbf{C}_1 \cdot \mathbf{B}_1 + \mathsf{noise}$ and $\mathbf{s}_2 = \mathbf{B}_1^{-1}(\mathbf{C}_2^\top \cdot \mathbf{r}_2)$, $\mathbf{v}_2^\top = \mathbf{r}_2^\top \cdot \mathbf{C}_2 \cdot \mathbf{B}_2 + \mathsf{noise}$. In this case, the testing algorithm outputs 1 because $\mathbf{v}_1^\top \cdot \mathbf{s}_2 \approx \mathbf{r}_1^\top \cdot \mathbf{C}_1 \cdot \mathbf{C}_2^\top \cdot \mathbf{r}_2$ $= \mathbf{r}_2^\top \cdot \mathbf{C}_2 \cdot \mathbf{C}_1^\top \cdot \mathbf{r}_1 \approx \mathbf{v}_2^\top \cdot \mathbf{s}_1$. However, if both ciphertexts are encryptions of $\mathbf{0}$, then both $\mathbf{v}_1^\top \cdot \mathbf{s}_2$ and $\mathbf{v}_2^\top \cdot \mathbf{s}_1$ are uniformly random elements, and therefore, they are likely not close to each other. At a high level, this approach works because in a key cycle, the $\mathbf{B}_1$ in $\mathbf{v}_1$ and $\mathbf{B}_1^{-1}$ in $\mathbf{s}_2$ cancel each other (and similarly the matrix $\mathbf{B}_2$ and $\mathbf{B}_2^{-1}$ in $\mathbf{v}_2$ and $\mathbf{s}_1$ respectively).

   Unfortunately, the BHW approach cannot be directly used to handle longer cycles.

---

[1] For simplicity, we use the notation $\mathbf{A}^{-1}(\cdot)$ to represent the pre-image $\mathbf{S}$. In the formal description of our algorithms, we use the pre-image sampling algorithm $\mathsf{SamplePre}$.

Our Approach via Cascading Cancellations: For simplicity, let us consider the problem of constructing a 3-cycle tester (this can be easily extended to handle longer cycles). The starting point of our approach is the following simple observation: for $i = 1, 2, 3$, let $\mathbf{B}_i$ be matrices with trapdoors, and let $\mathbf{X}, \mathbf{C}_1, \mathbf{C}_2, \mathbf{C}_3$ be arbitrary matrices. Consider the matrices $\mathbf{M}_1 = \mathbf{B}_3^{-1}(\mathbf{C}_1 \cdot \mathbf{X})$, $\mathbf{M}_2 = \mathbf{B}_1^{-1}(\mathbf{C}_2 \cdot \mathbf{B}_2)$ and $\mathbf{M}_3 = \mathbf{B}_2^{-1}(\mathbf{C}_3 \cdot \mathbf{B}_3)$. Then $\mathbf{B}_1 \cdot \mathbf{M}_2 \cdot \mathbf{M}_3 \cdot \mathbf{M}_1 = \mathbf{C}_2 \cdot \mathbf{C}_3 \cdot \mathbf{C}_1 \cdot \mathbf{X}$. The matrix $\mathbf{B}_1$ starts the 'chain reaction' by canceling $\mathbf{B}_1^{-1}$ in $\mathbf{M}_1$, and after each matrix multiplication, the product is a canceling matrix for the next one in the sequence.

In fact, this observation can be easily extended to have noisy matrices: for $i = 1, 2, 3$, let $\mathbf{B}_i$ be matrices with trapdoors, $\mathbf{C}_i$ matrices with low norm entries, and $\mathbf{X}$ any arbitrary matrix. Consider the matrices $\mathbf{M}_1 = \mathbf{B}_3^{-1}(\mathbf{C}_1 \cdot \mathbf{X} + \mathsf{noise})$, $\mathbf{M}_2 = \mathbf{B}_1^{-1}(\mathbf{C}_2 \cdot \mathbf{B}_2 + \mathsf{noise})$ and $\mathbf{M}_3 = \mathbf{B}_2^{-1}(\mathbf{C}_3 \cdot \mathbf{B}_3 + \mathsf{noise})$. Then $\mathbf{B}_1 \cdot \mathbf{M}_2 \cdot \mathbf{M}_3 \cdot \mathbf{M}_1 \approx \mathbf{C}_2 \cdot \mathbf{C}_3 \cdot \mathbf{C}_1 \cdot \mathbf{X}$. This observation inspires us to try the following approach: each ciphertext consists of two low norm matrices such that a key cycle gives us two parallel chains with the same end product matrix. Before discussing this approach in more detail, we will present an extension of the cycle tester framework which will help simplify our presentation.

Extending the BHW k-cycle Tester Framework: We introduce an extension of the BHW cycle tester framework, which we call the *Leader-Follower k-cycle tester* framework. This framework has a setup algorithm for outputting the parameters, two different key generation and encryption algorithms, and finally a tester algorithm. Looking ahead, in our separation, one of the public keys/ciphertexts has a special role, and they are generated using the 'leader' key generation/encryption algorithms, while the remaining are generated using the 'follower' key generation/encryption algorithms. For correctness, we require that the test algorithm outputs 1 if the k ciphertexts form an encryption cycle, else it outputs 0. For security, both the leader and follower encryption schemes must satisfy IND-CPA security. One can establish a simple reduction from our Leader-Follower framework to the BHW cycle-tester framework.

First Attempt via Two Parallel Chains: As an initial attempt, we present a Leader-Follower 3-cycle tester where any message/secret key consists of two strings, each of which can be used to sample a lattice trapdoor. To begin, we will describe the follower key generation/encryption algorithms.

The follower key generation algorithm chooses two strings $\mathbf{x}_1, \mathbf{x}_2$ and sets $(\mathbf{x}_1, \mathbf{x}_2)$ as the secret key. To compute the public key, it first chooses two matrices $\mathbf{B}_1, \mathbf{B}_2$ with trapdoors (using strings $\mathbf{x}_1$ and $\mathbf{x}_2$ respectively as randomness). The public key simply consists of the matrices $\mathbf{B}_1, \mathbf{B}_2$. The corresponding encryption algorithm uses the message $\mathsf{msg} = (\mathbf{y}_1, \mathbf{y}_2)$ to sample matrices $\mathbf{Z}_1, \mathbf{Z}_2$ together with the respective trapdoors. Next, it chooses a low norm matrix $\mathbf{C}$ and outputs $\mathbf{S}_1 = \mathbf{Z}_1^{-1}(\mathbf{C} \cdot \mathbf{B}_1 + \mathsf{noise})$, $\mathbf{S}_2 = \mathbf{Z}_2^{-1}(\mathbf{C} \cdot \mathbf{B}_2 + \mathsf{noise})$ as the ciphertext.

The leader key generation algorithm is a bit more involved. The secret key is chosen as in the follower key generation, and the public key has an additional component: a uniformly random matrix $\mathbf{X}$. As in the follower encryption algorithm, the leader encryption algorithm chooses matrices $\mathbf{Z}_1, \mathbf{Z}_2$ and their trapdoors. Next, it chooses low norm matrices $\mathbf{C}_1$ and outputs $\mathbf{S}_1 = \mathbf{Z}_1^{-1}(\mathbf{C} \cdot \mathbf{X} + \mathsf{noise})$, $\mathbf{S}_2 = \mathbf{Z}_2^{-1}(\mathbf{C} \cdot \mathbf{X} + \mathsf{noise})$.

The testing algorithm, on input three ciphertexts $(\mathbf{S}_{11}, \mathbf{S}_{12})$, $(\mathbf{S}_{21}, \mathbf{S}_{22})$, $(\mathbf{S}_{31}, \mathbf{S}_{32})$ and three public keys $(\mathbf{B}_{11}, \mathbf{B}_{12}, \mathbf{X})$, $(\mathbf{B}_{21}, \mathbf{B}_{22})$, $(\mathbf{B}_{31}, \mathbf{B}_{32})$, checks if $\mathbf{B}_{11} \cdot \mathbf{S}_{21} \cdot \mathbf{S}_{31} \cdot \mathbf{S}_{11} \approx \mathbf{B}_{12} \cdot \mathbf{S}_{22} \cdot \mathbf{S}_{32} \cdot \mathbf{S}_{12}$. The testing algorithm works as desired, because if $\mathbf{C}_i$ is the random matrix used for computing $\mathbf{S}_{ij}$ and the three ciphertexts form an encryption cycle, then both the expressions are approximately $\mathbf{C}_2 \cdot \mathbf{C}_3 \cdot \mathbf{C}_1 \cdot \mathbf{X}$.

For IND-CPA security of follower key generation/encryptions, note that by the LWE assumption, both $\mathbf{C} \cdot \mathbf{B}_1 + \mathsf{noise}$ and $\mathbf{C} \cdot \mathbf{B}_2 + \mathsf{noise}$ are indistinguishable from truly random matrices. As a result, $\mathbf{S}_1$ and $\mathbf{S}_2$ hide the randomness used to choose $\mathbf{Z}_1$ and $\mathbf{Z}_2$.

Next, let us consider IND-CPA security of leader key generation/encryptions. Unfortunately, this part is problematic, because the matrices $\mathbf{Z}_1^{-1}(\mathbf{C} \cdot \mathbf{X} + \mathsf{noise})$ and $\mathbf{Z}_2^{-1}(\mathbf{C} \cdot \mathbf{X} + \mathsf{noise})$ clearly reveal information about $\mathbf{Z}_1$ and $\mathbf{Z}_2$ (for example, one can check if $\mathbf{Z}_1 = \mathbf{Z}_2$). To address this problem, we first increase the number of parallel chains to a suitably large number (say ℓ), and have ℓ matrices $\mathbf{X}_1, \ldots, \mathbf{X}_\ell$ as part of the leader public key. These matrices satisfy the following relation: there exist a $\{-1, 1\}$ coefficient vector $\mathbf{x}$ such that $x_i \cdot \mathbf{X}_i = 0$. This vector $\mathbf{x}$ must be hidden from the IND-CPA adversary; however, the test algorithm must be able to somehow use this vector to cancel out the $\mathbf{X}_i$ matrices.

Our Solution: Our final solution is similar to the approach outlined above. The messages and secret keys consist of ℓ strings, each of which can be used as the randomness for lattice trapdoor generation. The follower key generation and encryption algorithms are similar to the ones described above, except that now there are ℓ public matrices $\mathbf{B}_1, \ldots, \mathbf{B}_\ell$, and the ciphertext consists of ℓ low norm matrices $\mathbf{S}_1, \ldots, \mathbf{S}_\ell$, where $\mathbf{S}_i = \mathbf{Z}_i^{-1}(\mathbf{C} \cdot \mathbf{B}_i + \mathsf{noise})$.

The main differences are in the leader key generation algorithm. The leader key generation algorithm first uses the secret key to sample ℓ matrices $\mathbf{B}_1, \ldots, \mathbf{B}_\ell$ along with their trapdoors. Next, it chooses an ℓ length string $\mathbf{x}$, chooses $\ell - 1$ matrices $\mathbf{X}_1, \ldots, \mathbf{X}_{\ell-1}$, and sets X_ℓ such that $\sum_i x_i \cdot \mathbf{X}_i = 0$. The public key consists of the matrices $(x_1 \cdot \mathbf{B}_1, \ldots, x_\ell \cdot \mathbf{B}_\ell, \mathbf{X}_1, \ldots, \mathbf{X}_\ell)$ (note the x_i coefficients attached to each $\mathbf{B}_i$). To encrypt a message, one chooses matrices $\mathbf{Z}_1, \ldots, \mathbf{Z}_\ell$ with trapdoors using the message strings as randomness. Then it chooses a matrix $\mathbf{C}$ and outputs $\mathbf{S}_i = \mathbf{Z}_i^{-1}(\mathbf{C} \cdot \mathbf{X}_i + \mathsf{noise})$ as the ciphertext. To argue IND-CPA security, note that the matrices $\mathbf{X}_i$ look uniformly random (since the vector $\mathbf{x}$ is hidden from the adversary). As a result, the matrices $\mathbf{C} \cdot \mathbf{X}_i + \mathsf{noise}$ look like ℓ uniformly random matrices, and therefore the adversary does not learn any information about the $\mathbf{Z}_i$ matrices.

The test algorithm is similar to what was described in the previous solution, except at the end, the algorithm computes the sum of the final products, and checks if it is of low norm. The correctness of the test algorithm for $k = 3$ can be verified easily from the table below. Let $\mathsf{pk}_1 = (x_1 \cdot \mathbf{B}_{11}, \ldots, x_\ell \cdot \mathbf{B}_{i\ell}, \mathbf{X}_1, \ldots, \mathbf{X}_\ell)$, and let $\mathsf{ct}_1, \mathsf{ct}_2, \mathsf{ct}_3$ be the three ciphertexts that form a 3-cycle.

$\mathbf{pk}_1$ matrices	$x_1 \cdot \mathbf{B}_{11}$	$\dots$	$x_\ell \cdot \mathbf{B}_{1\ell}$
$\mathbf{ct}_2$ matrices	$\mathbf{S}_{21} = \mathbf{B}_{11}^{-1}(\mathbf{C}_2 \cdot \mathbf{B}_{21} + \text{noise})$	$\dots$	$\mathbf{S}_{2\ell} = \mathbf{B}_{1\ell}^{-1}(\mathbf{C}_2 \cdot \mathbf{B}_{2\ell} + \text{noise})$
$\mathbf{ct}_3$ matrices	$\mathbf{S}_{31} = \mathbf{B}_{21}^{-1}(\mathbf{C}_3 \cdot \mathbf{B}_{31} + \text{noise})$	$\dots$	$\mathbf{S}_{3\ell} = \mathbf{B}_{2\ell}^{-1}(\mathbf{C}_3 \cdot \mathbf{B}_{3\ell} + \text{noise})$
$\mathbf{ct}_1$ matrices	$\mathbf{S}_{11} = \mathbf{B}_{31}^{-1}(\mathbf{C}_1 \cdot \mathbf{X}_1 + \text{noise})$	$\dots$	$\mathbf{S}_{1\ell} = \mathbf{B}_{3\ell}^{-1}(\mathbf{C}_1 \cdot \mathbf{X}_\ell + \text{noise})$
Product	$\approx x_1 \cdot \mathbf{C}_2 \cdot \mathbf{C}_3 \cdot \mathbf{C}_1 \cdot \mathbf{X}_1$	$\dots$	$\approx x_\ell \cdot \mathbf{C}_2 \cdot \mathbf{C}_3 \cdot \mathbf{C}_1 \cdot \mathbf{X}_\ell$

Clearly, the sum of the products is low norm, and therefore the testing algorithm outputs 1 if the input is a key cycle. For a non-key cycle, each of these products is a uniformly random matrix, and therefore, with high probability, their sum has large norm. This concludes our scheme.

2   Preliminaries

Notations: We will use lowercase bold letters for vectors (e.g. $\mathbf{v}$) and uppercase bold letters for matrices (e.g. $\mathbf{A}$). For any finite set S, $x \leftarrow S$ denotes a uniformly random element x from the set S. Similarly, for any distribution $\mathcal{D}$, $x \leftarrow \mathcal{D}$ denotes an element x drawn from distribution $\mathcal{D}$. The distribution $\mathcal{D}^n$ is used to represent a distribution over vectors of n components, where each component is drawn independently from the distribution $\mathcal{D}$. Ber_p denotes the Bernoulli distribution over $\{0, 1\}$, where $\Pr_{x \leftarrow \text{Ber}_p}[x = 1] = p$.

Given a randomized algorithm $A(\cdot)$, the notation $A(\cdot; \cdot)$ is used to explicitly describe the randomness used by A (e.g. $A(x; r)$ denotes computation on input x using randomness r).

Randomness Extraction: We will use the following theorem, which follows from the Leftover Hash Lemma.

Theorem 1. *Let $m > (n+1) \log_2 q + \omega(\log n)$ and q a prime. Then the statistical distance between the following distributions is negligible in n.*

$$\{(\mathbf{A}, \mathbf{A} \cdot \mathbf{r}) : \mathbf{A} \leftarrow \mathbb{Z}_q^{n \times m}, \mathbf{r} \leftarrow \{-1, 1\}^m\} \approx \{(\mathbf{A}, \mathbf{u}) : \mathbf{A} \leftarrow \mathbb{Z}_q^{n \times m}, \mathbf{u} \leftarrow \mathbb{Z}_q^m\}.$$

2.1   Lattice Preliminaries

Given positive integers n, m, q and a matrix $A \in \mathbb{Z}_q^{n \times m}$, we let $\Lambda_q^\perp(A)$ denote the lattice $\{x \in \mathbb{Z}^m : Ax = 0 \mod q\}$. For $u \in \mathbb{Z}_q^n$, we let $\Lambda_q^u(A)$ denote the set $\{x \in \mathbb{Z}^m : Ax = u \mod q\}$.

Discrete Gaussians. Let σ be any positive real number. The Gaussian distribution $\mathcal{D}_\sigma$ with parameter σ is defined by the probability distribution function $\rho_\sigma(\mathbf{x}) = \exp(-\pi \cdot ||\mathbf{x}||^2/\sigma^2)$. For any set $\mathcal{L} \subset \mathbb{R}^m$, define $\rho_\sigma(\mathcal{L}) = \sum_{\mathbf{x} \in \mathcal{L}} \rho_\sigma(\mathbf{x})$. The discrete Gaussian distribution $\mathcal{D}_{\mathcal{L}, \sigma}$ over $\mathcal{L}$ with parameter σ is defined by the probability distribution function $\rho_{\mathcal{L}, \sigma}(\mathbf{x}) = \rho_\sigma(\mathbf{x})/\rho_\sigma(\mathcal{L})$ for all $\mathbf{x} \in \mathcal{L}$.

The following lemma (Lemma 4.4 of [20, 25]) shows that if the parameter σ of a discrete Gaussian distribution is small, then any vector drawn from this distribution will be short (with high probability).

Lemma 1. *Let m, n, q be positive integers with $m > n, q \geq 2$. Let $\mathbf{A} \leftarrow \mathbb{Z}_q^{n \times m}$ be a uniformly random matrix of dimensions $n \times m$, $\sigma = \tilde{\Omega}(n)$ and $\mathcal{L} = \Lambda_q^{\perp}(\mathbf{A})$. Then*

$$\Pr[\|\mathbf{x}\| > \sqrt{m} \cdot \sigma : \mathbf{x} \leftarrow \mathcal{D}_{\mathcal{L},\sigma}] \leq negl(n).$$

Learning with Errors (LWE). The Learning with Errors (LWE) problem was introduced by Regev [28]. The LWE problem has four parameters: the dimension of the lattice n, the number of samples m, the modulus q and the error distribution $\chi(n)$.

Assumption 1 (Learning with Errors). *Let n, m and q be positive integers and χ a noise distribution on $\mathbb{Z}$. The Learning with Errors assumption (n, m, q, χ)-LWE, parameterized by n, m, q, χ, states that the following distributions are computationally indistinguishable:*

$$\left\{ (\mathbf{A}, \mathbf{s}^{\top} \cdot \mathbf{A} + \mathbf{e}) : \begin{array}{c} \mathbf{A} \leftarrow \mathbb{Z}_q^{n \times m}, \\ \mathbf{s} \leftarrow \mathbb{Z}_q^n, \mathbf{e} \leftarrow \chi^m \end{array} \right\} \approx_c \left\{ (\mathbf{A}, \mathbf{u}) : \begin{array}{c} \mathbf{A} \leftarrow \mathbb{Z}_q^{n \times m}, \\ \mathbf{u} \leftarrow \mathbb{Z}_q^m \end{array} \right\}$$

Under a quantum reduction, Regev [28] showed that for certain noise distributions, LWE is as hard as worst case lattice problems such as the decisional approximate shortest vector problem (GapSVP) and approximate shortest independent vectors problem (SIVP). The following theorem statement is from Peikert's survey [27].

Theorem 2 [28]. *For any $m \leq \mathsf{poly(n)}$, any $q \leq 2^{\mathsf{poly(n)}}$, and any discretized Gaussian error distribution χ of parameter $\alpha \cdot q \geq 2 \cdot \sqrt{n}$, solving (n, m, q, χ)-LWE is as hard as quantumly solving GapSVP_{γ} and SIVP_{γ} on arbitrary n-dimensional lattices, for some $\gamma = \tilde{O}(n/\alpha)$.*

Later works [12,26] showed classical reductions from LWE to GapSVP_{γ}. Given the current state of art in lattice algorithms, GapSVP_{γ} and SIVP_{γ} are believed to be hard for $\gamma = \tilde{O}(2^{n^{\epsilon}})$, and therefore (n, m, q, χ)-LWE is believed to be hard for Gaussian error distributions χ with parameter $2^{-n^{\epsilon}} \cdot q \cdot \mathsf{poly(n)}$.

LWE with Short Secrets. In this work, we will be using a variant of the LWE problem called *LWE with Short Secrets*. In this variant, introduced by Applebaum et al. [5], the secret vector is also chosen from the noise distribution χ. They showed that this variant is as hard as LWE for sufficiently large number of samples m.

Assumption 2 (LWE with Short Secrets). *Let n, m and q be positive integers and χ a noise distribution on $\mathbb{Z}$. The LWE with Short Secrets assumption (n, m, q, χ)-LWE-ss, parameterized by n, m, q, χ, states that the following distributions are computationally indistinguishable[2]:*

$$\left\{ (\mathbf{A}, \mathbf{S} \cdot \mathbf{A} + \mathbf{E}) : \begin{array}{c} \mathbf{A} \leftarrow \mathbb{Z}_q^{n \times m}, \\ \mathbf{S} \leftarrow \chi^{n \times n}, \mathbf{E} \leftarrow \chi^{n \times m} \end{array} \right\} \approx_c \left\{ (\mathbf{A}, \mathbf{U}) : \begin{array}{c} \mathbf{A} \leftarrow \mathbb{Z}_q^{n \times m}, \\ \mathbf{U} \leftarrow \mathbb{Z}_q^{n \times m} \end{array} \right\}.$$

---

[2] Applebaum et al. showed that $\{(\mathbf{A}, \mathbf{s}^{\top} \cdot \mathbf{A} + \mathbf{e}) : \mathbf{A} \leftarrow \mathbb{Z}_q^{n \times m}, \mathbf{s} \leftarrow \chi^n, \mathbf{e} \leftarrow \chi^m\} \approx_c$ $\{(\mathbf{A}, \mathbf{u}) : \mathbf{A} \leftarrow \mathbb{Z}_q^{n \times m}, \mathbf{u} \leftarrow \mathbb{Z}_q^m\}$, assuming LWE is hard. However, by a simple hybrid argument, we can replace vectors $\mathbf{s}, \mathbf{e}, \mathbf{u}$ with matrices $\mathbf{S}, \mathbf{E}, \mathbf{U}$ of appropriate dimensions.

Lattices with Trapdoors. Lattices with trapdoors are lattices that are statistically indistinguishable from randomly chosen lattices, but have certain 'trapdoors' that allow efficient solutions to hard lattice problems.

Definition 1. *A trapdoor lattice sampler consists of algorithms* TrapGen *and* SamplePre *with the following syntax and properties:*

- TrapGen$(1^n, 1^m, q) \to (\mathbf{A}, T_{\mathbf{A}})$: *The lattice generation algorithm is a randomized algorithm that takes as input the matrix dimensions n, m, modulus q and $\ell_{\mathrm{TG}}(n)$ bits of randomness, and outputs a matrix $\mathbf{A} \in \mathbb{Z}_q^{n \times m}$ together with a trapdoor $T_{\mathbf{A}}$.*
- SamplePre$(\mathbf{A}, T_{\mathbf{A}}, \mathbf{u}, \sigma) \to \mathbf{s}$: *The presampling algorithm takes as input a matrix $\mathbf{A}$, trapdoor $T_{\mathbf{A}}$, a vector $\mathbf{u} \in \mathbb{Z}_q^n$ and a parameter $\sigma \in \mathbb{R}$ (which determines the length of the output vectors). It outputs a vector $\mathbf{s} \in \mathbb{Z}_q^m$.*

   These algorithms must satisfy the following properties:

1. Correct Presampling: *For any string $\mathbf{y} \in \{0,1\}^{\ell_{\mathrm{TG}}}$, vector $\mathbf{u}$ and parameter σ, let $(\mathbf{A}, T_{\mathbf{A}}) \leftarrow$ TrapGen$(1^n, 1^m; \mathbf{y})$, $\mathbf{s} \leftarrow$ SamplePre$(\mathbf{A}, T_{\mathbf{A}}, \mathbf{u}, \sigma)$. Then $\mathbf{A} \cdot \mathbf{s} = \mathbf{u}$ and $\|\mathbf{s}\|_\infty \leq \sqrt{m} \cdot \sigma$.*
2. Well Distributedness of Matrix: *The following distributions are statistically indistinguishable:*

$$\{\mathbf{A} : (\mathbf{A}, T_{\mathbf{A}}) \leftarrow \mathsf{TrapGen}(1^n, 1^m)\} \approx_s \{\mathbf{A} : \mathbf{A} \leftarrow \mathbb{Z}_q^{n \times m}\}.$$

3. Well Distributedness of Preimage: *For any string $\mathbf{y} \in \{0,1\}^{\ell_{\mathrm{TG}}}$, let $(\mathbf{A}, T_{\mathbf{A}}) =$ TrapGen$(1^n, 1^m; \mathbf{y})$. Then if $\sigma = \omega(\sqrt{n \cdot \log q \cdot \log m})$, the following distributions are statistically indistinguishable:*

$$\{\mathbf{s} : \mathbf{u} \leftarrow \mathbb{Z}_q^n, \mathbf{s} \leftarrow \mathsf{SamplePre}(\mathbf{A}, T_{\mathbf{A}}, \mathbf{u}, \sigma)\} \approx_s \mathcal{D}_{\mathbb{Z}^m, \sigma}.$$

   Note that the first and third properties must be satisfied for all strings $y \in \{0,1\}^{\ell_{\mathrm{TG}}}$. These properties are satisfied by the gadget-based trapdoor lattice sampler of [24].

3   Circular Security and Our Framework for Generating Circular Separations

In this section, we define the notion of circular security for public key encryption schemes, and then discuss frameworks for obtaining separation between circular security and IND-CPA security. Let $\mathcal{PKE} = (\mathsf{Setup}, \mathsf{KeyGen}, \mathsf{Enc}, \mathsf{Dec})$ be a public key encryption scheme. A k-encryption cycle consists of k encryptions, where the ith encryption is an encryption of the $(i-1)^{th}$ secret key using the ith public key. Intuitively, the scheme is k-circular secure if no adversary can distinguish between an encryption cycle and k encryptions of zeros.

Definition 2 (k-Circular Security). *Let* $\mathcal{PKE} = (\mathsf{Setup}, \mathsf{KeyGen}, \mathsf{Enc}, \mathsf{Dec})$ *be a public key cryptosystem. The scheme is said to k-circular secure if for all PPT adversaries $\mathcal{A}$, the following expression is at most* $\mathsf{negl}(\lambda)$.

$$\left| \Pr\left[1 \leftarrow \mathcal{A}(\{(\mathsf{pk}_i, \mathsf{ct}_i)\}_i) \; : \; \begin{array}{l} \mathsf{pp} \leftarrow \mathsf{Setup}(1^\lambda); (\mathsf{pk}_i, \mathsf{sk}_i) \leftarrow \mathsf{KeyGen}(\mathsf{pp}); \\ \mathsf{ct}_i \leftarrow \mathsf{Enc}(\mathsf{pk}_i, \mathsf{sk}_{i-1}) \end{array} \right] \right.$$

$$\left. - \Pr\left[1 \leftarrow \mathcal{A}(\{(\mathsf{pk}_i, \mathsf{ct}_i)\}_i) \; : \; \begin{array}{l} \mathsf{pp} \leftarrow \mathsf{Setup}(1^\lambda); (\mathsf{pk}_i, \mathsf{sk}_i) \leftarrow \mathsf{KeyGen}(\mathsf{pp}); \\ \mathsf{ct}_i \leftarrow \mathsf{Enc}(\mathsf{pk}_i, 0^{|\mathsf{sk}_{i-1}|}) \end{array} \right] \right|.$$

The above definition is derived from the Key-Dependent Message (KDM) security notion of Black et al. [8]. A weaker security notion, proposed by Cash et al. [14] requires the adversary to output the secret key when given an encryption cycle. Koppula et al. [21] showed that if there exists an adversary that can distinguish between an encryption cycle and encryptions of zeros, then there exists an adversary that can recover the entire secret key given an encryption cycle. Therefore, in this work, we focus on Definition 2.

3.1   The BHW Cycle Tester Framework

In a recent work, Bishop et al. [7] introduced a generic framework for creating circular security counterexamples. In this *cycle tester* framework, there are four algorithms - Setup, KeyGen, Encrypt and *Test*. The setup algorithm outputs the public parameters, the key generation algorithm uses the public parameters to output a public key/secret key pair. The encryption algorithm takes a public key and message as input, and outputs a ciphertext. Finally, the testing algorithm takes as input k public keys and k ciphertexts, and outputs 1 if the k encryptions form an encryption cycle, else it outputs 0. *Note that in this framework, there is no decryption algorithm.* The security requirement is identical to the IND-CPA security game. The following description is taken from [7].

Definition 3 (k-Cycle Tester). *A cycle tester $\Gamma = (\mathsf{Setup}, \mathsf{KeyGen}, \mathsf{Enc}, \mathsf{Test})$ for message space $\mathcal{M}$ and secret key space $\mathcal{S}$ is a tuple of algorithms specified as follows:*

- $\mathsf{Setup}(1^\lambda, 1^k) \to \mathsf{pp}$. *The setup algorithm takes as input the security parameter λ and the length of cycle k. It outputs the public parameters pp.*
- $\mathsf{KeyGen}(\mathsf{pp}) \to (\mathsf{pk}, \mathsf{sk})$. *The key generation algorithm takes as input the public parameters pp and outputs a public key pk and secret key $\mathsf{sk} \in \mathcal{S}$.*
- $\mathsf{Enc}(\mathsf{pk}, m \in \mathcal{M}) \to C$. *The encryption algorithm takes as input a public key pk and a message $m \in \mathcal{M}$ and outputs a ciphertext C.*
- $\mathsf{Test}(\mathbf{pk}, \mathbf{ct}) \to \{0, 1\}$. *On input $\mathbf{pk} = (\mathsf{pk}_1, \ldots, \mathsf{pk}_k)$ and $\mathbf{ct} = (\mathsf{ct}_1, \ldots, \mathsf{ct}_k)$, the testing algorithm outputs a bit in $\{0, 1\}$.*

The algorithms must satisfy the following properties.

1. *(Testing Correctness) There exists a polynomial $p(\cdot)$ such that for all security parameters λ, the Test algorithm's advantage (given by the following expression) is at least $1/p(\lambda)$.*

$$\Pr\left[1 \leftarrow \mathsf{Test}(\mathbf{pk}, \mathbf{ct}) \; : \; \begin{array}{l} \mathsf{pp} \leftarrow \mathsf{Setup}(1^\lambda); (\mathsf{pk}_i, \mathsf{sk}_i) \leftarrow \mathsf{KeyGen}(\mathsf{pp}); \\ \mathsf{ct}_i \leftarrow \mathsf{Enc}(\mathsf{pk}, \mathsf{sk}_{i-1}) \end{array}\right]$$

$$- \Pr\left[1 \leftarrow \mathsf{Test}(\mathbf{pk}, \mathbf{ct}) \; : \; \begin{array}{l} \mathsf{pp} \leftarrow \mathsf{Setup}(1^\lambda); (\mathsf{pk}_i, \mathsf{sk}_i) \leftarrow \mathsf{KeyGen}(1^\lambda); \\ \mathsf{ct}_i \leftarrow \mathsf{Enc}(\mathsf{pk}_i, 0^{|\mathsf{sk}_{i-1}|}) \end{array}\right]$$

2. *(IND-CPA Security) Let $\Pi = (\mathsf{Setup}, \mathsf{KeyGen}, \mathsf{Enc}, \cdot)$ be an encryption scheme with empty decryption algorithm. The scheme Π must satisfy the IND-CPA security definition.*

Bishop et al. [7] showed that in order to construct a separation between IND-CPA and k-circular security, it suffices to construct a k-cycle tester (as defined in Definition 3).

Theorem 3 (CPA Separation from Cycle Testers, [7]). *If there exists an IND-CPA-secure encryption scheme Π for message space $M = (M_1 \times M_2)$ and secret key space $S_1 \subseteq M_1$ and an k-cycle tester Γ for message space M_2 and secret key space $S_2 \subseteq M_2$, then there exists an IND-CPA-secure encryption scheme Π' for message space $M = (M_1 \times M_2)$ and secret key space $S = (S_1 \times S_2)$ that is k-circular insecure.*

3.2   Our Leader-Follower Tester Framework

In this section, we propose an adaptation of the BHW cycle tester framework that we call *Leader-Follower Tester*. In this modification, the key generation and encryption have two modes - leader and follower. The tester algorithm takes k public keys and ciphertexts: the first public key (resp. ciphertext) is a 'leader' public key (resp. ciphertext). The remaining are 'follower' public keys/ciphertexts. It outputs 1 if the ciphertexts form a cycle, else it outputs 0. First, we will formally define the syntax/properties of this modification, and then show how this implies the cycle tester framework of [7].

Definition 4 (k-Leader-Follower Tester). *A Leader-Follower cycle tester Γ = (Setup, KeyGen-L, KeyGen-F, Enc-L, Enc-F, Test) for message space M and secret key space S is a tuple of algorithms specified as follows:*

- Setup$(1^\lambda, 1^k) \to$ pp. *The setup algorithm takes as input the security parameter n and length of cycle k, and outputs public parameters pp.*
- KeyGen-L$(\mathsf{pp}) \to (\mathsf{pk}, \mathsf{sk})$. *The leader key generation algorithm takes as input the public parameters pp, and outputs a public key pk and secret key $\mathsf{sk} \in S$.*
- KeyGen-F$(\mathsf{pp}) \to (\mathsf{pk}, \mathsf{sk})$. *The follower key generation algorithm takes as input the public parameters pp, and outputs a public key pk and secret key $\mathsf{sk} \in S$.*
- Enc-L$(\mathsf{pk}, m \in M) \to C$. *The leader encryption algorithm takes as input a leader public key pk and a message $m \in M$ and outputs a ciphertext C.*

- Enc-F(pk, $m \in M$) $\to C$. *The follower encryption algorithm takes as input a follower public key* pk *and a message* $m \in M$ *and outputs a ciphertext* C.
- Test(**pk**, **ct**) $\to \{0, 1\}$. *The test algorithm takes as input a public key vector* **pk** $= (\mathsf{pk}_1, \dots, \mathsf{pk}_k)$ *and a ciphertext vector* **ct** $= (\mathsf{ct}_1, \dots, \mathsf{ct}_k)$. *Of these, the first public key and ciphertext are of leader type, while the remaining are of follower type. The testing algorithm outputs a bit in* $\{0, 1\}$.

The algorithms must satisfy the following properties.

1. *(Testing Correctness) There exists a polynomial* $p(\cdot)$ *such that for all security parameters* λ, *the Test algorithm's advantage (given by the following expression) is at least* $1/p(\lambda)$.

$$\Pr \left[1 \leftarrow \mathsf{Test}(\{\mathsf{pk}_i, \mathsf{ct}_i\}) : \begin{array}{l} \mathsf{pp} \leftarrow \mathsf{Setup}(1^\lambda, 1^k); (\mathsf{pk}_1, \mathsf{sk}_1) \leftarrow \mathsf{KeyGen\text{-}L}(\mathsf{pp}); \\ \mathsf{ct}_1 \leftarrow \mathsf{Enc\text{-}L}(\mathsf{pk}_1, \mathsf{sk}_k); (\mathsf{pk}_i, \mathsf{sk}_i) \leftarrow \mathsf{KeyGen\text{-}F}(\mathsf{pp}); \\ \mathsf{ct}_i \leftarrow \mathsf{Enc\text{-}F}(\mathsf{pk}, \mathsf{sk}_{i-1}) \end{array} \right]$$

$$- \Pr \left[1 \leftarrow \mathsf{Test}(\{\mathsf{pk}_i, \mathsf{ct}_i\}) : \begin{array}{l} \mathsf{pp} \leftarrow \mathsf{Setup}(1^\lambda, 1^k); (\mathsf{pk}_1, \mathsf{sk}_1) \leftarrow \mathsf{KeyGen\text{-}L}(\mathsf{pp}); \\ \mathsf{ct}_1 \leftarrow \mathsf{Enc\text{-}L}(\mathsf{pk}_1, \mathsf{sk}_k); (\mathsf{pk}_i, \mathsf{sk}_i) \leftarrow \mathsf{KeyGen\text{-}F}(\mathsf{pp}); \\ \mathsf{ct}_i \leftarrow \mathsf{Enc\text{-}F}(\mathsf{pk}_i, 0^{|\mathsf{sk}_{i-1}|}) \end{array} \right]$$

2. *(IND-CPA Security for Both Modes) Let* $\Pi\text{-}L = (\mathsf{Setup}, \mathsf{KeyGen\text{-}L}\mathsf{Enc\text{-}L}, \cdot)$ *and* $\Pi\text{-}F = (\mathsf{Setup}, \mathsf{KeyGen\text{-}F}, \mathsf{Enc\text{-}F}, \cdot)$ *be two encryption schemes with empty decryption algorithm. We require that both* $\Pi\text{-}L$ *and* $\Pi\text{-}F$ *must satisfy* IND-CPA *security.*

We will now show that the Leader-Follower Tester defined above implies the tester framework of [7] (Definition 3).

Lemma 2. *Suppose there exists a* k-*Leader-Follower-Tester (*Setup, KeyGen-L, Enc-L, KeyGen-F, Enc-F, Test*) as defined in Definition 4. Then there exists a* k-*Tester (*Setup′, KeyGen′, Enc′, Test′*) that satisfies Definition 3.*

Proof. The proof of this lemma is fairly straightforward: the setup algorithm first chooses a bit $b \leftarrow \mathsf{Ber}_{1/k}$. If $b = 1$, it runs KeyGen-L and sets the mode to be 'leader', else it runs KeyGen-F and sets the mode to be 'follower'. The encryption algorithm, based on the mode, either uses Enc-L or Enc-F. The detailed proof can be found in the full version of this paper.

The purpose of introducing this modification is that it simplifies the description of our construction (see Sect. 4). In our construction, one of the k public keys/ciphertexts is used to 'tie' the ends together, and therefore is referred to as the Leader. A similar structure can be found in the counterexample shown by [21].

4 Separation for k-Circular Security

In this section, we will describe a Leader-Follower cycle tester $\mathcal{E} = (\mathsf{Setup}, \mathsf{KeyGen\text{-}L}, \mathsf{Enc\text{-}L}, \mathsf{KeyGen\text{-}F}, \mathsf{Enc\text{-}F}, \mathsf{Test})$ such that it satisfies the properties described in Definition 4. Recall, $\ell_{\mathrm{TG}}(n)$ denote the number of bits of randomness required by the TrapGen algorithm. For simplicity of description, we will drop the dependence on n.

Fix any $\epsilon < 1/2$. Our scheme has following algorithms:

- Setup($1^\lambda, 1^k$): The setup algorithm chooses the following parameters: matrix dimensions n, m, LWE modulus q, parameter σ for the Gaussian noise distribution χ, and an additional parameter ℓ. These parameters will be functions of λ, k and ϵ. We require the parameters to satisfy the following relations:
  - (n, m) are the dimensions of matrices output by TrapGen, therefore $m = \Omega(n \cdot \log q)$.
  - $q = 2^{n^\epsilon}, \sigma = \mathsf{poly}(n)$ and $\chi = \mathcal{D}_\sigma$ (for LWE noise/modulus ratio to be less than $\mathsf{poly}(n)/2^{n^\epsilon}$)
  - $\ell = \Omega(n \cdot \log q)$ (for Leftover Hash Lemma 1)
  - $\ell \cdot (\ell \cdot m \cdot n \cdot \sigma)^k < q/8$ (for the correctness of our Test algorithm)
    One instantiation which works is as follows: fix some constant $c > 0$, let $n = ((c+\epsilon) \cdot k)^{1/\epsilon} \cdot \lambda$, $m = 6n \cdot \log q$, $\sigma = n^c$ for some large enough constant c. Then setting $q = 2^{n^\epsilon}$, $\ell = 2n \log q$ satisfies the above relations.

  The message space of our scheme (which is also the space of secret keys) is $(\{0, 1\}^{\ell_{\mathrm{TG}}})^\ell$.

- KeyGen-L(pp): The leader key generation algorithm first chooses $\mathbf{y}_1, \ldots \mathbf{y}_\ell \leftarrow \{0, 1\}^{\ell_{\mathrm{TG}}}$. For $i \le \ell$, the algorithm computes $(\mathbf{B}_i, T_{\mathbf{B}_i}) = \mathsf{TrapGen}(1^n; \mathbf{y}_i)$. Next it chooses a string $\mathbf{x} \in \{-1, 1\}^\ell$ by choosing uniformly random bits $x_i \leftarrow \{-1, 1\}$ for $i \le \ell - 1$ and setting $x_\ell = 1$. The first part of the public key consists of matrices $\mathbf{D}_i$ defined as follows:

$$\mathbf{D}_i = x_i \cdot \mathbf{B}_i \in \mathbb{Z}_q^{n \times m} \quad \text{for all } i \le \ell$$

  Next, it selects random vectors $\mathbf{h}_i \in \mathbb{Z}_q^n$ for $i < \ell$ and lets $\mathbf{h}_\ell = -\sum_{i<\ell} x_i \cdot \mathbf{h}_i$. The second part of the public key consists of the vectors $\{\mathbf{h}_i\}_i$. The secret key is $\mathsf{sk} = \{\mathbf{y}_i\}_{i \le \ell}$ and the public key is $\mathsf{pk} = (\{\mathbf{D}_i\}_{i \le \ell}, \{\mathbf{h}_i\}_{i \le \ell})$.

- Enc-L(pk, msg): Let $\mathsf{pk} = \{\mathbf{D}_i\}, \{\mathbf{h}_i\}$ and $\mathsf{msg} = (\mathbf{m}_1, \ldots, \mathbf{m}_\ell)$. The leader encryption algorithm computes $(\mathbf{Z}_i, T_{\mathbf{Z}_i}) = \mathsf{TrapGen}(1^n; \mathbf{m}_i)$ for $i \le \ell$. Next, it chooses matrix $\mathbf{C} \leftarrow \chi^{n \times n}$, error vector $\mathbf{e}_i \leftarrow \chi^n$ for $i \le \ell$, and sets $\mathbf{f}_i = \mathbf{C} \cdot \mathbf{h}_i + \mathbf{e}_i$. Finally, it computes $\mathbf{s}_i \leftarrow \mathsf{SamplePre}(\mathbf{Z}_i, T_{\mathbf{Z}_i}, \sigma, \mathbf{f}_i)$. The ciphertext is set to be $\mathsf{ct} = (\mathbf{s}_1, \ldots, \mathbf{s}_\ell)$.

- KeyGen-F(pp): The follower setup algorithm takes as input the security parameter 1^n. It first chooses ℓ uniformly random binary vectors of length ℓ_{TG}; that is, it chooses $\mathbf{y}_i \leftarrow \{0, 1\}^{\ell_{\mathrm{TG}}}$ for $i \le \ell$. Next, it computes $(\mathbf{B}_i, T_{\mathbf{B}_i}) = \mathsf{TrapGen}(1^n; \mathbf{y}_i)$.
  The algorithm outputs secret key $\mathsf{sk} = \{\mathbf{y}_i\}_{i \le \ell}$ and public key $\mathsf{pk} = \{\mathbf{B}_i\}_{i \le \ell}$.

– Enc-F(pk, msg): Let msg $= (\mathbf{m}_1, \ldots, \mathbf{m}_\ell)$. The follower encryption algorithm computes $(\mathbf{Z}_i, T_{\mathbf{Z}_i}) = \mathsf{TrapGen}(1^n; \mathbf{m}_i)$ for $i \leq \ell$. Next, it chooses matrix $\mathbf{C} \leftarrow \chi^{n \times n}$, error matrix $\mathbf{E}_i \leftarrow \chi^{n \times m}$ and sets $\mathbf{F}_i = \mathbf{C} \cdot \mathbf{B}_i + \mathbf{E}_i$. Finally, it computes $\mathbf{S}_i \leftarrow \mathsf{SamplePre}(\mathbf{Z}_i, T_{\mathbf{Z}_i}, \sigma, \mathbf{F}_i)$.
The ciphertext is set to be $\mathsf{ct} = (\mathbf{S}_1, \ldots, \mathbf{S}_\ell)$.

– Test$((\mathsf{pk}^{(1)}, \ldots, \mathsf{pk}^{(k)}), (\mathsf{ct}^{(1)}, \ldots, \mathsf{ct}^{(k)}))$: Let $\mathsf{pk}^{(1)} = (\{\mathbf{D}_i^{(1)}\}, \{\mathbf{h}_i\})$, $\mathsf{ct}^{(1)} = (\mathbf{s}_1^{(1)}, \ldots, \mathbf{s}_\ell^{(1)})$ and $\mathsf{ct}^{(j)} = (\mathbf{S}_1^{(j)}, \ldots, \mathbf{S}_\ell^{(j)})$ for $2 \leq j \leq k$.
The test algorithm computes

$$\sigma = \sum_{i \in [\ell]} \mathbf{D}_i^{(1)} \cdot \left(\prod_{2 \leq j \leq k} \mathbf{S}_i^{(j)} \right) \cdot \mathbf{s}_i^{(1)}.$$

It tests if $\sigma \in [-q/8, q/8]$ and outputs 1 if so to indicate a cycle. Otherwise it outputs 0.

4.1 Proof of Correctness

First, we will show that the Test algorithm distinguishes between a cycle and encryptions of zeros with overwhelming probability. For this, we need to set up some notations. Let $\mathsf{Bd} = n \cdot \sigma$. From Lemma 1, it follows that if $\mathbf{x} \leftarrow \chi^n$, then $\|\mathbf{x}\|_\infty \leq \mathsf{Bd}$ with overwhelming probability. Let $\mathsf{pk}^{(1)} = (\{\mathbf{D}_i\}, \{\mathbf{h}_i\})$ where $\mathbf{D}_i = x_i \cdot \mathbf{B}_i^{(1)}$.

Next, the follower public keys are $\mathsf{pk}^{(p)} = \{\mathbf{B}_i^p\}$ for $2 \leq p \leq k$ and $T_{\mathbf{B}^p}$ denote the trapdoor corresponding to matrix $\mathbf{B}_i^p$ for $p \leq k, i \leq \ell$.

We will first analyse the case where the ciphertexts are encryptions of a cycle. Let $\mathsf{ct}^{(1)} = (\mathbf{s}_1, \ldots, \mathbf{s}_\ell)$. Here, $\mathbf{f}_i = \mathbf{C}^{(1)} \cdot \mathbf{h}_i + \mathbf{e}_i$ and $\mathbf{s}_i = \mathsf{SamplePre}(\mathbf{B}_i^{(k)}, T_{\mathbf{B}_i^{(k)}}, \sigma, \mathbf{f}_i)$.
Next, for $2 \leq p \leq k$, let $\mathbf{F}_i^{(p)} = \mathbf{C}^{(p)} \cdot \mathbf{B}_i^{(p)} + \mathbf{E}_i^{(p)}$ and $\mathbf{S}_i^{(p)} = \mathsf{SamplePre}(\mathbf{B}_i^{(p-1)}, T_{\mathbf{B}^{(p-1)}}, \sigma, \mathbf{F}_i^{(p)})$. Let $\Delta_{i,p^*} = \mathbf{D}_i \cdot (\prod_{p=2}^{p^*} \mathbf{S}_i^{(p)})$ and $\Delta'_{i,p^*} = x_i \cdot (\prod_{p=2}^{p^*} \mathbf{C}^{(p)}) \cdot \mathbf{B}_i^{(p^*)}$

Claim 1. For any $i \leq \ell, p^* \in [2, k]$,

$$\|\Delta_{i,p^*} - \Delta'_{i,p^*}\|_\infty \leq (\ell \cdot m \cdot \mathsf{Bd})^{p^*-1}.$$

Proof. The proof of this theorem involves a simple induction argument on p^*. First, the base case: $p^* = 2$. In this case, $\Delta_{i,p^*} = \mathbf{D}_i \cdot \mathbf{S}_i^{(2)} = x_i \cdot \mathbf{C}^{(2)} \cdot \mathbf{B}_i^{(2)} + x_i \cdot \mathbf{E}_i^{(2)}$. Hence $\|\Delta_{i,2} - \Delta'_{i,2}\|_\infty \leq \ell \cdot m \cdot \mathsf{Bd}$.

Suppose this holds true for all indices less than p^*. Now, $\Delta_{i,p^*} = \Delta_{i,p^*-1} \cdot \mathbf{S}_i^{(p^*)}$, and let $\Delta_{i,p^*-1} = \Delta'_{i,p^*-1} + \mathsf{Err}_{i,p^*-1}$, where $\|\mathsf{Err}_{i,p^*-1}\|_\infty \leq (\ell \cdot m \cdot \mathsf{Bd})^{p^*-2}$.

$$\Delta_{i,p^*} = \Delta'_{i,p^*-1} \cdot \mathbf{S}_i^{(p^*)} + \mathsf{Err}_{i,p^*-1} \cdot \mathbf{S}_i^{(p^*)}$$

$$= \Delta'_{i,p^*} + x_i \cdot \left(\prod_{p=2}^{p^*-1} \mathbf{C}^{(p)} \right) \cdot \mathbf{E}_i^{(p^*)} + \mathsf{Err}_{i,p^*-1} \cdot \mathbf{S}_i^{(p^*)}$$

Let $\mathsf{Err}_{i,p^*} = x_i \cdot (\prod_{p=2}^{p^*-1} \mathbf{C}^{(p)}) \cdot \mathbf{E}_i^{(p^*)} + \mathsf{Err}_{p^*-1} \cdot \mathbf{S}_i^{(p^*)}$.

$$\|\mathsf{Err}_{i,p^*}\|_\infty \leq (\ell \cdot n \cdot \mathsf{Bd})^{p^*-2} \cdot (\ell \cdot m \cdot \mathsf{Bd}) + \|\mathsf{Err}_{p^*-1}\|_\infty \cdot (m \cdot \mathsf{Bd})$$
$$\leq (\ell \cdot n \cdot \mathsf{Bd})^{p^*-2} \cdot (\ell \cdot m \cdot \mathsf{Bd}) + (\ell \cdot m \cdot \mathsf{Bd})^{p^*-2} \cdot (m \cdot \mathsf{Bd})$$
$$\leq (\ell \cdot m \cdot \mathsf{Bd})^{p^*-1}$$

Finally, let us now consider the term $\Delta_{i,k} \cdot \mathbf{s}_i$. By a similar analysis as above, we can show that $\Delta_{i,k} \cdot \mathbf{s}_i = x_i \cdot (\prod_{p=2}^{k} \mathbf{C}^{(p)}) \cdot \mathbf{C}^{(1)} \cdot \mathbf{h}_i + \mathsf{Error}_i$ where $\|\mathsf{Error}_i\|_\infty \leq (\ell \cdot m \cdot \mathsf{Bd})^k$. As a result,

$$\|\sum_i \Delta_{i,k} \cdot \mathbf{s}_i\|_\infty = \|\sum_i x_i \cdot \mathbf{h}_i\|_\infty + \sum_i \|\mathsf{Error}_i\|_\infty \leq \ell \cdot (\ell \cdot m \cdot \mathsf{Bd})^k.$$

Given our choice of parameters, $\ell \cdot (\ell \cdot m \cdot \mathsf{Bd})^k < q/8$, and as a result, the Test algorithm outputs 1.

On the other hand, if the k cycle consists of encryptions of $\mathbf{0}$, then for all $i \leq \ell$, $\mathbf{D}_i \cdot \prod_{p=2}^{k} \mathbf{S}_i^{(p)} \cdot \mathbf{s}_i$ is a uniformly random vector in $\mathbb{Z}_q^n$, and therefore the test algorithm outputs 1 with negligible probability.

4.2   Proof of INDCPA Security

In this section, we will show that the construction described above is IND-CPA secure as per Definition 4. Recall, the IND-CPA security definition for leader-based encryption schemes requires two separate IND-CPA proofs for both leader and follower modes.

INDCPA Security for Leader Mode. First, we will prove IND-CPA security for Leader mode. For this, we will define a sequence of hybrid experiments, and then show that the hybrids are computationally indistinguishable. The first hybrid will correspond to the IND-CPA security game, while the final hybrid will be one where the adversary has 0 advantage. Due to space constraints, we will describe the first hybrid in full detail, and then onwards, we will only present the step that is modified.

Hyb_0: This corresponds to the IND-CPA security game.

1. Setup Phase:
   (a) The challenger first chooses $\mathbf{y}_i \leftarrow \{0,1\}^{\ell_{\mathrm{TG}}}$ for $i \leq \ell$ and computes $(\mathbf{B}_i, T_{\mathbf{B}_i}) = \mathsf{TrapGen}(1^n; \mathbf{y}_i)$.
   (b) Next, it chooses $x_i \leftarrow \{-1, 1\}$ for $i < \ell$, sets $x_\ell = 1$.
   (c) It chooses $\mathbf{h}_i \leftarrow \mathbb{Z}_q^n$ for $i < \ell$, sets $\mathbf{h}_\ell = -\sum_{i<\ell} x_i \cdot \mathbf{h}_i$.
   (d) Finally, the challenger sends $(\{x_i \cdot \mathbf{B}_i\}, \{\mathbf{h}_i\})$ to the adversary.
2. Challenge Phase
   (a) The adversary sends two messages $\mathsf{msg}_0, \mathsf{msg}_1$. The challenger chooses matrix $\mathbf{C} \leftarrow \chi^{n \times n}$, error vector $\mathbf{e}_i \leftarrow \chi^n$ for $i \leq \ell$ and sets $\mathbf{f}_i = \mathbf{C} \cdot \mathbf{h}_i + \mathbf{e}_i$ for $i \leq \ell$.

(b) Next, it chooses $b \leftarrow \{0,1\}$. Let $\mathsf{msg}_b = (\mathbf{m}_1, \ldots, \mathbf{m}_\ell)$. The challenger computes $(\mathbf{Z}_i, T_{\mathbf{Z}_i}) = \mathsf{TrapGen}(1^n; \mathbf{m}_i)$.

(c) Using $T_{\mathbf{Z}_i}$, the challenger computes $\mathbf{s}_i \leftarrow \mathsf{SamplePre}(T_{\mathbf{Z}_i}, \mathbf{f}_i)$ for all $i \leq \ell$. It sends $\mathsf{ct}^* = (\{\mathbf{s}_i\})$.

3. Guess: The adversary sends its guess b' and wins if $b = b'$.

Hyb_1: In this game, the challenger chooses $\mathbf{B}_i$ uniformly at random, and outputs $\{\mathbf{B}_i\}$ as part of public key, instead of $\{x_i \cdot \mathbf{B}_i\}$.

1. Setup Phase:
   (a) The challenger first chooses $\mathbf{B}_i \leftarrow \mathbb{Z}_q^{n \times m}$.
   (d) Finally, the challenger sends $(\{\mathbf{B}_i\}, \{\mathbf{h}_i\})$ to the adversary.

Hyb_2: In this game, the challenger chooses $\mathbf{h}_\ell$ uniformly at random instead of setting it as $-\sum x_i \mathbf{h}_i$. Therefore, from this game onwards, the challenger does not need to choose x_i for $i < \ell$.

1. Setup Phase:
   (c) It chooses $\mathbf{h}_i \leftarrow \mathbb{Z}_q^n$ for $i \leq \ell$.

Hyb_3: In this game, the challenger modifies the challenge phase. It chooses uniformly random vectors $\mathbf{f}_i \leftarrow \mathbb{Z}_q^n$.

2. Challenge Phase
   (a) The adversary sends two messages $\mathsf{msg}_0, \mathsf{msg}_1$.
       The challenger chooses $\mathbf{f}_i \leftarrow \mathbb{Z}_q^n$ for all $i \leq \ell$.

Hyb_4: In this game, the challenger chooses $\mathbf{s}_i$ from the Discrete Gaussian distribution $\mathcal{D}_{\mathbb{Z}^m, \sigma}$ with parameter σ. Note that in this hybrid, the adversary has 0 advantage.

2. Challenge Phase
   (a) Next, the challenger chooses bit $b \leftarrow \{0,1\}$ and $\mathbf{s}_i \leftarrow \mathcal{D}_\sigma$. It sends $\mathsf{ct}^* = \{\mathbf{s}_i\}$.

Analysis: We will now show that any PPT adversary has nearly identical advantage in the hybrid experiments described above. Let $\mathsf{Adv}_{\mathcal{A}}^i$ denote the advantage of adversary $\mathcal{A}$ in experiment Hyb_i.

Claim 2. *For any adversary* $\mathcal{A}$, $\mathsf{Adv}_{\mathcal{A}}^0 - \mathsf{Adv}_{\mathcal{A}}^1 \leq negl(n)$.

Proof. We will show that the statistical distance between the distributions of public keys in Hyb_0 and Hyb_1 is negligible in the security parameter n. Note that the only difference between the two hybrids is the distribution of $\mathbf{B}_i$ for $i \leq \ell$.

From the well-distributedness property of TrapGen, we know that the following distributions have negligible statistical distance:

$$\{\mathbf{B}_i \ : \ (\mathbf{B}_i, T_{\mathbf{B}_i}) \leftarrow \mathsf{TrapGen}(1^n)\} \approx \{\mathbf{B}_i \ : \ \mathbf{B}_i \leftarrow \mathbb{Z}_q^{n \times m}\}.$$

Next, note that the following distributions are identical:

$$\{(x_i, x_i \cdot \mathbf{B}_i) \ : \ x_i \leftarrow \{-1, 1\}, \mathbf{B}_i \leftarrow \mathbb{Z}_q^{n \times m}\} \equiv \{(x_i, \mathbf{B}_i) \ : \ x_i \leftarrow \{-1, 1\}, \mathbf{B}_i \leftarrow \mathbb{Z}_q^{n \times m}\}$$

Therefore, we can conclude that

$$\left\{ (x_i, x_i \cdot \mathbf{B}_i) \ : \ \begin{matrix} x_i \leftarrow \{-1, 1\}, \\ (\mathbf{B}_i, T_{\mathbf{B}_i}) \leftarrow \mathsf{TrapGen}(1^n) \end{matrix} \right\} \approx \left\{ (x_i, \mathbf{B}_i) \ : \ \begin{matrix} x_i \leftarrow \{-1, 1\}, \\ \mathbf{B}_i \leftarrow \mathbb{Z}_q^{n \times m} \end{matrix} \right\}.$$

As a result, the public key distributions in Hyb_0 and Hyb_1 are statistically indistinguishable.

Claim 3. *For any adversary $\mathcal{A}$, $\mathsf{Adv}_{\mathcal{A}}^1 - \mathsf{Adv}_{\mathcal{A}}^2 \leq \mathsf{negl}(n)$.*

Proof. The only difference between hybrid experiments Hyb_1 and Hyb_2 is in the choice of $\mathbf{h}_\ell$. In Hyb_1, $\mathbf{h}_\ell = -\sum_i x_i \mathbf{h}_i$, while in Hyb_2, it is chosen uniformly at random. Here, we will use the Leftover Hash Lemma (Theorem 1). Since $\ell > (n+1) \log_2 q + \omega(\log n)$, it follows that

$$\{(\mathbf{A} = [\mathbf{h}_1 | \ldots | \mathbf{h}_{\ell-1}], \mathbf{h}_\ell = -\mathbf{A} \cdot \mathbf{r}) \ : \ \mathbf{h}_i \leftarrow \mathbb{Z}_q^n \text{ for all } i \leq \ell - 1, \mathbf{r} \leftarrow \mathbb{Z}_q^{\ell-1}\}$$

$$\approx$$

$$\{(\mathbf{A} = [\mathbf{h}_1 | \ldots | \mathbf{h}_{\ell-1}], \mathbf{h}_\ell) \ : \ \mathbf{h}_i \leftarrow \mathbb{Z}_q^n \text{ for all } i \leq \ell\}$$

Claim 4. *Assuming (n, ℓ, q, χ)-LWE-ss (Assumption 2), for any PPT adversary A, $\mathsf{Adv}_{\mathcal{A}}^2 - \mathsf{Adv}_{\mathcal{A}}^3 \leq \mathsf{negl}(n)$.*

Proof. The only difference in Hyb_2 and Hyb_3 is the manner in which $\mathbf{f}_i$ are computed. In Hyb_2, the challenger chooses $\mathbf{C} \leftarrow \chi^{n \times n}$, $\mathbf{e}_i \leftarrow \chi^n$ and sets $\mathbf{f}_i = \mathbf{C} \cdot \mathbf{h}_i + \mathbf{e}_i$ for all $i \leq \ell$. In Hyb_3, $\mathbf{f}_i$ are chosen uniformly at random from $\mathbb{Z}_q^n$.

Suppose there exists an adversary $\mathcal{A}$ such that $\mathsf{Adv}_{\mathcal{A}}^2 - \mathsf{Adv}_{\mathcal{A}}^3$ is non-negligible in n. Then there exists a reduction algorithm $\mathcal{B}$ that can use $\mathcal{A}$ to break Assumption 2 with non-negligible advantage. First, $\mathcal{B}$ receives as LWE challenge two $n \times \ell$ matrices $(\mathbf{H}, \mathbf{F})$. It chooses ℓ matrices $\mathbf{B}_i \leftarrow \mathbb{Z}_q^{n \times m}$, sets $\mathbf{h}_i$ as the ith column of $\mathbf{H}$ and sends $\{\mathbf{B}_i, \mathbf{h}_i\}$ as the public key.

On receiving the challenge messages $\mathsf{msg}_0, \mathsf{msg}_1$, $\mathcal{B}$ uses $\mathbf{F}$ to compute the challenge ciphertext. It first chooses $b \leftarrow \{0, 1\}$, computes $(\mathbf{Z}_i, T_{\mathbf{Z}_i})$ using msg_b and sets $\mathbf{f}_i$ to be the ith column of $\mathbf{F}$. Next, it computes $\mathbf{s}_i \leftarrow \mathsf{SamplePre}(\mathbf{Z}_i, T_{\mathbf{Z}_i}, \sigma, \mathbf{f}_i)$ and sends the vectors $\{\mathbf{s}_i\}$ as the ciphertext.

Finally, the adversary sends the guess b'. If $b = b'$, $\mathcal{B}$ guesses that $\mathbf{F}$ is an LWE matrix, else it guesses that $\mathbf{F}$ is uniformly random.

Clearly, if $\mathbf{F} = \mathbf{C} \cdot \mathbf{H} + \mathbf{E}$ for some $\mathbf{C} \leftarrow \chi^{n \times n}$, $\mathbf{E} \leftarrow \chi^{n \times \ell}$, then B simulates Hyb_2, and if $\mathbf{F}$ is uniformly random, then this corresponds to Hyb_3. This concludes our proof.

Claim 5. *Assuming the well-distributedness property of* $(\mathsf{TrapGen}, \mathsf{SamplePre})$ *(Definition 1), for any adversary* $\mathcal{A}$, $\mathsf{Adv}_{\mathcal{A}}^3 - \mathsf{Adv}_{\mathcal{A}}^4 \leq negl(n)$.

Proof. This follows directly from the well-distributedness property of $(\mathsf{TrapGen}, \mathsf{SamplePre})$ algorithms, because the vectors $\{\mathbf{f}_i\}_i$ are chosen uniformly at random from $\mathbb{Z}_q^n$. Therefore, the well-distributedness property states that for all random coins $\mathbf{y}$, $\{\mathbf{s}_i : (\mathbf{M}, T_{\mathbf{M}}) \leftarrow \mathsf{TrapGen}(1^n; \mathbf{y}), \mathbf{s}_i \leftarrow \mathsf{SamplePre}(\mathbf{M}, T_{\mathbf{M}}, \sigma, \mathbf{f}_i)\} \approx_s \mathcal{D}_{\mathbb{Z}^m, \sigma}$.

Using the above claims, we can show that $\mathsf{Adv}_{\mathcal{A}}^0 - \mathsf{Adv}_{\mathcal{A}}^5 \leq negl(n)$, and therefore, the scheme is IND-CPA secure for Leader setup.

4.3 INDCPA Security for Follower Mode

This case is similar to the Leader mode, therefore we will only describe the intermediate hybrids, and refer to the corresponding proofs from the section above.

Hyb_0: This corresponds to the IND-CPA security game.

1. Setup Phase:
   (a) The challenger first chooses $\mathbf{y}_i \leftarrow \{0,1\}^{\ell_{\mathrm{TG}}}$ for $i \leq \ell$. Next, it computes $(\mathbf{B}_i, T_{\mathbf{B}_i}) = \mathsf{TrapGen}(1^n; \mathbf{y}_i)$.
   The challenger sends $\{\mathbf{B}_i\}_i$ to the adversary.
2. Challenge Phase
   (a) The adversary sends two messages $\mathsf{msg}_0, \mathsf{msg}_1$. The challenger first chooses $\mathbf{C} \leftarrow \chi^{n \times n}$, $\mathbf{E}_i \leftarrow \chi^{n \times m}$ and sets $\mathbf{F}_i = \mathbf{C} \cdot \mathbf{B}_i + \mathbf{E}_i$.
   (b) Next, it chooses $b \leftarrow \{0,1\}$. Let $\mathsf{msg}_b = (\mathbf{m}_1, \ldots, \mathbf{m}_\ell)$. The challenger computes $(\mathbf{Z}_i, T_{\mathbf{Z}_i}) = \mathsf{TrapGen}(1^n; \mathbf{m}_i)$.
   (c) Using $T_{\mathbf{Z}_i}$, the challenger computes $\mathbf{S}_i \leftarrow \mathsf{SamplePre}(\mathbf{Z}_i, T_{\mathbf{Z}_i}, \sigma, \mathbf{F}_i)$ for all $i \leq \ell$. It sends $\mathsf{ct}^* = \{\mathbf{S}_i\}_i$ as the challenge ciphertext.
3. Guess: The adversary sends its guess b' and wins if $b = b'$.

Hyb_1: In this hybrid, the challenger uses truly random matrices $\mathbf{B}_i$.

1. Setup Phase:
   (a) The challenger chooses $\mathbf{B}_i \leftarrow \mathbb{Z}_q^{n \times m}$ for $i \leq \ell$ and sends $\{\mathbf{B}_i\}_i$ to the adversary.

Hyb_2: In this hybrid, the challenger uses truly random matrices $\mathbf{F}_i$ to compute the ciphertext.

2. Challenge Phase
   (a) The adversary sends two messages $\mathsf{msg}_0, \mathsf{msg}_1$. The challenger first chooses $\mathbf{F}_i \leftarrow \mathbb{Z}_q^{n \times m}$ for all $i \leq \ell$.

Hyb_3: In this hybrid, the challenger chooses the matrices $\mathbf{S}_i$ with entries from the discrete Gaussian distribution $\mathcal{D}_{\mathbb{Z}^m, \sigma}^m$. Therefore, in this game, any adversary has 0 advantage.

2. Challenge Phase
   (b) Next, it chooses $\mathbf{S}_i \leftarrow \mathcal{D}_{\mathbb{Z}^m, \sigma}^m$ for all $i \leq \ell$. It sends $\mathsf{ct}^* = \{\mathbf{S}_i\}_i$.

Analysis: As mentioned above, the proofs for this section will be very similar to the ones in Sect. 4.2.

Claim 6. *For any PPT adversary* $\mathcal{A}$, $\mathsf{Adv}_{\mathcal{A}}^0 - \mathsf{Adv}_{\mathcal{A}}^1 \leq negl(n)$.

The proof of this claim is identical to the proof of Claim 2.

Claim 7. *Assuming* $(n, m \cdot \ell, q, \chi)$-$\mathsf{LWE}$-$\mathsf{ss}$ *(Assumption 2), for any PPT adversary* $\mathcal{A}$, $\mathsf{Adv}_{\mathcal{A}}^1 - \mathsf{Adv}_{\mathcal{A}}^2 \leq negl(n)$.

The proof of this claim is similar to the proof of Claim 4.

Claim 8. *Assuming the well-distributedness property of* (SamplePre, TrapGen) *algorithms (Definition 1), for any PPT adversary* $\mathcal{A}$, $\mathsf{Adv}_{\mathcal{A}}^3 - \mathsf{Adv}_{\mathcal{A}}^4 \leq negl(n)$.

This proof is identical to the proof of Claim 5.

References

1. Acar, T., Belenkiy, M., Bellare, M., Cash, D.: Cryptographic agility and its relation to circular encryption. In: Gilbert, H. (ed.) EUROCRYPT 2010. LNCS, vol. 6110, pp. 403–422. Springer, Heidelberg (2010)
2. Adão, P., Bana, G., Herzog, J., Scedrov, A.: Soundness and completeness of formal encryption: the cases of key cycles and partial information leakage. J. Comput. Secur. **17**(5), 737–797 (2009)
3. Alperin-Sheriff, J., Peikert, C.: Circular and KDM security for identity-based encryption. In: Fischlin, M., Buchmann, J., Manulis, M. (eds.) PKC 2012. LNCS, vol. 7293, pp. 334–352. Springer, Heidelberg (2012)
4. Applebaum, B.: Key-dependent message security: generic amplification and completeness. In: Paterson, K.G. (ed.) EUROCRYPT 2011. LNCS, vol. 6632, pp. 527–546. Springer, Heidelberg (2011)
5. Applebaum, B., Cash, D., Peikert, C., Sahai, A.: Fast cryptographic primitives and circular-secure encryption based on hard learning problems. In: Halevi, S. (ed.) CRYPTO 2009. LNCS, vol. 5677, pp. 595–618. Springer, Heidelberg (2009)
6. Barak, B., Haitner, I., Hofheinz, D., Ishai, Y.: Bounded key-dependent message security. In: Gilbert, H. (ed.) EUROCRYPT 2010. LNCS, vol. 6110, pp. 423–444. Springer, Heidelberg (2010)

7. Bishop, A., Hohenberger, S., Waters, B.: New circular security counterexamples from decision linear and learning with errors. In: Iwata, T., Cheon, J.H. (eds.) ASIACRYPT 2015. LNCS, vol. 9453, pp. 776–800. Springer, Heidelberg (2015)

8. Black, J., Rogaway, P., Shrimpton, T.: Encryption-scheme security in the presence of key-dependent messages. In: Nyberg, K., Heys, H.M. (eds.) SAC 2002. LNCS, vol. 2595, pp. 62–75. Springer, Heidelberg (2003)

9. Boneh, D., Halevi, S., Hamburg, M., Ostrovsky, R.: Circular-secure encryption from decision Diffie-Hellman. In: Wagner, D. (ed.) CRYPTO 2008. LNCS, vol. 5157, pp. 108–125. Springer, Heidelberg (2008)

10. Brakerski, Z., Goldwasser, S.: Circular and leakage resilient public-key encryption under subgroup indistinguishability (or: Quadratic residuosity strikes back). IACR Cryptology ePrint Archive 2010, 226 (2010)

11. Brakerski, Z., Goldwasser, S., Kalai, Y.T.: Black-box circular-secure encryption beyond affine functions. In: Ishai, Y. (ed.) TCC 2011. LNCS, vol. 6597, pp. 201–218. Springer, Heidelberg (2011)

12. Brakerski, Z., Langlois, A., Peikert, C., Regev, O., Stehlé, D.: Classical hardness of learning with errors. In: Symposium on Theory of Computing Conference, STOC 2013, Palo Alto, CA, USA, 1–4 June 2013, pp. 575–584 (2013)

13. Camenisch, J., Lysyanskaya, A.: An efficient system for non-transferable anonymous credentials with optional anonymity revocation. IACR Cryptology ePrint Archive 2001, 19 (2001)

14. Cash, D., Green, M., Hohenberger, S.: New definitions and separations for circular security. In: Fischlin, M., Buchmann, J., Manulis, M. (eds.) PKC 2012. LNCS, vol. 7293, pp. 540–557. Springer, Heidelberg (2012)

15. Cheon, J.H., Han, K., Lee, C., Ryu, H., Stehlé, D.: Cryptanalysis of the multilinear map over the integers. In: Oswald, E., Fischlin, M. (eds.) EUROCRYPT 2015, Part I. LNCS, vol. 9056, pp. 3–12. Springer, Heidelberg (2015)

16. Coron, J.-S., et al.: Zeroizing without low-level zeroes: new MMAP attacks and their limitations. In: Gennaro, R., Robshaw, M. (eds.) CRYPTO 2015, Part I. LNCS, vol. 9216, pp. 247–266. Springer, Heidelberg (2015)

17. Garg, S., Gentry, C., Halevi, S.: Candidate multilinear maps from ideal lattices. In: Johansson, T., Nguyen, P.Q. (eds.) EUROCRYPT 2013. LNCS, vol. 7881, pp. 1–17. Springer, Heidelberg (2013)

18. Garg, S., Gentry, C., Halevi, S., Raykova, M., Sahai, A., Waters, B.: Candidate indistinguishability obfuscation and functional encryption for all circuits. In: FOCS (2013)

19. Gentry, C.: Fully homomorphic encryption using ideal lattices. In: Proceedings of the 41st Annual ACM Symposium on Theory of Computing, STOC 2009, Bethesda, MD, USA, 31 May – 2 June 2009, pp. 169–178 (2009)

20. Gentry, C., Peikert, C., Vaikuntanathan, V.: Trapdoors for hard lattices and new cryptographic constructions. In: Proceedings of the 40th Annual ACM Symposium on Theory of Computing, Victoria, British Columbia, Canada, 17–20 May 2008, pp. 197–206 (2008)

21. Koppula, V., Ramchen, K., Waters, B.: Separations in circular security for arbitrary length key cycles. In: Dodis, Y., Nielsen, J.B. (eds.) TCC 2015, Part II. LNCS, vol. 9015, pp. 378–400. Springer, Heidelberg (2015)

22. Laud, P.: Encryption cycles and two views of cryptography. In: NORDSEC 2002 - Proceedings of the 7th Nordic Workshop on Secure IT Systems (Karlstad University Studies 2002:31), pp. 85–100 (2002)

23. Marcedone, A., Orlandi, C.: Obfuscation $\Rightarrow$ (IND-CPA security $\nRightarrow$ circular security). In: Abdalla, M., De Prisco, R. (eds.) SCN 2014. LNCS, vol. 8642, pp. 77–90. Springer, Heidelberg (2014)
24. Micciancio, D., Peikert, C.: Trapdoors for lattices: simpler, tighter, faster, smaller. In: Pointcheval, D., Johansson, T. (eds.) EUROCRYPT 2012. LNCS, vol. 7237, pp. 700–718. Springer, Heidelberg (2012)
25. Micciancio, D., Regev, O.: Worst-case to average-case reductions based on gaussian measures. SIAM J. Comput. **37**(1), 267–302 (2007)
26. Peikert, C.: Public-key cryptosystems from the worst-case shortest vector problem: extended abstract. In: Proceedings of the 41st Annual ACM Symposium on Theory of Computing, STOC 2009, Bethesda, MD, USA, 31 May – 2 June 2009, pp. 333–342 (2009)
27. Peikert, C.: A decade of lattice cryptography. Cryptology ePrint Archive, Report 2015/939 (2015). http://eprint.iacr.org/
28. Regev, O.: On lattices, learning with errors, random linear codes, and cryptography. In: Proceedings of the 37th Annual ACM Symposium on Theory of Computing, Baltimore, MD, USA, 22–24 May 2005, pp. 84–93 (2005)

Author Index